Proceedings of the 41st Annual Conference of the
Association for Computer Aided Design in Architecture

TOWARD CRITICAL COMPUTATION

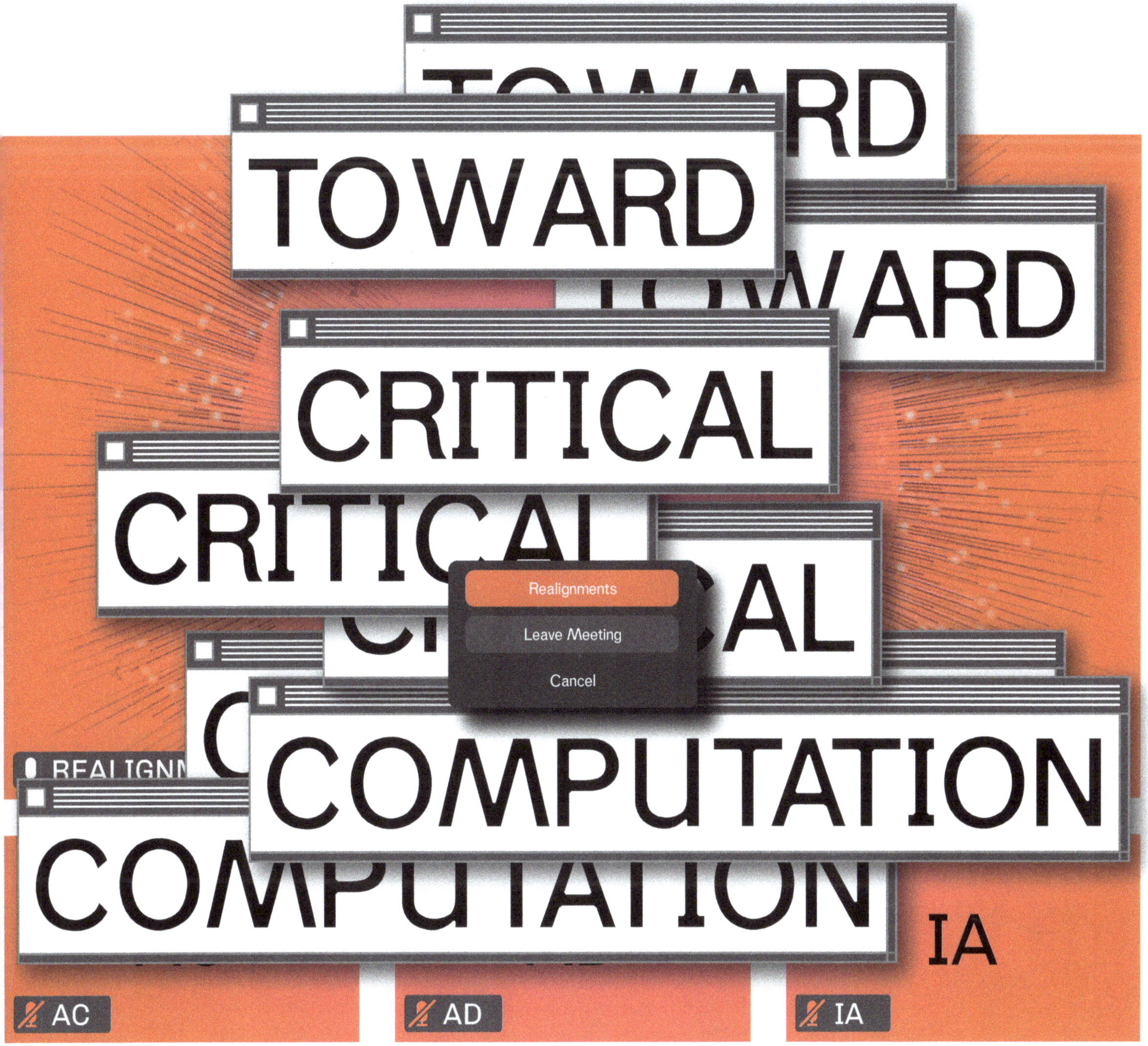

Editors
Kathrin Dörfler, Stefana Parascho, Jane Scott,
Biayna Bogosian, Behnaz Farahi, June A. Grant,
Jose Luis García del Castillo y López, Vernelle A. A. Noel

CO	NT
🎤̸ CO	

Realignments

Leave Meeting

Cancel

EN	TS
🎤̸ EN	🎤̸ TS

 Mute Stop Video Security Participants Chat Share Screen Polls Record Live Transcript Breakout Rooms Reactions More End

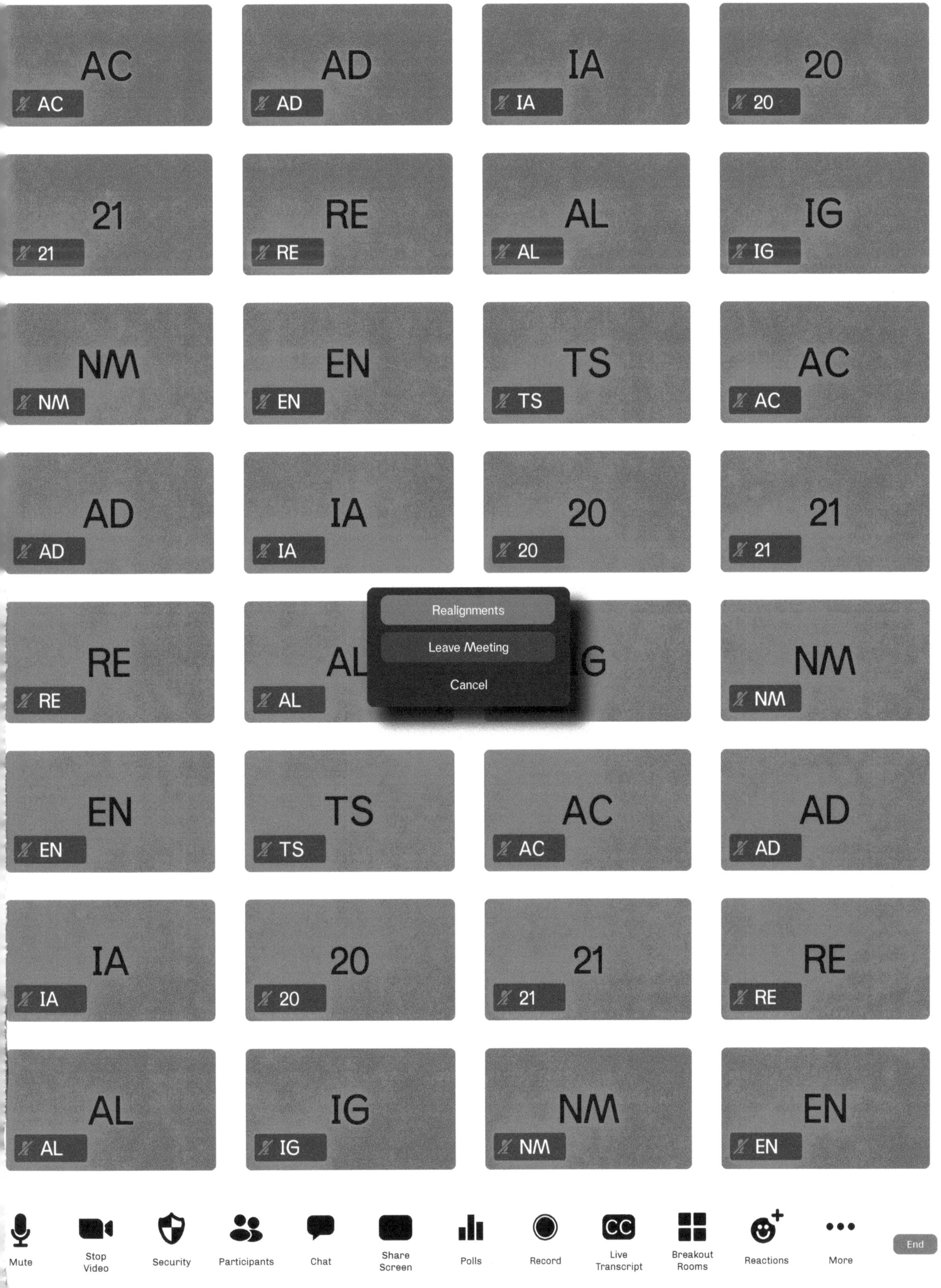

**REALIGNMENTS: TOWARD CRITICAL COMPUTATION
PROCEEDINGS OF THE 41ST ANNUAL CONFERENCE OF THE
ASSOCIATION FOR COMPUTER AIDED DESIGN IN ARCHITECTURE**

Editors
Kathrin Dörfler, Stefana Parascho, Jane Scott, Biayna Bogosian, Behnaz Farahi,
Jose Luis García del Castillo y López, June A. Grant, Vernelle A. A. Noel

Copy Editing
Gabi Sarhos

Graphic Design and Layout
Ian Besler

Printer
Ingram Spark

© Copyright 2022
Association for Computer Aided Design in Architecture (ACADIA)

All rights reserved by individual project authors who are solely responsible for their content.

No part of this work covered by copyright may be reproduced or used in any form, or by any means graphic, electronic or mechanical, including recording, taping or information storage and retrieval systems without prior permission from the copyright owner.

ISBN 979-8-9860805-6-7

INTRODUCTION

11   **Jenny E. Sabin**
Foreword: Collaborative Interfaces for Critical Computation

19   **Biayna Bogosian, Kathrin Dörfler, Behnaz Farahi, Jose Luis García del Castillo y López, June A. Grant, Vernelle A. A. Noel, Stefana Parascho, Jane Scott**
Introduction to the Proceedings

23   **Kathrin Dörfler, Stefana Parascho, Jane Scott**
Introduction to the Peer-Reviewed Papers

RESPONSES TO THE PANDEMIC

28   **Lavender Tessmer, Skylar Tibbits**
Personalized Knit Masks

38   **Mengda Wang, Ava Fatah gen. Schieck, Petros Koustsolampros**
Exploring the Role of Spatial Configuration and Human Behavior on the Spread of the Epidemic

48   **Alicia Nahmad Vazquez, Li Chen**
Automated Generation of Custom Fit PPE Inserts

58   **Zain Karsan**
TinyZ

CRTITICAL COMPUTATION

70   **Claire McAndrew, Gilles Retsin, Mollie Claypool, Clara Jaschke, Kevin Saey, Danaë Parissi**
House Block

76   **Rebecca Smith**
Passive Listening and Evidence Collection

82   **Behnaz Farahi**
Critical Computation

92   **Nate Imai, Matthew Conway**
Data Waltz

100   **Mona Ghandi, Mohamed Ismail, Marcus Blaisdell**
Parasympathy

AI-ASSISTED DESIGN

112   **Ridvan Kahraman, Zhetao Dong, Kurt Drachenberg, Katja Rinderspacher, Christoph Zechmeister, Ozgur S. Oguz, Achim Menges**
Augmenting Design

122   **Kathy Velikov, Matias del Campo, Lucas Denit, Kazi Najeeb Hasan, Ruxin Xie, Brent Boyce**
Design Engine

134   **Mikhael Johanes, Jeffrey Huang**
Deep Learning Isovist

146 **Cami Quinteros, Hesham Shawqy, Gabriella Rossi, Iliana Papadopoulou, David Leon**
Imaginary Vessels

156 **Hyojin Kwon, Adam Sherman**
Crooked Captures

DATA, BIAS, & ETHICS

160 **Shicong Cao, Hao Zheng**
A POI-Based Machine Learning Method in Predicting Health

170 **Achilleas Xydis, Nathanaël Perraudin, Romana Rust, Beverly Ann Lytle, Fabio Gramazio, Matthias Kohler**
Data-Driven Acoustic Design of Diffuse Soundfields

182 **Qi Yang, Jesus G. Cruz-Garza, Saleh Kalantari**
MindSculpt

194 **Klara Mundilova, Erik Demaine, Riccardo Foschi, Robby Kraft, Rupert Maleczek, Tomohiro Tachi**
Lotus: A Curved Folding Design Tool for Grasshopper

204 **Adam Marcus**
Arbor: Tectonic Contingencies and Ecological Engagement

MIXED REALITY

212 **David Gillespie, Zehao Qin, Francis Aish**
An Extended Reality Collaborative Design System

222 **Leslie Lok, Asbiel Samaniego, Lawson Spencer**
Timber De-Standardized

232 **Garvin Goepel, Kristof Crolla**
Augmented Feedback

238 **Hassan Anifowose, Wei Yan, Manish Dixit**
BIM LOD + Virtual Reality

246 **Nick Safley**
Reconnecting...

NEW MATERIALS/FABRICATION

258 **Edyta Augustynowicz, Maria Smigielska, Daniel Nikles, Thomas Wehrle, Heinz Wagner**
Parametric Design and Multirobotic Fabrication of Wood Facades

270 **Niccolò Dambrosio, Christoph Zechmeister, Rebeca Duque Estrada, Fabian Kannenberg, Marta Gil Pérez, Christoph Schlopschnat, Katja Rinderspacher, Jan Knippers, Achim Menges**
Maison Fibre

280 **Denitsa Koleva, Eda Özdemir, Vaia Tsiokou, Karola Dierichs**
Designing Matter

292 **Kyle Schumann, Katie MacDonald**
Pillow Forming

302 **Nancy Diniz, Frank Melendez**
Inoculated Matter

SUSTAINABLE CONSTRUCTION

308 **Gabriella Rossi, Ruxandra Chiujdea, Claudia Colmo, Chada ElAlami, Paul Nicholas, Martin Tamke, Mette Ramsgaard Thomsen**
A Material Monitoring Framework

318 **Alireza Borhani, Negar Kalantar**
Nesting Fabrication

328 **Mostafa Akbari, Yao Lu, Masoud Akbarzadeh**
From Design to the Fabrication of Shellular Funicular Structures

340 **Yu Zhang, Liz Tatarintseva, Tom Clewlow, Ed Clark, Gianni Botsford, Kristina Shea**
Mortarless Compressed Earth Block Dwellings

346   **Matthew Gordon, Roberto Vargas Calvo**
Digital Deconstruction and Design Strategies from Demolition Waste

REDEFINING SPATIAL TECTONICS

354   **Yulun Liu, Yao Lu, Masoud Akbarzadeh**
Kerf Bending and Zipper in Spatial Timber Tectonics

362   **Nicholas Bruscia**
Surface Disclination Topology in Self-Reactive Shell Structures

372   **Donghwi Kang, Nicholas Hoban, Maria Yablonina**
Discrete Quasicrystal Assembly

380   **Zhenxiang Huang, Yu-Chou Chiang, Jenny E. Sabin**
Automating Bi-Stable Auxetic Patterns for Polyhedral Surface

392   **Jason Carlow**
Al Janah Pavilion

NEW MATERIALS AND ADDITIVE PROCESSES

400   **Mackenzie Bruce, Gabrielle Clune, Ruxin Xie, Salma Mozaffari, Arash Adel**
Cocoon: 3D Printed Clay Formwork for Concrete Casting

410   **Mania Aghaei Meibodi, Ryan Craney, Wes McGee**
Robotic Pellet Extrusion: 3D Printing and Integral Computational Design

420   **Aya Shaker, Noor Khader, Lex Reiter, Ana Anton**
3D Printed Concrete Tectonics

428   **Philipp Rennen, Noor Khader, Norman Hack, Harald Kloft**
A Hybrid Additive Manufacturing Approach

438   **Ana Goidea, Mariana Popescu, David Andréen**
Meristem Wall: An Exploration of 3D-Printed Architecture

444   **Assia Crawford**
Mitochondrial Matrix

454   **Kimball Kaiser, Maryam Aljomairi**
DTS Printer: Spatial Inkjet Printing

DISTRIBUTED DESIGN AND FABRICATION

462   **Maria Yablonina, James Coleman**
Small Robots and Big Projects

470   **Grzegorz Łochnicki, Nicolas Kubail Kalousdian, Samuel Leder, Mathias Maierhofer, Dylan Wood, Achim Menges**
Co-Designing Material-Robot Construction Behaviors

480   **Tom Shaked, Amir Degani**
Shepherd

EMERGING DESIGN APPROACHES

492   **Olivia Römert, Malgorzata A. Zboinska**
Aligning the Analog, Digital, and Hyperreal

502   **Anna Mytcul**
ARchitect

512   **Zidong Liu**
Topological Networks Using a Sequential Method

520   **Katarina Richter-Lunn, Jose L. García del Castillo y López**
Affective Prosthesis

530   **Arash Adel, Edyta Augustynowicz, Thomas Wehrle**
Robotic Timber Construction

FIELD NOTES

540   **Shelby Doyle, Nick Senske**
Computational Access

546 **Cyle King, Jacob Gasper**
Process / Product

554 **Zain Karsan**
IN HOUSE: A Remote Making Studio

564 **Emily Pellicano, Carlo Sturken**
GPT–OA; Generative Pretrained Treatise–On Architecture

572 **Ricardo Cesar Rodrigues, Fábio A Alzate-Martinez, Daniel Escobar, Mayur Mistry**
Rendering Conceptual Design Ideas with Artificial Intelligence

KEYNOTES

578 **Amelia Jones, Lesley-Ann Noel, Krzysztof Wodiczko, Virginia San Fratello, Ronald Rael**
Critical Computation: Participation, Intersectionality & Emancipatory Design

586 **Benjamin Bratton, Sarah Williams, Lauren Lee McCarthy, Caitlin Mueller**
Designing with AI, Data, Bias & Ethics

594 **Paola Antonelli, Justin Garrett Moore, Lydia Kallipoliti, Mariana Popescu**
Design Imperatives in Social & Environmental Crisis

600 **Marta Novak, Dori Tunstall, Charlotte Materre-Barthes, Jenny Sabin**
Emerging Trends in Response to Critical Computation

AWARDS

608 **Axel Kilian**
Teaching Award of Excellence

612 **Alvin Huang**
Digital Practice Award of Excellence

616 **Wolf dPrix**
Lifetime Achievement Award

622 **Brian Slocum**
Society Award for Leadership

WORKSHOPS

632 **Daniel Bolojan, Shermeen Yousif, Emmanouil Vermisso**
Latent Morphologies: Disentangling Design Spaces

633 **Sigrid Brell-Cokcan, Johannes Braumann, Karl Singline, Sven Stumm, Etahn Kerber**
Distributed Collaborations - KUKAcrc Cloud Remote Control

634 **Chien-hua Huang, Zach Beale**
Collaborative AI – Human + AI Form

635 **Jeffrery Anderson, Ahmad Tabbakah**
Augmented Architectural Details

636 **Galo Canizares**
Building Web-Based Drawing Instruments

637 **Özgüç Bertuğ Çapunaman, Benay Gürsoy, Cemal Koray Bingöl**
Co-Crochet Computing Stitches for Collective and Distributed Crocheting

638 **Jonathan Dessi-Olive, Omid Oliyan, Ali Seyedahmadian**
Enhancing Fungi-Based Composite Materials With Computational Design and Robotic Fabrication

639 **Mariana Popescu, Robin Oval**
Knitted Growth: Scaffolds for Living Root Spans

640 **Daniela Atencio, Nicolas Turchi**
Physics Towards Critical Assemblies

641 **Runjia Tian, Zhaoyang Luos, Linhai Shen**
The Generative Game

642   **Rafael Pastrana, Isabel Moreira de Oliveira, Patrick Ole Ohlbrock, Pierluigi D'Acunto**
Form-Finding Staircases with COMPAS CEM

643   **Stefana Parascho, Edvard P.G. Bruun, Gonzalo Casas, Beverly Lytle**
Remote Robotic Assemblies

644   **Christoph Becker, Lilli Smith, Zach Kron**
Generative Design and Analysis in Early-Stage Planning with Spacemaker

CULTURAL HISTORY PROJECT

650   **Shelby Elizabeth Doyle, Melissa Goldman, Biayna Bogosian**
Reflections on the Past 40+ Years of ACADIA

658   **Philip Beesley, Karen Kensek, Branko Kolarevic, Mahesh Daas, Nancy Cheng, Jason Kelly Johnson, Kathy Velikov**
40th Anniversary Toast

ACADIA CREDITS

666   **Conference Chairs**
668   **Session Chairs**
673   **ACADIA Organization**
675   **Conference Management**
677   **Peer Review Committee**
681   **ACADIA 2021 Sponsors**

FO	RE
🔇 FO	

Realignments
Leave Meeting
Cancel

WO	RD
🔇 WO	🔇 RD

 Mute Stop Video Security Participants Chat Share Screen Polls Record Live Transcript Breakout Rooms Reactions More End

Collaborative Interfaces for Critical Computation

Jenny E. Sabin

ACADIA President 2021-2022
Arthur L. and Isabel B. Wiesenberger Professor in Architecture, Cornell University
Associate Dean for Design Initiatives, College of Architecture, Art, and Planning
Director, Sabin Lab
Principal, Jenny Sabin Studio

Realignments: Toward Critical Computation, ACADIA's 2021 conference marking the 40th anniversary of the ACADIA organization, emerged from the extraordinary and ongoing systemic challenges posed by the COVID-19 pandemic, climate catastrophe, and social justice crises. Since early March of 2020, our lives have been transformed in the ways that we work, live, rest, and engage with our communities. Our conception of "normal" has been altered, where time and space often conflate. For some of us, we have been able to go back to our labs, studios, classrooms, and offices, masked and socially distanced, while others have been confined to home offices and online platforms such as Zoom. While vaccination has brought relief, safety, and some sense of normalcy, we remain in a state of constant preparedness as new variants present new challenges. Although we all hoped to host an in-person ACADIA conference in 2021, it quickly became apparent that this would not be possible. As ACADIA's primary activity is the exchange of knowledge through an annual conference (mandated in the organization's by-laws), in March 2021 we realized we could not hold an in-person conference and thus postponed our planned conference. The ACADIA Board of Directors, while all facing their own uncertainties in their work and life, rallied together to re-envision a new conference theme and call for a second fully virtual and ACADIA Board-run conference. We realized that we could build upon the important topics raised in the 2020 ACADIA conference, recharge the kinds of discussions we have at ACADIA, and expand the ACADIA community.

In that uncertain time, eight individuals from the Board stepped up and agreed to take on the incredible amount of work to plan, organize, and run the 2021 conference. On behalf of the ACADIA community, I would like to sincerely thank Biayna Bogosian, Kathrin Dörfler, Behnaz Farahi, June Grant, Jose Luis Garcia del Castillio y López, Vernelle A. A. Noel, Stefana Parascho, and Jane Scott for volunteering their time to take on the incredible amount of work that it took to critically conceptualize, organize, and run ACADIA's 41st conference event, the *2021 Realignments: Toward Critical Computation* conference. In addition to the team, I want to recognize and thank Cameron Nelson for their dedication and tireless efforts as conference manager. Starting with the incredibly successful international workshops in September, followed by four days of presentations and powerful discussions, the conference is a testament to the resilience, energy, collaborative spirit, optimism, and commitment of these extraordinary individuals. In this conference the co-chairs raised our ambitions not only in terms of critical content, but also the conference organization. I would also like to sincerely thank ACADIA's former President and current Vice President, Kathy Velikov, for her dedication and guidance to the organization of this year's conference. Thank you also to the entire Board for stepping up and leading new initiatives and taking on the workload of committee work that this conference brought—from jurying the submission formats, to reviewing and awarding the new grants and paper awards programs. This was also an ambitious conference in terms of programming—it has been an immense organizational effort, and each session and keynote panel have been excellent, engaging, and so enriching and impactful.

Last year, the conference marked an important shift that this year's conference continues. A critical, thought-provoking, and inspiring conference appropriately titled *Distributed Proximities*, challenged us all to deepen our thinking, creative practices and engagement with ecology and ethics; access and bias as it pertains to algorithms and big data; automation, agency, labor and practice; computational design and issues of health and wellness; and new forms of collaboration to identify next steps and to in the conference chairs' words, highlight and reflect upon the "…resilience and ingenuity of the computational design community in the face of crisis." This year also marked the 40th anniversary of the ACADIA organization. ACADIA was founded in 1981 by some of the pioneers in the field of design computation including Bill Mitchell, Chuck Eastman, and Chris Yessios. A 40th Anniversary event took place at the conclusion of the conference to reflect upon and celebrate the 40+ years of ACADIA's history. Tremendous thanks go to Board members Shelby Doyle and Melissa Goldman for all their efforts towards this initiative, which included taking on the daunting task of amassing the entire archive of ACADIA proceedings and making them free and accessible to all on the ACADIA website.

Realignments: Toward Critical Computation continues to build upon and add to the important critical topics and questions that the 2020 conference raised, introducing conversations, people, and perspectives in the keynotes as a new and important trajectory to the ACADIA conference format. In the first keynote panel, Amelia Jones underscored how neutrality is a problematic concept in design and that the whiteness of tech and algorithm design can perpetuate systemic racism and sexism as it frequently does not prioritize diverse users. She challenged us to consider how design can be driven by issues of access, inclusivity, and equity. To do this, she offered that we need relational design strategies that foreground relationality and participation and to engage and acknowledge the embodiment of the receiver and to refuse the neutrality that has dominated art and design in the Global North. Instead, we must embrace connectedness and the engagement of broader communities. She offered some explicit methodologies for doing so, including cybernetics and generative strategies that are imbued with feedback loops where the user and interaction are part of the design process. Dr. Lesley-Ann Noel challenged us to center the voices of those who are directly impacted by the outcomes of the design process. We must redistribute and share power through co- and collective design.

In the second keynote panel, Benjamin Bratton argued that AI is used for both smart and stupid things, and its ultimate purposes may be largely undiscovered. Further, Bratton underscored the presence of a "sensing layer" and its promise to act back upon and govern itself. The pandemic has many failed areas within the sensing layer. For example, to be excluded from testing and therefore excluded from being sensed is to not count, which equals to exclusion. Sara Williams also argued for co-design and data-action, using data for the public good and to consider data for empowerment.

And in the third keynote panel, Paola Antonelli called for a more restorative attitude in design. She asked us to consider empathetic design and how it can help us be better communicators. Antonelli stated that "Design, at its best, is imagination." In contrast to this statement, Justin Garrett Moore described a manifest of a ship carrying enslaved people, in its meticulous calculation and representation, as imagination unchecked.

In our fourth keynote panel, Marta Nowak, Dori Tunstall, and Charlotte Malterre-Barthes presented critical presentations highlighting entanglements of technology, power, and society with reflections on future trajectories for critical computation in design education, applied research, and impactful practice. Dori Tunstall's work challenges us to decolonize design through the development of a better underlining consciousness with technology and the question, "What does it mean to bring an indigenousness consciousness to technology?" Tunstall argues that the values of design and tech have been and are still largely colonial, white supremacist, and patriarchal. Marta Nowak offers methods and tools for a more empathetic paradigm between humans and machines, one that considers the body as site through the addition of prosthetics to extend the capabilities of humans and our interactions with the environment and architecture. And finally, Charlotte Malterre-Barthes calls for a global moratorium on new construction and extractive processes and economies, citing the racial and social damage that exploitation brings. She argues for an architecture without extraction, one that promotes a labor of care where design offices pivot towards emancipated practices and consider technology as a caring friend.

We hope that now that we've opened the doors to these important questions and topics that these conversations persist, grow, and continue to contribute to the richness of the ACADIA community. The conference theme inspired high quality and exciting peer reviewed papers and projects and positive response to the new submission initiatives such as the Field Notes. Field Notes, which was a category that was introduced by the ACADIA 2020 Chairs, provides an opportunity to look "under the hood" at the work that we do, and from a different, and perhaps more personal and humanized perspective. The conference hosted 13 workshops, received 160 paper submissions, 42 projects, and 13 Field Notes. Thank you to this year's scientific committee composed of 171 peer reviewers. Our scientific committee ensures that the quality and caliber of the work at ACADIA is high and rigorous. This year our acceptance rate for the papers was 26 per cent and was 28 per cent for the projects. Thank you to the conference technical chairs, Jane Scott, Kathrin Dörfler, and Stefana Parascho for running such a rigorous peer review process. The technical submissions resulted in twelve sessions. We would also like to thank the session chairs who built on the conference theme and facilitated rich conversations with the authors.

As a non-profit, ACADIA's activities are only supported by membership, conference registration, and donations. For the second year, we wanted to radically reduce the cost of conference registration, and so we have depended on our sponsors for their generous financial support. Thank you to our Platinum Sponsor, Autodesk, whose donation provided free registration to all students, including doctoral students; Gold Sponsors, Facebook and Zaha Hadid Architects; our Silver sponsor, Grimshaw; our Bronze sponsors, Aruklura, ABB, as well as HKS Line, Architect Machines, and Oro editions. And a special thanks goes out to our development officer Matias del Campo for working with all our sponsors and to our communications officer Melissa Goldman and vice officer Shelby

Construction waste…

Big data…

Migration…

Metals smelting and processing…

The Great Pacific Garbage Patch…

New forms of communication…

Augmented reality and artificial intelligence…

Extreme environments…

Urban air quality…

Catastrophic natural disasters…

Climate change…

And global atrocities experienced right now and in the past…

… All of these complex topics and more demand that we embrace change, diversify and expand our community, and promote collaborative interfaces across disciplines and practice.

> The conference hosted 13 workshops...

> Received 160 paper submissions...

> 42 projects...

> And 13 field notes...

> Over 897 students from all over the world have been able to register for the conference for free...

> 58 students have been able to attend the workshops for free...

> ACADIA awarded 5 exceptional individuals...

> Our typical in-person conferences attract 300-350 participants...

> This year, we had 1082 people register and this number continues to increase as people register to access the online content asynchronously...

> This year, of attendees who chose to report their gender, we saw a significant increase to 45.47 per cent female attendance...

> Typically, 70-75 per cent of our registrants are from the US, Canada, and Mexico...

> This year, attendance outside of North America has gone from 25 to 44 per cent and we are seeing a much higher percentage of global registrants...

Doyle for the media outreach and management. Thanks to our sponsors, and in particular our longtime support from Autodesk, this year we were able to offer a third year of new programming of scholarships and awards. We worked closely with Autodesk to direct their funding toward increasing accessibility and diversity at the conference. I would also like to thank our Diversity Committee chaired by June Grant and our Vice President, Kathy Velikov, for their important and hard work on these outreach efforts.

Over 897 students from all over the world have been able to register for the conference for free, which demonstrates ACADIA's commitment to democratizing knowledge and access. The sponsorship from Autodesk also enabled us to initiate a series of more targeted grants and awards for a second year. We initiated a special grant for students from Mexico and for international students to attend the workshops. In 2020, we also initiated a partnership with NOMA—the National Organization of Minority Architects. And thanks to the work of June Grant, one of our newly elected Board members and President of the San Francisco NOMA Chapter, Acadia and Autodesk were able to award grants to attend the workshops for NOMA students and professionals, as well as conference registration for NOMA professionals and academics; 58 students have been able to attend the workshops for free. Thank you to the ACADIA scholarships committee chaired by Jane Scott with Vernelle Noel, Melissa Goldman, and Viola Ago and to the conference co-chairs for helping to manage and administer all these grants. ACADIA and Autodesk also continued the second year of a computational design sub-award to the NOMA annual awards program, and this year we are happy to announce that the recipients of the NOMAS students award went to the to the Savannah College of Art and Design and the University of Wisconsin, Milwaukee. We are looking forward to continuing to strengthen our partnership with NOMA in the coming years and to new collaborations that will occur between our communities.

The ACADIA 2021 Awards program highlights extraordinary individuals and exceptional research and projects through paper awards. This year, ACADIA awarded 5 awards to 5 exceptional individuals. Their outstanding talks are highlighted in these proceedings and available for viewing online to registered participants. The awards are Lifetime Achievement Award to Wolf dPrix, Innovative

Research Award of Excellence to Caitlin Mueller, Society Award for Leadership to Brian Slocum, Digital Practice Award of Excellence to Alvin Huang, and Teaching Award of Excellence to Axel Kilian. I want to thank the awards committee chaired by Maria Yablonina along with Biayna Bogosian and Kory Bieg for their leadership and effort towards the ACADIA awards process. This year, we also initiated new paper awards. The ACADIA Annual Conference Awards for peer-review submissions have three categories: Best Paper, Best Project, and the Vanguard Award (for paper submissions). A selected jury, composed of individuals from the Technical Committee, the Scientific Committee, and members of the ACADIA community, deliberated on the award selections. Thank you to the scientific committee chaired by Tsz Yan Ng for their hard work in coordinating the awards process.

ACADIA continued year two of a new initiative and collaboration with our sibling CAAD organizations that we launched last year. Working with the Presidents and colleagues of our sibling CAAD organizations we organized the second World CAAD PhD workshop. The aim of the World CAAD PhD Workshop is to introduce junior researchers at the PhD stage of different schools and different research cultures within the global CAAD community. The workshop offers students an opportunity to receive constructive feedback from prominent researchers and academics of the CAAD community and provides students with an occasion to position their research within the global CAAD research arena. Each sibling organization is represented at the workshop by three PhD student delegates, who were chosen through a competitive submission process, as well as experts from each community. This year, we also held an ACADIA 2021 PhD Workshop at the beginning of the conference. Thank you to the 2021 ACADIA delegates and to Board member Jane Scott for spearheading and coordinating these important efforts.

Lastly, we are excited to announce a new outreach program called the ACADIA Mentorship Program. This program will pair established mentors of the ACADIA community with mentees aspiring to publish their first paper at ACADIA and is specifically aimed at increasing access and diversifying our community of authors. Thank you to Board member Christoph Klemmt and the outreach committee for spearheading this important initiative.

The 2021 conference built upon last year's successful conference in unprecedented ways, including increasing attendance and international reach, which has positively impacted gender parity. Our typical in-person conferences attract 300-350 participants. This year, we had 1082 people register and this number continues to increase as people register to access the online content asynchronously. We also saw a strong interest from the student community with 897 student registrations. For a second year in a row, the open and accessible format of the conference has shifted the gender balance of our attendees, where typically female attendees have remained at 29 to 30 per cent in recent years. This year, of attendees who chose to report their gender, we saw a significant increase to 45.47 per cent female attendance. Given the online format of the conference, we have also seen the balance between North American and international attendees shift. Typically, 70-75 per cent of our registrants are from the U.S., Canada, and Mexico. This year, attendance outside of North America has gone from 25 to 44 per cent and we are seeing a much higher percentage of global registrants, including some new parts of the world in our top 15 registrant locations, such as Cairo, Tehran, and Tel Aviv.

As we celebrate the 40th anniversary of the ACADIA organization, it is also an important time to reflect, celebrate, critically assess, and generate new collaborative interfaces for critical computation. From the first decade of the organization, when there was a strong focus on pedagogy, CAAD tools, and design computing where topics such as "Introducing Computer-Aided Design into the Architecture Curriculum" framed the conference proceedings and discourse in the field, to now, ACADIA has maintained a commitment to evolve and at the same time maintain a foundation of rigor and excellence. A beleaguered planet still in the depths of a pandemic and an urgent climate emergency demands that computational design as a field and practice cultivate new collaborative models to comprehend key social, environmental, and technological issues. To begin to make necessary change, we need stronger integration and collaboration between research and practice, integration, and realignments that are reflected in the ACADIA 2021 conference. Construction waste, big data, migration, metals smelting and processing, The Great Pacific Garbage Patch, new forms of communication, augmented reality and artificial intelligence, extreme environments, urban air quality, catastrophic natural disasters, climate change, and global atrocities experienced right now and in the past: all of these complex topics and more demand that we embrace change, diversify and expand our community, and promote collaborative interfaces across disciplines and practice. At the same time, the COVID-19 crisis is forcing a reconsideration of workspaces, residences, and other occupied structures while also revealing ongoing systemic racism and persistent issues impacting diversity, equity, and inclusion. The diverse, experimental, and innovative array of ACADIA proceedings this year through papers and projects, keynote panels, field notes, and beyond reflect the complexity and urgency of our time and that we must come together to address these crises with radical new models for design research and collaboration across disciplines, alternative forms of practice, and communities. Our work has only just begun.

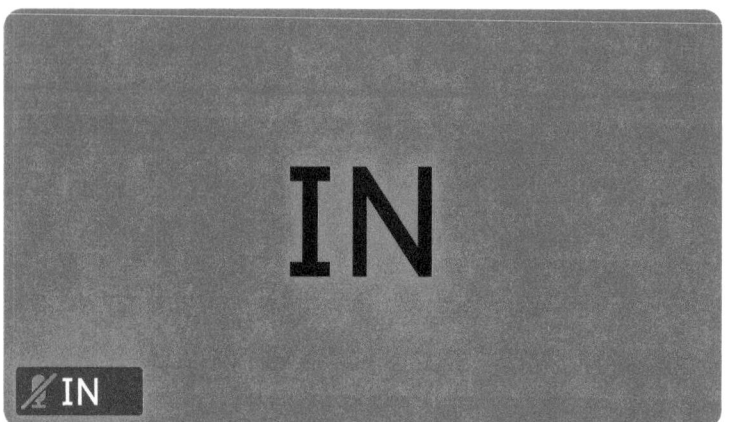

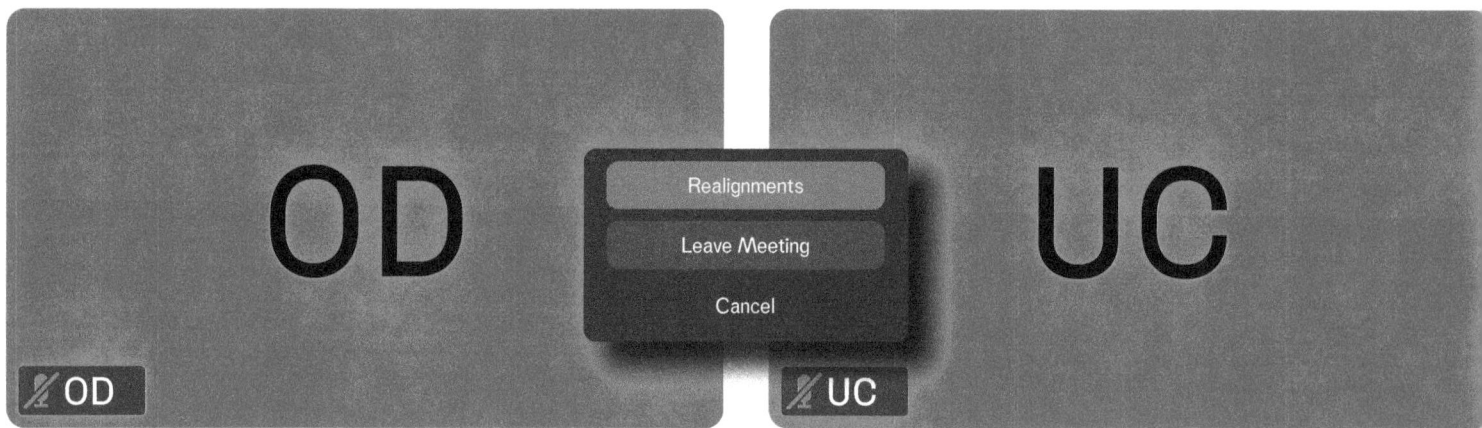
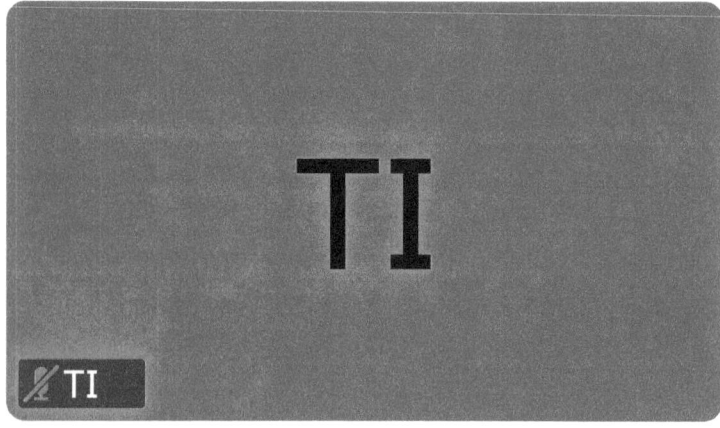

Introduction to the Proceedings of the 41ˢᵗ Annual ACADIA Conference

Biayna Bogosian
Florida International University

Kathrin Dörfler
Technical University of Munich

Behnaz Farahi
California State University Long Beach

Jose Luis García del Castillo y López
Harvard Graduate School of Design

June A. Grant
Design Principal, blinkLAB architecture

Vernelle A. A. Noel
Georgia Tech

Stefana Parascho
École Polytechnique Fédérale de Lausanne

Jane Scott
Newcastle University

At the convergence of social, political and ecological crises and a global pandemic, ACADIA 2021 took place as an online conference for the second time in a row. Entitled "Realignments: Toward Critical Computation", the goal of ACADIA 2021 was therefore to follow up on the issues raised in 2020, back then at the dawn of the pandemic, and to reflect on how the community continues to respond to the global crises.

As designers and researchers engaged in practices around computational systems we aim for creative solutions and emphasize the innovations that benefit societies and demonstrate the ingenuity of the design community. However, left unchecked, these systems we rely on can also exacerbate issues of inequality, bias, access and perpetuate methods and histories that may harm rather than foster positive change. ACADIA 2021 therefore considered how to realign our practices to allow for alternative and constructive ways of knowledge and world making.

With these entanglements of technology, power, and society as a backdrop, *ACADIA 2021 Realignments: Toward Critical Computation*, asked us to question our current practices and priorities to address the urgency of the now. This conference provided a platform to engage with conversations, tools and methodologies that included knowledge and communities currently missing to enable realignments toward inclusive and critical practices in architecture across different scales. How can the computational design community critically address questions of emancipation, intersectionality and our computational publics?

Whilst the online format creates its own barriers, limiting attention spans, exacerbating fatigue, and enforcing distance and formality, moving online has enabled the ACADIA community to extend its international reach. TThe online format has enabled us to begin breaking down hierarchies established by barriers like cost, time, and distance. The format allowed us to redefine outreach, and empower a new generation of researchers and designers.

The online conference included authors' presentations of technical papers, projects and field notes, alongside keynote conversations, award presentations and workshops.

This year's series of keynote conversations organized around specific prompts related to the conference theme *REALIGNMENTS: Toward Critical Computation*. The format aimed to encourage dialog, discussion, and debate around topics that span critical computation to practices and pedagogies of design computation, including: Critical Computation, Ecology, Environmental Crisis, AI, Data, Bias, & Ethics, Culture & Access, Labor & Practice, and Speculation & Critique.

As every year, the conference included the ACADIA Awards of Excellence, which recognizes consistent contributions and impact on the field of architectural computing. These include: Teaching, Digital Practice, Innovative Research, Lifetime Achievement and Society awards.

As part of the online conference, experienced instructors from around the world offered and led a series of online workshops. We curated these workshops to uniquely address the conference's theme of distributed modes of inclusion and collaboration.

A new initiative successfully introduced as part of the ACADIA 2021 conference was a dedicated PhD workshop set up to support early-stage researchers in the development of their doctoral research. This workshop was an opportunity for doctoral students to present and discuss their dissertation proposals in the field of computational design, fabrication, and construction in the built environment. The workshop aimed to promote interaction between doctoral students and specialists in the field to support collaboration and network development.

Through this year's proceedings, *ACADIA 2021 Realignments: Toward Critical Computation* captures the diversity and ambition of the ACADIA community at this critical moment in time. It builds on urgent discussions initiated in 2020, which highlighted how researchers and practitioners can come together to investigate new models and modes of research that will lead to a critical understanding of computation.

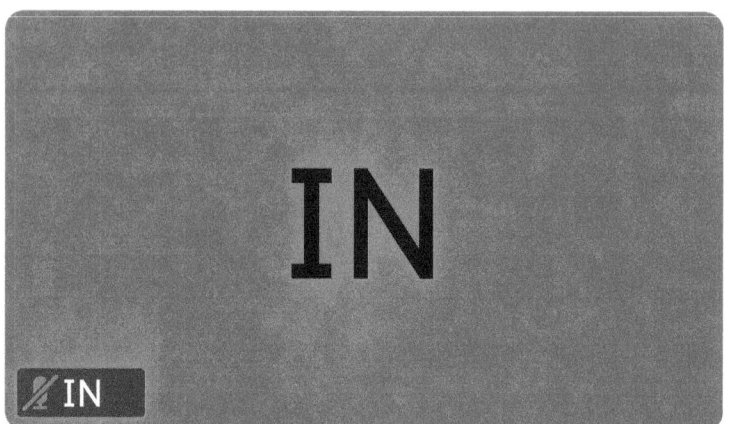

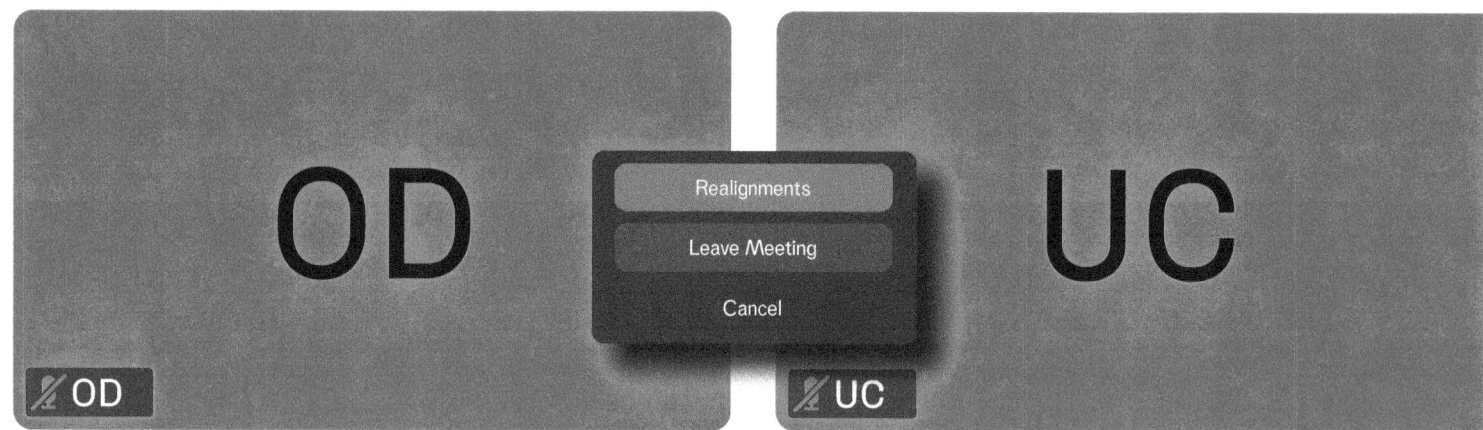
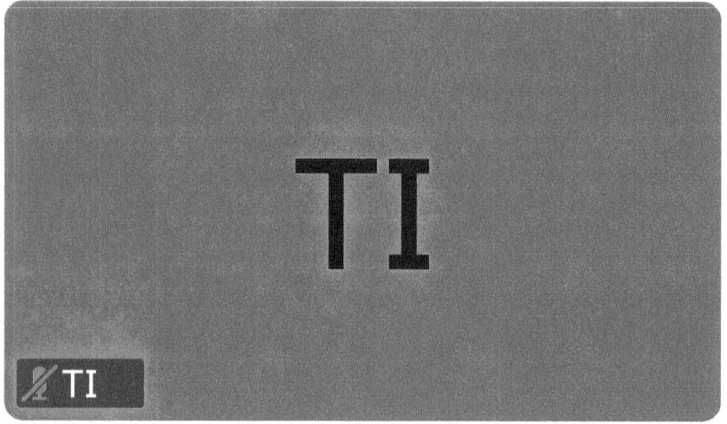

Introduction to the Peer-Reviewed Papers of the 41st Annual ACADIA Conference

Kathrin Dörfler
Technical University of Munich

Stefana Parascho
École Polytechnique Fédérale de Lausanne

Jane Scott
Newcastle University

This year's Call for Papers, Projects and Field Notes aimed at encouraging discussions leading to a critical examination of our field, as well as disseminating technical research results. The call took the occasion of ACADIA's 40th Anniversary to invite contributions of recent and emerging work in computational design, fabrication, and construction in the built environment. We particularly encouraged conceptualizations of emancipatory computation, computational diversity, and experiments in collaboration, participation, and empowerment. As such, we sought contributions with resonance across technical tools and conceptual approaches, inviting submissions of recent research, practice, and theory that aligned with but were not limited to integrations of the following approaches and methods and tools:

Approaches:

- Critical Computation, Participatory, and Emancipatory Design
- Computational Publics
- Intersectionality in Computational Design
- Interdisciplinarity, and Experiments in Collaboration
- Data, Bias, and Ethics
- Politics of Access
- Environmental Crisis
- Architectures of Care
- Design Emergency
- Cities, Health, and Policies

Methods and Tools:

- Artificial Intelligence, Machine Learning, and Statistical Approaches
- Robotic Construction, and Operations
- Distributed Fabrication, and Inter/Intra Operability
- Adaptation, Responsiveness, and Interactivity
- New Materials, and Construction Practices
- Spatial Computing, Sensing, and VR/AR/MR
- Environmental, and Regenerative Design
- Low vs. High Tech, and Analog vs. Digital
- Emerging Pedagogies

We intentionally stepped back from leading the conversation or suggesting answers and directions through top down dissemination, but offered the conference participants and paper authors a way of engaging in discussions beyond their specific paper contents so that new questions, directions and topics could emerge. In order to achieve this, the sessions were framed around both technical and critical themes, and with session chairs carefully chosen to promote discussion beyond the technicalities of their contributions.

The paper sessions were grouped thematically to give a sense of the underlying narrative that emerged from the papers submitted. It is a testament to the responsiveness of our community that this proceedings begins with the session Responses to the Pandemic, introducing four outstanding contributions that sought to engage directly with the immediate circumstances of COVID-19. Whilst the pandemic is made explicit within the contributions of this first session, the subtext of remote working, reconfigured studios and new distributed networks is implicit in the methods and approaches across many of this year's Technical Papers. With the intention of finding a dedicated space between theory-informed practice and practice building theory, Session 2 initiated a dialogue of Critical Computation. Of particular note were questions of surveillance, data and power, suggesting new narratives for exploration as well as the challenges of the current systems. While some of the themes return every year such as New Materials and Fabrication, Sustainable Construction, Mixed Reality and AI Assisted Design, newly introduced themes concerned Redefining Spatial Tectonics, New Materials and Additive Processes and Emerging Design Approaches. In addition, the conversations that were initiated at ACADIA 2021 Distributed Proximities have continued to resonate with the community, exemplified by sessions focused on Data, Bias and Ethics and Distributed Design and Fabrication and a dedicated Field Notes session.

With Realignments: Toward Critical Computation, we wanted to provide an opportunity for both critical and technical perspectives to engage in dialogue, and to explore how the mix and match of different tools and approaches can advance and transform research in our field.

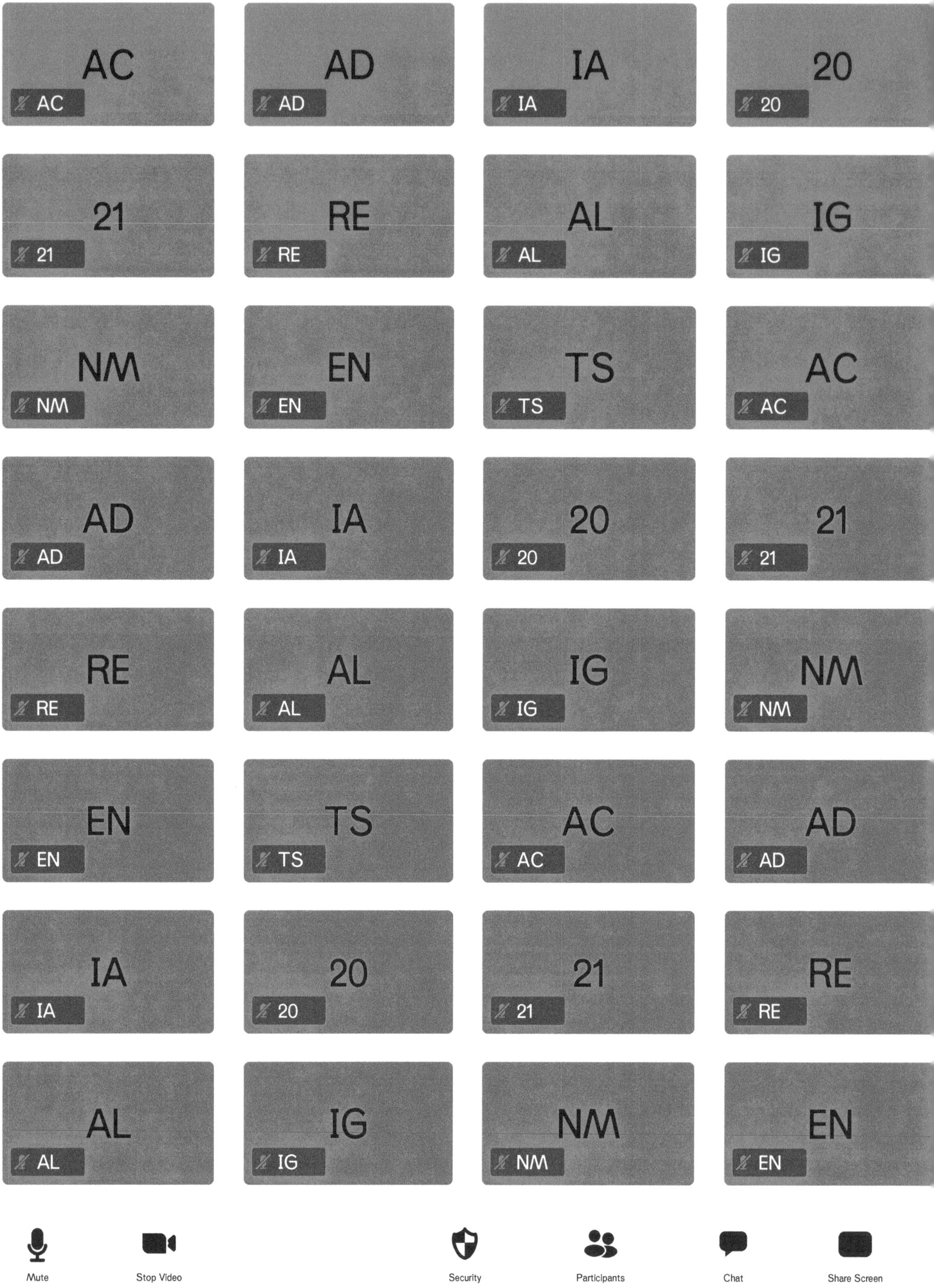

AC	AD	IA	20
21	RE	AL	IG
NM	EN	TS	AC
AD	IA	20	21
RE	AL	IG	NM
EN	TS	AC	AD
IA	20	21	RE
AL	IG	NM	EN

 Mute Stop Video Security Participants Chat Share Screen

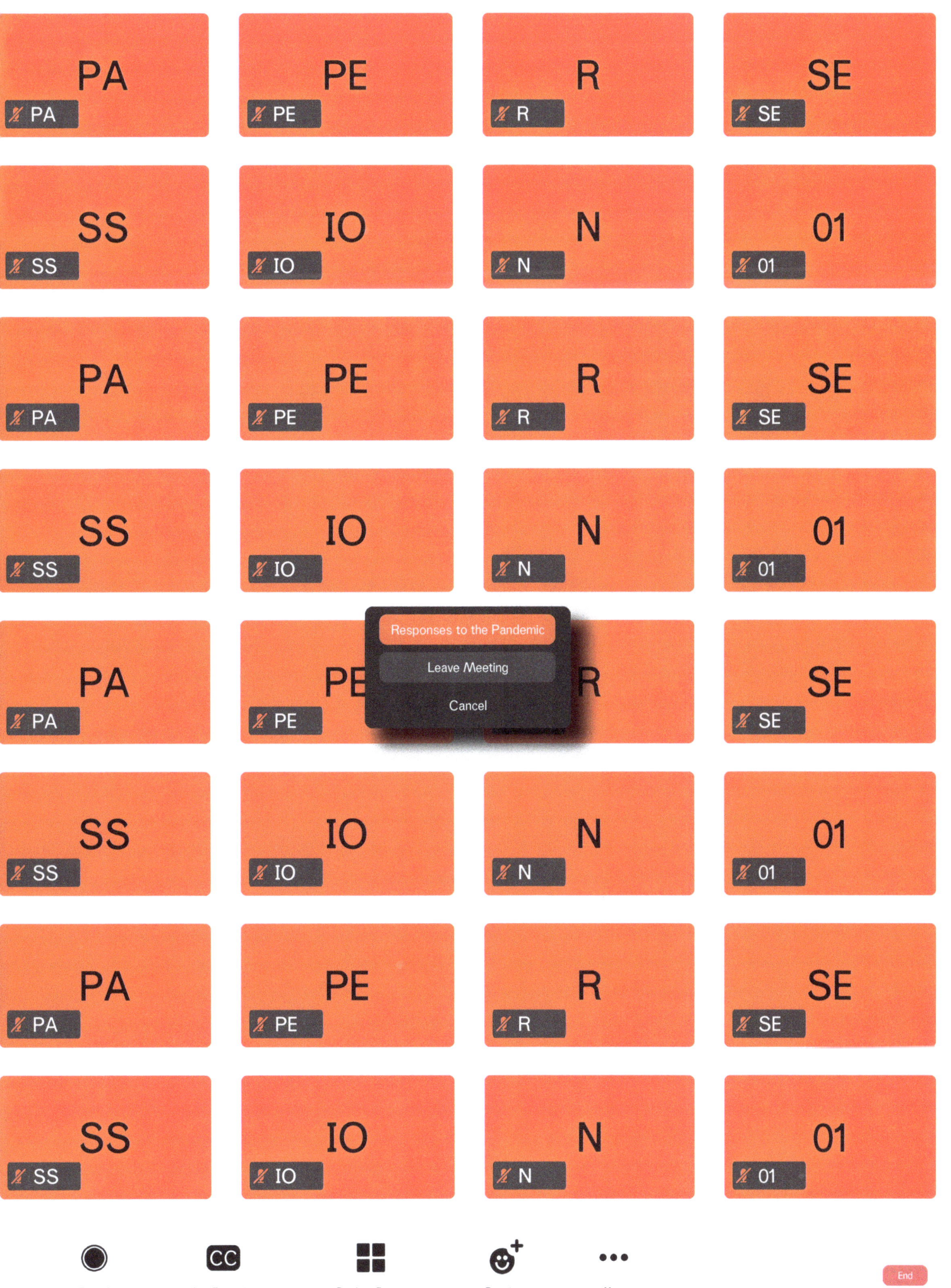

Personalized Knit Masks

Lavender Tessmer
Massachusetts
Institute of Technology

Skylar Tibbits
Massachusetts
Institute of Technology

Programmable Shape Change for Customized Fit

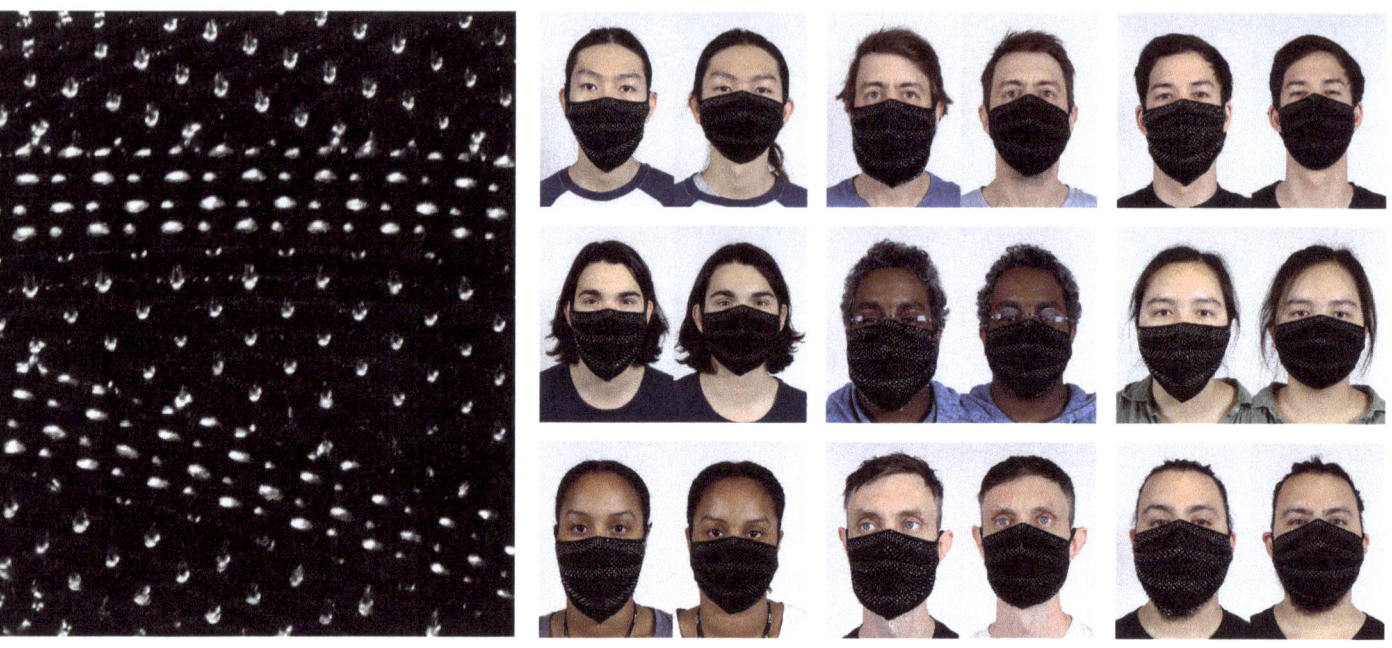

1 Knit fabric with zones of temperature-responsive fibers (left); knit mask, before and after the personalized robotic tailoring process (right)

ABSTRACT

In this paper we outline a new workflow for textile customization through the design and fabrication of knit shape-changing masks that contain multi-material fibers to create programmable transformation. We have created a process for producing standardized and scalable textile goods using a flatbed industrial CNC knitting machine, which are then "tailored" to an individual's body measurements through a system of programmable textiles, custom multi-material fiber, and robotic heat activation. Hybridizing the efficiency of standardized textile production with unique geometric variation, the proposed strategy centers on the shape-change behavior of fibers and precise knit structures to produce personalized textiles. This work focuses on the face mask as an example of a now-ubiquitous textile good that is often ill fitting and yet can now be highly tailored to an individual's personal fit and comfort. This paper outlines the materials, knit fabric development, mask design, digital workflow, and fabrication steps for producing truly customized masks for an individual's unique facial geometry.

INTRODUCTION

Mass-customization often comes at the expense of scalability and efficiency. Existing workflows for mass customization often rely on design-to-machine processes to produce parts that are geometrically unique, resulting in increased complexity. A customized production approach is especially relevant in the apparel industry offering the potential for better fit, and more comfortable and individualized clothing. In particular, industrial knitting machines have significant potential as a production tool for customizable garments due to the highly reconfigurable CNC control of fabric shapes and patterns. However, there are significant challenges for producing on-demand customized knit geometries that relate to the complexity of material behavior and digital workflows for machine output.

Current explorations in mass-customized garments and accessories in the apparel industry have focused almost exclusively on producing customizable garment patterns in software that can be sent directly to a textile machine or garment construction process (Apeagyei and Otieno 2007). With a similar strategy, computer scientists have made progress in developing these automated workflows for on-demand production of knit geometries, but these approaches often acknowledge numerous barriers to realizing a scalable and seamless process for generating unique and dimensionally accurate pattern shapes (Narayanan et al. 2018; Wu, Swan, and Yuksel 2019). This software-based approach to customization has significant challenges due to the low dimensional stability of knit fabric, the inability of the automated process to adapt to the nuances of material, the difficulty of predicting the behavior of a textile, as well as the interaction of forces applied during knitting and its effect on the final geometry.

This paper explores a new process for garment customization that does not rely solely on software-based pattern generation, but instead focuses on material properties and tailored activation, demonstrating an alternative workflow for customization and production. Balancing repeatability with the potential for variety, the proposed solution leverages the efficiency of scalable garment production using industrial knitting machines coupled with programmable material fibers to produce fully customized objects through temperature activation (Figure 1a). To test this process, we have fabricated a set of standardized knit masks that are pre-programmed with the ability to change shape for a personalized fit (Figure 1b). The customization of fit is achieved using heat post-processing with a robotic toolpath that is translated from each wearer's individualized face measurements (Figure 2). This approach allows for the full speed and efficiency of industrial knitting machines and standardized files that are widely used in the apparel industry total, while still enabling a fully tailored and personalized textile product through a post-process material transformation.

As many people adapt to the new and widespread wearable accessory of face masks, it has become clear that masks often do not fit correctly or cannot accommodate the wide range of face sizes and shapes. Often ill-fitting and uncomfortable, mass-produced mask designs highlight a forced standardization in contrast to the diversity of individual human shape and proportion. Our approach to creating shape-changing customized masks presents a unique opportunity to develop and evaluate new methods of producing personalized items.

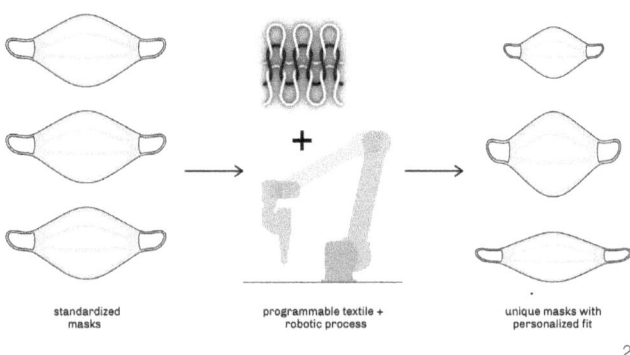

This project builds on previous research in the Self-Assembly Lab at the Massachusetts Institute of Technology. This work includes 4D printing that uses contrasting material properties along with intricate 3-dimensional geometry to promote precise shape-transformation in a printed part when subjected to moisture (Papadopoulou, Laucks, and

2   Concept: robotic process and programmable textile are used to transform standardized masks into personalized masks.

Tibbits 2017). More recent work translates these strategies to textiles, demonstrating a proof of concept for a heat-activated "robotic tailoring" technique for a multi-material knit sweater and a climate-adaptive garment that changes porosity and thickness based on the environment (Tessmer et al. 2019). Continuing this research, this project aims to further develop the earlier projects with a new computational workflow to link personal fit measurements to precise shape change and transformation.

BACKGROUND

This work aims to synthesize a number of different research areas into a novel strategy that combines apparel design, knit fabric, custom-made fibers, mask design, and robotic fabrication. Among these areas are manufacturing workflows for mass-customizable garments, industrial knitting, the behavior of knit structures, material research in heat-responsive fibers, fabric heat-setting techniques, and customizable robotic toolpaths.

In the area of fashion apparel, researchers have identified methods of evaluating fit variables for customized garments, as well as verified consumer desire for personalized fit (Bellemare 2018). Other work in this area has established a production process that works between customer size data, garment design, and knitting production. Through this work, researchers have identified a need for addressing the dimensional stability of knit fabrics in the custom-manufacturing workflow by incorporating a thermosetting process that adapts to the dimensions of the individual input data (Buecher et al. 2018). Likewise, our workflow incorporates a heat post-processing technique with individual size data and industrial flat-bed knitting. However, we expand on this approach by introducing a heat setting process that incorporates multi-material fibers and an individualized heat application.

Among the methods of industrial textile production, which include weaving, braiding, and others, knitting offers key advantages for leveraging the small-scale configuration of fibers to generate large-scale behaviors in fabric. Knit fabrics can integrate stitch-by-stitch material changes and fiber configurations. Furthermore, local properties at the scale of a single stitch can accumulate into overall behaviors in larger zones of fabric. Conventional flat-bed industrial knitting provides numerous ways to arrange multiple materials in small areas on the face and the cross section of the fabric. Industrial knitting is a scalable and widespread manufacturing process used to produce a wide range of items which includes garments, accessories, and shoes.

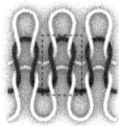 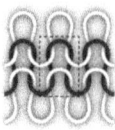 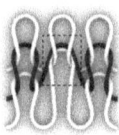 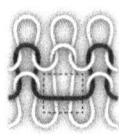

a. front stitch   b. back stitch   c. tuck stitch   d. float stitch

3

Knit fabrics are composed of rows of interlocking loops formed by needle mechanisms in the knitting machine (Spencer 2001). Each needle can perform a set of different actions that result in different loop types, programmed as a set of symbols. Common types of needle actions include front stitch, back stitch, tuck, and float (Figure 3). "Knit structure" refers to the pattern in which these stitch types are combined and deployed in the overall fabrics. The knit structure has a significant effect on the overall behavior of the fabric, including its appearance, texture, and ability to stretch, and in this case can serve to constrain the direction of shape transformation in the textile.

There are a number of examples of existing knitting research that apply knit structure to achieve fabric behaviors for different applications and functions. In architecture and design, existing projects have utilized knit structure to manipulate fabric to produce folding, actuation, and variation in elasticity with a wide range of effects (Ahlquist and Menges 2013; Baranovskaya 2016; Scott 2013; Pavko-Čuden and Rant 2017). In engineering, researchers have combined knit structure with shape memory materials to produce curling mechanisms and to adapt to the shape and motion of the human body (Abel, Luntz, and Brei 2013; Eschen et al. 2018). Our proposed approach uses both knit structure in combination with custom-made multi-material fibers and a robotic activation process to control shape change in fabric.

To produce fabric transformation, this project employs various heat responsive "active" fibers. This includes a shrinkage force and a bulking behavior using a thermoplastic fiber and a bicomponent fiber. Thermoplastic fiber shrinkage in knit fabrics is well-documented, and is characterized in its effect on washing, drying, or manufacturing of knit fabrics and garments (Perera and Lanarolle 2020). Bicomponent fibers combine two materials with contrasting properties that are extruded and fused together in cross section. These fibers typically produce a curling behavior in response to a moisture or heat stimulus that affects the texture of the fabric (Rwei, Lin, and Su 2005; Prahsarn et al. 2013). In this project, we leverage these fiber properties as a feature for controlling shape change, utilizing a combination of both types of fiber behaviors in response to heat.

Heat setting is a widely used technique in industrial textile production. It can be used to improve dimensional stability of knit garments, or to produce interesting textural effects (Haar 2011; Perera, Lanarolle, and Jayasundara 2019). Our technique expands the role and purpose of heat setting, applying a unique and precise process to achieve full shape customization for individual masks.

Finally, robotic toolpaths for customization and repeatability have been extensively used in architectural research (Gramazio, Kohler, and Willmann 2014). Particularly relevant are the robotic means of transforming standardized material into geometrically differentiated objects in response to input data. Our strategy virtualizes the connection between unique instances of input data and the customized outcome, performed here by the interaction between precise application of heat (based on an individual's facial measurements) with the active fiber material and programmed knit fabrics.

METHOD

Our approach to creating customized knit textiles involves several steps: the selection of active and inactive materials; the integration of these materials into knit structures that are able to transform directionally; the characterization of the directional fabrics' shape change relative to the robotic parameters; the design of a mask that integrates different knit structures to accommodate unique face shapes; and a computational workflow that generates a robotic tool path for tailoring the mask to a specific individual's measurements.

Materials

The knit fabric of the mask contains a combination of inactive and active materials. The first active material, Grilon K85, is a thermoplastic fiber that exhibits a contracting force when heated to a temperature range of 95 - 120°C. Grilon is a self-fusing polyamide copolymer that is commonly used in industrial textile settings, often to create fused seams and edges (EMS-CHEMIE Holding AG, n.d.). The material has a high linear contraction in response to heat (Figure 4a, Figure 4b). Each of the knit structures within the mask utilizes two strands of 380 denier Grilon K85 (760 denier total). Grilon is plied with one strand of a second

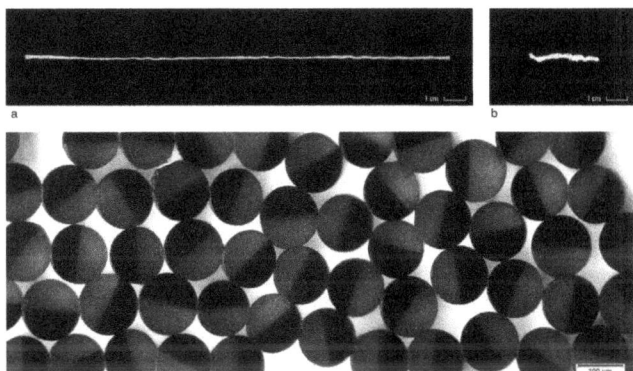

active material: a custom-made 72-filament bicomponent fiber which contains two materials: linear low-density polyethylene (LLDPE) and polyamide-6 (PA6) in its cross section (Figure 4c). When both types of active fiber are heated together in the fabric surface, they produce a permanent, heat-sensitive contraction with a soft surface texture. In the knit fabric, the active fibers work together with an "inactive" synthetic staple yarn which consists of 72% viscose and 28% polyester. Small amounts of a synthetic elastic yarn are used in the mask's periphery, forming the edges and ear loops.

Knit Structures

Linear contraction of the active fibers can be transferred to directional contraction in the fabric through the configuration of two materials together in a knit structure. Using these materials, we develop a set of knit structures where the inactive material serves to elongate non-looping lengths of active material. By inserting different patterns of active material into the inactive material, it is possible to control

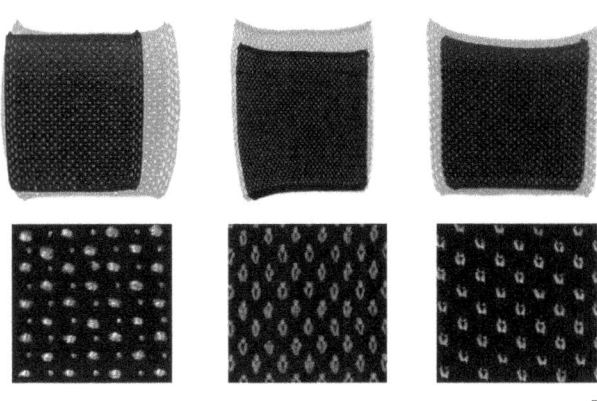

3   Examples of knit stitch types, showing basic interlocking loop structures (a, b) and non-loop-forming stitches (c, d)

4   Thermoplastic shrinkage of Grilon fiber before (a) and after (b) exposure to 95-120°C temperature range; cross section of bicomponent fibers containing LLDPE and PA6 (c)

5   Detail of knit structures with horizontal (left), vertical (center), and diagonal (right) behaviors, showing overall fabric contraction after heating

the direction of dimensional change in an area of the fabric. We employ three transforming knit structures in the mask, each with a different directional behavior: weft inlay to produce vertical contraction, held stitches to produce vertical contraction, and diagonal floats to produce diagonal contraction (Figure 5a, Figure 5b, Figure 5c). In each case, the active material forms a directional pattern within the inactive material, influencing the physical transformation. These structures are designed with symmetry on the front and back fabric faces to limit unwanted curling during the heating process, as well as sufficient thickness to feel durable and protective.

Three Stitch Patterns of Active Fiber

1) Horizontal contraction: A weft inlay is a method of inserting a continuous fiber into a horizontal course of stitches; the inserted fiber does not form loops and maintains an entirely width-wise orientation. When an active fiber is placed as the inlay material, the result is a horizontal contraction of the entire fabric surface with very minimal change in the vertical direction (Figure 6a).

2) Vertical contraction: Our strategy for generating a proportionally higher vertical contraction is to drastically elongate the loops of front stitches following an alternating pattern. This results in a predominantly lengthwise orientation of uninterrupted linear active fiber. When the fabric is heated, the longer stitches of active material contract predominantly in a lengthwise direction (Figure 6b).

3) Diagonal contraction: The third strategy for producing controlled dimensional change in knit fabric is through the placement of diagonal floats. This results in a secondary network of active fibers that pull the fabric inwards in both directions. Since the floats are oriented diagonally in two

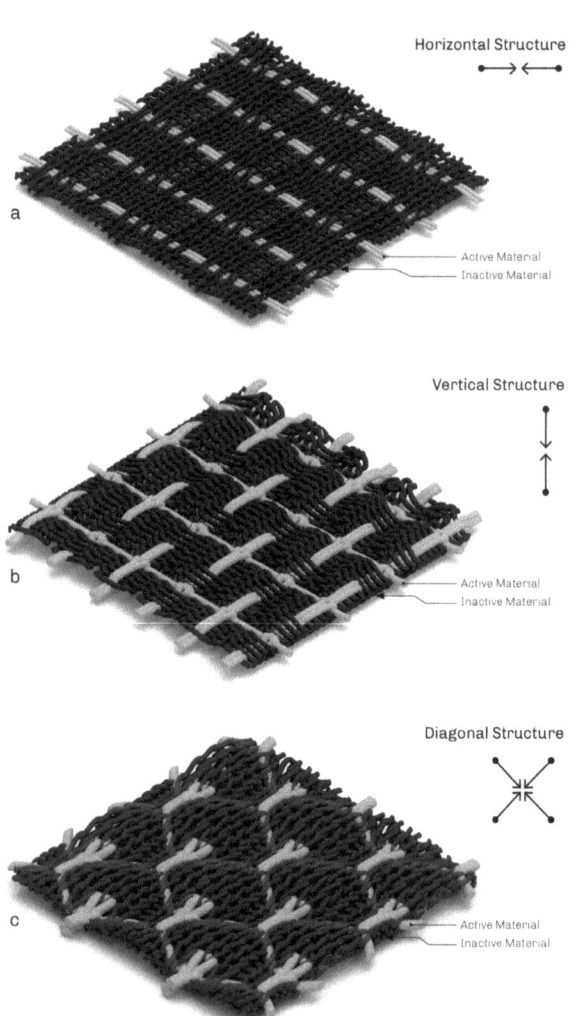

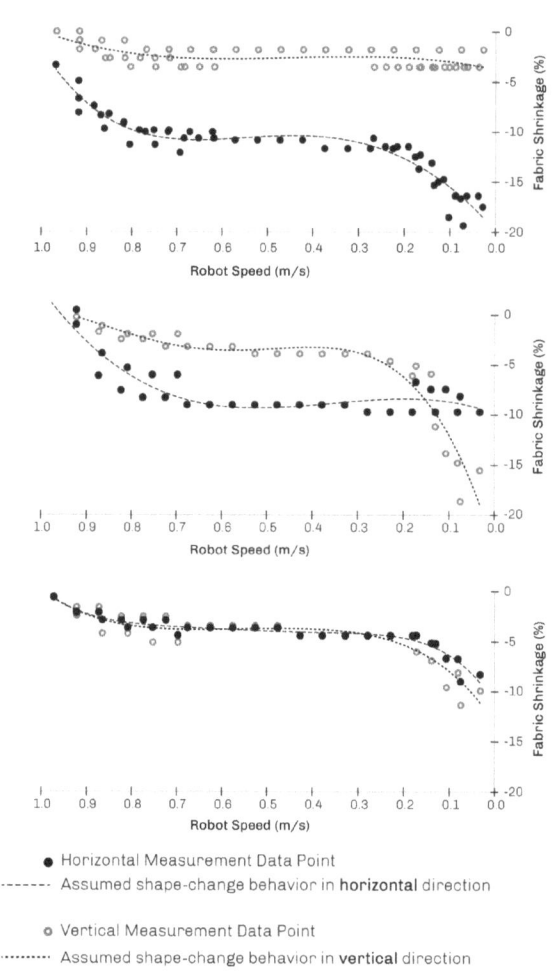

6   Illustration of the active and inactive fiber locations in each of the three transforming knit structures

7   Characterization of horizontal (top), vertical (middle), and diagonal (bottom), active knit structures with robot speed and size measurement data. The measurements show that slower robot speeds correlate to increased transformation of the fabric; each knit structure exhibits a different shrinkage behavior in the horizontal and vertical directions.

directions, the horizontal-to-vertical proportion of contraction is relatively even compared to the previous strategies (Figure 6c).

Robotic Heat Application and Shape Change

To test how each of the knit structures behaves when heated, a heat gun with a 5mm nozzle was mounted to a 6-axis robotic arm and was used to apply heat to a set of fabric samples following precise toolpaths. The temperature of the heat gun was set at 137°C and was stepped across the fabric in 2mm increments, traveling 12mm from the fabric surface. The sample was fixed to the table surface and covered with a wire mesh screen to prevent unwanted motion of the fabric in the vertical axis. After each full pass of the robot, the horizontal and vertical size of each sample was measured and recorded. The speed of the robot was varied during each test which exposed the fabric to different durations of heat; all other parameters remained the same. After tests were conducted on a range of different robot speeds, the measurement data was recorded to establish a predictable relationship between robot speed and percentage of shrinkage. The tests demonstrate three differing shrinkage behaviors corresponding to each knit structure which informs the design of the mask (Figure 7).

Mask Design and Fabrication

Standardized, mass-produced masks are typically available in different size gradations (S, M, L, etc.) or contain built-in strategies such as elasticity, adjustability, or expandability for adapting to multiple face sizes or shapes. Many of these mask designs inherently exclude certain face shapes or apply a single presumed set of facial proportions across a set of standardized sizes. As an alternative to this, the proposed strategy uses a self-measurement system that can record facial differences and embed them in the design of the mask to actively morph to the unique geometry of someone's face, through the strategic placement of active material and knit structures. The mask design is based on seven facial measurements that seek to identify the key dimensional relationships and proportions with facial features including the nose, mouth, chin, and ear (Figure 8):

Horizontal face measurements:

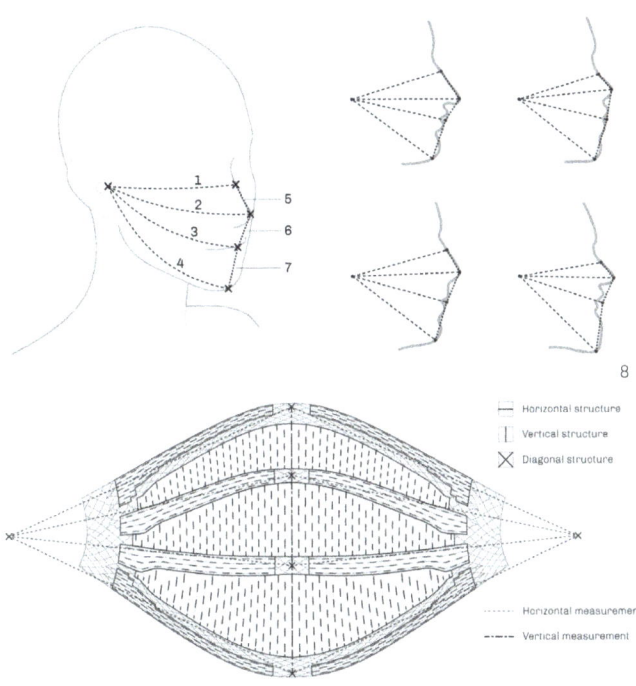

8 Location of seven facial measurements used to develop the mask design that can adapt to different facial sizes and proportions. The measurements are recorded as inputs to generate the geometry and corresponding speeds of the robotic tool path.

9 Mask design with layout of horizontal, vertical, and diagonal knit structure areas. Each area corresponds with a facial measurement location.

- 1: Nose bridge to ear
- 2: Nose tip to ear
- 3: Mouth to ear
- 4: Chin to ear

Vertical face measurements:
- 5: Nose bridge to nose tip
- 6: Nose tip to mouth
- 7: Mouth to chin

The measurement system acts as a scaffold for locating and organizing the three knit structures, and the regions of different fabric behaviors are applied as a tool for adjusting each of the seven input dimensions. The horizontally acting zones can adjust the size and relative proportions of the nose bridge, nose, mouth, and chin while the vertically acting areas can adjust the spacings between them. The regions with diagonal structures absorb the contrast between the vertical and horizontal areas while providing

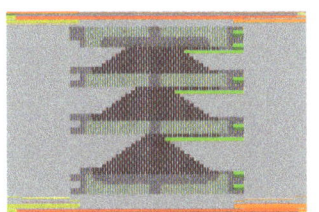

10 Programming and production of masks on a STOLL CNC knitting machine

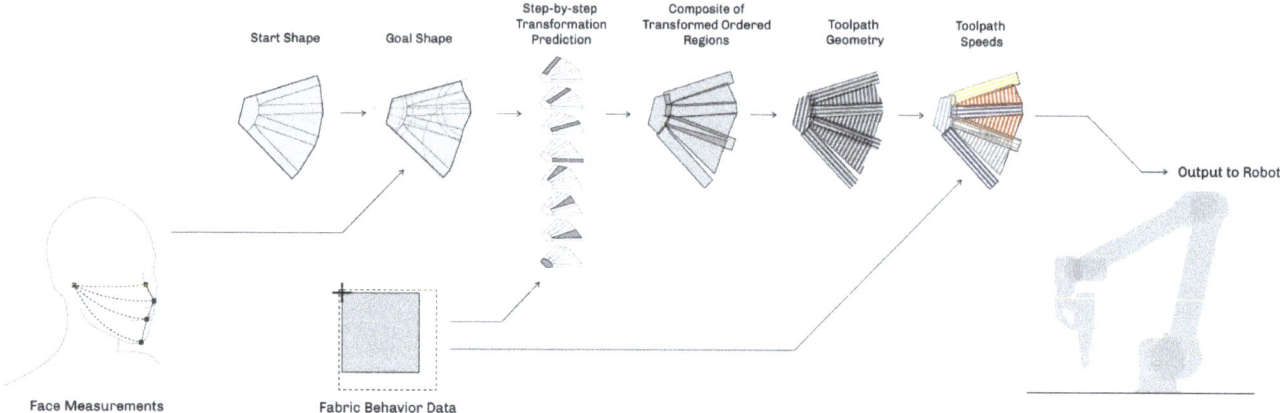

11 Digital workflow integrating facial measurements, fabric behavior data, and tool path output

stability to the sides of the mask (Figure 9). The initial size of the mask is calibrated to encapsulate an approximate range of standard sizes from XS to L, enabling a maximum potential of 60mm of shape change for each of the horizontal regions and 40mm of total shape change for vertical regions.

The knit masks are produced on a STOLL CMS HP-W TT Sport Industrial knitting machine, which is programmed using the M1PLUS software (version 7.2.037) interface (Stoll, n.d.). After knitting, the masks are hand-washed once in cold water, folded in half, and dried flat (Figure 10).

Toolpath Script

A script using Grasshopper 3d and the Scorpion plugin translates between the individual measurements and the standardized mask design to generate and output a robotic tool path with varied speeds that correspond to the necessary amount of shape change for each set of input measurements (Rutten, *Grasshopper 3d*, V. 1.0.0; *Scorpion* V. 0.2). In addition to producing the tool path, the script predicts the amount of shape change during the robotic process while the fabric moves and shrinks as it is being activated. Using the previous swatch testing measurements, the script builds an approximation of the original and transformed states and outputs the toolpath geometry with assigned speed parameters for the robot (Figure 11).

First, the geometry of the initial standardized shape is constructed and organized into a sequence of zones that corresponds to eight different areas of the mask. Each of the zones contains a single knit structure, the first seven of which correspond directly to the seven facial measurements. Zones one through four contain horizontal structures; zones five through seven contain vertically contracting structures, and the final zone contains the diagonally acting structure near the ear loops (Figure 12b). Next, the individual input measurements are used to construct the intended geometry of the finished mask; the difference is calculated between the initial mask size and measured inputs from the individual. Both the original and transformed representations of the mask are constructed using the same geometric "scaffold", approximating both existing and potential conditions using measured behavior that is known about the three knit structures.

A composite of both sets of measurements is then assembled to estimate the zone-by-zone motion during heating, producing a series of geometries that represent each step of the mask's shape during the heating process (Figure 12).

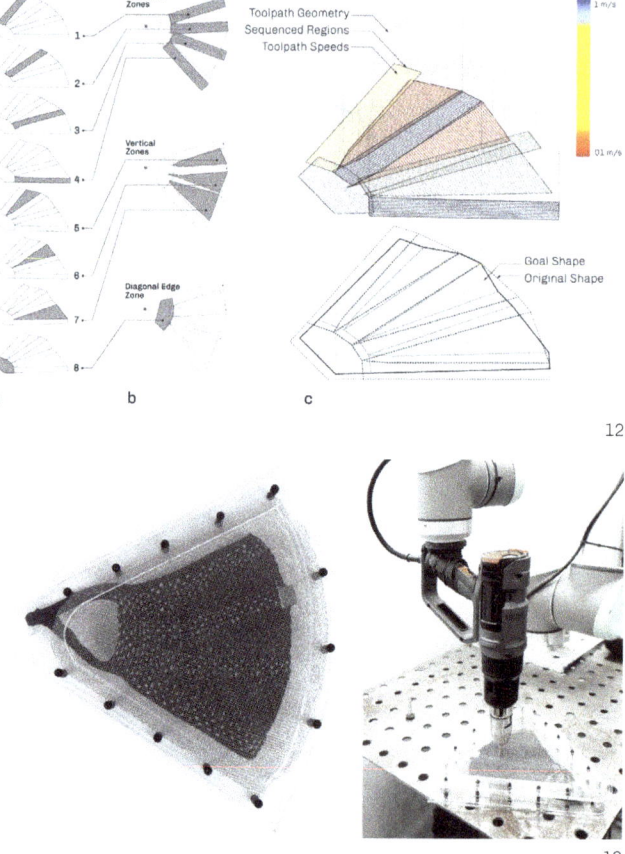

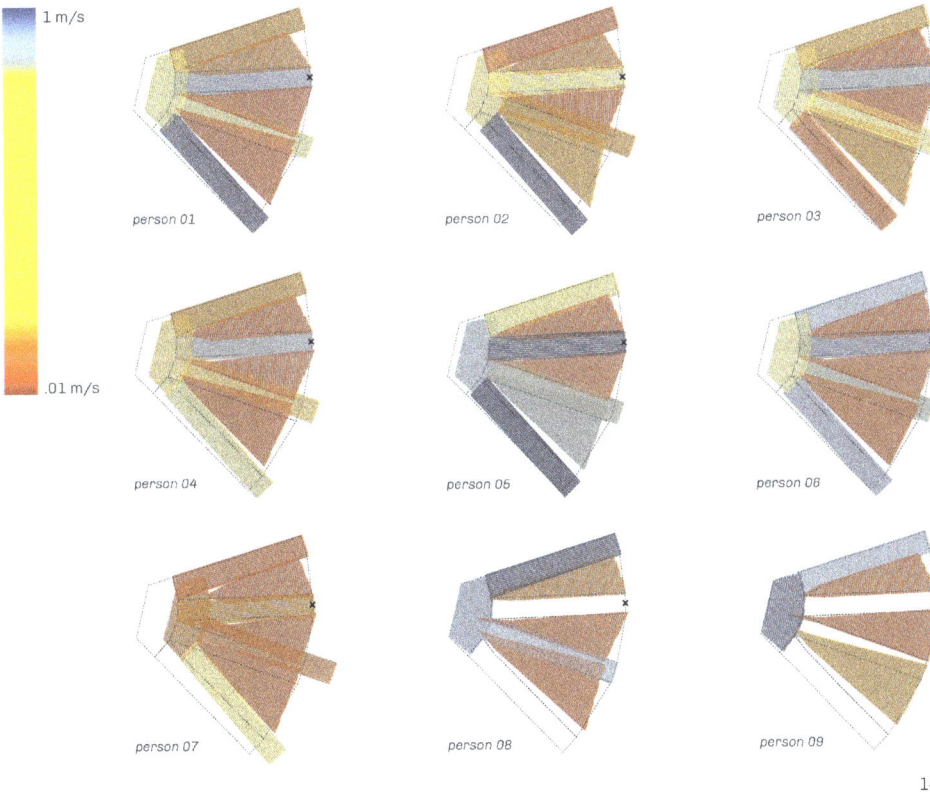

12 Sequence and approximation of zone-by-zone transformation (a); areas of directional fabric transformation used to form a composite geometry (b); application of stepping tool path geometry and robotic speeds to the activation areas (c)

13 Mask-holding frame (left) and robotic heating setup (right)

14 Set of unique tool path geometries and robot speeds corresponding to the unique facial measurements of nine participants

15 Examples of 2D scans of the original and transformed masks from three participants overlaid with facial measurement guidelines

Before an area is heated, it is represented by its initial shape and location; after it is heated, its size and location are updated into a transformed shape along with any zones that are affected by the change. The transformed shapes are generated using the measured data—represented as functions that describe the behavior of each knit structure—from the earlier swatch tests. A 2mm stepping tool path is then assigned to each zone, and a robot speed is determined based on the amount of intended transformation using the swatch test data (Figure 12c). If no transformation is necessary for a particular zone, it is omitted from the path.

Robotic Activation

When heat is applied with the robot, the mask is affixed inside a holding device and anchored at a single point at the nose area. The holding device holds the mask in a precise location on the table surface and is covered on the front and back with a screen to prevent unwanted out-of-plane motion during heating (Figure 13). After the tool path is run once, the mask is flipped, and a mirrored version of the process is applied on the reverse side. The mask is then turned inside out, and the entire process is repeated so that both sides of fabric are exposed to heat.

RESULTS

The proposed process was tested using the measurements of nine people. Each individual measured their own face, and a unique tool path was generated for each mask (Figure 14). The masks were evaluated by 2D scanning and comparing the original with an overlay of the transformed mask (Figure 15). To further test the fit of the masks, each individual was photographed wearing the mask before and after the heating process is performed (Figure 16).

This process resulted in the successful production of nine differentiated masks, and in every case the transformed fit was a significant improvement from the initial fit. The participants also reported that the masks were comfortable and fit much better than a standard mask, subjectively demonstrating effectiveness of the workflow. The masks were able to adapt to a wide range of face sizes and shapes represented by the participant group.

However, some aspects of the shape-change process were more successful than others, and there are a number of challenges to overcome in future work. The vertically acting knit structure is limited by inherent constraints of weft knitting, but there could be ways to improve its effectiveness

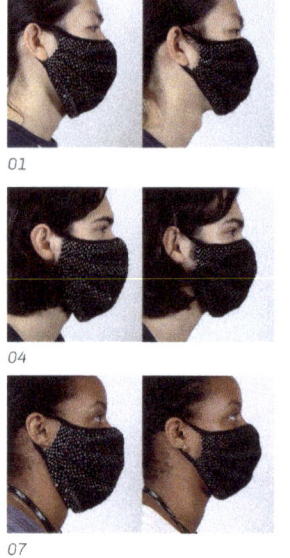
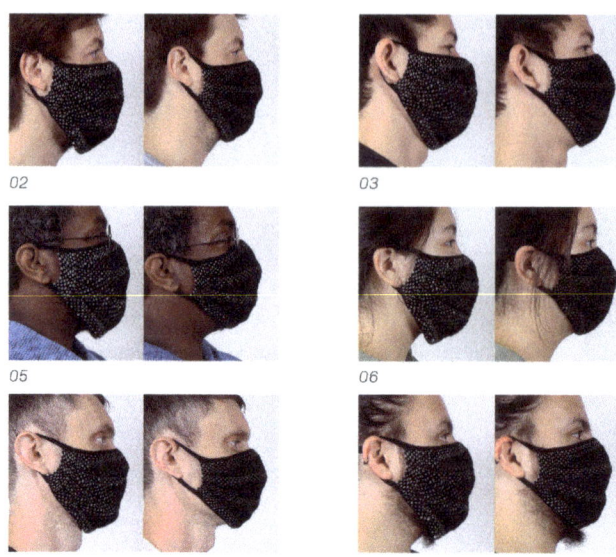

16 Before (left) and after (right) front views of standardized masks transformed into personalized masks for nine participants

with further knit structure development. Many of the input measurements contained a concave condition where a person's mouth is located (visible in person 7 and 8 in Figure 17), and the mask design was unable to fully adapt; this is most likely due to surrounding forces in the fabric that were not present during the initial characterization of knit structures. During the robotic heating process, the flat activation method was successful in reducing complexity during heat application, but it introduced new limitations that constrained the amount of possible transformation. This could be reevaluated in future iterations by testing three-dimensional frameworks.

Since the sample size of unique individuals was small, there is the potential for a more comprehensive method of evaluating how much facial variation can be accommodated by the mask design. To address this, our future work aims to increase the amount of shrinkage to at least -30% from the current -20% to improve on the current 60mm maximum transformation of the mask. We also aim to improve the process of evaluating the success of the fit after the mask is activated to account for face motion and interaction with other facial accessories such as glasses.

CONCLUSION

Through the use of programmable material, this project demonstrates a new approach for creating mass customized textile products, enabling the transformation from scalable and standardized production into truly personalized products. This process aims to differentiate the manufacturing process from the customization tool, allowing for a scalable production method while also creating a streamlined method for personalization with the user. We have been able to achieve this through the careful design and integration of a computational workflow, customized multi-material fibers, precise knit structure and a robotic post-process activation.

ACKNOWLEDGEMENTS

The authors would like to acknowledge support from the Advanced Functional Fabrics of America, Oye Ajewole and Gihan Amarasiriwardena (Ministry of Supply), Ziyue Wendy Wu, the MIT Department of Architecture, and Hills Inc.

REFERENCES

Abel, Julianna, Jonathan Luntz, and Diann Brei. 2013. "Hierarchical Architecture of Active Knits." *Smart Materials and Structures* 22 (12): 125001.

Ahlquist, Sean, and Achim Menges. Oct. 2013. "Frameworks for Computational Design of Textile Micro-Architectures and Material Behavior in Forming Complex Force-Active Structures." In *ACADIA 13: Adaptive Architecture, Proceedings of the 33rd Annual Conference of the Association for Computer Aided Design in Architecture (ACADIA)*. Cambridge, Ontario. 281–292.

Apeagyei, Phoebe R., and Rose Otieno. 2007. "Usability of Pattern Customising Technology in the Achievement and Testing of Fit for Mass Customisation." *Journal of Fashion Marketing and Management* 11 (3): 349–65.

Baranovskaya, Yuliya. Aug. 2016. "Knitflatable Architecture - Pneumatically Activated Preprogrammed Knitted Textiles." In Vol. 1 of *Complexity & Simplicity - Proceedings of the 34th ECAADe Conference*, edited by A. Herneoja, T. Österlund and P. Markkanen. Oulu, Finland. 571–580.

Bellemare, Jocelyn. 2018. "Fashion Apparel Industry 4.0 and Smart Mass Customization Approach for Clothing Product Design." In *Customization 4.0: Springer Proceedings in Business and*

Economics, edited by S. Hankammer, K. Nielsen, F. T. Piller, G. Schuh, and N. Wang. Springer: Cham. 619–33.

Buecher, Daniel, Yves-Simon Gloy, Bernhard Schmenk, and Thomas Gries. 2018. "Individual On-Demand Produced Clothing: Ultrafast Fashion Production System." In *Customization 4.0: Springer Proceedings in Business and Economics*, edited by S. Hankammer, K. Nielsen, F. T. Piller, G. Schuh, and N. Wang, Springer: Cham. 635–44.

EMS-CHEMIE Holding AG. n.d. "EMS-GRILTECH - Grilon." Accessed October 5, 2021. https://www.emsgriltech.com/en/products-applications/products/grilon/.

Eschen, Kevin, Julianna Abel, Rachael Granberry, and Brad Holschuh. Sept. 2018. "Active-Contracting Variable-Stiffness Fabrics for Self-Fitting Wearables." In *Proceedings of the ASME Conference on Smart Materials, Adaptive Structures, and Intelligent Systems*.

Gramazio, Fabio, Matthias Kohler, and Jan Willmann. 2014. *The Robotic Touch: How Robots Change Architecture*. Zurich: Park Books.

Haar, Sherry J. 2011. "Studio Practices for Shaping and Heat-Setting Synthetic Fabrics." *International Journal of Fashion Design, Technology and Education* 4 (1): 31–41.

Knittel, Chelsea, Diana Nicholas, Reva Street, Caroline Schauer, and Genevieve Dion. 2015. "Self-Folding Textiles through Manipulation of Knit Stitch Architecture." *Fibers* 3 (4): 575–87.

Luo, Jin, Fumei Wang, Guangbiao Xu, and Hoe Hin Chuah. 2011. "Effects of Fiber Crimp Configurations on the Face Texture of Knitted Fabrics Made with PTT/PET Bicomponent Fibers." *Journal of Engineered Fibers and Fabrics* 6 (1): 155892501100600.

Stoll, n.d. *M1PLUS*, V. 7.2.037. Stoll. https://www.stoll.com/en/software/m1plusr/.

Narayanan, Vidya, Lea Albaugh, Jessica Hodgins, Stelian Coros, and James McCann. 2018. "Automatic Machine Knitting of 3D Meshes." *ACM Transactions on Graphics* 37 (3): 35:1-35:15.

Pavko-Čuden, Alenka, and Darja Rant. 2017. "Multifunctional Foldable Knitted Structures: Fundamentals, Advances and Applications." In *Textiles for Advanced Applications*, edited by B. Kumar and S. Thakur. Rijeka: InTech. https://doi.org/10.5772/intechopen.69292.

Perera, Ayomi Enoka, Gamini Lanarolle, and Ravindi Jayasundara. 2019. "Effects of Panel Parameters and Heat Setting Temperature on Thermal Shrinkage of Heat Cured Polyester Plain Knitted Fabric Panels Statistical Modeling Approach." In *2019 Moratuwa Engineering Research Conference (MERCon)*. 722–26.

Perera, Henadeera A. A. E. and Wilathgamuwage D. G. Lanarolle. 2020. "Comparative Study on the Thermal Shrinkage Behaviour of Polyester Yarn and Its Plain Knitted Fabrics." *The Journal of The Textile Institute* 111 (12): 1755–65.

Papadopoulou, Athina, Jared Laucks, and Skylar Tibbits. 2017. "Auxetic Materials in Design and Architecture." *Nature Reviews Materials* 2 (12): 1–3. https://doi.org/10.1038/natrevmats.2017.78.

Prahsarn, Chureerat, Wattana Klinsukhon, Nanjaporn Roungpaisan, and Natee Srisawat. 2013. "Self-Crimped Bicomponent Fibers Containing Polypropylene/Ethylene Octene Copolymer." *Materials Letters* 91 (January): 232–34.

Rutten, David. 2018. *Grasshopper 3d*. V. 1.0.0007. Robert McNeel & Associates. https://www.rhino3d.com/.

Rwei, S.P., Y.T. Lin, and Y.Y. Su. 2005. "Study of Self-Crimp Polyester Fibers." *Polymer Engineering & Science* 45 (6): 838–45.

Scorpion Robotics. 2017. *Scorpion* (plugin). V. 0.2 . https://www.food4rhino.com/en/app/scorpion.

Scott, Jane. 2013. "Hierarchy in Knitted Forms: Environmentally Responsive Textiles for Architecture." In *ACADIA 13: Adaptive Architecture [Proceedings of the 33rd Annual Conference of the Association for Computer Aided Design in Architecture (ACADIA)*, Cambridge. Ontario. 361–366.

Spencer, David J. 2001. *Knitting Technology: A Comprehensive Handbook and Practical Guide*. CRC Press.

Tessmer, Lavender, Carmel Dunlap, Bjorn Sparrman, Schendy Kernizan, Jared Laucks, and Skylar Tibbits. 2019. "Active Textile Tailoring." In *ACM SIGGRAPH 2019 Emerging Technologies*. 2 vols. New York, NY, USA: Association for Computing Machinery.

Wu, Kui, Hannah Swan, and Cem Yuksel. 2019. "Knittable Stitch Meshes." *ACM Transactions on Graphics* 38 (1): 1–13.

IMAGE CREDITS
All drawings and images by the authors.

Lavender Tessmer is a PhD student in Design and Computation in the Department of Architecture at the Massachusetts Institute of Technology.

Skylar Tibbits is an Associate Professor of Design Research and director of the Self-Assembly Lab in the Department of Architecture at the Massachusetts Institute of Technology.

Exploring the role of spatial configuration and human behavior on the spread of the epidemic

A study of factors that affect Covid-19 spreading in the city

Mengda Wang
The Bartlett School
of Architecture/University
College London

Ava Fatah gen. Schieck
The Bartlett School
of Architecture/University
College London

Petros Koutsolampros
The Bartlett School
of Architecture/University
College London

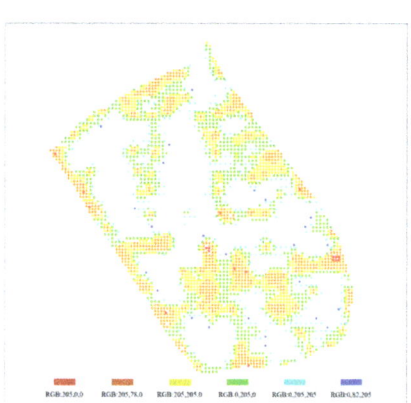
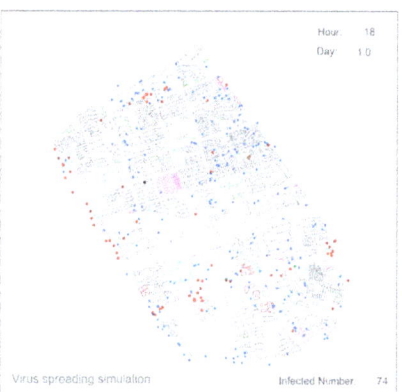
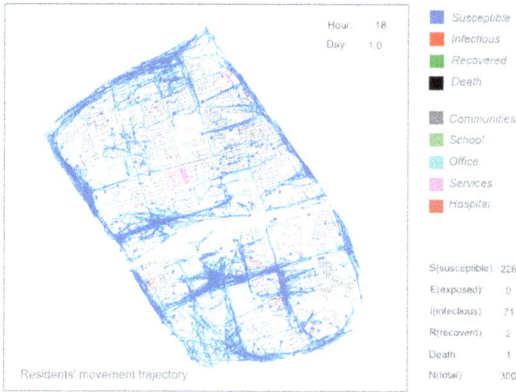
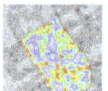

1

1   Introduction of simulation experiments and movement trajectory

ABSTRACT

This research explores how exterior public space—defined through the configuration of the city—and human behavior affect the spread of disease. In order to understand the virus spreading mechanism and influencing factors of the epidemic which accompany residents' movement, this study attempts to reproduce the process of virus spreading in city areas through computer simulation.

The simulation can be divided into residents' movement simulation and the virus spreading simulation. First, the Agent-based model (ABM) can effectively simulate the behavior of the individual and crowd; real location data—uploaded by residents via mobile phone applications—is used as a behavioral driving force for the agent's movement. Second, a mathematical model of infectious diseases is constructed based on SIR (SEIR) Compartmental models in epidemiology.

Finally, by analyzing the simulation results of the agent's movement in the city and the virus spreading under different conditions, the influence of multiple factors of city configuration and human behavior on the virus spreading process is explored, and the effectiveness of countermeasures such as social distancing and lockdown are further demonstrated.

INTRODUCTION

The spread of the COVID-19 epidemic has brought threats to human society, causing huge public health risks and property losses. Because the virus is mainly transmitted through the mouth and nose, the infection between individual contact (such as coughing and talking) can cause possible infection (WHO 2020). It is of great significance to understand the transmission mechanism of the virus in different spaces. At the same time, it helps to formulate targeted policies to protect public health.

In the face of COVID-19, there are still gaps in using Agent-based modelling (ABM)—supported by real data— for infectious disease analysis. Technically speaking, the use of relatively new methods combined with ABM analysis of real data has practical significance. In terms of research content, the relationship between the spread of the virus and city space using COVID-19 as an example, and the relationship between the virus and human behavior (such as social distancing) need to be explored. In this study, ABM is created and combined with real mobile phone data to simulate infectious diseases in city space, while addressing research into and analysis of specific issues. Using this method can lead to better understanding the mechanisms and influencing factors of a new type of infectious disease. It should, however, be noted that this study does not address biological factors.

The simulation model consists of two parts: the first is a movement model mainly driven by real residents' location data. In addition, the movement model is inspired by Isovist and Boids algorithms, and rules are designed to realize different movement behaviors. The second model is an epidemic model, using SIR as the main framework and using COVID-19 data as a reference for research.

BACKGROUND

In recent years, with the development of computer technology, the use of computer simulation technology in public health related research has been increasingly applied to the spread of diseases (Ajelli et al. 2010). And agent-based models (ABM) supported by data are particularly important for infectious disease research; they have higher flexibility (Bonabeau 2002), a factor important in the evaluations of the model in this study.

ABM allows individuals to be distinguished, which is the basis of epidemiological research (Koopman and Lynch 1999). In addition, this model can simulate the spread of diseases based on the connection (contact) between independent individuals (Sattenspiel 2009). So this model is more intuitive than population-based differential equation models (Chen et al. 2015), which clearly demonstrates causality. Because an equation-based method cannot directly match the data to the individual level, it is impossible to achieve high-precision simulation. Therefore, the ABM method brings more details and high reliability.

Chen et al. (2015) introduced a large number of theories and cases of infectious disease analysis modelling, including examples based on agent-based model analysis. Friesen and McLeod (2014) established an infectious disease model based on smartphone trajectory data, and conducted simulation analysis at the provincial and town levels. This is due to the advantage that agent-based model is more suitable for combining individual real data (individual behavior is driven by data).

METHODS

The research process can be briefly described as follows: a) First, select a public space (exterior space) and create the built environment required for simulation, which includes borders, obstacles (buildings), and other elements; b) Create an ABM simulation model that can be divided into two parts: movement model and epidemic model. Of the two, the movement model is mainly driven by residents' dynamic location data and gives individuals more behavioral capabilities; the epidemic model is built based on the mathematical models of SIR and SEIR, which intergrates the real data of COVID-19; c) To simulate virus and disease transmission under different conditions, adjust various behavior parameters and spatial parameters; d) Compare and analyze the influence of spatial configuration and personal behavior on virus transmission, and propose effective epidemic prevention and control strategies based on the simulation results, while verifying the effectiveness of existing strategies.

Build Environment: Location and Elements

When using an Agent-based model to simulate in an urban public space, the size of the area is directly related to the input samples and time, so reducing the activity space of the agent can effectively reduce the cost of the experiment. In addition, the selected area should be representative, and the area should have clear boundaries and a variety of public buildings (including traffic buildings) (Figure 2).

After determining the location, it is necessary to collect city information data to complete the establishment of the built environment. This research is based on two-dimensional virus spreading simulation, so the build requires only the flat plane content of the exterior space environment in the city including: boundaries, buildings (obstacles), streets. The built environment data is based on an open city GIS database (Figure 3).

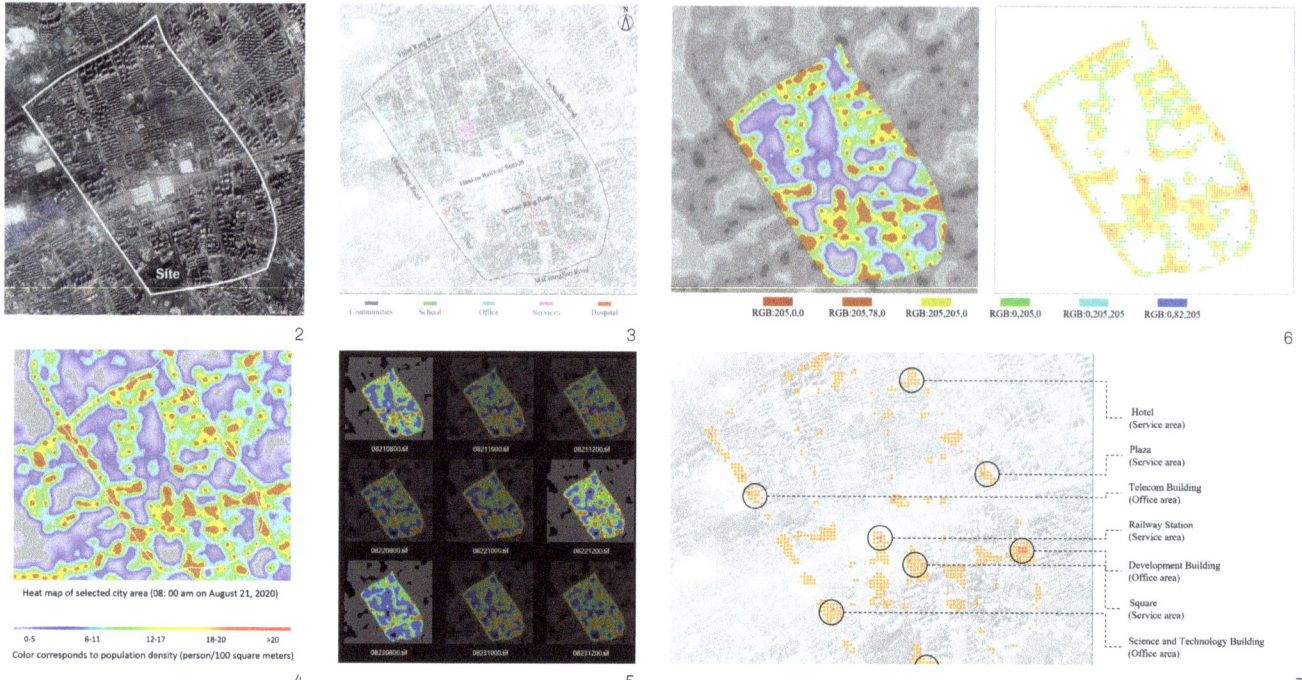

Data Processing

The most important driver of simulation model is location data. The basic movement behavior of the agent is determined according to heat map function provided by the mobile app, which yields dynamic location data (picture format) of residents in a specific area. The regional heat map data shows the population location and heat information of the real-time urban area, as shown in Figure 4.

In order to meet the simulation requirements, the heat map data for the three days from August 21 to August 22, 2020 were collected for the identified area (Figure 5). These three days, from Friday to Sunday, from weekday to weekend, represent different city phases of being. In addition, due to the large data sets collected from people during the day, night data is not considered in this collection. The heat map used in the final simulation is the data from 8:00 a.m. to 00:00 a.m. on the next day, and the collection interval is two hours. Although the heat map represents a transient situation, in this simulation, each heat map represents a two-hour period.

Subsequently, the heat map data is processed to accomplish two tasks: a) complete the initialization of the movement model; b) determine the driving mechanism of the agent's movement. Specifically, the first step is to establish a data set of resident coordinates through heat map pixels, and determine the initial position of the agent in the area through the data set (pixels are individual spatial coordinates). In order to make the agent's movement driven by the data, in the second step, it is necessary to make the agent gain power. Through the heat map, we can calculate the initial individual position and its potential vector to attractors, such as a individual (pixel) moving to the nearest densely populated area (attractors) in the next time period (Figure 6).

Take the heat map from 08:00 a.m. to 10:00 a.m. on August 21, 2020 as an example. By analyzing the pixels of the heat map, the pixels can be extracted proportionally as individual coordinates based on the population density in the legend; high-heat areas are used as the "destination," and the individual is attracted to move toward the nearby "destination" in the movement module of ABM. What needs to be clear is that the attractor as the "destination" of the movement is calculated from the heat map of the next period (10:00 a.m. - 12:00 p.m.) (Figure 7).

Movement Model

The movement model only considers two-dimensional movement. Individuals in the movement model have basic attributes and behaviors. The attributes include the position of the individual, the direction of movement and the speed of the movement. These three factors are manifestations of individual behavior in the simulation.

The behavior of an individual is the rule of its movement and the reason for determining the direction of movement. The behaviors are Navigation, Group, Separate, Observation, and Quarantine, which are shown in Figures 8 and 9.

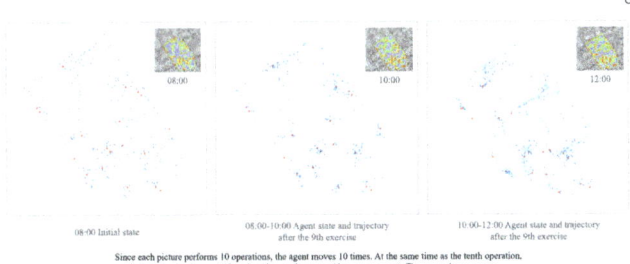

8

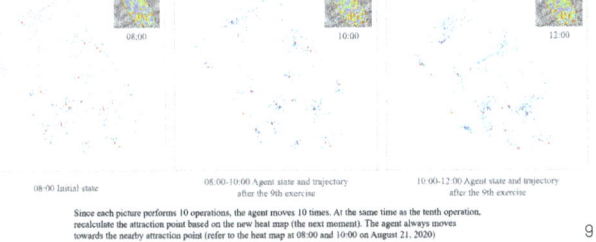

9

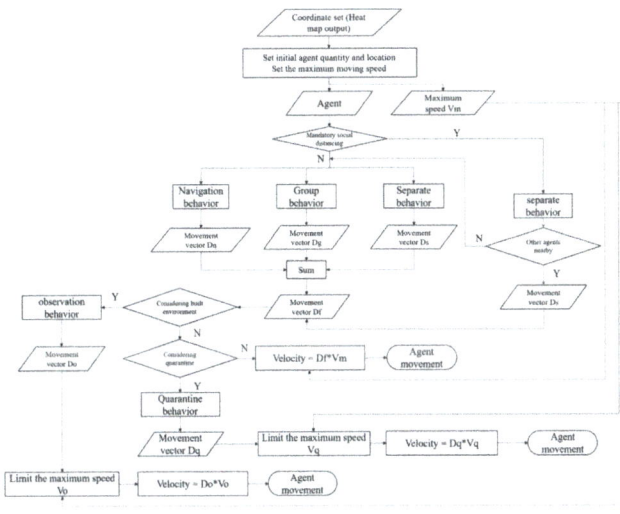

11

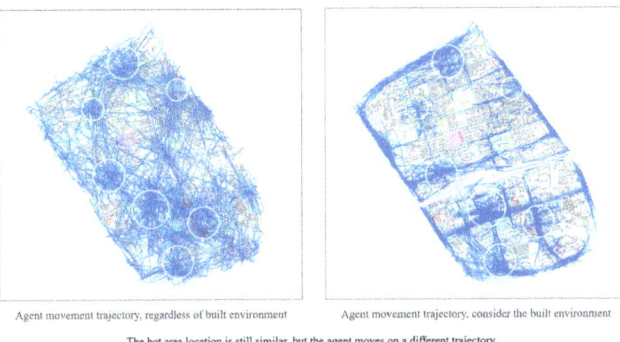

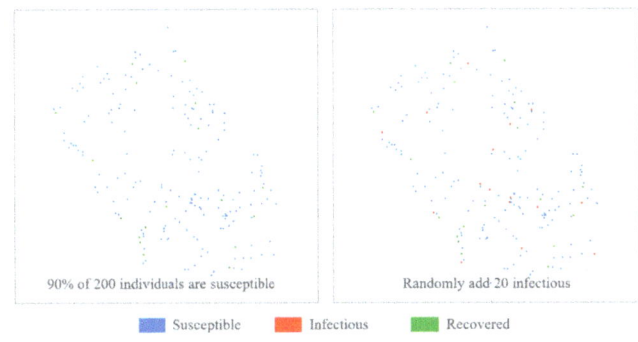

10

12

When considering the built environment of the city, the behavior of the agent will change greatly (Figure 10). If the built environment is not considered, the agent will ignore the urban obstacles and move directly to the attraction point. In this research, the built environment should be considered when analyzing the city space, while it should not be considered when analyzing human behavior. The specific flow chart for the executed movement model is shown in Figure 11.

Epidemic Model

The SIR and SEIR models are a population-based models. Maintaining the number of existing individuals is the premise of this type of model (Beckley 2013). That is to say, natural births and deaths are not considered. In this model, a "label" is added to each agent. This label is the individual state in the epidemic model, such as susceptible, infectious, and recovered, and the sum of the number of individuals in all states remains constant. In the initialization process, the entire population can be divided into two parts, susceptible and recovered, in proportion to the needs. The recovered here can be understood as part of the population that has acquired virus immunity. In addition, initialization can also

2   Center area of city

3   Selected city area: built environment

4   Heat map of selected city area

5   Part of heat maps for three day from August 21 to August 22, 2020

6   Establish a coordinate set by different colors

7   Pixels in high-heat areas are used as "destination"

8   Behaviors in movement model

9   Simulation of individual navigation behavior

10 The trajectory of the individual in two different situations

11 Flow chart of individual behavior

12 Individual infection status initialization

introduce viruses in a random manner, that is, transform part of susceptible into infectious (Figure 12).

The parameters that directly affect the epidemic model include R0 (which represents the transmission capacity of the virus), the average time required for recovery, mortality rate, the time required for a death case, and SEIR's unique incubation period time, etc. All parameter descriptions, values, and sources are shown in Figure 13.

Parameter	Description	Value	Calculation	Reference
Model Type	Choose SIR or SEIR as the basic mathematical model	SIR/SEIR	-	Kermack and McKendrick, 1927
R0	Basic Reproduction Number: used to describe the infectiousness of a disease	4	-	Description: Kermack and McKendrick, 1927; Value: Flaxman et al., 2020
Recover Time	Average duration between infectious and recovered	2-8 weeks	Each movement (each frame) represents 12 minutes (refer to 4.2.1). Based on the 14-day recovery duration, the **recovery rate** within 12 minutes can be calculated to be 0.0794%(12 minutes/minutes in 14 days)	World Health Organization, 2020
Death Time	Average duration between infectious and death	6-41 days, average 14 days	Based on the 6-day death time, the **death rate** per 12 minutes can be calculated as 0.0102% (the death rate per minute * 12 minutes)	Wang et al., 2020
Death Rate	mortality rate worldwide	3.3% (September 9, 2020) 5.5% (April 7, 2020)		Johns Hopkins University, 2020
Exposed Time	Duration of incubation period	1-14 days	If the exposed period is 2 days, the **exposed rate** per 12 minutes can be calculated to be 0.556% (12 minutes/minutes in 2 days)	Anna Carthaus, 2020
Average Contact	Average number of contacts in one day	5 physical contacts	Contacts are 0.0556 every 12 minutes	Davies et al., 2020
β	Infection Coefficient	-	R0 * Recover Rate	Kermack and McKendrick, 1927; Coburn et al. 2009
Infection Rate	Probability of individual infection	-	β / Average contact	Kermack and McKendrick, 1927
Infection Distance	Effective distance of infection: uncovered cough can lead to droplets travelling up to 4.5 meters	4.5 meters	-	Jayaweera et al., 2020

13

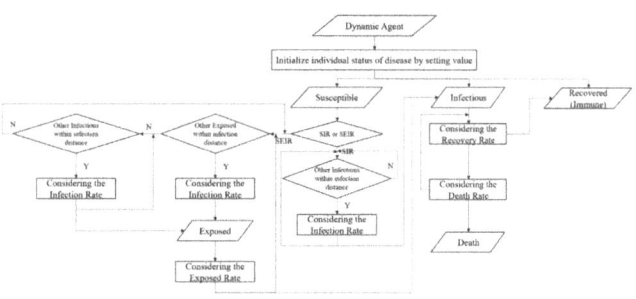

14

Category		Parameter	Description	Value
Location Data		Initial coordinate set	Calculated based on the heat map. Potential initial position of the agent.	-
		Attractor	As the "destination" of the agent, it is calculated based on the high heat area of the heat map. Determine the number of attraction points as needed.	Quantity: 50
				Factor: 0.5
Movement Model	Individual and Behavior	Number of Agents	Randomly select from the Initial coordinate set.	Set as needed (200-1000)
		Location of Agents		From Initial coordinate set
		Maximum Speed of Agent	According to the average human moving speed	Max speed: 1.5 m/s
				Stationary factor: 10
		Group Parameter	Agent aggregation behavior	Group Distance: 7m
				Factor: 0.5
				Probability: 50%
		Social Distancing Parameter	Keep a certain distance between agents	Social Distance: 1.2
				Factor: 0.5
	Built Environment and Space	Observation Parameter	Use Line of Sight to observe the built environment and move	Observation Range: 20
				Observation Detail: 3
		Quarantine Parameter	Movement area limitation	Set as needed
Epidemic Model		Initial susceptible number	-	Set as needed
		Initial infectious number	-	Set as needed
		Other Parameter(R0 etc.)	-	See table 4.2
Simulation Control		Simulation environment	-	City area, Jianghan District, Wuhan
		Simulation Duration	Set each heat map to calculate 10 times. Since each heat map represents 2 hours, each calculation (agent movement) represents real time 12 minutes	Set as needed (7 days)

15

13 Parameter initialization based on COVID-19

14 Flow chart of the epidemic model

15 Default parameter setting

The epidemic model is inherited from the movement model and is relatively independent. Specifically, the agent's movement is only affected by its attributes and behaviors, while the infectious disease model shows the results of disease transmission after the agent moves, and the disease transmission itself does not affect the agent's movement mode. The model tracks the location and state of each agent, and changes the agent's "label" (disease state) under certain conditions. The flow chart of the epidemic model is shown in Figure 14.

Simulation Model

The Movement model realizes the simulation of individual movement behaviour. The epidemic model judges and changes the individual's disease state based on the individual's movement. Based on the above two models, a complete ABM-based infectious disease simulation model is realized.

RESULTS AND REFLECTION

Due to the large number of model variables, it is necessary to set default parameters in simulation experiments for different factors. The parameters here consider people's daily behavior (normal movement, stay, social distance) and normal operation of public spaces (not considering lockdown, etc.), as shown in Figure 15.

The Impact of City Spatial Configuration

When analyzing the impact of different space configurations on the spread of the virus, the typical space of the area is divided into different parts. In terms of functional characteristics and spatial forms, these typical spaces are newly built communities, old communities, service and commercial areas, office areas, school areas, and hospital areas. The reason for classifying according to function is that different function spaces have different urban area characteristics (form, organization, density). This experiment is not about the relationship between space function and virus transmission, but the influence of space configuration embodied by this function on virus transmission.

These six typical spaces have similarities and differences in their shape, size, and density. Based on the relevant information of the area, these spatial characteristics are shown in Figures 16 and 17.

The above data is only for this area, and it is believed that the same type of space has similar spatial characteristics. However, this space is only a typical case and does not have universal applicability. Based on these criteria for these six different types of spatial configurations, when only the same type of built environment is considered, simulations are performed separately. The simulation results and the line graph of the infection cases are shown in Figure 18.

In summary, from a larger scale (city area), the concentrated spatial layout may reduce the total amount of virus infection. The net-shaped space (multiple cross paths) will also slow down the spread of the virus to a certain extent and reduce the amount of infection. The linear and fishbone-shaped space (single path) will accelerate the spread of the virus in the early stage. The larger space area (more buildings and complex urban environment) will also reduce the cases of virus transmission, which may be related to the limitation of space on the speed of individuals. From a small scale (community) point of view, there are more cases of infection in areas with higher density, and the spreading process is faster, which may be related to the limitation of local space on the range of human movement.

City Decentralization Strategy

In this project, the agent's "destination" (attractor) is obtained through the calculation of the heat map. It reflects the attraction of the hot areas of the heat map to individuals. Fewer city centers will lead to the tendency of individuals to centralize and also mean larger-scale gatherings. On the contrary, more urban centers are a decentralized approach to a certain extent. Based on this, it is necessary to conduct experimental demonstrations and analyze the impact of decentralization on the spread of the virus. The influence of different numbers of city centers on the spread of the virus is analyzed (Figure 19).

It can be seen from the simulation results (Figure 16) that the number of urban centers directly affects the speed of early virus transmission, and fewer urban centers means faster infections. The virus transmission rate is basically inversely proportional to the number of centers. But when approaching a certain number, its spreading rate remains basically unchanged. It can be seen that the number of multiple centers may lead to relatively more infections, but the overall difference is not obvious.

Interaction of Sight and City Space

In this study, vision has two important parameters: Field of View and Detail (the degree of detail of observation). These two parameters directly affect the interaction between individual vision and the city space. When analyzing vision alone, we set up several different combinations of vision parameters to analyze how it affects the movement of individuals and thus the spread of the virus (Figure 20).

From the above simulation analysis, it can be seen that as the field of view increases and the observation details

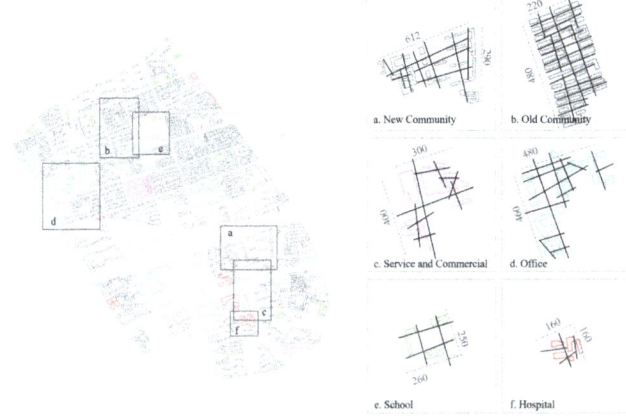

16

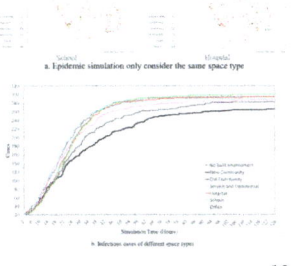

17

16 Six typical spaces in cities

17 Characteristics and attributes of six typical spaces

18 Virus spreading simulation in six typical spaces

19 Virus spreading simulation with multiple center

20 Simulation results with different visual parameters

18

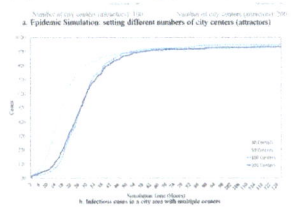

19   20

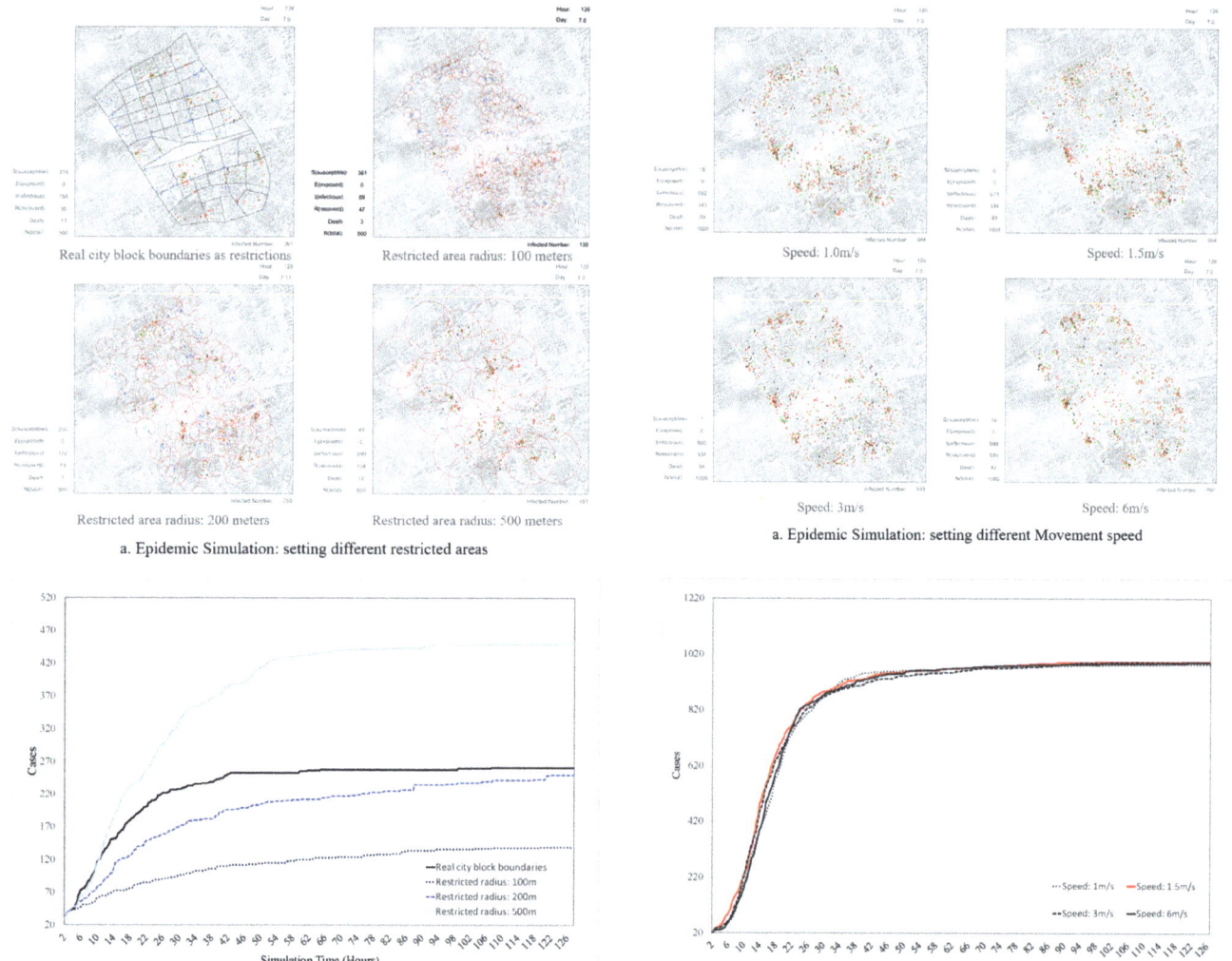

21

22

increase, the overall infection cases show an increasing trend, and the infection rate also increases to a certain extent.

This shows that under the condition of having a wider field of view (greater than 120°), individuals will explore the city to a greater degree, which may cause faster and more infections. With a smaller field of view, individuals are more likely to be restricted by space obstacles, with smaller moving range and speed, and less infection.

Partial Lockdown and Quarantine in City Area

City lockdown is a restriction policy implemented on a larger scale (city level), while the very small-scale spatial restriction belongs to personal quarantine. An isolation strategy that allows individuals to move within a certain range can be adopted, which not only guarantees the freedom of individual movement to a certain extent, but also effectively limits the spread of the virus.

This study chose two space restriction strategies. One is based on the real geographic environment and limited by blocks; the other is space limitation based on size, which analyzes the impact of the size of the limited space on the spread of the virus. The results are as Figure 21 shows.

As the radius of the restricted area decreases, the speed of virus transmission and the amount of infection have decreased significantly. In general, the blockade strategy based directly on community boundaries will cause the virus to spread quickly in early stage, but the total amount of infection will not increase significantly. The blockade of areas less than 200 meters can effectively slow down the spread of the virus and greatly reduce the total amount of infection.

The Impact of Movement Behavior

Speed, as one of the most important parameters of movement, describes the state of a person to a certain extent.

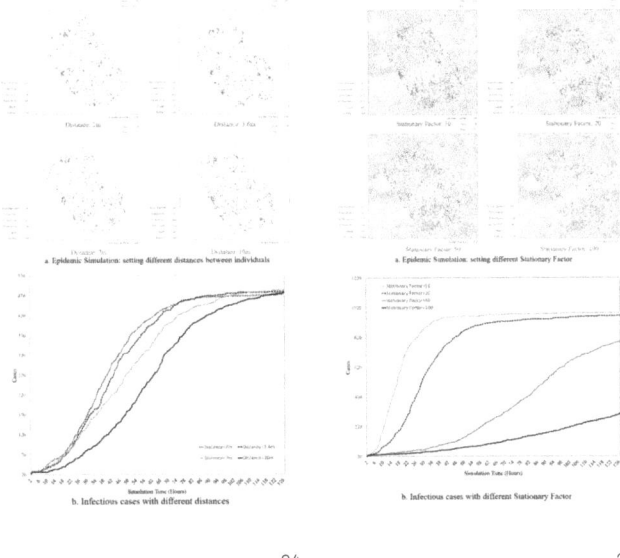

21 Virus spreading simulation with restricted areas

22 Simulation result of movement with different speed

23 Simulation with different stationary factors

24 Simulation result with different distances

Speed changes will also affect the speed of virus spread. The movement speed of young people is 3.31 mph to 3.37 mph (1.48 m/s to 1.51 m/s) (TranSafety 1997). Therefore, when considering lower movement speeds, the parameters are specified at 1 m/s. In addition, considering the individual jogging situation, Keller (1996) pointed out that the average running speed of the individual is 3.0 to 3.4 m/s (female) and 3.4 to 3.8 m/s (male). Therefore, a running speed of 3 m/s is also considered in this experiment.

The simulation results according to different speeds are shown in Figure 22.

On the whole, the movement of individuals at different speeds did not result in significant differences in the spread of the virus. The possible reason is that, without considering the built environment, the speed limit will not affect the individual's movement mode and movement path. Speed does not change the chance of individual contact, so the infection results are similar.

The Impact of Stationary Behaviour

In human behavior stationary activity, such as waiting for the bus, rest, etc. plays an important role. In this experiment, a larger stay parameter will make the individual have a higher probability of staying behavior. The results are shown in Figure 23.

A larger Stationary Factor will significantly slow down the spread of the virus. When the Stationary Factor is 10 and 20, the virus spreads quickly in the early stage. At the 50th hour, its infection has reached 962 and 865, which are close to the peak. However, when the Stationary parameter is 50 and 100, the spread of the virus is significantly reduced. At 50 hours, the infection amount is 148 and 80, respectively.

It shows that staying behavior has a small impact on the final total amount of infection, and can only reduce the number of infection cases within a limited time, but it will greatly delay the time of virus spreading.

The Impact of Social Distancing Behaviour

In the face of COVID-19, social distancing methods are also adopted to deal with the spread of the virus. For example, the UK NHS (National Health Service 2020) recommends a social distance of 2 meters, while the WHO (World Health Organization 2020) recommends a social distance of 1 meter.

This experiment analyzes the differences between different social distance recommendations during the COVID-19 epidemic. The social distance control here is a non-mandatory control, which means after entering the virus transmission distance an individual who finds that the distance is too close can then reacts to increase the distance.

The simulation results of different social distances are shown in Figure 24.

From the results, we can observe that different social distances will obviously affect the speed of virus transmission. The distance becomes larger, the effect becomes better. But the total infections are almost the same. However, the reason may be that this simulation is based on a fixed area while the population remains the same; that is, individuals are moving in a closed space. Therefore, the spread in this kind of enclosed space only affects the speed of the infection and cannot avoid the spread of the virus.

CONCLUSION

After simulation and experiment, the conclusions are as follows:

- Discussing virus spreading in city space should start from two scales: at large scale (city level), the space with decentralized characteristics (network form, distributed layout) shown to slow down the speed of virus transmission; concentrated spatial Features (linear form, centralized layout, high density) will reduce the total amount of infection. From the perspective of small scale (community level), spaces with concentrated characteristics (centralized layout, linear, high-density, etc.) will cause faster and more infections, and vice versa.

- Small-scale space restrictions will accelerate the speed in early stage of virus transmission. This kind of regional restriction measure is actually aimed at controlling the total amount of infection to avoid spreading on a larger scale.

- "Multi-center" strategy of dispersing the population can delay the spread of the virus, but the effectiveness of this effect is also within a certain range of "multi-center" numbers. But it should be noted that this method may bring slightly more infections.

- The wider the view, the more adequate the exploration of urban space. This kind of exploration is essentially different from the movement based on the heat map. It is representative of "Aimless" movement behavior similar to roaming. It will cause faster and larger infections. However, when the field of view is small, that is, when performing "purposeful" behaviors that are mainly driven by heat maps, the early virus spreading speed is faster, but the total amount of infection is small.

- The change of individual movement speed basically does not affect the virus transmission process. This speed refers to normal movement, not low speed or stationary. This is because the individual's behavioral trajectory has not undergone a fundamental change. In a fixed area, individuals will still gather and cause the virus to spread, which has no essential relationship with individuals taking different vehicles (not considering the closed transmission of vehicles).

- Higher staying probability and low-speed movement would significantly slow down the spread of the virus, but did not change the number of spreads. This approach will only delay the spread of the virus, not stop it.

- Social distancing strategy can effectively control the spread of the virus. As the distance between individuals increases, the spread of the virus becomes slower. However, the total amount of infection will not change.

This is because the experiment adopts a "non-mandatory" strategy.

The study itself has certain limitations. In terms of research methods, the simulation is based on the heat map, rather than on real trajectory data. And in this model, many behaviors are assigned to the individual, which leads to differences between the individual's movement and real behavior, further contributing to uncertainty in the results. In terms of research objects, this research is based on a specific city area, which is limited to 2D and exterior space. All conclusions can only be used as a reference rather than universal application. At the same time, the population-based infectious disease model does not consider changes in the number of people in the area, which is an ideal simulation. Since there is no precise mathematical simultion description and biological analysis for COVID-19 input for the infectious disease model, there will be certain limitations in its spreading mechanism. In terms of research content, personal behavior only focuses on human movement behavior, and does not involve strategies such as wearing masks and hands washing.

To sum up, this study is based on real data, applying agent-based model, to create a correlating model of individual movement and epidemics. Conducting a large number of simulation experiments based on this model, reseachers aimed to study the virus spreading process in different situations. Researchers then analyzed the influence of city factors, defined by spatial configuration and individual movement behaviors on the spread of the virus. These research results can be used as a reference for dealing with the spread of the epidemic.

REFERENCES

Ajelli, M., B. Gonçalves, D. Balcan, V. Colizza, H. Hu, J.J. Ramasco, S. Merler and A. Vespignani. 2010. "Comparing large-scale computational approaches to epidemic modeling: agent-based versus structured metapopulation models." *BMC Infectious Diseases* 10 (1): 190.

Beckley, R., C. Weatherspoon, M. Alexander, M. Chandler, A. Johnson, and G. S. Bhatt. 2013. "Modeling epidemics with differential equations." Class project Tennessee State University and Philander Smith University.

Bonabeau, E. 2002. "Agent-based modeling: Methods and techniques for simulating human systems." *Proceedings of the National Academy of Sciences* 99 (suppl 3): 7280–7287.

Chen, D., B. Moulin, and J. Wu, eds. 2015. *Analyzing and Modeling Spatial and Temporal Dynamics of Infectious Diseases*. Hoboken, New Jersey: John Wiley & Sons, Inc.

Friesen, M., and R. McLeod. 2014. "Smartphone trajectories as data sources for agent-based infection-spread modeling." In *Analyzing and Modeling Spatial and Temporal Dynamics of Infectious Diseases*, edited by D. Chen, B. Moulin and J. Wu. Hoboken, NJ: John Wiley & Sons, Inc. 443-472.

Keller, T.S., A.M. Weisberger, J.L. Ray, S.S. Hasan, R.G. Shiavi, and D.M. Spengler. 1996. "Relationship between vertical ground reaction force and speed during walking, slow jogging, and running." *Clinical Biomechanics* 11 (5): 253–259.

Koopman, J. S. and J. W. Lynch. 1999. "Individual causal models and population system models in epidemiology." *American Journal of Public Health* 89 (8): 1170–1174.

National Health Service (NHS). 2020. "Social distancing: what you need to do." Accessed August 20, 2020. https://www.nhs.uk/conditions/coronavirus-covid-19/social-distancing/what-you-need-to-do/.

Sattenspiel, L. and A. Lloyd. 2009. *The Geographic Spread of Infectious Diseases: Models and Applications*, vol. 5. Princeton: Princeton University Press.

TranSafety, Inc. 1997. "Study compares older and younger pedestrian walking speeds." *Road Management & Engineering Journal*.

World Health Organization (WHO). 2020. Q&A on coronaviruses. Accessed May 14, 2020.

World Health Organization (WHO). 2020. "Coronavirus disease (COVID-19) advice for the public." Accessed August 20, 2020. https://www.who.int/emergencies/diseases/novel-coronavirus-2019/advice-for-public.

World Health Organization. 2020. "Report of the WHO-China Joint Mission on Coronavirus Disease 2019 (COVID-19) February 2020." Archived (PDF) from the original on 29 February 2020. Accessed March 21, 2020. https://www.who.int/publications/i/item/report-of-the-who-china-joint-mission-on-coronavirus-disease-2019-(covid-19).

IMAGE CREDITS

All drawings and images by the authors.

Mengda Wang is an architect and architectural algorithm designer. He hold an MSc degree in Architectural Computation with distinction in the Bartlett School of Architecture, University College London. He works in digital product R&D for the world's top 500 enterprises. His research interests are architectural modeling algorithms, agent-based modelling, particle systems and machine learning.

Ava Fatah gen. Schieck is an Educator, architect and researcher. She is Associate Professor in Media Architecture and Urban Digital Interaction, the director of the PhD Programme "Architectural Space and Computation," the Departmental Graduate Tutor for the PhD Programme Architecture and Digital Theory, and leads the studio "Body as Interface | City as Interface" for the Architectural Computation Program at the Bartlett School of Architecture. Her research is practice based resulting in a unique 'living lab,' environment (2010, UK). The research focuses on the area of Human Building Interaction (Architecture, Interaction Design, and Ubiquitous Computing) developed since 2001 through teaching and research positions at The Bartlett (UCL). She has published over 100 peer-reviewed publications in the areas of Media, Architecture, and Human Computer Interaction and has acted as the programme chair for the Media Architecture Biennale (since 2012).

Petros Koutsolampros is an architect by training and a specialist in spatial analysis and computation. He holds an architecture engineering degree from the National Technical University in Athens and an MSc in Adaptive Architecture and Computation at The Bartlett, UCL. His PhD focused on the effects of spatial configuration on workplace movement and interaction. Petros currently works for Space Syntax Ltd. and also teaches various technical subjects at UCL, primarily the MSc "Architectural Computation" and MSc "Space Syntax," and "Architecture and Cities" courses. His teaching and research interests include spatial analysis and visualisation, virtual and augmented-reality applications, and agent-based modelling.

Automated Generation of Custom Fit PPE Inserts

For Respiratory Masks Using
2D and 3D Anthropometric Data

Alicia Nahmad Vazquez
University of Calgary, SAPL

Li Chen
Generative Technology Limited

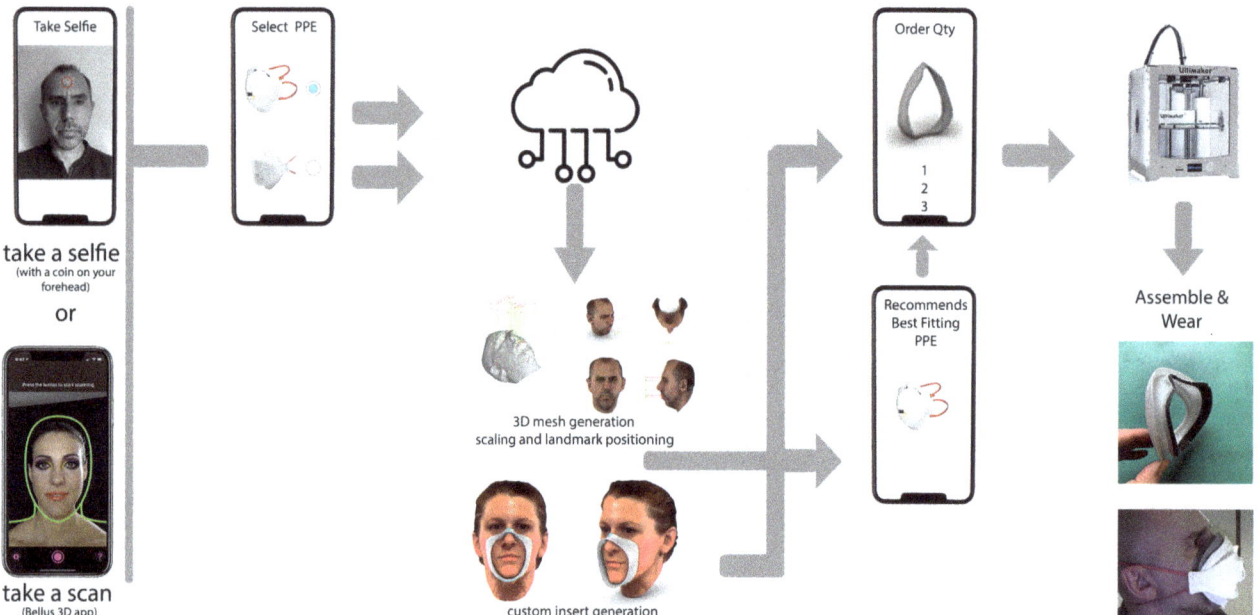

1   Mask app user journey

ABSTRACT

This research presents a machine learning-based interactive design method for the creation of customized inserts that improve the fit of the PPE 3M 1863 and 3M 8833 respiratory face masks. These two models are the most commonly used by doctors and professionals during the recent COVID-19 pandemic. The proper fit of masks is crucial for their performance. Characteristics and fit of current leading market brands were analyzed to develop a parametric design software workflow that resulted in a 3D printed insert customized to specific facial features and the mask that will be used. The insert provides a perfect fit for the respirator mask. Statistical face meshes were generated from an anthropometric database, and 3D facial scans and photos were taken from two hundred doctors and nurses on an NHS Trust hospital. The software workflow can start from either a 2D image of the face (picture) or a 3D mesh taken from a scanning device. The platform uses machine learning and a parametric design workflow based on key performance facial parameters to output the insert between the face and the 3M masks. It also generates the 3D printing file, which can be processed onsite at the hospital. The 2D image approach and the 3D scan approach used to initialize the system were digitally compared, and the resultant inserts were physically tested by twenty frontline personnel in an NHS Trust hospital. Finally, we demonstrate the criticality of proper fit on masks for doctors and nurses and the versatility of our approach augmenting an already tested product through customized digital design and fabrication.

INTRODUCTION

In line with the "Realignments" topic of this year's conference, this paper uses the digital and material knowledge and skills of digital fabrication, parametric design, and machine learning to develop a research to fabrication project that addresses the issues of comfort and fitting of respiratory masks. Both are critical factors that require a mass-customization design approach. The lack of customized respiratory masks affects doctors and frontline personnel fighting the pandemic. We realign our design skills to address matters of urgency.

During the recent COVID-19 pandemic, the 3M 1863 and 3M 8833 became the main masks used by doctors and frontline workers. Fit and comfort level are critical factors, and mask technical characteristics and protection levels have been perfected through extensive research and testing. However, mask performance is dependent on proper fit. Typical respiratory mask fit is based on a flexible metal clip that goes on top of the nose. The clip is located in the top panel for the 1863 mask and the top of the 8833 mask. The nose clip must be pressed upon the nose once the mask has been pre-formed over the face. After pulling the headbands over the head, the nose clip needs to be pinched again until it is comfortable and positive pressure is felt inside the piece. During the fit (i.e. pinching it only with one hand), errors can result in less effective respirator performance (3M 2016). The respiratory mask fitting problem is twofold. On the one hand, although the mask design allows fitting a wide range of face shapes, only two sizes—M and L—are available to fit an extensive range of faces. Moreover, mask sizes are based on studies of American male and female soldiers, which may not correspond to the majority of the population which they are intended to fit (Zhuang and Bradtmiller 2005). Comfort becomes critical when masks have to be worn every day for prolonged periods, as has been the case during the pandemic.

On the other hand, it is easy to make errors during the fitting process due to the lack of comfort, compromising performance. The leakage of aerosol pathogens through the gaps between the face and the respiratory masks has been a well-documented and simulated problem during the recent pandemic. Leakage compromised the device's effectiveness whilst increasing the risk for the wearer (Hariharan et al. 2021). The design quality of mass-produced masks does not correspond to their performance. The design is uncomfortable and does not fit all faces properly, leading to leaks and discomfort. Satisfactory solutions for the design of customized 3D respiratory masks are lacking.

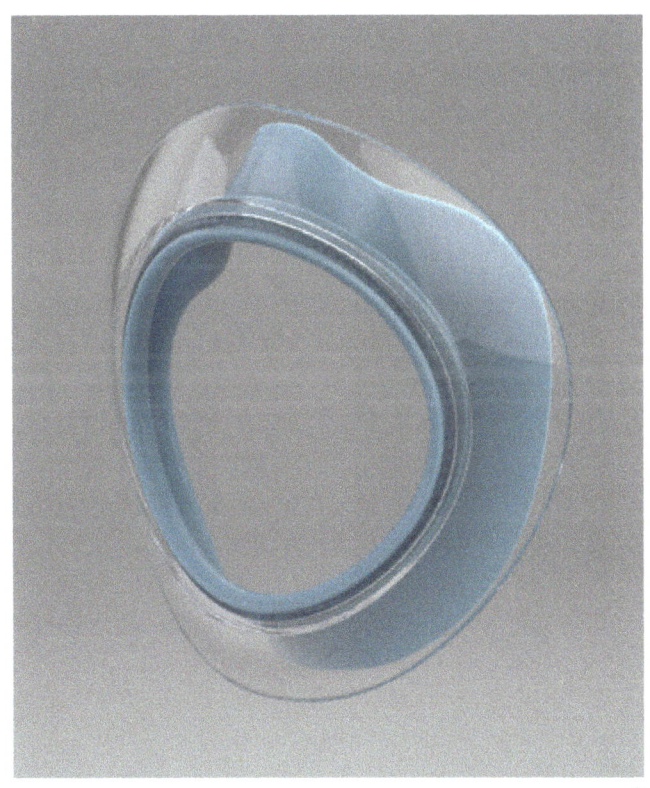

2

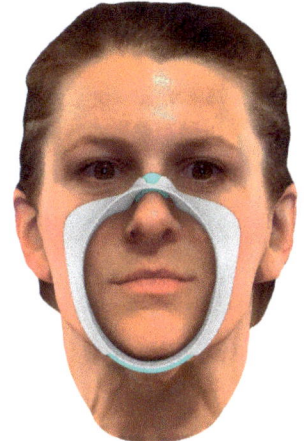 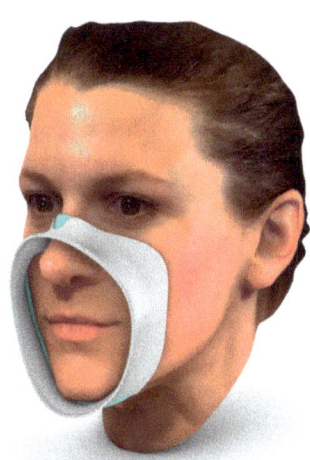

3

2 Bespoke hybrid soft/rigid PPE interface: the interior is a 2D profile cut from foam material that fits in a groove printed in the interior of the insert (mask shell not shown)

3 3D printed PPE interface with foam liners (mask shell not shown)

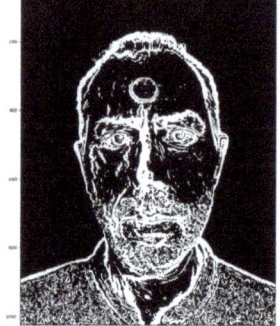
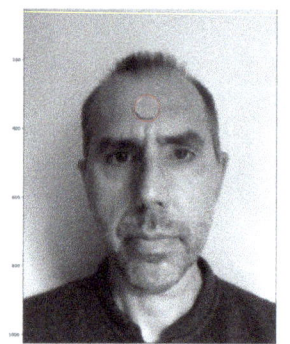

4

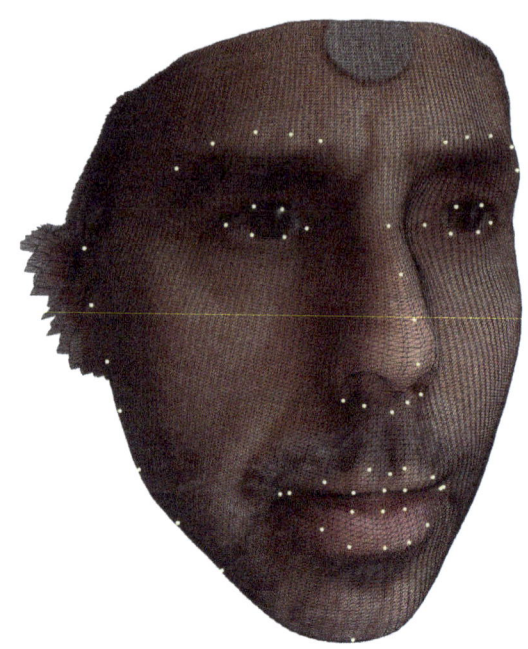

6

5

The rapid development of 3D scanning technologies allows designers to collect large amounts of data economically, quickly, and efficiently. Popular mobile phones such as the iPhone come equipped with state-of-the-art depth sensors, and high-quality 3D reconstruction software available for them, like Bellus scanning technologies. Personalized design based on anthropomorphic data is a trend in the fashion and accessories industries that enhances the value added to a product whilst satisfying individual customer requirements (Zheng et al. 2007; Tseng et al. 2014).

Two different approaches can be identified using anthropomorphic data on product design: design enhancement and design customization. Design enhancement uses anthropomorphic knowledge to improve product performance. Designers, in this case, determine key design parameters based on the information extracted from anthropometric data (Sanders and McCormick 1998; Loker et al. 2005). Examples of this are the tools designed for bra fitting by Loker et al. (2005). Design customization focuses on a variety of characteristics of the individual wearer. They scan parts or the whole shape of the customer and adjust the design automatically or semi-automatically through the use of CAD tools (Istook and Hwang 2001; Tognola et al. 2004; Chu et al. 2015). This method has been widely used in the fashion industry (Istook and Hwang 2001). It has also been deployed to design medical applications such as ear canal impressions (Tognola et al. 2004). Tognola et al. use 3D laser scanning to reconstruct ear canals. The 3D mesh reconstructed file is then passed to 3D printing machines to print the moulds and ensure a perfect fit of the device.

Shape transformation methods have also been explored as tools for customization using statistical shape analysis (Unal et al. 2008). In this method, relationships are established between two classes of shapes, which generate a shape of one class when given a shape of the other class. Finally, the United States National Institute for Occupational Safety and Health (NIOSH) developed an anthropometric database detailing the face size distributions of respirator users. They established a fit test panel by the face dimensions and landmark positions measured from 3997 subjects. As a result of this database, a new five-category sizing system was proposed using PCA or principal component analysis (Zhuang et al. 2007). However, a common problem across all of these studies was the high cost of 3D scanning devices and their inaccessibility to the general public. Customized products are expensive, limiting mass-customization to a selected few. The COVID pandemic started at a moment in which scanning devices with high resolution, machine learning, parametric modelling techniques, and rapid prototyping machines are widely available. These tools provide an alternative to this problem extending the feasibility of having customized products related to the human face. This research presents a workflow to design and fabricate a customized mask insert without the need for conventional scanning technology that requires highly expensive equipment and skilled professionals. The aim is to make workflow into an app accessible to medical professionals at their places of work and together with an on-site 3D printer, enable them to have a customized fit. This paper presents the development and testing of the workflow before encapsulating it into the app.

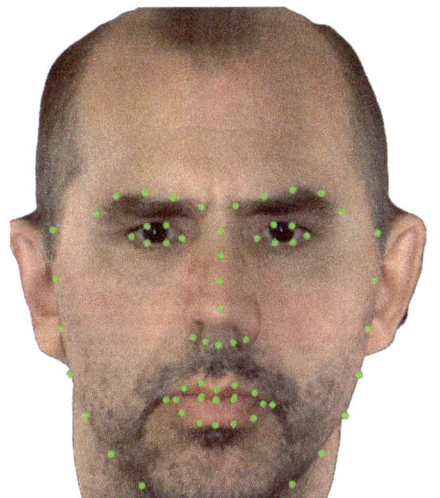
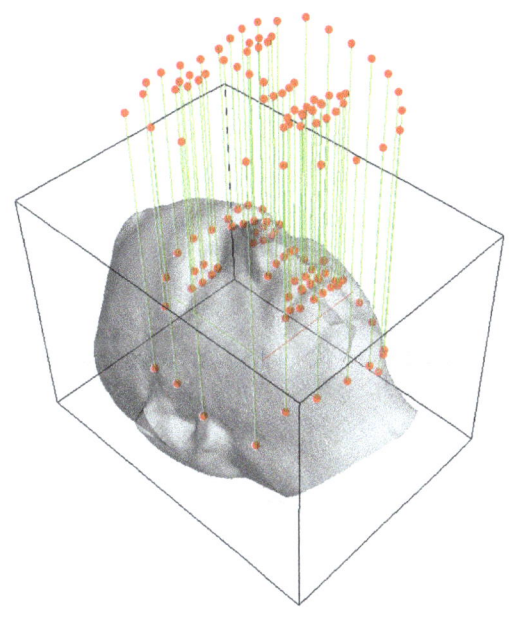

4 Annotated landmarks

5 3D face reconstruction from 2D image with landmarks

6 10p coin used for scaling as an object of identifiable radii, placed on the forehead

7 (Left) 2D landmark detection; (Right) Landmarks projected into their 3D locations

Unlike other research projects that focus on generative methods for the entire customized mask respirators—masks that can then be fabricated in some cases by 3D printing (Chu et al. 2015), or by folding (Nejur and Szentesi-nejur 2021)—this project focuses on developing a digital workflow that can parametrically model a custom fit insert to be worn between the face and the mask. The approach allows incorporating all the respiratory technical advances of well-tested masks whilst improving the fit through customized parametric design and 3D printing technology. Focusing on an insert reduces the timeframe of testing and certifying a new respirator mask allowing for the fast deployment of mass-customized, well fitting masks across hospital facilites in a short period of time. Furthermore, the design software and workflow address the two respiratory masks most commonly available in hospitals, presenting the potential to reach more doctors and personnel faster and avoid being dependent on a single product's supply chain. This was an issue exceedingly pressing in the early days of the pandemic when masks were in short supply.

PROJECT

The research developed in this paper aims to augment the fitting of medical-grade respiratory masks rather than redesigning them. It develops a digital workflow to translate 2D images of a user's face into a 3D model to perform a virtual fitting process and parametrically model a custom fit PPE interface worn between the face and the mask respirator. The research outcome is a software application that can be successfully used to custom fit multiple respiratory masks through the design and fabrication of the insert. The developed workflow can be deployed in two ways. Firstly, a simple 2D photograph of a face is taken with any mobile device, which in turn uses machine learning to convert 2D data to 3D data, detecting facial landmarks, positioning the respiratory mask, and generating the insert for a perfect fit. Alternatively, the workflow can also be initiated via a 3D scan of a face using a Bellus scanner or Bellus app. The same steps of landmark detection, mask positioning, and insert generation runs after the 3D mesh is either created or retrieved. The application output in both cases is an .obj file for 3D printing the insert that can be immediately processed onsite at the hospital for an instant fit.

DETAILED DESCRIPTION

Data for this project was collected in the form of 2D photographs and 3D scans of doctors and nurses from the Brighton and Sussex University Hospital NHS Trust. Two hundred doctors were scanned and photographed, which allowed for a diverse sample that included a variation on facial features.

A digital pipeline (Figure 1) and a UI were designed for the doctors to generate their insert and print it on-site. The software starts with an input: either a 2D photograph taken from a phone camera or a 3D scan utilizing the Bellus scanner and similar on the iPhone mobile device. In the 2D to 3D approach, the second step utilizes a machine learning model PRNet (Feng et al. 2018) to reconstruct the mesh and place key landmarks around the face. In the scanning approach, the second step uses machine learning to detect facial features and place landmarks around the face. From this step onwards both workflows work similarly: the landmarks are used as generative points to parametrically

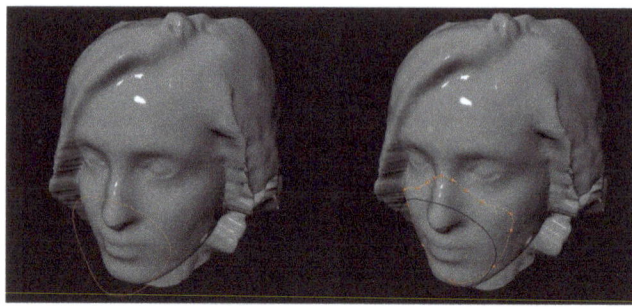

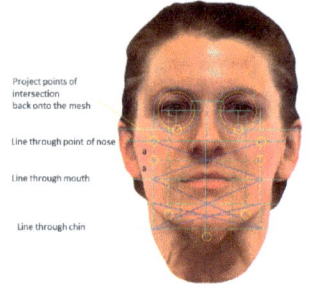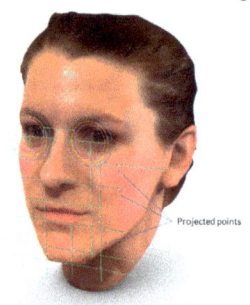

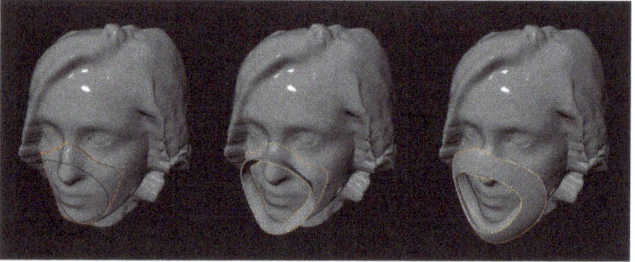

8 (Left) Profile in front of face; (Right) Soft profile based on landmarks

9 Setup of critical geometry based on facial landmarks analysis and generative system setup

10 (Left) Curve subdivision; (Middle and Right) Loft and hard shell

create a hard shell insert—near the face—based on face contours and connecting the face to the mask. The insert has a groove to receive a soft 'U' shaped insert between the customized hard shell and the face (Figure 2). The mask fit of the insert is simulated using a physics solver in blender. The designed insert—based on facial landmarks—provides a hybrid soft/hard shell interface between the face and the 3M masks that ensures a perfect fit but also keeps the user comfortable (Figure 3). The detailed workflow is described in the next section.

STEP BY STEP WORKFLOW
2D to 3D

The research develops two pipelines for the generation of the hard insert. The first one begins from a 2D image, the second one from a 3D scanned mesh.

PRNet (Feng et al. 2018) is used for generation of a 3D face model from a 2D image. It utilizes a standardized 3D face mesh model with a set of facial landmarks as anchor points (Figure 4). The machine learning model is trained to firstly pull these anchor points to the corresponding 2D locations of the input image, resembling the facial topology two-dimensionally, and then regressing the depth of each vertex on the mesh to their approximated 3D coordinates. Thus, the resulting output is not only the reconstructed 3D face model, but also its 3D facial landmarks (Figure 5), which can be use directly for the parametric generation of the mask geometry.

With 2D-to-3D output being an approximation of facial structure in a local coordinate system, real-world reference is required to obtain the actual scale of the face. To achieve this, objects like a credit card or a ten pence British coin can be included in the photo of the face. Such objects can then be easily detected using mature computer vision tools like OpenCV, providing a real-world scale reference (Figure 6).

3D Workflow

Another viable and developed workflow is to obtain accurate 3D faces scans through professional 3D scanning equipment like the ones provided by Bellus Technology, or directly through mobile devices with a depth sensor such as iPhone X. Once the accurate 3D scan is obtained, the software will extract the 3D locations and place facial landmarks.

Since the facial landmark detection of 2D images is already a mature machine-learning technology, it was decided to take this advantage and first detect landmarks two-dimensionally with a front-view render of the face scan (Figure 7, Left), and then project these landmarks back to their 3D locations (Figure 7, Right). For 2D landmarks detection a pre-trained machine learning model from dlib was used, and for 3D projection compas (Mele 2017-2021) was used. It should be noted that there are also methods to infer landmarks directly from 3D models through analysing the mesh topology and surface curvature, which can potentially create more accurate results. However, this direction was not further explored due to the time limit and possibly much higher computing cost on the production stage.

A fixed profile is orientated in front of the face (Figure 8, Left), which will act as the connection to the hard-shell part of the insert. Then, based on the landmark analysis and locations (Figure 9), a group of control points are taken and used to create the soft profile (Figure 8, Right). The next step is to further subdivide the soft profile (Figure 10, Left).

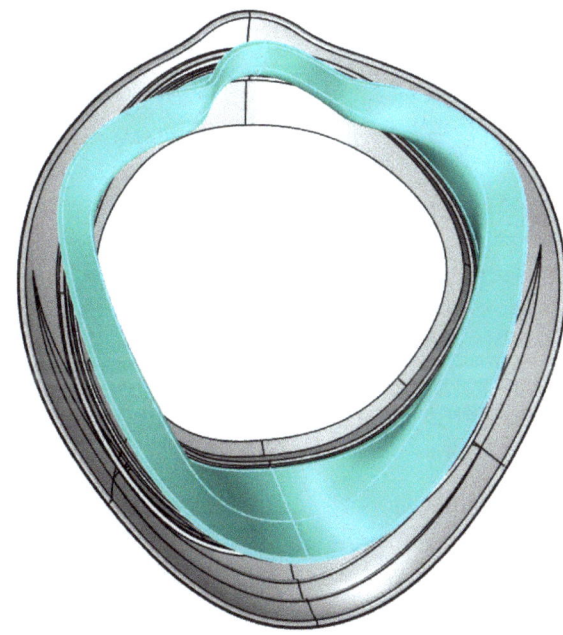

11 Grooves on the hard insert allowing for different positions of the soft insert

The intermediate subdivision points are projected to the face to achieve a better conformation between the insert and the face geometry. Finally, the 3D geometry is generated by lofting between the two profiles using 3D Bézier curves (Figure 10, Middle & Right). After these steps are completed, the insert gets 3D printed and attached to the mask for a perfect fit—without fogging, wrinkling around the face, or nose clip, etc. (Figure 13).

DATABASE

A database was capture from ninety-seven doctors and front line personnel from the Brighton and Sussex University Hospital NHS Trust. Two sets of four photographs were taken from each doctor: set one consisted of two frontal and two profile images (left and right) without a mask; and set two consisted of the same images of the subject wearing one of the two 3M masks used in this study. Additionally, two head scans using Bellus technology were taken from each doctor (with mask and without a mask).

All doctors willingly participated in this project. The participants gave their consent to have their pictures and scans taken and used for the context of this research. Nevertheless, consent hasn't been sought at this stage towards their photographs being made public.

DIGITAL PIPELINE TESTS & VALIDATION

Both developed workflows (2D image-based and scan-based) were tested with samples from the database. The 3D scan workflow was run through the ninety-seven doctors in the database to generate inserts (Figure 14). By running the digital pipeline on the different doctors, the process was refined. It allowed identification of facial landmarks in which significant variance can result in differences crucial to the fit of the insert. Additional considerations were incorporated for these cases. The two sets of facial landmarks that their variance beyond range affect the insert are:

1. *Landmark at the tip of the chin.* When the chin is pointing up, the point at the chin becomes problematic when generating the customized as values are negative (Figure 15). To calculate that point, a different methodology was devised. A line was dropped from the center of the nose towards the chin and measured. If the result was on the negative range with the chin pointing upwards, then the point was angled back to adapt the insert. Otherwise, initial chin points are used.

2. *Distance between the nose and the edge of the eye.* Considered when the nose was extremely small and did not pop-out of the face or a high cheek-to-nose ratio. In these cases, the insert became very shallow as it had to spread over a larger area (cheeks) and did not have the required depth (Figure 16). A new constraint had to be introduced to move the outer edge of the insert further out from the face hence generating a larger insert that allowed for situations when the variance between landmarks was out of range.

After these two discoveries—emerging from testing the workflow digitally with the ninety-seven doctors on the dataset and analyzing the resulting 3D files—the digital pipeline was physically tested with ten doctors. Due to COVID constraints no more physical tests could be performed at this stage. The 2D photograph-to-mesh method was used and inserts were 3D-printed. The doctors wore the insert for the week between the November 23 and 27, 2020. During this period, data was collected through interviews regarding the fit and comfort level of the insert. Specific questions regarding fogging of glasses, breathability, and fit were considered.

PHYSICAL TESTS

Ten doctors from the Brighton and Sussex University Hospital NHS Trust tested the inserts for five days as part of this project. Doctor pictures and scans were run through the digital pipeline to generate the customized insert. The workflow allowed for a 2D insert, independently cut from soft foam, and the shell to be printed on a regular Ultimaker machine using PLA.

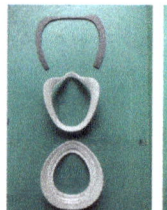   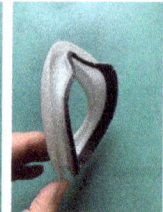

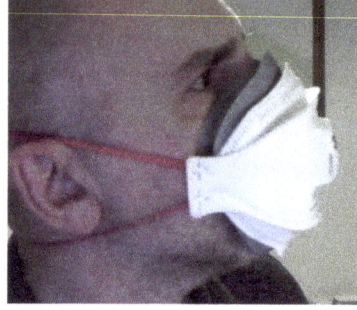 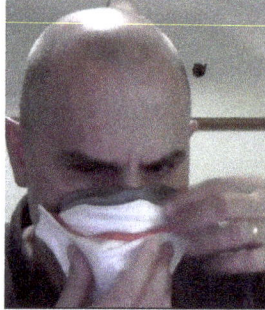

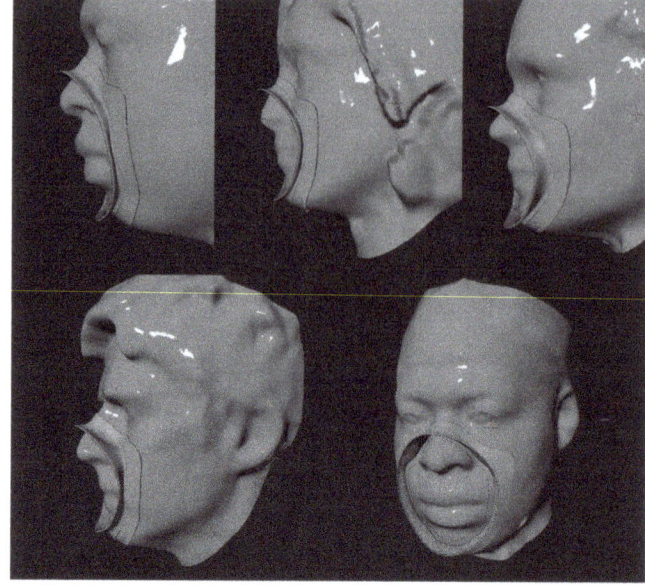

12 Photos of early prototype exploring the soft and hard shell concept
13 Fitting of 3M respiratory mask using customized insert
14 Successful generation of customized PPE inserts on various doctors and nurses from the dataset
15 Variance on the chin landmark becoming problematic

In the case of two inserts, there was a minimal gap on the area below the eyes between the face and the insert, which caused air to flow, meaning the insert was not sealing correctly (Figure 17). The insert was not used. The remaining eight doctors used the customized insert with a 3M mask for five days.

The feedback collected through interviews was mainly positive. Doctors appreciate the customized fit of the insert and found comfortable the presence of a soft layer between their face and the mask differently from the traditional clip that the masks have on the nose area. It was also appreciated that doctors who wear glasses did not experience fogging problems that they would have otherwise.

One issue that was not considered in the original design and emerged from the testing phase was the use in an OR situation. Doctors need to wear specific goggles inside the OR, which could not fit comfortably when the insert was in place. A solution for this is to bring the insert and mask further out from the face. The fit would be lower, and there would be sufficient space for the goggles.

The two inserts that did not fit properly required more points to be introduced during the generative process to allow for a smoother curvature between the nose and the cheek. This issue seems related to point two on the digital pipeline test.

CHALLENGES
Scaling Deviation

In the 2D-to-3D workflow, we used circle detection to mark a physical coin's radius as a scaling reference. The accuracy of detection could be affected by factors such as poor lighting and low color contrast between skin and coin. The deviation could also be amplified if the coin was too small. Initially, the scaling was visualized with a credit card next to the face. However, people tend to hold the credit card at an angle, causing inaccurate scaling of the face and affecting downstream design processes. Coins adhere to the forehead without any additional effort. Three coins—the most commonly used in the UK—were measured and defined to scale the photographs. Doctors were asked to place a coin 5p or 10p on their forehead before taking a photograph. The 20p and 50p coins are not circular and could be used. Coins had a great level of accuracy in scaling faces and features and proved to be and adequate and easily accessible object. After scaling, facial measures were compared with the measures of the participants faces to calibrate the dimensions. The coins proved an efficient scaling tool.

Further, comparisons were made between the 2D to 3D method using the coin scaling process and the 3D Bellus scan (Figure 18). The aim was to measure the deviations of the points and landmarks. The comparison between both methods showed that the 2D-to-3D method remains like a 2.5D image. The features work correctly but lack depth, specifically in cases where the nose is more prominent. The result is a mesh more similar to an engraving than an actual 3D representation. To compare the deviation between two methods, we created a simple Rhino Grasshopper script to project equally distanced points from mask profile onto each face models and measured

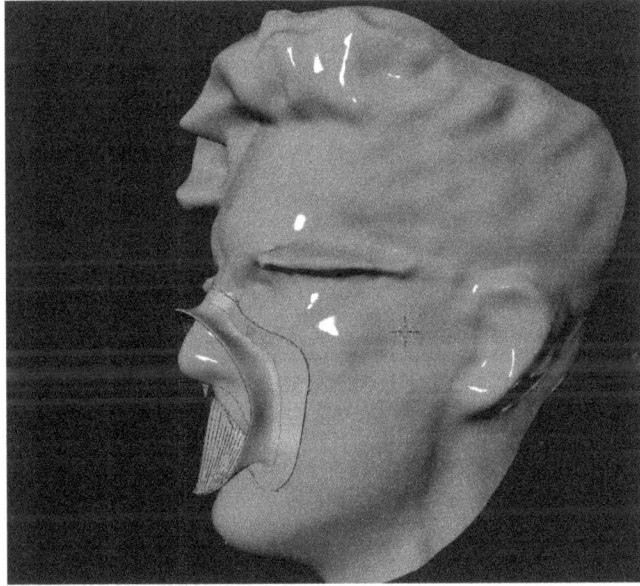

15

deviations between each set of these points (Figure 19). As a result, the maximum deviation between both methods was 10mm with a 4mm deviation between the tips of the nose.

Interestingly, the insert fits better on the 2D-to-3D face than on the 3D mesh resultant from scanning. A possible explanation is the 2D-to-3D machine learning algorithm produces a more 'generic' 2.5D face that makes fitting easier. The 3D scan contains all the face information so the fit is worse, though more faithful to the real face.

STREAMLINING PROCESS INTO REDISTRIBUTABLE SOFTWARE

Whilst the design, research, and prototyping for this project was carried out in commercial CAD software like Rhinoceros with its parametric design tool Grasshopper, the proprietary nature of the software means that it is impossible to automate the workflow as a production pipeline or for it to be repackaged into redistributable applications. Thus, in the first version of the application, a wide range of open-source projects with compatible licenses for the necessary functionalities were utilized. The more important ones are:

Pipeline A - 2D images to 3D:
- 2D image to 3D model: PRNet (MIT license)
- Circle Detection: OpenCV (Apache 2 License)

Pipeline B - with 3D scans:
- Face scan Rendering: Vispy (BSD license)
- Landmark Detection: Dlib (Boost Software License)
- 3D landmark projection: COMPAS (MIT license)

Shared by both Pipelines:
- 3D projection: COMPAS (MIT license)
- Parametric modelling of mask: Blender (GNU General Public License)

The final application is delivered as a Linux Docker container image, supporting cross-platform usages while using Docker as a subsystem. This application can used as a standalone program that takes in picture or 3D scan and automatically output watertight 3D printable mask insert files. It can also be deployed on a server environment to process images and scans received directly from a front-app mobile app or webpage in a Internet browser, sending out 3D printable files into a production queue. Both setups do not require any intervention of trained professionals (Figure 20).

IMPACT AND OUTCOMES

Based on the feedback from the doctors, minor modifications will be added to the digital pipeline for a new round of physical testing. The project aims to have an app that can be deployed across NHS hospitals with their corresponding 3D printers and rubber profiles to guarantee the perfect fit of each mask to the doctors and other frontline personnel.

CONCLUSIONS

Doctors, nurses, and customers have a strong need for a customized alternative to respiratory masks for comfort during prolonged use and protect them from infectious diseases. Most existing design customization methods for respiratory masks focus on redeveloping the mask. This approach requires expensive 3D scanning software and multiple testing to reach already approved market leaders' performance levels and certifications. The research presented in this paper focuses on developing an insert that augments the comfort and fit of existing masks whilst taking advantage of their properties.

The developed and presented digital pipelines was successful in both workflows—1) starting from a 2D photograph and converting it into a 3D mesh through the use of machine learning, and 2) starting from a 3D scan— in creating inserts that are customized and comfortable to wear for doctors with a normal approved 1863 or 8833 3M mask. The workflow from an image or a point cloud generates an .obj 3D printable file of the insert in around 30 seconds. Thus, inserts are 3D printed and can be worn by the doctors in two hours, ensuring a perfect, comfortable fit whilst increasing the performance of the mask.

The parametric analysis and setup of facial parameters were successful in generating well-positioned and well-fitting inserts. However, the resultant facial landmark

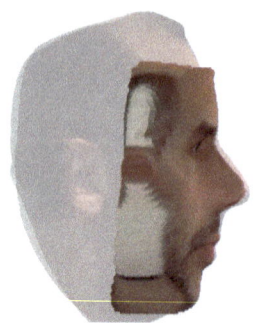
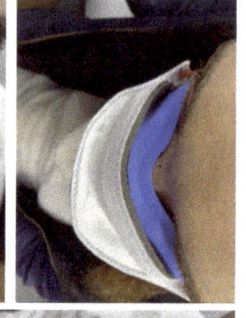
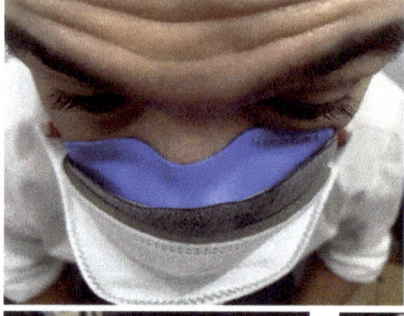
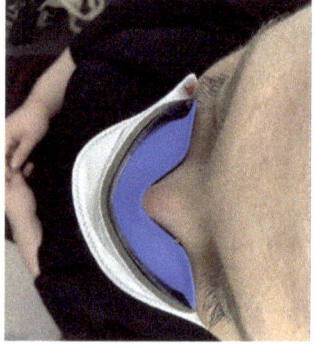
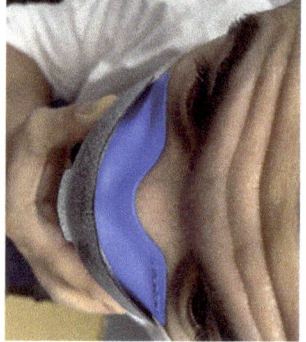

16 Variance on the nose to cheek ratio between users required revisions of the generative process

17 Different results of physical tests with doctors

18 Comparison of final mesh between 2D-to-3D method and 3D scan.

research highlighted specific cases that require large ranges due to the amount of variance within them. These cases were successfully resolved by allowing more flexibility on the parameters of the digital fitting. The generative process has been tested with a random sample of ninety-seven doctors and nurses of the Brighton and Sussex University Hospital NHS Trust. The selection has sufficient variation in facial features, ethnicities, and gender representation composition and dimensions for this project to be considered successful. However, doctors from a different geographical region might show variance in facial landmarks that are not accounted for in current generative process.

ACKNOWLEDGMENTS

This work was developed in collaboration with Rogue Product UK and Emteqlabs, funded by Innovate UK.

REFERENCES

3M. 2016. "Respiratory Protection for Health Care Workers: Fitting a 3MTM 1863 Unvalved FFP3 Respirator." 3M HSO guidelines.

Chu, Chih Hsing, Szu Hao Huang, Chih Kai Yang, and Chun Yang Tseng. 2015. "Design Customization of Respiratory Mask Based on 3D Face Anthropometric Data." *International Journal of Precision Engineering and Manufacturing* 16 (3): 487–94. https://doi.org/10.1007/s12541-015-0066-5.

Feng, Yao, Fan Wu, Xiaohu Shao, Yanfeng Wang, and Xi Zhou. 2018. "Joint 3d Face Reconstruction and Dense Alignment with Position Map Regression Network." In *Computer Vision, ECCV 2018*. Lecture Notes in Computer Science, vol. 11218. Cham: Springer. 557–74. https://doi.org/10.1007/978-3-030-01264-9_33.

Hariharan, Prasanna, Neha Sharma, Suvajyoti Guha, Rupak K Banerjee, Gavin D'Souza, and Matthew R. Myers. 2021. "A Computational Model for Predicting Changes in Infection Dynamics Due to Leakage through N95 Respirators." *Scientific Reports* 11 (1): 10690. https://doi.org/10.1038/s41598-021-89604-7.

Istook, Cynthia L, and Su-Jeong Hwang. 2001. "3D Body Scanning Systems with Application to the Apparel Industry." *Journal of Fashion Marketing and Management* 5 (2): 120–32. https://doi.org/10.1108/EUM0000000007283.

King, Davis. 2021. "dlib-models." https://github.com/davisking/dlib-models#shape_predictor_68_face_landmarksdatbz2.

Loker, Suzanne, Susan Ashdown, and Katherine Schoenfelder. 2005. "Size-Specific Analysis of Body Scan Data to Improve Apparel Fit." *Journal of Textile and Apparel, Technology and Management* 4 (3): 1–15.

Nejur, Andrei, and Szende Szentesi-Nejur. 2021. "The F8ld Mask." In *Projections; Proceedings of the 26th CAADRIA Conference*, vol. 1. Hong Kong and Online. 503–12.

Sanders, M. S., and E. J. McCormick. 1998. "Human Factors in Engineering and Design." *Industrial Robot* 25 (2): 153. https://doi.org/10.1108/ir.1998.25.2.153.2.

Tognola, Gabriella, Marta Parazzini, Cesare Svelto, Manuela Galli, Paolo Ravazzani, and Ferdinando Grandori. 2004. "Design of Hearing Aid Shells by Three Dimensional Laser Scanning and Mesh Reconstruction." *Journal of Biomedical Optics* 9 (4): 835–43. https://doi.org/10.1117/1.1756595.

Tseng, Chun Yang, I. Jan Wang, and Chih Hsing Chu. 2014. "Parametric Modeling of 3D Human Faces Using Anthropometric

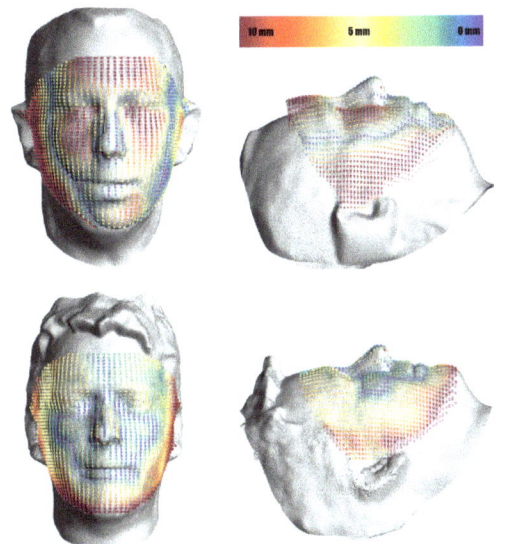
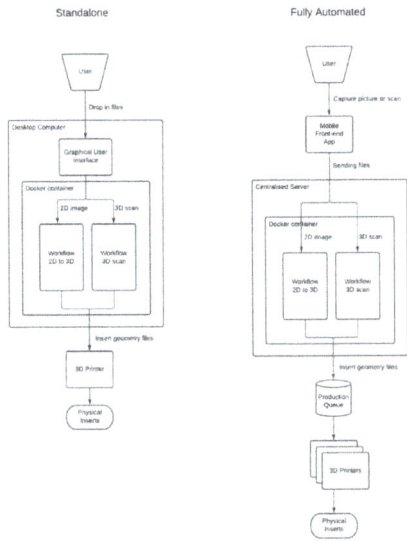

19 Comparison between 2D to 3D workflow and 3D scan based workflow

20 Software architecture

Data." In *2014 IEEE International Conference on Industrial Engineering and Engineering Management*. 491–95. https://doi.org/10.1109/IEEM.2014.7058686.

Unal, Gozde, Delphine Nain, Greg Slabaugh, and Tong Fang. "Customized Design of Hearing Aids Using Statistical Shape Learning." 2008. In *Lecture Notes in Computer Science: Artificial Intelligence and Lecture Notes in Bioinformatics*, 5241 LNCS, no. PART 1 (2008): 518–26. https://doi.org/10.1007/978-3-540-85988-8_62.

Van Mele, Tom et al. "COMPAS: A Framework for Computational Research in Architecture and Structures.", n.d. https://doi.org/10.5281/zenodo.2594510.

Zheng, Rong, Winnie Yu, and Jintu Fan. "Development of a New Chinese Bra Sizing System Based on Breast Anthropometric Measurements." 2007. *International Journal of Industrial Ergonomics* 37 (8): 697–705.

Zhuang, Ziqing, and Bruce Bradtmiller. 2005. "Head-and-Face Anthropometric Survey of U.S. Respirator Users." *Journal of Occupational and Environmental Hygiene* 2 (11): 567–76. https://doi.org/10.1080/15459620500324727.

Zhuang, Ziqing, Bruce Bradtmiller, and Ronald E Shaffer. 2007. "New Respirator Fit Test Panels Representing the Current U.S. Civilian Work Force." *Journal of Occupational and Environmental Hygiene* 4 (9): 647–59. https://doi.org/10.1080/15459620701497538.

Alicia Nahmad Vazquez is the founder of Architecture Extrapolated (R-Ex) and Assistant Professor at the University of Calgary SAPL. She is also co-director of the Laboratory for Integrative Design. As a research-based practicing architect, Alicia explores materials and digital design and fabrication technologies along with the digitization of building trades and the wisdom of traditional building cultures. Her projects include the construction of award-winning 'Knit-Candela' and diverse collaborations with practice and academic institutions. She holds a PhD from Cardiff University and a MArch from the AADRL.

Li Chen is a software engineer specialised in computational geometries and machine learning for computer vision. He is the director of Generative Technology Limited, a Hong Kong based technology company for customized software solutions utilizing generative design and machine learning.

TinyZ

Zain Karsan
Massachusetts
Institute of Technology

A Desktop CNC Machine to Enable Remote Digital Fabrication

1 TinyZ 4-axis Pumpkin Machining Experiment

ABSTRACT

The circumstances of the pandemic have resulted in the closure of workshops and Fab Labs and put physical making on hold for fabrication-based design courses. However, with digital fabrication having become a crucial component of design education, involving the critical transition from design ideas represented digitally to being realized physically, alternative approaches needed to be found. Remote making can be enabled by the potentials of small-scale modular machines, which due to their low cost, are easily distributable and can be shipped to each student in a design studio. The use of at-home fabrication offers new possibilities for project-adaptive prototyping tools.

Desktop scaled fabrication tools designed to reach a distributed audience abound in industry, academia, and amongst DIY-ers. Drawing from these precedents, a desktop milling machine called the TinyZ was developed to support digital fabrication in an architectural studio held at MIT in the Spring of 2021. The machine was designed to be an easily reconfigurable rapid prototyping tool intended to adapt to evolving design processes.

The TinyZ Kit introduced students to the basics of machine building, electronics, and computer numerically controlled (CNC) programming. The outcome of the studio showed the potential for different home labs to develop specializations and to collaborate by outsourcing, offering a way for students to work together remotely. Finally, the work of the studio demonstrated that new material processes developed remotely could return to fab labs and extend the capacities of shared maker spaces.

INTRODUCTION

The TinyZ is a desktop milling machine developed to support remote making during the pandemic. At a time when students at the department of architecture at MIT were unable to use machines and tools in collective maker spaces and workshops throughout campus, the TinyZ allowed students to re-engage digital fabrication and physical making at home, with readily available materials.

Digital fabrication represents a critical portion of design education in architecture schools, and access to CNC machines enables students to fluidly transition from digital representations to material artifacts. Thus, students engage in a process of learning based on tangible experimentation akin to the heuristic method or guided inquiry in education (Rayner-Canham 1995).

As a pedagogical experiment, a graduate architecture studio entitled IN HOUSE was devised in the spring of 2021 to make use of the TinyZ machine and prompt students to develop their own fabrication labs, termed Home Labs. In the process of assembling their machine kit, students were introduced to basic concepts of machine building and CNC programming, and developed protocols to guide their work with the machine and in their remote contexts. Here, distributed manufacturing becomes the setup for an emerging pedagogy that investigates fabrication tools down to their constituent components, introducing students not only to machine operation, but also machine assembly, maintenance, and tuning. Subsequent hacking and operation involve low-tech and high-tech strategies that range from simple mechanisms like wire tourniquets for part fixturing to multi-axis machining and paste extrusion.

BACKGROUND

The proliferation of digital fabrication and the maker movement originates in part with the work of the Center for Bits and Atoms (CBA) led by Neil Gershenfeld and the FAB conferences held to support institutions in opening Fab Labs. Fab Labs, or more generally termed Maker Spaces are places where designers and fabricators can work with advanced fabrication equipment to make (almost) anything (Gershenfeld 2012). A continuing trend among STEM curricula is to incorporate physical making and digital fabrication to amplify the communication of fundamental engineering concepts (Berry et. al 2010). Thus, Fab Labs are a common part of universities, with workshops supporting different disciplines and departments to perform experiments and enable students and researchers to prototype designs (Wong and Partridge 2016).

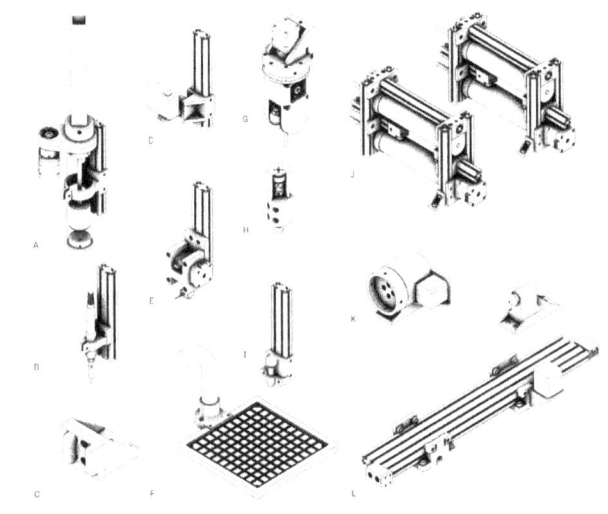

2 Hacks
(A) Extruder
(B) Melting Tool
(C) Mortiser Vise
(D) Go Pro
(E) Hot End
(F) Vacuum Hold Down
(G) 5-Axis Aggregate
(H) Spindle
(I) Drag-Knife
(J) Continuous Feed Roller
(K) Horizontal Rotary Indexer
(L) Axis Extension

With the advent of the pandemic came also rethinking of digital fabrication and physical making as critical components in design and STEM curricula. An array of approaches has emerged to support remote making and access to digital fabrication, outlined by Peek and Jacobs (2021), from simulation of CNC Machines to furnishing students with low-cost 3D printers and vinyl cutters. Through interviews with internal and external instructors, the researchers agree that at-home fabrication resources dramatically improve students' tacit knowledge of the material process and machine operation.

One of the key capacities of any Fab Lab is self-replication, to make machines that can make machines, which is an ongoing project at the CBA. The Rep-Rap, for example is one such machine that belongs to a growing family of small-scale rapid prototyping tools that can make rapid prototyping tools. However, this prospect remained largely overlooked amongst the responses to fabrication-based curricula; instead the strategy was to source low cost commercially available tools. The project of self-replication, or FAB 2.0, was tested in an academic context for the 2020 class "How to Make Almost Anything," where students were equipped with a small PCB milling machine designed and produced by the CBA lab. The course successfully demonstrated the implementation of a custom fabrication tool to enable remote making, in this case, PCB Manufacture. The course demonstrated that while this approach does not yield a commercially competitive fabrication tool, there is tremendous value in developing a customized pedagogical device to communicate particular concepts to students.

Table 1: Survey of rapid prototyping tools

   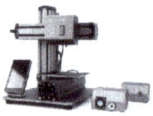  

	TinyZ	Prusa Mini	Genmitsu 3018	SnapMaker	Shapeoko	Ender 3
Process	Milling	FDM Printing	Milling	Milling Engraving FDM Printing	Routing	FDM Printing
Work Volume	5" x 5" x 4"	7" x 7" x 7"	11.8" x 7.1" x 1.8"	3.5" x 3.5" x 2"	16" x 16" x 3"	8.7" x 8.7" x 9.8"
Cost	$640	$399	$349	$699	$1320	$236

Learning from this precedent, the TinyZ machine aspires to extend the project of self-replication in the context of an architectural studio, with an open ended and modular machine platform capable of reconfiguration based on a student's material experiments. To that end, the market for open ended desktop CNC tools remains relatively modest. Table 1 shows a selection of desktop CNC tools that enable varying scales of making and material processes.

While these machines are produced at competitive cost, the intent is generally for the designer to work within the bounds of the machine. The TinyZ encourages its reconfiguration in a similar manner to the modular machines project formulated by Peek and Coleman (2017). The Snap Maker is a commercially available modular machine; however, the set of possible reconfigurations is relatively limited. Here, reconfigurable slides are fastened to one another, and control software is modified to match predefined hardware configurations.

The design goal for the TinyZ was both to develop an open-ended platform for machine reconfiguration while at the same time performing as a robust machine tool with which students could design. The machine has a work volume of roughly 5" x 5" x 4" and features a Dremel 3000 Rotary Tool as an end effector. The machine is largely composed of standard aluminum extrusions, off the shelf components, and a few custom components making it amenable to reconfiguration and modifications, termed hacks.

This approach was tested in the context of an architecture studio, wherein the TinyZ was used to enable design investigations. The prompt of the studio was simple, to design and construct an architectural installation larger than the TinyZ machine itself. This prompt encouraged students to engage questions of part to whole, of joinery and assembly, and ultimately of inhabitation.

METHODS

Design of the TinyZ

The TinyZ is a modular desktop-scaled milling machine capable of being reconfigured to evolve with student design processes. Myriad low-cost machines exist with varying degrees of flexibility, work volume and material capacity, refer to Table 1 for a selection. The design of the machine was therefore a primary site of concern especially in preparation for its use in the context of an architectural design studio. The machine was considered, on the one hand, as a fabrication tool and, on the other, as a pedagogical device to encourage machine iteration and hardware hacking.

Hardware

The design of the TinyZ leveraged the modularity and standardization of 2020 aluminum profiles which enables any surface of the machine to become a mounting feature for

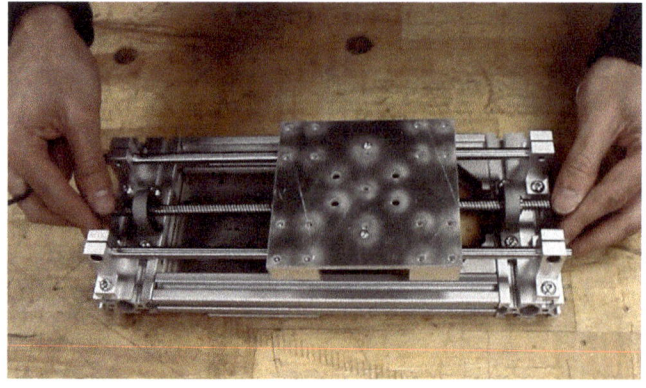

3

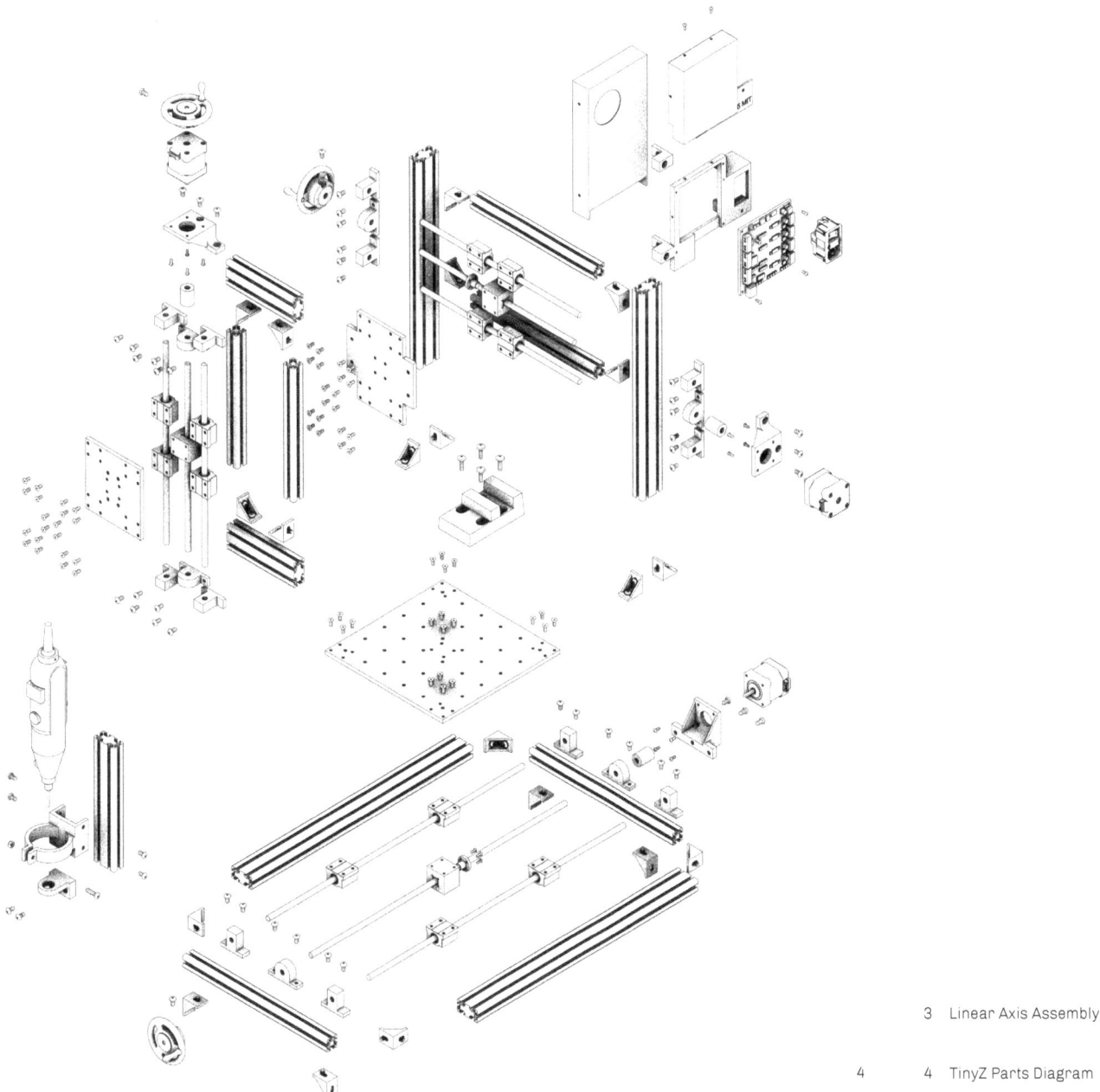

3   Linear Axis Assembly

4   TinyZ Parts Diagram

additional components devised by students. To that end, the machine has redundant aluminum framing in an effort to minimize custom components and amplify opportunities for subsequent hardware hacks.

The X-, Y- and Z-axes are lead screw-driven with NEMA17 84oz.in bipolar stepper motors providing the actuation. These motors are mounted with 3D printed parts with flexible shaft collars to attach motor output shafts to the lead screw.

The ability of the machine to be used manually was an important pedagogical decision that meant incorporating handwheels to actuate each axis. These are mounted directly to the lead screws of the X- and Y-axis, with the Z-axis featuring a dual shafted stepper motor to facilitate both manual and stepper driven motion. The goal is to enable students to build an intuition for their material experiments by manually advancing the end effector. Communicating the fundamentals of digital fabrication and CNC machining was effectively achieved by incrementally building up technical sophistication as shown in Figure 5. The process begins by using the Dremel Rotary Tool by hand, then attaching the tool to the Z-axis of the machine and manipulating the tool manually with turn wheels, and finally approaching CNC programming with predefined feed rates based on previous manual experiments.

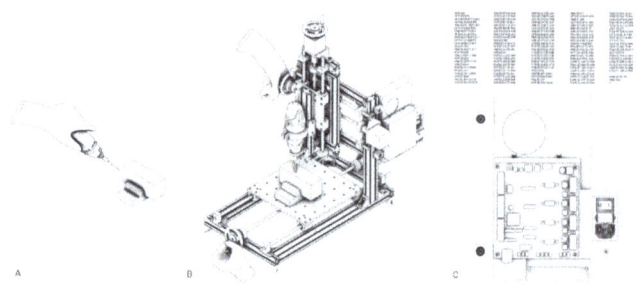

A   B   C

7

8

Each axis is composed of the same parts with differing proportions, a water jet aluminum plate with countersunk M4 holes is used to attach four linear bearings and a lead nut. This assembly is fastened to a frame composed of aluminum 2020 with shaft clamps and pillow blocks. Each axis is fastened to the next with some combination of water jet cut aluminum plates or brackets as shown in Figure 6.

The X-axis bed features a 1.5"x1.5" grid of M5 tapped holes for fixturing material. A small jeweler's vise was included in the TinyZ kit to use for part fixturing as shown in Figure 7. Additionally, students devised their own work holding strategies involving wire tourniquets, Vee-Blocks, or 3D printed attachments to reorient the jeweler's vise.

Software and Electronic Parts

A TinyG v8 Board from Synthetos was used as a controller to coordinate the motion of each stepper motor. This board was powered by a 350W Power Supply Unit with a 10A Fuse and Rocker Switch assembly, shown in Figure 8. The board features micro-stepping, current control of each motor, a host of digital outputs and inputs, and a total of 4 motor

9

10

drivers to accept different axis configurations. Wiring the TinyG involves locating coil pairs in each bipolar stepper motor and connecting them to terminal blocks on the TinyG board. The configuration of the TinyG is the next step to align hardware with software.

A range of open-source software solutions exist to communicate G Code to the TinyG, but CNCJs (Wu 2015) was found to be the most intuitive and robust. The configuration of the TinyG occurred through the serial console within the CNCJs interface, as well as jogging commands, and CNC program previewing and cycling, shown in Figure 9.

Mastercam 2019 (CNC Software Inc. 2019) was used primarily to generate G Code for milling operations. Mastercam is an industrial CAM package used to program machines in workshops across the MIT campus, from 2-axis to multi-axis operations. The fundamentals of Mastercam programming were communicated in the studio such that students could program their TinyZ machines, but could also apply their understanding of the software to work with larger industrial machines on campus.

End Effector

The base TinyZ Kit includes a Dremel rotary tool, which can hold an 1/8" diameter end mill. 3D printed components

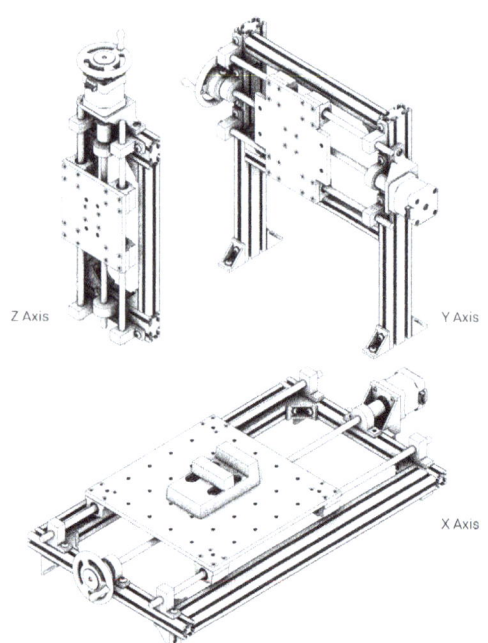

Z Axis   Y Axis

X Axis

6

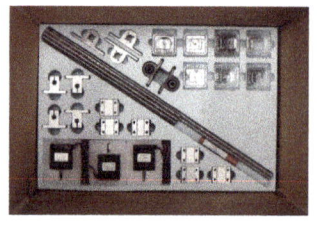

11

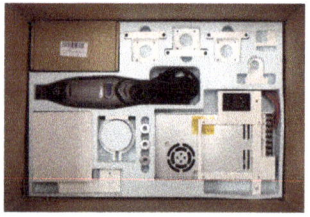

12

were devised to mount the Dremel to the Z-axis of the TinyZ, making use of coarse plastic threads on the Dremel itself, shown in Figure 10. Two tools were provided, both high speed steel tools with a 3/4" length of cut and 1" overall length, a ball end mill and flat end mill. These tools were tested by the author, machining patterns in expanded polystyrene foam, construction grade pine, and 6061 aluminum. Stiffer materials caused higher cutting forces and vibration to the machine, resulting in longer run times and more frequent machine maintenance. Foam and basswood became recommended materials for machining, and longer ball end mills were provided as student testing began scaling up to take advantage of the machine's 4" Z-travel.

Packaging and Cost

The TinyZ kit parts were packaged in machined extruded polystyrene, and organized to facilitate assembly by locating similar parts in sequence or in clusters with related parts, shown in Figure 11 and 12. The cost divides into roughly $260 in hardware components and $240 in electronic components. The remaining cost of the kit, roughly $160 comes from the choice of end effector, tooling, and accessories like shop vacuum and jeweler's vise

Hardware Hacks and Extendability

Throughout the studio, different hardware hacks were devised to extend the TinyZ and further enable material investigations. Hacking represents an important part of the culture of some academic institutions, where creative uses and misuses of technology can produce disruptive innovations. Within the context of the studio, hacking involved creative modifications to the machine to further enable material experimentation. These included changes to end effectors, axis configurations, and accessory components.

Extruder

A paste extruder was devised at the kickoff of the studio to extend investigations into additive manufacturing. The extruder was made largely of 3D printed components with some standard parts from vendors like McMaster-Carr. The extruder includes a 3D printed plastic plunger driven by a lead screw, similar to that used to drive the linear axes of the machine. A gear reduced stepper motor with a belt and pulley is used to actuate the lead screw and advance the plunger. Coordination of the extruder occurs by mapping the gear motor to the fourth motor slot on the TinyG Board, essentially treating the extruder as an additional axis. Used this way, the motion of the X-, Y-, and Z-axes are treated synchronously with the A or extruder axis. A flow rate variable was used to augment the travel of the extruder's plunger according to the distance travelled by the X-, Y-, and Z-axes. A range of open-source 3D printing software

6 Learning sequence (A) hand tool (B) manual milling (C) CNC milling
7 TinyZ subassembly diagram
8 X-axis bed and jeweler's vise
9 Power supply unit and controller assembly
10 CNCJs controller interface
11 Dremel 3000 mounting
12, 13 TinyZ packaging
14 Parts for extruder assembly
15 Student experiments
 (A) Manual Extrusion;
 (B, C) Drooping Ceramics;
 (D) Fluted Extrusion;
 (E, F) Biophilic Substrate;
 (G) Food Extrusion;
 (H, I) Extruding and Frying

14

were used to generate Gcode, however the most flexible solution was found to be Silkworm (Holloway and Mamou-Mani 2012), a plugin for Grasshopper and Rhinoceros (McNeel 2020). The path planning strategy here involves discretizing curves or polylines into simple line segments, and applying an A-axis travel to be performed across each segment, thereby achieving flow control across a 3D path. This end effector was used by students to print clay ceramics, biophilic substrates for plants, and food. The strategies for path planning with the extruder involved hybrid manual and machine control where the turn wheels were used to manipulate the extruder in space to coil ceramic filament across domestic objects, shown in Figure 15B and C. Iterating with fluted nozzle dies offered different formal readings that also had performative implications

15

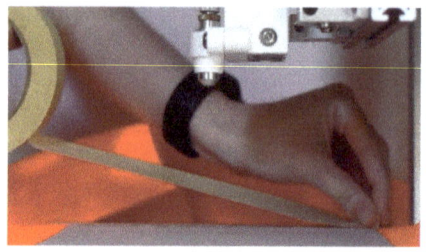
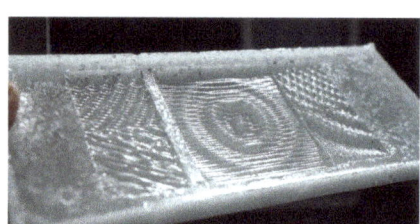

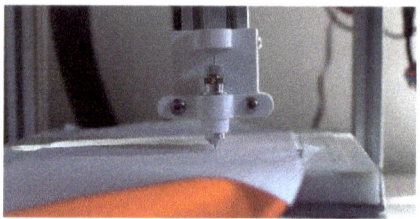

16 17 18

for biophilic substrates, shown in Figure 15 D, E and F. The extruder was also used to print food, where cutting boards and hot plates were placed directly onto the X-axis of the TinyZ, and various mixtures were used to print fried finger foods shown in Figure 15 G, H and I.

Heat
An engraving tool was used to selectively melt and remelt wax blanks. In addition to subtractive processing of wax, the student used a soldering iron with various tool tips to apply heat strategically to previously machined surfaces shown in Figure 16. Rather than using the router tool as a strategy for surface finishing, the approach was to relegate routing as a stock removal strategy, then use heat to selectively melt and produce a fluid surface finish. Process stacking allowed the student different formal readings of her material.

Part Fixturing
This student intended to machine the ends of long basswood stock. The hardware hack here was relatively simple. By turning the machine on its side and manufacturing a bracket for work holding, the student was able to turn the TinyZ desktop mill into a horizontal mortiser as shown in Figure 17. This machine was brought back to the workshop as restrictions relaxed. The change of location to the workshop enabled the student to gain access to larger blanks of material, more space to work, and adequate ventilation. Here, the TinyZ mortiser enabled machining of the material in a way that the workshop could not effectively support. More involved part fixturing strategies included the development of a vacuum bed for drag knife cutting shown in Figure 18. This hardware hack involved a High Density Polyethylene (HDPE) plenum and Low Density Fiberboard (LDF) sacrificial layer stacked atop the X-axis. Each TinyZ kit included a micro shop vacuum for dust collection, but because fabric cutting produces no dust, the vacuum was ported into the HDPE plenum to achieve vacuum fixturing.

Multi-Axis Configuration
Two students worked materials using multi-axis operations which involved significant changes to the X-axis bed. This hardware hack incorporated a horizontal rotary indexer mounted to an extended X-axis stage shown in Figure 19. The lead screw driven stage was replaced with a six-foot 2080 aluminum extrusion, actuated with a timing belt with a pulley mounted to the output shaft of the X-axis stepper motor, typical of the setup in most 3D printers.

Parts for the headstock and tailstock were 3D printed, with an additional stepper motor driving the rotary axis shown in Figure 20. A direct drive rotary axis was eventually replaced with a belt driven version which proved more robust and capable in machining operations. Both the long X-axis and the rotary axis were equipped with turn wheels to enable manual machining. The fourth motor slot of the TinyG was used here to control the rotary indexer.

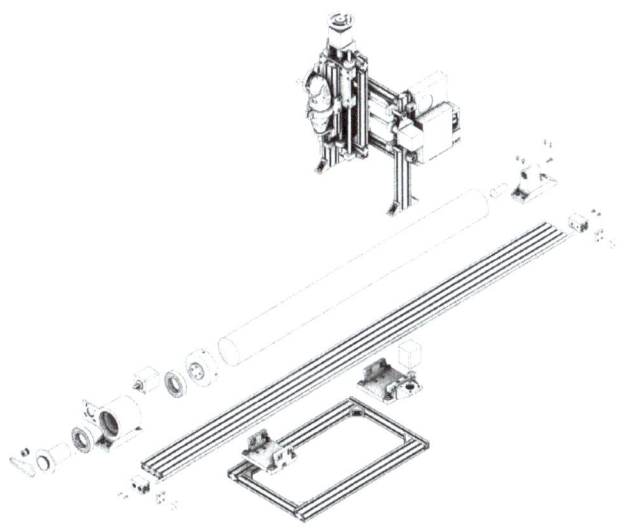

16 Daisy Zhang wax melting tool mounting, in operation, and surface finish experiments

17 Mengqiao Zhao's TinyZ turned sideways and used as a horizontal mortiser, a machine capability the architecture shops do not currently have

18 Taylor Boes preparing her drag knife setup to cut fabric, using a custom X Axis vacuum plate and cutting tool

19 TinyZ 4-axis diagram

20 Customized X-axis and tailstock

21 Jitske Swagemakers machining twigs using a horizontal rotary positioner

22 Flexural detail for bearing preload

A crucial detail of this hardware hack involved preloading bearings against the aluminum stage to constrain movement and reduce backlash in the X-axis shown in Figure 22. Flexures are used to achieve preload, based on designs of similar machines from the CBA "Machines That Make" project (2006).

RESULTS AND REFLECTION

The TinyZ desktop mill was an experiment in remote making in the Spring of 2021. Over the course of the semester, students built their machine from a kit, learned to operate the machine and eventually to modify it, augmenting the TinyZ in different ways to enable experiments with materials. The studio reflected two contentious issues raised for instructors and students by the pandemic. The first was related to physical making, specifically digital fabrication and the increased barriers to access that resulted from shop closures and shelter in place orders. The second related to inhabitation, which had also presented a challenge to students and researchers, namely how to make spaces for making in their domestic environments. Thus, the protocols devised at the start of the semester challenged students to think through the impact that their material experiments would have upon their apartments and dorms, and to strategize rules to guide their work at home to be practicable and sustainable, and not disruptive to the people living there or destructive to property.

The Home Lab therefore began as a scaled down exercise in creating a fab lab, and the subsequent material testing that occurred throughout the semester created challenges to the setups that encouraged students to acknowledge their home lab protocol as a continuous work in progress. This exercise is achieved in part because of the flexibility of the TinyZ machine to be rapidly reconfigured. In some way, this tool is a collection of standard materials held together with very few custom components. For example, one project made use of the stepper motors and controller, fabricating new mounting fixtures with a 3D printer and turning the TinyZ into a large cable-actuated delta robot shown in Figure 24 E and F.

As a pedagogical device, then, the investment of time and effort into a customizable and modular machine proved valuable not merely as an introduction to machine operation, but also to machine design. The end effectors developed by students throughout the class are an indication of the flexibility of the system.

The implications of hardware hacking have some demonstrable promise beyond the pandemic. An example previously described was the horizontal mortiser, where the TinyZ was rotated and a custom fixture was devised to enable the student to machine the ends of very long wooden parts shown in Figure 17. The machine was brought back

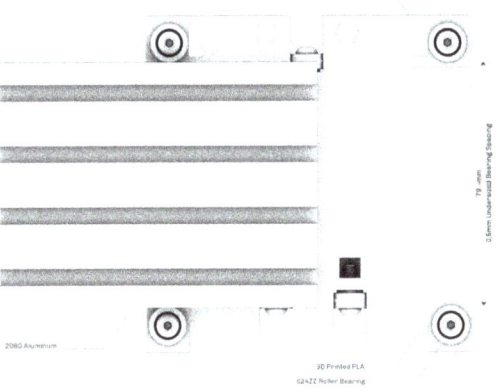

23 Fab Labs and the Home Lab Ecosystem

24 (A) Tristan Searight experimented with edible extrusions, printing his own food in isolation.
(B) Carolyn Tam experimented with growing materials, tending to her installation throughout the semester.
(C) Taylor Boes developed a sound attenuating veil to augment the partition wall separating her room from common areas.
(D) Mengiao Zhao treated the TinyZ as her roommate giving the machine mobility and sight.
(E) & (F) Gil Sunshine developed a room-scaled Delta Bot that intervenes in various sites on MIT Campus.

to the shop to perform the work due to the space requirements that working with these parts demanded. The hybrid nature of some of the projects meant that the workshop could support the home lab; however, it was also true that machines developed at home labs could support the workshop. It was encouraging to see that an augmented desktop scale machine could perform tasks more effectively than larger industrial machines that existed at the workshop. The implication being that processes developed at Home Labs could support and enrich collective maker spaces as restrictions are lifted and access returns.

Another implication of the Home Lab is the possibility of collaboration from one specialized lab to another. At some point in the semester, a student with an extruder postulated an improved extruder design that required a complex part, but this part could not be fabricated at their lab. However, another student who had a 4-axis TinyZ could fabricate the part and send it to support the other. Individual Home Labs could potentially outsource parts to each other, allowing hardware hacks to cross-pollinate and for students to leverage each other's specialties, converging and springboarding off one another. This moment was encouraging because it demonstrated a way for collaboration to occur during a pandemic. The potentials for collaboration abound where a dispersed network of home labs can exchange knowledge and expertise, outsource material for processing to each other, exchange machine parts and hardware hacks, and exchange projects as different processes are required.

CONCLUSION

In response to shop closures and shelter in place mandates, a low cost and easily distributable rapid prototyping tool was developed called the TinyZ. This machine was tested in the context of an architectural studio that enabled students to design through digital fabrication, material experiments, and physical making.

The TinyZ machine was, on the one hand, devised to lower the barriers of entry to working with CNC fabrication techniques. On the other hand, the machine served as a pedagogical tool to focus the discourse surrounding fabrication on inhabitation, which was similarly put into question by the circumstances of the pandemic. To accomplish these goals, the machine was designed to be easily reconfigurable, and hackable, with the majority of its constituent parts sourced from low-cost manufacturers. The TinyZ enabled students to reclaim a kind of agency over their domestic environments, devising installations that intervened not only as a built construction upon their interior spaces but also upon their lifestyle. The students' projects included the design of food, Figure 24A, biophilic substrates to host plants Figure 24B, a sound attenuating veil between roommates Figure 24C, and the TinyZ machine itself as a roommate Figure 24D. The range of projects demonstrate the capacity for digital fabrication to be mobilized to answer design questions that do not end merely in something constructed, but rather something that is lived with.

Through the use of at-home fabrication tools, students could enact changes in personal lifestyle. This was the result of students designing with dramatically scaled down machines and readily reconfigurable machine tools. Therefore, the research studio is evidence of a potential shift of digital fabrication from centralized large scale maker spaces toward dispersed home labs. This shift suggests that frequently reconfigured prototyping tools will lead to a more profound integration of design process and fabrication resulting ultimately in a "soft" approach to fabrication that is fluid and flexible. Low cost and reconfigurable prototyping tools will make digital fabrication more accessible and enable project specific adaptation, encouraging designers and practitioners to be at home with machines.

ACKNOWLEDGEMENTS

Many thanks to the efforts of students who approached the studio with aplomb: Mengqiao Zhao, Gil Sunshine, Jitske Swagemakers, Carolyn Tam, Tristan Searight, Taylor Boes, Florence Ma, Daisy Zhang, and Zhifei Xu. Thanks also to Christopher Dewart for the use of n51 as a base of operations.

24

REFERENCES

Benabdallah, Gabrielle, Sam Bourgault, Nadya Peek, and Jennifer Jacobs. May 2021 "Remote Learners, Home Makers: How Digital Fabrication was Taught Online During a Pandemic." In *Yokohama '21: ACM CHI Conference on Human Factors in Computing Systems*, edited by P. Bjorn and S. Drucker. Yokohama, Japan: ACM.

Berry III, R. Q., G. Bull, C. Browning, C. D. Thomas, K. Starkweather and J. H. Aylor. 2010. "Preliminary considerations regarding use of digital fabrication to incorporate engineering design principles in elementary mathematics education." *Contemporary Issues in Technology and Teacher Education* 10 (2): 167–172.

Center for Bits and Atoms. "Machines That Make." Accessed June 1, 2021. http://mtm.cba.mit.edu.

CNC Software Inc. *Mastercam*. V. 2019. CNC Software Inc. Windows PC. 2019.

Fox, Michael and Miles Kemp. 2009. *Interactive Architecture*. New York: Princeton Architectural Press.

Gershenfeld, Neil. 2012. "How to Make Almost Anything: The Digital Fabrication Revolution." *Foreign Affairs* 91 (6): 43–57.

Holloway, Adam and Arthur Mamou-Mani. *Silkworm*. V. 0.1.1. https://github.com/ProjectSilkworm/Silkworm. Windows PC. 2012.

McNeel, Robert. *Rhinoceros*. V. 6. Robert McNeel & Associates. Windows PC. 2018.

Peek, Nadya, James Coleman, Ilan Moyer and Neil Gershenfeld. May 2017. "Cardboard Machine Kit: Modules for the Rapid Prototyping of Rapid Prototyping Machines." In *CHI '21: Proceedings of the 2017 CHI Conference on Human Factors in Computing Systems*, edited by C. Lampe, M. C. Schraefel, J. P. Hourcade, C. Appert and D. Wigdor. 3657–3668. Denver, Colorado USA: ACM.

Rayner-Canham, Geoff A. and Marlene Rayner-Canham. 2015. "The Heuristic Method, Precursor of Guided Inquiry: Henry Amrstrong and British Girls' Schools, 1890-1920." *Journal of Chemical Education* 92 (3): 463–466.

Wong, Anne and Helen Partridge. 2016. "Making as Learning: Makerspaces in Universities." *Australian Academic & Research Libraries* 47 (3): 143-159.

Wu, Cheton. *CNCJs*. V. 1.9.22. https://github.com/cncjs/cncjs. Mac OS. 2015.

IMAGE CREDITS

Figure 15B,C, 17, 24D by Mengiao Zhao
Figure 15D,E,F, 24B by Carolyn Tam
Figure 15G,H,I, 24A by Tristan Searight
Figure 16 by Daisy Zhang
Figure 18, 24C by Taylor Boes
Figure 20, 21 by Jitske Swagemakers
Figure 24E,F by Gil Sunshine
All other drawings and images by the author

Zain Karsan is a teaching fellow and technical instructor at the architecture workshops at MIT. He holds a Master of Architecture from MIT and a Bachelor of Architecture from the University of Waterloo. His research focuses on the relationship between machine tools, material culture, and the architectural imaginary.

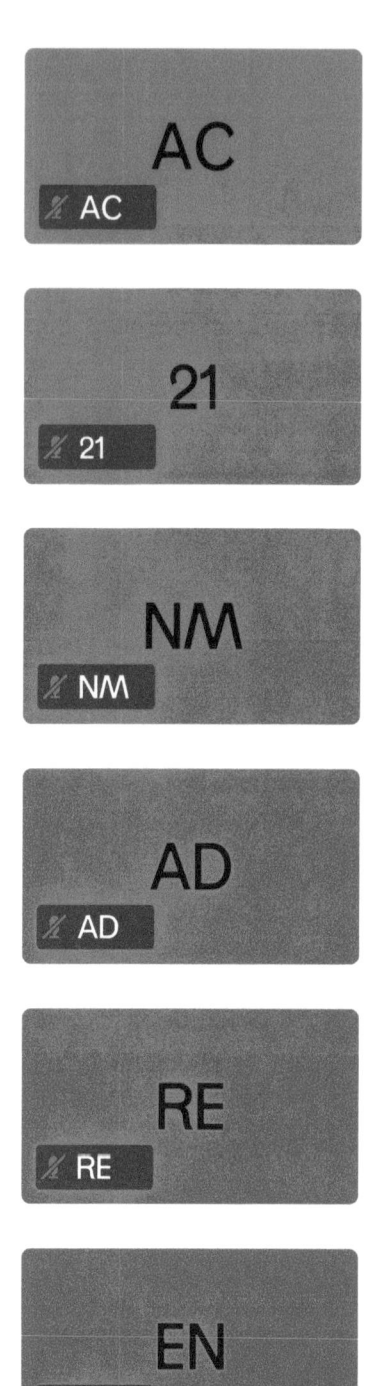
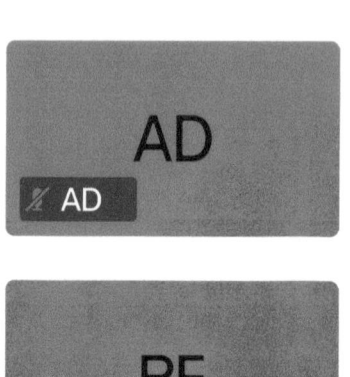
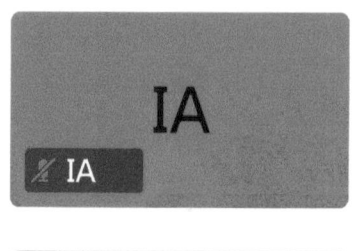

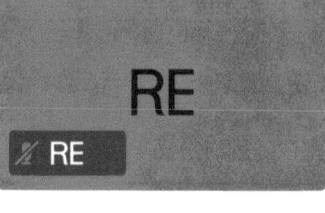
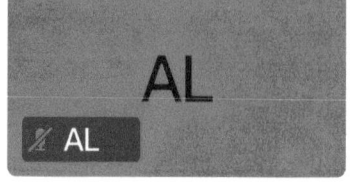
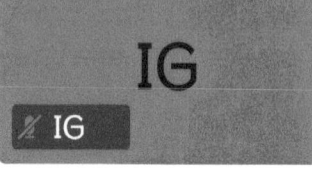

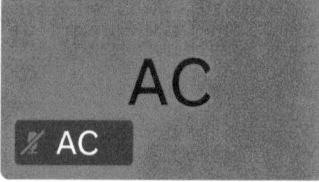
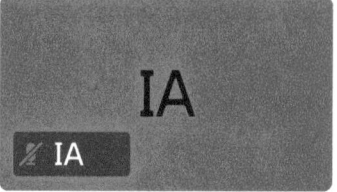

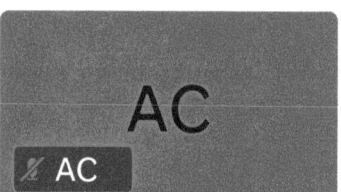

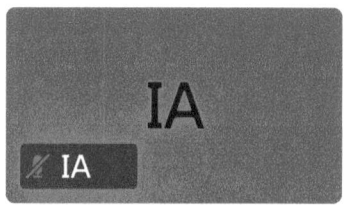

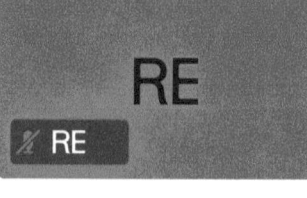
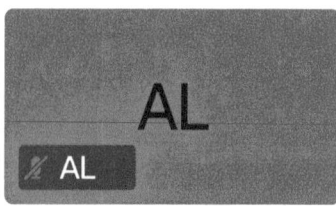

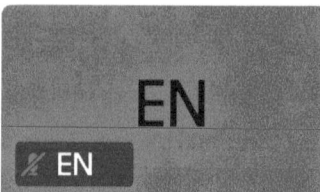

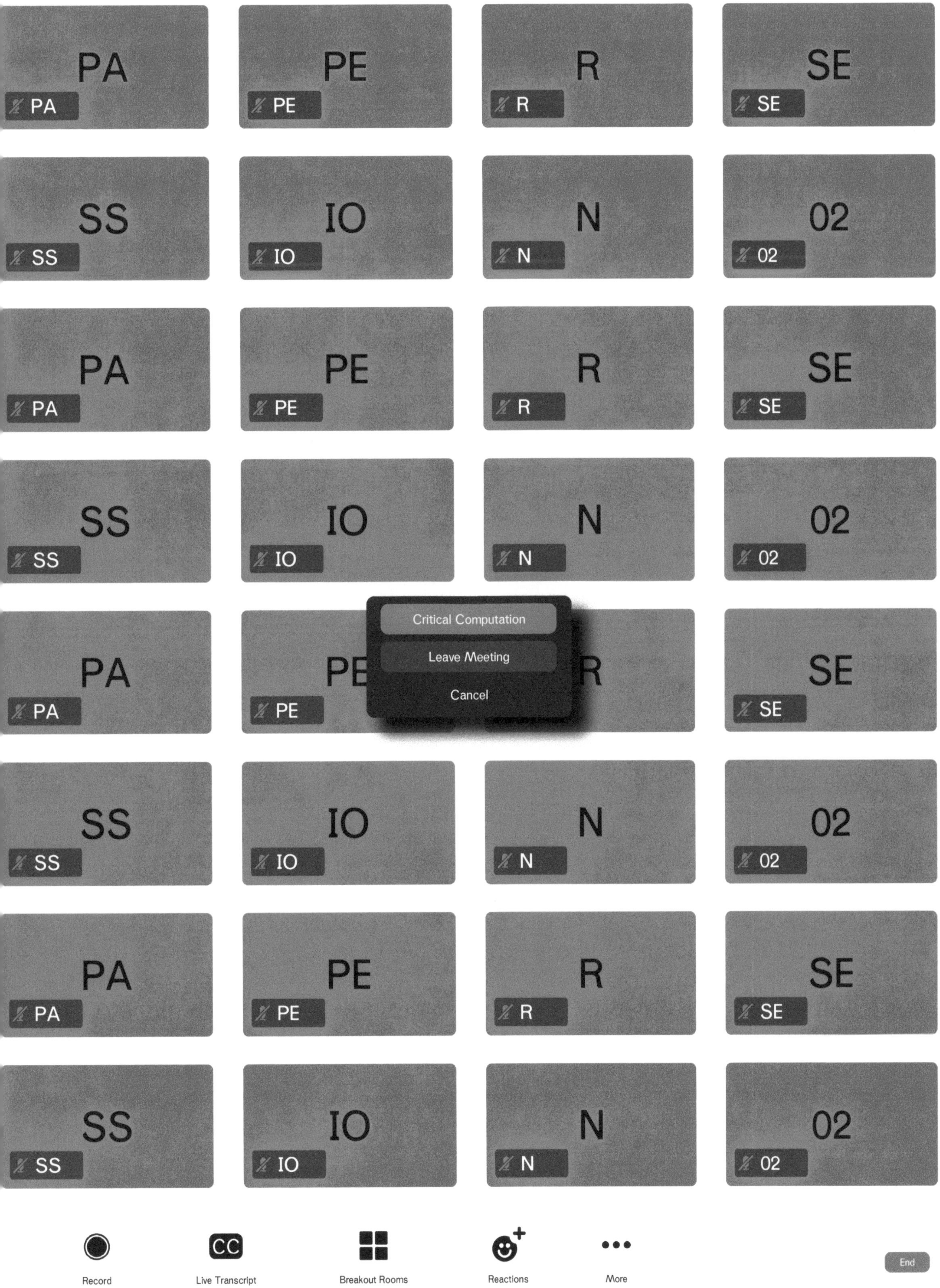
PA PE R SE
SS IO N 02
PA PE R SE
SS IO N 02
PA PE R SE
SS IO N 02
PA PE R SE
SS IO N 02
Critical Computation
Leave Meeting
Cancel
Record
CC
Live Transcript
Breakout Rooms
Reactions
More
End

House Block

Claire McAndrew
AUAR Labs

Clara Jaschke
AUAR Labs

Gilles Retsin
AUAR Labs

Kevin Saey
AUAR Labs

Mollie Claypool
AUAR Labs

Danaë Parissi
AUAR Labs

1   *House Block*, prototype of a two-storey dwelling unit in Clapton, East London (James Harris, 2021)

An Experiment in Discrete Automation

House Block was a temporary housing prototype in East London, UK from April to May 2021. The project constituted the most recent in a series of experiments developing Automated Architecture (AUAR) Labs' discrete framework for housing production, one which repositions the architect as curator of a system and enables participants to engage with active agency. Recognizing that there is a knowledge gap to be addressed for this reconfiguration of practices to take form, this project centred on making automation and its potential for local communities tangible. This sits within broader calls advocating for a more material alignment of inclusive design with makers and 21st Century making in practice (see, for example, Luck 2018).

House Block was designed and built using AUAR's discrete housing system consisting of a kit of parts, known as Block Type A. Each block was CNC milled from a single sheet of plywood, assembled by hand, and then post-tensioned on site. Constructed from 270 identical blocks, there are no predefined geometric types or hierarchy between parts. The discrete enables an open-ended, adaptive system where each block can be used as a column, floor slab, wall, or stair—allowing for disconnection, reconfiguration, and reassembly (Retsin 2019). The democratisation of design and production that defines the discrete creates points for alternative value systems to enter, for critical realignments in architectural production.

PRODUCTION NOTES

Architect: AUAR
Status: Built
Site Area: 52 sq. m.
Location: Clapton, London, UK
Date: 2021

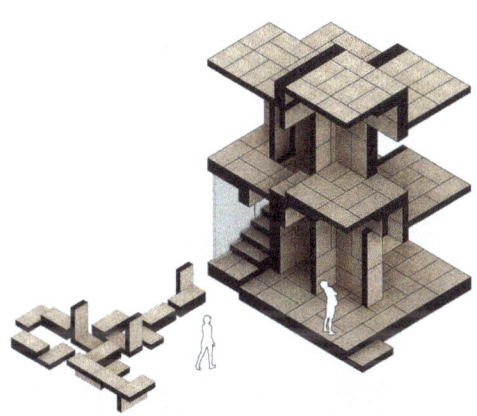

2   House Block 'Takeover' - A Catalyst for Community Making

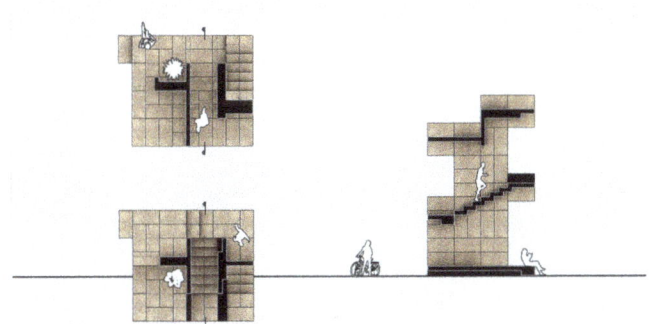

3   *House Block* plans & section

4   Concept rendering of *House Block* further development

The two-storey, 52m² structure was built on the site of a former playground in Clapton over a three-week period. It was delivered in partnership with London Borough of Hackney, as part of an ongoing discussion and shared interest in alternative methods of affordable housing delivery and development of small plots. *House Block* made tangible the social and economic possibilities of decentralized modes of digital fabrication and reduced building syntax. In partnership with New City College's Hackney campus, a work experience programme invited students who would normally not engage with automation to explore how it may benefit their practice. Through the assembly of 320 blocks, seventy construction, carpentry, multi-skills students/apprentices participated in dialogue about the potential of localised automation (such as CNC milling and robotic assembly) with short production chains that upskill local labour and create new kinds of jobs for more inclusive, just futures.

A series of 'takeovers'—small-scale, creative, experimental, explorations—designed and built by four project partners, The Building Centre, Gonzalo Herrero Delicado + Pati Santos, Hackney Wick Underground and L U C I N E, extended critical understandings of automation. Activated by responses to the housing prototype, the takeovers enabled live experimentation with *House Block*. Fifty blocks supported the delivery of weekly takeovers over the course of one month. Local residents participated in their production, a process of gaining lifelong skills and agency, along with experiential learning of part-to-whole relationships within a local context. A public programme of talks and events provided further entry points, reaching over 200 members of the public on the potential of discrete automation for the commons, domesticity, sustainability, new housing strategies and policies.

House Block is one experiment in becoming an activator for the discrete. The discrete depends on more than academic knowledge and designerly ways of knowing. It demands participation to redefine established positions and the development of practical skill sets to access digital modes of production (Claypool 2019). *House Block* has offered

5　Interior View (James Harris, 2021)

6   House Block Side View (James Harris, 2021).

7   *House Block* as seen from one of the site entrances (James Harris, 2021)

8   *House Block* Front Elevation (James Harris, 2021)

a tangible demonstration of the viability of rethinking long production chains in architecture and using 'digital materials' (Popescu and Gershenfeld 2009) at scale. Our modes of engagement have offered a lens into the value of decentralized processes—elevating horizontal forms of collaboration in a local setting for more equitable, affordable, and sustainable housing realities. Whilst advocating for lived experience as a form of expertise, this project illustrated the freedom part-to-whole relations bring in generating conversations and design ideas for new modes of living that work for "everyone's unique difference" (Luck 2018, 116). It offers access to an alternate housing reality, a demonstration of the necessity of material engagement for the transformation of architectural design and construction practices.

ACKNOWLEDGMENTS

Project Partner: London Borough of Hackney

Collaborators: The Building Centre, The Good Thing + Gonzalo Herrero, Hackney Wick Underground, L U C I N E, Studio Wayne McGregor, New City College, Waltham Forest Future Creatives Programme, Valentina Soana + MArch Architectural Design (RC2) at The Bartlett School of Architecture.

Structural Engineer: Manja Van de Worp, YIP Engineering

Funder: UCL Trellis: Community Partnership Building Events, EPSRC Impact Acceleration Account Follow-On Funding, UCL East Community Engagement Seed Fund, UCL Innovation & Enterprise HEIF.

REFERENCES

Claypool, Mollie. 2019. "Our Automated Future: A Discrete Framework for the Production of Housing." *Architectural Design* 89 (2): 46–53.

Luck, Rachael. 2018. "Inclusive Design and Making in Practice: Bringing Bodily Experience in Close Contact With Making." *Design Studies* 54: 96–119.

9   Detail of House Block's Upper Floor (James Harris, 2021)

Popescu, George A. and Neil Gershenfeld. 2009. "Digital Materials." *MIT FabCentral*. Draft publication. http://fab.cba.mit.edu/classes/961.09/04.13/DM.draft.pdf

Retsin, Gilles. 2019. "Bits and Pieces: Digital Assemblies: From Craft to Automation." *Architectural Design* 89 (2): 38–45.

IMAGE CREDITS

Figure 1, 5-11: © James Harris, 2021

All other drawings and images by the authors

Claire McAndrew is a social scientist that works across architecture and practices of care. She focuses on new frameworks for participation—drawing upon contemporary theory, research and debate around architecture, technology, community, and public engagement in the production of the built environment. Her writing and practice consider the use of co-design methodologies to shift social practices, inform public policy, and enact more care-full capacities. Claire is Director of Public Engagement and Co-Director of AUAR Labs at The Bartlett School of Architecture, UCL.

Gilles Retsin is an architect, educator and thinker working at the intersection of computation, fabrication, and architecture. His work has been internationally recognised through awards, lectures and exhibitions at major cultural institutions such as the Museum of Art and Design in New York, Royal Academy in London and Centre Pompidou in Paris. He edited an issue of *Architectural Design (AD)* (Wiley 2019) and co-authored *Robotic Building: Architecture in the Age of Automation* (Detail Edition 2019). Gilles is Associate Professor in Architecture, Co-Director of AUAR Labs and Programme Director of MArch Architectural Design at The Bartlett School of Architecture, UCL.

10 Construction of House Block in Clapton, East London (James Harris, 2021).

11 Building a House Block 'Takeover' (James Harris, 2021).

Mollie Claypool is a leading architecture theorist focused on issues of social justice highlighted by increasing automation in architecture and design production, and the potential of automation in architecture and the built environment to provide more socially engaged and environmentally sustainable ways of designing and building. Mollie is co-author of *Robotic Building: Architecture in the Age of Automation* (Detail Edition 2019) and author of the SPACE10 report "The Digital in Architecture: Then, Now and in the Future" (2019). She is also Associate Professor in Architecture at The Bartlett School of Architecture, UCL. At The Bartlett she is Co-Director of AUAR Labs.

Clara Jaschke is an architectural designer and researcher specializing in digital tools and technology. She holds a Master in Architecture from the University of Innsbruck as well as a MRes in Architecture & Digital Theory from the Bartlett School of Architecture, UCL, where she is now a Teaching Fellow in addition to being a Researcher at AUAR.

Kevin Saey is an architect and researcher in automation, digital fabrication and computational design with a background in game design. Kevin studied Digital Arts and Entertainment at University College West Flanders, MSci in Architecture at KU Leuven and MArch Architecture at The Bartlett School of Architecture, UCL.

Danaë Parissi is an architect interested in bringing together the traditional discipline of architecture with cutting edge computational theory to explore new interdisciplinary paths of design. She has completed her integrated MArch at the NTUA School of Architecture and an MSc in Architectural Computation at The Bartlett School of Architecture, UCL.

Passive Listening and Evidence Collection

The Social Stakes of Acoustical Sensing

Rebecca Smith
University of Michigan

1   From left: Wildlife Acoustics sensing unit; gunshot detection sensor; image of the Audobon Society's birdsong identification app.

In this paper, I present the commercial, urban-scale gunshot detection system ShotSpotter in comparison with a range of ecological acoustical sensing examples which monitor animal vocalizations. Gunshot detection sensors are used to alert law enforcement that a gunshot has occurred and to collect evidence. They are intertwined with processes of criminalization, in which the individual, rather than the collective, is targeted for punishment. Ecological sensors are used as a "passive" practice of information gathering which seeks to understand the health of a given ecosystem through monitoring population demographics, and to document the collective harms of anthropogenic change (Stowell and Sueur 2020). In both examples, the ability of sensing infrastructures to "join up and speed up" (Gabrys 2019, 1) is increasing with the use of machine learning to identify patterns and objects: a new form of expertise through which the differential agendas of these systems are implemented and made visible. I trace these agendas as they manifest through varied components: the spatial distribution of hardware in the existing urban environment and/or landscape; the soft l processes that organize and translate the data; the visualization of acoustical sensing data; the commercial factors surrounding the production of material components; and the apps, platforms, and other forms of media through which information is made available to different stakeholders. I take an interpretive and qualitative approach to the analysis of these systems as cultural artifacts (Winner 1980) to demonstrate how the political and social stakes of the technology are embedded all through them.

INTRODUCTION

The dynamic membrane of an audio sensor takes different forms, but at a basic level it functions by converting ambient environmental sound energy into analog signal, which is then translated into digital information. While this moment of conversion is common to both of the sensing configurations discussed in this paper, they are vastly different in their overall agendas. Gunshot detection sensors are used for evidence collection, and for dispatching police personnel. They are intertwined with processes of criminalization in which the individual is targeted for punishment; a process decontextualized from larger collective responsibility for the conditions that give rise to gun violence in the first place. By contrast, the sensing arrays of ecological acoustics are used as a "passive" practice of information gathering, one that seeks to understand and document the health of a given ecosystem through monitoring population demographics; a framing in which the harms of anthropogenic change are understood as a collective societal responsibility.

In this paper I analyze these two systems comparatively, by considering signal flow through them, to ask how do the differential agendas of passive listening versus evidence collection manifest across their varied spatial, informational, representational, technological and social components? In the context of the ACADIA conference, I am placing this work in the lineage of recent papers such as Shelby Doyle, Leslie Forehand, and Nick Senske's, "Computational Feminism," which proposes computational feminism as having an 'ethical narrative' or the ability to address equity and serve as a disruptive force (Doyle, Shelby and Senske 2017); and also Maya Przblyzki's "Critical Computational Literacy," which argues for computation as part of the 'material assembly' with which architects work, an element to be engaged with critically and ethically (Przbylski 2018). My approach in this paper is slightly different: rather than an investigation of an architectural computational process, I follow an analytical approach towards urban technology as-found. This is presented with an understanding of the conceptual latitude and critical sensibilities found within architectural computation (particularly in the research or experimental setting), and the foundationally different ways that technologies of sensing and machine learning might be approached from within it.

SHOTSPOTTER AND ECOLOGICAL ACOUSTICS

ShotSpotter gunshot detection sensing arrays are leased by local municipalities throughout the United States, where the company dominates the gunshot detection technology industry. ShotSpotter sensing arrays are designed with two purposes: to alert law enforcement that a shooting has occurred, and to aid in the collection of evidence for investigation into the shooting [1]. Individual sensing nodes are generally visible, and occasionally—although not always—the resulting data is made public (for example, the map shown in Figure 4). But what happens in-between—the signal flow and processes of translation—is a proprietary, urban-scale black box. Through this intermediate step, the gunshot event is cauterized, made exchangeable, and reduced to a datapoint. This data is then described in the language of objectivity and fact, divorced from its larger social and political context.

Similarly to urban infrastructures and built environment policies of the past in areas like housing and transportation, ShotSpotter arrays are the result of successful political lobbying for funding at the federal level (Schwenk 2020). Like other forms of surveillance technology, the relationship between ShotSpotter, crime rates, and the actual occurrence of violence is hard to decipher; more the product of political discourse than any clear metric of the actual impact that the system may or may not be having. Several municipalities have already made public statements that they don't feel the system is effective, or have cancelled their contracts altogether (Swilley 2017). The siting of ShotSpotter is also determined by larger social and historical conditions that are not made explicit in the discourse surrounding its implementation: the disinvestment that makes high-vacancy cities susceptible to the appeal of a system like ShotSpotter (which, in the context of shrinking tax bases promises to do more with less), the larger context of structural racism and widespread inequality in the United States that contributes to the epidemic of gun violence and the conditions that give rise to it, particularly in Black and brown communities; and the powerful position of the gun industry itself within the American context.

In contrast to the evidence collection model of gunshot detection, ecological acousticians advocate for "passive listening"—an open-ended engagement with the entity and physical environment under study (Howe et al. 2019). Ecological acousticians site their arrays based on the interrelationships of the animal they are studying and the greater ecology of the location. Instead of the comprehensive and totalizing coverage of gunshot detection arrays, single sensors are placed in a location that animals are likely to pass through or occupy, sometimes in relation to physical or environmental features [2]. This produces a territory or catchment area similar perhaps to a core sample, in contrast to the abstracted zone of undifferentiated coverage produced by the gunshot detection array.

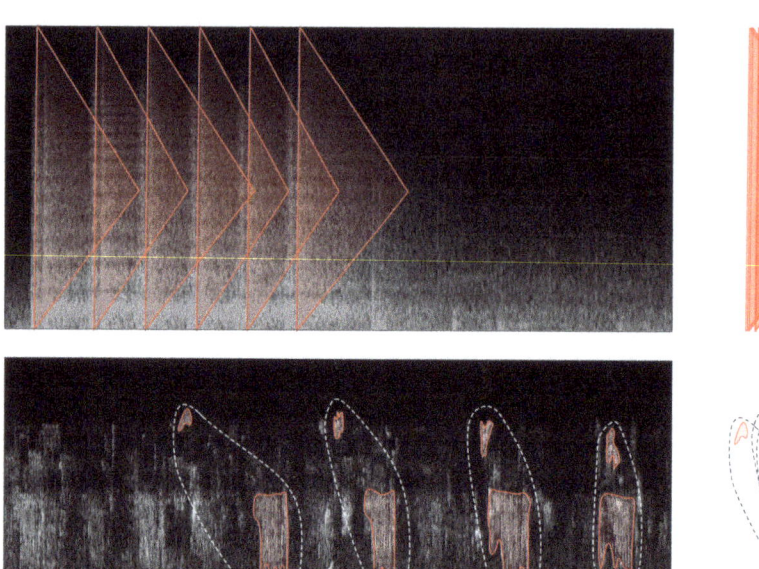
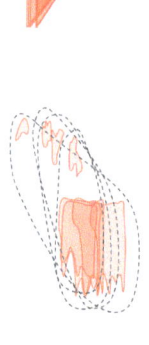

2   Spectral visualizations of gunshots (top) and birdsong (bottom) with diagrammatic representation of how classification occurs.

Within these workflows, the placement of the hardware, and the configuration of the software and scripts used to process the resulting data are determined by research design principles and intended to gather a representative sample, which can give insight into the health of the ecosystem overall (Sugai et al. 2020). The data is described as simultaneously a measure of the individual animal, as well as its relationship to the environment and context: this data is an inherently ecological entity (Stowell and Sueur 2020). Unlike ShotSpotter, where the contextualism of factors like policing practices, structural racism, and the legacy of urban disinvestment are omitted within a narrative of clinical technological objectivity, the practice of ecological acoustics makes contextuality explicit as an intentional component of research design.

SPECTRAL VISUALIZATION AND MACHINE LEARNING

Within both sensing configurations, a key component is the use of spectral images to visualize and classify acoustical data. Figure 2 illustrates two examples of this: the top image shows a series of gunshots, while the bottom shows an example of birdsong. This form of visualization shows the spectrum of acoustical energy over time, in which the figure of the sound emerges against the background as an area of density. Within gunshot detection workflows, the objective is to determine a simple binary: the sound either is—or is not—a gunshot, based on how quickly the transient moves from loud to quiet. Once defined as a gunshot, a larger event-chain is triggered, prompting a combined human-technological response. In contrast, birdsong is defined by relationships between recurring clusters of pattern. Traditionally, this has not happened through an abstracted measurement, such as the slope in the gunshot, but instead by the recognition and classification of patterns, a process undertaken by researchers or other workers, who also listen to the actual recordings.

Machine learning is increasingly being used to perform these same operations. Critically, in both gunshot detection and ecological acoustics, the visualization of acoustical data remains a key component, as this allows for the application of convolutional neural networks to the image matrices. Within ecological acoustics, machine learning is used to disambiguate relevant patterns in increasingly complex environments. This allows for the monitoring of habitats that are the most precarious in relation to anthropogenic change and for research to move beyond border conditions to include more fine-grain locations like traffic intersections—locations with more significant background noise which present an additional degree of complexity.

Similarly to the distributed spatial locations of the sensing units, these visualizations analyze temporal moments at set intervals rather than taking the surveillant approach of constant and comprehensive monitoring. Although there are commercially available products available for this process, researchers are increasingly developing

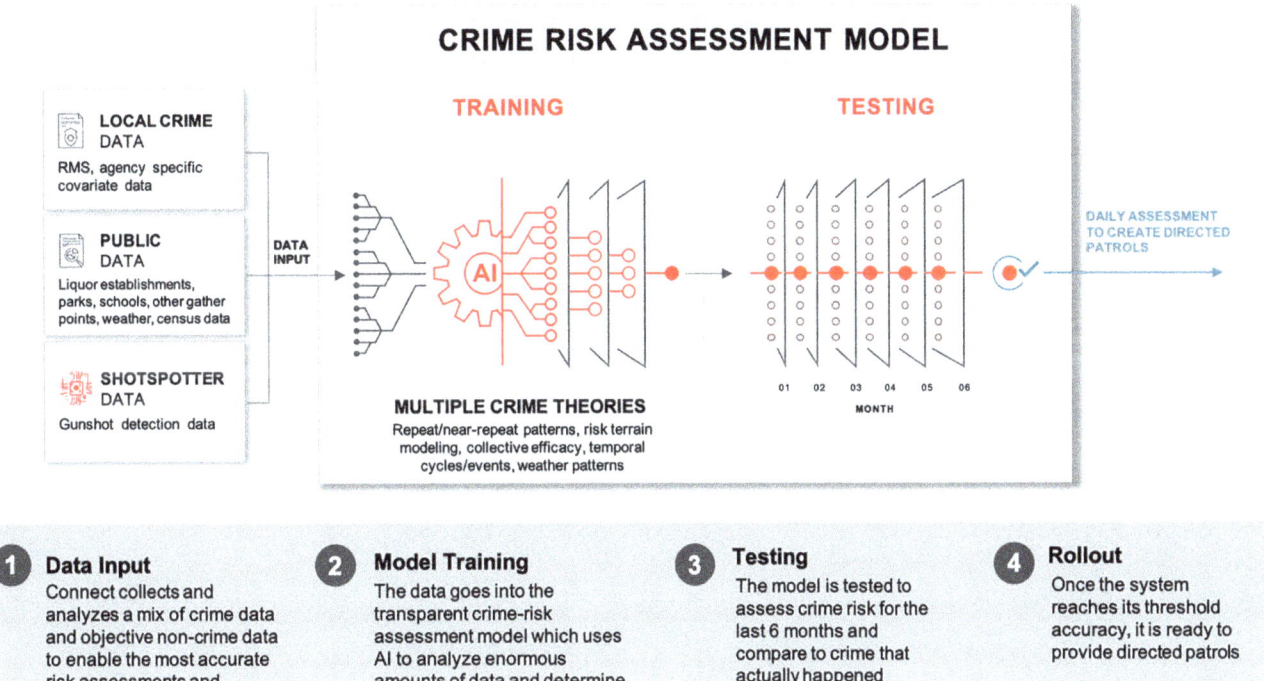

3 ShotSpotter promotional material describing the role that machine learning plays in its patrol management suite of products.

their own alternatives. They argue this is necessary to allow transparency around machine learning methods, models, and workflows, and that stronger and more fully contextualized data will result. These workflows retain openly described elements of human labor: for example, outsourcing the work of image annotation for model training to skilled and experienced birding enthusiasts. This culture of volunteer engagement is oriented around a sense of shared contribution to the task of understanding the health of the overall ecosystem. These partnerships are cultivated through apps, websites, and other platforms which both crowdsource data and make it available to the public (Aodha et al. 2018).

Machine learning is also integrated into ShotSpotter's workflows. The company regularly registers new patents for the interpretation of spectral images [3]. Shortly after the death of George Floyd in May of 2020, the company underwent a redesign, and now features machine learning as a component in a reform-oriented wave of promotional materials, many of which are explicitly designed to ease community interactions around the implementation of the company's products. The company's scope of operation has also expanded, from the temporal event-chain of gunshot response, into the larger domain of patrol management. Throughout this suite of products, machine learning is presented as a technology of accountability and objectivity, using common tropes of technological efficiency in which the fairness of the larger system is never questioned.

There is a disconnect, however, between how the image of machine learning is promoted throughout ShotSpotter's carefully curated public relations strategies, and the company's closed stance regarding the actual details of these workflows and the data that results. This is seen in the apps and platforms that accompany its systems, which are accessible only to registered users (typically law enforcement personnel). The company also tightly controls the datasets which result from its arrays, and has publicly promoted its interest in monetizing them as the foundation of their business model moving forwards. Even when made available, visualizations of the ShotSpotter data create a self-perpetuating cycle of rationalization; a representational narrative that characterizes the communities in which the technology is placed as violent, and therefore, deserving of its continued and expanded uses.

DISCUSSION

The larger agendas and political stakes of these sensing assemblages can be seen throughout their varied material, informational, and social components. These examples illustrate that our systems of measurement, visualization, and representation matter: they function as world-making devices that reproduce societal beliefs (Daston and Galison 2007; Haraway 1988). Both anthropogenic change and gun violence are large-scale social phenomena, ones with multiple and complex causes, which create significant harm for communities of many scales. But the systems that measure, record, and represent information about

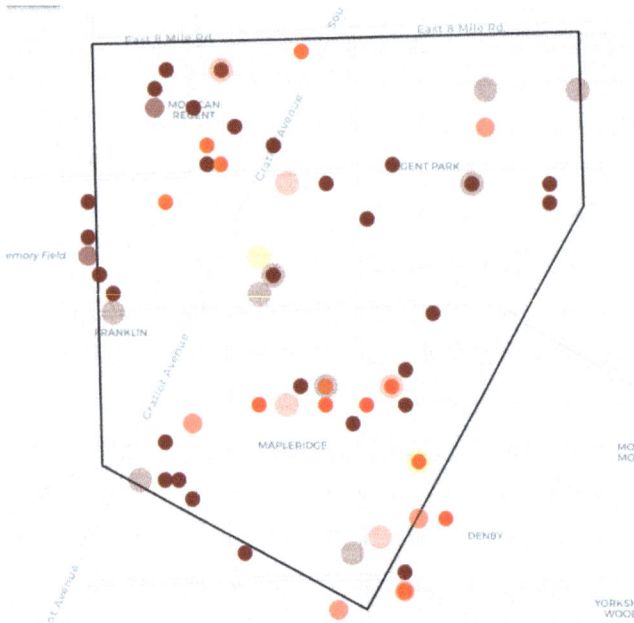

4   Visualization of gunshot data gathered through Detroit ShotSpotter array, December 31, 2015. From The Detroit News, February 24, 2016. Map by Tom Gromak.

them have drastically different methods of assigning responsibility. The technologies and workflows of ecological acoustics articulate an ecology of passive listening, stressing collective responsibility and interrelationships. In contrast, technologies of gunshot detection seek to isolate and punish, to reduce the scale of collective responsibility to a single event by extracting the it from the larger social context.

If sensing has always functioned to "join up and speed up" (Gabrys 2019, 1), the increasing adoption of machine learning into these workflows both amplifies the epistemic values embedded within them, and also makes these orientations and underlying dispositions more legible. This is particularly critical because, unlike some other forms of surveillance technology, gunshot detection can deceptively give the illusion of fairness and objectivity. As there is no obviously faulty algorithm to take as an entry point, moving away from "fixing" a distinct component or "proving" that one piece is broken requires an understanding that something much larger is fundamentally wrong; a recognition of more comprehensive, multidimensional and systemic failures.

It is this multidimensionality, then, that I am suggesting might be approached through a critical computational sensibility. This is not to say that theoretical approaches or experimental practices are a "solution" for the very real issues presented in this paper, but instead, to suggest that critical computation in architecture, specifically given its disciplinary adjacencies to both the social dimensions of urbanism and the technological specifics of computation, might have a contribution to make; a position from which to begin to address some of this complexity. It might offer a stance beyond a simple fix or correction, offering instead a way to address the mutually constitutive relationship between the technological, the social, the material, and the political. But perhaps this can only be achieved through real and critical engagement with the social and political context of the contemporary moment. So then, a final closing question, or direction for future research might be, what does this look like?

Jutta Weber, writing of the rise of feminist STS in the last decades of the 21st century, identifies myriad ways that the political context—from the experience of individual female researchers and scientists, to the larger influence of the feminist and other social movements—revised and shaped foundational beliefs about what did and did not constitute knowledge within the domain of scientific research (Weber 2006). This encompassed multiple scales, from the personal experiences of individual female researchers and thinkers, to the larger feminist movements of the time, and other movements such as anti-nuclear activism. It was not enough, for example, to simply avoid research contributing to military or nuclear endeavors; researchers also took to heart a foundational questioning of binaries—such as male and female, nature and technology, subject and object—and integrated these paradigmatic shifts into their formulations of knowledge.

In both of the sensing array examples presented here, machine learning increases the capacity to quite literally take up space and time, to reinterpret and territorialize both through processes of visualization; to amplify what Jennifer Gabrys describes as sensing's innate ability to "speed up and join up" (Gabrys 2019, 1). If one takeaway is that surveillance technology should be replaced by ecologies of care—that ideas of safety need to move beyond false metrics like crime data and gunshot occurrences—this paper might ask, in the tradition of both feminist STS and critical computation, can the specific technological components identified here, themselves, perhaps become sites or drivers for action?

A very different conception of both space and time, for example, is offered by activist and writer Adrienne Marie Browne, who discusses themes of temporality, connection, and ecology in her concept of emergent strategy. Browne uses metaphors of emergence to advocate for deep relational building as a force for social change, perhaps most often referenced in her encouragement to build an "inch wide" and a "mile deep" (Brown 2017, 16). In the context of digital urbanism, Shannon Mattern has recently written about practices of care and repair—small-scale actions that take time—as both responses to our current sociopolitical context, as well as a counterbalance to the scope and scale of our technological infrastructures (Mattern 2021). Both of these perspectives influence a stance that asks, beyond just fixing things or seeing the world in terms of problems that we need to solve, how do we allow ourselves, and our technological practices, to be modulated or hybridized by the times we are living in, and by our current social and political contexts?

NOTES

1. ShotSpotter. Accessed July 2021. http://www.shotspotter.com.
2. For example, the work of Dr. Lindsay Swierk. Accessed July 2021. https://www.wildlifeacoustics.com/customer-stories/understanding-amphibian-responses-to-noisy-suburban-habitats.
3. "ShotSpotter's Latest U.S. Patent Enables Major Advancement in Machine Learning Accuracy for it's Gunshot Detection Technology." Press Release. Accessed July 2021. https://www.shotspotter.com/press-releases/shotspotters-latest-u-s-patent-enables-major-advancement-in-machine-learning-accuracy-for-its-gunshot-detection-technology/.

REFERENCES

Daston, Lorraine and Peter Galison. 2007. *Objectivity*. Cambridge, Mass.: Zone Books, distributed by the MIT Press.

Doyle, Shelby, Leslie Forehand, and Nick Senske. 2017. "Computational Feminism: Searching for Cyborgs." In *ACADIA 2017: Disciplines & Disruption; Proceedings of the 37th Annual Conference of the Association for Computer Aided Design in Architecture*. Cambridge, Mass: ACADIA. 232–237.

Gabrys, Jennifer. 2019. *How to Do Things with Sensors*. Minneapolis: University of Minnesota Press. https://doi.org/10.5749/j.ctvpbnq7k.

Haraway, Donna. 1988. "Situated Knowledges: The Science Question in Feminism and the Privilege of Partial Perspective." *Feminist Studies* 14 (3): 575–99. https://doi.org/10.2307/3178066.

Howe, Bruce M., Jennifer Miksis-Olds, Eric Rehm, Hanne Sagen, Peter F. Worcester, and Georgios Haralabus. 2019. "Observing the Oceans Acoustically." *Frontiers in Marine Science* 6: 426. https://doi.org/10.3389/fmars.2019.00426.

Mac Aodha, Oisin, Rory Gibb, Kate E. Barlow, Ella Browning, Michael Firman, Robin Freeman, Briana Harder, et al. 2018. "Bat Detective-Deep Learning Tools for Bat Acoustic Signal Detection." *PLoS Computational Biology* 14 (3): e1005995–e1005995. https://doi.org/10.1371/journal.pcbi.1005995.

Mattern, Shannon. 2021. *A City Is Not A Computer*. Princeton: Princeton University Press.

Moreira Sugai, L.S., Camille Desjonquères, Thiago Sanna Freire Silva, and Diego Llusia. 2020. "A Roadmap for Survey Designs in Terrestrial Acoustic Monitoring." *Remote Sensing in Ecology and Conservation* 6 (3): 220–35. https://doi.org/10.1002/rse2.131.

Przbylski, Maya. 2018. "Critical Computational Literacy: A Call for the Development of Socially Aware, Ethically Minded Research Within ACADIA." In *ACADIA 2018 Recalibration: On Imprecision and Infidelity; Proceedings of the 38th Annual Conference of the Association for Computer Aided Design in Architecture*. Mexico City: ACADIA. 9–10.

Schwenk, Katya. 2020. "Operation Legend Is Bringing Surveillance Tech to Cities." *The Intercept*. September 13, 2020. Accessed November 21, 2021. https://theintercept.com/2020/09/13/police-surveillance-technology-operation-legend/.

Stowell, Dan and Jérôme Sueur. 2020. "Ecoacoustics: Acoustic Sensing for Biodiversity Monitoring at Scale." *Remote Sensing in Ecology and Conservation* 6 (3): 217–19. https://doi.org/10.1002/rse2.174.

Swilley, Kristin. 2017. "Some Cities Are Ditching ShotSpotter, But Cincinnati Police Say It's Working." WCPO Channel 9: Cincinnati. Accessed July 15, 2021. https://www.wcpo.com/news/local-news/hamilton-county/cincinnati/some-cities-are-ditching-shotspotter-but-cincinnati-police-say-its-working.

Weber, Jutta. 2006. "From Science and Technology to Feminist Technoscience." In *Handbook of Gender and Women's Studies*, edited by Kathy Davis, Mary Evans, and Judith Lorber. London: Sage.

Winner, Langdon. 1980. "Do Artifacts Have Politics?" *Daedalus* 109 (1): 121–36.

IMAGE CREDITS

Figure 1: (left) Wildlife Acoustics, accessed July 2021, http://www.wildlifeacoustics.com; (center) Image by the author; (right) Cornell Lab of Ornithology, accessed July 2021, www.birdnet.cornell.edu.

Figure 3: Crime Risk Assessment Model, accessed July 2021, http://www.ShotSpotter.com.

Figure 4: Map by Tom Gromak, from "ShotSpotter Recorded Gunfire Incidents In Detroit," *Detroit News*, February 24, 2016.

All other drawings and images by the authors.

Rebecca Smith is a Detroit-based researcher, designer, and educator who works at the intersection of digital technology and urbanism. She is currently a PhD candidate at the University of Michigan.

Critical Computation

The Gaze and Surveillance Feminism

Behnaz Farahi
CSULB Department of Design
Human-Experience Design
Interaction (HXDI)

1 LAUREN is an art installation that addresses the role of human labor in the future of automation and surveillance (Lauren Lee McCarthy, 2020)

ABSTRACT

Can computation be critical or will various forms of bias always be found embedded in computational systems? Could surveillances act as a form of resistance? This paper provides a theoretical reflection on these questions and explores the notion of critical computation. It addresses the discourse of the gaze, and surveillance feminism, using some critical computational projects by way of illustration. This paper argues that critical computation integrates two strands of theory and practice in a seamless way. The theory originates from the tradition of critical theory and reveals the underlying algorithmic biases behind pervasive technologies, such as the scholarly work of Ruha Benjamin, Slavoj Zizek, and Yuval Harari. The practice uses the technology itself in a critical approach as a way to reflect our privacy or as a strategy to undermine various forms of power structure and to promote forms of resistance such as creative works of Diller Scofidio + Renfro, Lauren Lee McCarthy, and my own practice.

This paper first provides a brief theoretical context to the notion of critical computation. Then by differentiating between technological determinism and intersectional affordance, it aims to provide a lens through which to study surveillance computation. This paper attempts to avoid any form of technological determinism. Rather than rehashing arguments as to whether computation and in particular surveillance is inherently good or bad, it aims to take an "intersectional feminist affordance" approach to show what constitutes the gaze and surveillance, and to consider what strategies of resistance might prove to be effective in art and design practices.

INTRODUCTION

We live in an age where we are immersed in a world of machine learning algorithms, computer vision, and biometric sensors. But what are the ethical implications of this? Might these computational systems not replicate—somewhat disturbingly—our underlying biases and serve to reinforce the marginalization of those who typically have been excluded? Is AI necessarily complicit in replicating human prejudice? Or are there ways of deploying AI as a means of exposing human bias? I would like to highlight strategies that could be deployed for using computation critically to help us overcome bias. This could be viewed either as a form of critical lens to unfold algorithmic biases embedded in pervasive technologies or as a critical approach to the implementation of computation in art and design practices.

CRITICAL COMPUTATION

Can computation be critical? Could we look at computation through a critical lens? What is critical thinking and how could it be applied to computation? Could critical computation become a tool for reflection or an instrument for interrogating the underlying algorithms of ubiquitous technologies?

Critical thinking implies going beyond the obvious in order to interrogate the roots of an issue with all its nuances. Critical thinking questions issues which have been taken for granted, or what often goes unquestioned. This way of thinking has its own roots in Critical Theory, a school of thought initiated in the early twentieth century by a group of scholars at the Institute for Social Research in Frankfurt, Germany known as 'the Frankfurt School'. Some of the influential thinkers in this school of thought, such as Theodor Adorno, Walter Benjamin, Herbert Marcuse, and Juergen Habermas, were addressing questions of social change. As Sensoy and DiAngelo put it, "Their work was guided by the belief that society should work toward the ideals of equality and social betterment" (Sensoy and DiAngelo 2011, 25). One of the reasons for the emergence of Critical Theory was to respond to the overemphasis of the scientific method (referred as "positivism") by raising questions about "whose" rationality and "whose" presumed objectivity underlies scientific methods. What is interesting here is that in the 1960s this school of thought came to be absorbed in North America in the context of "antiwar, feminist, gay rights, Black power, Indigenous Peoples, and other emerging social justice movements" (Sensoy and DiAngelo 2011, 26). In simple terms, Critical Theory engages with the study of society from the perspective of the marginalized, especially in relation to indigenous, racialized, and postcolonial peoples, whose voices have been suppressed due to structural oppression. But could we apply such a critical approach to computational systems?

Ruha Benjamin, a sociologist aligned with the Critical Theory tradition and a critical race theorist, explores algorithmic biases in her book *Race After Technology* (Benjamin 2019). She is critical of the often glossy promises of tech companies, and seeks to question the divisions of class and race that they exacerbate. She thinks that they are also responsible for perpetuating a form of eugenic practice especially through the presence of AI systems. She invites us to ground our computational designs in a deeper historical and social context, so we don't unwittingly introduce racial and social inequality. Benjamin explains that many computational algorithms "hide, speed up and deepen discrimination while appearing neutral and even benevolent compared to the racism of a previous era" (Benjamin 2019, 45). She argues that technology can hide the ongoing nature of social domination and allow it to penetrate every aspect of our lives under the guise of progress. But where did the problem originate? Is computation complicit in various forms of social domination or is our contemporary culture the real source of this problem?

For example, could we not say that the problem is actually already 'hidden' as a result of our contemporary culture, particularly in the North American and European context, where we 'politely' hide many forms of social inequalities? For instance, to hide racism, many western companies, organizations and institutions have been embracing the importance of diversity. However, under the disguise of institutionalization of diversity, we see an increasing amount of racism and sexism. In her book, *On Being Included: Racism and Diversity in Institutional Life*, Sara Ahmed addresses the gap between the symbolic commitment to diversity and the experience of those who embody diversity (Ahmed 2012). As she puts it, "Diversity as an ego ideal conceals experiences of racism, which means that multiculturalism is a fantasy which supports the hegemony of whiteness" (Ahmed 2008). In her book she offers a critique of what happens when diversity is offered as a 'solution'. But—equally—isn't this exactly the same problem that we have when we project too much onto computation or in fact see technology as a solution? By seeing the potential in AI to promise a better vetting and recruiting system, or to solve our housing crises, could it not be said that we are subscribing to a form of technological determinism?

TECHNOLOGICAL DETERMINISM VS. INTERSECTIONAL AFFORDANCE

Technological determinism is a reductionist theory elaborated by American sociologist Thorstein Veblen, which states that the "improvement of society's cultural values and social structure is driven by the technology it possesses."[1] In other words, technology has the agency to shape and alter our behaviors and is a fundamental force behind patterns of social organization. As a result, we ascribe too much agency to computational systems to either save or destroy our civilization. Such a view can often be found expressed in relation to surveillance technologies, pattern recognition, and computer vision technologies.

In contrast to such a deterministic view, the theory of 'affordance' offers us a far more nuanced understanding of how a tool can be used, beyond the limitations of technological determinism. The theory of affordances was initially introduced by James Gibson in his book *The Ecological Approach to Visual Perception* (1979). He distinguished between the intrinsic properties of things and the potential actions that they invite. Gibson illustrated his idea with an example, "a glass wall affords seeing through but not walking through, whereas a cloth curtain affords going through but not seeing through" (Gibson 1979, 187). In this example, glass displays its own intrinsic properties that determine its rigidity and stiffness. But it is the capacity of the material, which allows us to see through it, but not walk through it. It is important to emphasize that the affordances of an object are related to its potential use in coupling with an organism's capabilities. As Jelle Van Dijk explains, "James Gibson's affordances can be seen as the way the world shows up as affording some action, with a certain sensorimotor coupling in place" (Dijk 2018, 7). The theory of affordances applies to tools especially. A tool, after all, is no use, if we cannot use it. Likewise a tool might 'lend itself' to certain operations, but not to others. A tool, however, has no agency. It cannot force us to use it in a certain way. Nonetheless, with an appropriate tool, the capacity of human beings can be enhanced considerably. As Neil Leach (Leach 2016, 350) explains:

> The theory of affordances suggests that there is a particular action or set of actions that is afforded by a tool or object. Thus a knob might afford pulling—or possibly pushing—while a cord might afford pulling. This is not to say that the tool or object has agency as such. In other words the tool or object does not have the capacity to actually 'invite' or 'prevent' certain actions. Rather it simply 'affords' certain operations that it is incumbent on the user to recognize, dependent in part on a set of pre-existing associations that have been made with that tool or object. Moreover, certain tools afford certain operations, but do not preclude other operations. For example, we might perhaps affix a nail with a screwdriver – albeit less efficiently – if we do not have a hammer at hand.

While notion of 'affordance' is extremely helpful in the context of computational thinking and in fact, it considers how an action shaped by different physical bodies and abilities (Franchak and Adolph 2013; Kytta 2004), it does not engage with social dimension of one's being. For instance, we can see that people are frequently discriminated based on their identity. As such a room could not afford entering for all people if access is limited. In other words, beyond physical affordance, there are aspect of social affordance which defines the use of technology. The feminist concept of "intersectionality" coined by Kimberly Crenshaw in 1991 could give us more insight on how we could think of intersectional affordance where different aspects of our identities (e.g., race/ethnicity, socioeconomic status, gender identity) cannot be dissociated from and necessarily interact with one another. As Paxton (Paxton et al. 2019) puts it:

> For ecological psychology, intersectionality provides an opportunity for us to situate the entire person in their full context. Rather than reducing a person's existence to merely their physical environment, a social ecological psychology that is sensitive to intersectional identity would capture the person's full history and experience as they move through their physical and social worlds.

SURVEILANCE, AI, AND BIAS

One area where the question of technological determinism is apparent is in the use of surveillance. In her book, *The Age of Surveillance Capitalism: The Fight for a Human Future at the New Frontier of Power*, Shoshana Zuboff, subscribes to a form of technological determinism, and argues that surveillance technologies benefit capitalist corporates and their economy, such as those in the Silicon Valley. She believes that these technologies are influencing our behavior, as our data is being bought and sold, and "the production of goods and services is subordinated to a new 'means of behavioral modification'" (Zuboff 2019). To her this new version of Big Brother is operating in the interest of surveillance capitalism, free from any democratic oversight.

Yuval Noah Harari, however, has a more nuanced perspective on surveillance. In the battle against the coronavirus epidemic, he warns us that surveillance is shifting from "over the skin" to "under the skin," while at the same time it could also help us to overcome this global crisis. As he puts it, "with coronavirus, the focus of interest shifts. Now the government wants to know the temperature of your finger and the blood-pressure under its skin" (Harari 2020). While surveillance and monitoring system could help us to control and sustain the spread of the virus, he warns us that the trend to monitoring citizens could last even after the epidemic is over and become an exploitive tool for totalitarian government. Instead of these technologies being in control of the government or any security authorities, he thinks that they should be used to empower citizens. In other words, in order to avoid politicians abusing such systems, we might need special organizations where the data is not shared with other authorities. But, more importantly, Harari invites us to see how surveillance can be used not only by governments to monitor individuals, but also by individuals to monitor governments. Such a view could be seen in the video recording captured by a witness of the brutal murder of George Floyd by the police. This evidence has made authorities accountable for their brutal action and give a new spin to the Black Lives Matter movement in the United States.

In his article, "We are already controlled by the digital giants, but Huawei's expansion will usher in China-style surveillance," Slovenian philosopher Slavoj Žižek reminds us about the current battle for power on the digital network, the main control mechanism over a citizen's life. He believes that "the digital network that regulates the functioning of our societies as well as their control mechanisms is the ultimate figure of the technical grid that sustains power today" (Žižek 2019). He similarly seems to be warning us that rather than blaming surveillance itself we should be more concerned about 'who' is controlling the data. In one of his interviews Žižek comments, "I don't think big technological companies simply form one evil block against ordinary people. What I fear more is that this rule of barbarian technocracy will get combined with some new brutal populism."[2]

Let us here differentiate, then, between the notion of who 'controls' and the actual 'surveillance'. Instead of arguing whether surveillance is good or bad in and of itself, let's explore who is in control, who is scrutinized, why and at what cost. From a critical perspective, it is important to note here that when we are addressing surveillance, questions of race, gender and class are crucial. For instance, many religious groups such as Muslims have been monitored disproportionally (Kanji 2020) or often

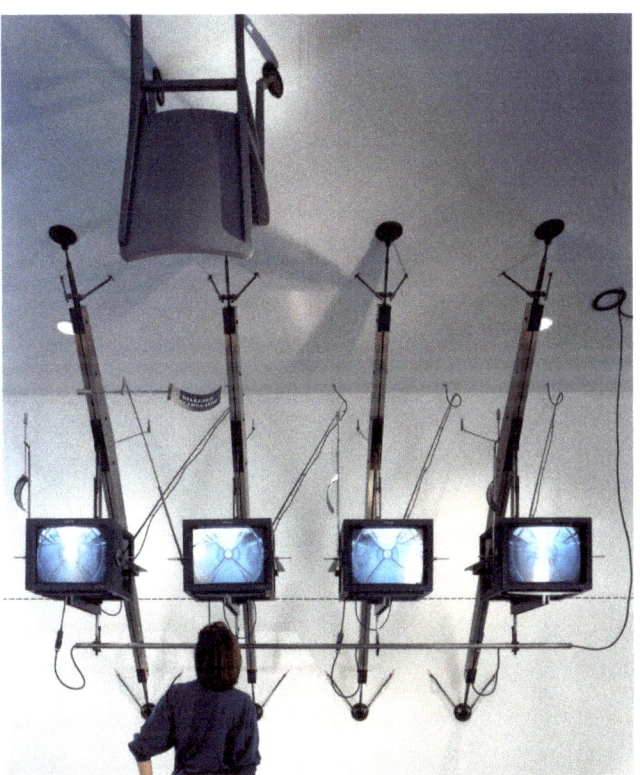

2   Para-site, The Museum of Modern Art, New York (Diller and Scofidio, 1989)

the privacy of refugees and immigrants has been invaded (Lahav 2003, 89) as though it is unimportant. In the collection of essays on feminist surveillance studies, the authors expose the ways in which surveillance practices are mostly tied to systematic forms of discrimination and normalization of whiteness, from full-body airport scanners, mainstream media reports about honor killings, depiction of women's bodies on the media, to surveillance that aims at curbing the trafficking of women and sex work. In their book Dubrofsky and Magnet argue that surveillance studies must be seen in the context of feminist theories: "A feminist praxis is not limited to gender issues, but rather sees gender as part and parcel of a number of contingent issues, such as race, sexuality, class, and able- and disabled-bodiedness, insisting that these cannot be viewed in isolation" (Dubrofsky and Magnet 2015, 4).

In the context of "intersectional affordance" could surveillance be used to subvert forms of power domination? In order to study surveillance, I hope to address what it means to be observed. To illustrate this argument, I am also going to draw upon two critical art/design projects.

THE FEELING OF BEING OBSERVED

Research into the effect of being observed can be traced back to a series of experiments that began at Western

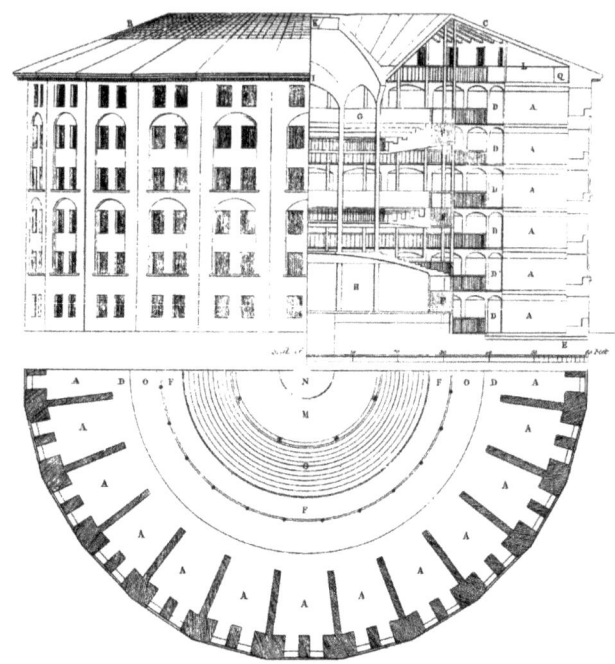

3   Bentham's Panopticon (1791), plan, section, and elevation drawn by Wiley Reveley (Source: Wikimedia Commons)

Electric's Hawthorne Works factory in Cicero, near Chicago in 1924. Hawthorne Works was a factory complex consisting of a hospital, ballpark, library, and much more, which employed about 40,000 workers. The management at Hawthorne Works sponsored the National Research Council to work with MIT to show that illumination increased productivity (Bastian 2013). Surprisingly the results showed that a worker's productivity would improve when any changes were made, by either increasing or decreasing the lighting level. This seemed to suggest that increase in productivity had nothing to do with the lighting level, but depended on the workers receiving motivational attention from being observed (Cox 2000, 158). In other words, the Hawthorne Effect (or observer effect), "is a type of reactivity in which individuals modify an aspect of their behavior in response to their awareness of being observed" (McCarney et al. 2007, 2). After the results were published, there were several critical attempts to debunk the findings of this experiment particularly as the results were not replicated in other attempts to repeat the same experiment. As Jones notes, "little or no evidence of Hawthorne effects was found" (Jones 1992).

Despite criticism of the Hawthorne Effect within the scientific community, psychologist Edward Titchener, went on to conduct further research on the feelings of being watched. In his paper, "The Feeling of Being Stared At," he explains that, "if one thinks hard of one's knee, or foot, e.g., one will obtain a surprisingly intensive and insistent mass of cutaneous and organic sensations of which one was previously unconscious" (Titchener 1898, 897). This is because human beings and many other creatures have evolved a dedicated neural architecture to detect facial features such as gaze detection that serves as an important evolutionary tool for purposes such as detecting the potential threat from a predator or enemy.

In fact, human beings are so sensitive to being watched that even just a drawing or a photograph of a pair of eyes can influence their behaviors. A group of researchers from the University of Newcastle in the UK found that even displaying images of watching eyes and related verbal messages could reduce the number of bicycles stolen on the campus. As they explain, "the effectiveness of this extremely cheap and simple intervention suggests that there can be considerable crime-reduction benefits to engaging the psychology of surveillance, even in the absence of surveillance itself" (Angwin 2014, 44).

The feeling of being watched could also lead to more socially desirable behavior. In 1979, more than 360 Halloween trick-or-treaters unknowingly participated in an experiment conducted by Beaman et al. The results showed that when children see their own reflection in a mirror

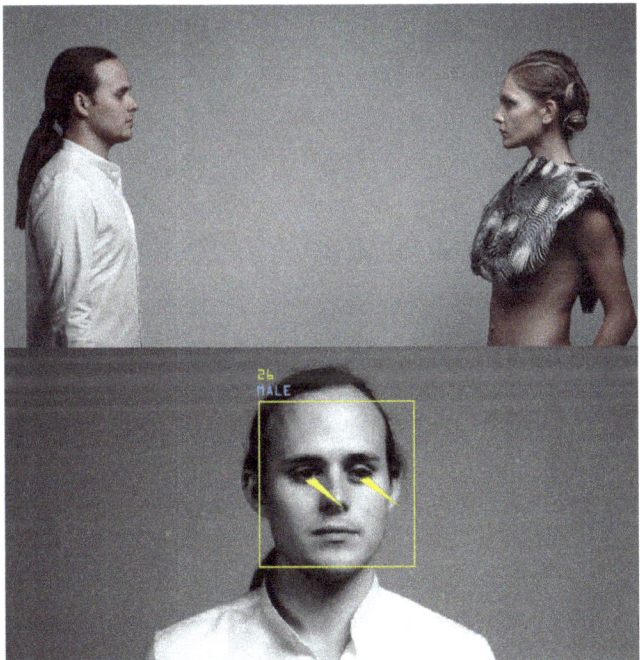

4   *Caress of the Gaze*, a 3D-printed, gaze-actuated wearable (Behnaz Farahi, 2015)

5   A small camera embedded inside the *Caress of the Gaze* wearable (Behnaz Farahi, 2015)

placed behind the candy bowl, their transgressive behavior is reduced (Beaman et al. 1979). Moreover, in their article, "Engineering Human Cooperation," Burnham and Hare argue that people make more cooperative choices in computer games when they sense they are being watched by images of a robot on their computer screen (Burnham et al. 2007).

The effect of being observed on social behavior is quite intuitive. When we know we are being observed we tend to act differently and present our best behavior. This could be achieved by the presence of a human observer, or simply by an image of eyes or the use of surveillance camera. In recent years, camera technologies have improved greatly and are now being used in a variety of applications whether outdoors or indoors, and have even been used on wearables. These computer vision technologies have the capacity to sense information, such as gestures, posture, voice, gaze, and facial expressions. Moreover, any hint of their presence will make us feel that we are being watched. It is crucial to understand both the conscious and unconscious behavioral changes related to the feeling of being watched.

One example of the critical implementation of surveillance in architectural space could be seen in the *Para-site* (1989) project of Liz Diller and Ricardo Scofidio (DS+R).

The installation consists of seven surveillance cameras mounted on metal rods above the entranceways and escalators at the Museum of Modern Art. They transmit the videos from three remote sites to the main gallery where viewers see the activities in the museum with a perplexing effect, while mirrors and upside-down chairs disorient. DS+R describe Para-site:

> Just as the social parasite entertains its host to earn welcome at the dinner table, the installation offers the entertainment value of voyeurism to a public unwittingly drawn into an interrogation of vision; just as the technological parasite creates interference in an information network, the installation interrupts the systems of the museum to interrogate it."[3]

THE PANOPTICON

Reflecting on 'value of voyeurism' and feeling of being observed, the panopticon is one of the most important historical precedents. The panopticon refers to a type of institutional building and a system of control envisioned by the English philosopher, Jeremy Bentham, in the 18th century. The panopticon has a central tower in which the guard sits, and the cells are arranged radially, so that from the tower the guard is afforded a view all around— as the name 'panopticon' implies—into each of the cells. Meanwhile, the openings in the tower itself, through blinds

6   After the installation of a series of custom-designed networked smart devices, Lauren watches inhabitants 24/7 and controls all aspects of their home (Lauren Lee McCarthy, 2020).

and other devices, prevent the inmates in the cells from knowing whether or not the guard is watching them. Thus the inmates remain under the perpetual control of the gaze of the guard.

French philosopher Michel Foucault revives interest in the panopticon in his 1975 book, *Discipline and Punish*, and uses it to illustrate how such a model could be applied to disciplinary societies in order to control their citizens. He describes the prisoner of a panopticon as being at the receiving end of asymmetrical surveillance. As a consequence, the inmate polices himself or herself for fear of punishment. As Foucault notes: "Hence the major effect of the Panopticon: to induce in the inmate a state of conscious and permanent visibility that assures the automatic functioning of power" (Foucault 1977).

What is fascinating about this example is that the gaze of the watcher is internalized to such an extent that each prisoner becomes his/her own watchperson. "Foucault emphasized the productive potential of surveillance as a technology of statecraft—one by which the state produces the form of state scrutiny is not only the province of external forms of police come to police themselves" (Dubrofsky and Magnet 2015, 6). Similarly, we could argue that—more generally—it is this process that allows rules and regulations to be internalized so as to inform our actions, behaviors, and even beliefs. Moreover, the manner in which we naturalize and internalize rules, it could be claimed, causes society to be less willing to contest unjust laws and the dominant, accepted outlook. Similar to how the inmates are not aware as to whether or not they are being watched, we are not aware that we are being controlled through naturalized rules rooted in our culture.

This internalized asymmetrical power structure could be seen in the notion of 'male gaze.' In her essay on "Visual Pleasure and Narrative Cinema," Laura Mulvey exposes the asymmetry of social and political power relations between men and women (Mulvey 1975). Mulvey claims that the male gaze serves to depict women as the object of pleasure for the heterosexual male viewer. Besides the fact that women are regularly subject to sexual harassment whenever they enter public space through various forms of 'looked-at-ness,' women have absorbed all this unconsciously as a form of internalized male gaze. What if women were to subvert this through the power of their gaze? Could we draw upon computer vision technologies to allow women to know when onlookers are staring at them?

Caress of the Gaze, developed by the author, is a 3D printed cape augmented with facial and gaze tracking technology and smart materials that responds to the onlooker's gaze.[4] This project engages with broader social questions such as the male gaze on the female body. A facial tracking algorithm in this piece detects age, gender, and gaze of the onlooker. While we know that gender is performative and doesn't depend on pure representation, the movement of a garment based on the viewer's gaze could unfold a new set of social meanings. If you are the wearer, you know which part of your body is being looked at, and if you are an onlooker, you know that your action has been noticed. This project shows how different strategies could be used to undermine the patriarchal system by developing forms of resistance using surveillance technologies. In other words, surveillance itself can be undermined by surveillance technologies.

In her art project *Lauren*, Lauren Lee McCarthy addresses the question of surveillance and privacy in a different way. People can sign up for this installation and have custom-made devices installed in their apartments, including cameras, microphones, switches, door locks, faucets, and other electronic devices. She then watches them remotely 24/7 and controls all aspects of their home In this, she literally becomes a human version of Amazon Alexa, a smart home intelligence for people, as she observes them, anticipates and fulfils their needs. During this work a form of bond between Lauren (the observer) and inhabitants (the observee) emerges which "falls in the ambiguous space between human-machine and human-human." She describes it as follows, "Lauren is a meditation on the smart home, the tensions between intimacy and privacy, convenience and agency they present, and the role of human labor in the future of automation."[5] Beyond making a clear judgement as to whether the surveillance is good or bad—as we allow more smart devices into the most intimate spaces of our homes—Lauren's work takes a more nuanced perspective by allowing the participants/viewers to make up their own mind.

CONCLUSION

Winston Churchill once said, "We shape our buildings, and afterwards our buildings shape us,"[6] which later was paraphrased in the 1960s by Marshall McLuhan with a more topical "we shape our tools and thereafter our tools shape us."[7] But doesn't this suggest a form of technological determinism? After all, there is no inherent application for any tool. A knife could be used as a murder weapon or just as a tool for cutting an apple. Similar arguments could be made about surveillance. The questions we should ask are 'in what context', 'how' and 'for whom' are we going to use it?

We should be aware of the potential problems with surveillance and AI in general. For once an issue is recognized as a problem, it becomes a different kind of problem. It becomes a problem not by which we are trapped, but rather with which we can deal. Bias is certainly a problem that originates with human beings and is replicated and even exacerbated in AI. As we are training the artificial algorithms to learn to see the world, it is important to ask ourselves, what does it mean to be 'seen' by a machine? And through the lens of what kind data have they been trained to look out at the world?

There are many arguments on the dark and bright side of surveillance. The question of surveillance, it seems, is not so straightforward after all. This article aims to suggest an intersectional affordance approach to the study of surveillance in which many facets of surveillance such as race, gender, class and the feeling of being observed should be studied altogether. We should be open to a more nuanced approach towards surveillance, where surveillance is not seen as so resolutely negative or positive (as in measures to help us track COVID, or when certain religious groups' privacy has been invaded). Lauren McCarthy's installation, *Lauren*, is a perfect example of this nuanced perspective.

NOTES

1. Techopedia, "Technodeterminism," accessed May 15, 2021, https://www.techopedia.com/definition/28194/technodeterminism.
2. Slavoj Žižek, "Coronavirus, surveillance, and the End of Capitalism," excerpted by The Radical Revolution, April 9, 2020, YouTube video, excerpt of full interview with Renata Ávila and DiEM25 TV interview, streamed live on Mar 31, 2020, https://youtu.be/UYc7eJ_Txq0.
3. Diller, Scofidio + Renfro, "Para-site," accessed May 15, 2021. https://dsrny.com/project/para-site.
4. Behnaz Farahi, "Caress of the Gaze," accessed May 15, 2021, http://www.behnazfarahi.com/caress-of-the-gaze/.
5. Lauren Lee McCarthy, "Lauren," accessed May 15, 2021. https://lauren-mccarthy.com/LAUREN.
6. Churchill and the Commons Chamber, accessed March 2022, https://www.parliament.uk/about/living-heritage/building/palace/architecture/palacestructure/churchill/.
7. Jukka Jouhki and Pertti Hurme. 2017. "We Shape Our Tools, and Thereafter Our Tools Shape Us." *Human Technology* 13. https://doi.org/10.17011/ht/urn.201711104209.

REFERENCES

Ahmed, Sara. 2008. "Liberal Multiculturalism is the Hegemony – Its an Empirical Fact' – A response to Slavoj Žižek." *Darkmatter*. February 19, 2008. Accessed May 15, 2021. http://www.darkmatter101.org/site/2008/02/19/%E2%80%98liberal-multiculturalism-is-the-hegemony-%E2%80%93-its-an-empirical-fact%E2%80%99-a-response-to-slavoj-zizek/.

Ahmed, Sara. 2012. *On Being Included: Racism and Diversity in Institutional Life*. Durham; London: Duke University Press Books.

Angwin, Julia. 2015. *Dragnet Nation: A Quest for Privacy, Security, and Freedom in a World of Relentless Surveillance*. Reprint ed. New York: St. Martin's Griffin.

Bastian, Hilda. 2013. "The Hawthorne effect: An old scientists' tale lingering 'in the gunsmoke of academic snipers'." *Absolutely Maybe* (blog), *Scientific American*. July 26, 2013. Accessed May 15, 2021. https://blogs.scientificamerican.com/absolutely-maybe/the-hawthorne-effect-an-old-scientistse28099-tale-lingering-e2809cin-the-gunsmoke-of-academic-sniperse2809d/.

Beaman, Arthur L., B. Klentz, E. Diener and S. Svanum. 1979. "Self-awareness and transgression in children: two field studies." *Journal of Personality and Social Psychology* 37 (10): 1835–46.

Benjamin, Ruha. 2019. *Race After Technology: Abolitionist Tools for the New Jim Code*. 1st ed. Medford, MA: Polity.

Burnham, T.C. and B. Hare. 2007. "Engineering Human Cooperation: Does Involuntary Neural Activation Increase Public Goods Contributions?" *Human Nature* 18: 88-108. https://doi.org/10.1007/s12110-007-9012-2.

Cox, Erika. 2000. *Psychology for AS Level*. Oxford: Oxford University Press.

Dubrofsky, Rachel E. and Shoshana Amielle Magnet, eds. 2015. *Feminist Surveillance Studies*. Durham: Duke University Press Books.

Foucault, Michel. 1977. *Discipline and Punish: The Birth of the Prison*. Translated by Alan Sheridan. 1st American ed. New York: Pantheon Books.

Gibson, James J. 1979. *The Ecological Approach to Visual Perception*. 1st ed. New Jersey: Lawrence Erlbaum Assoc. Inc.

Harari, Yuval Noah. 2020. "The world after coronavirus." *Financial Times*. March 20, 2020. Accessed May 15, 2021. https://www.ft.com/content/19d90308-6858-11ea-a3c9-1fe6fedcca75.

Jones, Stephen R. 1992. "Was there a Hawthorne effect?" *American Journal of Sociology* 98 (3): 451–468. Accessed May 15, 2021. https://psycnet.apa.org/record/1993-16087-001.

Kanji, Azeezah. 2020. "Islamophobia in Canada; Submission to the UN Special Rapporteur on Freedom of Religion or Belief." *United Nations Office of the High Commissioner for Human Rights*. November 30, 2020. Accessed May 15, 2021. https://www.ohchr.org/Documents/Issues/Religion/Islamophobia-AntiMuslim/Civil%20Society%20or%20Individuals/Noor-ICLMG-ISSA.pdf

Lahav, Gallya. 2003. "Migration and Security: The role of non-state actors and civil liberties in liberal democracies." October 15-16, 2003. Paper prepared for the Second Coordination Meeting on International Migration, Department of Economic and Social Affairs, United Nations, Population Division, New York.

Leach, Neil. 2016. "Digital Tool Thinking: Object Oriented Ontology versus New Materialism." In *ACADIA 16: Posthuman Frontiers; Data, Designers and Cognitive Machines [Proceedings of the 36th Annual Conference of the Association for Computer Aided Design in Architecture*, edited by K. Velikov, S. Ahlquist, M. del Campo and G. Thün. Ann Arbor: ACADIA. 344–351.

McCarney, R., J. Warner, S. Iliffe, R. van Haselen, M. Griffin and P. Fisher. 2007. "The Hawthorne Effect: A randomised, controlled trial." *BMC Med Res Methodol*. 7: 30. https://doi:10.1186/1471-2288-7-30.

Mulvey, Laura. 1975. "Visual Pleasure and Narrative Cinema." *Screen* 16 (3): 6–18. https://doi.org/10.1093/screen/16.3.6.

Paxton, A., J. J. C. Blau, and M. Weston. 2019. "The case for intersectionality in ecological psychology." In *Studies in Perception and Action: Proceedings from the Twentieth International Conference on Perception and Action*, edited by L. van Dijk and R. Withagen. Enschede: Ipskamp Printing.

Prasad TVSNV. 2020. "Pattern Recognition: The basis of Human and Machine Learning." *Analytics Vidhya*. December 11, 2020. Accessed May 15, 2021. https://www.analyticsvidhya.com/blog/2020/12/patterns-recognition-the-basis-of-human-and-machine-learning/.

Sensoy, Ozlem, Robin DiAngelo and James A. Banks. 2011. *Is Everyone Really Equal?: An Introduction to Key Concepts in Social Justice Education. Multicultural Education Series.* New York: Teachers College Press.

Titchener, E. B. 1898. "The 'Feeling of Being Stared At'." *Science* 8 (208): 895–897. https://doi: 10.1126/science.8.208.895.

Van Dijk, Jelle. 2018. "Designing for Embodied Being-in-the-World: A Critical Analysis of the Concept of Embodiment in the Design of Hybrids." *Multimodal Technologies and Interaction* 2 (1): 7. https://doi.org/10.3390/mti2010007.

Zizek, Slavoj. 2019. "We are already controlled by the digital giants, but Huawei's expansion will usher in China-style surveillance." *The Independent*. May 14, 2019. Accessed May 15, 2021. https://www.independent.co.uk/voices/huawei-5g-china-surveillance-social-credit-google-facebook-assange-a8912891.html.

Zuboff, Shoshana. 2019. *The Age of Surveillance Capitalism: The Fight for a Human Future at the New Frontier of Power.* 1st ed. New York: PublicAffairs.

IMAGE CREDITS
Figure 1 & 6: © Lauren Lee McCarthy
Figure 2: © Diller Scofidio + Renfro
Figure 3: © Wikimedia Commons
Figure 4 & 5: © Behnaz Farahi

Behnaz Farahi Trained as an architect, Behnaz Farahi is an award-winning designer and critical maker based in Los Angeles. She holds a PhD in Interdisciplinary Media Arts and Practice from USC School of Cinematic Arts and is currently Assistant Professor of design at CaliforniaState University, Long Beach. She is a co-editor of an issue of AD, '3D Printed Body Architecture' (2017), and of 'Interactive Futures' (forthcoming). www.behnazfarahi.com.

Data Waltz

Nate Imai
Texas Tech University

Matthew Conway
UCLA + USC

Interacting with Architecture through the Internet of Things

1 The installation connects local and remote users through their real-time interactions with Wikipedia.

ABSTRACT

This paper explores the impacts of the Internet of Things (IoT) on the field of interactive architecture and the ways this novel technology enables realignments toward inclusive and critical practices in the design of computational systems across different scales. Specifically, it examines how the integration of IoT in the design of architectural surfaces can encourage interaction between local and remote users and increase accessibility amongst contributors. Beginning with a survey of media facades and the superimposition of architectural surfaces with projected images, the paper outlines a historical relationship between buildings and the public realm through advancements in technology.

The paper next reveals ways in which IoT can transform the field of interactive architecture through the documentation and analysis of a project that stages an encounter between local and remote Wikipedia contributors. The installation creates a feedback loop for engaging Wikipedia in real-time, allowing visitors to follow and produce content from their interactions with the gallery's physical environment. Light, sound, and fabric contextualize the direction and volume of real-time user-generated event data in relation to the gallery's location, creating an interface that allows participants to dance with dynamic bodies of knowledge.

By incorporating IoT with the field of interactive architecture, this project creates a framework for designing computational systems responsive to multiple scales and expanding our understanding of computational publics.

INTRODUCTION

The growth of the Internet of Things (IoT) has created novel opportunities for interacting with the built environment. IoT's value lies in connecting "physical-first products and items to each other as well as connecting them to digital-first devices" (Greengard 2015). This technology has the potential to radically change the field of interactive architecture by transferring "intelligence and knowledge about location and state of things" to objects from a bottom-up rather than top-down approach (Achten 2015). By leveraging real-time data to allow users to engage in "two-way dynamic conversations," the scope of interactive architecture can be expanded beyond local interactions to engage a broader range of people and information (Fox and Kemp 2009).

A new generation of Wi-Fi-enabled "single-purpose microcontrollers for handling discrete sensing or actuation activities and larger server-like mini-computers that serve as network brokers" allows IoT data to be distributed across larger networks and scales (Zarzycki 2018). This development is in contrast with older models of interactive architecture, such as the Kunsthaus Graz BIX Facade by realities:united, in which centralized command and control systems limit opportunities to scale and dynamically respond to contextual conditions. Furthermore, the emergence of the Web 2.0 "based on collaboration, interactivity, crowdsourcing, and user-generated content" permits bidirectional flows of information through application programming interfaces (APIs) (Carpo 2017).

IoT's capacity to impact interactive architecture is twofold: 1) it can connect local and remote users in real-time through distributed interface components, and 2) it can increase accessibility and participation from a range of communities through mobile devices accessible to the general public. This paper examines the methods and results of an installation that uses IoT to connect people through Wikipedia's digital commons. By superimposing spatiotemporal data upon a set of constructed surfaces, this project demonstrates interactive architecture's capacity to engage local and global networks through IoT.

BACKGROUND

In "Face and Screen: Toward a Genealogy of the Media Façade," Craig Buckley establishes a historical relationship between architecture and display technologies through the framework of the media facade (Buckley 2019). Buckley outlines three distinct phases that characterize building facades as a medium. The first phase is highlighted by Leon Battista Alberti's Palazzo Rucellai that creates a schism between buildings' interior and exterior through facades

2

3

4

2 Signage within the space instructs visitors on how to make updates to Wikipedia on their mobile devices.

3 The composition of fabric, light, and sound create a soft boundary for engaging Wikipedia within the gallery.

4 QR code link to the 360 video of the gallery installation (https://vimeo.com/572931557).

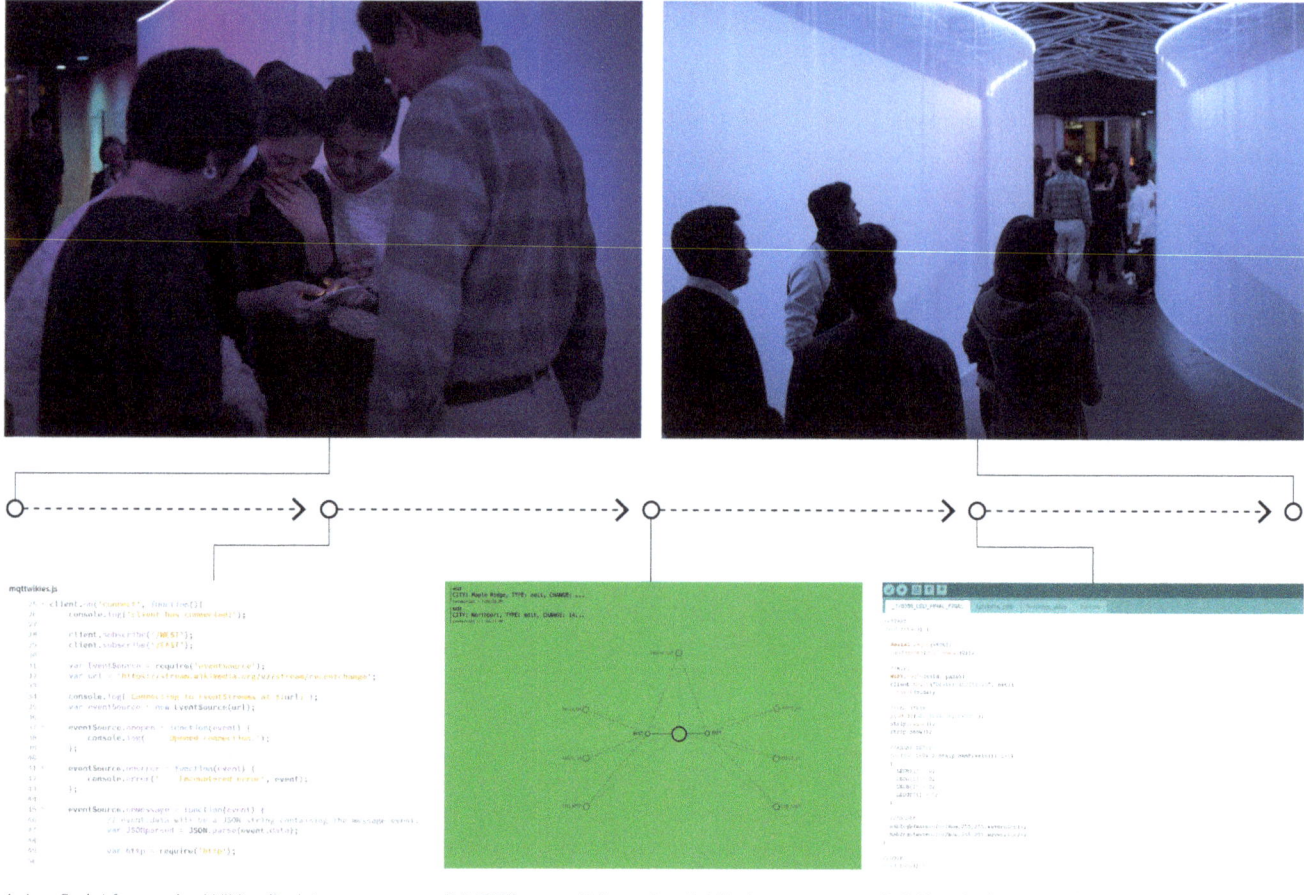

1. JavaScript for parsing Wikipedia data   2. MQTT connectivity protocol shiftr.io   3. C/C++ Arduino sketch

5

that perform as a public "screen" (Buckley 2019). The next phase is characterized by the commercialization of facades through their incorporation of billboards and electrical signage. The final phase is the contemporary media facade exemplified by the Centre Pompidou's original competition design that uses electronic lighting systems to create a surface that changes in meaning. In describing these phases Buckley identifies facades as an architectural medium for engaging the public realm through advancements in technology.

In *Kissing Architecture*, Sylvia Lavin frames the intersection of architectural surfaces and emergent forms of media as an opportunity to reconsider "medium specificity in architecture today" (Lavin 2011). Using the term "superarchitecture," which is "produced specifically by the collision and superimposition of the architectural surface with the projected image," Lavin argues the synthesis of architecture and multimedia projections can create a "surplus" and "added value" (Lavin 2011). Two impacts of superarchitecture are its ability to transform the way space is occupied and its capacity to connect people across multiple scales.

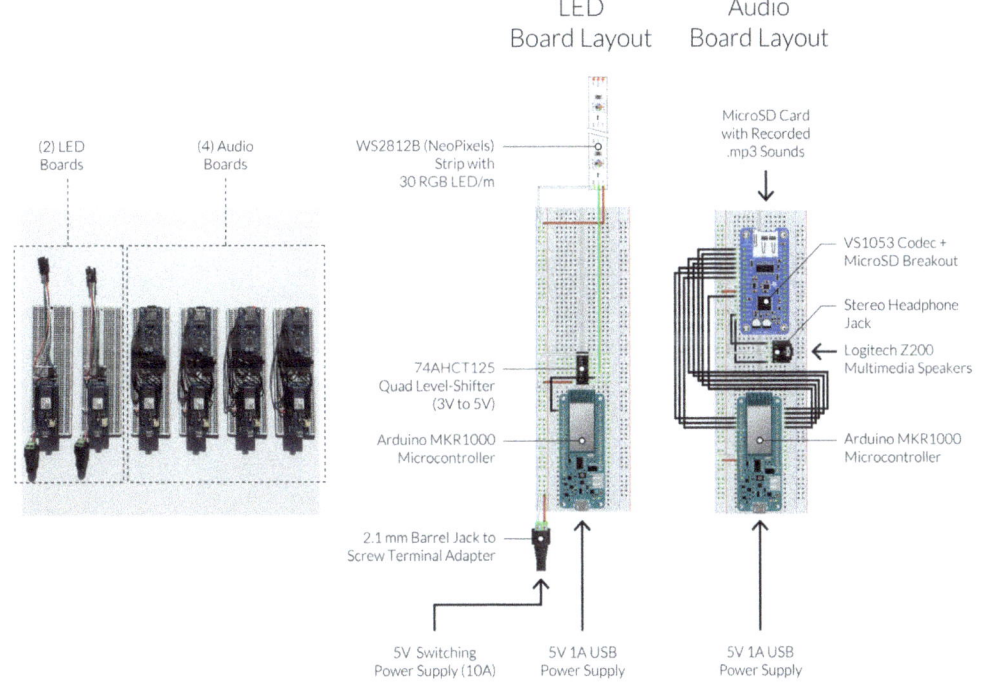

5   Wikipedia data is converted into lights and sounds via a three-step process: 1) a JavaScript code parses data from Wikipedia's recent changes stream; 2) the MQTT platform shiftr.io shares the parsed data with Wi-Fi-enabled Arduino microcontrollers; and 3) the Arduino microcontrollers running C/C++ sketches control addressable LED lights and audio speakers. The code is open source and can be found in the following repository: https://github.com/archgame/DataWaltz.

6   LED and audio board layout diagrams.

Building upon previous work, we are interested in exploring the relationship between IoT and interactive architecture to increase accessibility and connectivity across different scales. By creating immersive interfaces that connect both local and global networked systems, we can activate architectural space in novel ways and add value to the built environment through users' interactions.

METHODS

Data Waltz is designed as a feedback loop for engaging Wikipedia in real-time. Installed at the Woodbury University Hollywood Outpost (WUHO) Gallery in Los Angeles, this project makes tangible the geolocation and size of web updates through materials, light, and sound. The use of IoT to create a distributed interface for connecting local and global communities places new evolutionary pressure on the field of interactive architecture to change and adapt to an increasingly dynamic and networked society of users.

Wikipedia was selected from among other social media networks because of its unique organizational structure that promotes dialogue between remote contributors across different scales. Freely accessible in over 323 languages and editable by any user, the online platform represents a diverse set of voices from a wide spectrum of communities (Wikipedia 2021).

One important development in this project is the ability for visitors to interact with the installation by using their own web-enabled mobile devices while following the contributions of others across the globe (Figure 2). Red, green, and white pulses of light indicate article edits, new articles, and locally based updates respectively. The width and location of pulses within the gallery correspond to the size and azimuth direction of updates taking place globally. Notes arranged on a pentatonic scale sonically communicate the size and type of Wikipedia event data completing an arrangement of light and sound to form an urban relationship between the gallery, the city, and online contributors worldwide (Figure 3).

Translating Wikipedia Updates to LED Lights and MP3 Sounds

Addressable LED lights and pre-recorded MP3 sounds are used to communicate the size and type of real-time Wikipedia contributions within the gallery space. The method of translating Wikipedia updates is a three-part process: 1) a JavaScript code parses data from Wikipedia's recent changes stream, 2) an MQ Telemetry Transport (MQTT) platform shares the parsed data with Wi-Fi-enabled Arduino microcontrollers, and 3) Arduino microcontrollers operate LED lights and speakers to transmit the parsed data as light and sound (Figure 5). An azimuthal map projection encoded in the JavaScript code locates the lights and sounds in relation to the center of the gallery to create an interface that allows visitors to understand the direction from which live Wikipedia updates originate.

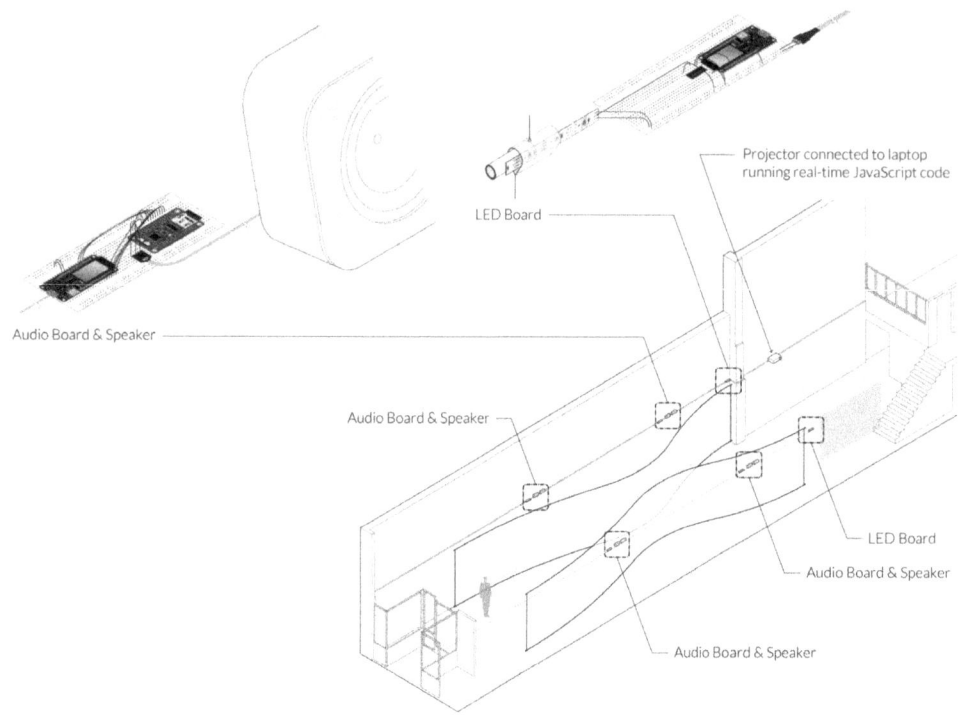

7 (2) LED boards drive (2) LED strips 480 pixels in length, and (4) audio boards control (4) sets of Logitech Z200 speakers within the gallery. A projector connected to a laptop displays the JavaScript code running in real-time.

8 The design responds to the building's orientation to create a compass for locating updates in relation to the gallery.

9 An azimuthal map projection determines the direction from which Wikipedia updates originate.

Parsing Wikipedia using JavaScript

All article changes are publicly viewable as a JavaScript Object Notification (JSON) and displayed in real-time on Wikipedia's "recent changes" stream (Wikipedia 2021). A JavaScript code is used to parse the following data: user information, type of edit, length of edit, and the name of the server accessed. IP information is available for anonymous contributors and the JavaScript code connects with an IP geolocation API to locate these contributors' city, latitude, and longitude location. The JavaScript code is executed on an internet-connected computer using the JavaScript runtime environment Node.js (Node.js, n.d.).

Using MQTT to Connect JavaScript with C/C++ Arduino Devices

An MQTT machine to machine (M2M) Internet of Things connectivity protocol is used to relay the parsed Wikipedia data to Wi-Fi-enabled Arduino controllers running C/C++ (MQTT, n.d.). The IoT prototyping platform shiftr.io was selected for its capacity to connect JavaScript and Arduino devices in real-time (shiftr.io, n.d.). This platform permits seven devices to be connected simultaneously: one internet-connected computer running the JavaScript code, two Wi-Fi-enabled Arduino microcontrollers operating LED lights, and four Wi-Fi-enabled Arduino microcontrollers operating audio speakers.

Controlling Addressable LED Lights and Audio Speakers

Arduino MKR1000 microcontrollers are used for their Wi-Fi-capacity and compact size (Figure 6). Two MKR1000s control two strips of WS2812B "NeoPixel" addressable LED lights, each strip running 480 pixels in length. Four MKR1000s control four sets of Logitech Z200 audio speakers using a VS1053 codec that plays thirty-one prerecorded custom MP3 audio files arranged on a pentatonic scale (Figure 7). C/C++ sketches written using Arduino's Integrated Development Environment (IDE) locate pulses of light and sound that correspond to the direction from which Wikipedia edits originate. Because these components are pieces forming a larger whole, the installation can grow, shrink, and be redistributed based on its spatial setting.

Azimuthal Map Projection

The design responds to the gallery's narrow 15'-0" wide by 90'-0" long dimensions and north-to-south orientation to form a spatial compass that allows visitors to understand where contributions are originating from (Figure 8). Drawing inspiration from Islam's usage of azimuth maps for locating the Qibla—the direction of the Ka'ba in Mecca—the JavaScript code unfolds the earth in relation to the gallery's center point to locate updates within the space (Figure 9) (Almakky and Snyder 1996).

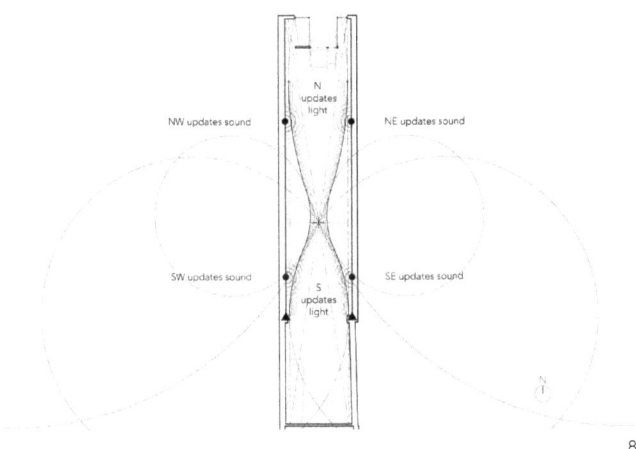
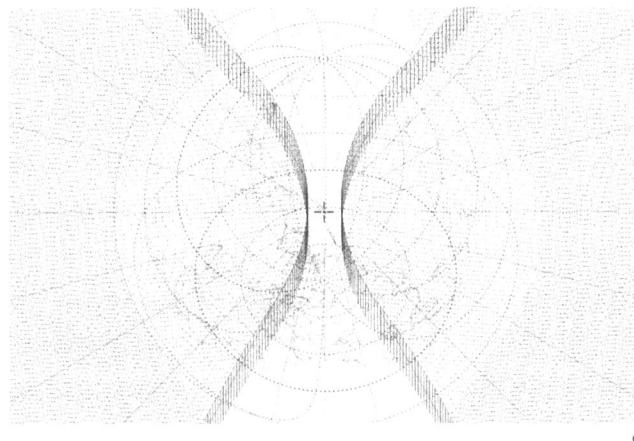

8

9

The installation allows visitors to see themselves within the constellation of contributions made by the Wikipedia community. Two lengths of sheer fabric 8'-9" high and 53'-0" long suspended on 1/2" diameter manually-bent EMT pipe create a soft boundary for framing visitors' interaction with each other and defines the gallery's center. A mounted projector showing the JavaScript code running in real-time displays updates as text (Figure 10). Updates taking place within the city trigger a sound sequence played by all speakers and a pulse of white light that originates from the gallery's center to communicate edits taking place locally (Figure 11). The effect is a choreography of fabric, lights, and sound that allows visitors to interact with Wikipedia and connects the physical gallery with the online network.

RESULTS

Data Waltz was installed for a duration of three weeks. During this period the project demonstrated IoT's capacity to impact interactive architecture in two specific ways. The first is IoT's ability to connect local and remote users through distributed interface components that include locally installed microcontrollers and contributors' internet-connected devices located across the globe. The second is IoT's ability to increase accessibility and participation from a range of communities through the incorporation of prevalent mobile devices available to the public.

One technical outcome to emerge is a method for translating real-time API web data into microcontroller outputs using a JavaScript to MQTT to C/C++ workflow. MQTT connectivity protocols are beneficial for two reasons: 1) they allow devices using different computer languages to communicate with one another, and 2) they can be monitored remotely using website interfaces that display connections in real-time. Although the physical installation is no longer active, the software for this interface remains accessible online as an open-source project (GitHub 2021).

As the images and video emphasize, there was a notable phenomenological effect created within the space. While similar in impact to other precedents outlined in this paper, this project differentiates itself by abstracting the distance between local and remote users through metrics translated into the mediums of light and sound.

CONCLUSION

This project is interested in continuing to explore the relationship between IoT and interactive architecture. One new trajectory for the research is applying the interface to other online networks that connect local and remote users through spatiotemporal data. Engaging with public health and weather data for example will allow users to better understand and participate with larger networks of people and information through their interactions with the built environment.

IoT has the potential to radically transform the field of interactive architecture by expanding its scope and reach. The integration of IoT with interactive architecture should not be framed as a purely technological or socially driven development, but as the creation of a "feedback loop" for addressing new challenges in our built environment systemically (Moe and Smith 2012). Through IoT, interactive architecture can better include missing knowledges and communities and enable realignments toward inclusive and critical practices in the design of computational systems across different scales.

10

11

ACKNOWLEDGEMENTS

Data Waltz was designed in collaboration by Nate Imai, Matthew Conway, Rachel Lee, and Max Wong as a part of the 2017-2018 WUHO Fellowship Program.

Special thanks to the following, whose contributions made this project possible:

Ingalill Wahlroos-Ritter, Heather Peterson, Galina Kraus, Tim Ottman, ACROBOTIC, Rose Brand, Hatnote, Asher Blum, Lamont Burnley, Angel Escobar, Hosam Fatani, David Hermosillo, Ulysses Hermosillo, Marta Huo, Arda Kilickan, Kevin Lugo-Negrete, Adrian Rios, Micol Romano, Karla Sandoval, Melissa Uyuni, Rodney Yasmeh, and Jordan Conway.

REFERENCES

Achten, Henri. 2015. "Closing the Loop for Interactive Architecture - Internet of Things, Cloud Computing, and Wearables." In *RealTime; Proceedings of the 33rd eCAADe Conference*, vol. 2, edited by B. Martens, G. Wurzer, T. Grasl, W.E. Lorenz, and R. Schaffranek. Vienna: eCAADe. 623–632.

Almakky, Ghazy A. and John P. Snyder. 1996. "Calculating an Azimuth from One Location to Another A Case Study in Determining the Qibla to Makkah." *Cartographica* 33 (2): 29–36.

Buckley, Craig. 2019. "Face and Screen: Toward a Genealogy of the Media Façade." In *Screen Genealogies: From Optical Device to Environmental Medium*, edited by C. Buckley, R. Campe and F. Casetti. Amsterdam: Amsterdam University Press. 73-114.

Carpo, Mario. 2017. *The Second Digital Turn: Design Beyond Intelligence. Writing Architecture Series*. Cambridge: MIT Press.

Fox, Michael and Miles Kemp. 2009. *Interactive Architecture*. New York: Princeton Architectural Press.

GitHub. 2021. "DataWaltz." Last modified June 8, 2019. https://github.com/archgame/DataWaltz.

Greengard, Samuel. 2015. *The Internet of Things*. Cambridge: MIT Press.

Lavin, Sylvia. 2011. *Kissing Architecture*. Princeton: Princeton University Press.

Moe, Kiel and Ryan E. Smith. 2012. *Building Systems: Design, Technology, and Society*. New York: Routledge.

MQTT. n.d. "MQTT." Accessed July 1, 2021. https://mqtt.org/.

Node.js. n.d. "About Node.js." Accessed July 1, 2021. https://nodejs.org/en/about/.

shiftr.io. n.d. "shiftr.io." Accessed July 1, 2021. https://www.shiftr.io/.

teamLab. n.d. "What a Loving, and Beautiful World - ArtScience Museum." Accessed July 1, 2021. https://www.teamlab.art/w/wlbw-artsciensemuseum/.

Wikipedia. 2021. "API:RecentChanges." Last modified May 24, 2021. https://www.mediawiki.org/wiki/API:RecentChanges.

Wikipedia. 2021. "Wikipedia." Last modified October 6, 2021. https://en.wikipedia.org/wiki/Wikipedia.

Zarzycki, Andrzej. 2018. "Strategies for the Integration of Smart Technologies into Buildings and Construction Assemblies." In *Computing for a Better Tomorrow, Proceedings of the 36th eCAADe Conference*, vol. 1, edited by A. Kepczynska-Walczak and S. Bialkowski. Lodz: eCAADe. 19–21.

10 The real-time projection of the JavaScript code in juxtaposition with LED lights, sounds and fabric allow the installation to be experienced through multiple mediums and senses.

11 Pulses of white light originating from the center signal updates taking place within the city.

12 The integration of IoT with interactive architecture can increase connectivity between local and remote users and increase accessibility and participation from a range of communities.

IMAGE CREDITS

Figure 1-3, 5, 10-12: © Mikey Tnasuttimonkol, 2017
Figure 4: Video by Cory Seeger
Figure 7-9: Drawings by Rachel Lee
All other drawings and images by the authors.

Nate Imai is an Assistant Professor at the Texas Tech University College of Architecture. He was the 2018-2019 Tennessee Architecture Fellow at the University of Tennessee College of Architecture + Design and has previously taught at Woodbury University's School of Architecture. He holds a Master in Architecture (MArch) from Harvard University Graduate School of Design and a Bachelor of Arts in Architectural Studies from UCLA Architecture & Urban Design. Imai's work includes built projects in the United States and Japan.

Matthew Conway is a Computational Design Leader with experience in a range of AEC softwares, technologies, and programming languages with a focus on parametric design, geometric approaches, interactive and immersive technologies, and integrated practice. He is faculty at UCLA Architecture and Urban Design and USC School of Architecture where he teaches technology seminars at the graduate level. He received a Master in Architecture (MArch) from Harvard University Graduate School of Design. His research work focuses on programming literacy and exploring architectural design through contemporary digital methods such as scripting, animation, and game engines.

Parasympathy

A Space of Empathy and Active Compassion

Mona Ghandi
Washington State University

Mohamed Ismail
Washington State University

Marcus Blaisdell
Washington State University

1 A visitor's emotional interaction with the installation. Detected changes in biomarkers change the color and form of the installation to drive the occupant's emotions toward a pleasant memory or feeling.

Affective Computing, Artificial Intelligence, Inclusion and Social Justice in the Built Environment

What if emotions, unspoken feelings, fears, and desires could manifest as architectural elements to reflect the experiences and feelings of a community, perhaps even feeling empathy? How could architecture be a more active contributor to our social and psychological wellbeing?

Parasympathy is an interactive spatial experience operating as an extension of visitors' minds. By integrating Artificial Intelligence (AI), wearable technologies, affective computing (Picard 1995; Picard 2003), and neuroscience, this project blurs the lines between the physical, digital, and biological spheres and empowers users' brains to solicit positive changes from their spaces based on their real-time biophysical reactions and emotions.

The objective is to deploy these technologies in support of the wellbeing of the community especially when related to social matters such as inclusion and social justice in our built environment. Consequently, this project places the users' emotions at the very center of its space by performing real-time responses to the emotional state of the individuals within the space.

The project leveraged AI as extended intelligence and relied on the human brain for information-processing, employing wearable technology, and sensory environments

PRODUCTION NOTES

Director: Mona Ghandi
Design: Morphogenesis Lab
Fabrication: Morphogenesis Lab
Type: Interactive Installation
Method: Artificial Intelligence
Sensory Environment
Wearable Technology
Adaptive Architecture
Cyber-Physical Space
Status: Built
Exhibition: Lewis-Clark State College Center for Arts & History
Dimensions: 16 ft x 8 ft
Date: 2020

 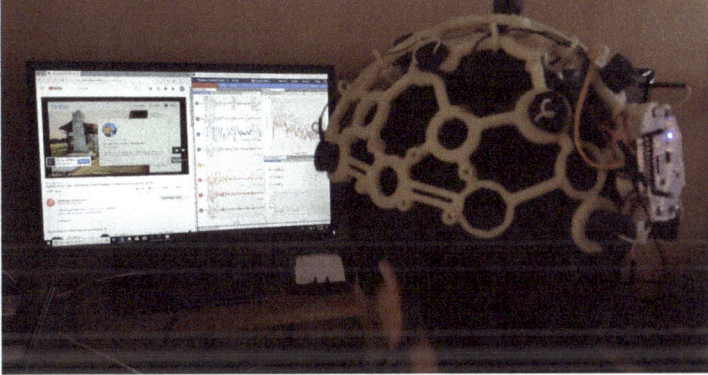

2   Neurological data collection using OPEN BCI EEG headset for the Neural Network Machine Learning

to foster a process in which synapses in the brain triggered responses in the installation, ultimately modulating emotion.[1]

The employed method is unique in its use of wearable technology (i.e., Empatica E4 and Open BCI EEG) as prostheses to collect data and integrate AI for real-time emotion detection and communication with an intelligent interactive installation, synchronizing changes in the space to the emotional data received. Acknowledging the role of Machine Learning (ML) in performance-based design since the mid-1990s (Tamke, Nicholas, and Zwierzycki 2018), here ML is used for detecting emotions and translating them into data used in design to achieve rather inclusive spaces.

3   Color and pattern created by one module

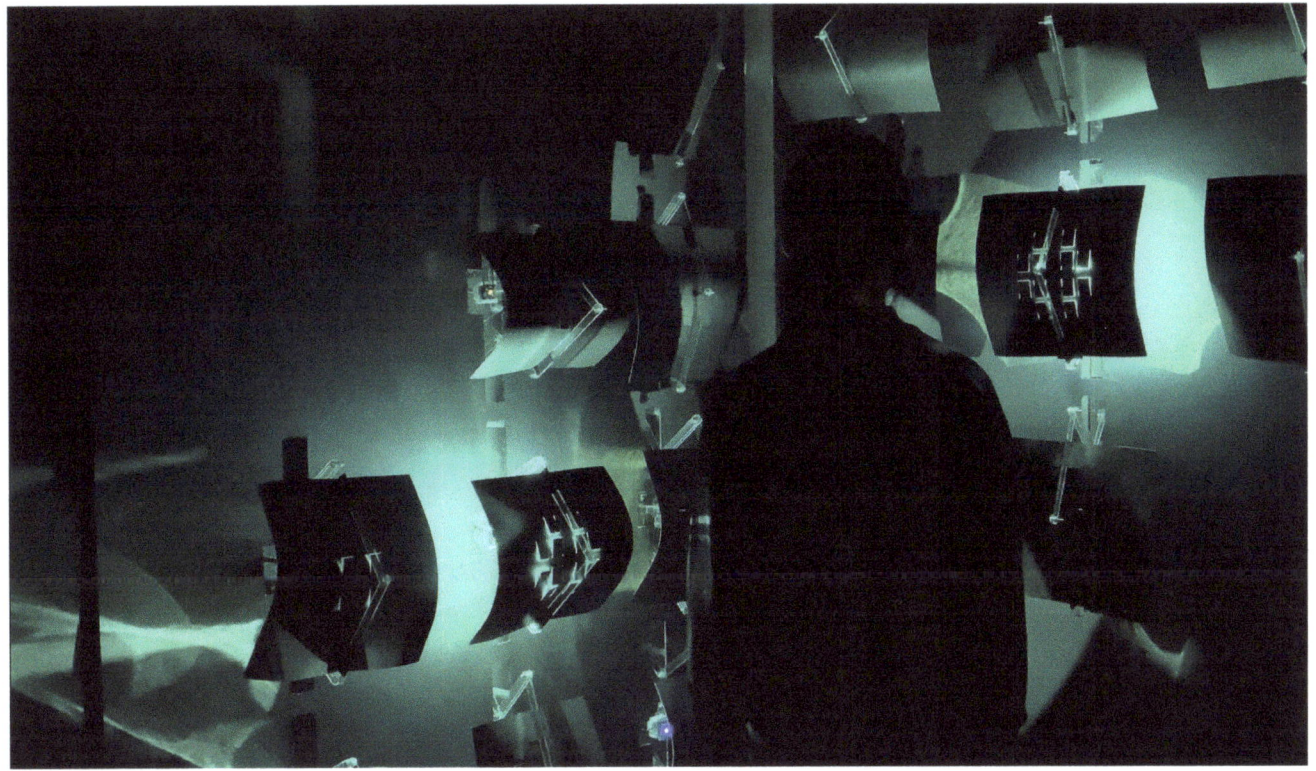

4   The empathetic interaction between a visitor and the installation. Visitors were given physical indications of their emotional and physiological states.

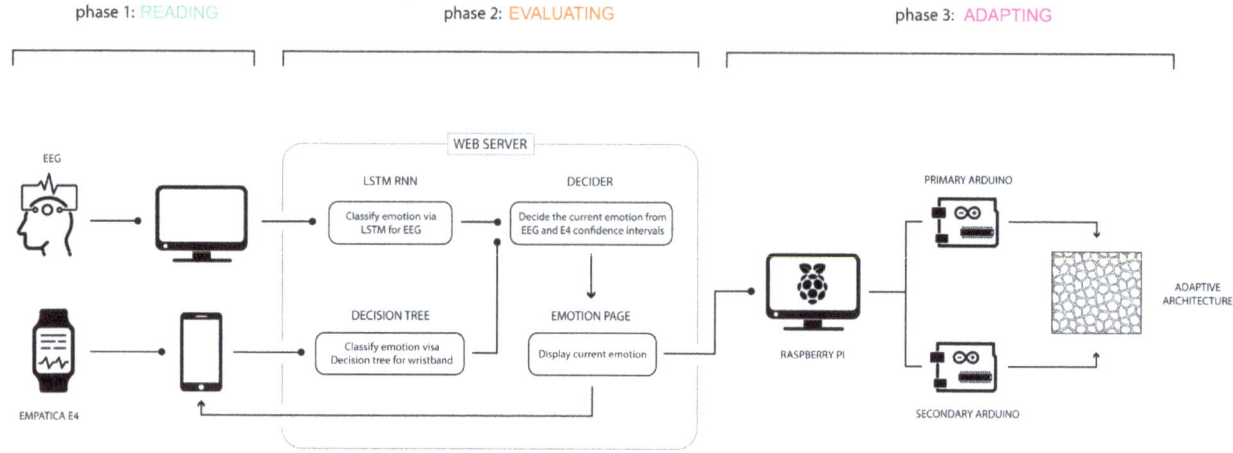

5   Flowchart represents the research method with three phases: Reading, Evaluating, and Adapting

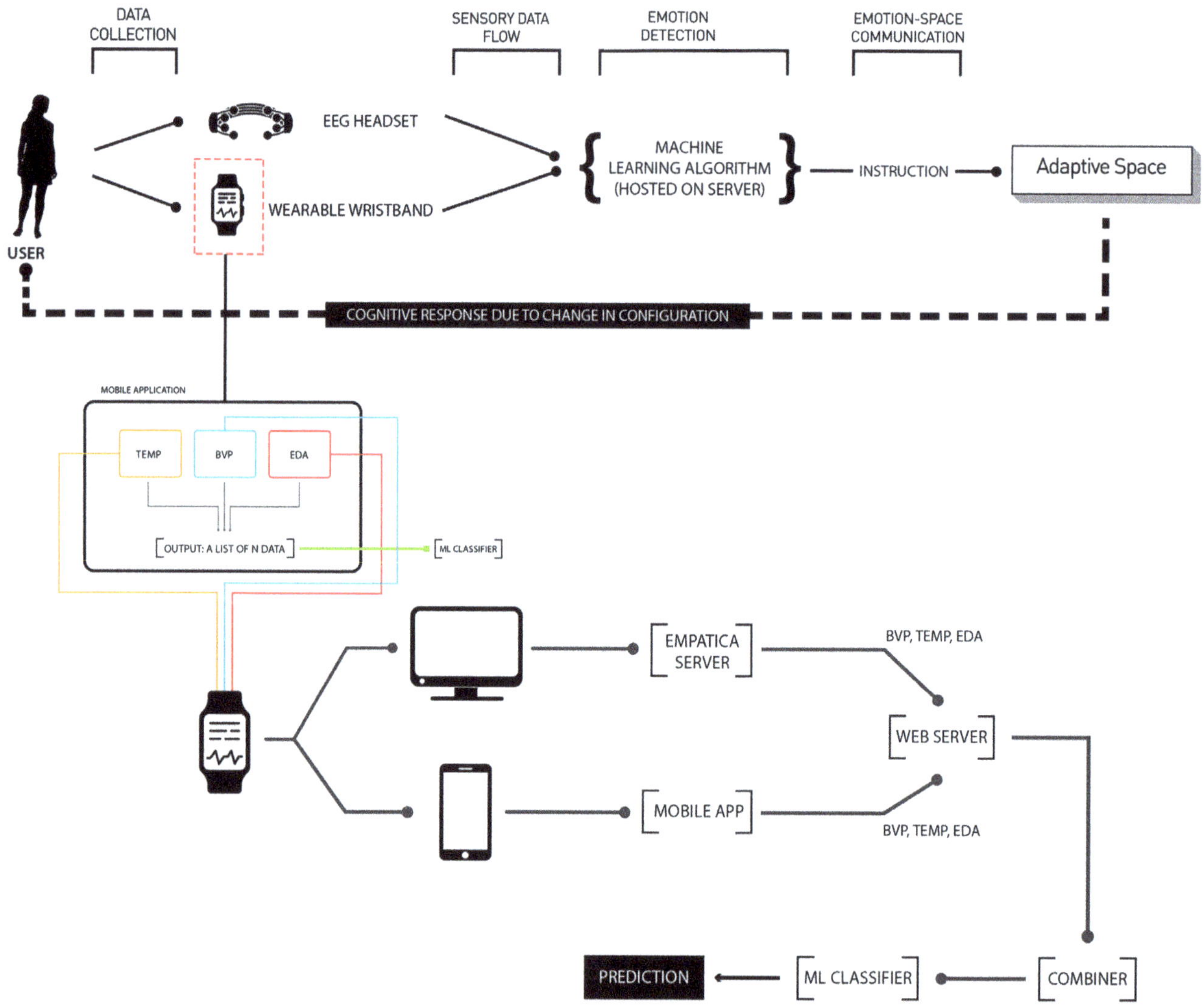

6   Flowchart showing emotional state detection based on biological data

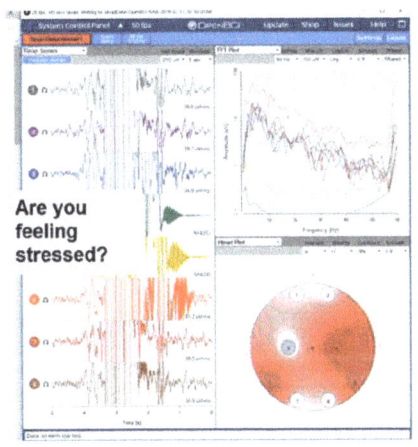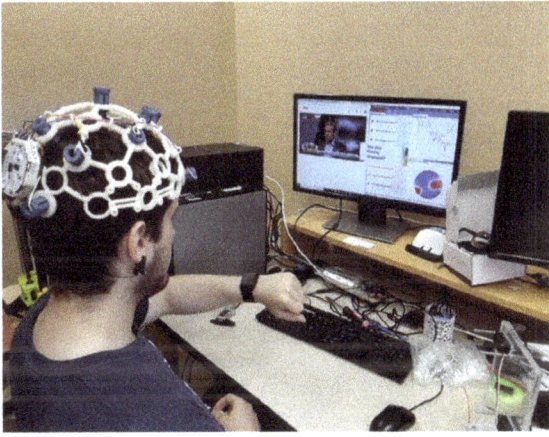

7   Neurological data collection using EEG and wristband, mapping the data into a graph for data analysis and Active Learning to improve the machine learning prediction precision with ground truth data.

OpenBCI EEG Raw Data

OpenBCI Emotion Labels

8   Stored data in a comma-separated text file as raw data and labeled emotion based on the biological and neurological data.

9   Brain and body interaction with the installation using smart wristband and EEG. Detected emotion from neurophysiological data applies changes in the form and colour of the installation.

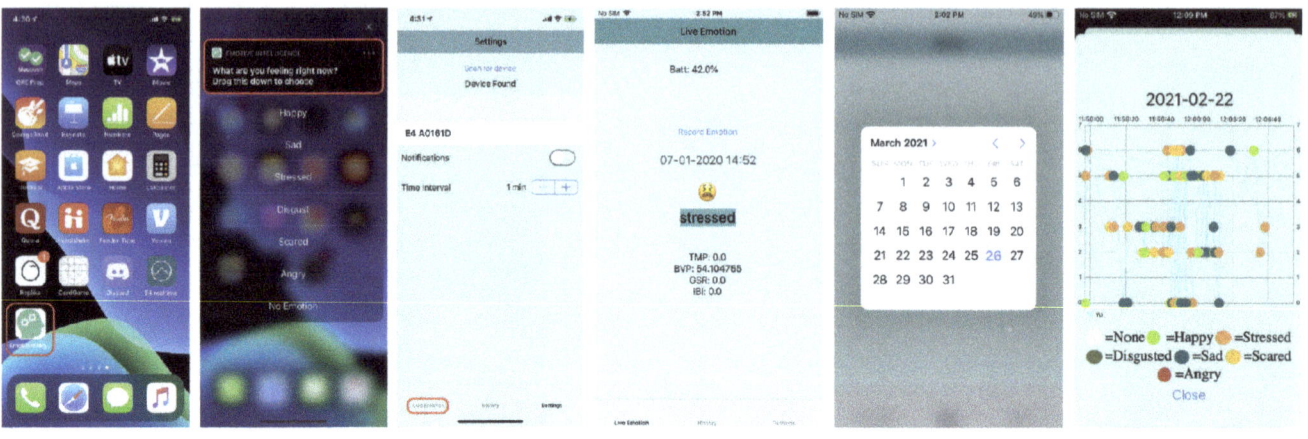

1. Launch the Emotive Intelligence app
2. When in learning mode, as the app makes predictions on the users emotional state it will periodically prompt them to either
3. Once the app is paired to the wristband, the user can view their predicted emotion by clicking on the "live emotion" page.
4. The app makes a prediction of the users emotional sate and displays it to the user. the current sensor levels are provided for reference.
5. The user can select the history page and choose a date that they would like to view their data in graphical format.
6. The user can view their recorded emotions for a given date in a chart.

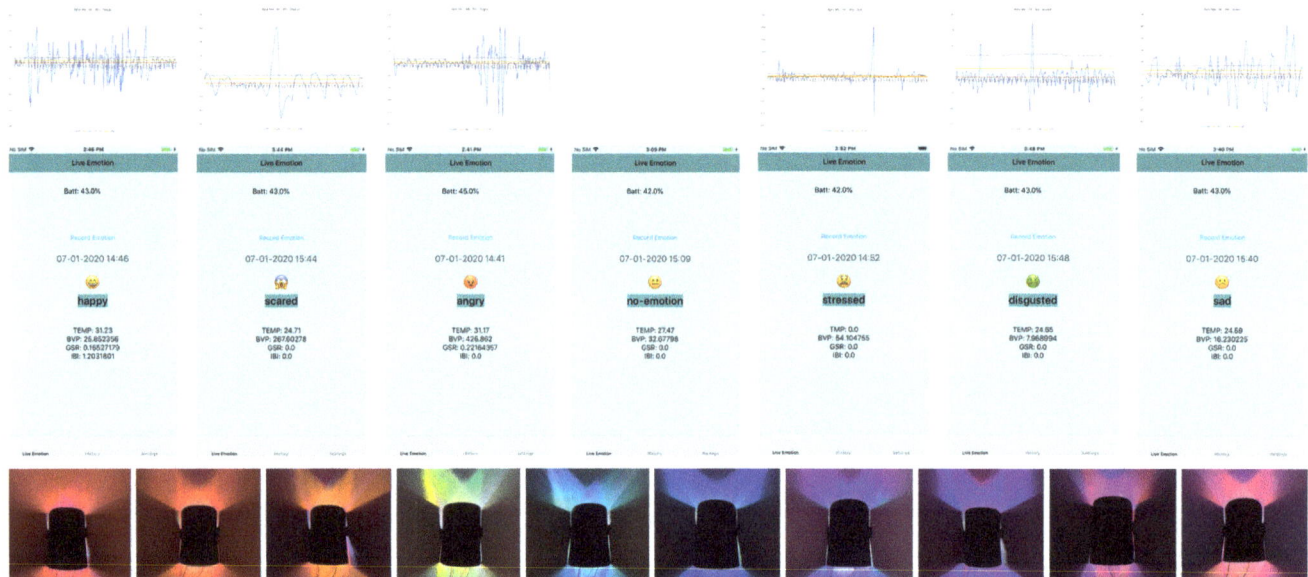

11 Mapping biodata to the seven distinct emotions (top), the phone app interface showing the emotions and biological data (middle), installation color and form changing based on detected emotions (bottom).

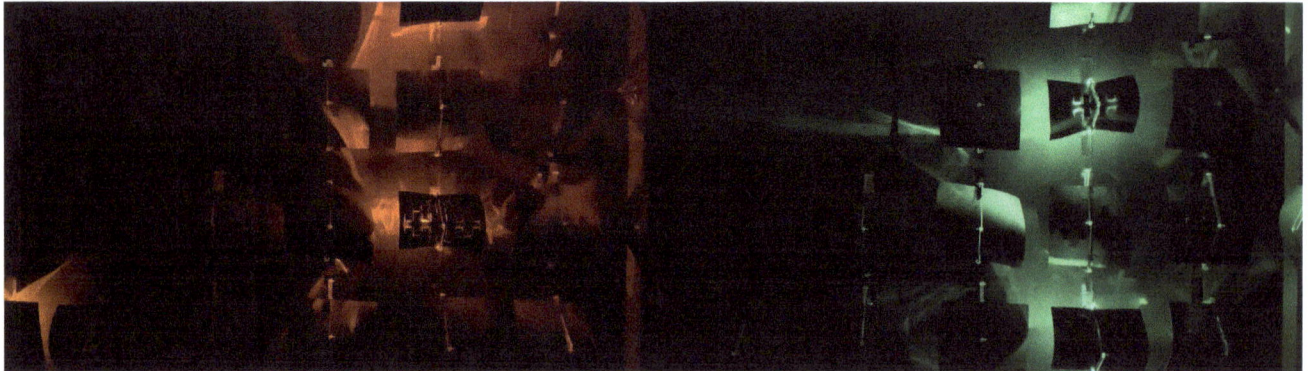

12 Installation calibration to actively respond to visitors' biological and emotional data and changing patterns and colors to create an ambience that would improve the users' emotions.

13 Full-scale installation in the gallery

Parasympathy is made of a series of kinetic reflective tiles folded and fluctuated in a calculated rhythm, producing a spectacle of color and patterns akin to the northern lights. The effect was contingent upon the involvement of its users; using a smart wristband, biophysical data (i.e. heart rate, skin electricity, blood volume, and temperature) was gathered and analyzed by our ML algorithm and translated into emotion categories. The installation was then calibrated to actively respond to these data by changing patterns and colors to create an ambient that would improve the users' emotions. For example, if the stress was detected, space morphed, and colors shifted to calming bright colors, such as blue. Based on earlier color studies, we assigned different colors to different emotions to navigate and index the moods of the users.

The installation was comprised of four 4' x 8' panels positioned at 90° in the corner of a gallery, drawing users to the center of the space to amplify the sense of immersion. Each panel consisted of a grid of retractable tiles that acted as the kinetic component of the installation. Nestled within each module was a colored LED light that activated in concert with the module's movement and detected emotion. A Raspberry Pi computer queried the web server every six seconds to read the last predicted emotional state and check for changes, calibrating the rhythm and sequence of the mechanisms. While becoming aware of their mental state through the wristband and cell phone provided, users found that the therapeutic immersion in color and light increased their sense of self-awareness of their key role in activating the installation. Users learn their emotional and physiological states and thus acquire a tool to enhance, mitigate, or simply become aware of their emotions.

By using each individual's biosignature as a noticeable trace and by breaking away from the traditional designer-centric concept, this user-centric installation is a medium that actively responds to the mood of the community, helping to promote communication for those who often go unheard. This installation demonstrates responsible

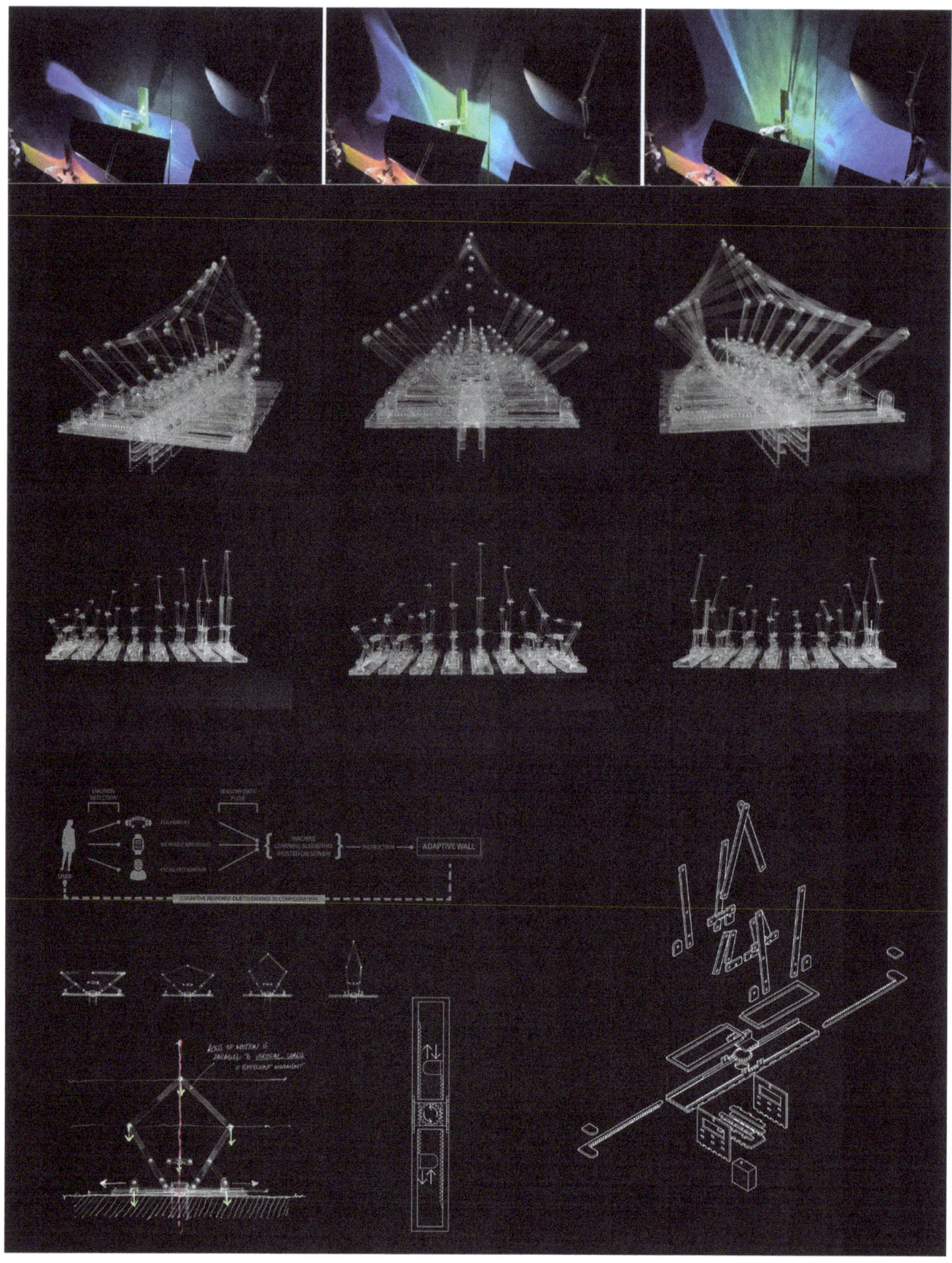

14 Light reflection and mechanical actuation occurring in sync, producing oscillating waves of color (top). Details of the retractable mechanism. Each mechanism retracts using a servo and a gear system run by an Arduino (bottom).

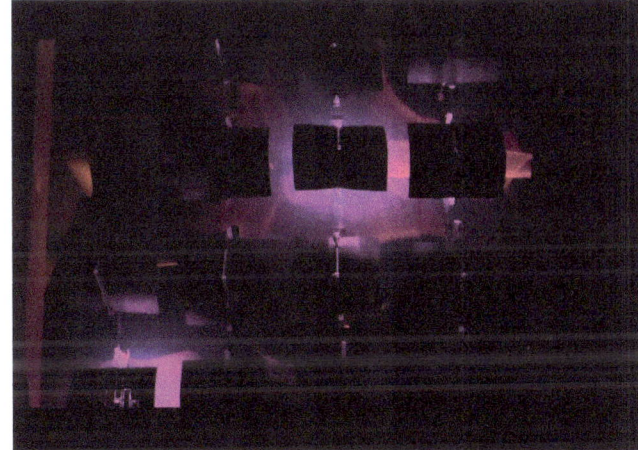

15 A selection of views showing the immersive interaction of visitors with the installation exhibiting changes in the patterns and colors of the installation, which result in distinctive atmospheric conditions of the space.

uses of emerging technology that can promote social awareness and enhance the agency of the democratic populace and equitable design.[2]

It contributes to research on cyber-physical design and the interaction of technology and empathy. This project had a singular objective, to reconcile the relationship between humans and architecture and redefine it as one of emotional empathy and active compassion.

ACKNOWLEDGMENTS

This project was funded by WSU Office of Research New Faculty SEED grant. Special thanks to Sal Bagaveyev, Aisha Marcos, Ruri Adams, and Nicole Liu for their valuable contributions to this project. The video can be viewed here: https://www.youtube.com/watch?v=qrUnVqfXy8Y.

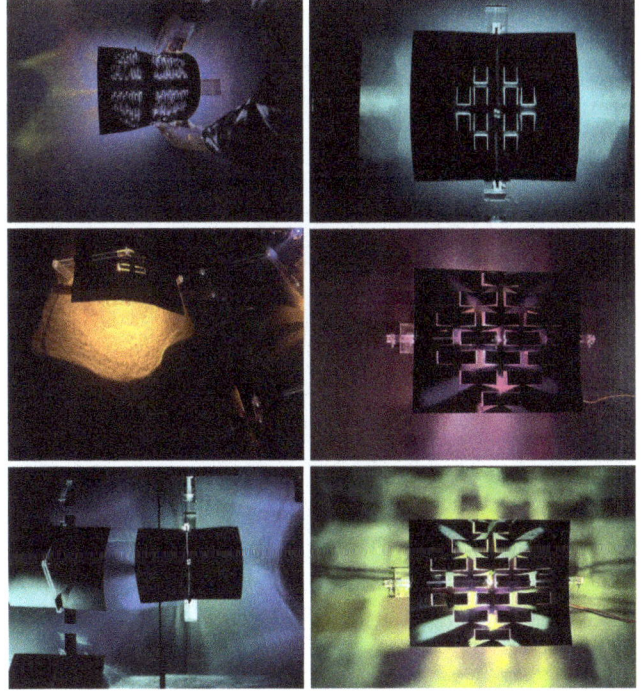

16 Primitive studies showing the effects of surface reflectivity and concavity on color and pattern generation. Nestled within each module was a colored LED light that activated in concert with the module's movement and detected emotion.

17 The results included: 1) an ML algorithm with the capacity to detect emotions from biodata, 2) a mobile app that served as an interface for active learning and emotion mapping, 3) a website for storing data and personalizing settings, and 4) a smart interactive installation capable of real-time response to people's emotions.

NOTES

1. Emotions are the result of activities in the brain and the use of neurophysiological data to determine a person's emotional state is an active research area (Donavan et al. 2018; Gilligan 2016; Hui and Sherratt 2018; Lui et al. 2018; Ollander 2015; Ordóñez and Roggen 2016; Picard and Healey 1997; Schwarz 1990; Zheng et al. 2014).

2. The application of this cognition-emotion-space interaction system has the potential to be utilized as a method of remedial therapy, provide augmented assistant living for people with physical and mental disabilities and elderlies ultimately empowering them to regain control over their environments and live more equal and independent lifestyles.

REFERENCES

Donavan, R. and et al. 2018. "SenseCare: Using Automatic Emotional Analysis to Provide Effective Tools for Supporting." In *2018 IEEE International Conference on Bioinformatics and Biomedicine (BIBM)*. Madrid, Spain: IEEE. 2682–2687.

Gilligan, T. and B. Akis. 2016. "Emotion AI , Real-Time Emotion Detection using CNN." Accessed October 6, 2021. http://web.stanford.edu/class/cs231a/prev_projects_2016/emotion-ai-real.pdf

Hui, T. and R. S. Sherratt. 2018. "Coverage of Emotion Recognition for Common Wearable Biosensors." *Biosensors* 8 (2): 30. https://doi.org/10.3390/bios8020030.

Liu, J., H. Meng, M. Li et al. 2018. "Emotion detection from EEG recordings based on supervised and unsupervised dimension reduction." *Concurrency and Computation: Practice and Experience* 30 (23): e4446. https://doi.org/10.1002/cpe.4446.

Ollander, S. 2015. "Wearable Sensor Data Fusion for Human Stress Estimation." Master diss., Linköping University.

Ordóñez, F. J. and D. Roggen. 2016. "Deep Convolutional and LSTM Recurrent Neural Networks for Multimodal Wearable Activity Recognition." *Sensors* 16 (1): 115. https://doi.org/10.3390/s16010115.

Picard, R. W. 1995. "Affective Computing." M.I.T Media Laboratory Perceptual Computing Section Technical Report No. 321.

Picard, R. W. 2003. "Affective computing: challenges." International Journal of Human-Computer Studies 59 (1–2): 55–64. https://doi.org/10.1016/s1071-5819(03)00052-1.

Picard, R. W. and J. Healey. 1997. "Affective wearables." *Personal Technologies* 1: 231–240. https://doi.org/10.1007/BF01682026.

Schwarz, N. 1990. "Feelings as information: Informational and motivational functions of affective states." In *Handbook of Motivation and Cognition: Foundations of Social Behaviour*, edited by E. T. Higgins ET and R. Sorrentino R. New York: Guilford Press. 527–556.

Tamke, M., P. Nicholas and M. Zwierzycki. 2018. "Machine learning for architectural design: Practices and infrastructure."

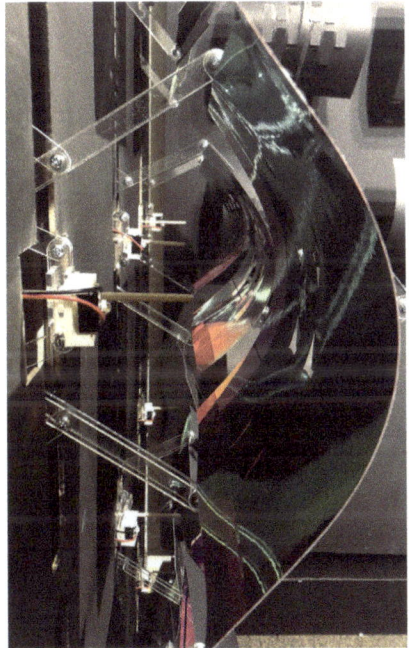 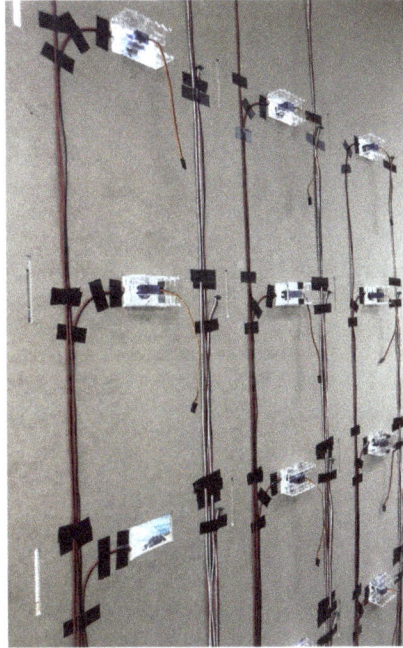 

18 Details of LEDs circuitry, 3D printed LED fixture to host a heatsink, and reflective vinyl (left), the position of the servos controlling each tile (middle), and Arduino, wiring, transistor, and resistor system of one board (right).

International Journal of Architectural Computing 16 (2): 123–143. https://doi.org/10.1177/1478077118778580.

Zheng, W. L., J. Zhu, Y. Peng and B. Lu. 2014. "EEG-based emotion classification using deep belief networks." *2014 IEEE International Conference on Multimedia and Expo (ICME)*. 1-6. https://doi.org/10.1109/ICME.2014.6890166 2014.

IMAGE CREDITS

All drawings and images by the Morphogenesis Lab, © Morphogenesis Lab.

Mona Ghandi is an Assistant Professor of architecture and the director of the Morphogenesis Lab at Washington State University. Her interdisciplinary research focuses on the Architecture of Emotive Intelligence, which examines the role of Artificial Intelligence, machine learning, wearable technologies, robotics, and adaptive architecture to improve well-being, equity, and sustainability in buildings. Her work has been recognized with several awards such as Architizer A+Award, World Architecture Award, and the Vilcek Foundation Prize for creative promises in Architecture. Her work was exhibited in different national and international exhibitions such as the Venice Biennale of Architecture, Bellevue Arts Museum, Melbourne Design to name a few. https://www.morphogenesislab.com/.

Mohamed Ismail is an architectural designer and graduate of the School of Design and Construction at Washington State University. Following his undergraduate studies, Mohamed joined the Morphogenesis Lab where he worked as a research assistant under the direction of Mona Ghandi. Mohamed's responsibilities as the head of the architecture team include the research, design, and fabrication of architectural prototypes. He works closely with colleagues in the fields of computer science and engineering to synthesize an interdisciplinary approach to design. This cross-pollination of knowledge and expertise has challenged him to advance the role of architectural possibilities.

Marcus Blaisdell is a Machine Learning Data Scientist with a BS in Computer Science from Washington State University. He has conducted research in emotion prediction with biometrics in Morphogenesis Lab at WSU and has experiences in smart spaces, physical computing, and Human-Computer Interaction. His interests are in Data Science, Artificial Intelligence, and Robotics.

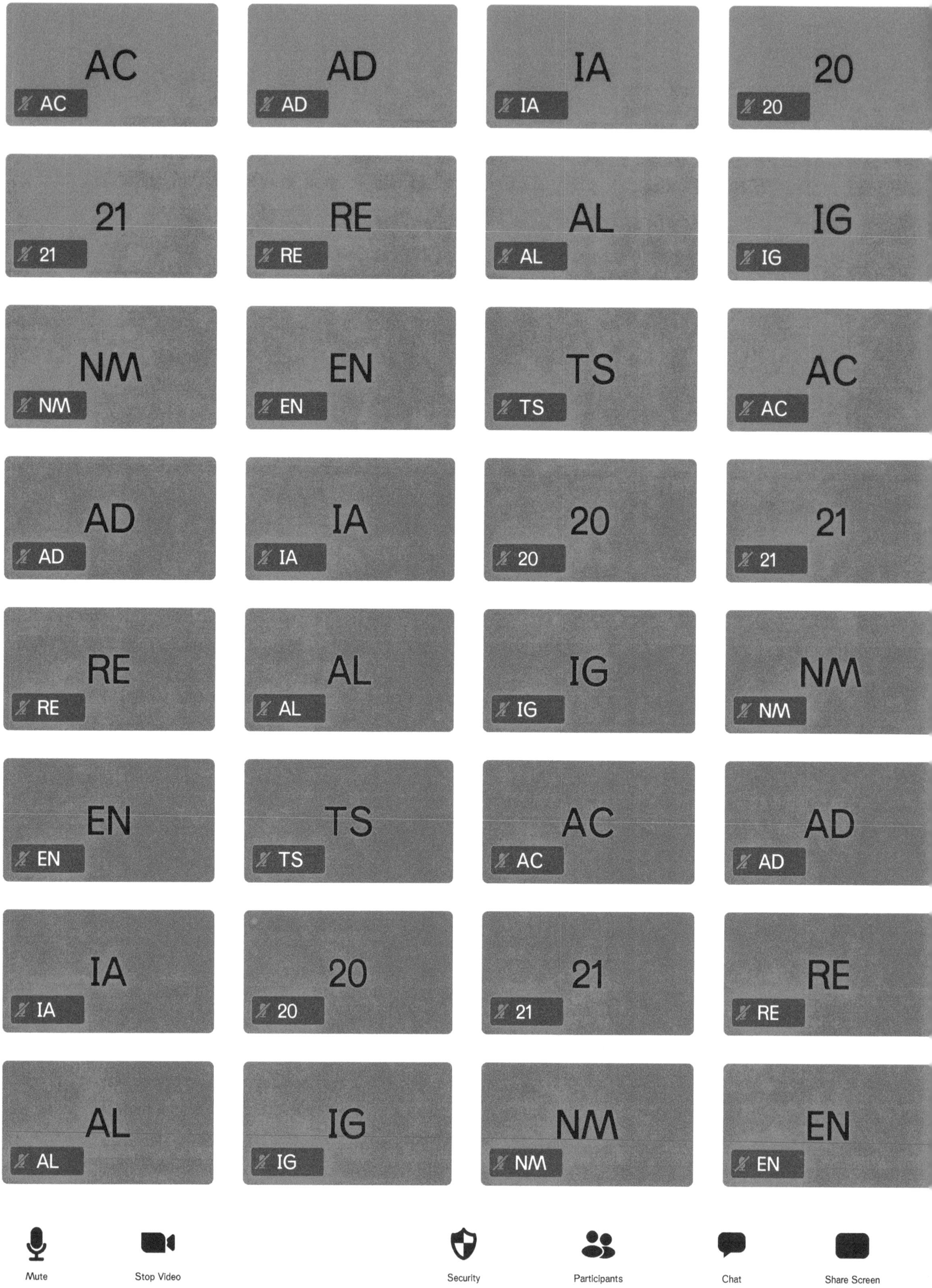

AC	AD	IA	20
21	RE	AL	IG
NM	EN	TS	AC
AD	IA	20	21
RE	AL	IG	NM
EN	TS	AC	AD
IA	20	21	RE
AL	IG	NM	EN

 Mute Stop Video Security Participants Chat Share Screen

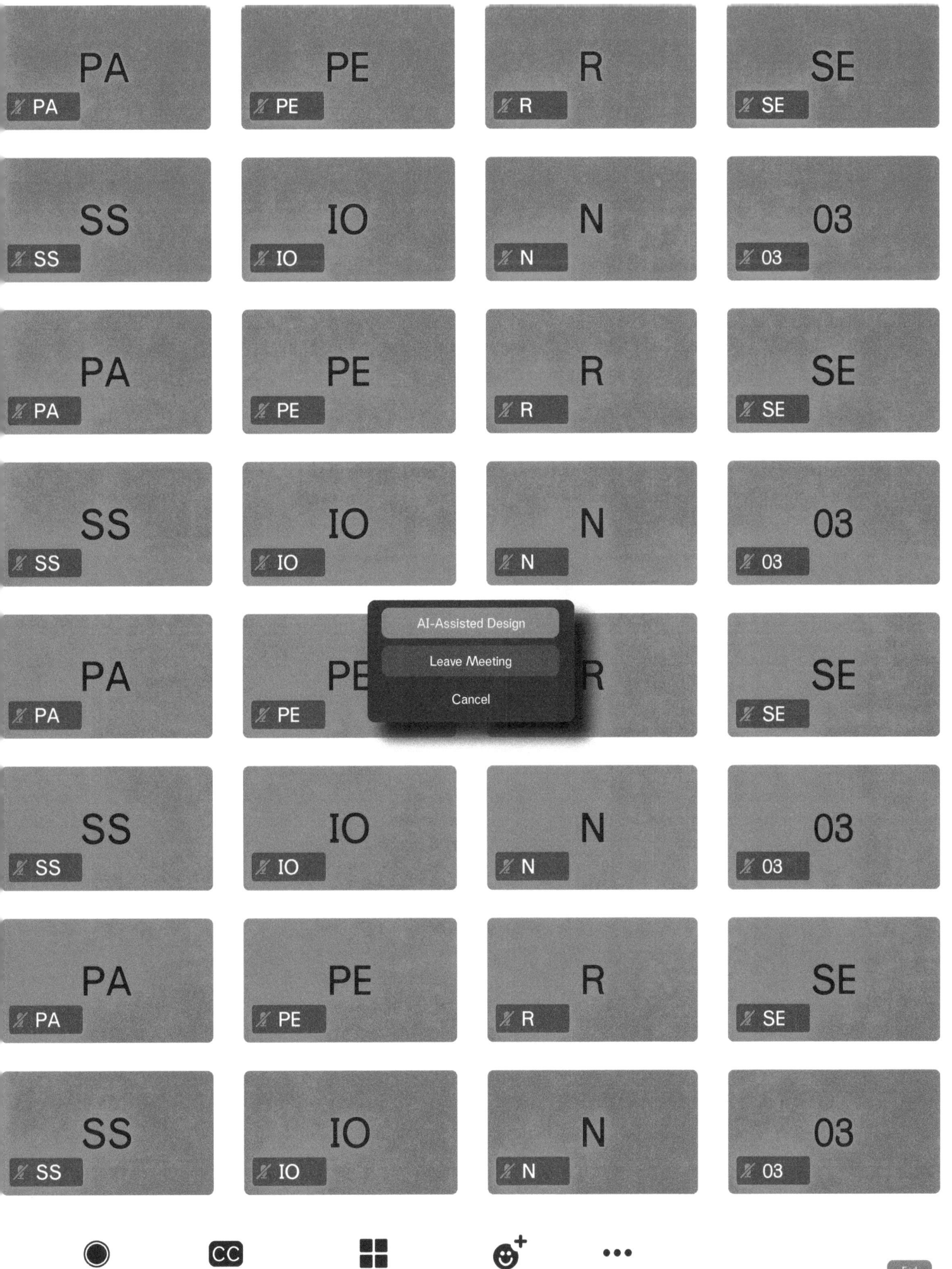

Augmenting Design

Solving design problems using generative deep learning frameworks with multiple objectives

Ridvan Kahraman
ITECH University of Stuttgart

Zhetao Dong
ITECH University of Stuttgart

Kurt Drachenberg
ITECH University of Stuttgart

Katja Rinderspacher
ICD University of Stuttgart

Christoph Zechmeister
ICD University of Stuttgart

Ozgur S. Oguz
IPVS University of Stuttgart

Achim Menges
ICD University of Stuttgart

1   A set of chairs generated using the multiple-objective variational autoencoder (MoVAE) network

ABSTRACT

In recent years, generative machine learning methods such as variational autoencoders (VAEs) and generative adversarial networks (GANs) have opened up new avenues of exploration for architects and designers. The presented work explores how these methods can be expanded by incorporating multiple abstract criteria directly into the formulation of the algorithm that negotiates these complex criteria and proposes a fitting design. It draws inspiration from the works of several design theorists who have developed such goal-oriented approaches to design, and sets up multiple-objective VAE and GAN frameworks with this idea in mind. The research demonstrates that by incorporating multiple constraints using auxiliary discriminator networks, the developed algorithms are able to generate innovative solutions to two example problems: the design of 2D digits, and the design of 3D voxel chairs. By speculating and examining the role of the designer in data based generative computational design workflows, the research aims to provide an approach for solving design tasks in the age of big data.

INTRODUCTION

As artificial intelligence becomes more prevalent, more researchers are exploring the possibilities of how AI can be used in the fields of architecture and design. Some approaches aim to combine inputs from different sources to create unexpected and creative results (Özel and Ennemoser 2019). Others use the predictive capabilities of machine learning algorithms as a tool to inform early design processes (Zhang et al. 2018), or use generative models as design drivers for generating ideas (Chaillou 2019).

Generative models are able to generate realistic images, text, or objects, but they so far infamously suffer from a "lack of control over the generative process, or the poor understanding of the learned representation" (Plumerault, Borgne, and Hudelot 2020). The challenge lies in establishing computational workflows that push the algorithms towards creativity and originality, while still enabling a level of control that incorporates real world constraints or specific designer inputs. This research aims to address these issues by introducing additional abstract constraints to established generative machine learning algorithms in the form of auxiliary neural networks. These networks are trained separately to classify designs according to given criteria, and introduced to the generative network to influence its generation according to the designer's preference. Two established generative deep learning methods have been chosen as examples: variational autoencoder - VAE (Kingma and Welling 2013), and generative adversarial networks - GANs (I. J. Goodfellow et al. 2014). The research extends these existing methods by integrating additional networks to modify the learned latent space: the network's learned representation of the design object.

In the field of computer science, such methods are often used to replicate the dataset's essential features as accurately and detailed as possible. Taking these methods to the design context requires a shift of focus to achieving novelty, inspiration, and integration into a design philosophy and method. The background section serves to outline the theoretical underpinnings, while the results and conclusion sections discuss outcomes that move beyond perfected replication, aiming to strike a balance between novelty and designer control.

At a meta-level, the work aims to push machine learning related research in architecture and design towards incorporating real world constraints within a semiautonomous computational framework, with the belief that if we are to solve real world design problems, we must work towards incorporating mechanisms into our algorithms that respond to specific demands and pressures. Workflows such as the one presented here can guide the designer in the search for design solutions that address these complex issues.

BACKGROUND

Design as a Complex, Multiple-Objective Problem

Understanding of the theory of design is essential for developing computational workflows, as the design of the networks and algorithms themselves are connected to the understanding of the tasks that they solve (Goodfellow, Bengio, and Aaron 2016). What procedures or universalities are common to each design situation and how might they be understood as performative cognitive machines which interact and symbiotically synthesize a design prototype? While in the beginning, designers often do not have a clear idea of what their solution to a presented design problem may become, it is generally agreed that design can be viewed as a goal-based process. For Hubert Simon, the science of design is "concerned with how things ought to be, with devising artifacts to attain goals" (Simon 1988). Gero describes design as "a goal-oriented, constrained, decision-making, exploration, and learning activity that operates within a context that depends on the designer's perception of the context" (Gero 1990). These ideas suggest that a method which is able to integrate abstract concepts as design goals is crucial to successfully solve design problems in a computational workflow.

In addition to being goal-based, design processes are chaotic and their evaluation criteria for success are often not well defined (Dillenburger 2016). Therefore, proposed methodologies must be able to sufficiently represent the inherent complexity involved. Alexander represents a design instance with nodes and linkages, each node being a design variable, and link being the correlation between given nodes (Alexander 1964). This representation is useful for the computational translation of a design problem, and is very similar in character to a neural network. The difficult task of changing and negotiating design variables is thus relegated to the computer, which is especially advantageous in problems with high complexity. Even though the reliance on data in identifying these connections biases the solutions towards existing examples, the networks still can use their knowledge about the designed object in unexpected ways to realize their own creativity.

Creativity and Originality in Design Problems

Human invention comes from the ability to appropriate existing patterns of information into new ones. Hume notes that this creative power of the mind amounts to no more than the faculty of compounding, transposing, augmenting, or diminishing the materials afforded us by the senses and experience (Hume 2007). He indicates that creativity emerges from the recombination of data already in memory either as long term or immediate feedback from experimentation. How well a designer is able to combine these aspects will illuminate the possible designs that can be achieved. As Alexander notes, the "act of design is governed entirely by the patterns [the designer] has in [their] mind at the moment" and their ability to combine these patterns (Alexander 1964).

New design prototypes are conceived from a conceptual schema "representing a class of generalized heterogeneous grouping of alike design cases that provides the basis for the start and continuation of a design (Gero 1990). The ability to analyze the existing examples provides a foundation for envisioning something new and uncanny to input knowledge, thus the envisioning of a new pattern based on precedents. If no past examples can be used directly, something new must be created. However, it is a rare possibility to proceed without precedent. The rich data environment of the world today and in the possible future lends itself to the creative possibilities of intelligent machines, as their abilities of mixing and re-combining existing patterns opens up new avenues of creativity that we, as human designers, might not have imagined.

The Changing Role of the Designer

Özel and Ennemoser have questioned human agency in machine learning based generative design methods. Similar to the presented work, the designer was tasked with the curation of the data and converting the final outputs from 2D concept to a 3D model, while the computer was tasked with an interdisciplinary design endeavor (Özel and Ennemoser 2019). This kind of workflow requires the designer to understand the design process holistically, and to view it as a system of information flow. In such workflows, in addition to the task of contextualizing and evaluating the design, the designer has to choose which criteria must be satisfied in order to be successful. Figure 2 represents the abstract overview of the presented research.

Leaving the mathematical definition of the design object to the computer has the additional advantage that in data-based methods, the output is not simply an object, but a network within which a latent space is embedded. This network can be viewed as a meta-object, one that contains multiple possibilities that answer the designer's call for fitting designs (Roman 2013; DeLanda 2012). In parametric design, a design topology is created, which the designer can explore by changing the parameters. A similar low

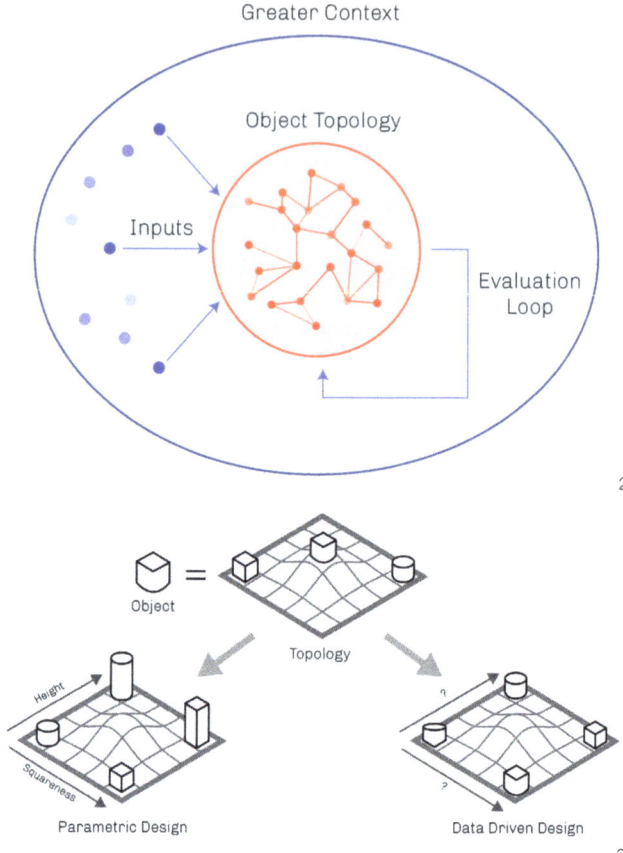

2   A sample design process, divided into the human (blue) and computer (red) tasks

3   Topological space of a designed object. On the left is a topological space of two dimensions: height and squareness. On the right is an example of a data-inferred topological space or latent space, where the dimensions are not defined by the designer, but rather by the computer through the analysis of data.

dimensional topological space is generated in data based generative methods, only this time through the analysis of data (Figure 3). This has the effect of changing the role of the designer from a creator of parameters or rules, to a creator of context. The designer's ability to evaluate assumptions and to perceive the scope of the task will dictate the possibilities that can be achieved.

Related Work

A GAN that generates 3D objects has been proposed by Wu et al., which demonstrates the capabilities of this framework in 3D (Wu et al. 2016). Meanwhile, Bidgoli and Veloso have extended the autoencoder network developed by Achlioptas et al. and Brock et al. to the architectural realm by adding a way in which the human designer can interact with the network to generate 3D chairs (Bidgoli and Veloso 2018). This study showcases the VAE method's ability to generate objects by interpolating existing data. However, even though some latent space vectors do correspond to understandable design parameters, they do not allow the designer to preselect specific criteria to modify the generation. This gap points to the need to establish a reliable framework in which data-driven generative methods can be easily controlled and understood by the designer.

Ideas to address this need have been proposed for GANs: conditional GANs that take additional input data that influences the generation (Mirza and Osindero 2014; Park, Yoo, and Kwak 2018), or style-based GANs that allow a level of control over the 'style' of the generation (Karras, Laine, and Aila 2018). While these approaches do increase control, the goals that designers usually operate with are abstract and relate to the specific design object. In a design context, an Auxiliary Classifier GAN (AC-GAN) was used to generate façade patterns that fit daylight performance criteria (Odena, Olah, and Shlens 2017; Shaghaghian and Yan 2019). Another approach where abstract goals such as 'monumental' or 'delicate' were incorporated into a GAN setup was proposed by Liu et al. (Liu, Liao, and Srivastava 2019). The setup learns how to generate variables for a parametric model that will result in objects fitting designer criteria, although limited by the parametric model's design space.

This study addresses the issue of enabling designer control by incorporating conditions chosen by the designer as auxiliary discriminator networks to a base generator. The base generator creates the initial object topology, which is then modified according to the designer's needs. To our knowledge, multiple objective conditioning in GANs or VAEs has not been used for this purpose before. However, there have been works that incorporate a similar strategy in different fields for different purposes: increasing the stability of GANs (Durugkar, Gemp, and Mahadevan 2016; Albuquerque et al. 2019), increasing the quality of generations (Wang et al. 2019), and for tackling other problems where multiple tasks are involved (Ferstl, Neff, and McDonnell 2019). On the VAE side, a similar idea to the MoVAE presented in this paper was proposed where a pre-trained classifier network is added to the original VAE for improving the quality of generated image samples (Lamb, Dumoulin, and Courville 2016).

METHODS

Extending from their predecessors, the two frameworks, multi-objective VAE (MoVAE) and multi-objective GAN (MoGAN), propose the post-integration of a flexible amount of customizable auxiliary discriminator networks to allow multi-objective control of the generation. In both cases, the first step is to train the generative networks, VAE or GAN, with their established methods. The auxiliary discriminator networks are trained separately to evaluate the data according to given criteria. The combined network is then trained based on the designer's preference. This final training step takes only a few minutes, so the designer can select new criteria and quickly retrain the network for new results.

The two frameworks were tested on a 2D task and a 3D task respectively. The 2D task addresses the creation of a new digit, and is designed as a simple case study. The 3D task explores the design of a chair that fits into multiple design criteria. The chair as a design object provides enough complexity and variety for interesting results to emerge, while being visually evaluable. Tensorflow version

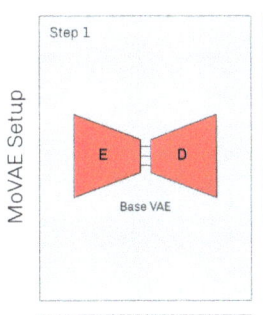
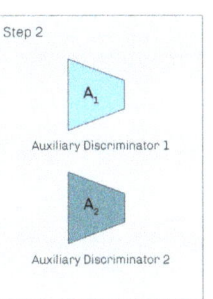
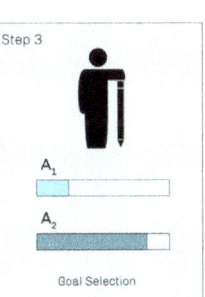
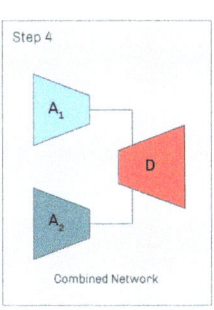
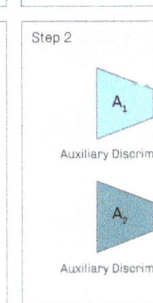
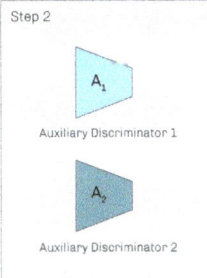
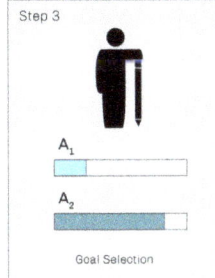
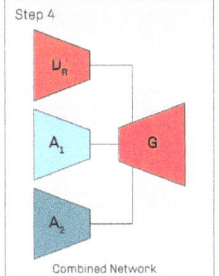

4 The training steps of MoVAE and MoGAN frameworks.

MoVAE:
1) The base VAE is trained.
2) The auxiliary discriminators are trained.
3) The designer selects the desired criteria.
4) The decoder part of the network, D, is further trained with the MoVAE loss function that takes inputs from $A_1, A_2 ... A_k$.

MoGAN:
1) The base GAN is trained.
2 and 3) Identical to MoVAE.
4) The generator part of the GAN, G, is further trained with the additional discriminators $A_1, A_2 ... A_k$, with the modified loss function.

2.5 in the Google Colaboratory environment is used to train the networks. The chairs are visualized in Rhinoceros 3D (Abadi et al. 2016).

MoVAE

VAEs are known for their ability to learn a low dimensional representation of the dataset, and their capability for interpolations in this lower dimension representation. The Multiple-objective VAE (MoVAE) is trained in the following sequence shown in Figure 4. The base VAE (Kingma and Welling 2014) is trained as a traditional variational autoencoder. The auxiliary discriminator networks are trained as classification or regression models, depending on the training data. Their goal is to learn how to identify if the input data fits into a given criteria by looking at labeled data. For example, they can be trained to identify if a digit is a '2', or if a voxelized chair is stable or not.

The trained auxiliary discriminators are then added to the network to incorporate additional goals. In the case of the digit task, the goal digit is given to the network as a number input for each criteria. For the chair generation, the goal is given as a real number input between 0 and 1 for each criteria. For the MoVAE training, the networks are combined using the loss function:

$$L_{MoVAE} = L_{VAE} + \sum_{k=1}^{K} \alpha_k L_{Dk}$$

Where L_{VAE} = base VAE loss, K= number of auxiliary discriminators, = weight coefficient, L_{Dk}= auxiliary discriminator loss. The base VAE loss is the VAE loss from Step 1. The alpha value is a coefficient that is used to adjust the strength of each auxiliary discriminator. This coefficient is necessary to make sure that the L_{Dk} terms have a similar degree of influence as the L_{VAE} term.

MoGAN

GANs show a powerful generative ability by having two networks competing with each other: one network learns to generate results that fools the other to mistake the generated results with those already existing dataset, and the other network learns to differentiate the generated results from the dataset. The Multiple-objective GAN (MoGAN) is trained in the sequence shown in Figure 4. For the base GAN, the traditional GAN training documented in the original paper is used (Goodfellow et al. 2014). The auxiliary discriminators are trained in the same way as MoVAE. Once the goals are chosen, the following loss function is used for combining the networks:

$$L_{MoGAN} = L_{DR} + \sum_{k=1}^{K} \alpha_k L_{Dk}$$

Where L_{DR} = realistic discriminator loss (same as Step 1). Similar to the MoVAE loss function, the alpha value is a coefficient that is used to adjust the strength of each auxiliary discriminator, and is necessary to make sure that the L_{Dk} terms have a similar degree of influence as the L_{DR} term.

RESULTS AND REFLECTION

Creating a New Digit

For the first proof of concept study, a MoVAE and MoGAN were trained to generate a digit that fits into the criteria of being a '0' and a '2'. Once the combined network is trained (step 4), the objectives can easily be altered and the network can be retrained in a few minutes to produce results that fit different criteria. The MNIST dataset is used for the training (LeCun, Cortes, and Burges 1999). An appendix detailing the training and the networks is provided as supporting documentation in the Notes section [1].

MoVAE

As elaborated in the methods section, the first step is the training of a standard VAE. The outputs of this VAE are shown in Figure 5A, where the dataset is well replicated. The auxiliary discriminator networks trained in the next step are able to identify digits correctly with an accuracy of around 98%. After completing the final step where the networks are combined and retrained, the entire latent space morphs into a selection of either twos, zeros,

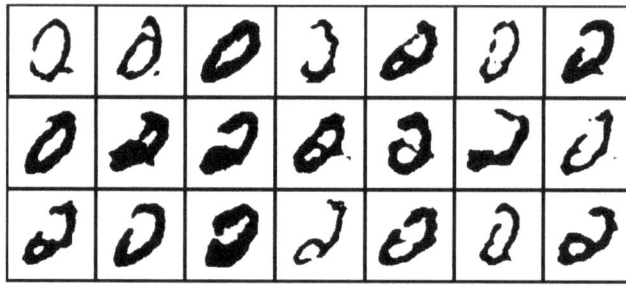

5 Results of digit MoVAE. A) Random digits from the latent space of base VAE. B) Digits sampled from the same noise vectors as A, but after further training using the MoVAE setup. C) MoVAE results that were rated highly for both criteria.

6 Digits generated using the MoGAN framework. It is trained with auxiliary discriminators to generate digits that look like a '0' and a '2' at the same time.

or a digit in between twos and zeros. The network succeeds in finding the closest '0' or '2' to any digit from the dataset (compare Figure 5A and 5B). Therefore, a '2' that evolved from a '1' still has a distinct slender shape, but has become identifiable as a '2'. Figure 5C displays the creative potential of the method, where the results that are highly rated by both criteria are selected.

MoGAN

While the previous experiment used the MoVAE, this experiment attempts to achieve a similar result using MoGAN. Figure 6 shows the results that score highly in both criteria. The MoGAN is successful in producing novel digit patterns by negotiating the same two criteria. Similar to the MoVAE results, this is a case where the human and computer judgments converge: human designers also identify these patterns as a '2' or a '0', even though the network generates solutions that solve the challenge in different ways. The solution space for the two methods look similar, however, there are interesting patterns specific to this method such as the 'zero with a tail' pattern seen in some designs.

Chair Design Using MoVAE

The goal of this exercise is to use multi-objective generative networks to design three dimensional chairs that fit into pre-selected criteria. As GANs are significantly harder to train in the 3D domain, it is extremely difficult to train a stable MoGAN for chair design. The previous section demonstrates that the MoGAN framework can generate solutions that are different from those generated by MoVAE. However, due to stability issues regarding training 3D GANs, the MoVAE setup proved to be a better choice for this task and will be the focus of this section.

The ShapeNet dataset was used for this exercise (Chang et al. 2015). Three design criteria are defined for the generation of the chairs: stability, function, and aesthetic preference (Figure 7). Due to the lack of available labeled data, ratings of each criterion had to be conducted manually. The three criteria are consciously chosen to test the network's ability to assess a range of objective and subjective ratings. The stability factor as an objective rating can be calculated using a simple formula assessing the moment of overturning. The aesthetic preference and function ratings as rather subjective criteria were defined by a single researcher as a categorical variable ranging from 0 to 10. The limitations of using a single researcher's rating is apparent, nevertheless, averaging preferences of more researchers was not attempted at this stage of the research to allow an easier evaluation of the outcomes. Future studies should employ ratings of multiple sources to incorporate a wide range of aesthetic preferences. More detail on the labeling process can be found in the Notes [1] section.

To demonstrate how each condition influences the generative process in the MoVAE framework, one auxiliary discriminator at a time was integrated. Figure 8 shows a chair generated from the same noise vector by the original base VAE, and MoVAE. In these examples, $L_{VAE}= 0$, as the influence of the base generator for evaluating single discriminators was not included. These results show that each auxiliary network is able to influence the generation

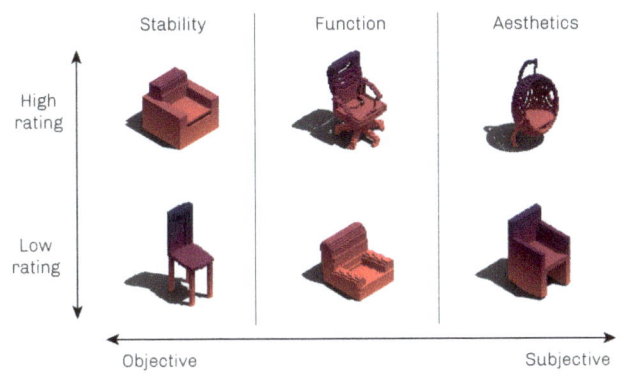

7

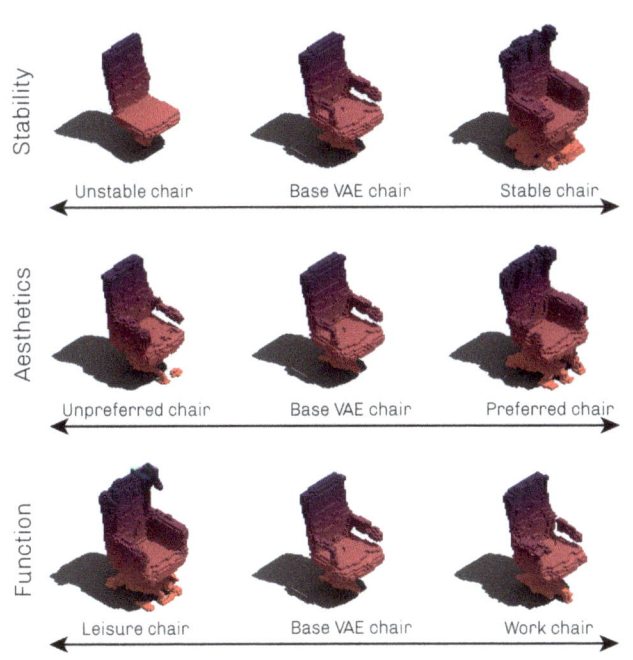

8

7   Three classes of ratings were purposefully chosen to test the network's ability to grasp concepts with different levels of subjectiveness and abstraction: stability, function, and aesthetic preference.

8   The chairs in the middle are generated from a single noise vector using the base VAE. The multitude of chairs around it are chairs generated using the same noise vector after being further trained with singular criteria. Each row shows the effect of a singular auxiliary discriminator on the base VAE after further training.

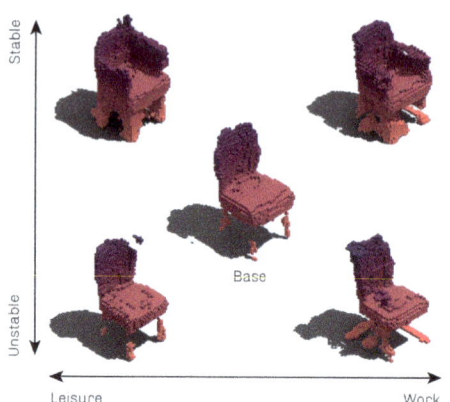 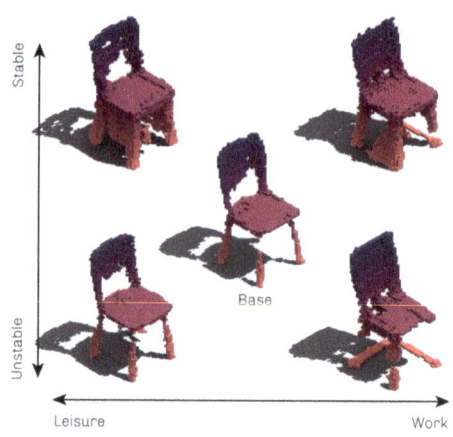

9 Two-objective chair designs using the MoVAE. In both charts, the chair in the middle is generated by the base VAE. The X-axis is the objective given to the function discriminator, while the Y-axis is the objective given to the stability discriminator.

10 On the left column are the outputs from the MoVAE with different designer requests. On the right are the three chairs from the dataset with the least different L2-norms compared to the generated output.

and grant the networks an understanding of abstract concepts such as stability, aesthetic preference, or function. For example, the output from MoVAE is modified to be more of a leisure chair with the addition of more couch-like features. Its stability is visibly altered as more voxels are either added or subtracted.

The results using multiple discriminators simultaneously are displayed in Figure 9. The outputs demonstrate that the network is able to negotiate between the criteria and increase or decrease the auxiliary discriminator outputs accordingly. The auxiliary discriminator determining leisure/work level of the chair made an impact on the generated results, as some of the characteristics of work chairs such as thinner backrests and office chair style legs are evident. The stability discriminator also influenced the generation by adding a level of bulkiness to the output. MoVAE networks with all objectives (stability, aesthetic preference, leisure/work) turned on at the same time were also trained. Adding the aesthetic preference discriminator changes the outcome of the generations, however, the reasons for these changes are not as easily identifiable as those for the other discriminators.

The 'originality' of the generated chairs were evaluated by assessing how similar the new chairs are to existing chairs in the dataset (similar to Wu et al. 2016) (Figure 10). The L2-norms of the generated outputs are compared to that of all the geometries in the dataset and the three that are least different are selected. Even though the generated chairs resemble the dataset, the network demonstrates its ability to generate novel geometries by mixing patterns and features found in different data. The final generations display design elements that do not exist as wholes in the dataset.

Other examples taken from various trained networks are presented in Figure 11. The latent space contains some unexpected results that diverge from anything that is found in the dataset. These 'over-constructed' chairs raise the question of how far the 'chairness' of a chair can be stretched before the concept loses its meaning, and how humans and computers stretch this meaning in different ways. While the presence of a seat is perhaps the most important feature in a chair for humans, a computer might prefer additional armrests when pushed to its limits (Figure 11H), or decide to remove the seat altogether for stability and aesthetic preference (Figure 11E). Taking both figures into account, it can be said that it takes conflicting criteria to really force the MoVAE to innovate outside the existing data-scape. In this case, the sacrifice in the clarity and 'chairness' of the chair is a necessary price to pay. As designers we want to operate at the boundary of what is possible. By striking a balance between originality and precedent, the algorithm operates at this boundary. It is this kind of intelligence that the machines will make use of, in order to assist the human designer as they accept their role as creators of context.

CONCLUSION

The potential to use neural network workflows such as

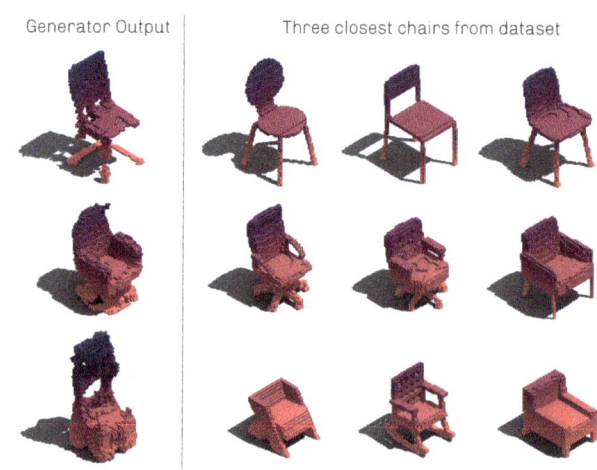

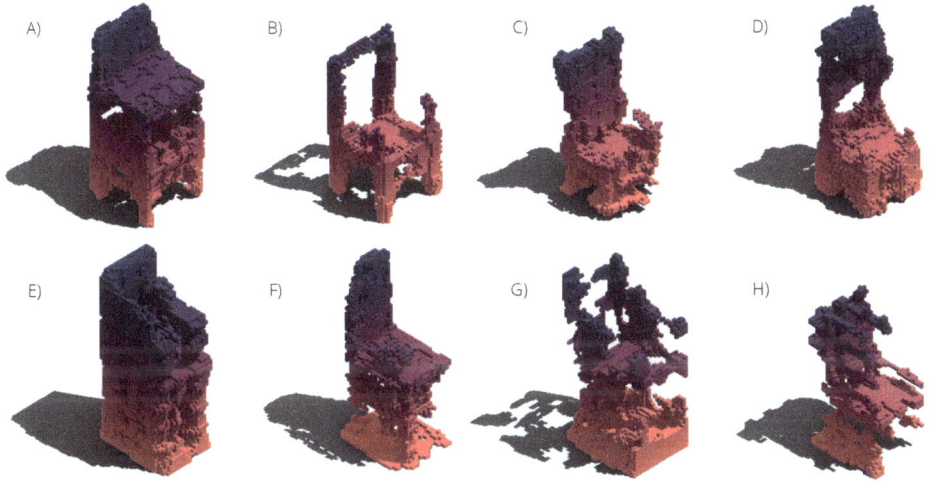

11 Some interesting examples of the chairs found in the latent space of various networks trained for the following criteria:
A) High stability, leisure chair.
B) Low stability, leisure chair.
C) High stability, work chair.
D) Low stability, work chair.
E) High stability, aesthetically unpreferred, leisure chair.
F) Low stability, aesthetically preferred, work chair.
G) High stability, leisure chair.
H) Low stability, aesthetically unpreferred, work chair.

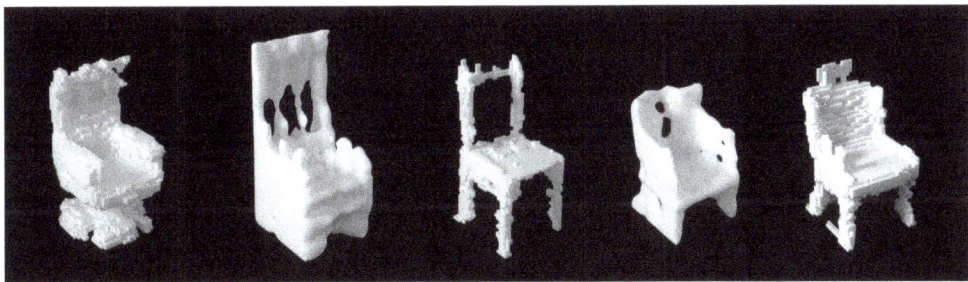

12 Some 3D printed chairs that were designed by the networks. For the second and fourth chairs, isosurfaces were created from voxel representations using the Chromodoris plugin for Grasshopper (Newnham 2016).

MoVAE and MoGAN to solve complex, multi-objective design problems was demonstrated in this research. It illustrates the ability of neural networks to generate novel patterns in outputs by drawing from existing data. It further exemplifies how such a data-driven approach can allow designers to externalize the task of parameterization to focus on curating the data and using their holistic understanding of the design process. The series of studies are intended as a proof of concept of using MoVAE and MoGAN to model the generative process in design problems. The usefulness of neural networks' ability to understand abstract criteria is demonstrated, and a method to incorporate these criteria into the formulation of the design object is established.

Generating such vast quantities of various novel outputs as demonstrated in the results is unthinkable for human designers in such short time frames. With the proposed frameworks, neural networks quickly conduct trial and error processes as they learn from the data and generate a vast array of designs. However, these methods are not immune to problems. As mentioned earlier, the instability of GANs, the increased dimensionality and the limited amount of available data resulted in the failure to generate satisfactory 3D chairs using the MoGAN framework.

As mentioned in the Methods section, gathering and manipulating multiple sources for the ratings should be included in future research as they would allow a tailored analysis of how the machine is able to interpret the convergence of multiple customized preferences. As another future direction, a 2D generation problem with more complex criteria can be added to the hard and easy difficulty problems that were tackled in this study. Finally, making these tools widely accessible to designers is another important future direction. While the development of an interface where the designer can interact with the computer in a more intuitive manner remained out of scope in this work, this subject needs to be addressed in order to make these tools accessible to designers outside the computational design domain.

NOTES

1. Github page for relevant codes and appendix: https://github.com/krz2020/.

ACKNOWLEDGEMENTS

This research project was realized as a Master's thesis in the Integrative Technologies and Architectural Design Research MSc Programme (ITECH) at the University of Stuttgart, Germany, led by the Institute of Computational Design and Construction (ICD) and the Institute of Building Structures and Structural Design (itke). The research has been partially supported by the Deutsche Forschungsgemeinschaft (DFG, German Research Foundation) under Germany's Excellence Strategy, EXC 2120/1-390831618.

REFERENCES

Abadi, M., P. Barham, J. Chen, Z. Chen, A. Davis, J. Dean, M. Devin, et al. 2016. "TensorFlow: A System for Large-Scale Machine Learning." In *12th USENIX Symposium on Operating Systems Design and Implementation (OSDI 16)*. 265–83.

Achlioptas, P., O. Diamanti, I. Mitliagkas, and L. Guibas. 2018. "Learning Representations and Generative Models for 3D Point Clouds." *35th International Conference on Machine Learning (ICML)*. http://arxiv.org/abs/1707.02392.

Albuquerque, I., J. Monteiro, T. Doan, B. Considine, T. Falk, and I. Mitliagkas. 2019. "Multi-Objective Training of Generative Adversarial Networks with Multiple Discriminators." In *Proceedings of the 36th International Conference on Machine Learning*. PMLR 97: 202–211.

Alexander, C. 1964. *Notes on the Synthesis of Form*. Vol. 5. Cambridge, MA: Harvard University Press.

Bidgoli, A. and P. Veloso. 2018. "DeepCloud: The Application of a Data-Driven, Generative Model in Design." In *ACADIA 18: Recalibration, Proceedings of the 38th Annual Conference of the Association for Computer Aided Design in Architecture*. 176–85.

Brock, A., T. Lim, J. M. Ritchie, and N. Weston. 2016. "Generative and Discriminative Voxel Modeling with Convolutional Neural Networks." *ArXiv* Preprint. https://arxiv.org/abs/1608.04236.

Chaillou, S. 2019. "ArchiGAN: A Generative Stack for Apartment Building Design." *NVIDIA Developer* (blog). July 17, 2019. https://devblogs.nvidia.com/archigan-generative-stack-apartment-building-design/.

Chang, A. X. et al. 2015 . "ShapeNet: An Information-Rich 3D Model Repository." *ArXiv* Preprint. https://arxiv.org/abs/1512.03012.

DeLanda, M. 2012. "Genetic Algorithms in Art." In *ACADIA 12: Synthetic Digital Ecologies, Proceedings of the 32nd Annual Conference of the Association for Computer Aided Design in Architecture*. 25–31.

Dillenburger, B. 2016. "Raumindex - Ein Datenbasiertes Entwurfs-instrument." PhD diss., ETH Zürich, 2016. https://doi.org/10.3929/ethz-b-000161426.

Durugkar, I., I. Gemp, and S. Mahadevan. 2016. "Generative Multi-Adversarial Networks." *ArXiv* Preprint. https://arxiv.org/abs/1611.01673.

Ferstl, Y., M. Neff, and R. McDonnell. 2019. "Multi-Objective Adversarial Gesture Generation." In *Proceedings MIG '19: Motion, Interaction and Games*. New York: Association for Computing Machinery. 1–10. https://doi.org/10.1145/3359566.3360053.

Gero, J. S. 1990. "Design Prototypes: A Knowledge Representation Schema for Design." *AI Magazine* 11 (4): 12.

Goodfellow, I., Y. Bengio, and C. Aaron. 2016. *Deep Learning*. Cambridge, MA: MIT Press.

Goodfellow, I. J., J. Pouget-Abadie, M. Mirza, B. Xu, D. Warde-Farley, S. Ozair, A. Courville, and Y. Bengio. 2014. "Generative Adversarial Networks." *ArXiv*. http://arxiv.org/abs/1406.2661.

Hume, D. 2007. An Enquiry Concerning Human Understanding. Oxford World's Classics, edited by P. F. Millican. Oxford; New York: Oxford University Press.

Karras, T., S. Laine, and T. Aila. 2018. "A Style-Based Generator Architecture for Generative Adversarial Networks." *ArXiv*. http://arxiv.org/abs/1812.04948.

Kingma, D. P. and M. Welling. 2013. "Auto-Encoding Variational Bayes." *ArXiv* Preprint. https://arxiv.org/abs/1312.6114.

Lamb, A., V. Dumoulin, and A. Courville. 2016. "Discriminative Regularization for Generative Models." *ArXiv*. http://arxiv.org/abs/1602.03220.

LeCun, Y., C. Cortes, and C. J. C. Burges. 1999. "The MNIST Database of Handwritten Digits." http://yann.lecun.com/exdb/mnist/.

Liu, H., L. Liao, and A. Srivastava. 2019. "An Anonymous Composition: Design Optimization Through Machine Learning Algorithm." In *ACADIA 19: Ubiquity and Autonomy, Proceedings of the 39th Annual Conference of the Association for Computer Aided Design in Architecture*.

Mirza, M. and S. Osindero. 2014. "Conditional Generative Adversarial Nets." ArXiv Preprint. https://arxiv.org/abs/1411.1784.

Newnham, C. *Chromodoris*. V. 0.0.9.1. 2016.

Odena, A., C. Olah, and J. Shlens. 2017. "Conditional Image Synthesis with Auxiliary Classifier GANs." In *Proceedings of the 34th International Conference on Machine Learning*. PMLR 70: 2642–2651.

Özel, G. and B. Ennemoser. 2019. "Interdisciplinary AI: A Machine Learning System for Streamlining External Aesthetic and Cultural Influences in Architecture." In *ACADIA 19: Ubiquity and Autonomy: Proceedings of the 39th Annual Conference of the Association for Computer Aided Design in Architecture*. 380–91.

Park, H., Y. J. Yoo, and N. Kwak. 2018. "MC-GAN: Multi-Conditional Generative Adversarial Network for Image Synthesis." *ArXiv*. http://arxiv.org/abs/1805.01123.

Plumerault, A., H. Le Borgne, and C. Hudelot. 2020. "Controlling Generative Models with Continuous Factors of Variations." *ArXiv*. http://arxiv.org/abs/2001.10238.

Roman, M. 2013. "Four Chairs and All the Others - Eigenchair." In *Proceedings of the 31st ECAADe Conference*, vol. 2. 405–14.

Shaghaghian, Z. and W. Yan. 2019. "Application of Deep Learning in Generating Desired Design Options: Experiments Using Synthetic Training Dataset." *ArXiv* Preprint. https://arxiv.org/abs/2001.05849.

Simon, H. A. 1988. "The Science of Design: Creating the Artificial." *Design Issues* 4 (1/2): 67-82.

Wang, H., N. Schor, R. Hu, H. Huang, D. Cohen-Or, and H. Huang. 2019. "Global-to-Local Generative Model for 3D Shapes." *ACM Transactions on Graphics* 37 (6): 1–10. https://doi.org/10.1145/3272127.3275025.

Wu, J., C. Zhang, T. Xue, W. T. Freeman, and J. B. Tenenbaum. 2016. "Learning a Probabilistic Latent Space of Object Shapes via 3D Generative-Adversarial Modeling." In *Proceedings of the 30th International Conference on Neural Information Processing Systems*. 82–90.

Zhang, Y., A. Grignard, A. Aubuchon, K. Lyons, and K. Larson. 2018. "Machine Learning for Real-Time Urban Metrics and Design Recommendations." In *ACADIA 18: Recalibration, Proceedings of the 38th Annual Conference of the Association for Computer Aided Design in Architecture*. 196–205.

IMAGE CREDITS
All drawings and images by the authors.

Rıdvan Kahraman is a materials scientist and computational designer, with a B.Sc. and M.Sc. in Materials Science and Engineering and an M.Sc. in Integrative Technologies and Architectural Design Research. He is interested in applying ideas from mathematics and the sciences in architecture and urban design.

Zhetao Dong is a researcher, designer and programmer, with a Master of Science in Integrative Technologies and Architectural Design Research (ITECH). He has an interdisciplinary background in architectural design, computer science and psychology. His expertise is developing artificial intelligence projects to process data, aid design and facilitate robotic fabrication.

Kurt Drachenberg is a Computational Designer and Civil Engineer with a B.Sc. in Civil Engineering from the University of Alberta and a M.Sc. in ITECH from the University of Stuttgart with a special interest in history, theory, and philosophy. Kurt has a background set of skills and experiences in structural design and analysis, artificial neural networks, coding, and visual scripting.

Katja Rinderspacher is a research associate at the Institute for Computational Design and Construction (ICD) and the coordinator of the ITECH M.Sc. Program at University of Stuttgart. She holds a. M.Arch. degree with honors from Pratt Institute and is a registered architect in Germany. Katja has gained professional experience as an architect and project manager in offices in the US, Switzerland and Germany. Her current research focuses on the design and fabrication with indeterminate fabrication processes for high-resolution surface structures.

Christoph Zechmeister is a research associate at the Institute for Computational Design and Construction (ICD) at the University of Stuttgart. He holds a Master of Science from Vienna University of Technology as well as a postgraduate Master of Advanced Studies in Architecture and Information from ETH Zürich. Before joining the ICD, Christoph worked as a Junior Architectural Designer at UNStudio, Amsterdam, as well as in multiple offices in and around Zürich, Switzerland. Christoph is currently involved in developing high performance lightweight structures.

Ozgur S Oguz received the B.Sc. and M.Sc. (summa cum laude) degrees in computer science from Koc University, Istanbul, Turkey, and the Ph.D. degree (summa cum laude) from the Department of Electrical and Computer Engineering, Technical University of Munich, Munich, Germany, in 2018. He is currently a Post-Doctoral Researcher with the Machine Learning and Robotics Laboratory, University of Stuttgart, Stuttgart, Germany. His research interests are developing autonomous systems that are able to reason about their states of knowledge, take sequential decisions to realize a goal, and simultaneously learn to improve their causal physical reasoning and manipulation skills.

Achim Menges is a registered architect in Frankfurt and professor at the University of Stuttgart, where he is the founding director of the Institute for Computational Design and Construction (ICD) and the director of the Cluster of Excellence on Integrative Computational Design and Construction for Architecture (IntCDC). In addition, he has been Visiting Professor in Architecture at Harvard University's Graduate School of Design and held multiple other visiting professorships in Europe and the United States. He graduated with honors from the Architectural Association, AA School of Architecture in London.

Design Engine

Generative Multi-Objective Performance Design Scenarios

Kathy Velikov
University of Michigan

Matias del Campo
University of Michigan

Lucas Denit
University of Michigan

Kazi Najeeb Hasan
University of Michigan

Ruxin Xie
University of Michigan

Brent Boyce
Guardian Glass

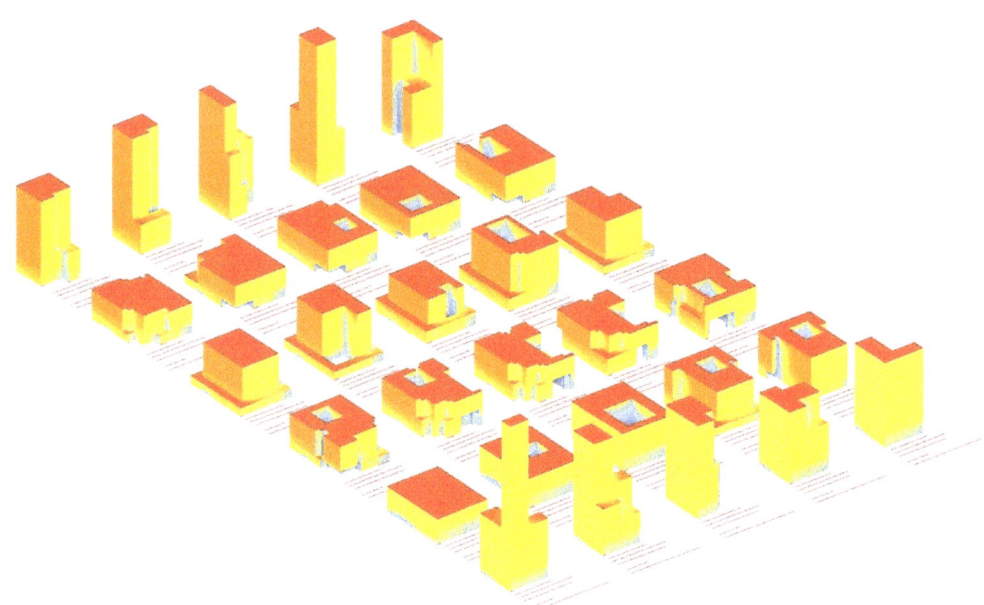

1   Families of generated solutions for a single site show a range of characteristics based on compromises and trade-offs in optimization across multiple fitness parameters.

ABSTRACT

Generative design offers the possibility to heuristically explore data-driven design iterations during the design process. This enables performance-informed feedback and the possibility for exploring viable options with stakeholders earlier in the design process. Since architectural design is a complex, nonlinear process that requires trade-offs and compromises among multiple requirements, many of which are in conflict with each other, a multi-objective solver provides a spectrum of possible solutions without converging on a single optimized individual. This enables a more informed design possibility space that is open to collaborative decision-making. This paper describes the development of a custom multi-objective generative design workflow to visualize families of possible future building typologies with a focus on the impact of site, form, envelope performance, and glazing. Three future design scenarios are generated for three urban U.S. locations projected to grow and where progressive environmental performance stretch codes have been adopted. Drivers—such as plausible site, procurement, financing, value chain, and construction typology inform possibilities for built form, envelope technologies, and performance in relation to local codes, environment, and occupant health—are transformed into design inputs through urban, spatial, and environmental simulation tools for a "building design generator," or a multi-objective optimizer tool that produces an array of possible building massing and schematic envelope design options. The paper concludes with pointing out some of the gaps in data of current evaluation tools, the need for interoperability across platforms, and this points to multiple trajectories of future research in this area.

INTRODUCTION

Building design decision-making is an inherently complex process that requires the consideration of myriad factors balanced relative to each other. This process is characterized by a combination of design possibility afforded by the site, program, and budget; the vision of the designer; informational input from various stakeholders ranging from engineers to contractors and material suppliers, to clients and users, to building authorities; and tacit domain knowledge on the part of the architectural, engineering, and contractor team. Iterative development in form and aesthetics during the early design stages typically means that building performance software has primarily been used for later-stage analysis and reporting versus early-stage development and feedback-based design decision-making. Recent software advances are enabling earlier heuristic performance assessment for design aspects such as massing, structure, daylighting, shading, and climate analysis. Built-in rapid simulation engines, such as the recently launched ClimateStudio's Radiance processor (Solemma 2021) and accessible user interfaces are enabling faster evaluation to be undertaken by designers, versus the time-consuming brute force methods of the previous generation of simulation software. The majority of simulation software, however, is still typically designed to evaluate and optimize one to two primary parameters at a time, such as structural performance, or daylighting and thermal analysis. This makes it difficult to be able to rapidly assess the trade-offs and balancing across parameters that the design decision-making process typically entails. Recently, computational design researchers and software developers have begun to turn toward evolutionary solvers as a way to simulate and inform complex decision-making for building planning and design performance. Evolutionary solvers are able to generate families of solutions with characteristics based on trade-offs across sometimes contradictory fitness parameters. What is compelling about evolutionary solvers is that the spectrum of possible individuals is open-ended, in that it leaves decisions to be made by the design team, rather than converging on single optimal solutions (Figure 1). This points to a realignment in future design workflows that become more inclusive of knowledge domains across different scales, and ultimately to a more holistic approach in energy and carbon reduction for buildings.

This paper describes the experimental development of a multi-objective generative design workflow and subsequent design scenario generation development that combines site-level massing with daylighting, ventilation, and basic energy analysis. The research was undertaken within the context of research in alliance with Guardian Glass called "The Design Ecologies of Glass." This exploratory research aims to investigate the most significant decision-making drivers, paradigm shifts, and transformations within the ecology of building design, delivery, and real estate that will impact new building envelopes and future building typologies, with a focus on the potential impact of high-performance glazing. The work that follows describes one thrust of the larger research efforts wherein the team developed a custom multi-objective generative design workflow in order to speculate on and demonstrate what factors could drive future building typologies and, in turn, building envelope design and performance.

Three different design scenarios were generated in three urban U.S. locations projected to grow in the future and where progressive building performance stretch codes (i.e. building performance codes that surpass the national energy code) have been adopted: Austin (Texas), Minneapolis (Minnesota), and Washington, DC. The scenarios speculate on a plausible site, procurement, financing, value chain, and construction typology in exploring possibilities for built form, envelope technologies, and performance in relation to local codes, environment, health, and other occupant drivers. These drivers were transformed into design inputs through urban, spatial and environmental simulation tools in the form of a "building design generator," or a multi-objective optimizer tool that produces an array of possible building massing and schematic envelope design options. Each scenario describes and illustrates a family of possible building types for each city, and then focuses on one to develop in more detail. The building design scenarios remain at a highly schematic level, their value being in the description of the process and assembly of parameters and decision-making drivers, as opposed to the particular design outcome.

STATE OF THE ART: EVOLUTIONARY SOLVERS

The team chose to explore evolutionary solvers for the scenario studies for two primary reasons. The first was that our previous research has indicated that this type of multi-objective decision-making will become increasingly prevalent in early design development, especially as designers will be asked to meet multiple energy performance and financial pro-forma goals simultaneously during preliminary design (Figure 2). The second is that an evolutionary computational model has the capability to reduce idiosyncratic design bias on the part of the authors, and our hypothesis was that developing a tool to generate design possibilities would allow us to rapidly explore more "generic" options as well as the possibility that the genetic algorithm might be able to "find" solutions that may be unexpected or counterintuitive.

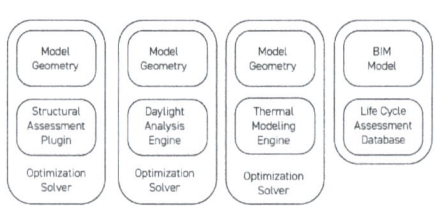
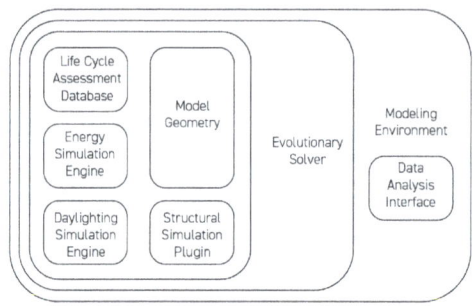

2 Diagram illustrating the current paradigm of an incremental approach to simulation and optimization (left) and a schematic diagram illustrating the proposed method, with an integrated workflow and simultaneous optimization (right).

An evolutionary solver is a computational tool that can be used to reach conclusions for complex problems. The concept was first developed by John H. Holland in his *Adaptation in Natural and Artificial Systems* (Holland 1975) and advanced by John Frazer in his seminal book *An Evolutionary Architecture* (Frazer 1995). The underlying logic of evolutionary algorithms is inspired by processes of biological evolution, where the inheritance of genes and genomes, and the randomness of mutation, drive the evolution of a species toward a form that is best-suited to a particular ecological niche. This process can be simulated for any number of tasks, by assigning variables to a model and then iterating through different instances of that model. In theory there is no limit to the complexity of these models, although in practice, primarily due to limits in computational power, it can become extremely difficult to account for the accelerating complexity of a model as more variables are introduced (Rutten 2013).

Evolutionary solvers might be classified as "single-objective" or "multi-objective," with the key differences between these methods being how input data is prepared and how the optimized results are returned to the user. In a single-objective optimization tool, all fitness objectives are combined into a fitness function returning a single value. This is an equation combining all the optimization objectives along with all necessary weighting coefficients and any additional variables, and produces a single, compound fitness value that accounts for all the genome parameters, and typically converges on a single, optimized value. Because a single-objective solver typically returns a single solution, or several very similar solutions, there is a major dependence on the quality and accuracy of the model to return useful results. This results in scenarios where it is plausible that data bias, faulty assumptions, or unconsidered variables might skew the results with less opportunity for the human authors to make informed decisions about the results.

In contrast, a multi-objective optimization solver sidesteps the challenge of authoring a single fitness function by accommodating several fitness values. The solver searches the design space for the most "fit" genotypes, beginning with a pool of random combinations and re-evaluating over

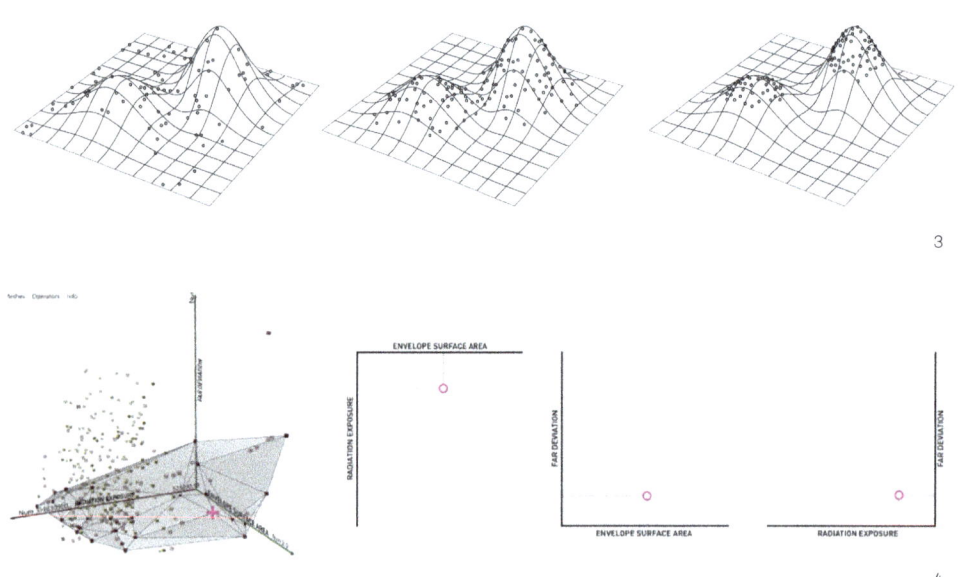

3 A simple representation of a two-parameter search-space evolution, where optimal solutions have a high Z-height as a function of their X- and Y-parameters.

4 The search space visualization from the evolutionary solver, with a single instance selected from the Pareto front (illustrated with a grey mesh on the scatter plot). This instance is also plotted against three axes to illustrate its relative values as a set of trade-offs.

subsequent generations to converge towards a spectrum of increasingly optimal solutions, some of which may optimize one variable at the expense of all others, and some of which might attempt to satisfy all criteria to a lower standard (Figure 3). This multi-objective optimization is based on a concept taken from economics known as Pareto efficiency or the Pareto front (or frontier). In theory, all the solutions that exist along the Pareto front are equally optimal, absent any external subjective preferences (Vierlinger 2013). Because the solver returns a range of solutions along this optimal boundary, it requires an external decision to evaluate which of the collection is the most appropriate for the specific situation, and as such returns agency back to the designer as decision-maker (Figure 4).

Enabled by cloud computing and the maturation of machine learning algorithms, driven by increasingly complex performance requirements as well as the need to engage stakeholders in the early process of design development, parametric generative design software is emerging as a way for designers and developers to explore data-driven options for design possibilities while optimizing for multiple project requirements and constraints (Wilson 2019). In the past several years, a number of generative design engines have emerged, with several geared toward site development and early-stage design space. These include the Scout software developed by architecture firm Kohn Pedersen Fox (https://ui.kpf.com/scout), Autodesk's Project Refinery Beta (www.autodesk.com/campaigns/refinery-beta), Gensler's Generative Design Tool for Modular Buildings (www.gensler.com/gri/generative-design-tool-for-modular-buildings), the startup generative site design engines Spacemaker (www.spacemakerai.com/), Hypar (www.hypar.io), and Digital Blue Foam (www.digitalbluefoam.com), and Sidewalk Labs' recently launched generative urban design software called Delve (www.sidewalklabs.com/products/delve). Delve, for example, uses machine learning to generate and assess the impact of "millions of design possibilities for a given project" by optimizing across financial, environmental, and quality of life indicators so that urban development teams can "discover" best performing site design options for projects (Ikhena 2020). Given that all of these evolutionary solvers were still in early stages of commercial software development at the time this work was started in early 2020, and since few incorporated energy modeling and advanced envelope design parameters, the team developed its own custom tool and interface that could combine building massing and typology with schematic envelope, climate, and daylighting performance.

METHODS: SCENARIO AND EVOLUTIONARY SOLVER DEVELOPMENT

Design Scenario Selection and Parameters

A scenario-based design research approach was adopted for this study. Scenario planning "is a disciplined method for imagining possible futures" that is capable of exploring combined impacts of multiple uncertainties and various factors, are able to capture elements that "cannot be formally such as new regulations, value shifts, or innovation," and "organizes those possibilities into narratives that are easier to grasp and use than great volumes of data" (Schoemacher 1995, 25-27). This method is valuable in its ability to generate an array of future possibilities, stemming from the identification of basic future trends and uncertainties, as well as their background parameters. This makes for an accommodative research method that is conscious against possible bias toward any one or particular set of drivers. The primary goal of the scenarios is not to predict the future but instead to challenge mindsets and assumptions about how the future might unfold in addition to sponsoring curiosity, learning, and anticipatory decision-making.

Buildings are specific to their particular site, urban, and regulatory context and order to develop the scenarios, specific sites in U.S. cities were chosen. Possible urban locations for testing out the design impacts of the concerned drivers were identified by overlaying data on growing residential and commercial construction markets in the U.S. with that of municipalities with notable comprehensive urban development policies, specifically those concerned with advanced building energy efficiency codes and health standards. As such, three cities, Austin, Minneapolis, and Washington, DC were selected, with a twenty year future horizon of 2040. These cities were also chosen for their respective variability in climatic and geographic conditions relative to each other, and for different transformations in the context of climate change. The team chose possible development site options by identifying underutilized or undeveloped land in growth areas identified by the cities' respective urban masterplans. Key parameters guiding the design development across the three sites include the building program, procurement and ownership model, developer and design consultancy, market segment, and construction technology. A host of other drivers also participated in the design process and are summarized in Figure 5. All speculative buildings are either in the mid- or high-rise category, and all anticipate some level of pre-manufactured component-based construction. The latter construction method was prioritized based on research and indicators that prefabricated, modular, and factory-built construction is projected to

	Austin	Minneapolis	Washington DC
Program	Affordable Housing with office and ground floor food and/or retail.	Housing integrated with urban farming & ground floor groceries/amenities.	Office with ground floor food and recreation.
Ownership Method	Developer lease to multiple private & corporate entities. Afforable housing & student housing renters. Commercial at grade.	Combination of developer owned and operated, and built to suit for single renters and families including senior housing and commercial/office at grade.	Built on specification/potentially owned by large private corporation and leased to federal agencies and businesses.
Developer Type	Small scale, local developer & architecture firm (architect-developer)	Medium scale, regional developer & property management firm.	Large scale, international developer, builder, & property mangement firm.
Design Firm Type	Small scale, local developer & architecture firm (architect-developer)	Nationally acclaimed medium scale architecture & landscaping firm specializing in cross laminated timber.	Globally acclaimed studio based architecture firm.
Market Segment	Mid level market rate & affordable rental.	Mid level market rate & affordable rental.	High end.
Key Construction and Envelope Technologies	Prefabricated, mass customized volumetric modular housing construction with factory installed glazing	Prefabricated, mass customized unitized panel, modular construction.	Mass timber columns and concrete floor plate hybrid construction with prefabricated unitized windows.
Key Structural Material	Lightweight timber, concrete podium, glass	Hoeavy cross laminated timber, glass	Heavy timber, concrete, glass
Height	100'	100'-150'	120'
FAR	5	6.5	6.5
Window Type	Triple Glazing	Triple Glazing	VIG
WWR (Window to Wall Ratio)	30%	40%	70%
Total Envelope Area	135,000 sf	90,000 sf	200,000 sf

5   Summary of scenario design drivers for Austin, Minneapolis, and Washington DC

grow, especially in multi-unit residential construction prioritizing low-cost and high performance project delivery (Bertram et al. 2019).

Evolutionary Solver Tool Development
This research prompted the development of a demonstration workflow to explore the three design scenarios. The workflow is built in the Rhino/Grasshopper CAD environment for its high degree of customization and a rich ecosystem of plugin tools for both design optimization and building performance simulation. The project demanded a degree of architectural elaboration and formal differentiation, so while the fundamental logic of the tool remained consistent across all three site proposals, the generative logic of massing geometry was tuned to meet the proposed building typologies, site scales, and proposed construction methods. After generating a building massing iteration, the tool processes the geometry for quantitative evaluation using three primary evaluation tools. The method uses the Ladybug and Honeybee (Roudsari et al. 2013) plugins for climate and thermal behavior simulation, which connect Grasshopper 3D to EnergyPlus, Radiance, Daysim, and OpenStudio. These were selected over other available environmental tools for their high degree of customization and flexibility in configuration, robust support resources, and quality of visual output. The tool uses the Octopus multi-objective solver for the evolutionary optimization of massing schemes (Verlinger 2013). Quantitative information about a selected scheme was aggregated and displayed in a visualization dashboard within the Rhino modeling environment using Proving Ground's Conduit software (Miller 2020). The immediate feedback about the relative benefits of a particular scheme is well suited to the early-stage, scenario-based comparative study that the research anticipates will be an increasingly important step in data-informed design practice (Figure 6).

This project posed several challenges in terms of computation requirements as existing tools take different approaches. Some solutions simply run intensive simulations for each iteration and accept the cost of high computing time. Others sometimes pre-compute values and then call on a stored database during optimization as a way to accelerate the process. A third approach is to use metrics based only on model geometry, rather than additional simulation data generated with a plugin engine

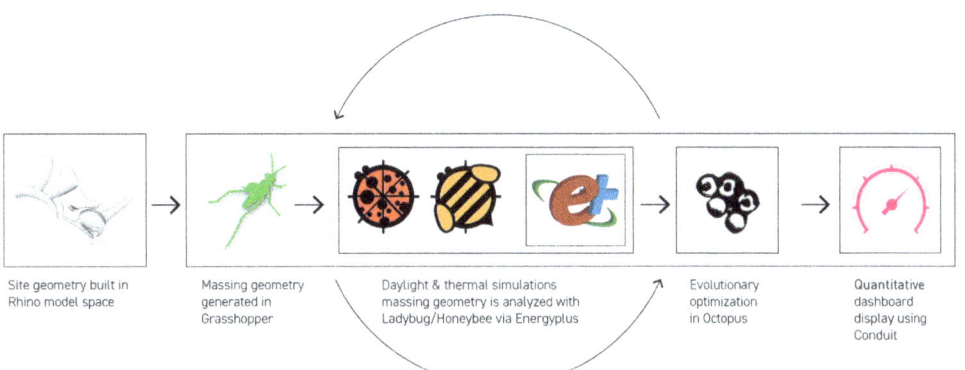

6   Multi-optimization tool workflow. Base geometry from the Rhino workspace is fed into Grasshopper, where it is analyzed with Ladybug/Honeybee via EnergyPlus .idf files. These simulations are logged in the Octopus solver, which directs the process to repeat toward a convergence along a Pareto Front. The information about these optimized solutions is displayed in the workspace via a visualization dashboard using Conduit.

(using distances of points to planes, tallying vector intersections, etc). Our own demonstration tool applies each of these methods in different cases when appropriate: some solar simulations are calculated in real time, some thermal simulation values are calculated ahead of time and stored in a data set, and some measurements are produced based on internal model geometry.

Rather than produce a conclusive "optimal" solution for each design site, we present several solutions within the Pareto optimal family to demonstrate how a range of solutions can supply designers with quantitative analysis about the subjective design decisions being made. The fitness goals for each site were selected to be adversarial and demand trade-offs from one another, in order to better visualize the constraints of the solution search space (Wilson 2019). The fitness goals that were incorporated in the models varied slightly for each site relative to local climate and heating versus cooling priority, and included building density (FAR), orientation and views, massing, solar exposure, daylighting, and envelope area. They are listed in more detail in the results and discussion for each scenario.

DESIGN SCENARIO RESULTS AND DISCUSSION
Austin Design Scenario

The chosen scenario design site is located adjacent to the campus of the University of Texas at Austin, suitable for mixed-use housing and commercial use. The development scenario chosen is a fully vertically integrated design, development, construction and property management company leveraging volumetric, prefabricated modular residential unit construction through advanced construction technologies towards mass customization of residential units. The choice of modular construction has definitive impacts on the possible solutions produced by the generative design tool due to the inherent constraints of the modules.

The design scenario for Austin explores design options with a high surface area and plenty of opportunity for self-shading and natural ventilation, responding through physical means to the hot and dry climate of southern Texas (Figure 7). The selected fitness objectives for this site are:

- Maximize adequate daylight
- Minimize glare
- Maximize naturally ventilated thermal comfort in summer months
- Maximize units with views of campus landmarks

The input parameters allow for a randomization of massing strategy based on a gridded aggregating method, where prefabricated modular units can be stacked around the perimeter of the podium, as well as subtracted to create "punches" through the facade to encourage wind-driven ventilation to the inner courtyards. The window-to-wall

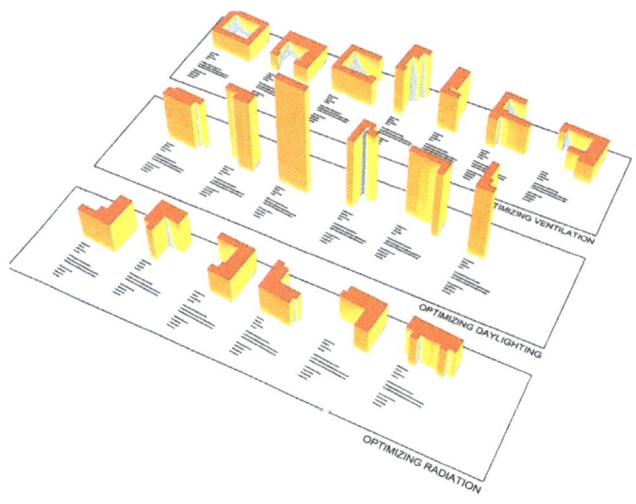

7   Families of individual massing prioritizing different performance optimization: the most viable options evolve from compromises across parameters

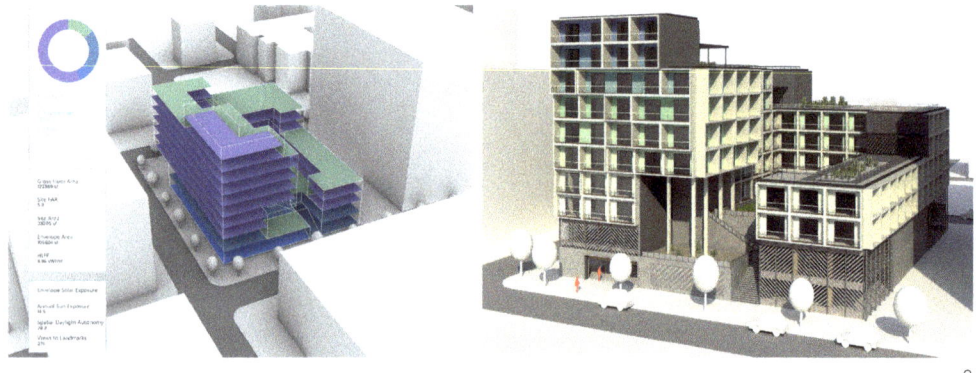

8   These four individuals from the pareto front illustrate some of the variations and trends within the solution set for the Austin scenario

9   Solution Z chosen for further development in Austin: a color coded floor plate/wireframe shows different programs corresponding to the output data in the toolbar at left (left); and schematic design development of self-shaded envelope, interior courtyards and terraces (right)

ratio of the design is 30%, with factory-installed U0.1 triple-pane glass with operable units to maximize the opportunity for natural ventilation in the hot and dry Austin climate. The courtyards and stacking strategy also provide ample accessible outdoor space for residents. Given the modular prefabricated construction technique, this model developed and analyzed each separate unit type for daylight and thermal comfort and this data could then be called up during the optimization sequence and fed into the algorithm as proxy data based on the parameters of a given solution. Each unit type was also assigned a variable shade/balcony component. Units with a high solar exposure were assigned deep shading geometry, while units with low solar exposure were assigned shallow shading geometry. We also proposed a simple system to randomize facade offsets to contribute to the effects of self-shading as a method to reduce incident solar radiation at the glazed facade surfaces. The results from this study demonstrate the competing goals (Figure 8). Solutions with high numbers of north-facing units have high scores for daylight quality (solution "W"), while solutions with primarily east-facing windows have poor daylight quality but clear views of the University campus (solution "X"). Solutions with a high proportion of single loaded units tend to have the best scores for natural ventilation, particularly when the massing is oriented to allow access from the prevailing southern wind. Solution "Y" has a very high Spatial Daylight Autonomy (SDA) score, at the expense of also having high glare. Solution "Z" is a compromise across goals, with low envelope solar exposure, some views toward the campus, and a balance between sufficient daylight and overlit glare. This was chosen to be developed to a higher resolution (Figure 9).

Minneapolis Design Scenario

The chosen scenario design site is 170,000sf brownfield at the urban periphery and targeted by the city's master plan as a future medium to high density transit-oriented development area. Following the Minneapolis 2040 development guidelines (www.minneapolis2040.com/), we explored a catalogue of building typologies and urban block configurations. Here, the construction typology selected is heavy timber highrise with panelized prefabricated envelopes that can be craned into place and assembled with quick single-pass construction practices. The selected fitness objectives for this scenario are:

- Minimize total Heat Loss Form Factor
- Maximize street level solar access
- Maximize greenhouse solar access

This design scenario has generally tried to propose buildings with a high level of compactness as a method for reducing winter heat loss. This can be quantified as Heat Loss Form Factor, or the ratio of floor area to thermal envelope area. We explore this project as a promising application for speculative U0.07 high performance triple-pane glazing, and that with this high-quality envelope the minimally-conditioned greenhouse spaces might prove occupiable as winter gardens even in the cold Minnesota climate. The residential volumes of the project have a window-to-wall ratio of 40%, in line with the Minneapolis development guideline.

Different building types each host a different greenhouse typology: some greenhouses sit on the main podium, some are nested onto terraces, and some sit within the primary mass of a tower. The script isolates the geometry of the greenhouse spaces and evaluates their solar exposure independently of the rest of the building geometry as a way to test the balance of useful solar radiation (Figure 10). Overall, it was found that many of the solutions with the best scores for sufficient solar exposure at street level were the solutions with low scores for useful greenhouse radiation, primarily because of the terrace configurations (solutions "W" and "X"). Buildings which terrace towards the south and provide sufficient space for useful greenhouses will be taller at their northern face and will block more sun at the pedestrian zones to their north. Conversely, a building which terraces down to the north will offer more sunlight to the pedestrian zone, but this massing strategy will introduce unproductive self-shading at the spaces which require solar radiation to be most effective and efficient (solution "Y"). These sacrifices and compromises are visible in the selected solutions presented in Figure 10. The iteration at the urban-scale aggregation which best demonstrated compromise between fitness goals (solution "Z") has been developed further (Figure 11).

Washington, DC Design Scenario

The chosen scenario design site in Washington, DC is located within the downtown commercial urban core in close proximity to the Convention Center and other prominent commercial and federal landmarks. The development scenario selected is a large commercial real estate developer, manufacturer, and property operator and the construction typology is hybrid construction with prefabricated unitized glass facades. Given the primarily commercial and office uses, the proposals for this scenario integrate WELL standards (www.wellcertified.com/) for holistic, human-oriented building performance into a quantitative framework for optimization. The fitness objectives we selected for this study are:

- Minimizeworkspace area within 25' of windows (WELL Standard 61.2)
- Maximize exterior terrace area (WELL Standard 100.1)
- Maximize exterior view quality
- Minimize building envelope area

The building mass generations are produced through a carving/subtraction logic as a way to both introduce daylight into the depth of the building and as a way to generate outdoor terraces above the ground plane. This project has a high window-to-wall ratio of 70%, which is accommodated with the use of speculative U0.05 super high performing vacuum-insulated glazing units. This envelope system enables a reduction in lighting demand

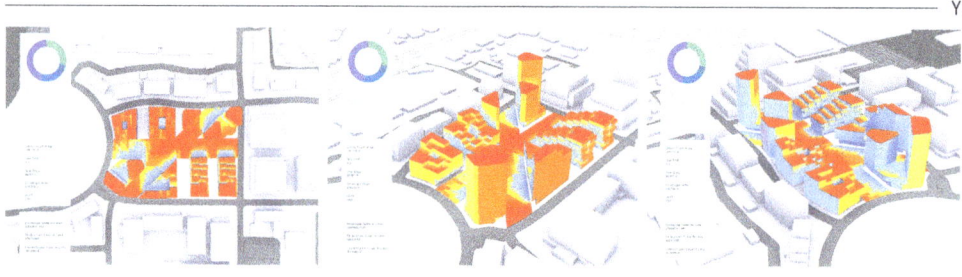
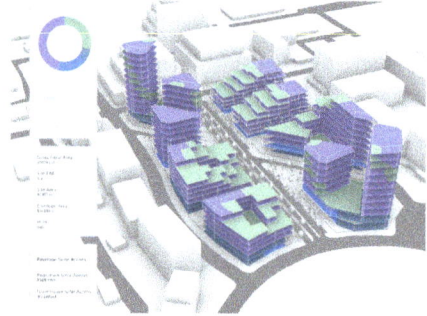

10 These four solutions generated from the design engine show the preferences of some solutions and some of the possible compromises for the Minneapolis scenario

11 Solution Z chosen for further development in Minneapolis: the solver produces color coded surfaces to rapidly assess massing and program distribution (right); based on this, schematic facades are developed, driven by logics of heavy timber structure and panelized modular construction (right)

without an increase in cooling energy expense. The facade system proposes a typical cassette dimension across the facade, within which the configuration of the glazing lites is variable depending on the orientation of the unit and its expected solar exposure. Facade units facing to the east and west tend to feature vertical shading to reduce morning and afternoon glare, while units facing South have more horizontal shading to protect from midday overhead solar gain. Between these two conditions is a spectrum of possible configurations which produces a gradient of glazing units and contributes to a more coherent architectural aesthetic.

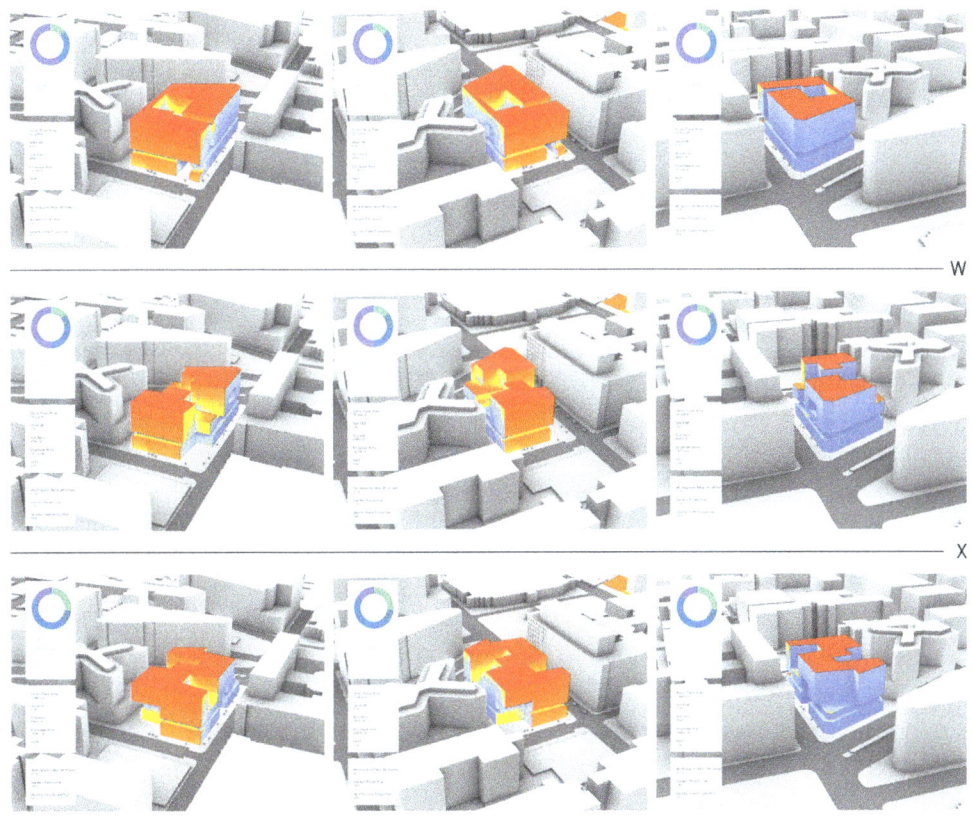

12 These four possible solutions generated by the solver show the preferences of some solutions and some of the possible compromises for the Washington, DC scenario

13 Floor plate view mode of the solver illustrates the spatial relationships of different program areas, paired with quantifiable feedback metrics (left); schematic envelope development elaborates on the high performance variable cassette glazing units that create a continuous visual aesthetic across the facade (right)

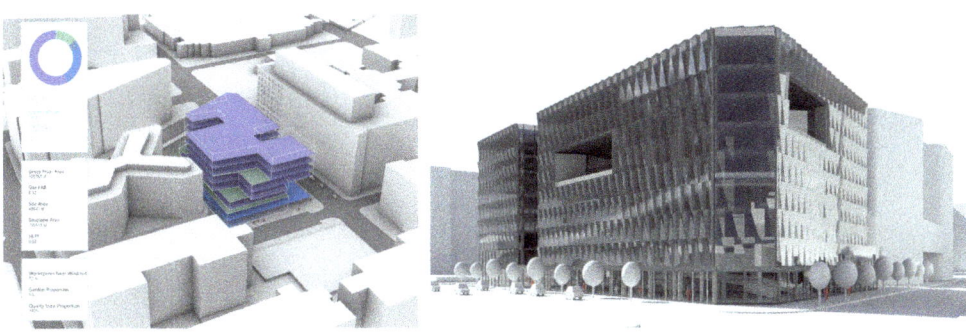

The results from the solver tool suggest several formal trends (Figure 12). Solutions which prioritize compactness typically do so at the expense of view quality and window access (solution "W"). Solutions with a high score for window access and view quality tend to feature more, smaller cuts to the building mass with a high envelope area (solutions "X" and "Y"). The solution selected for the DC site (solution "Z") balances these criteria; the scenario attempts to balance the compactness of the building's massing to reduce conditioning demand, with the need to break up the building mass for both daylight access and as a response to an architectural need for thoughtful and responsible urban design (Figure 13). Carved voids and terraces introduce outdoor space through the building while also helping to improve daylight access to the interior and aesthetic effects are produced by the variable cassette window geometry.

REFLECTIONS AND CONCLUSION

The multi-objective design decision-making workflow and the design scenarios simulated through this research aim to simultaneously explore future tools for design decision-making and, using these tools, develop a framework for speculation for how future building typologies might be impacted by advances in envelope technology, combined with advances in design and construction value chains, in the context of increasingly stringent local codes for energy performance and human health and well-being. The development of a "pre-alpha" generative multi-objective design solver allowed our team to not only explore how various drivers might impact building design typologies and high performance envelope decision-making in the future, but also to understand the gaps in knowledge and data for the future of generative solvers themselves.

Of these various drivers assessed within the project, the most impactful with regard to building typology and built form continues to in fact be the building site, closely followed by the structural and construction system used. With regard to the former, compact urban sites provide far more constraints on formal possibilities. In the latter, prefabricated, modular, and mass timber construction systems—which are anticipated to grow due to their ability to deliver high energy performance and construction quality with decreased construction time and cost—have clear implications on building form possibilities due to their dimensional and tectonic logics. These potential differences and their opportunities are demonstrated across the three scenarios, which each model different degrees of prefabricated and modular construction systems.

Construction cost, a significant major driver in building design, was not able to be explored in the scenarios due to the distance in the future, the volatility in construction markets and supply chains being currently experienced, and the very specific and fragmented domain knowledge of the construction industry that has not been able to produce reliable data on construction costs. A future evolutionary solver tied to real procurement and cost data from the contractor's side would represent a revolution in the building design process. That said, the cost of the building envelope, whether glass or other, will continue to be one of the highest costs of a building, and thus minimizing the envelope to floor area ratio will continue to be a major driver for clients and developers. While computational design and performance simulation can generate and evaluate complex building geometries with greater ease and efficiency, simpler geometries are anticipated to continue to be preferred from an energy efficiency and cost aspect, still prioritizing rectilinear and modular built forms.

Overall, this research anticipates that the future of design decision-making will increasingly depend on complex informational models. These models are however only as accurate as the information and data they contain, and at the moment there are enormous gaps in data available to designers. This ranges from data on material performance, to construction costs, to life-cycle assessment, to post-occupancy systems performance. For example, at the outset of the project, the team had also hoped to include embodied carbon, or whole-building life-cycle assessment (WBLCA) as a parameter in the multi-optimization software. However, further investigation of possible tools revealed gaps in reliable data for current tools (Herrero-Garcia 2020), as well as a specific lack of data for some of the major materials our team was investigating such as emerging triple-glazed and VIG envelope units. The more accurate and detailed information for various building systems can be, the more effectively it can be integrated into decision making and performance evaluation platforms. Provided the right information, generative tools allow the interrogation of multiple models in a rapid sequence. Parametric tools designed to test the material properties of a project allow designers to find appropriate solutions for any given typology. In conclusion it can be stated that the continuous development of highly flexible design tool sets that adapt quickly to as wide a possible variety of given conditions allow design teams to respond to future environmental and societal needs.

ACKNOWLEDGEMENTS

This work was funded through a Taubman Guardian Research Alliance Grant that ran from 2019-2021. The authors would like to thank the extended project team, who included Geoffrey Lewis, Carol Menassa, Vineet Kamat, Gregory Keolian, and Harsheen Kaur from University of Michigan, Sheldon Davis and Brian Schulz from Guardian Glass, and external professional advisors Thom Culp, Marc Simmons, Anthony Mosellie, and Derek Basinger.

REFERENCES

Bertram, Nick, et al. "Modular Construction: From Projects to Products." McKinsey & Company, 2019. Accessed October 14, 2020. https://www.mckinsey.com/~/media/mckinsey/business%20functions/operations/our%20insights/modular%20construction%20from%20projects%20to%20products%20new/modular-construction-from-projects-to-products-full-report-new.pdf.

Frazer, John. 1995. *An Evolutionary Architecture*. London: Architectural Association.

Herrero-Garcia, Victoria. 2020. "Whole-building life cycle assessment: Comparison of available tools." *Technology | Architecture + Design* 4 (2): 248–252.

Holland, John H. 1975. *Adaptation in Natural and Artificial Systems*. Ann Arbor: University of Michigan.

Ikhena, Okalo. 2020. "Announcing Delve: Discovering Radically Better Urban Designs." *Sidewalk Labs, Insights*. Accessed November 25, 2020. https://bit.ly/2VqNfBC

Miller, Nathan. *Conduit*. v.2008.11.19.0. PG Apps. 2020. https://apps.provingground.io/conduit/.

Rutten, David. 2013. "Galapagos: On the Logic and Limits of Generic Solvers." *Architectural Design* 83 (2): 132–135.

Roudsari, Mostapha, Chris Mackay, et al. *Ladybug Tools*. V. 1.1.0. Ladybug Tools. 2021. https://www.ladybug.tools/.

Schoemaker, Paul J. H. 1995. "Scenario Planning: A Tool for Strategic Thinking." *Sloan Management Review* 36: 25–40.

Solemma LLC. 2021. "Why is ClimateStudio So Fast." Accessed October 2, 2021. https://www.solemma.com/climatestudio/speed.

Vierlinger, Robert. *Octopus*. V.0.4. Grasshopper for Rhino6 plugin. Windows. 2018. https://www.food4rhino.com/en/app/octopus.

Vierlinger, Robert. 2013. "Multi-Objective Design Interface." MSc diss., University of Applied Arts Vienna. https://doi.org/10.13140/RG.2.1.3401.0324.

Wilson, Luc, Jason Danforth, Carlos Cerezo Davila, and Dee Harvey. 2019. "How to Generate a Thousand Master Plans: A Computational Framework for Urban Design." In *SIMAUD '19: Proceedings of the Symposium on Simulation for Architecture and Urban Design*. 113–120.

IMAGE CREDITS

All drawings and images by the authors.

Kathy Velikov is Professor at the University of Michigan Taubman College of Architecture and Urban Planning, and Vice President of ACADIA. She is a licensed Architect and founding partner of the research-based practice rvtr (www.rvtr.com), which serves as a platform for exploration and experimentation in the intertwinements between architecture and the environment. Her work ranges from material experimentation and physical prototyping, to mapping and analysis, to speculative design propositions.

Matias del Campo is a registered Architect, Designer and Educator. He is Associate Professor at the University of Michigan Taubman College for Architecture and Urban Planning, and director of the AR2IL–The Architecture and Artificial Intelligence Laboratory. He is the co-founder of the architecture practice SPAN (www.span-arch.org), a globally acting practice best known for their application of contemporary technologies in architecture design.

Lucas Denit is a graduate of the University of Michigan's MArch and MSDMT programs and a Designer at Lake|Flato Architects.

Kazi Najeeb Hasan is a graduate from the University of Michigan's MArch and Real Estate Development Certification programs. He is currently engaged with Volta Homes—a vertically aligned offsite homebuilding startup—as a founding team member and business analyst, where his work is situated between business and branding strategies, operational process standardization, product development, and customer experience.

Ruxin Xie is a graduate from the University of Michigan's MArch and MSDMT programs and a Technical Designer at Gensler. She focuses on the prototyping method towards the synergy between architectural design experience and digital technology.

Brent Boyce is Director of Advanced Products at Guardian Glass, charged with identifying future opportunities that will arise out of shifts in the architectural, engineering, and construction industry. He has applied his background in thin film physics to lead development of over fifty new low-E and technical coated glass products.

Deep Learning Isovist

Unsupervised Spatial Encoding in Architecture

Mikhael Johanes
École Polytechnique Fédérale de Lausanne

Jeffrey Huang
École Polytechnique Fédérale de Lausanne

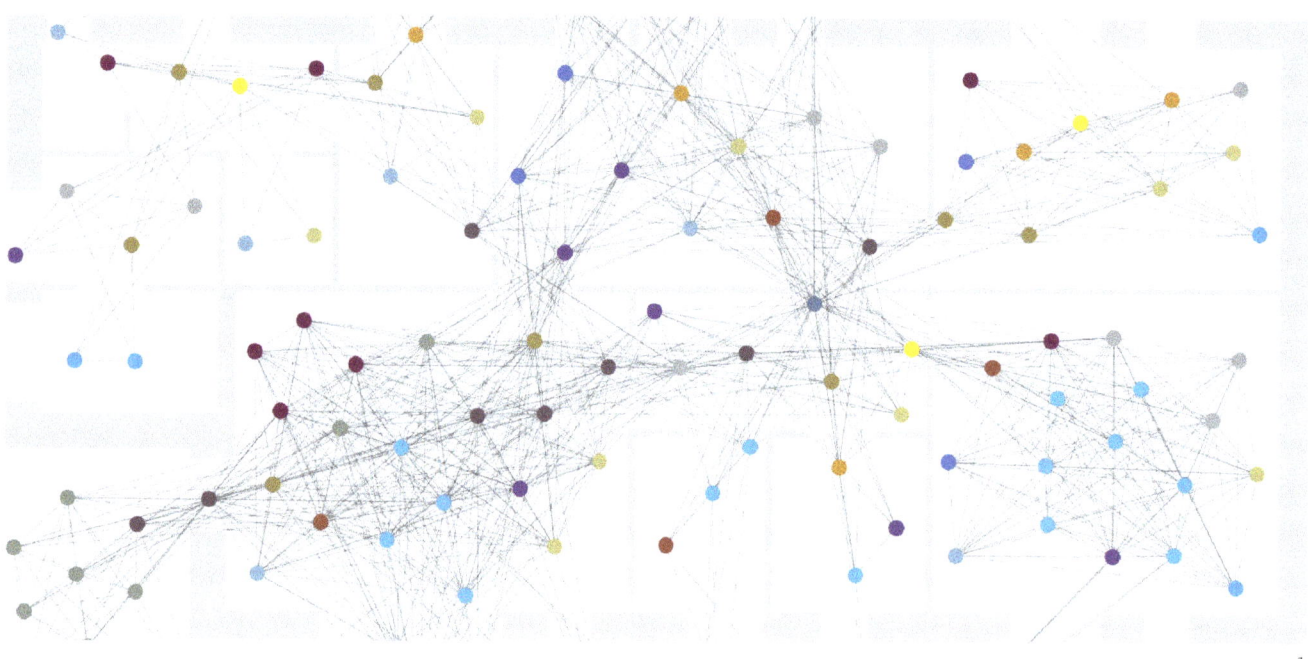

1 Spatial fingerprints using using isovist and deep learning techniques

2 Isovist in an environment

ABSTRACT

Understanding the qualitative aspect of space is essential in architectural design. However, the development of computational design tools has lacked features to comprehend architectural quality that involves perceptual and phenomenological aspects of space. The advancement in machine learning opens up a new opportunity to understand spatial qualities as a data-driven approach and utilize the gained information to infer or derive the qualitative aspect of architectural space. This paper presents an experimental unsupervised encoding framework to learn the qualitative features of architectural space by using isovist and deep learning techniques. It combines stochastic isovist sampling with Variational Autoencoder (VAE) model and clustering method to learn and extract spatial patterns from thousands of floor plan data. The developed framework will enable the encoding of architectural spatial qualities into quantifiable features to improve the computability of spatial qualities in architectural design.

INTRODUCTION

Spatial Recognition in Architecture

While the notion of space and spatial qualities has been central in architectural design, the development of architectural computational tools has lacked features to comprehend such abstract spatial qualities and weigh itself into a more concrete geometric representation of the architecture (Bhatt, Schulz and Huang 2012). The recent development in spatial recognition and informatics has attempted to address the problem and provides a foundational agenda to create a human-centered computational design. By borrowing machine vision and spatial recognition techniques, we propose a framework to make architectural space machine learnable and searchable, thus assisting the computability of architectural space.

This research investigates computational encoding techniques to identify, extract, and compare architectural spatial qualities using isovist representation and deep learning methods. Isovist is defined as a set of all points visible from a given point in an environment as a representation of perception of space (Benedikt 1979). Isovist representation offers a reading of machine learnable architectural patterns that encompass phenomenological and morphological aspects of space. Deep learning algorithms are developed to identify the hypothetical statistical structures from the data, called manifolds, to automatically perform actions or inferences (Goodfellow, Bengio, and Courville 2016). It is assumed that the spatial properties can exist in such manifolds for machines to learn and utilize the information to infer or derive particular architectural qualities without the need for high-level formalization.

Deep Learning Isovist

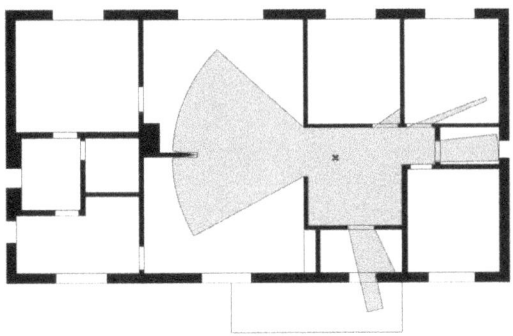

2

Although research on utilizing isovist in architectural and urban analytics has been well developed for a couple of decades, deep learning techniques for isovist encoding are still relatively rare. A notable study is a work of (Leduc, Chaillou, and Ouard 2011), which couples isovists field with digital signal processing technique, Fast Fourier Transform (FFT), to express isovist as a sum of its periodic components. As such, a clustering algorithm could be performed to classify a pedestrian surrounding space based on the discrete Fourier analysis of isovist data. Nevertheless, the authors suggest that the proposed encoding method perhaps will not yield much differentiation for a more complex environment and propose extending the research with a more significant number of sampling and a more complete Fourier analysis function.

The advancement of machine learning opens up a new episode of spatial analysis in architecture. One of few attempts in architectural research is the use of generated labeled 2D depth map images of 3D isovist from a set of spatial prototypes for training a deep convolutional neural network classifier using purely supervised learning (Peng et al. 2017). The classifier can then identify several predefined spatial patterns in a particular area of analyzed architecture. Supervised learning assumes assigned classes from which the neural networks are trained to automate the classification. On the other hand, unsupervised learning aims to discover the emerging classes that reflect similarities and differences in the data. Unsupervised classifiers in isovist spatial analysis can lead to new typologies that integrate the perceptual, phenomenological, and programmatic aspects of space (Derix and Jagannath 2014).

Some architectural relevant experiments of unsupervised machine learning of isovist are found in the domain of machine vision to provide a shared machine-human spatial for autonomous mobile robots. Sedlmeier and Feld (2018) construct and apply an unsupervised clustering algorithm to 6-dimensional feature vectors of isovist descriptor, as described in Benedikt (1979), and extract meaningful spatial structures from a building floor plan such as rooms, corridors, and doorways. An unsupervised deep learning application in isovist data can be found in Feld et al. (2018), which uses VAE and isovist representation to learn the semantically meaningful encoding of spatial-temporal trajectory data. The works above demonstrate the possibility of learning meaningful spatial patterns from isovist representation using deep learning methods. However, the data used in the experiment was still minimal; expanding the study using a larger dataset will leverage the capacity of deep learning models for architectural analysis. This research aims to extend the research of deep learning of isovists in extracting the significant qualitative aspects of architectural design by establishing a workflow that consists of the sampling, learning, and clustering of isovist data.

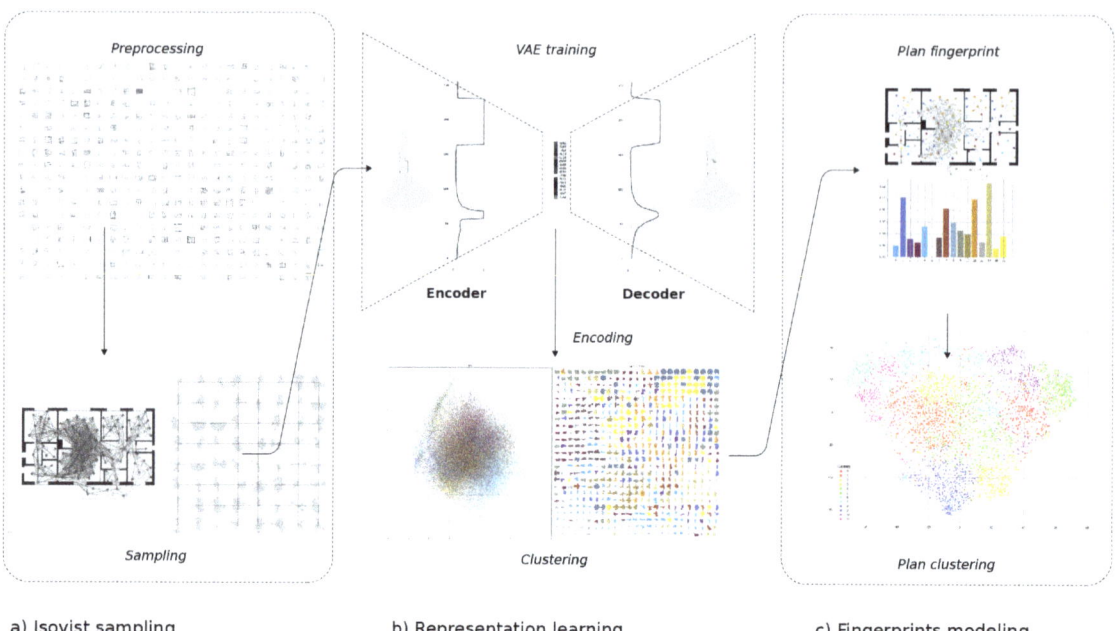

3   The workflow of unsupervised deep learning for architectural spatial pattern recognition

METHODS:
REPRESENTATION LEARNING OF ISOVIST

This study developed a framework for unsupervised encoding of architectural space by learning the spatial representation of isovists using deep learning techniques (Figure 3). The framework aims to extract meaningful spatial patterns from the given dataset. The experiments employ various tools for respective stages of (a) isovist sampling, (b) representation learning, and (c) fingerprint modeling. The dataset preparation and isovist sampling are conducted using custom Grasshopper scripts. The consequent stages are conducted in a Python local environment. A custom TensorFlow implementation of the VAE deep learning algorithm is used to learn and encode the isovists into a 16-dimensional feature vector. Scikit-learn implementation of -means clustering is used to cluster and classify the encoded isovists and floor plan data (Pedregosa et al. 2011).

Isovist Sampling

As experimental data, a publicly available dataset of approximately five thousand floor plan drawings of house apartments from the Finland region is used (Kalervo et al. 2019). The dataset is comprised of scaled, annotated floor plan drawings in SVG format. The wall and floor information are extracted from the dataset, divided into 85% training set, 10% evaluation set, and 5% test set. A custom Grasshopper script is used to sample the isovists from the defined floor area stochastically. The isovist size and sampling density determine the amount of information sampled from the given data on an architectural scale. In this case, we assume an isovist radius of 5m and density of one sample point per 1 m^2 as pragmatic human scale (Turner et al. 2001). For transforming the isovist samples into machine learnable features, discretization of isovist in polar coordinate with a regular angle (Figure 4) is used to create a function of radial distances. In this experiment, the isovists are sampled into 256 radials that give enough spatial resolution in architectural scale (Leduc, Chaillou, and Ouard 2011).

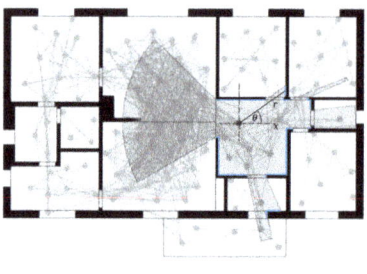 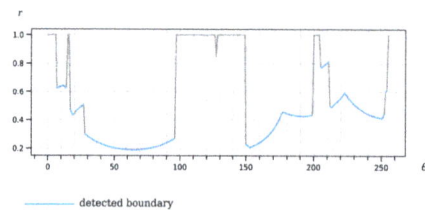

4   Isovist sampling

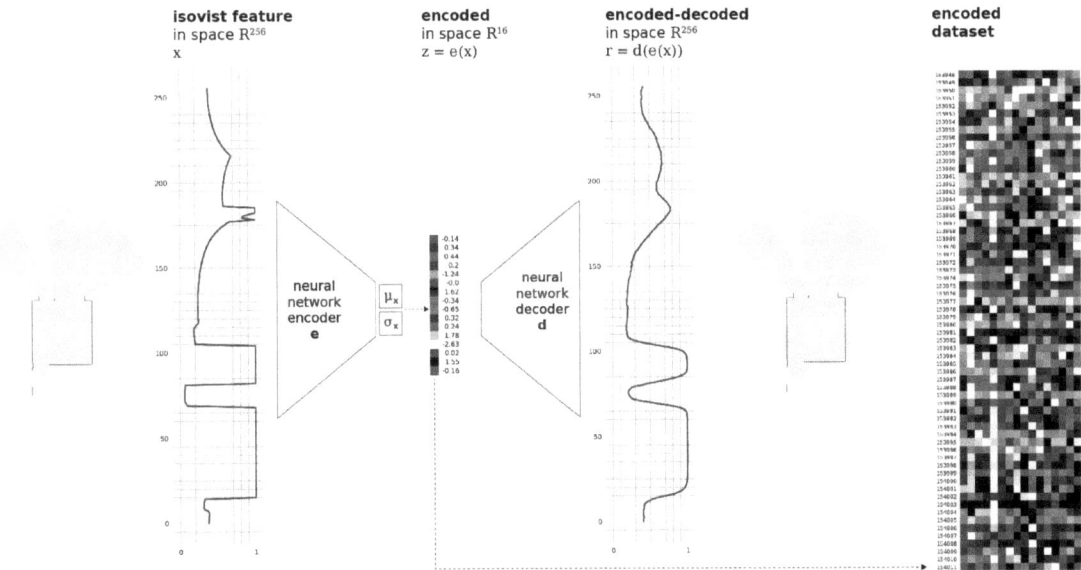

5   Encoding Isovist Features.

Representational Learning

Unsupervised machine learning algorithms extract the spatial information from the isovist features and cluster the captured spatial patterns. The results of the clustering process will be used as a semantic index for isovist sampling and provide a base for spatial fingerprints modeling.

The first step of representation learning is the extraction of spatial patterns from isovist samples using autoencoders. Autoencoder is an unsupervised artificial neural network designed to learn the encoding of input data in a lower-dimensional internal representation and reconstruct it back to match the input data (Goodfellow et al. 2016). Autoencoders are designed to learn a specific internal representation from the data by enforcing particular constraints during encoding and decoding data. As such, autoencoders are forced to learn the features needed the most to reconstruct the input, and are thus potentially effective in learning meaningful features from the data. Variational Autoencoders (VAE) is a more advanced implementation of autoencoders that assume continuous probabilistic distribution in the internal representational space, thus improving the learning results' generalizability (Kingma and Welling 2014). In this experiment, the VAE model is trained to learn the compact representation of isovists. After training, the encoder of the trained VAE is thus being used to encode the isovists into 16-dimensional feature vectors (Figure 5).

K-means clustering algorithm is used to cluster the encoded isovist. The number of clusters is determined by using an elbow method by increasing the number of clusters while measuring the variation within the clusters. The elbow point of the curve occurs when having more clusters will not significantly decrease the variation of data in the clusters, thus indicating the optimum number of clusters. This experiment uses an automatic curve's knee detection to determine the number of clusters (Satopaa et al. 2011). As the algorithm groups similar data to a cluster, the resulting clusters capture recurring spatial patterns from the floor plans and provide semantic vocabulary for spatial indexing.

Fingerprints Modeling

The clustering results from the previous stage are used as the index to label each floor plan in the dataset, providing spatial semantic that can be used to compare different floor plans (Figure 6). There are two strategies of spatial fingerprints modeling that are envisioned from the resulting isovist clusters. The first strategy is to create a feature vector for each plan by using the weighted mean of the corresponding cluster to represent the general morphological characteristic of the plan (Figure 6b). Comparing the floor plans can be done by measuring the distance between the feature vectors. This simple strategy will not accurately result in finding similar floor plans but rather provide analogical results that expand to different floor plans that potentially share similar morphological characteristics. A clustering of floor plans has been done by using

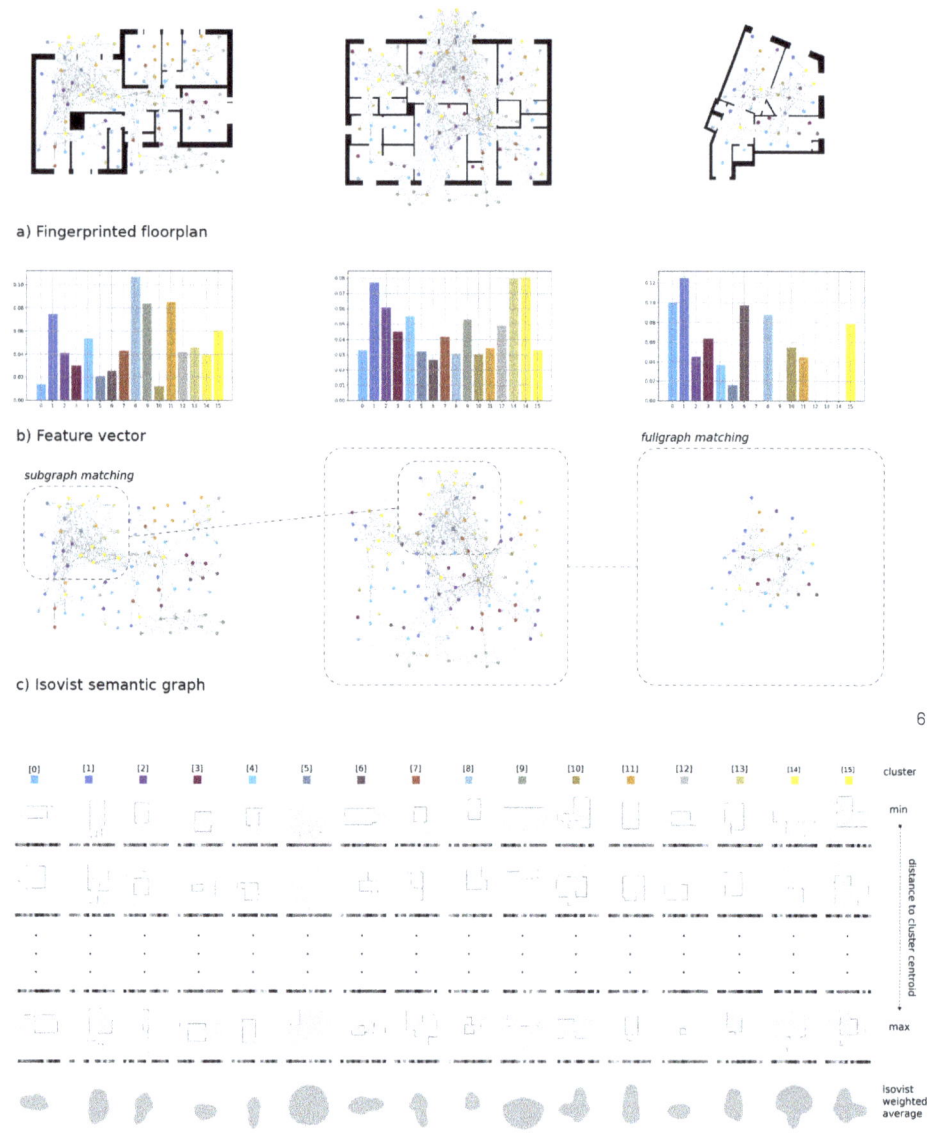

6 Spatial fingerprinting: the isovist classes are used as a feature vector and semantic node of the topological graph.

7 Clustering of isovists based on its encoding

these feature vectors; the results will be discussed in the following section.

The second strategy is to combine the isovist semantic with the graph of isovist co-visibility (Figure 6c). Both the topological and morphological information of the plan will be incorporated into one fingerprint model. The possibility of decomposing and clustering the graph into subgraphs allows the discovery of spatial patterns from different granularity in which isovist semantic nodes will be essential to introduce a morphological aspect to graph-based spatial fingerprints. A set of graphs with semantic from isovist clusters has been extracted from the data. Further work will develop on the embedded information to identify the spatial pattern by using graph matching (Langenhan et al. 2013), graph clustering (Schaeffer 2007), and graph embedding (Goyal and Ferrara 2018).

RESULTS AND DISCUSSION: TOWARD MACHINE UNDERSTANDING OF SPACE

Encoded Isovists

From the VAE encoding and K-means clustering process, 16 unlabeled clusters of isovist are identified as spatial prototypes (Figure 7). These spatial prototypes represent the basic spatial elements that correspond to certain spatial morphology and qualities. The clusters will provide a semantic basis for spatial classification and recognition in the next stage of research. By decomposing the morphological features of architecture, these spatial prototypes can be used as a building block for generative modeling that uses spatial qualities as its defining parameters. Further work will expand the generative potential of these decomposed spatial elements and leveraging its emerging semantic classification.

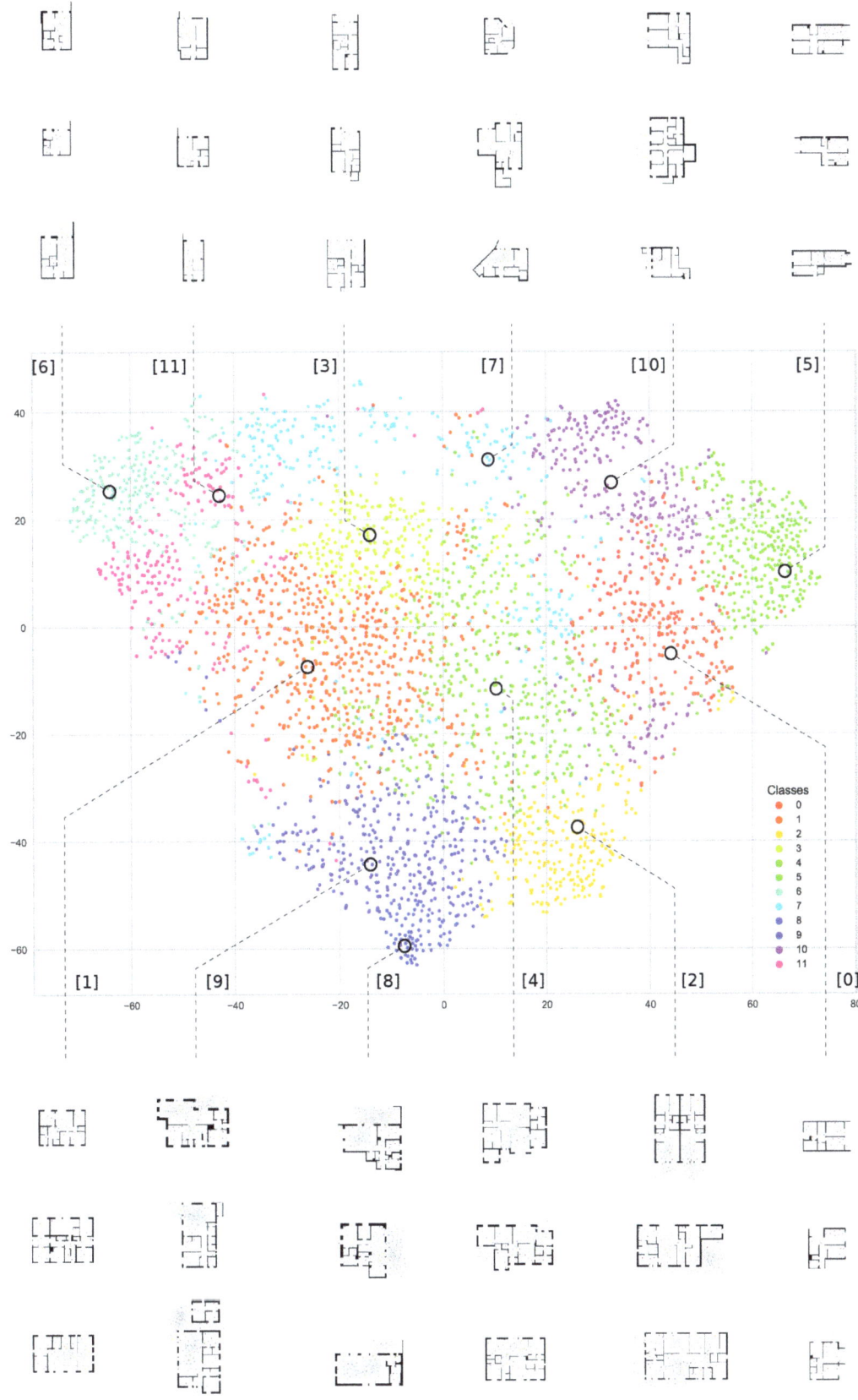

8   T-SNE visualization of floor plan clusters based on its feature vector

Data-driven Typology

The floor plans are classified by using the K-means clustering algorithm of the weighted mean of occurrences for each spatial prototype as floorplan fingerprints in which the number of clusters is determined by the elbow method. The visual observation of the plan shows some recurring spatial pattern that characterizes each cluster, such as orientation, the geometry of the spaces, and configuration complexity (Figure 8). The clustering result indicates the possibility of unsupervised architectural spatial recognition from isovist representation without prior semantic labeling. This approach could offer a computational approach to allow an empirical, data-driven understanding of architectural spatial patterns. Furthermore, the extracted pattern could provide a robust spatial fingerprint that relies less on human-made semantic labels that reduce labor in organizing architectural spatial information.

CONCLUSION AND FUTURE WORK

This study proposes a framework for unsupervised encoding of architectural space by learning the spatial representation of isovists using deep learning techniques. The work contributes to developing and investigating computational encoding techniques to identify, extract, and compare architectural spatial patterns that reflect space's phenomenological and morphological features. The framework, therefore, provides a basis for further investigation of perceptual space that is machine learnable and searchable. The results obtained from the experiment show the potential use of the developed techniques to extract different architectural spatial patterns without prescribed semantic labels. Therefore, the framework aims to improve the computability of spatial qualities to be integrated into architecture design and analysis.

Work in progress consists of incorporating semantic nodes from isovist clusters with their topological information to compare spatial quality in different granularity. By combining the topological structure of isovist co-visibility, graph-related techniques such as graph clustering, graph matching, and graph embedding can be implemented to improve the classification and spatial pattern recognition task. Since the VAE algorithm only learns the recurring spatial pattern from isovist data, the salient and unique properties are left out in the encoding process. Some research in deep learning has addressed this issue, and future implementation is envisioned. Possible extension of the developed framework to 3-dimensional architectural space is also anticipated in the future work of the study.

REFERENCES

Benedikt, Michael L. 1979. "To Take Hold of Space: Isovists and Isovist Fields." *Environment and Planning B: Planning and Design* 6 (1): 47–65.

Bhatt, Mehul, Carl Schultz, and Minqian Huang. 2012. "The Shape of Empty Space: Human-Centred Cognitive Foundations in Computing for Spatial Design." In *2012 IEEE Symposium on Visual Languages and Human-Centric Computing (VL/HCC)*. 33–40. https://doi.org/10.1109/VLHCC.2012.6344477.

Derix, Christian and Prarthana Jagannath. 2014. "Digital Intuition – Autonomous Classifiers for Spatial Analysis and Empirical Design." *The Journal of Space Syntax* 5 (2): 190–215.

Feld, Sebastian, Steffen Illium, Andreas Sedlmeier, and Lenz Belzner. 2018. "Trajectory Annotation Using Sequences of Spatial Perception." In *SIGSPATIAL '18: Proceedings of the 26th ACM SIGSPATIAL International Conference on Advances in Geographic Information Systems*. New York, NY, USA: ACM. 329–38. https://doi.org/10.1145/3274895.3274968.

Goodfellow, Ian, Yoshua Bengio, and Aaron Courville. 2016. *Deep Learning; Adaptive Computation and Machine Learning*. Cambridge, Massachusetts: The MIT Press.

Goyal, Palash and Emilio Ferrara. 2018. "Graph Embedding Techniques, Applications, and Performance: A Survey." *Knowledge-Based Systems* 151 (July): 78–94. https://doi.org/10.1016/j.knosys.2018.03.022.

Kalervo, Ahti, Juha Ylioinas, Markus Häikiö, Antti Karhu, and Juho Kannala. 2019. "CubiCasa5K: A Dataset and an Improved Multi-Task Model for Floorplan Image Analysis." *ArXiv*, April 3, 2019. http://arxiv.org/abs/1904.01920.

Kingma, Diederik P. and Max Welling. 2014. "Auto-Encoding Variational Bayes." *ArXiv*, May 1, 2014. http://arxiv.org/abs/1312.6114.

Langenhan, Christoph, Markus Weber, Marcus Liwicki, Frank Petzold, and Andreas Dengel. 2013. "Graph-Based Retrieval of Building Information Models for Supporting the Early Design Stages." *Advanced Engineering Informatics* 27 (4): 413–26. https://doi.org/10.1016/j.aei.2013.04.005.

Leduc, Thomas, Francoise Chaillou, and Thomas Ouard. 2011. "Towards a 'Typification' of the Pedestrian Surrounding Space: Analysis of the Isovist Using Digital Processing Method." In *Advancing Geoinformation Science for a Changing World. Lecture Notes in Geoinformation and Cartography.* Berlin; Heidelberg: Springer. 275–92. https://doi.org/10.1007/978-3-642-19789-5_14.

Pedregosa, Fabian, Gaël Varoquaux, Alexandre Gramfort, Vincent Michel, Bertrand Thirion, Olivier Grisel, Mathieu Blondel, Peter Prettenhofer, Ron Weiss, and Vincent Dubourg. 2011. "Scikit-Learn: Machine Learning in Python." *The Journal of Machine Learning Research* 12: 2825–30.

Peng, Wenzhe, Fan Zhang, and Takehiko Nagakura. 2017. "Machines' Perception of Space: Employing 3D Isovist Methods and a Convolutional Neural Network in Architectural Space Classification." In *ACADIA 17: Discipline & Disruption, Proceedings of the 37th Annual Conference of the Association for Computer Aided Design in Architecture.* Cambridge, MA: ACADIA. 474–81.

Satopaa, Ville, Jeannie Albrecht, David Irwin, and Barath Raghavan. 2011. "Finding a 'Kneedle' in a Haystack: Detecting Knee Points in System Behavior." In *2011 31st International Conference on Distributed Computing Systems Workshops.* Minneapolis, MN, USA: IEEE. 166–71. https://doi.org/10.1109/ICDCSW.2011.20.

Schaeffer, Satu Elisa. 2007. "Graph Clustering." *Computer Science Review* 1 (1): 27–64. https://doi.org/10.1016/j.cosrev.2007.05.001.

Sedlmeier, Andreas, and Sebastian Feld. 2018. "Learning Indoor Space Perception." *Journal of Location Based Services* 12 (3–4): 179–214. https://doi.org/10.1080/17489725.2018.1539255.

Turner, Alasdair, Maria Doxa, David O'sullivan, and Alan Penn. 2001. "From Isovists to Visibility Graphs: A Methodology for the Analysis of Architectural Space." *Environment and Planning B: Planning and Design* 28 (1): 103–21.

IMAGE CREDITS

All drawings and images by the authors.

Mikhael Johanes is a Doctoral Assistant at Media x Design Lab EPFL. His works examine the computational medium with its generative and analytical possibilities in architectural design. Since he finished his Master of Architecture in Universitas Indonesia 2013, he has continued his career as a lecturer and research staff in the same University. He is part of the editorial team of Interiority and ARSNET journal, and has also published in several international journals.

Jeffrey Huang is the Director of the Institute of Architecture at EPFL, Head of the Media x Design Lab, and a Full Professor in Architecture and Computer Science at EPFL. His research examines the convergence of physical and digital architecture. His recent work on Artificial Design (Design Brain) is featured at the Seoul Biennale of Architecture and Urbanism 2021.

Imaginary Vessels

Machine learning digital dataset creation for robotic pottery fabrication

Cami Quinteros*
UMass Amherst/ IAAC

Hesham Shawqy*
IAAC

Gabriella Rossi
CITA/Royal Danish Academy

Iliana Papadopoulou
UCL/IAAC

David Leon
IAAC

* Authors contributed equally to the research

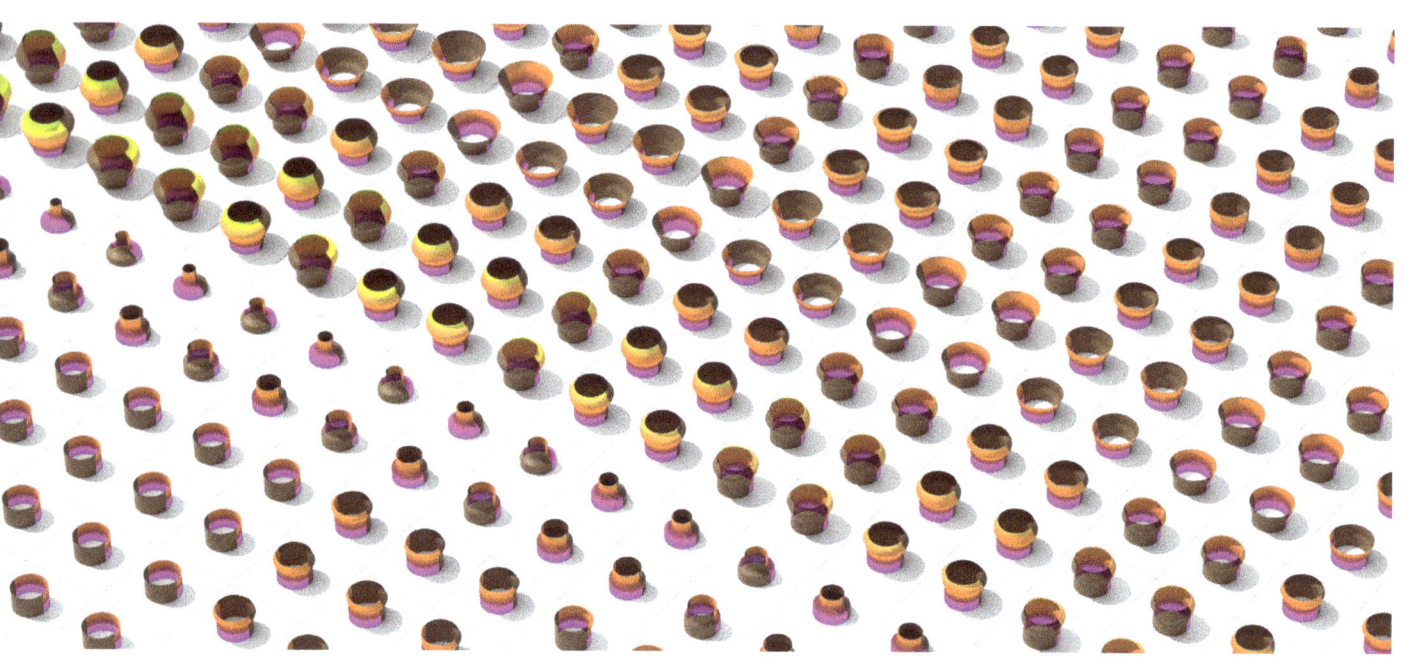

1

ABSTRACT

Clay is one of the foundational materials in art and architecture, traced in the development of mud walls and adobe structures, and showcased in utilitarian and ornamental pottery. Wheel throwing is the process of shaping clay mainly into symmetrical objects, a complex craft in which the master potter has the knowledge and skill to manipulate the clay into the final design of various physical objects. This project explores how machine learning can be used to translate the richness and complexity of wheel throwing for digital fabrication.

In this paper we present a surrogate digital dataset for robotic fabrication and geometric prediction used to train neural networks and provide a bridge between digital fabrication and handcraft. We report on the parametric model that abstracts wheel throwing as the interaction between a rotating mass and a given set of forces, as well as on data wrangling methods, dataset composition considerations, and training methodology. We present two models, one in which geometry is predicted based on a given set of forces, and a second in which forces are predicted based on a given geometry. Lastly, we give a critical assessment of the predictions of both networks and discuss future steps.

INTRODUCTION

Clay is a natural material that has been used in architecture for centuries due to its plasticity, formal variability, and structural strength gained after firing. Reinforced concrete became the building material of choice during the modernist era, hence historicizing clay bricks; nonetheless, clay has resurfaced in contemporary practice with renewed design potential attached to the ceramic module (Keuning 2007). It is, however, not only in modularity but also in material plasticity that clay offers a solution to contemporary designs of complex geometric expression (Sabin 2010). With the rise in use of digital tools for fabrication, different methods to interact with clay have been developed—from extruded 3D printed buildings, as seen in the work of Rael and San Fratello's Casa Covida (Rael and San Fratello 2020) or Mario Cucinella and WASP's TECLA housing prototype (Cucinella 2021), to optimization of brick modules through wire cutting (Andreani and Bechthold 2014), or pairing of oscillating wire cutting with adaptive pick-and-place production for brick panel assemblies (Rossi et al. 2021). Many of these projects showcase hardware and software advances to print with clay and all the incongruences that this process presents. Sabin goes as far as to explore the outer edges of 3D printing by changing from orthotropic layer deposition to woven-like extrusions that cure on pre-formed plaster molds. In so doing, Sabin explores a live feedback loop that allows the user/designer to actively interact with the printing path, and therefore introduce "human error" or a "maker's mark" (Sabin et al. 2018).

Our project expands on the idea of the maker's mark outside of the world of 3D printing. Rather than programming a robotic arm to interact in an additive or subtractive manner, our project aims to recreate the clay throwing process, based on the analysis and translation of a pottery wheel into a parametric model. This makes the creation of continuous smooth surfaces that carry the mark of the maker. By challenging the typical 3D printing operation, we free forms from design constraints that lack advanced material intelligence (Chronis et al. 2017) and create a path that can best interact with the material and all its strengths. In doing so, we provide a critical response to the shortcomings of clay 3D printing and aim to enhance knowledge inherited from craftspeople. As a point of departure, our project first considers the material properties of clay, as taught to us by master potters, and then develops a method of emulation of their craft.

Wheel throwing utilizes rotational kinetic energy in combination with manual forces, an action in which various physical phenomena have an impact. Parameters considered include the velocity of the wheel, the moment of inertia as the clay distributes its mass around the central axis, and the distinct and multiple manual pressures applied with one or two hands. As master potters pull upwards or push sideways, altering the thickness of the surface, they compact the clay changing its molecular arrangement and, in so doing, improve its tectonic performance (Roux and Corbetta 1989). Nordmoen's humanMADE robot uses Machine Learning, a genetic algorithm, and a single silicone human-like finger to generate novel designs on the pottery wheel (Nordmoen 2021). Nordmoen's project is focused on questioning the notions of unique and handmade in opposition to machine-made, limiting its scope to the replacement of human creativity. In contrast, our project builds on human creativity, analyzing it, and developing the initial stages of a manufacturing process that can expand into architectural application. A primary shortcoming of human-made pottery is in the scale of the manufacturing—for example, controlling the thickness of the pot's walls becomes exponentially more difficult as the pot grows in width and height, and the width and height of the pot is limited by the potter's arm length. A new method of robotic fabrication will provide a path to overcome these shortcomings.

Our project considers multiple forces interacting, both in the axial movement of the wheel and in the combined forces of both hands, resulting in the simulation of geometric and tectonic transformation of clay. An initial attempt to collect data through film and hand tracking with the OpenCV library yielded two hours of recordings and ten original physical pots. However, this method of data collection proved to be inadequate since the potter at times would obstruct the view for hand tracking, and hands became covered in clay which further obfuscated the process. By translating the wheel thrower's action into a parametric model, the potential for creating synthetic datasets with numerous samples occurs, as it is needed for neural network training. Furthermore, a digital surrogate dataset allows for the creation of samples without human constraints such as pot size, wall thickness, or the need to physically make all the samples and digitize them *a posteriori*. Time and physical scale constraints are removed from the data collection process, and a feedback loop calibrating the dataset to reality could become feasible by continuing the project into the robotic fabrication stage.

In addition, a digital dataset provides the possibility for the neural network to learn what forces are needed to generate a given geometry, and what geometry may be generated from a given set of forces. With the surrogate dataset, we can train models at multiple scales to predict geometries for robotic fabrication parameters, and in so

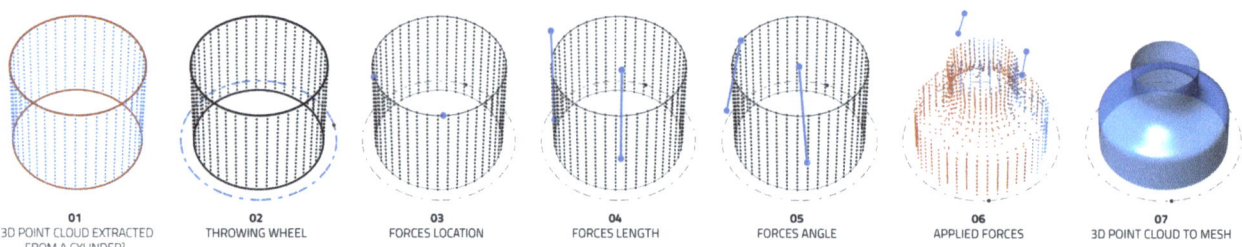

2 Step-by-step explanation of the parametric model

doing, we are not proposing to replace human craft but rather to augment it through digital fabrication.

Machine learning has proven capable of translating complex fabrication systems and crafts into robotic toolpaths, as shown by Rossi and Nicholas's work with the English Wheel (Rossi and Nicholas 2018) as well as the work of Brugnaro and Hanna with wood chisels (Brugnaro and Hanna 2017). This research project pushes the state of the art of training a neural network for robotic fabrication by using a surrogate digital model rather than directly analyzing the motions of a master craftsperson or digitizing produced samples. In general, synthetic datasets are useful for machine learning tasks, particularly in the fields of architecture, where there is a lack of available datasets, and the material constrains of existing ones make the process unsustainable and time-consuming. In contrast, this proposal's research of a parametric model produces a synthetic dataset with controllable sample complexity and data distribution. In creating a digital parameter space, data collection is easily accessible and is geometrically rich and varied, ridding ourselves of data augmentation techniques and their implied biases.

DATASET DESIGN
Time-based Parametric Model

A Grasshopper script was developed to abstract and simulate the clay throwing process. A cylinder (the clay) rotates along a central axis (the wheel) and forces are applied (the hands of the potter). The script makes use of action loops using the Anemone plugin for Grasshopper, and it is organized so that the forces can be given a location, length, and rotation angle as inputs. The centered cylinder has an original height of 15cm, and a diameter of 10cm, becoming a constant starting value, and it is transformed by the forces as it rotates along the central axis (Figure 2). The resulting digital wheel is an iterative time-based simulation, where each wheel rotation further modifies the original geometry as the forces interact with the cylinder.

In the initial development phase, the script showed the step-by-step transformation of each pot, and the user interacted with the pot by changing the position and angle of the hands through Grasshopper number sliders. This live user input allowed to validate the accuracy of the script by visually examining the result in the form of a pot. Once the script was tested for accuracy, the 'hand' positions were automated by providing a series of multipliers that would alter the angle and force applied to the lump of clay at each iteration, and a total of 9,500 unique pots were generated over 24 hours, with total of 19 parameters were extracted from each as seen in Figures 3, 4 and 5.

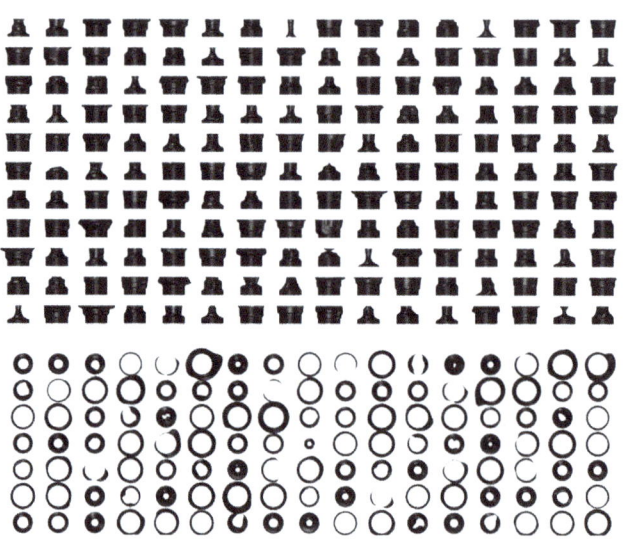

3 Elevation view (top) and corresponding top view (bottom) of a subsample of pots generated using the parametric script.

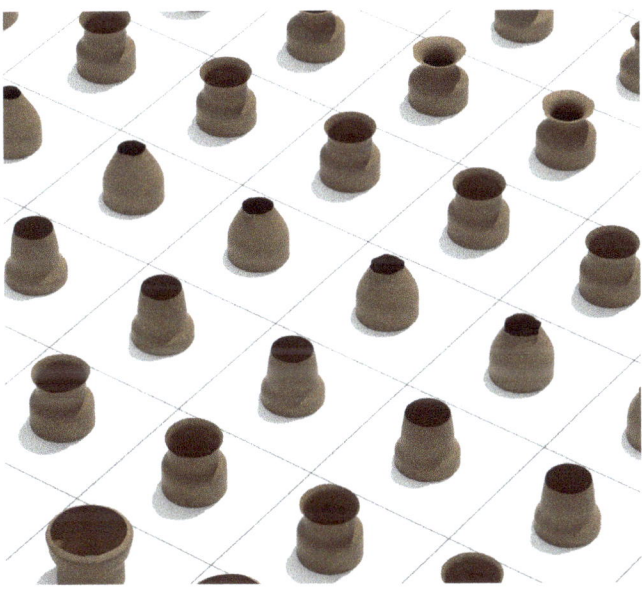

4 Orthographic view of pots generated using the parametric script.

Analytical Dataset Composition

The recorded parameters were divided into two categories: (A) Robotic Fabrication Parameters, and (B) Geometry Parameters as shown in Figure 5. Robotic Fabrication Parameters (Figure 5A) included information on the central axis rotation iterations and speed, as well as force values, rotation angles, length, and friction values. Geometry Parameters (Figure 5B) are specific points at the cross-section line of each individual pot from which the complete geometry can be reconstructed. In dividing the parameters into two categories, we designed a two-avenue ML training, whereby we either predict the resultant geometry based on a given set of forces, or alternatively predict the forces needed to create a desired cross-section.

Our dataset had multiple iterations, and each ran through Principal Component Analysis (PCA) and Exploratory Data Analysis (EDA). The original dataset included 11 robotic fabrication parameters and 60 geometry parameters. We decreased the geometry parameters by assigning constant values on the x- and z-axis of our cross-sections, and only measuring deviation from zero on the y-axis. Moreover, the robotic parameters were only kept insofar as they were deemed necessary for robotic fabrication. The final dataset consisted only of 10 robotic fabrication parameters and 9 geometry parameters.

Geometry Parameters Data Encoding

The data encoding for the geometry parameters required simplicity and accuracy, since the success of the neural network predictions is reliant on them. On the first try, each cross-section point was given an x, y, and z value. This tripled the number of values needed per point and made it difficult for the neural network to recognize patterns for prediction. The solution was to place the original cross-section at point 0,0,0 and maintain constant values for the x and z coordinates of each point as seen on Figure 6.

This way, the transformed cross-section of each pot was encoded as the distance between the original and final points. This method of data encoding allowed for ease of training, as well as a simple and precise method to compare predicted versus true values visually and numerically.

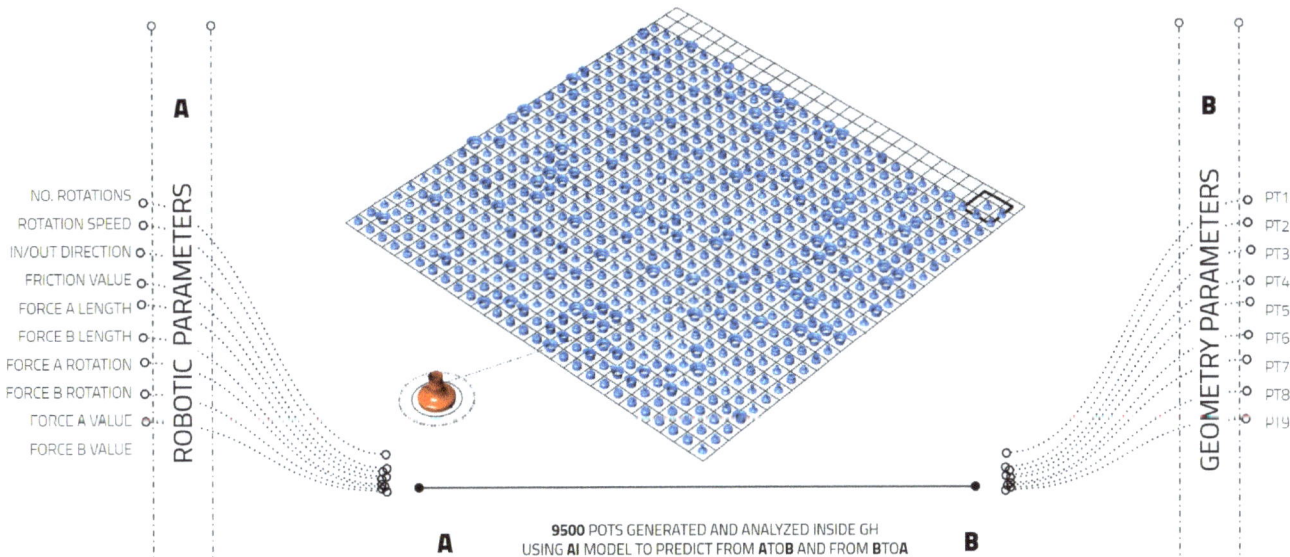

5 Recorded parameters and time-based script parameter space

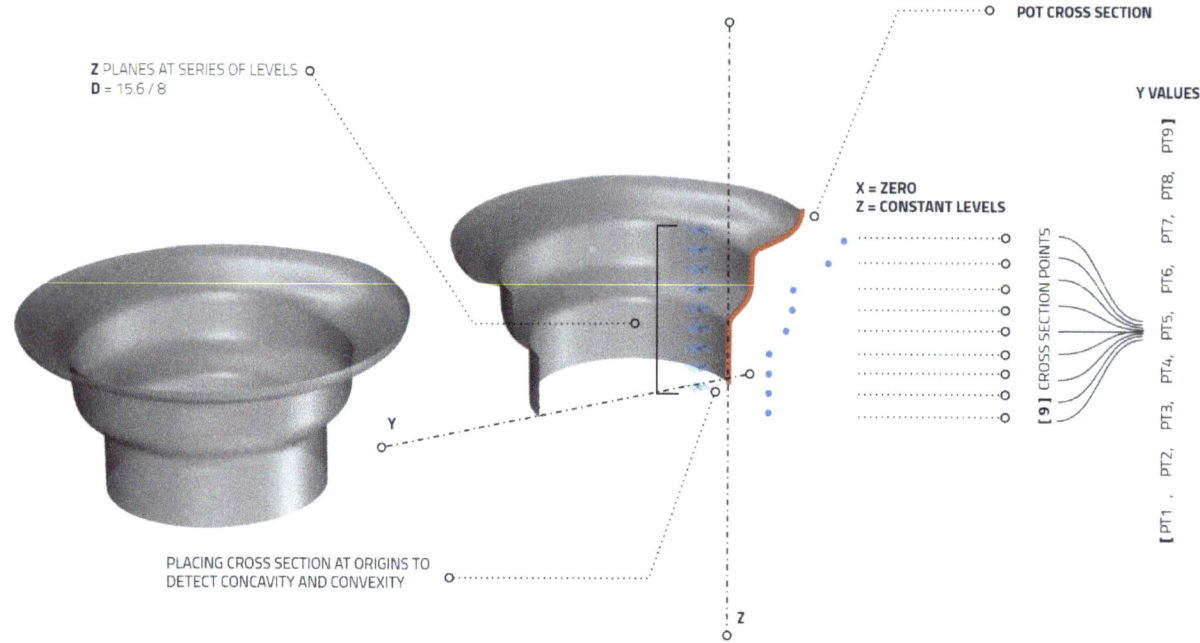

6   Graphic explanation of Geometry Parameters data encoding, where X and Z are constant values, and Y is a measure of displacement.

PREDICTING GEOMETRY PARAMETERS FROM ROBOTIC FABRICATION PARAMETERS

Neural Network Architecture

The first neural network model trained used (A) Robotic Parameters to predict (B) Geometry Parameters and was named the AtoB predictor. As with the work of Brugnaro and Hanna, this presented a regression problem with back-propagation learning (Brugnaro and Hanna 2017).

The deep neural network is a double funnel of 10-18-52-27-18-9 neuron architecture, predicting 9 parameters based on 10. The hidden layers use ReLu activation, while the input and output layers are Sigmoid activated. Important to note is the use of two types of scaling methods for our data from the SciKit-Learn library, MinMax Scaler for the Geometry Parameters (in which we did not expect many outliers), and Standard Scaler for the Robotic Fabrication parameters, which had a normal distribution. This model proved efficient, as it required only 20 epochs to train at an above 90% success rate.

Data Pipelines

The neural network training was outsourced to a free cloud-based GPU in Google Colab, providing an easy collaborative environment. After training, the model was migrated to a local computer to be integrated into Rhino/Grasshopper. The methodology included the coordination between the native Grasshopper component Hops, Visual Studio Code, a development environment for Python, and a local server. With this workflow, the Hops component becomes the host of a trained AI in the Grasshopper

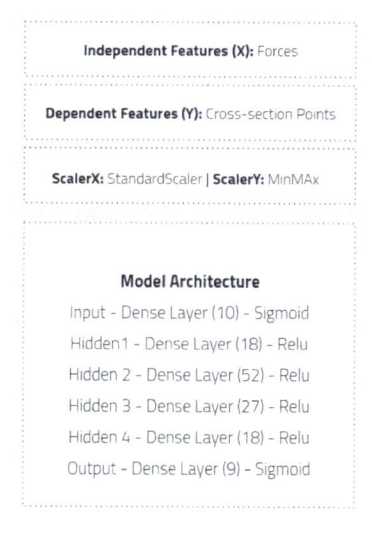

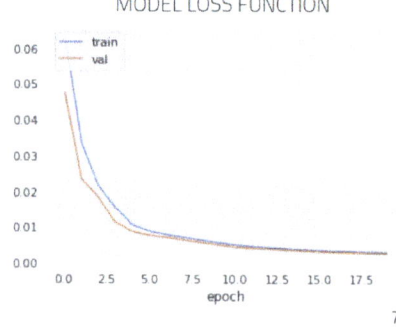

7

environment to which number sliders can be connected, representing (A) Robotic Parameters, transforming the design space into a de facto control panel for prediction of (B) Geometry Parameters and digital fabrication.

7 Neural network architecture and model loss function during training on free cloud-based GPU

8 Set of 500 pot subsamples selected to analyze deviation values. The blue line represents the ground truth cross section, while the red represents the predicted one.

Result Evaluation

Having migrated the AI into the parametric space, visual geometric comparison of predicted versus true values was possible by superimposing the cross-section of the resultant geometries and their deviation (Figure 7).

This visual comparison is especially important in the design field since the usual tools of loss and accuracy graphs for neural networks do not show the margins of error for each individual pot (Figure 8). Imperfection is an inherent part of craft and, in this case, algorithmic deviation can provide space for exploration. It is in this visual evaluation of results that we see the potential to use neural networks as a bridge between handcrafts and digital fabrication. The AI model recalibrates at every iteration, without having the final configuration explicitly preprogrammed, converging towards an approximation that is deemed good enough, yet not error-free (Rossi and Nicholas 2018).

A subset of 500 pots were selected to analyze deviation values. While the neural network validation accuracy showed 0.99, and validation loss showed 0.0029, the deviation values in the visual comparison ranged from 1.78 to 23.7 cm with an average value of 7.16 cm.

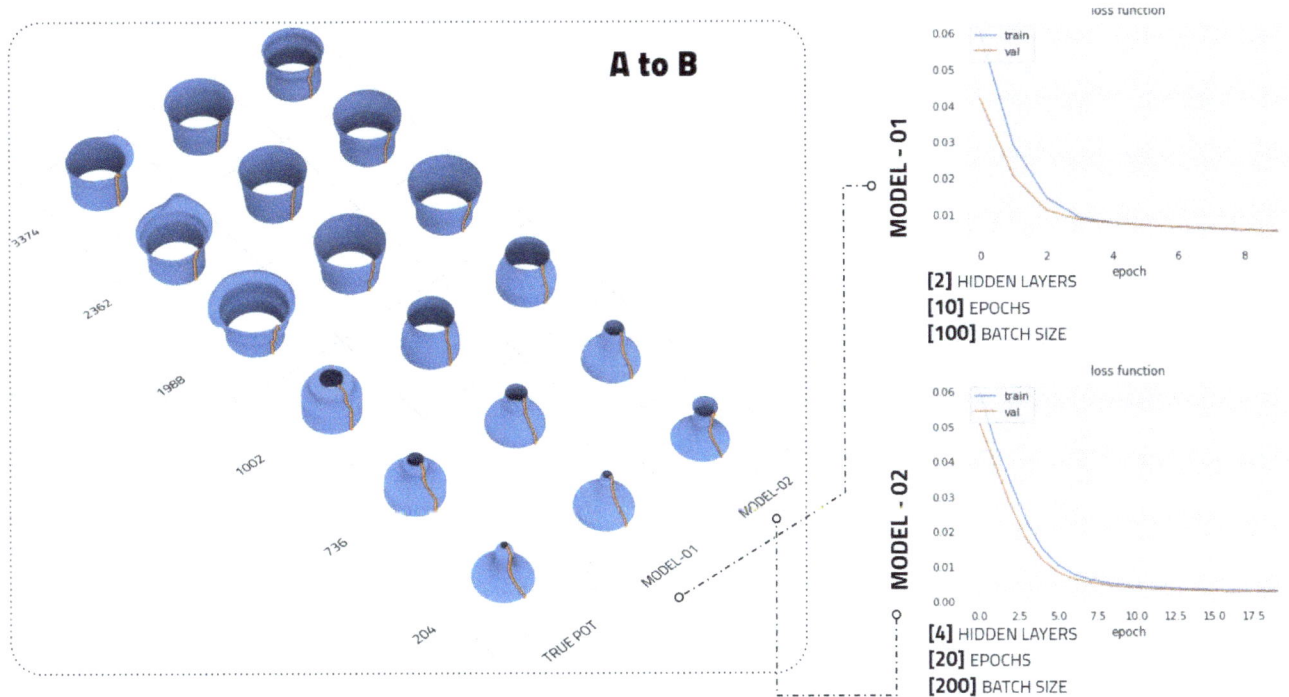

9 Subsample of pots selected to analyze deviation values

AI-ASSISTED DESIGN   REALIGNMENTS   147

PREDICTING ROBOTIC FABRICATION PARAMETERS FROM GEOMETRY PARAMETERS

Neural Network Architecture

The second neural network model trained used (B) Geometry Parameters to predict (A) Robotic Fabrication Parameters, the exact inverse of the first model, and was named the BtoA predictor. Again, an ANN regression model was used with backpropagation. However, this second model proved to be more complex.

The PCA analysis showed that we did not need as many as 9 parameters to run the prediction; however, lowering the number of cross-section points would lower the resolution of the resultant geometry. Instead of sacrificing resolution, the decision was to increase our training iterations as well as further complicate the neuron architecture.

The neural architecture became a deeper double funnel of 9-18-27-52-27-18-10, where the hidden layers were once again ReLu activated. The input layer was Sigmoid activated, while the output layer was Elu activated. For our training, the epoch iteration had to be increased between 1,000 and 2,000, as well as increase the batch size. Overall, this second model proved more difficult to arrive to a satisfactory precision.

Data Pipelines

Using the same methodology as with the first model, we ran the training on a cloud-based GPU and then migrated to a local computer. The Grasshopper-native Hops component for this model became a host for the AI prediction in the same way as for the first model. In addition, however, this second Hops component was a summary of the parametric script originally created, and it negated the need for non-native Grasshopper components or plug-ins.

Result Evaluation

Following the same method, a subset of 500 pots was selected, the cross-sections of the predicted versus ground truth pots were overlapped, and the deviations measured (Figure 11). On this second model, the validation loss was 0.56, and the validation accuracy was 0.3272. Graphic deviation values ranged from 2.18 to 16.53 cm, with an average deviation of 7.95 cm, a 0.8 cm difference from the first model. The deviation between prediction and ground truth was present but acceptable. Greater deviation values were observed towards the top of each pot, which can be due to a dataset bias where a majority of pots showed a wider top.

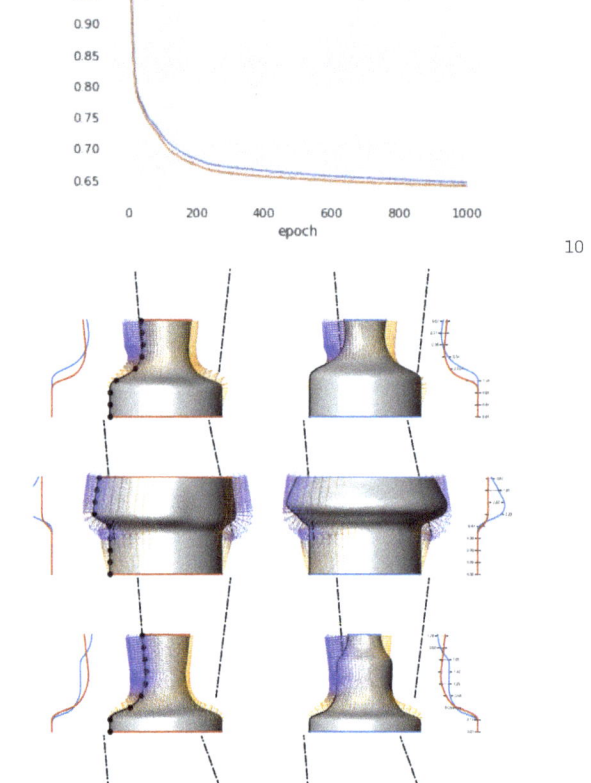

10 Neural network architecture and model loss function during training on free cloud-based GPU

11 Set of 500 pot subsamples selected to analyze deviation values: the blue line represents the ground truth cross section, while the red represents the predicted one

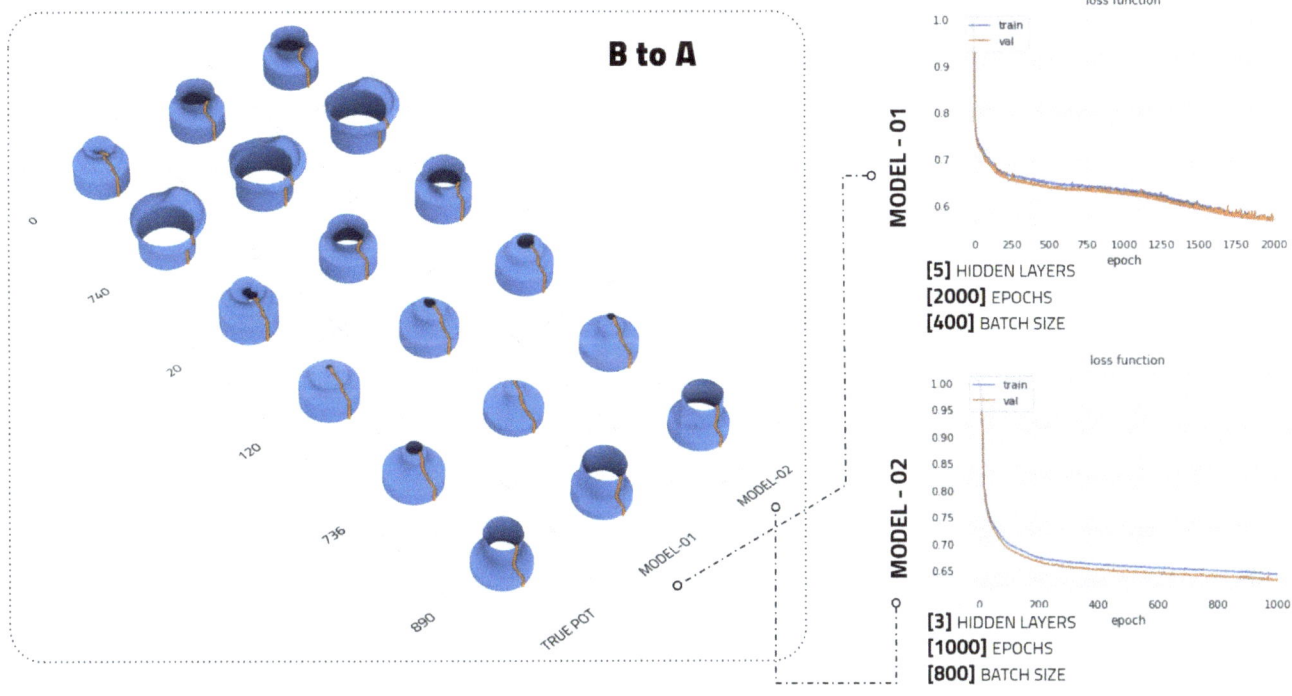

12 Subsample of pots selected to analyze deviation values

CONCLUSIONS AND FUTURE WORK

In this paper we presented an initial attempt at using surrogate digital parametric scripts for the modelling of craft processes. The development of synthetic datasets became the bases for novel manufacturing methods without the need of time consuming and expensive physical experimentation, thus creating a path of accessibility for future designers. The development of a synthetic dataset has allowed us to ideate a feasible path towards robotic fabrication that can explored as a next step.

By using a time-based parametric model that generates a synthetic dataset to abstract the process of clay throwing, we engage with the renewed design potential attached to the clay module and further develop the use of material intelligence associated to clay's molecular composition. While our model certainly carries built-in imprecisions and biases, improvements can be made by physically testing and calibrating through fabrication experiments.

We also presented a conceptual workflow for the encoding of geometric data into datasets for machine learning. While the average deviation in the geometric results for predictions AtoB and BtoA was around 7 cm, we believe that future steps for improvement could be made by including further variety within our dataset. It is possible that the deviation values are caused by the machine learning algorithm memorizing and giving preference to one kind of pot over the other, specifically as related to the top of each pot. We also foresee that additional features might be necessary to represent the effects of different clay types on the form finding process.

Learning from handcrafts using machine learning requires rigorous analysis, experimentation, and an expandable dataset. Surrogate digital datasets are a way to make use of machine learning for complex fabrication processes, allowing us to expand, contract, and test various datasets with our neural networks. They also allow control via the equal distribution of representative features, a crucial aspect to the success of the predictive model as well as making thousands of samples at no physical cost of time or resources. New methods of performance evaluation beyond loss values graphs are needed for proper evaluation of design-based predictions; therefore, we presented an integrated computational workflow for visual evaluation of the predicted imaginary pots that focuses more on qualitative assessment rather than the numerical.

Finally, we believe that the key to enabling further uses of ML in architecture is dataset availability, combined with open-source code. This positions computational design as a tool of accessibility for designers and enables collaborative workflows.

ACKNOWLEDGEMENTS

This project was developed as part of the seminar "Computer Vision Machine Learning and Data Encoding" taught by Gabriella Rossi and assisted by Iliana Papadopoulou as part of the AI and Architecture module of the Master in Advanced Computation in Architecture and Design at the Institute of Advanced Architecture of Catalonia.

REFERENCES

Andreani, Stefano and Martin Bechthold. 2014. "[R]Evolving Brick: Geometry and Performance Innocation in Ceramic Building Systems through Design Robotics." In *Fabricate 2014; Negotiating Design & Making*, edited by F. Gramazio, M. Kohler, and S. Langenberg. London: UCL Press. 182–91. https://doi.org/10.2307/j.ctt1tp3c5w.26.

Brugnaro, Giulio and Sean Hanna. 2017. "Adaptive Robotic Training Methods for Subtractive Manufacturing." In *ACADIA 17: Disciplines and Disruption, Proceedings of the 37th Annual Conference of the Association for Computer Aided Design in Architecture.* 164–69.

Chronis, Angelos, Alexander Dubor, Edouard Cabay, and Mostapha Sadeghipour Roudsari. 2017. "Integration of CFD in Computational Design: An Evaluation of the Current State of the Art." *Proceedings of the 35th ECAADe Conference*, vol. 1. 601–10.

Cucinella, Mario. 2021. "TECLA, An experimental 3D-printed dwelling made of local earth." Mario Cucinella Architects (website). Accessed December 6, 2021. https://www.mcarchitects.it/project/tecla-2.

Keuning, D. 2007. "Beautiful Brick Architecture." In *Brick: The Book*, edited by P. Brandon and M. Betts. London: Chapman and Hall. 251–61.

Nordmoen, Charlotte. 2021. "HumanMADE." Accessed December 6, 2021. http://www.cnordmoen.com/humanmade.

Rael, Ronald and Virginia San Fratello. 2020. "Casa Covida." Accessed December 6, 2021. https://www.rael-sanfratello.com/made/casa-covida.

Rossi, Gabriella and Nicholas Paul. 2018. "Re/Learning the Wheel." In *ACADIA 18, Recalibration, On Imprecision and Infidelity, Proceedings of the 38th Annual Conference of the Association for Computer Aided Design in Architecture.* 146–155.

Rossi, Gabriella, James Walker, Asbjorn Sondergaard, Isak Worre Foged, Anke Pasold, and Jacob Hilmer. 2021. "Oscillating Wire Cutting and Robotic Assembly of Bespoke Acoustic Tile Systems." *Construction Robotics* 5: 63–72.

Roux, Valentine and Daniela Corbetta. 1989. *The Potter's Wheel: Craft Specialization and Technical Competence*. New Delhi: Oxford & IBH Publishing.

Sabin, Jenny, Bennett Norman, David Rosenwasser, Jungyang Leo Lui, and Jeremy Bilottin. 2018. "Robosense 2.0: Robotic Sensing and Architectural Ceramic Fabrication." In *ACADIA 18: Recalibration, On Imprecision and Infidelity; Proceedings of the 38th Annual Conference of the Association for Computer Aided Design in Architecture.* 276–85.

Sabin, Jenny. 2010. "Digital Ceramics: Crafts-Based Media for Novel Material Expression & Information Mediation at the Architectural Scale." In *ACADIA 10: Life in:Formation; Proceedings of the 30th Annual Conference of the Association for Computer Aided Design in Architecture.* 174–81.

IMAGE CREDITS

All images by Cami Quinteros and Hesham Shawqy.

Cami Quinteros is an architect born in Lima, Peru currently living in Massachusetts, USA. Their academic trajectory has ranged from fine arts, in particular sculpture and installation art, to historic preservation, and are currently involved in advanced computational design. Cami has a particular interest in the use of compostable materials—such as mud, wood, or slate—and their application to complex architectural forms, as well as an interest in human mobility and migratory patterns.

Hesham Shawqy is a architect and researcher focusing on computational design and artificial intelligence in the field of architecture and urban development. He received his Master's degree in Advanced Computation for Architecture & Design in 2021. His research interests focus on digitizing handcrafts using machine learning and robotic fabrication. He is currently working as a researcher and teaching assistant at IAAC.

Gabriella Rossi is a PhD Fellow at CITA, Royal Danish Academy. Her research focuses on machine learning as an emerging modelling paradigm and how it can change architectural practice. Specific accent is placed on architectural datasets, harvested or simulated, and potential applications for complex material behavior and design performance. She has experience with developing novel fabrication methods using Industrial Robotics and digital tools.

Iliana Papadopoulou is an architect focusing on computational design, computer graphics, and interactive technologies. She holds a diploma of Architectural Engineering from AUTh, and she's currently studying and working as a postgraduate teaching assistant at The Bartlett School of Architecture in the MSc Architectural Computation program.

David Leon is an architect with focus on the research, training and development of computational tools for the AEC industry. He currently works for McNeel Europe providing support for third-party developers pushing the boundaries of interoperability and computational design globally. He is currently the director of the MACAD master's program at IAAC.

Crooked Captures

Hyojin Kwon
Harvard Graduate School of Design

Adam Sherman
Technical University of Vienna

1   *Crooked Captures* Exhibition. M Gallery, CICA Museum, Seoul, Korea. January 27-31, 2021 (Photo by Mingi Lee, 2021)

With flashy renderings dominating news feeds and high-flying drones filming from otherwise inaccessible vantage points, our encounters with the built environment increasingly involve perspectival views, but not necessarily those experienced firsthand. As tools for image production and consumption evolve, so too will methods for studying historical precedents.

Crooked Captures treats this proliferation of digital images as fertile ground for photogrammetric explorations into how two-dimensional imaging techniques can influence three-dimensional form. While photogrammetry, the process of determining spatial measurements of physical objects from photographic inputs, has been an area of investigation for almost two centuries, the technique's potential has blossomed with increased access to high quality cameras. Typical photogrammetric applications couple high-fidelity scanning and computing to produce faithful digital copies of physical artifacts and scenes for measuring and surveying. Leading photogrammetry software packages promise *accuracy* and *precision*, touting the exact replication of physical forms in digital space—so-called *reality capture*—as an indisputable virtue.

Crooked Captures is a multimedia exhibition composed of three-dimensional architectural scenes processed with Autodesk ReCap, one such photogrammetric software package. The project utilized frames taken from drone footage of New York City found on the internet as raw material for a two step process. First, image processing operations performed on

2   Process: Image Input; drone footage widely accessible on the internet is used as the input source material for processing

3   Process: Pixel Parsing; pixels within a selected range of brightness, hue, and saturation are filtered to select a given building's facade

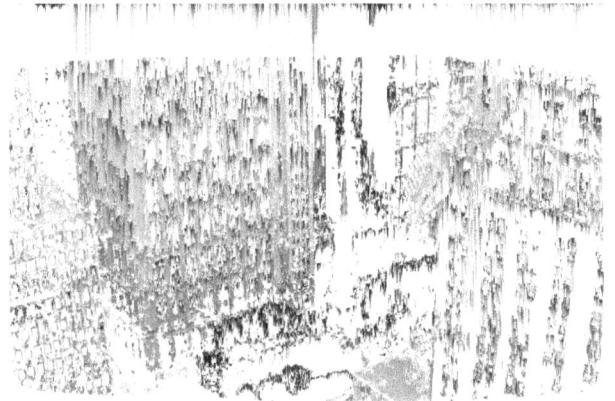

4   Process: Pixel Sorting; pixels in each vertical column are sorted by brightness value

5   Process: Image Stitching; the original and new images are merged by replacing the original pixels with sorted pixels

the video stills extracted, emphasized, and exaggerated qualities only attainable through computation. Second, respatialization of the images through photogrammetric processing generated three-dimensional scenes that the manipulated frames could possibly depict. The result is a series of anamorphic landscapes in which recognizable images give way to mutations revealed through orbits and zooms.

These computer vision methods produce readings and misreadings that become a basis for abstraction as blurs, streaks, bumps, and folds arise from purely computational operations. While the human eyes and brain may work in tandem to identify each individual component when viewing an architectural photograph, the computer "sees" these input images completely differently as it parses through flattened pixel matrices. Here, distinctions between discrete elements fade away as components melt into each other and buildings blend into their sites. This process reduces precedents to repositories of spatial data at the mercy of tools that capture, store, and distribute information.

Circumstance, therefore, determines much of the interpretation, as contingent qualities of color, light, and perspective allow representation to supersede the lost notion of an *original*. By remaining willfully ignorant of the as-built physical geometry of precedents, the project reinforces the subjectivity inherent in the perception of architectural form.

Technologies supposedly simulating exact virtual replicas of the physical world have long provided opportunities to speculate on glitches and misreadings, as in the unsteady recordings of a static GPS receiver in Laura Kurgan's *You Are Here: Information Drift* (Kurgan 1994). Within this history, the uncanniness of photogrammetric distortions arises from the complexity of visual landscapes that are increasingly realistic and inhabitable, yet still decidedly warped. In *Crooked Captures*, intentional modification of the input data set hindered recognition of the original physical artifacts, making the photogrammetric operation less a reconstruction and more a construction. These assemblies emerge not from the orthographic marks of architectural drawings but rather from the gradients of pixel values

6   American Telephone & Telegraph Building. 3D Scan constructed from manipulated images from found drone footage (the Dronalist 2020c)

7   Distorted perspective

8   461 Fifth Avenue. 3D Scan constructed from manipulated images from found drone footage (the Dronalist 2020a)

9   130 William. 3D Scan constructed from manipulated images from found drone footage (Kalucci 2020)

10-11   Legible Perspective: Each piece contains relatively coherent views among the otherwise deformed scenes, achieving an anamorphic effect as the photogrammetric software attempts to identify the original vantage points

across digital videos (as well as in the biases solidified within the programming of photogrammetric software).

While many design computation software packages promise greater precision and control over new geometries, *Crooked Captures* surrenders some authorial agency as it looks past the limits of the designer's own mind and searches instead for architecture as a product of unplanned computational processes. The works seen here anticipate new methods of production of both images and tectonics, anticipating new relationships between human experience and computer vision in physical and digital artifacts alike.

ACKNOWLEDGMENTS

Thanks to the Autodesk Residency Program, who provided us with software and fabrication services for the realization of this project.

REFERENCES

Autodesk. *ReCap*. V. 7.0. Autodesk. PC. 2020.

the Dronalist. "Drone Midtown Manhattan 4k." Premiered online March 21, 2020. YouTube video, 6:36. https://youtu.be/8F50Ctq3rA0.

the Dronalist. "Seaport District, Downtown NYC 4k Drone sunset." Premiered online February 19, 2020. YouTube video, 2:35. https://youtu.be/GN0wAVD2vyg.

the Dronalist. "1+ Hour Downtown NYC Drone." Premiered online December 2, 2020. YouTube video, 1:08:36. https://youtu.be/CL60t4WyhvI.

Kalucci. "Aerial New York - One Hour Relaxation Music - 4K Drone Footage." Premiered online Mar 21, 2020. YouTube video, 1:00:00. https://youtu.be/iEmfFT1CoqQ.

Kurgan, Laura. *You Are Here: Information Drift*. Installation, Storefront for Art and Architecture, New York. March 12, 1994. http://storefrontnews.org/programming/you-are-here-information-drift/.

IMAGE CREDITS

Figure 1, 14-17: © Mingi Lee, January 27, 2021
All other drawings and images by the authors.

12 Distorted perspective

13 Distorted perspective

14 - 17 Exhibition Photographs: The project was exhibited at a gallery to share with a larger audience (Photo by Mingi Lee, 2021)

18 Physical Model: The three-dimensional scene was materialized into a physical model via color 3D printing

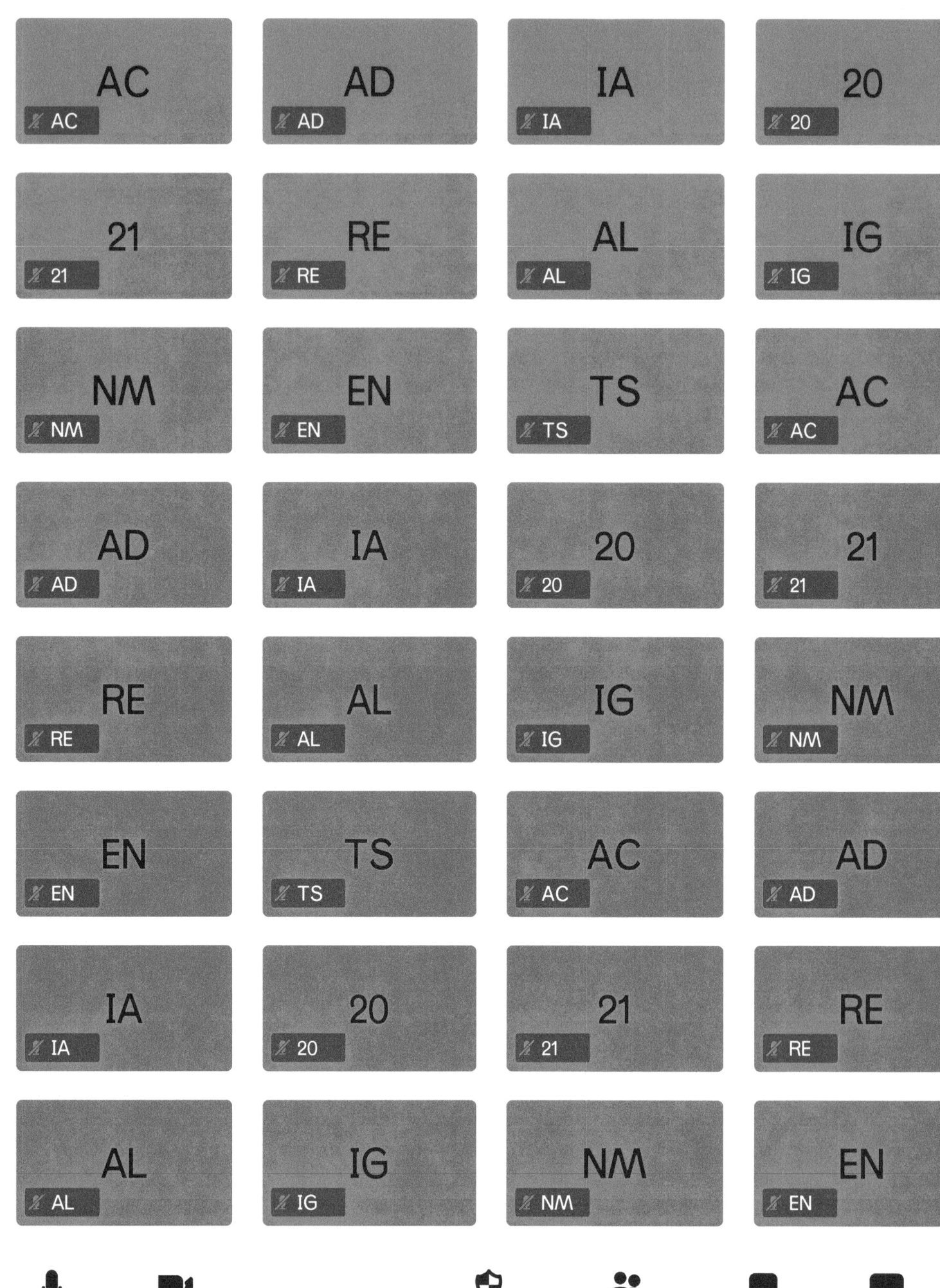

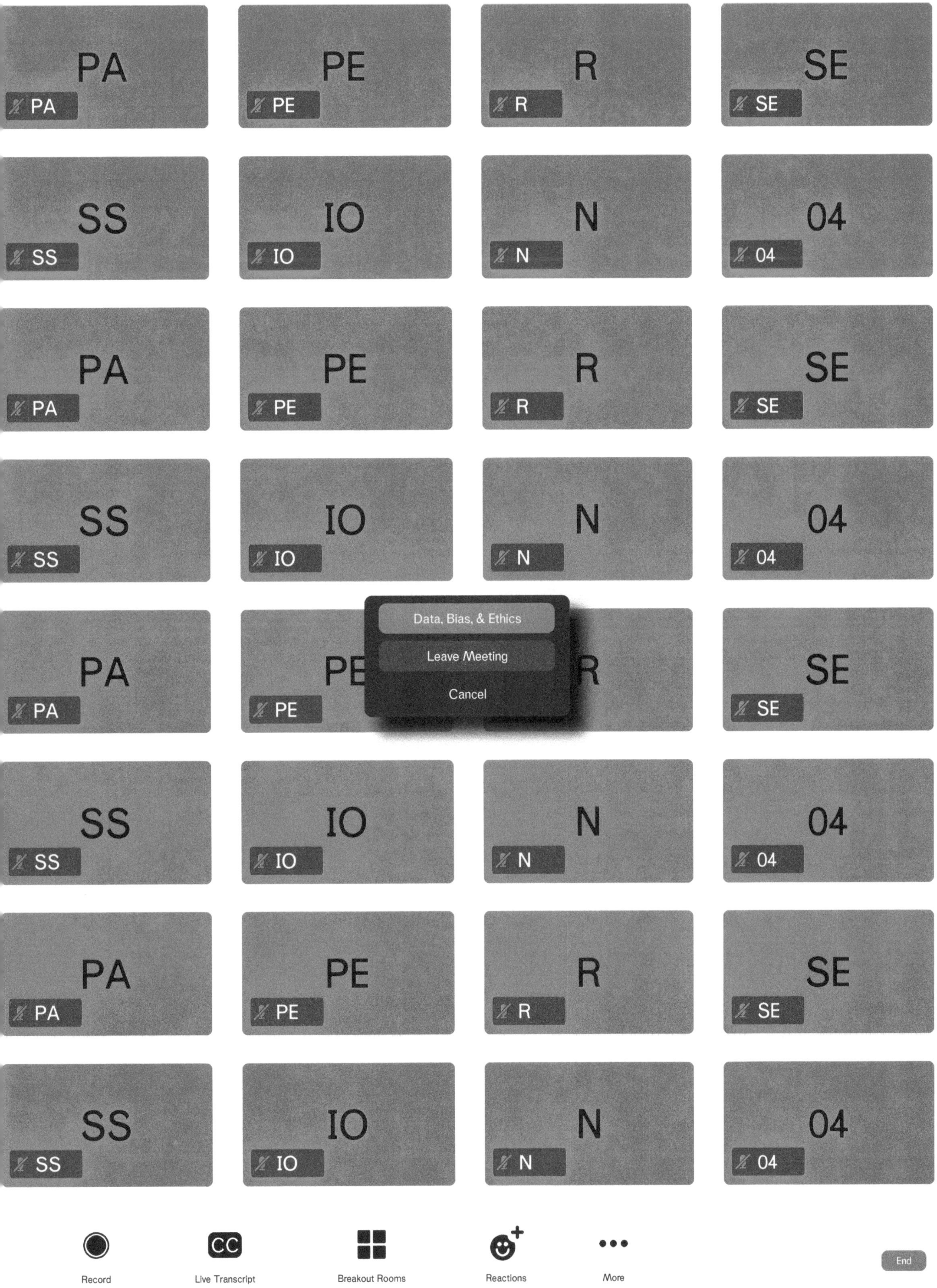
PA
PE
R
SE
SS
IO
N
04
PA
PE
R
SE
SS
IO
N
04
PA
PE
R
SE
SS
IO
N
04
PA
PE
R
SE
SS
IO
N
04
Data, Bias, & Ethics
Leave Meeting
Cancel
Record
Live Transcript
Breakout Rooms
Reactions
More
End

A POI-Based Machine Learning Method in Predicting Health

Predicting Residents' Health Status and Implications for Healthy City Planning

Shicong Cao
Heinle, Wischer und Partner Freie Architekten

Hao Zheng
University of Pennsylvania
Stuart Weitzman School of Design

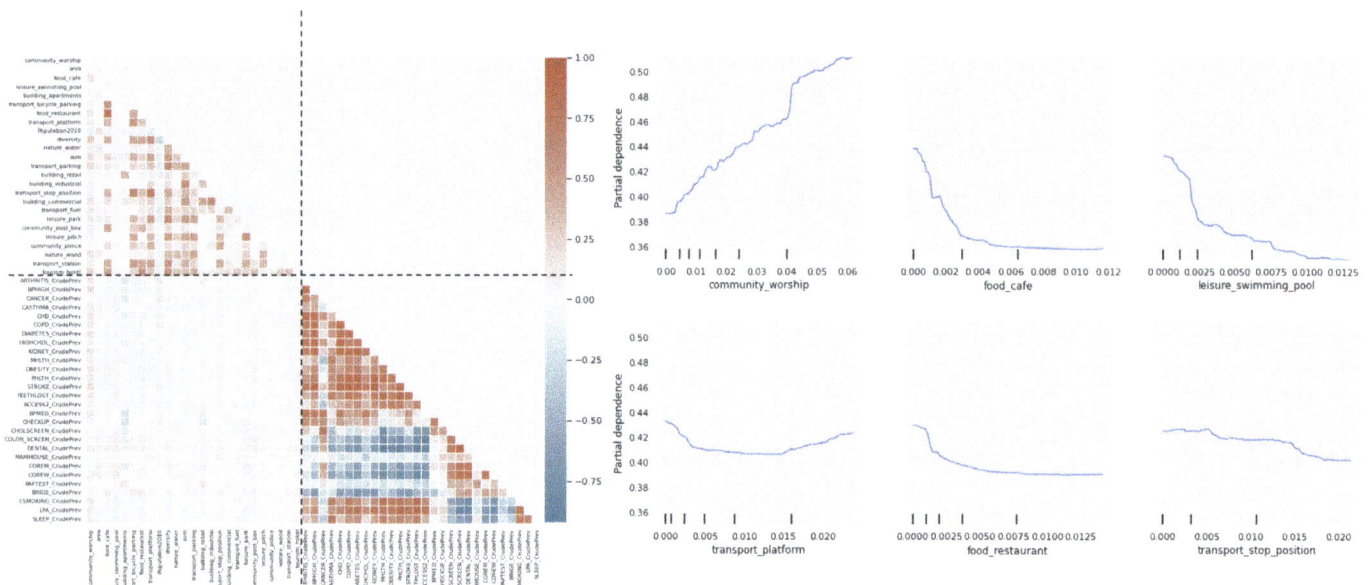

Above left: Correlation between selected features and output variables

Above right: Partial dependence plot of major input variables and prevalence of obesity

ABSTRACT

This research aims to explore the quantitative relationship between urban planning decisions and the health status of residents. By modeling the Point of Interest (POI) data and the geographic distribution of health-related outcomes, the research explores the critical factors in urban planning that could influence the health status of residents. It also informs decision-making regarding a healthier built environment and opens up possibilities for other data-driven methods. The data source constitutes two data sets, the POI data from OpenStreetMap, and the CDC dataset PLACES: Local Data for Better Health. After the data is collected and joined spatially, a machine learning method is used to select the most critical urban features in predicting the health outcomes of residents. Several machine learning models are trained and compared. With the chosen model, the prediction is evaluated on the test dataset and mapped geographically. The relations between factors are explored and interpreted. Finally, to understand the implications for urban design, the impact of modified POI data on the prediction of residents' health status is calculated and compared. This research proves the possibility of predicting residents' health from urban conditions with machine learning methods. The result verifies existing healthy urban design theories from a different perspective. This approach shows vast potential that data could in future assist decision-making to achieve a healthier built environment.

INTRODUCTION
Health and POI Data
Healthy living is a goal of many 21st century cities. The World Health Organization's Healthy Cities project has identified urban planning principles supporting health and example cities whose development can be learned from (Duhl and Sanchez 1999). The definition of health in the constitution of the WHO is a state of complete physical, mental and social wellbeing and not merely the absence of disease or infirmity (Kelley 2008). Health is, therefore, a social issue that needs to be addressed systemically and not only from a medical care point of view.

Studies show that about 60% of people's health depends on their lifestyle and the environment (Schroeder 2007). In contrast, only 3% of United States health expenditure goes into public health activities. The U.S. has a higher healthcare expenditure as a percentage of GDP than any other developed country but the lowest healthcare performance (Battisto and Wilhelm 2019). This makes us question the effectiveness of the policy to shape our healthcare system through the lens of medical care rather than through that of more encompassing public health.

Big data analytics, artificial intelligence, and other emerging technologies make it possible to understand and evaluate the effect of the built environment on residents' health quantitatively. The available data source, data processing procedures, and interaction technologies are poised to revolutionize urban management (Engin et al. 2020). By learning from data, solutions that benefit the community in the long term could be discovered and communicated. Data-driven process empowers the local community with the evidence they need to make better decisions for their city and neighborhoods.

Research Background
The link between the built environment and health has long been acknowledged. Research shows that there is interdependence between environments and individual behavior (Macintyre, Ellaway, and Cummins 2002). Primarily there are three domains where urban planning can most effectively focus support for health and wellbeing—physical activity, community interaction, and healthy eating—since these domains address some of the significant risk factors for chronic diseases (Kent and Thompson 2014). But the limitation of the traditional urban research method is relying on qualitative methods, sometimes without hard evidence.

To estimate the obesity rate, one of the machine learning models, Convolutional Neural Network (CNN), has been used to analyze the satellite image (Newton et al. 2020). Analysis of the convolutional layers suggests which visual features are more critical for a low obesity rate. The limitations of the imagery method are the amount of computing it requires and the obscurity of the conclusion due to the restriction of the dataset and the black box effect of the algorithm. Street view imagery is also used as a source of data to measure visual walkability (Zhou et al. 2019). This approach considers the human perception of the built environment. The amount of data processing and redundancy could be the problem.

The use of OpenStreetMap (OSM) data to generate socio-economic indicators and urban crime risk has been studied and testified (Feldmeyer et al. 2020; Cichosz 2020). The data processing method can be used for reference, and it showcases that POI data can be a good indicator of urban conditions and activities. Urban POI data analysis can also be integrated with other methods of data collection. POI data, location-based service positioning data, and street view images are used in conjunction to measure greenway suitability and give suggestions on greenway networks planning (Tang et al. 2020).

Objectives
This research aims to explore using open-source city Point of Interest (POI) data to predict the health status of the residents using machine learning methods. By modeling the Point of Interest (POI) data and the geographic distribution of health-related outcomes, the research evaluates the key factors in urban planning that support the residents' health and wellbeing. The data processing and modeling methods inform decision-making to support a healthier built environment and open up possibilities for other data-driven methods.

METHODOLOGY
The workflow of this research follows five steps. In the first step, POI data from OpenStreetMap for the study area were collected and spatially joined with the health-related outcomes data within the census tract boundary. Second, Principal Component Analysis (PCA) and correlation analysis were conducted after some initial data cleaning. Third, a set of machine learning models were trained, and feature importance was calculated for feature selection. Fourth, the selected features were used to train machine learning models, and the results evaluated. Finally, the effect of modification of input variables on output variables was interpreted.

Data Source
The data source constitutes two data sets, the POI data

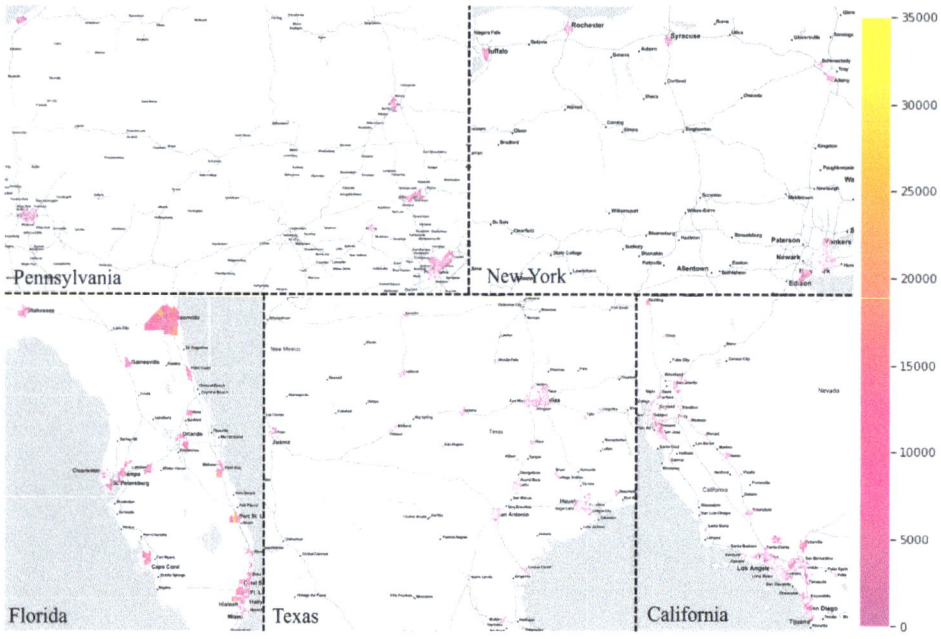

1 Sample area and spatial distribution of population

from OpenStreetMap, and 500 Cities Project dataset from the CDC. The test regions are within the five most populous states of the United States—California, Texas, Florida, New York, and Pennsylvania—and are a compromise between data availability, handling capacity, and statistical accuracy.

- 500 Cities: Local Data for Better Health. 500 Cities is a project that provides city and census tract-level small area estimates for chronic disease risk factors, health outcomes, and clinical preventive services use for the largest five hundred cities in the United States (CDC n.d. "500 Cities Project"). The dataset is generated with an innovative peer-reviewed multilevel regression and poststratification (MRP) approach that links geocoded health surveys and high spatial resolution population demographic and socioeconomic data. The twenty-eight measures include four unhealthy behaviors, fourteen health outcomes, and ten prevention practices. The measures include major risk behaviors that lead to illness, suffering and early death related to chronic diseases and conditions, as well as the conditions and diseases that are the most common, costly, and preventable of all health problems (CDC n.d., "Places"). The population size of each census tract is also included as a column in the dataset as well as the census tract boundaries. These small area estimates allowed cities and local health departments to understand better the geographic distribution of health-related variables in their jurisdictions and assisted them in planning public health interventions.
- OpenStreetMap is an open-source database with volunteers mapping geographic elements of the world. It represents physical features on the ground using tags attached to its basic data structures (its nodes, ways, and relations) (Feldmeyer et al. 2020). The research uses Overpass API to query the database by tags to get the geographic location of certain features. Fifty features were initially selected—based on the relation with physical wellbeing and the abundance of data points—and can be categorized into food, healthcare, transportation, community service, leisure, tourism, building, nature, and shop. Since the OpenStreetMap is user-generated data, there are various levels of completeness and accuracy of features. Overall, the data quality in the U.S. is good enough for POI analysis (Idham Muttaqien 2017; Barrington-Leigh and Millard-Ball 2017).

The two datasets are spatially joined within the boundary of each census tract. There are in total 12588 rows; namely, 12588 census tracts were sampled. As Figure 1 shows, the sample area consists of the major cities within each state. Most census tracts have a moderate population while some have a substantial population.

Two columns are added—a column describing the total count of POI and another column describing the total number of POI categories available within each census tract—to get a better understanding of the density and diversity of POI points.

Feature Extraction
POI Data Exploration and Preparation
Some initial data exploration showed that there are rows

with zero or minimal data points. The lower quartile of rows with a total number of POI less than 12 were dropped from the dataset since it provided little information. The reason for lacking data points could be due to the completeness of OSM or fewer activities within the area. The model performance improved after dropping this part of the data.

A boxplot (Figure 2, upper left) shows that most census tracts have a moderate number of POI while there are also many outliers with a large number of POI, representing larger and denser census tracts.

Considering the different sizes and densities of the census tracts, we divide the POI count by population and area and get the POI counts per capita and per unit area. A principal component analysis (PCA) shows different patterns of cumulative explained variance for each data manipulation method (Figure 2, upper right, lower left and right). For the original total POI count data, a few features can explain most of the variance. The curve is flatter for the per-unit-area data, which means that variance is more spread out. The per-capita data has the best balance between redundancy and bias. It is more suitable for feature selection. It also makes more sense to evaluate the number of POIs serving the designated population.

Health Data Exploration and Preparation

A boxplot (Figure 3) of the test data shows that the health data have different ranges, including some outliers. Since there are no explicit patterns, we use Tukey's rule to remove outliers and set the outer range to three-interquartile range (IQR). The outliers are assigned either to the upper fence value or lower fence value.

A mapping of the health outcome shows that there are some spatial patterns. As in Figure 4, the prevalence of coronary heart disease is higher in Florida, where there is a higher percentage of older population. The prevalence is lower in New York City, where there are more young people (United States Census Bureau n.d.).

Correlation Analysis

A Pearson correlation coefficient heat map, as on Figure 5, showed correlations among input variables both within and among categories. The category of food has a strong correlation within itself. It also correlates with public transportation, bike parking, hotels, and shops. The healthcare category does not connect with other features. The transportation category has some correlation within itself and with leisure and tourism categories. Leisure has some correlation within itself except for the swimming pool criteria; as a category, it correlates with

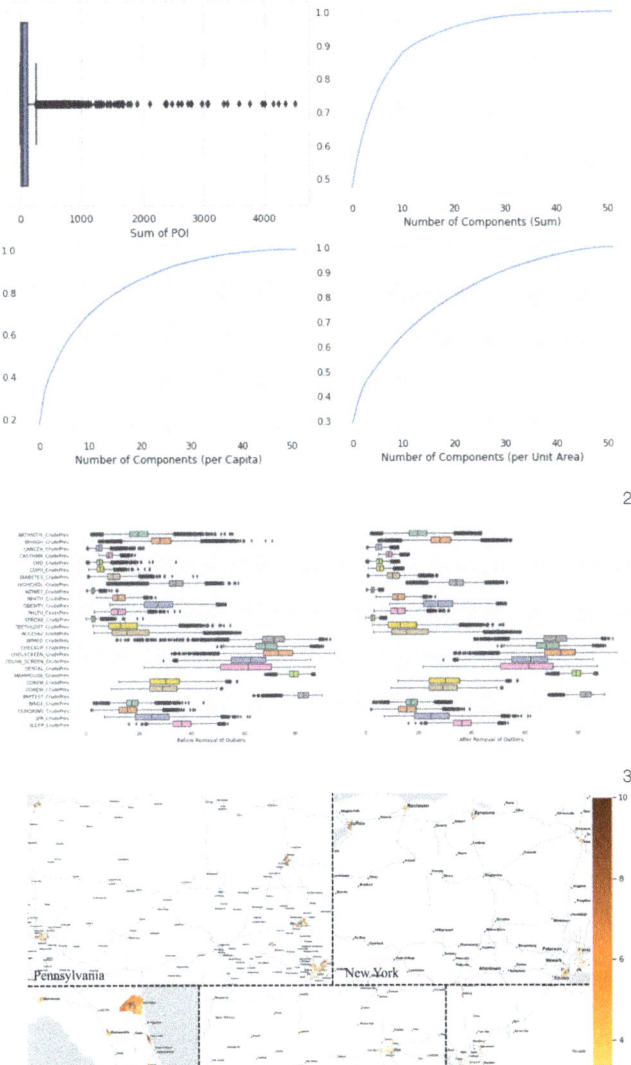

2 Upper left: Distribution of total POI counts; Upper right: Cumulative explained variance for POI counts; Lower left: Cumulative explained variance for POI counts divided by population; Lower right: Cumulative explained variance for POI counts divided by area.

3 Distribution of health data before and after removal of outliers

4 Spatial distribution of prevalence of CHD

transportation, tourism, and moderately with nature. There is some correlation between commercial, industrial, retail, and apartment buildings, and the building category does not correlate with other categories. Water correlates with wetland within the nature category. Shops correlate with food services and moderately with transportation.

Hierarchical clustering groups similar objects into clusters and helps us to understand the relationships among

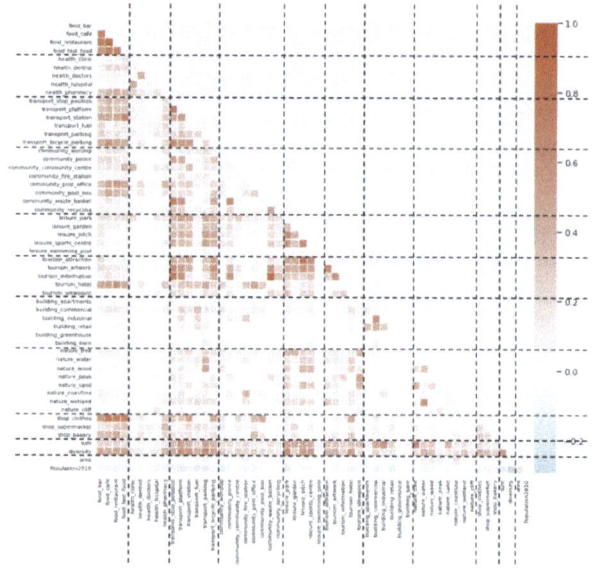

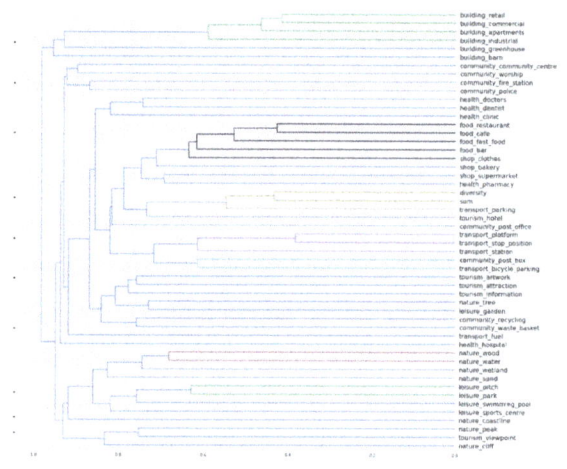

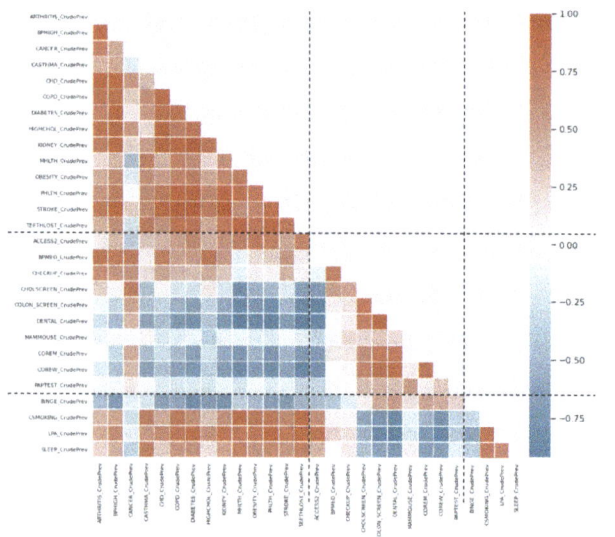

5   Correlation heatmap of input variables

6   Hierarchical clustering of input variables

7   Correlation analysis of output variables

features. As shown in Figure 6, there is a similarity between building_retail and building_commercial. They also belong to the same cluster with building_apartment and building_industrial, indicating the building density of an area. Food_restaurant and food_cafe have similarities, and they also belong to the same cluster with other food services and shop_clothes, which might indicate the prosperity of retail in the area. Diversity, sum, and transportation_parking belong to the same cluster, indicating the overall abundance of POI. Transport_platform, transport_stop position, and transport_station are within the same cluster, indicating the accessibility by public transportation. There is also a similarity between community_post_box and transport_bicycle_parking. Nature_wood and nature_water belong to the same cluster, which might indicate natural resources. Leisure_pitch and leisure_park have similarities that might indicate the recreation and outdoor activities within an area.

Among the output variables, there is overall a strong correlation, as shown in Figure 7. Bad health outcomes correlate positively within themselves and unhealthy behaviors and negatively with prevention measures. However, cancer seems to have a negative correlation with some bad health outcomes while positive with preventions. Within the prevention group, taking blood pressure medicine and routine checkup in previous years positively correlates with harmful health outcomes, while others correlate negatively. Within the unhealthy behaviors group, binge drinking correlates negatively with bad health outcomes and positively with preventions, which might indicate a correlation of better health and socioeconomic status (CDC 2012).

Machine Learning

An initial model training was conducted for model selection. We implemented six machine learning models, as follows. A Random Prediction model is implemented with random values within the test data range are generated. Then a Linear Regression is conducted as a basic statistical prediction. A Decision Tree model is a simple non-linear machine learning algorithm, where the data is continuously split according to a specific parameter. The Random Forest model is an ensemble learning method that operates by constructing many decision trees at training time. K-nearest Neighbors algorithm is a non-parametric classification method that returns the average values of k-nearest neighbors. Artificial Neural Network (ANN) is a deep learning method that digitally mimics the human brain to predict values. A 5-layer neural network is used in the research, and a training step of 4000 achieves the best accuracy.

Model	Mean Absolute Error (Before/After)	Median Accuracy (%) (Before/After)
Random Prediction	0.2567/0.2572	77.20/77.09
Linear Regressor	0.1154/0.1151	90.62/90.53
Decision Tree Regressor	0.1359/0.1368	89.51/89.425
Random Forest Regressor	0.0961/0.0961	92.33/92.34
K-Neighbors Regressor	0.1129/0.1111	90.85/90.99
Artificial Neural Network	0.1003/0.0987	91.98/92.26

Table 1   Model Performance before and after feature selection

Since the data were standardized into the range of 0 to 1, the mean absolute error could represent the performance of the model. The median accuracy was calculated for each model with the test dataset. An average accuracy rate for the 28 predictions is calculated as the output.

Furthermore, as there are correlations among features, a permutation method is used for calculating the feature importance of the Random Forest model instead of the built-in impurity method of sklearn. The idea is to permute the values of each feature and measure how much the permutation decreases the accuracy of the model. Therefore, the respective importance of correlated features will not be shared with one another, and scores will not be reduced (Strobl et al. 2007).

We selected the features that had an importance score over 0.005 and retrained the model from the result of the feature importance calculation. The performance of the model improved slightly. We also did an R-squared score analysis for each of the output variables to evaluate to what extent the model's variance can be explained. Moreover, a partial dependence plot was used to understand the relation between deciding features and the output variable. The plot was implemented with sklearn plot_partial_dependence. We also mapped the prediction accuracy to see if there is any spatial pattern for prediction accuracy. Lastly, input variables were changed by a certain percentage, and the effect on the output variable was measured and compared.

RESULT AND DISCUSSION
Model Evaluation
As can be seen from Table 1, before feature selection, the Random Forest model has the lowest mean absolute error and highest median accuracy.

After feature selection, the ANN model's performance has increased the most because of removing irrelevant features and avoiding overfitting. The performance of K-Neighbors Regressor increased by a little. The version of Random Forest Regressor stays the same and still has the best performance among all the models. The performance of Decision Tree Regressor, Linear Regressor, and Random Prediction decreased by a little, probably because of fewer input variables.

A plot of different models' performance on each output variable (Figure 8) shows that machine learning models

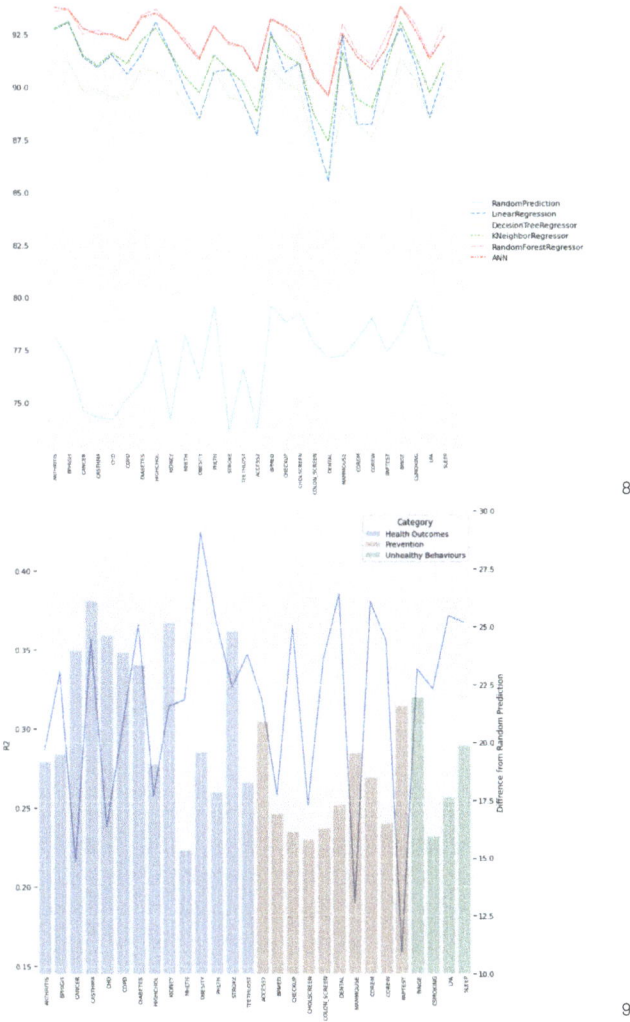

8   Model Performance on All the Output Variables

9   Outperformance over Random Prediction and R-squared Score of Random Forest Model on All the Output Variables

perform better than random prediction for all health outcomes. The Random Forest model has the best performance on all the output variables.

The overall R-squared score for Random Forest Model is 0.3088, which means that the model could explain about one-third of the variance of the health outcomes. This result is in line with the perception that part of people's health depends on their lifestyle behaviors and environmental exposure. As on Figure 9, a plot of the R-squared score on each output variable shows that the model could best explain the variance for the prevalence of obesity, which has an R-squared score of 0.4239.

As in the bar plot part of the right image of Figure 9, we calculated the percentage of the increased model performance of the Random Forest model compared to random prediction. The model has better prediction accuracy for health outcomes than prevention and unhealthy behaviors, especially for the prevalence of cancer, asthma, coronary heart disease (CHD), chronic obstructive pulmonary disease (COPD), diabetes, chronic kidney disease, and stroke.

A mapping of the difference of prediction for CHD from ground truth shows that most of the prediction is close to the ground truth, as is shown in Figure 10. There is no clear pattern of overestimation or underestimation spatially.

Feature Importance

As shown in Figure 11, the permutation feature importance of the Random Forest model shows that community_worship has by far the highest importance, followed by area, food_cafe, leisure_swimming pool. The next tier includes building_apartments, transport_bicycle_parking, food_restaurant, transport_platform, population, POI diversity, natural_water, and POI sum. The other features that influence the model accuracy are transport_parking, building_retail, building_industrial, transport_stop_position, building_commercial, transport_fuel, leisure_park, community_post_box, leisure pitch, community_police, and natural_wood.

By adding the feature importance score of each category, as shown in Figure 12, we can see that community service has the highest cumulative feature importance, followed by food, leisure, transport, building, and nature. In contrast, tourism, health, and shop categories have little impact on the model's accuracy. Also, apart from places of worship, the rest of the community category does not have a lot of impact on the model's accuracy.

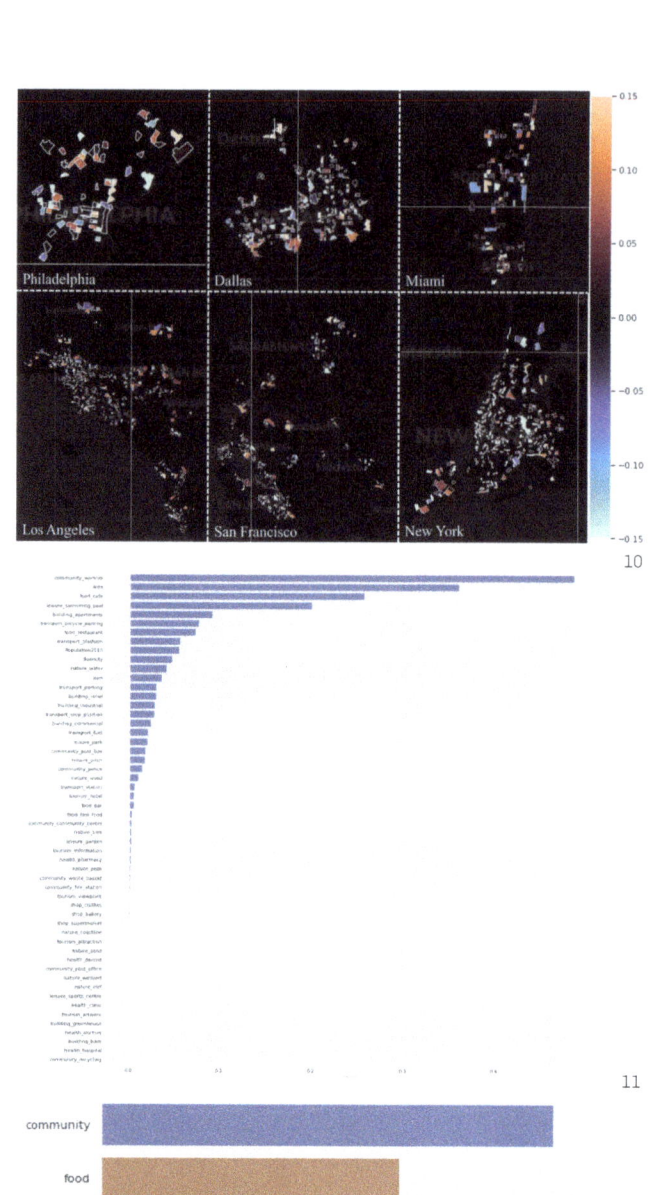

10

11

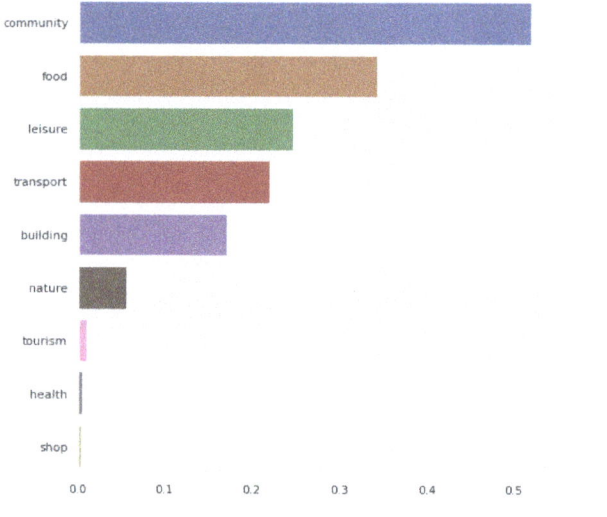

12

10 Spatial distribution of prediction difference from ground truth for CHD

11 Feature importance of random forest regressor for CHD

12 Accumulated feature importance for eight POI categories

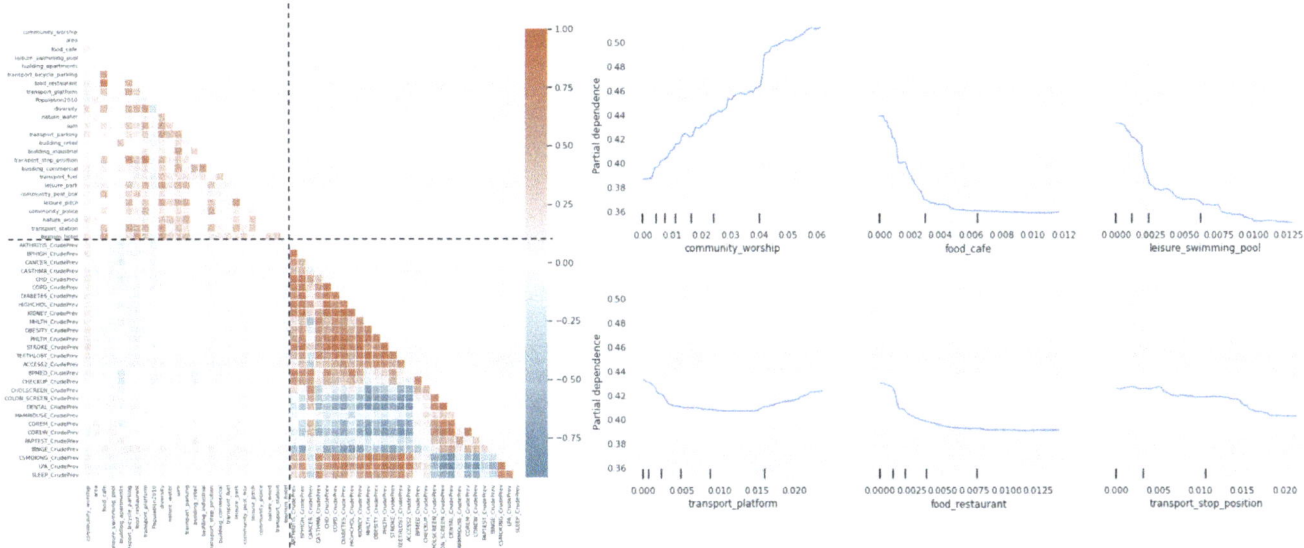

13 Left: Correlation between selected features and output variables; Right: Partial dependence plot of major input variables and prevalence of obesity

The four most important categories support the theory that physical activity, community interaction, and healthy eating could best support health and wellbeing (Kent and Thompson 2014). Building density and use have some influence on health outcomes. Because the sample area is primarily within the urban context, nature's influence is relatively small. It is counterintuitive to see that the health category has little prediction power on residents' health.

Partial Dependence Analysis

The heatmap of selected input data and output data (Figure 13, left) shows no apparent correlation between the input and output data. A slight positive correlation between community worship per capita and bad health outcomes compared to many other features has a slight negative correlation with bad health outcomes. Previous research shows that religions are relevant for poverty. More churches in the community might indicate lower socioeconomic status, which is relevant for worse health outcomes, especially as a source of normativity and motivation. The twofold influence is particularly exercised in the three areas of business and finance, politics and culture, education, and health (Sedmak 2019).

A partial dependence plot (Figure 13, right) depicts how the predictions partially depend on values of the input variables of interest (Wright 2018). In this case, we plot the relationship of the prediction of obesity and the six features with the highest feature importance scores. The prediction prevalence of obesity increases as places of worship per capita rise and decreases as the number of cafés, swimming pools, and restaurants increases. However, not many data points with large values are available, so the reliability of the estimates decreases in those areas. The prediction decreases as the number of stop positions increases, implying that public transportation has a health benefit. The prediction decreases and increases as the number of platforms increases, indicating a difference in urban landscape related to railways.

Modification of POI data and the Change of Prediction

Then we modified several POI features and made the prediction again. By decreasing the number of places of worship, we got a new prediction for the prevalence of obesity on the test set. As shown in Figure 14, most predictions go down while some predictions go up, which reflects the non-linearity of the model. On average, the prediction of the prevalence of obesity decreased by 0.29%. By increasing the number of cafés by 10 percent, the average prediction decreased by 0.05%, and by increasing the number of swimming pools by 10%, the average prediction dropped by 0.09%. The effect of increasing the amount of stop positions was to move the prediction down by 0.03%.

The result is in line with that of the partial dependence plot. The influence of the number of places of worship is still greatest, followed by the number of swimming pools, cafés, and stop positions. We think that it is a result of both the socioeconomic status and the lifestyle that these features indicate (Gelormino et al. 2015). Although there is no direct causality established, it is noticeable that these factors are not related to the medical care system as usually expected. Urban planners could use this machine learning prediction as a tool to evaluate quantitatively which measures potentially have the best return on investment. It is also an effective way to create a consensus among participants on what aspects of the built environment could be improved to support residents' health.

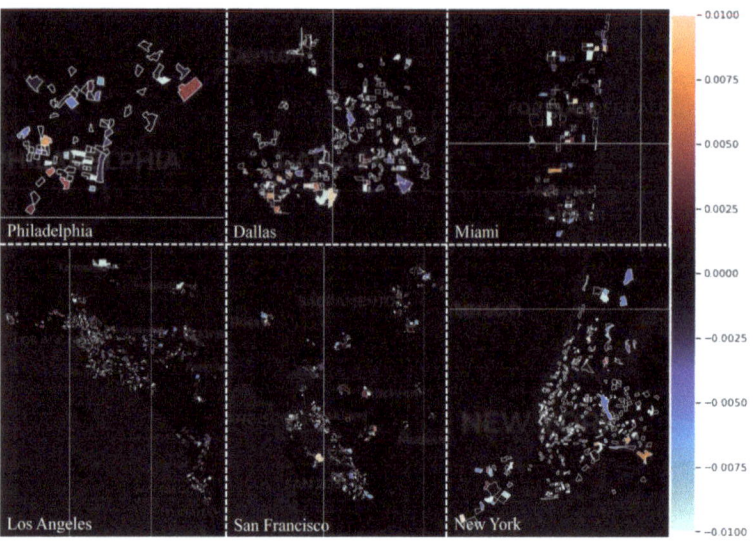

14 Change of prediction after decreasing the number of worship places per capita by 10%

14

CONCLUSION AND DISCUSSION

This research explores a method to predict the health outcomes of residents by using the POI data available from OSM, thus exploring the linkage between the built environment and the health of residents. Different machine learning methods were used and evaluated. The result shows that the Random Forest model strikes the best balance between prediction accuracy and ease of implementation. The interpretation of the machine learning model suggests some critical features that potentially influence residents' health. The result supports as well as complements the existing healthy urban design theory.

This research showcased the possibility that big data analysis could benefit the urban design process. Urban planners could use this method to discover and support ideas that improve the health status of residents. With the increasing availability of data and computational resources, there is considerable potential that data could change how a healthy city is designed and become common ground for decision-making.

For the next step of research, a more comprehensive analysis of different urban characteristics would be helpful, for instance, the building code, density, vehicular and pedestrian traffic environments, and medical state benefits. Comparing other countries would also give a more complete picture of the most important factors for residents' health within different contexts.

One of the challenges of this research is the data source. Although the quality of OpenStreetMap data in the U.S. is relatively good, it is not complete everywhere (Idham Muttaqien 2017). The reliability of POI data can affect the final result. In the future, if we have more comprehensive and real-time data, predictions could be more accurate. Another challenge is to understand the correlation and causality between built environment and health. The built environment results from socioeconomic factors, and it also shapes the way people live (Williams et al. 2009). It is essential to have a case-by-case understanding of how a specific feature influences the health status of residents. As more relationships between the built environment and health may be established through data, there is a challenge and an opportunity for traditional urban planning practice. By making data more explicable, the decision-making process could become more transparent. The community would be empowered to make healthier decisions for themselves in the long run.

ACKNOWLEDGEMENTS

The code of the research is available at: https://github.com/shicong0720/Finding-Healthy-Community-Design-Criteria-with-AI

REFERENCES

Barrington-Leigh, Christopher, and Adam Millard-Ball. 2017. "The World's User-Generated Road Map Is More than 80% Complete." *PLoS ONE* 12 (8): 1–20. https://doi.org/10.1371/journal.pone.0180698.

Battisto, Dina, and Jacob J. Wilhelm. 2019. *Architecture and Health: Guiding Principles for Practice.* London: Routledge.

Centers for Disease Control and Prevention (CDC). 2012. "Vital Signs: Binge Drinking Prevalence, Frequency, and Intensity among Adults - United States, 2010." *Morbidity and Mortality Weekly Report* 61 (1): 14–19. http://www.ncbi.nlm.nih.gov/pubmed/22237031.

Centers for Disease Control and Prevention (CDC). n.d. "Places: Local Data for Better Health." Accessed 15 September 2021. https://www.cdc.gov/places/methodology/index.html.

Centers for Disease Control and Prevention (CDC). n.d. "500 Cities Project: 2016 to 2019." Accessed 15 September 2021. https://www.cdc.gov/places/about/500-cities-2016-2019/index.html.

Cichosz, Paweł. 2020. "Urban Crime Risk Prediction Using Point of Interest Data." *ISPRS International Journal of Geo-Information* 9 (7). https://doi.org/10.3390/ijgi9070459.

Duhl, L.J, A.K. Sanchez, and World Health Organization. Regional Office for Europe. 1999. *Healthy Cities and the City Planning Process: A Background Document on Links between Health and Urban Planning.* Copenhagen: WHO Regional Office for Europe.

Engin, Zeynep, Justin van Dijk, Tian Lan, Paul A. Longley, Philip Treleaven, Michael Batty, and Alan Penn. 2020. "Data-Driven Urban Management: Mapping the Landscape." *Journal of Urban Management* 9 (2): 140–50. https://doi.org/10.1016/j.jum.2019.12.001.

Feldmeyer, Daniel, Claude Meisch, Holger Sauter, and Joern Birkmann. 2020. "Using OpenStreetMap Data and Machine Learning to Generate Socio-Economic Indicators." *ISPRS International Journal of Geo-Information* 9 (9): 1–16. https://doi.org/10.3390/ijgi9090498.

Gelormino, Elena, Giulia Melis, Cristina Marietta, and Giuseppe Costa. 2015. "From Built Environment to Health Inequalities : An Explanatory Framework Based on Evidence." *PMEDR* 2: 737–45. https://doi.org/10.1016/j.pmedr.2015.08.019.

Idham Muttaqien, Bani. 2017. "Assessing the Credibility of Volunteered Geographic Information: The Case of OpenStreetMap." MS diss., University of Twente. https://webapps.itc.utwente.nl/librarywww/papers_2017/msc/gfm/muttaqien.pdf.

Kelley, Lee. 2008. "The World Health Organization (WHO)." The World Health Organization (WHO), no. July 1994: 1–157. https://doi.org/10.4324/9780203029732.

Macintyre, Sally, Anne Ellaway, and Steven Cummins. 2002. "Place Effects on Health: How Can We Conceptualise, Operationalise and Measure Them?" *Social Science and Medicine* 55 (1): 125–39. https://doi.org/10.1016/S0277-9536(01)00214-3.

Newton, David, Dan Piatkowski, Wesley Marshall, and Atharva Tendle. 2020. "Deep Learning Methods for Urban Analysis and Health; Estimation of Obesity." In *ECAADe 2020: Health and Materials in Architecture and Cities.* Vol. 1. 297–304.

Schroeder, Steven A. 2007. "We Can Do Better — Improving the Health of the American People." *New England Journal of Medicine* 357 (12): 1221–28. https://doi.org/10.1056/NEJMsa073350.

Sedmak, Clemens. 2019. "Evidence-Based Dialogue: The Relationship between Religion and Poverty through the Lens of Randomized Controlled Trials." *Palgrave Communications* 5 (1): 1–7. https://doi.org/10.1057/s41599-019-0215-z.

Strobl, Carolin, Anne Laure Boulesteix, Achim Zeileis, and Torsten Hothorn. 2007. "Bias in Random Forest Variable Importance Measures: Illustrations, Sources and a Solution." *BMC Bioinformatics* 8. https://doi.org/10.1186/1471-2105-8-25.

Tang, Ziyi, Yu Ye, Zhidian Jiang, Chaowei Fu, Rong Huang, and Dong Yao. 2020. "A Data-Informed Analytical Approach to Human-Scale Greenway Planning: Integrating Multi-Sourced Urban Data with Machine Learning Algorithms." *Urban Forestry and Urban Greening* 56 (January). https://doi.org/10.1016/j.ufug.2020.126871.

United States Census Bureau. n.d. "Quick Facts." n.d. Accessed 15 September 2021. https://www.census.gov/quickfacts/fact/table/US/PST045219.

Williams, Oli, Teresa Lavin, C. Higgins, Margaret Kelaher, Deborah J. Warr, Theonie Tacticos, Terrence D. Hill, et al. 2009. "Health Impacts of the Built Environment." *Institute of Public Health in Ireland.* Vol. 15.

Wright, Ray. 2018. "Interpreting Black-Box Machine Learning Models Using Partial Dependence and Individual Conditional Expectation Plots," 1950–2018. Accessed 15 September 2021. https://www.sas.com/content/dam/SAS/support/en/sas-global-forum-proceedings/2018/1950-2018.pdf.

Zhou, Hao, Shenjing He, Yuyang Cai, Miao Wang, and Shiliang Su. 2019. "Social Inequalities in Neighborhood Visual Walkability: Using Street View Imagery and Deep Learning Technologies to Facilitate Healthy City Planning." *Sustainable Cities and Society* 50 (129): 101605. https://doi.org/10.1016/j.scs.2019.101605.

IMAGE CREDITS
All drawings and images by the authors.

Shicong Cao is a registered architect and researcher currently working at Heinle Wischer und Partner in Berlin. Her research interests focus on artificial intelligence and the healthy built environment. She holds a Master's degree in Architecture + Health from Clemson University, a Master's degree in Architecture from Politecnico di Milano and a bachelor's degree in Civil Engineering from Tongji University.

Hao Zheng is a PhD Candidate at the University of Pennsylvania, Stuart Weitzman School of Design, specializing in machine learning, digital fabrication, mixed reality, and generative design. He holds a Master of Architecture from the University of California, Berkeley, and Bachelor of Architecture and Arts degrees from Shanghai Jiao Tong University.

Data-Driven Acoustic Design of Diffuse Soundfields

Self-Organizing Maps as an Exploratory Design Tool for Big Data

Achilleas Xydis
ETH Zurich

Nathanaël Perraudin
Swiss Data Science Center

Romana Rust
ETH Zurich

Beverly Ann Lytle
ETH Zurich

Fabio Gramazio
ETH Zurich

Matthias Kohler
ETH Zurich

1 3D visualization of a portion of GIR Dataset's diffusive surfaces demonstrating possible geometric variations. For scale reference, a human is placed at the bottom left part of the image.

ABSTRACT

The paper demonstrates a novel approach to performance-driven acoustic design of architectural diffusive surfaces. It uses unsupervised machine learning techniques to analyze and explore the GIR Dataset, an extensive collection of real impulse responses and acoustically diffusive surfaces. The presented approach enables designers to explore many alternative acoustically-informed material patterns with various diffusive properties without requiring expert knowledge in acoustics. The paper introduces the computational pipeline, describes the used methods, and presents two use-cases in the form of design experiments. Finally, the paper discusses the challenges of developing such a method, its advantages, limitations, and future work.

INTRODUCTION

During the design phase, architects examine a wide range of alternative design ideas. Early-stage design decisions significantly impact the final design's performance, whereas late-stage design modifications can rarely compensate for poor early-stage choices. In fundamental building components such as structure or façade, performance is an integral design driver, which is included early on. Usually, this is done in close collaboration with experts in an iterative process where a design is analyzed, evaluated, and adjusted to meet the desired performance criteria. This process has become a standard practice for most architectural projects because architects are trained to understand these topics (e.g., structural design). However, room acoustics are rarely included in the early design process, even though they significantly impact our perception of space and well-being. Apart from cases where sound quality is critical (e.g., concert halls, auditoriums), acoustics are either not included in the design process or come as an afterthought relying on standardized solutions in the form of absorbent or diffusive panels.

The acoustic quality of a room is determined by its geometry and the structure or pattern of its surfaces. Slight manipulations of the surface geometry could yield significant gains in acoustic quality (Cox and D'Antonio 2004). Currently, through computational design and digital fabrication, architects already design, visualize, and fabricate surfaces with complex geometries. While these geometries are not designed with acoustics in mind, they could act as sound diffusers, enhancing the room's acoustic qualities. Diffusive surfaces reflect sound in multiple directions and, by doing so, reduce echoes, standing waves, and sound coloration while promoting spaciousness. Suppose acoustics were included as a design criterion. In that case, these complex geometries could be an integral part of architectural elements and combine multiple acoustic properties, targeting the acoustic needs of their immediate surrounding.

To employ acoustic performance as a design driver, we must be able to quantify and interpret the acoustic effects of our geometric design choices. In a classical design process, architects have no starting point for an acoustically performative design of surfaces as they lack expert knowledge. Different computer simulation software (Odeon,[1] Pachyderm,[2] CATT-acoustics[3]) can be used to analyze and characterize the acoustic performance of digitally designed geometries. Nevertheless, this paradigm relies on the premise that the user is knowledgeable in room acoustics and knows what adjustments need to be made to achieve the desired goal. As a result, architects are discouraged from using such software to evaluate their design, especially early on. Furthermore, no CAD nor acoustic simulation software exists that proposes a geometrical solution to an acoustical question. Further effort is needed to increase acoustic performance awareness in architecture and provide architects with simple and accessible workflows for designing diffusive surfaces.

Machine Learning (ML) has enabled significant breakthroughs in automated data processing and pattern recognition within various fields (Voulodimos et al. 2018; Arulkumaran et al. 2017; Niu et al. 2019). Architecture and engineering have also seen an increase in research on how to employ ML techniques in performance-based design (Tamke, Nicholas, and Zwierzycki 2018), style transferring (Steinfeld 2017; Gatys, Ecker, and Bethge 2015), and clustering (Saldana Ochoa et al. 2020). In ML techniques, the quality and size of the dataset heavily influence the ML model's final quality (Mirvis 2002). Although a larger dataset is desirable—as it makes for a more confident prediction—the larger the dataset is, the more challenging it is to navigate, especially for non-expert users. Given its success in other fields, ML is also used in acoustics research, mainly as a predictive tool. Datasets built for this purpose could also be explored as a knowledge base of known acoustic properties. This paper combines data clustering techniques with a large dataset of geometries and impulse responses. It provides an exploratory design tool for diffusive surfaces, bringing acoustic performance-based evaluation earlier into the design stage.

BACKGROUND
Acoustics

In recent years, significant research has been carried out on acoustic performance-based design. Shtrepi et al. (2020) presented a design process that provides architects and designers with rapid visual feedback on the acoustic performance of diffusive surfaces. Peters (B. Peters 2015; Isak Worre Foged 2014) demonstrated methods that allow tuning acoustic performance while geometry and materials change. Badino et al. (Badino, Shtrepi, and Astolfi 2020) presented the state-of-the-art of acoustic performance-based design application in practice using nineteen built projects. Most of these projects were conducted by big architectural firms in collaboration with expert acoustic consulting groups but were only geared towards spaces intended for music performance. Several computational tools exist that enable the design and optimization of acoustically diffusive surfaces. However, their primary focus is phase grating surfaces (stepped diffusers, quadratic residue diffusers, primitive root diffusers) (Cox and D'Antonio 2000), based on sound diffusers introduced by

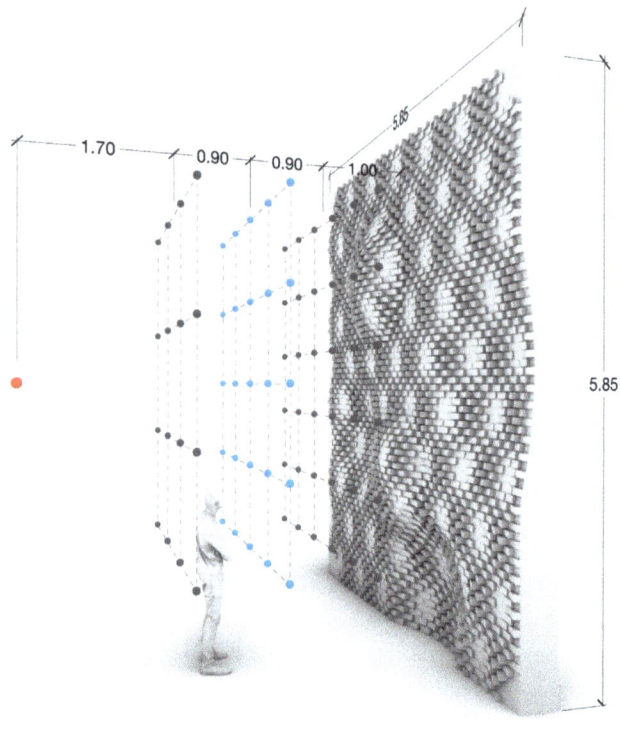

2

Schroeder (Schroeder 1975). Although these tools simplify the acoustic design process, the generated diffusers have particular and limited geometries, a substantial thickness, and a dedicated placement according to acoustic criteria. These factors make them unattractive and difficult to integrate into an architectural design that is not purely focused on music performance.

Machine Learning

The main ML applications in room acoustics have focused on characterization, information extraction, or classification. For example, ML has been used to extract the Reverberation Time and the Early Decay Time of a room from music signals (Gamper and Tashev 2018) and the room volume (Genovese et al. 2019). Peters et al. and Papayiannis et al. (N. Peters, Lei, and Friedland 2012; Papayiannis, Evers, and Naylor 2020) presented methods to identify the room type from an audio recording.

Most of the contributions above used supervised learning, which generally requires large amounts of labeled data. Data such as impulse response, absorption and scattering coefficient, early decay time, and many more are primarily quantitative in nature, therefore, hard to evaluate by non-acousticians. Moreover, architectural design is often focused on qualitative measures that depend on the application context and the designer's personal preferences. Alternately, unsupervised learning allows the extraction of information from data even when no labels are available.

For example, dimensionality reduction organizes high-dimensional data samples in a lower-dimensional space—also known as embedding—by clustering similar samples together. A high-dimensional space contains data samples with multiple attributes; for example, an image with a resolution of just 100 by 100 pixels is a 10000-dimensional sample if we view each pixel within the image as an attribute. Classical dimensionality-reduction techniques include Principal Component Analysis (Wold, Esbensen, and Geladi 1987), t-SNE (Van der Maaten and Hinton 2008), or Self-Organizing Maps (SOM) (Kohonen 1982). SOMs have been successfully employed in several fields such as environmental studies (Gorgoglione et al. 2021), cancer research (Mazin et al. 2021), chemistry (Motevalli, Sun, and Barnard 2020), structural design (Saldana Ochoa et al. 2020), and architectural design (Kobayashi 2006). SOMs are particularly useful in this context. They use unsupervised training to create a nonlinear data transformation of a high-dimensional space to a low-dimensional space (usually a two-dimensional map) while preserving the topological relationships of the original high-dimensional space (Moosavi 2017). Topology preservation implies that if two data points are close in the high-dimensional space, they must also be near each other in the new low-dimensional space and, therefore, belong to the same cluster. This reduction in complexity makes it possible for designers to associate a qualitative measure on the embedding.

Dataset

As mentioned in the introduction, the success of ML techniques relies heavily on the quality and size of the dataset they use. Several acoustic datasets exist containing room impulse responses (IRs), but their main application is in the field of speech enhancement and speech recognition: AIR,[4] BUT ReverdDB,[5] RWCP[6] (Jeub, Schafer, and Vary 2009; Szoke et al. 2019; Nakamura et al. 2000); acoustic environment characterization, ACE Corpus[7] (Eaton et al. 2017); or for smart-home applications, DIRHA[8] (Ravanelli et al. 2015). Furthermore, these datasets do not contain any three-dimensional geometrical data. The open-sourced GIR Dataset[9] (Xydis et al. 2021), an extensive collection of three-dimensional diffuse surfaces and their corresponding real impulse responses, was recently released. It can be used for ML applications to predict the acoustic properties of three-dimensional surfaces.

METHODS

As highlighted in sections 1 and 2, acoustic performance criteria are mainly considered in projects where spaces host music performances. Furthermore, current methods mainly focus on design optimization and heavily rely on expert knowledge in acoustics. This research presents

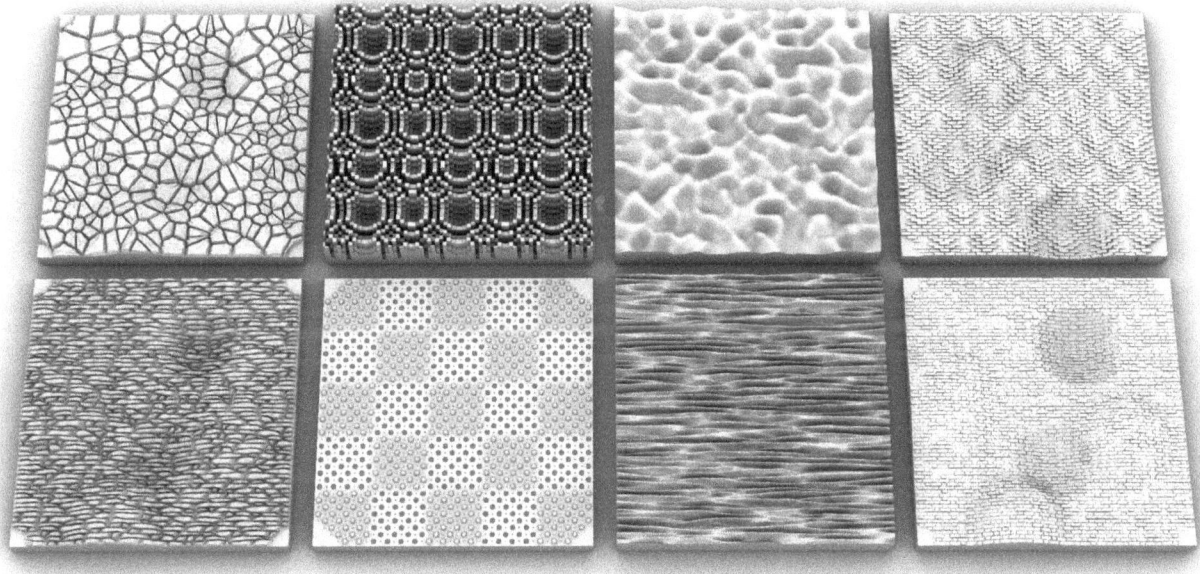

a workflow that enables architects to explore several possible design solutions, given specific acoustic performance criteria (energy per frequency band). It uses the GIR Dataset for its unique set of three-dimensional surfaces and the high number of real IRs per surface. Machine learning techniques and specifically SOMs are used to cluster the surfaces based on acoustic performance criteria.

The GIR Dataset
The GIR Dataset contains 873,496 real impulse responses from 296 surfaces (2951 per surface), spread in three layers (Figure 2). Layer_0 contains 36 measurements in a 6x6 grid and is the closest to the surface at a distance of 1 meter. Layer_1 and layer_2 contain 25 and 16 measurements in a 5x5 and 4x4 grid at a distance of 1.9 meters and 2.8 meters, respectively. The IRs were captured inside a semi-anechoic room and time-windowed only to contain the first reflections. The surfaces of the dataset resemble architectural material systems and are arranged in nine typologies such as brick walls, stone walls, and more (Figure 3). The geometry of each surface is composed of a microstructure and a macrostructure. The first defines the typology and the second its overall shape. Several typology-specific material and construction characteristics are coded in the geometry generation algorithm and used to create different material patterns. The brick dimensions, its rotation along the *z-axis*, its shift along the macrostructure's normal vector, and the width and depth of the mortar are used for the brick typologies.

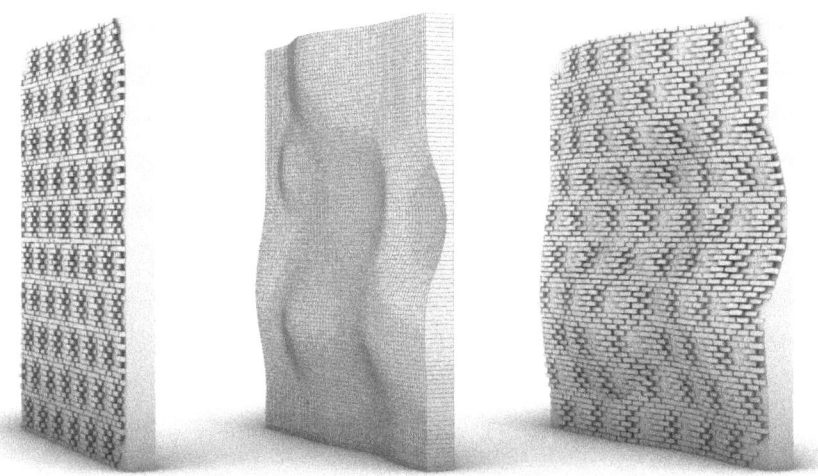

2 The measuring grid in front of a surface. Red represents the source position and blue the selected receivers' layer. The source is located in the center of the surface and 4.5 meters away from it. The receivers' layer (layer_1) is approximately 1.9 meters away from the surface.

3 A sample of different surface typologies. From top left to bottom right: Polygonal rubble stones, PRD diffuser, IDL, Stretcher bond bricks, Coursed ashlar stones, Primitives, IDL, Flemish bond bricks.

4 Micro-macrostructure. Left: A surface with only a microstructure (Stretcher bond bricks). Middle: A surface with only a macrostructure. Right: A surface that combines the micro- and macrostructure.

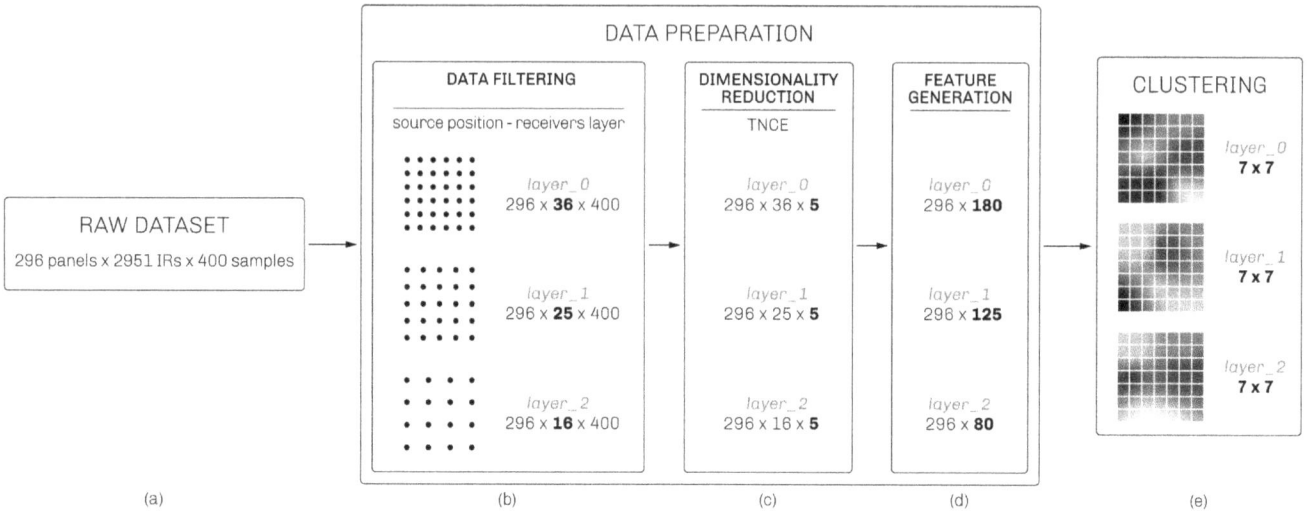

5 Data preparation pipeline

The number of stones per square meter is used for the stone typologies, along with the surface roughness and the joint depth between them. The macrostructure enhances the low-frequency diffusion by significantly increasing the depth variation (Figure 4).

PROPOSED DESIGN WORKFLOW

The proposed design workflow contains three main steps: data preparation, clustering, and design exploration (Figure 5). We use the impulse responses of the GIR Dataset to compute the primary performance criteria for our method. The large size of the dataset dictates the need for data reduction strategies. We use the open-sourced MiniSom Python library (Vettigli 2018) and create several custom data visualization algorithms for the clustering step. These algorithms provide an easy and understandable way to visualize complex and high-dimensional data and validate the quality and performance of several steps of our workflow. Finally, we describe using the trained SOM to explore design options based on given acoustic criteria.

Data Preparation

The principal challenge when constructing a low-dimensional embedding using a SOM is the size of a sample. Given that each of the patterns contains 2951 impulse responses of 400 float numbers, the total size of the raw feature vector is 1,180,400 (Figure 5a). We use three steps to reduce this large dimension. First, a source position is selected from the measuring grid, which yields 83 measurements spread across the three grid layers (Figure 5b). In the second step, we use the post-processing pipeline from Rust et al. (2021) to build low-dimensional feature vectors from the selected IRs. This pipeline removes the direct sound from the IR, retaining only the reflected sound coming from the surface. A custom-designed band-pass filter is used to split the above-mentioned acoustic descriptors into five frequency bands, with center frequencies at 250Hz, 500Hz, 1kHz, 2kHz, and 4kHz.[10] As a last step, we use the provided functions to convert the IR into total normalized cumulative energy[11] (TNCE). This step effectively reduces each IR's 400 samples to 5 numbers (Figure 5c). Finally, we concatenate the features of each pattern and obtain feature vectors of size 36x5, 25x5, 16x5 for layers 0, 1, and 2, respectively (Figure 5d).

Clustering

Clustering operations aim to group various design options into sets with similar features (in this case, TNCE). Analogous to the clustering methods used by (Saldana Ochoa et al. 2020; Fuhrimann et al. 2018), this paper proposes a method to cluster multiple design options based on their acoustic performance. Therefore, one can expect similar acoustic performance for all the designs of the same cluster. Such clustering can be used as a data-driven catalog that enables designers to explore the available design space based on acoustic criteria. The SOM algorithm organizes all the patterns on a two-dimensional plane. Figure 6 shows the embedding of 296 patterns based on TNCE values. As highlighted with the colored outline, the macrostructure is one of the most discriminative features for the SOM.

Design Exploration

Using the two-dimensional SOM described in the previous subsection, designers can get a fast and precise overview of possible design options. Each cell of the SOM contains a group of design options clustered based on the acoustic performance feature selected by the designer (e.g., TNCE). The hypothetical examples described below are used to illustrate the proposed design workflow.

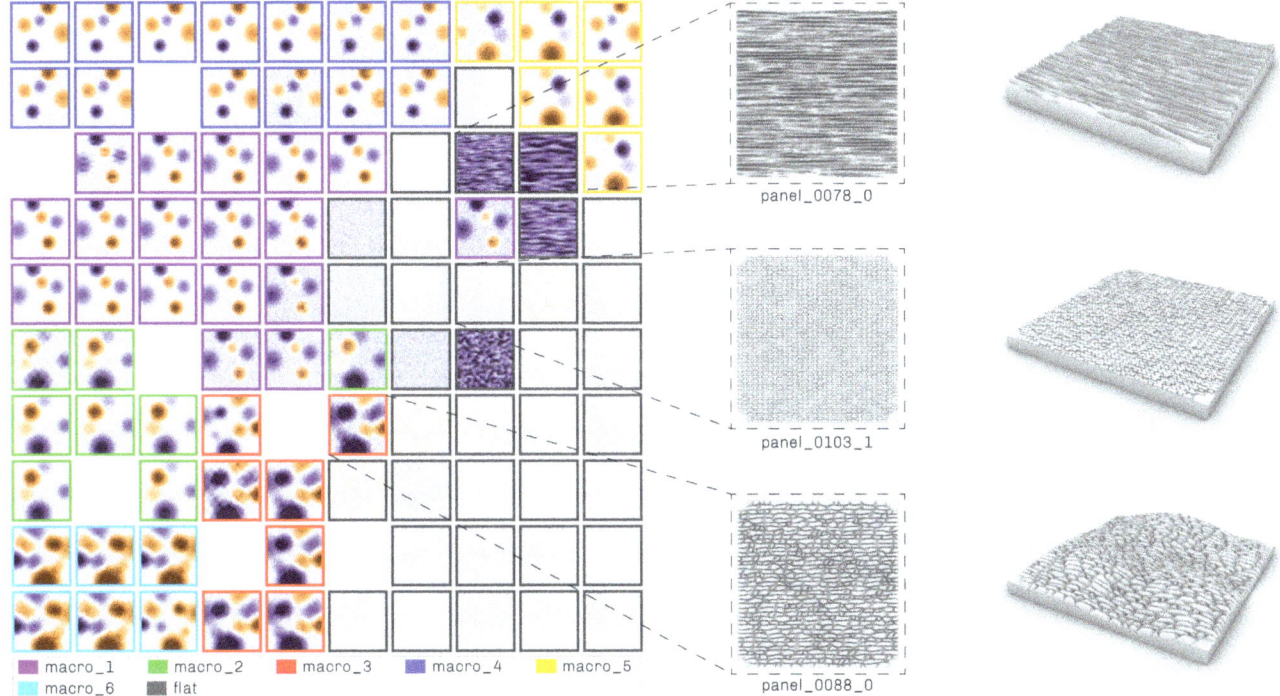

6 An SOM of 296 surfaces based on the TNCE values of layer_1. The displayed surfaces are colored from violet to orange to represent their depth and the colored outline indicates their macrostructure.

We imagine a generic meeting room where one of its walls may be freely designed to improve the room's acoustical properties. For our performance criterion, we choose the TNCE values of layer_1 because they are located very close to the center of the room. Because the IRs contain only early reflection information (see the GIR Data Set), the TNCE values also contain only the energy from these early reflections. Although the form of the room does not influence our method, for simplicity, the meeting room has a shoebox shape measuring 5 meters wide, 6 meters long, and 4 meters high. We consider the reflected energy of a flat surface as our 100 percent reference (maximum specular reflection). The criterion is the reflected energy of the desired surface, represented as a ratio of the flat surface's energy. Values higher than 100 percent represent amplification and lower values energy reduction.

Scenario A does not have a specific material system in mind, but scenario B assumes designers have already decided on a material system, specifically, a brick wall. These different decisions result in two different sets of panels for the SOM training. Scenario A uses all the dataset typologies, resulting in 279 surfaces, and scenario B only the brick wall typologies, resulting in 146 surfaces. For the SOM training, the MiniSom library requires us to provide values for the following arguments: map dimensions (x, y number of neurons), training iterations, the neighborhood function, the sigma, and the learning rate[12] (Table 1). Sigma defines the spread of the neighborhood function in number of neighbors. The appropriate value for sigma varies by map dimensions. When the sigma value is too small, the samples cluster near the center of the map; when it is too large, the map exhibits several large empty areas towards the center (Hearty 2016). The learning rate defines only the initial value of the learning rate for the SOM. With every training iteration, the learning rate adjusts according to the following function:

learning rate(t) = learning rate / (1 + t / (0.5 x iterations))

We iterated over different training values to achieve an optimum embedding (Figures 7 and 8). A SOM with many neurons has enough space to arrange the data samples. When multiple very similar samples exist, the SOM

	neurons		iterations	neighborhood function	sigma	learning rate	training time
	x	y					
scenario A	10	10	100,000	Gaussian	0.8	1.5	28 sec
scenario B	7	7	100,000	Gaussian	1.0	2.5	22 sec

Table 1: SOM training values

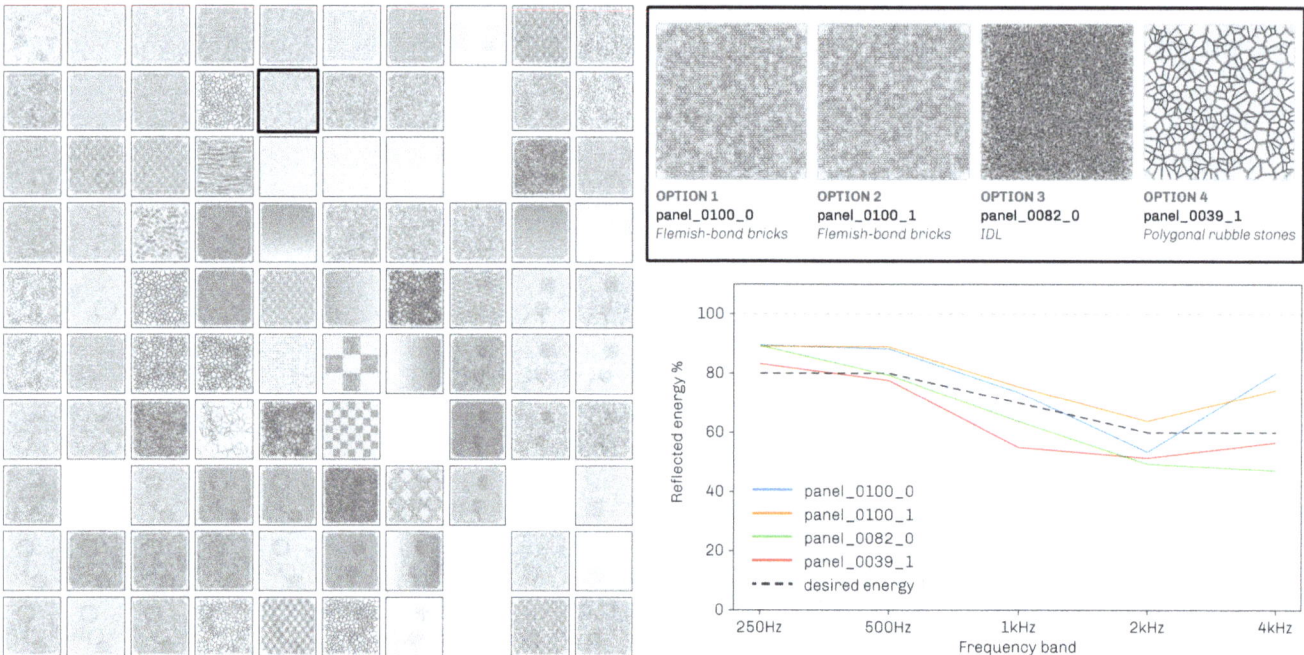

7   Left: The 10x10 SOM for Scenario A. The black outline indicate the best matching cell acording to the desired energy values.
    Right-top: The surfaces of the best matching cell. Right-bottom: Energy ratios of all matching surfaces.

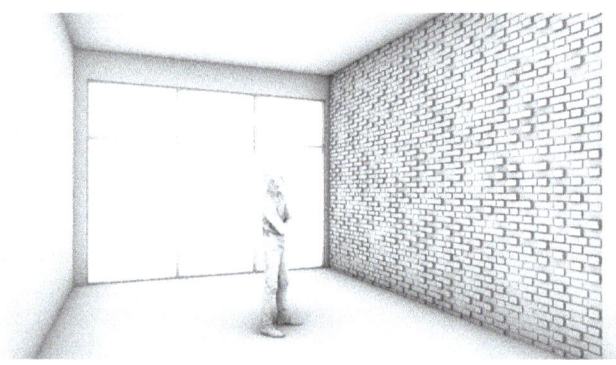

8   3D Visualization of option 1 from scenario A (panel_0100_0)

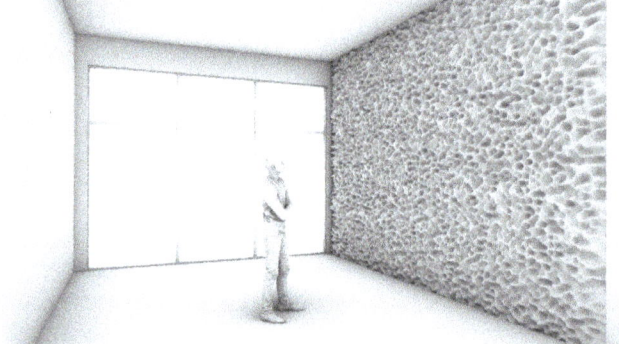

9   3D Visualization of option 3 from scenario A (panel_0082_0)

algorithm places these samples in the same cell; thus, the resulting embedding can have several empty cells. On the other hand, a SOM with a very small number of neurons may not have enough space to arrange the samples. This constraint will force the algorithm to place less similar samples on the same cell, resulting in a less representative data embedding.

Scenario A

This scenario aims to design a surface that, compared to a flat reflective surface, lowers the specularly reflected energy in the whole spectrum, emphasizing the mid- and high-frequency bands. This emphasis will make the room sound softer by reducing the often harsh high-frequency specular reflections. Combined with the overall reduction in reflected energy, the person speaking will sound more clear. To achieve the desired energy goal, we input the following values: [80, 80, 70, 60, 60], and the SOM cell with the closest matching values is displayed (Figure 7). Selecting the cell reveals all the surfaces with similar values in descending order, from the closest to the least matching option. Nevertheless, because of how the SOM clustering algorithm works, even the least matching option is very close to our desired acoustic criterion. Figure 7 shows the energy ratios of all matching surfaces compared to the desired energies and their close-up views. Option 1 (panel_0100_0) and option 3 (panel_0082_0) are also visualized inside the room to evaluate them based on aesthetic qualities. At this point, the architect decides which surface best suits their design idea.

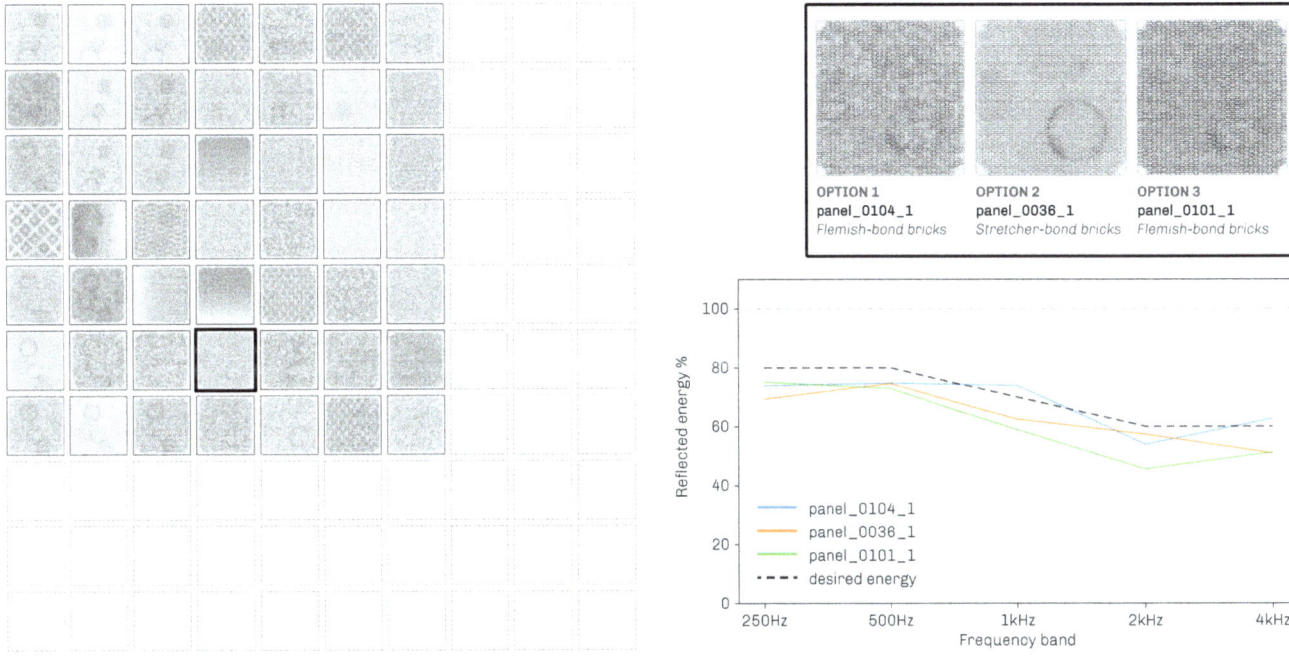

10 Left: The 7x7 SOM for Scenario A. The black outline indicates the best matching cell acording to the desired energy values.
Right-top: The surfaces of the best matching cell. Right-bottom: Energy ratios of all matching surfaces.

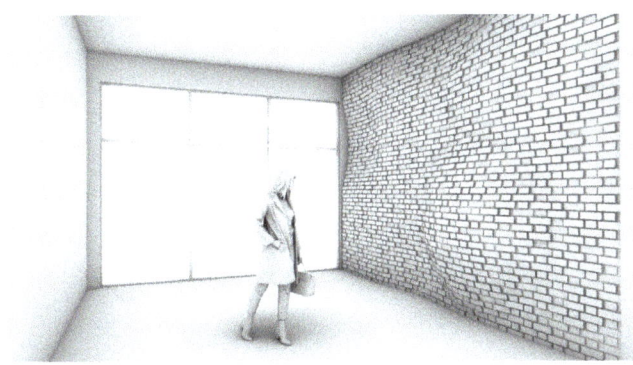

11 3D visualization of option 1 from scenario B (panel_0104_1)

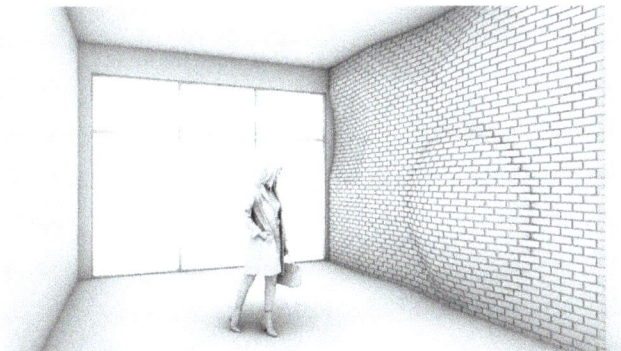

12 3D Visualization of option 2 from scenario B (panel_0036_1)

Scenario B

Like the previous scenario, the performance criterion is again the TNCE values of layer_1. Figure 10 shows the cell with the best matching values, the close-up views of the associated surfaces and their energy ratios. In this case, the SOM cell contains only three surfaces. Contrary to scenario A, these surfaces happen to have a macrostructure, making them more spatially expressive. Options 1 and 3 are from the same typology and have very similar designs and energy values; therefore, we focus on options 1 and 2. Panel_0036_1 lowers the energy by five to ten percent more than the desired energy goal in all frequency bands. Although panel_0104_1 also lowers the energy a little more than the set goal in the two lowest frequency bands (250Hz, 500Hz), it matches the desired goal in the 1kHz and 4kHz frequency bands (see Figure 10). Therefore, option 1 better matches our desired acoustic performance criteria.

RESULTS AND DISCUSSION

We have proposed a novel and fast workflow for a performance-driven acoustic design of diffusive surfaces. We described its components and how each of them contributes to the entire workflow. We have demonstrated its application with two design experiments. These experiments showed that thanks to its visual and intuitive implementation, users need little acoustic expert knowledge to specify and explore early design options compared to traditional room acoustic surface design processes. When no predefined typology is chosen, the design proposals could include several different typologies.

This approach could inspire or drive the designer's choices and could also be used as a basis for discussion and further refinement with acoustic experts. Compared to sometimes days of computing time when using numerical modeling algorithms (B. Peters 2015), our method needs only 20 to 30 seconds[13] to train the SOM depending on the dataset size (See Table 1). Then, computing the closest matching designs requires less than a second. Although the presented workflow is based on the GIR Dataset and a panel's precise TNCE values, one could use any other acoustical descriptor from the GIR Dataset (impulse response, frequency response, cumulative energy, and more). Furthermore, the presented methodology is not limited to the GIR Dataset. It can be adapted and applied to any other acoustic dataset..

The two design scenarios have shown that both flat-like (scenario A) and spatially varied surfaces (scenario B) are considered options. Flat-like surfaces are more likely to have uniform TNCE values across points of the same layer, and are more likely, therefore, to be the closest matching sample in the SOM. Nevertheless, the design workflow is not limited to a single set of desired energy values. We can assign different values to each layer, assign individual values to each grid point of a specific layer, and finally, assign a few values at desired locations and let the algorithm interpolate the in-between values. The fewer sets of energy values one uses as a performance criterion, the more likely it will result in a flat-like design.

Limitations

Although the presented design workflow proposes material patterns based on desired acoustic performance criteria, these patterns can only be from the GIR dataset. Nevertheless, the dataset can be expanded to include more patterns for a specific typology or introduce an entirely new typology. Furthermore, because the measurements were not according to the ISO standard, they cannot be used to derive standard acoustical descriptors such as absorption and scattering coefficients. Therefore, the clustering can only be done using the descriptors provided by Rust et al. (2021) (e.g., cumulative energy, normalized cumulative energy, tonal normalized cumulative energy). Nevertheless, we believe that total energy values (TNCE), split into five filter bands, are metrics most users can understand or quickly get familiar with.

Future Work

The proposed design workflow provides initial ideas or inspiration for a more acoustically informed design direction. However, choosing the desired acoustical parameters for the different frequency bands may still require some basic understanding of acoustics or initial consultation with an acoustics expert. Therefore, predefined acoustic use-cases should be implemented. These cases will translate qualitative intentions into quantitative parameters. Currently, the design workflow can be used via a Jupyter notebook, and it is available as an open-source code in *https://renkulab.io/gitlab/ddad/ddad-renku/*. The interface can be further streamlined and possibly integrated as a tool within existing CAD software or a stand-alone web-based application.

AUTHORS CONTRIBUTIONS

Conceptualization, A.X.; methodology, A.X.; coding A.X., N.P.; code optimization, B.L.; writing-original draft, A.X.; writing-review and editing, A.X., R.R., F.G., M.K.; figures and visualization, A.X.; supervision, R.R., F.G., M.K.

ACKNOWLEDGMENTS

This research stemmed out of a collaborative and multidisciplinary project between Gramazio Kohler Research at ETH Zurich, the Swiss Data Science Center, the Laboratory for Acoustics/Noise Control at EMPA, and STRAUSS ELEKTROAKUSTIK GMBH. Therefore, the authors would like to thank Dr. Fernando Perez-Cruz, Dr. Kurt Heutschi, Kurt Eggenschwiler, and Jurgen Strauss for their inputs. Furthermore, we would like to thank Dr. Nikola Marinčić for his valuable inputs on self-organizing maps and Gonzalo Casas for always being keen (hopefully) on helping on Python related topics.

NOTES

All website were accessed on October 8th, 2021.

1. http://www.odeon.dk.
2. http://www.orase.org.
3. http://www.catt.se.
4. https://www.iks.rwth-aachen.de/forschung/tools-downloads/databases/aachen-impulse-response-database/
5. https://speech.fit.vutbr.cz/software/but-speech-fit-reverb-database.
6. http://research.nii.ac.jp/src/en/RWCP-SSD.html.
7. https://acecorpus.ee.ic.ac.uk/.
8. http://dirha.fbk.eu/English-PHdev.
9. https://renkulab.io/projects/ddad/gir-dataset/.
10. The geometries and frequencies in the data set are in 1:10 scale.
11. The TNCE is the last value from the NCE list, representing the total energy arrived at the receiver position.
12. Further documentation can be found on MiniSom's Github repository. https://github.com/JustGlowing/minisom.
13. On a 2.9GHz 6-core Intel i9 cpu and 32GB 2400MHz DDR4 RAM.

REFERENCES

Arulkumaran, Kai, Marc Peter Deisenroth, Miles Brundage, and Anil Anthony Bharath. 2017. "Deep Reinforcement Learning: A Brief Survey." *IEEE Signal Processing Magazine* 34 (6): 26–38.

Badino, Elena, Louena Shtrepi, and Arianna Astolfi. 2020. "Acoustic Performance-Based Design: A Brief Overview of the Opportunities and Limits in Current Practice." *Acoustics* 2 (2): 246–78.

Cox, Trevor J., and Peter D'Antonio. 2000. "Acoustic Phase Gratings for Reduced Specular Reflection." *Applied Acoustics* 60 (2): 167–86.

Cox, Trevor J., and Peter D'Antonio. 2004. *Acoustic Absorbers and Diffusers: Theory, Design and Application*. Boca Raton: CRC Press.

Eaton, James, Nikolay D. Gaubitch, Alastair H. Moore, and Patrick A. Naylor. 2017. "Acoustic Characterization of Environments (ACE) Challenge Results Technical Report." *ArXiv*:1606.03365.

Fuhrimann, Lukas, Vahid Moosavi, Patrick Ole Ohlbrock, and Pierluigi D'acunto. 2018. "Data-Driven Design: Exploring New Structural Forms Using Machine Learning and Graphic Statics." *Proceedings of IASS Annual Symposia* 2018 (2): 1–8.

Gamper, Hannes, and Ivan J. Tashev. 2018. "Blind Reverberation Time Estimation Using a Convolutional Neural Network." In *Proceedings of the 16th International Workshop on Acoustic Signal Enhancement (IWAENC)*. Tokyo, Japan. 136–140.

Gatys, Leon A., Alexander S. Ecker, and Matthias Bethge. 2015. "A Neural Algorithm of Artistic Style." *ArXiv*:1508.06576.

Genovese, Andrea F., Hannes Gamper, Ville Pulkki, Nikunj Raghuvanshi, and Ivan J. Tashev. 2019. "Blind Room Volume Estimation from Single-Channel Noisy Speech." In *ICASSP 2019; 2019 IEEE International Conference on Acoustics, Speech and Signal Processing (ICASSP)*. Brighton, UK. 231–35.

Gorgoglione, Angela, Alberto Castro, Vito Iacobellis, and Andrea Gioia. 2021. "A Comparison of Linear and Non-Linear Machine Learning Techniques (PCA and SOM) for Characterizing Urban Nutrient Runoff." *Sustainability* 13 (4): 2054.

Hearty, John. 2016. *Advanced Machine Learning with Python*. Birmingham: Packt Publishing Ltd.

Isak Worre Foged, Anke Pasold, and Mads Brath Jensen. 2014. "Evolution of an Instrumental Architecture." In *Fusion; Proceedings of the 32nd ECAADe Conference*, vol 2., edited by Emine M. Thompson. Northumbria University, Department of Architecture and Built Environment. Newcastle upon Tyne, UK: ECAADe. 365–372.

Jeub, Marco, Magnus Schafer, and Peter Vary. 2009. "A Binaural Room Impulse Response Database for the Evaluation of Dereverberation Algorithms." In *2009 16th International Conference on Digital Signal Processing*. Santorini, Greece. 1–8.

Kobayashi, Yosihiro. 2006. "Self-Organizing Map and Axial Spatial Arrangement: Topological Mapping of Alternative Designs." In *Synthetic Landscapes; Proceedings of the 25th Annual Conference of the Association for Computer-Aided Design in Architecture (ACADIA)*. 342–355.

Kohonen, Teuvo. 1982. "Self-Organized Formation of Topologically Correct Feature Maps." *Biological Cybernetics* 43 (1): 59–69.

Mazin, Asim, Samuel H. Hawkins, Olya Stringfield, Jasreman Dhillon, Brandon J. Manley, Daniel K. Jeong, and Natarajan Raghunand. 2021. "Identification of Sarcomatoid Differentiation in Renal Cell Carcinoma by Machine Learning on Multiparametric MRI." *Scientific Reports* 11 (1): 3785.

Mirvis, Stuart E. 2002. "Garbage in, Garbage out (How Purportedly Great ML Models Can Be Screwed up by Bad Data)." *Applied Radiology* 31 (2): 5.

Moosavi, Vahid. 2017. "Contextual Mapping: Visualization of High-Dimensional Spatial Patterns in a Single Geo-Map." *Computers, Environment and Urban Systems* 61 (January): 1–12.

Motevalli, Benyamin, Baichuan Sun, and Amanda S. Barnard. 2020. "Understanding and Predicting the Cause of Defects in Graphene Oxide Nanostructures Using Machine Learning." *The Journal of Physical Chemistry C* 124 (13): 7404–13.

Nakamura, Satoshi, Kazuo Hiyane, Futoshi Asano, Takanobu Nishiura, and Takeshi Yamada. 2000. "Acoustical Sound Database in Real Environments for Sound Scene Understanding and Hands-Free Speech Recognition." In *Proceedings of the Second International Conference on Language Resources and Evaluation (LREC'00)*. Athens, Greece: European Language Resources Association (ELRA).

Niu, Haiqiang, Zaixiao Gong, Emma Ozanich, Peter Gerstoft, Haibin Wang, and Zhenglin Li. 2019. "Deep-Learning Source Localization Using Multi-Frequency Magnitude-Only Data." *The Journal of the Acoustical Society of America* 146 (1): 211–22.

Papayiannis, Constantinos, Christine Evers, and Patrick A. Naylor. 2020. "End-to-End Classification of Reverberant Rooms Using DNNs." *IEEE/ACM Transactions on Audio, Speech, and Language Processing* 28: 3010–17.

Peters, Brady. 2015. "Integrating Acoustic Simulation in Architectural Design Workflows: The FabPod Meeting Room Prototype." *Simulation* 91 (9): 787–808.

Peters, Nils, Howard Lei, and Gerald Friedland. 2012. "Name That Room: Room Identification Using Acoustic Features in a Recording." In *MM '12: Proceedings of the 20th ACM International Conference on Multimedia*. New York: ACM. 841–44.

Ravanelli, Mirco, Luca Cristoforetti, Roberto Gretter, Marco Pellin, Alessandro Sosi, and Maurizio Omologo. 2015. "The DIRHA-ENGLISH Corpus and Related Tasks for Distant-Speech Recognition in Domestic Environments." In *2015 IEEE Workshop on Automatic Speech Recognition and Understanding (ASRU)*. 275–82.

Rust, Romana, Achilleas Xydis, Kurt Heutschi, Nathanael Perraudin, Gonzalo Casas, Chaoyu Du, Jürgen Strauss, et al. 2021. "A Data Acquisition Setup for Data Driven Acoustic Design." *Building Acoustics* 28 (4): 345–360. https://doi.org/10.1177/1351010X20986901.

Saldana Ochoa, Karla, Patrick Ole Ohlbrock, Pierluigi D'Acunto, and Vahid Moosavi. 2020. "Beyond Typologies, Beyond Optimization: Exploring Novel Structural Forms at the Interface of Human and Machine Intelligence." *International Journal of Architectural Computing* 19 (3): 466–490. https://doi.org/10.1177/1478077120943062.

Schroeder, Manfred R. 1975. "Diffuse Sound Reflection by Maximum-length Sequences." *Journal of the Acoustical Society of America* 57 (1): 149.

Shtrepi, Louena, Tomás Mendéz Echenagucia, Elena Badino, and Arianna Astolfi. 2020. "A Performance-Based Optimization Approach for Diffusive Surface Topology Design." *Building Acoustics* 28 (3): 231–247. https://doi.org/10.1177/1351010X20967821.

Steinfeld, Kyle. 2017. "Dreams May Come." In *ACADIA 17: Disciplines & Disruption; Proceedings of the 37th Annual Conference of the Association for Computer Aided Design in Architecture (ACADIA)*. Cambridge, MA. 590–99.

Szoke, Igor, Miroslav Skacel, Ladislav Mosner, Jakub Paliesek, and Jan "Honza" Cernocky. 2019. "Building and Evaluation of a Real Room Impulse Response Dataset." *IEEE Journal of Selected Topics in Signal Processing* 13 (4): 863–76.

Tamke, Martin, Paul Nicholas, and Mateusz Zwierzycki. 2018. "Machine Learning for Architectural Design: Practices and Infrastructure." *International Journal of Architectural Computing* 16 (June): 123–43.

Van der Maaten, Laurens, and Geoffrey Hinton. 2008. "Visualizing Data Using T-SNE." *Journal of Machine Learning Research* 9 (11): 2579–2605.

Vettigli, Giuseppe. 2018. *MiniSom: Minimalistic and NumPy-Based Implementation of the Self Organizing Map*. Accessed January 15, 2022. https://github.com/JustGlowing/minisom.

Vouldimos, Athanasios, Nikolaos Doulamis, Anastasios Doulamis, and Eftychios Protopapadakis. 2018. "Deep Learning for Computer Vision: A Brief Review." *Computational Intelligence and Neuroscience* 2018 (February). https://doi.org/10.1155/2018/7068349.

Wold, Svante, Kim Esbensen, and Paul Geladi. 1987. "Principal Component Analysis." *Chemometrics and Intelligent Laboratory Systems, Proceedings of the Multivariate Statistical Workshop for Geologists and Geochemists* 2 (1): 37–52.

Xydis, Achilleas, Nathanaël Perraudin, Romana Rust, Kurt Heutschi, Gonzalo Casas, Oksana Riba Grognuz, Kurt Eggenschwiler, Matthias Kohler, and Fernando Perez-Cruz. 2021. *GIR Dataset: A Geometry and Real Impulse Response Dataset* [Data set]. Zenodo. https://doi.org/10.5281/zenodo.5288744.

IMAGE CREDITS

All drawings and images by the authors.

Achilleas Xydis is an architect and current a doctoral researcher at the Chair of Architecture and Digital Fabrication (Gramazio Kohler Research) at ETH Zurich. He received his Diploma in Architecture from the University of Patras in 2010. In 2013 he completed the post-graduate Master of Advanced Studies in Architecture and Information at ETH Zurich. He focuses on combining machine learning techniques with architectural acoustics.

Nathanaël Perraudin finished his Master in electrical engineering at the Ecole Fédérale de Lausanne (EPFL) and worked as a researcher in the Acoustic Research Institute (ARI) in Vienna. In 2013, he came back to EPFL for a PhD. He specialized in signal processing, graph theory, machine learning, data science, and audio processing. After graduating in 2017, he began working as a senior data scientist at the Swiss Data Science Center (SDSC), focusing on deep neural networks and generative models.

Romana Rust is a computational architect and senior researcher at Gramazio Kohler Research, ETH Zurich within the Design++ initiative: Centre for Augmented Computational Design in AEC. She is the co-coordinator of the Immersive Design Lab, a lab for collaborative research and teaching in the field of extended reality and machine learning in architecture and construction. Her particular interest is the development of innovative computational methods that integrate multiple design objectives such as geometry, acoustics, materiality, and robotic fabrication.

Beverly Ann Lytle is a software engineer at the Chair for Architecture and Digital Fabrication at ETH Zurich, where she contributes to COMPAS, an open-source python framework for AEC. She received her master's degree at Ohio State University and her PhD at ETH Zurich, both in mathematics.

Fabio Gramazio and Matthias Kohler are professors of Architecture and Digital Fabrication at ETH Zurich. In 2000, they founded the architecture practice Gramazio & Kohler, which realized numerous award-winning projects. Opening the world's first architectural robotic laboratory at ETH Zurich, Gramazio & Kohler's research has been formative in the field of digital architecture, setting precedence and de facto creating a new research field merging advanced architectural design and additive fabrication processes through the customized use of industrial robots.

MindSculpt

Using a Brain–Computer Interface to Enable Designers to Create Diverse Geometries by Thinking

Qi Yang
DAIL Lab, Department
of Design + Environmental
Analysis, Cornell University

Jesus G. Cruz-Garza
DAIL Lab, Cornell University

Saleh Kalantari
DAIL Lab, Cornell University

ABSTRACT

MindSculpt enables users to generate a wide range of hybrid geometries in Grasshopper in real-time simply by thinking about those geometries. This design tool combines a non-invasive brain-computer interface (BCI) with the parametric design platform Grasshopper, creating an intuitive design workflow that shortens the latency between ideation and implementation compared to traditional computer-aided design tools based on mouse-and-keyboard paradigms. The project arises from transdisciplinary research between neuroscience and architecture, with the goal of building a cyber-human collaborative tool that is capable of leveraging the complex and fluid nature of thinking in the design process. MindSculpt applies a supervised machine-learning approach, based on the support vector machine model (SVM), to identify patterns of brain waves that occur in EEG data when participants mentally rotate four different solid geometries. The researchers tested MindSculpt with participants who had no prior experience in design, and found that the tool was enjoyable to use and could contribute to design ideation and artistic endeavors.

INTRODUCTION

Abraham Maslow is credited with formulating the "law of the instrument," which is commonly summed up with the quotation: "If all you have is a hammer, everything looks like a nail" (Maslow 1966, 15). This phrase reminds us that the tools we use in design will inevitably affect our perception and workflow, and thus contribute to shaping the resulting design products. In the age before powerful computers, designers such as Antoni Gaudí and Frei Otto developed physical prototyping tools to help them better visualize the complex forms taking shape in their minds. The primary purpose of these tools was to provide timely feedback so that the designers could evaluate their visions and make improvements. As the computer era advanced, design workflows based on the WIMP paradigm (Windows, Icons, Menus, and Pointers) became the new standard (Caetano et al. 2020; Shankar et al. 2014). These tools were meant to enable a more intuitive technological interface for people without programming backgrounds (Sutherland 1964). However, multiple empirical studies have indicated that using WIMP-based computer-aided design tools can impede design ideation and the conceptual design process. Compared to old-fashioned sketching, WIMP-based design has been shown to discourage the generation of diverse ideas (Stones et al. 2007; Alcaide-Marzal et al. 2013), to mislead designers to focus excessively on details and neglect the big picture (Ibrahim et al. 2010; Charlesworth and Chris 2007), and promote premature design fixation and constrain designers' capabilities due to the unintuitive interface (Robertson and Radcliffe 2009).

Some researchers have put forth plausible theories and principles for tools that can better support design creativity (Hutchins et al. 1985; Shneiderman et al. 2006). For example, according to Hutchins's direct manipulation theory, design tools should seek to minimize the "gulf of execution," which is the mental gap or delay that occurs when computer users have to translate their intentions into linear and sequential machine-understandable commands. A significant gulf of execution will lead to high latency between idea generation and realization, interrupting the intuitive process of visualization and design feedback. One of the main problems with WIMP-based design tools is that they have relatively significant gulfs of execution, particularly for users who are not yet fully fluent in the relevant software and input technologies.

To address the deficiencies of WIMP-based computer interfaces during the design ideation phase, we developed MindSculpt. This new design tool directly translates the user's imagination of different forms into high-fidelity, real-time digital prototypes in common design software platforms, without the need for pointing and clicking (Figure 1). MindSculpt is based on a machine learning algorithm that has been trained to identify brain signals associated with the user's visualization of different solid geometries. This technology is a work in progress, and the output is currently quite fluid and, in some cases, difficult to control, mirroring the complex nature of thinking itself. Overall, the feedback from our test participants has been quite positive, indicating that MindSculpt has potential for application in design ideation.

BACKGROUND

Brain-computer interfaces (BCI) are capable of translating a user's brain activities into messages or commands (Lotte et al. 2018). Previous studies have demonstrated the possibility of classifying EEG signals associated with the mental visualization of complex geometries and shapes (Esfahani et al. 2012; Rai and Akshay 2016). Furthermore, the mental rotation of objects has become a common task used in the BCI field to demonstrate the technology's potential for good binary classification scenarios (Friedrich et al. 2013; Jeunet et al. 2015). Using mental rotation may accommodate a wider range of task difficulties and offer users more flexibility when using BCI (Gardony et al. 2017).

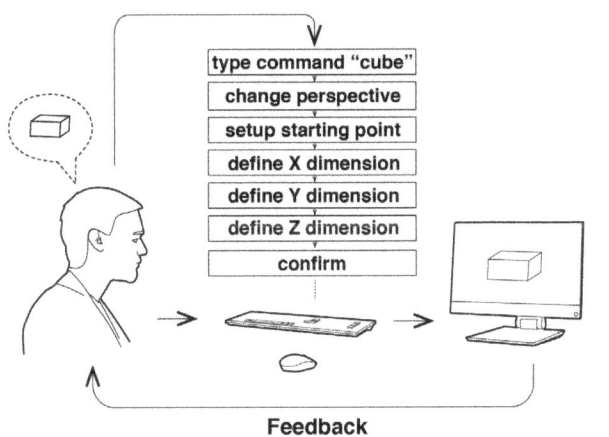
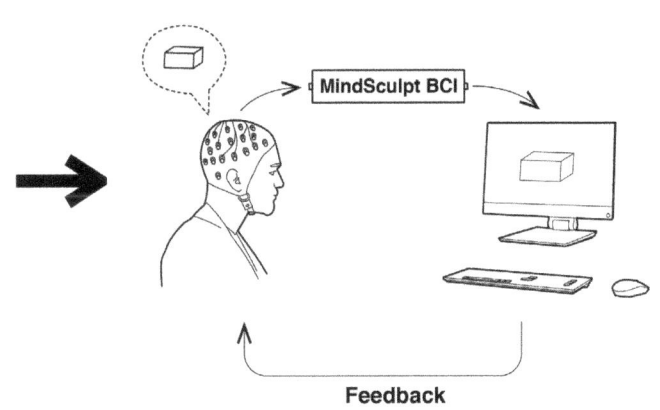

1 Traditional CAD tool feedback loop vs. the MindSculpt feedback loop

BCI applications include writing by selecting letters on a screen (Farwell and Donchin 1988; Yi et al. 2012; Yin et al. 2013), or imagining handwriting movements to create sentences (Willet et al. 2021). Users have been also able to control a computer cursor on a screen (Blankertz et al. 2007; Wolpaw and Birbaumer 2002). Although the majority of BCI applications are oriented toward clinical scenarios, especially for physically disabled people (Abiri et al. 2019), the use of BCI technology in art and design has also been explored (Nijholt 2019, 104-111). Users used BCI to control a group of robots in an art installation (Ulrike 1991). Through motor imagery, BCI users had been shown to compete in an electronic ping-pong game (Babiloni et al. 2007) and 'Connect Four' (Maby et al. 2012) using only EEG-measured brain activity. Other studies explored mental activities acquired by BCI as a metric to evaluate the design or environment in the Generative Design and Smart Environment frameworks (Barsan-Pipu 2019; Cutellic 2018; Yu-Chun 2006). More recently, artistic and social explorations of BCI in music performance (Mullen et al. 2015; Rosenboom and Mullen 2019), digital drawings (Wulff-Abramsson et al. 2019; Kübler et al. 2019), digital modeling (Shankar et al. 2014) and responsive built environments (Todd et al. 2019; Kovacevic et al. 2015) have motivated the field to engage in new forms of brain-interface applications.

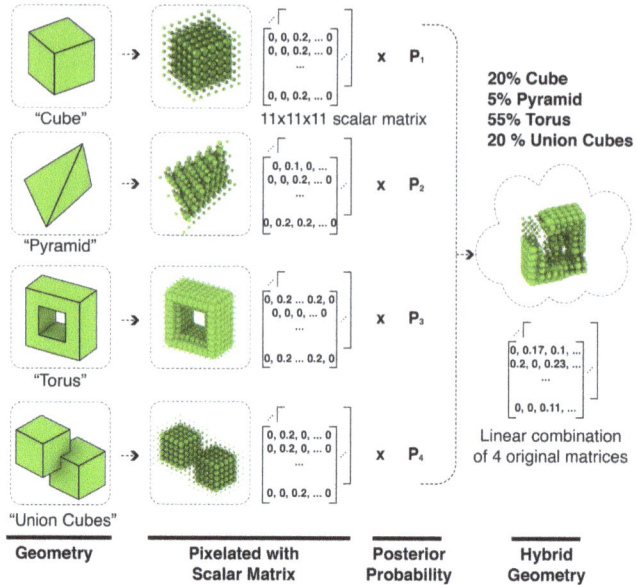

3 Geometry generation using osterior probability allows for the creation of hybrid figures

In many of those studies based on non-invasive BCI, the applications created intriguing external visualizations of brain signals and have the capability to customize the relationship between the user's mental imagery and the output. However, the type of brain activity linked to various BCI outputs was often tangential or even counterintuitive.

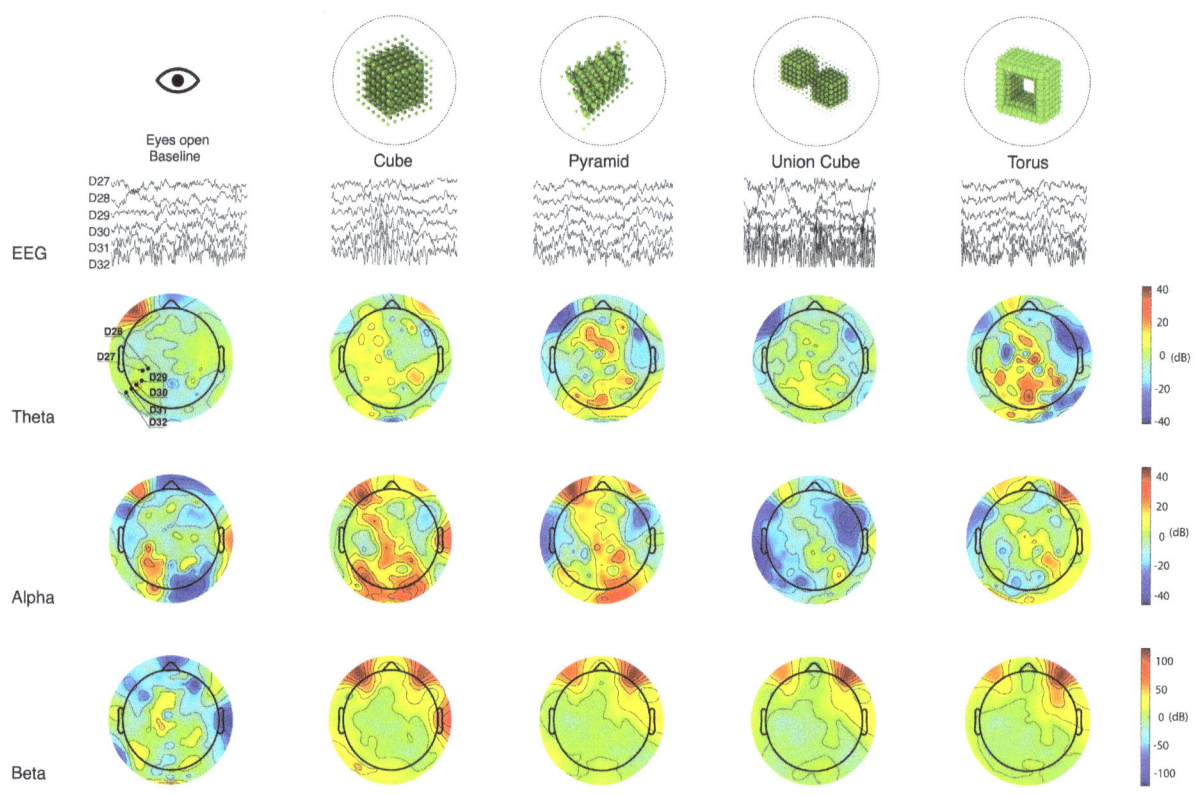

2 Different brain patterns associated with the mental rotation of each geometry for Participant 1

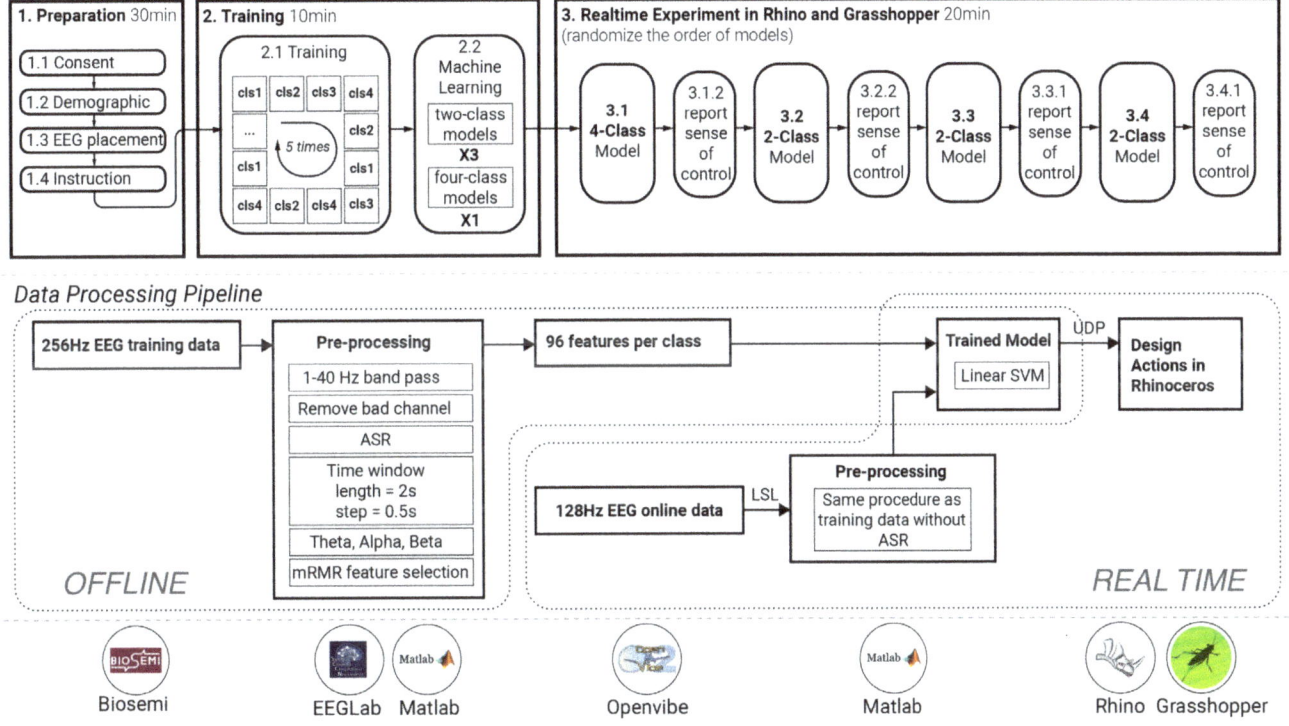

4 Experiment protocol and data processing pipeline of MindSculpt

One system has mapped motor activities such as eyeblinks to commands such as "draw a circle" (Shankar et al. 2014). One system made icons flicker to decode where the user attended to (Kübler et al. 2019). Another exploration used brain activity to evaluate and influence an autonomously developing design rather than to exert active, dynamic control (Barsan-Pipu 2019; Yu-Chun 2006). Here, we provide an experience to the user where a BCI is used to create an intuitive and instant feedback loop between the ideation of geometric forms and their external representation.

The MindSculpt project used a machine learning approach to identify neural signals associated to the mental rotation of different solid geometries, and then represented those same geometries on the screen via common design software programs (Figure 2). Discrete classification paradigms seemed insufficient to satisfy the need for design exploration during the ideation and conceptual design phases. The approach that we adopted for MindSculpt used the posterior probability score as the output of the machine learning model, and pixelated the 3D geometries based on the probability score. As a result, MindSculpt was able to provide a wide spectrum of complex geometries beyond the four basic shapes. When the user hesitated between a cube and a pyramid during the ideation process, MindSculpt would generate a hybrid geometry that was in-between a cube and a pyramid (Figure 3).

METHOD

During the MindSculpt project the researchers first determined the best approach to creating the BCI, and then conducted a pilot study with non-designer participants to test the system.

Hardware and Software Setup

The current set-up of MindSculpt included an EEG headset, Openvibe (Renard et al. 2010), Matlab, and Grasshopper (Figure 4). EEG data were collected using a non-invasive, 128-channel, gel-based Actiview System (BioSemi Inc. n.d.) with Ag/AgCl active electrodes. Openvibe is an open-source BCI platform that can acquire real-time EEG data and transfer the results to Matlab through the Lab Streaming Layer (Kothe et al. 2014). Matlab was used to pre-process the EEG data, train the participant-specific machine-learning model, predict the subsequent geometries that users visualized, and send the target class posterior probability scores to Grasshopper to visualize the outcomes in real-time.

MindSculpt Workflow

Using MindSculpt was a three-stage process (Figure 4): preparation, training, and real-time experimentation. Previous studies have shown that BCI self-regulation involves learning psychological factors—including cognitive states experienced by the users in a given session (e.g. motivation, concentration, flow)—and experiencing external factors at the time of measurement (Roc et al. 2020). For distinct participants, these factors could affect the EEG-data mental imagery representation of the same abstract object during different sessions. Therefore, we built a personalized ML model for each participant (Shah et al. 2021; Wang et al. 2021). The training session was meant to calibrate the BCI for different users and optimize the performance.

During the training, each participant viewed a video in which each of the four pre-selected geometries appeared on the computer screen for 10 seconds in a randomized sequence (50 seconds per class in total). Those geometries were a cube, a pyramid, a square torus, and two cubes combined diagonally ("union cubes"). When any geometry appeared, users were asked to perform a rotation of the image in their minds, which helped to create a strong mental visualization of the figure (Friedrich et al. 2013). This process continued until each figure had been shown five times.

As the participants viewed and visualized these solid geometries, EEG data were recorded at 256 Hz that enabled enough resolution for robust band-power estimation (Thomson 1982). Raw EEG data were preprocessed to bandpass the preferred frequency band (1~40 Hz), and remove bad channels with poor contact with the scalp. We applied artifact subspace reconstruction (Mullen et al. 2013) to remove noisy channels and artifactual power bursts (sd threshold = 15). Then, we visually inspected and removed EEG spikes caused by motion artifacts such as blinking, clenching, and body movements because they would cause confusion for the ML model to classify mental rotation of different geometries.

Linear classifiers such as Linear Discriminant Analysis (LDA) and Support Vector Machines (SVM) are the most popular types of classifiers for EEG-based BCIs (Lotte et al. 2018). Although deep learning approaches have shown promising results in EEG signal classification (Craik et al. 2019), SVMs often outperform other classifiers (Lotte et al. 2018; Nahmias et al. 2020).

Power-spectral density band-power features in the theta, alpha, and beta bands have shown to have robust discriminative power in motor-imagery classification (Herman et al. 2008; Oikonomou et al. 2017; Ravindran et al. 2019; Cruz-Garza et al. 2020). BCIs using linear SVMs for classification have been successfully implemented to classify between imagined movement (Song et al. 2013; Nahmias et al. 2020; You et al. 2020; Astrand et al. 2021), including multi-class motor intent detection (Yi et al. 2013; Wang et al. 2012).

Spectral band-power based BCIs are effective to extract relevant motor intent information, and reduce complexity for real-time BCI applications (Bhattacharyya et al. 2021). Furthermore, band-power based features have been implemented for EEG signal classification in contextually rich real-world environments (Kontston et al. 2015; Kovacevic et al. 2015; Todd et al. 2019).

Sensorimotor rhythms are associated with preparation, control, and executing volitional motor intent, particularly close to brain areas associated to such cognitive processes. The alpha (8-12 Hz) at central electrodes, named my-activity when recorded from central electrodes, is typically found to desynchronize (less power) when a motor action is imagined or executed. The beta band (13-30 Hz) is associated with higher order cognitive processes during motor imagery (Pfurtscheller et al. 2000; Fu et al. 2019). The alpha and beta bands in frontal-parietal electrodes are also predictive of video-game play in participants in freely-behaving settings (Ravindran et al. 2019). Patterns of connectivity between frontal and posterior electrodes have been associated to complex motor intention task that involve the human creative process, such as creative writing (Cruz-Garza et al. 2020) and observation of art pieces in a real-world museum (Kontston et al. 2015). In virtual reality environments the theta band has been recently found to be involved in and salient form recognition (Kalantari et al. 2021) from parietal regions, and aesthetic assessment of shapes (Banaei et al. 2017) from frontal regions, which makes it an interesting candidate to explore in form-making BCI systems.

Based on those rationales, band-power features were obtained at 2 second time windows for the theta (4-7 Hz), alpha (8-12 Hz) and beta (13-30 Hz) bands (Kontston et al. 2015; Cruz-Garza et al. 2017).

Band-power features were obtained for 2 second data windows, with 0.5 second overlap, resulting in in 96 data points per class. The frequency bands selected were the theta, alpha, and beta. We used Minimum Redundancy Maximum Relevancy (mRMR) (Peng et al. 2005) algorithm to select the most salient features out of 384 features (Figure 2). We defined the minimum number of selected features as 16 to increase the robustness of the subsequent real-time performance because the data of any channel could be affected by unpredicted events such as fierce movements from the user in real time. To obtain robust classification models, the EEG data from the training session was divided into a 5-fold cross-validation scheme—where 80% of the data was selected for training the model, and 20% for validation—and repeated 5 times. The best-performing

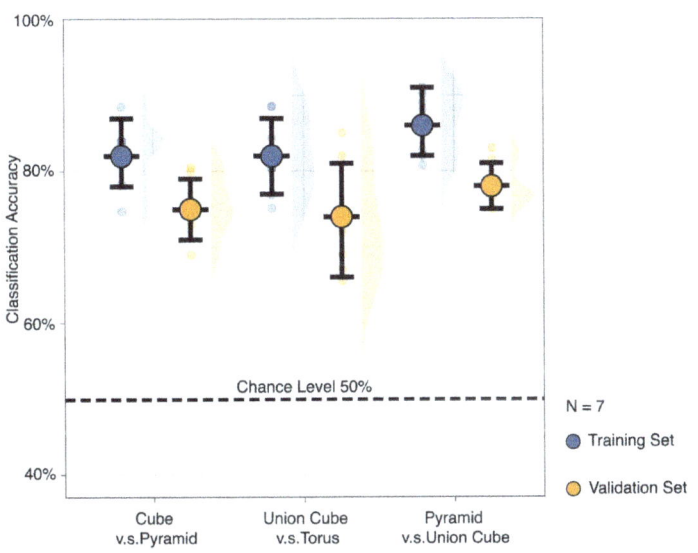
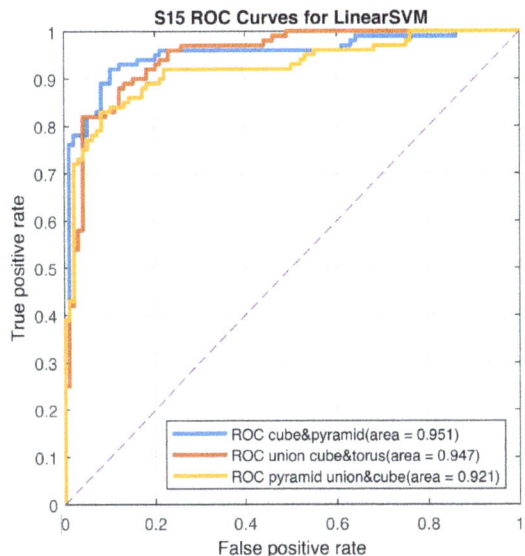

5 Classification, Validation Accuracy, and ROC curves for Three 2-class Models

(validation accuracy) cross-validation model was used for subsequent analyses. The resulting participant-specific linear SVM model was used for the real-time test set classification subsequently.

At this point, the participant could begin using MindSculpt to visualize hybrid shapes in real-time. During our pilot study, participants were asked to visualize two-class combinations of the sample geometries. These combinations included [cube + pyramid], [union cube + torus], and [pyramid + union cube]. The participants were also asked to visualize one four-class combination, in which all of the sample geometries were combined to create a final shape. They were not only told to create these hybrid geometries however they wished, but also encouraged to consider mental rotations and mergers of the sample shapes and to work on each hybrid geometry for at least 3 minutes. The trained machine-learning model, in turn, would assign weights to those four pre-defined geometries that were directly proportional to the posterior probability that a data sample belonged to each class. The hybrid geometry derived from the EEG data was represented on a computer screen by Grasshopper in real-time (Figure 3). Users were able to press a button to pause the reception of brain data and save the displayed geometry if they liked the design. They were then able to use the same button to resume the data reception and real-time visualization to continue exploring their mental control of the system. After the experiment, the researchers conducted a brief, semi-structured debriefing interview to obtain each participants' feedback about the MindSculpt system performance.

Participant

For the feasibility pilot study, we recruited seven participants using a convenience sampling method (word-of-mouth and announcements on departmental e-mail lists). None of the participants had prior experience working or studying in design fields, nor did they have prior BCI experiences. All of them were undergraduate university students. Each participant gave informed written consent before participating in the experiment, and the overall study protocol was approved by the Institutional Review Board prior to the start of research activities. All of the experiment sessions took place at the same physical location.

RESULT

For the seven participants in the pilot study, we reached 82% mean classification accuracy for the training set 2-class models (SD = 4%, n = 7), and 78% mean off-line validation accuracy (SD = 3%, n = 7) by iteratively using 4 trials to train a Linear SVM classifier and 1 trial as the validation set (Figure 5).

We also asked participants to self-report their sense of control over the represented hybrid geometry on a scale from 1 (low) to 10 (high). Five out of seven participants' data were collected. Although some participants reported a high sense of control when using MindSculpt, we observed a wide range of perceived performance for the 2-class models (min = 1.67, max = 7.3, M = 4.93, SD = 2.29, n = 5). The high variance of self-reported performance in this study matched the observations from previous BCI research (Jeunet et al. 2015).

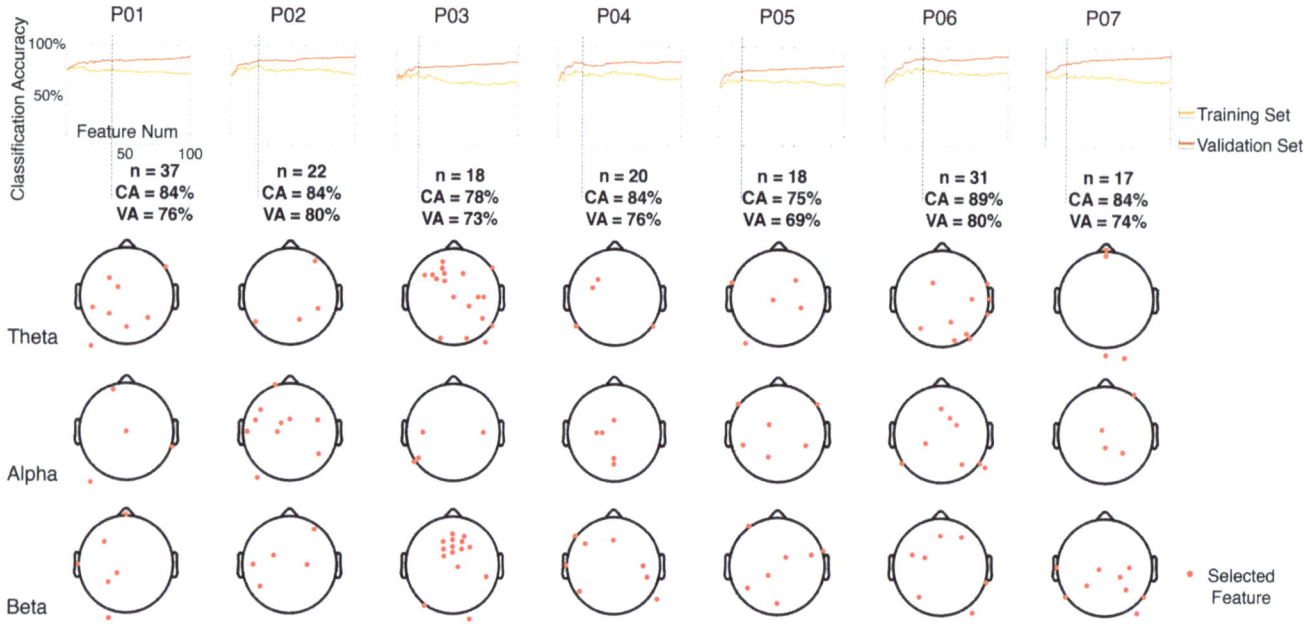

6 Feature selection for all participants based on mRMR of Cube vs. Pyramid model

More alpha power was observed in frontal and posterior regions of the scalp for the simple geometries (cube, pyramid) and suppression for more complex geometries (union, torus) (Figure 2). The observations resonated with previous findings (Michel et al. 1994; Riečanský et al. 2010; Gardony et al. 2017). In terms of selected EEG features, 23 features were selected on average for all 2-class models (SD = 6.83, n = 21). The feature selected showed that the validation accuracy reached a plateau (or decreased) after 17 to 23 features selected. More features would potentially impair real-time performance. Selected features for each model varied among different participants (Figure 6), but comprised a network of frontal, central, and parietal regions.

The most relevant features for classification, as obtained with mRMR, showed distinct patterns of channel relevance when participants were mental rotating geometries with different complexity (Figure 6). In the alpha band (8-12 Hz), central regions (close to motor regions) showed consistently most relevance and least redundancy to predict the mental imagery, although the spatial location of the electrodes was broad. The theta (4-7 Hz) and Beta (15-30 Hz) frequency bands also provided relevant information for classification performance, including relevant channels in midline frontal and parietal areas (Gardony et al. 2017).

During the experiments, we noticed that almost all of the participants were willing to spend much more time than the required 3 minutes to explore each hybrid shape.

Three participants reported that they were motivated to try different thoughts other than mere mental rotation and combination to generate geometries. For example, one participant explored thinking about rectangular objects and tested if a cube could be stretched accordingly.

In the interviews, all participants described the tool as "inventive" and "interesting." They said that they felt motivated to explore different shapes, and reported that the instant real-time feedback was helpful for their thinking process. However, one participant mentioned that they had difficulty concentrating on their internal visualization while also watching the changing real-time feedback on the screen. One participant indicated that he preferred to close his eyes and think, then open them briefly to check the outcomes. Five participants reported difficulty and high mental workload in continuously maintaining focus on the geometric shapes. When asked explicitly about the relationship between the MindSculpt BCI and conventional WIMP-based computer-aided design, three participants indicated that we should use both techniques in the future. When asked about the limitations of MindSculpt, the most common concern was the BCI's lack of accuracy in clearly capturing the user's mental model on the screen. When the EEG-based representation failed to match participants' intentions for what seemed like a long period of time, three participants experienced feelings of confusion, frustration, or annoyance. We even observed one participant self-blamed for not thinking hard enough, although the participant is not at fault.

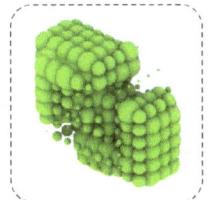 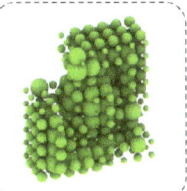 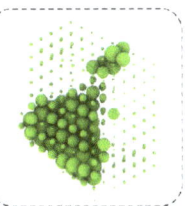 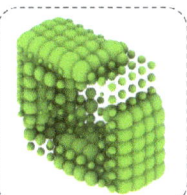 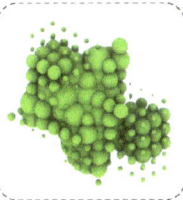 

Cube	= 25%	Cube	= 25%	Cube	= 0%	Cube	= 0%	Cube	= 10%	Cube	= 30%
Pyramid	= 10%	Pyramid	= 0%	Pyramid	= 85%	Pyramid	= 20%	Pyramid	= 60%	Pyramid	= 0%
Torus	= 0%	Torus	= 55%	Torus	= 15%	Torus	= 60%	Torus	= 0%	Torus	= 0%
Union Cubes	= 65%	Union Cubes	= 20%	Union Cubes	= 0%	Union Cubes	= 20%	Union Cubes	= 30%	Union Cubes	= 70%

7   Wide spectrum of geometries generated by MindSculpt

DISCUSSION AND CONCLUSION

In this paper, we introduce BCI to the architectural design field and explore its potentials in design ideation by developing MindSculpt. MindSculpt aims to shorten the feedback loop from design ideation to design implementation, so that practitioners can get instant, high-fidelity feedback and external visualization and recording of their ideas. The study involves recording participants' brain activity using a 128-channel EEG headset and developing a machine-language classifier to reflect the users' mental rotation of solid geometries. Adding to the previous study that uses mental imagery tasks to classify different geometries (Esfahani et al. 2012), MindSculpt demonstrates a novel alternative to classify geometries with different complexity by asking user to conduct mental rotation tasks. To test the performance of the interface, a feasibility study with 7 participants is conducted. The results indicates that this approach to human-computer interaction has potential, though the reliability of the system degenerates when users are confused by the instruction and not sure what to imagine, or distracted during the training session.

The current proof-of-concept study indicates the value of the MindSculpt approach, and lays the groundwork for approaching more complex design scenarios and longer-term interactions between users and the system. We believe that MindSculpt opens new avenues for the application of BCI to design research and practice.

Limitation

MindSculpt is limited by the prescriptive geometry generation algorithm. While linear combinations of four geometries provided many fascinating outputs (Figure 7), increasing the number of base shapes would allow the system to much more closely approximate the many possible geometries that an architectural designer would want to explore. More fundamental works are needed to determine how much variation in a user's brain patterns can be expected from day to day, and whether or not it may be possible for MindSculpt to accommodate such variations and adapt closer to a user over time. On the user experience side, setting up the EEG headset and conducting training session each time when the user intends to use the BCI can be time-consuming, and current MindSculpt offered a relatively low refresh rate (2 Hz) in Grasshopper due to the latency of real-time data processing, which may interfere the fluent interaction. Finally, the current pilot study relied strongly on user feedback to evaluate the viability of the tool. While the strong positive reception and interest in using MindSculpt is a valuable indication of the system's potential, more work is needed to consider objective measures such as the visual similarity between user sketches, quantity and diversity of the MindSculpt output, as well as the practical effectiveness of integrating MindSculpt into actual design workflows.

Future Directions

To expand the possible design space of MindSculpt, we expect more fundamental studies that help us understand the association between the physiological indicators and design intentions including but not limited to EEG.

While MindSculpt provides a new BCI modality in design, it is not likely that the traditional mouse and keyboard paradigm, or traditional sketching and prototyping, will soon disappear. One likely path forward in design will be the combination of BCI with other interfaces. Thus, future work in this area will need to evaluate the optimal uses of BCI during the design process. What design tasks are best implemented through BCI, and how can the use and output of these interfaces best be integrated with other aspects of the workflow? For example, BCI could be used to choose between a "Boolean Union" vs. "Boolean Difference" relationship while other modalities are used to conduct "selection" or "changing position" commands.

Making mistakes and encountering situations when design outcomes fail to match the original intention is common during design processes. When this happens, designers should be able to readily correct those mistakes or discover new directions embedded within them. The participants in the current study expressed some frustration when the outcome of the MindSculpt system diverged from their intent and could not be readily brought into line. Future work in this area will benefit from investigating the issue of error correction and non-BCI feedback in the system. A related issue is the steep learning curve of the BCI tools, which should be addressed by the development of clear guidebooks and interactive tutorials (Lotte and Camille 2015).

Future studies in this area may also want to attend more closely to the paradigms of passive, reactive, and active BCI (Nijholt 2019, 5-7). Reactive BCI requires users to attend to a stimulus (for example, a flickering segment of the displayed model) to select and alter that area. This type of BCI has higher accuracy but involves less voluntary thinking. Passive BCI monitors users' brain waves and triggers specific actions based on metrics such as stress levels. Though those paradigms are not precisely "doing by active thinking," they may have some potential for design applications. For example, one may imagine using passive BCI to evaluate and calibrate design options generated by active BCI so that increasing user frustration would trigger a shift in the system's responses. As for active BCI, brain signals associated with various body movements of even emotional feelings could be used to create inspiring design processes. Future researchers could also focus more overtly on the value of BCI-based tools for designers who have motor disabilities or other limitations in using conventional WIMP-based computer interfaces. Attending specifically to this population will provide opportunities to evaluate and improve the overall utility of the technology and will help to expand its adoption.

REFERENCES

Abiri, Reza, Soheil Borhani, Eric W. Sellers, Yang Jiang, and Xiaopeng Zhao. 2019. "A Comprehensive Review of EEG-based Brain–computer Interface Paradigms." *Journal of Neural Engineering* 16 (1): 011001.

Alcaide-Marzal, Jorge, José Antonio Diego-Más, Sabina Asensio-Cuesta, and Betina Piqueras-Fiszman. 2013. "An Exploratory Study on the Use of Digital Sculpting in Conceptual Product Design." *Design Studies* 34 (2): 264–284.

Astrand, Elaine, Jeanette Plantin, Susanne Palmcrantz, and Jonatan Tidare. 2021. "EEG non-stationarity across multiple sessions during a Motor Imagery-BCI intervention: Two post stroke case series." In *2021 10th International IEEE/EMBS Conference on Neural Engineering (NER)*. IEEE. 817–821.

Babiloni, F., F. Cincotti, M. Marciani, S. Salinari, L. Astolfi, A. Tocci, F. Aloise, F. De Vico Fallani, S. Bufalari, and D. Mattia. 2007. "The Estimation of Cortical Activity for Brain-Computer Interface: Applications in a Domotic Context." *Computational Intelligence and Neuroscience* 2007: 9165. https://doi.org/10.1155/2007/91651

Banaei, Maryam, Javad Hatami, Abbas Yazdanfar, and Klaus Gramann. 2017. "Walking through Architectural Spaces: The Impact of Interior Forms on Human Brain Dynamics." *Frontiers in Human Neuroscience* 11: 477.

Barsan-Pipu, Claudiu. 2019. "Artificial Intelligence Applied to Brain-Computer Interfacing with Eye-Tracking for Computer-Aided Conceptual Architectural Design in Virtual Reality Using Neurofeedback." In *Proceedings of the 2019 DigitalFUTURES: The 1st International Conference on Computational Design and Robotic Fabrication*, edited by Philip F. Yuan, Yi Min Xie, Jiawei Yao, and Chao Yan. Springer, Singapore. 124–135.

Bhattacharyya, Sumanta, Swatilekha Das, Arijit Das, Rajesh Dey, and RudraSankar Dhar. 2021. "Neuro-feedback system for real-time BCI decision prediction" *Microsystem Technologies* 27: 3725–3734. https://doi.org/10.1007/s00542-020-05146-4.

BioSemi Inc. n.d. Accessed 06 July 2021. https://www.biosemi.com/active_electrode.htm.

Blankertz, B., G. Dornhege, M. Krauledat, K-R. Müller, and G. Curio. 2007. "The non-invasive Berlin BrainComputer Interface: Fast acquisition of effective performance in untrained subjects." *NeuroImage* 37 (2): 539–550.

Caetano, Inês, and António Leitão. 2020. "Architecture Meets Computation: An Overview of the Evolution of Computational Design Approaches in Architecture." *Architectural Science Review* 63 (2): 165–174.

Charlesworth, Chris. 2007. "Student Use of Virtual and Physical Modelling in Design Development: An Experiment in 3D Design Education." *The Design Journal* 10 (1): 35–45.

Craik, Alexander, Yongtian He, and Jose L. Contreras-Vidal. 2019. "Deep learning for electroencephalogram (EEG) classification tasks: A Review." *Journal of Neural Engineering* 16 (3): 031001.

Cruz-Garza, Jesus G., Akshay Sujatha Ravindran, Anastasiya E. Kopteva, Cristina Rivera Garza, and Jose L. Contreras-Vidal. 2020. "Characterization of the stages of creative writing with mobile EEG using Generalized Partial Directed Coherence." *Frontiers in Human Neuroscience* 14: 533.

Cutellic, Pierre. 2018. "UCHRON: An Event-Based Generative Design Software Implementing Fast Discriminative Cognitive Responses from Visual ERP BCI." In *Computing for a Better Tomorrow; Proceedings of the 36th eCAADe Conference*, vol. 2, edited by A. Kepczynska-Walczak and S. Bialkowski. 131–138.

Dong, Enzeng, Changhai Li, Liting Li, Shengzhi Du, Abdelkader Nasreddine Belkacem, and Chao Chen. 2017. "Classification of Multi-class Motor Imagery with a Novel Hierarchical SVM Algorithm for Brain–computer Interfaces." *Medical & Biological Engineering & Computing* 55 (10): 1809–1818.

Esfahani, Ehsan Tarkesh, and V. Sundararajan. 2012. "Classification of Primitive Shapes Using Brain–computer Interfaces." *Computer-Aided Design* 44 (10): 1011–1019.

Farwell, Lawrence Ashley, and Emanuel Donchin. 1988. "Talking off the top of your head: Toward a mental prosthesis utilizing event-related brain potentials." *Electroencephalography and Clinical Neurophysiology* 70 (6): 510–523.

Friedrich, Elisabeth V.C., Christa Neuper, and Reinhold Scherer. 2013. "Whatever Works: A Systematic User-centered Training Protocol to Optimize Brain-computer Interfacing Individually." *PLoS ONE* 8 (9): e76214.

Fu, Rongrong, Yongsheng Tian, Tiantian Bao, Zong Meng, and Peiming Shi. 2019. "Improvement motor imagery EEG classification based on regularized linear discriminant analysis." *Journal of Medical Systems* 43 (6): 1–13.

Gardony, Aaron L., Marianna D. Eddy, Tad T. Brunyé, and Holly A. Taylor. 2017. "Cognitive Strategies in the Mental Rotation Task Revealed by EEG Spectral Power." *Brain and Cognition* 118: 1–18.

Herman, Pawel, Girijesh Prasad, Thomas Martin McGinnity, and Damien Coyle. 2008. "Comparative analysis of spectral approaches to feature extraction for EEG-based motor imagery classification." *IEEE Transactions on Neural Systems and Rehabilitation Engineering* 16 (4): 317–326.

Huang, Yu-Chun. 2006. "A space make you lively: A brain-computer interface approach to smart space." In *CAADRIA 2006; Proceedings of the 11th International Conference on Computer Aided Architectural Design Research in Asia*. 303–312.

Hutchins, Edwin L., James D. Hollan, and Donald A. Norman. 1985. "Direct Manipulation Interfaces." *Human–Computer Interaction* 1 (4): 311–338.

Ibrahim, Rahinah, and Farzad Pour Rahimian. 2010. "Comparison of CAD and Manual Sketching Tools for Teaching Architectural Design." *Automation in Construction* 19 (8): 978–987.

Jeunet, Camille, Bernard N'Kaoua, Sriram Subramanian, Martin Hachet, and Fabien Lotte. 2015. "Predicting Mental Imagery-based BCI Performance from Personality, Cognitive Profile and Neurophysiological Patterns." *PLoS ONE* 10 (12): e0143962.
Kontson, Kimberly, Murad Megjhani, Justin A. Brantley, Jesus Gabriel Cruz-Garza, Sho Nakagome, Dario Robleto, Michelle White, Eugene Civillico, and Jose Luis Contreras-Vidal. 2015. "'Your Brain on Art': Emergent cortical dynamics during aesthetic experiences." *Frontiers in Human Neuroscience* 9: 626.

Kerous, Bojan, Filip Skola, and Fotis Liarokapis. 2018. "EEG-based BCI and Video Games: A Progress Report." *Virtual Reality* 22 (2): 119–135.

Kothe, C., D. Medine, and M. Grivich. 2014. "Lab Streaming Layer." Accessed 19 August 2020. https://github.com/sccn/labstreaminglayer.

Kovacevic, Natasha, Petra Ritter, William Tays, Sylvain Moreno, and Anthony Randal McIntosh. 2015. "'My Virtual Dream:' Collective Neurofeedback in an Immersive Art Environment." *PLoS ONE* 10 (7): e0130129.

Kübler, Andrea, and Loic Botrel. 2019. "The Making of Brain Painting—From the Idea to Daily Life Use by People in the Locked-in State." In *Brain Art: Brain-Computer Interfaces for Artistic Expression*, edited by A. Nijholt, 409–431. Cham: Springer.

Li, Yueqing, Jincheol Woo, and Chang S. Nam. 2012. "A preliminary research on P300-based BCI application for people with motor disabilities." *Proceedings of the Human Factors and Ergonomics Society Annual Meeting* 56 (1). Los Angeles: SAGE Publications.

Lotte, Fabien, Chang S. Nam, and Anton Nijholt. 2018. "Introduction: Evolution of Brain-computer Interfaces." In *Brain-Computer Interfaces Handbook: Technological and Theoretical Advances*, edited by F. Lotte, C.S. Nam, and A. Nijholt, 1–11. Taylor & Francis (CRC Press).

Lotte, Fabien, and Camille Jeunet. 2015. "Towards Improved BCI based on Human Learning Principles." In *The 3rd International Winter Conference on Brain-Computer Interface*. IEEE. 1–4.

Maby, E., M. Perrin, O. Bertrand, G. Sanchez, and J. Mattout. 2012. "BCI could make old two-player games even more fun: A proof of concept with 'connect Four,'" *Adv. Human-Computer Interact.* 2012: 1. https://doi.org/10.1155/2012/124728.

Maslow, Abraham H. 1966. *The Psychology of Science a Reconnaissance*, 15. New York; London: Harper & Row.

Michel, Christoph M., Lloyd Kaufman, and Samuel J. Williamson. 1994. "Duration of EEG and MEG Alpha Suppression Increases with Angle in a Mental Rotation Task." *Journal of Cognitive Neuroscience* 6 (2): 139–150.

Mullen, T., C. Kothe, Y.M. Chi, A. Ojeda, T. Kerth, S. Makeig, G. Cauwenberghs, and T.P. Jung. 2013. July. "Real-time Modeling and 3D Visualization of Source Dynamics and Connectivity Using Wearable EEG." In *2013 35th Annual International Conference of the IEEE Engineering in Medicine and Biology Society (EMBC)*. IEEE. 2184–2187.

Nahmias, David O., Eugene F. Civillico, and Kimberly L. Kontson. 2020. "Deep learning and feature based medication classifications from EEG in a large clinical data set." *Scientific Reports* 10 (1): 1–11.

Nijholt, Anton, ed. 2019. *Brain Art: Brain-Computer Interfaces for Artistic Expression*. Cham: Springer.

Oikonomou, Vangelis P., Kostas Georgiadis, George Liaros, Spiros Nikolopoulos, and Ioannis Kompatsiaris. 2017. "A comparison study on EEG signal processing techniques using motor imagery EEG data." In *2017 IEEE 30th international symposium on computer-based medical systems (CBMS)*. IEEE. 781–786.

Peng, Hanchuan, Fuhui Long, and Chris Ding. 2005. "Feature Selection based on Mutual Information Criteria of Max-dependency, Max-relevance, and Min-redundancy." *IEEE Transactions on Pattern Analysis and Machine Intelligence* 27 (8): 1226–1238.

Pfurtscheller, Gert, Christa Neuper, Clemens Brunner, and F. Lopes Da Silva. 2005. "Beta rebound after different types of motor imagery in man." *Neuroscience Letters* 378 (3): 156–159.

Rai, Rahul, and Akshay V. Deshpande. 2016. "Fragmentary Shape Recognition: A BCI study." *Computer-Aided Design* 71: 51–64.

Ravindran, Akshay Sujatha, Aryan Mobiny, Jesus G. Cruz-Garza, Andrew Paek, Anastasiya Kopteva, and José L. Contreras Vidal. 2019. "Assaying neural activity of children during video game play in public spaces: a deep learning approach." *Journal of Neural Engineering* 16 (3): 036028.

Renard, Yann, Fabien Lotte, Guillaume Gibert, Marco Congedo, Emmanuel Maby, Vincent Delannoy, Olivier Bertrand, and Anatole Lécuyer. 2010. "Openvibe: An Open-source Software Platform to Design, Test, and Use Brain–computer Interfaces in Real and Virtual Environments." *Presence* 19 (1): 35–53.

Riečanský, Igor, and Stanislav Katina. 2010. "Induced EEG Alpha Oscillations are Related to Mental Rotation Ability: the Evidence for Neural Efficiency and Serial Processing." *Neuroscience Letters* 482 (2): 133–136.

Robertson, B.F., and D.F. Radcliffe. 2009. "Impact of CAD tools on Creative Problem Solving in Engineering Design." *Computer-Aided Design* 41 (3): 136–146.

Roc, Aline, Léa Pillette, Jelena Mladenovic, Camille Benaroch, Bernard N'Kaoua, Camille Jeunet, and Fabien Lotte. 2020. "A Review of User Training Methods in Brain Computer Interfaces based on Mental Tasks." *Journal of Neural Engineering* 18 (1): 011002.

Rosenboom, D. and T. Mullen. 2019. "More than one—Artistic explorations with multi-agent BCIs." In *Brain Art: Brain-Computer Interfaces for Artistic Expression*, edited by A. Nijholt, 117–143. Cham: Springer.

Rounds, James D., Jesus Gabriel Cruz-Garza, and Saleh Kalantari. 2020. "Using posterior eeg theta band to assess the effects of architectural designs on landmark recognition in an urban setting." *Frontiers in Human Neuroscience* 14: 537.

Shah, Rutvik V., Gillian Grennan, Mariam Zafar-Khan, Fahad Alim, Sujit Dey, Dhakshin Ramanathan, and Jyoti Mishra. 2021. "Personalized Machine Learning of Depressed Mood Using Wearables." *Translational Psychiatry* 11 (1): 1–18.

Shankar, S. Sree, and Rahul Rai. 2014. "Human Factors Study on the Usage of BCI Headset for 3D CAD Modeling." *Computer-Aided Design* 54: 51–55.

Saha, Simanto, and Mathias Baumert. 2020. "Intra-and Inter-subject Variability in EEG-based Sensorimotor Brain Computer Interface: A Review." *Frontiers in Computational Neuroscience* 13: 87.

Shneiderman, Ben, Gerhard Fischer, Mary Czerwinski, Mitch Resnick, Brad Myers, Linda Candy, Ernest Edmonds et al. 2006. "Creativity Support Tools: Report from a US National Science Foundation Sponsored Workshop." *International Journal of Human-Computer Interaction* 20 (2): 61–77.

Song, Xiaomu, Suk-Chung Yoon, and Viraga Perera. 2013. "Adaptive common spatial pattern for single-trial EEG classification in multi-subject BCI." In *2013 6th International IEEE/EMBS Conference on Neural Engineering (NER)*. IEEE. 411–414.

Stones, Catherine, and Tom Cassidy. 2007. "Comparing Synthesis Strategies of Novice Graphic Designers Using Digital and Traditional Design Tools." *Design Studies* 28 (1): 59–72.

Sutherland, Ivan E. 1964. "Sketchpad a Man-machine Graphical Communication System." *Simulation* 2 (5): R-3.

Todd, Eric, Jesus G. Cruz-Garza, Austin Moreau, James Templeton, and Jose Luis Contreras-Vidal. 2019. "Self-conscience/physical Memory: An Immersive, Kinetic Art Installation Driven by Real-Time and Archival EEG Signals." In *Brain Art: Brain-Computer Interfaces for Artistic Expression*, edited by A. Nijholt, 309–323. Cham: Springer.

Ulrike, Gabriel. 1993. "Terrain 01." Media Art Net, Publishing Organization. Accessed 6 July 2021.. http://www.medienkunstnetz.de/works/terrain/.

Wang, Deng, Duoqian Miao, and Gunnar Blohm. 2012. "Multi-class motor imagery EEG decoding for brain-computer interfaces." *Frontiers in Neuroscience* 6: 151.

Wang, Jian-Guo, Hui-Min Shao, Yuan Yao, Jian-Long Liu, and Shi-Wei Ma. 2021. "A Personalized Feature Extraction and Classification Method for Motor Imagery Recognition." *Mobile Networks and Applications* 26: 1359–1371.

Willett, Francis R., Donald T. Avansino, Leigh R. Hochberg, Jaimie M. Henderson, and Krishna V. Shenoy. 2021. "High-performance brain-to-text communication via handwriting." *Nature* 593 (7858): 249–254.

Wolpaw, J. R., N. Birbaumer et al. 2002. "Brain-computer interfaces for communication and control." *Clinical Neurophysiology* 113 (6): 767–791.

Wulff-Abramsson, Andreas, Adam Lopez, and Luis Antonio Mercado Cerda. 2019. "Paint with Brainwaves—A Step Towards a Low Brain Effort Active BCI Painting Prototype." In *Mobile Brain-Body Imaging and the Neuroscience of Art, Innovation and Creativity*, 183–188. Springer Series on Bio- and Neurosystems, vol. 10, Cham: Springer.

Yi, Weibo, Shuang Qiu, Hongzhi Qi, Lixin Zhang, Baikun Wan, and Dong Ming. 2013. "EEG feature comparison and classification of simple and compound limb motor imagery." *Journal of Neuroengineering and Rehabilitation* 10 (1): 1–12.

Yin, Erwei, Zongtan Zhou, Jun Jiang, Fanglin Chen, Yadong Liu, and Dewen Hu. 2013. "A speedy hybrid BCI spelling approach combining P300 and SSVEP." *IEEE Transactions on Biomedical Engineering* 61 (2): 473–483.

You, Yang, Wanzhong Chen, and Tao Zhang. 2020. "Motor imagery EEG classification based on flexible analytic wavelet transform." *Biomedical Signal Processing and Control* 62: 102069.

IMAGE CREDITS

All drawings and images by the authors.

Qi Yang is a PhD student at Cornell University with research interests in cognitive tools, behavior modeling, and computational architectural design.

Jesus G. Cruz-Garza is a postdoctoral researcher at Cornell University with research interests in neuroengineering, machine learning, brain-machine interface and neuroaesthetics.

Saleh Kalantari is an assistant professor in Cornell University's Department of Design and Environmental Analysis. He is the director of the Design and Augmented Intelligence Lab (DAIL) at Cornell, where his research group investigates human–technology partnerships in the design process and the resulting opportunities for innovation and creativity.

Lotus: A curved folding design tool for Grasshopper

Klara Mundilova
MIT

Erik Demaine*
MIT

Riccardo Foschi*
University of Bologna

Robby Kraft*
University of Innsbruck

Rupert Maleczek*
University of Innsbruck

Tomohiro Tachi*
University of Tokyo

* Authors contributed equally to the research.

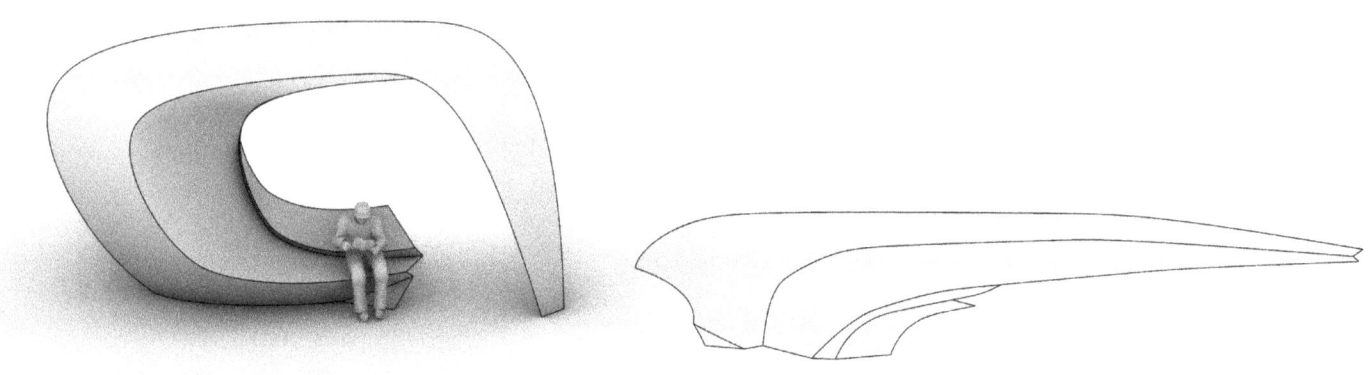

1   Roofed bench and its development designed with *Lotus*

ABSTRACT

Curved-crease origami design is a novel area of research with applications in fields such as architecture, design, engineering, and fabrication ranging between micro and macro scales. However, the design of such models is still a difficult task which requires preserving isometry between the 3D form and 2D unfolded state. This paper introduces a new software tool for Rhino/Grasshopper for interactive computational curved-crease origami design. Using a rule-line based approach, this tool has two functions: rigid-ruling bending of a flat sheet, and a patch-by-patch additive construction method for cylindrical and conical surfaces along curved creases.

1 INTRODUCTION

Developable surfaces are surfaces that can be obtained by bending a planar sheet without stretching or tearing. Such surfaces have various applications due to their aesthetics, ease of fabrication, and cost efficiency (Lawrence 2011). Examples include structural elements (Lienhard 2014), architecture facades (Glaeser and Gruber 2007), boat hull making (Chalfant 1997), and furniture design.

Developable surfaces can be folded along curves, resulting in so-called *curved-crease origami*. Shapes with curved creases provide structural stiffness, have less material offcut, and are easier to transport in their flat state (Demaine et al. 2015, Maleczek et al. 2019, Bhooshan et al. 2014).

Several computational algorithms have been proposed to deal with developable surfaces and simulate curved folding. Kilian et al. (2018) digitize scanned curved crease folding through an optimization-based approach. Solomon et al. (2012) guide interactive modelling of discretized developable surfaces with a discrete mean curvature bending energy. Developability constraints are imposed on the control points of a spline surface by Tang et al. (2016). Rabinovich et al. (2018, 2019) use discrete geodesic orthogonal nets to approximate developable surfaces.

However, methods for computationally designing curved crease models are still limited. In particular, for designers including architects, it is crucial to be able to design parametrically, allowing for adjustments as discussed in Foschi et al. (2021). Mitani (2011) implements a parametric software that explores shapes that can be obtained by planar reflection; however, the family of curved folding geometry is limited. Maleczek et al. (2021) construct curved crease shapes by rounding the edges of a polyhedral surface and adding curved creases around the polyhedron's vertices.

In this paper, we present a novel digital toolbox called *Lotus* for Rhino/Grasshopper, based on the methods described by Mundilova (2019), see Figure 2. Our tool allows the user to design developable surfaces and implement certain types of curved folding in a parametric design environment. Since the toolbox is provided as a set of components for Grasshopper, the users can freely arrange and concatenate the components to obtain a large family of shapes with multiple curved creases. The algorithms have the benefit of being constructive and do not require any optimization. Thus, the parametric inputs lead to fast feedback suited for interactive design of complicated shapes with multiple curved creases. *Lotus* aims to be accessible to users without a strong knowledge in computational origami

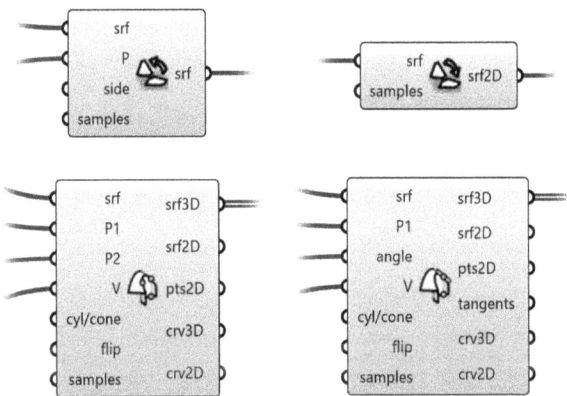

2 *Lotus* Grasshopper components: *SlidingDevelopable* (top left), *UnrollDevelopable* (top right), *PatchToCylCone(3pt)* (bottom left), *PatchToCylCone(tan)* (bottom right)

theory, and it has been tested in multiple architecture courses.

The first component in *Lotus* assists in the design of developable surfaces by pulling a planar surface into 3D. Given one planar surface patch and one point in 3D above the plane, the component bends the surface isometrically with rigid rulings to intersect the point while one curved boundary remains in the original plane. As the input point moves around, the constrained edge of the solution surface has the effect of sliding back and forth in the plane, giving the name *SlidingDevelopable*. *SlidingDevelopable* is universal, that is, it can locally construct any developable surface including cylinders, cones, and general tangent developables. This construction allows users to design developable surfaces by rigid ruling bending of their developments. Consequently, if the input surface does not contain singularities (edge-of-regression), neither will the constructed surface. The implementation of *SlidingDevelopable* is discussed in Section 3.2.

The second set of components are for folding developable surfaces. These take a developable surface as an input and compute the curved crease that connects the given surface with a new cylinder or a cone patch. The geometry of the second surface is specified by the cylinder's profile curve base plane or cone apex point. Two components offer unique interfaces to Mundilova (2019); the user may either prescribe two points on the crease curve or one point and one incident tangent direction. These components go by the names *PatchToCylCone(3pt)* and *PatchToCylCone(tan)*. The background theory is described in Section 2, and the implementation is described in Section 3.3.

The final component, *UnrollDevelopable*, computes the development of a developable surface, a subroutine of the

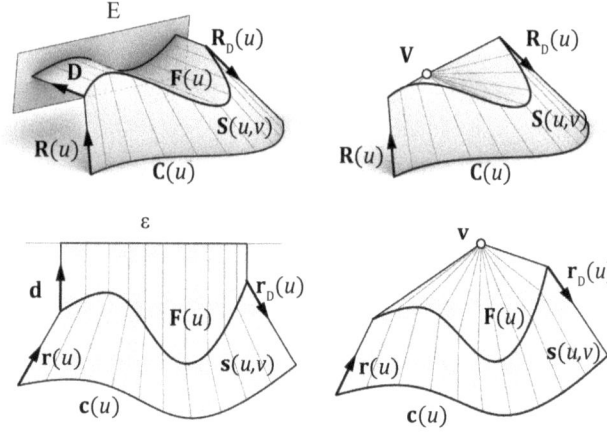

3 Illustration of the notation in Section 2; Left: Patch-to-cylinder construction; Right: Patch-to-cone construction

folding algorithms, which was found to be an independently useful component.

The combination of sliding developables and the patch-to-patch folding constructions enables a wide variety of curved folded design. By serially connecting patch-to-patch folding steps, we obtain parametric design of curved-crease folding with multiple creases. In Section 4, we show case studies of versatile designs using our tool.

2 THEORETICAL BACKGROUND

2.1 Developable Surfaces

Developable surfaces are surfaces that can be obtained by bending a planar sheet without stretching or tearing. A developable surface is a combination of one or more of these basic types of surfaces: planes, cylinders, cones, and tangent developables (Pottmann et al. 2010, Chapter 15). Geometrically, these surfaces contain a family of lines, the so-called rule lines, along which the surface tangent planes are the same. Consequently, developable surfaces allow a parametrization $S(u,t)$ where parameter t follows the rule lines with normalized direction $R(u)$ that are attached to a curve $C(u)$ on the surface, that is,

$$S(u,t) = C(u) + t\, R(u).$$

If the surface is a cylinder, the rule line directions are parallel and thus $R(u)$ is constant. If the surface is a cone, the curve $C(u)$ may degenerate to a point, the apex of the cone.

The development of a developable surface is its planar counterpart, obtained by unbending the surface without stretching or tearing. Similarly, it can be parametrized by

$$s(u,t) = c(u) + t\, r(u),$$

where the small letters indicate the respective 2D counterparts. In the following, we will assume without loss of generality that $|R(u)| = |r(u)| = 1$, and that the speed of the parametrization of the curves is the same.

2.2 Patch-to-cylinder and Patch-to-cone Construction

A curved crease is a curve in 3D space that joins two developable surfaces and maintains that the combined surface is also developable. From a construction approach, given two independently developable surface patches, in general, joining them together along a curve will typically not result in a developable seam; the development will contain a gap or an overlap.

The special case where the two developments do fit together is referred to as a curved-crease folding, as the folded model can be obtained from a flat sheet. In this section, we briefly review patch-to-patch construction concepts by Mundilova (2019), which computes crease curves on a given first developable patch such that the computed second surface patch is either a cylinder or a cone constructed from a cylinder's profile curve base plane or cone apex in 3D and 2D.

In both cylinder and cone cases, we assume that the crease curve $F(u)$ and its developed counterpart $f(u)$ can be parametrized as

$$F(u) = C(u) + l(u)\, R(u) \quad \text{and} \quad f(u) = c(u) + l(u)\, r(u),$$

that is, we can reach a point on the crease curve by firstly following the curve on the surface and then going $l(u)$ units in the ruling direction (see Figure 3). The distance $l(u)$ depends on the type of the second surface. In the following, we discuss the ruling-by-ruling computation of $l(u)$ from distance constraints.

Patches to cylinders. Let S be the parametrization of a first surface. Recall that a profile curve $P(u)$ of a cylinder lies in a plane perpendicular to the cylinder's ruling direction. As unfolding preserves the angles between curves on the surface, the development $p(u)$ of a profile curve is a line orthogonal to the developed ruling direction. We specify the second surface by the base plane E and base line ε of the cylinder's profile curve or line in the development, respectively (see Figure 3). Let $D \cdot X = K$ be the Hessian normal form of E with unitized normal vector D, and $d \cdot x = k$ the normal form of ε with unitized normal vector d. We assume that the profile curve of the cylinder can be parametrized by

$$P(u) = F(u) + l_p(u) \, D \quad \text{and} \quad p(u) = f(u) + l_p(u) \, d,$$

with an initially unknown distance function $l_p(u)$, the distance between the crease curve and the base planes. Constraining $P(u)$ and $p(u)$ to E and ε results in two linear constraints for the unknown functions $l(u)$ and $l_p(u)$, that is,

$$K = P(u) \cdot D = C(u) \cdot D + l(u) \, R(u) \cdot D + l_p(u),$$
$$k = p(u) \cdot d = c(u) \cdot d + l(u) \, r(u) \cdot d + l_p(u),$$

solving for the two unknown functions $l(u)$ and $l_p(u)$ yields

$$l(u) = \frac{c(u) \cdot d - C(u) \cdot D - (k - K)}{r(u) \cdot d - R(u) \cdot D}$$

$$l_p(u) = \frac{(c(u) \cdot d - k) \, r(u) \cdot d - (C(u) \cdot D - K) \, R(u) \cdot D}{r(u) \cdot d - R(u) \cdot D}.$$

Patches to cones. Let **S** be the parametrization of a first surface and let the second surface be a cone specified by the normalized 3D and 2D apices **V** and **v** (see Figure 3). As the lengths on a developable surface are preserved when unfolding, a point on the crease curve $F(u)$ has the same distance to the 3D apex **V** as its developed counterpart $f(u)$ to the 2D apex **v**, that is,

$$| F(u) - V |^2 = | f(u) - v |^2.$$

The quadratic terms in this quadratic constraint on the length function $l(u)$ cancel out and solving for $l(u)$ results in

$$l(u) = \frac{1}{2} \frac{|v - c(u)|^2 - |V - C(u)|^2}{r(u) \cdot (v - c(u)) - R(u) \cdot (V - C(u))}.$$

2.3 Valid Surface Patch Combination

The length function $l(u)$ above gives the location of the crease curve as the intersection of developable surfaces. The intersection of two surfaces and pairwise split results in four patches (see Figure 4). There are two combinations of two patches which create a developable combination of surfaces, resulting in a model that can be constructed by folding as it has no overlap in the development. This combination is found by selecting one patch from either side of the crease curve, choosing one from each surface. A natural extension of Maleczeck et al. (2020) gives a mathematical description. Namely, the valid patch combination is encoded in the denominator of the length function,

$$D(u) = r(u) \cdot d - R(u) \cdot D \quad \text{or} \quad D(u) = r(u) \cdot (v - c(u)) - R(u) \cdot (V - C(u))$$

for cylinder and cone, respectively. Specifically, if we want to combine the second surface patch containing the profile

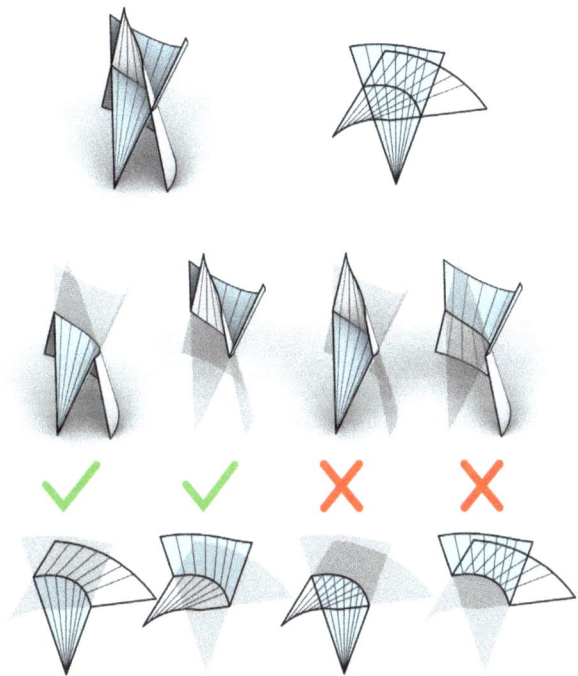

4   Intersecting developable surfaces along a curved crease and their development (top); two developable and two non-developable surface patch combinations (bottom)

curve or the apex, we rebuild the compatible first surface by extending the 3D and 2D crease curves in $R_D(u)$ and $r_D(u)$ directions, where

$$R_D(u) = -\text{sign}(D(u)) \, R(u) \quad \text{and} \quad r_D(u) = -\text{sign}(D(u)) \, r(u),$$

as illustrated in Figure 3.

3 IMPLEMENTATION

Lotus was written in C# and leverages the geometry primitives in the RhinoCommon library. Each component in *Lotus* requires the input surface to be a developable NURBS surface of degree $1 \times n$ where $n \geq 1$, with $2 \times m$ control points where $m \geq 3$ (quads do not have a defined rule line direction). Note that *Lotus* can also take polyhedral surface patches as input.

3.1 Discretization

Each component begins by discretizing the input surface into a strip of (almost) planar quads or triangles. We evaluate the rule lines of the developable surface and connect them appropriately to an (almost planar) quad or triangle mesh that will be used for the subsequent computations. The rule line sampling density for the discretized mesh can optionally be adjusted by the user (see Figure 5).

Furthermore, if the input surface is not planar, we compute a development by laying out the quads and triangles in the

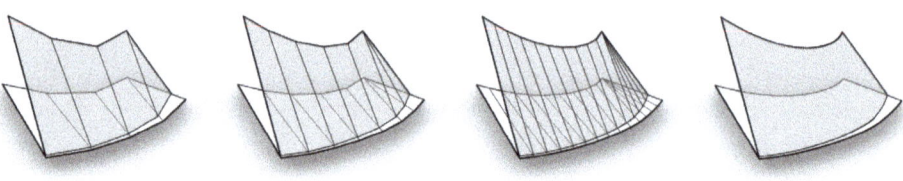

5 Different levels of discretization of a developable surface and their developments

6 Sliding developable construction process

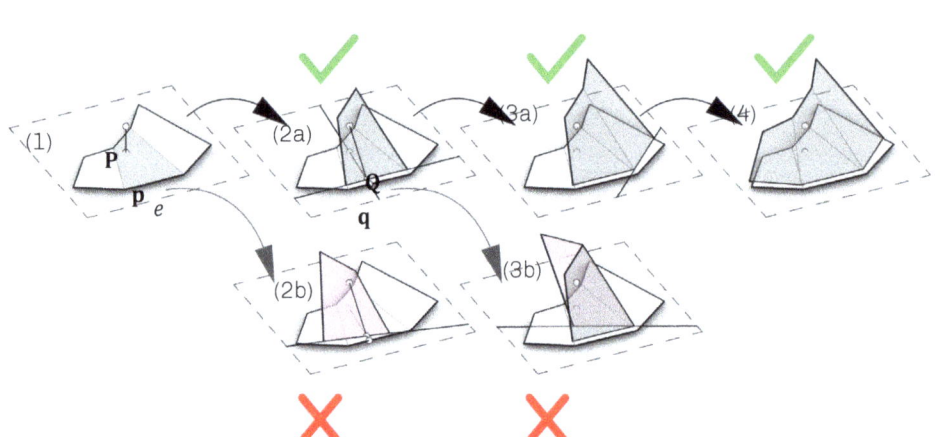

plane. For convenience, *Lotus* provides a custom unrolling component, *UnrollDevelopable*, specifically tailored for developable surfaces.

3.2 Sliding Developables

The *SlidingDevelopable* component takes as input a 3D point **P** and a planar ruled surface **s** (see Figure 6). Such a surface can be generated in Rhino by extruding a planar curve in-plane or lofting two curves that are in the same plane. In the following, let **s** be positioned in the *xy*-plane. The input point **P** is expected to be positioned in such a way that its projection to the *xy*-plane **p** lies inside **s**. The component constructs a surface **S** whose development is **s**, such that one of the surface's curved boundaries lies in the *xy*-plane and **P**'s development as a point on **S** is **p**. Only the curved boundaries of the surface patch can be constrained to lie in the plane. Generally, there are two such curves, but in the case of a cone it's possible that one curve degenerates to a point. In case of two curved boundaries, the *SlidingDevelopable* component provides a Boolean input to select one. Note that the desired **S** may only exist for points in a small neighborhood of **s**, see Section 3.4.

In this algorithm, illustrated in Figure 6, we first determine which quad or triangle contains **p** by orthogonally projecting into the plane (1). We then place this quad or triangle in 3D space by bringing **p** to **P** while anchoring the selected boundary edge *e* in the *xy*-plane. We reduce the number of ways to position the incident face to two by selecting a point **q** on the boundary *e* of **s** and set **Q** to be the point on the segment **pq** with distance dist(**p**,**q**) to **P** (2a and 2b). Finally, from these two solutions we pick the one in which the boundary direction is closer to the corresponding direction of **s** (2a). After we solve one face, we can find a solution for its neighbor face(s). The common ruling of a positioned face and its neighbor acts as an axis of rotation for the placement of the neighbor face. In general, there will be two configurations that place the appropriate boundary edge in the *xy*-plane (3a and 3b); we pick the configuration in which the planes enclose a larger dihedral angle (3a). If there is no valid configuration, **P** is too far from the input surface and the algorithm returns only a partial result. We repeat this process to both sides of the initial face (4).

In our suggested use, we start with point **P** near the center of the surface, incident to it, and slowly move the point upwards orthogonal from the surface. This is the area least likely to have issues and will usually return a solution. We explore the space of valid solutions by moving the point both away from the surface, and sideways towards each of the edges of the surface.

3.3 Patch-to-patch Implementation

Lotus offers two components, *PatchToCylCone(3pt)* and *PatchToCylCone(tan)*, which implement the patch-to-patch construction method of Mundilova (2019) as outlined in Sections 2.2 and 2.3. Both components construct the crease between a user-specified general developable surface patch and a cylinder or cone, and upon success return the generated surfaces and their development.

Component *PatchToCylCone(3pt)* has three input points: P_1, P_2 and **V** (see Figures 7 and 8). The first two points are expected to lie on the surface, and if a crease curve is returned, it will pass through P_1 and P_2. If the second surface is a cylinder the 3D ruling direction is P_1V, and V will lie on the base plane x of the profile curve of the cylinder. If the second surface is a cone, the 3D apex position will be V.

Component *PatchToCylCone(tan)* has two input points, P_1 and V, and one input angle α (see Figure 9). The point P_1 is expected to lie on the surface, and if a crease curve is returned, it will pass through this point. Furthermore, the tangent of the crease curve and the ruling of the input surface will enclose the user specified angle. The functionality of V is analogous to the *PatchToCylCone(3pt)* component.

Both components contain a Boolean toggle to allow switching between valid patch combinations, as shown in Figure 4.

Each component's unique set of inputs finds the 3D and 2D profile curve's base plane or cone apices necessary for the formula in Section 2, but because the components only operate on the 3D model without knowledge of the 2D developed state, the inputs vary from the inputs in the equations. Each component's approach to obtaining the 2D profile curve's base plane or cone apex is discussed in detail in the following paragraphs.

Both algorithms begin by discretizing the input surface and unrolling it to the *xy*-plane. We compute p_1 and if applicable, p_2, the locations of P_1 and P_2 on the unrolled surface. We find the location of the 2D profile curve's base line ε or apex position v by computing the 2D counterpart v of V. In case of the patch-to-cylinder construction, the base line ε has normal vector p_1v and contains v. In case of the patch-to-cone construction, v is the location of the developed apex.

PatchToCylCone(3pt) – Cylinder. As the distances between points on surfaces in 3D and their developed counterparts must be the same, we have

$$\mathrm{dist}(P_1, E) = \mathrm{dist}(p_1, \varepsilon) \quad \text{and} \quad \mathrm{dist}(P_2, E) = \mathrm{dist}(p_2, \varepsilon).$$

Thus, ε is a common tangent of the two circles c_1 and c_2 centered at p_1 and p_2 with radii $\mathrm{dist}(P_1, E)$ and $\mathrm{dist}(P_2, E)$, respectively. In general, there are two suitable solutions for the outer tangents, candidates of ε on opposite sides of p_1 and p_2, which determine two solutions for v as their base point of p_1. We choose ε such that $\{p_1, p_2, v\}$ are positioned counterclockwise.

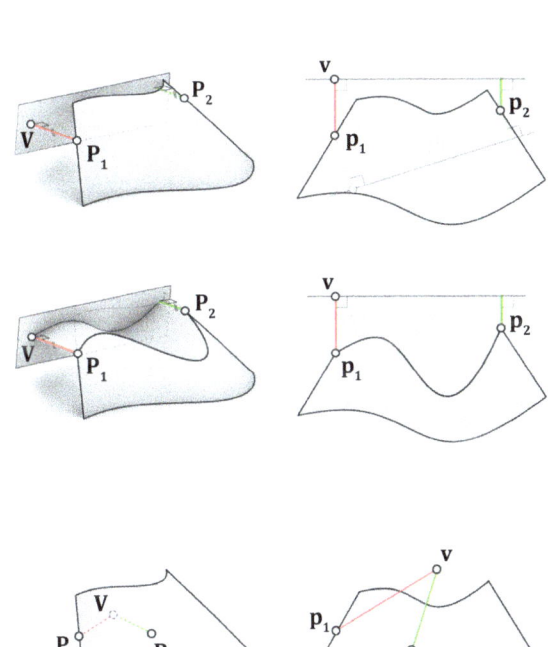

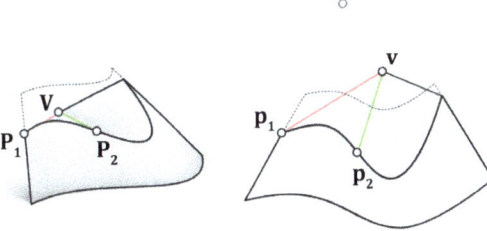

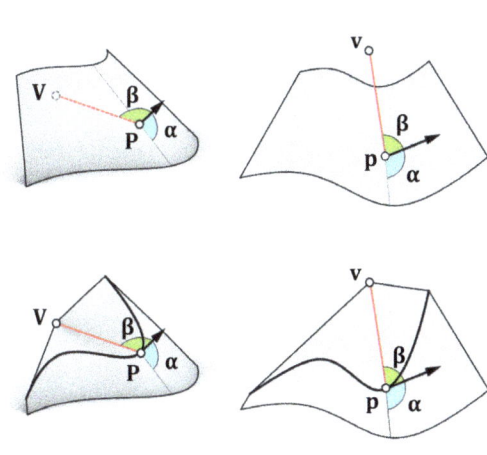

7 Input of the *PatchToCylCone(3pt) - Cylinder* component

8 Input of the *PatchToCylCone(3pt) - Cone* component

9 Input of the *PatchToCylCone(tan)* component

PatchToCylCone(3pt) – Cone. As the distances between points on surfaces in 3D and their developed counterparts must be the same, the location of **v** is constrained by two distances,

$$\mathrm{dist}(\mathbf{P}_1,\mathbf{V}) = \mathrm{dist}(\mathbf{p}_1,\mathbf{v}) \quad \text{and} \quad \mathrm{dist}(\mathbf{P}_2,\mathbf{V}) = \mathrm{dist}(\mathbf{p}_2,\mathbf{v}).$$

Thus, **v** is an intersection point of two circles c_1 and c_2 centered at $\mathbf{p}_1$ and $\mathbf{p}_2$ with radii $\mathrm{dist}(\mathbf{P}_1,\mathbf{V})$ and $\mathrm{dist}(\mathbf{P}_2,\mathbf{V})$, respectively. In general, there are two solutions for **v** on opposite sides of $\mathbf{P}_1$ and $\mathbf{P}_2$. Again, we choose the solution such that $\{\mathbf{p}_1,\mathbf{p}_2,\mathbf{v}\}$ are positioned counterclockwise.

PatchToCylCone(tan). After finding the location of $\mathbf{p}_1$, we compute its 3D and 2D incident rulings. We then reconstruct the 3D tangent line in the incident tangent plane and 2D tangent line as we are given the counterclockwise angle that they enclose with the ruling. Let α be the angle between $\mathbf{P}_1\mathbf{V}$ and the 3D tangent. As developing a surface preserves angles and distances, we find **v** as the endpoint of the line segment of length $\mathrm{dist}(\mathbf{P}_1,\mathbf{V})$ enclosing the counterclockwise angle β with the 2D tangent line.

Finally, after *PatchToCylCone(3pt)* or *PatchToCylCone(tan)* determines the 2D ruling direction or apex position, we apply the formulas in Section 2.2 rule line by rule line. First, we build the second surface containing the cylinder's profile curve and cone apex. We then build the compatible first surface according to Section 2.3.

Note, that the crease curve can exceed curved boundaries of the input surface (see Figure 10). In addition, it is possible for the profile curve of the second surface to intersect the crease curve (see Figure 7). In both cases, the corresponding surfaces need to be appropriately extended. On the other hand, some inputs can result in crease curves with large amplitudes, potentially exceeding the cones apex, such that extending the surfaces would be undesirable. In this case, the plug-in returns without result. If the input surfaces are smooth, we interpolate the appropriate surface pair's sampled points to return smooth surfaces.

Note that we choose **v** such that $\{\mathbf{p}_1,\mathbf{p}_2,\mathbf{v}\}$ are positioned counterclockwise. Exchanging the order of $\mathbf{P}_1$ and $\mathbf{P}_2$ changes **v** to the other location.

3.4 Limitations

Lotus requires sensible input from the user, as not all input configurations provide a desired result. Common unrealistic inputs are as follows:

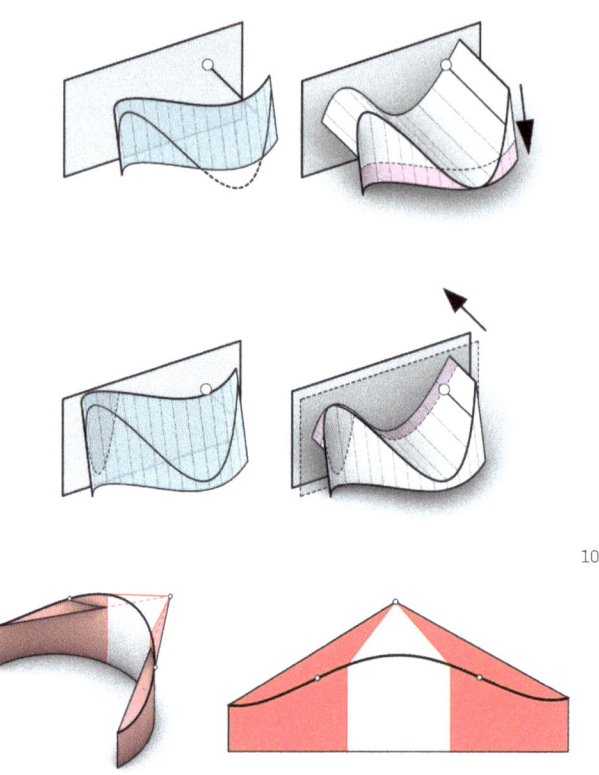

10

11

SlidingDevelopable: If the input point is further than the valid range from the base plane, a 3D configuration does not exist. Beyond the trivial case where the distance is longer than the ruling length, there is a geometric limit resulting from the shape of the slide curve of the input patch. For explicit limits see Mundilova (2019).

PatchToCylCone(3pt): For some configurations of the input points, finding the location of ε or **v** might not be possible. In the cylindrical case, if the input is such that one of the two circles c_1 and c_2 is contained in the other, computing the common outer tangent ε will fail. In the conical case, if the input is such that the circles c_1 and c_2 do not intersect, computing their intersection **v** will fail.

PatchToCylCone(tan): For input angles α close to 0 mod π, the crease curves tangent direction will be close to the incident ruling direction. This may cause large absolute values of the length function and large extensions of the surface.

Furthermore, *Lotus* does not check whether surfaces intersect (see Figure 11). Once a model is completed, it falls to the user to double check that no surfaces are intersecting.

4 DESIGN STUDY

Finally, we present some usage examples of our method and show works of students and colleagues using *Lotus*.

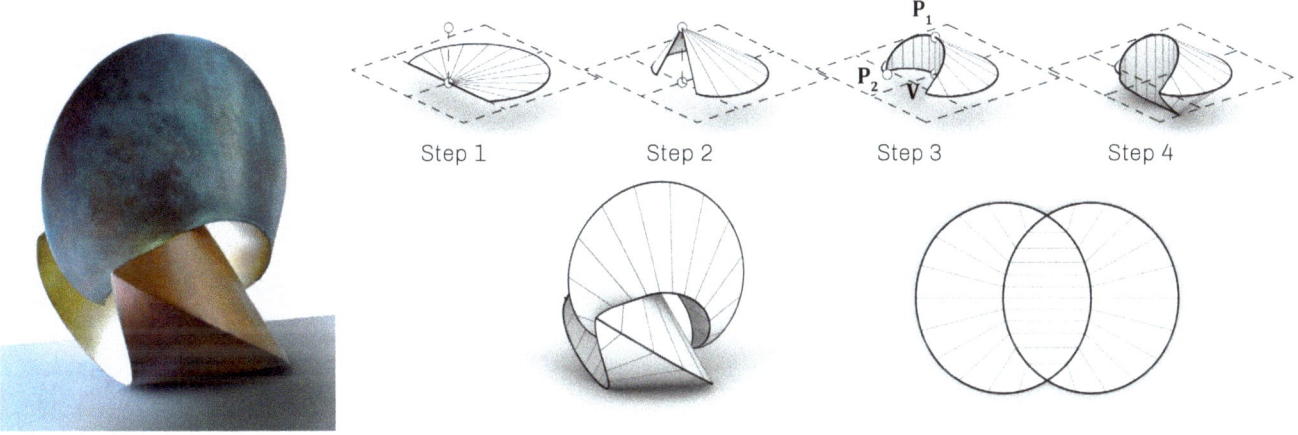

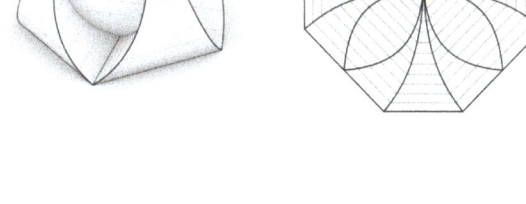

10 Crease curve exceeds boundary of input surface (top) and crease curve intersects profile curve of cylinder (bottom); in both cases, the surfaces need to be appropriately extended

11 Illustration of intersecting surfaces generated by the *Lotus* component

12 Left: Susan Latham's sculpture "Attraction" (2008); Right: Step-by-step reconstruction of Susan Latham's sculpture with *Lotus*

13 Jeanine Mosely's bud reconstructed with *Lotus*

4.1 Reconstruction of Susan Latham's Vesica Piscis

Susan Latham's sculpture "Attraction" (Santa Fe, 2008) consists of two folded Vesica Piscis. One of the shapes is obtained by folding flat material in the shape of two overlapping disks, which are positioned such that the center of one lies on the boundary of the other. These two overlapping disks are folded along their inner circle segments, such that the outer circle segments can be glued together. Mundilova and Wills (2018) prove that the resulting shape consists of two cones and a cylinder. The developed apex of the cones coincides with the center of the overlapping disks, and the cylinder's ruling direction is perpendicular to the line along which one circle would be rotated onto the other. We can reconstruct the Vesica Piscis with *Lotus* as follows (Figure 12):

- Step 1: Construct a unit circle segment in the *xy*-plane centered at $\{½,0,0\}$ and trim it with the *y*-axis. Extrude the circular arc to a planar cone with apex $\{0,0,0\}$.
- Step 2: Position a point at $\{0,0,¾\}$ and use it together with the planar cone as input for the *SlidingDevelopable component*. This results in a 3D cone.
- Step 3: Evaluate the cone at its *uv*-parameter $\{²/_3,½,0\}$ to obtain $\mathbf{P}_1$. Construct $\mathbf{V} = \mathbf{P}_1 + \{0,0,-½\}$.
  - *PatchToCylCone(3pt)*: Evaluate the cone at $\{0,1,0\}$ to obtain $\mathbf{P}_2$. Use *PatchToCylCone(3pt)* to obtain a cylinder with this input.

- *PatchToCylCone(tan)*: Alternatively, use *PatchToCylCone(tan)* with angle $-π/2$ to obtain a cylinder with this input.
- Step 4: Mirror the cylinder and cone on the *xy*-plane to obtain the closed folded Vesica Piscis shape.

4.2 Mosely's Bud

Jeanine Mosely's "Square-Based Origami Bud" is an intricate design whose curved crease elements can be approximated by a series of cylinders (Lang 2018). We reconstruct this shape, shown in Figure 13, by twice iterating *PatchToCylCone(3pt)*'s cylinder-to-cylinder construction to compute an eighth of the design. We chose the input points so that the first and third cylinders' profile curve base planes enclose an angle of $π/4$. By reflection on these planes, we obtain a closed 3D shape.

4.3 Living Room Table

The table's structure in Figure 14 was designed by Duks Koschitz (Professor at Pratt Institute, New York) by using *PatchToCylCone(3pt)* twice consecutively. Beginning with the upper cylindrical surface, based on an elastica in section, the sides were added by a patch-to-cone construction and were then connected to another cylindrical surface. Both cylindrical surfaces have parallel rulings that result in a common symmetry plane for the generation of the hollow tube. While the upper surface is joined as one continuous surface, the dove-tail-shaped joint is located on

14 Living room table designed by Duks Koschitz with *Lotus*

15 Left: Two variations of a pavillion designed by Lucy Czarnecka with *Lotus*; Right: Paper model and development

the bottom surface. The straight edges are folded over to give the overall shape more stiffness.

4.4 Curved-crease Pavilion

During a course taught by Maleczek and Mundilova at the University of Innsbruck, students had the opportunity to develop designs with a beta version of the *Lotus* plug-in. Lucy Czarnecka used the plug-in to develop a curved folded pavilion on a pentagonal base with curved walls that has inward-folded openings and a roof with five lenses. After digitalization of her paper studies (see Figure 15), she adapted her design so that development of the structure could be fabricated from a single edge-connected component. The design uses 19 consecutive creases for the roof with five lenses, all computed with the *PatchToCylCone(3pt)* component. The parametric setup allowed her to experiment with the design and develop variations.

CONCLUSION

In conclusion, we have shown that *Lotus* is a ready to use plug-in for Rhino/Grasshopper that provides tools for generating parametric curved folding. *Lotus* handles both discretized and non-discretized geometries and offers a parametric interface for interactive design. We explained its implementation and functionality, and highlighted its versatility and user-friendliness for a general audience without a deep knowledge of curved folding. Looking ahead, this component can be extended to incorporate details about materials and fabrication, such as material thickness.

ACKNOWLEDGEMENTS

We thank Tony Wills and Susan Latham for the initial inspiration and the participants of the Structural Origami Gathering 2020 for their input and feedback. We thank Duks Koschitz and Lucy Czarnecka for their designs. Robby Kraft, Rupert Malezcek, and Klara Mundilova were supported by Project "Fold to Bend" / FFG Austria, Project #864731. Tomohiro Tachi is supported by JST Presto JPMJPR1927.

REFERENCES

Bhooshan, Shajay, El-Sayed Mustafa, and Chandra Surajansh. 2014. "Design-Friendly Strategies for Computational Form-Finding of Curved Folded Geometries: A Case Study." In *Proceedings of the Symposium on Simulation for Architecture & Urban Design*. San Diego: Society for Computer Simulation International. 117–124.

Chalfant, Julie. 1997. "Analysis and design of developable surfaces for shipbuilding". MSc Thesis, Massachusetts Institute of Technology.

Demaine, Erik, Martin Demaine, Duks Koschitz, Tomohiro Tachi. 2015. "A review on curved creases in art, design and mathematics." *Symmetry: Culture and Science* 26 (2): 145–161.

Foschi, Riccardo, Robby Kraft, Rupert Maleczek, Klara Mundilova and Tomohiro Tachi. 2021. "How to use parametric curved folding design methods – a case study and comparison." To appear in *Proceedings of the IASS Annual Symposium* 2020/21.

Glaeser, Georg and F. Gruber. 2007. "Developable surfaces in contemporary architecture." *Journal of Mathematics and the Arts* 1 (3): 59–71.

Kilian, Martin, Simon Flöry, Zhonggui Chen, Niloy Mitra, Alla Sheffer, and Helmut Pottmann. 2008. "Curved folding". *ACM Transactions on Graphics (TOG)* 27 (3): 1–9.

Koschitz, Duks. 2016. "Designing with curved creases." In *Advances in Architectural Geometry 2016*, edited by Sigird Adrianssens, Fabio Gramazio, Matthias Kohler, Achim Menges, and Mark Pauly. Herausgeber: Hochschulverlag, AG ETH Zürich vdf. 82–103.

Lang, Robert. 2018. *Twists, Tilings and Tessellations: Mathematical Methods for Geometric Origami*. Natick, Mass.: A K Peters, Limited/CRC Press..

Lawrence, Snezana. 2011. "Developable surfaces: Their history and application." *Nexus Network Journal* 13 (3): 701–714.

Lienhard, Julian. "Bending-active structures: form-finding strategies using elastic deformation in static and kinetic systems and the structural potentials therein." PhD thesis, Stuttgart, 2014.

Maleczek, Rupert, Gabriel Stern, Astrid Metzler and Clemens Preisinger. 2019. "Large Scale Curved Folding Mechanisms." In *Impact: Design with all Senses; Proceedings of the Design Modelling Symposium, Berlin 2019*. Berlin: Springer. 539–553.

Maleczek, Rupert, Klara Mundilova, and Tomohiro Tachi. 2021. "Curved Crease Edge Rounding of Polyhedral Surfaces." In *Advances in Architectural Geometry 2020*. Paris: Ponts et Chaussées. 131–153.

Mitani, Jun, and T. Igarashi. 2011. "Interactive Design of Planar Curved Folding by Reflection." In *Proceedings of the Pacific Conference on Computer Graphis and Applications*. 77–81.

Mundilova, Klara. 2019. "On mathematical paper-folding: Sliding developables and parametrizations of folds into cylinders and cones." *Computer-Aided Design* 10: 34–41.

Mundilova, Klara, and Tony Wills. 2018. "Folding the Vesica Piscis." In *Proceedings of the 21st Annual Conference of BRIDGES: Mathematics, Music, Art, Architecture, Culture*. 535–538.

Pottmann, Helmut, Andras Asperl, Michael Hofer, and Axel Kilian. 2011. *Architectural Geometry*. Springer & Bentley Institute Press.

Rabinovich, Michael, Tim Hoffmann, Olga Sorkine-Hornung. 2018. "Discrete geodesic nets for modeling developable surfaces." *ACM Trans. Graph.* 37 (2): 16:1–16:17.

Rabinovich, Michael, Tim Hoffmann, Olga Sorkine-Hornung. 2019. "Modeling curved folding with freeform deformations." *ACM Trans. Graph.* 38 (6): 170:1–170:12.

Solomon, Justin, Etienne Vouga, Max Wardetzky and Eitan Grinspun. 2012. "Flexible developable surfaces." *Computer Graphics Forum* 31 (5): 1567–1576.

Tang, Cheng-Cheng, Po Bo, Johannes Wallner, and Helmut Pottmann. 2016. "Interactive design of developable surfaces." *ACM Trans. Graph.* 35 (2): 12:1–12:12.

IMAGE CREDITS

Figure 12 (left): Susan Latham, 2008.

Klara Mundilova is a PhD student in Computer Science at Massachusetts Institute of Technology. She is exploring the geometry of curved crease origami and developing software to design such foldings.

Erik Demaine is a Professor of Computer Science and Engineering at Massachusetts Institute of Technology. His research interests range from algorithms, to data structures for improving web searches to the geometry of understanding how proteins fold, to the computational difficulty of playing games.

Riccardo Foschi is a research fellow and adjunct professor in the field of architectural representation and real time rendering in the University of Bologna. His main research interests are virtual representation, virtual reconstruction of never built or lost architectures, applied digital origami, parametric and algorithmic modeling, descriptive geometry, and surveys of cultural heritage.

Robby Kraft is an origami artist, programmer, and instructor in New York City, currently working abroad as a researcher at the University of Innsbruck.

Rupert Maleczek is an architect, researcher and digital consultant, currently working as Senior Scientist at the institute of structure and design (i.sd) at the University Innsbruck. In his multidisciplinary work, he explores the relation between form, structure, performance, materiality, and the digital, as well as physical production. The aim of his research is the understanding of complex relations to enhance control over them through controlled simplification.

Tomohiro Tachi is an associate professor in Arts and Sciences at the University of Tokyo. He keeps exploring three-dimensional and kinematic forms through computation and developed origami software tools "rigid origami simulator," "origamizer," and "freeform origami," which are available from his website. His research interests include origami, structural morphology, computational design, and fabrication.

Arbor: Tectonic Contingencies and Ecological Engagement

Adam Marcus
Variable Projects /
California College of the Arts

1   View of *Arbor*

Introduction

Arbor is a data spatialization of the urban forest of Palo Alto, California. The sculptural installation consists of 120 ribs arranged radially within King Plaza fronting Palo Alto City Hall. It uses the database of over 45,000 public trees in the city's Open Data Portal (City of Palo Alto, n.d.) as the basis for a collective, three-dimensional map of one aspect of the city's ecology. The installation performs like a compass, with each rib corresponding directionally to a respective "pie slice" of territory raiding outwards from City Hall. The trees are represented by bumps on the outer edge of each rib, so the zones with more trees result in ribs with more relief. The ribs are arranged in a circle, gradually changing in height, profile, and color to create a dynamic form that is different from each side.

Data Spatialization and Environmental Representation

Arbor looks to historical examples of optical devices that operate radially, such as the zoetrope and the cyclorama, both of which use radial geometry to create novel and immersive environmental representations. Just as these devices use conventions of optical abstraction such as perspective, trompe-l'oeil, and depth of field to represent spatial environments (Crary 1990), *Arbor* employs techniques of data-driven analysis, design computation, and digital fabrication to produce an alternative representation of the city—one premised on its inventory of trees. This approach—data spatialization—expands computation's purview beyond technical concerns like efficiency and performance optimization to include more qualitative capacities of public engagement and ecological

PRODUCTION NOTES

Architect: Adam Marcus /
             Variable Projects
Client: City of Palo Alto
Status: Built
Location: Palo Alto, California, USA
Date: 2021

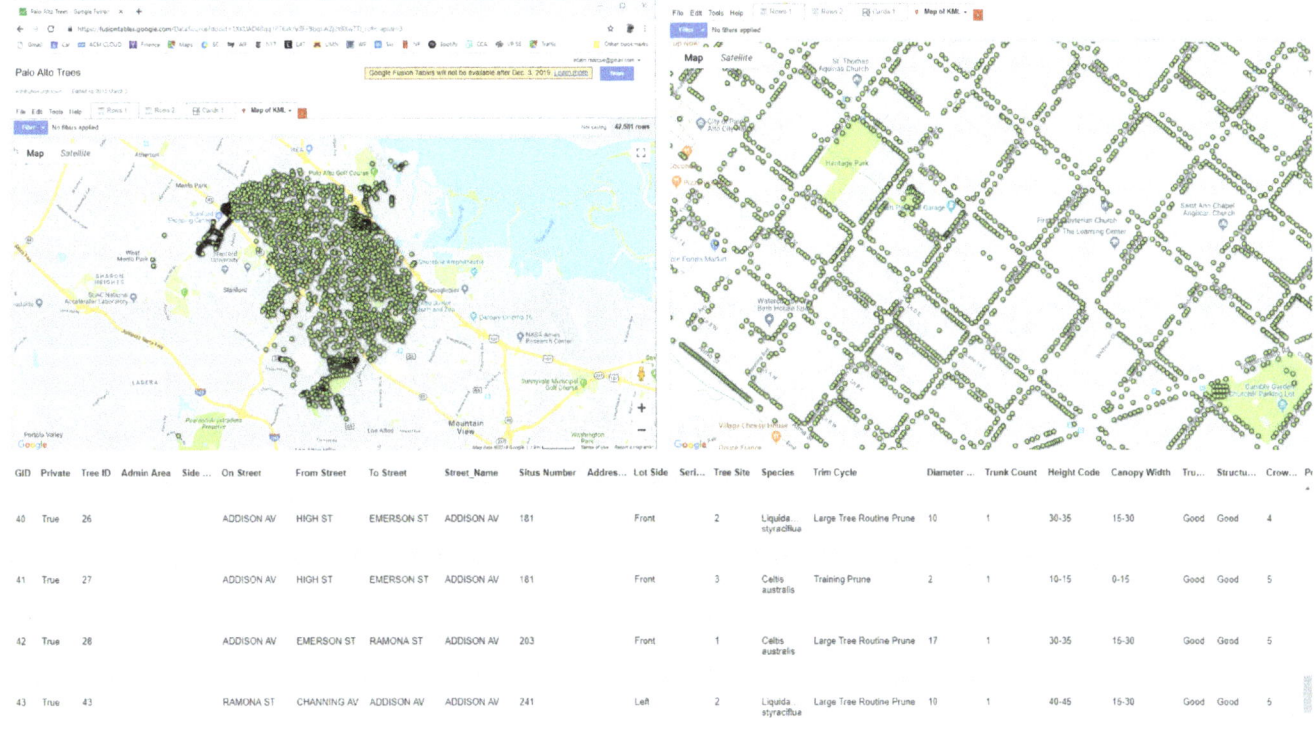

2 Screenshots of the Palo Alto's Open Data Portal, which contains a geolocation database of every tree in the public realm

3 Mapping study translating the tree location data to a three-dimensional design environment

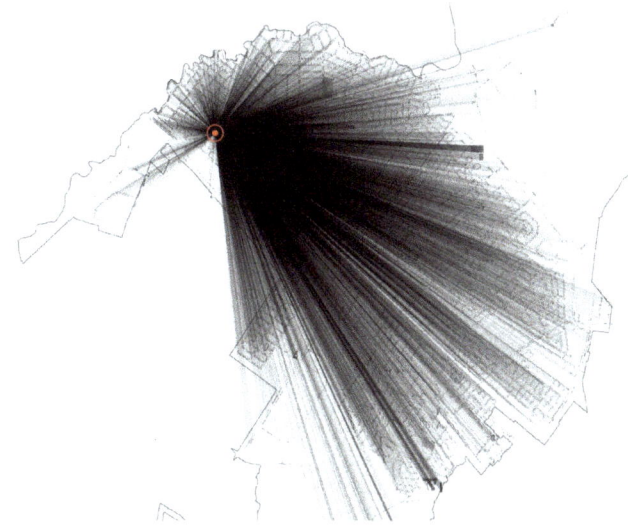

4 Lines connect the project site, City Hall, with each of the 45,000+ public trees in the city

awareness (Marcus 2014; Marcus, Dean, Kim, and Reichert 2017).

Design Process

The Open Data Portal contains granular data for each of the 45,000+ trees in the city's public realm; parameters include geolocation coordinates, species, trunk diameter, trunk height, and canopy diameter, among many others (Figure 2).

The design process began by developing a custom script using Grasshopper (Rutten 2019) to parse this data and translate it to a parametric design environment that allowed for creative iteration in producing three-dimensional form and tectonic assemblies. The model positions the site, King Plaza, within the field of trees, represented as points located at the coordinates derived from the tree geolocation data (Figure 3, Figure 4). Using City Hall as a

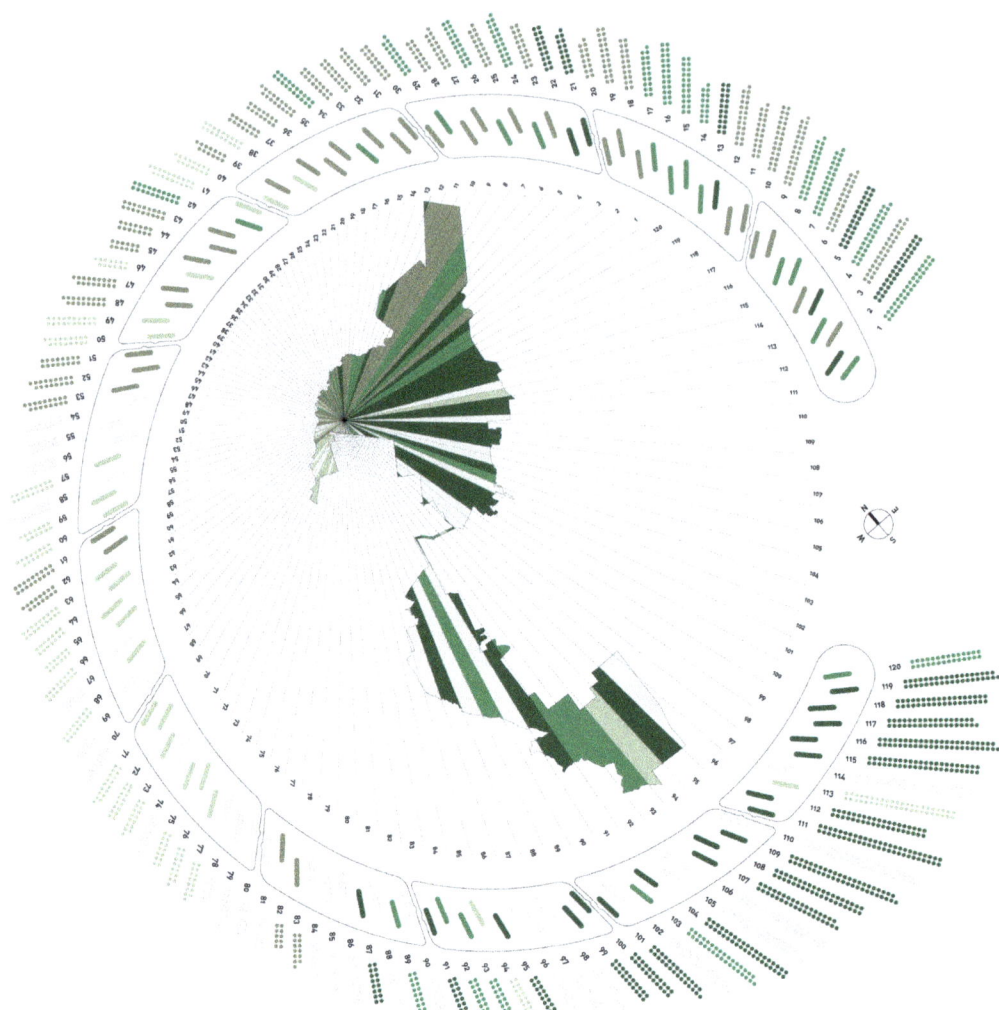

5   This diagram explains how tree location data from the city's Open Data Portal was translated into three-dimensional form. The map of Palo Alto shows the city divided into radial slices, with the center located at City Hall. The slices, numbered from 1 to 120, correspond to the ribs in the installation, which are also numbered according to the plan drawing on the outer edge of the diagram. The circular dots for each rib measure the trees in that particular area of the city. Each dot corresponds to approximately 15 trees. The more trees, the more dots, and the more bumps in that particular rib.

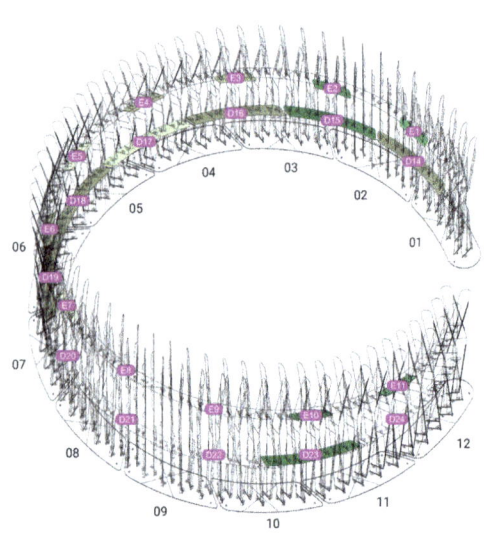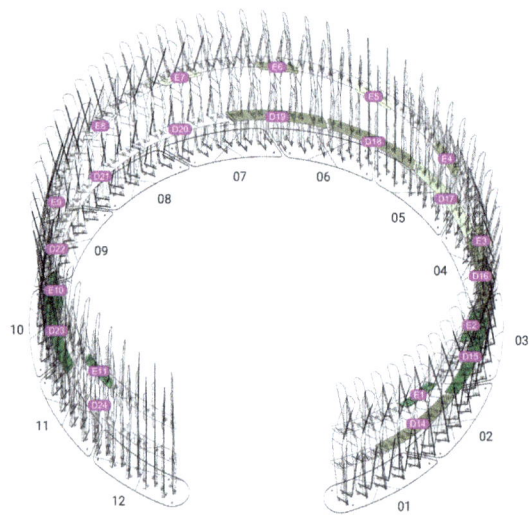

6   Assembly drawings showing the different component types

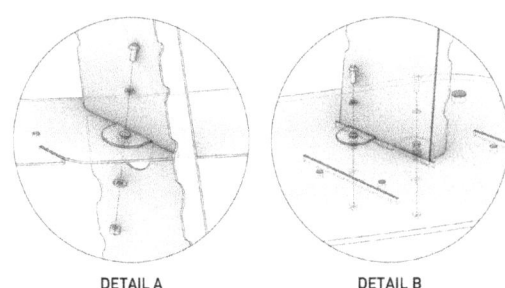
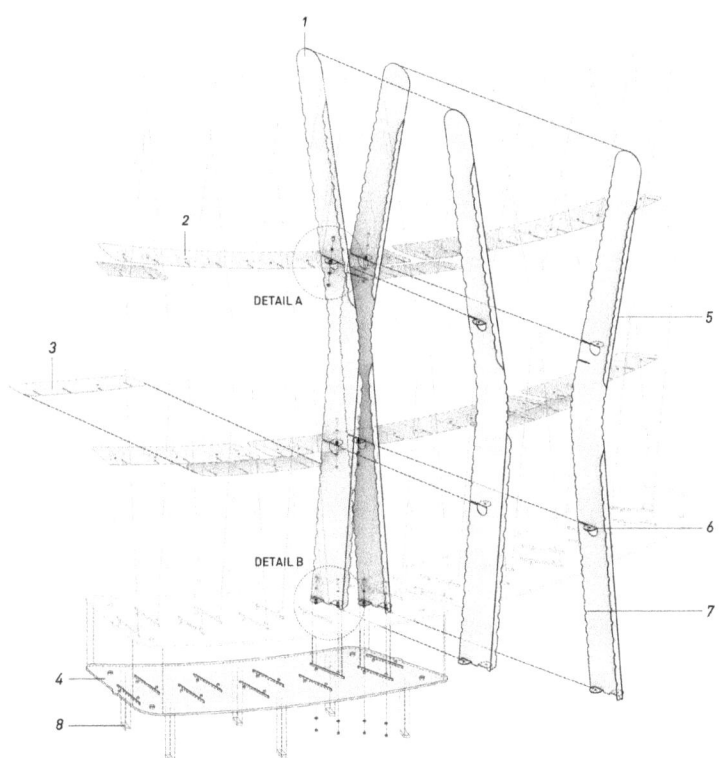

1— PRIMARY (VERTICAL) RIB - 16 GA. STEEL
2— SECONDARY (HORIZONTAL) RIB - 16 GA. STEEL
3— TERTIARY (HORIZONTAL) RIB - 16 GA. STEEL
4— BASE PLATE - 1/4" STEEL
5— INTEGRAL FLANGE @ TOP AND BOTTOM OF EACH PRIMARY RIB
6— INTEGRAL TAB @ EACH CONNECTION POINT
7— BUMPS ALONG RIBS CORRESPOND TO TREE DATA
8— POLYURETHANE RUBBER BUMPER, ADHERED TO BASE PLATE

7   Exploded axonometric and connection detail drawings

center point, 120 lines extend radially, creating 120 slices of territory. The quantity of trees within each slice corresponds directly to the quantity of bumps on each of the structure's 120 vertical ribs; the more trees in a particular region, the more bumps on that particular rib (Figure 5). The script allows for fine tuning of the scale and number of bumps on each rib, maintaining the proportional distribution sourced from the tree data.

Tectonic Contingencies

The installation is designed to produce maximum spatial effect with a minimum amount of material that nonetheless maintains structural integrity as a long-term public artwork. 120 primary ribs fabricated from laser-cut, powder-coated 16 ga. steel with integral stiffening flanges are supported by 1/4 in. steel base plates, and connected by a series of staggered horizontal 16 ga. ribs (Figure 6). All connection points consist of a tab bent from the vertical rib that is mechanically fastened to the horizontal rib (Figure 7, Figure 9). A recursive scripting operation negotiates the joint locations with both the global variation in height across the vertical ribs and the variation of the data-driven relief pattern cut into each rib; the script calculates the joint location, generates the tab and slot geometry, ensures that the joints conform to structural constraints, accounts for fabrication and assembly tolerances that were determined by full-scale prototypes, and automatically produces the two-dimensional fabrication files. This algorithmic negotiation produces a tectonics of contingency between the ecological data and the material assembly, whereby the form, texture, and effect of the installation are resultant from both but irreducible to neither.

Computational Publics and Ecological Engagement

As an experiment with techniques of data spatialization, *Arbor* demonstrates one way to leverage computational processes for purposes of public and ecological engagement. Visitors to *Arbor* can enter the installation, use the didactic diagram mounted on the ground in the center to locate themselves spatially and in relationship to their homes, and understand how one of the city's most significant communal ecological resources—the urban forest—is distributed throughout the public realm. In a more abstract sense, the installation itself evokes a grove of trees, changing throughout the day and in different lighting conditions, producing dappled shadow patterns, and even sometimes gently swaying in the wind. As computational processes become ever more powerful and data-driven tools become increasingly associated with extractive and exploitative protocols (Benjamin 2019, Crawford 2021, Zuboff 2020), *Arbor* reminds us that design technology can take on more positive, public-oriented capacities that increase engagement with our broader environment.

8 Interior view of *Arbor*

9 Detail view of *Arbor*

10 Detail view of *Arbor*

11 Detail view of *Arbor*

ACKNOWLEDGMENTS
Design: Adam Marcus, Pete Pham
Steel Fabrication: Seaport Stainless
Powder Coating: Richmond Metal Painting
Structural Consultant: Taylor Brady / Hohbach-Lewin
Assembly Team: Adam Marcus, Pete Pham, Nadya Chuprina, Joe Saxe, David Bentley

REFERENCES
Benjamin, Ruha. 2019. *Race After Technology: Abolitionist Tools for the New Jim Code*. Cambridge: Polity.

City of Palo Alto. n.d. "Open Data Portal." Accessed January 5, 2019. https://data.cityofpaloalto.org/.

Crary, Jonathan. 1990. *Techniques of the Observer: On Vision and Modernity in the Nineteenth Century*. Cambridge: MIT Press.

Crawford, Kate. 2021. *Atlas of AI: Power, Politics, and the Planetary Costs of Artificial Intelligence*. New Haven: Yale University Press.

Marcus, Adam. 2014. "Centennial Chromagraph: Data Spatialization and Computational Craft." In *ACADIA 14: Design Agency; Proceedings of the 34th Annual Conference of the Association for Computer Aided Design in Architecture*, edited by D. Gerber, A. Huang, and J. Sanchez. Los Angeles: ACADIA. 167–176.

Marcus, Adam, Daniel Dean, John Kim, and Molly Reichert. 2017. "Meander: Data Spatialization and the Mississippi River." In *2015 TxA Emerging Design + Technology Conference Proceedings*, edited by K. Bieg. Austin: TxA. 100–121.

Rutten, David. *Grasshopper*. V. 1.0. McNeel. PC. 2019.

Zuboff, Shoshana. 2020. *The Age of Surveillance Capitalism: The Fight for a Human Future at the New Frontier of Power*. New York: PublicAffairs.

IMAGE CREDITS
Figures 1, 3-12: © Adam Marcus / Variable Projects

Figure 2: Palo Alto Open Data Portal

12 View of *Arbor*

Adam Marcus is a licensed and registered architect and educator. He directs Variable Projects, an independent architecture practice in Oakland, and he is a partner in Futures North, a Minneapolis-based public art collaborative dedicated to exploring the aesthetics of data. Adam is Associate Professor of Architecture at California College of the Arts in San Francisco, where he teaches design studios in computational design and digital fabrication, co-directs the Architectural Ecologies Lab, and collaborates with CCA's Digital Craft Lab. From 2011 to 2013, Adam was Cass Gilbert Assistant Professor at University of Minnesota School of Architecture, where he chaired the symposium "Digital Provocations: Emerging Computational Approaches to Pedagogy & Practice." He has also taught at Columbia University and the Architectural Association's Visiting School Los Angeles. Adam is a graduate of Brown University and Columbia University's Graduate School of Architecture, Planning and Preservation.

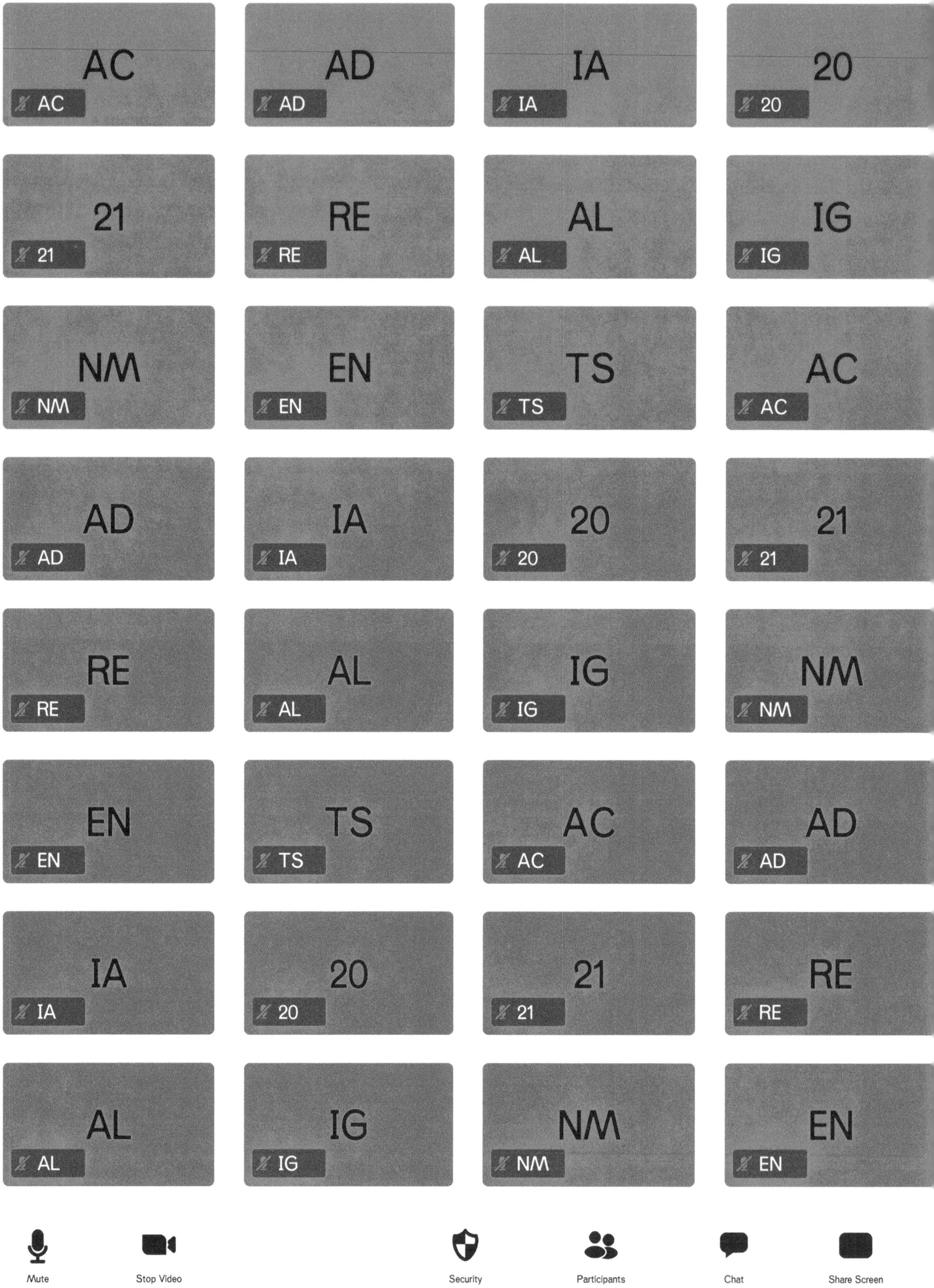
AC
AD
IA
20
21
RE
AL
IG
NM
EN
TS
AC
AD
IA
20
21
RE
AL
IG
NM
EN
TS
AC
AD
IA
20
21
RE
AL
IG
NM
EN
Mute
Stop Video
Security
Participants
Chat
Share Screen

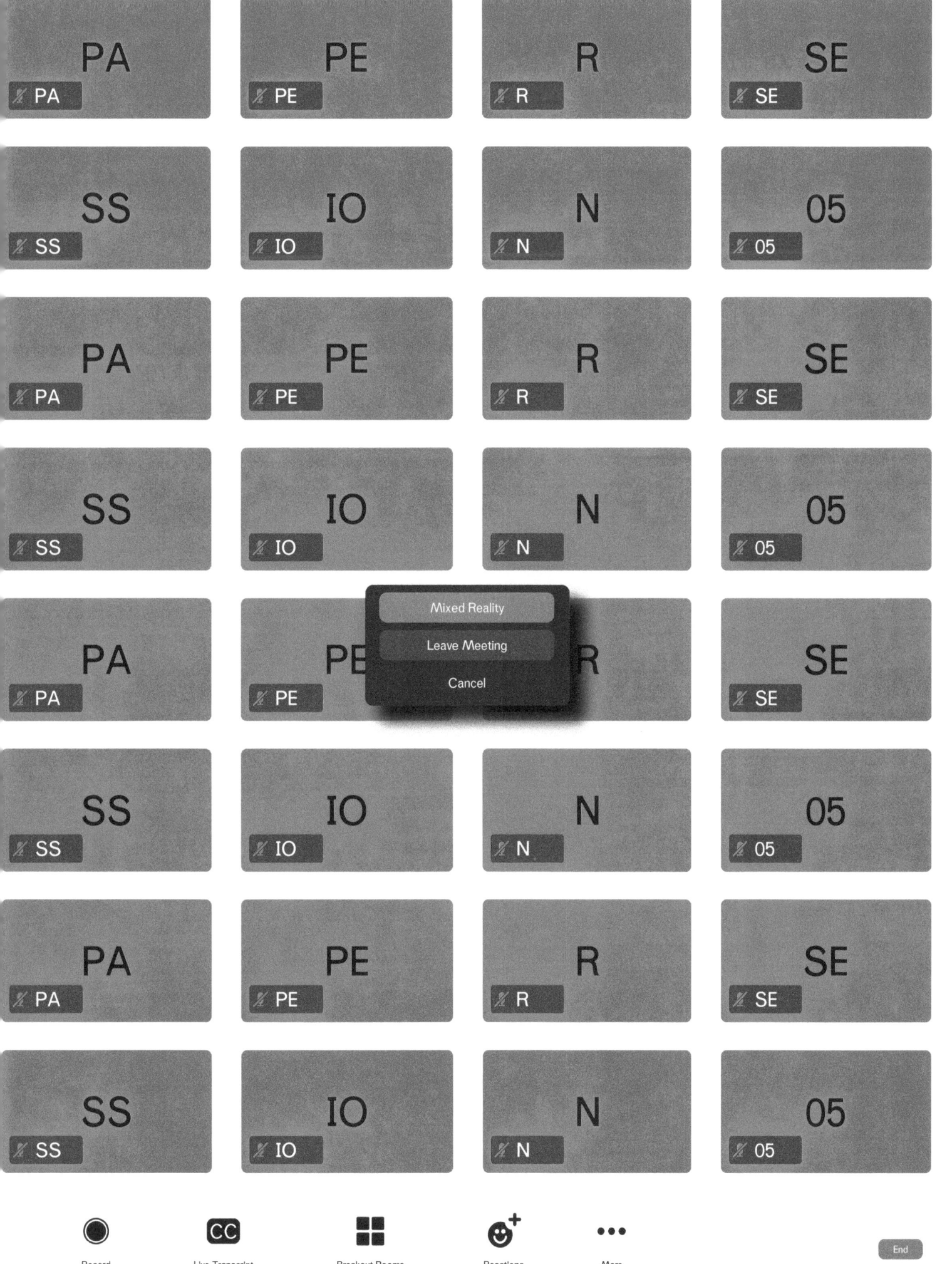

An Extended Reality Collaborative Design System

In-situ Design Reviews in Uncontrolled Environments

David Gillespie*
Applied R&D
Foster + Partners

Zehao Qin*
Applied R&D
Foster + Partners

Francis Aish*
Applied R&D
Foster + Partners

*Authors contributed equally to the research

1   Extended reality collaborative design system in use on site

ABSTRACT

This paper presents a new system that enables an eXtended Reality (XR) collaborative design review process, by augmenting an existing physical mockup or environment with virtual models at 1:1 scale in situ. By using this new hybrid approach, existing context can be extended with minimal or no base physical structure through a simulated VR/AR environment to facilitate stakeholder design collaboration in a manner that was previously either cost prohibitive or technically unfeasible. Through combining real and virtual in this way, the sense of realism can be enhanced, increasing engagement and participation in the design process. An approach to apply AR/VR to uncontrolled environments is described, allowing it to overcome challenges such as tracking and mapping, and allowing users to walk around freely in situ.

Two examples are presented where the system has been used in live project environments, one as a design tool for client review and engagement, and the other as part of a public planning process.

INTRODUCTION

Physical and virtual mockups are two of the most powerful ways of understanding and experiencing buildings before they are built. Whilst both can provide part of the sensory experience of a real building, neither can fully replicate it. Physical mockups can provide a very strong and intuitive tactile experience but are generally static and expensive. Virtual mockups can provide strong visual and auditory experiences and the ability to rapidly evaluate multiple options at minimal cost. However, they lack the intuitive navigation and human collaboration of the real world.

The aim of this investigation was to create a collaborative, intuitive and multi-user XR platform for design and planning reviews that allows the creation of realistic spatial experiences of design proposals in both large controlled indoor environments and large uncontrolled outdoor environments.

STATE OF THE ART

The rising need of participatory and collaborative design reviews sets demands on how communication between stakeholders, architects, managers, and end-users should be conducted and facilitated.

Physical mock-ups and prototyping are common practice, can provide significant experiential insights and readily engage a diverse set of project stakeholders. However, considerable time, effort, and cost is spent in their construction. Given finite project resources, only a limited size and number of these may be possible. Adding to this, mock-ups are usually discarded after they have served their purpose, contributing to single-use construction waste.

Current technological approaches are often based on existing tools that do not have straightforward solutions to improve communication in design reviews. However, collaborative and participatory virtual environments are being explored and deployed to overcome several problems that are critical to the creation and progression in a social creative design review (Arias et al. 2000, Fischer et al. 2005, Xue et al. 2012, Sunesson et al. 2008, Faliu et al. 2018). The use of XR systems has also been discussed for extending the capability of the shared virtual environments and improving the communication between users in design processes (Xue et al. 2012, Du et al. 2018).

XR has been widely explored in the AEC industry as an evaluation tool to support stakeholders, architects, and end-users in the evaluation of the proposed design to feed back into different design phases. Several authors claim that immersive virtual environments can be used as an approach for end-users to evaluate and validate the proposed design with the sense of presence in physical mock-ups (Bardram et al. 2002, Dunston et al. 2007, Eastman et al. 2008, Loyola et al. 2019) and improve the communication between architects, engineers, and stakeholders (Maldovan, Messner and Faddoul 2006, Wagner et al. 2013). Current XR systems mainly rely on two types of tracking technologies: outside-in and inside-out tracking systems. Outside-in XR Head Mounted Displays (HMD) are tracked by external stationary sensors or fixed infrastructure; inside-out XR HMDs are based on Simultaneous Localization And Mapping (SLAM) and are less constrained by fixed infrastructure.

AR has been examined in outdoor and indoor environments with the advance of the inside-out tracking and improved the design process (Azuma et al. 1999, Moeslund et al. 2003, Kieferle and Wössner 2003, Piekarski, Smith and Thomas 2004, Chung et al. 2009). However, AR devices have their limitations in the AEC industry. For instance, handheld tablet AR devices have their limitations on the size of the display; AR HMDs focus on superimposing digital content on the display with a limited field of view, and their displays are often insufficiently bright to work in outdoor settings due to the external lighting conditions.

Current application scenarios for VR primarily focus on providing end-users with a user-perspective experience of design alternatives in small scale rooms and controlled environments. Several authors have claimed that the use of VR technologies has often been a product of the outside-in tracking technologies required for room-scale VR to be undertaken, which is a problem when transferring the methods to outdoor public projects (Schneider et al. 2013, Loyola et al. 2019). Although large indoor VR environments for commercial and academic uses are possible (UNC Tracker Project 1997, VOID 2016), these require dedicated spaces and infrastructure that are not feasible in uncontrolled outdoor environments, where ongoing activities would conflict with fixed VR tracking infrastructure,and would be cost-prohibitive for AEC use cases.

A promising but less explored realm is the use of VR applications to review design proposals in uncontrolled outdoor environments, allowing users to freely explore the projects on the real site. Although, inside-out VR HMDs generally have been designed to work best in small scale environments and suffer from tracking issues, they differentiate themselves from the outside-in tracking by the need of fixed infrastructure. Nevertheless, most of the applications of VR in the outdoor environments are not applied to changing altitudes and inclinations and do not have out-of-the-box solutions to register and align virtual content to the real world. Several solutions for navigating users in VR to avoid collisions with objects and wormholes have been introduced. For instance, an inside-out tracking system could simultaneously retrofit a virtual environment to the physical world and dynamically redirect the user towards the chosen destination by mounting extra sensors on the headset, which allows users to navigate in large spaces when fully immersed (Marwecki et al. 2018, Cheng et al. 2019, Yang et al. 2019). Other navigation techniques allow users to intuitively navigate in VR and AR (Psarras et al. 2019, Spatial.IO 2021), or use customized hardware to give the multi-sensory impression of motion in virtual environments (Knight and Brown 2001).

METHODS
User Experience
The aims of this research were to put people and user experience at the forefront in determining technical solutions to apply to project scenarios and to enable the most objective decisions possible.

The best fit of technology to user experience was established through a cyclical evaluation process with design teams who, from their project experience, are best placed with an intuitive understanding of the various stakeholder requirements. Through this engagement, the following requirements were identified:

- **Freedom:** The ability to walk anywhere in the whole space at scale as if the design were physically in front of them.
- **Collaborative Environment:** The ability to engage with other users to facilitate discourse.
- **Objective:** The process should enable objective decisions to be made rather than the process itself becoming the focus.
- **Intuitive:** Individuals should intuitively know how to interact with the experience and movement/interaction should feel instinctive.
- **Fidelity:** The visual rendering quality and display resolution should be maximized to create a realistic, immersive and engaging experience.
- **Safety:** The process should be safe, and everyone should be comfortable using the technology. This was particularly important given the range of stakeholders, various site conditions and COVID-19 implications.

2   Warehouse site conditions (pre mock-up construction)

3   Tower site conditions

Project Requirements
Two live projects influenced the development of these tools, which whilst both being large scale environments, had contrasting environments and project requirements (Figure 4).

Tower Lobby Mock-Up
This project extended a full-scale mockup of a proposed building lobby in a warehouse measuring approximately 100x30m (Figure 2). Constructed in timber, the goal was to give the client an understanding of the scale of the space, to allow the entrance arrangements to be physically tested, and to evaluate stair design options (of which only half of each option were constructed). The space was digitally extended to allow the full-size lobby to be experienced complete with lighting, materials finishes, and switchable stair options.

Public Realm Planning Project
The goal of this project was to demonstrate the impact of a tower and its public realm intervention by overlaying it digitally on top of the existing site, measuring approximately

	Lobby Mock Up	Tower and Public Realm Planning Project
Dimensions	• 100m x 30m warehouse interior	• 100m x 100m public plaza
Environmental conditions	• Interior space • Daylit top and sides • Direct sunlight at open ends • Construction site task lighting	• Outdoor public plaza • Uncontrolled lighting ranging from bright direct sun to overcast/dull • Variable weather
Site setup constraints	• Tight project timeline • Active construction site until client review • Setup and testing occurred in parallel with construction activities • Minimal set up time available	• Access to site over extended period • Minimal set-up time due to varying weather • Random varying obstacles (furniture, people)
Proposed vs. built	• BIM model as ground truth (reality) • Ground truth consistent with proposed design (with exception of set extension and stair options) • Deviations from the design model and what was built arose from construction tolerances/error and design changes from mock up being used as design development tool	• 3D laser scan of site as ground truth (reality) • Ground truth differs from proposed design; proposal makes fundamental changes to site • Deviations from reality and proposed design needed to be mapped against physical site conditions and physical differences compensated for
Tracking system constraints	• An external tracking system whilst possible, would have required a complex setup to cover full extents of space • Project timeline required a system to be setup within a day	• An external tracking system was not feasible as the systems required would need to have been installed on adjacent buildings, across streets etc., which would adversely impact the wider public realm
Hazards and risk	• Construction site • Physical stairs and stair option changes	• Physical urban context • People • Traffic • Level changes (eg. pavement edges, steps) • Obstructions (eg. furniture) • COVID-19
Additional requirements	• Digital extension of existing mock-up • Comparison of physical and digital design options	• Comparison of real with proposed • Collaborative presentation and review

4 Summary Project Requirements

5 VR backpack and hardware

6 Spatial alignment and reality compensation

7 Multi-user interaction

100x100m (Figure 3). It was to be presented by the architect to stakeholders as part of a planning inquiry, allowing them to engage with sensory nature of the context, experiencing the project as if it were built. The project team sought to also use this as a comparative tool, comparing the existing to the proposed by live swapping the experience and reality.

IMPLEMENTATION

The system allows for a broad set of hardware devices and platforms. To create an optimal user experience, several AR/VR technologies were evaluated based on capabilities and user-experience feedback. These included:

- **Tablet AR:** Use of a tablet such as an Apple iPad to overlay digital content to a live camera view from the device's camera.
- **Optical see-through AR Headset:** Use of an AR headset to overlay digital content directly on top of an individual's direct view of reality.
- **Video see-through AR Headset:** Use of an immersive VR headset to overlay digital content on top of a live video feed of reality approximating an individual's view of reality.
- **VR Headset:** Use of an immersive VR headset to display replace an individual's field of view with a virtual environment.

Tablet AR, whilst accessible, did not offer the desired immersion. Optical AR headsets were promising, but they were limited by field of view or used optical displays best suited to darker indoor environments. The video see-through AR headsets available at the time also required external tracking systems, limiting the scale of space a user could freely navigate. A fully VR solution was adopted as it also solved situations where physical levels differed from the virtual, and adjustments to a user's virtual location were required where a disconnect from reality would be evident using an AR approach.

Coupling a backpack VR PC and a high-resolution VR headset with inside-out tracking offered the best balance of graphical fidelity, immersion, and field of view for the project goals (Figure 5). It allowed for freedom to physically walk through a physical environment without requiring an external tracking system or a cable tied back to a PC in a fixed location. The high-resolution display allowed stakeholders to understand the fine detail of the projects, which was favorably received by the design team. A standalone VR headset was also evaluated; however, the advantage of a lighter hardware trade-off was not offset by the graphical fidelity of the tethered VR headset.

Because of its accessibility, Tablet AR was developed in parallel, and while it was not as immersive, it formed a fallback to address any concerns over user reticence in wearing a VR backpack and headset.

Both versions used a custom in-house system called Glaucon, which was developed on top of a games engine. Marker tracking libraries, additional cameras, and custom mounts were combined to implement the extended VR tracking.

Spatial Alignment and Reality Compensation

In order for users to experience the virtual space on site together in the correct location, a method for spatial alignment was required. This was achieved by first creating a ground truth model of reality that aligned with the design models, allowing for correspondence to be established between physical reality, ground truth modelled reality, and the virtual design proposal.

Building upon this correspondence, a practical way to align a user in physical reality to the corresponding position in virtual space was required. Several techniques were investigated that used known consistent locations between reality and virtual design, such as known start positions, VR motion controllers for anchor points, and camera marker tracking. Camera marker tracking was settled upon, as it was deemed to require the least user input, achieved an acceptable level of precision, and was least prone to user error. From a user experience perspective, camera marker tracking required users to simply look at a specific object in their physical environment to trigger the alignment. Given the errors inherent in the VR tracking systems, the marker alignment process created a simple way that user misalignments could be rectified during the review process.

Instances where proposed floor levels differed from actual reality created discrepancies between where the user's eye position was in reality and where their desired view should have been in the virtual design model. To address this, a reality compensation system was developed that allowed for localized offset of a user's height during marker alignment. This in effect created zones of particular heights in the space; where a user could align at a position at one level, walk across the space to another space and arrive at an incorrect height, a simple marker realignment in that zone would fix it. Other techniques for dynamically adapting a user's height as they walked over the space were tested; however, in most cases additional data on the users' height/pose was required (such as from foot/ankle tracking), for instance, to sufficiently differentiate between a user crouching down with no reality compensation offset and a user who had already been offset by 1m. Therefore, the approach taken to combine marker alignment and reality compensation was deemed the best overall balance given the available technology (Figure 6).

Whilst not automatic, from a user experience perspective, the point where a user would need to pause to realign was integrated to facilitate design discourse. Working with the design team, markers were located across the environment in such a way as to ensure that these realignments occurred in natural conversational spots encouraging conversation whilst complimenting technical need.

Tracking in Uncontrolled Environments and Precision

Correct user alignment in both physical and virtual environments is important when using VR as a comparative tool: switching between current reality (headset off) and proposed reality (with headset on). This was particularly important in the Public Realm project where incorrect alignment could make the difference between experiencing the design as it would be on site or being incorrectly situated and unable to make the necessary value judgement.

Because inside-out tracking systems had inherent drift—and where lighting conditions could also adversely affect this—camera markers were positioned relative to key viewpoints so that spatial alignment and key comparison points could coincide. This allowed a sufficiently accurate correspondence and opportunity for realignment for users to make design review decisions.

Multi-User Interaction

Multi-user interaction was implemented as a representational and presentation tool, and to aid gesture-based physical communication between users. Care was taken to ensure user head orientation mapped to those of their avatars facilitating convincing user interaction and abstracted body language between physical and virtual (Figure 1).

Spatial alignment between users allowed one to physically and virtually follow another walking around the space. Navigation mapped the action of physically walking to movement in VR, and simple pointing using hand controllers allowed parts of the design to be highlighted using a virtual pointer. As users were in the same physical typically in close proximity to one another, there was no need to provide an audio communication system. A display screen was provided on the back of each VR backpack to aid inclusion for non-VR users. The experience was also shared remotely, both using standard online collaboration tools and its own remote collaboration functionality (Figure 7).

Hazards and Risk

Hazards in the physical environment such as abrupt level changes and obstructions pose a heightened risk in VR, where users may not be aware of their surroundings and in situ, unplanned, and dynamic obstacles also play a factor. Given the risk profile of site hazards to the participating parties, a human "bodyguard" was considered the best approach accompanying each VR user around the space. This was essential where reality differed substantially from the virtual space, such as with the virtual stair options in the warehouse mock-up, on curbside edges adjacent to a road, or where the real and proposed levels differed.

A COVID risk assessment was also conducted incorporating hardware redundancy to reduce risk of exposure from shared devices.

RESULTS AND DISCUSSION

User Experience and Feedback

The system was successfully tested by at least thirty users from a broad range of demographics and experiences of XR systems. Due to most of these tests taking place as part of larger formal design reviews, it was not possible to perform a systematic assessment or use questionnaires to evaluate users' experiences. However, the general feedback from users has been extremely positive, with many of them describing it as the most realistic VR experience they had ever used, and very strongly encouraging their colleagues to try the system. Factors that are considered to support that feedback are described below.

Ergonomics and Immersion

The VR backpack system did not prove overly onerous for users to wear, and the quality of experience strongly outweighed any physical constraints. Streaming technology was considered, but at the time, wired HMDs had higher resolution than wireless ones. They were very few reports of motion discomfort compared to other more static VR systems, and almost all users found walking in VR to be an intuitive experience. Some users wore the system for extended periods (up to 30 minutes) without prompting and typically walked considerable distances. The combination of virtual visual, physical tactile, and augmented auditory experiences was found to be very immersive.

Collaboration and Participation

The multi-user collaborative aspects of the system proved effective, with users interacting and communicating in a natural way.

Avatar representation was evaluated with end users to understand their preferences on visual style, interpersonal interaction, and intuitive design. A range of visual representations were considered from photorealistic to fully abstracted. A human representation that allowed realistic body language to be expressed—whilst abstracted to a point that it was not distracting itself—was received most favorably. Several approaches to capturing human motion were tested, with a head driven system providing a good balance between quantity of data (to infer a reasonable pose for effective communication) and ease of use by the end user. Through use of head and shoulder body movement inferred from physical HMD transforms to maintain consistent eye lines between real and virtual users, together with the ability to physical point at objects, a correspondence was established between the real and the virtual that facilitated an effective means of communication.

Switching between real and proposed environments by lowering and raising the HMDs was found to be a very useful comparative tool. Because of the strong correspondence between real and virtual—buildings and people were accurately aligned across both, and social cues such as body language and conversation were preserved— the natural flow of the collaborative experience was maintained.

The system also created opportunities for more stakeholder participation—other parties in the reviews and planning inquiry could see the users' experiences via a screen on the rear of the backpacks or use iPads. The overlay of real and virtual allowed challenge and response to design decisions and issues. The system was generally considered by users to be accurate and objective .

Use of VR in situ at 1:1 scale also brought to the fore potential accessibility issues, and this reinforced the authors' ongoing investigation into the use of VR and AR for more inclusive design. The tablet AR system was less immersive but proved very useful during lockdown to keep the planning process going, and enabled participation for users who felt uncomfortable using full VR. Overall, participants in the planning process considered the system very useful and indicated they though it should be used more widely in future. It was also observed that the system had great potential as a storytelling tool and as a means for designers to truly understand and communicate their design.

CONCLUSION

The use of collaborative XR at 1:1 scale in situ has found to be both feasible and highly effective in real world projects. The combination of physical mockups and environments, augmented with realistic digital models, provides spatial experiences that are more immersive than their individual parts. At a broader level, the use of multi-user XR design collaboration systems can enhance inclusivity and participation in the design and planning processes.

Future work could include increasing the robustness of outdoor tracking, enhancing multi-user collaboration through further research in avatar representation and user interaction, and integrated design and analysis tools. Inclusion both from a technical (VR and non-VR user) and socioeconomic perspective could be improved by future hardware that would be cheaper, faster, lighter, and wireless compared to current hardware. The overall system could be further extended to gather objective measures of diverse user experience, and thus could better inform both participatory design processes and built design outcomes.

ACKNOWLEDGEMENTS

We would like to thank our colleagues, clients, and collaborators for their contributions to the development and evaluation of this system, in particular Gamma Basra and Ewan Couper from our Visualisation team, and Zoe De Simone, formerly of our team, for her review and feedback on this paper.

REFERENCES

Arias, Ernesto, Hal Eden, Gerhard Fischer, Andrew Gorman, and Eric Scharff. 2000. "Transcending the individual human mind—creating shared understanding through collaborative design." *ACM Transactions on Computer - Human Interaction* 7 (1): 84–113.

Azuma, Ronald, Jong Weon Lee, Bolan Jiang, Jun Park, Suya You, and Ulrich Neumann. 1999. "Tracking in unprepared environments for augmented reality systems." *Computers & Graphics* 23 (6): 787–793.

Bardram, Jakob E., Claus Bossen, Andreas Lykke-Olesen, Rune Nielsen, and Kim Halskov. 2002. "Virtual video prototyping of pervasive healthcare systems." *DIS '02: Proceedings of the 4th conference on Designing interactive systems: processes, practices, methods, and techniques*. London, England: ACM. 167–177.

Cheng, Lung-Pan, Eyal Ofek, Christian Holz, and Andrew D. Wilson. 2019. "VRoamer: Generating On-The-Fly VR Experiences While Walking inside Larger, Unknown Real-World Building Environments." *2019 IEEE Conference on Virtual Reality and 3D User Interfaces (VR)*. 359–366.

Chung, Daniel Hii Jun, Zhou Steven Zhiying, Jayashree Karlekar, Miriam Schneider, and Weiquan Lu. 2009. "Outdoor mobile augmented reality for past and future on-site architectural visualizations." *Joining Languages, Cultures and Visions - CAADFutures 2009; Proceedings of the 13th International CAAD Futures Conference*. 557–571.

Du, Jing, Yangming Shi, Zhengbo Zou, and Dong Zhao. 2018. "CoVR: Cloud-Based Multiuser Virtual Reality Headset System for Project Communication of Remote Users." *Journal of Construction Engineering and Management* 144 (2): 1–19. doi:10.1061/(ASCE)CO.1943-7862.0001426.

Dunston, Phillip S., Laura L. Arns, James D. Mcglothlin, Gregory C. Lasker, and Adam G. Kushner. 2007. "An Immersive Virtual Reality Mock-Up For Design Review of Hospital Patient Rooms." *7th International Conference on Construction Applications of Virtual Reality*. PA, USA: University Park.

Eastman, Chuck, Paul Teicholz, Rafael Sacks, and Kathleen Liston. 2008. *BIM Handbook: A Guide to Building Information Modeling for Owners, Managers, Designers, Engineers and Contractors*. Hoboken, NJ: Wiley.

Faliu, Barnabé, Alena Siarheyeva, Ruding Lou, and Frédéric Merienne. 2018. "Design and prototyping of an interactive virtual environment to foster citizen participation and creativity in urban design." *27th International Conference on Information Systems Development*. Lund, Sweden. 1–13.

Fischer, Gerhard, Elisa Giaccardi, Hal Eden, Masanori Sugimoto, and Yunwen Ye. 2005. "Beyond binary choices: Integrating individual and social creativity." *International Journal of Human-Computer Studies* 63 (4-5): 482–512.

Kieferle, J., and U. Wössner. 2003. "Combining Realities - Designing with Augmented and Virtual Reality." *Digital Design, 21th eCAADe Conference Proceedings*. Graz, Austria. 29–32.

Knight, Michael, and Andre Brown. 2001. "Towards a natural and appropriate Architectural Virtual Reality: the nAVRgate project." *Proceedings of the Ninth International Conference on Computer Aided Architectural Design Futures*. Eindhoven, The Netherlands: Eindhoven University of Technology. 139–149.

Loyola, Mauricio, Bruno Rossi, Constanza Montiel, and Max Daiber. 2019. "Use of Virtual Reality in Participatory Design." *Architecture in the Age of the 4th Industrial Revolution; Proceedings of the 37th eCAADe and 23rd SIGraDi Conference*, vol. 2. Porto, Portugal. 449–454.

Maldovan, Kurt D., John I. Messner, and Mera Faddoul. 2006. "Framework for Reviewing Mockups in an Immersive Environment." *CONVR 2006: 6th International Conference on Construction Applications of Virtual Reality*. Orlando, Florida. 6.

Marwecki, Sebastian, Maximilian Brehm, Lukas Wagner, Lung-Pan Cheng, Florian F. Mueller, and Patrick Baudisch. 2018. "VirtualSpace - Overloading Physical Space with Multiple Virtual Reality Users." *Proceedings of the 2018 CHI Conference on Human Factors in Computing Systems*. 1–10.

Moeslund, Thomas B., Moritz Stoerring, Wolfgang Broll, Francis Aish, Yong Liu, and Erik Granum. 2003. "The ARTHUR System: An Augmented Round Table." *Journal of Virtual Reality and Broadcasting* 34.

Piekarski, Wayne, Ross Smith, and Bruce H. Thomas. 2004. "Designing Backpacks for High Fidelity Mobile Outdoor Augmented Reality." *3rd IEEE and ACM International Symposium on Mixed and Augmented Reality*. Arlington, VA, USA. 2–5. Accessed July 2021. https://www.hitlabnz.org/.

Psarras, Stamatios, Marcin Kosicki, Khaled Elashry, Sherif Tarabishy, Martha Tsigikari, Adam Davis, and Francis Aish. 2019. "SandBOX - An Intuitive Conceptual Design System." In *Impact: Design With All Senses*, edited by C. Gengnagel, O. Baverel, J. Burry, T.M. Ramsgaard, and S. Weinzierl. Cham: Springer. 625–635.

Schneider, Sven, Saskia Kuliga, Christoph Hölscher, Ruth Conroy-Dalton, André Kunert, Alexander Kulik, and Dirk Donath. 2013. "Educating architecture students to design buildings from the inside out." In *Proceedings of the 9th international space syntax symposium*. Seoul.

Spatial.IO. 2021. "Spatial - Virtual spaces that bring us together." Accessed July 12, 2021. https://spatial.io/.

Sunesson, Kaj, Carl Martin Allwood, Dan Paulin, Ilona Heldal, Mattias Roupe, Mikael Johansson, and Borje Westerdahl. 2008. "Virtual reality as a new tool in the city planning process." *Tsinghua Science and Technology (TUP)* 13 (S1): 255–260.

UNC Tracker Project. 1997. UNC-CH Tracker research group. Accessed July 2021. https://www.cs.unc.edu/~tracker/.

VOID. 2016. "Welcome to the VOID." Accessed July 2021. https://www.polygon.com/features/2016/5/5/11597482/the-void-virtual-reality-magician-tracy-hickman.

Wagner, Rosana, Sandra Dutra Piovesan, Profa. Dra. Liliana Maria Passerino, and José de Lima. 2013. "Using 3D virtual learning environments in new perspective of education." *12th International Conference on Information Technology Based Higher Education and Training (ITHET)*. IEEE. 1–6.

Xue, Xiaolong, Qiping Shen, Hongqin Fan, Heng Li, and Shichao Fan. 2012. "IT supported collaborative work in A/E/C projects: A ten-year review." *Automation in Construction* 21 (1): 1–9.

Yang, Jackie (Junrui), Christian Holz, Eyal Ofek, and Andrew D. Wilson. 2019. "DreamWalker: Substituting Real-World Walking Experiences with a Virtual Reality." *UIST '19: Proceedings of the 32nd Annual ACM Symposium on User Interface Software and Technology*. New Orleans LA, USA: Association for Computing Machinery. 1093.

David Gillespie is an Associate Partner at Foster + Partners. His background spans architecture, construction, BIM, computational design, XR, and realtime technologies. Joining Foster + Partners in 2007, he has worked on projects across all scales, implementing computation design processes to aid inter disciplinary integration, coordination and developing progressive worklows and developing BIM processes to aid delivery of complex projects, such Glasgow's Hydro Arena. He currently leads XR development within the team which incorporates work on computer graphic technologies, augmenting design, construction and the experience of the built environment and developing associated tools and proceesses. David is a member of the Royal Institute of British Architects and has lectured on the integration of computational design in the design and construction process, BIM and XR.

Zehao Qin is a Design Systems Analyst in the Applied Research + Development group at Foster + Partners. He is a specialist in a wide range of areas, including human-computer interaction (HCI), XR and performance-driven design. His interest lies in developing algorithmic design and scripting techniques to produce tools and applying thorough and proactive research of design systems and prototypes to design. He completed his MSc Architectural Computation at the Bartlett, UCL and BFA Public Art at Central Academy of Fine Arts, graduating in both as top of his class with distinction. He also taught and lectured on the subjects of parametric and algorithmic design to graduate students at the Bartlett, led a series of algorithmic design workshops and published papers in conferences.

Francis Aish was a Partner and Head of Applied Research and Development at Foster + Partners. He studied Aerospace Systems Engineering at the University of Southampton, and is currently completing an Engineering Doctorate at University College London. After graduation, he spent two years in the Advanced Technology division of Nortel Networks, investigating the application of VR to the design, management and marketing of telecommunications networks. He joined Foster + Partners in 1999, responsible for the research and development of systems to model and solve complex, multi-disciplinary design problems. In the course of this work he was involved in over 200 projects and competitions, including the Swiss Re HQ in London, the Smithsonian Institution, and Beijing International Airport. He also conducted collaborative research with leading universities and companies, and published academic papers on computational design, as well as lecturing widely on the subject in Europe, Asia, and North America.

Timber De-Standardized

A Mixed Reality Framework for the Assembly
of Irregular Tree Log Structures

Leslie Lok
Cornell University

Asbiel Samaniego
Cornell University

Lawson Spenser
Cornell University

1   Left to Right: Physical prototype at 1:1 scale, digital model, and structural feedback model visualized with mixed reality tools

ABSTRACT

Timber De-Standardized is a framework that salvages irregular and regular shaped tree logs by utilizing a mixed reality (MR) interface for the design, fabrication, and assembly of a structurally viable tree log assembly. The process engages users through a direct, hands-on design approach to iteratively modify and design irregular geometry at full scale within an immersive MR environment without altering the original material. A digital archive of 3D scanned logs are the building elements from which users, designing in the MR environment, can digitally harvest (though slicing) and place the elements into a digitally constructed whole. The constructed whole is structurally analyzed and optimized through recursive feedback loops to preserve the users' predetermined design. This iterative toggling between the physical and virtual emancipates the use of irregular tree log structures while informing and prioritizing users' design intent. To test this approach, a scaled prototype was developed and fabricated in MR.

By creating a framework that links a holographic digital design to a physical catalog of material, the interactive workflow provides greater design agency to users as co-creators in processing material parts. This participation enables users to have a direct impact on the design of discretized tree logs that would otherwise have been discarded in standardized manufacturing. This paper presents an approach in which complex tree log structures can be made without the use of robotic fabrication tools. This workflow opens new opportunities for design in which users can freely configure structures with non-standardized elements within an intuitive MR environment.

INTRODUCTION
Environmental Observation

Presently active forest management and responsible harvesting practices have been employed to reduce the carbon footprint of timber, one of the most common and sustainable materials for building construction. However, the manufacturing and production process of timber products can affect and pollute its surrounding environments and ecosystems (Adhikari and Ozarska 2018) such as sawmilling industries, the use and production of glue toxins, the process of transportation, etc. Furthermore, the manufacturing process to produce dimensional lumber and plywood products requires the standardization of logs into straight members (Belly et al. 2019). These manufacturing processes also create enormous amounts of waste in the form of tree branches, off-cuts, sawdust, bark, and juvenile trees. As Adhikari and Ozarska (2018) reported in Nigeria: "Out of every 1 m3 of tree that is cut and removed from the forest, about 50% goes to waste in the form of damaged residuals, followed by abandoned logs (3.75%), stumps (10%), branches (33.75%), and butt trimmings (2.5%)."

The waste produced from tree harvesting is primarily comprised of irregular shaped logs and branches that cannot be standardized for manufacturing and consequently cannot be used for standardized applications. Therefore, the use of non-standard tree geometry has the potential to mitigate the carbon impact of industrial dimensional lumber manufacturing processes by salvaging the wood material that is presently considered to be waste. Though the irregularity of the waste produced from tree harvesting is incompatible with standard applications, the irregularity of the wasted tree geometry is easily workable in 3D user interfaces of augmented and mixed reality platforms.

A Framework

Timber De-Standardized is a framework that demonstrates how non-standard materials in the form of irregular tree logs can be salvaged and repurposed to create a tree log assembly through the integration of 3D scanning, mixed reality, and several iterative stages of structural analysis and optimization. Designing with non-standard materials is fundamentally more complex than designing with standard materials because each element is discrete. However, designing in virtual and augmented realities (3D user interfaces) does not privilege one geometry over another; each element is equally compliant because the interaction between geometries is strictly in accordance with the user's intuitive gestures. The intuitive familiarity of working in a 3D user interface (UI), as compared to a 2D UI, has the opportunity of making the UI more accessible to a broader audience of non-expert and expert users alike. Salvaging non-standard tree logs necessitates the cataloging of such elements with a 3D scanner in order to link the digital mesh derived from the 3D scan with each discrete physical member. *Timber De-Standardized* creates a platform that utilizes what would otherwise be considered as discarded tree logs to democratize design through the employment of a mixed reality environment. Mixed Reality (MR) enables the interaction between human, computer, and one's immediate place through immersion of both physical and digital interfaces (Microsoft 2020). In order to test the framework, a MR-designed spatial tree log assembly is analyzed and optimized to reduce structural deformation as a scaled prototype that was originally proposed for a conceptual housing project. The MR environment initiates the use of virtual instructions for the fabrication and assembly of the prototype. For the purpose of contextualizing and testing this framework, the research goals were to (1) establish design and structural parameters specific to working with irregular tree logs, (2) develop an interactive workflow for users to manipulate the tree logs for design and fabrication between the mixed-reality, digital, and physical environments, and (3) to develop an optimization process that would prioritize the users' design intent. *Timber De-Standardized* is a framework that utilizes MR to initiate a platform that aims to democratize design agency through several iterative design and structural analysis stages followed by the provision of direct fabrication and assembly instruction, all of which are linked to a physical catalog of salvaged non-standard tree logs.

STATE OF THE ART
Digital Log Constructions

Recent projects such as the *Wood Chip Barn* (Self et al. 2016), *Limb* (Von Buelow et al. 2018), *Branch Formations* (Alner et al. 2021), and *torinosu* (Eri Sumitomo Architects 2020) explore the potential of repurposing irregular tree geometries for structural applications in architecture. *Wood Chip Barn* at Hooke Park was revelatory in its workflow between scanning trees, selecting trees, and indexing 'Y' shaped members to design a truss with selected members, structurally evaluating these members, and finally to fabricating a *Tree Fork Truss* (Mollica and Self 2016). *Limb* continued the exploration of the tree bifurcations as it relates to shell structures, (Von Buelow et al. 2018), while *Branch Formations* also studies similar tree bifurcations, this time with regard to structures derived from a tetrahedral based recursive aggregation algorithm (Allner et al. 2019). Because of the irregularity of the non-standard members selected for design, each of these three projects relies on robotic fabrication for the construction of their final assemblies. Still reliant on the process of 3D scanning, Aki Hamada and Eri Sumitomo led a team on the repurposing of irregular tree logs into a reciprocal frame in *torinosu* (Eri Sumitomo Architects 2020). Instead of relying on robotic fabrication, the team utilized an augmented reality interface for the manual fabrication and assembly of *torinosu*. Augmented reality (AR) is a composite view of the real world and a virtual scene that is digitally generated for the perspective of the user (Nakata et al. 2012). In recent years, AR has become an increasingly popular tool for architectural practices to project virtual instruction into a real environment (Song et al. 2021).

A Familiar Environment

AR has become an increasingly popular area of research in architectural based design practices (Song et al. 2021), as it has shown to enhance education and increase student participation (Arici et al. 2019). AR offers a number of affordances as a user interface (Bowman et al. 2006) and as a tool for digital construction (Abe et al. 2017). As a user interface (UI), there is a greater degree of 'unnatural' interaction through a 2D interface as opposed to a 3D interface when orienting, transforming, or otherwise manipulating 3D information (Alaikseyeu et al. 2012). In other words, 3D information can be more fully represented and communicated through a 3D UI (Alaikseyeu et al. 2012). With AR, virtual models made digitally can be displayed and seen in physical space through the live view of a computer device (Nakata et al. 2012). Similar to AR, MR is the intersection between virtual and physical realities with which a haptic stimulus is also presented (Skarbez et al. 2021). This becomes a powerful design tool as it gives the user control of how digital models are placed, oriented, and perceived with respect to the physicality of their environment.

These tools create new approaches in design and fabrication where, what would otherwise be considered complex digital models, can be made without robotic-based fabrication techniques through the instruction provided in a 3D UI. This concept was explored in the *Woven Steel Pavilion* (2018) (Jahn et al. 2019) during a workshop at the 2018 CAADRIA conference in Beijing, in which researchers from Fologram tested their mixed reality designed interface as a tool for metal rods bending instruction. The users were able to accurately bend each rod in several multi-axial directions as informed by the MR environment (Jan et al. 2019). As inferred through this project and other projects such as *torinosu*, MR becomes highly relevant in the manipulation, reorientation, and communication of the multi-axial nature of irregular tree logs through the 3D UI visual instruction.

Through the hybridity of human input and MR technologies, design can become an iterative and generative tool for engaged users. When informed by immediate structural simulations, the processes of fabrication and assembly can become a localized educational instruction manual for the users. With the added stages of structural simulation to a MR environment, *Timber De-Standardized* enables a participatory and iterative approach by providing the user with immediate structural feedback throughout the design process. The further active participation in such a framework enhances the design agency of the participant from user to co-creator of custom non-standard log components. The spatialization and tangibility of an AR environment (similar to the proposed MR setup) empowers users to interact with 3D information through an authentic learning environment (Cheng and Tsai 2012).

METHODS

Design Approach

The prototype is part of a three-story housing complex that explores innovative uses of ash wood. The housing complex is composed of three twisted rectangular volumes that enclose a central courtyard. The voids created from the corners of the twisted volumes infer the bounding geometry from which the tree log assembly is derived. The voids also serve as the continuation of the adjacent circulation corridors and facilitate the movement of the inhabitants across the complex. The prototype for *Timber De-Standardized* is a portion of one of the voids at the entry to the housing complex (Figure 2). The volume is a triangular pyramid with a circulation corridor at each level. In this conceptual housing complex, each level of the structure is laterally supported with fixed connections at the ground level. However, for the purpose of the prototype, the structure is self-supported and the base connections are not fixed. These aspects of the housing complex are the driving geometry that informs the tree log assembly designed in MR from irregularly shaped logs.

Workflow

Timber De-Standardized is a framework that enables an iterative process of design through several recursive

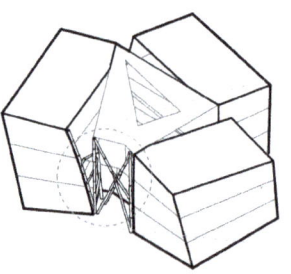
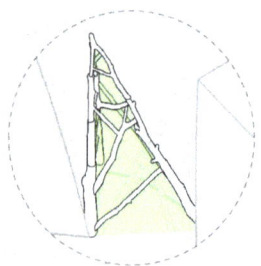

2   Housing complex with structure

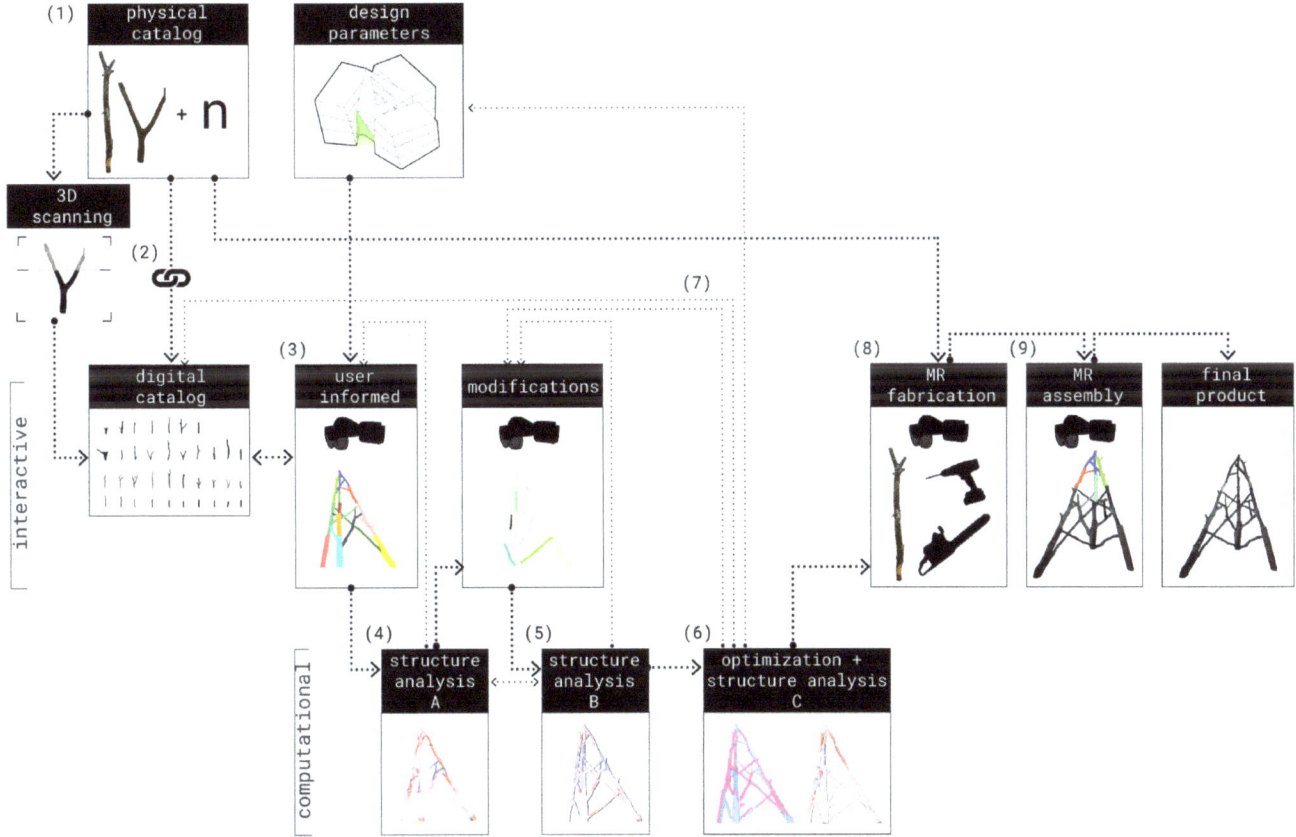

3   Workflow diagram

stages of structural analysis in a MR environment (Figure 3). For the purpose of this prototype and the demonstration of this framework, irregular tree logs were (1) sourced from a local groundskeeping entity, and (2) cataloged and indexed using a 3D scanner. Next, the tree logs were used as the (3) 'building blocks' to design a tree log assembly in the MR environment. The resulting structure was immediately (4) analyzed for its structural viability as the first structural loop. From this, (5) the second structural loop, in which the user can continue to modify the design in MR according to the results observed from the structural analysis feedback loop. Once the user completes the modifications, (6) the third structural analysis 'jitters' the model to find the optimum orientation and alignment of the members placed by the user, then (7) the resulting structure is re-indexed. The automatic re-indexing (8) instructs the user on how to proceed with the following fabrication steps (cutting the members and drilling holes for their joinery) in the MR environment. Finally, (9) the MR environment continues to be utilized as each member is assembled. The final structure is the derivative of an iterative process of MR design, structural simulations and a low-tech fabrication and assembly process enabled by the MR environment. This framework serves as a participatory and educational platform for the design and construction of structures sourced from irregular tree logs.

Cataloging

As previously mentioned, irregular, discarded tree logs were sourced from an already existing local stockpile. A wide variety of profiles from the stockpile were selected and scanned using Scanner - Structure SDK, a mobile application for iPad and iPhone by Occipital, Inc. with the associated accessory hardware, Structure. Each tree log was indexed independently as a digital mesh. The centerlines from each tree log (Figure 4) is derived by first contouring the log to get a series of cross sections that were connected together at their center. In the case of tree logs with one or more bifurcations, duplicate lines were removed. Each log, its centerlines, and respective cross sections were cataloged with a unique file name, so that any one or several files could be referenced into a single library file (Figure 5). Using Fologram, a mixed

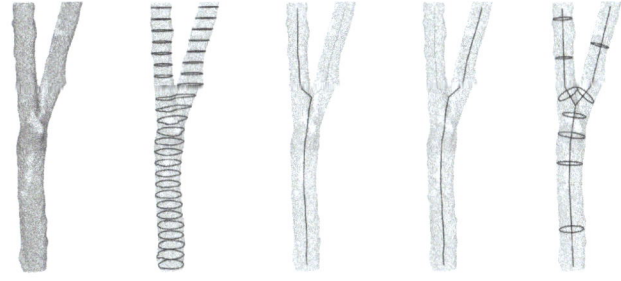

4   Computational analysis of individual log

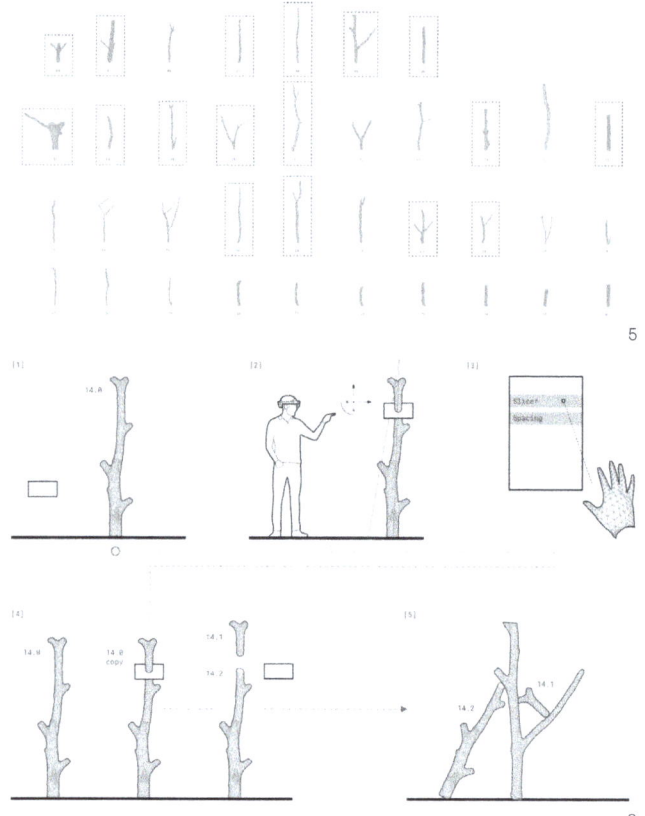

5

6

reality software for mobile devices and Rhino 3D and Grasshopper 3D (digital modeling softwares by Robert McNeel & Associates), with HoloLens, a mixed reality pair of smart glasses by Microsoft, the library file is the platform through which the user can select tree branches from the digital catalog to modify and manipulate within the MR environment.

Mixed Reality Design

The Fologram environment with HoloLens easily allows the user to move, orient, and position the scanned logs in space as desired. This enables the user to design and prototype at scale, while giving the user an accurate sense for the actual size of the building components in relation to physical space. Once the digital tree meshes are positioned in the MR environment, the user can use the 'slicer,' a plane for slicing the digital tree log meshes, to determine the cut locations for each log. In this process, after each 'slice' operation, a copy of the whole tree log mesh is created in order to re-index the part to the whole during the fabrication process. This immersive process engages the user work iteratively according to their own intent and design exploration (Figure 6). The user can modify and make unique parts without altering any of the discretized physical tree logs until the final design is set.

When designing in the MR environment, it was necessary to illustrate the design parameters of one of the voids from the conceptual housing project to guide the user in their design. As such, floor slabs, structural guides, and a corridor were visibly represented and served as the guiding principles for the user (Figure 7).

Structural Feedback Loop_A (Real-Time Visualization)

The structure was designed in Fologram with HoloLens according to the design parameters defined by the user. However, this method of working does not presently allow for the centerlines of the mesh to automatically snap together. In order to test the structural viability of the augmented reality designed structure, each tree mesh was evaluated as shell structures, while the assumed dowel joint connections between each tree mesh were evaluated as beam elements using Karamba 3D (Preisinger and Heimrath 2014), a parametric engineering plugin for Grasshopper 3D and Rhino 3D. The location of the dowel joint connection was determined by both the size and location of the 'slicer'. In the first structural loop, Structure Loop_A, the user can see the structural parameters as well as deformation and utilization in the Fologram environment.

Structural Feedback Loop_B (User Modification)

Based upon the information visually communicated to the user, additional members can be manually added by the

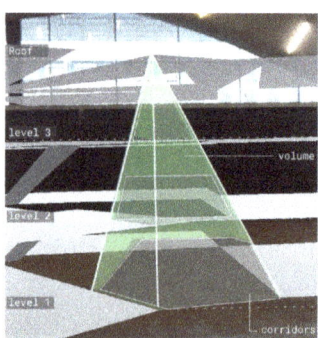
Parameters informing the placement of log members

Log selection and placement according to parameters

Passage entry framed by logs

5 Whole log catalog

6 Mixed reality interactive workflow diagram

7 Initial bounding parameters

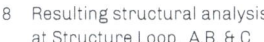

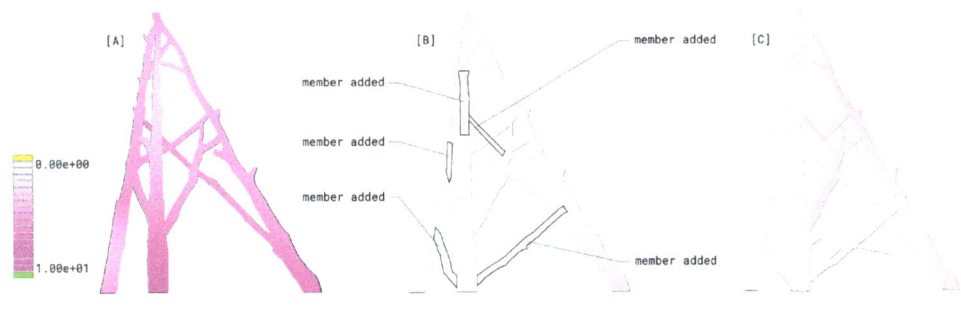
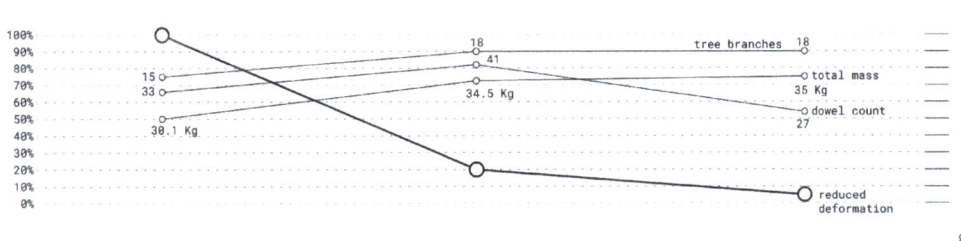

8 Resulting structural analysis at Structure Loop_A, B, & C

9 Structure Loop_C

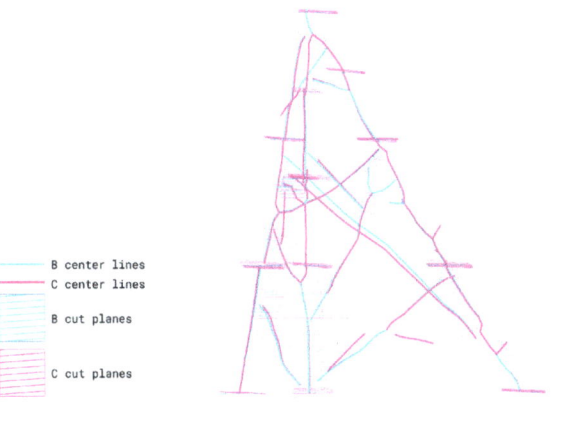
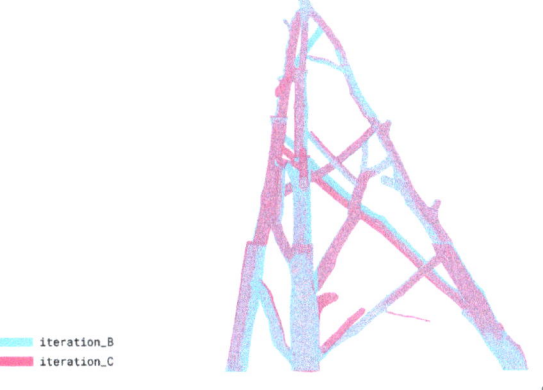

user using the MR environment to reduce the total deformation of the model. This is the second structural loop (Structural Loop_B). In this prototype, the user added five members and removed two from the previous iteration. Figure 8 illustrates the reduction of the deformation between Structure Loops A, B, and C.

Structural Feedback Loop_C (Realignment)

Upon completing the second iteration of design, the geometry from the model is then optimized with Galapagos, an evolutionary solver for Grasshopper 3D, by realigning and rotating each member around its local x, y, and z axes to reduce the total deformation of the model. Additionally, the 'slicers' used to split the digital tree meshes in the MR environment are able to transition and reorient in their local x, y, z axes. Each connection derived from the 'slicer' is referenced as a beam element in the Karamba 3D model. Structure Loop_C utilizes the Galapagos solver to adjust variables with the goal of reducing the total deformation in the model. Figure 9 illustrates the differences between Structure Loop_B and Structure Loop_C.

Re-Indexing

The final transformation of each tree mesh is recorded and applied to each of the elements associated with the tree mesh, i.e. center lines, cross sections, layer, and object attributes, etc. This method allows for the reorientation of each part to its original whole according to the orientation of the derivative optimization before the parts were split by the 'slicer.' The members are also referenced to the associated members in the physical catalog. This allows the user to see the updated cut location for each tree branch as well as the location to drill the holes after each tree branch is cut (Figure 10).

Fabrication

The fabrication for each individual log required strategic planning and the use of standard power tools as guided by the HoloLens. In order to align the digital tree log meshes with their respective physical tree log, it was crucial to orient the digital tree log mesh to the natural resting position of the physical tree log in Rhino 3D.

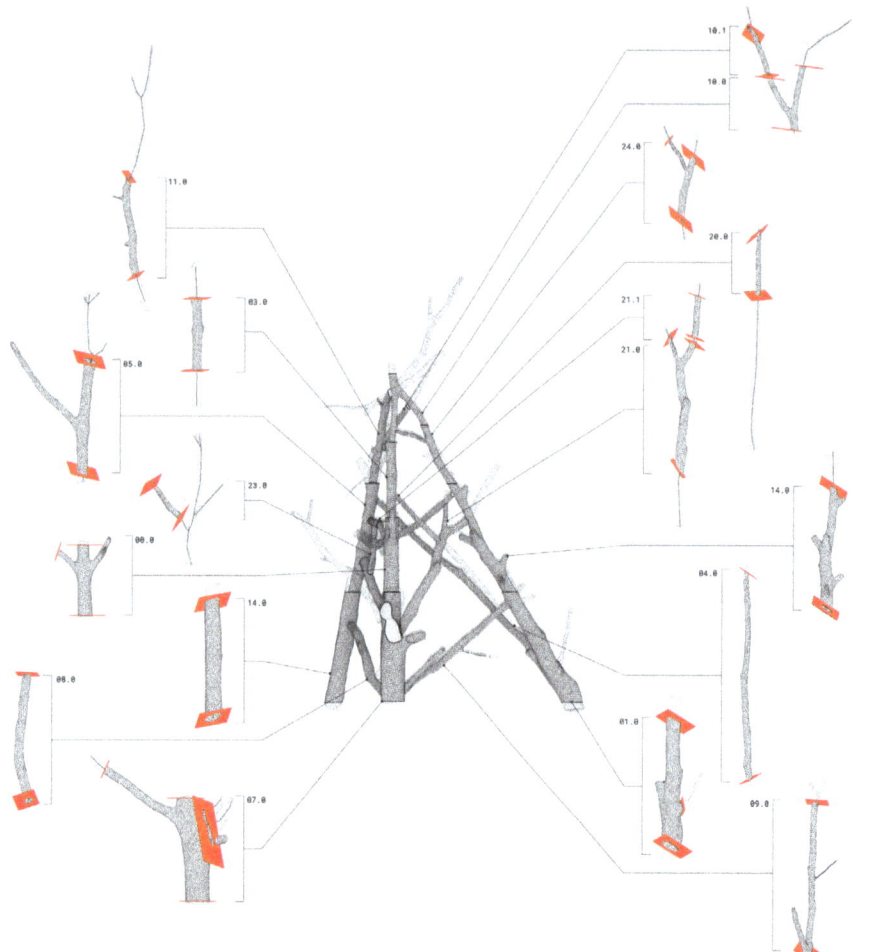
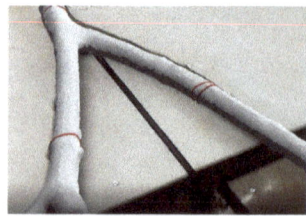

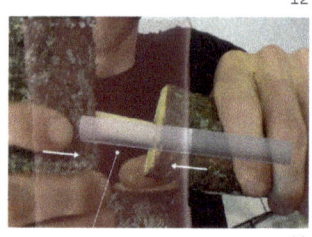

10 Re-indexing
11 Digital log simulated in natural resting position on a flat surface to overlay with the physical log
12 Fabrication of joint locations
13 Dowelled joints connecting two members

For this task, Kangaroo (Piker 2013), a built-in physics simulation plugin for Grasshopper 3D, was utilized to simulate the logs' natural resting position on a flat surface, which could then be viewed in the MR environment. One by one, the digital logs were positioned on a table and were then overlaid with the physical logs until both profiles matched (Figure 11). With the assistance of the MR environment, section cuts were visibly displayed on the logs. For safety purposes it was decided to first mark and trace these section cuts with a marker and then have the logs cut with a chainsaw. The same process would be followed for the location of the joints. However, because the joints used for this prototype are dowelled connections made from dowel rods, there were instances where the connections between logs occurred at different angles. To accommodate this, the MR environment was used to visualize the location and angle in which the drill bit had to enter (Figure 12).

Assembly
After pre-drilling each hole for the dowel-based connections, the large, upright log at the base was selected to be the first member to guide the assembly of the rest of the members. The prototype uses dowelled joints (Figure 14) with each member being either an end-to-end connection or a side-to-end connection. Two different dowel dimensions are used in the prototype. Most members use two dowels per connection to mitigate any type of local x or y rotation. The wearer of the HoloLens instructed the other two users as to the assembly of the members and often directly positioned the members to ensure the angles of each member were accurate to the digital model (Figure 15). The use of a MR environment was necessary because of the complex, multi-axis nature of each non-standard element.

RESULTS AND DISCUSSION
The Final Structure
The built prototype demonstrates the possibilities of working and designing with irregularly shaped logs at the intersection between MR and structural optimization. The prototype serves as a proof of concept highlighting key design decisions made by the user. This is evident in the log selection, the unique log segments created by slicing locations, and the placement of each individual member with the thicker logs placed towards the bottom and the thinner ones placed up top.

14 Prototype assembly

The final prototype is a self-supporting structure, measuring 9 feet in height and covering an area of 11 square feet. The structure uses a total of 16 salvaged tree logs cut into 18 members (Figure 15) with 28 dowel connections. The time required to complete the assembly was seven hours.

Reflections
Throughout the three phases of this project, several findings were made regarding the MR workflow. There are multiple advantages to the interactive design process: the MR technology provides intuitive commands that render the interaction with digital models almost physical. This analog/digital approach towards design enables the user to freely design and physically explore structural designs according to their intuition, without the actual weight of the log. Unlike designing through a 2D UI, the MR environment allows the user to truly understand the scale and context in which the design will be situated. As for the fabrication process, MR provides a platform where irregular shaped logs can be cut and drilled fairly accurately according to the information provided by the 3D UI.

The framework performed well for the purpose of this prototype, however, due to the limitations of the MR interface, the constant switch between HoloLens and the computer (for the purpose of referencing various logs) significantly slowed the design process. Another challenging aspect that emerged from working in MR was the issue of holographic drift, which tended to complicate the alignment and precision between the physical and digital tree logs and the resulting cut lines and drill locations. Advanced drift correction has recently been implemented in Fologram's 2021 sibling program Twinbuild and could be tested in future experiments.

Furthermore, the MR workflow discussed in this paper assumes that users have a basic foundational knowledge in design and construction. Due to the intuitive nature of the workflow, almost anyone can operate and design within the mixed reality setting. However, someone with the appropriate knowledge and understanding of structure is better suited to work and design with the material.

Future Investigations
The next series of investigations will focus on improving the precision of the joint connections within the mixed reality framework. By building upon the interactive workflow, the joinery-making process could be further developed beyond simple dowel connections.

The scale of the prototype was manageable and parts were easy to assemble. However, moving forward, working with larger logs could present additional logistical challenges during the assembly process, as large machinery will be required to move members into position. The downside to scaling up is that part of the direct user implementation will be relegated to machinery coordination. Moving up in scale presents additional design parameters that were not factored into the structural analysis of this prototype.

CONCLUSION
Key Contributions
The MR workflow realigns environmental, technological, and social aspects within the construction industry. Natural logs that were once considered waste can now be repurposed and utilized for new opportunities in design and construction. Along with these new opportunities is the reduction of the carbon footprint of the manufacturing process. With the use of mixed-reality tools, the design, fabrication, and assembly of complex irregular structures can be accessible without the use of robotic infrastructure. The process engages the user through a direct, hands-on design approach to iteratively manipulate, slice, modify, and assemble their customized non-standard kit-of-parts at full scale within an immersive environment. This approach realigns users' interaction and design agency with a computationally and digitally driven protocol. The uniqueness of the MR workflow is that it preserves the design aesthetic set by the user despite the structural optimization phase. The end result is a structurally viable, spatial, non-standard, log assembly that maintains and facilitates the design intent originally predetermined by the user (Figure 16).

Future Value/Potential
Due to the flexibility and adaptability of the framework, *Timber De-Standardized* can be utilized to design custom structural systems. Furthermore, it presents the construction and environmental viability of using non-standard building material. The approach can be expanded to

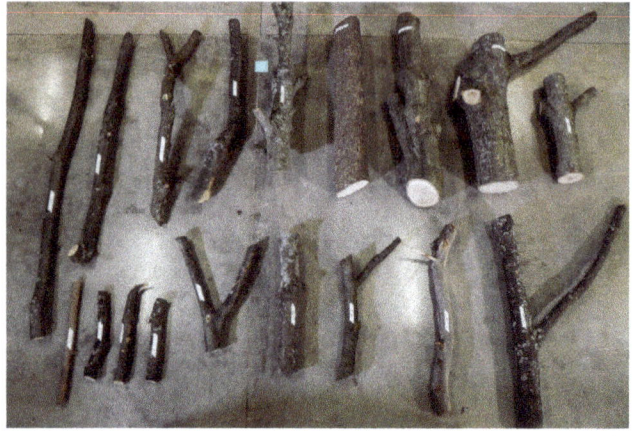
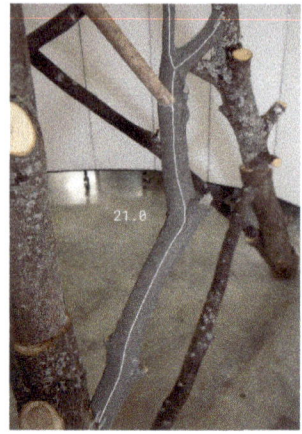

15 Log pieces used

16 Log labeled with centerline to visualize its multi-axis nature

innovate irregular building envelopes made from wood or the use of small round wood elements to create viable conventional structural systems such as trusses or space frames. Additionally, the MR workflow provides an intimate interactive platform between users and the building material. The hybridized workflow takes the physical act of making and combines it with immersive technological tools, to create a computational platform that can be used and engaged by almost anyone. The hands-on workflow within the MR environment provides a new experience, one that gives the user a greater sense of design authorship. Moving forward, there is potential in exploring how the workflow can engage multiple users to coordinate and work together as a team in the MR environment. A team dynamic would restructure and introduce new variables to the current framework.

Timber De-Standardized realigns how we interpret and utilize timber in our current design and construction models. More importantly, it presents a viable minimum infrastructure solution to our environmental crisis regarding the manufacturing and processing of lumber based products.

ACKNOWLEDGEMENTS

For this project, RUBI Lab received funding support from Cornell AAP Department of Architecture and material from the Cornell University Grounds Department. The initial design of the irregular log structure was part of a robotic log housing investigation developed by the RUBI Lab team including research assistant Byungchan Ahn. The authors would like to thank the AAP facilities and shop team for providing tools and equipment and RUBI Lab member Lily Jantarachota for assembly assistance.

REFERENCES

Abe, U-ichi, Kensuke Hotta, Akito Hotta, Yosuke Takami, Hikaru Ikeda, and Yasushi Ikeda. 2017. "Digital Construction: Demonstration of Interactive Assembly Using Smart Discrete Papers with RFID and AR Codes." In *Proceedings of the 22nd International Conference of the Association for Computer-Aided Architectural Design Research in Asia (CAADRIA)*, edited by P. Janssen, P. Loh, A. Raonic, and A. Schnabel. Hong Kong: The Association for Computer-Aided Architectural Design Research in Asia (CAADRIA). 75–85.

Adhikari, Shankar, and Barbara Ozarska. 2018. "Minimizing Environmental Impacts of Timber Products through the Production Process 'From Sawmill to Final Products.'" *Environmental Systems Research* 7 (1): 6. https://doi.org/10.1186/s40068-018-0109-x.

Aliakseyeu, Dzmitry, Sriram Subramanian, Jean-Bernard Martens, and Matthias Rauterberg. 2002. "Interaction Techniques for Navigation through and Manipulation of 2D and 3D Data." In *EGVE '02: Proceedings of the Workshop on Virtual Environments 2002*. Goslar, DEU: Eurographics Association. 179–88.

Allner, L., C. Kaltenbrunner, D. Kröhnert, and P. Reinsberg. 2021. Conceptual Joining: Wood Structures from Detail to Utopia/ Holzstrukturen im Experiment. Boston/Basel: Birkhäuser. 256.

Allner, Lukas, Daniela Kroehnert, and Andrea Rossi. 2020. "Mediating Irregularity: Towards a Design Method for Spatial Structures Utilizing Naturally Grown Forked Branches." In *Impact: Design With All Senses*, edited by Christoph Gengnagel, Olivier Baverel, Jane Burry, Mette Ramsgaard Thomsen, and Stefan Weinzierl. Cham: Springer International Publishing. 433–45.

Arici, Faruk, Pelin Yildirim, Şeyma Caliklar, and Rabia M. Yilmaz. 2019. "Research Trends in the Use of Augmented Reality in Science Education: Content and Bibliometric Mapping Analysis." *Computers & Education* 142 (December): 103647. https://doi.org/10.1016/j.compedu.2019.103647.

Belley, Denis, Isabelle Duchesne, Steve Vallerand, Julie Barrette, and Michel Beaudoin. 2019. "Computed Tomography (CT) Scanning of Internal Log Attributes Prior to Sawing Increases Lumber Value in White Spruce (Picea Glauca) and Jack Pine (Pinus Banksiana)."

Canadian Journal of Forest Research 49 (12): 1516–24. https://doi.org/10.1139/cjfr-2018-0409.

Bowman, Doug, Jian Chen, Chadwick Wingrave, John Lucas, Andrew Ray, Nicholas Polys, Qing Li, et al. 2006. "New Directions in 3D User Interfaces." *International Journal of Virtual Reality* 5 (2): 3–14. https://doi.org/10.20870/IJVR.2006.5.2.2683.

Cheng, Kun-Hung, and Chin-Chung Tsai. 2013. "Affordances of Augmented Reality in Science Learning: Suggestions for Future Research." *Journal of Science Education and Technology* 22 (4): 449–62. https://doi.org/10.1007/s10956-012-9405-9.

Jahn, Gwyllim, Cameron Newnham, Nick Berg, and Matthew Beanland. 2019. "Making in Mixed Reality." In *ACADIA '18: Proceedings of the 2018 Association for Computer Aided Design in Architecture (ACADIA)*, edited by Pablo Kobayashi, Brian Slocum, Philllip Anzalone, Andrew Jon Wit, Marcella Del Signore, Jorge Ramirez, Marcela Delgado, Pablo Iriate, Irma Soler, and Jose Luis Guiterrez Brezmes. Mexico City: Association for Computer Aided Design in Architecture.

Microsoft. 2020. "What Is Mixed Reality? - Mixed Reality." Accessed September 26, 2020. https://docs.microsoft.com/en-us/windows/mixed-reality/discover/mixed-reality.

Mollica, Zachary, and Martin Self. 2016. "Tree Fork Truss: Geometric Strategies for Exploiting Inherent Material Form." In *Developable Surfaces, Polyhedral Surface, Timber Plate Shells, 3D-Scanning, Digital Fabrication, Membrane, Meshing, Computational Design, Freeform*, edited by Sigrid Adriaenssens, Fabio Gramazio, Matthias Kohler, Achim Menges, and Mark Pauly. Zürich: vdf Hochschulverlag AG an der ETH Zürich. https://doi.org/10.3218/3778-4.

Nakata, Norio, Naoki Suzuki, Asaki Hattori, Naoya Hirai, Yukio Miyamoto, and Kunihiko Fukuda. 2012. "Informatics in Radiology: Intuitive User Interface for 3D Image Manipulation Using Augmented Reality and a Smartphone as a Remote Control." *RadioGraphics* 32 (4): E169–74. https://doi.org/10.1148/rg.324115086.

Piker, Daniel. 2013. "Kangaroo: Form Finding with Computational Physics." *Architectural Design* 83 (2). https://doi.org/10.1002/ad.1569.

Preisinger, Clemens, and Moritz Heimrath. 2014. "Karamba—A Toolkit for Parametric Structural Design." *Structural Engineering International* 24 (May): 217–22. https://doi.org/10.2749/101686614X13830790993483.

Self, M., E. Vercruysse, A. Menges, B. Sheil, R. Glynn, and M. Skavara. 2017. "Infinite Variations, Radical Strategies." In *Fabricate 2017*. London: UCL Press. 30–35. https://doi.org/10.2307/j.ctt1n7qkg7.8

Skarbez, Richard, Missie Smith, and Mary Whitton. 2021. "Revisiting Milgram and Kishino's Reality-Virtuality Continuum." *Frontiers in Virtual Reality* 2 (March). https://doi.org/10.3389/frvir.2021.647997.

Song, Yang, Richard Koeck, and Shan Luo. 2021. "Review and Analysis of Augmented Reality (AR) Literature for Digital Fabrication in Architecture." *Automation in Construction* 128 (August): 103762. https://doi.org/10.1016/j.autcon.2021.103762.

Suzuki, Takaharu, Hikaru Ikeda, Issei Takeuchi, Fumiya Matsunaga, Eri Sumitomo, and Ikeda Yasushi. 2020. "Holonavi: A Study on User Interface for Assembly Guidance System with Mixed Reality in a Timber Craft of Architecture." In *Proceedings of the 25th International Conference on Computer-Aided Architectural Design Research in Asia (CAADRIA)*, edited by Dominik Holzer, Walaiporn Nakapan, Anastasia Globa, and Immanuel Koh, vol. 1. Bangkok, Thailand: The Association for Computer-Aided Architectural Design Research in Asia (CAADRIA). 691–700.

Eri Sumitomo Architects. 2020. "Torinosu." Accessed June 19, 2021. https://www.erisumitomo.com/%E8%A4%87%E8%A3%BD-house-in-hayama.

Von Buelow, Peter, Omid Oliyan Torghabehi, Steven Mankouche, and Kasey Vliet. 2018. Combining Parametric Form Generation and Design Exploration to Produce a Wooden Reticulated Shell Using Natural Tree Crotches. In *Proceedings of IASS Annual Symposia, IASS 2018 Boston Symposium: Timber Spatial Structures*. Boston: International Association for Shell and Spatial Structures (IASS). 1–8(8).

Leslie Lok is Assistant Professor at the Cornell University Department of Architecture and directs the Rural-Urban Building Innovation (RUBI) Lab. Her research and teaching explore the intersection of technology, novel material methods, and urbanization to experiment with hybridized design and construction processes. Lok is also a co-founder at HANNAH, an experimental design practice that explores the implementation of innovative forms of construction such as additive concrete manufacturing and robotic wood construction at full scale.

Asbiel Samaniego is a research associate at the Rural-Urban Building Innovation Lab and a MS. MDC candidate at Cornell University Department of Architecture.

Lawson Spencer is a research associate at Cornell AAP and a Master of Architecture candidate at Cornell University.

Augmented Feedback

Garvin Goepel
CUHK

Kristof Crolla
HKU

A case study in Mixed-Reality as a tool for assembly and real-time feedback in bamboo construction

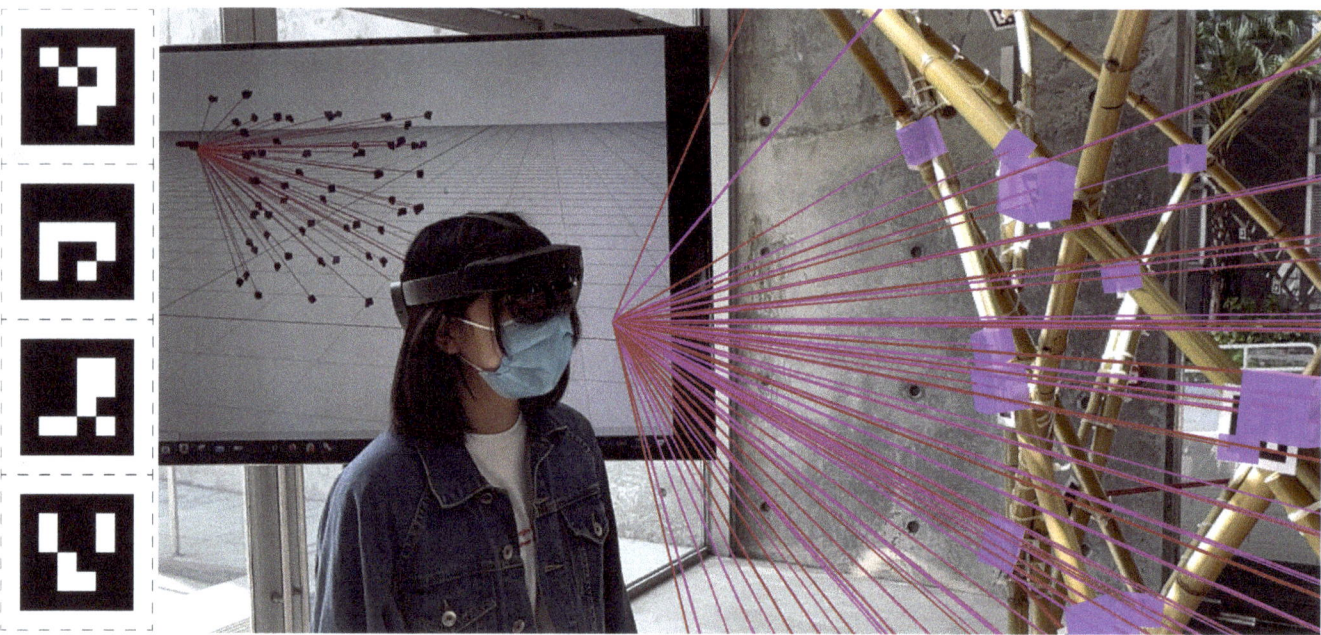

1   Scanning Aruco markers on the installation. Live stream between Rhino and the HMD to bring the tracked points back into a 3D environment.

ABSTRACT

Augmented Reality (AR) has the potential to create a paradigm shift in the production of architecture. This paper discusses the assembly and evaluation of a bamboo prototype installation aided by holographic instructions. The case study is situated within the framework of AR-driven computational design implementation methods that incorporate feedback loops between the as-built and the digital model. The prototype construction aims to contribute to the ongoing international debate on architectural applications of digital technology and computational design tools and on the impact these have on craftsmanship and architecture fabrication. The case study uses AR-aided construction techniques to augment existing bamboo craftsmanship in order to expand its practically feasible design solution space. Participating laypersons were challenged to work at the interface of technology and material culture and engage with both latest AR systems and century-old bamboo craft. This paper reflects on how AR tracking can be used to create a constant feedback loop between as-built installations and digitally designed source models and how this allows for the real-time assessment of design fidelity and deviations. The case study illustrates that this is especially advantageous when working with naturally varying materials, like bamboo, whose properties and behavior cannot be digitally simulated straightforwardly and accurately. The paper concludes by discussing how augmented feedback loops within the fabrication cycle can facilitate real-time refinement of digital simulation tools with the potential to save time, cost, and material. The augmentation of skills available onsite facilitates the democratization of non-standard architectural design production.

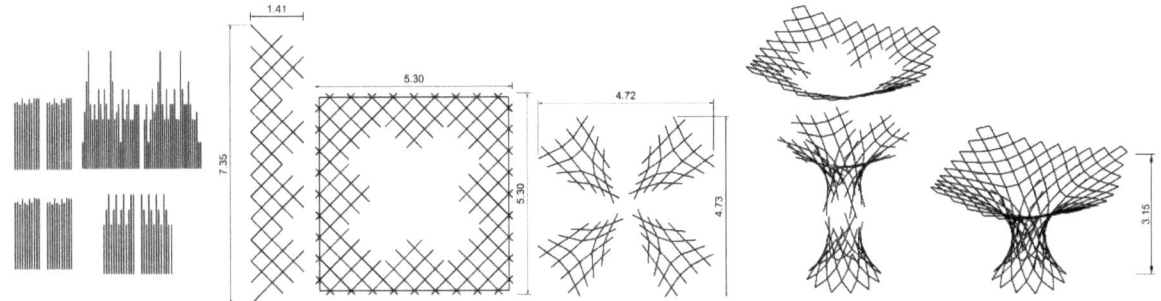

2   Fabrication Sequence; Left: Cut to length,
    Middle: Layout and connection of grids, Right: Assembly of parts

INTRODUCTION

AR in Fabrication and Construction

Integration of digital and robotic fabrication techniques in typical production chains and construction processes is often hindered by a lack of dexterity—otherwise found in human interaction—that limits their ability to adapt to the often-unstructured nature of onsite construction tasks (Kyjanek et al. 2019). In addition, a strong dichotomy exists between the increased architectural design agency offered by digital tools today and the affordances given by many construction contexts, especially building environments in developing countries with limited available means (Crolla 2018). Mixed-Reality has the ability to bridge this gap by taking physical measurements for on-site alignment to overlap holographic information on top of the physical fabrication area (Fazel and Izadi 2018). AR holographic instructions can reduce construction time, increase skill development, and increase the participations of non-experts in construction processes (Jahn et al. 2019). Conventional 2D drawings and physical templates can be replaced by an overlay of holographic instructions with clear, intuitive, and contextualized on-site information (Jahn 2020). In doing so, augmented construction techniques enable the democratization of skill through simple and intuitive holographic instructions of people that do not require professional training (Goepel 2019).

There is an exponential increase in the implementation of AR in fabrication and construction disciplines, with a strong focus not only on AR for 3D holographic instruction for assembly and fabrication, but also for AR as an immersive design tool (Song, Koeck and Luo 2021). Projects from peers have showcased the implementation of AR instructions in, among others, the assembly of complex brick walls (Jahn 2020; Mitterberger et al. 2020), the fabrication and assembly of a steam bend plywood (Jahn, Wit and Pazzi 2019; Jahn et al. 2019), a bent steel structure (Jahn et al. 2018), and a bamboo split art installation (Goepel and Crolla 2020).

AR immersive design techniques have been used to alter design shapes in real-time by the control of points through an AR interface (Betti, Aziz and Ron 2019), to influence digital simulation through gesture recognition and hand gestures (Hahm et al. 2019), and to adjust to dynamic structural performance simulations by tracking the movement of a structure's control points (Forren et al 2019). Mitterberger et al. (2020) applied an "eye-in-hand" system (Schmalstieg and Höllerer 2016) with an object tracking and registration system that estimates the deviations between the as-planned and the as-built model in real-time for the assembly of a fair-faced brickwork facade. Qi et al. (2021) investigated a feedback driven adaptive workflow by using computer vision algorithms through sensory scanned data by a depth camera to compensate for deviations in a bamboo pole structure by 3D printing joint connections on-the-fly.

This study aims to expand the practically feasible design solution space of existing bamboo craftsmanship by augmenting onsite skill through the use of AR-aided construction techniques. Rather than shifting construction complexity to automated pre-manufacturing, which reduces the onsite need for human labour skill, this setup enables digital laypersons to participate in complex building processes.

Bending-active Bamboo Gridshells

Bending-active bamboo gridshell construction systems cannot rely on hyper-precision, but must adapt to inevitable and system-specific indeterminacies that occur onsite. The choice of natural, unprocessed bamboo as a material implies working with unpredictability as each culm has naturally varying properties that directly affect the structural behavior when bent. These factors include variance due to each culm's species, age, and moisture content and its diameter, which changes along its length.

This paper discusses a strategy for the practical, onsite handling of such variance through a demonstrator

3

4

5

case-study which was developed in preparation of a larger future bamboo installation based on this tectonic system. The study tests the bamboo's perceived bending behavior and evaluates it in relation to its digital simulation. AR is used to inform both the construction sequence and the evaluative feedback system.

METHODS

The prototype's design consisted of a number of equidistant bamboo grids—with no shear rigidity—that were prefabricated on the ground and were then interconnected and bent into the final shape (Figure 2). The system included four equilateral pentagonal cells that connect the column to the top grid. These could be fabricated flat on the ground but had to be pre-assembled in a bent state (Figure 3). The fabrication sequence relied on three-dimensional AR instructions for cutting culms, aligning the grid, assembling the components, and for the tracing of the as-built structure for comparison with the digital source model. Fologram (Jahn, Newnham and van den Berg 2020), a third-party plugin for McNeel's NURBS modeller Rhinoceros 3D (Robert McNeel & Associates 2021), was used to stream this digital information directly from the procedural modeller Grasshopper (Rutten 2021) to the multiple HoloLenses and mobile devices used onsite to visualize the fabrication instructions.

The holographic projections, overlaid onto the builders' field of vision, acted like three-dimensional stencils that had to be transferred from the digital into the physical space. For the grid assemblies, different colors indicated the different pole layers and their orientations (Figures 4, 5).

The prefabricated components were then manually bent and connected to the top grid. This step required the top grid and column to deform into shape. Participants wearing the Head-Mounted AR Devices (HMD)—the "holographic instructors"—aligned the components in the correct position while assistants without HMDs held the poles together and fixed them with standard zip ties. During construction, an image target was used to align the holographic projection of the digital model with the prototype as it defined location of the digital file's origin in physical space.

Once completed, a series of Aruco trackers, measuring 45mm by 45mm, were fixed to the structure at the intersection points of the bamboo poles (Figure 1). These trackers could be detected by the HoloLens in real time, producing a virtual 3D point cloud in Rhino3D and placed in reference to the model space's origin, which was defined in the real world using an image target tracker (Figure 1). Once detected by the HoloLens, a semi-transparent, purple holographic cube was generated on each tracked Aruco marker to visually inform the participant wearing the HMD about the success of the detection. Another participant simultaneously followed the progress live in a Rhino3D viewport to help to identify untracked markers. This allowed the automated digital remodelling of the physical structure and the consequent study of deviations between both (Figure 6).

RESULTS

Successful Holographic Fabrication

The case study's main objective was to demonstrate it is practically feasible to expand the design solution space of

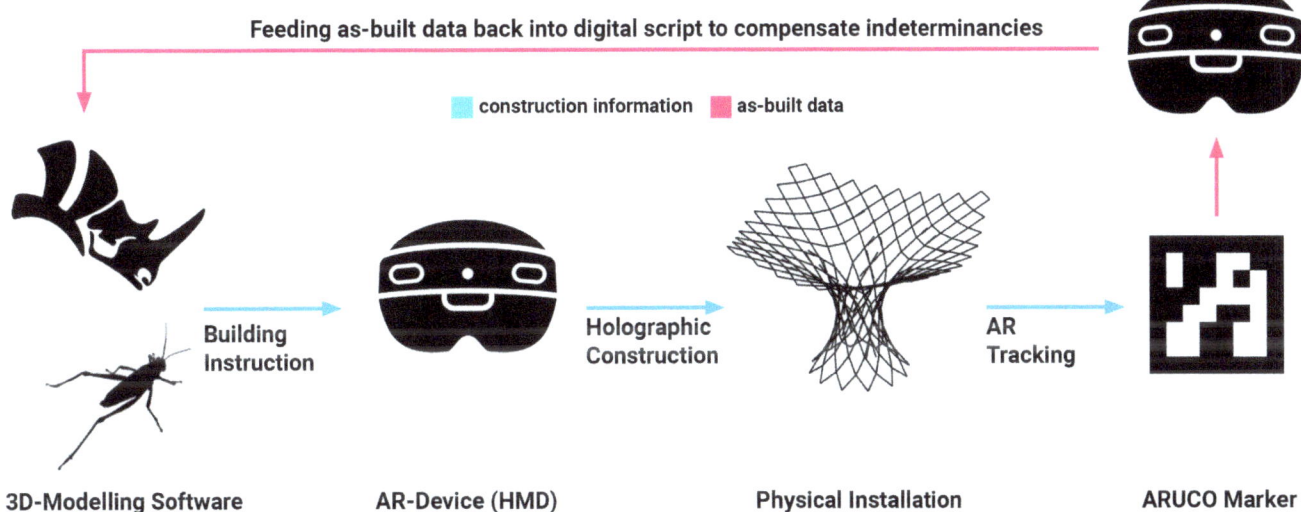

existing bamboo craftsmanship by augmenting onsite skill through the use of AR-aided construction techniques.

The installation was completed within two working days by fifteen participants. Participants took turns to use the three available Microsoft HoloLens HMDs. The "holographic instructors" collaborated with participants without holographic guidance. Participants stated that it was easy to grasp the design and scale of the project from the beginning, since the holographic display would continuously provide guidance in the working space. As no onsite measuring was needed, components could be cut and assembled more easily and quicker than when using traditional analog methods.

While the participants reported that the holographic guidance for the fabrication of the flat grids was intuitive, helpful, and clear, the more complex bent components were challenging to align, as the physical bent outcome did not always align with the digital simulated bending behaviour. Here, the AR 3D holographic guidelines helped to assemble the parts by placing the holographic projection in reference next to the physical model, rather than exactly on top of the to-be-assembled physical components.

Challenging Creation of Holographic Feedback

The detection of the Aruco markers by the HoloLens required the HMD to be in close proximity to the markers and needed the participant to move their head in front of the trackers multiple times. Problematic shifts and offsets frequently occurred during this tracking, resulting in the necessary resetting of the AR origin by reorienting the digital file in physical space with the image target.

3 Assembly of bent components
4 Grid layout and marking through holographic guidelines
5 Laid out grid with holographic guidelines
6 Feedback loop

Additionally, the number of Aruco markers that can be tracked simultaneously within one single digital file is predefined and limited by the Fologram software to 100 in order to reduce error between the codes. As we needed more, tracked points had to be internalized in the digital file so that identical ID Aruco markers could be reused for the scanning of additional intersection points. Doubly tracked points could, however, be easily identified. Here we detected small deviations of identical ID points, caused by shifts that occurred due to offsets in the tracking. These could be fixed by rescanning following a reorientation as per above.

The main deviation between the as-built and digital model was found in the middle area of the column where the tectonic system was most flexible and the predicted possible deviations were the largest. An additional tension rope was added and wrapped around the column to tighten its radius to reduce flexible movement there. This allowed calibration of the overall shape to more closely match the digital file.

Smaller deviations were detected at individual intersection points. These deviations, however, did not cause any issues, as they were absorbed by the integrated system and fell within the overall tolerance/allowance. These deviations were caused by multiple factors: Aruco markers could have been placed slightly off the intersection point, zip ties could inhibit the correct placement of the Aruco markers,

7   Final installation at the School of Architecture atrium at CUHK

CONCLUSION

This case study demonstrated how, for bamboo architecture design and construction, the integration of AR in construction can lead to the augmentation of existing craftsmanship and the expansion of its associated practically feasible design solution space. It additionally showed how an augmented feedback loop within the fabrication cycle can help to refine digital simulation tools in real-time.

ACKNOWLEDGEMENTS

Workshop Team: Garvin Goepel (CUHK), Dr. Kristof Crolla (LEAD, HKU) - Workshop Participants: Adwin Shing Fung KEUN, Anson Yik Hei CHEUNG, Fergal Yau Wai TSE, Nomy Jia Ning YU, Oscar Chun Yan WONG, Vincent Chun Ki YAN, Katie Hoi Tung LEUNG, Vanessa Heimen LIN, May Hyunjoo KIM, Rachel Gayoung KIM, Silviane Sum Yin LO, Seiyoung JOUNG, Klementine Ka Ting WEI, Erika Anna FERNANDES, Carson Yan Shun WONG.

This workshop was supported by a grant from the Research Grants Council of the Hong Kong Special Administrative Region China (Project Ref No. 14604618)

REFERENCES

Betti, Giovanni, Saqib Aziz, and Gili Ron. 2019. "Pop up factory: collaborative design in mixed reality, interactive live installation of the makeCity festival." In *Proceeding of the 37th Education and Research in Computer Aided Architectural Design in Europe (eCAADe)*. Berlin, 2019. 115–124.

Crolla, Kristof. 2018. "Building simplexity: The 'more or less' of post-digital architecture practice." PhD diss., RMIT University.

Fazel, Alireza and Abbasali Izadi. 2018. "An interactive augmented reality tool for constructing free-form modular surfaces." *Automation in Construction* 85: 135–145

Forren, James, Makenzie Ramadan, and Sebastien Sarrazin. 2019. "Action Over Form: Combining Off-loom Weaving and Augmented Reality in a Non-specification Model of Design." In *Artificial Realities: Virtual as an Aesthetic Medium in Architecture Ideation Symposium*. Lisbon Architecture Triennale, Portugal.

Goepel, Garvin, and Kristof Crolla. 2020. "Augmented Reality-based Collaboration; ARgan, a bamboo art installation case study." In *RE: Anthropocene, Design in the Age of Humans; Proceedings of the 25th CAADRIA Conference*, vol. 2. Bangkok, 2020. 313–322.

Goepel, Garvin and Kristof Crolla. 2021. "Secret Whispers and Transmogrifications: A case study in online teaching of Augmented Reality technology for collaborative design production" In *Proceedings of the 26th CAADRIA Conference*. Hong Kong, 2021. 21–30.

and pole intersection points could have slipped and shifted slightly throughout the assembly, etc. The digital simulation, however, maintained a perfectly equidistant grid throughout the deformation.

Studying the as-built deviation can inform designers to adjust digital parameters as to more accurately simulate real-world behaviors. These parameters can be changed in real-time during the feedback loop, which enables the adjustment of the digital base file to these indeterminacies found onsite.

A test to extract useful as-built 3D mesh data from the HoloLens spatial mapping failed: the Hololens' current scanning resolution does not provide sufficient detail and resolution to allow for the tracking of poles of diameters of 3cm. Our case study, however, did not require such an object detecting system, as the orientation in space of each pole can easily be derived from the intersection points. The Aruco marker tracking method was computationally fast in comparing physical points to virtual points as opposed to comparing scanned mesh geometries with digital simulated curves or meshes.

Goepel, Garvin. 2019. "Augmented Construction - Impact and opportunity of Mixed Reality integration in Architectural Design Implementation." In *ACADIA '19, Ubiquity and Autonomy: Proceedings of the 39th Annual Conference of the Association for Computer Aided Design in Architecture*. Austin, Texas. 430–437.

Hahm, Soomeen, Abel Maciel, Eri Sumitiomo, and Alvaro Lopez Rodriguez. 2019. "FlowMorph - exploring the human-material interaction in digitally augmented craftsmanship." In *Intelligent & Informed; Proceedings of the 24th CAADRIA Conference*, vol. 1. Wellington, New Zealand, 2019. 553–562.

Jahn, Gwyllim, Cameron Newnham and Nick van den Berg. *Fologram*. V 2020.3.5.0. Robert McNeel & Associates. PC. 2020.

Jahn, Gwyllim. 2020. "Constructing curved benches from cut bricks in mixed reality." Fologram community post. Uploaded September 2, 2020. https://community.fologram.com/t/constructing-curved-benches-from-cut-bricks-in-mixed-reality/434.

Jahn, Gwyllim, Cameron Newnham, Nicholas van den Berg, Melissa Iraheta, and Jackson Wells. 2019. "Holographic Construction." *Design Modelling Symposium*. Springer, Berlin. 314–324.

Jahn, Gwyllim, Cameron Newnham, Soomeen Hahm, and Igor Pantic. 2019. "Steampunk Pavilion." Fologram. Uploaded October 12, 2019. Vimeo video, 0:02:59. https://vimeo.com/365917769.

Jahn, Gwyllim, Andrew Wit, and James Pazzi. 2019. "[BENT] Holographic handcraft in large-scale steam-bent timber structures." In *ACADIA '19, Ubiquity and Autonomy: Proceedings of the 39th Annual Conference of the Association for Computer Aided Design in Architecture*. Austin, Texas. 438–447.

Jahn, Gwyllim, Cameron Newnham, Nicholas van den Berg, and Matthew Beanland. 2018. "Making in Mixed Reality: Holographic design, fabrication, assembly and analysis of woven steel structures." In *ACADIA '18, Recalibration, On Imprecision and Infidelity Structures: Proceedings of the 38th Annual Conference of the Association for Computer Aided Design in Architecture*. Mexico City, 2018. 88–97.

Kyjanek, Ondrej, Bahar Al Bahar, Lauren Vasey, Benedikt Wannemacher, and Achim Menges. 2019. "Implementation of an augmented reality AR workflow for human robot collaboration in timber prefabrication." In *Proceedings of the 36th International Symposium on Automation and Robotics in Construction (ISARC)*, vol. 36. Banff, Canada. 1223–1230.

Mitterberger, Daniela, Kathrin Dörfler, Timothy Sandy, Foteini Salveridou, Marco Hutter, Fabio Gramazio, and Matthias Kohler. 2020. "Augmented bricklaying; Human-machine interaction for in situ assembly of complex brickwork using object-aware augmented reality." *Construction Robotics* 4: 151–161.

Robert McNeel & Associates. *Rhinoceros*. V.6.0. Robert McNeel & Associates. PC. 2021.

Rutten, David. *Grasshopper3D*. V1.0.0007. Robert McNeel & Associates. PC. 2021

Schmalstieg, Dieter, and Tobias Höllerer. 2016. *Augmented Reality: Principles and Practice*. Boston: Addison-Wesley.

Song, Yang, Richard Koeck, and Shan Luo. 2021. "Review and analysis of augmented reality (AR) literature for digital fabrication in architecture." *Automation in Construction* 128: 103762.

Qi, Yue, Ruqing Zhong, Benjamin Kaiser, Long Nguyen, Hans Jakob Wagner, Alexander Verl, and Achim Menges. 2021. "Working with Uncertainties: An Adaptive Fabrication Workflow for Bamboo Structures." In *Proceedings of the 2020 DigitalFUTURES, 2nd International Conference on Computational Design and Robotic Fabrication (CDRF)*. Springer: Singapore. 265–279.

IMAGE CREDITS
All drawings and images by the authors.

Garvin Goepel is a PhD researcher at the Chinese University of Hong Kong (CUHK) specialized in Augmented Reality (AR) implementation in fabrication and design processes. He believes that AR has the ability to enhance human capacities to participate in complex processes through simplified instructions—instead of surrendering human skill to automation in manufacturing. His research advances studies in collaborative holographic-driven construction, expands opportunities for technology-infused craftsmanship, and reflects on workflows that replace conventional paper drawing-based communication with holographic instruction.

Dr. Kristof Crolla is a Hong Kong-based architect who combines his architectural practice, Laboratory for Explorative Architecture & Design Ltd. (LEAD), with his position as Associate Professor at the University of Hong Kong (HKU) Departments of Architecture and Engineering, where he also directs the Bachelor of Arts & Sciences in Design+ program. His work has received numerous design, research, and teaching awards and accolades, including the RMIT Vice-Chancellor's Prize for Research Impact - Higher Degree by Research. He is best known for the projects "Golden Moon" and "ZCB Bamboo Pavilion," for which he received the World Architecture Festival Small Project of the Year 2016 award.

BIM LOD + Virtual Reality

Using Game Engine for Visualization in
Architectural & Construction Education

Hassan Anifowose
Texas A&M University

Wei Yan
Texas A&M University

Manish Dixit
Texas A&M University

ABSTRACT

Architectural education faces limitations due to the tactile approach to learning in classrooms supplemented by existing 2D and 3D tools. At a higher level, Virtual Reality provides a potential for delivering more information to individuals engaged in design learning. This paper investigates a hypothesis establishing grounds towards new research in Building Information Modeling (BIM) and Virtual Reality (VR). The hypothesis is projected to determine best practices for content creation and tactile object virtual interaction, which potentially can improve learning in architectural and construction education with a less costly approach and ease of access to well-known buildings. We explored this hypothesis in a step-by-step game design demonstration in VR by showcasing the exploration of the Farnsworth House and reproducing assemblage of the same with different game levels of difficulty that correspond with varying BIM levels of development (LODs). The game design prototype equally provides an entry way and learning style for users with or without a formal architectural or construction education seeking to understand design tectonics within diverse or cross-disciplinary study cases. This paper shows that developing geometric abstract concepts of design pedagogy, using varying LODs for game content and levels—while utilizing newly developed features such as snap-to-grid, snap-to-position and snap-to-angle to improve user engagement during assemblage—may provide deeper learning objectives for architectural precedent study.

INTRODUCTION

The integration of interactive building anatomy modeling into the construction education system improves learning while enabling effective knowledge transfer to learners (Park et al. 2016). The COVID-19 pandemic proved education's overreliance on the classroom to be inadequate. Adaptation to future demands is overdue. Researchers indicated in previous works that interactive gaming improves learning and motivation (Kharvari and Höhl 2019), and it is also acknowledged that BIM provides the framework for developing 3D geometric data that can be used in the generation of levels and maps in games (Yan, Culp, and Graf 2011). Virtual Reality (VR) on the other hand is technology that is fast gaining adoption, however, with limitations based on extra work required to visualize BIM models (Zaker and Coloma 2018).

We used an empirical research method by developing a simple demonstration to test the hypothesis. The overall objective is to study how building anatomy modeling process combined with varying levels of development (LODs) help the process of learning architectural precedents. Evidence shows that using this anatomical approach—varying LODs provided in geometric form—holds a potential to improve interaction and overall learning. This demonstration shows the significance of new features for improved engagement in VR.

STATE OF THE ART

VR provides full scale perception of spaces while accurately representing materiality in deeper levels of immersion when compared to desktop screens and projection systems (Angulo 2015). The common use of VR is to enhance walking experiences inside and around a virtual structure; however, it has been used to evaluate form and design of a building (Abdelhameed 2013).

Increasing the feeling of presence has been determined to improve overall VR experiences, which must be accompanied with an increased level of realism within VR (Du et al. 2016). Outside environment realism, there are limitations of accessibility to data from BIM environments, requiring improvement (Kieferle and Woessner 2015). Our research towards gamification aims to help users create and manipulate objects within the VR environment thereby limiting dependence on the BIM environment for new objects generated in a game scene. Designing and providing content and interface design, alongside with coding and programming, are some of the identified issues to overcome (Maghool, Moeini, and Arefazar 2018).

VR + Architectural Education

The exploration of existing architectural projects, precedents, or case studies is at the core of architectural education. Design students are generally expected to provide clarity of understanding of such precedents' abstract principles and tectonics before tendering design proposals during studio sessions. Visiting architectural precedents can become very costly for students and institutions depending on various factors. Additionally, there are no layers of interactivity involved in local visits, and therefore concept retention varies between students (Bourdakis and Charitos 1999; Kharvari and Höhl 2019).

Learning within Virtual Environments (VEs) can be developed to create a foundation for learning about Virtual environment design. Pushing learning in VE further within tutored environments, is proven to "benefit young architects on their first step towards understanding the essence of architecture" (Kamath, Dongale, and Kamat 2012). The speed of learning in architectural education has been attributed to students' ability to detect their errors faster and learn from them (Dvořák et al. 2005; Abdelhameed 2013). Moloney indicated that "the advantages of working in a real-time environment where early design iterations can be tested from multiple points of view" has been established alongside limitations of communicating while wearing a VR device (Moloney and Harvey 2004; Milovanovic et al. 2017).

VR + Construction Education

The study of building composition in construction education is limited by teaching strategies. The Interactive Building Anatomy Modeling (IBAM) research has indicated that an anatomical approach can deliver better learning outcomes in construction education via assemblage and gamified techniques (Park et al. 2016; P. Wang et al. 2018). This motivates our research idea for a virtual precedent study environment combined with BIM to provide an avenue where students apply their education to solve real problems. However, various settings for VEs require testing in order to determine which situation provides flexibility as well as improve virtual construction learning (Ghosh 2012). Gamifying construction learning is one method of continuously engaging the younger population (Generation Y) who have difficulties in engaging with traditional educational methods (Goedert et al. 2011).

The combination of building models in VR alongside a game process for learning may hold considerable impact on learning when applied to education. Wang, Newton, and Lowe (2015) indicated that learning in VR provides an experiential outcome which is obvious in the learner's behavior.

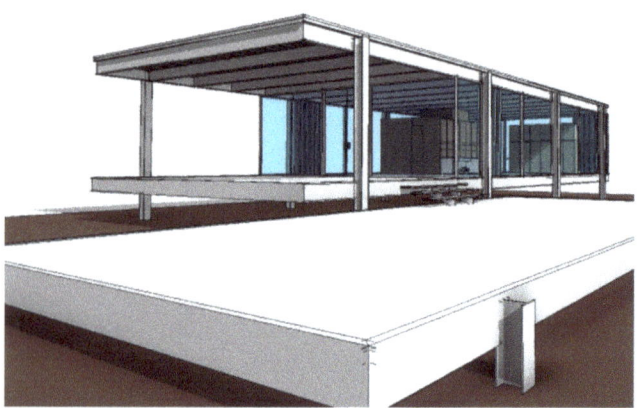

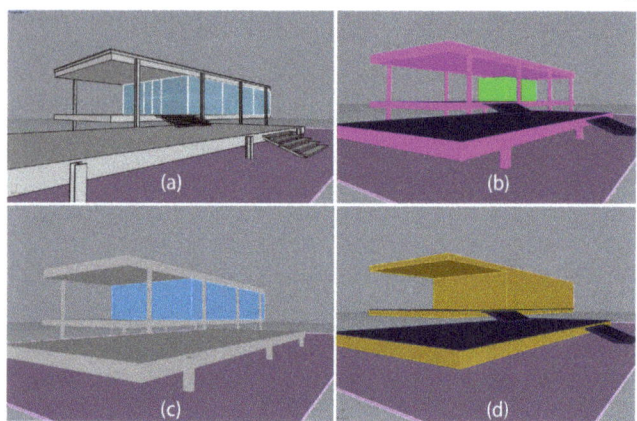

1 Completed detailed model of the building in Autodesk Revit

2 Abstract concept modeling of the Farnsworth House showing top left to bottom right, (a) full geometry, (b) structure, (c) lighting, and (d) circulation respectively

3 Precedent exploration in First Person View in the VR environment

4 Stairs: LOD 200 (left) and LOD 300 (right)

METHODS

Gamification for Studying Architectural Precedents

Various classifications for Virtual Reality (VR) in Education have been established (Motejlek and Alpay 2019). These are: purpose of the application, technology of delivery, user interaction, user experience, system interaction, and gamification types. Gamification is largely unexplored for developing learning experiences of architectural precedents. Beyond the conventional methods, this paper showcases new techniques for exploring architectural pedagogy through a new lens, i.e. Virtual Reality. To achieve cross-disciplinary simplified learning for users without prior design knowledge, we utilized varying levels of development (LODs). This study seeks to answer the question "What are best game design practices recommended to enhance playability in a BIM and VR game development?"

Pre-Game – Prototype Development

A prototype development provides an extensive tool for teaching varying LODs from concept design to final detailed design. Users explore a collaborative virtual environment and embark on design activities thereafter. For this study, we used the Farnsworth House.

Stage 1/Modeling: Detailed modeling of the building was completed in Autodesk Revit as a basis for studying (Figure 1). Abstract concepts are modeled in Rhinoceros and merged in Unity Game Engine (Unity 2021b) using the SteamVR (Valve 2021) plugin for gameplay and interactions.

Figure 2 shows different concept schemes that provide learning opportunities about underlying design pedagogy.

Stage 2/Precedent Exploration: Users are introduced to the architectural precedent and allowed 10 minutes for exploration (Figure 3).

Thereafter, users proceed to the game area for an assemblage exercise to test their knowledge.

Stage 3/Studio Setup, Practice Areas: In the VR environment, we developed a studio setup comprising an in-game projector screen that played a video narrating the architectural precedent while the game player explored the precedent and environment including interacting with abstract concept models in one case. The practice area in Level 1 allows the player to get accustomed to object interaction and manipulation techniques required for the game challenge in Level 2.

Stage 4/Gamification & Level Development: Borrowing

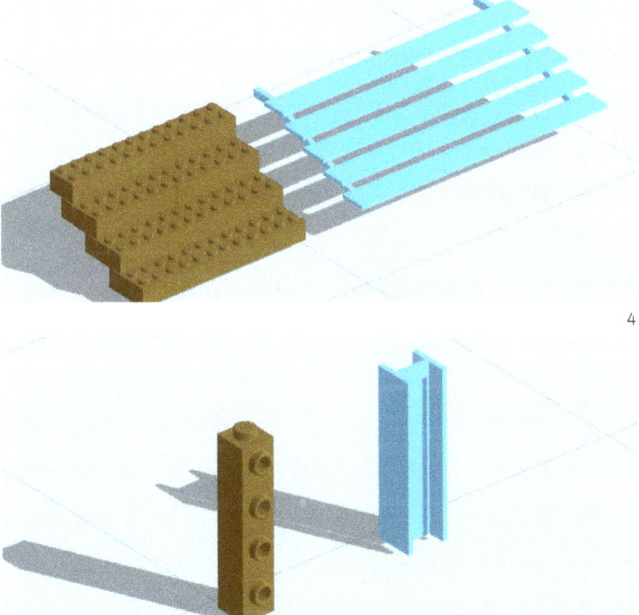

5   Steel Stanchion: LOD 200 (left) and LOD 300 (right)

6   Floor Slab: LOD 200 (left) and LOD 300 (right)

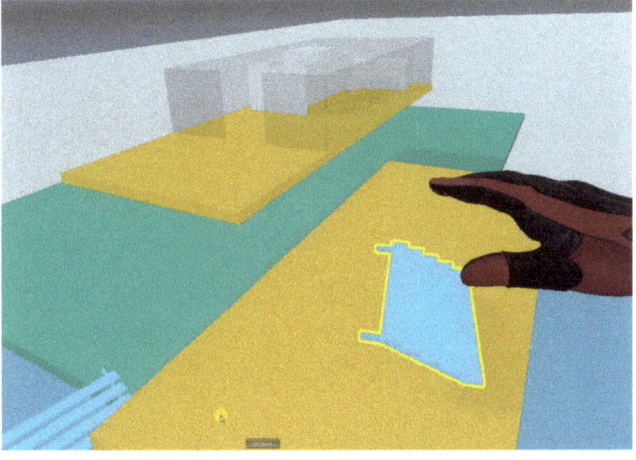

7   Completed Game Level 1 with LOD 200

8   Game Challenge in progress for LOD 300

from universally established BIM Level of Developments standard classifications (TrueCADD 2021), game levels were designed to correspond with increasing LOD. For the purpose of this research, we have classified the building construction components into game artifacts via two classes of LODs. The lowest LOD used is LOD 200 (approximate geometry represented by LEGO®), while the highest LOD used is LOD 300 (precise geometry of the building). LEGO® was used because of its geometric simplicity and popularity for low detail geometry assemblage. The classifications are created specifically for this game based on geometric complexity and similarity to universal BIM LOD definitions. Users are presented with each game level to complete assemblage tasks after the precedent explorations and pre-exercise stages. The assemblage task comprises of components with varying level of development (Figure 4, Figure 5, Figure 6).

Game Level 1 features the player, assembling the building with LOD 200 construction components generated from Mecabricks (Mecabricks 2020), a web-based application with LEGO® parts (Figure 7). Upon completion, the player proceeds to Level 2 where they complete the assemblage with LOD 300 components (Figure 9), which represent precise geometry.

In-Game – Implementation and Prototype Testing

Our in-lab demonstration involved two players. We received immediate feedback on playability, implemented features and overall comfort within the VR environment. Player 1 explored the game with a studio environment setup, an in-game video narrative, and was presented with the assemblage task. Player 2 explored the game with all the

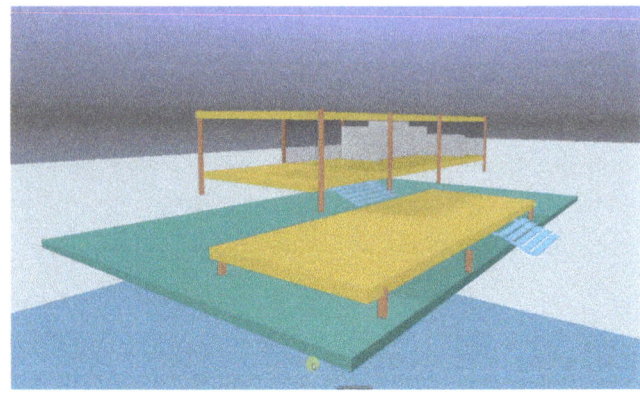

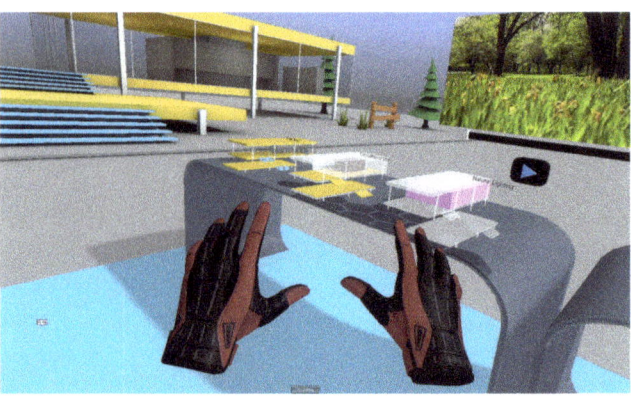

9   Completed Game Level 2 with LOD 300

10 Game scene showing the base model and geometric abstract concepts (Circulation-to-use and Natural Lighting) in the exploration mode

features from Player 1's experience including interacting with the geometric abstract concepts (Figure 10). A new VR object-snapping feature developed for the game allowed both players to complete the tasks with the same levels of difficulty. Each player spent 10 minutes exploring the precedent and 10 minutes assembling the parts in the exercise area. The game was live demonstrated to a team of researchers at a college symposium (Anifowose 2020).

RESULTS AND DISCUSSION

Feedback was collected from the two demonstration players based on four categories namely player experience, user behavior, level of object interaction, and knowledge retention. We observed that the game players were enthusiastic to learn about the architectural precedent. Player 2 had significant shorter time to reproduce the assemblage while also reporting a higher level understanding and motivation for learning and reproducing the building than Player 1. This indicates that having abstract concepts in precedent study virtual environments other than only the building geometry provides an opportunity for better knowledge retention. Players also indicated increased depth of understanding with the voice/video narration during the exploration stage. Especially important given the ongoing travel restrictions due to the pandemic, the BIM+VR prototype developed in this research—using virtual game development modalities—provides a less costly approach and improved accessibility to well-known architectural precedents with a potential for learning and engagement. Institutions may adopt this strategy to provide rich learning objectives for both design-related and cross-disciplinary studies.

To improve user interaction and overall engagement, new game features were created that increased the intuition level for interacting with the assemblage components and increased the users' abilities to complete the assemblage tasks. By developing an object-spawning system (Figure 11) using the Bolt Visual Scripting tool (Unity 2021a), the dependence on importing heavy BIM files is drastically reduced, thereby increasing adoption possibilities of this game's approach to learning in Virtual Reality. This tool was used in developing the embedded features saving time beyond the conventional C# scripting tools. This demonstration enabled us establish a case scenario for application within the design and construction education environment. As shown in (Figure 12), the BIM LOD + VR game research is grouped into four major stages of work. The first two stages and a portion of the third stage are reported in this paper.

CONCLUSION

Virtual Reality can potentially eliminate the need for local and international travel in architectural studies. This paper investigated the possibilities of using BIM+VR in a gamified LOD test case as a way of increasing effective learning objectives in architectural forms and tectonics. Feedback from the demonstration indicated that higher understanding of the main architectural precedent could be gained by studying the developed abstract concepts in geometric form with varying LODs inside the game before attempting the assemblage task. In diverse cases where the user has no previous knowledge of architecture or construction, the BIM+VR approach provides a learning alternative beyond reading technical drawings, a task typically difficult for the majority potential users in other fields of study. Developed features such as snap-to-grid, snap-to-angle, and snap-to-position are also necessary to improve user engagement during the assemblage phase of the game. By generalizing results, we can confirm that this approach could provide a deeper knowledge of architectural precedents while simulating a design studio and modeling environment inside virtual reality. Limitations

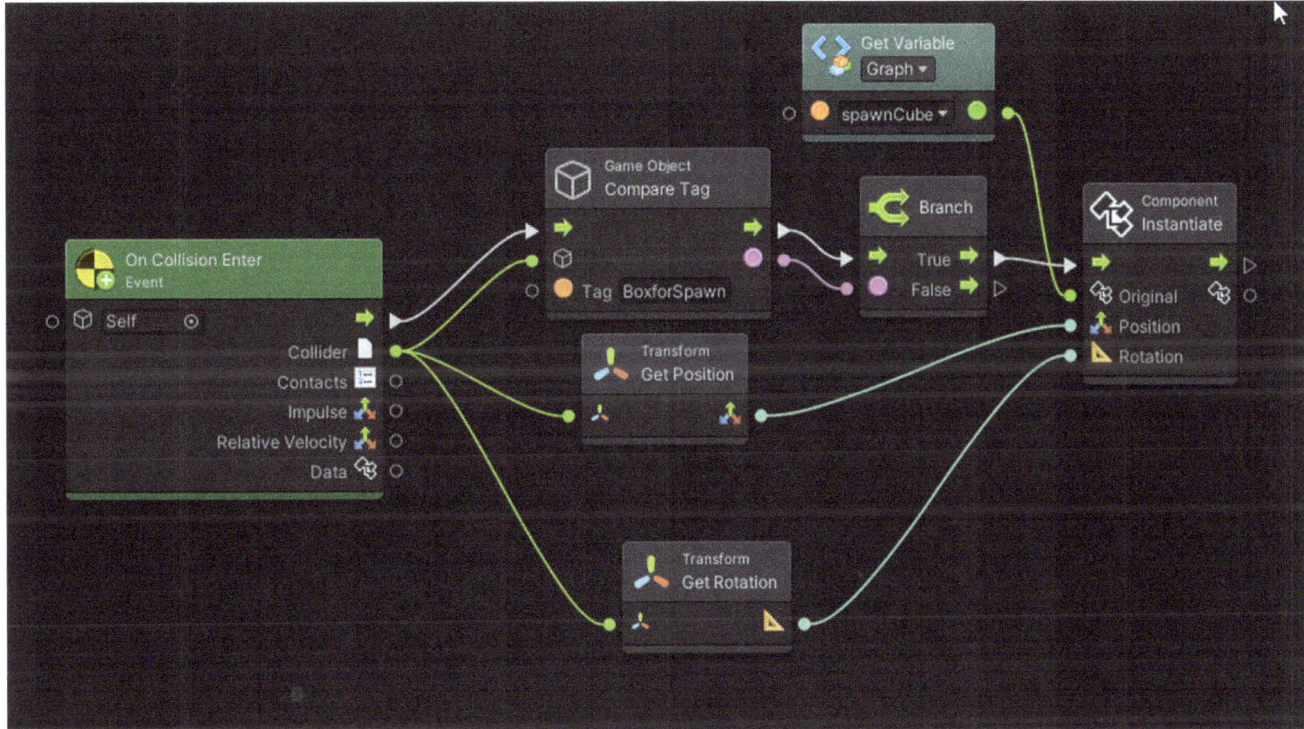

11 Object spawning system for game artifacts using Bolt Visual Scripting

experienced included difficulty in generating custom scripts and retention of BIM geometry data inside the Unity software while converting objects from native Revit to game objects. More challenges were faced in developing mechanics for improved game performance. This was overcome by developing object spawning systems.

A larger scale experiment is currently being developed for construction education to enable researchers to answer questions as to (1) how effective learning can occur using BIM and VR with varying LODs and also to (2) determine which gamification features enhance interaction while contributing to effective learning. This will be carried out as a full study with a larger group of human subjects. With the findings in this research, gamification is expected to make its way into Architectural and Construction Education both in the classroom and in the field.

REFERENCES

Abdelhameed, Wael A. 2013. "Virtual Reality Use in Architectural Design Studios: A Case of Studying Structure and Construction." *Procedia Computer Science* 25: 220-230. https://doi.org/10.1016/j.procs.2013.11.027.

Angulo, Antonieta. 2015. "Rediscovering Virtual Reality in the Education of Architectural Design: The immersive simulation of spatial experiences." *Ambiances; International Journal of Sensory Environment, Architecture and Urban Space* 1. https://doi.org/10.4000/ambiances.594.

Anifowose, Hassan. 2020. "BIM LOD + VR - Integrated codes, abstract concepts, interactions," Uploaded September 24, 2020. YouTube video, 00:07:37. https://youtu.be/_le_EpDeS9s.

Bourdakis, Vassilis, and Dimitrios Charitos. 1999. "Virtual Environment Design-Defining a New Direction for Architectural Education." In *Architectural Computing from Turing to 2000; eCAADe Conference Proceedings*. Liverpool (UK), 1999. 403–409.

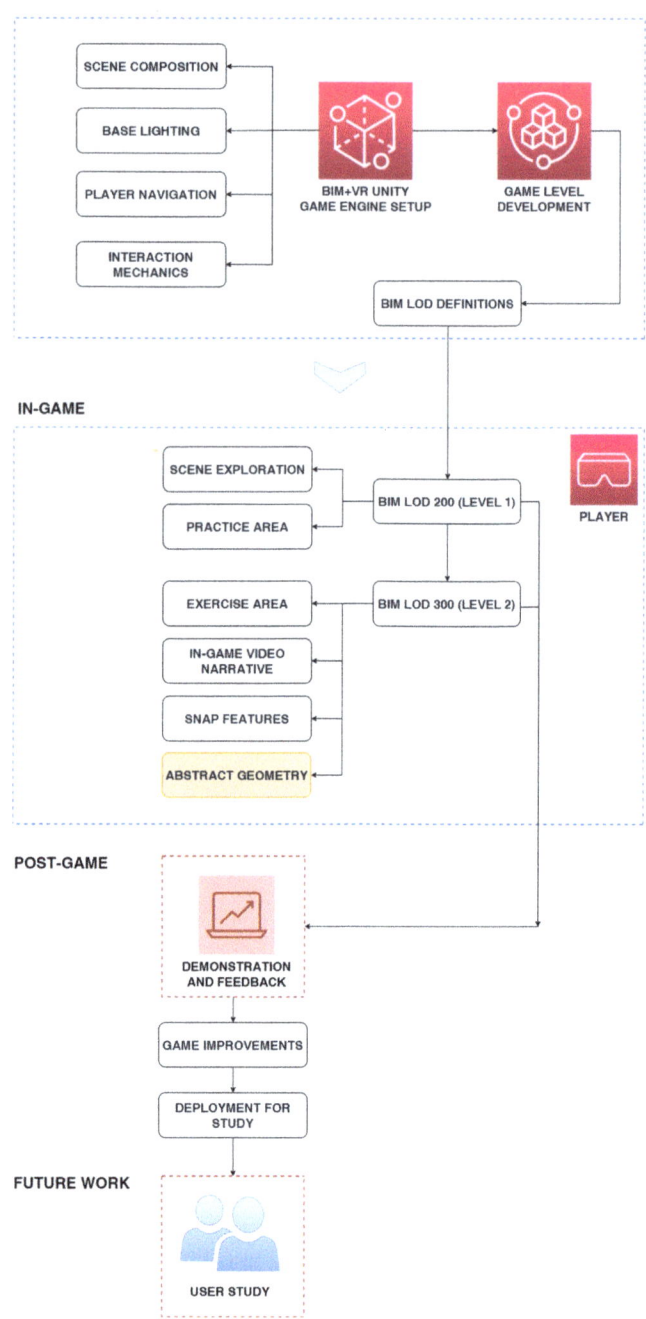

12 Diagram showing 4 stages of this research and features of each stage

Du, Jing, Yangming Shi, Chao Mei, John Quarles, and Wei Yan. 2016. "Communication by Interaction: A Multiplayer VR Environment for Building Walkthroughs." In *Proceedings of the 2016 Construction Research Congress*. San Juan, Puerto Rico, 2016. Reston, VA: ASCE. 2281–2290. https://doi.org/10.1061/9780784479827.227.

Dvořák, J., V. Hamata, J. Skácilik, and B. Beneš. 2005. "Boosting up architectural design education with virtual reality." In *CEMVR '05: Central European Multimedia and Virtual Reality Conference*. 95–200.

Ghosh, Arundhati. 2012. "Virtual Construction + Collaboration Lab: Setting a new paradigm for BIM education." In *2012 ASEE Annual Conference & Exposition*. San Antonio, Texas. https://doi.org/10.18260/1-2--2221.

Goedert, James, Yong Cho, Mahadevan Subramaniam, Haifeng Guo, and Ling Xiao. 2011. "A framework for Virtual Interactive Construction Education (VICE)." *Automation in Construction* 20 (1): 76–87. https://doi.org/10.1016/j.autcon.2010.07.002.

Kamath, R.S., T.D. Dongale, and R.K. Kamat. 2012. "Development of Virtual Reality Tool for Creative Learning in Architectural Education." *International Journal of Quality Assurance in Engineering and Technology Education (IJQAETE)* 2 (4): 16–24. https://doi.org/10.4018/ijqaete.2012100102.

Kharvari, F., and W. Höhl. 2019. "The Role of Serious Gaming using Virtual Reality Applications for 3D Architectural Visualization." In *2019 11th International Conference on Virtual Worlds and Games for Serious Applications (VS-Games)*. Vienna, 2019. 1–2. https://doi.org/10.1109/VS-Games.2019.8864576.

Kieferle, Joachim, and Uwe Woessner. 2015. "BIM Interactive - About combining BIM and Virtual Reality; A Bidirectional Interaction Method for BIM Models in Different Environment." In *Real Time: Proceedings of the 33rd eCAADe Conference*, vol. 1. Vienna, 2015. 69–75.

Maghool, Sayyed Amir Hossain, Seyed Hossein Iradj Moeini, and Yasaman Arefazar. 2018. "An Educational Application based on Virtual Reality Technology for Learning Architectural Details: Challenges and Benefits." *ArchNet-IJAR: International Journal of Architectural Research* 12 (3): 246–272.

Mecabricks. 2020. "Mecabricks Workshop Platform." Accessed July 20, 2021. https://www.mecabricks.com/en/workshop.

Milovanovic, Julie, Guillaume Moreau, Daniel Siret, and Francis Miguet. 2017. "Virtual and Augmented Reality in Architectural Design and Education An Immersive Multimodal Platform to Support Architectural Pedagogy." In *Future Trajectories of Computation in Design, 17th International Conference, CAAD Futures*. Istanbul, 2017. 513–532.

Moloney, Jules, and Lawrence Harvey. 2004. "Visualization and 'Auralization' of Architectural Design in a Game Engine based Collaborative Virtual Environment." In *IV 2004, Proceedings of the Information Visualisation, Eighth International Conference*. IEEE Computer Society, USA.

Motejlek, Jiri, and Esat Alpay. 2019. "A Taxonomy for Virtual and Augmented Reality in Education." *arXiv* preprint. arXiv:1906.12051.

Park, Chan Sik, Quang Tuan Le, Akeem Pedro, and Chung Rok Lim. 2016. "Interactive building anatomy modeling for experiential building construction education." *Journal of Professional Issues in Engineering Education* 142 (3): 04015019.

TrueCADD. 2021. "BIM Level of Development - LOD 100, 200, 300, 350, 400, 500 Experts." TrueCADD. Last modified February 16, 2021. Accessed February 16, 2021. https://www.trace-software.com/blog/the-level-of-detail-and-the-level-of-development-in-the-bim-environment/.

Unity. 2021a. "Bolt Visual Scripting." Accessed July 15, 2021. https://assetstore.unity.com/packages/tools/visual-scripting/bolt-163802.

Unity. 2021b. "Unity Software." Accessed 9th July 2021. https://unity.com/download.

Valve. 2021. "Steam VR Plugin." Accessed July 16, 2021. https://assetstore.unity.com/packages/tools/integration/steamvr-plugin-32647.

Wang, Peng, Peng Wu, Jun Wang, Hung-Lin Chi, Xiangyu Wang. 2018. "A critical review of the use of virtual reality in construction engineering education and training." *International Journal of Environmental Research and Public Health* 15 (6): 1204. https://doi.org/10.3390/ijerph15061204.

Wang, Rui, Sidney Newton, and Russell Lowe. 2015. "Experiential Learning Styles in the Age of a Virtual Rurrogate." *ArchNet-IJAR: International Journal of Architectural Research* 9 (3): 93–110.

Yan, Wei, Charles Culp, and Robert Graf. 2011. "Integrating BIM and Gaming for Real-time Interactive Architectural Visualization." *Automation in Construction* 20 (4): 446–458. https://doi.org/10.1016/j.autcon.2010.11.013.

Zaker, Reza, and Eloi Coloma. 2018. "Virtual Reality-integrated Workflow in BIM-enabled Projects Collaboration and Design Review: A Case Study." *Visualization in Engineering* 6 (1). https://doi.org/10.1186/s40327-018-0065-6.

IMAGE CREDITS
All drawings and images by the authors.

Hassan Anifowose is an Architect and 3D visualization expert with several years of experience in design, construction and entrepreneurship. Hassan is a PhD candidate at Texas A&M University, where he is constantly working on the edge, with interests spanning across a wide range of subjects including user experience, game design, and human-building interaction. He has won eight scholarships during his PhD tenure, including the AIA's David Lakamp Scholarship. Hassan is presently exploring new frontiers of BIM and virtual reality applications for improved project performance on construction projects and aims to improve productivity in the construction industry with technology.

Wei Yan is the Mattia Flabiano III AIA/Page Southerland Endowed Professor of Architecture at Texas A&M University, with expertise in computational methods in eesign, Building Information Modeling, augmented reality, etc. He has led research projects funded by the National Science Foundation, the National Endowment for the Humanities, etc. He received the Best Paper Prize in Design Computing and Cognition '06. He was named a Texas A&M Presidential Impact Fellow in 2017. Yan was educated at the University of California, Berkeley (PhD in Architecture and MS in Computer Science), ETH Zurich, and Tianjin University.

Manish Dixit PhD, LEED AP is Associate Professor in Construction Science at Texas A&M University with over ten years of experience in the design and commercial construction industry. His research interests include life-cycle energy and environmental modeling, green building materials, embodied energy modeling, 3D printing in construction, Building Information Modeling (BIM) and facilities performance assessment. He served as National Participant in Annex57 of the International Energy Agency (IEA) and was endorsed as an expert in embodied energy analysis field by the U.S. Department of Energy (USDOE). He represents the nation in Annex 72 (IEA), focusing on life cycle energy assessment and optimization of buildings.

Reconnecting...

Nick Safley
Kent State University

Reconceptualizing Details with 3D Scanning

1   Point cloud scan of fountain detail at the Querini Stampalia

2   "Der Knoten" from Gottfried Semper, *Der Stil*, 1860

ABSTRACT

This design research reimagines the architectural detail in a postdigital framework and proposes digital methods to work upon discrete tectonics. Drawing upon Marco Frascari's writing "The Tell-the-Tale Detail" (Frascari 1996), the study aims to reimagine tectonic thinking for focused attention after the digital turn. Today, computational tools are powerful enough to perform operations more similar to physical tools than in the earlier digital era. These tools create a "digital materiality," where architects can manipulate digital information in parallel and overlapping ways to physical corollaries (Abrons and Fure 2018, 185-195). To date work in this area has focused on materiality specifically. This project reinterprets tectonics using texture map editing and point cloud information, particularly reconceptualizing jointing using images. Smartphone-based 3D digital scanning was used to captured details from a series of Carlo Scarpa's influential works, isolating these details from their physical sites and focusing attention upon individual tectonic moments. As digital scans, these details problematize the rhetoric of smoothness and seamlessness prevalent in digital architecture as they are discretely construed loci yet composed of digital meshes (Jones 2014, 29-42). Once removed from their contexts, reconnecting the digital scans into compositions of "compound details" necessitated a series of new mechanisms for constructing and construing not native to the material world. Using Photoshop editing of texture-mapped images, digital texturing of meshes, and interpretation of the initial material constructions, new joints within and between these the digital scanned details were created to reframe the original detail for the post-digital.

WHY STUDY DETAILS NOW?

Details occupy a murky territory between isolated cultural objects and the vast continuum of the material in the world. In the pioneering treatise, *The Four Elements of Architecture*, Gottfried Semper insisted that the threading, twisting, and knotting of linear fibers were among the most ancient of human arts, from which all else was derived, including both building and textiles. In Semper's work, the most fundamental element of both building and textiles was the knot (Ingold 2013). For Semper, a knot was both part of and separate from the world, existing as part of the chord (material world) but not simply due to the chord's materiality. Knots are neither autonomously authored nor a result of inherent material forces. In the postdigital era, architects question the networked continuity, smoothness, and focusless logic of much of the digital age and its practitioners, who found parallels to the knot and chord in Semper's writing. The renewed need for cultural expression and perceptual focus drives skepticism of the detail's demise at the hand of computational tools and data management via the computer. By reengaging the historic pre-digital notions of tectonic, we can situate details in ways which neither reject nor fetishize the material or digital realms.

Practically, when we can effortlessly create and manage vast amounts of data outside of our mind via the computer, we must curate, organize, and then reconnect isolated chunks if we aim to focus our limited attention. Today, ubiquitous digitality makes this necessary for both the discipline of architecture and, more importantly, for larger cultural practices. What would otherwise be an ocean of information, overwhelming in volume and too big for comprehension, needs to be isolated for discussion and practice. By creating working methods to focus attention and relate areas of concentration, one can strategically bound information to solicit new meaning. Contrary to early postmodern narratives of meaning, this method of reconnecting asks for attention as cognitive participatory labor, not stable embodiment or a prescribed narrative.

BACKGROUND

As within Semper's illustrations of knotted chords shown in Figure 2, details can be isolated moments. In both—the illustrations of knots and mundane technical drawings of details—only parts of the material world are shown, removed via convention or technique from their surroundings. In technical drawings, notation and graphic location relate details to one another, but creating such isolation requires different tools and techniques in physical reality. The questions of how to isolate and focus attention were at the heart of Marco Frascari's "The Tell-the-Tale Detail,"

2

where the locus of detailing was in material jointing (Frascari 1996, 498-513). In his historical description of this process, candles and flashlights were the tools used to focus a sensing subject who visited the building site in the darkness of night. While Scarpa's designs were holistic pieces of architecture, he found it necessary to study, during design and construction, the details in isolation from the larger work. Details viewed in perceptual isolation function much more like miniature worlds and stand in stark contrast to arrays of parametric connections standard in digital design work. For Frascari—in a process he terms "the mind's eye"—each detail required a suspension of contextual continuity and asked for the observer's mental participation relating an individual detail to the details and organization of the same built work.

While this analysis will focus on how we can adapt Frascari's thinking to a postdigital context, it is worth outlining other interpretations of details. Edward Ford is one of the only scholars writing upon this topic, and in his work *The Architectural Detail*, he outlines four types, or ways, of thinking about details to which we will add two terms. Ford categorizes details as either *motific*, a representation of construction and an expression of material joints, or as *autonomous* from the larger work as a whole (Ford 2011). To this list, architect Nader Tehrani adds *grain* or the directional organization of a surface to address materiality in computational terms. Tehrani's addition moves away from the discrete detail, opening the tectonic to progressive geometric shifts and gradients required in field conditions (Tehrani 2017). Grain defines details through geometric variation and not typology. His definition is the legacy of the digital project in material fabrication

3 4

and furthers Ford's analysis to conceptualize the computational surface itself.

At the time of Frascari's writing, in the early 1980s, he sought to frame the detail as the generator of design, yet he also identified the detail as ill-defined. He proposed the definition of details to be the union of construction and construing. Construction, in this definition, results from the *logos of techne* or thoughts of making. Construing, on the other hand, results from the *techne of logos* or the making of thoughts. Through his definition, the detail could become the vehicle for communication and narrative. Today, we remain rightfully skeptical of the detail's capacity to signify direct meaning. Yet, the relationship between construction, as an act of material organization, and construing, as the appearance of the material once organized, remains a helpful structure. In light of evolving digital culture, the proliferation of images, and easy movement of visual information between the digital and physical world, one new detail type needs to be added to Frascari's list. This detail type directly addresses the space between our updated version of constructing and construing. Details in the work that follows address digital tools, techniques, and construction in an effort to find places to create joints in the digital substrate—what contemporary thinkers Adam Fure and Ellie Abrons term "postdigital materiality" (Abrons and Fure 2018, 185-195). This detail type could be referred to as the *image detail*, which exists not purely as a material joint, organization of a surface condition, or *grain*, but uses image-based digital information and high-fidelity geometric approximations of material to construe the qualities of physical material unrelated to the underlying construction. To politely borrow Frascari's term, the construction and construing do not have the same "logos" (reason), instead privileging the autonomous image as that which is construed, with material construction purely in its service.

METHODS

Scarpa made a practice of visiting the building site during the night for verification with a flashlight. He thereby controlled the execution and the expression of the individual details. In normal daylight, it would indeed be impossible to focus on details in such a selective manner. It is also a procedure whereby the indirect vision becomes a critical process in design decisions.

> The flashlight is a tool by which is achieved an analog of both the process of vision and the eye's movement in its perception field (with only one spot in focus and the eye darting around).
> —Marco Frascari, "The Tell-the-Tale Detail"

Today, isolation of details from architectural works is possible in a similar way using digital 3D scanning. This process mimics the use of light in a dark environment to control the extents of individual vision. Dense scans comprised of captured color data points are organized in Cartesian space, closely matching the geometric extents of the original detail. The data that comprises the scans constitute information curated not just by the human eye, but in collaboration with the digital scanning technology. Further, the connecting of details does not simply need to rely upon indirect vision, the spatial and mental connection of details experience individually, to find connections. The collections of details can be shared readily with others in geographically distant locations and viewed collaboratively.

Lastly, the flattening of Scarpa's initial details into a common "digital material" comprised of mesh point locations and color information can be used to speculate upon new means of connecting detail previously distinct from one another.

Scanning

Digital scanning, in this project, utilizes inexpensive photogrammetry via the iPhone app Capture (Standard Cyborg 2019) to created isolated details from built projects by Carlo Scarpa. The scanning process is analogous to the flashlight-based methods that Scarpa used in his construction verification visits. The small phone-based scanner allowed the capture of projects that would have otherwise required a special permit or occured only within institutions that allow photography. This lack of access does not affect the rhetoric of the project, but it is worth mentioning as an influence on the details curated. Scans of the details were created by slowly moving the phone's camera over and around the area of interest while watching the phone screen to check the fidelity of the scan. The Capture app generates point clouds, each with a Cartesian location in model space that has a central *0,0,0* origin. This technical fact has ideological implications as the detail, once scanned, is quite literally the center of its universe. Color information is encoded into the point cloud also and is later used to generate a texture-mapped mesh.

Conceptually, each scan's creation parallels Scarpa's projects. Each of his details and each of the scans function alone as discrete information and also in conjunction with other details to create the larger whole of the project. Marco Frascari describes the part to the whole relation in Scarpa's work:

> While it is whole, Scarpa's architecture cannot be characterized as complete. An architectural whole is seen as a phenomenon composed of details unified by a structuring 'device,' a structuring principle.
> —Marco Frascari, "The Tell-the-Tale Detail"

In digital scans, discrete parts, themselves composed of points, work in the service of a larger image of the scanned detail. This process is nearly identical to the way Frascari posits that details work in relation to the built whole in Scarpa's work. This conception of the part to whole relationships does not neatly fit neatly within the either Ford's or Tehrani's expanded list of detail categories. Here details are semi-autonomous, working discreetly to construe an image without specific meaning themselves. The computational modeling software used to collect and combine the scans functions as a new platform, whereby the scanned

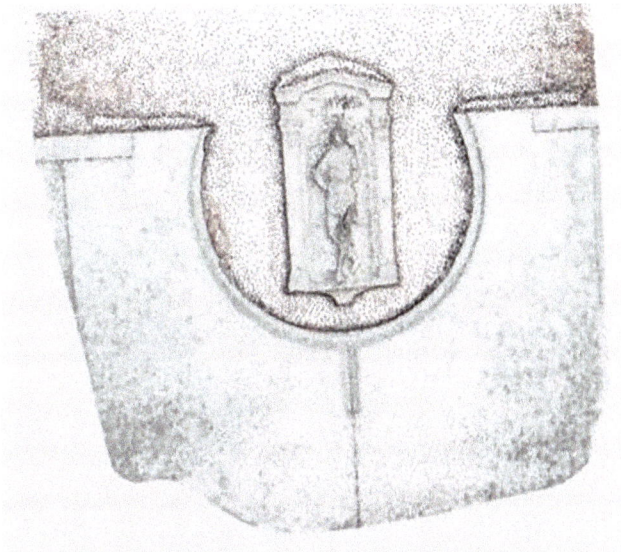

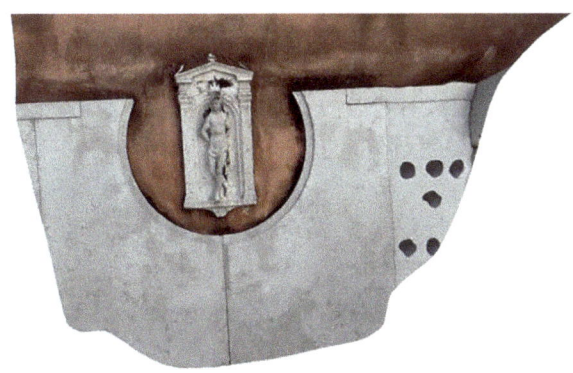

5

3 Scanning Process using smart phone at Carlo Scarpa's San Sebastiano Gate

4 Enlarged cut stone stepped detail of San Sebastiano Gate

5 Top to Bottom: Point Cloud, Digital Mesh Geometry, and Mesh Scan with Color Information as Texture Map

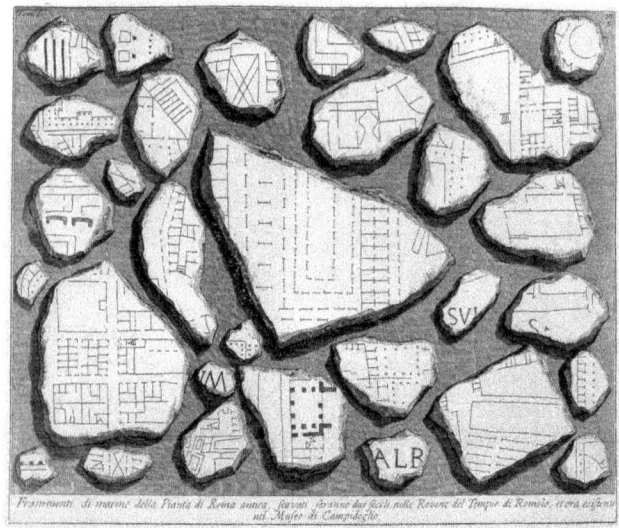
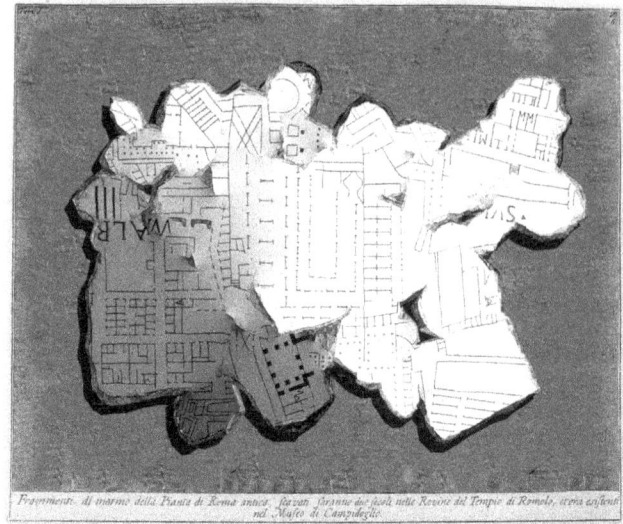

6

7

details are combined independent of the human body or the physical the built environment. Using the computer to act upon scanned details allows architects to work collectively and collaboratively with other human and nonhuman subjects to create architectural wholes.

Collecting

Collecting requires the capability to remove something from its context. Historically architects have used collections to frame theoretical arguments and expose those unable to travel far distances to seminal works of architecture. During the nineteenth century, drawings and plaster casting were a relatively expedient means for circulating works of architecture to those who could not physically inhabit or experience it firsthand. At the end of the nineteenth century, photography began to replace drawings, but the three-dimensional quality of plaster castings, albeit free of color, contained information not compressible or transferable in two-dimensional media (Lending 2017). As such, the plaster cast occupies a unique position in architectural history and discipline. One to one in scale, plaster casts of entire projects were not practical or technically possible due to the realities of transportation and physical access to built works. Details sized so they could be placed in crates, handled, and shipped made their way from the world's far corners into the collections of museums located in European centers of colonial power and the United States. Collecting these details involved complex political, economic, and ethical relationships, which have generated a body of scholarship and may inform this project in the future. For this research, we will narrowly view this casting practice as the collection of detail fragments and purely as a technical process of curation. Once collected, many of these casts found themselves on display in spatial proximity unimaginable in situ, and their intention as pedagogical tools aside, the capriccio-like collections were immersive material compositions. The current practice of handheld 3D scanning addressed in this research has given the collection process an ease similar to that of photography, and one might imagine the educational intent of earlier plaster casting taking place not in physical galleries but in an online viewable database or search engine.

This project uses a method of pulling apart and collecting built work by its details, like detail casts in the Carnegie Gallery at the Pittsburgh Museum of Art, or other architectural plaster galleries. New negotiations between the individual fragments replace polite voids in the museum. Figure 7 expresses this, conceptually joining with the work of another Venetian, Piranesi, who interestingly also explored details, as fragments of larger works, in darkness by candlelight. The curatorial display of the scans also finds a counterpoint to plaster casts. Historically, casting was a way to appropriate distant history and cultures from which one created rooms to display the power of the collecting nation. As a method to be updated via new technology and worldviews, collecting details in a digital format can be a means of reimagining the material world through an expanded conception of tectonics and not simply as a colonial endeavor.

Reconnecting

The projects whose details were collected come from recent history and belong to a single architect's body of work. Counter to museums' historical collections, the collected details are not displayed as detail fragments encrusting a preexisting room, which serves as a container, frame, and method of unification. Instead, the scanned details—a digital parallel to portable plaster castings—are combined into compound detail objects without an accessible interior.

In this way, the newly designed compound details are their own context, and each collection of details from Scarpa's work are only reconnected with scans from the same initial project.

More specifically, collections of scanned details are composed and organized into new architectural wholes utilizing qualities found in the digital point cloud scan. Qualities—such as similar color, point cloud shape, point cloud density, mesh direction, and fidelity of the 3D scan—organize the new objects as compound details. Once located, the scanned details were collapsed into one dataset and the resulting point cloud was meshed via the open-source software MeshLab (Cignoni 2009). Automatically created meshes and numerically averaged color information fill in the gaps between what began as separate scans. The spaces between scans are exemplary of the image detail discussed earlier, connecting the digital construction of the mesh into a seamless object. What differentiates the image detail from surface-based connections is that the new object is construed as a collection of materials from the initial material details but without clear edges or physical joints.

This research project combines scans from three projects into a series of objects, though scans from a much larger body of Scarpa's work exist for future research. The Castelvecchio in Verona, Scarpa's largest work, is included as it has the most range of detail scale, bridging spatial or formal joints like stairs, with overtly material joints like doorknobs. The Querini Stampalia is quite different from Castelvecchio and is primarily composed of surface linings of an existing building. The scans here were much smaller, more discrete, and composed of surfaces, putting more pressure upon the composition and adjacency between them to define an enclosed volume. The last project used for this research is the Brion Tomb, whose archipelago of small pavilions juxtaposes the myriad of stepped concrete motific details. From all the projects scanned, these three

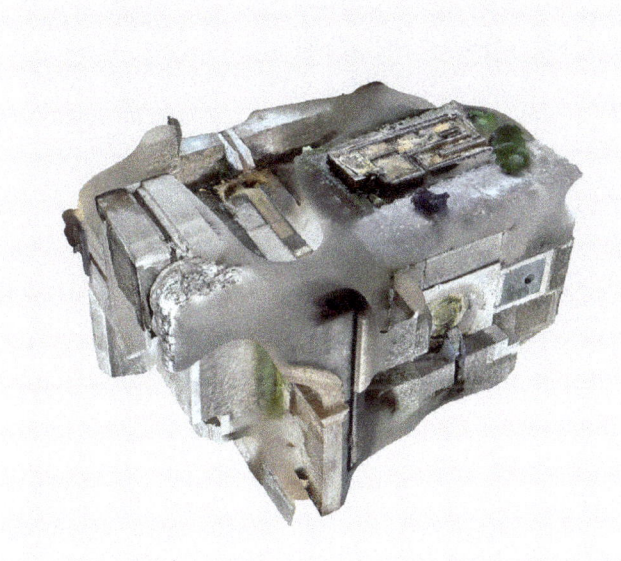

6   Fragments of the Marble Plan of Ancient Rome, Giovanni Battista Piranesi, *Le antichita romane (The Roman Antiquities)*, 1756

7   Digital image (created using the photomerge command in Photoshop) shows seamlessly joining fragments

8   Compound detail created from scans of the Castelvecchio

9   Compound detail created from scans of the Brion Tomb

10 Compound detail created from scans of the Querini Stampalia

most exemplify the narrative of Frascari. Once the sites are visited and scanned, his narrative and rhetoric are problematized by the reality of the material construction. In many cases that which is being construed places material in the complete service of form, something Edward Ford explains:

> ...the reality of Scarpa's work is that if it lacks a conceptual structure, it has no shortage of formal unity; the excess of typical Scarpa motifs is, if anything, oppressive in its unity in works such as the Bank of Verona in Verona Italy (1973), or the Brion Cemetery. It cannot be said that there is no system to the use of these motifs. The step-like echelon motif at Brion is used at the ends of all planes of concrete and stone. It is not the absence of a system that is problematic here; it is that the system is primarily a decorative one indifferent to material and function, and used to excess.
> —Edward R. Ford, *The Architectural Detail*, 124

While Ford seems to be ethically chastising Scarpa, his comment is a possible direction for contemporary work to operate. Scarpa's motifs presuppose collections of things that simply look alike, and the image as a formal architectural device.

Material joints in both the Brion Tomb and the 3D scans of material details unify the appearance of multiple materials negotiated using physical joints into single plastic material, either concrete or the "digital material" of the scans. Now that 3D printing in various materials and with integral color allows for the recreation of the digital scans in physical reality, the space between Scarpa and contemporary digital techniques nearly collapses. Today, architects can operate on the gaps and joints construed between isolated materials themselves, as these materials do not necessarily need to match the construction. Images transgress the boundaries of detail fragments; joints become expanded in their scope and can become objects or areas that negotiate between parts through image yet continuous as material. We can organize details loosely and create negotiated material compositions in what otherwise has remained a gap or joint.

> Despite the motific character of the Scarpa building(s) the details take on an autonomous character. They are autonomous narratives of construction in buildings otherwise unconcerned with construction.
> —Edward R. Ford, *The Architectural Detail*, 255

A second design exercise of this research explores a more traditional notion of jointing as the discontinuous negotiation between material parts. As seen in Figure 13, a single

11 Stepped motif in the endwall of the Brion Tomb; material layers are implied in the plastic material

12 Texture map of digital mesh detail scan in Figure 13

13 Combined mesh detail of two fountains at the Querini Stampalia, with edited texture map applied and joints created around color areas

scanned detail is broken into parts unrelated to its initial material configuration. The texture map of the digital mesh, which contains the color information related to the material in the initial detail scanned, is overlaid with a simple tile pattern. This subdivision of the image into four parts maps to the 3D scan in unpredictable ways due to the automated methods the meshing software used to texture map. These areas, shown as brightly colored patches, are then treated as differing materials and joints created between the color patches. In both example processes, the texture map as an image is the site of tectonic operation. In the color jointing example, an autonomous narrative of detail construction is used in the scanned detail, which is unconcerned with construction.

RESULTS AND REFLECTION

Scanning Scarpa's works, removing, isolating, and collecting details is a colonial endeavor to find an optimistic outlook for digital technologies. Like plaster casting, 3D scanning as technology and technique envelopes and approximates material joints from the initial material construction as an image joint with no material necessity and pure expression. The scan is not a copy but an approximation with image qualities consistent enough to allow correlation. Material joints from the initial detail are only approximated and sometimes captured in the texture-mapped image. In the texture mapping of detail scans, we find the most significant difference between plaster casting and scanning. While both techniques create forms in media with a high degree of granularity, for plaster, the granules are actual plaster granules, and for scans, it is the digital mesh resolution. Color information can only be hand-painted onto plaster casts, while 3D scans work more like photographs displayed on the 3D mesh surface. In the .jpeg file, tools within Adobe Photoshop (Adobe 2019) operate upon the texture map, and the results suggest material tectonics, joints, and organizations not native to traditional tools and their related materials.

Interestingly, the automated 3D scan process generates seams between the image fragments accessible in the mesh texture map. Gradient fills that negotiate pixel information, surface texture joints that do not cohere to the initial material logic, and finishes created via visual pixel manipulation challenge traditional jointing tectonics. With these and other tools, new tectonic thinking can connect or reconnect anything once digitized.

Once details are scanned or more generally isolated from their context, we can create a blueprint for new architecture. This process is familiar to the discipline of analytique drawings, where details were collected onto a single drawing as delineations by the architect to be interpreted and executed by skilled craftsmen. Further, the spaces between the details are omitted, to be either collaborated upon during construction or interpreted, truly making the architect the team leader of a skilled collective. The real and virtual categories are much more overlapping in the postdigital computational environment, if not a false binary. Details can be collaboratively joined in a similar process to that of a knowledgeable craftsperson as an automatic or inferred process originating in the computer program code.

12

13

14 Mesh connection joining a series of scanned details from the Querini Stampalia

Craft in construction stands in contrast to the drawing or mediated construction documentation, which assumes little, or at least less, knowledge of the person constructing the detail. A shared commonality, or culture, is assumed between architect and craftsperson. A similar relationship with the person or people who developed the computer code can be considered in the latter, as both utilize an assumption of a known but collaborative outcome. Encoded software operations can here function more like motifs or images, placing them in closer parallel to the knowledge of the preindustrial craftsperson than of draftsperson.

So, what then is the architect's position and role in this sort of design practice? Firstly, automation of connections still requires collaboration, not necessarily the type of craftsperson Scarpa notoriously collaborated with, but instead their contemporary corollaries, programmers writing code and fabricators controlling digital making processes. Second, in controlling the desired perception of the design, the architect is still responsible for its organization. In the postdigital era, the perception of the designed object is also not limited to that of the human body directly experiencing the physical material world. Instead, at stake is the entirety of the mediation of the designed object. The variety and scope of this meditation form the new responsibility of the architect, no matter the sort of collaboration, craftsperson or digital code. The location and organization of the constituent parts in this proposal closely parallel that of Frascari's phenomenological discussion of Scarpa.

Collocation is the composing by place, that is, the functional placement of the details. The function in this case not only is limited to the practical and structural dimensions but it embodies, as well, historical and aesthetic dimensions. The placing of details, then, is deeply related to the other two requirements: numbers and analogies. The detail in this manner is not defined by scale, but, rather, the scale is the tool for controlling it. The geometrical and mathematical construction of the architectural detail is in no sense a technical question. The matter should be regarded as falling within the philosophical problem of the foundation of architecture or geometry, and ultimately within the theories of perception.
—Marco Frascari, The Tell-the-Tale Detail

Curation, location, and organization then become the design choices of the postdigital architect. Proposed is not a totalizing or total design process desired in some early modern practices and reappeared in a high degree of digital customization fabrication enabled in the early 2000s. We do not design to produce totalizing environments but instead discrete objects that solicit collaboration during construction and during their mediation. There is no desire to recreate the overwhelming conditions Adolf Loos outlined for his "poor little rich man," who had no space to imagine. Our task is to create a collaborative relationship in which the designed object solicits interaction with the viewing subject. Objects made for this project reframe the title of Frascari's essay. Here tale might better be spelled tail. The narrative character of the detail is superseded by the detail as a character, as a thing with an agency in and of itself. This definition wags the agency of the detail from being a repository of human speech, embodied or otherwise, to functioning as a quasi-subject, authored by the architect, but needing the viewer's consistent labor in trying to make an array of narrative tales. This process imbues the object with life, or something like it, as the viewing subject's attention literally places it there. This process is further pushed with the automated connections between details found in the compound detail objects. Each is created, not by the architect, but by the relationships the architect places the details in that allow for collaboration. These designed objects are composites of vital moments, both in their scanned details, those of the computer's labor, and from the meaning creating subject.

CONCLUSION

While the research for this grant was conducted in the summer of 2019, its fruition took much longer. The global Covid-19 pandemic placed this research on the back burner as teaching and navigating social isolation dominated. Strangely, there could have been no better time to focus upon topics of isolation and how computational tools can—and do—allow for created connections. Details have a life of their own when removed from their in situ location but cannot construe in isolation. Reconnecting and refiguring the material world as a social one via digital tools means that we can use the past to imagine an optimistic future with the material world. A postdigital tectonic hopes that we find a social relationship with one another and social relationship in equal part with the material in which work. Like knots, details as material constructions exist within the larger ecology of the planet, and at the same time, a locus of attention for tectonic thinking. A postdigital flattening of the hierarchy between the material world and the digital realm expands our focus further. Colloquially, this gives us more rope to hang ourselves as we are able to manage large sums of information without specific locations of commonality. The postdigital detail uses our perception in conjunction with technical devices to remove areas of focus from the larger world to construe digital constructions of a common world.

ACKNOWLEDGMENTS

This research was completed with the generous support of the Kent State University College of Architecture and Environmental Design 2019 Faculty Travel Fellowship.

REFERENCES

Abrons, Ellie, and Adam Fure. 2018. "Postdigital Materiality." In *Lineament: Material, Representation, and the Physical Figure in Architectural Production*, edited by G. Peter Borden and M. Meredith. New York: Routledge. 185–195.

Adobe Inc. *Adobe Photoshop CC 2019*. V. 20.0.6. Adobe.com. 2019.

Cignoni, P., M. Callieri, M. Corsini, M. Dellepiane, F. Ganovelli, G. Ranzuglia. *MeshLab: Open Source Mesh Processing Tool*. V. 2020.07. ISTI - CNR Research Center. GitHub. 2020.

Ford, Edward R. 2011. *The Architectural Detail*. New York: Princeton Architectural Press.

Frascari, Marco. 1996. "The Tell-the-Tale Detail." In *Theorizing a New Agenda for Architecture, an Anthology of Architectural Theory 1965-1995*, edited by Kate Nesbitt. New York: Princeton Architectural Press. 498–513.

Ingold, Tim. 2013. "Of Blocks and Knots: Architecture as Weaving." *Architectural Review*, October 23, 2013. Accessed July 20, 2020. http://www.architectural-review.com/essays/of-blocks-and-knots-architecture-as-weaving.

Jones, Wes. 2014. "Can Tectonics Grasp Smoothness?" *Log* 30 (Winter 2014): 29–42. http://www.jstor.org/stable/43631731.

Lending, Mari. 2017. "Monuments in Flux." In *Plaster Monuments: Architecture and the Power of Reproduction*, edited by M. Kending, Princeton, NJ: Princeton University Press. 1–29.

Piranesi, Giovanni Battista. 1756. "Map of ancient Rome and Forma Urbis." Etching. In *The Roman Antiquities*, vol. 1, plate V.

Semper, Gottfried. 1860. "Die textile Kunst für sich betrachtet und in Beziehung zur Baukunst, Illustration of a Knot." Pencil Drawings. University of Heidelberg Historical Digital Holdings. Accessed July 20, 2020. https://digi.ub.uni-heidelberg.de/diglit/semper1860/0226.

Tehrani, Nader. 2017. "The Tectonic Grain." The Ohio State University, Knowlton School of Architecture, Columbus, OH. Filmed Autumn 2017. YouTube video, 1:33:40. https://www.youtube.com/watch?v=BkJK0rAFYXM.

Standard Cyborg. *Capture: 3D Scan Anything*. V. 1.2.5. Standard Cyborg. iPhone. 2019.

IMAGE CREDITS

Figures 2 and 6: Creative Commons CC0, Public Domain, https://creativecommons.org/publicdomain/zero/1.0/.

All other drawings and images by the authors.

Nick Safley is a designer whose work focuses on architectural character, materiality, and methods of fabrication. His work takes place at the intersection of form, material, and the technologies used to negotiate between these categories. At Kent State he teaches courses on construction and coordinates the fourth year Integrated Design Studio. Prior to teaching, Safley worked professionally for the LADG, Utile, and NADAAA.

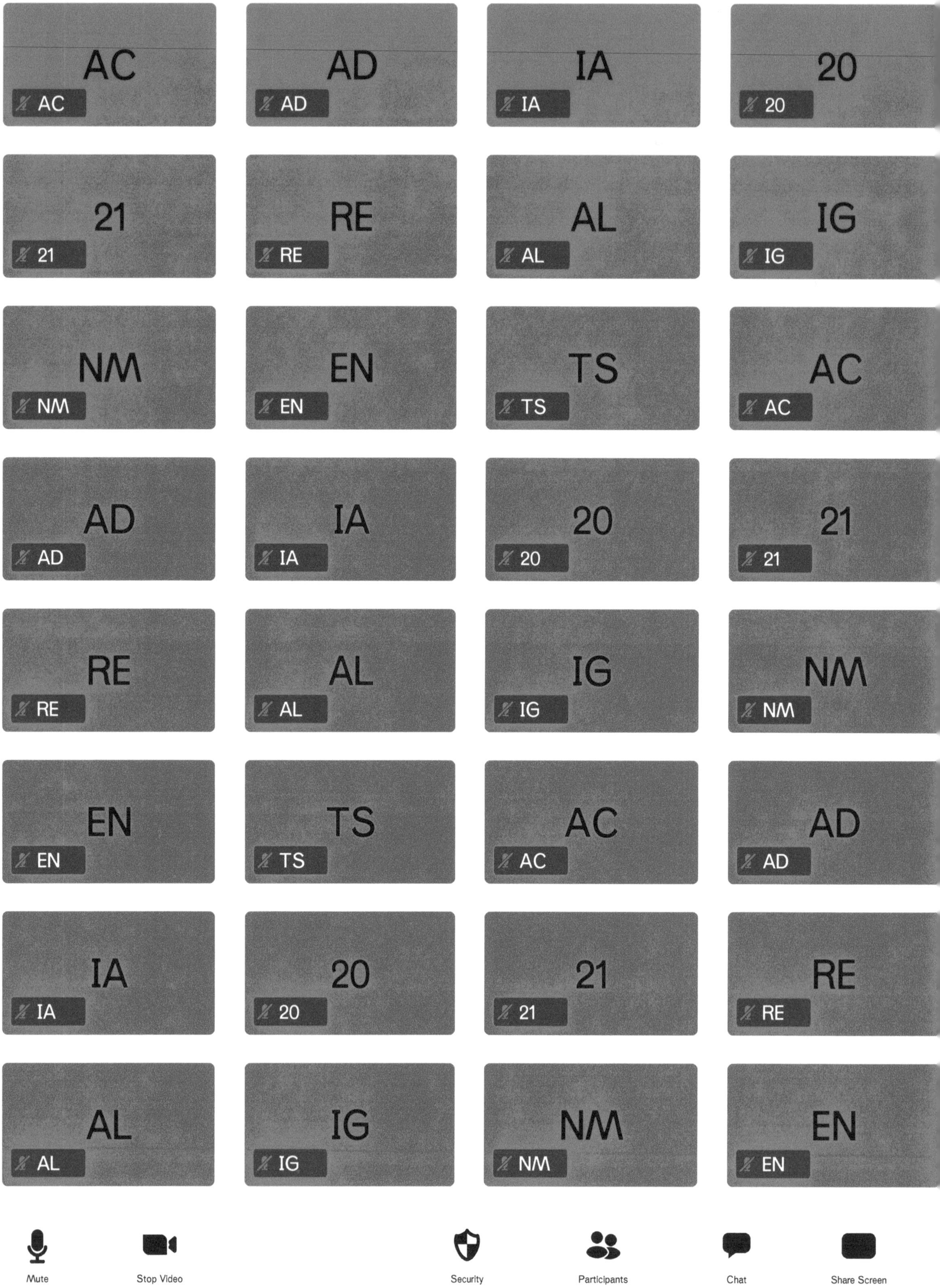
AC
AC
AD
AD
IA
IA
20
20
21
21
RE
RE
AL
AL
IG
IG
NM
NM
EN
EN
TS
TS
AC
AC
AD
AD
IA
IA
20
20
21
21
RE
RE
AL
AL
IG
IG
NM
NM
EN
EN
TS
TS
AC
AC
AD
AD
IA
IA
20
20
21
21
RE
RE
AL
AL
IG
IG
NM
NM
EN
EN
Mute
Stop Video
Security
Participants
Chat
Share Screen

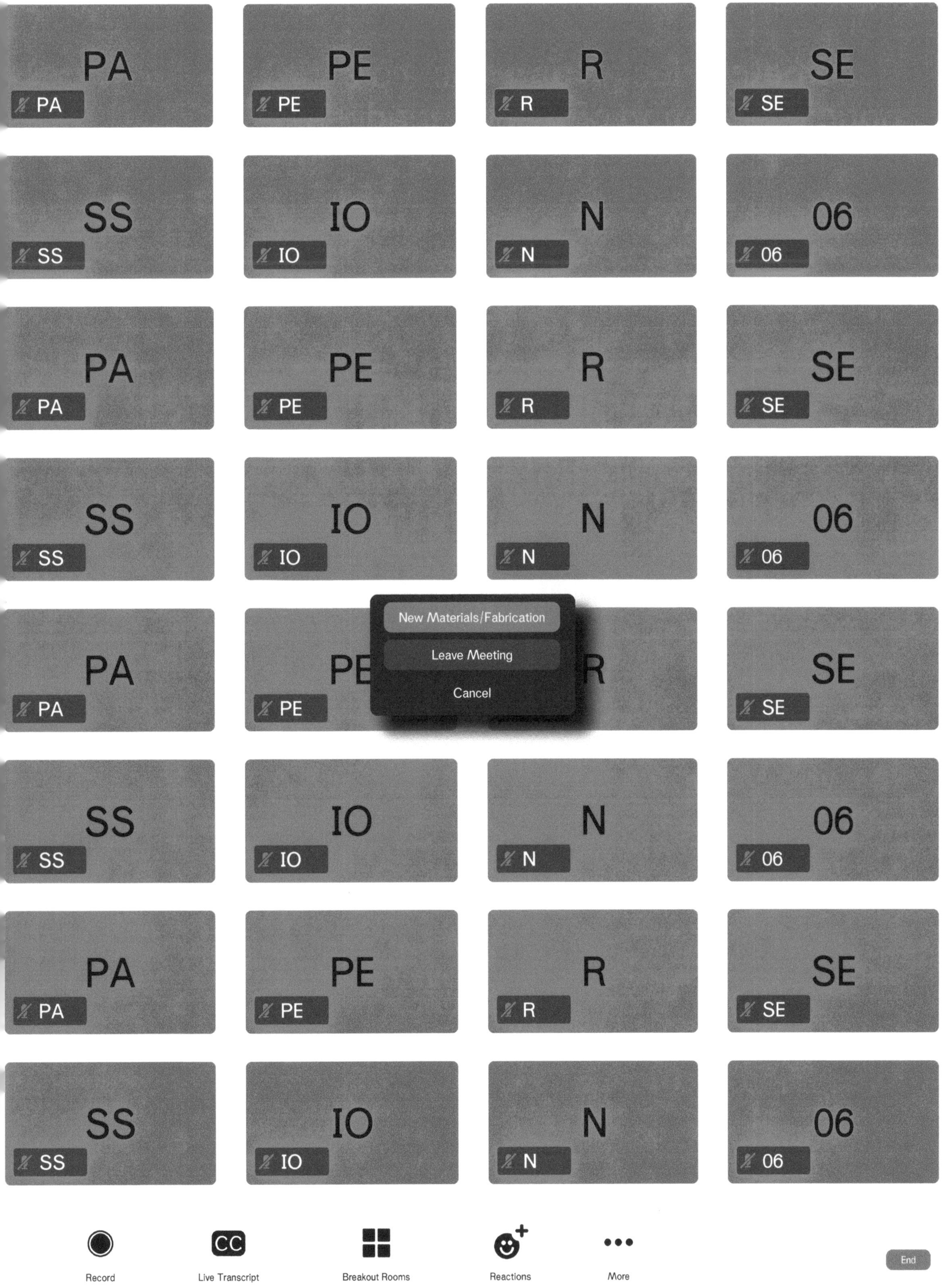

PA	PE	R	SE
SS	IO	N	06
PA	PE	R	SE
SS	IO	N	06
PA	PE	R	SE
SS	IO	N	06
PA	PE	R	SE
SS	IO	N	06

New Materials/Fabrication
Leave Meeting
Cancel

 Record   Live Transcript   Breakout Rooms   Reactions   More End

Parametric design and multirobotic fabrication of wood facades

Edyta Augustynowicz*
ERNE AG Holzbau/IID, FHNW

Maria Smigielska*
IID, HGK, FHNW

Daniel Nikles
IID, HGK, FHNW

Thomas Wehrle
ERNE AG Holzbau

Heinz Wagner
IID, HGK, FHNW

* Authors contributed equally to the research

1 Final project demonstrator, 2020, perspective view

ABSTRACT

The paper describes the findings of the applied research project by Institute Integrative Design (currently ICDP) HGK FHNW and ERNE AG Holzbau to design and manufacture prefabricated wooden façades in the collaborative design manner between architects and industry. As such, it is an attempt to respond to the current interdisciplinary split in construction, which blocks innovation and promotes standardized inefficient building solutions. Within this project, we apply three innovations in the industrial setup that result in the integrated design-to-production process of individualized, cost-efficient, and well-crafted façades. The collaborative design approach is a method in which architect, engineer, and manufacturer start exchange on the early stage of the project during the collaborative design workshops. Digital design and fabrication tools enable architects to generate a large scope of façade variations within the production feasibility of the manufacturer, and engineers to prepare files for robotic production. Novel multi-robot fabrication processes, developed with the industrial partner, allow for complex façade assembly. This paper introduces the concept of digital craftsmanship, manifested in a mixed fabrication system, which intelligently combines automated and manual production to obtain economic feasibility and highest aesthetic quality. Finally, we describe the design and fabrication of the project demonstrator consisting of four intricate façades on a modular office building, inspired by local traditional solutions, which validate the developed methods and highlight the architectural potential of the presented approach.

INTRODUCTION

Although integration within the building industry has been discussed for years (Kolarevic 2009), it has still not become a common practice. Currently design, engineering, and manufacturing function as disconnected disciplines with individual interests that are often misaligned. This situation leaves architects to pursue design with limited fabrication understanding, which in turn leads to higher planning costs, fabricators' disregard of architects' intentions, and incline towards standardized solutions (Correa, Krieg, and Meyboom 2019). As a result, contemporary cities offer mostly serialized, unattractive, and anonymous urban frontages. Within this project, we aimed at improving our built environment on a larger scale by facilitating interdisciplinary communication within the construction process and combining novel digital technologies with local traditional woodcraft techniques.

We believe that the convergence of design and manufacturing, as well as constant interdisciplinary feedback, can trigger the true potentials of technological innovation. Thus, one of our goals was to strengthen this link through the use of digital means. In this paper we present the outcomes of an applied research project between IID FHNW and ERNE AG Holzbau that introduces three innovations in the industrial setup—(1) the collaborative design approach, (2) digital design and fabrication tools, and (3) novel multi-robot fabrication processes—that result in a fluid process to design and fabricate wooden façades that are individualized, well-crafted, and cost-efficient. Combining an integrated design-to-fabrication model based on the mastery of craft (Correa, Krieg, and Meyboom 2019) with bringing the architect closer to making, this solution introduces the concept of digital craftsmanship as a new value in the industrial realm. Finally, the paper describes the implementation of this process in a constructed demonstrator: four bespoke façades on the temporary office building (Figure 1).

STATE OF THE ART

Robotics in Timber Construction

Examples are broad where academic research applies experimental robotics in complex timber structures. At the ETH Zürich, it was explored and applied with 2-dimensional stacking process on the ITA roof (Apolinarska et al. 2016) and spatially assembled constructions (Zock et al. 2014; Helm et al. 2012). Following those examples, several other academic institutions engaged in this topic (Williams and Cherrey 2016; Søndergaard et al. 2016; Robeller and Weinand 2016; Johns and Foley 2014; Schwinn, Krieg, and Menges 2013; Dank and Freissling 2013). The non-industrial approach to working with non-engineered lumber is represented by the Wood Chip Barn project (Mollica and Self 2016), where robotic fabrication and computational process informed by real-time data led to optimized tectonics that maximized the natural structural properties of wood. The multi-robotic assembly of wood constructions remains a relatively unexplored field. During recent years this topic was researched in fields like metal folding (Saunders and Epps 2016), hot wire cutting (Rust et al. 2016), or filament winding (Parascho et al. 2015), and led to either increased production speed, bigger size of the built object, or increased design possibilities. Further developments in robotic cooperation were applied in support-free spatial structures made of metal rods (Parascho et al. 2017), bricks (Parascho et al. 2020), or timber (Thoma et al. 2019). These projects resulted in reduced waste and cost and time of production by the renouncement of substructure but required complex path-planning. The proposed solution makes use of two robots—working within clearly defined regions to eliminate collision—during the automated assembly process.

Despite the intense academic research, the building-scale robotic timber construction is almost solely applied to prefabrication of standardized building elements. Given the impressive amount of developed knowledge at universities, the next step should be to integrate it into a real construction workflow.

Design and Production of Prefabricated Wooden Façades

Even though the façade is the most prominent element of a building (Koolhaas 2018), constructing a non-standard building envelope is a highly demanding task. While standardized building parts are assembled mostly automatically, wooden prefabricated façades are still constructed manually (Ackermann 2018). This time-consuming process generates high costs and often limits the design possibilities to simple, modularized solutions. Moreover, the lack of communication between architects and manufacturers requires additional detailing by the façade planner, engineer, or manufacturer, which further prolongs the planning phase. Overall the disciplinary disconnection, as well as low levels of automation in industrial production, leads to either higher manufacturing costs or inclination towards standardized façade solutions.

Collaborative Design in the Construction Industry

Collaboration in the design process between various fields is seen as a mainstay of efficiency improvements (Mehrjerdi 2009). That is why concepts such as integrated design, integrated practice, and integrated project delivery (IPD) have recently gained more prominence in an architectural practice that faces growing complexity within the building industry (Kolarevic 2009). Most efforts in that

2 Variety of geometries created with generative design tools: a) plate-based façades (parameters: width, length, bottom shape, and direction of plates), b) beam-based façades (parameters: gap size, sequential pattern, angle, length, and width of the slats), c) mixed façades (parameters: number of layers, pattern, rhythm, and typology of elements)

domain are focused on building information modelling (BIM), which, however, do not bridge effectively different perspectives of various stakeholders (Çıdık, Boyd, and Thurairajah 2017) and require digital-integration-driven organizational change (Dossick and Neff 2011; Harty 2008; Whyte and Lobo 2010; Whyte 2011) that is often hard to control (Çıdık, Boyd, and Thurairajah 2017). Moreover, basic BIM software does not enhance the design and production of customized building components, and additional tools have to be introduced (Wang 2021). An alternative approach is a custom digital-to-fabrication platform developed by a project team for one specific project. Such highly sophisticated solutions are mainly applied in academically driven projects, such as the BUGA Wood Pavilion (Alvarez et al. 2019), and are usually impossible to be scaled up to broader industrial applications. The main challenge of integration is thus to avoid closed systems and to keep integrative tendencies as open as possible, conceptually, and operationally (Kolarevic 2009).

METHODS
Generative Design and Fabrication Tools

The main objective for the development of the digital design and fabrication tools was to provide maximal flexibility for architects to design and for engineers to produce a wide variety of façades with minimal necessary changes in the code needed. These tools were developed independently as they had different users and purposes. The integrated design tool (Tool 1) allows generating various façade geometries according to the company's internal standards and the material properties of wood. It is a basis for a discussion between an architect and a manufacturer during collaborative design workshops, when a skilled computational designer adapts it to the aesthetic preferences of the architect. The fabrication tool (Tool 2) is operated by trained engineers. It translates the B-rep (boundary representation) geometry of the façade to fabrication data for two robotic units. The base geometry for Tool 2 can originate from Tool 1 or a different design software, as no special data encoding of individual elements is required. As such architects work in their preferred way without limiting their design freedom. Both tools were written in the Grasshopper plugin (Rutten 2020) for Rhinoceros (Robert McNeel & Associates 2020) with custom Python components. Tool 2 additionally uses KUKA|Prc plugin (Braumann 2020) for planning of robotic paths.

Tool 1: Generative Design Tool

The Generative Design Tool allows architects to playfully explore the design options of various façade types (Figure 2). The external wall panels and windows are fed to the script as planar polygonal curves. Users can control façade geometry with a plethora of geometrical, numerical, and choice-based parameters. Those define the general characteristics of the façades (e.g. angle of the slats, sequential pattern of the beams, framing of the windows, and wrapping of the slats around the openings), relationship between the elements (e.g. gap size between the elements, split of the beams along the length), or geometrical properties of individual elements (e.g. width, depth, bottom angle). Firstly, the script generates axes of beams and then 3D B-rep geometry, which includes bottom and top angled cuts of the slats enabling uninterrupted water flow. This geometry is subject to further deformations, depending on individual aesthetical vision of architects and chosen fabrication process (e.g. milling). The script provides direct feedback about the estimated production time and cost of the designed façade per m^2 based on the number of elements, volume of the material, and average time needed for a robot to assemble each element. This feature enhances transparency in communication with manufacturers and enables architects to make more informed design decisions.

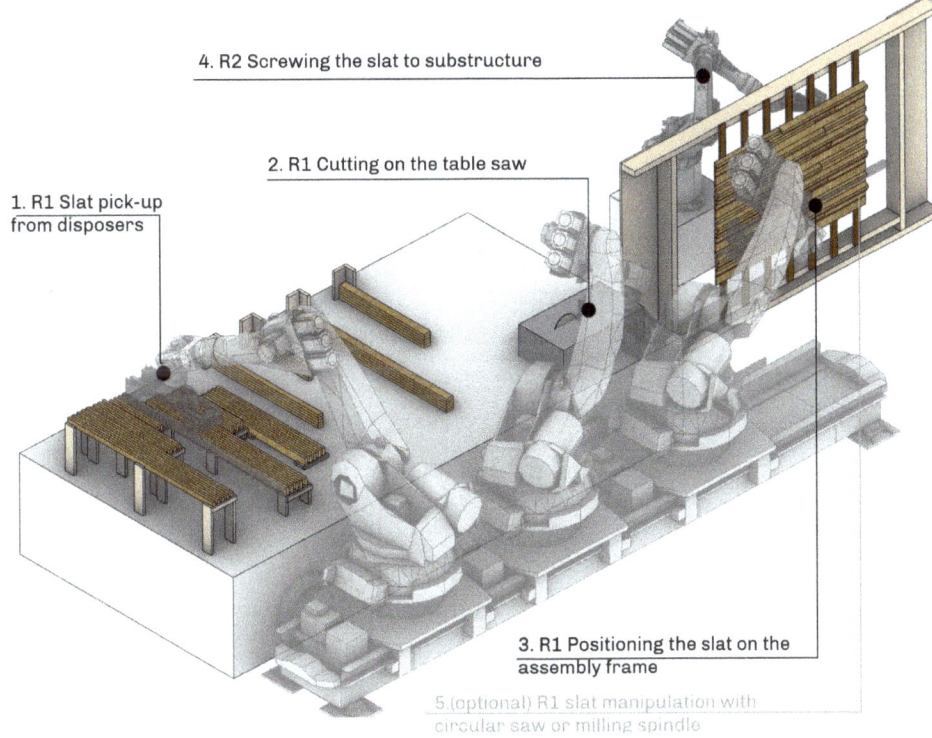

1. R1 Slat pick-up from disposers
2. R1 Cutting on the table saw
3. R1 Positioning the slat on the assembly frame
4. R2 Screwing the slat to substructure
5. (optional) R1 slat manipulation with circular saw or milling spindle

3 Fabrication process: Kuka KR240 (R1) (1) picks a timber slat from one of the disposers, (2) cuts it on a table saw, and (3) positions it on the façade panel, and (4) Kuka KR6 (R2) screws the elements to the preassembled substructure from behind. (Optional) R1 slat manipulation with circular saw or milling spindle.

4 Different assembly strategies for different façade types: a) shingle façade, assembly from the bottom right corner; b) board-on-board façade, assembly in layers from the top left corner. The different colors symbolize two grippers: blue lines - vacuum gripper; magenta lines - asymmetric gripper.

Tool 2: Fabrication Tool

The fabrication tool is a highly flexible prototypical software to prepare files for automated robotic processing and assembly of a wide variety of wooden façades. Based on the geometrical input of a façade panel and settings of robotic toolheads, the fabrication tool generates commands for two robots in the form of .src files. Communication between the robots is described in the next section. The software automates various operations, as illustrated in Figure 3. To avoid collisions between the gripper and the already fixed façade slats, the assembly order plays an essential role and depends on the spatial relationship between individual elements, façade type, and user-defined order strategy (Figure 4). The fabrication-relevant attributes for each façade element are the gripping plane, both-end cut planes, positioning planes on the façade panel, and screw planes.

The picking position of a slat from a correct disposer is solved automatically based on the element bounding box and predefined dimensions of the beams in the disposers. The cutting planes can be perpendicular, angled, or parallel to the slat's axis. For short elements that are cut on both ends, the program automatically incorporates the regripping of the elements in-between the cuts to avoid collision between the gripper and a saw.

According to the company's internal standards, all the screws are inserted from behind the panel to avoid any visible connections. The screw planes are calculated in regard to material and engineering requirements of individual façades. Depending on the façade type, the screws can be placed either on the intersection between slats and substructure (beam façades) or between overlapping elements (shingles or mixed structures). For slats that are wider than 60mm, at least two screws on each intersection are required to prevent elements from bending. The number of screws per element is dependent on its length and width but cannot be smaller than two.

Multi-robotic Fabrication

Robotic setup at the industrial partner's facility consists of one robotic arm KR240 on the rail equipped with a milling spindle, circular saw, and a custom tool for assembling the standard façade panels. To expand the design possibilities, we established innovative multi-robot fabrication processes for complex façade assembly. The new setup consists of the vertically positioned wooden assembly frame of the size 2600mm x 1800mm (WxH) that can be

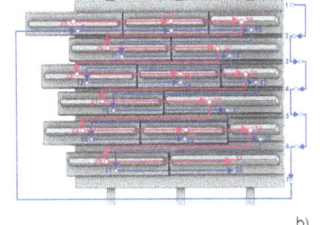

a)   b)

4

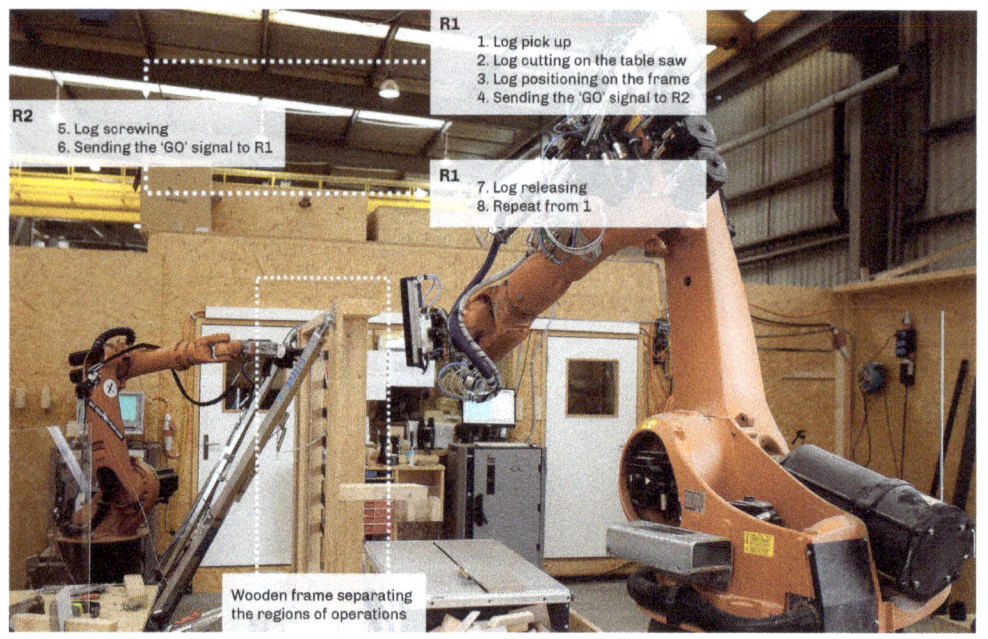

5   Robotic setup with a wooden assembly frame separating regions of operation for two robots and generalized sequencing and operations between R1 and R2 robots.

6   Asymmetric gripper: construction and functionality.

7   Manual operations performed before robotic assembly: a) assembly of the substructure, b) feeding of the log dispenser.

accessed from two sides by each robotic arm (Figure 5). Two robots work simultaneously within clearly defined regions that exclude their collision during the automated assembly process. KR240 (R1) is responsible for multiple operations of picking, handling, processing, positioning, and temporarily stabilizing of façade wood elements, while the statically positioned KR6 (R2) robot screws them to substructure on the other side of the frame. The size of the frame is a result of limited reachability of the R2 robot and the estimated limits of the façade panel weight, which should not exceed 50kg when carried by two persons. We employed additional hardware, including a stand-alone table saw with 100mm blade radius and manually adjustable tilt angle, and a robotic screw gun handling the screws between 35mm and 50mm lengths. Newly developed wood picking tools consist of vacuum gripper for plates of the tested width (60mm to 400mm) and length (200mm to 1700mm), maximal tested weight (10kg), and an asymmetric gripper allowing to pick up beam elements of varied widths (25mm to 115mm), lengths (300mm to 2500mm) and depths (20mm to 90mm). The slat's size diversity was obtained through special tool construction with five guided drive actuators placed in two orientations: two vertical ones pressing the beam towards substructure to guarantee the tight screw connection, and three horizontal ones allowing for gripping and straightening of wood beams pressed against an 8mm steel plate (Figure 6). The use of such an asymmetric construction allows to minimize the gaps between the beams to 10mm and for the diverse depths of slats in one panel, which would be impossible with the use of parallel grippers. Apart from the mentioned equipment, we developed two types of customized dispensers depending on the proportion of logs. Plates and wider beams were organized in vertical stacks of maximum number of ten to minimize the material accumulated error while gripping. On the other hand, the narrow and deep beams required separated rails to prevent them from tipping over and to additionally help to keep the beams straight. Depending on the façade type, different amounts of dispensers were used: two for the shingle façade, seven for the board-on-board façade, and ten for the beam-based façade.

8　Workshops with architects at the beginning of the project. Fifteen invited architects from eleven offices designed façades in an experimental manner, and the workshop consisted of lecture, discussions, and hands-on digital and physical experiments with small-scale wood-based façade systems.

The robots' communication is based on the I/O signal exchange to hand over the sequences grouped by consecutive slats starting with R1 and the following with operations by R2 (Figure 5). We introduced a double signal exchange to prevent the unexpected skipping of the signal and undesired jumping to the next sequence.

The developed fabrication system was a combination of a majority of automated procedures with a few manual operations for prep- and post-processes. These included log surface treatment, material preparation, removing the façade panel from the frame, feeding the log dispensers, and installing the new substructure elements on the frame (Figures 7a,b). This mixed fabrication, in which a workspace is partly shared by a craftsperson and a robot allowing for contact when the robot is not operational, presents a medium level of human-robot collaboration and was driven by industrial safety and efficiency requirements. At the same time, it maximized the flexibility of the overall process.

Collaborative Approach

We base our research on the concept of collaborative customization (Salvador, Martin de Holan, and Piller 2009), in which the needs of the customers inform the manufacturing processes. Secondly, we also rely on collaborative design strategies (Achten 2002) which aim at enhancing each participant's contribution by incorporating their inputs at the early design stage. Our collaborative approach is manifested in three areas:

- Researchers, Industrial Partner, and Architects: In order to understand architects' expectations towards prefabricated façade systems, we started the research with a workshop with eleven architectural offices (Figure 8) as well as conducted an online survey. The findings guided subsequent research developments.
- Researchers and Wood Engineers: The exchange with the company's engineers during the project guided the development of the digital design tool in regard to structural, material, and construction requirements.
- Industrial Partner and Architects: Each new project starts with collaborative design workshops, where architects' expectations are enhanced with custom parametric design tools and confronted with the company's engineering guidelines. This early exchange allows for the development of custom designs variants that are informed with manufacturing capabilities of ERNE and are optimized in terms of material use and price.

Digital Craftsmanship

The project explores how the concept of craftsmanship enhanced through digital means can be reintroduced in the timber industrial setup. It can be observed on multiple levels. Firstly, the design space developed during the project is related to the local woodcraft tradition. Three most typical façade families—beam-based, shingle, and board-on-board systems—have been aesthetically reinterpreted and digitally encoded while their original logic was preserved.

Secondly, each of the developed systems required a different set of production and assembly methods. During the project, we adjusted and expanded the manufacturing infrastructure of the industrial partner by building several custom robotic tools and processes that allowed for production of a wide variety of façades. These were designed as a set of diverse and independent components that can be rearranged to answer different production and assembly requirements. To that end, the developed fabrication setup follows the logic of "flexible automation" (Löfving et al. 2018), where the machines can be quickly and easily re-tasked for multiple façade families.

Thirdly, the design and fabrication were carried out in parallel through experimentation and prototyping. Such a model is close to the traditional notion of craft, where the result is born out of a dialogue-based interaction between maker and objects through material and tools (Gramazio, Kohler, and Langenberg 2014).

An additional level of flexibility was obtained by deliberately giving up on full automation and "lights-out-factories" scenarios, which in many industrial cases result in poor system performance (Endsley 1997; Endsley and Kiris 1995). The developed system intelligently shares different tasks of the production chain among automated and manual procedures. While the robotic processes improve precision, efficiency, and speed of construction, the material quality and surface finishes are ensured by manual skills of the ERNE's carpenters. This quality control takes place at the very beginning and end of the production chain without interrupting the automated robotic tasks to obtain high production efficiency and aesthetic quality at the same time.

Lastly, the proposed process is a first step to bring architects closer to making. By using integrative digital tools and through a close collaboration with an industrial partner, they expand the palette of fabrication solutions and have the chance to take a more active role in the construction process.

RESULTS AND DISCUSSION
Project Demonstrator
Project demonstrator are four unique façades on a 1-storey-high temporary office building, with a total area of 220m², constructed in November 2020 (Figure 10). These intricate claddings are reinterpreting three most common

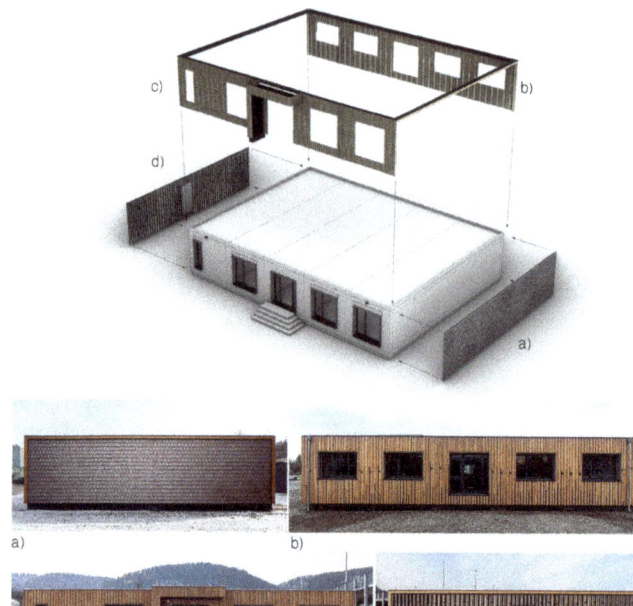

10 Architectural concept of demonstrator. Four unique façades, where front and back create the architectural frame for two side façade

traditional Swiss façade types (Figure 9) presenting different construction logic and challenges.

Shingles
Shingle façade is characterized by the dynamic, attractor-based pattern applied to the bottom of the plates (Figure 12). A robot equipped with a vacuum gripper assembled 1,341 unique elements, cut from the plates of 165mm x 500mm x 20mm and split to 24 panels of maximal size of 1.8m x 1.4m. Its design required an individual double-sided cut for each plank on the table saw with a 45° tilted blade. The front cut had the aesthetic purpose to expose the irregular pattern of the façade, while the back cut ensured that the rain flowed smoothly on the panel and was rooted in the internal quality standards of the company. Each shingle was screwed from behind in four points, two to the substructure and two to the shingles beneath. The fabrication process is depicted in Figure 11. To maximize the efficiency, we focused on reduction of processing time

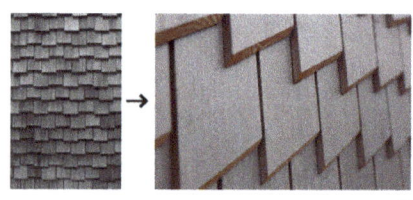
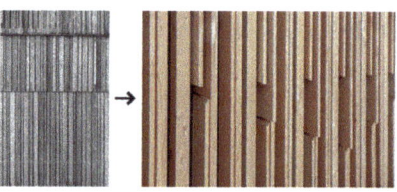
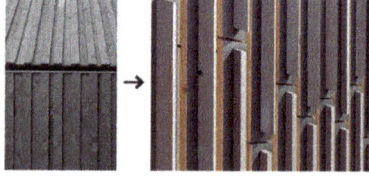

9 Three façade types inspired by local traditional carpentry solutions: a) Shingles, b) Beam-based façade, c) Board-on-board system

per element. An average of 70 sec was achieved by optimizing the robotic path, shortening the distance between the disposers and production frames, and placing the substructure manually.

Bea-based Façade

Beam-based façades systems were applied to the front and back elevation of the demonstrator (Figure 13) and constructed with 38 and 32 panels, respectively. Each panel consisted of four different slat dimensions 25mm x 50mm, 50mm x 25mm, 75mm x 25mm, 25mm x 75mm. Each type was pre-ordered with three different length values to optimize the amount of material waste during robotic cutting. Therefore, this system required the most complex disposer organization.

The façade was assembled with an asymmetric gripper, which picked the slat from one of ten disposers, adjusted its length on the table saw, and positioned it on the façade. During the screwing, the asymmetric gripper was slid to force the element to the correct shape. Consequently, the difference along the length between the slats is within 1mm to 1.5mm margin, which is comparable to tolerances in manually produced façades.

Horizontal organization of elements was based on a 10-digit-long pattern. The irregular cuts along the slats and vertical pattern were controlled with multiple closed curves defining the regions with different applied effects to choose from: creating a gap between the elements, removing it or changing its size.

Board-on-board Façade

The last constructed façade was a mixed, plate-beam system, inspired by traditional board-on-board solutions (Figure 14). It consisted of three layers of plates and beams that were screwed on the top of each other and divided into 22 panels of maximal size of 2m x 1.5m. The plates were of two geometrical types (150mm x 1600mm x 22mm and 180mm x 1000mm x 20mm) and slats (40mm x 80mm x 1000mm) had three types of angled cut from the back, creating a dynamic shading pattern. This complex façade required working with two grippers: vacuum gripper for the two bottom layers, and asymmetric gripper for the top beams. Each element on the middle and top layers was individually cut by the robot to the correct length on the table saw. Due to the complicated overlapping between the panels, the pattern of the façade and panel division was partially driven by the assembly requirements. To simplify

  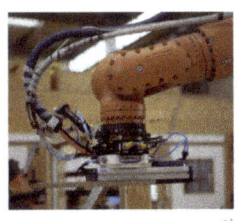 

  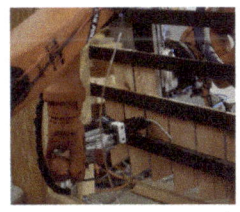 

11 Fabrication of shingle façade: a) manual assembly of substructure on the frame, b) placing plates in disposer, c) gripping plates with vacuum gripper, d) double cut of plates with a table saw, e) placing of the element on the frame, f) screwing from the back by KR6, g) manual finishing.

12 Close up of shingle façade depicting a dynamic pattern of bottom cuts

13 Close up of beam-based facade

14 Close up of board-on-board facade

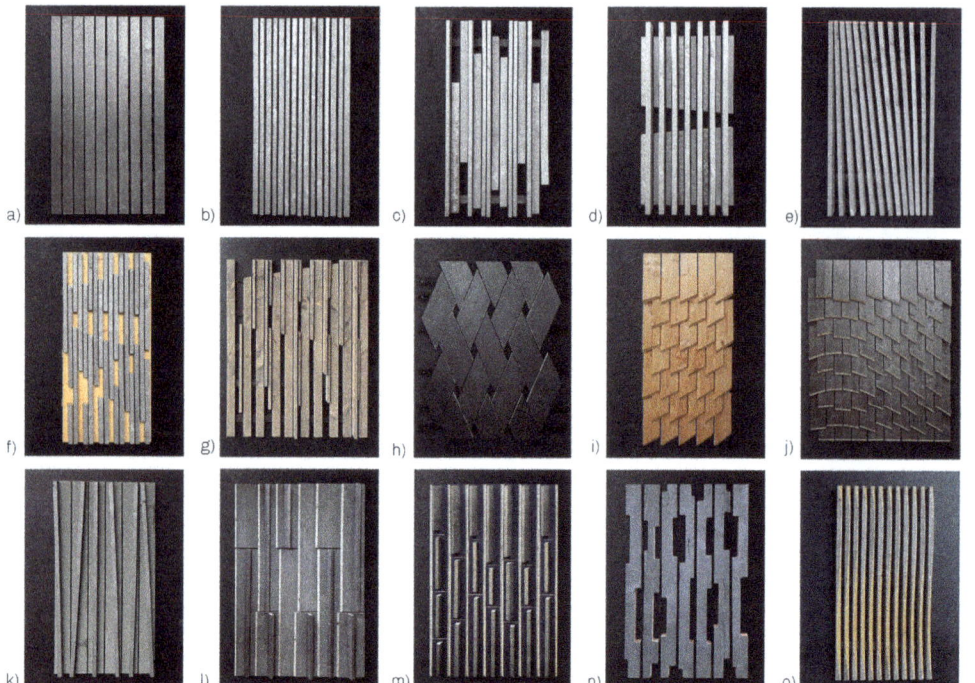

15 Physical mock-ups representing design potential of the developed process: a) uneven gaps between the slats, b) different widths, c) non-repetitive pattern, d) different depths of elements, e) different orientations, f) non-repetitive division along the slats, g) non-repetitive width, length and depth, h) plate-based façades, i,j) shingle façades, k,l,m) multi-layered façades, n,o) façades with milled pattern

and optimize the production time, we decided to perform most of the repetitive geometrical manipulation on the individual elements manually before the actual robotic fabrication. As a result, the average time per panel was reduced to 35 minutes (Table 1). Therefore, this façade proved to be the most cost-efficient in our demonstrator but also the most prone to human error.

Other Applications

Besides the office façades, we have produced other physical artefacts, like 15 samples of 1:1 façade mock-ups (Figure 15), presenting various design and assembly variants, G-Box depicting seamless connection between vertical beam panels and two mock-ups for the exhibition purposes, each of the size 1.2m x 0.8m x 2m.

Discussion

The proposed system is a response to the growing need for flexibility and adaptability in the construction industry, while keeping the high quality of the final product. Where manual manufacturing technique would either be impossible to implement or require custom-made formworks and complicated individual treatment of each element, the described computational and fabrication methods proved

	Plate-based façade	Beam-based façade	Board on board façade		
Assembly method	Vacuum gripper (VG)	Asymmetric gripper (AG)	Vacuum/Asymmetric gripper (VG/AG)		
Material preparation and placement in disposer	Manual	Manual	Manual		
Robotic pick up	VG	AG	VG	VG	AG
Robotic cut	2x	1x	-	1x	1x
Robotic place	VG	AG	VG	VG	AG
Tool changing	-	-	-	VG-> AG	-
Quality checkup and panel removal from the frame	Manual	Manual	Manual		
Avr. Time /panel	1:30:00	1:10:00	00:37:00		
Avr. no of elements/ panel	54	44	31		
Avr. time/ element	00:01:10	00:01:10	00:01:03		
Avr. time /element brutto (including additional processes)	00:01:40	00:01:35	00:01:10		

Table 1   The comparison of the production processes and average manufacturing time of panels and elements in each façade type

to be effective in designing and fabrication of a large variety of custom wooden façade panels that are both well-crafted and cost-efficient. This distinguishes the developed system from individualized expensive façade solutions available on the market.

The collaborative work between architects and industry was adjusted to the architects' demands formulated at the beginning of the research. The design tool was initially dedicated for designers to actively use within their architectural practice as a stand-alone-software. However, we decided to keep it on the level of a flexible prototype handled by a computational specialist during an interdisciplinary design workshops at the beginning of each project, as architects expressed no interest in learning new design tools. It was also impossible to generalize the vast design space while maintaining a clear and understandable user interface. Although this solution allows architects to gain understanding of the production process and make more informed design decisions, it requires further improvements to enable them more independent design control.

The production based on the medium level of human-robot collaboration proved to be valid in the timber industry. Higher level of collaboration would be impossible to achieve without compromising either safety, efficiency, or flexibility of the overall system.

Due to its applied character, this research project guarantees the direct implementation of its content into the company's infrastructure and thus, might improve the aesthetic quality of our urban environment on a larger scale.

CONCLUSIONS
Contributions
The project's strongest contribution lies in the introduction of design-driven flexible automation strategies embedded in the industrial setup. This is facilitated through (1) flexible digital design tool, (2) flexible robotic asymmetric gripper allowing for tight assembly of wooden slats of varied depths, and (3) overall reconfigurable fabrication setup. Additionally, the findings of the initial workshop with architects directly guided the research directions, which only brought the results closer to original intentions—allowing architects more design freedom.

Outlook
The developed fabrication system and digital design tools have been incorporated into ERNE infrastructure and the proposed process has been integrated into the company's communication plans for 2021. The innovative content of this research led the company to start a new production line for robotically assembled building elements in 2021-2022 and to purchase a second robot on a linear axis, which will enable production of larger façade elements (up to 3.5m x 1.5m) on a lower cost.

Additionally, the process is applied on a large-scale project with the local architectural office. The collaboration started in April 2020 and will result with 265,5m^2 of frontal façade of a five storey residential building. The assembly is planned for spring 2022.

ACKNOWLEDGEMENTS
This research was co-funded by Innosuisse, Swiss Innovation Agency, under project code CoDeFa: Collaborative Design of Prefabricated Facade Systems. The authors would like to sincerely thank all collaborators from both involved institutions. ERNE: Oliver Ackermann for support with robotic production of facades, Heiri Treier for detailed facade planning, and Thomas Reiner for wood engineering. FHNW: Ralf Michel for academic advisory.

REFERENCES
Achten, H.H. 2002. "Requirements for Collaborative Design in Architecture." In *Proceedings of the 6th Design & Decision Support Systems in Architecture & Urban Planning Conference*, edited by H.J.P. Timmermans and B. de Vries. Eindhoven. 1–13.

Ackermann, Oliver. 2018. "Methods of Parametric Design and Future Implementation." BA diss., University of Applied Sciences and Arts Northwestern Switzerland (Basel), 2018. Accessed March 20, 2022. https://www.hypermagazine.ch/wp-content/blogs.dir/1/files/2018/09/Oliver_Ackermann_Diplomdokumentation.pdf.

Alvarez, Martin, Hans Jakob Wagner, Abel Groenewolt, Oliver David Krieg, Ondrej Kyjanek, Daniel Sonntag, Simon Bechert, Lotte Aldinger, Achim Menges, and Jan Knippers. 2019. "The Buga Wood Pavilion Integrative Interdisciplinary Advancements of Digital Timber Architecture." In *ACADIA '19: Ubiquity and Autonomy; Proceedings of the 39th Annual Conference of the Association for Computer Aided Design in Architecture*. Austin, Texas. 490–500.

Apolinarska, Aleksandra Anna, Ralph Bärtschi, Reto Furrer, Fabio Gramazio, and Kohler Matthias. 2016. "Mastering the 'Sequential Roof' - Computational Methods for Integrating Design, Structural Analysis and Robotic Fabrication." In *Advances in Architectural Geometry*, edited by S. Adriaenssens, F. Gramazio, M. Kohler, A. Menges, and M. Pauly. vdf Hochschulverlag AG an der ETH Zürich. 240–58.

Braumann, Johannes. *KUKA|prc*. V. 20200508. Association for Robots in Architecture. Windows. 2020.

Çıdık, Mustafa Selçuk, David Boyd, and Niraj Thurairajah. 2017.

"Ordering in Disguise: Digital Integration in Built-Environment Practices." *Building Research and Information* 45 (6). https://doi.org/10.1080/09613218.2017.1309767.

Correa, David, Oliver David Krieg, and Anna Lisa Meyboom. 2019. "Beyond Form Definition: Material Informed Digital Fabrication in Timber Construction." In *Digital Wood Design*. Lecture Notes in Civil Engineering, vol. 24. Cham: Springer. 61–92. https://doi.org/10.1007/978-3-030-03676-8_2.

Dank, Richard, and Christian Freissling. 2012. "The Framed Pavilion; Modeling and producing complex systems in architectural education." In *Rob | Arch 2012; Robotic Fabrication in Architecture, Art and Design*. Vienna: Springer. 238–247. https://doi.org/10.1007/978-3-7091-1465-0_28.

Dossick, Carrie Sturts, and Gina Neff. 2011. "Messy Talk and Clean Technology: Communication, Problem-Solving and Collaboration Using Building Information Modelling." *Engineering Project Organization Journal* 1 (2): 83–93. https://doi.org/10.1080/21573727.2011.569929.

Endsley, Mica R. 1997. "Level of Automation: Integrating Humans and Automated Systems." In *Proceedings of the Human Factors and Ergonomics Society* 41 (October 1997): 200–204. https://doi.org/10.1177/107118139704100146.

Endsley, Mica R., and Esin O. Kiris. 1995. "The Out-of-the-Loop Performance Problem and Level of Control in Automation." *Human Factors* 37 (2): 381–94. https://doi.org/10.1518/001872095779064555.

Gramazio, Fabio, Matthias Kohler, and Silke Langenberg. 2014. "Mario Carpo in Conversation with Matthias Kohler." In *Fabricate 2014: Negotiating Design & Making*, edited by F. Gramazio, M. Kohler, and S. Langenberg. London: UCL Press. 12–21. https://doi.org/10.2307/j.ctt1tp3c5w.5.

Harty, Chris. 2008. "Implementing Innovation in Construction: Contexts, Relative Boundedness and Actor-Network Theory." *Construction Management and Economics* 26 (10): 1029–41. https://doi.org/10.1080/01446190802298413.

Helm, Volker, Selen Ercan, Fabio Gramazio, and Matthias Kohler. 2012. "Mobile Robotic Fabrication on Construction Sites: DimRob." In *2012 IEEE/RSJ International Conference on Intelligent Robots and Systems*. 4335–4341. https://doi.org/10.1109/IROS.2012.6385617.

Johns, Ryan Luke, and Nicholas Foley. 2014. "Bandsaw Bands: Feature-Based Design and Fabrication of Nested Freeform Surfaces in Wood." In *Robotic Fabrication in Architecture, Art and Design 2014*, edited by W. Mcgee and M. Ponce de Leon. Switzerland: Springer International Publishing. 17–32.

Kolarevic, Branko. 2009. "Towards Integrative Design." *International Journal of Architectural Computing* 7 (3): 335–44. https://doi.org/10.1260/147807709789621248.

Koolhaas, Rem. 2018. *Elements of Architecture*. Cologne: Taschen.

Löfving, Malin, Peter Almström, Caroline Jarebrant, Boel Wadman, and Magnus Widfeldt. 2018. "Evaluation of Flexible Automation for Small Batch Production." *Procedia Manufacturing* 25: 177–184. https://doi.org/10.1016/j.promfg.2018.06.072.

Mehrjerdi, Yahia Zare. 2009. "Excellent Supply Chain Management." *Assembly Automation* 29 (1): 52–60. https://doi.org/10.1108/01445150910929866.

Mollica, Zachary, and Martin Self. 2016. "Tree Fork Truss: Geometric Strategies for Exploiting Inherent Material Form." In *Advances in Architectural Geometry 2016*, edited by S. Adriaenssens, F. Gramazio, M. Kohler, A. Menges, and M. Pauly. Zurich: vdf Hochschulverlag AG.

Parascho, Stefana, Augusto Gandia, Ammar Mirjan, Fabio Gramazio, and Matthias Kohler. 2017. "Cooperative Fabrication of Spatial Metal Structures." In *Fabricate 2017*. UCL Press. 24–29. https://doi.org/10.3929/ETHZ-B-000219566.

Parascho, Stefana, Isla Xi Han, Samantha Walker, Alessandro Beghini, Edvard P. G. Bruun, and Sigrid Adriaenssens. 2020. "Robotic Vault: A Cooperative Robotic Assembly Method for Brick Vault Construction." *Construction Robotics* 4: 117–126. https://doi.org/10.1007/s41693-020-00041-w.

Parascho, Stefana, Jan Knippers, Moritz Dörstelmann, Marshall Prado, and Achim Menges. 2015. "Modular Fibrous Morphologies: Computational Design, Simulation and Fabrication of Differentiated Fibre Composite Building Components." In *Advances in Architectural Geometry 2014*, edited by P. Block, J. Knippers, N. Mitra, and W. Wang W. Cham: Springer. https://doi.org/10.1007/978-3-319-11418-7_3.

Robeller, Christopher, and Yves Weinand. 2016. "Fabrication-Aware Design of Timber Folded Plate Shells with Double Through Tenon Joints." In *Robotic Fabrication in Architecture, Art and Design 2016*, edited by D. Reinhardt, R. Saunders, and J. Burry J. Cham: Springer. https://doi.org/10.1007/978-3-319-26378-6_12.

Robert McNeel & Associates. *Rhinoceros*. V6.SR26. Robert McNeel & Associates. Windows. 2020.

Rust, Romana, David Jenny, Fabio Gramazio, and Matthias Kohler. 2016. "Spatial Wire Cutting : Cooperative Robotic Cutting of Non-Ruled Surface Geometries for Bespoke Building Components." In *CAADRIA 2016: Living Systems and Micro-Utopias; Towards Continuous Designing, Proceedings of the 21st International Conference of the Association for Computer-Aided Architectural

Design Research in Asia, edited by S. Chien, S. Choo, M.A. Schnabel, W. Nakapan, M.J. Kim, S. Roudavski. 529–538.

Rutten, David. *Grasshopper in Rhinoceros 3D*. V 1.0.0007. Robert McNeel & Associates. Windows. 2020.

Salvador, Fabrizio, Pablo Martin de Holan, and Frank Piller. 2009. "Cracking the Code of Mass Customization." *MIT Sloan Management Review* 50 (3).

Saunders, Andrew, and Gregory Epps. 2016. "Robotic Lattice Smock." In *Rob | Arch 2016: Robotic Fabrication in Architecture, Art and Design*, edited by D. Reinhardt, R. Saunders, and J. Burry J. Cham: Springer. https://doi.org/10.1007/978-3-319-26378-6_6.

Schwinn, Tobias, Oliver David Krieg, and Achim Menges. 2013. "Robotically Fabricated Wood Plate Morphologies." In *Rob | Arch 2012: Robotic Fabrication in Architecture, Art and Design*, edited by S. Brell-Çokcan and J. Braumann. Vienna: Springer. https://doi.org/10.1007/978-3-7091-1465-0_4.

Søndergaard, Asbjørn, Oded Amir, Phillip Eversmann, Luka Piskorec, Florin Stan, Fabio Gramazio, and Matthias Kohler. 2016. "Topology Optimization and Robotic Fabrication of Advanced Timber Space-Frame Structures." In *Rob | Arch 2016: Robotic Fabrication in Architecture, Art and Design*, edited by D. Reinhardt, R. Saunders, and J. Burry. Cham: Springer. https://doi.org/10.1007/978-3-319-26378-6_14.

Thoma, Andreas, Arash Adel, Matthias Helmreich, Thomas Wehrle, Fabio Gramazio, and Matthias Kohler. 2019. "Robotic Fabrication of Bespoke Timber Frame Modules." In *Rob | Arch 2018: Robotic Fabrication in Architecture, Art and Design*, edited by J. Willmann, P. Block, M. Hutter, K. Byrne, and T. Schork. Cham: Springer International Publishing. https://doi.org/10.1007/978-3-319-92294-2_34.

Wang, Zichu Will. 2021. "Real Design Practice, Real Design Computation." *International Journal of Architectural Computing* 19 (1): 104–15. https://doi.org/10.1177/1478077120958165.

Whyte, Jennifer. 2011. "Managing Digital Coordination of Design: Emerging Hybrid Practices in an Institutionalized Project Setting." *Engineering Project Organization Journal* 1 (3): 159–68. https://doi.org/10.1080/21573727.2011.597743.

Whyte, Jennifer, and Sunila Lobo. 2010. "Coordination and Control in Project-Based Work: Digital Objects and Infrastructures for Delivery." *Construction Management and Economics* 28 (6): 557–67. https://doi.org/10.1080/01446193.2010.486838.

Williams, Nicholas, and John Cherrey. 2016. "Crafting Robustness: Rapidly Fabricating Ruled Surface Acoustic Panels." In *Rob | Arch 2016: Robotic Fabrication in Architecture, Art and Design*, edited by D. Reinhardt, R. Saunders, and J. Burry. Cham: Springer. https://doi.org/10.1007/978-3-319-26378-6_23.

Zock, P., E. Bachmann, F. Gramazio, M. Kohler, T Kohlhammer, M. Knauss, C. Sigrist, and S. Sitzmann. 2014. "Additive Robotergestützte Herstellung Komplexer Holzstrukturen." In *46 Tagungsband Fortbildungskurs Holzverbindungen Mit Klebstoffen Für Die Bauanwendung*. Weinfelden: Swiss Wood Innovation Network (S-WIN). 197–208. http://www.gramaziokohler.com/data/publikationen/1105.pdf.

Edyta Augustynowicz is an architect, researcher, and computational designer, specializing in complex timber structures. She is currently affiliated with ERNE AG Holzbau. She holds a Masters in Architecture and Urban Design from the TU Poznan and a MAS degree in CAAD from the ETH Zurich. Her previous experience include Digital Technology Group at Herzog & de Meuron, Block Research Group at the ETH Zurich, and Institute Integrative Design, HGK, FHNW in Basel.

Maria Smigielska is an architect and researcher with diverse experience in academia and practice. Maria holds MSc degree in Architecture (TU Poznan) and MAS in Architecture and Information (ETH Zurich). She is currently affiliated with the chair of Digital Building Technologies, ETH Zurich, and ICDP HGK FHNW Basel. Maria formerly worked with Baierbischoferberger Architects, Creative Robotics UfG Linz, Digital Knowledge at ENSAPM.

Daniel Nikles is a post-industrial designer and robotics researcher in the field of art and design. He holds BA degree from the Institute HyperWerk and MA in Design from HGK FHNW Basel, with the thesis employing industrial robots for designers. Trained as an architectural draftsman, currently he is working as a researcher for the Institute HyperWerk FHNW and as project lead robotics at ERNE AG Holzbau.

Thomas Wehrle is CTO at ERNE AG Holzbau. He is responsible for the development of digitalization and introduction of robotic production at the company. For his achievements with implementation of robotics in collaboration with ETH Zurich, he was awarded in 2017 with the European Innovation Award for the Forestry and Wood Industries. In parallel he is tutor and researcher at Hochschule Luzern, Berner Fachhochschule, and ETH Zurich. He holds a Master degree from ZHAW School of Engineering.

Heinz Wagner is the former Head of the Institute Integrative Design|Masterstudio (currently ICDP) at HGK, FHNW. He contributed to the creation of the Master's program at HGK. He studied Economics, Sociology and Law at the University of Basel, completed the ABU diploma course at the Swiss Federal Institute for Vocational Education and Training in Zollikofen and a postgraduate course in Architecture/Timber house construction at ETH Zurich.

Maison Fibre

Design and development of an FRP-Timber hybrid building system for multi-story applications in architecture

Niccolo Dambrosio
ICD University of Stuttgart

Christoph Zechmeister
ICD University of Stuttgart

Rebeca Duque Estrada
ICD University of Stuttgart

Fabian Kannenberg
ICD University of Stuttgart

Marta Gil Peréz
ITKE University of Stuttgart

Christoph Schlopschnat
ICD University of Stuttgart

Katja Rinderspacher
ICD University of Stuttgart

Jan Knippers
ITKE University of Stuttgart

Achim Menges
ICD Universitz of Stuttgart

0   Interior space of the *Maison Fibre*

ABSTRACT

This research demonstrates the development of a hybrid FRP-timber wall and slab system for multi-story structures. Bespoke computational tools and robotic fabrication processes allow for adaptive placement of material according to specific local requirements of the structure thus representing a resource-efficient alternative to established modes of construction. This constitutes a departure from pre-digital, material-intensive building methods based on isotropic materials towards genuinely digital building systems using lightweight, hybrid composite elements. Design and fabrication methods build upon previous research on lightweight fiber structures conducted at the University of Stuttgart and expand it towards inhabitable, multi-story building systems. Interdisciplinary design collaboration based on reciprocal computational feedback allows for the concurrent consideration of architectural, structural, fabrication, and material constraints. The robotic coreless filament winding process only uses minimal, modular formwork and allows for the efficient production of morphologically differentiated building components. The research results were demonstrated through Maison Fibre, developed for the 17th Architecture Biennale in Venice. Situated at the Venice Arsenale, the installation is composed of thirty plate-like elements and depicts a modular, further-extensible scheme. While this first implementation of a hybrid multi-story building system relies on established glass and carbon fiber composites, the methods can be extended towards a wider range of materials ranging from ultra-high-performance mineral fiber systems to renewable natural fibers.

INTRODUCTION

Urbanization and growth of the global population represent two major demographic mega-trends. Estimates and projections of urbanization indicate that the future growth of the human population can be accounted for almost entirely by a growing number of city dwellers. By mid-century, around 68 per cent of the world's population will be living in urban areas (United Nations 2019). The need for new buildings and infrastructures is already posing one of the most important ecological and social challenges to society. Construction has become one of the most materially intense and environmentally detrimental human activities: as it is currently operating, this sector can be held responsible for solid waste production, greenhouse gas emissions at global scale, high-energy utilization, external and internal pollution, profound environmental impact and resource exhaustion (Ortiz, Castells, and Sonnemann 2009; Melchert 2007; Zimmermann, Althaus, and Haas 2005). The manufacturing, transportation and use of all construction materials for buildings resulted in energy and process CO_2 emissions of 3.5 Gt in 2019, or 10% of all energy sector emissions (IEA 2020). Moreover, driven by a rapid increase in demand from various sectors of economy, with the construction sector ahead, between 1980 and 2010 the global production of steel nearly doubled, while the cement production more than tripled (Van Ruijven et al. 2016). The current building paradigm, which prioritizes simple and standardized construction processes over saving material and resources, seems no longer sustainable. Computational design methods, workflows and tools, employed to design with specific classes of materials, as in this research fibrous composite materials, could offer a viable alternative to disrupt such paradigm. Composite materials, with particular regard to glass and carbon fiber reinforced polymers (G/CFRP), occur as a potential backup to the more conventional structural materials, such as concrete and steel. In the case of CFRP, this is mainly due to their high strength-to-weight ratio, low thermal expansion, and high fatigue and corrosion resistance (Fitzer 1985). Their use in architecture and construction is not entirely novel, dating back to the second half of the 20th century (Bakis et al. 2002), an early iconic example being Matti Suuronen's Futuro houses (Menges and Knippers 2015). G/CFRP applications also span from large-scale structural elements, as in the BMW Guggenheim Lab (Schittich 2014), to façade tiles as in the SF Moma (Sterrett and Piantavigna 2018). However, considering that commonly used manufacturing processes (molding processes and pultrusion) rely on special molds or dies, and involve significant amounts of material cut-offs and production waste (Bader 2002), G/CFRP's adoption in the field of construction remained limited and discontinuous over time as it proved to be not always economically viable for the production of large-scale one-off components like the ones often encountered in buildings. Alternative approaches to the use of composite materials in architecture and construction, based on the study of fibrous systems in nature, have been investigated by the Institute for Computation al Design and Construction (ICD) and the Institute for Building Structures and Structural Design (ITKE) of the University of Stuttgart. Since the logic of CFRP and GFRP is quite closely related to that of natural fibrous systems (Menges and Knippers 2015), integrative computational design and simulation methods (Reichert et al. 2014) and novel manufacturing processes for G/CFRP—which facilitate automated tailored fiber placement in terms of quantity, direction, and orientation while also keeping the production costs relatively low (La Magna, Waimer, and Knippers 2014)—were developed by ICD and ITKE researchers. Such methods were then applied to the realization of demonstrators that would showcase principles of locally adapted fiber layups in a single layer shell structure (Knippers et al. 2015), as well as in a composite segmented shell with highly differentiated components (Parascho et al. 2015). Further developments have subsequently led to the introduction of specific design features allowing an interface with other parts of a building (Prado et al. 2017) and the development of an full architectural scale novel fiber composite building system (Dambrosio et al. 2019). The following sections illustrate the design and development of the *Maison Fibre* project. Realized in occasion of the 2021 Venice Architecture Biennale and made possible by the use of new integrative co-design processes, this project represents the first built example that utilizes a novel FRP-timber hybrid building system (Figure 1) for multi-story inhabitable applications in architecture, which starts challenging current paradigms in construction.

METHODS

System Conceptualization and Demonstrator

The FRP-timber hybrid system presented in this paper aims to create a novel building system for multi-story applications in architecture consisting of horizontal structural elements and vertical load-bearing wall elements. The horizontal elements or slabs combine the advantageous aspects of coreless wound FRP components—lightweight, high potential for adaptability, and minimal production waste (Dambrosio et al. 2019)—with a thin LVL plate. By creating a walkable surface, the LVL takes the live load, providing compression resistance to the system, and transfers these forces to the fiber structure that caters to structural depth. In order to keep the timber as thin as possible, the supports or connections between LVL and

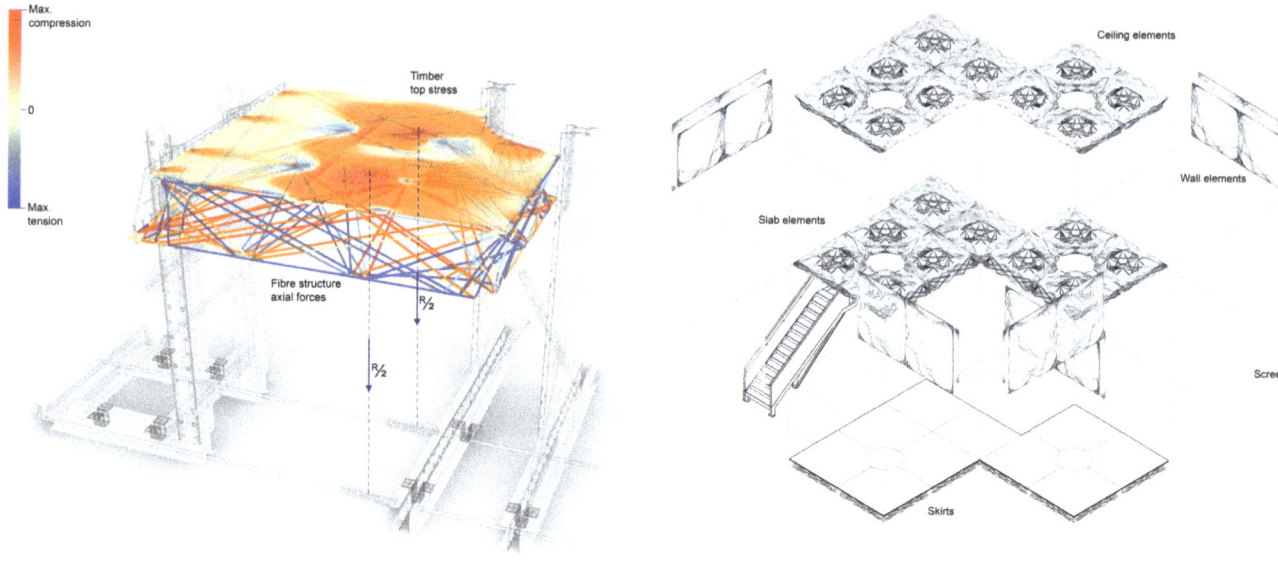

fiber structure are designed for the maximum span that the target thickness of the timber can withstand. This became one of the driving design factors for the fiber structure and allowed the hybrid system to be lighter and performative.

The research challenges associated with such building system are:
- Development of a suitable fiber layup for the horizontal structural elements that can successfully interface with the LVL plate.
- Development of fiber layup for the vertical load-bearing perimeter elements that can effectively connect the slabs and generate a stiff box-like structure.
- Structural modeling, testing, and validation and establishing a data exchange protocol and subsequent feedback loop with the design modeling platforms.
- Implementation of a relevant fabrication setup with winding frame that enables the production of every designed fiber component at reasonable costs.
- Interfaces and connection details development.

Such challenges were thoroughly investigated by the research team in the design and implementation of a specific case study: the *Maison Fibre*.

Developed for the 2021 Venice Architecture Biennale and located in the Corderie of the Arsenale, the installation is composed of thirty plate-like components, divided into ten slab, ten ceiling, and ten wall elements. Additionally, four screens were designed and fabricated to provide enclosure on the first floor, and thirteen skirts were used on the ground floor as an envelope to the base, also alluding to the virtually possible vertical repetition of the system in a truly multi-story configuration (Figure 2).

Integrative Design Approach—Interdisciplinary Collaboration

The fiber layup of each component is developed in reciprocal relation to the structural simulation. The structural performance of a fiber element is a decisive aspect of the design process and is directly linked to the component's geometry and the fiber layup. Important factors for structural performance are the fiber interaction of individual fiber layers—depending on the amount of pressure the currently laid fiber exerts on previously deposited fibers and their subsequent bonding strength in order to create structurally active, surface like formations—and the cross-section of fiber bundles and its buckling length in lattice-like structures. To control and evaluate those factors, computational design methods act as interfaces and enable interdisciplinary data exchange. The employed digital workflow simultaneously considers the logic of the winding sequence, the component aesthetics, structural analysis, and robotic fabrication constraints. Bespoke computational tools enable direct data exchange and rapid feedback from finite element method (FEM) simulations derived from surface and beam based structural models, guiding and evaluating the design development of all fiber layers (Figure 3).

In addition, this workflow was supported by the full-scale structural testing representing both the timber and the fibers of a single hybrid module. The evaluation of each iteration of the test informed this computational interdisciplinary loop, resulting in the optimization of the fiber layup and the reduction of the amount of material in less loaded areas (Gil Peréz et al. 2021). A purpose-built interface containing all fabrication relevant data in tabular form allowed for platform- and manufacturer-independent

Maison Fibre Dambrosio, Zechmeister, Duque Estrada, Kannenberg, Gil Peréz, Schlopschnat, Rinderspacher, Knippers, Menges

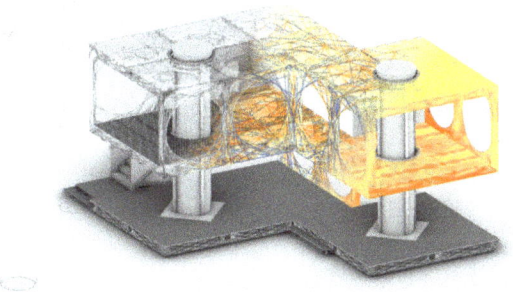

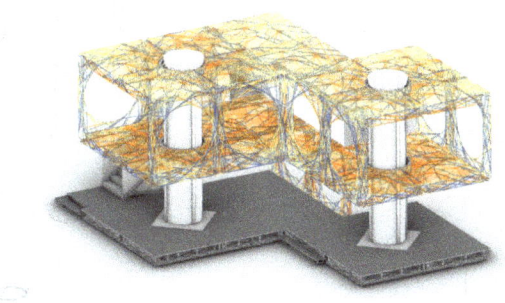

1 Diagram of stress distribution in a prototypical FRP-timber hybrid slab element
2 Exploded view of construction elements making up the full installation
3 Model showcasing the integrative design approach
4 Fiber finite element method (FEM) beam model

exchange of production data and facilitated the cyclical interaction with industry partners during prototyping and production.

Structural Design

The structural design of the *Maison Fibre* installation was realized with a multi-level approach as the final fiber layup of the components is not defined from the early stages of design (Gil Pérez et al. 2020). After a few preliminary models to evaluate the overall stability of the system, the FEM became more detailed with the evolution of the structural design and the component prototyping. Once the volumetric geometry of the components was better known, it was used to model walls and slabs as surface elements. The distribution of forces and stresses within the components, as well as the connection forces between them, were better studied with this iteration. In the last level of modelling, the carbon fiber layup of each wall and slab component was represented by beam elements (Figure 4). The models were checked for buckling, deflection, and maximum forces in the fibers in an iterative process between design and structural analysis (Gil Pérez et al. 2021). In parallel, to verify the structural safety and further optimize the composite components design, a full structural test of one of the slabs was designed. The test represented the real boundary conditions of the final installation for a single slab component element including both fiber composite and timber plate (Figure 1). A total of three iterations were performed resulting in the optimization of both fabrication and structural design.

Fabrication Setup

The fibrous wall, slab, and ceiling elements were manufactured using the robotic coreless filament winding process developed by the project team. This process employs a robotic arm to lay fibers on a very minimal linear framework equipped with winding pins. The initially set fibers become the mold in the winding process, allowing the elimination of the core or mandrel, and therefore minimizing the production waste to almost zero. The process is enhanced by the ability to tailor material orientation and quantity to structural, geometrical, and aesthetic requirements. This feature allows for locally load-adapted design and alignment of the fibers, thus enabling an extraordinary lightweight construction: the code-compliant, load-bearing fiber structure of the upper floor weighs just 9.9 kg/m². The wall elements are even lighter. An eight-axis robotic fabrication system was used to manufacture the fiber components, consisting of a six-axis industrial robot arm and an external, two-axis positioner. The two-axis positioner extends the kinematic system of the robotic setup and carries the winding frame, enabling optimal reachability of all anchor points during the fabrication process. Bundles of six glass fiber rovings and eight carbon fiber rovings are impregnated with epoxy resin during manufacturing in a drum-type resin bath. A mechanical tension control system maintains constant fiber tension throughout the fabrication process and ensures consistent material quality (Figure 5).

System Flexibility—Modular Winding Frame

Slab, ceiling, and wall components were designed to specific load cases and followed two main levels of differentiation. At the geometrical level, they were defined by the boundary conditions resulted from component-component interface and/or component-existing building interface. The slabs were divided into three boundary types, the ceilings in two types, while the walls only had one. At the material configuration level, each component was designed to perform its structural function through a precise and locally load-adapted design and alignment of the fibers. As for screens and skirts—since they mainly serve an enclosure function rather than a structural one—both geometrical boundaries and syntax were defined to fit their positioning and address the fiber continuity mostly from a visual aspect, therefore exhibiting a low level of differentiation. The proposed modular system and both

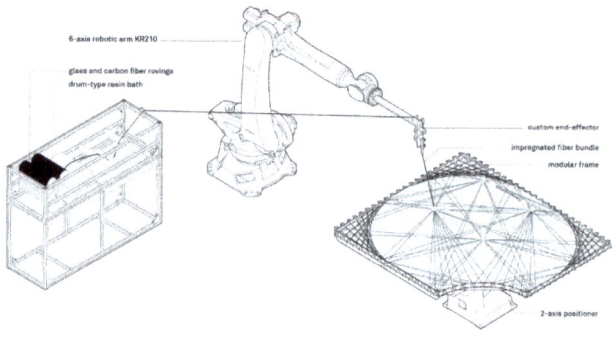

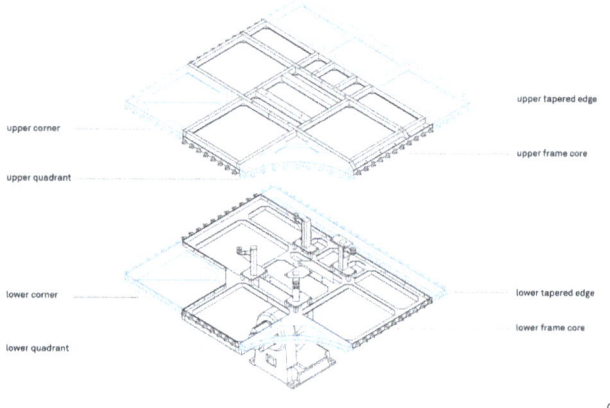

5   Fabrication setup diagram, with winding frame configured for the first winding stage

6   Modular winding frame, with highlighted add-ons that allow the fabrication of all the typologies of elements with minimal and reversible adaptations of the frame itself

7   Winding steps of a slab element: a) Anchor points along the component's perimeter and infield points, b) radial reinforcement syntax, c) hyperboloidal inner ring syntax, d0 hyperboloidal outer ring syntax

levels of differentiation were directly reflected in the frame development. In order to fabricate six different component types (three slab types, two ceiling types and one wall type), a modular frame was designed, allowing the use of one core and few attachments. Since the screens needed a specific pin orientation, with anchors normal to the winding surface, an additional lighter frame was developed for their fabrication. The main winding frame core was composed of waterjet cut steel plates and rectangular steel profiles forming a T-shape. Specific edges presented connection plates in which different attachments could be connected. According to the specific boundary conditions of each component type, the attachments were designed to provide the necessary interface between components. For example, in the case of wall-slab interface, the 45-degree chamfered edge that allows walls and slabs to meet was translated (in terms of frame configuration) into overlapping long and short edges. The orientation of the tapered edge was also considered, facing up or down, to differentiate the slab from ceiling elements. For the creation of the interface between slab elements and steel vertical supports, the quadrant winding frame attachment was used to generate the relevant quadrant cut out into the fiber elements, which would then enable the connection to the steel ring surrounding the existing columns. Inner anchor points were placed in the same position for all the slab and ceiling elements. Different syntaxes informed the positioning of double or single anchors, with different washer sizes. Considering the presence of inner winding pins and their necessary reachability during the initial winding stages, the frame was designed as two parts to be combined at a specific moment during fabrication, in a multi-stage winding process. The lower level contained the boundary edges with external winding pins, consisting of a bolt-sleeve-washer system, and vertical profiles responsible for holding the inner anchor points and for supporting the upper level. The upper level was configured as a mirrored version of the lower one (Figure 6).

Component Development and Manufacturing

The slab components represent significant innovation in coreless filament winding, as they for the first time form inhabitable fiber surfaces. Building up on previous research, the acquired knowledge to create modular structural building components (Prado et al. 2017) was expanded towards a high performance, ultra-lightweight, FRP-timber hybrid component. While carbon and glass fibers efficiently transfer the occurring loads, maintaining high material efficiency, a 27mm thin sheet of laminated veneer lumber (LVL) provides an inhabitable surface and adds to the component's structural capacity. The slab components consist of two substantially different fiber layup typologies: a hyperboloidal inner and outer ring, each consisting of glass fiber body, carbon fiber reinforcement, and a radial reinforcement layup. The radial reinforcement is wound first, and its primary task is to support the timber at its inner area and to transfer forces to the outer ring fibers. It is composed of four different fibers sub-layers that establish structural carbon fiber bundles of about 12 mm and incorporate a set of four infield points, acting as support for the LVL decking sheet. The inner and outer ring create the components characteristic aperture and enclosure along its perimeter (Figure 7). There are three different slab types—their geometry was specifically designed for the use case at the Arsenale in Venice and allows for the integration of the Arsenale's historic columns into the installation by introducing quadrant cut outs into

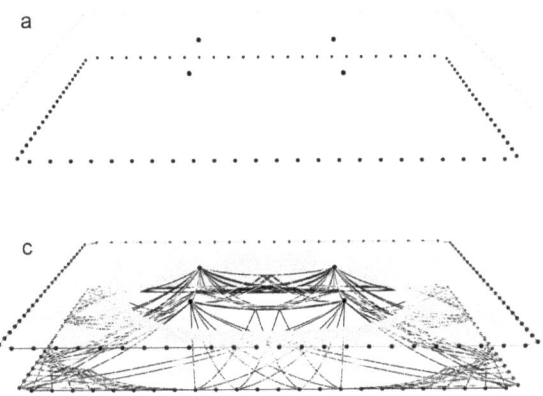
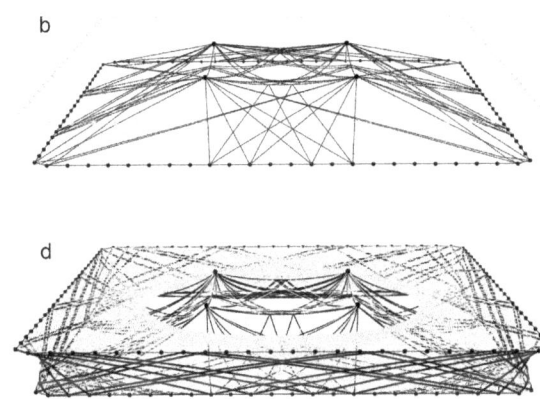

two slab types. To facilitate connections with wall elements, the connecting edges of slab and wall components exhibit 45-degree chamfers. This avoids the introduction of more infield points and maintains a visually clean edge. Owing to the flexibility of the fiber syntax and the minimal boundary geometry of the winding frames without depending on any mandrel or mould, the geometrically different slab types could be economically manufactured with only minimal adaptations of the fabrication equipment. In total, 747m of carbon, and 890m of glass fiber bundles were used for the production of one slab element, which results in a weight of only 63kg.

The ceiling elements depict the installation's enclosure towards the top and structurally act as horizontal bracing. Similar to the slab elements they are FRP-timber hybrid components and follow similar logic in terms of fiber syntax configuration. Due to the reduced structural requirements of the ceiling elements, the material expenditure can be significantly reduced compared to the slab elements. By adjusting the material amount to the specific local structural needs, a high degree of material differentiation is achieved, reducing redundancy and thus saving material and weight.

The wall elements are used for the vertical connection of the slab components, and together with the screens they form a spatial closure on both floors of *Maison Fibre*. There are ten wall elements in total, two on the lower floor and eight on the upper floor, and they are formed into modular, axially symmetrical pairs. Regarding its design, a single wall element consists of a total of fourteen fiber layers in different fiber bundle thicknesses, depending individually on the components' simulated structural performance. To ensure a frictional overlay of the fibers, wide-span and narrow-span circular glass fiber rings for each wall component were developed. Both rings are used as scaffolds for long-spanned structural carbon fibers. These take the connection points from the interface with the carbon fiber layers of the ceiling and floor components and connect them along both radial rings, emphasising a direct force-flow between top and bottom slab elements. With a dead weight of 61kg for the walls of the lower floor, and just 48kg for the components of the upper floor, the wall elements are among the lightest high-performance components of the entire installation.

To provide enclosure and safety, the walls and the two open sides on the first floor were equipped with glass fiber screens. The screens of the wall components were integrated as their first layer, while the screens for the open sides were manufactured and attached separately, as two pieces that connect in the center of the opening. To minimize the visual impact and provide for uniformity all around the structure, a mostly vertical pattern was chosen.

Interfaces

Maison Fibre is conceived as a self-standing, structurally autonomous installation. The inclusion of the Arsenale's historical columns into the volume of the installation permitted the maximization of the floor area while highlighting the flexibility of the proposed building system and its capability to adapt to given pre-existing conditions. Moreover, no contact or load transfers between *Maison Fibre* and the historical structures of the Arsenale were allowed. Therefore, four vertical steel U-profiles were placed around each column, horizontally braced by circular quadrant steel elements, forming a cage-like structure. The quadrant steel elements were designed to nest into the quadrant cut out of slab and ceiling elements, allowing them to be mechanically fastened to the fiber components. These vertical steel supports were then responsible for redirecting the loads coming from the FRP-timber structure to the ground. Loads are then evenly spread on the Arsenale floor through a steel frame that extends underneath the podium of the *Maison Fibre*.

8

To connect the ceiling-to-wall and wall-to-slab components and transfer the forces between them, custom steel connection interfaces were developed. These bespoke steel parts were connected to the anchor points of the load-bearing carbon bundles. While the lateral connection between neighboring slab components could be achieved by directly bolting their anchor points together, the ceiling-to-wall and wall-to-slab connections required custom steel brackets. These brackets also providing assembly tolerances and reachability. The timber plate that sits on top of the slab was connected to the fiber component with 90-degree custom steel brackets on the perimeter of the component. The central area of the plate was directly bolted to the infield points. Resting on top of the slabs and ceiling, the plates allowed the distribution of loads from the timber into the fibers (Figure 8).

Due to the alignment of their anchor vectors, the ceiling components that connected to the walls received 45-degree brackets for the timber on the inner perimeter, while the components themselves could be directly connected on the outer edge. This alignment of the anchor vectors is beneficial for the fiber direction, as collisions between the fibers and the edges of the winding frame during the fabrication process can cause uncontrolled thinning of the fiber bundles' sections, and consideration of the anchor orientation can improve the structural capacity of the FRP components (Gil Pérez et al. 2019).

DEMONSTRATOR AND CONCLUSION
Principles for material-saving and resource-efficient design, enabled by computational co-design methodologies, translate into a unique architectural experience in the *Maison Fibre*. With its base orthogonal grid rotated 45 degrees in the relation to the main axis of the Arsenale's Corderie, the structure opens up towards the visitors, inviting them to experience an inhabitable space made solely of fibers (Figure 9).

When observing the structure from afar, the accentuated vertical orientation of the fibers of the screens constituting the façade provides for a perceivably unifying element. While approaching the installation, the same screens start to visually vanish, allowing visibility through the filigree structure and offering glimpses of the first-floor spatial qualities. At the ground level, looking up towards the slab, the force flow of the structure is emphasized by the dense black carbon fiber bundles departing from the columns, thinning out towards the perimeter, and seamlessly flowing into the wall and ceiling, where they become denser once again in proximity of the columns.

A custom-designed stair takes the visitor to the first floor. Once there, one is immersed in a diaphanous fibrous space that communicates aspects of extreme lightness and dematerialization. For safety in the sections of the installation that are not enclosed by walls, a dark monolithic volume continues the balustrade of the stairs (Figure 10).

The structure covers a floor area of around 63 square meters, repeated over three levels separated by a clear distance of 2.4 m. Between the supports it achieves a free span of 5.6 m. Calculated for a live load of 4 kN/sqm in accordance with the regulation for public exhibition spaces (BSI 2011), the primary load bearing structure is made from thirty bespoke elements, twenty of which are FRP-timber hybrid with an average weight of 23.7 kg/sqm (9.9 kg/sqm considering the composite portion only), which makes it approximately twenty times lighter than a generic reinforced concrete slab.

OUTLOOK
The project demonstrates how co-design strategies, considering architectural, structural engineering, and fabrication instances, can lead to the development of a novel FRP-timber hybrid building system oriented towards an efficient use of resources. The combination of fibrous composite materials with LVL plates seems also very promising for the expansion of the scope of such hybrid system: from slab-based towards more long-span applications, as well as for the realization of strong yet lightweight modules that can be used to build extensions of the existing building stock. Moreover, the compactness of the fabrication setup

Maison Fibre Dambrosio, Zechmeister, Duque Estrada, Kannenberg, Gil Peréz, Schlopschnat, Rinderspacher, Knippers, Menges

8   LVL plates assembly on the fiber components. In the image it is possible to notice the spoke structure supporting the LVL plates

9   Exterior view of Maison Fibre from the main approaching axis

together with that of the material stocks, open up scenarios where small production units could be deployed directly on site, reducing costs and emissions—deriving from transportation of massive construction elements—and the overall impact of the building. Lastly, the building method investigated in the project shows potential for being used for a wide range of materials. While the project still uses glass and carbon fibers, a possible expansion of the material spectrum is already emerging, ranging from mineral fibers that can withstand extreme temperature stresses, to natural fibers that with an annual growth cycle render construction methods even more resource-efficient.

ACKNOWLEDGMENTS

The authors would like to express their gratitude to the entire scientific development and robotic fabrication team, including students and support staff who helped to complete this project. The research has been partially supported by the Deutsche Forschungsgemeinschaft (DFG, German Research Foundation) under Germany's Excellence Strategy – EXC 2120/1 – 390831618. The research has also been supported by University of Stuttgart, Cluster of Excellence IntCDC, Ministry of Science and Research and the Arts, Baden-Württemberg.

Project Team // ICD – Institute for Computational Design and Construction, University of Stuttgart, Prof. Achim Menges, Niccolo Dambrosio, Katja Rinderspacher, Christoph Zechmeister, Rebeca Duque Estrada, Fabian Kannenberg, Christoph Schlopschnat // ITKE – Institute of Building Structures and Structural Design, University of Stuttgart, Prof. Jan Knippers, Nikolas Früh, Marta Gil Pérez, Riccardo La Magna // FibR GmbH, Stuttgart, Moritz Dörstelmann, Ondrej Kyjanek, Philipp Essers, Philipp Gülke //

With additional support of GETTYLAB, Teijin Carbon Europe GmbH, Elisabetta Cane with Bipaled s.r.l. – Annalisa Pastore, Trimble Solutions Germany GmbH.

REFERENCES

Bader, Michael G. 2002. "Selection of Composite Materials and Manufacturing Routes for Cost-Effective Performance." *Composites Part A: Applied Science and Manufacturing* 33 (7): 913–34. https://doi.org/10.1016/s1359-835x(02)00044-1.

Bakis, C. E., Lawrence C. Bank, V. L. Brown, E. Cosenza, J. F. Davalos, J. J. Lesko, A. Machida, S. H. Rizkalla, and T. C. Triantafillou. 2002. "Fiber-Reinforced Polymer Composites for Construction: State-of-the-Art Review." *Journal of Composites for Construction* 6 (2): 73–87. https://doi.org/10.1061/(asce)1090-0268(2002)6:2(73).

British Standards Institution (BSI). 2011. *Eurocode 1: Actions on Structures*. London: BSI.

Dambrosio, Niccolo, Christoph Zechmeister, Serban Bodea, Valentin Koslowski, Marta Gil Pérez, Bas Rongen, Jan Knippers, and Achim Menges. 2019. "Buga Fibre Pavilion: Towards an Architectural Application of Novel Fiber Composite Building Systems." In *ACADIA 2019: Ubiquity and Autonomy; Proceedings of the 39th Annual Conference of the Association for Computer Aided Design in Architecture*, edited by K. Bieg, D. Briscoe, and C. Odom. Austin, Texas. 140–149.

International Energy Agency (IEA). 2020. "Energy Technology Perspectives 2020." World Energy Outlook Special Report on Sustainable Recovery, IEA, Paris. Accessed June 20, 2021. https://iea.blob.core.windows.net/

10 Interior view of the *Maison Fibre*

assets/7f8aed40-89af-4348-be19-c8a67df0b9ea/Energy_Technology_Perspectives_2020_PDF.pdf.

Fitzer, E. 1985. "Technical Status and Future Prospects of Carbon Fibres and Their Application in Composites with Polymer Matrix (CFRPs)." In *Carbon Fibres and Their Composites*, edited by E. Fitzer. Berlin, Heidelberg: Springer. 3–45. https://doi.org/10.1007/978-3-642-70725-4_1.

Gil Pérez, Marta, Niccolò Dambrosio, Bas Rongen, Achim Menges, and Jan Knippers. 2019. "Structural optimization of coreless filament wound components connection system through orientation of anchor points in the winding frames." In *Proceedings IASS Annual Symposia 2019: FormForce*, edited by C. Lazaro, K.U. Bletzinger, and E. Onate. International Association for Shell and Spatial Structures (IASS). 1381–1388.

Gil Pérez, Marta, Bas Rongen, Valentin Koslowski, and Jan Knippers. 2020. "Structural Design, Optimization and Detailing of the BUGA Fibre Pavilion." *International Journal of Space Structures* 35 (4): 147–59. https://doi.org/10.1177/0956059920961778.

Gil Pérez, Marta, Bas Rongen, Valentin Koslowski, and Jan Knippers. 2021. "Structural design assisted by testing for modular coreless filament-wound composites: The BUGA Fibre Pavilion." *Construction and Building Materials* 301 (2021): 124303. https://doi.org/10.1016/j.conbuildmat.2021.124303.

Knippers, Jan, Riccardo La Magna, Achim Menges, Steffen Reichert, Tobias Schwinn, and Frédéric Waimer. 2015. "ICD/ITKE Research Pavilion 2012: Coreless Filament Winding Based on the Morphological Principles of an Arthropod Exoskeleton."
Architectural Design 85 (5): 48–53. https://doi.org/10.1002/ad.1953.

La Magna, Riccardo, Frédéric Waimer, and Jan Knippers. 2016. "Coreless Winding and Assembled Core: Novel Fabrication Approaches for FRP Based Components in Building Construction." *Construction and Building Materials* 127 (2016): 1009–16. https://doi.org/10.1016/j.conbuildmat.2016.01.015.

Melchert, Luciana. 2007. "The Dutch Sustainable Building Policy: A Model for Developing Countries?" *Building and Environment* 42 (2): 893–901. https://doi.org/10.1016/j.buildenv.2005.10.007.

Menges, Achim, and Jan Knippers. 2015. "Fibrous Tectonics." *Architectural Design* 85 (5): 40–47. https://doi.org/10.1002/ad.1952.

Ortiz, Oscar, Francesc Castells, and Guido Sonnemann. 2009. "Sustainability in the Construction Industry: A Review of Recent Developments Based on LCA." *Construction and Building Materials* 23 (1): 28–39. https://doi.org/10.1016/j.conbuildmat.2007.11.012.

Parascho, Stefana, Jan Knippers, Moritz Dörstelmann, Marshall Prado, and Achim Menges. 2014. "Modular Fibrous Morphologies: Computational Design, Simulation and Fabrication of Differentiated Fibre Composite Building Components." In *Advances in Architectural Geometry 2014*. Cham: Springer. 29–45. https://doi.org/10.1007/978-3-319-11418-7_3.

Prado, Marshall, Moritz Dörstelmann, Achim Menges, James Solly, and Jan Knippers. 2017. "Elytra Filament Pavilion: Robotic Filament Winding for Structural Composite Building Systems." In *Fabricate 2017*, edited by A. Menges, B. Sheil, R. Glynn, and M. Skavara. London: UCL Press. 224–31.

Reichert, Steffen, Tobias Schwinn, Riccardo La Magna, Frédéric Waimer, Jan Knippers, and Achim Menges. 2014. "Fibrous Structures: An Integrative Approach to Design Computation, Simulation and Fabrication for Lightweight, Glass and Carbon Fibre Composite Structures in Architecture Based on Biomimetic Design Principles." *Computer-Aided Design* 52 (July 2014): 27–39. https://doi.org/10.1016/j.cad.2014.02.005.

Schittich, Christian. 2014. "Details Around the Corner." *Architectural Design* 84 (4): 36–43. https://doi.org/10.1002/ad.1779.

Sterrett, Jill, and Roberta Piantavigna. 2018. "Building an Environmentally Sustainable San Francisco Museum of Modern Art." *Studies in Conservation* 63 (sup1): 242–50. https://doi.org/10.1080/00393630.2018.1481324.

Maison Fibre Dambrosio, Zechmeister, Duque Estrada, Kannenberg, Gil Peréz, Schlopschnat, Rinderspacher, Knippers, Menges

United Nations, Department of Economic and Social Affairs, Population Division (DESA). 2019. *World Urbanization Prospects: The 2018 Revision (ST/ESA/SER.A/420)*. New York: United Nations.

van Ruijven, Bas J., Detlef P. van Vuuren, Willem Boskaljon, Maarten L. Neelis, Deger Saygin, and Martin K. Patel. 2016. "Long-Term Model-Based Projections of Energy Use and CO2 Emissions from the Global Steel and Cement Industries." *Resources, Conservation and Recycling* 112 (Sept. 2016): 15–36. https://doi.org/10.1016/j.resconrec.2016.04.016.

Zimmermann, M., H.-J. Althaus, and A. Haas. 2015. "Benchmarks for Sustainable Construction." *Energy and Buildings* 37 (11): 1147–57. https://doi.org/10.1016/j.enbuild.2005.06.017.

Niccolo Dambrosio is a research associate at the Institute for Computational Design and Construction at the University of Stuttgart. He holds a Master in Architecture degree from the Polytechnic University of Bari as well as an advanced degree as a Master of Architecture from the Graduate School of Design at Harvard University. Niccolò's current research focuses on the use of anisotropic fiber composite materials in additive production processes for high performance, lightweight structures.

Christoph Zechmeister is a research associate at the Institute for Computational Design (ICD) at the University of Stuttgart. He holds a Master of Science from Vienna University of Technology as well as a postgraduate Master of Advanced Studies in Architecture and Information from ETH Zürich. Before joining the ICD, Christoph worked as a Junior Architectural Designer at UNStudio, Amsterdam, as well as in multiple offices in and around Zürich, Switzerland. Christoph is currently involved in developing high performance, lightweight structures.

Rebeca Duque Estrada is a research associate and doctoral candidate at the Institute for Computational Design and Construction (ICD) at the University of Stuttgart, where she works with fiber composite materials and robotic fabrication in the development of high-performance lightweight structures.

Fabian Kannenberg is a research associate at the Institute for Computational Design and Construction (ICD) at the University of Stuttgart, where he also received his Master of Science in Architecture from the Integrative Technologies and Architectural Design Research (ITECH) program. His current work at ICD focuses on computational design methods for high-performance fiber composite building systems.

Marta Gil Pérez is a research associate at the ITKE Institute of Building Structures and Structural Design, University of Stuttgart. She obtained a Master in architecture at ETSAM, University Politécnica de Madrid (Spain). Afterwards, she was granted a full scholarship for a Master in Structural Engineering at Seoul National University (South Korea). For three years she joined the firm C.S. Structural Engineering in South Korea participating in the structural design of projects in South Korea and the Middle East. Currently her research focuses on the structural design and building system development of robotically fabricated coreless wound fibre composite structures.

Christoph Schlopschnat is a research associate at the Institute for Computational Design and Construction (ICD) at the University of Stuttgart, Germany, where his research focuses on advanced computational design and novel fabrication technologies of lightweight material systems and its integration in the built environment. He holds a Dipl.-Ing. degree in architecture from University of Innsbruck, Austria and was awarded the Dean's Award of the Faculty of Architecture for his thesis on textile-reinforced, 3d-printed concrete structures in a multi-robotic framework.

Katja Rinderspacher is a research associate at the Institute for Computational Design and Construction (ICD) and the coordinator of the ITECH M.Sc. Program at University of Stuttgart. She holds a M.Arch. degree with honors from Pratt Institute and is a registered architect in Germany. Katja has gained professional experience as an architect and project manager in offices in the U.S., Switzerland, and Germany. Her current research focuses on the design and fabrication with indeterminate fabrication processes for high-resolution surface structures.

Jan Knippers is a structural engineer and since 2000 head of the Institute for Building Structures and Structural Design (ITKE) at the University of Stuttgart. His interest is in innovative and resource-efficient structures created at the intersection of research and development and practice. In 2001, he co-founded Knippers Helbig Advanced Engineering. In 2018, he founded Jan Knippers Ingenieure in Stuttgart to give more personal attention to innovative projects from concept to completion. The current focus of the practise is on the design and construction of novel fibre composite and timber structures.

Achim Menges is a registered architect in Frankfurt and professor at the University of Stuttgart, where he is the founding director of the Institute for Computational Design and Construction (ICD) and the director of the Cluster of Excellence on Integrative Computational Design and Construction for Architecture (IntCDC). In addition, he has been Visiting Professor in Architecture at Harvard University's Graduate School of Design and held multiple other visiting professorships in Europe and the United States. He graduated with honours from the Architectural Association, AA School of Architecture in London.

Designing Matter

Autonomously Shape-changing Granular Materials in Architecture

Denitsa Koleva
Eda Özdemir
Vaia Tsiokou
Karola Dierichs

Institute for Computational Design and Construction, University of Stuttgart

1   The final prototype: (left) aggregated and (right) disaggregated. While the interlocked structure remains stable at room temperature, it can be entirely resolved by actively heating it up with a heat gun.

ABSTRACT

Autonomously shape-changing granular materials are investigated as architectural construction materials. They allow the embedding of different mechanical behaviors in the same material system through the design of their component particles. Granular materials are defined as large numbers of individual elements larger than a micron. Since they are not bound to each other, only contact forces act between them. The design of individual particles affects the behavior of a granular substance composed of such materials. The design process involves the definition of the form and materiality of the particle in relation to the desired function of the granular material. If shape-change materials are deployed in the making of the particles, the granular material can have more than one designed behavior, for example, both liquid and solid phases. Autonomously shape-changing granular materials have seldom been explored in either architecture or granular physics. Thus their exploration is both a relevant and a novel contribution to the field of granular architectures in specific and computational architectural design in general.

This article outlines the field of autonomously shape-changing granular materials and embeds them in the current state. Experimental and simulation methods for the development of shape-changing particles and granular materials are introduced. A case study on the development and testing of autonomously shape-changing particles made from a bimetal is also presented. Further research is outlined with respect to the practical, methodological, and conceptual development of an autonomously shape-changing designed granular material.

INTRODUCTION

Autonomously shape-changing granular materials are designed for architectural construction. Apart from being entirely reversible and reconfigurable, they allow the integration of more than one mechanical behavior in a single material system, such as the ability to be pourable during construction and to interlock once a structure is set in place. Therefore, they are highly pertinent as a sustainable approach to architectural construction (Figure 1).

Granular materials consist of extremely large numbers of elements, that is, particles larger than 1 micron. There are no permanent bonds between particles, and they interact only through contact forces (Law and Rennie 2015; Andreotti et al. 2013; Duran 2000; de Gennes 1999, 1998; Jaeger et al. 1996a, 1996b). Sand, gravel, or snow are examples of such material systems (Law and Rennie 2015; Andreotti et al. 2013; de Gennes 1999, 1998; Jaeger et al. 1996a, 1996b).

Designing matter or materials denotes the interrelation of form and function in a material across all length of scales (Reis et al. 2015). The area of designed materials is explored across several disciplines such as architecture, materials science, and engineering, and one of the key claims is the abolition of machines in favor of autonomous matter (Reis et al. 2015; Fratzl et al. 2013).

Designed granular materials are examples of such substances, of which the form and materiality of a particle or group of particles is defined to achieve the desired behavior (Dierichs and Menges 2016, 2017; Miskin 2016; Keller and Jaeger 2016; Jaeger 2015; Reis et al. 2015; Athanassiadis et al. 2014; Miskin and Jaeger 2013, 2014; Hensel et al. 2010; Hensel and Menges 2006a, 2006b, 2006c, 2008a, 2008b). If a shape-change material is used in particle fabrication, more than one behavior can be achieved in a granular material consisting of such particles (Dierichs et al. 2017; Dierichs and Menges 2016; Keller and Jaeger 2016; Jaeger 2015). Autonomy is understood as the absence of external—mostly electrically driven—machines for the system assembly and disassembly. This autonomy in assembly and disassembly is driven by the shape-change of the individual particles. This is embedded in the larger field of research on autonomous construction systems, particularly those based on large numbers of small and frequently simple robots or units (Petersen et al. 2019; Tibbits 2017; Andreen et al. 2016).

Granular materials have properties distinct from solid, liquid, or gaseous states (Jaeger et al. 1996a, 1996b). The transition between such states is entirely reversible. Owing to these characteristics, they are both recyclable and reconfigurable (Aejmelaeus-Lindström et al. 2016; Keller and Jaeger 2016; Dierichs and Menges 2012, 2016; Hensel et al. 2010; Hensel and Menges 2008a, 2008b). Custom-designed particles in a granular material widen the spectrum of possible characteristics and allows the development of granular materials specifically suitable for architectural construction (Keller and Jaeger 2016; Dierichs and Menges 2015, 2016; Jaeger 2015; Hensel et al. 2010; Hensel and Menges 2006a, 2006b, 2006c).

CURRENT STATE

The current-state review of the designed granular materials includes research conducted for both the architecture and granular physics. Both non-variable and variable—also called shape-changing—particle geometries, are considered. While variable particle geometries directly relate to the research presented here, non-variable ones give relevant information on possible behaviors that can be obtained by a specific particle shape and that can then be considered in the design of a shape-change. This review evaluates the initial versions of shape-changing particles and possible non-variable particle geometries that can be integrated into variable geometries. The following review was previously presented in greater depth (Dierichs 2020). Initial projects on designed granular materials were conducted by Kentaro Tsubaki at Cranbrook Academy of Art, Eiichi Matsuda at Diploma Unit 4 at the Architectural Association School of Architecture (AA) as well as Anne Hawkins and Catie Newell in a GPA studio project at Rice University (Hensel et al. 2010; Hawkins and Newell 2008; Matsuda 2008; Tsubaki 2008, 2009, 2012; Hensel and Menges 2006b, 2006d, 2008a). Tsubaki's approach was further pursued by Richard Peterson at Tulane University (Peterson 2014); and the approach by Matsuda was also taken on in the AA's Emergent Technologies and Design program by Kyle Schertzing and Selim Bayer (Hensel et al. 2010). Gramazio Kohler Research at the Swiss Federal Institute of Technology (ETH) Zurich and the Self-Assembly Lab at the Massachusetts Institute of Technology (MIT) developed research on granular materials, more specifically on "Jammed Architectural Structures." Granular materials from industrially produced particles were also investigated in the early stages of this project (Aejmelaeus-Lindström et al. 2017a; Aejmelaeus-Lindström et al. 2017b; Aejmelaeus-Lindstrom et al. 2016; Fauconneau et al. 2016). Another project by Gramazio Kohler Research at the Swiss Federal Institute of Technology (ETH) Zurich used custom-made particles to investigate structures made of projectiles (Dörfler 2018; Dörfler et al. 2014; Piskorec 2014; Gramazio and Kohler 2014a, 2014b). At the Institute for Computational Design and Construction, the research field of "Granular Architectures" was established, and a wide range of particle shapes were tested, ranging from convex over non-convex to double non-convex—hook-shaped—geometries (Dierichs 2020; Dierichs and Menges 2016). In this study, some initial shape-changing particles for architectural applications were developed (Dierichs et al. 2017; Dierichs and Menges 2016, 2015).

The following review of the designed granular materials used in physics that are relevant for architecture is based on previous reviews by Heinrich M. Jaeger and collaborators, Scott V. Franklin and collaborators and Robert P. Behringer and collaborators (Keller and Jaeger 2016; Murphy et al. 2016; Zhao et al. 2016; Jaeger 2015; Athanassiadis et al. 2014; Franklin 2014; Gravish et al. 2012). A detailed version of this review is provided in the doctoral thesis by the last author (Dierichs 2020).

(i) Convex geometries range from spherical to prolate shapes. Albert P. Philipse of the Van't Hoff Laboratory for Physical and Colloid Chemistry at Utrecht University investigated granular materials which are composed of rod-like particles (Philipse 1996). Joshua Blouwolff and Seth Fraden from the Complex Fluids Group at Brandeis University investigated the coordination number of granular cylinders (Blouwolff and Fraden 2006). Alan Wouterse, Stefan Luding and Albert P. Philipse conducted research on granular materials composed of rods at Van 't Hoff Laboratory for Physical and Colloid Chemistry at Utrecht University and at Multi Scale Mechanics at the Universiteit Twente (Wouterse et al. 2009). Melissa Trepanier and Scott V. Franklin of the Rochester Institute of Technology explored cylinders composed of granular materials consisting of rods (Trepanier and Franklin 2010). The Jaeger Lab at the University of Chicago and the Sibley School of Mechanical and Aerospace Engineering at Cornell University explored the impact of particle shapes—also including convex geometries—on the behavior of the respective granular materials under confining pressures (Athanassiadis et al. 2014; Murphy et al. 2019; Murphy and Jaeger 2018).

(ii) Non-convex geometries have tended to be explored in the form of particles with arm extensions. In this group of particle shapes, a collaboration between the MoSCoS School of Mathematics and Physics at the University of Queensland, CSIRO Exploration and Mining, the Golder Geomechanics Center at the University of Queensland, and the Grupo de Simulación de Sistemas Físicos at the Universidad Nacional de Colombia aimed to establish a molecular dynamics (MD) model that allows the characterization of granular materials consisting of such particles (Galindo-Torres et al. 2009). Another collaborative project of LCD, SP2MI, and UPMC explored the transport properties of granular materials consisting of "spiky" particles (Malinouskaya et al. 2009). Non-convex geometries have also been explored in the aforementioned study by the Jaeger Lab at the University of Chicago and the Sibley School of Mechanical and Aerospace Engineering at Cornell University (Athanassiadis et al. 2014; Murphy et al. 2019; Murphy and Jaeger 2018). A research group composed of the Department of Mechanics Engineering at the South China University of Technology and of the Department of Mechanics and Aerospace Engineering at Peking University investigated "maximally dense random packing (MDRP)" of "intersecting spherocylinders" (Meng et al. 2017; Meng et al. 2016). In another cooperation with the Electric Power Research Institute of Guangdong Power Grid Cooperation, the same team of South China University of Technology focused on particle shapes fabricated by deforming or assembling rods (Meng et al. 2018). A collaboration between the Laboratory of Energy Science and Engineering, the Department of Mechanical and Process Engineering and the Institute of Energy and Process Engineering at the ETH Zurich and the Swiss Federal Laboratories for Materials Science and Technology (Empa) investigated the correlation between the morphology of 2D and 3D packings of non-convex particles and the contact forces in such granular materials (Conzelmann et al. 2020).

(iii) Double non-convex geometries include hook-shaped particles. Among other geometries, they were explored in the form of "u-shaped" particles in a collaboration of the School of Physics and the School of Mechanical Engineering at Georgia Institute of Technology and the Department of Physics at Rochester Institute of Technology (Gravish et al. 2012). Scott V. Franklin in the School of Physics at the Rochester Institute of Technology continued this research into such particles (Franklin 2014). At the Jaeger Lab at the University of Chicago, Kieran A. Murphy and collaborators investigated granular materials consisting of a group of particle shapes which were termed "Z," "U," and "Z90" based on the rotation of two arm extensions in a backbone (Murphy et al. 2017a; Murphy et al. 2017b; Murphy et al. 2016).

Another approach to the design of granular materials is the "inverse" modeling of particle geometries using evolutionary algorithms (EAs) (Miskin 2016; Miskin and Jaeger 2013, 2014).

Shape-changing particles appear to be comparatively seldom explored. A group from the Martin Fisher School of Physics, Brandeis University observed the collective motion of molecular motors in microtubule filaments (Sanchez

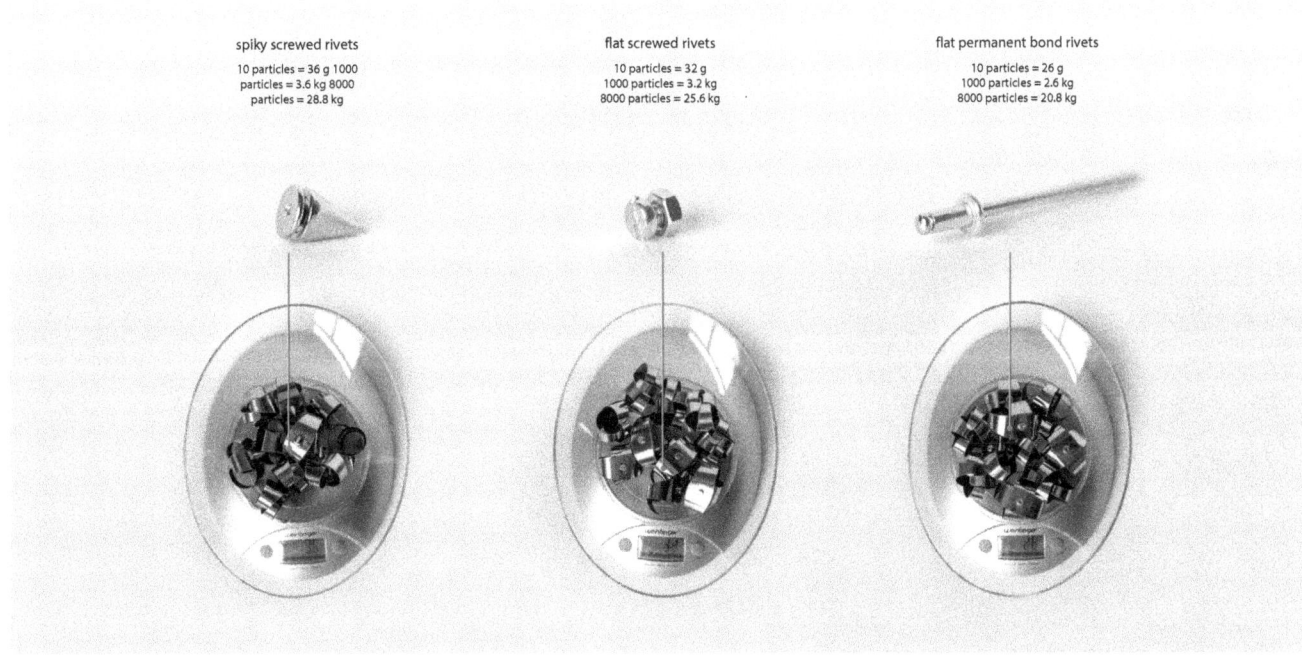

2   Weight comparison for different rivets. Three different types of rivets were considered: two screwed rivet types—spiky and flat—and flat permanent bond rivets. Flat permanent bond rivets had the lowest weight and were chosen to avoid increasing the loading through self-weight of the structure.

et al. 2012). The departments of Biomaterials and Colloid Chemistry at the Max Planck Institute of Colloids and Interfaces, the Helmholtz-Zentrum Berlin for Soft Matter and Functional Materials, the Department of Physical Chemistry at the University of Chemistry and Technology in Prague and the Institute of Physics at Humboldt-Universität zu Berlin explored the entanglement of "porous poly(ionic liquid) (PIL) membrane actuators" (Zhao et al. 2015). A team from the Institute for Multiscale Simulations (MSS) at the Friedrich-Alexander-Universtität Erlangen-Nürnberg explored the use of "Vibrots" as a granular material, translating linear motion into rotational motion, and helping simulate collective motion patterns (Scholz et al. 2021; Scholz et al. 2018; Scholz et al. 2016).

A wide gamut of particle shapes for designed granular materials—used in the architecture and granular physics fields—have been explored, ranging from convex to non-convex and double non-convex. These shapes are mostly non-variable in geometry, and variable, that is shape-changing, geometries have been relatively little explored with examples that investigate interlocking and motion though a change in shape.

The novel contribution of this article is an in-depth investigation into the possible geometries of autonomously shape-changing particles used in architecture. In addition, bimetal is introduced as a suitable material for the fabrication of particles because it has a range of actuation beyond regular temperature fluctuations and high mechanical properties, as well as material homogeneity for reproducible results. Furthermore, a full-scale architectural demonstrator is presented to validate the research.

METHODS

The methods applied include a simulation of the individual particles, as well as experiments with individual particles and many particles forming a granular material. The following section provides an overview of the respective layouts of these simulations and experiments.

Simulation

The geometries of individual particles were developed using a parametric modeling interface. The particles were designed in a non-actuated closed state. Their behavior during actuation was simulated using a dynamic relaxation in a parametric modeling environment. To approximate the geometry of the actuated open particle, an equal force is applied on its mesh in the direction of the low-expansion bimetal side. To verify the accuracy of the simulations, the results were compared with the behavior of the actuated particles during heating and cooling.

Experimental Methods
Particles

The particles were digitally designed and fabricated using a waterjet cutter. Their behavior during actuation were simulated, and the geometries were optimized to maximize the achieved shape-change. The particles are made of bimetal and assembled using various types of rivets in a series of

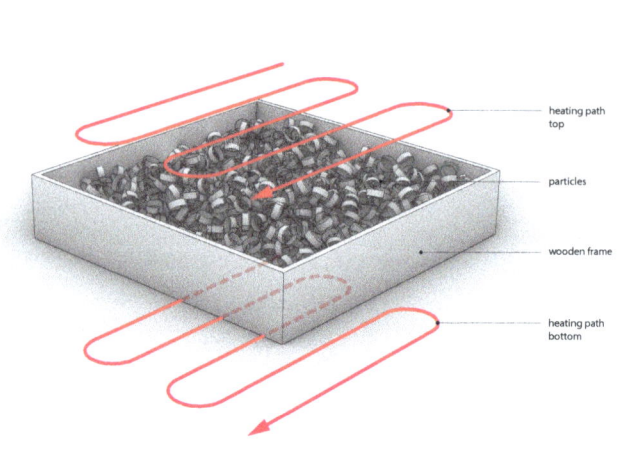
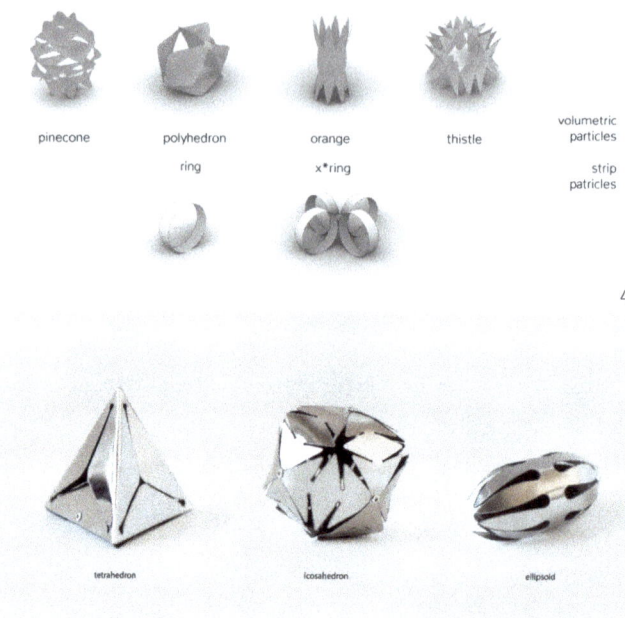

different experiments. Rivets were used in the final demonstrator to minimize the weight of the structure (Figure 2). The particle sets were heated simultaneously from the top and bottom through a grate and cooled.

Particle Mixtures
Setup
A 50 cm × 50 cm × 10 cm hollow wooden frame was built and fixed on a perforated metal sheet. This setup was anchored between two tables to leave the bottom and top of the frame free of solid surfaces during the heating process. The camera was placed on top of the wooden frame aligned to its center, and the camera lens was parallel to the frame to minimize distortion (Figure 3).

Pattern
Three patterns were developed involving single (S), assembled (A), and mixed (M) particles—called pattern S, pattern A, and pattern M, as follows. Pattern M comprised single and assembled particles at a ratio 1 to 1. For each pattern, the test was conducted over 12 iterations. The same particles were repeatedly used to evaluate the life cycle of the material empirically.

Test Protocol
Both sides were simultaneously heated from the top and bottom. Therefore, all the particles open simultaneously and interlock. To form a solid surface, it is crucial that the particles are heated equally.

For each test, particles were left to equalize at 24.5 °C for 10 minutes. They were then brought to the testing room, manually poured into a wooden frame and evenly distributed on the surface. A double-sided heating strategy was developed to ensure that all particles were heated at the same time and opened up as much as possible to achieve maximum interlocking. The heating process was executed with two people moving the heat guns simultaneously, corresponding to the same point: one at the bottom of the frame, the other on top. Starting from the bottom-left corner of the frame, the heating was carried out in five rows over a course of 5 minutes. Each row took 1 minute, and the rows were completed in opposite alternating directions for continuous heating. Once the heating process was finished, the interlocking was recorded for 10 minutes, and the aggregation was removed from the frame for evaluation.

RESULTS
The results are presented with respect to the particle design, particle fabrication, particle mixtures, and final demonstrator. To conclude, these results are discussed with respect to the key parameters that affect the design and construction processes they allow..

Particle Design
The design parameters for a single particle are closely related to the behavior of the particle system. The system has two states: liquid and solid. Therefore, the particle also has two states: open and closed.

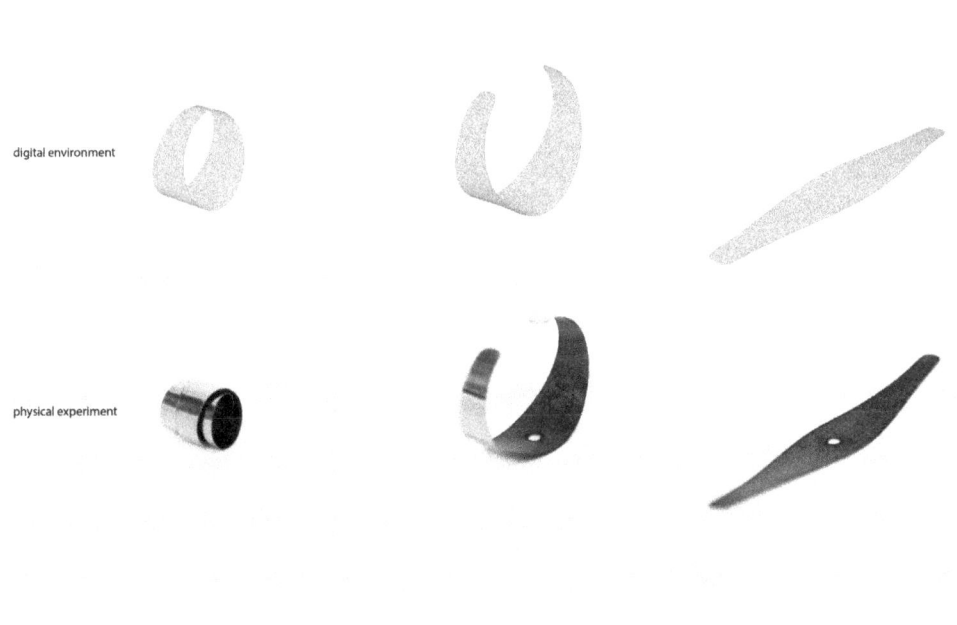

3   Experimental setup of particle mix testing. The particles are poured on top of a perforated metal sheet and then heated simultaneously from top and bottom in sinusoidal curves.

4   Overview of particle geometries. All particle geometries were simulated and tested experimentally using bimetals cut with a water jet.

5   Volumetric particles, the geometries of which are based on a tetrahedron, an icosahedron, and an ellipsoid. These basic geometries are used to design shape-changing particles with parts that unfold to interlock with the neighboring particles.

6   Simulation and image of a single strip particle. A single strip particle and an assembled strip particle are simulated and tested experimentally. The experiments in this case serve to validate the simulation.

The closed position is the "normal" state. The open position is the "transitional" state. In both liquid and solid system states, the particles are closed. They open for a short period of time, actuated by a temperature change in the environment, and allow the system to transition from one state to another. In the liquid state, the particles have not yet been actuated. After heating for the first time inside a frame or mold, they interlock and form a solid structure. To disassemble the structure and turn it back to the liquid state, the particles need to be reheated without using a mold. Reheating them inside the mold can strengthen and weaken the structure, depending on the heating technique.

In the following sections, both the material and different geometric approaches to a particle design are outlined.

Material
An important criterion for choosing the metal alloy—with MnCu18Ni10 as the high-expansion side and FeNi36 as the low-expansion side—was its operating temperature of maximum 350°C. This allows for controlled actuation, where the material is not affected by sun exposure or small fluctuations in the room temperature. Although the material thickness of 0.115 mm is one of the biggest limitations for the size of the particle, the material has to remain as thin as possible to allow the desired shape-change. Another limitation of the size of the particle is the width of the 195 mm coil.

Particle Geometry
The following experiments were carried out to determine the geometry and size of a single particle (Figure 4).

Volumetric particles: Three volumetric particle geometries were designed for an experiment on the initial particle behavior: a tetrahedron, an icosahedron, and an ellipsoid (Figure 5). Each particle had a volumetric shape with cuts on its sides to allow it to open and close. To ease and speed up the assembly process, their geometries were made out of as few flat components as possible and were joined by aluminum rivets.

The length and angle of the cuts are the two most important parameters for manipulating the behavior of the particles. The longer and thinner the cuts, the more the particles opened, and the better they interlocked with their neighbors.

A second set of tests was carried out to determine whether a set of 10 ellipsoid particles could form an aggregation. The particles were poured into a glass cylinder and heated for 3 minutes. After cooling for 5 minutes, the cylinder was removed. This cycle was repeated 10 times. In all repetitions, the particles failed to form a single aggregation and instead formed clusters. Particles in the middle of the aggregations were often crushed during heating because they did not have sufficient space to expand inside the cylinder.

Strip particles: Strip particles were developed for another test series. Their geometry allows them to change their

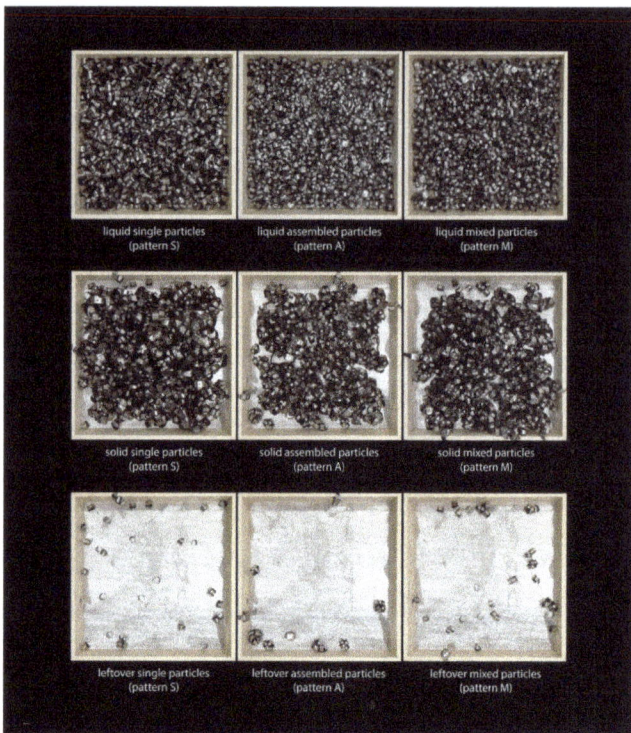

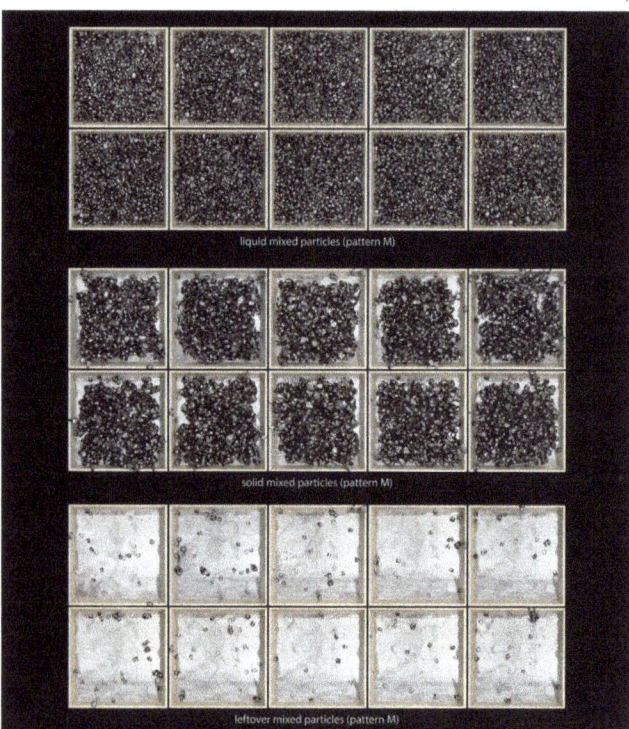

7 Aggregation sequence of all three patterns showing a heating sequence. The top row shows the loose particles after pouring into the wood-frame. The middle row shows the aggregated and cooled down particles. The bottom row shows the particles remaining in the formwork after the aggregated particles have been lifted out.

8 All ten probes of the mixed pattern (pattern M). Although all probes interlocked, the individual probes display a high degree of difference owing to the fact that granular materials are chance-based systems where contacts between particles are randomly formed.

shape from a convex three-dimensional ring—the closed state—to a non-convex strip (Figure 6). The length and width of the strips were the two most important parameters for achieving a good interlocking behavior. In a set of 10 repetitions, 10 particles with a flat strip length of 100 mm × 10 mm were poured and heated for 3 minutes in a glass container. After cooling for 5 minutes, the container was removed. In all test repetitions, the particles formed tight knots. To disassemble them, the knots were reheated for 1 minute.

Strip particle size study: Assembled strip particles were developed for another experiment. Its goal was to determine the appropriate particle size based on the material thickness. Each particle was made out of two single strip particles that were perpendicularly attached to each other at their midpoint using a rivet. They changed shape from a convex three-dimensional ball—a closed state, into a non-convex X-shape—an open state. To speed up the fabrication and assembly process and reduce the cost, the particles had to be as large as possible. Four sets of different sizes were tested: 20, 30, 40, and 50 mm in diameter. Each consisted of five particles. The larger the particle, the less stable it was. Particles from the 40 and 50 mm sets were too weak to form a solid aggregation. The 20 and 30 mm specimens exhibited promising interlocking behavior and formed well-interconnected aggregations. Because the 30 mm particles were the largest to show a promising interlocking behavior, they were chosen for further experiments.

Volumetric versus strip particles: Although volumetric particles interact with their neighbors, the connections they form are usually too weak. This interlocking problem can be solved by adding thinner cuts to their surfaces. However, this would slow down the fabrication and assembly processes and increase the cost.

Single and assembled strip particles exhibited much greater potential for forming stable, well-interconnected aggregations on a large scale. They were also faster to cut and assemble than volumetric particles.

Particle Fabrication

The single and assembled strip particles were built in two and three steps, respectively. The strips used for both types were identical and cut using a waterjet cutter. Single particles consisted of a single strip rolled into the form of a ring. Assembled rings were made of two single rings that were attached perpendicularly to each other at their midpoint using a rivet. The strips were rolled without using any heat to ensure that they were always in a closed—convex—state when cold, and were opened up to their strip—non-convex—state when heated. The assembly of single and assembled particles take 15 s and 50 s, respectively.

Particle Mixtures

Single (pattern S): The average weights of the single and assembled particles were calculated as 1.177 and 2.605 g/particle, respectively. The weights of both were approximated by measuring 50 particles of each genre on a precision scale. Before activation—that means in a liquid state—all patterns formed a loose surface of approximately 3 cm in thickness within the 50 cm × 50 cm wooden frame with a volume of 7500 cm³. Based on this stipulation, 1274 single particles of 1500 g in total, 384 assembled particles of 1000 g in total, and 600 mixed particles of 1135 g in total were used for patterns S, A, and M, respectively.

Particle aggregation expanded during heating and contracted during cooling. The large number of single particles resulted in denser surfaces than those formed by the assembled and mixed particles. Owing to the density, the particles were not evenly connected throughout the surface. Pockets of loose particles or small clumps were formed within the surface. After removing the aggregation from the frame, particles that failed to connect to the structure were found at the bottom (Figure 7).

Assembled (pattern A): Assembled particles formed surfaces that were less dense than those formed by the single particles. Although they did not create pockets, they formed clusters of particles and areas that were disconnected from one another. Therefore, their aggregations were weaker and easier to break off than the aggregations of single particles. In some of the experiments, the clusters were completely detached from one another and failed to form a single surface (Figure 7).

Mixed (pattern M): Mixed particles were most likely to form a stable, well-interconnected surface. Their aggregations were not as dense as the aggregations formed by the single particles and had fewer particle pockets. The assembled particles in the mixture allowed for a lower density, whereas the single particles helped to form better connections (Figures 7 and 8).

FINAL DEMONSTRATOR

The final prototype with a length of 3.20 m, a thickness of 0.06–0.15 m, and a width of 1.0 m was built using a mixture of approximately 5000 single and 5000 assembled particles. Eight separate components were assembled. Each was heated from the top and bottom simultaneously using a

9 Full view (A) and details (B) of final demonstrator. The final prototype was suspended from steel cables. Compared to previous architectural structures made from designed particles which can display a wall thickness of up to 0.30–1.00 m, it is relatively thin with a thickness of between 0.06 and 0.15 m.

wooden frame as a mold with dimensions of 50 cm × 50 cm × 10 cm. The interlocked particles were placed next to each other to form the final design. The spaces between them are filled with loose particles. The same method of heating from the top and bottom was used to reheat the edges of each surface and loose particles in between. After cooling, the surfaces formed a larger aggregation that was lifted up to a height of 2 m using a cable system. To disassemble it, the structure was heated from the bottom, allowing the particles to detach separately (Figure 9).

SUMMARY AND DISCUSSION

The main parameters influencing the design and behavior of autonomously shape-changing particles made out of bimetals were the material thickness, size of the coil, and width and length of the particle parts that were free to move. A mixture of assembled and strip particles is the most suitable for creating a stable, well-connected surface. The assembled particles created a volume, and the single strip particles made strong connections. Another crucial parameter is the initial thickness of the liquid particles before actuation. To assemble or disassemble a surface, all particles are heated simultaneously. Heat can also be strategically used in the assembly process to actuate parts of the structure that need to be changed, as well as to attach or detach the components.

Autonomously shape-changing granular materials allow the builder to make both local and global changes to the system. The aggregations they form can be split into separate components—particle clusters—and joined back together through local changes. Global changes switch the state of the entire system from solid to liquid. Therefore, global actuation is used to either assemble or disassemble the entire particle system. Although other granular materials allow the construction of different aggregations using the same set of particles, it is often impossible to partially disassemble them and change the structure. Splitting them into components that can later be rejoined is also extremely difficult. The shape-changing characteristic of a material enables the building of structures that are temporary, completely reversible, and easily changeable.

CONCLUSION

Future research into autonomously shape-changing granular materials needs to address developments at the practical, methodological, and conceptual levels.

Practical developments will focus on the development of tools that will ease the particle and system fabrication and assembly processes. A robot arm can be used to distribute the particles and create areas with different densities and thicknesses throughout the system. This allows for better control of the properties of the structure. Although the robot can precisely distribute particles in the mold, it is impossible—and also not necessary—to control the exact position of each particle during actuation. Therefore, the border between different areas in the system is a thick strip of mixed particles.

Methodological developments will focus on creating digital tools to accurately simulate not only the behavior of a single particle, but also the entire particle system while being poured, actuated, and cooled, helping create a library of possible particle geometries corresponding to different material behaviors. It will also facilitate the process of finding the most suitable particle mix for the desired system behavior.

Conceptual developments will investigate the notion of an autonomous material and its relevance to the construction industry.

ACKNOWLEDGEMENTS

KD acknowledges the Terra Incognita project funding by the University of Stuttgart / Deutsche Forschungsgemeinschaft (DFG, German Research Foundation) and the support by the Deutsche Forschungsgemeinschaft (DFG, German Research Foundation) under Germany's Excellence Strategy – EXC 2120/1 – 390831618. The authors are very grateful to Professor Heinrich M. Jaeger, Ph.D., of the Jaeger Lab, University of Chicago and Professor Dr. Thorsten Pöschel of the Institute for Multiscale Simulation, Friedrich Alexander Universität Erlangen Nürnberg for their scientific advice on the project. Manufacturing was conducted by Auerhammer Metallwerk GmbH and Roy Hohlfeld Wasserstrahlschneiden. The authors would also like to thank Editage (www.editage.com) for English language editing.

REFERENCES

Aejmelaeus-Lindström, P., A. Mirjan, F. Gramazio, M. Kohler, S. Kernizan, B. Sparrman, J. Laucks, and S. Tibbits. 2017a. "Granular Jamming of Loadbearing and Reversible Structures: Rock Print and Rock Wall." *Architectural Design* 87 (4): 82–87. https://doi.org/10.1002/ad.2199.

Aejmelaeus-Lindström, P., A. Thoma, A. Mirjan, V. Helm, S. Tibbits, F. Gramazio, and M. Kohler. 2017b. "Rock Print: An Architectural Installation of Granular Matter." In *Active Matter*, edited by S. Tibbits. Cambridge, Massachusetts: The MIT Press. 291–300. https://doi.org/10.7551/mitpress/11236.003.0045.

Aejmelaeus-Lindström, P., J. Willmann, S. Tibbits, F. Gramazio, and M. Kohler. 2016. "Jammed Architectural Structures: Towards Large-Scale Reversible Construction." *Granular Matter* 18 (2): 28. https://doi.org/10.1007/s10035-016-0628-y.

Andreen, D., P. Jenning, N. Napp, and K. Petersen. 2016. "Emergent Structures Assembled by Large Swarms of Simple Robots." In *ACADIA '16: Posthuman Frontiers: Data, Designers, and Cognitive Machines; Proceedings of the 36th Annual Conference of the Association for Computer Aided Design in Architecture*, edited by K. Velikov, S. Manninger, M. del Campo, S. Ahlquist, and G. Thün.

Andreotti, B., Y. Forterre, and O. Pouliquen. 2013. *Granular Media: Between Fluid and Solid*. Cambridge University Press.

Athanassiadis, A.G., M.Z. Miskin, P. Kaplan, N. Rodenberg, S.H. Lee, J. Merritt et al. 2014. "Particle Shape Effects on the Stress Response of Granular Packings." *Soft Matter* 10 (1): 48–59. https://doi.org/10.1039/C3SM52047A.

Blouwolff, J., and S. Fraden. 2006. "The Coordination Number of Granular Cylinders." *Europhysics Letters* 76 (6): 1095–1101. https://doi.org/10.1209/epl/i2006-10376-1.

Conzelmann, N.A., A. Penn, M.N. Partl, Frank J. Clemens, L.D. Poulikakos, and C.R. Müller. 2020. "Link between packing morphology and the distribution of contact forces and stresses in packings of highly nonconvex particles." *Phys. Rev. E* 102 (6): 62902. https://doi.org/10.1103/PhysRevE.102.062902.

de Gennes, P.-G. 1998. "Reflections on the mechanics of granular matter." *Physica A: Statistical Mechanics and its Applications* 261 (3): 267–293. https://doi.org/10.1016/S0378-4371(98)00438-5.

de Gennes, P.-G. 1999. "Granular matter: A tentative view." *Review of Modern Physics* 71 (2): S374-S382. https://doi.org/10.1103/RevModPhys.71.S374.

Dierichs, K. 2020. "Granular Architectures: Granular Materials as 'Designer Matter' in Architecture." Dr.-Ing. thesis, Research Reports Institute for Computational Design and Construction 2, University of Stuttgart.

Dierichs, K., and A. Menges. 2012. "Aggregate Structures: Material and Machine Computation of Designed Granular Substances." *Architectural Design* 82 (2): 74–81. https://doi.org/10.1002/ad.1382.

Dierichs, K., and A. Menges. 2015. "Granular Morphologies: Programming Material Behaviour with Designed Aggregates." *Architectural Design* 85 (5): 86–91. https://doi.org/10.1002/ad.1959.

Dierichs, K., and A. Menges. 2016. "Towards an Aggregate Architecture: Designed Granular Systems as Programmable Matter in Architecture." *Granular Matter* 18 (2): 25. https://doi.org/10.1007/s10035-016-0631-3.

Dierichs, K., and A. Menges. 2017. "Granular Construction: Designed Particles for Macro-Scale Architectural Structures." *Architectural Design* 87 (4): 88–93. https://doi.org/10.1002/ad.2200.

Dierichs, K., D. Wood, D. Correa, and A. Menges. 2017. "Smart Granular Materials: Prototypes for Hygroscopically Actuated Shape-Changing Particles." In *ACADIA '17: Disciplines & Disruption; Proceedings of the 37th Annual Conference of the Association for Computer Aided Design in Architecture. Association for Computer Aided Design in Architecture*, edited by T. Nagakura, S. Tibbits, C. Mueller, and M. Ibanez. MIT School of Architecture and Planning, Cambridge, Massachusetts. 222–231. http://papers.cumincad.org/cgi-bin/works/paper/acadia17_222.

Dörfler, K. 2018. "Strategies for Robotic in Situ Fabrication." Dr. sc. thesis, ETH Zurich. https://doi.org/10.3929/ethz-b-000328683.

Dörfler, K., S. Ernst, L. Piskorec, J. Willmann, V. Helm, F. Gramazio, and M. Kohler. 2014. "Remote material deposition: Exploration of reciprocal digital and material computational capacities." In *What's the Matter: Materiality and Materialism at the Age of Computation; International Conference, COAC, ETSAB, ETSAV*, edited by M. Voyatzaki. Barcelona, Spain: European Network of Heads of Schools of Architecture (ENHSA). 361–377.

Duran, J. 2000. *Sands, Powders, and Grains: An Introduction to the Physics of Granular Materials*. 1st ed. Partially Ordered Systems. New York: Springer-Verlag.

Fauconneau, M., F.K. Wittel, and H.J. Herrmann. 2016. "Continuous Wire Reinforcement for Jammed Granular Architecture." *Granular Matter* 18 (2): 27. https://doi.org/10.1007/s10035-016-0630-4.

Franklin, S.V. 2014. "Extensional Rheology of Entangled Granular Materials." *Europhysics Letters* 106 (5): 58004. https://doi.org/10.1209/0295-5075/106/58004.

Fratzl, P., J. Dunlop, and R. Weinkamer, eds. 2013. *Materials Design Inspired by Nature: Function Through Inner Architecture*. Cambridge, UK: Royal Society of Chemistry Publishing. https://doi.org/10.1039/9781849737555.

Galindo-Torres, S.A., F. Alonso-Marroquín, Y. Wang, D. Pedroso, and J.D. Munoz Castano. 2009. "Molecular Dynamics Simulation of Complex Particles in Three Dimensions and the Study of Friction Due to Nonconvexity." *Physical Review E* 79 (6): 60301. https://doi.org/10.1103/PhysRevE.79.060301.

Gramazio, F., and M. Kohler. 2014a. "Remote Material Deposition." Accessed December 29, 2015. http://gramaziokohler.arch.ethz.ch/web/e/lehre/277.html.

Gramazio, F., and M. Kohler. 2014b. "Remote Material Deposition Installation." Accessed December 29, 2015. http://gramaziokohler.arch.ethz.ch/web/e/lehre/276.html.

Gravish, N., S.V. Franklin, D.L. Hu, and D.I. Goldman. 2012. "Entangled Granular Media." *Physical Review Letters* 108 (20), 208001. https://doi.org/10.1103/PhysRevLett.108.208001.

Hawkins, A., and C. Newell. 2008. "Aggregat gefertigter Partikel 02." *Archplus* 41 (188): 82–85.

Hensel, M., and A. Menges, eds. 2006a. "Anne Hawkins and Catie Newell, Aggregates 02, 2004." Project Description. In *Morpho-Ecologies: Towards a Discourse of Heterogeneous Space in Architecture*. London: AA Publications. 274–283.

Hensel, M., and A. Menges, eds. 2006b. "Eiichi Matsuda, Aggregates 01, 2003-2004." Project description. In *Morpho-Ecologies: Towards a Discourse of Heterogeneous Space in Architecture*. London: AA Publications. 262–271.

Hensel, M., and A. Menges. 2006c. "Material Systems; Introduction, 20 Proto-Architectures, Research and Design Projects." In *Morpho-Ecologies: Towards a Discourse of Heterogeneous Space in Architecture*, edited by M. Hensel and A. Menges. London: AA Publications. 62–67.

Hensel, M., and A. Menges. 2006d. "Morpho-Ecologies: Towards an Inclusive Discourse on Heterogeneous Architectures." In *Morpho-Ecologies: Towards a Discourse of Heterogeneous Space in Architecture*, edited by M. Hensel and A. Menges. London: AA Publications. 16–61.

Hensel, M., and A. Menges. 2008a. "Aggregates." *Architectural Design* 78 (2): 80–87. https://doi.org/10.1002/ad.645.

Hensel, M., and A. Menges. 2008b. "Materialsysteme 05: Aggregate." *Archplus* 41 (188): 76–77.

Hensel, M., A. Menges, and M. Weinstock. 2010. "Aggregates." In *Emergent Technologies and Design: Towards a Biological Paradigm for Architecture*, edited by M. Hensel, A. Menges, and M. Weinstock. Abingdon, New York: Routledge. 227–241.

Jaeger, H.M. 2015. "Celebrating Soft Matter's 10th Anniversary: Toward Jamming by Design." *Soft Matter* 11 (1): 12–27. https://doi.org/10.1039/C4SM01923G.

Jaeger, H.M., S.R. Nagel, and R.P. Behringer. 1996a. "Granular Solids, Liquids, and Gases." *Reviews of Modern Physics* 68 (4): 1259–1273. https://doi.org/10.1103/RevModPhys.68.1259.

Jaeger, H.M., S.R. Nagel, and R.P. Behringer. 1996b. "The Physics of Granular Materials." *Physics Today* 49 (4): 32–38. https://doi.org/10.1063/1.881494.

Keller, S., and H.M. Jaeger. 2016. "Aleatory Architectures." *Granular Matter* 18 (2): 29. https://doi.org/10.1007/s10035-016-0629-x.

Law, J., and R. Rennie. 2015. "Granular Material." In *A Dictionary of Physics*, 7th ed. e-book, edited by J. Law and R. Rennie. Oxford University Press. https://doi.org/10.1093/acref/9780198714743.001.0001.

Matsuda, E. 2008. "Aggregat Gefertigter Partikel 01." *Archplus* 41 (188): 80–81.

Malinouskaya, I., V.V. Mourzenko, J.-F. Thovert and P.M. Adler. 2009. "Random packings of spiky particles: geometry and transport properties." *Phys. Rev. E* 80 (1): 11304. https://doi.org/10.1103/PhysRevE.80.011304.

Meng, L., Y. Jiao, and S. Li. 2016. "Maximally Dense Random Packings of Spherocylinders." *Powder Technology* 292: 176–185. https://doi.org/10.1016/j.powtec.2016.01.036.

Meng, L., S. Li, and X. Yao. 2017. "Maximally Dense Random Packings of Intersecting Spherocylinders with Central Symmetry." *Powder Technology* 314: 49–58. https://doi.org/10.1016/j.powtec.2016.07.059.

Meng, L., C. Wang, and X. Yao. 2018. "Non-Convex Shape Effects on the Dense Random Packing Properties of Assembled Rods." *Physica A: Statistical Mechanics and its Applications* 490: 212–221. https://doi.org/10.1016/j.physa.2017.08.026.

Miskin, M.Z. 2016. *The Automated Design of Materials Far from Equilibrium*. Springer Theses. Cham: Springer International Publishing. https://doi.org/10.1007/978-3-319-24621-5.

Miskin, M.Z., and H.M. Jaeger. 2013. "Adapting Granular Materials through Artificial Evolution." *Nature Materials* 12: 326–331. https://doi.org/10.1038/nmat3543.

Miskin, M.Z., and H.M. Jaeger. 2014. "Evolving Design Rules for the Inverse Granular Packing Problem." *Soft Matter* 10 (21): 3708–3715. https://doi.org/10.1039/C4SM00539B.

Murphy, K. L. Roth, D. Peterman, and H. Jaeger. 2017a. "Aleatory Construction Based on Jamming: Stability through Self-Confinement." *Architectural Design* 87 (4): 74–81. https://doi.org/10.1002/ad.2198.

Murphy, K.A., K.A. Dahmen, and H.M. Jaeger. 2019. "Transforming Mesoscale Granular Plasticity through Particle Shape." *Physical Review X* 9 (1): 11014. https://doi.org/10.1103/PhysRevX.9.011014.

Murphy, K.A., and H.M. Jaeger. 2018. "Designed to Fail: Granular Plasticity and Particle Shape." In *Proceedings of the IUTAM Symposium Architectured Materials Mechanics*, edited by T. Siegmund and F. Barthelat, 17–19. Chicago: Purdue University Libraries Scholarly Publishing Services. https://docs.lib.purdue.edu/iutam/presentations/abstracts/53.

Murphy, K.A., N. Reiser, D. Choksy, C.E. Singer, and H.M. Jaeger. 2016. "Freestanding Loadbearing Structures with Z-Shaped Particles." *Granular Matter* 18 (2): 26. https://doi.org/10.1007/s10035-015-0600-2.

Murphy, K.A., L.K. Roth, and H.M. Jaeger. 2017b. "Adaptive Granular Matter." In *Active Matter*, edited by S. Tibbits, 287–289. Cambridge, Mass.: The MIT Press. https://doi.org/10.7551/mitpress/11236.003.0044.

Petersen, K.H., N. Napp, R. Stuart-Smith, D. Rus, and M. Kovac. 2019. "A Review of Collective Robotic Construction." *Science Robotics* 4 (28), eaau8479. https://doi.org/10.1126/scirobotics.aau8479.

Peterson, R. 2014. "Tumbling Units: The Tectonics of Indeterminate Extension." Supervised by Kentaro Tsubaki. Tulane University. Accessed Sept 28, 2016. http://winwooddesignworks.com/tumbling-units-the-tectonics-of-indeterminate-extension.

Philipse, A.P. 1996. "The Random Contact Equation and its Implications for (Colloidal) Rods in Packings, Suspensions, and Anisotropic Powders." *Langmuir* 12 (5): 1127–1133. https://doi.org/10.1021/la950671o.

Piskorec, L. 2014. "Remote Material Deposition." *Space 562, Envisaging the Reality of Digital Fabrication: 3D Printing and Robotics* (2014): 65.

Reis, P.M., H.M. Jaeger, and M. van Hecke. 2015. "Designer Matter: A Perspective." *Extreme Mechanics Letters* 5: 25–29. https://doi.org/10.1016/j.eml.2015.09.004.

Sanchez, T., D.T.N. Chen, S.J. DeCamp, M. Heymann, and Z. Dogic. 2012. "Spontaneous Motion in Hierarchically Assembled Active Matter." *Nature* 491 (7424): 431. https://doi.org/10.1038/nature11591.

Scholz, C., S. D'Silva, and T. Pöschel. 2016. "Ratcheting and Tumbling Motion of Vibrots." *New Journal of Physics* 18 (12): 123001. https://doi.org/10.1088/1367-2630/18/12/123001.

Scholz, C., M. Engel, and T. Pöschel. 2018. "Rotating Robots Move Collectively and Self-Organize." *Nature Communications* 9 (1): 931. https://doi.org/10.1038/s41467-018-03154-7.

Scholz, C., A. Ldov, T. Pöschel, M. Engel, and H. Löwen. 2021. "Surfactants and Rotelles in Active Chiral Fluids." *Science Advances* 7 (16). https://doi.org/10.1126/sciadv.abf8998.

Tibbits, S., ed. 2017. *Autonomous Assembly: Designing for a New Era of Collective Construction*. Architectural Design 87 (4). Oxford: John Wiley & Sons.

Trepanier, M., and S.V. Franklin. 2010. "Column Collapse of Granular Rods." *Physical Review E* 82 (1): 11308. https://doi.org/10.1103/PhysRevE.82.011308.

Tsubaki, K. 2008. "Tumbling units: tectonics of indeterminate extension." In *Material Matters: Making Architecture: 2008 West Fall Conference Proceedings*, edited by G.P. Borden and M. Meredith. Washington DC: ACSA Press. 270–277.

Tsubaki, K. 2009. "Tumbling units: tectonics of indeterminate extension." In *The Value of Design: Design is at the Core of What We Teach and Practice: Papers from the Association of Collegiate Schools of Architecture 97th Annual Meeting*, edited by P. Crisman and M. Gillem. Washington DC: ACSA Press. 292–98.

Tsubaki, K. 2012. "Tumbling units: tectonics of indeterminate extension." In *Matter: Material Processes in Architectural Production*, edited by G.P. Borden and M. Meredith. Abingdon, New York: Routledge; Taylor & Francis Group. 187–203.

Wouterse, A., S. Luding and A.P. Philipse. 2009. "On contact numbers in random rod packings." *Granular Matter* 11 (3): 169–77. https://doi.org/10.1007/s10035-009-0126-6.

Zhao, Q., J. Heyda, J. Dzubiella, K. Täuber, J.W.C. Dunlop, and J. Yuan. 2015. "Sensing Solvents with Ultrasensitive Porous Poly(Ionic Liquid) Actuators." *Advanced Materials* 27 (18): 2913–2917. https://doi.org/10.1002/adma.201500533.

Zhao, Y., K. Liu, M. Zheng, J. Barés, K. Dierichs, A. Menges, and R.P. Behringer. 2016. "Packings of 3D Stars: Stability and Structure." *Granular Matter* 18 (2): 24. https://doi.org/10.1007/s10035-016-0606-4.

IMAGE CREDITS
All drawings and images by the authors.

Denitsa Koleva holds an MSc degree from the Integrative Technologies and Architectural Design Research (ITECH) program from the University of Stuttgart.

Eda Özdemir holds an MSc degree from the Integrative Technologies and Architectural Design Research (ITECH) program from the University of Stuttgart.

Vaia Tsiokou holds an MSc degree from the Integrative Technologies and Architectural Design Research (ITECH) program from the University of Stuttgart.

Karola Dierichs holds the Material and Code professorship at weißensee school of art and design berlin in the Cluster of Excellence Matters of Activity at Humboldt-Universität zu Berlin. Previously, she was a research associate at the Institute for Computational Design and Construction (ICD) within the Cluster of Excellence Integrative Computational Design and Construction for Architecture (IntCDC).

Pillow Forming

Digital Fabrication of Complex Surfaces
through Actuated Modular Pneumatics

Kyle Schumann*
University of Virginia /
After Architecture

Katie MacDonald*
University of Virginia /
After Architecture

*Authors contributed
equally to the research

1   Apparatus with visible forming
    surface and pneumatic control
    system

ABSTRACT

Recent decades have seen the development of increasingly powerful digital modeling and fabrication tools applied to the creation of molds or formwork for cast or formed materials. Many of these processes are highly customizable but resource intensive, singular in geometry, and disposable. This paper introduces pillow forming as a customizable, reusable forming system aimed at minimizing the resource intensity of construction and capable of producing both standardized and unique curved molded panels. The apparatus consists of a field of pneumatic pillows that inflate to form a complex curved surface with which various materials can be formed or cast. The design and construction of the system is discussed, including the modular inflation system, pneumatic and electronic control systems, control software run through Rhinoceros, Grasshopper, and Arduino, as well as the standard operation procedure. The system is demonstrated through the production of *Homegrown*, an architectural installation built of pillow-formed biomaterial aggregate. Various limitations and opportunities of the system are discussed and analyzed, and opportunities for future development and applications in sustainable construction are posited.

INTRODUCTION

Computer-aided industrial manufacturing has seen the proliferation and refinement of automated approaches to both subtractive and additive manufacturing, resulting in the viability of constructing customizable, complex forms in a variety of materials. This paper addresses the limitations of current widespread molding processes by proposing an automated method for reformable molds. The fabrication of molds typically relies on subtractive processes such as CNC milling, foam cutting, and carving of EPS, MDF, and other petroleum- and wood-based materials, with each mold used to produce only a single unique geometry. This results in waste throughout the fabrication process: offcuts are produced in the creation of each mold, multiple molds must be produced for non-repeating geometries, and because mold geometries are fixed, they cannot be reused for subsequent construction projects, leading to their disposal. This waste of valuable resources and the environmental impacts associated with disposal create a need for molding techniques which are reusable rather than single-use.

The need for reusable molding materials has been explored for both the production of repeating and customizable forms, resulting in a number of new formwork materials: fabric (West 2016; Chandler and Keable 2009); sand (Hollis 2021); earth (Ensamble Studio 2010; Tippet Rise Art Center n.d.); and even ice (Sitnikov et al. 2019). Each of these techniques leverages materials which are both easier to obtain locally and to reform than typical petroleum- and wood-based materials, reducing the environmental impact of these processes and offering new aesthetic effects in the cast output.

Building on such work, pillow forming, the ground-up, modular forming system described in this paper, advances variable over repetitious form allowing for the production of serial unique forms through a malleable process—the injection and removal of air—therefore eliminating the waste intrinsic to standard means of mold production (Figure 1). Because the inflation process is reversible, repeatable, and adaptable to various surface geometries, many different forms can be molded without the need for multiple molds, effectively avoiding the creation of disposable molds and the offcuts associated with their production. A key advancement of this system is its modularity: pillow forming is composed of a field of pneumatic units (pillows) that are individually operated. The modular nature of the system means that it is scalable, overcoming physical limitations such as CNC bed size, robotic arm reach, and others presented by typical mold production equipment. Pillow-formed wall or ceiling panels can also be assembled edge-to-edge, producing continuous surfaces many times larger than the apparatus itself, without using any additional material except that which constitutes the finished product.

STATE OF THE ART

Aiming to address the challenge of creating customized formwork, the project responds to and builds upon past works in three categories: actuated surfaces, malleable forming, and adaptable formwork.

Actuated Surfaces

The modular surface, composed of a grid of individually operated units, is a strategy explored in shape changing interfaces such as inFORM, developed at the MIT Tangible Media Group (Leithinger et al. 2015). In this project, a surface is physically modeled by individually actuated pixels. The calibration of movement of each pixel to create a single continuous global surface is achieved through the control software. While inFORM is applied as a physical display surface, its creation of an actuated, reformable pixel-based system shares similarities to the customizable molding system described in this paper.

Other methods for controlled actuated surfaces have been explored using textiles to program various surface deformations and curvatures. Knitflatable and PneumaKnit each explore the programing of geometry into knitwork which is inflated, producing controlled forms (Baranovskaya et al. 2016; Ahlquist et al. 2017). In these projects, emphasis is placed on the design of a knit pattern that inflates into a specific three-dimensional geometry, while the actuation itself consists of only three fixed states (inflated, partially inflated, deflated) rather than variable geometries. More similar in approach to inFORM are Barkow Leibinger's (2014) Kinetic Wall and Asif Khan's (Khan 2014) MegaFaces. Both were exhibited as large scale architectural surfaces in 2014 and employ actuated pistons to map three-dimensional data across reformable fabric surfaces.

Malleable Forming

Recent efforts are expanding in the area of malleable forming, in which material is not subtracted (CNC milling), or added (3D printing), but physically manipulated, often through some combination of pressure and/or heat.

Incremental sheet metal forming creates three-dimensional surfaces through the application of pressure along a toolpath, gradually deforming a growing area of the sheet metal. Such processes have been applied using robotic arms and a sturdy clamping apparatus (Kalo and Newsum 2014). While such examples of incremental sheet

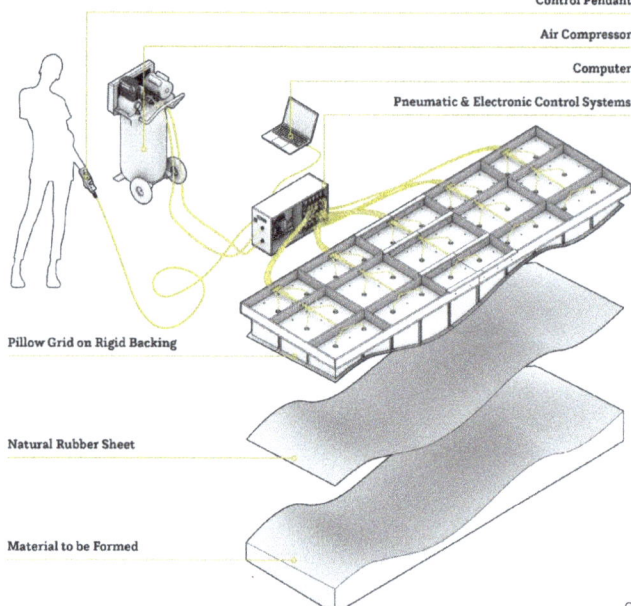

positive or negative air pressure deforms it up or down. It is possible to apply the thermoforming method to create reformable molds, though like incremental sheet metal forming, the degree to which the sheet can be reformed would likely be limited.

Adaptable Formwork

Various precedents exist in which molds are created from reformable or reconfigurable materials.

Pneumatic formwork has been explored in several projects at ICD Stuttgart, in which inflatable volumes are used to create voids within an assembly of granular aggregate units, or as a scaffold on which a robotically applied fiber composite surface can be constructed (Rusenova et al. 2016; Vasey et al. 2015). Like pillow forming, these projects involve pneumatic surfaces against which material is molded or placed; however, the inflated forms are fixed in geometry, and must be reconfigured manually rather than through programmable actuation.

In several P_Wall projects by Matsys, an adaptable formwork is created by positioning a series of dowels in a field condition and draping them with a stretchable fabric surface onto which plaster is cast (Kudless 2011). The molds are manually reconfigurable, and the resultant forms are a negotiation between dowel placement and the sagging of fabric under the weight of the plaster.

Zero Waste Free-Form Formwork, developed at ETH Zürich similarly explores reconfigurable formwork for material molding, but works to overcome material irregularities with a robotically positioned surface (Oesterle et al. 2012). A robotic arm is used to manipulate a series of linear pistons arrayed to scaffold a flexible surface. This surface is used to produce a wax mold for concrete casting in a lost wax process. The repositionable surface and recycling of wax similarly reduce material waste in the mold making process. While the ambitions of pillow forming are perhaps most closely aligned with this technique, a principal difference in approach taken with pillow forming is a shift to a standalone system that does not rely on a robotic arm to accurately position the surface, enabling mold surfaces larger than the arm's reach to be created, and the use of the system in the field away from industrial robots.

forming are conceived as final outputs, it is feasible that the resulting surfaces could be used as molds. However, the incremental toolpath necessary would require a long production time, and there would likely be a limit on the degree to which the sheet could be reformed depending on material ductility.

Several projects reimagine plastic thermoforming operations, abandoning the traditional physical mold for a pressure chamber under the surface to be thermoformed (Schumann and Johns 2019; Mueller et al. 2019). Selective robotic heating makes the surface malleable, while

Zero Waste Free-Form Formwork proposes a closed-loop material cycle, in which the wax used to produce one mold is melted down and formed into the next. Similarly, IceFormwork creates cast concrete elements using blocks of ice CNC milled in a subtractive process as molds (Sitnikov et al. 2019). The creation of an ice mold that can be cast

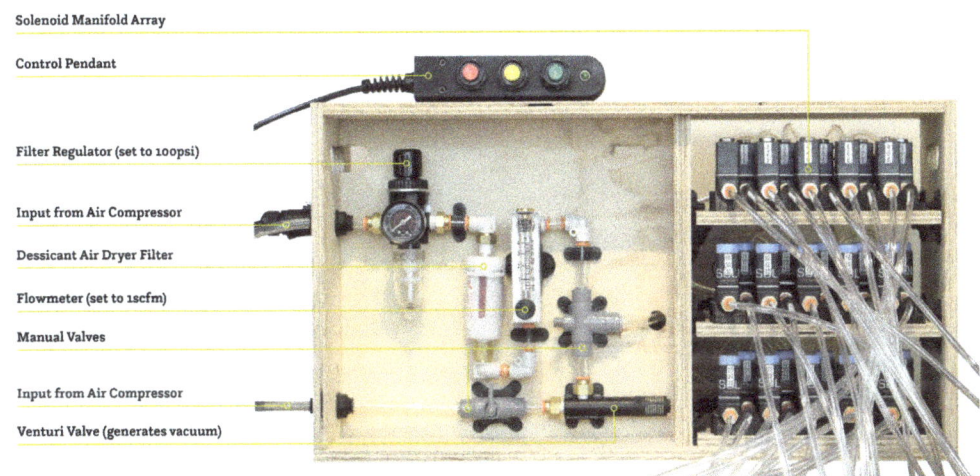

2   Pillow forming apparatus

3   Pillows are connected along the top edges to create a single continuous surface

4   Pneumatic control system

5   Electronic control system

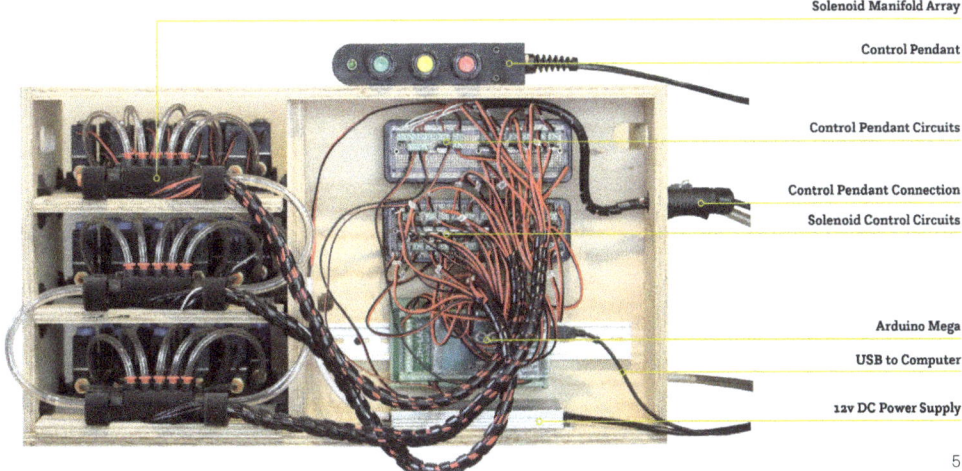

upon, melted, refrozen, milled, and cast upon again demonstrates the possibilities for a formwork materiality that is easy to source, demold, and reuse. Pillow forming also proposes a closed-loop material cycle, but instead eliminates the need for mold material to be recycled by casting directly on an actuated surface.

METHODS

Pillow forming is demonstrated through the construction of a fully functional apparatus used to fabricate an architectural installation. The apparatus consists of a physical inflation system, pneumatic control system, electronic control system, and control software (Figure 2). These systems are described herein, along with the typical operation procedure.

Physical Inflation System

The physical inflation system consists of a rigid plywood surface reinforced with a plywood waffle grid on which a grid of one foot cubic inflatable pillows is mounted. The pillows are made of clear vinyl sheet, which is reinforced through mechanically sewn seams and made airtight with a vinyl adhesive applied over these seams. Each pillow can be inflated independently, expanding upwards as it fills with air. Each pillow has an extra strip of material along the perimeter of the top and bottom faces, with which the pillows are attached to each other and to the rigid plywood backing. Attaching the pillows to each other at the corners and midpoints of the top edges allows for a continuous surface to be maintained, regardless of discrepancies between inflation heights of neighboring pillows (Figure 3). The final large scale apparatus measures 3 feet by 10 feet and consists of thirty individual pillows.

Pneumatic Control System

The pneumatic system consists of a standard air compressor, attached with two air hoses to a custom pneumatic control system. Air coming from the compressor is first run through a filter regulator and a desiccant air dryer filter to remove moisture and prevent condensation from building up in the hoses and pillows (Figure 4). It then runs through a flowmeter with a range of 0.4 to 4 standard cubic

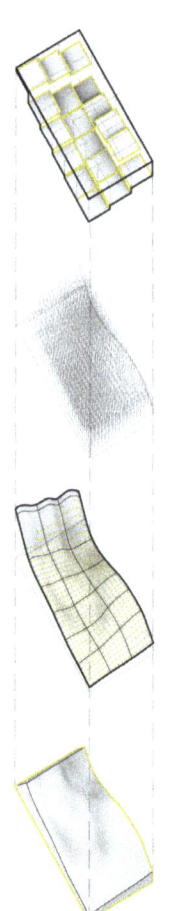

Average Heights
Average heights within each 12 inch region. These heights will inform inflation times for each pillow.

Measure Heights
Calculate heights of each point on surface, measured from horizontal plane at fixed height.

Divide Surface
Extend surface to boundary. Create 12 inch square grid. Locate one point every 2 inches within these regions.

Wall Panel Geometries
Panels are oriented horizontally for fabrication with flat outer face down.

6

7

6 Analyzing the digital surface to determine average pillow heights

7 Overlay of single pillow inflation tests for calibrating inflation timing

A pair of manually operated valves allows the user to switch between inflation and deflation modes. When switched to deflation mode, the compressed air is run directly through a venturi valve, generating vacuum and allowing the pillows to be emptied at the end of each inflation cycle.

Electronic Control System

The machine is controlled with an Arduino Mega (Banzi et al 2020), which provides enough digital output pins to wire all thirty solenoid valves, one to control each pillow (Figure 5). The solenoid valves are 12v, so a separate power source is used to convert 120v AC to 12v DC. The Arduino operates at 5v, powered via USB from a computer, which is also used to prepare and upload the control software and monitor the system during use.

A control pendant houses three push-button switches, the operation of which is described in the following section, as well as a small indicator light which will be illuminated if the Arduino is powered and successfully running the uploaded program.

Control Software

The machine is simple to operate, beginning with a Rhinoceros (Robert McNeel & Associates 2018) model. Once the user has digitally modeled the desired surface, it is input through a Grasshopper (Rutten 2018) script that analyzes the surface and converts the surface height to inflation times for each individual pillow.

In Grasshopper, the input surface is divided into a one-foot square grid, corresponding with the size of the pillows. Simply reading the height at the center point of each square produced less than desirable accuracy. Instead, the average height of each pillow is calculated by measuring the height at intervals every two inches and dividing by the total number of measurements (Figure 6).

Once the average pillow height is found, it is converted to an inflation time for each pillow. To determine this conversion, a series of inflation tests were conducted using a single pillow (Figure 7). With an input flow rate of 1 scfm, the pillow was repeatedly inflated, increasing the inflation time by one second intervals. It was determined that the pillow height increased at a fairly constant rate of one inch per second of inflation time, resulting in a straightforward conversion of pillow height to inflation time: average height (in inches) multiplied by 1000 to achieve the desired units (milliseconds).

feet per minute (scfm) that allows the airflow to be precisely controlled. This is important for calibrating pillow inflation, as discussed further below.

Next, a series of manifolds splits the line into thirty individual air supply hoses, each operated by an electronic solenoid valve controlling a single pillow. Push-to-connect fittings are used throughout to enable easy replacement or reconfiguration of parts.

These millisecond time values for each pillow are compiled into Arduino-ready code using a series of concatenate components in Grasshopper, producing a text panel which is copy-pasted directly into an Arduino script. This can be accomplished automatically, but the manual step of copying the values over affords the user an opportunity to verify that the input surface was read correctly and that there are no null or otherwise extreme values. Once the values are pasted into the prepared Arduino code, it is uploaded to the machine via USB cable.

Operation

The machine is operated through a pair of aforementioned manual valves to switch between positive and negative pressure air supply, and a control pendant with three physical push-button switches that begin the inflation sequence (green button), run a vacuum sequence to deflate all pillows (yellow button), or cancel the currently running operation (red button).

A digital readout is provided on the computer through the Arduino serial monitor. This readout provides reminders for the operator to manually switch the air supply from positive to negative pressure when beginning the inflation or vacuum sequences and provides values for inflation/vacuum times for each pillow and a total time value at the conclusion of each sequence.

When the operator has uploaded Arduino code and is ready to begin, the green button is pressed to initiate the inflation sequence, and a reminder is provided via Arduino serial monitor to manually switch to the positive pressure air supply. Once this is done, a second press of the green button begins inflation, with times provided on screen as each pillow inflates, until the inflation sequence is complete. The process for deflation is very similar but using the yellow button, and the red button can be used at any time to stop the current process—this functions as a hard reset on the Arduino, which maintains the current software upload but the interrupted sequence needs to be restarted from the beginning.

Once the inflation sequence is complete, the surface is ready to be used for forming or casting, as discussed in the following section. Since the solenoid valves operating the pillows are closed when unpowered, the apparatus will hold its form for many hours, but may eventually deform due to any small air leaks that may exist in the seams of the pillows. This is true even if it is disconnected from the air supply, electrical power, or the control computer, any of which can be done to make it easier to manipulate the apparatus as needed for forming or casting.

RESULTS AND DISCUSSION

Pillow forming was successfully demonstrated in the translation of digitally modeled undulating surfaces into physical molds and formed architectural wall panels through a temporary installation, *Homegrown*. The work revealed several opportunities for improvement in future iterations, as discussed below.

Application

Pillow forming is conceived as a low-waste, reusable mold for sustainable construction. After prototyping, the application of the machine is demonstrated at full scale by forming a biomaterial wall panel assembly realized for the Knoxville Museum of Art (Figure 8). The design of the installation focuses on how addition rather than subtraction can eliminate the production of offcuts—in this case, reimagining how biomaterials like wood are reduced at harvest, at the factory, and on the construction site into an appropriate unit size, that then becomes a unit within a larger assembly. In contrast, the installation uses invasive plant species and landscaping waste as its raw material and aggregates discrete parts into additive formed panels.

The installation consists of four wall panels shaping an exterior room measuring 10 feet by 10 feet. Each panel is flat on the exterior surface and undulating on the interior surface—the side made via pillow forming. The panels are composed of a selection of biomaterial including the invasive species kudzu and golden bamboo, as well as forestry waste including sticks, twigs, and pine needles (Figure 9).

The plant materials are coated in a liquid bio-based adhesive binder and arranged on a flat surface. Once the material is laid out, the code is prepared using a digital surface geometry, and the pillow forming apparatus is inflated, inverted, and pressed down onto the panel, compressing the material and conforming it to its final surface geometry. A series of wood legs are placed around the perimeter of the plywood base of the forming apparatus to maintain overall thickness as determined in the digital model. Weights can be added on top as needed to compress the cast material down to this point. A thin layer of natural rubber is placed between the cast material and the vinyl pillows to ensure that they are not adhered to the final product, and that they stay clean to be reused again for the next panel. The natural rubber sheet likewise remains clean and reusable.

Several of the wall panels needed for the installation exceeded the size of the 3 foot by 10 foot apparatus,

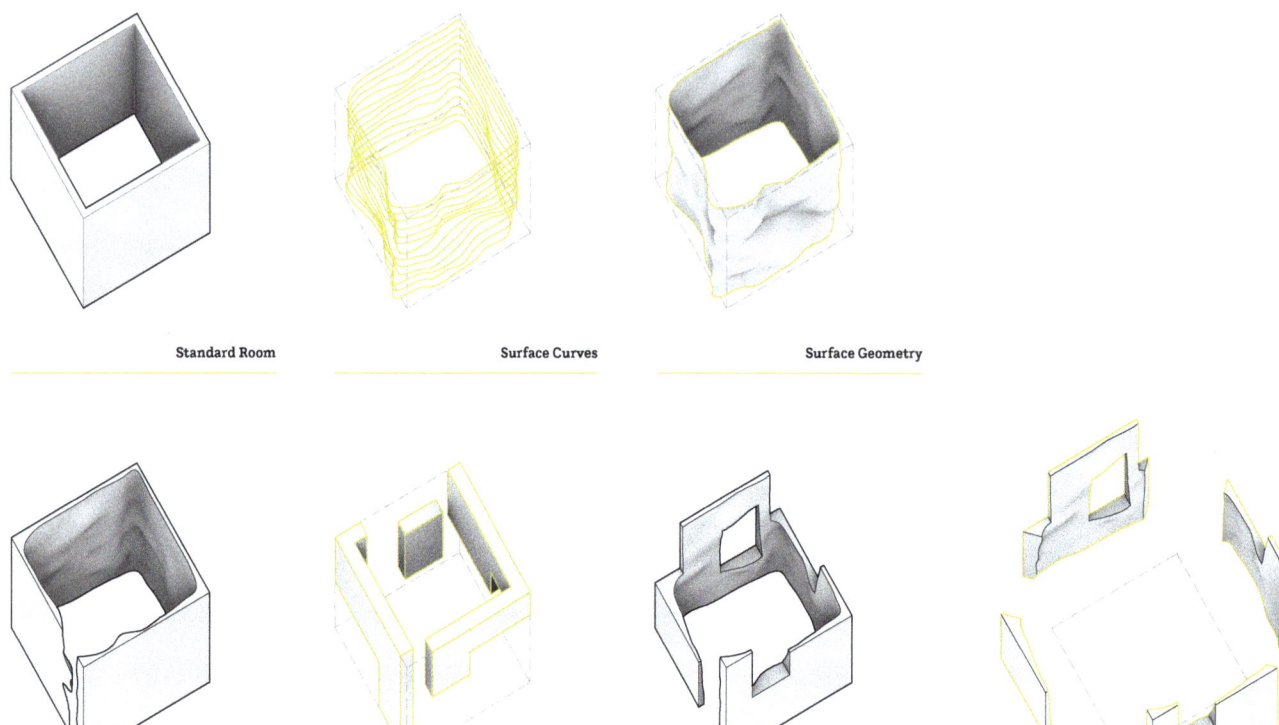

8   Installation modeling process.

measuring 10 feet wide by 6 feet tall (Figure 10). Therefore, they required two sequential forming operations. The line at the center of the panel where the edge of the apparatus meets in the two forming operations is essentially invisible in the completed wall, demonstrating that two distinct edges can be made accurately with separate inflations to produce a continuous surface larger than the area of the arrayed pillows.

Inflation Sequence

The pillows need to be inflated with particular sequencing. Before the inflation sequence is run, the pillows are entirely deflated with vacuum, and since the solenoid valves are closed, they are held fully deflated. If the first pillow is inflated to its full amount immediately, the neighboring deflated pillows will prevent it from expanding, and the resulting tension can rip the pillows apart at their connections. To prevent this, the pillows need to be inflated evenly a bit at a time. This can be accomplished by opening all solenoids at once and closing them as needed, but the rate of air flow into the pillows will be different depending on how many solenoids are open at any given time. Instead, the inflation time for each pillow is divided by 3, and the pillows are all inflated to one third of their final height, then another third, then fully inflated. This produces a more gradual rising between adjacent pillows and excessive stress is mitigated.

Surface Limitations

The surface physically formed by the pillow forming apparatus can accommodate a range of possible curvatures that are limited by the ways in which any given pillow geometry can move. The surfaces are essentially 2.5-dimensional in definition, as no cantilevers or fold overs are possible. The surface must be smooth and continuous given the attachment of the pillows to one another, producing a single surface. If the pillows were to be selectively detached at their tops, steps in the surface could be achievable. The resolution or degree of undulation possible in the surface is related to the area of the pillows, constructed in a 1 foot by 1 foot grid. Narrower pillows with smaller surface areas can be used to produce more tightly undulating surfaces, as discussed in the following section.

Pillow Variations

Variations in the design and aggregation of the pneumatic

9   Detail of a pillow formed surface in biomaterial aggregate.

pillows have the potential to alter the range of possible inflated surfaces. Changing the current square grid of pillows to a triangular, hexagonal, or other grid could smooth transitions across the surface in different ways. Altering the height of the fully inflated pillow, or its internal structure—several stacked or interconnected bladders instead of one large volume, for example—would affect how the pillows inflate and rise and how much weight they can support. Such iterations will be developed and tested in future work.

CONCLUSION

The ability to create a complex physical surface from a digital model accurately and with a reusable system creates opportunities for fabrication in a variety of materials. While the pillow forming apparatus is tested in the production of a small installation, scaling the molding process toward a more traditional architectural application such as facade paneling is conceivable. In the demonstrated application, material is formed through compression by the pillow forming apparatus. Future work will explore alternative material strategies, including surface-applied materials such as fiberglass and shotcrete, and cast materials such as concrete and plaster. While cast materials will present a few challenges in terms of adding temporary sides around the pillow forming apparatus to create a sealed, castable mold, the ability to accurately form complex three-dimensional surfaces, either repeatable or unique, suggests that it would be possible to mold panels that can be added together to create large and continuous surfaces. Such applications might include custom wall or ceiling paneling or unique tilt-up concrete wall slabs.

Pillow forming advances techniques of adaptable formwork, toward the larger aim of minimizing the resource intensity of construction—in particular, custom mold production. While the materials against which molded materials are formed have typically been costly and petroleum- or wood-based, the forming system described in this paper demonstrates the potential of air as an economical material that does not need to be harvested, refined, stored, or transported. In contrast to the fixed, heavy, disposable molds typical of construction, the pillow forming apparatus is light, reformable, and highly customizable.

10 *Homegrown*, a pillow formed installation with biomaterial aggregate

ACKNOWLEDGEMENTS

The work presented in this paper was produced with funding from the University of Tennessee through the College of Architecture + Design's Tennessee Architecture Fellowship and The Office of Undergraduate Research's Research Assistant Award. The authors would like to thank students Rachel Crosslin, Zherti Jasa, and Rose Gowder for their work in the 'Material Misbehavior' advanced studio taught by the authors, Kevin Saslawaky for his work as research assistant, and Dean Jason Young for his support of the work. The work was further developed at the University of Virginia. The technology presented has been filed under U.S. Provisional Patent Serial No. 63/253,693, "Pillow Forming: Digital Fabrication of Complex Surfaces through Actuated Modular Pneumatics," filed October 8, 2021.

REFERENCES

Ahlquist, Sean, Wes McGee, and Shahida Sharmin. 2017. "PneumaKnit: Actuated Architectures Through Wale- and Course-Wise Tubular Knit-Constrained Pneumatic Systems." In *ACADIA '17: Disciplines & Disruption; Proceedings of the 37th Annual Conference of the Association for Computer Aided Design in Architecture*, edited by T. Nagakura, S. Tibbits, and C. Mueller. 38–51. Cambridge, Mass.: ACADIA.

Banzi, Massimo, David Cuartielles, Tom Igoe, Gianluca Martino, and David Mellis. *Arduino Software* (IDE). V. 1.8.13. Arduino.cc. Windows 10. 2020.

Baranovskaya, Yuliya, Marshall Prado, Moritz Dörstelmann, and Achim Menges. 2016. "Knitflatable Architecture: Pneumatically Activated Preprogrammed Knitted Textiles." In *Complexity & Simplicity; Proceedings of the 34th eCAADe Conference*, vol. 1, edited by A. Herneoja, T. Österlund, and P. Markkanen. 571–580. Oulu, Finland: eCAADe.

Barkow Leibinger. 2014. "Kinetic Wall." Venice Biennale Installation. Project web page. Accessed July 17, 2021. https://barkowleibinger.com/archive/view/kinetic_wall.

Chandler, Alan, and Rowland Keable. 2009. "Achieving carbon neutral structures through pure tension: Using a fabric formwork to construct rammed earth columns and walls." In *Proceedings of the 11th International Conference on Non-conventional Materials and Technologies (NOCMAT)*. Bath, United Kingdom: NOCMAT.

Ensamble Studio. 2010. "The Truffle. Costa da Morte, 2010." Project web page. Accessed July 17, 2021. https://www.ensamble.info/thetruffle.

Hollis, Sophie Aliece. 2021. "Studio Anne Holtrop's 35 Green Corner builds texture with sand-cast concrete and aluminum." The Architect's Newspaper, October 15, 2021. Accessed Nov. 1, 2021. https://www.archpaper.com/2021/10/facades-35-green-corner-studio-anne-holtrop-muharraq-sand-cast-concrete/.

Kalo, Ammar, and Michael Jake Newsum. 2014. "An Investigation of Robotic Incremental Sheet Metal Forming as a Method for Prototyping Parametric Architectural Skins." In *Robotic Fabrication in Architecture, Art and Design 2014*, edited by W. McGee and M. Ponce de Leon. 33–49. Switzerland: Springer International Publishing.

Kahn, Asif. 2014. "MegaFaces." Project web page. Accessed July 17, 2021. http://www.asif-khan.com/project/sochi-winter-olympics-2014/.

Kudless, Andrew. 2011. "Bodies in Formation: The material evolution of flexible formworks." In *ACADIA '11: Integration through Computation; Proceedings of the 31st Annual Conference of the Association for Computer Aided Design in Architecture*, edited by J. S. Johnson, B. Kolarevic, V. Parlac, and J. M. Taron. 98–105. Banff, Alberta: ACADIA.

Leithinger, Daniel, Sean Follmer, Alex Olwal, and Hiroshi Ishii. 2015. "Shape Displays: Spatial Interaction with Dynamic Physical Form." *IEEE Computer Graphics and Applications* 35 (5): 5–11.

Mueller, Stefanie, Anna Seufert, Huaishu Peng, Robert Kovacs, Kevin Reuss, François Guimbretière, and Patrick Baudisch. 2019. "FormFab: Continuous Interactive Fabrication." In *Proceedings of the Thirteenth International Conference on Tangible, Embedded, and Embodied Interactions (TEI 2019)*. Association for Computing Machinery (ACM). 315-323. Tempe, Arizona: ACM.

Oesterle, Silvan, Axel Vansteenkiste, Ammar Mirjan, Jan Willmann, Fabio Gramazio, and Matthias Kohler. 2012. "Zero Waste Free-Form Formwork." In *Proceedings of the Second International Conference on Flexible Formwork (ICFF)*. 258–267. Bath: BRE CICM University of Bath.

Robert McNeel & Associates. *Rhinoceros 3D*. V. 6. Robert McNeel & Associates. Windows 10. 2018.

Rusenova, Gergana, Karola Dierichs, Ehsan Baharlou, and Achim Menges. 2016. "Feedback- and Data-driven Design for Aggregate Architectures: Analyses of Data Collections for Physical and Numerical Prototypes of Designed Granular Materials." In *ACADIA '16: Posthuman Frontiers; Data, Designers, and Cognitive Machines. Proceedings of the 36th Annual Conference of the Association for Computer Aided Design in Architecture*, edited by K. Velikov, S. Ahlquist, and M. del Campo. 62–72. Ann Arbor: ACADIA.

Rutten, David. *Grasshopper 3D*. Robert McNeel & Associates. Windows 10. 2018.

Schumann, Kyle, and Ryan Luke Johns. 2019. "Airforming: Adaptive Robotic Molding of Freeform Surfaces through Incremental Heat and Variable Pressure." In *Intelligent & Informed; Proceedings of the 24th International Conference of the Association for Computer-Aided Architectural Design Research in Asia (CAADRIA)*, vol. 1. 33–42. Victoria University of Wellington, New Zealand.

Sitnikov, Vasily, Peter Eigenraam, Panagiotis Papanastasis, and Stephan Wassermann-Fry. 2019. "IceFormwork for Cast HPFRC Elements: Process-Oriented Design of a Light-Weight High-Performance Fiber-Reinforced Concrete (HPFRC) Rain-Screen Façade." In *ACADIA '19: Ubiquity and Autonomy; Proceedings of the 39th Annual Conference of the Association for Computer Aided Design in Architecture*, edited by K. Bieg, D. Briscoe and C. Odom. 616–627. Austin: ACADIA.

Tippet Rise Art Center. n.d. "Structures of Landscape, Ensamble Studio." Accessed July 17, 2021. https://tippetrise.org/films/structures-of-landscape-ensamble-studio.

Vasey, Lauren, Ehsan Baharlou, Moritz Dörstelmann, Valentin Koslowski, Marshall Prado, Gundula Schieber, Achim Menges, and Jan Knippers. 2015. "Behavioral Design and Adaptive Robotic Fabrication of a Fiber Composite Compression Shell with Pneumatic Formwork." In *ACADIA '15: Computational Ecologies; Design in the Anthropocene. Proceedings of the 35th Annual Conference of the Association for Computer Aided Design in Architecture*, edited by C. Perry and L. Combs. 297–309. Cincinnati: ACADIA.

West, Mark. 2016. *The Fabric Formwork Book: Methods for Building New Architectural and Structural Forms in Concrete*. London; New York: Routledge.

IMAGE CREDITS

All drawings and images by the authors.

Kyle Schumann is cofounder of After Architecture and Assistant Professor of Architecture at the University of Virginia. He was previously the 2019-2020 Tennessee Architecture Fellow at the University of Tennessee and is the recipient of the Robert A.M. Stern Architects Travel Fellowship. Schumann seeks to advance the accessibility of digital fabrication techniques, leveraging democratized technologies in his teaching and research as well as inventing and building low-cost ground-up fabrication systems. His work spans analog processes in woodworking, metalworking, casting, ceramics, and textile production, to advanced and novel digital fabrication technologies, robotics, and machine visioning systems.

Katie MacDonald is cofounder of After Architecture and Assistant Professor of Architecture at the University of Virginia. She held the 2019-2020 Tennessee Architecture Fellowship at the University of Tennessee. MacDonald's work reconsiders construction standards and materials in light of new lifecycle questions and overextended supply chains. MacDonald prototypes new biomaterial assemblies, with the aim of creating building material systems that sequester carbon and reduce construction's contribution to the environmental crisis.

Inoculated Matter

3D Printing with Mycelium and Upcycled Waste

Nancy Diniz
Central Saint Martins, UAL

Frank Melendez
City College of New York, CUNY

1   Designed objects 3D printed with upcyled waste and mycelium (© bioMATTERS LLC, 2021)

INOCULATED MATTER looks towards new possibilities for designing and making architectural elements with living organisms, upcycled waste, and 3D printing technologies. This research project, which is currently ongoing and has been developed over the past two years, includes a series of multi-scalar mycelium bio-composites, as a means of redefining material, water, and energy in the face of changing scales of manufacturing and resource cycles. Our methods build on previous methods of fungal based, bio-based, materials [1] and 3D printed assembles for microbial transformation [2], with a different focus on bio-techniques for fungal growth in waste substrates in order to advance inoculated mycelium waste based extrusion pastes to 3D print product design objects, furniture, and interior design modular systems (Figures 1, 3). By upcycling domestic and industrial waste (e.g. coffee grounds, sawdust, cardboard, etc.), our team has developed 3D printable pastes that can be inoculated and used as a substrate for growing mycelium, the vegetative part of fungi (Figures 2, 4). The elements are computationally designed and 3D printed, allowing for the production of complex, customized parts, that are difficult to achieve using traditional mold-making and casting methods (Figures 5, 6). The mycelium hyphae networks grow within and on the paste, creating a naturally strong bond and solidifying the 3D printed layers into a cohesive whole. Microscopy and computational simulation techniques are used to better understand the behaviors and characteristics of the hyphae networks (Figures 7, 8). The mycelium is grown for approximately seven days in a sterile environment to avoid contamination, at which point it is removed to stop its growth, followed by a drying and baking process, to remove any additional moisture. The dried mycelium surface

2   Various substrates and mycelium strains (© bioMATTERS LLC, 2021)

reflects a quality similar to a soft, velvet-like texture, producing a unique and inviting haptic experience. These designed objects serve as a proof-of-concept for 3D printing parts that can be grown with waste materials and natural binders. These elements, while strong and durable, will eventually biodegrade, promoting circular economies and ecological design strategies, as opposed to synthetic, non-biodegradable products and material systems.

In line with the 2021 ACADIA conference theme, *INOCULATED MATTER* reflects a realignment in practice, a paradigm shift in making and fabricating architecture in collaboration with living systems. This realignment opens up new possibilities for architectural design, fabrication, and speculation, and challenges traditional and conventional methods of design and making in a post-human world to advance interspecies communication. The project seeks opportunities for architectural design to shift from 'systems thinking' to 'living systems thinking'. The project questions cultural attitudes towards debris, by valuing waste contributing to circular material cycles and ecologically positive material processes. It prompts rethinking methods in how individuals, neighborhoods, and cities address waste, and seeks to improve sustainable practices, raise ecological awareness, and promote ecological 'tuning' [3]. By working in tandem with living organisms and taking measure to integrate upcycled waste, we can take measures to reduce the amount of materials that end up in landfills and find alternative ways to replace certain objects that are made out of plastics and other environmentally harmful materials. *INOCULATED MATTER* seeks to address these topics through applied research, and the development of techniques and workflows that demonstrate alternative biologically based methods for designing and making architecture.

3 3D printed product, suspended lamp, made with upcylced waste and mycelium (© bioMATTERS LLC, 2021)

4 3D printed objects made with upcylced waste and mycelium, and various substrates and mycelium strains (© bioMATTERS LLC, 2021)

NOTES

1. F. V. W. Appels. 2020. "The use of fungal mycelium for the production of bio-based materials." PhD thesis, Utrecht University, Utrecht, NL.
2. A. Goidea, D. Floudas, and D. Andreen. 2020. "Pulp Faction, 3D printed material assemblies through microbial biotransformation." In *Fabricate 2020: Making Resilient Architecture Conference*, edited by J. Burry, J. Sabon, B. Sheil, and M. Skavara. London: UCL Press.
3. T. Morton. 2019. *Being Ecological*. London, UK: MIT Press.

IMAGE CREDITS

Figures 1-8: © bioMATTERS LLC, 2021

Nancy Diniz is a registered architect and educator. She is a co-founder of Augmented Architectures and bioMATTERS LLC, based in New York City and London. Parallel to her practice she is the Course Director of the MA in Biodesign at Central Saint Martins, University of the Arts London. Her research and practice involve working with living systems and computational design, and pertain to topics including biomaterials, bio and digital fabrication, virtual reality, and data visualization. Nancy is a co-editor of *Data, Matter, Design* (Routledge 2020). Her work has been exhibited internationally at various venues including Lisbon Architecture Triennale, London Design Festival, Istanbul Design Biennale, EYEBEAM, The Today Art Museum, MAAT Museum and GAA Foundation. She is the recipient of several grants and fellowships namely from New York State Council on the Arts/Storefront, MacDowell Colony, EYEBEAM, Seoul Art Space Geumcheon, and The Foundation for Science and Technology, Portugal. Nancy is a member of the Design and Living Systems Lab at Central Saint Martins UAL. She has held academic positions in the US, UK, China, Italy, and Portugal.

5   3D printed waste-based extrusion paste (© bioMATTERS LLC, 2021)

6   3D printed waste-based extrusion paste (© bioMATTERS LLC, 2021)

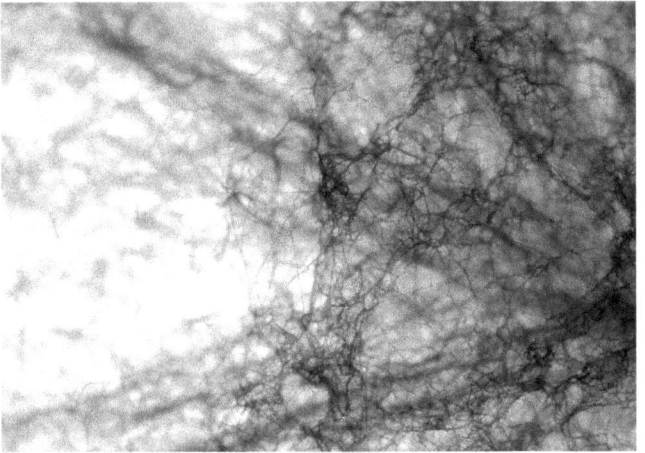

7   Microscopic image of hyphae network (© bioMATTERS LLC, 2021)

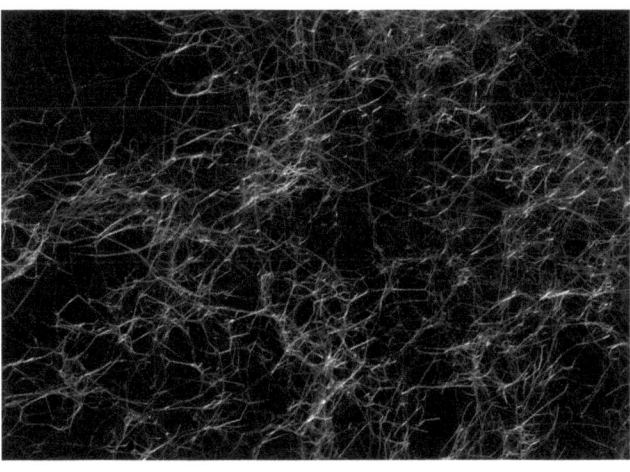

8   Computational simulation of hyphae network (© bioMATTERS LLC, 2021)

Frank Melendez is an architectural designer, educator, and researcher. He is a partner at Augmented Architectures and bioMATTERS LLC, based in New York City and London. Parallel to his practice he is an Associate Professor at the Spitzer School of Architecture, City College of New York, CUNY. His practice, teaching, and research focus on the advancement of architectural design through the integration of emerging digital technologies within the built environment. This work engages topics pertaining to architectural drawing, computation, ecology, digital fabrication, bio and synthetic materials, physical computing, and robotics. Frank is the author of *Drawing from the Model* (Wiley 2019) and a co-editor of *Data, Matter, Design* (Routledge 2020). He has held academic appointments at Carnegie Mellon University and Louisiana State University, and his work has been supported through grants, fellowships, and memberships including, the New York State Council of the Arts (NYSCA) / Van Alen Institute, the MacDowell Colony, and NEW INC.

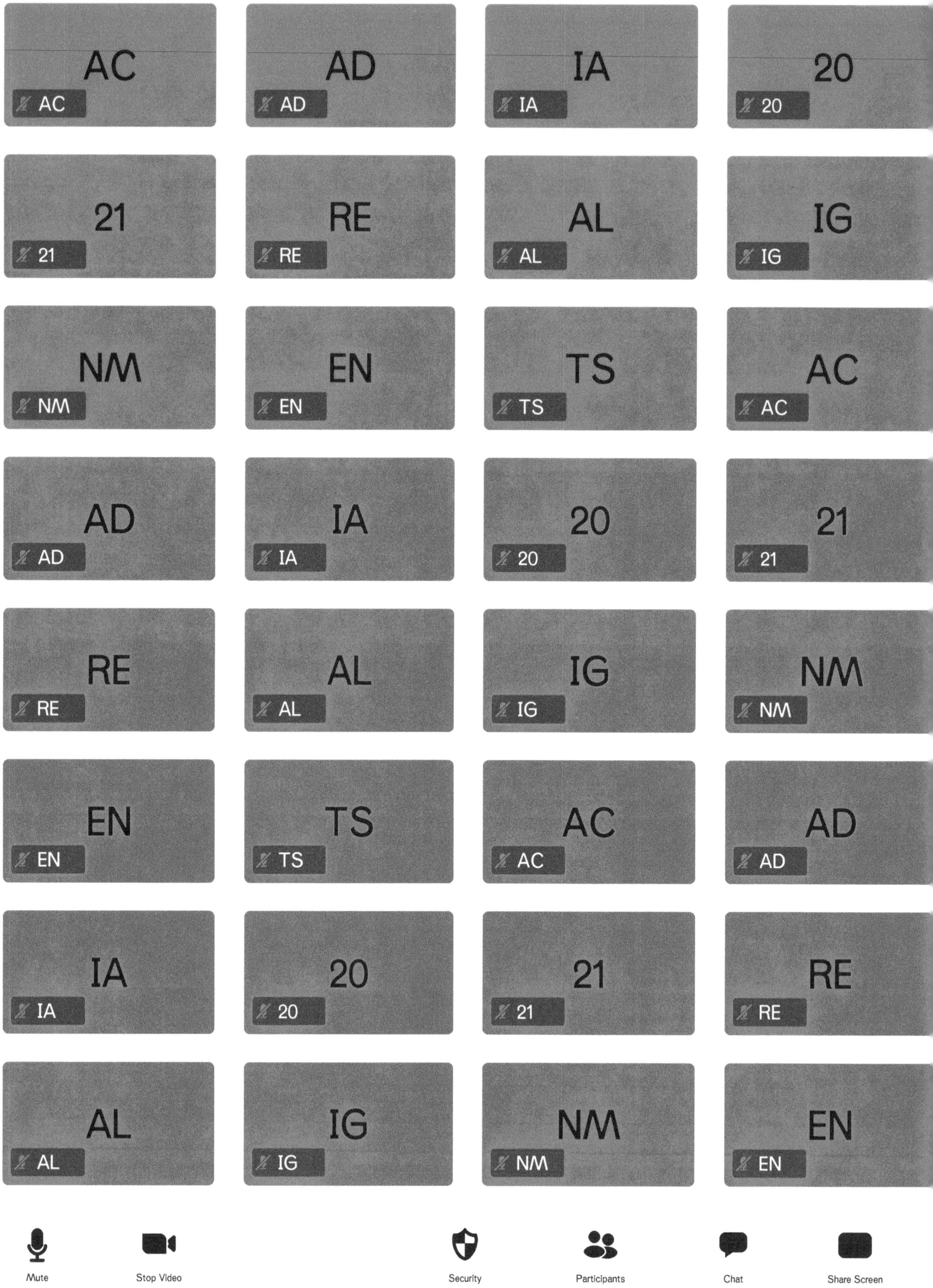

AC	AD	IA	20
21	RE	AL	IG
NM	EN	TS	AC
AD	IA	20	21
RE	AL	IG	NM
EN	TS	AC	AD
IA	20	21	RE
AL	IG	NM	EN

 Mute   Stop Video   Security   Participants   Chat   Share Screen

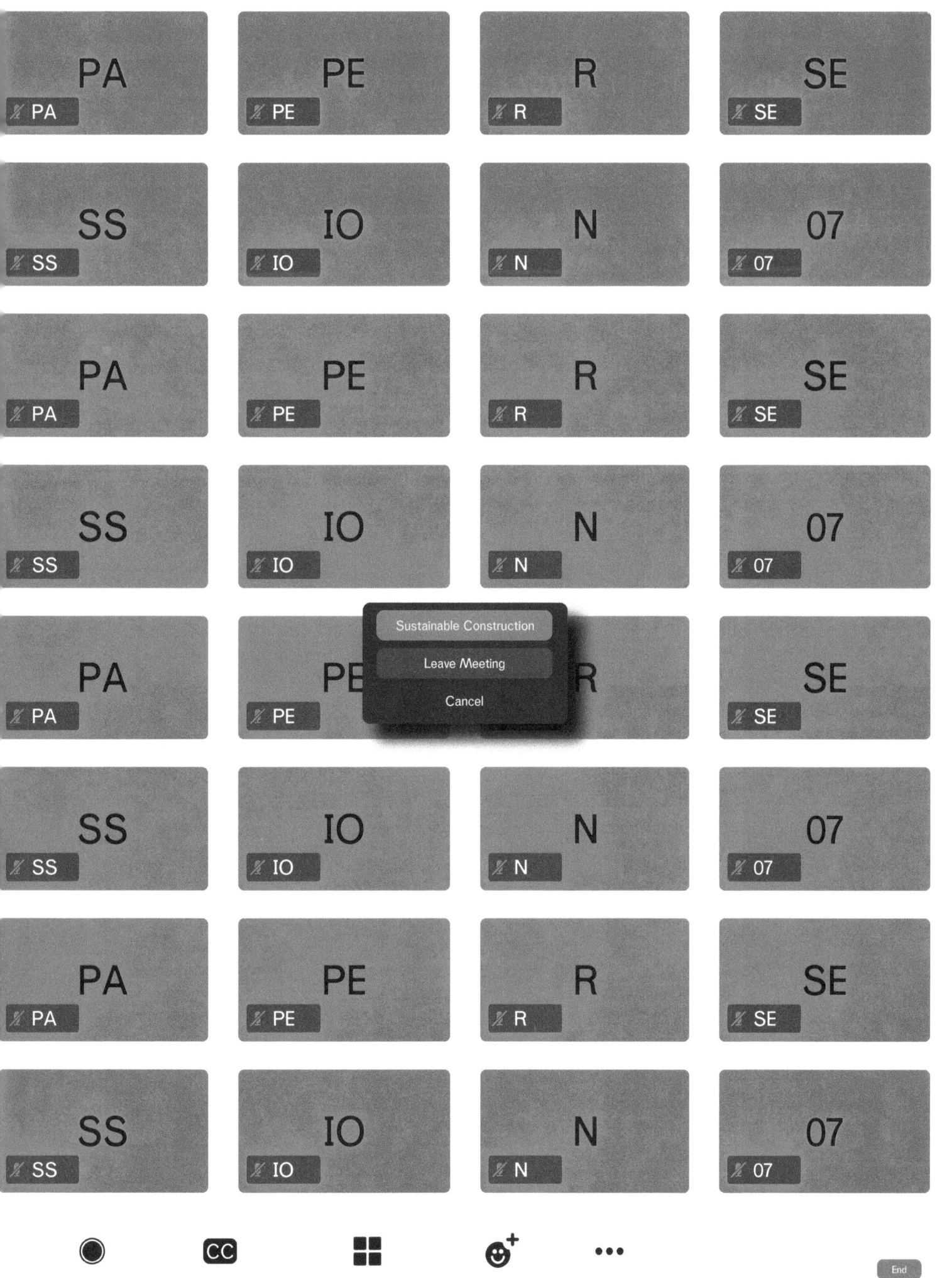

A Material Monitoring Framework

Tracking the curing of 3d printed cellulose-based biopolymers

Gabriella Rossi
Ruxandra Chiujdea
Claudia Colmo
Chada ElAlami
Paul Nicholas
Martin Tamke
Mette Ramsgaard Thomsen

CITA/Royal Danish Academy

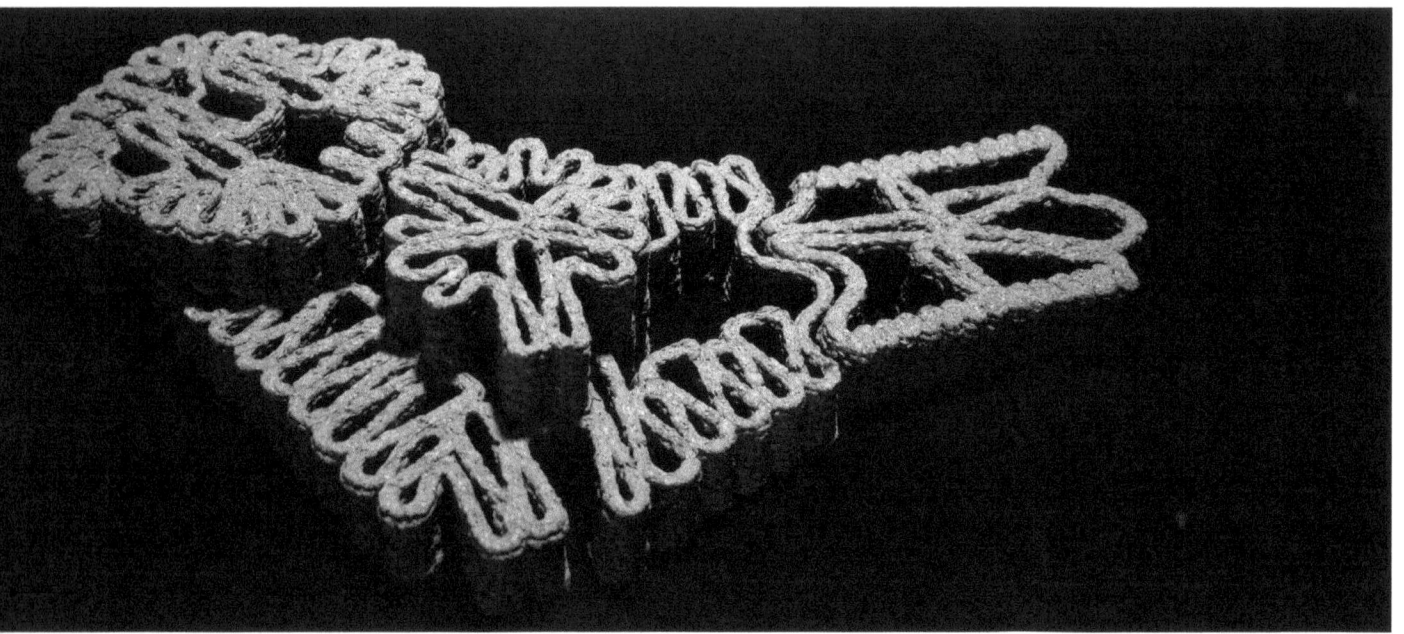

1   Behavior control of cellulose-based biopolymer 3d printed components offers potential for architectural applications

ABSTRACT

Through 3d printing, cellulose-based biopolymers undergo a two-staged hybrid fabrication process, where initial rapid forming is followed by a slower secondary stage of curing. During this curing large quantities of water are evaporated from the material which results in anisotropic deformations. In order to harness the potential of 3d printing biopolymers for architectural applications, it is necessary to understand this extended timeline of material activity and its implications on critical architectural factors related to overall element shrinkage, positional change of joints, and overall assembly tolerance. This paper presents a flexible multi-modal sensing framework for the understanding of complex material behavior of 3d printed cellulose biopolymers during their transient curing process.

We report on the building of a Sensor Rig that, using a mix of image-based, marker-based, and pin-based protocols for data collection, interfaces multiple aspects of the curing of our cellulose-slurry print experiments. Our method uses timestamps as a common parameter to interface various modes of curing monitoring through multi-dimensional time slices. In this way, we are able to uncover underlying correlations and affects between the different phenomena occuring during curing. We report on the developed data pipelines enabling the Monitoring Framework and its associated software and hardware implementation. Through graphical Exploratory Data Analysis (EDA) of three print experiments, we demonstrate that geometry is the main driver for behavior control. This finding is key to future architectural-scale explorations.

INTRODUCTION

Industrialization has focused architectural production on materials, extracted from the geosphere, that present standardized behaviors and hold poor end-of-life possibilities. Defining ways of working with biospheric materials that are inherently regenerative and biodegradable are important steps for future sustainable building practice. Cellulose, our material base, and one of many abundant, inexpensive and renewable bio-based materials (Pattinson and Hart 2017) can be mixed with different binders, fiber reinforcement, and fillers into 3d printable slurries. However, 3d printed biopolymers undergo radical changes in their material performance and geometry as they transition from the wet to dried state. This paper reports on the Predicting Response research project investigating how the inherent heterogeneity, temporal plasticity, and anisotropy of biopolymers can be characterized in order to build advanced computational models that predict their behavior in turn enabling their usage as materials for the built environment. This paper reports on the development of the methodological steppingstones to enable such modelling practices: the making of a Monitoring Framework and Sensing Rig to characterize material behavior during fabrication.

Data-driven Understanding of Biomaterials

Technological advancements increasingly allow architects to interface fabrication and design to the material properties of their artefacts (Ramsgaard Thomsen 2012). By combining this material design approach to new methods of data collection, we can develop new forms of analysis and modelling (Del Signore et al. 2021). Sensor-based data-driven principles in computational design research have focused on three information-rich domains where data gathering is tractable. The first is in urban and city studies where big data is gathered using city-scale sensor networks or live satellite images (Kang et al. 2020; Najari et al. 2020). The second is cyber-physical robotic fabrication, where sensors are used to measure real-time parameters punctually or via the cloud and adapt the robotic motion on the fly (Brugnaro et al. 2016; Vasey et al. 2018, Rossi and Nicholas 2018; Nicholas et al. 2020). The third is the case of monitoring material behavior, where time-based thermal sensor measurements of both samples and their environment as well as image-based captures to monitor visual properties are registered (Faircloth et al. 2018a).

Our research sits within a growing interest in additive manufacturing of biopolymer composites. While casting might provide a high-volume production method of modular elements, 3d printing allows for a larger design space not only through geometrical freedom which has performance and formal drivers (Goidea et al. 2020), but also allow for material recipe grading (Mogas-Soldevila et al. 2015), and on the fly online adaptation. However, this potential does not come without complex manufacturing considerations: 3d printing of cellulose-based slurries is a two-staged hybrid fabrication system presenting an initial rapid forming, followed by a slower curing and hardening phase. The curing phase is characterized by the evaporation of the slurry's water content, allowing the slurry to acquire strength properties but resulting in large-scale shrinkage and warpage of the printed geometry. While geometric considerations and manipulations of curvature and edge condition matching (Goidea et al. 2020) or corrective models that overcompensate the printing toolpath to obtain as close dried target geometry as possible can be used to obtain consistent printed results (Dritsas et al. 2018; Sanandiya et al. 2018), these applications limit their investigation of the element to a binary initial "wet" and final "dry" state. We go beyond the state of the art by studying the transient curing process at different scales and across different phenomena using multidimensional feature vectors across time. By using a multi-modal sensing regime on this extended timeline of material activity, our aim is to better characterize the complexity and interactions between the curing environment and the resulting transformation thereby enabling a stronger control of the behavior of slurry biomaterials (Ramsgaard Thomsen et al. 2021), a more precise component assembly and paving the way for predictive behaviour models using Machine Learning algorithms.

In this paper we present a Material Monitoring Framework (Figure 2) for tracking the curing of 3d printed cellulose biopolymer. We first contextualize our sensing approach with respect to industry state-of-the-art curing monitoring techniques. We describe the development of our framework using IOT: a Sensor Rig which combines multiple sensors, markers and image devices, as well as data management protocols and cloud-based data storage. Our method utilizes the timestamp to bring together various modes of curing monitoring and uncover underlying correlations and affects between different phenomena, thanks to the multiple data point structure of time slices. Using Exploratory Data Analysis (EDA), we report on three printing experiments and showcase how within a controlled curing environment, geometry is the main driver for the complex behavior of cellulose-slurry curing. This finding, central to architectural and design consideration, will be our main guide for the next steps of our research.

CONTEMPORARY SENSING PRACTICE

Architectural Sensor-Material practice is deeply embedded

in contemporary construction contexts. Powered by the rise of IOT, sensors are used for punctual quality control in pre-fabrication (Stavropoulos et al. 2013) where they are specifically designed for low-variety large-volume sensing. Another usage of sensors is lifetime tracking and safety monitoring of completed structures (Zhang et al. 2021) or of construction processes such as concrete casting. Commercially available systems, such as Sensohive's concrete monitoring [1] or the Built2spec [2] sensors can measure temperature, humidity, and CO^2 concentration. To give insights about the curing state. These products are limited in terms of high costs, and data analysis and synthesis customization. They are often sacrificial to the structure they monitor and cannot be reused. A low-cost cloud-based plug-and-play sensor network for architectural application has been developed by KieranTimberlake named "Pointelist" [3]. It is employed to sense the building envelope giving a real time feedback loop of the indoor level of comfort. Composed of a Wi-Fi integrated node and a Raspberry Pi microcomputer, the system uploads data from different humidity and temperature sensors, and thermal imaging to a web interface where data can be analyzed. In the Building 661 case (Faircloth et al. 2018b) the data made it possible to predict the thermal behavior and the expected material properties of the non-insulated facade. This example confirms the viability of our data-driven approach for material monitoring.

METHOD

Studying the curing of 3d printed cellulose slurry requires monitoring of multiple phenomena across the duration of the curing process which, depending on the size of the print and the thickness of the print beads, lasts 7 to 15 days. In our research context, the Sensing Rig cannot be sacrificial but rather reusable, since we need to repeatedly monitor the curing of many material probes and analyze and compare the collected data. The Sensor Rig also needs to be replicable and able to accommodate an open-ended range of sensors. As the 3d prints are fragile when still wet, the sensors should be as non-intrusive as possible. These criteria necessitate remote and vision-based sensors. We utilize timestamp tagging as a tool to interface and compare data incoming from different sources. This creates a multiple-data-point time slice which we use to uncover data patterns and correlate phenomena driving the curing process.

Sensing Hardware

The Monitoring Framework includes the monitoring of the following phenomena: (1) weight (sensor: 5 kg bar-type load cells HX711); (2) temperature of material surface for tracking of evaporation (sensor: measurement and false

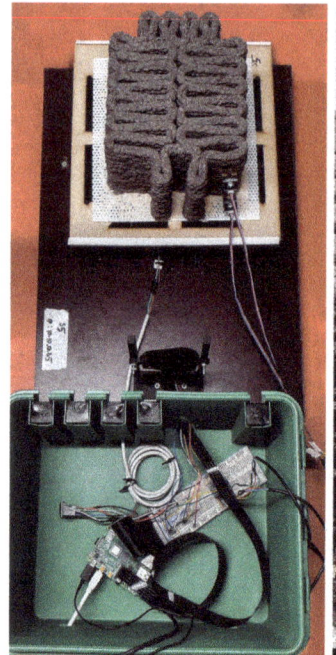

2

color imaging with FLIR Lepton 3.5 radiometric camera); (3) humidity in material interior (sensor: SEN00114DFROBOT resistive soil moisture sensors); (4) overall geometry (sensor: Intel RealSense LiDAR); (5) position of specific points in space (sensor: Motion capture with Optitrack PrimeX13); and (6) Texture, Color and Look (sensor: RGB imaging with PiCamera). This print local sensing is further framed by environmental sensing (sensor: relative humidity and temperature of the environment with AM2315 Sensor).

The Sensing Rig (Figure 3) sits on a board for easy repositioning. The scale is mounted using a 3d printed spacer to correct the strain. A 300 mm x 300 mm wooden frame holds the printed sample. The two camera tracking systems (piCamera and thermal camera) are fixed at 150 mm distance from the sensing plate. Material humidity sensors (SEN00114DFROBOT) are inserted into the sample to track the gradual evaporation of water. To sense changes in geometry, we have explored the possibility of integrating an Intel RealSense LiDAR scanner to record a pointcloud that can be transformed into mesh in a CAD environment; however, the usage of punctual markers to track the motion of specific points of interest proved better. We use eight Optitrack PrimeX13 cameras set up around the drying station, for this task. The cameras are calibrated using Wanding on Motive [4] to 0.1 mm tolerance. Camera feed masking ensures the stability of the reflective marker tracking by avoiding inference from reflective matter in the setup such as the humidity sensors. We use Motive's loopback streaming for marker position broadcasting. Simultaneously, we are monitoring the temperature and

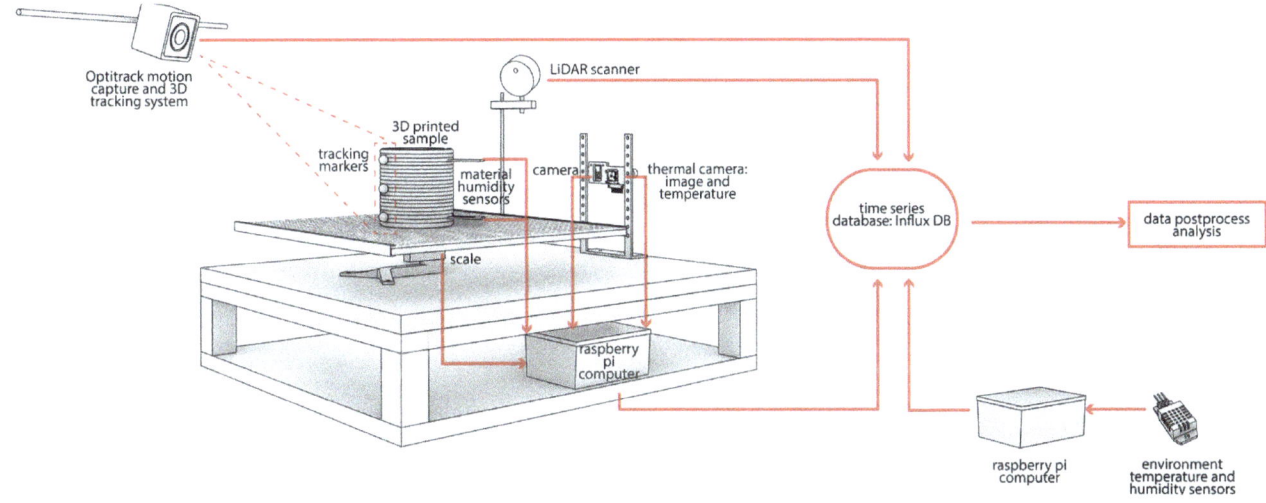

2   Material Monitoring Framework for tracking 3d printed cellulose: closeup on our Sensing Rig; weight and moisture sensors and image and thermal cameras are connected to a RaspberryPi, while the Optitrack reflective markers are tracked by eight surrounding cameras

3   Diagram showcasing the multiple devices that constitute the Sensing Rig and the data flux between it and the cloud

humidity around the sensing station using two Raspberri Pis placed at opposite sides of the room.

Data Pipeline

We develop custom upload and download data pipelines to fit with our custom sensing hardware setup, and to accommodate for different data types from multiple sources. We use the cloud-based service of InfluxDB [5] an open-source time-series database for storing our monitoring data. Their python-based API client [6] defines the Line protocol for data upload: a measurement is assigned to a timestamp and can be referenced using tags and fields, which allow for ease of query and analysis later. Measurements are taken every five minutes. For each experiment, an umbrella tag is created with subfields. The Optitrack motion-capture data is streamed using our custom version of the NatNet C++ client [7], and we upload xyz positions for every marker at 0.1 mm precision. Different measurements from the RPi sensing hardware are formatted into the Line protocol and uploaded tagged by measure, under the same experiment tag. Environment monitoring of the room temperature and humidity condition is uploaded as a separate tag. Finally, RGB color image and a false-color thermal image are time stamped and uploaded to a Google Drive storage using their Python API [8]. Once the curing is complete, the data is parsed for analysis. By using the influxDB API, we run custom queries and compile a Pandas dataframe, which allows the processing of data into meaningful information. Normalization is an important step of data processing: for instance, the timestamp is translated into curing time, the Optitrack marker coordinates are translated into distance from the initial wet point, and the weight of the sample is translated into weight loss percentage. With this method, we create a dataframe for each experiment and use graphical Exploratory Data Analysis to understand the curing process.

CURING EXPERIMENTS SETUP

In-house variations of the cellulose slurry recipe developed at CITA (Chiujdea and Nicholas 2020) have shown the impact of varying proportions and ingredients on print behavior. The recipe used for all prints here reported optimizes nozzle flow and inter-layer adhesion, developed by our DTU partners (Rech et al. 2021a; Rech et al. 2021b). It blends cellulose floc, wood flour, glycerol, xanthan gum, calcium chloride and water (72% of the total weight). The material is packed into an acrylic tube which feeds to our custom extruder end effector. Due to the material's viscosity, a pneumatic pressure of 2 bar is sufficient to move the material to the printing head. Here a motor driven screw ensures consistent flow of material during extrusion. Using an ABB140 arm, we print at 25 mm/s speed, with a 9 mm nozzle and 4 mm layer height. The samples are printed on a perforated mesh which aids their later curing process (Figure 4).

The goal of the experiments is to isolate, for a given recipe and curing environment, the role of the print geometry in driving the curing process. We report on three prints below:

- Experiment 1 (Figure 5a) examines a cylinder as a baseline to the behavior of the material. The hypothesis

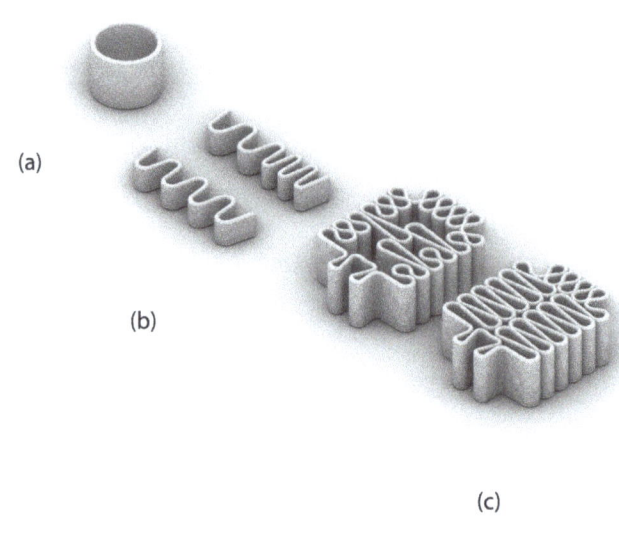

4   Print setup with ABB 140: using the perforated mesh as a print bed allows air to circulate through the sample during curing, thus avoiding molding

5   Digital printing toolpath of the three experiments; the toolpaths are spiralized to guarantee print stability

is that all material is equidistant from the center, and therefore, equally exposed to evaporation. Our sample is 100 mm in diameter and 74 mm high. It is monitored using a weight scale, two humidity sensors (one near top layer, one near bottom layer), and a 3x3 grid of Optitrack markers and thermal imaging.

- Experiment 2 (Figure 5b) aims to understand how material distribution affects evaporation, and therefore curing behavior, where larger exposure to evaporative surface would lead to faster curing. It comprises two "sine wave" samples 240 mm long: a constant 55 mm wave-length sine wave and a varying sine wave with 35 mm and 55 mm wavelengths, both 50 mm wide amplitude.
- Experiment 3 (Figure 5c) expands this understanding to volumes, with particular focus on the joints and assemblies. The hypothesis is that the density of the infill would differentiate the curing of the joint. We print two component samples of size 250 x 200 x 80mm fitted with the same male joint. Both present the same outer surface area (87,000 mm^2) but a varying inner evaporative surface area: a dense component with 240,000 mm^2 inner surface and a light component with 264,000 mm2 inner surface. The experiment questions how evaporation affects the deformation of the component and its assembly joint.

RESULTS

Experiment 1 "The Cylinder"

We represent the findings of our multimodal sensing through graphical Exploratory Data Analysis (EDA), see Figure 6, which shows that the cylinder print reaches a drying plateau on day 5. The sample loses approximately 60% of its weight. As water evaporates, we observe a drop in humidity registered by the moisture sensors. In the graph we can see that there is a one day delay between the sharp drop in humidity at the top sensor (at 15% weight loss) and at the bottom sensor (at 30% weight loss), indicating that the print dries from top to bottom. However, the slight increase in the bottom sensor reading indicates that water travels downwards during the print process due to gravity. This hypothesis is confirmed by the thermal imagery: the sample is initially all blue (10 degrees) and slowly transitions downwards into red as it dries and reaches room temperature (20 degrees). We also observe that 95% of the shrinkage occurs over the first 48 hours of curing. The travel distance of the Optitrack markers, as well as their travel speed is higher for the markers placed at the top of the sample (20 mm which in this case represent 30% of the total sample height), and relatively small for those placed at the bottom (approximately 2 mm). We can therefore formalize a baseline curing behavior: with all material equally exposed to an evaporative surface, a mostly vertical shrinkage occurs, with water travelling through the sample due to gravity.

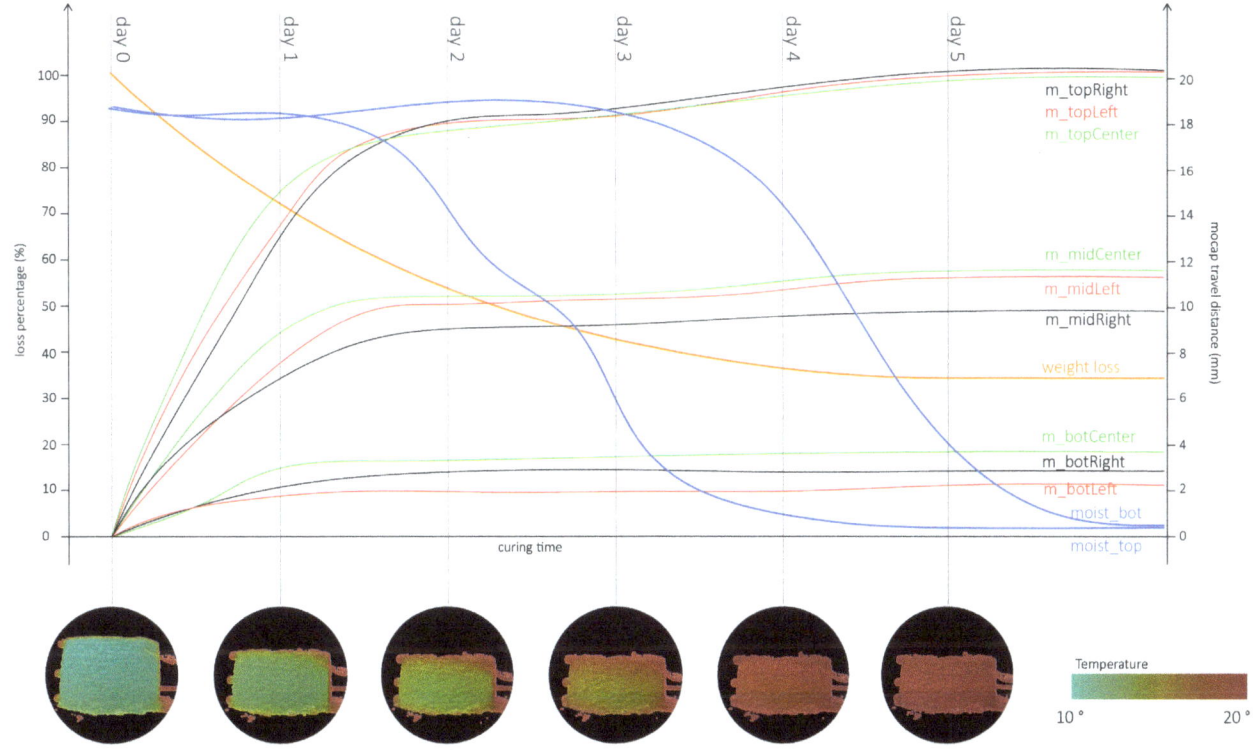

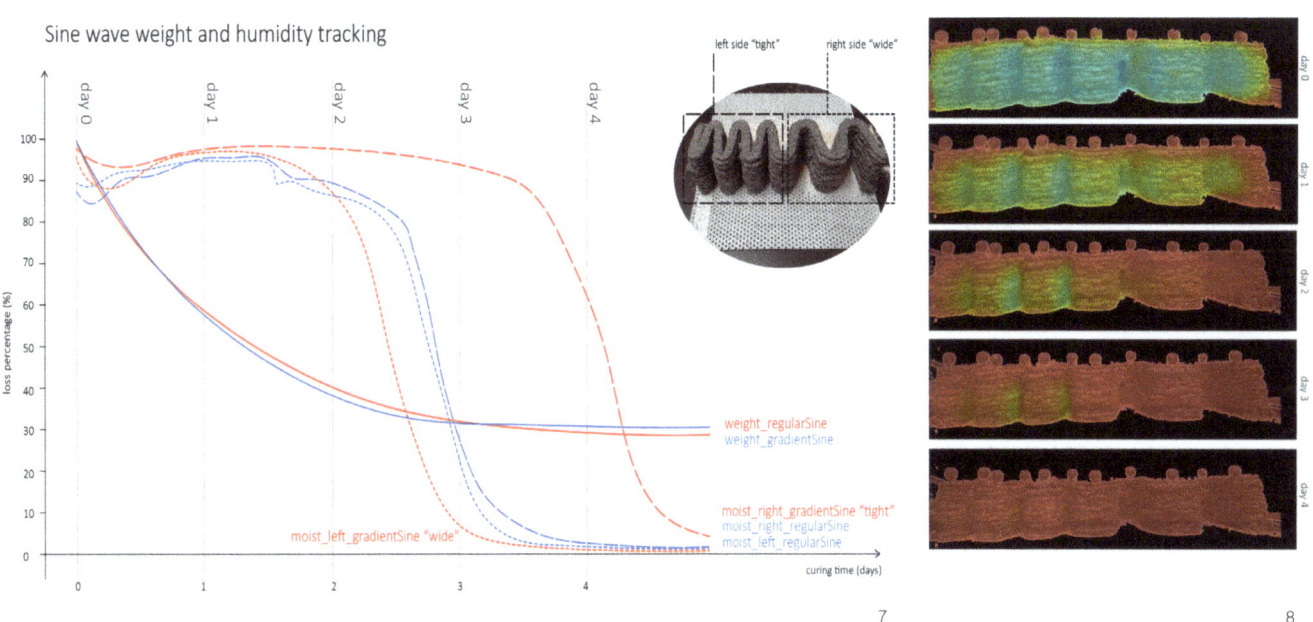

6   Graphical EDA of Experiment 1: the curing behaviour of a 100 mm diameter cylinder; Multi-Sensor data is overlaid in the graph, and thermo-imagery vignettes showcase the curing state every day at 13:00

7   Graphical EDA of Experiment 2: the simultaneous curing of a regular and a graded sine wave prints; the key legend shows the placement of moisture sensors of the sample

8   Monitoring of the gradient sine wave drying using thermal imaging; notice the left "tight" side of the print drying later than the right "wide" side

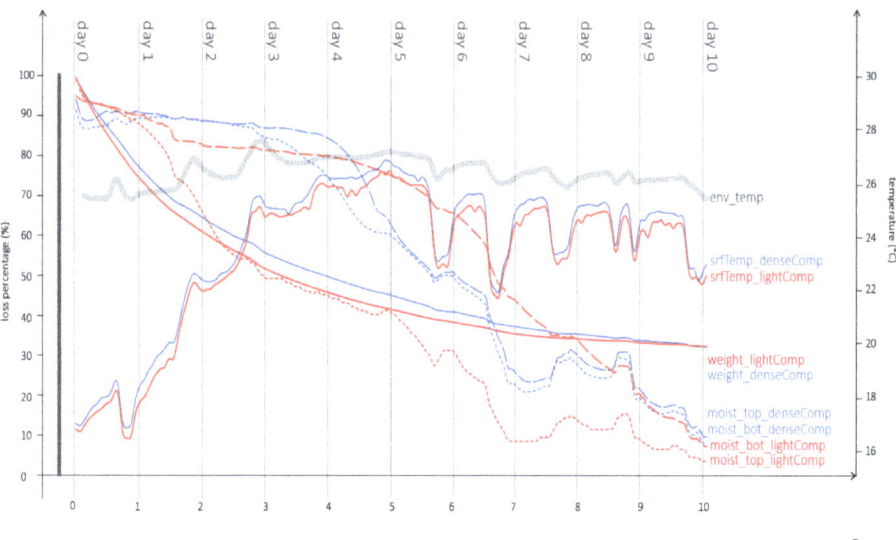

9 Graphical EDA of Experiment 3: measurements in red relate to the light component, in blue relate to the dense component, and in gray to the curing environment in which the components are left to dry

10 Plotting of the Optitrack Marker displacement across time for the light (a) and dense (b) components reveals the speed and amplitude of the shrinkage; in (c)(d) we spatially represent the geometric deformation across five curing moments; markers on the dense component (d) dry at unequal rate if compared to the ones on the light component (c)

Experiment 2 "The Sine Wave"

We compare the curing process of the two sine waves samples over five days (Figure 7). We observe that while both samples lose water at the same rate, the localization of this humidity loss is different. The humidity sensor readings show that in the case of the regular wave geometry, both left and right humidity sensors register a moisture loss on day 2, but in the case of the varying wave geometry, the humidity sensor on the wide wave side registers the drop early on day 2, but the one on the tight wave side only registers it on day 4. This geometry-driven differential curing behavior hypothesis is confirmed by the thermal imagery captures over the five days (Figure 8), where we see that on day 2 the wide side of the sine wave presents a surface temperature of 20 degrees—meaning it is dried—while the cavities of the tight part of the wave are still at 14 degrees—meaning they are still humid. We also observe that the vertical shrinkage is more accelerated on the wide side than on the tight side. We conclude that evaporative surface exposure—in this case the sine wave period—allows the acceleration or slow-down of the drying.

Experiment 3 "The Modular Component"

The two volumetric components are monitored over ten days. Through graphical EDA (Figure 9), we observe that the light component, the top moisture sensor registers a humidity drop already on day 1, while the bottom one only registers it on day 5. This is different from the dense sample, where both sensors start to register a drop between day 3 and 4. This implies that the airiness of the infill allows water to leave the material faster. Examining the minimum temperature at the joint, which has been placed facing the thermal camera, we can see that for both samples it rises steadily from 17 degrees to 24 at the same rate, while following the temperature cycles within the lab space. The central part of the joint remains colder on the dense component than the airy one. We also observe that the dense sample presents less fluctuations. We assume that this is because it has 30% more material mass and can therefore better retain the heat. Analyzing the Optitrack markers via graphical EDA (Figure 10 a/b) shows that drying does not occur equally across the samples. While Experiments 1 and 2 dried explicitly from top to bottom, in Experiment 3 we observe that the denser the infill, the more the drying will also be from edges to center. We can see that in the case of the light component (Figure 10a), all the marker points at the top layer start to move at the same time and at similar rates, no matter if it is the joint or the central or lateral part of the lattice. However, for the dense component (Figure 10b), it is the joint that starts shrinking first, followed by the lattice edges, and finally the lattice center starts to shrink as well. This is due to the way the infill structure forms evaporative cavities. Our flexible data pipelines allows us to visualize this data in different platforms: in addition to graphical EDA, we can also reconstruct and visualize the Optitrack travel data in Grasshopper/Rhino (Robert McNeel & Associates 2021) environment to understand the spatial implications of the curing process. In Figure 10 c/d we show the location of the Optitrack markers over 5 timestamps: 0, 500, 1000, 1500 and 4000 minutes after curing. This representation confirms that the motion of Optitrack markers, and therefore the shrinkage of the components, does not occur equi-temporally.

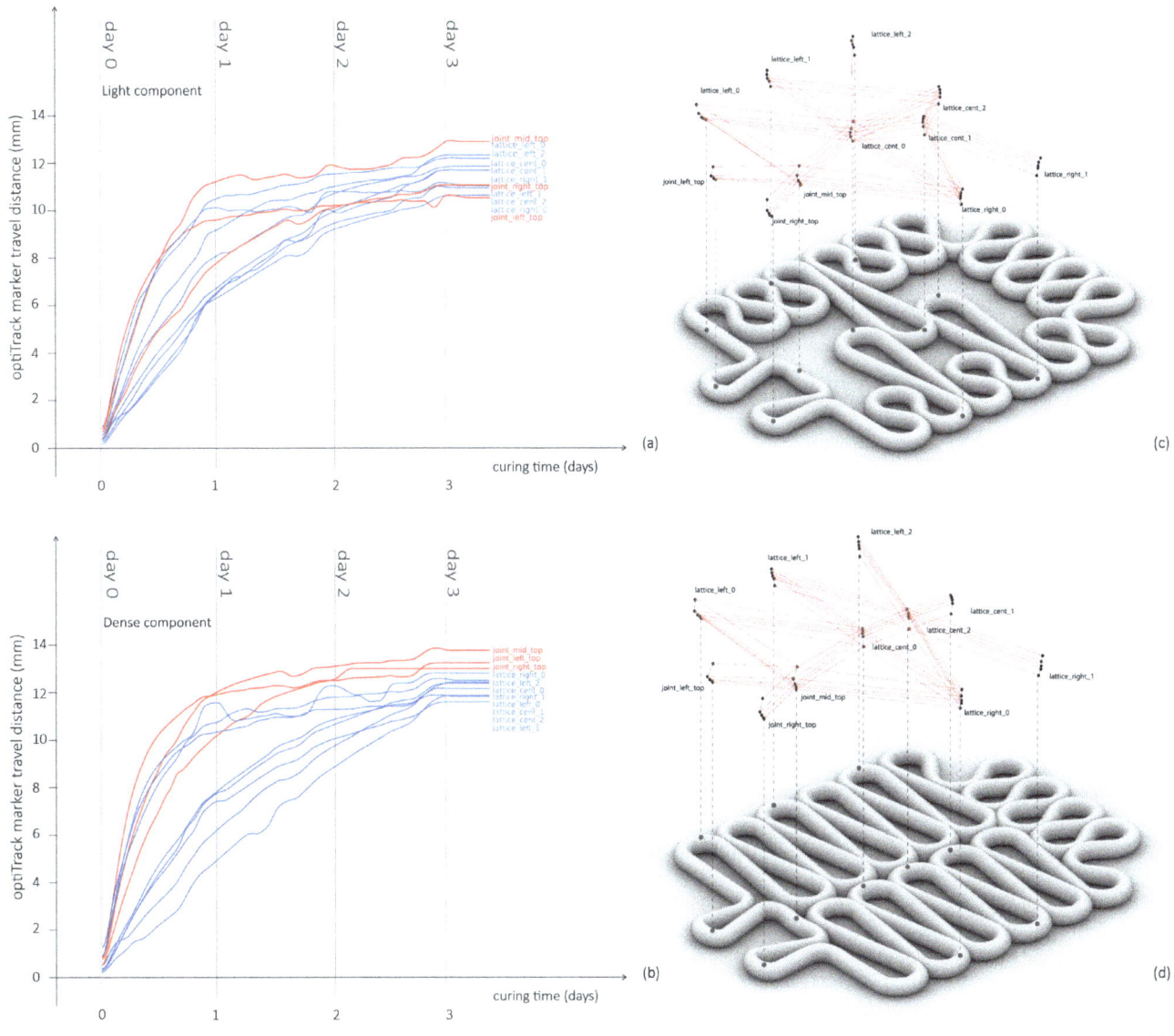

DISCUSSION & CONCLUSION

In this paper, we present the development and testing of a Material Monitoring Workflow and Sensing Rig, which provides deeper insight into the complex curing behavior of 3d printed cellulose-based slurries. Our Sensing Rig hardware has been developed following an accessible approach that can be easily replicated using readily available components and open-source protocols. While initial setup and calibration of these components and data wrangling processes can be laborious, since sensors are prone to data outliers and spikes, overall, the data quality produced by our rig is sufficient for our research, where the change of measured data over time is more important than the absolute precision of a single data point. Our framework's decentralized architecture with local micro-computing power and reliable cloud database provides a stable base for continuous sensing of many probes, where sensing can continue, while changing numbers of rigs are added, maintained, and removed. Additionally the sensors on each rig are plug-and-play and their configuration can be tailored to specific experiments needs.

The experiments we have described, through their associated sense data, have revealed important aspects of material behavior during curing that have crucial implications on critical architectural factors driving design and fabrication, related to element shrinkage, positional changes of joints and fits which would affect a final assembly. We demonstrate that the curing process follows a specific non-linear timescale across the samples and occurs at a changing rate. Printed elements shrink in height (-25%) more than in print plane. The impact of gravity on the drying, as water moves downwards across the layers, is minor and may improve interlayer bonding. Moreover, while the drying phase extends up to ten days, the fact that nearly all water loss and shrinkage occurs within the first three days implies that a printed object could be positioned and left to finish drying in place after this initial stage, as its geometry and density then remain relatively stable.

Our experiments have also revealed that geometry is the main driver for surface evaporation. Open geometries that expose more surface area to airflow dry more quickly and more evenly, therefore, will be less affected by warping. By intentionally grading the airiness of the sample, one can therefore dictate the drying "sequence" of the piece, and control not only the final tolerances of the piece, but also chose to dry more robust or more fragile parts before others. This can also be used strategically to control interlocking joint tolerances, which would guarantee the feasibility of an architectural assembly using these 3d printed components. Initial material characterization is currently underway, and points to the possibility of this material being used for interior partitioning, furniture scale, and thermal and sound insulation. Taken together, these insights point to ways in which a cellulose-based biopolymer could be graded, or a production environment configured, to manage, counteract, or speed the natural curing process, and reveal opportunities for steering and control that are missed by a binary 'before and after' approach. Further experiments will be made to unpack this fabrication and assembly strategy unique to two-staged hybrid fabrication processes, with focus on male/female joint warpage driven by the infil of the components, and explore design strategies that negotiate material distribution between the need for airyness versus structural stability.

Finally, from a conceptual point of view, we report that the timestamp is a successful method of combining multiple data streams from multiple sensor sources. Through minor data wrangling and normalization operations, we can use graphical EDA mapping of monitoring data to understand and theorize the behavior of these components. Formatting data as a time series also offers advantages for the next stages of this research, where we use statistical analysis tools such as Principal Component Analysis to backward engineer the design of our Sensor Rig and choice of sensed phenomena. Lastly, next stages of this research will also take advantage of the developed formatting pipeline, and build a library of curing samples over time, to interface with Machine Learning models and predict time-varying data readings on our next printed samples.

ACKNOWLEDGEMENTS

This work is funded by Independant Research Fund Denmark (DFF) PROJECT NUMBER 9131-00034B "Predicing Response," in collaboration with Anders Egede Daugaard and Arianna Rech (Denmark Technical University) and John Harding (University of Reading).

NOTES

All website references listed below were accessed October 2021.
1. https://maturix.com/knowledge-center/.
2. https://cordis.europa.eu/project/id/637221.
3. Kieran Timberlake, n.d. *Pointelist*. https://www.pointelist.com/.
4. Natural point. *Motive*, v2.2.0, 2019. https://optitrack.com/software/motive/.
5. InfluxData, *influxDB 2.0 Python Client*, 2019. https://github.com/influxdata/influxdb-client-python.
6. NaturalPoint, *NatNet SDK*, 2021. https://optitrack.com/software/natnet-sdk/.
7. Google, *Drive API*, V.3, 2021. https://developers.google.com/drive/api/v3/quickstart/python.

REFERENCES

Brugnaro, G., E. Baharlou, L. Vasey, and A. Menges. 2016. "Robotic softness: An adaptive robotic fabrication process for woven structures." In *ACADIA '16: Posthuman Frontiers: Data, Designers, and Cognitive Machines; Proceedings of the 36th Annual Conference of the Association for Computer Aided Design in Architecture*. 154–163.

Chiujdea, Ruxandra, and Paul Nicholas. 2020. "Design and 3D printing methodologies for cellulose-based composite materials." In *Anthropologic: Architecture and Fabrication in the Cognitive Age; Proceedings of the 38th eCAADe Conference*, vol. 1., edited by L. Werner and D. Koering. TU Berlin, Berlin, Germany. 547–554.

Del Signore, Marcella, Nancy Diniz, and Frank Melendez. 2021. "Information Matters." In *Data, Matter, Design: Strategies in Computational Design*, edited by Frank Melendez, Nancy Diniz, and Marcella Del Signore. New York: Routledge.

Dritsas, Stylianos, Samuel E. P. Halim, Yadunund Vijay, Naresh G. Sanandiya, and Javier G. Fernandez. 2018. "Digital Fabrication with Natural Composites: Design and Development towards Sustainable Manufacturing." *Construction Robotics* 2 (1–4): 41–51. https://doi.org/10.1007/s41693-018-0011-0.

Faircloth, Billie, Ryan Welch, Yuliya Sinke, Martin Tamke, Paul Nicholas, Phil Ayres, Erica Eherenbard, and Mette Ramsgaard Thomsen. 2018. "Coupled Modeling and Monitoring of Phase Change Phenomena in Architectural Practice." In *Symposium on Simulation for Architecture & Urban Design 2018*. 81–88.

Faircloth, Billie, Ryan Welch, Martin Tamke, Paul Nicholas, Phil Ayres, Yulia Sinke, Brandon Cuffy, and Mette Ramsgaard Thomsen. 2018. "Multiscale Modeling Frameworks for Architecture: Designing the Unseen and Invisible with Phase Change Materials." *International Journal of Architectural Computing* 16 (2): 104–22.

Goidea, Ana, Dimitrios Floudas, and David Andréen. 2020. "Pulp Faction: 3d Printed Material Assemblies through Microbial Biotransformation." In *Fabricate 2020*. London: UCL Press. 42–49.

Kang, Yuhao, Fan Zhang, Wenzhe Peng, Song Gao, Jinmeng Rao, Fabio Duarte, and Carlo Ratti. 2020. "Understanding House Price Appreciation Using Multi-Source Big Geo-Data and Machine Learning." *Land Use Policy* 111: 104919. https://doi.org/10.1016/j.landusepol.2020.104919.

Mogas-Soldevila, L., J. Duro-Royo, D. Lizardo, M. Kayser, S. Sharma, S. Keating, J. Klein, C. Inamura, and N. Oxman. 2015. "Designing the Ocean Pavilion: Biomaterial Templating of Structural, Manufacturing, and Environmental Performance," In *Proceedings of IASS 2015 Amsterdam Symposium: Future Visions – Digital Architecture and Design*. 1–13.

Najari, Arman, Diego Pajarito, and Areti Markopoulou. 2020. "Data Modeling of Cities, a Machine Learning Application." In *SimAUD '20: Proceedings of the 11th Annual Symposium on Simulation for Architecture and Urban Design*. 1–8.

Nicholas, Paul, Gabriella Rossi, Ella Williams, Michael Bennett, and Tim Schork. 2020. "Integrating Real-Time Multi-Resolution Scanning and Machine Learning for Conformal Robotic 3D Printing in Architecture." *International Journal of Architectural Computing* 18 (4): 371–84. https://doi.org/10.1177/1478077120948203.

Pattinson, Sebastian W., and A. John Hart. 2017. "Additive Manufacturing of Cellulosic Materials with Robust Mechanics and Antimicrobial Functionality." *Advanced Materials Technologies* 2 (4): 1600084. https://doi.org/10.1002/admt.201600084.

Ramsgaard Thomsen, Mette ,and Ayelet Karmon. 2012. "Informing Material Specification." In *Digital Aptitudes: + Other Openings*. Boston: ACSA. 677–83.

Ramsgaard Thomsen, Mette, Paul Nicholas, Martin Tamke, and Tom Svilans. 2021. "A New Material Vision." In *Data, Matter, Design: Strategies in Computational Design*, edited by Frank Melendez, Nancy Diniz, and Marcella Del Signore. New York: Routledge.

Rech, Arianna, Paul Nicholas, Mette Ramsgaard Thomsen, and Anders E. Dougaard. 2021a. "Waste-based biopolymer slurry for 3d printing with recycled materials." In *Nordic Polymer Days 2021*. https://nrs.blob.core.windows.net/pdfs/nrc-2-41381210-885e-4e8b-9ea3-054d3ec822b4.pdf.

Rech, Arianna, Ruxandra Chiujdea, Claudia Colmo, Gabriella Rossi, Paul Nicholas, Martin Tamke, Mette Ramsgaard Thomsen, and Anders Egede Daugaard. 2021b. "Predicting Response: Waste-based biopolymer slurry recipe for 3d-printing (CelluloseFloc_v01)." Data set. Zenodo. https://doi.org/10.5281/zenodo.5557218.

Robert McNeel & Associates, *Rhinoceros 3D & Grasshopper*. V.7, SR13. Windows. 2021.

Rossi, Gabriella, and Paul Nicholas. 2018. "Modelling a Complex Fabrication System: New Design Tools for Doubly Curved Metal Surfaces Fabricated Using the English Wheel." In *Proceedings of ECAADe 2018*. 811–20.

Sanandiya, Naresh D., Yadunund Vijay, Marina Dimopoulou, Stylianos Dritsas, and Javier G. Fernandez. 2018. "Large-Scale Additive Manufacturing with Bioinspired Cellulosic Materials." *Scientific Reports* 8 (1): 8642. https://doi.org/10.1038/s41598-018-26985-2.

Stavropoulos, P., D. Chantzis, C. Doukas, A. Papacharalampopoulos, and G. Chryssolouris. 2013. "Monitoring and Control of Manufacturing Processes: A Review." *Procedia CIRP* 8: 421–25. https://doi.org/10.1016/j.procir.2013.06.127.

Vasey, Lauren, Ehsan Baharlou, Moritz Dörstelmann, Valentin Koslowski, Marshall Prado, Gundula Schieber, Achim Menges, and Jan Knippers. 2015. "Behavioral Design and Adaptive Robotic Fabrication of a Fiber Composite Compression Shell with Pneumatic Formwork." In *ACADIA '15: Computational Ecologies: Design in the Anthropocene; Proceedings of the 35th Annual Conference of the Association for Computer Aided Design in Architecture*. 297–309.

Zhang, Lixiao, Guoyang Qiu, and Zhishou Chen. 2021. "Structural Health Monitoring Methods of Cables in Cable-Stayed Bridge: A Review." *Measurement* 168 (January): 108343.

IMAGE CREDITS

Figures 1-10: © Photographs and drawings by the authors; CITA / Royal Danish Academy

Gabriella Rossi is a PhD fellow at CITA.

Ruxandra Chuijdea, Claudia Colmo, and Chada el Alami are research assistants at CITA.

Paul Nicholas is Associate Professor at the Royal Danish Academy, senior researcher at CITA, and leads the international masters program Computation in Architecture.

Martin Tamke is Associate Professor at the Royal Danish Academy and senior researcher at CITA.

Mette Ramsgaard Thomsen is Professor of Digital Technology at the Royal Danish Academy and leads CITA.

Nesting Fabrication

Alireza Borhani
California College of the Arts

Negar Kalantar
California College of the Arts

Using Minimum Material & Production Time
to Deliver Nearly Zero-Waste Construction
of Freeform Assemblies

ABSTRACT

Positioned at the intersection of the computational modes of design and production, this research explains the principles and applications of a novel construction-informed geometric system called nesting fabrication. Applying the nesting fabrication method, the authors reimage the construction of complex forms by proposing geometric arrangements that lessen material waste and reduce production time, transportation cost, and storage space requirements. Through this method, appearance and performance characteristics are contingent on fabrication constraints and material behavior. In this study, the focus is on developing design rules for this method and investigating the main parameters involved in dividing the global geometry of a complex volume into stackable components when the first component in the stack gives shape to the second. The authors introduce three different strategies for nesting fabrication: 2D, 2.5D, and 3D nesting. Which of these strategies can be used depends on the geometrical needs of the design and available tools and materials. Next, by revisiting different fabrication approaches, the authors introduce readers to the possibility of large-scale objects with considerable overhangs without the need for nearly any temporary support structures (Figure 1). After establishing a workflow starting with the identification of geometric rules of nesting and ending with fabrication limits, this work showcases the proposed workflow through a series of case studies, demonstrating the feasibility of the suggested method and its capacity to integrate production constraints into the design process.

1   By using the nesting fabrication method, the arrangement of fourteen custom-designed modules created a self-supporting arch without the need for mortar or complementary mechanical connectors

INTRODUCTION

More than just facilitating the creation of unorthodox buildings and atypical structures, digital fabrication techniques minimize material waste and production time due to the level of sustainability and efficiency they provide. However, this expectation is not fully satisfied when the moderation of waste plays a subordinate role to fulfilling the demand for freeform architecture (Lavery 2013; Craveiro et al. 2019; Abdelmohsen and Hassab 2020). Moreover, in some cases, the links among material waste, capital expenditure, and affordability are not particularly clear. For the last decade, to strengthen fabrication efficacy and sustainability and optimize material usage and production processes, the authors have endeavored to find new opportunities for creating tectonic links between construction and design.

BACKGROUND

Nesting is widely employed in the industry to optimize material usage in 2D and 3D environments (Nee et al. 1986; Lam and Sze 2007; Lutterset al. 2012; Struckmeier and León 2019). For example, if they fit well together, the term "nesting" refer to a cluster of 2D drawings used for cutting in sheet metal manufacturing (Lutters et al. 2012). To improve the productivity and cost-effectiveness of 3D printing, especially in binder jetting and laser sintering techniques, it is common to nest as many complex parts together as possible per build. Typically, the printer build volume is filled via two nesting techniques: bounding box and parts geometry. In this research, though the presented method is not entirely associated with the definition and practices of nesting currently available in the industry, the authors use the term because its main intention is minimizing fabrication waste when creating compact arrangements of parts. Here, nesting implies a geometrical process and fabrication technique for creating successive parts that shape a stackable volume with freeform geometry. Representationally, the nesting method can be compared to a set of Russian nesting dolls of reducing size, positioned one inside the other (Jacobson 2017) (Figure 2). Despite its relevance and applicability to the construction of complex forms, in architecture, design cases related to nesting are rare. Examples include Kudless's Zero/Fold Screen completed at the University of Calgary in 2010 (Kudless 2010; Forward and Taron 2019). And the 2011 method of making freeform shells from precast stackable components presented by Cepaitis and team (Cepaitis et al. 2011; Enrique et al. 2016). Rene van Zuuk Architects (2016) presented an igloo-shaped house named "RE-Settle". He used a hot wire cutter machine to cut a block of expanded polystyrene while creating nested rings.

NESTING FABRICATION METHOD

In this research, the term "nesting" indicates the geometric process of transforming a freeform volume into stackable components, allowing them to be fabricated, assembled, and transported with ease. In a nested assembly, the whole volume is first subdivided into identical yet interrelated pieces. Then, when reorganized, these neighboring pieces nest together, either side by side (horizontally) or on top of one another (vertically), forming a stack of snugly fitting parts (Figure 3). In such cases, every piece in a stack is part of a larger whole, as one side is compatible with the other side of the next. This work introduces three nesting methods that leverage three parametric definitions: 2D, 2.5D, and 3D.

2D Nesting Method

2D nesting is a design and fabrication strategy for creating undulating walls (or ceilings) out of parallel rib-like components. During CNC or laser cutting, the components are cut from flat sheets with minimum waste. Here, the rib curvatures are interrelated, as are the geometries of the front and back sides of the walls (Figure 4). Prior attention to the sheet dimensions plays a significant role in minimizing material use. By specifying only a few input parameters such as the material's thickness, number of ribs, and preferred gap between them, an in-house parametric definition in Grasshopper can generate a 3D model and the required 2D cut files.

To minimize material waste, all neighboring ribs must be lined up next to one another, with the front side of one rib serving as the rear side of the previous. Consequently, the cutting layout is composed of several adjacent profiles with shared edges and no gaps in between. Other than the material cutoff due to the cutting tool's width, nothing is taken from the adjacent pieces when cut. Since just three cutting operations produce every two ribs, the matching time is also lessened.

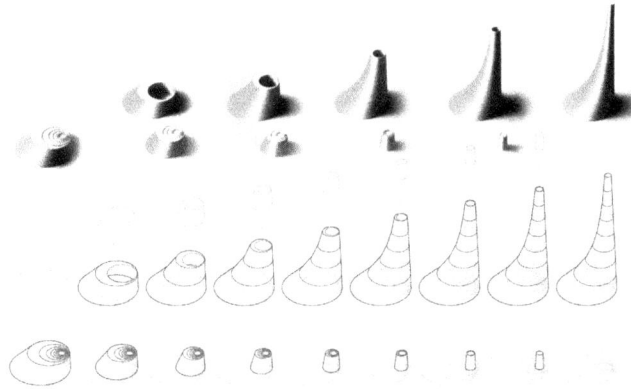

2   Placement of one ring of a nested volume inside another to obtain a compact stack and deliver nearly zero-waste construction of freeform volumes.

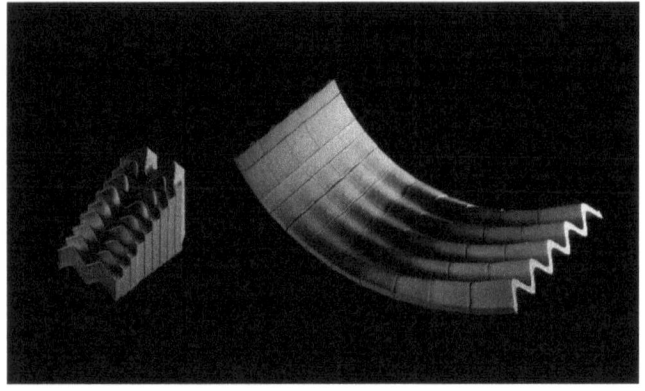 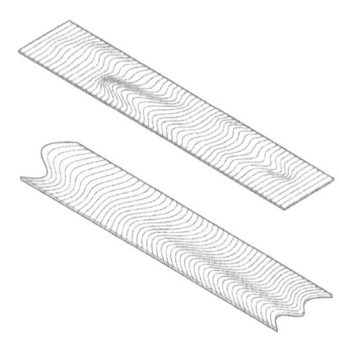

2.5D Nesting Method

The subdivided components of an object should nest inside one another to form a stack. As a result, the components must be organized consecutively, one after the other, with the inside of one component concurrent with the outside of the next (Figure 5). Not only should the first nested component shape the second, it should also be possible to slide the second component out from the first without collision. Objects with both open and closed geometries can be nested via the 2.5D nesting method. Different opportunities and limitations must be considered when dealing with these types of objects. The opportunities and limitations associated with each type are discussed below:

Closed Geometry—An object with a closed geometry has a continuous closed enclosure with space in between. When dividing the object via the 2.5D nesting method, a stack of closed ring-like components is achieved. Since the stack rings maintain their continuity, they offer certain advantages, such as using less storage area (due to the maximum stack density) and easy and speedy assembly. To retain all of the components in a centric stack while also providing proper thickness for each ring, it is essential to deal with a volume with tilted surface bodies. Steeper angles for the surfaces provide more thickening opportunities for the stacked components. However, the necessity of having angled surfaces may result in some limitations. For instance, a general limit associated with these object types is dissimilar thicknesses of the components in the stack. To include the components in a single stack, the Gaussian curvature of the object surface should be constant, with either negative or positive values. Some objects may have regions of both negative and positive Gaussian curvatures changed by a curve of points with a zero Gaussian curvature (Goldman 2005). When the curvature direction changes, it is necessary to split the object from those points. Such a division results in more than one stack to eliminate the collision possibility of the components in areas with different Gaussian curvatures (Figure 6).

Open Geometry—An object with an open geometry has an expandable enclosure, without taking up any enclosed space from its environment. Contrasting noticeably with the former category, when an object with an open geometry is nested, its stack is no longer centric. Therefore, the overall thickness of the components can be the same, producing the best custom-fit pieces with the desired depth. Also, all components can be made in one uninterrupted stack, regardless of the Gaussian curvature value of the object surface angles (Figures 7, 8). More importantly, unlike the

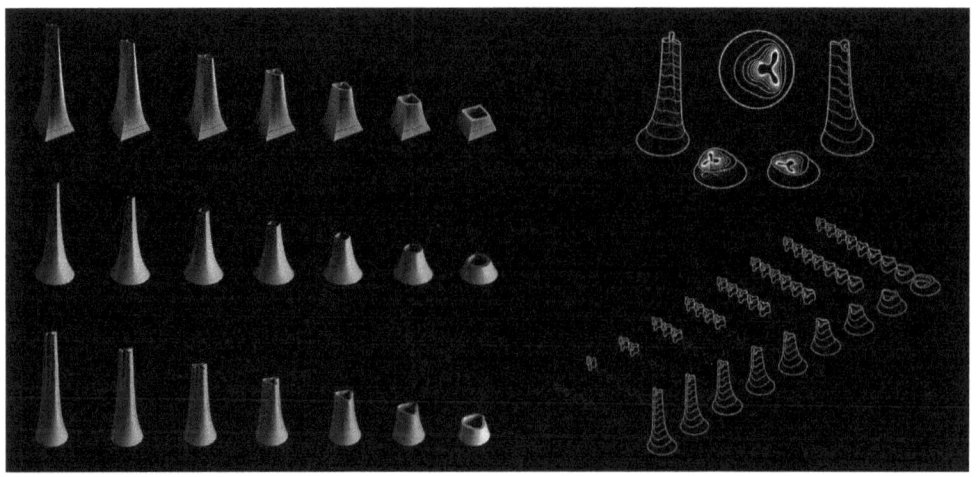

3 Breaking the main geometry into stackable components

4 2D Nesting method: producing an undulating volume with no waste when the front of the first rib and the back of the second one are identical

5 2.5D Nesting method when the inner part of the lower volume is congruent with the outer side of the upper

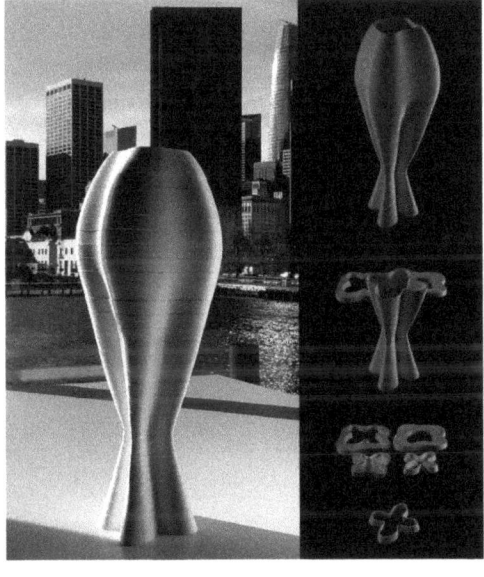
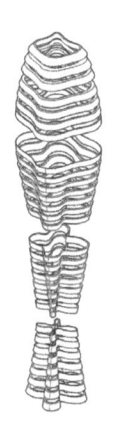
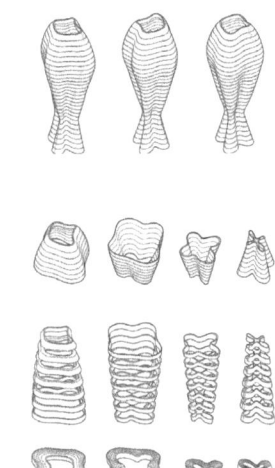

6   2.5D Nesting method of a volume with a closed geometry: producing more than one stack due to the curvature direction changes

7   2.5D Nesting method of a volume with an open geometry: producing all components in one continuous stack

8   2.5D Nesting method of a freeform volume with an open geometry

6

previous category, it is unnecessary to have an object with slanted surfaces.

3D Nesting Method

3D nesting was developed to construct geometrically complex large-scale building elements without trade-offs. Relying on producing a series of three dimensional parts with complementary geometries, a freeform volume is tessellated into stackable components for ease of assembly, transportation, and storage, or to allow for a specific fabrication process (Figure 9).

The governing principles of nesting permit components with different shapes to be fabricated alongside or on top of one another. In this fabrication method, it is possible to make more compact stacks when dealing with more complex components (Figure 10). Depending on the fabrication tools and materials, the 3D design process for a nested object requires more attention.

In the 2.5D nesting method, the volume is sliced with parallel planes. Thus, the components have two parallel sides when assembled. In the 3D nesting method, non-parallel planes subdivide the volume to produce wedge-shaped or voussoir pieces sitting flank to flank (Figure 11). Since the beveled shape of voussoirs helps them fit together to transfer the thrust of mass from one component to another, the stability of the structure and angle of the cut planes are interdependent. Depending on the geometry, the volume can be tessellated via an array of cutting planes with different angles and orientations. These planes might be convergent, divergent, rotated, twisted, or slanted to adapt to the volume geometry. To create a compact stack, it is essential to consider the spatial relationship between these planes as they pull apart or compress.

POSSIBLE 2.5D & 3D FABRICATION NESTING METHODS

The main advantage of the nesting method is to generate smaller and lighter components during transportation, storage, and assembly, as opposed to dealing with monolithic pieces. Consequently, large-size tools are not required to produce the components. This may impact the necessary space dedicated to fabrication and reduce the total

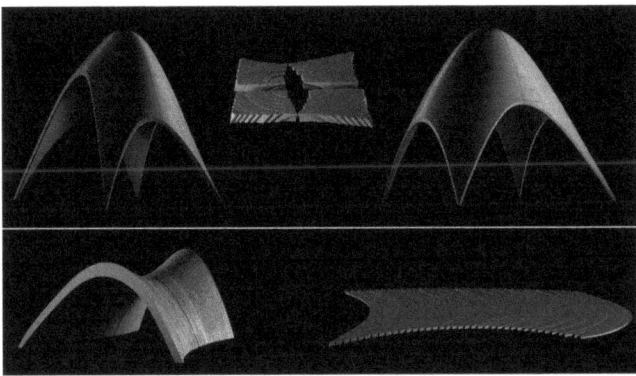
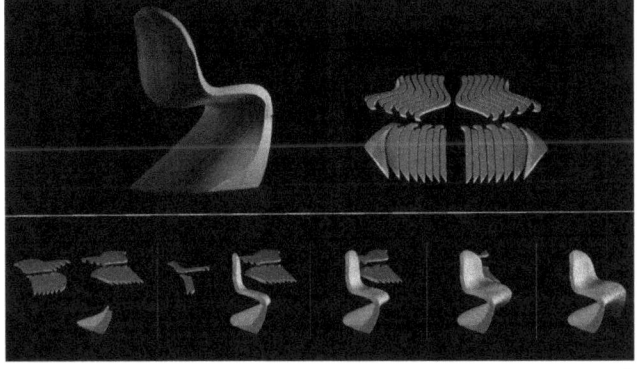

7

8

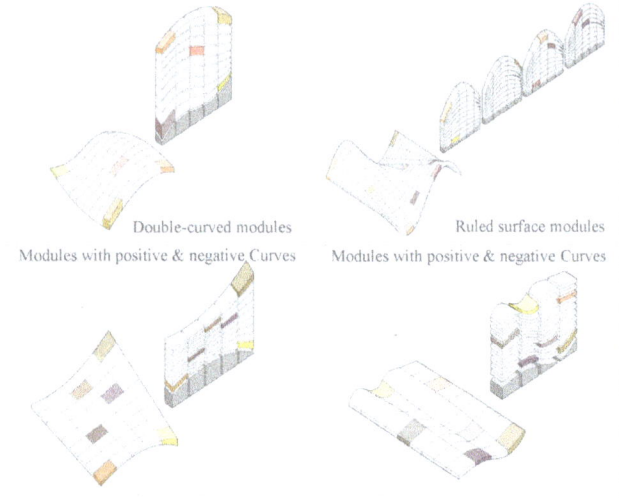

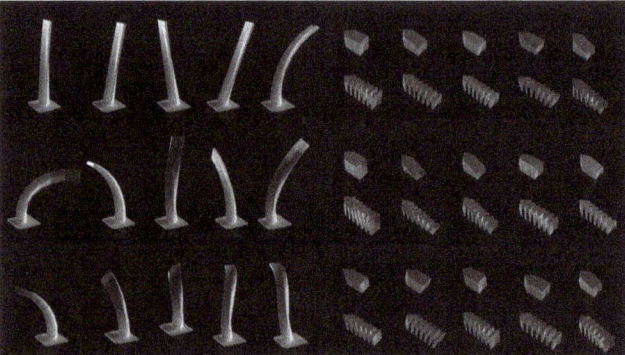

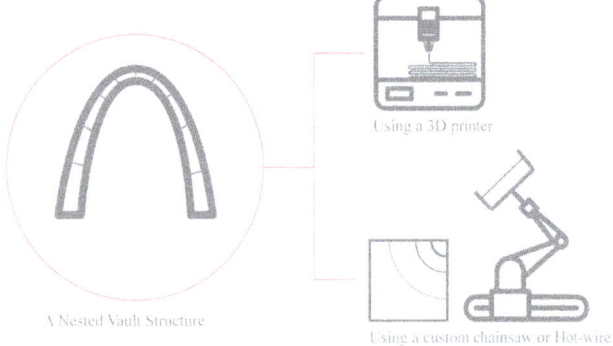

investment. Also, using smaller tools can enhance fabrication tolerance. Different fabrication strategies and tools can be exploited depending on the positioning of nested components within the stack and their global geometry. Here, two fabrication strategies are reviewed: subtractive and additive fabrication approaches (Figure 12).

Subtractive Strategies

A single block of material can be sliced via sequential cuts in predefined directions to make a nested assembly (Figure 13). For this process, a variety of materials can be considered such as stone, wood, steel, or foam. Different tailored cutting techniques can be utilized in conjunction with a robotic arm, including waterjet, diamond wire saw, wire EDM, chainsaw, plasma arc, and hotwire cutter. The cutting techniques applied define the level of surface complexity generated in the slicing process. Since most of the tools mentioned above belong to the family of line-form cutting tools, they slice the block into different ruled surfaces via multiple straight lines. In general, such ruled surfaces reduce the complexity of the geometry of the original surfaces. Since the ruled geometry of cutting surfaces impacts the shape of the final assembly, proper attention should be given to the difference between the initial and attainable cutting geometries. A CNC router can transform

9   3D Nesting method: the inner and outer components support one another during fabrication, transportation, and storage

10 Producing wedge-shaped or voussoir pieces to transfer loads

11 Showcasing the potential of the 3D Nesting method to create any freeform surfaces, including twisted, bowed, slanted geometries; showcasing the potential of the 3D Nesting method to create any freeform surfaces, including twisted, bowed, slanted geometries

12 Two possible fabrication strategies to produce nested volumes

13 Showcasing the potential of line-form cutting tools to create freeform shells out of a block of material with no waste

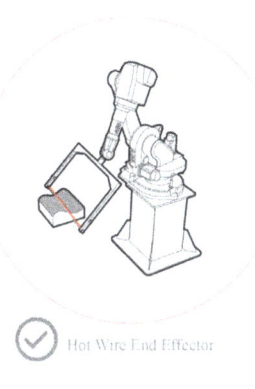

14

a stock of material into a desired shape, but it produces significant waste and dust and takes a great deal of time. Compared to the CNC milling scenario in which plenty of the material is cut off, the amount of waste is minimum when slicing because line-form cutting tools remove the least amount of kerf in the material. Another advantage of this method is production speed; two pieces are made with just three cutting passes. The proposed method can be considered an alternative to CNC machines when more flexibility, speed, and reachability are demanded (Figure 14). When moving along cutting paths, cutting tools should go all the way through the block. However, depending on the tools and material thickness, once the depth of the cut increases, the possibility of undesirable results should be considered (e.g., undercutting, thermal distortion). Also, based on the

15

16

shape of the pieces and specifications of each cutting tool, creating lead-in and lead-out cutting at the entry and exit of the cutting paths may be required before cutting.

To examine the feasibility of the method mentioned above, in 2018, two self-standing modular interlocking vaulted structures were built using robotic hotwire cutting techniques (Figure 15). The modules were produced out of large expanded polystyrene blocks, with fifteen cuts per structure. The size of each foam block was 48" x 48" x 36" for each structure (Figure 16). Consisting of a 1:1 scale section, all of the modules were cut with a minimum of material waste when all edges were concurrent within the block boundary. The interlocking modules were assembled without any scaffolding or specific tools. In approximately twenty minutes, five people completed the entire assembly process. In this project, a track-mounted ABB 4600 industrial robot arm was used to control the hotwire end effector's cutting sequence. Fabrication of both structures with a single line-form cutting tool revealed how nesting principles could reduce material expenditure while also enhancing production time (Figure 17). The principle of nesting can be extended to inform a cost-effective formwork strategy for casting concrete shells. For instance, a block of foam can be subdivided into pieces with differing geometries via successive cuts. Later, these stacked pieces can be split up with a specific space in between to create the proper void for casting. Applying the nesting method to the formwork design allows for maximum flexibility and minimal cost.

14 Comparing CNC milling and line-form cutting to save more resources
15 Using the 3D Nesting method to create two self-standing interlocking structures with nearly no waste
16 After taking down the components of vaulted structures, they nested together to occupy minimum space for shipping and storage.
17 Instead of using CNC milling, the constraints of a line-form cutter were intentionally chosen to reduce the material waste and increase the production time

Additive Strategies

There are still challenges to 3D printing large-scale buildings. Some stem from the behavior of printed materials. The rheological behavior of concrete freshly mixed for 3D printing should be adjusted to be easy to pump and extrude. Providing enough fluidity to obtain a printable mix contradicts the stability of concrete to withstand the weight of the material layers above (Roussel 2018). Despite the increasing popularity of concrete 3D printing, it remains difficult to use cementitious materials in their fresh state for printing overhanging objects with no support (Wei et al. 2019). Due to the lack of effortless approaches to making temporary support structures out of concrete and removing them during post-processing, the challenges to transforming any freeform design (with features suspended in midair) into a desired concrete volume have yet to be fully overcome. Consequently, the application of concrete printing is mainly restricted to creating extruded-like volumes with self-supporting walls that do not lean too much from the vertical when additional scaffolding or wooden planks are not used to hold subsequent printed layers in place. Even in Rudenko's 3D Printed Concrete Castle, the walls' maximum slope angle was designed to create a row of connected walls for a closed structure (Molitch-Hou 2014). In this case, every two adjacent overhangs support one another to avoid structural collapse.

Pushing the boundaries of what is possible with large-scale 3D printing, the nesting fabrication method can significantly reduce or even eliminate the need to produce support materials for a single type of building material when a series of complementary pieces are printed in stacks. Using this method allows different components of a nested stack to be printed side by side or on top of one another. Depending on the geometry of the nesting, one of the following approaches is possible.

PRINT on PRINT—When the nested components make a vertical pile, the lower parts in the stack serve as a substrate for the upper (Figure 18). If needed, a temporary support structure is printed to hold up the first component. After 3D scanning the top printed surface, the printer head begins moving over the surface to make the next component (Figure 19). In the case of robotic printing, the components are not restricted to a planar material deposition with uniform or discrete layers and the staircase effect (Aniruddha 2019; Mansoori et al. 2018, Farahbakhsh et al. 2020). Besides saving a significant amount of support material, this approach substantially reduces the build time, as compared to separately printing each component. Also, this approach can lessen the post-processing phases to eliminate the need for supports.

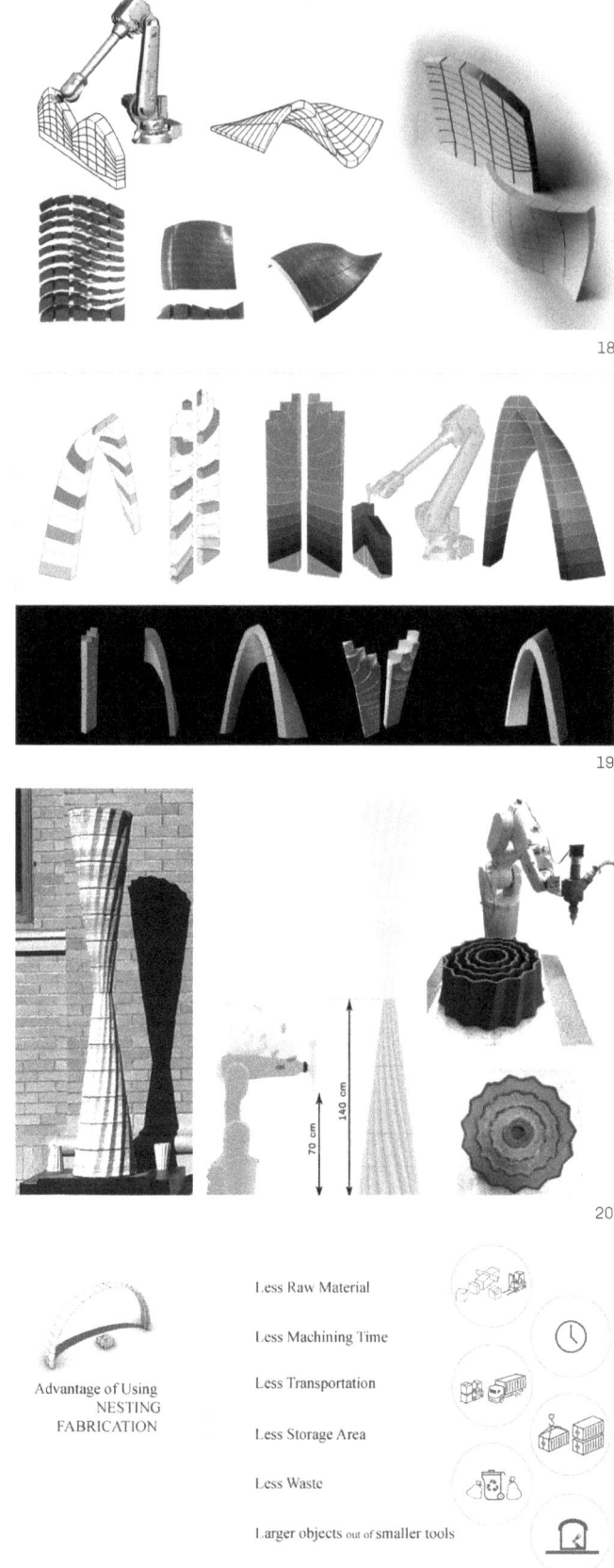

18 Eliminating support materials for large-scale 3D printed components
19 PRINT on PRINT: When printing on top of a previous piece, the lower piece is used as a bottom support or substrate for the next piece produced; therefore, the geometry of the upper piece in the stack is supported by the lower
20 PRINT in PRINT: When printing side by side, the surface of each piece supports the adjacent printed parts, 3D printed by Mehdi Farahbakhsh
21 Advantages of the proposed nesting fabrication method

PRINT in PRINT—When a series of nested components sit horizontally side by side, the inner parts can serve as a support structure for the outer, and vice versa. Thus, it is important to understand the respective impacts of the inside-out and outside-in printing options for components with steep overhanging angles. When support is not needed, the components in the stack are simultaneously printed to increase the speed of production.

Printing nested stacks can expand the potential of clay and concrete printing. In 2020, with the PRINT in PRINT fabrication approach in mind, one tower-like volume was made out of nested concrete parts that came together in two horizontal stacks, with corresponding inner and outer surfaces (Figure 20). In this project, nesting principles not only informed the character of the final artifact, but were also used to establish a novel strategy of making removable formwork for concrete casting. After finalizing the compact geometry of the nested parts in the stack, only the outer surfaces of these parts were 3D printed with clay. In this case, the printed clay shells served as formwork, later to be filled with concrete mix. After pouring and solidification of the concrete mix, the clay shells were then removed. In this way, the nested components were built to form the concrete pieces. In this project, the HAL Robotics plugin for Grasshopper and RobotStudio were used to generate and simulate the printing toolpath and robot movement. Proper perforations can be printed inside each component in a stack, after accounting for the passage of post-tensioning steel wires (Anton et al. 2019). Once temporarily assembled, the tendons can be stressed to keep the component held tightly together. Once the wires are anchored to the structure base, each nested component pushes firmly against the surface of adjacent pieces and uniformly conducts loads.

RESULT

The nesting method is a rational approach to transforming fabrication restrictions into geometric propositions. In a nested assembly, the geometry of the constituent components drives formal choices. Still, the geometric principles of nesting can be tuned to the intrinsic characteristics of the tools and materials being used, depending on the most favorable dimensional tolerances. As a playground for the exploration of nesting design, the authors have created several structures in different forms and on a variety of scales. The constituent components of these volumes stack in a nested fashion, mainly to be 3D printed. The nesting method provides more opportunities to create large freeform objects with substantial overhangs that need nearly no temporary support when printed or subtracted (Figure 21). Moreover, the advantages of the proposed method are three-fold:

- The method saves a significant amount of material that would otherwise be wasted making temporary structures. Such a material reduction is a step towards more sustainable construction.
- In additive manufacturing, this method liberates the final print from the time-consuming and labor-intensive process of removing support material.
- The method helps boost production speed, since the fabrication tools directly make the primary parts without spending extra time making supports. Thus, the method improves the tools' lifespan and minimizes required maintenance costs, while also reducing unnecessary movements of mechanical parts. Moreover, since the expected components in the stack are not scattered in different locations, the tools can perform in a clustered zone. This will increase the printing speed while reducing the printer dimensions and setup time.

REFLECTION

In a 3D nested assembly, components' wedge-shaped surfaces impact how the thrust of one voussoir counteracts the thrust of its neighbors. Thus, to force the voussoirs together instead of apart, the proposed design's global geometry and structural behavior should be incorporated in the component shapes and dimensions, taking advantage of the compressive strength of components in their final assembly. In the future, to fulfill specific fabrication requirements (for instance, clay printing) and negotiate among various formal and structural needs, an integrated parametric model will be developed to evaluate the whole geometry of the nested design, as well as the dimension of components and their final stacks. Also, during structural analysis, the minimum thickness of components should be calculated to achieve the least possible material consumption and structure weight. During the final assembly, nested components may require temporary scaffolds from below until they are set in place (Figure 22). Using an interlocking method for connecting the pieces is beneficial because it minimizes the use of scaffolds and reduces waste generated on the construction site. Based on the authors' previous experience fabricating two full-scale vaulted structures with topological interlocking components, it is feasible to eliminate the use of any temporary scaffolds or formworks. To complete this line of research, in the future, the parametric model will be further developed to generate the required interlocking parts with proper mechanical properties, buttressing the thrusts and conducting them into the structural supports. The series of cases shown here employs the nesting method to present how a given freeform volume can be fabricated via stackable components and with minimal waste. In the future, the authors will make more prototypes on a larger scale to bring together tool parameters, production speed, material optimization, shipping, and assembly concerns, with the eventual goal of more sustainable construction.

CONCLUSION

By establishing a profound connection between formal, structural, and manufacturing advances, the nesting fabrication method, as a strategy for material efficiency, addresses challenges in the construction of large-scale freeform volumes. Employing simultaneous innovation and an integrated approach to both geometry and manufacturing, the proposed method will improve construction productivity by reducing fabrication time and cost on a broad scale. More importantly, this method creates a new realm of possibilities for the design and parametric modeling of complex forms. Although the nesting fabrication method is more intrinsic and correlative to fabrication constraints, it encompasses more than just technological translations of material and tool restraints into a form. As a new explorative method in form-finding, this process emphasizes fabrication as an expressive way to generate architectonic forms when opening new possibilities for craft. Traversing from pragmatic to geometrical concerns, the approach discussed here offers an integrated approach supporting functional, structural, and environmental matters important when turning material, technical, assembly, and transportation systems into geometric parameters. The nesting method explores and develops the tectonics by which built forms are realized. By incorporating the competence of material and time, as well as assembly concerns such as tolerance and seams, the logic of nesting becomes a part of the structure's aesthetic and tectonic qualities, and not something to be disregarded. These qualities are not indifferent to the processes that deliver them.

ACKNOWLEDGEMENTS

The authors would like to special thank Jiries Alali and Mehdi Farahbakhsh, who provided insight and expertise to advance this research. Also, the authors extend their thanks to the Autodesk Technology Centers in Boston and San Francisco.

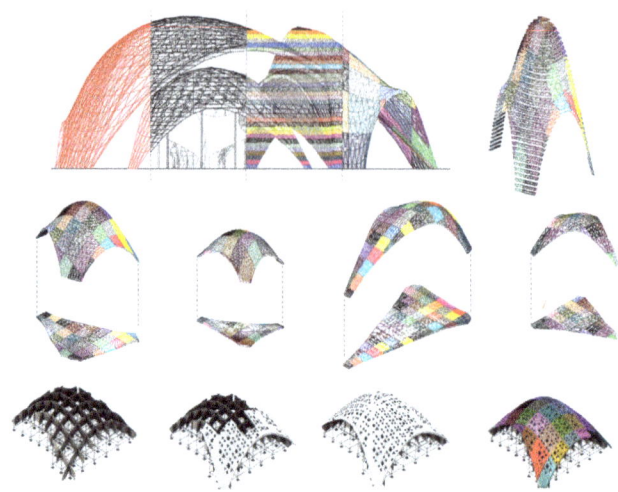

22 Showcasing the best way to erect the nested components of a pavilion and assemble them while minimizing the use of scaffolds. CCA Advanced Studio, Fall 20: taught by the authers, Designed by Venessa Davidenko.

REFERENCES

Abdelmohsen, Sherif, and A. Hassab. 2020. "A Computational Approach for the Mass Customization of Materially Informed Double Curved." In Proceedings of the 25th Conference on Computer-Aided Architectural Design Research in Asia (CAADRIA). Bangkok, Thailand. 163–172.

Aniruddha V. Shembekar, Yeo Jung Yoon, Alec Kanyuck, and Satyandra K. Gupta. 2019. "Generating Robot Trajectories for Conformal 3D Printing Using Nonplanar Layer." *Journal of Computing and Information Science in Engineering* 19(3): 031011. https://doi.org/10.1115/1.4043013.

Anton, Ana, Angela Yoo, Patrick Bedarf, Lex Reiter, Timothy Wangler and Benjamin Dillenburger. 2019. "Vertical Modulations Computational design for concrete 3D printed columns." In *ACADIA 19: Ubiquity and Autonomy; Proceedings of the 39th Annual Conference of the Association for Computer Aided Design in Architecture*. Austin, Texas. 596–605.

Craveiro, Flávio, José Pinto Duarte, Helena Bartolo, and Paulo Jorge Bartolo. 2019. "Additive manufacturing as an enabling technology for digital construction: A perspective on Construction 4.0." *Automation in Construction* 103: 251–267.

Cepaitis, Povilas, L. Enrique, D. Ordoñez, and C. Piles. 2011. "Towards waste-free concrete fabrication without conventional molds." Accessed June 10, 2021. https://www.lafargeholcim-foundation.org/media/news/awards/towards-waste-free-concrete-fabrication-without-convention-al-mol.

Enrique, Lluis, Povilas Cepaitis, Diego Ordoñez, and Carlos Piles. 2016. "CASTonCAST: Architectural freeform shapes from precast stackable components." *VLC Arquitectura* 3 (1): 85–102.

Farahbakhsh, Mehdi, Negar Kalantar, and Zofia Rybkowski. 2020. "Impact of Robotic 3D Printing Process Parameters on Bond Strength." In *ACADIA 20: Distributed Proximities; Proceedings of the 40th Annual Conference of the Association of Computer Aided Design in Architecture*, vol. 1. 594–603.

Forward, Kristen, and Joshua Taron. 2019. "Waste Ornament: Augmenting the Visual Potency of Sustainable Facade Designs." In *ACADIA 19: Ubiquity and Autonomy; Proceedings of the 39th Annual Conference of the Association for Computer Aided Design in Architecture*. Austin, Texas. 90–99.

Goldman, Ron. 2005. "Curvature formulas for implicit curves and surfaces." *Computer Aided Geometric Design* 22 (7): 632–658.

Holmes, Gillian. "Matryoshka Doll." How Products Are Made. Advameg, Inc. Accessed June 10, 2021. http://www.madehow.com/Volume-6/Matryoshka-Doll.html.

Jacobson, Alec. 2017. "Generalized Matryoshka: Computational Design of Nesting Objects." *Computer Graphics Forum* 36 (5): 27–35.

Kudless, Andrew. 2010. "Zero/Fold Screen." Accessed June 10, 2021. https://www.matsys.design/zero-fold-screen/.

Lam, T. F., W. S. Sze, and S. T. Tam. 2007. "Nesting of Complex Sheet Metal Parts, Computer-Aided Design and Applications." *Computer-Aided Design and Applications* 4 (1-4): 169–179.

Lavery, Conor. 2013. "Spencer Dock Bridge." *Concrete International* 35 (2013): 28–31.

Lutters, Eric, D. ten Dam, and T. Faneker. 2012. "3D Nesting of Complex Shapes." In *Procedia CIRP: 45th CIRP Conference on Manufacturing Systems 2012*, vol. 3. 26–31. https://doi.org/10.1016/j.procir.2012.07.006.

Mansoori, Maryam, Negal Kalantar, and William Palmer. 2018. "Handmade by Machine : A Study on Layered Paste Deposition Methods in 3D Printing Geometric Sculptures." In *Hyperseeing: Proceedings of SMI'2018 Fabrication and Sculpting Event (FASE)*. Lisbon, Portugal.

Molitch-Hou, Michael. 2014. "Finally, It Stands! Andrey Rudenko's 3D Printed Concrete Castle." 3D Printing Industry. August 27, 2014. Accessed June 10, 2021. https://3dprintingindustry.com/news/finally-stands-andrey-rudenkos-3d-printed-concrete-castle-32097/.

Nee, Andrew Y. C., Kek Wee Seow, S. L. Long, and V. C. Venkatesh. 1986. "Designing Algorithm for Nesting Irregular Shapes With and Without Boundary Constraints." *CIRP Annals* 35 (1986): 107–110.

René van Zuuk Architects. 2016. "Re-Settle Studio." Website portfolio project. Accessed October 9, 2021. https://www.renevanzuuk.nl/re-settlestudio.

Roussel, Nicolas. 2018. "Rheological requirements for printable concretes." *Cement and Concrete Research* 488 (October): 76–85.

Struckmeier, Frederick, and Fernando Puente León. 2019. "Nesting in the sheet metal industry: Dealing with constraints of flatbed laser-cutting machines." *Procedia Manufacturing* 29: 575–582.

Tay, Yi Wei Daniel, Ming Yang Li, and Ming Jen Tan. 2019. "Effect of printing parameters in 3D concrete printing: Printing region and support structures." *Journal of Materials Processing Technology* 271: 261–270.

Wikipedia. n.d. "Matryoshka dolls." Accessed June 10, 2021. https://en.wikipedia.org/wiki/Matryoshka_doll.

IMAGE CREDITS

Figure 20: © Mehdi Farahbakhsh, 2020.
Figure 22: © Venessa Davidenko, 2020.
All other drawings and images by the authors.

Alireza Borhani is an innovator, architect, educator, and co-principal of the transLAB. His interdisciplinary experience has allowed him to expand his career into a broad scale and type of projects at the intersection of design computation, emerging material systems, additive manufacturing workflows, and robotics. At the forefront of kinematic structures, ranging from architectural-scale shelters to small products, Borhani has been immersed in the world of transformable and adaptive design for the past twenty years. At the California College of the Arts, Texas A&M, and Virginia Tech, Alireza has taught architecture studios, concurrent with research and practice, for over a decade.

Dr. Negar Kalantar is Associate Professor of Architecture and a Co-Director of the Digital Craft Lab at California College of the Arts (CCA) in San Francisco. Her cross-disciplinary research focuses on materials exploration, robotic and additive manufacturing technologies to engage architecture, and science and engineering as platforms for examining the critical role of design in global issues and built environments. Kalantar is the recipient of several awards and grants, including the Dornfeld Manufacturing Vision Award 2018, the National Science Foundation, Autodesk Technology Center Grant, and X-Grant 2018 from the Texas A&M President's Excellence Fund on developing sustainable material for 3D printed buildings.

From Design to the Fabrication of Shellular Funicular Structures

Mostafa Akbari
Yao Lu
Masoud Akbarzadeh

Polyhedral Structures Laboratory, Stuart Weitzman School of Design, University of Pennsylvania

(above)
A shellular funicular structure fabricated using tuck-folding technique.

ABSTRACT

Shellular Funicular Structures (SFSs) are single-layer, two-manifold structures with anticlastic curvature, designed in the context of graphic statics. They are considered as efficient structures applicable to many functions on different scales. Due to their complex geometry, design, and fabrication of SFSs are quite challenging, limiting their application in large scales. Furthermore, designing these structures for a predefined boundary condition, control, and manipulation of their geometry is not an easy task. Moreover, fabricating these geometries is mostly possible using additive manufacturing techniques, requiring a lot of supports in the printing process.

Cellular funicular structures (CFSs) as strut-based spatial structures can be easily designed and manipulated in the context of graphic statics. This paper introduces a computational algorithm for translating a Cellular Funicular Structure (CFS) to a Shellular Funicular Structure (SFS). Furthermore, it explains a fabrication method to build the structure out of a flat sheet of material using the origami/kirigami technique as an ideal choice because of its accessibility, processibility, low cost, and applicability to large scales. The paper concludes by displaying a structure that is designed and fabricated using this technique.

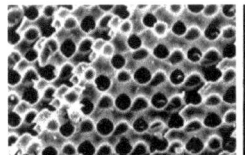 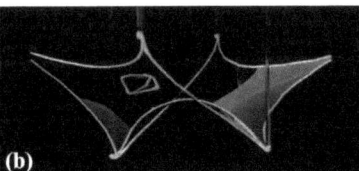  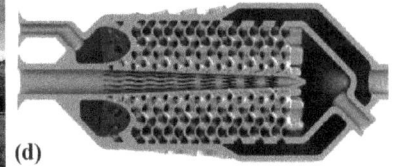

1 Shellular structures in nature, architecture and engineering: (a) cross-section through a sea urchin skeletal plate (Lai et al. 2007); (b) Frei Otto's physical form-finding using soap film (Hassel 2016); (c) bridge over the Basento by Sergio Musmeci (Milan 2020); (d) shellular structure's application in an aerospace turbine engine (Vlahinos and O'Hara 2020).

INTRODUCTION

Shellular Structures

Shellular (shell cellular) structures are a category of cellular structures in nature, composed of single, continuous, smooth-curved shells (Figure 1). Their geometry includes surfaces filling the space with the least amount of material, called minimal surfaces (Meeks and Perez 2011; Han et al. 2015). Minimal surfaces in nature (e.g., soap film) have inspired many architects and engineers to design lightweight structures on large scales for many years (Hassel 2016). Considering $k1$ and $k2$ as the principal curvatures of a minimal surface in each point, these surfaces have zero mean curvature ($H = (k1 + k2)/2 = 0$) and negative Gaussian curvature ($G = k1 \times k2 < 0$) (Hilbert and Cohn-Vossen 1990). Recent studies show that due to their specific morphology and high surface-to-volume ratio, these structures have better mechanical performance compared to the other cellular structures, such as strut-based cellular structures (Han et al. 2015; Han et al. 2017).

Graphic Statics

Graphic statics as an intuitive method of structural design has been used to design and analyze structures for many years (Maxwell 1864; Rankine 1864; Culmann 1866; Cremona 1890; Beghini et al. 2013). This method allows the designer to design a structure by exploring its form and force simultaneously. Using reciprocal diagrams, one is able to control the internal flow of force and external loading scenario while designing the structure. Three-dimensional graphic statics (3DGS), as an extension of 2D graphic statics, enables the designer to design three-dimensional and axially loaded structures in equilibrium in which no bending occurs (Akbarzadeh 2016; McRobie 2016; Konstantatou et al. 2017; D'acunto et al. 2017; Akbarzadeh et al. 2021). In this method, there is a clear relation between the form and force diagrams as a pair of reciprocal diagrams linked through simple geometric constraints. In 3DGS, a closed polyhedron can represent the equilibrium of a three-dimensional node (Figure 2a). Each edge e_i or force f_i in the form diagram is perpendicular to the corresponding face $f_i^†$ in the force diagram. The form diagram represents the geometry of the structure combined with the reaction forces and applied loads (Figure 2, bottom), while the force diagram represents the equilibrium of internal and external forces (Figure 2, top). In this paper, the form is denoted by Γ and the force diagram by Γ†. Furthermore, all the topological elements related to the force diagram are denoted by † superscript.

Topologically speaking, these diagrams consist of vertices v_i, edges e_i, faces f_i, and cells c_i. Each vertex, edge, face, and cell ($v_i^†, e_i^†, f_i^†, c_i^†$) in the force diagram corresponds to a cell, face, edge, and vertex (v_i, e_i, f_i, c_i) in the form diagram (Figure 2) (Akbarzadeh 2016). Since these diagrams are reciprocal and the faces of the force diagram (corresponding to the form diagram's edges) are planar, the form diagram's faces (corresponding to the force diagram's edges) are planar as well (Akbarzadeh 2016).

A closed and convex force diagram signifies a compression/tension-only structure in equilibrium. Furthermore, the magnitude of the force f_i in each strut member in the form diagram is proportional to the area of the corresponding face $A(f_i)$ in the force diagram (Figure 2a) (Akbarzadeh 2016). This theory can be simply proved based on the divergence theorem (Strokes 1901). Based on this theory, the sum of all area-weighted normals of a polyhedron is zero. As a form-finding technique, subdividing the internal space of the force diagram results in a variety of topologically different structures, designed for a defined boundary condition and loading scenario (Figure 2) (Akbarzadeh 2016; Akbari et al. 2019). Adding thickness to each edge of these form diagrams proportional to the area of its corresponding face in the force diagram results in a strut-based cellular funicular structure (CFC) (Figure 2).

Increasing the number of subdivisions in the force diagram results in a form diagram with smaller edges and distributed forces in the members (Figure 2a-f). This increase can finally result in the edges with near zero-length, approximating a surface as a form diagram. Specific subdivision techniques in 3DGS can approximate surfaces with synclastic or anticlastic curvatures as form diagrams (Akbari et al. 2019). Figure 3a displays a tetrahedron as a force diagram corresponding to a node in equilibrium with two forces upward and two downwards as a form diagram (a node with an anticlastic curvature). This tetrahedron is

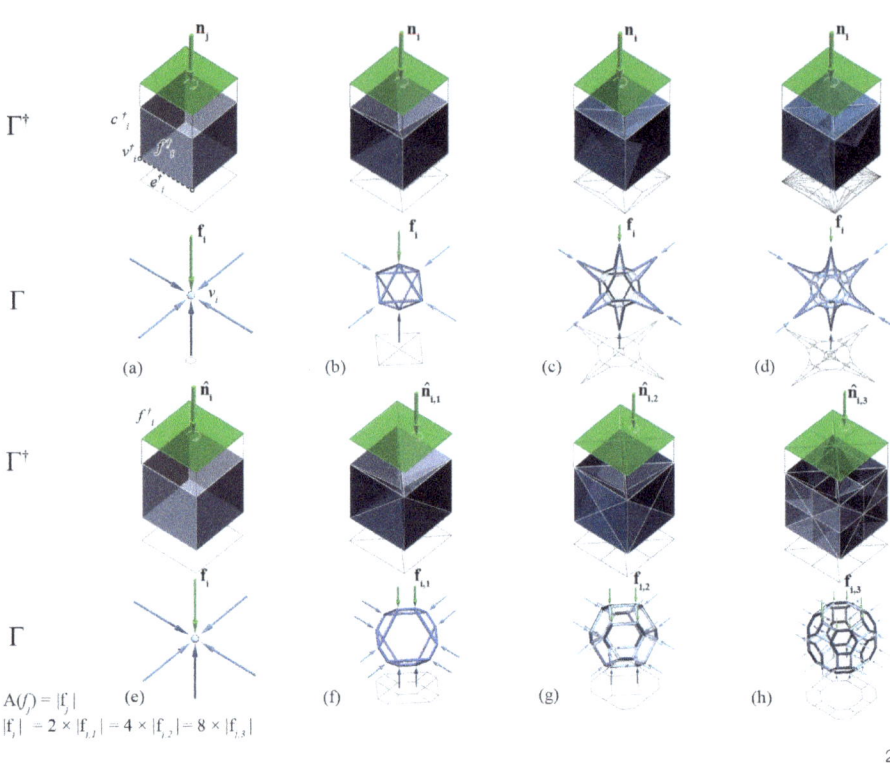

2   Different types of force diagram's subdivisions in 3DGS result in new form diagrams: (a, e) a closed cube as a force diagram represents a node in equilibrium; (a-d) internal subdivision of a force diagram without subdividing the global faces; (e-h) subdivision of a force diagram while subdividing the external faces and extruding inside

3   Different types of force diagram's iterative subdivision of a tetrahedron as a force diagram approximates a discrete surface with anticlastic curvature (a-d) as the form diagram: using this subdivision between specific labyrinth graphs, one can design a shellular funicular structure (e) and use the labyrinths as control handles to manipulate the structure (f)

4   Origamizing a polyhedral surface using tuck-folding technique

generated by connecting the end points of the two skew lines $l^\dagger_i$ and $l'^\dagger_i$. Dividing $l^\dagger_i$ and $l'^\dagger_i$ into equal segments and establishing a tetrahedron between each two skew segments from each line subdivides the force diagram into multiple tetrahedrons, resulting in a discrete anticlastic surface as a form diagram, as shown in Figure 3a-d (Akbari et al. 2019). Two lines $l^\dagger_i$ and $l'^\dagger_i$ play the role of subdivision axis in the force and the curvature axis in the form diagram (Figure 3d). These lines are parts of two connectivity graphs, named labyrinths, connecting two segregated regions, divided by the anticlastic surface (Fischer and Koch 1989). Using the anticlastic subdivision technique, one can design an anticlastic polyhedral surface, named a Shellular Funicular Structure (SFS). The labyrinths, as the subdivision axes in the form and the control handles in the force, facilitate the design process and manipulation of the SFS form-finding technique (Akbari et al. 2020).

Origamizing Polyhedral Surfaces Using Tuck-folding
Origami, as an art of folding a flat sheet of material to the desired shape without cutting or stretching, has been investigated for many years. On the other hand, Kirigami is the combination of cuts and folds on a flat sheet of material in order to result in the desired geometry (Castle et al. 2014). Tuck-folding as an origami method can design any freeform surface from flat sheet material by tucking and hiding the unwanted areas of the paper (Tachi 2009). In this process, the Origamizer software can be used to generate the folding/cutting pattern (Demaine and Tachi 2017). The input is a polyhedral mesh and the output is the folding pattern with edge tucking molecules (ETM) and vertex tucking molecules (VTM) added (Figure 4). Each ETM is a quadrilateral with a crease pattern, inserted between a pair of edges in the pattern, corresponding to an edge on the reference mesh (Figure 4). Each VTM is an N-gon surrounded by N edge-tucking molecules (ETM) corresponding to a vertex on the reference mesh. Despite the use of more materials, tuck-folding has many advantages over regular folds. It is easier to fold for complex geometries (especially with anticlastic curvatures) as the fold angle is embedded in the cutting patterns. Furthermore, the ETMs provide extra stiffness and stability to the folds compared to regular folds. This technique includes three steps, cutting to a disk, mapping surface polygons, and generating the crease/cut pattern (Tachi 2009), as follows.

- **Cutting to a disk:** To map the surface's polygons to a 2D plane, one needs to construct a polygonal schema, a polyhedral surface homeomorphic to a flat disk that covers the original surface (Figure 4).
- **Mapping the surface's polygons:** Next, the polyhedral mesh is decomposed into individual polygons and isometrically mapped to a 2D plane. In isometric mapping, the edge lengths and the polygon's angles will be preserved (Figure 4).
- **Designing the folding/cutting pattern:** Surface flattening is the problem of generating a 2D pattern from a given

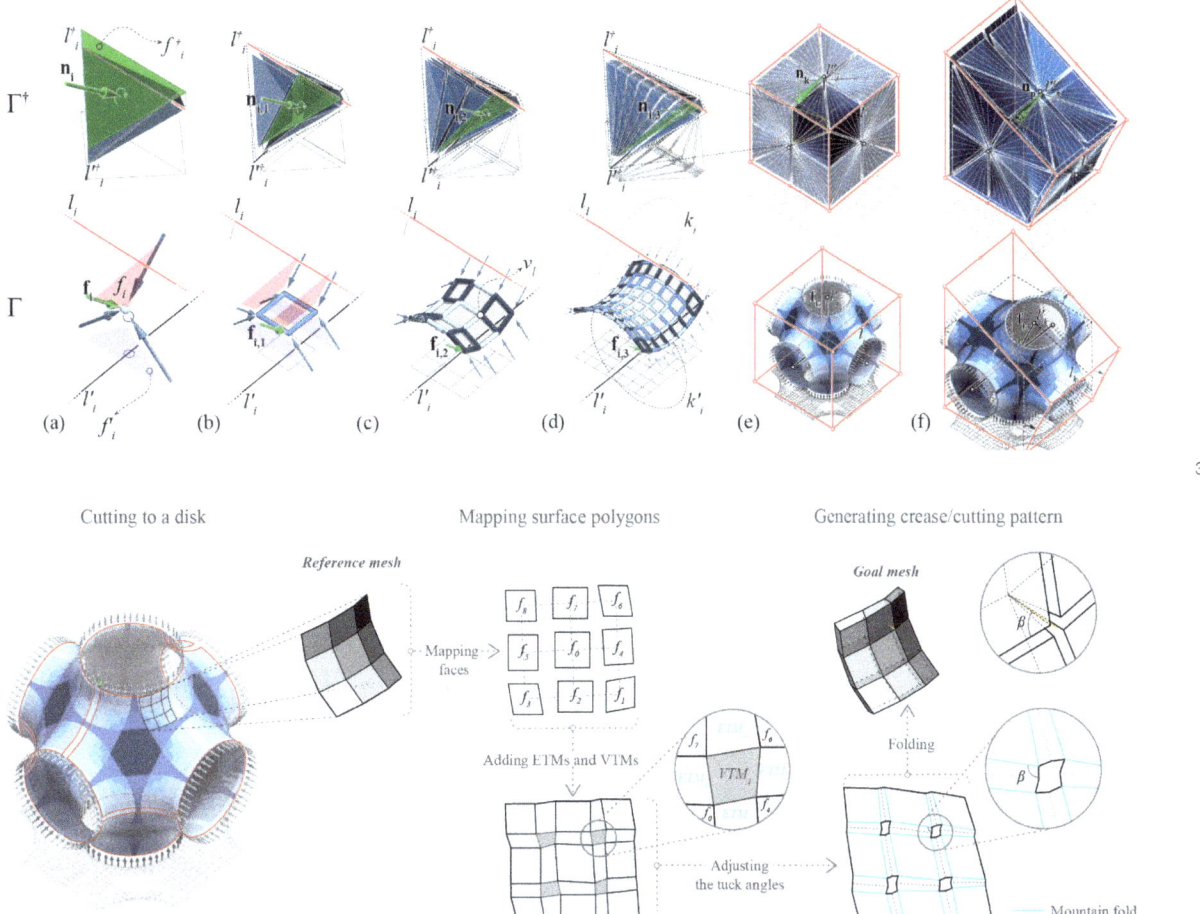

3

4

3D shape using an isometric mapping. After adding ETMs and VTMs between the mapped polygons, using specific equality and inequality conditions (Tachi 2009), the method ensures that each ETM is an isosceles trapezoid and each VTM is a closed convex polygon. By folding a valley crease defined in the middle of each ETM, the ETMs convert to an edge with a hidden tuck in the folding state. Considering the thickness and stiffness of the material, VTMs are cut out as holes, combining the kirigami and origami techniques to facilitate the folding process (Liu et al. 2019). In the goal mesh, the angle between the vertex axis and the edge (i.e., $β$) is equal to the angle between the VTM's and the ETM's edge (Figure 4). This angle is defined as the tuck angle which assures us that the curvature of the goal mesh in each vertex is equal to the one in the reference mesh.

Problem Statement

Although SFSs are considered as efficient structures applicable to different functions in many scales, due to their complex geometry, their process of computational modeling and fabrication are quite challenging, limiting their application in large scales. Furthermore, designing these structures for a predefined boundary condition, control and manipulation of their geometry are not easy tasks. Moreover, fabricating these geometries is mostly possible using additive manufacturing techniques, requiring many supports in the printing process. Recently, Akbari et al. (2019; 2020) proposed different form-finding techniques to design Shellular Funicular Structures in the context of graphic statics based on designing the labyrinths' graphs and applying anticlastic subdivisions in between. But these techniques lack a robust computational algorithm to translate a Cellular (CFS) to a Shellular Funicular Structure (SFS). Furthermore, due to the specific criteria that need to be satisfied in each step, designing the labyrinths' graphs as an intuitive process is not an easy task (Akbari et al. 2020). Moreover, due to their complex interwoven geometry, their fabrication process using digital manufacturing techniques on large scales is challenging. This paper seeks a robust computational method for the design and fabrication of these structures to facilitate their application in macro scales.

Objectives

The main objectives of this paper are to explain a

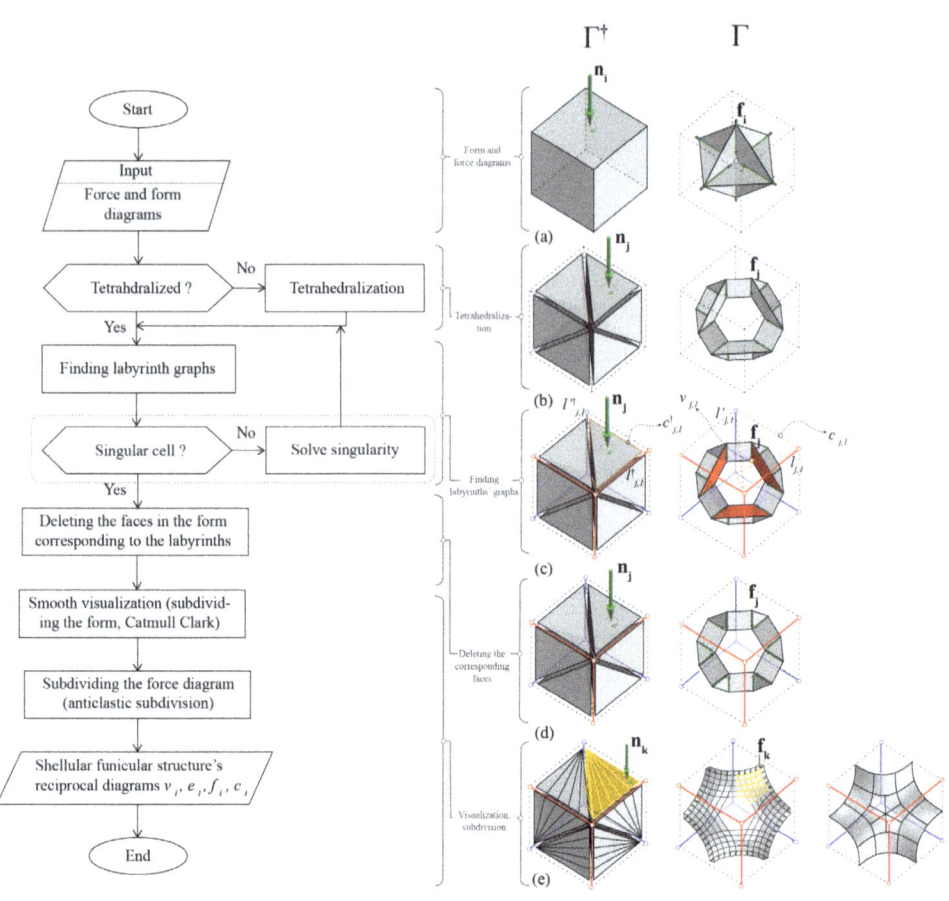

5 The form-finding flowchart (left) and the workflow (right), representing the algorithm for translating a CFS to SFS in the context of graphic statics.

6 Different tetrahedralization of a cube as a force diagram, e.g., into six (a, b) and five (c, d) tetrahedrons, results in different types of the labyrinths' graphs.

7 The process of finding a labyrinths' graph in a tetrahedralized force diagram (a-f) and 3 different types of graphs resulting from this algorithm (g, h, i).

computational algorithm for translating a Cellular Funicular Structure (CFS) to a Shellular Funicular Structure (SFS), and describing a fabrication method to build SFSs out of flat sheets of material using the origami/kirigami technique. As a proof of concept, a structure's geometry with its folding/cutting pattern is designed and fabricated using the SFS and tuck-folding techniques.

METHODS

The methodology for designing and fabricating SFSs is two-fold. The first section explains the computational algorithm behind the process of translating a CFS to an SFS and the second one describes the fabrication process, inspired by the tuck-folding technique.

Computational Design Process

To translate the geometry of a Cellular Funicular Structure (CFS) to a Shellular Funicular Structure (SFS), one needs to subdivide the corresponding force diagram, such that each vertex v_i connected to the group of edges e_i converts to an anticlastic patch, consisting of a group of vertices v_i, faces f_i and edges e_i with smaller lengths, approximating a discrete surface (Figure 3). Materializing the new form diagram by adding faces between the edges (instead of adding thickness to the edges) results in a Shellular Funicular Structure. It is important to notice that the labyrinths l_i and l_i' in the force diagram, overlay on the edges e_i and e_i', corresponding to the faces f_i and f_i' in the form diagram (Figure 3a, b). While materializing the form diagram in the SFS form-finding process, the faces in the form corresponding to the labyrinth edges in the force should not be materialized. Removing these faces results in a 2-manifold geometry (Figure 3).

Although one can use this technique to design an SFS, designing the labyrinths' graphs intuitively is not an easy task. This section describes a computational algorithm for translating a CFS to SFS. The main objective of this algorithm is to find proper three-dimensional labyrinths' graphs in the form and the force diagrams (inputs) and using them to design the shellular structure (Figure 5). This algorithm receives a form and the force diagram (Figure 5a) that are designed using *Polyframe* (Nejur and Akbarzadeh 2021). After converting the force diagram to a group of tetrahedrons (tetrahedralization, Figure 5b), the algorithm finds the possible labyrinths' graphs in the force diagram (Figure 5c), and remove the corresponding faces to these labyrinths in the form diagram to result in a 2-manifold geometry (Figure

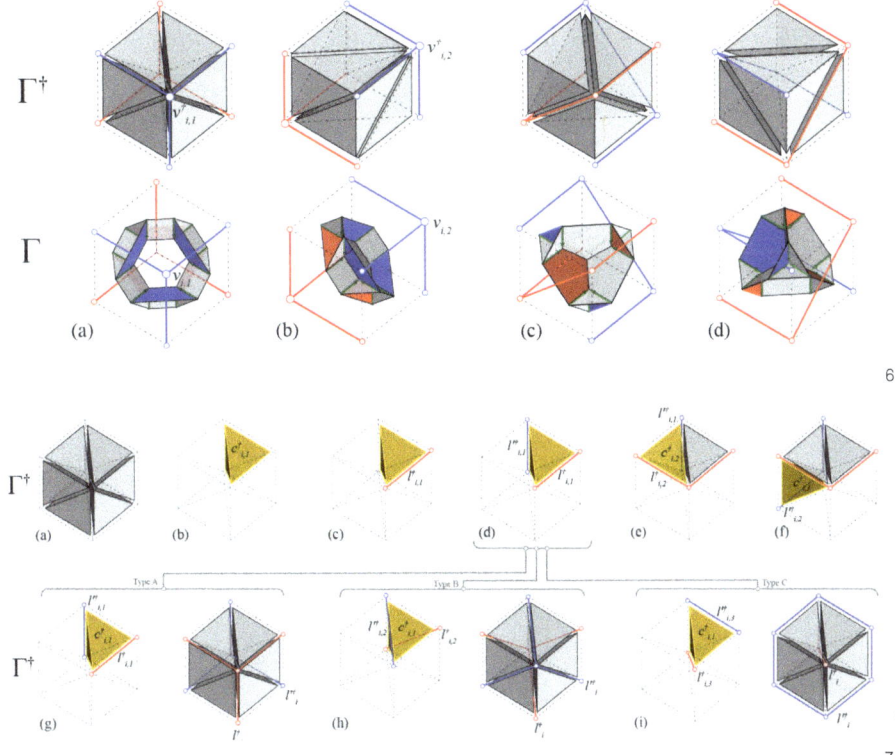

5d). Finally, the smooth version of the form diagram will be visualized using the Catmull-Clark subdivision (Catmull and Clark 1978), and the main shellular load-path will be constructed using the anticlastic subdivision in 3DGS (Figure 5e) (Akbari et al., 2019). In the SFS design process, there are three main principles that need to be considered:

- each pair of labyrinths in the form diagram (e.g., $l_{j,1}$ and $l'_{j,1}$) should form a tetrahedron in between (e.g., $c_{j,1}$), corresponding to an anticlastic node (e.g., $v_{j,1}$) in the form diagram (Figure 5c),
- the force diagram includes two sets of labyrinths ($l^†_{j,1}$ and $l'^†_{j,1}$) and each tetrahedron in the force diagram (e.g., $c^†_{j,1}$), should only include one labyrinth edge from each set which is in a skew position to the other (Figure 5c),
- each labyrinth's edge in the force diagram can be connected to the labyrinth edges from the same set, assuring that the resulting surface (form diagram) will divide the space into two subspaces (Figures 5d, e).

Tetrahedralization

According to the first principle mentioned above, each force diagram should only include tetrahedrons, resulting in 4-valency nodes (nodes that are connected to 4 edges) in the form diagram, representing an anticlastic curvature. The polyhedron tetrahedralization function decomposes a 3D polyhedron into a set of non-overlapping tetrahedra whose vertices are chosen from the vertices of the polyhedron (Toussaint et al. 1993). A convex polyhedron in 3D can always be tetrahedralized without adding new vertices (Steiner points) by connecting any vertex of the polyhedron to all the other vertices (Lennes 1911). Therefore, there might be multiple solutions for tetrahedralizing an input force diagram (Figure 6).

Unfortunately, the minimum-complexity triangulation problem is NP-complete (Below et al. 2000). Hence, in this algorithm, the authors provide part of the possible solutions, using the user's preference (Figure 6). In this process, each input vertex (e.g., $v^†_{i,1}$, $v^†_{i,2}$) will be connected to all the other vertices in the polyhedron, resulting in a group of tetrahedrons (Figures 6a, b). In some examples like a cube, there might be different techniques resulting in fewer tetrahedrons and new labyrinths' graphs (Figures 6c, d). Since this subject is beyond the scope of this paper, it is going to be studied further in future researches.

Finding the Labyrinths' Graph

After tetrahedralizing the force diagram, two edges from each tetrahedron, as the edges of the labyrinths' graphs, need to be selected. Figures 7a-f displays the process for finding the labyrinths' graphs in a cube partitioned into six tetrahedrons, as the force diagram.

This process starts with selecting a random cell $c^†_{i,1}$ (Figure 7b). Selecting one of the edges $l^†_{i,1}$ of this tetrahedron as a labyrinth's edge (Figure 7c) determines the skew edge $l'^†_{i,1}$

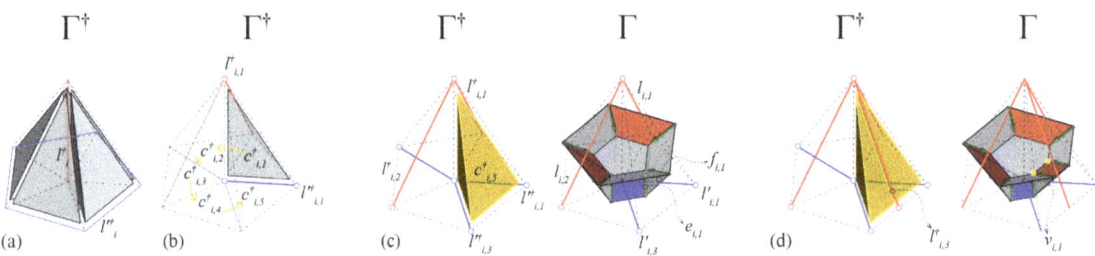

8   A force diagram comprising 5 tetrahedrons (a); an attempt to find the new labyrinths' graphs (b); finding the singular cell (c); and resolving the singularity by subdividing the cell (d)

as the other labyrinth's edge (Figure 7d). The edges $l'^\dagger_{i,1}$ and $l^\dagger_{i,1}$ have specific topological relations to each other and are called the dual of each other. The algorithm proceeds through the four neighbors of the tetrahedron cell $c^\dagger_{i,1}$. Each neighbor (e.g., $c^\dagger_{i,2}$) has a face with three edges and three vertices in common with the previous cell $c^\dagger_{i,1}$ (Figure 7e). Among these edges, only one edge is already recognized as a labyrinth edge (i.e., $l^\dagger_{i,1}$). If $l'^\dagger_{i,1}$ is the dual of $l^\dagger_{i,1}$ in cell $c^\dagger_{i,1}$, it is the dual of $l^\dagger_{i,1}$ in the cell $c^\dagger_{i,2}$ as well (Figure 7e). Similarly, the algorithm marches through all the cells in the force diagram and finds the dual labyrinth edges of the previous cell (Figure 7f), until all the cells are passed. In this process, each cell has a pair of labyrinth edges, and the force diagram has two labyrinth graphs $l^\dagger_i$ and $l'^\dagger_i$ (Figure 7g).

The resulted labyrinths' graphs are the first type of possible graphs for a force diagram. Since there is a possibility of choosing three different pairs of labyrinth edges in a tetrahedron out of 6 edges, choosing different pairs in the first cell $c^\dagger_{i,1}$ (Figure 7d) as the labyrinth edges will result in different labyrinths' graphs for the force diagram (Figures 7g, h, i). Hence, according to this algorithm, there are three geometrically different labyrinths' graphs for a force diagram.

Singularity

In some situations, due to the specific edge-cell connectivity (when a labyrinth edge is connected to the odd number of cells, e.g., Figure 8a), the algorithm explained above has only one solution (instead of three). Figure 8a displays a tetrahedralized force diagram with 5 cells and possible labyrinths' graphs $l^\dagger_i$ and $l'^\dagger_i$. Figure 8b shows an attempt to find the second pair of the possible graphs. In cell $c^\dagger_{i,1}$, two labyrinth edges $l^\dagger_{i,1}$ and $l'^\dagger_{i,1}$ are selected. After selecting the labyrinths' edges of to the next cells respectively ($c^\dagger_{i,2}$, $c^\dagger_{i,3}$, $c^\dagger_{i,4}$, $c^\dagger_{i,5}$), the algorithm faces the cell $c^\dagger_{i,5}$ with two edges $l^\dagger_{i,1}$ and $l'^\dagger_{i,3}$ from the same set of labyrinth's graph which is against the main principles of the SFS's technique (Figure 8c). Having two labyrinths in the cell $c^\dagger_{i,5}$ will eliminate two neighbor faces of the edge $e_{i,1}$ in the form diagram (faces are marked with blue), resulting in the edge $e_{i,1}$ connected to one face (Figure 8c). In this situation the cell $c^\dagger_{i,5}$ is called a singular cell. This issue happens when an edge in the force diagram is connected to an odd number of cells. To solve this issue, one may subdivide the singular cell into two cells and add the new edge as one of the labyrinths' edges (e.g., $l^\dagger_{i,3}$) (Figure 8d).

Eliminating the Extra Faces

Each of the labyrinths' graphs corresponds to a group of faces in the form diagram (Figures 9b, g, l). Eliminating these faces from the form diagram results in a 2-manifold discrete surface (Figures 9c, h, m).

Visualization and Subdivision

There are two ways to generate a smooth curved surface from the discrete surface resulted in the previous section. The first solution is to apply a Catmull-Clark subdivision, generating a bi-cubic uniform B-spline surface to visualize the smooth anticlastic surface and to compare the geometry of different structures (Figures 9d, i, n) (Catmull and Clark 1978). To result in a smooth form diagram in the context of graphic statics, and finding the corresponding load path, one needs to apply the anticlastic subdivision technique (explained in the Introduction) to each tetrahedron cell between each pair of labyrinths (Figures 9 e, j, o). Using this technique, one can intuitively design a Cellular Funicular Structure (CFS) for a specified boundary condition and loading scenario (e.g., a connection, a column, or a bridge) and translate it to its counterpart, a Shellular Funicular Structure (SFS) (Figure 10). It is worth mentioning that in graphical form finding methods, the self-weight of the structure is not considered. To consider the self-weight, one needs to assign mass to each to each node in the system and apply numerical form finding methods (e.g., mass spring or force density) to find the new form in equilibrium (Adriaenssens et al. 2014).

Fabrication Process

Shellular Funicular Structures (SFSs) comprise planar faces resembling a polyhedral surface (Introduction). Therefore, they are a genuine candidate to be fabricated based on Origami/kirigami techniques using flat sheet material. Tuck-folding can be used as a fabrication method to design the folding pattern of these structures. Using this technique, one can either fabricate a shellular system as a

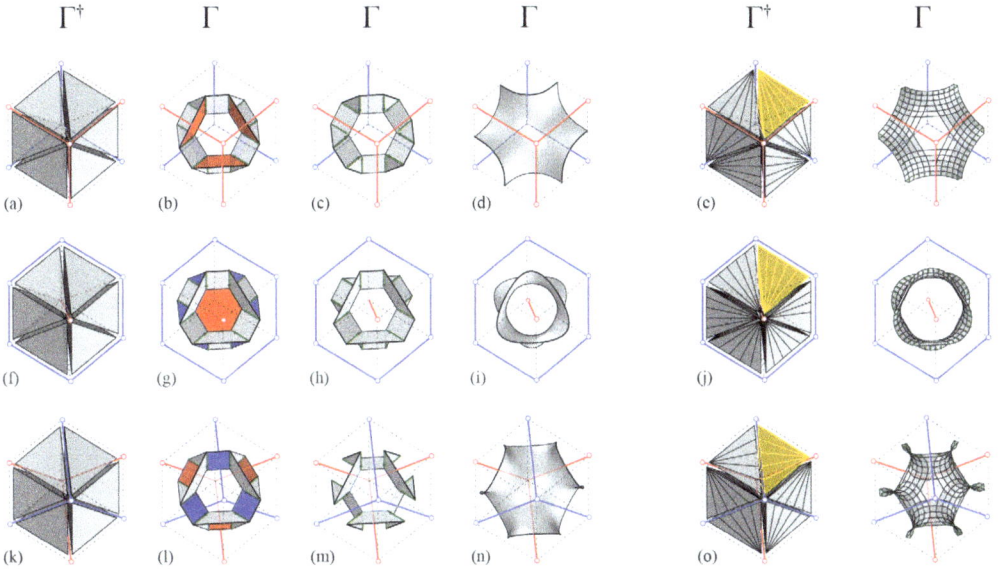

9   Three types of possible labyrinths' graphs for a force diagram (a, f, k) and their reciprocal form diagram with faces correspond to the labyrinths' edges (b, g, l); removing these faces results in a 2-manifold geometry (c, h, m) which can be converted to a smooth surface by either Catmull-Clark (d, i, n) or anti-clastic subdivision in 3DGS (e, j, o)

discrete single-layer structure, or a double layer formwork for pouring a structural material like concrete (Figures 11a, b). This method finds the folding pattern of a polyhedral surface out of a flat sheet of material, with hidden tucks instead of constructing the exact faces (Figure 11). Furthermore, the edge tucking molecules (ETMs) provide extra stiffness and stability to the folds compared to regular folds. This section explains the process of fabricating a shellular geometry with the tuck-folding method introduced in the introduction (Tachi 2009; Liu et al. 2019).

Fabricating a Shellular Funicular Structure Using the Tuck-folding Technique

A shellular geometry with a bounding box dimension of 620mm by 620mm by 310mm as a part of the shellular funicular bridge displayed in Figure 9 is generated using the proposed algorithm for a trial fabrication (Figure 12). This geometry is cut and flattened into four cutting patterns, and each pattern has a bounding rectangle dimension of 760mm by 700mm. For the purpose of reducing the shipping cost and the level of folding difficulty, each cutting pattern is further split into seven smaller parts (Figure 12).

After examining the durability, mechanical properties, and foldability of different materials, 0.5mm stainless steel sheet is selected as the material for fabrication, and an industrial laser cutter is used to cut steel sheets to the target patterns. To mark the fold lines and increase the ease of manual folding, dash line cuts are added to the cutting patterns. The design parameters of the dash line cuts include material density, cutting interval, and dash line width (Figure 13). Material density is defined by the percentage of material left on a folding hinge after dash line cutting; cutting interval means the distance between two neighboring dash line cuts; dash line width indicates the width of the dash line cuts, which confines the bending area along the fold lines. Lower material density, smaller cutting interval, and larger dash line width lead to easier folding, however, small cutting interval also impairs the material strength along the hinges. Based on a series of physical experiments (Figure 15), the cutting interval is determined at 1.5mm. The dash line width has more complex impacts on the folding behavior as too small a width may incur a larger size in the folded geometry, while excessive width increases the folding error. Those parameters are designed in such a way that all folds can be easily achieved by hand, and the dimensions of the folded geometry are as close as possible to the original (Figure 15). To secure the tuck-folding, a pair of circular cuts are also added to each edge tucking molecule which are later aligned and fastened using M3 screws and nuts after folding (Figure 14). When considered as a frictionless hinge and rigid face system, each folding pattern has only one degree of kinematical indeterminacy, meaning that all the folding lines need to be folded at the same time (Figure 16). After folding each part, the shellular structure is assembled and secured using screws and nuts (Figures 17, 18).

Results and Discussion

In this research, the authors proposed a robust computational algorithm to translate a CFS to a SFS. Using this algorithm, one can design any compression/tension-only

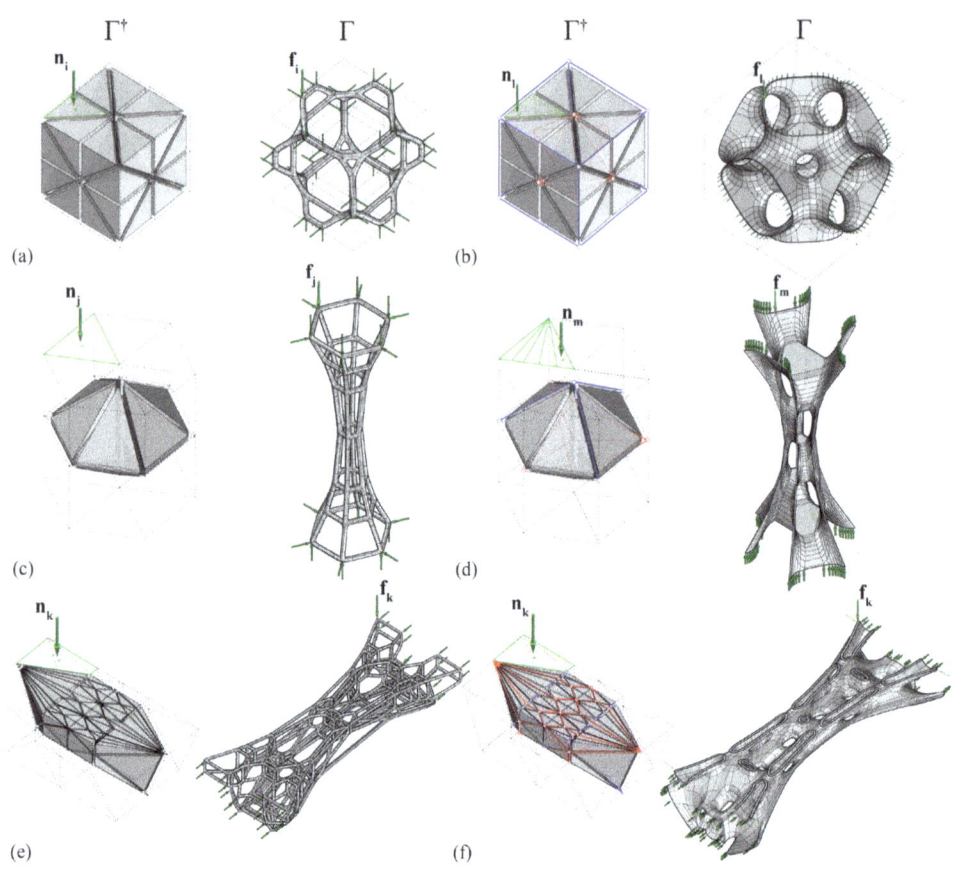

10 The examples of the translation of force and form diagrams of CFSs (a, c, e) to SFSs using the SFS's computational algorithm (b, d, f)

11 Fabricating an SFS using the tuck-folding technique as a discrete single-layer structure (a), or double-layer form-work for pouring concrete (b)

12 The shellular geometry for trial fabrication and the fold pattern developed using the Origamizer software (Demaine and Tachi 2017)

13 The design parameters of dash line cuts

14 Securing the tuck-folding using scews and nuts

shellular structure for a defined boundary condition, from meso-scale (e.g., columns and beams) to macro-scale (e.g., a building or a bridge). From the fabrication point of view, due to the interwoven geometry of SFSs which necessitates lots of supports in the printing process, additive manufacturing techniques may not be considered as an efficient process on a large scale. In the meantime, flat sheet materials are an ideal choice because of their accessibility, processibility, low cost, and applicability to large scales. Hence, the second part of the research focused on the fabrication process of SFSs out of flat metal sheets. This method facilitates the use of these structures on the large scale, as a single-layer structure or a double-layer formwork for pouring concrete (Figures 17, 18). Thanks to the proposed design methodology, the strctures that are designed using this technique can be considered as self-supporting structures, designed for specific loading scenarios. Moreover, adding thickness to the edges of the structures in the fabrication process results in a stiffer system, increasing the structural capacity of the formwork.

CONCLUSION

Using the proposed technique, shellular funicular structures can be designed and fabricated for macro-scale applications. Future researches include implementing interactive software for the design and manipulation of SFSs and investigating the effect of different tetrahedralization on the results. In the fabrication process, there is a need for improving the process of designing the folding pattern. Controlling the with widths of ETMs using the Origamizer software is not an easy task, resulting in a large waste of material. Adding ETMs with different widths to the structure (proportional to the force distribution in the system) results in an efficient system in terms of the material usage with constant stress distribution. It is important to notice that the structures that are designed using this technique are already in equilibrium for the defined boundary condition and adding non-uniform thickness in the materialization process only yields constant stresses in them.

ACKNOWLEDGEMENTS

The authors acknowledge support by National Science Foundation Future Eco Manufacturing Research Grant (NSF, FMRG-CMMI 2037097 and the National Science Foundation CAREER Award (NSF, CAREER-CMMI 1944691) to Dr. Masoud Akbarzadeh. Moreover, the authors are grateful for an inspiring discussion with Dr. Shu Yang from material science department at the University of Pennsylvania.

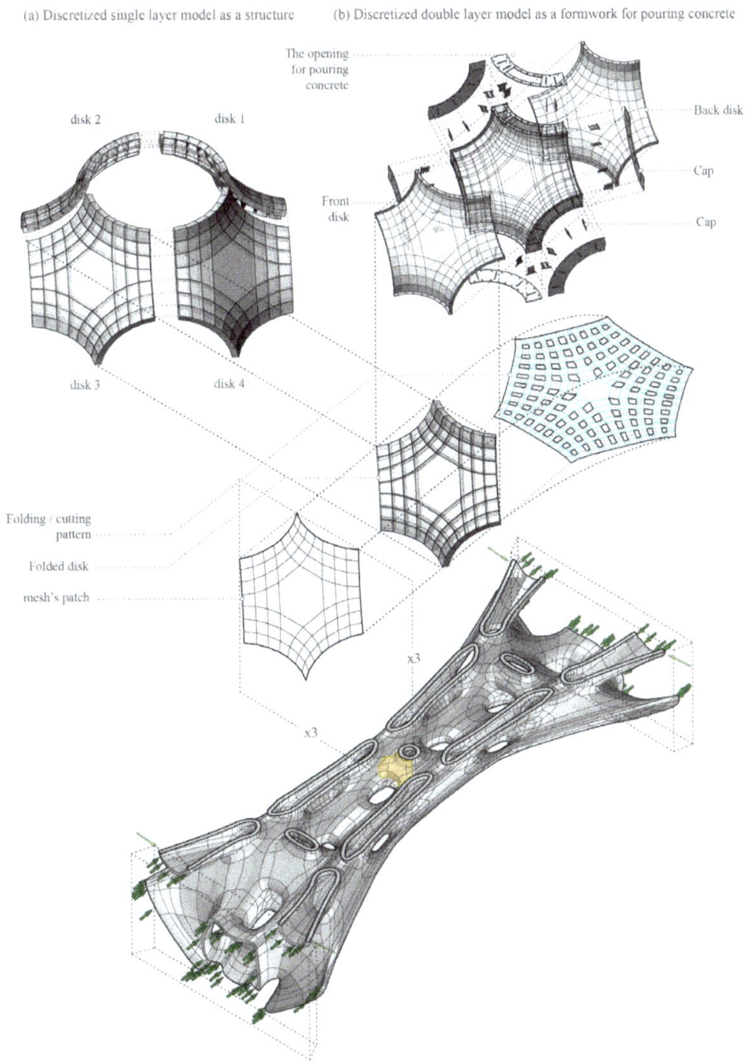

REFERENCES

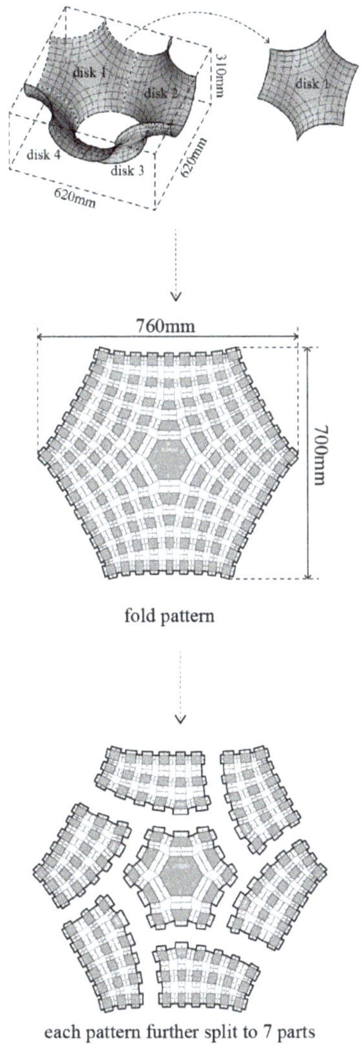

Adriaenssens, S., P. Block, D. Veenendaal, and C. Williams, eds. 2014. *Shell Structures for Architecture: Form Finding and Optimization*. London: Routledge.

Akbari, Mostafa, Masoud Akbarzadeh, and Mohammad Bolhassani. 2019. "From Polyhedral to Anticlastic Funicular Spatial Structures." In *Form and Force: Proceedings of IASS Annual Symposium 2019*. Barcelona, Spain.Akbari, Mostafa, Armin Mirabolghasemi, Hamid Akbarzadeh, and Masoud Akbarzadeh. 2020. "Geometry-based structural form-finding to design architected cellular solids." In *SCF '20: Symposium on Computational Fabrication*.

Akbarzadeh, M., M. Bolhassani, A. Tabatabaie, M. Akbari, A. Seyedahmadian, K. Papalexiou, and N. Designs. 2021. "Saltatur: Node-Based Assembly of Funicular Spatial Concrete." In *ACADIA 20: Distributed Proximities; Proceedings of the 40th Annual Conference of the Association for Computer-Aided Design in Architecture*. Online and Global. 108–113.

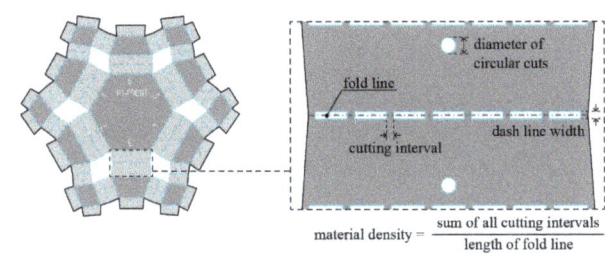

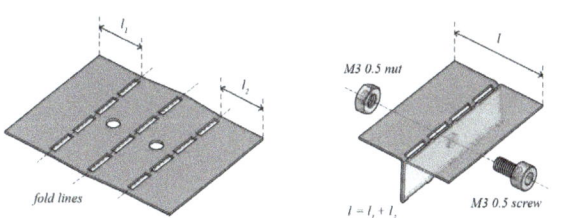

$$\text{material density} = \frac{\text{sum of all cutting intervals}}{\text{length of fold line}}$$

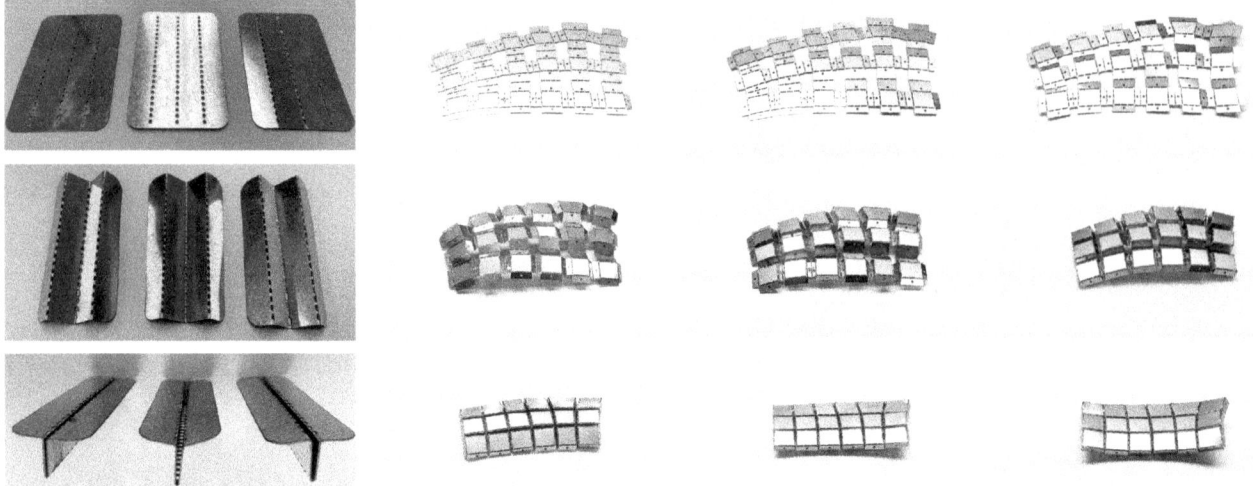

15 Folding a stainless steel sheet with different design parameters of dash line cuts
16 The folding process of a part of the physical model
17 The final assembled model (top view)
18 The final assembled model (front view)

Akbarzadeh, Masoud. 2016. "3D Graphical Statics Using Reciprocal Polyhedral Diagrams." PhD diss., ETH Zurich.

Beghini, Lauren L., Juan Carrion, Alessandro Beghini, Arkadiusz Mazurek, and William F. Baker. 2014. "Structural Optimization Using Graphic Statics." *Structural and Multidisciplinary Optimization* 49: 351–366.

Below, A., J. A. De Loera, and J. Richter-Gebert. 2000. "Finding Minimal Triangulations of Convex 3-polytopes is NP-hard." In *SODA '00: Proceedings of the Eleventh Annual ACM-SIAM Symposium on Discrete Algorithms*. 65–66.

Castle, T., Y. Cho, X. Gong, E. Jung, D.M. Sussman, S. Yang, and R.D. Kamien. 2014. "Making the Cut: Lattice Kirigami Rules." *Physical Review Letters* 113 (24): 245502.

Catmull, Edwin, and James Clark. 1978. "Recursively generated B-spline surfaces on arbitrary topological meshes." *Computer-Aided Design* 10 (6): 350-355.

Cremona, Luigi. 1890. *Graphical Statics: Two treatises on the Graphical Calculus and Reciprocal Figures in Graphical Statics*. Oxford: Clarendon Press.

Culmann, Karl. 1866. *Die Graphische Statik*. Zurich: Verlag Von Mayer & Zeller.

D'Acunto, Pierluigi, Jean-Philippe Jasienski, Patrick Ole Ohlbrock, and Corentin Fivet. 2017. "Vector-based 3d Graphic Statics: Transformations of Force Diagrams." In *Interfaces: Architecture. engineering.science; Proceedings of the IASS Annual Symposium 2017*. Hamburg, Germany.

Demaine, Erik D. and Tomohiro Tachi. 2017. "Origamizer: A Practical Algorithm for Folding Any Polyhedron." In *33rd International Symposium on Computational Geometry (SoCG 2017)*. Leibniz International Proceedings in Informatics.

Fischer, Werner, and Elke Koch. 1989. "Genera of minimal balance surfaces." *Acta Crystallographica Section A: Foundations of Crystallography* 45 (10): 726–732.

Han, Seung Chul, Jeong Myung Choi, Gang Liu, and Kiju Kang. 2017. "A microscopic shell structure with Schwarz's D-surface." *Scientific Reports* 7 (1): 1–8.

Han, Seung Chul, Jeong Woo Lee, and Kiju Kang. 2015. "A New Type of Low Density Material: Shellular." *Advanced Materials* 27 (37): 5506–11.

Hassel, Joshua V., director. *Frei Otto, Spanning the Future*. Tensile Evolution North America, 2016. 1:00:09. Accessed July 15, 2021. http://www.freiottofilm.com/.

Hilbert, David, and Stefan Cohn-Vossen, 1990. *Geometry and The Imagination*. New York: Chelsea Publishing Company.

Konstantatou, M. and Mcrobie, A., 2017, September. 3D Graphic statics and graphic kinematics for spatial structures. In *Proceedings of International Association for Shell and Spatial Structures (IASS) Annual Symposia* 2017 (12): 1–10.

Lennes, N. J. 1911, "Theorems on the simple finite polygon and polyhedron." *American Journal of Mathematics* 33: 37–62.

Liu, Jingyang, Grace Chuang, Hun Chun Sang, and Jenny E. Sabin. 2019. "Programmable Kirigami: Cutting and Folding in Science, Technology and Architecture." In *ASME 2019 International*

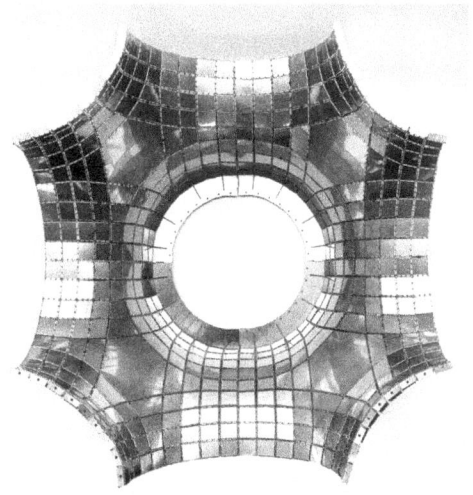

17

18

Design Engineering Technical Conferences and Computers and Information in Engineering Conference, vol. 5B. Anaheim, California.

Macquorn Rankine, W. J. 1864. "XVII. Principle of the equilibrium of polyhedral frames." *The London, Edinburgh, and Dublin Philosophical Magazine and Journal of Science* 27 (180): 92.

Maxwell, J. Clerk. 1864. "L. On the Calculation of the Equilibrium and Stiffness of Frames." *The London, Edinburgh, and Dublin Philosophical Magazine and Journal of Science* 27 (182): 294–299.

McRobie, A., 2016. "Maxwell and Rankine reciprocal diagrams via Minkowski sums for two-dimensional and three-dimensional trusses under load." *International Journal of Space Structures* 31 (2-4): 203–216.

Meeks, W. H., and J. Pérez. 2011. "The classical theory of minimal surfaces." *Bulletin of the American Mathematical Society* 48 (3): 325–407.

Milan, Laura. 2020. "Potenza, finalmente un progetto per il Ponte Musmeci." Teknoring (blog). December 29, 2020. Accessed July 15, 2021. https://www.teknoring.com/news/infrastrutture/potenza-progetto-ponte-musmeci/.

Nejur, A. and M. Akbarzadeh. 2021. "PolyFrame, Efficient Computation for 3D Graphic Statics." *Computer-Aided Design* 134: 103003.

Stokes, George Gabriel. 1901. *Mathematical and Physical Papers*, vol. 3. Cambridge: University Press.

Tachi, Tomohiro. 2009. "3D Origami Design based on Tucking Molecule." In *Origami 4*, edited by Robert J. Lang. New York: A K Peters/CRC Press. 259–272.

Tango, Alba Fermoso. 2012. "Equilibrio." sumaunahoramenos (blog). July 15, 2012. Accessed July 15, 2021. http://sumaunahoramenos.blogspot.com/2012/07/equilibrio.html.

Toussaint, Godfried T., Clark Verbrugge, Caoan Wang, and Binhai Zhu. 1993. "Tetrahedralization of simple and non-simple polyhedra." In *Proceedings of the 5th Canadian Conference on Computational Geometry*. 24–29.

Vlahinos, Maiki. and Ryan O'Hara. 2020. "Unlocking Advanced Heat Exchanger Design and Simulation with nTop Platform and ANSYS CFX." Accessed July 15 2021. https://www.aerospacemanufacturinganddesign.com/article/ntopology-heat-exchanger-design-simulation/.

IMAGE CREDITS
Figure 1: Image credits are listed in the caption.
All other drawings and images by the authors.

Mostafa Akbari is a PhD student at the Polyhedral Structures Laboratory, Weitzman School of Design, University of Pennsylvania, specializing in computational design and advanced manufacturing. His main research topic is Shellular Funicular Structures, a novel geometrical form finding method to approximate the geometry of minimal surface structures in the context of graphic statics.

Yao Lu is a PhD student at the Polyhedral Structures Laboratory, Weitzman School of Design, University of Pennsylvania. He is a design researcher with a great interest in generative design, robotic fabrication, 3D printing, and computer graphics.

Masoud Akbarzadeh is a designer with a unique academic background and experience in architectural design, computation, and structural engineering. He is an Assistant Professor of Architecture in Structures and Advanced Technologies and the Director of the Polyhedral Structures Laboratory (PSL) at the University of Pennsylvania.

Mortarless Compressed Earth Block Dwellings

A Low-Cost Sustainable Design and Fabrication Process

Yu Zhang
ETH Zurich

Liz Tatarintseva
Gianni Botsford Architects

Tom Clewlow
Arup

Ed Clark
Arup

Gianni Botsford
Gianni Botsford Architects

Kristina Shea
ETH Zurich

1 Single-story CEB Dwellings

This project develops a template design and an adaptive fabrication process for sustainable Compressed Earth Block (CEB) dwellings for low-income countries. Most existing projects (Wilton et al. 2019; WASP 2021) on sustainable dwellings involve high-tech equipment or skilled workers on-site. This project integrates digital technologies into the design and fabrication processes to reduce these requirements and make the design compatible with conventional construction methods that are actively adopted in low-income countries using minimum infrastructure, skilled labor, and investment.

The template design is a single-story CEB dwelling (Figures 1, 2, 3) consisting of a small number of interlocking brick types (Figure 4). The transitioning geometry, as shown in Figure 5, is optimized for structural efficiency as well as a practical architectural floorplan. The dwelling can either be built as a single-room house or form a multi-room house and complexes according to needs (Figure 6).

To enable a mortarless and stable assembly during construction, interlocking features, including left-right interlocking and top-bottom interlocking, are embedded into the brick design (Figure 4). The interlocking nature of the bricks acts as a guide to locate and orientate the bricks so that specialist construction knowledge and training are not required,

PRODUCTION NOTES

Status: In progress
Location: ETH Zurich
Date: 2021

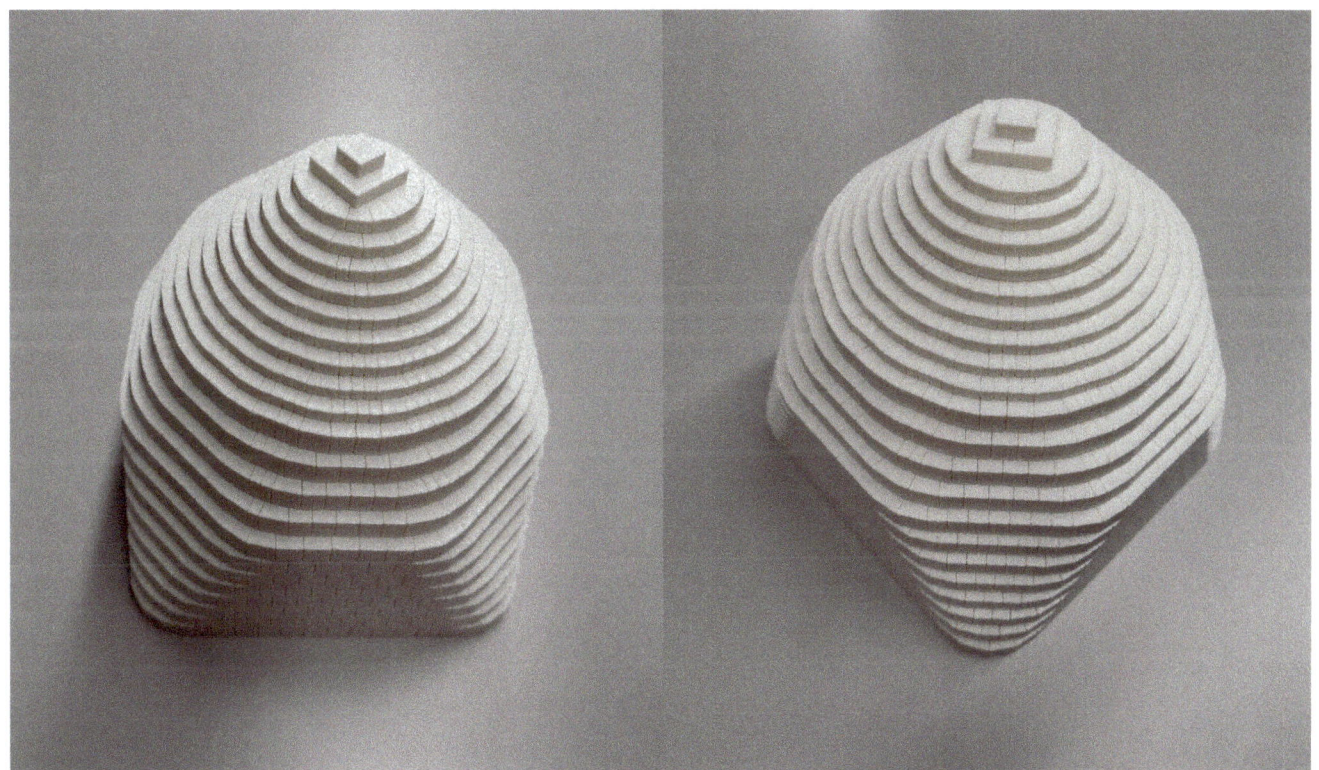

2 A 3D-printed dwelling model

3 A CEB dwelling consisting of eight interlocking brick types

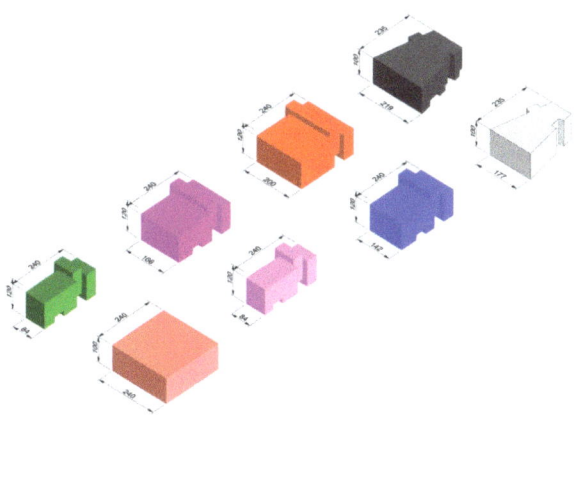

4 Eight brick types and their designs

Square to octagon zone   Octagon to hexadecagon zone   Hexadecagon zone   Rooftop

5   Zones of the CEB dwelling

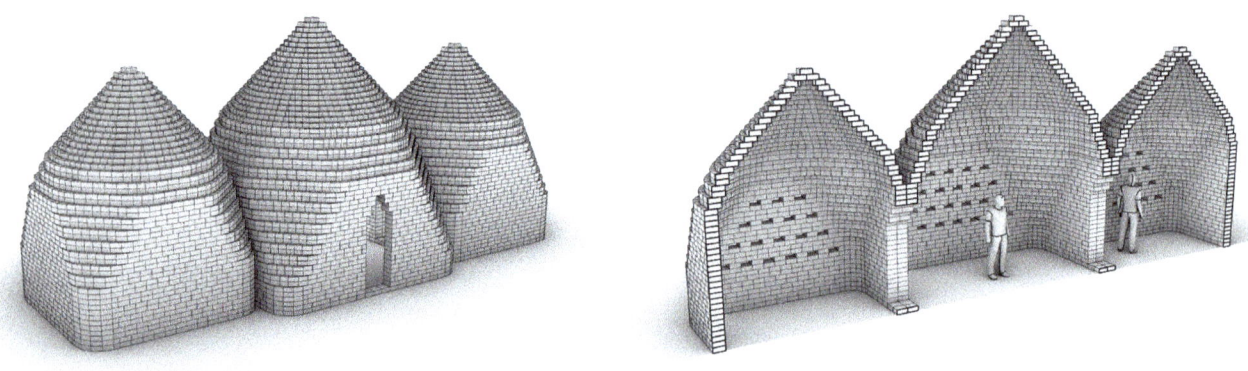

6   Dwelling units can be used in a flexible layout by interconnecting them to form multi-room houses

7 The compressed earth block (CEB) press with the default steel mold

enabling access for low-skilled and first-time builders. However, this interlocking brick design, together with the overall geometry design of the structure, makes it challenging to fit the bricks into the target geometry without any gaps or overlaps. This problem is solved by rigorously computing the trigonometric relationship and developing a parametric design tool that generates feasible designs, including block geometry and layout.

As shown in Figure 4, the brick dimensions, yielded from the aforementioned computational design, are not standardized. Thus, this project proposes a fabrication process for customized interlocking bricks that applies to most of the CEB press machines existing on the market. The press this project started with is the Auram 3000 manual CEB press, with the default steel mold Plain 240 (Figure 7). However, the proposed method can adapt to presses with or without replaceable steel mold/chamber by 3D-printing a set of

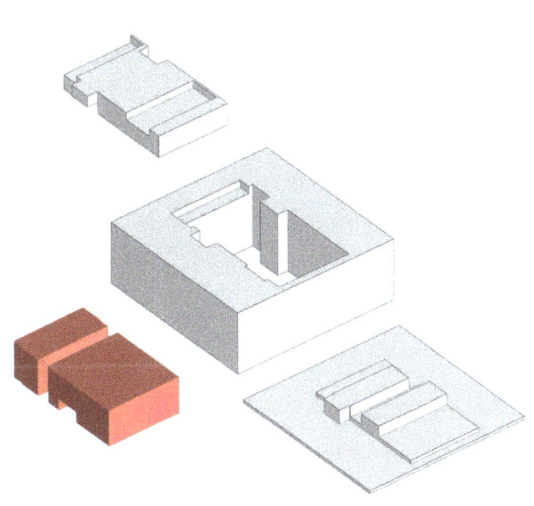

8 Example of a brick mold model and its 3D-printed copy

SUSTAINABLE CONSTRUCTION   REALIGNMENTS   343

9 The earth material is loaded into the 3D-printed mold before closing the lid of the press

10 An assembled section (a corner of the square to octagon zone) of the structure

embedded molds for the customized brick geometries (Figure 8), such that it refrains from any modifications to the original steel chamber/mold.

The brick fabrication process is similar to the conventional CEB process except that a 3D-printed mold is put into the steel chamber. Then the raw earth material is loaded into the 3D-printed mold (Figure 9) and compressed. After demoulding, the earth brick is placed on a rack for drying. To complete the sustainable approach, linseed oil is used for waterproofing and no further additives (e.g. cement or lime) are used, which allows for infinitely reusing the bricks or recycling the raw earth material. Figure 10 shows an assembled section of the structure.

By not modifying the press itself, this method drastically expands the application to almost any CEB press, especially low-cost presses in developing countries without replaceable chambers. Compared to other existing projects that use customized steel molds, the proposed method requires less equipment and lead time for mold making. Moreover, the 3D-printed molds are lightweight, portable, and economical. All the digital processes, i.e. computational design, 3D printing, and structural analysis, can be conducted offsite, which significantly reduces the requirements on the construction site, e.g. no need for skilled personnel, high-tech and high-cost machines, and electricity. The next step is to build a full-size demonstration dwelling in the correct climatic and social context.

ACKNOWLEDGMENTS

The authors would like to thank Giulio Antonutto (Arup, London) for his useful discussions. Funding for this research project is provided by the National Centre of Competence in Research (NCCR) Digital Fabrication at ETH Zurich in Switzerland and an Arup Global Research Challenge award, "Designing with Uncertainty: Digital Fabrication for Developing Countries."

REFERENCES

Wilton, Oliver, Matthew Barnett Howland, and Peter Scully. 2019. "The Role of Robotic Milling in the Research and Development of the Cork Construction Kit." In *FABRICATE 2020: Making Resilient Architecture*, edited by Burry Jane, Sabin Jenny, Sheil Bob, and Skavara Marilena. London: UCL Press. 36–41.

WASP. 2021. "TECLA." Accessed July 06, 2021. https://www.3dwasp.com/en/3d-printed-house-tecla/.

IMAGE CREDITS

All the drawings and images are by the authors.

Yu Zhang, SM, BEng is a PhD student at the Engineering Design and Computing Laboratory at ETH Zürich. She obtained her Master's degree at the Massachusetts Institute of Technology in 2017. She is passionate about additive manufacturing and uncertainty-based computational design and optimization methods.

Liz Tatarintseva, AADipl, ARB studied art and design at Central Saint Martins and graduated from the Architectural Association in 2014. Throughout her studies she has acquired a high level of proficiency in computational design which has led her to hold Grasshopper and digital media workshops at Cardiff University and the Bartlett. Liz worked for several London based practices including Gianni Botsford Architects and AL_A before establishing ao-ft in 2017 alongside Zachary Fluker.

Tom Clewlow, MEng, CEng, MIStructE is a structural engineer at Arup in London. Studied at The University of Cambridge (2010-2014). He joined Arup as a graduate in 2014 and has a passion pushing boundaries in the construction industry to enable more sustainable designs in the future.

Ed Clark, MEng, CEng, MICE, FIStructE is a structural engineer and Director of Arup in London. Studied at The University of Leeds (1992-1996). Joined Arup as a graduate in 1996. Technical tutor at the Bartlett School of Architecture, UCL (2011-2020). Currently a Trustee Board Member of the Institution of Structural engineers. Recipient of the 2011 IABSE Milne Medal for Structural Design Excellence.

Gianni Botsford, BA, (Hons) AADipl, RIBA is Director at Gianni Botsford Architects, London starting in 1996, studied at Kingston University (1982-1985) The Architectural Association, London (1994-1996), Architectural Association Research Fellow (1997), teacher Architectural Association 1998-1999, London Metropolitan University 2010-2012, The Welsh School of Architecture starting in 2018, and is Founder member Studio in the Woods, Lubetkin Prize 2008.

Kristina Shea, PhD is Full Professor for Engineering Design and Computing in Mechanical and Processing Engineering at ETH Zürich since 2012. She received her PhD in Mechanical Engineering from Carnegie Mellon University in 1997. Her lab's research merges engineering design, computation and fabrication to design and prototype creative engineering systems with new functionalities and an emphasis on sustainability and low-resource settings.

Digital Deconstruction and Design Strategies from Demolition Waste

Matthew Gordon
Institute for Advanced Architecture of Catalonia

Roberto Vargas Calvo
Institute for Advanced Architecture of Catalonia

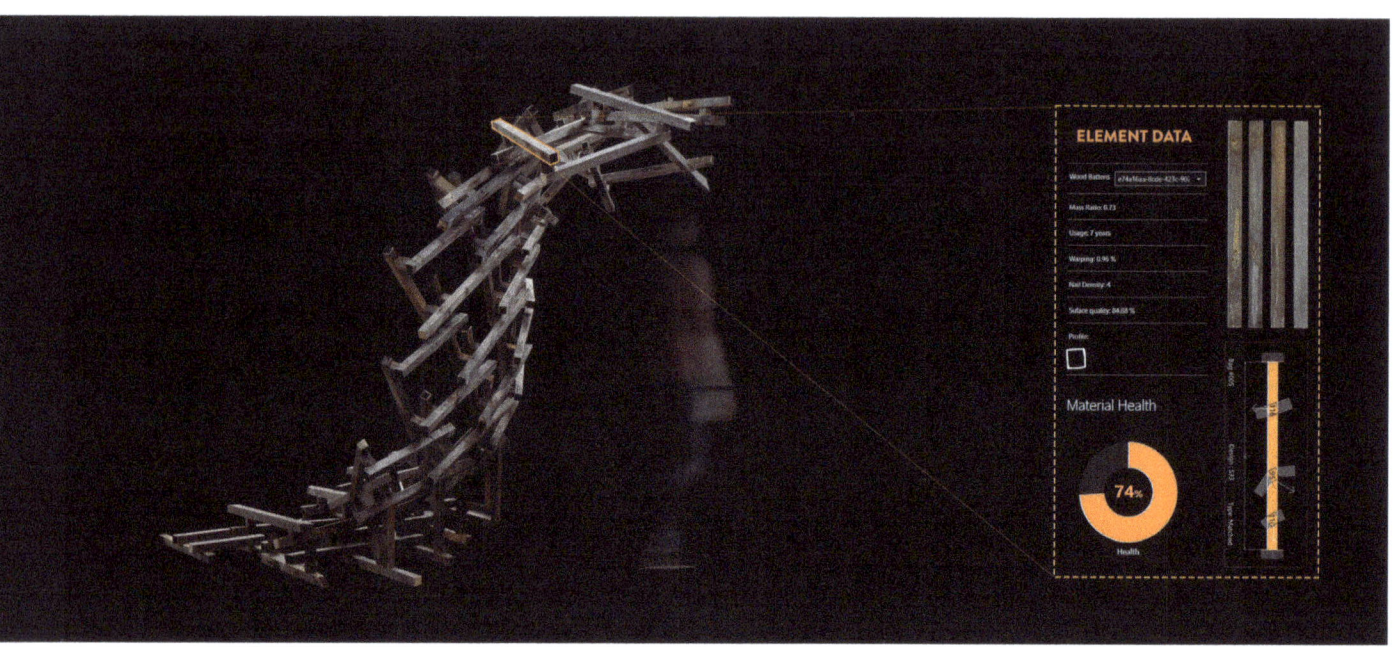

1   Optimized and data-integrated structure from reclaimed wood battens

A Digital Approach to Increase Reliability on Reclaimed Materials

The project develops pre- and post-demolition digital assessment protocols in order to better inform reclaimed material implementation in new projects. The application of the protocols are demonstrated in a pavilion constructed of reused timber (Figure 1). By facilitating the data capture, analysis, identification, and characterization of available secondary raw materials, and creating database systems for pre- and post-demolition sites, it promotes gains in high quality upcycled materials for new construction projects. Modern reality capture technologies allow for collecting high density and quality Construction and Demolition Waste (CDW) data, presenting the opportunity to also increase the reliability and trust in upcycled materials by data specifically structured to relevant actors.

PRODUCTION NOTES

Status: Built
Site Area: 97 sq. ft.
Location: Barcelona
Date: 2021

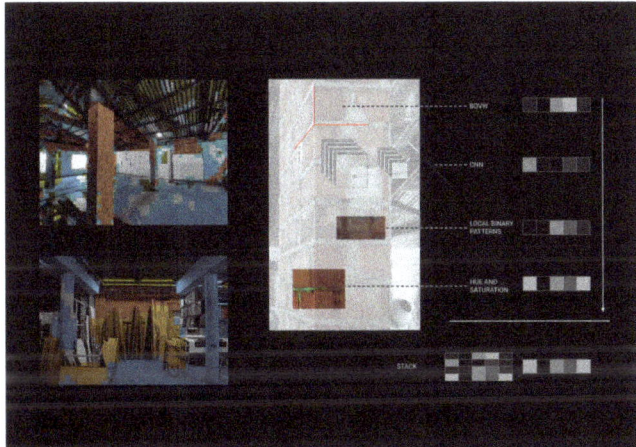

2   Stacking and integration of several material classification methods at different levels of detail.

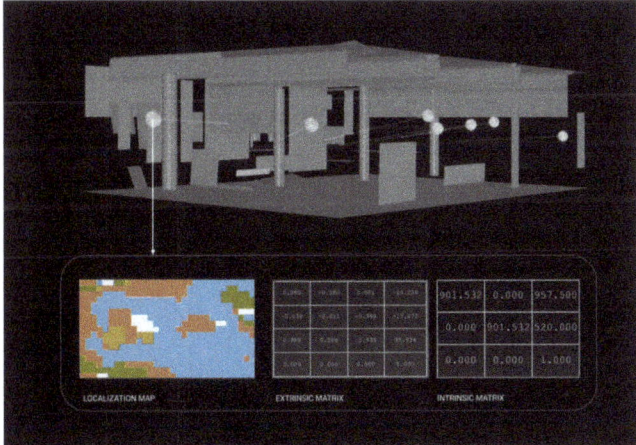

3   Integrating 2d material maps with 3d site geometry

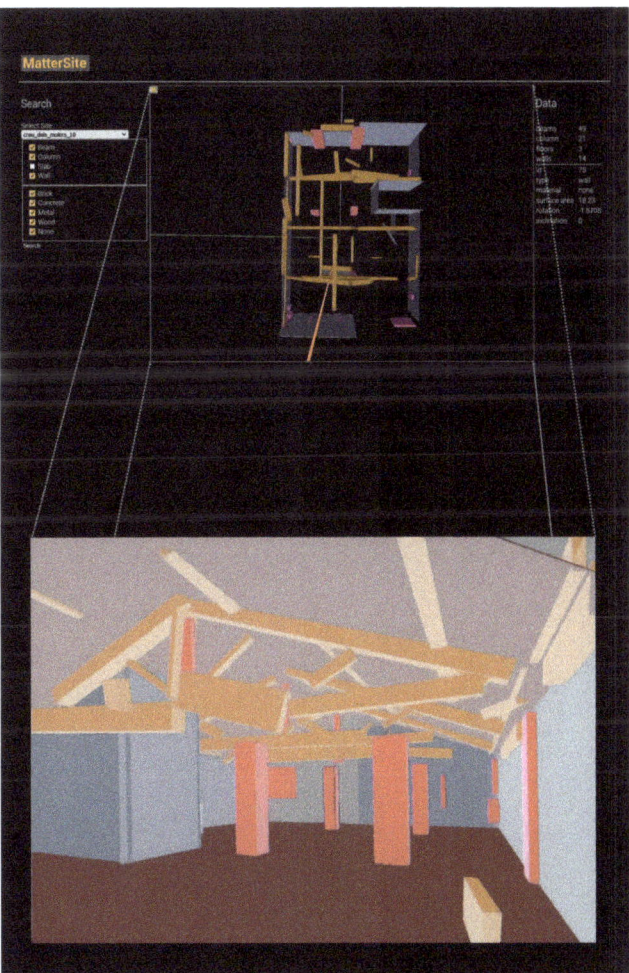

4   Top view of a reconstructed site in the web interface and perspective view of an interior

Pre-demolition Digitalization and Assessment

A digitization-based decision support system was created for managing pre-demolition sites, integrating digital reality-capture tools and processes (photogrammetry scanning, scan-to-BIM, and CV classification) that aid in defining the most sustainable and economical deconstruction and reuse strategy for a building.

This process is divided into five steps: building inspection, geometric reconstruction, material localization and classification (Figure 2), BIM integration (Figure 3), and planning interfaces (Figure 4). The results show that it is possible to automatically quantify and locate recoverable materials into material reports using imagery-based machine learning techniques. The resulting output assists the mapping of material flows and the potential for reuse and recycling of products from buildings undergoing renovation, redevelopment, or deconstruction into new projects. Once the results of the information gathered about each material is validated with a harvest expert, the value and security of upcycled material supply can potentially improve, reducing the time spent on manual classification and quantification.

Post-demolition Digitalization and Assessment

Once the materials are registered, it is necessary to fulfill the reentry of these recovered materials into new construction, wherein a "form follows availability" design strategy is explored, from a sparse quantity of reclaimed material. It develops a database from a trial post-demolition site, and the consequent extracted material (Figure 5) is cataloged and stored before being matched and used for a new demonstrator construction. Each recovered element is imaged, scanned, geometrically analyzed, and weighed to create a unique material health indicator (Figure 6-7). The associated retrieval tool can provide reliable information about each element for an architect or planner, increasing confidence in the usability and applicability, and allowing for pre-filtering for different levels of structural necessity (Figures 8, 9). The final step of the system matches designed components using multi-objective optimization (Figure 12) from a parameterized surface-following design

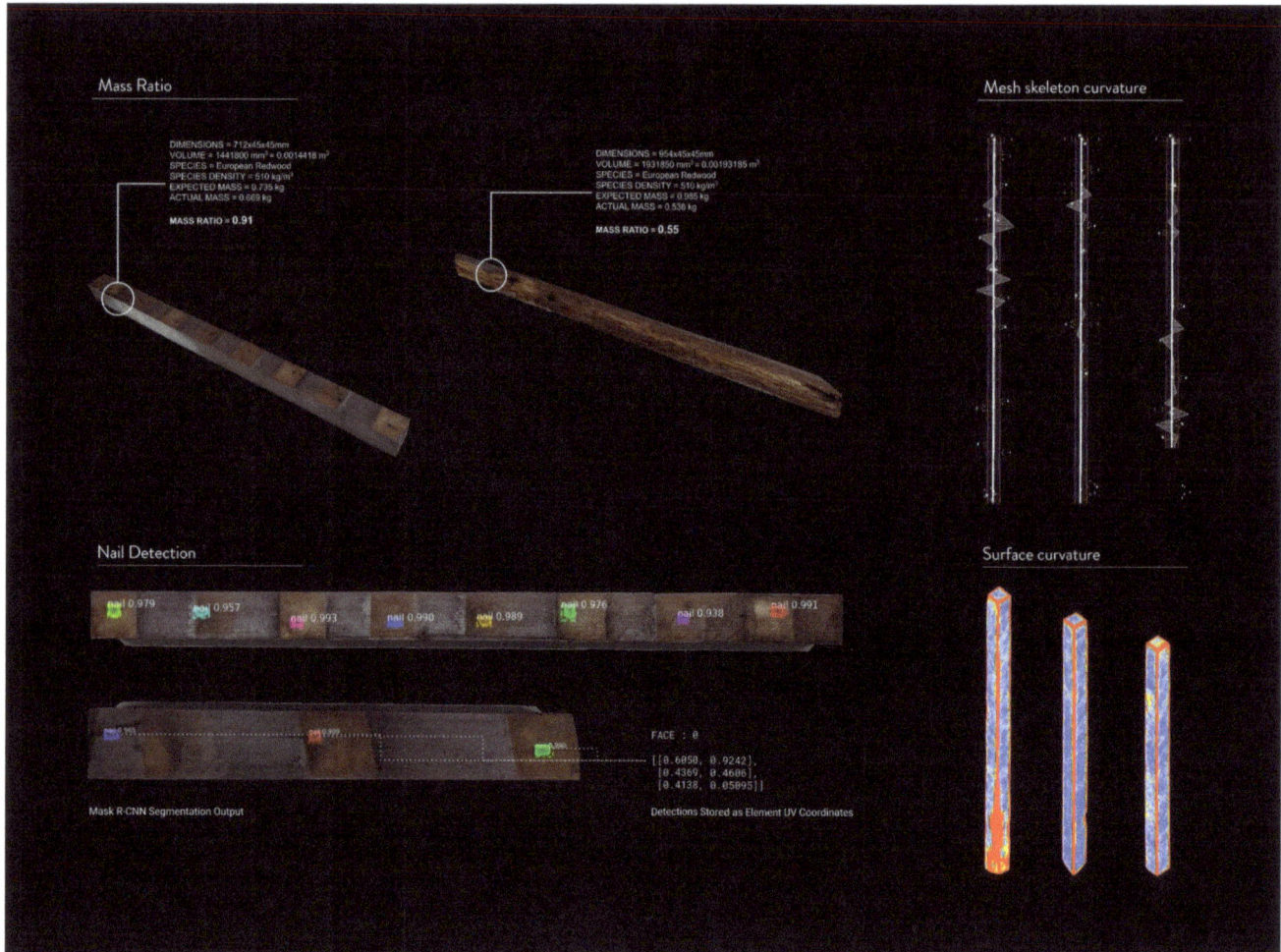

5   Post-demolition assessment on wood battens

system with relevant stored materials by their generative design requirements (Figure 10).

Design Demonstrator

Results from the demonstrator and proposed iterations aimed to optimize repurposed material utilization and fabrication efficiency (Figures 13, 16). By extracting multi-dimensional data on each wood batten and presenting their relevant indicators in a user-friendly interface, it is possible to create a dialogue between the designer and irregular shapes, augmenting the widespread use of reclaimed materials in structurally predictable assemblies. The predictability of the system is verified by a 10mm maximum deviation (Figure 15) between the as-designed and the as-built structure (Figure 14), where 85% of all the material was matched and found by the presented method.

Continuing, the research will demonstrate the potential for constructing big digitally availale datasets of reusable materials for facilitating material harvesting and promoting their incorporation in new construction. Expanding material and element types can be directed to reuse actors, promoting the circular economy in a short loop. Finally, integration with design tools will simplify the architects' process for designing with and specifying these materials, affecting the overall environmental costs of the industry at multiple levels.

ACKNOWLEDGMENTS

This research was developed during the Master in Robotics and Advanced Construction + Thesis (2019-2021) in the Institute for Advanced Architecture of Catalonia.

We would like to recognize the support we received from our tutors: Alexandre Dubor and Aldo Sollazzo; also from the faculty technology support: Raimund Krenmueller and Angel Muñoz. Finally our industrial advisors: Shajay Bhooshan and Stuart Maggs.

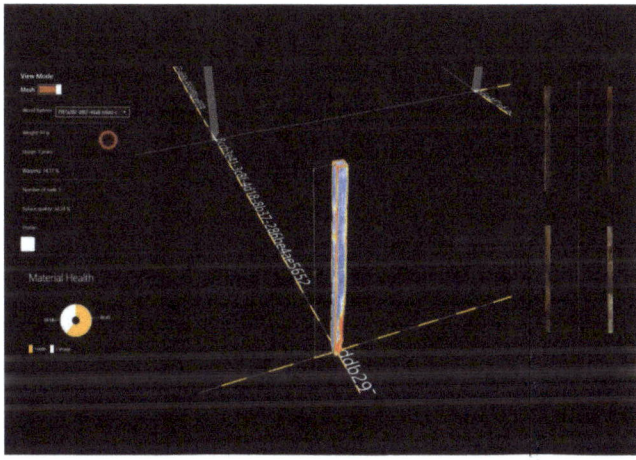
6 The user interface shows useful indicators characterizing the material

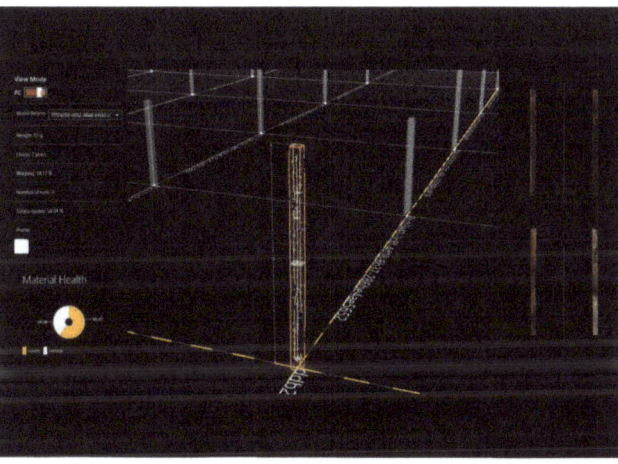
7 Perspective view of a digital item displaying the mesh skeleton curvature

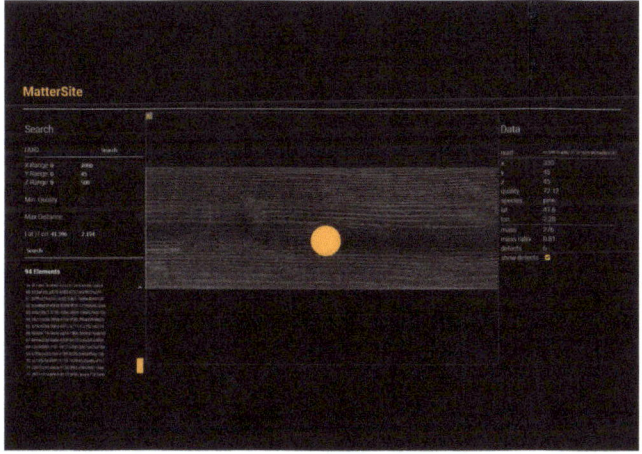
8 Web interface to visually inspect the digital inventory; zoomed view

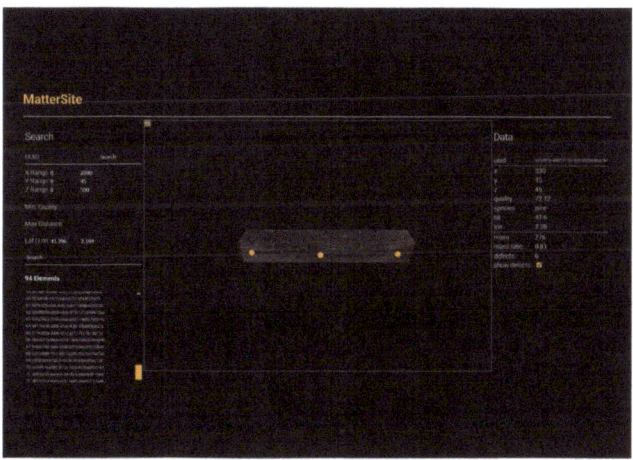
9 Web interface: orange circles indicate defects like nails

10 Length-matching flexibility embedded in the design logic

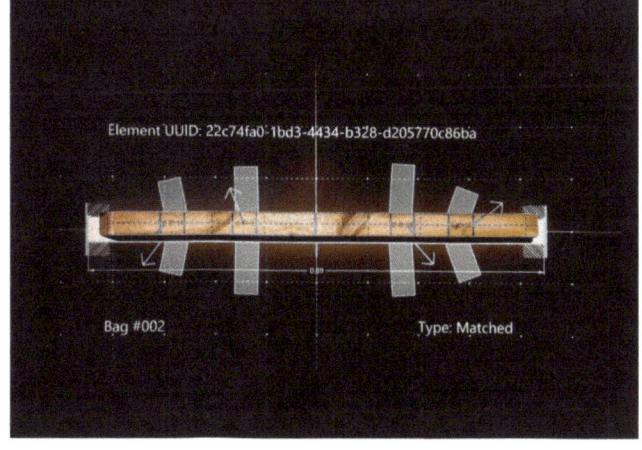
11 Digitally augmented construction assistance with fabrication information projected onto physical inventory

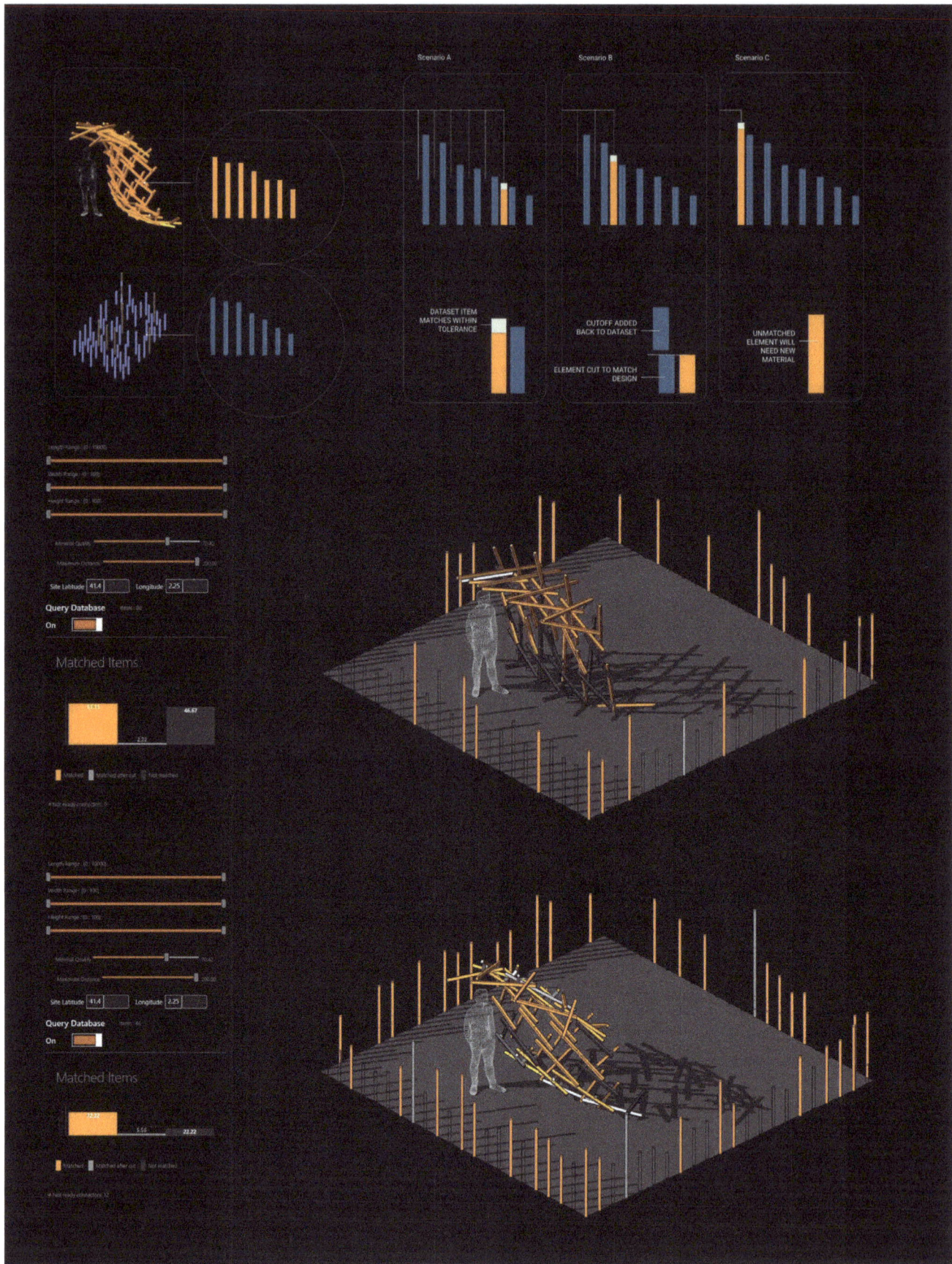

12 Matching strategy and design interface for optimizing the available materials; a graph on the left side indicates how much material is matched based on the design parameters

13 Connectors for fabrication, optimized for a robotic cutting process

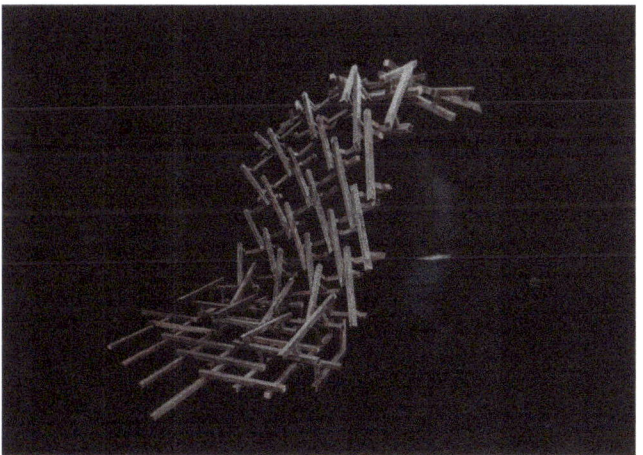

14 Photograph of final structure

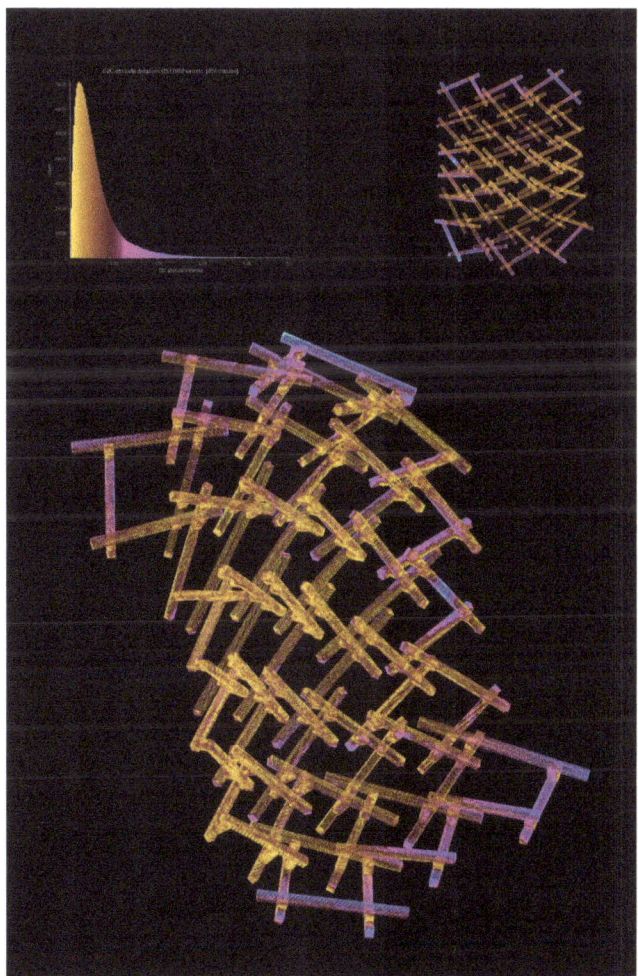

15 Views of structure digitally scanned after construction, indicating measured deviation from design geometry

---

Matthew Gordon is a designer and construction technologist, originally trained in architecture and holding a Masters in Robotics and Advanced Construction from IaaC, Barcelona. After working with IaaC and the University of Virginia Architectural Robotics Research Group focusing on new-building processes, his research ultimately turned to center on deconstruction and circular concepts, combining design and engineering to find new methods for transforming the impact of the built environment.

Roberto Vargas is a licensed architectural designer, based in Barcelona. He holds a Master's degree in Robotics and Advanced Construction + Thesis from the Institute for Advanced Architecture in Catalonia (IAAC), Barcelona. Roberto has developed his professional career with a multidisciplinary vision, in collaboration with academic and industry leaders, where he has expanded his skills in communication, research, computer vision, robotic fabrication, machine learning, parametric and circular design. Since the beginning of his design profession in Costa Rica, his driving goal has been the essential properties that sustainable and technological solutions can bring into the built environment.

16 Photograph of the robot cutting process for the connectors

AC	AD	IA	20
21	RE	AL	IG
NM	EN	TS	AC
AD	IA	20	21
RE	AL	IG	NM
EN	TS	AC	AD
IA	20	21	RE
AL	IG	NM	EN

 Mute   Stop Video   Security   Participants   Chat   Share Screen

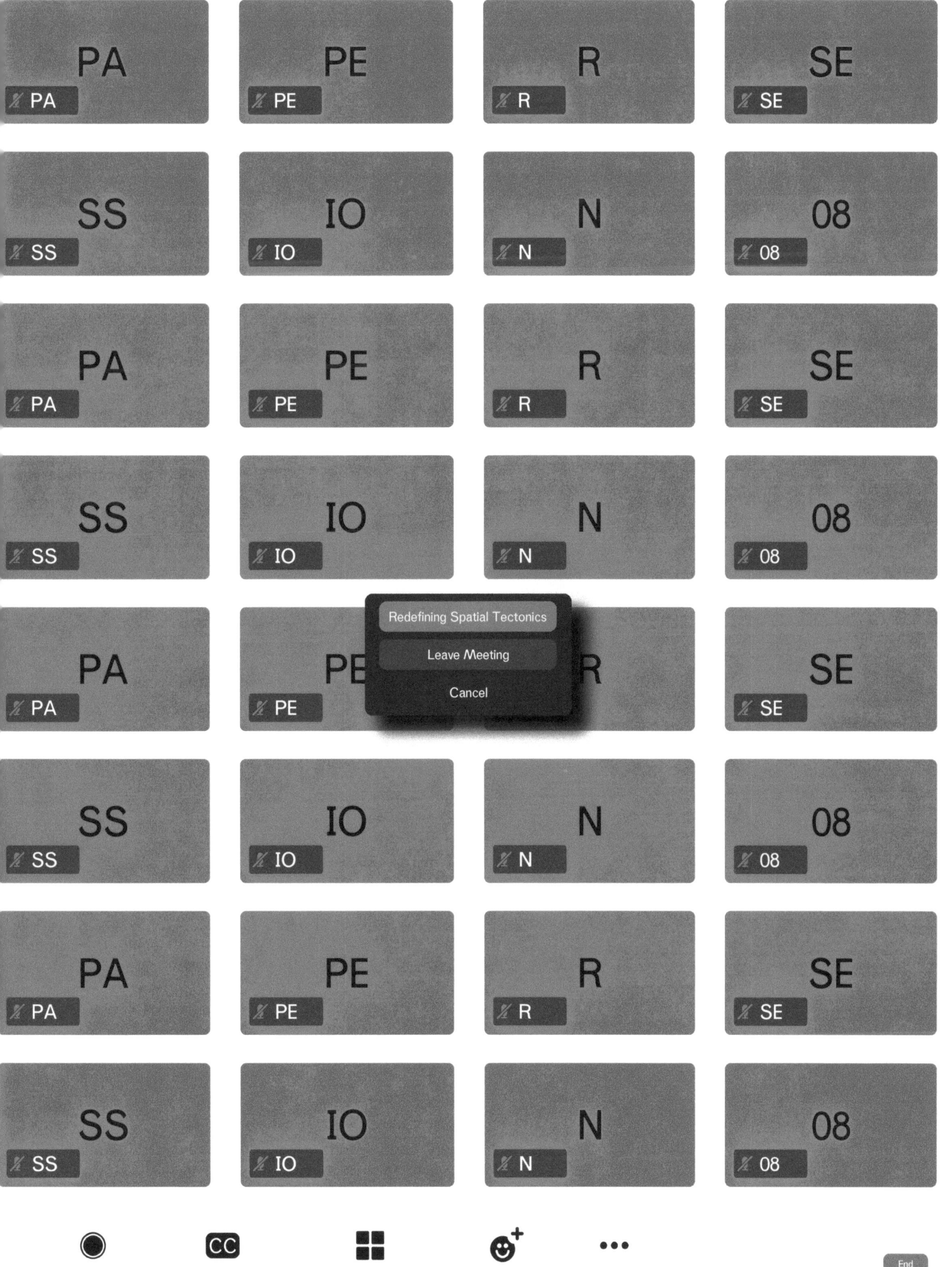

Kerf Bending and Zipper in Spatial Timber Tectonics

A Polyhedral Timber Space Frame System Manufacturable by 3-Axis CNC Milling Machine

Yulun Liu*
University of Pennsylvania

Yao Lu*
University of Pennsylvania

Masoud Akbarzadeh
University of Pennsylvania

*Authors contributed equally

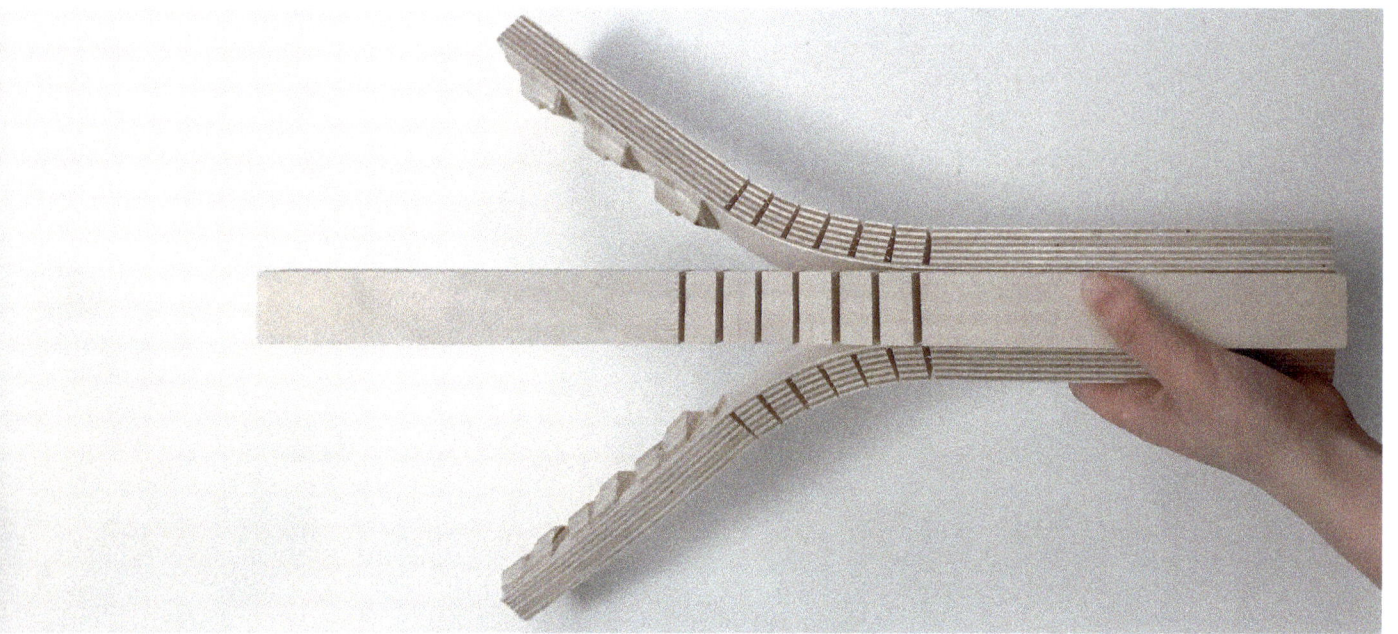

1 One branch of the bar shows Kerf-bending part and Zippered node

ABSTRACT

Space frames are widely used in spatial constructions as they are lightweight, rigid, and efficient. However, when it comes to the complex and irregular spaces frames, they can be difficult to fabricate because of the uniqueness of the nodes and bars. This paper presents a novel timber space frame system that can be easily manufactured using 3-axis CNC machines, and therefore increase the ease of the design and construction of complex space frames. The form-finding of the space frame is achieved with the help of polyhedral graphic statics (PGS), and the resulting form has inherent planarity that can be harnessed in the materialization of the structure. Inspired by the traditional wood tectonics, kerf bending and zippers are applied when devising the connection details. The design approach and computational process of this system are described, and a test fabrication of a single node is made via 3-axis CNC milling and both physically and numerically tested. The structural performance shows its potentials for applications in large-scale spatial structures.

INTRODUCTION

Space frames are widely used in spatial constructions as they are lightweight, rigid, and efficient. Despite the fact that the systematic development of specialized production methods and digital fabrication technology has enabled the construction of timber structures to reach a new level (Lennartz and Jacob-Freitag 2015), the fabrication of such spatial structure is still not easy due to the uniqueness of bars and nodes, and always involves robotic fabrication, multi-axis machining, 3D printing as well as casting. The diverse nodes of some branching wooden structural systems are produced by flip-milled on a CNC router (Lamere and Gunadi 2019), and other spatial wooden structures introduce steel sleeve anchor or tension rods to help construct the nodes (Momoeda 2017; Teeple 2015).

To reduce the complexity of the design and fabrication of space structures, this paper presents a novel timber space frame system that blurs the boundary between bars and nodes and can be milled simply by 3-axis CNC milling machines. This system incorporates the traditional manufacturing techniques of kerf bending and dovetailed joints and introduces a new connection using zippered wood. The form-finding process is achieved with the help of polyhedral graphic statics (PGS) as the inherent planarity of the resulted form is required for developing the structural details.

Kerf-bending and Zippered-wood

In practice, there are many techniques to bend a piece of wood into the desired curvature. One way to easily create the precise bending wood surface is the relief-cutting process known as kerfing (Mansoori et al. 2019). It is a well-established carpentry technique for producing curved wooden pieces in a wide range of applications. By cutting a series of deep notches (or kerfs) at the bending area allows for in-plane expansions and compressions perpendicular to the cutting lines (Figure 2). If the two edges of each kerf on the compression side are tightly attached, a mathematical relationship can be found among the angle of bending, the thickness of the wood, and the dimension of cutting slots.

The zippered wood connection was inspired by dovetailed joints which provide interlocking between the components while hiding the connection details. A series of 'pins' cut to extend from the end to interlock with a series of 'tails' from another board (Figure 3). This connection can not only accommodate the orientation change of the bars but also achieve the desired curvature and twisting (Satterfield et al. 2020).

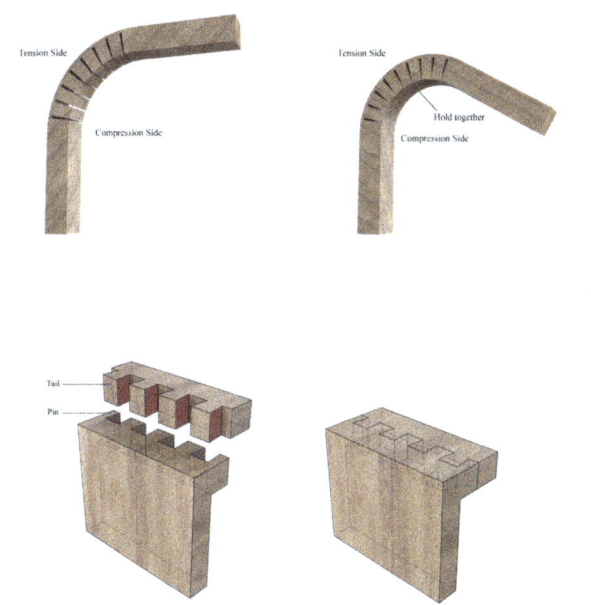

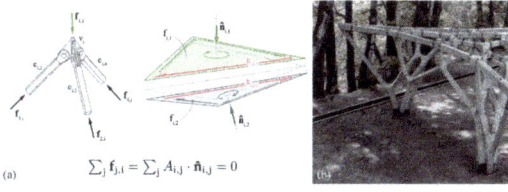

2 Kerf-bending technology allows the rigid wood to bend by the desired angle when the pair of edges of each kerf is tightly attached

3 Dovetail joint manufactured using classic technology, highlighted are fraction surfaces

4 (a) A spatial structural joint in equilibrium, and the reciprocal relationship between polyhedral form and force polyhedron (Akbarzadeh et al. 2019); (b) the first built structure designed by using 3D Graphic Statics methods (Akbarzadeh et al. 2017; Bolhassani et al. 2018)

Polyhedral Graphic Statics

Geometric structural design methods depend on 2D reciprocal diagrams are regarded as a compelling design tool that has long been studied and practiced. The geometric interrelation between force and form was initially proposed by Rankine (1864), and Maxwell (1870) formulated the topological and reciprocal relationship as reciprocal form and force diagrams. The development of 3D graphic statics (3DGS) further increased the ease of designing complex spatial structures. In the realm of 3DGS, the method uses polyhedral reciprocal diagrams, usually referred to as

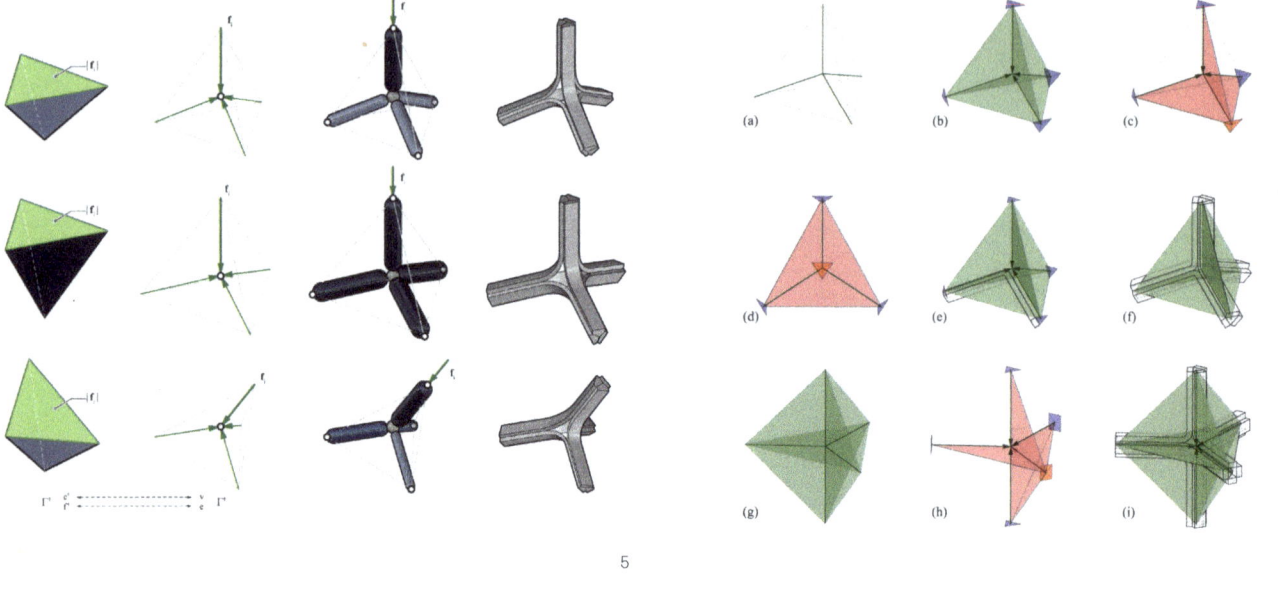

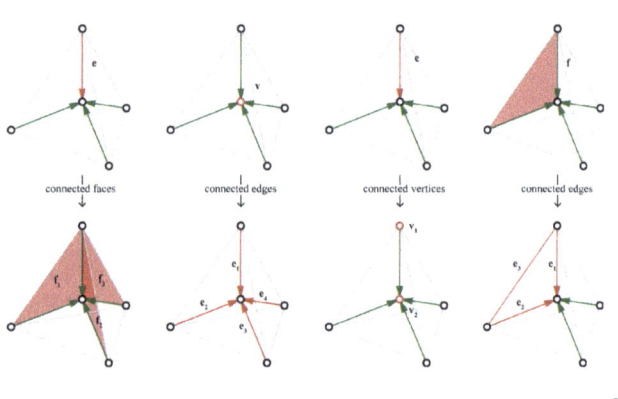

5 The form can be manipulated by changing the force diagram

6 Fast queries of geometric and topological information are made possible through the half-face data structure

7 (a) Spatial node with four bars; (b) edges and faces; (c) triangular section profile for one edge determined by connected three faces; (d) aligned view for the edge section; (e) side face perpendicular to the corresponding input face; (f) all side faces perpendicular to their corresponding input faces; (g) Spatial node with five bars; (h) tetragonal section profile for one edge determined by connected four faces; (i) final generated parts

polyhedral graphic statics (PGS), has recently been developed and extended (Akbarzadeh 2016; Akbarzadeh et al. 2015a; McRobie 2016). It provides methods to find the equilibrium of structure in 3D by enforcing the closeness of the global force polyhedron and nodal force polyhedrons, and the reciprocity is built by projecting form polyhedrons to force vertices, form faces to force edges, form edges to force faces, form vertices to force polyhedrons, where the force magnitude of each edge in the form polyhedron is equivalent to the area of the corresponding face in the force polyhedron (Figure 4). The inherent planarity of the polyhedral geometries has the potentials to be harnessed for efficient construction processes.

DESIGN METHODOLOGY

This research is an ongoing investigation in the design and construction of timber spatial structures designed by polyhedral graphic statics. The design approach built upon the reciprocal form and force diagrams of PGS will be explained in this section. The computational model is implemented using Rhino Python.

Form-finding through PGS

This timber space frame system is developed upon polyhedral forms as the intrinsic planarity helps develop the detail of kerf bending. PGS is used because it allows manipulating the form while being aware of the internal force distribution. PolyFrame (Nejur and Akbarzadeh 2018), an efficient computational PGS plugin for Rhinoceros (Robert McNeel & Associates 2020), together with its underlying half-face data structure (Nejur and Akbarzadeh 2021) are used throughout the design process.

To start the form-finding process, a set of closed polyhedrons is made as the force diagram, followed by the generation of its corresponding form diagram. The simplest frame with one node is used here to demonstrate the adjustment of the form as well as the organization of the geometrical and topological data (Figure 5). Through the half-face data structure, fast queries of needed information are made possible (Figure 6).

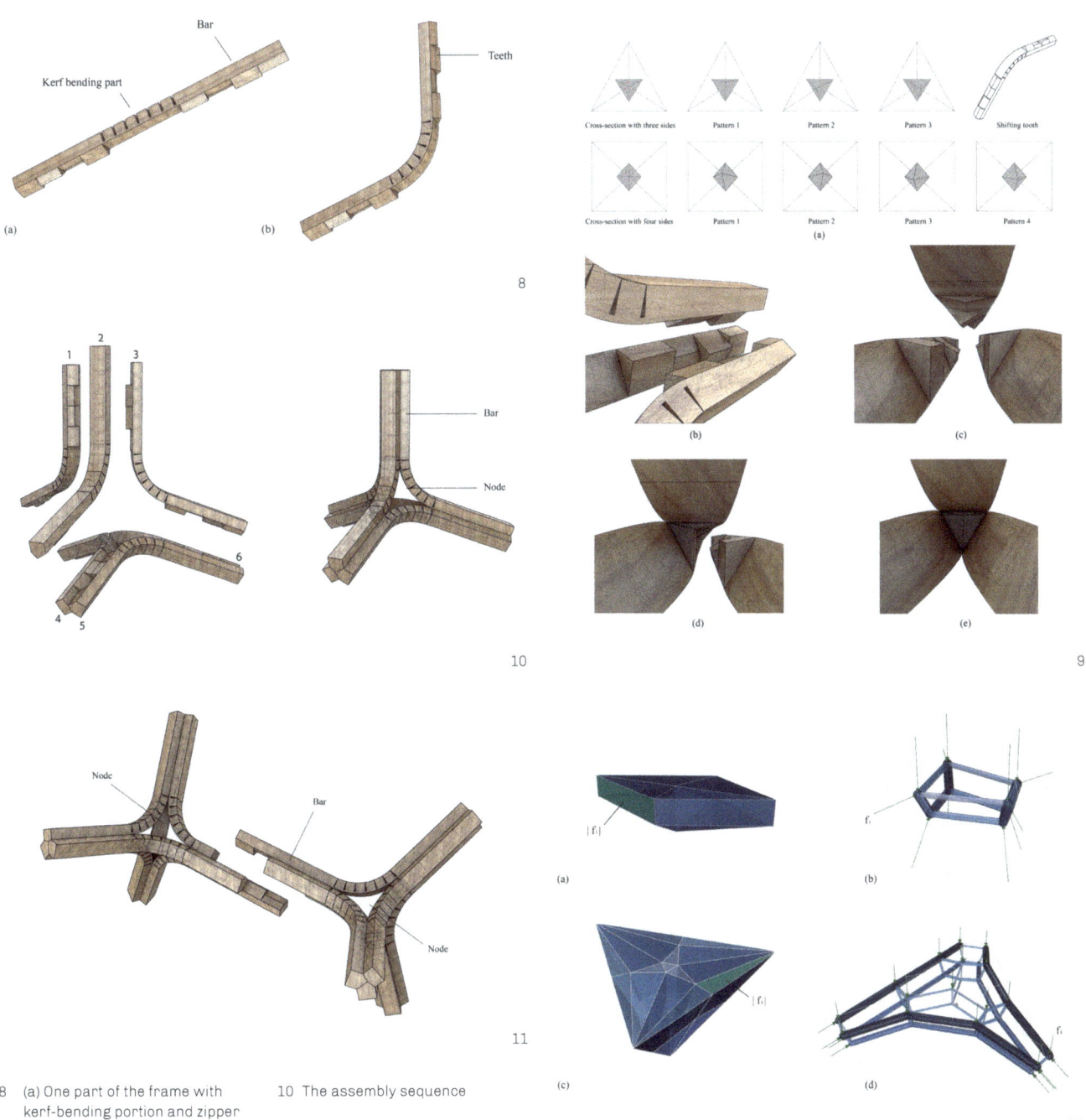

8 (a) One part of the frame with kerf-bending portion and zipper teeth; (b) the part after bending with the edge pair of each kerf tightly attached

9 (a) The logic of generating zippered 'tooth' and the mode of patterns for triangle and quadrangle cross-section; (b)(c)(d)(e) details for the three-directional zippered part

10 The assembly sequence

11 The connection between two neighboring joints

12 (a) The force diagram of a table; (b) the form diagram of the table; (c) the force diagram of a shell; (d) the form diagram of the shell

Materialization

After getting the form diagram (Figure 7a), the edges are materialized into timber bars. The cross-section of each bar is determined by both the connected faces of the corresponding edge and the magnitude of its force. Taking one edge from the form diagram, the number of sides of the profile is determined by the count of the connected faces; the direction of each side is determined by the normal of the corresponding face; the area of the profile is proportional to the internal force of the edge. For instance, if the edge connects to three faces, the profile will be triangular (Figures 7b–7d); if the edge connects to four faces, the profile will be quadrangular (Figures 7g–7i).

Later, a smooth singly curved blend is generated for each pair of bar sides that are connected and sharing the

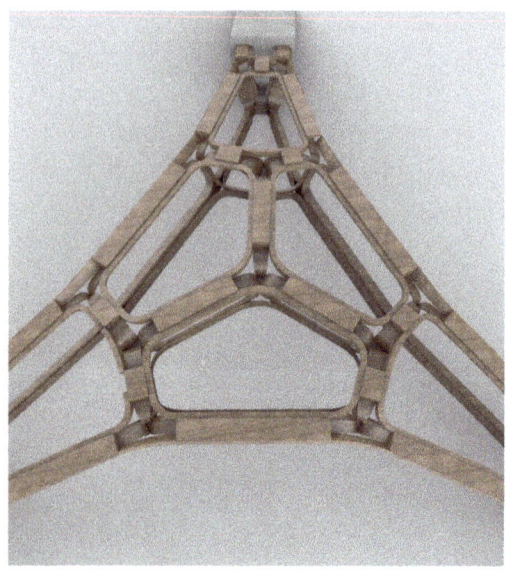
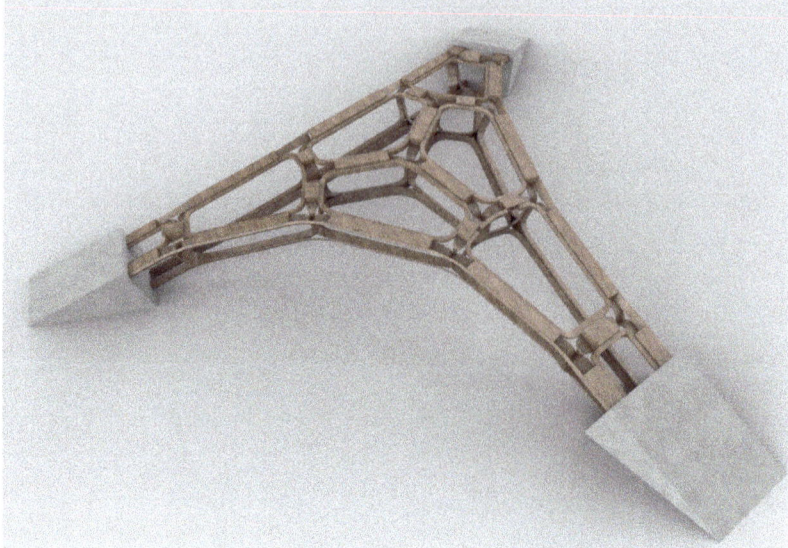

same face, forming the kerf bending node that joins all the connected bars (Figure 7e). Then, the side extrusions of each bar as well as the zipper teeth are made along the corresponding edge, whose details will be described later. Finally, all the geometries that are associated with each face will be grouped and fabricated as one part. As a result, this simple spatial structure with four bars and one node can be made by only six parts (Figure 7f). Since both the extrusion and kerf bending portions of each part are perpendicular to the same face, it can be unrolled and fabricated from flat material (Figure 8).

The zippered teeth are located on the interior side of the straight segment of each part, serving as the mechanism that interlocks all side parts that compose each bar.

As a further description of the interlocking zipper teeth, the generation of the tooth patterns for one bar is illustrated as follows (Figure 9). First, the length of the bar is evenly divided into a number of segments. Then, the profile of the bar is split into triangles by connecting its centroid to all corners. Next, the centroid of each triangle is successively used as the new splitting point for the tooth pattern

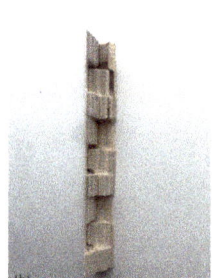

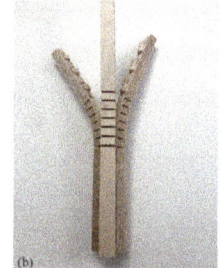
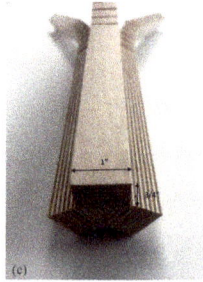

13 Renderings of a funicular shell and a desk

14 Assembly process of the teeth of a bar

15 (a) one part; (b) three parts after assembly; (c) cross-section of a bar

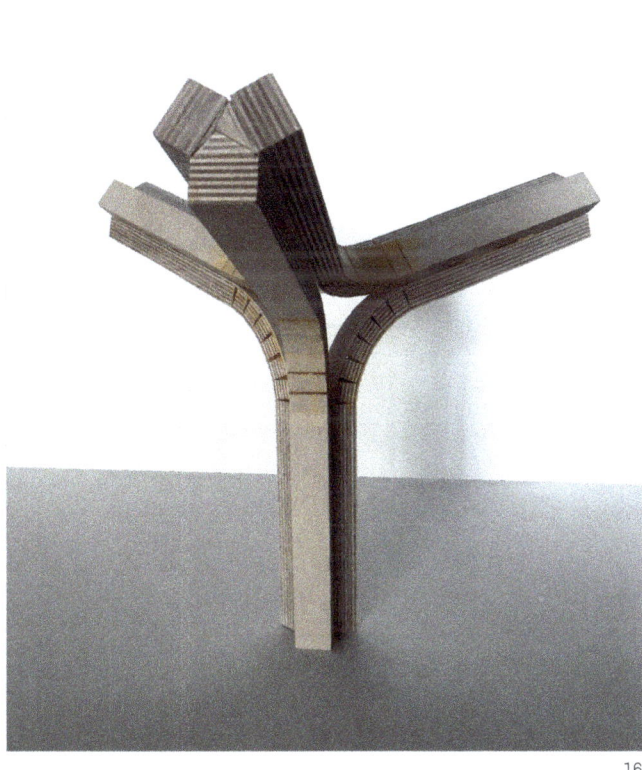
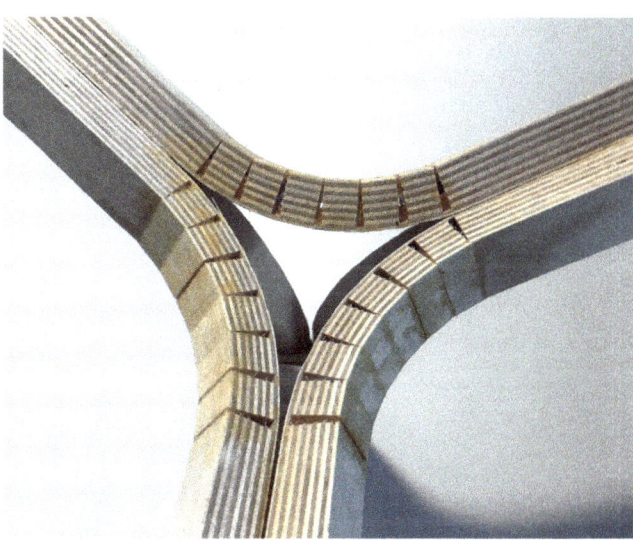
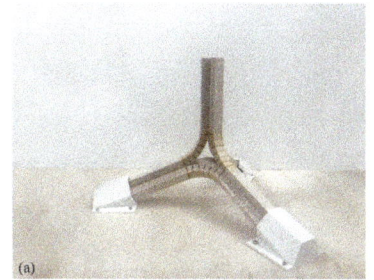
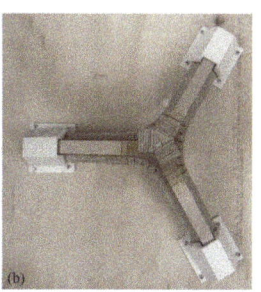

16 One whole joint after assembly

17 Details of the kerf-bending part

18 (a) Set-up for load test (b) top view of the set-up

of the segments, and consequently, the interlocking teeth are created.

For kerf cuttings, the cutting depth, cutting width, and kerf count are the three key parameters that determine the curvature of the bending. They need to be calculated accordingly such that the two exterior edges of each kerf are tightly pressed together after bending (Figure 9b).

The material properties and fabrication constraints are also considered parallel alongside the geometrical development of the space frame system. Timber is selected as the construction material because of its sustainability and ease of processing. All geometries of the structural parts are generated in the way that only the table saw and 3-axis CNC milling machine are needed for rapid fabrication, where the table saw cuts the kerfs and the 3-axis CNC milling machine carves the teeth. It is also manufacturable by a 3-axis CNC milling machine only with flip milling technique. The assembly of the frame is illustrated in Figure 10, and a bonding agent is required on the interface between the parts.

The Application in Multi-Node Polyhedral Forms

With a connection devised between every two neighboring nodes, the workflow described above can also be applied to complex polyhedral space frames with more bars and nodes. As shown in Figure 11, the two adjacent nodes share the same bar. By varying the contact location of the sides, an interlock is created between these nodes, and the contact area is increased. As case studies, a compression-only funicular shell and a table are generated using the workflow proposed above (Figures 12–13).

FABRICATION AND EXPERIMENTS

The strength and stiffness of the proposed system need to be carefully investigated as the bending part is made vulnerable due to the kerfs. To evaluate its structural performance, a frame prototype with four bars and one node is fabricated, assembled, and tested through both physical and numerical methods. The overall fabrication process would use the 3-axis CNC milling machine for the zippered shape and table saw for cutting the wood.

Fabrication

The simple frame is built within the 450x400x360 mm bounding box, and 3/4 inch thick plywood is used.

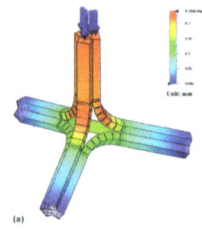 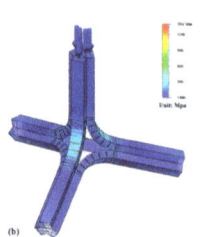 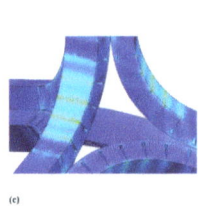

19 Physical applied load test for the prototype: (a) 50 pounds; (b) 70 pounds; (c) 90 pounds

20 (a) Displacement map; (b) stress distribution map; (c) detail of maximum stress point after bearing the force of 1KN

Although each part is manufacturable as a whole, the zipper teeth of each part are separately made and glued back in this prototype for a faster fabrication (Figure 14, Figure 15a). For the kerf-bending part, the slots are cut by a 1/8 inch table saw. For the zipper teeth, appropriate tolerances need to be left to ensure the tight interlocking. Figure 15a shows one part of the assembly, and Figures 15b and 15c show one assembled bar by interlocking and bonding three parts together. As explained before, all the exterior edges of each kerf are pressed together after the assembly, and bonding agents are applied to fix the curvature and strengthen the structure (Figures 16–17).

Physical Test

Due to the limited accessibility of testing machines, a simple loading test is conducted to study the strength and stiffness of the prototype. The supports are 3D-printed and screwed on a wood panel to confine the three bottom bars from any lateral movement (Figure 18). The self-weight of this prototype is 857 grams, and after applying a total load of 90 pounds on the top bar, there is no visible buckling or deformation, which shows the potentials of its load-bearing capacity (Figure 19). To further understand its structural performance and limit, a universal testing machine will be used for the next step to complete an accurate experiment.

Numerical Simulation

The Autodesk Fusion 360 is used to simulate the stress distribution and displacement of the prototype under heavier loads. A total load of 1 KN is applied to the top bar while the bottom three bars are fixed at the end. The result shows stress concentrating on the exterior side of the kerfs, where the maximum stress is 34.93 MPa and the maximum displacement is 0.006mm (Figure 20).

CONCLUSION

This paper introduces a novel timber space frame system by combining kerf bending with zippered wood, which can be easily manufactured by accessible tools like 3-axis CNC milling machines. The assembly is also made easy since no jig or locator is needed. This system greatly facilitates the design and fabrication of complex timber space frames and allows for rapid fabrication and construction. A computational pipeline is also developed based on PGS for fast modeling and user-friendly parameter control.

The outcome of both the physical test and numerical simulation show promising structural performance in terms of the little deformation under the heavy loads. It also contributes towards the applications of polyhedral graphic statics in materialization and construction.

ACKNOWLEDGEMENTS

This research was generously supported by the Polyhedral Structures Lab at Weitzman School of Design, University of Pennsylvania. We thank Masoud Akbarzadeh for his support and advising and Edyta Augustynowicz for offering valuable consultation and expertise.

REFERENCES

Akbarzadeh, M. 2016. "3D graphic statics using polyhedral reciprocal diagrams." PhD thesis, ETH Zurich.

Akbarzadeh, Masoud, Tom Van Mele, and Philippe Block. 2015a. "3D graphic statics: geometric construction of global equilibrium." In *Proceedings of IASS Annual Symposia*, vol. 21. Amsterdam, The Netherlands. 1–9.

Akbarzadeh, Masoud, Tom Van Mele, and Philippe Block. 2015b. "On the equilibrium of funicular polyhedral frames and convex polyhedral force diagrams." *Computer-Aided Design* 63: 118–128.

Akbarzadeh, Masoud, Mohammad Bolhassani, Andrei Nejur, Joseph Robert Yost, Cory Byrnes, Jens Schneider, Ulrich Knaack, and Chris Borg Costanzi. 2019. "The design of an ultra-transparent funicular glass structure." In *Structures Congress 2019: Blast, Impact Loading, and Research and Education*. Reston, VA: American Society of Civil Engineers. 405–413.

Akbarzadeh, Masoud, Mehrad Mahnia, Ramtin Taherian, and Amir Hossein Tabrizi. 2017. "Prefab, Concrete Polyhedral Frame:

Materializing 3D Graphic Statics." In *Proceedings of IASS Annual Symposia*, vol. 2017 (6). International Association for Shell and Spatial Structures (IASS). 1–10.

Bolhassani, Mohammad, Masoud Akbarzadeh, Mehrad Mahnia, and Ramtin Taherian. 2018. "On structural behavior of a funicular concrete polyhedral frame designed by 3D graphic statics." *Structures* 14: 56–68.

Capone, Mara, and Emanuela Lanzara. 2019. "Parametric Kerf Bending: Manufacturing Double Curvature Surfaces for Wooden Furniture Design." In *Digital Wood Design*, edited by F. Bianconi and M. Filippucci M. Lecture Notes in Civil Engineering, vol. 24. Cham: Springer. 415–439.

Lamere, Joel, and Cynthia Gunadi. 2019. "Lost House." In ACADIA '19: Ubiquity and Autonomy; Paper Proceedings of the 39th Annual Conference on the Association for Computer Aided Design in Architecture, edited by Kory Bieg, Danelle Briscoe, and Clay Odom. Mexico City: ACADIA. 140–145.

Lennartz, Marc Wilhelm, and Susanne Jacob-Freitag. 2015. *New Architecture in Wood: Forms and Structures*. Basel: Birkhäuser Press.

Mansoori, Maryam, et al. 2019. "Adaptive wooden architecture. Designing a wood composite with shape-memory behavior." In *Digital Wood Design*, edited by F. Bianconi and M. Filippucci M. Lecture Notes in Civil Engineering, vol. 24. Cham: Springer. 703–717.

Maxwell, J. Clerk. 1870. "I.—on reciprocal figures, frames, and diagrams of forces." *Earth and Environmental Science Transactions of the Royal Society of Edinburgh* 26.1: 1–40.

McRobie, Allan. 2016. "Maxwell and Rankine reciprocal diagrams via Minkowski sums for two-dimensional and three-dimensional trusses under load." *International Journal of Space Structures* 31.2-4: 203–216.

Momoeda, Yu. 2017. "Agri Chapel / Yu Momoeda Architecture Office." ArchDaily. Accessed July 8, 2021. https://www.archdaily.com/884875/agri-chapel-yu-momoeda-architecture-office.

Nejur, A. and M. Akbarzadeh. 2021. "PolyFrame, Efficient Computation for 3D Graphic Statics." *Computer Aided Design* 134: 103003.

Nejur, A. and A. Masoud. *PolyFrame Beta*, V. 0.1.9.3. Polyhedral Structures Laboratory. Rhino 6, 7 for Windows. 2018.

Rankine, W. J. M. 1864. "Principle of the Equilibrium of Polyhedral Frames." *Philosophical Magazine Series* 4 27 (108): 92.

Robert McNeel & Associates. *Rhinoceros 3D*. V. 6.0. Robert McNeel Associates. Windows. 2010.

Sebera, Vaclav, and Milan Šimek. 2014. "Finite element analysis of dovetail joint made with the use of cnc technology." *Acta universitatis agriculturae et silviculturae mendelianae brunensis* 58 (5): 321–328.

Satterfield, Blair, et al. "Bending the Line Zippered Wood Creating Non-Orthogonal Architectural Assemblies Using The Most Common Linear Building Component (The 2X4)." In Fabricate 2020: Making Resilient Architecture, edited by Jane Burry, Jenny Sabin, Bob Sheil, and Marilena Skavara. London: UCL Press. 58–65.

Teeple Architects. 2015. "Philip J. Currie Dinosaur Museum, Teeple Architects." ArchDaily. Accessed July 8, 2021. https://www.archdaily.com/618989/philip-j-currie-dinosaur-museum-teeple-architects.

IMAGE CREDITS
Figure 4: ©Masoud Akbarzadeh, 2017.
All other drawings and images by the authors.

Yulun Liu is an interdisciplinary designer with an interest in structural design and spatial efficiency. She is currently a Master of Architecture candidate at the Weitzman School of Design, University of Pennsylvania. She also holds a Master of Structural Engineering from Tianjin University.

Yao Lu is currently a PhD student at the Polyhedral Structures Laboratory, Weitzman School of Design, University of Pennsylvania. He is a design researcher with a great interest in generative design, robotic fabrication, 3D printing, and computer graphics. Before joining PSL, he graduated from Cornell University with a Master of Science in Matter Design Computation. He also holds a Master of Architecture and a Bachelor in Engineering degrees and from Tongji University.

Masoud Akbarzadeh is a designer with a unique academic background and experience in architectural design, computation, and structural engineering. He is Assistant Professor of Architecture in Structures and Advanced Technologies and the Director of the Polyhedral Structures Laboratory (PSL). He holds a DSc from the Institute of Technology in Architecture, ETH Zurich, where he was a Research Assistant in the Block Research Group. He holds two degrees from MIT: a Master of Science in Architecture Studies (Computation) and a MArch, the thesis for which earned him the renowned SOM award. He also has a degree in Earthquake Engineering and Dynamics of Structures from the Iran University of Science and Technology and a BS in Civil and Environmental Engineering. His main research topic is Three-Dimensional Graphical Statics, which is a novel geometric method of structural design in three dimensions.

Surface Disclination Topology in Self-Reactive Shell Structures

Nicholas Bruscia
University at Buffalo, State University of New York

1 Thin plywood prototypes of triangle and square based surface disclinations

ABSTRACT

This paper discusses recent developments on the geometric construction and fabrication techniques associated with large-scale surface disclinations. The basic concept of disclinations recognizes the role of "defects" in the composition of materials, the strategic placement of which shapes the material by inducing curvature from initially planar elements. By acknowledging the relationship between geometry and topology that governs disclination based form-finding and material prototyping, this work consciously explores its potential at the architectural scale. Basic geometric figures and their topological transformations are documented in the context of digital modeling and simulation, fabrication, and a specific material palette. Specifically, this work builds on recent efforts by focusing on three particular areas of investigation: a) enhancing the stability of surface disclinations with a synthetic fibrous layer, b) aggregation via periodic tilings, and c) harnessing snap-through instability to increase bending stiffness in thin surfaces.

INTRODUCTION

The work presented in this paper builds on methods that were developed in 2019 and subsequently explored in a second sponsored graduate research studio conducted early 2021, as a design-build educational exercise resulting in refined digital modeling and fabrication strategies and new knowledge associated with surface disclinations. Here, the term more broadly refers to the topological concept that introduces angle deflections in sheet materials resulting in out of plane buckling, while a more specific term, *wedge disclinations* (wd) refers to a particular variant that topologically induces elliptic or hyperbolic geometry in elastic surfaces and lattice structures (Harris 1977; Weeks 1985; Bruscia 2020).

The notation used here to describe this variant is [+/-x°] wd, referring to the overall sector (or wedge) of inserted material into the original shape as a rotation angle—otherwise described as "angular excess" (producing negative curvature) or "angular deficit" (producing positive curvature) (Terrones and Mackay 1993). An additional terminology used in this paper, "2-circle" or "3-square," describes a [-360°]wd or [-720°]wd respectively. This casually refers to the overall amount of surface area, while the degree notation refers more specifically to the inserted amount. The physical form-finding process for surface disclinations involves the rotational bending of pre-cut sheet materials whose edges reconnect back upon themselves. Strain is introduced to the sheet, and the sheet bends in reaction. The strain is then relieved and distributed evenly across the now stable surface (Harris 1977). Nothing is added, but the surface takes on a new form; in a sense, it is reacting against it's own natural tendency to become flat again. The term "self-reaction" is used here to describe this behavior.

Wedge disclinations are found in various sculptural and woven work. In architecture, Buckminster Fuller's geodesic dome is a two-dimensional hexagonal framework with several wedge disclinations occurring as well-placed pentagons (deWit 1971). Surface and lattice transformations caused by placing singularities within tri-axial meshes (Ayres et al. 2020) share common references in the natural world; the defect in the molecular arrangement of carbon molecules are present in thin periodic structures such as virus shells, the pattern of fingerprints, and in the pelts of striped animals (Harris 1977). Because carbon atoms are arranged into a hexagonal arrangement, it is possible to build complex molecular structures with curvature using disclinations, as demonstrated in the work on controlled shape change in graphite by Terrones (1993), Mackay (1985), and

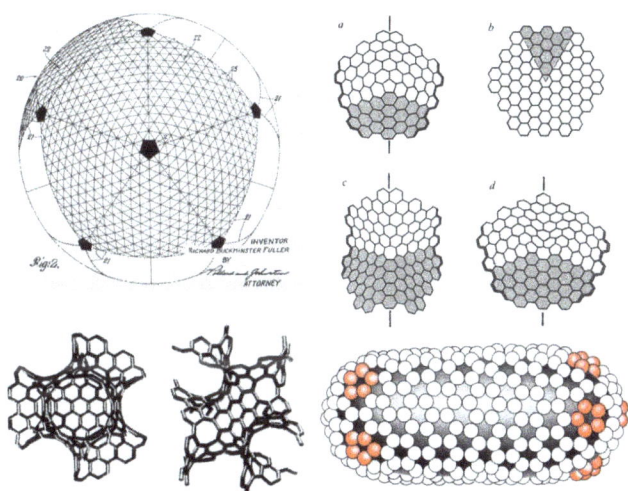

2

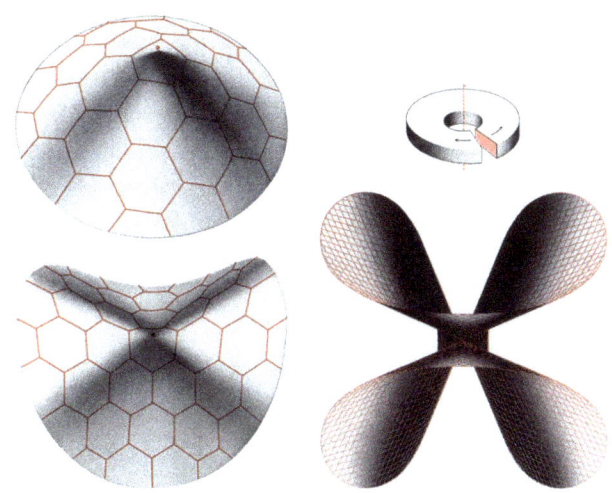

3

4

2   Disclinations in hexagonal lattices (clockwise rom top left): Buckminster Fuller (1954), Iijima et.al (1992), Terrones and Mackay (1993), W. Harris (1977)

3   Left: [+60°] and [-60°] wedge disclinations (wd) in thin sheets (W. Harris 1977); Right: [-360°] wedge disclination (wd) in a thin sheet

4   Left: [+180°]wd plywood prototype, and Right: [-288°] wd plywood prototype

5 Gauss-Bonnet bending of a 90° equilateral triangle (Harriss 2020)

6 Gauss-Bonnet bending through curved and serrated seams (Delp 2011)

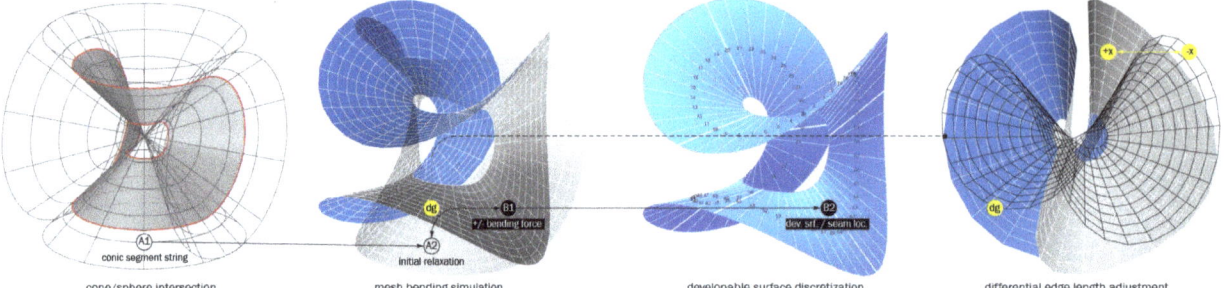

7 Basic geometric construction and bending simulation of a 2-circle [-x°]wd

Iijima et al. (1992) (Figure 2). The observation of the effect of disclinations in thin surfaces was illustrated by crystallographer William F. Harris (1977), from which this work adopts the general concept, terminology, and notation (Figure 3).

Previous work demonstrated that [+x°]wd produced a variety of conic shells with positive global curvature, and [-x°]wd gave rise to a variety of saddle-like shells with negative global curvature. Both were prototyped at a large scale; the [+x°]wd forming a tilted conic shell, and the [-x°]wd resulting in a 8-sided saddle-like form, found by combining two pentagons (Bruscia 2020) (Figure 4). Although the prototypes showed that topologically induced bending using the disclination method could scale, the shells were only moderately stable and were subject to shearing during the assembly process. The advancement of this work had three primary goals intended to address these shortcomings; a) surface reinforcement, b) expansion via tiling, and c) stability via snap-through buckling, all of which are discussed in the following sections.

STATE OF THE ART

Disclination topology may be described more formally with the Gauss-Bonnet theorem, and we find related work in mathematics and sculpture. Paper strip models are used to teach mathematical concepts such as discrete Gaussian curvature, face-defects, and angle deflections; the paper model as a tool to demonstrate them. For example, we can look to Ilhan Koman's developable sculptures to see how angle deflections describe local bending behavior in surfaces (Akleman and Chen 2006). Inversely (and serendipitously), we have discovered the connection in our work to these mathematical concepts by first exploring surface disclinations intuitively. Mathematician Edmund Harriss provides a clear explanation of Gauss-Bonnet: that the sum of the angles in a triangle is 180°, and by changing the corner angles of the triangle, it must bend into positive or negative curvature (Harriss 2020). For example, consider that the corner angles in an equilateral triangle are 60°. By increasing the corner angles to 90°, we increase the overall sum. Constructing this triangle from paper strips will force it to buckle with positive curvature, and eight of them constructed the same way will create an octahedral sphere (Harriss 2020) (Figure 5). Paper bending visually demonstrates this transition from Euclidian to elliptic and hyperbolic polygons by combining them into spherical platonic solids or minimal surface approximations (Akleman et al. 2010; Harriss 2020).

Another example is seen in the work of Kelly Delp and Bill Thurston (2011), whose technique pushes the curvature away from vertices and into the edges. Considering again the equilateral triangle, a 30° sector of a circle replaces each side, forming a 'fat octahedron' when the curved seams are forced to connect. Meandering profiles along the curved edges are added, forcing the parts to bend and interlock, evenly distributing the curvature across the seam and smoothing the polyhedron into a sphere (Figure 6). Simply forcing the profiles to come together without gaps automatically determines the localized bending and global curvature. The method is easily adapted to create models

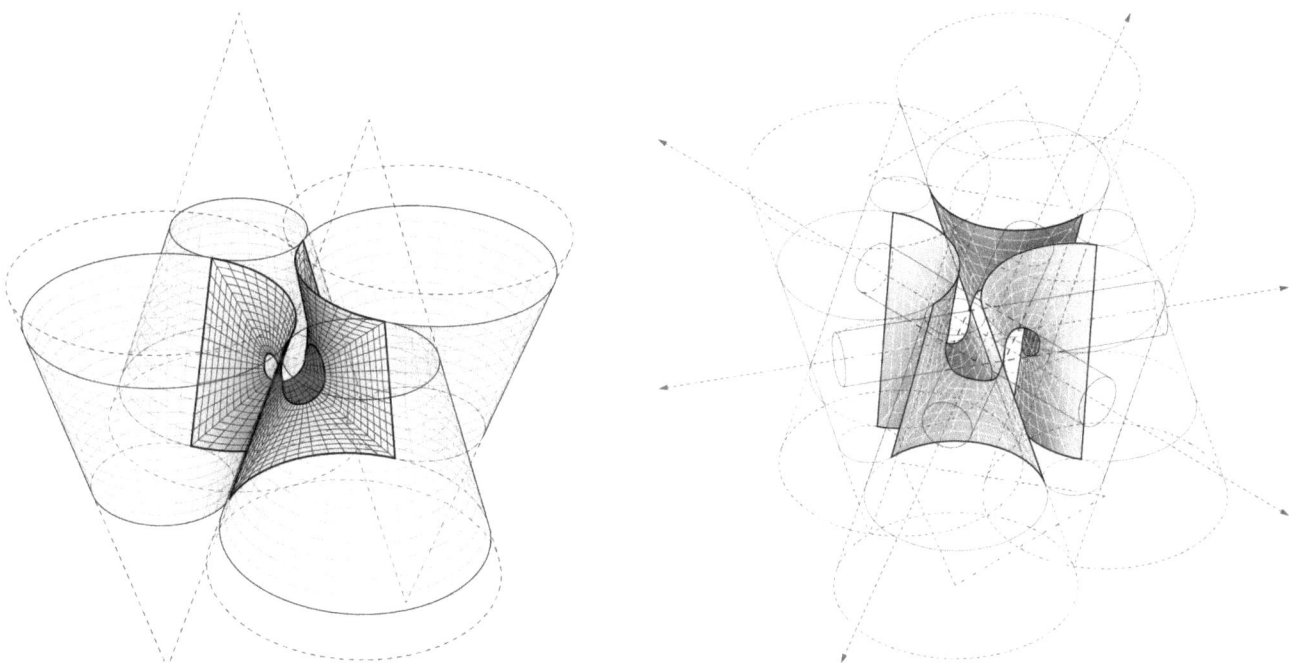

8   Geometric construction techniques for a 2-square [-x°]wd (left) and a 3-square [-x°]wd (right)

of negative curvature and combinations of both, as shown in a variety of their models (Delp and Thurston 2011). The serrated profiles presented in this paper were designed to evenly distribute bending forces across the seam, but they themselves do not induce the curvature. In the discussed prototypes, serrated seams function to combine smaller parts into structures with dimensions that exceed off-the-shelf materials. However, both the techniques introduced above, as well as the disclination techniques described below, result in surfaces that are formed automatically during assembly. The wooden surfaces are not steam-bent or forced into molds; they begin flat, and are shaped entirely by disclination topology and active bending forces.

DESIGN AND FABRICATION METHODS

While it can be argued that the [-x°]wd can itself be discussed as an efficient method for fabricating curved surfaces from developable sheet materials, the specific application of those surfaces remains one of the goals of future work. Instead, this work is concerned with the methods for developing sculptural shell structures that are intended only to demonstrate how the potential structural qualities of surface disclinations translate to a larger scale. Recent developments in the modeling and simulation of surface disclinations are discussed below, paying particular attention to the geometric construction of base mesh topologies.

Geometric Construction of Disclination Topology

Three techniques to digitally construct and simulate surface disclinations were studied previously (Bruscia 2020). Of the three, the "double-cone intersection" technique has shown to be the most versatile, especially when used in combination with the "expanded concentric polylines" technique for added control over the result (Figure 7). Initially devised to prove that the surfaces are made up of strings of conic sections and thus maintaining their developability on-screen, new variations of this technique can now accommodate a wider range of base geometric figures. The subsequent mesh bending simulation is useful to arrive at more "natural" digital representations.

Bending simulation is achieved using a physics simulation platform, for example Kangaroo2 (Piker 2013) which models physical behavior qualitatively on triangle meshes using a 3-point approach. The following sections focus on the geometric approximations of disclination topology, with the assumption that a variety of platforms may be used to apply bending simulation. The work in this paper was developed by using default algorithmic sequences for mesh bending in Kangaroo2 as a proof-of-concept. Once the mesh topology is modeled correctly, the concentric mesh edges are assigned a numerical input to dynamically change their length. The radiating mesh edges are forced to remain at their original lengths as the concentric edges shrink or grow, forcing the mesh buckling to increase or decrease. For example, a numeric range to adjust the amount of curvature could be [0.5 < 1 < 2.0] with 1 referring to the original length. The amount of material insertion based on the differential growth of mesh edges depends on the initial state of the mesh. Assuming that the geometric

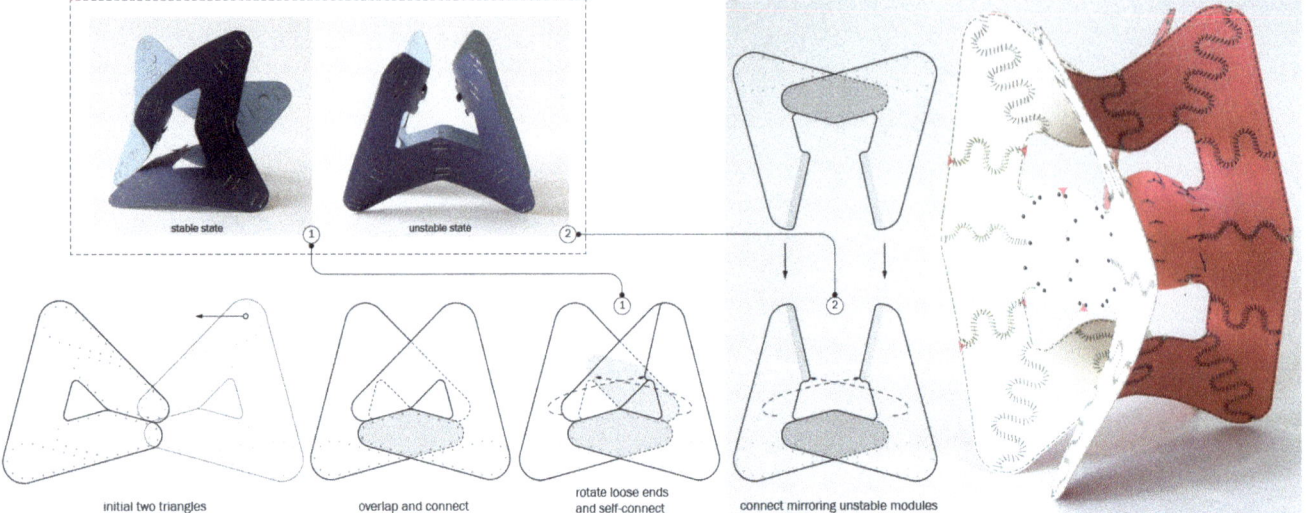

9 Counter-reactive stability by combining (2) opposing 2-triangle [-x°]wd modules forced to connect between their stable states

construction shown in Figure 4 produces a surface that is an exact approximation of the 2-circle [-360°]wd, the angular excess can be found by simply multiplying the edge length factor (x) by 360. If x=0.25, the angular excess is 90°. The following paragraphs include few examples to illustrate the relationship between the desired geometric profile and the digital modeling process.

Topology Based on the Circle and Ellipse

The simplest geometric construction intersects two spheres (one large, and one small) with two double-cones whose apices meet at the same point. The intersections form circles that can be split using the tangents between the cones. Selecting every other circle segment will string together the inner and outer profiles of a 2-circle [-360°]wd. The profiles may then be used to draw a ruled surface and/or a quad or triangle mesh between them. The exact process can also be used to model a 3-circle [-720°]wd by adding one more double-cone to the cube, such that a circular cone base is drawn along each of the cube's 6 faces. Geometrically constructing the topology for a surface disclination based on the ellipse generally follows the same process. Instead of a cube to organize the cones, the faces of rectangular prism organizes the cones which are intersected with ellipsoids instead of spheres. As an alternative to the rectangular prism, a hexagonal prism may be used to construct a 3-ellipse surface disclination with 3-fold symmetry.

Topology Based on the Square

Square-based surface disclinations follow the same topologically induced bending behavior as the circle and ellipse, but a different arrangement of cones is needed to model the 3-dimensional cubic profile. Both 2-square and 3-square structures begin with mirrored sets of vertical cones that touch along a straight tangent line (Figure 8). The method chosen to ensure tangency will determine input parameters that may be numerically controlled to dynamically change the outer edge tilt and overall proportions. For example, we begin with three circles whose center points rest at the vertices of an equilateral triangle, and whose radii are the distance between the triangle vertices and the triangle center point. Three cones are drawn from the base circles, whose intersections create the apices of three opposite cones. For both 2- and 3-square [-x°]wd, it is advantageous to cut the set of cones with horizontal planes that are equidistant from the cone apices. This will prevent the structure from self-intersecting, and spacing them apart to the same distance as one side of the equilateral triangle will ensure more cubic proportions. The tangent lines between the set of 6 cones are used for the outer, rectilinear edges of the surface. The center profile is drawn from a separate process, however, for instance the profile of a 3-square circle disclination may be used and centered among the set of vertical cones. Finally, the center and outer profiles are connected as meshes or NURBS surfaces for further development.

Topology Based on the Triangle

Triangle-based surface disclinations are unique because they tend toward a stable state that appears twisted, and in a sense, incomplete. Between states, when the bending forces are at their highest and the surface is trying to release its energy by rotating back, is when the 2-triangle [-x°]wd has the appearance of being stable due its perceived symmetry. After many attempts to discover how to repeat and control this behavior, it was eventually

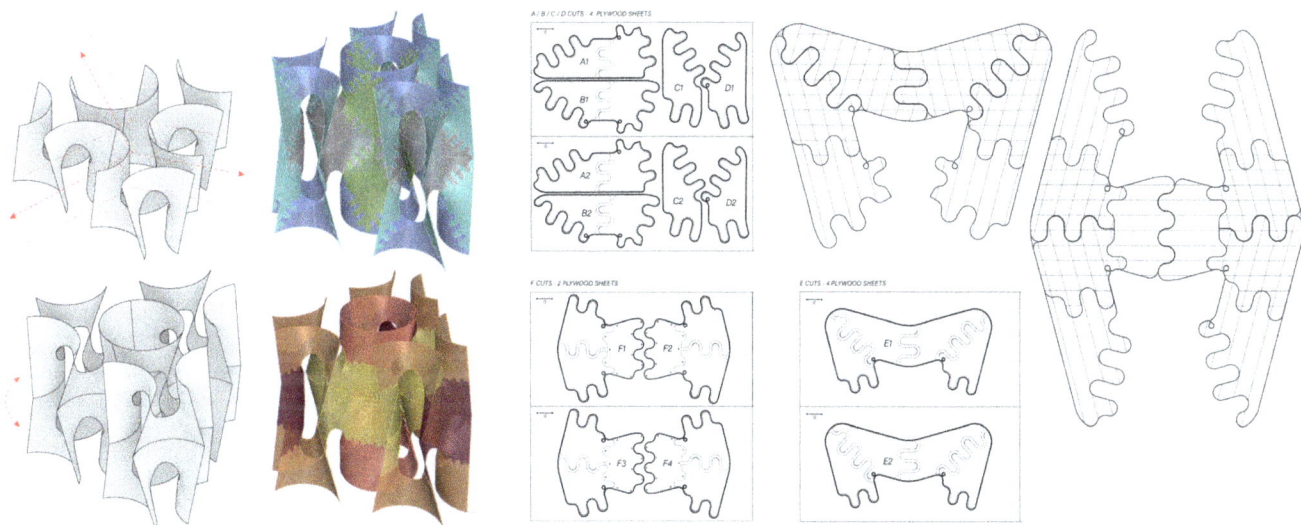

10 (Left) 3-square [-x°]wd modules blended into two continuous layers with staggered seams; (Right) CNC cut-sheets and grain orientation of the 2-module, 4-triangle [-x°]wd

harnessed as a strategy for increasing the overall stiffness by connecting two opposing surface disclinations at their mid-way and most unstable position, thereby maximizing the self-reaction (Figure 9). The simulation of such behavior remains an area of interest for future investigation as it could hold interesting potential in the design of thin and lightweight sheet materials in larger structures. The geometric construction borrows the same cone arrangement as the 2-square [-x°]wd, but with two major differences: to find the inner profile, folded planes are intersected with the cones, and to obtain the outer profile, planes are extended off of the cones from where they meet tangentially.

Fabrication and Assembly

Three large-scale prototypes were constructed from rotary cut 4ft x 8ft x 1/8in (3mm thick) 2-ply sheets of Okoume BS 1088 Lloyds approved marine-grade plywood, and rotary cut 4ft x 8ft x 1/8in 3-ply sheets of bendable barrel grain exterior grade Lauan plywood. To maintain bending continuity throughout the entire surface, each prototype consists of two layers of material for a total thickness of 1/4in (6mm). The 2-layer approach avoids awkward lapping such that all seams have the same visual hierarchy and each side of the surface is smooth (Figure 10). Deeply serrated edges allow the material to bend evenly, although a very small area of single-layer thickness remains where the seam terminates at the outer edge. A relief hole is placed in these locations to distribute the bending forces deeper into the 2-layer surface. The prototypes were assembled as a sequence of parts that were gradually bent and forced into position, automatically taking shape as parts are connected. The form emerges naturally as a result of the topologically induced bending, and the seams between parts are located so that each part can be cut from a single 4ft x 8ft sheet and with a desirable grain direction that is assigned based on whether it is advantageous achieve a tight bending radius or to locally increase the stiffness. Each of the three demonstrators represent a different base geometry, aggregation, and stiffening strategy, but all demonstrate the topologically induced bending process of [-x°]wd.

RESULTS AND DISCUSSION
Disclination based on the Ellipse

The 2- or 3-circle [-x°]wd is perhaps the most clear demonstration of disclination topology, but its stability varies depending on proportions, surface depth and overall dimensions, and seemingly requires the most additional stiffening after assembly. The studio sought to test the limits of this disclination variant, and a form finding study based on the ellipse led to the design of a free-standing ribbon that stands 9ft (3m) tall (Figure 11). The total surface area of material used is about 199 square feet, including both layers of 1/8in Okoume plywood for a total weight of approximately 56 lbs.

Smaller scale investigation revealed that the self-reactive bending could not provide enough stability for the structure to resist its own weight. This was addressed by adding a carbon fiber exoskeleton that increases the structural depth of the surface, allowing it more quickly spring back. The structure easily carries its own weight and wiggles confidently under minor loading, and has remained symmetrical without tilting or sagging several months after completion. The exoskeleton consists of ~1125 ft of 12k T700 carbon fiber roving impregnated with epoxy resin cured in room temperature. The fibers are looped around

11

12

792 nylon threaded rods that extend 20mm from each side of the surface, holding fibers approximately 7mm to 15mm away. Fiber roving may be added to further reinforce specific areas of the surface and the threaded connections can vary in length to create a gradient of deep and shallow areas along the surface. The lightweight CFRP roving ensured that the materials work as a uniform system, assisting and not overtaking the active bending.

Disclination based on the Square

Polygon-based surface disclinations are able to more easily aggregate due to their points and edges. Aligning modules by overlapping edges gives rise to periodic tilings of disclination modules, shifting the focus from the development of the singular object to that of larger structures. The studio aimed to provide a clear example by tiling the 3-square [-x°]wd disclination both horizontally and vertically. The 3-square disclination is particularly conducive to tiling due its 6 "sides"—three curves along the ground plane, and 3 above rotated 60°. 3-square modules may thus be arranged in a hexagonal tiling and easily stacked in the z-axis. Three modules centered at the vertices of an equilateral triangle will form a cylindrical space, however the aggregate surface is always continuous and open. The finished prototype stands at 8in tall and consists of ~430 square feet of material including a combination of 1/8in Okoume and barrel grain Lauan plywood fastened together with 3/16in (4mm) aluminum rivets and #10-32 SS bolts along demountable seams (Figure 13).

Given the overall footprint, this structure is extremely stable. However, a stiffened proof-of-concept was built using a single 2-square module. 3in wide carbon fiber tape was manually applied to the surfaces after bending into a completed assembly, consisting of 5.7oz IM2 unidirectional layers, topped with a layer of 3K carbon and fiberglass biaxial tape. The pattern was manually drawn onto the surface using a strip of PETG that was pinned between CNC-cut holes and laid out along the geodesic path.

Disclination based on the Triangle

Interestingly, the 2-triangle [-x°]wd is more conducive to tiling periodically when in the tensed state. This provided a key opportunity to harness the structural potential of disclination topology; in order for one 2-triangle module to hold its position, a second module must counter the natural tendency for it to rotate/twist. The second module is designed with identical geometry but with the tendency to rotate in the opposite direction. When connected, the two modules balance each other by providing an equal and opposite reaction, as observed in both small scale models and the larger prototype that blends the modules into a single, continuous surface. While the prototype maintained this mutually self-reactive behavior, the simulation requires a different approach to demonstrating the natural tendency toward a state that differs from its initial form. Some work has been done to experiment with adding 'bias' to the meshes by introducing diagonal edges in opposing directions, but as of the writing of this paper this remains a work-in-progress.

11 3-ellipse [-x°]wd prototype with standoff CFRP roving for added stability

12 2-module, 4-triangle [-x°]wd prototype

13 6-module, 18-square [-x°]wd tiling prototype, with 2-square [-x°]wd CFRP tape test

The approximately 8ft tall, 2-module (4-triangle) prototype was designed and built as a blended counter-reactive surface (Figure 12). Twenty-six parts assemble into the 2-layer plywood surface that is ~145.6 square feet of total material weighing ~41lbs. The parts and layers are manually sewn together simultaneously using braided Spectra twine through a CNC drilled hole pattern. This method has shown to be the most successful in maintaining layer adhesion without air gaps, while having enough flexibility in the joint to allow the layers to adjust while bending. In-depth analysis of fibrous joints applied to segmented timber shell structures is documented in Bechert et al. (2016) and Garufi et al. (2019).

Partially assembled portions exhibited the tendency to twist as expected and canceled each other out in the finished assembly. The structure is confidently stable and the internalized counter-reaction seems evident in the way it resists external forces pushing on it. While this prototype was intended to test the structural potential of counter-reactive bending, a portion of the surface was built separately as a proof-of-concept for additional fiber-reinforced stiffening, experimenting with 2in wide, 9oz T700 unidirectional carbon fiber tape topped with a layer of biaxial tape of the same specification.

The tendency for the surface to rest in multiple semi-stable states reveals a particularly interesting strategy for increasing the overall stiffness of the structure by using characteristics inherited from surface self-reaction. By attempting to understand what determines this behavior, we recognized that surface disclinations exhibit snap-through buckling, always maintaining a degree of increased stiffness between stable states. It has been observed throughout this research to date that physical models will behave differently under subtle changes, for example upon which side the material overlap occurs, assembly sequence, and accidental material "training" or fatigue that results from forcing the card stock to buckle and bend. This has not been a hindrance to larger scale prototyping however, and the counter-reactive assembly studied in the triangle-based prototype has inspired continued investigation into "functionalized instability" for increasing the structural capacity of very thin sheet materials (Schleicher et al. 2015).

CONCLUSION

In summary, the work outlined in this paper shows the continuation of research focused on developing new fabrication and form-finding techniques based on the topological and geometrical properties of surface disclinations. Yet to be explored at this scale previously, the results illustrate the feasibility of applying this particular form of topological bending at a large scale, and point toward paths forward that have both educational and material value. Specifically, the results demonstrate how the goals of this phase were addressed by demonstrating proof-of-concept approaches to: a) adding stability by integrating a lightweight carbon fiber reinforcement b) periodic tilings by blending disclinations into larger surfaces, and c) how snap-through instability may be harnessed to pre-tension disclination modules and

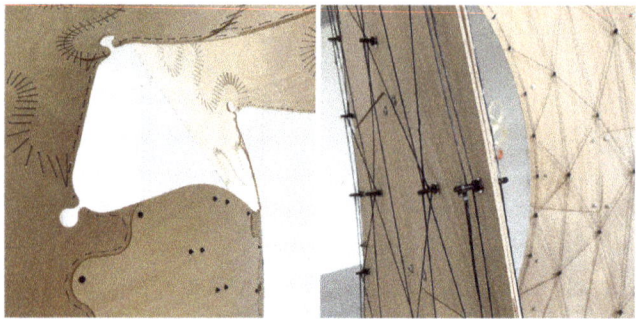

14 Sewn plywood layering and CRFP standoff details

15 Single, continuous surface of the square-based [-x°]wd tiling

16 Spatial pockets of the square-based [-x°]wd tiling

increase the overall bending stiffness in planar, off-the-shelf building materials.

In addition to the exercise of enacting disclinations from an architectural point of view, the work may contribute to a variety of contexts that are also seen here as areas of future improvement: (1) moldless shell fabrication wherever developable materials are advantageous, (2) the discourse on bending-active shell structures from thin sheet materials, (3) the geometric construction of base mesh topologies for computational analysis and simulation, and (4) design-based exploration of the relationship between mathematics, materials, and form. Linking the surface disclination fabrication technique with the Gauss-Bonnet formula is also a desirable area of further investigation, as it may open the door to a new design technique that expands the disclination repertoire to include spherical portions to the catalog of conic segments, and lead to a mathematical application of surface disclinations in architectural form rationalization and construction.

ACKNOWLEDGEMENTS

The work presented in this paper was supported by the Nohmura Foundation for Membrane Structure's Technology, and was conducted at the intersection between research and teaching together with students from the Situated Technologies Research group at the University at Buffalo, State University of New York. The author would like to express their gratitude towards Josh Barzideh, Tyler Beerse, Tom Cleary, Camilo Copete, Bhalendu Gautam, Sam Goembel, Marissa Hayden, Ollie He, Nick Hills, Jamie Jones Lovepreet Kaur, and Ben Starr for their inspiring work in the Spring 2021 STRG design studio. The author would also like the thank the peer reviewers for their constructive comments and suggestions.

REFERENCES

Akleman, Ergun and Jianer Chen. 2006. "Insight for Practical Subdivision Modeling with Discrete Gaussian-Bonnet Theorem." In *Proceedings of Geometry, Modeling and Processing 2006*, edited by M-S. Kim and K. Shimada. Pittsburg: Springer-Verlag Publishing. 287–298.

Akleman, Ergun, Jianer Chen, and Jonathan L. Gross. 2010. "Paper-Strip Sculptures." In *IEEE International Conference on Shape Modeling and Applications (SMI)*, edited by J-P. Pernot et.al., Aix-en-Provence. IEEE Computer Society. 236–240.

Ayres, Phil, Ji You-Wen, Jack Young, and Alison Grace Martin. 2020. "Meshing with Kagome Singularities." In *Advances in Architectural Geometry 2020*. Ecole de Ponts ParisTech and Université Gustave Eiffel. Accessed May 10, 2021. https://thinkshell.fr/advances-in-architectural-geometry-2020-on-line-library/.

Bechert, Simon, Jan Knippers, Oliver David Krieg, Achim Menges, Tobias Schwinn, and Daniel Sonntag. 2016. "Textile Fabrication Techniques for Timber Shells. Elastic Bending of Custom-Laminated Veneer for Segmented Shell Construction Systems." In *Advances in Architectural Geometry 2016*, edited by S. Adriaenssens et al. Zurich: vdf Hochschulverlag AG an der ETH Zurich. 154–169.

Bruscia, Nicholas. 2020. "Structural Papercuts: Scaling Disclinations in Self-Reactive Surfaces." In *ACADIA '20: Distributed Proximities: Proceedings of the 40th Annual Conference of the Association for Computer-Aided Design in Architecture*, edited by B. Slocum, V. Ago, S. Doyle, A. Marcus, M. Yablonina, M. del Campo. Online+Global. 536–545.

Delp, Kelly, and Bill Thurston. 2011. "Playing with Surfaces: Spheres, Monkey Pants, and Zippergons." In *Proceedings of the Bridges Conference 2011*. Coimbra: Tessellations Publishing. 1–8.

17 Selected paper form-finding models

deWit, Roland. 1971. "Relation between Dislocations and Disclinations." *Journal of Applied Physics* 42 (9): 3304–3308.

Garufi, Dominga, Hans Jakob Wagner, Simon Bechert, Tobias Schwinn, Dylan Marx Wood, Achim Menges, and Jan Knippers. 2019. "Fibrous Joints for Lightweight Segmented Timber Shells." In *Research Culture in Architecture: Cross-Disciplinary Collabortation*, edited by. C. Leopold, C. Robeller, and U. Weber. Basel: Birkhäuser. 53–63.

Harris, William F. 1977. "Disclinations." *Scientific American* 237 (6): 130–145.

Harriss, Edmund. 2020. "Gauss-Bonnet Sculpting." In *Proceedings of Bridges 2020: Mathematics, Art, Music, Architecture, Education, Culture*. Phoenix, Arizona: Tessellations Publishing. 137–144.

Iijima, Sumio, Toshinari Ichihashi, and Yoshinori Ando. 1992. "Pentagons, heptagons and negative curvature in graphite microtubule growth." *Nature* 356: 776-778

Mackay, Alan L. 1985. "Periodic Minimal Surfaces." *Nature* 314: 604–606.

Piker, Daniel. 2013. "Kangaroo - Form Finding with Computational Physics." *AD, Architectural Design* 83 (2): 136–137.

Schleicher, Simon, Andrew Rastetter, Riccardo La Magna, Andreas Schönbrunner, Nicolas Haberbosch, and Jan Knippers. 2015. "Form-Finding and Design Potentials of Bending-Active Plate Structures." In *Modelling Behaviour. Design Modelling Symposium 2015*, edited by M. Ramsgaard Thomsen et. al. Cham: Springer International Publishing. 53–63.

Terrones, Humberto, and Alan L. Mackay. 1993. "Hypothetical Curved Graphite." *Nanostructured Materials* 3: 319–329.

Weeks, Jeffrey R. 1985. *The Shape of Space. How to Visualize Surfaces and Three-Dimensional Manifolds.* New York: Marcel Dekker, Inc.

IMAGE CREDITS

Figure 2: Clockwise from top left: Buckminster Fuller 1945, Iijima et.al. 1992, Terrones and Mackay 1993, W. Harris 1977
Figure 3: W. Harris 1977
Figure 5: E. Harriss 2020
Figure 6: Delp and Thurston 2011
Figure 8: Left: Kaur 2021, Right: Hayden 2021
Figure 9: Hills, Kaur, Copete, Starr 2021
Figure 10: Left: Hayden, Cleary, Goembel, Barzideh 2021, Right: Hills, Kaur, Copete, Starr 2021
Figures 15 and 16: Hayden 2021
All other images by the author.

Nicholas Bruscia is Assistant Professor in the Department of Architecture at the University at Buffalo, State University of New York, where he is also a researcher in the Sustainable Manufacturing and Advanced Robotics Technology Community of Excellence (SMART CoE) and the Center for Architecture and Situated Technologies (CAST). An interest in architectural geometry applied to formal and structural elegance inspires his work with materials and fabrication processes.

Discrete Quasicrystal Assembly

Integrated Machine-Material System
for Discrete Assembly of Quasicrystal Blocks

Donghwi Chris Kang
University of Toronto

Nicholas Hoban
University of Toronto

Maria Yablonina
University of Toronto

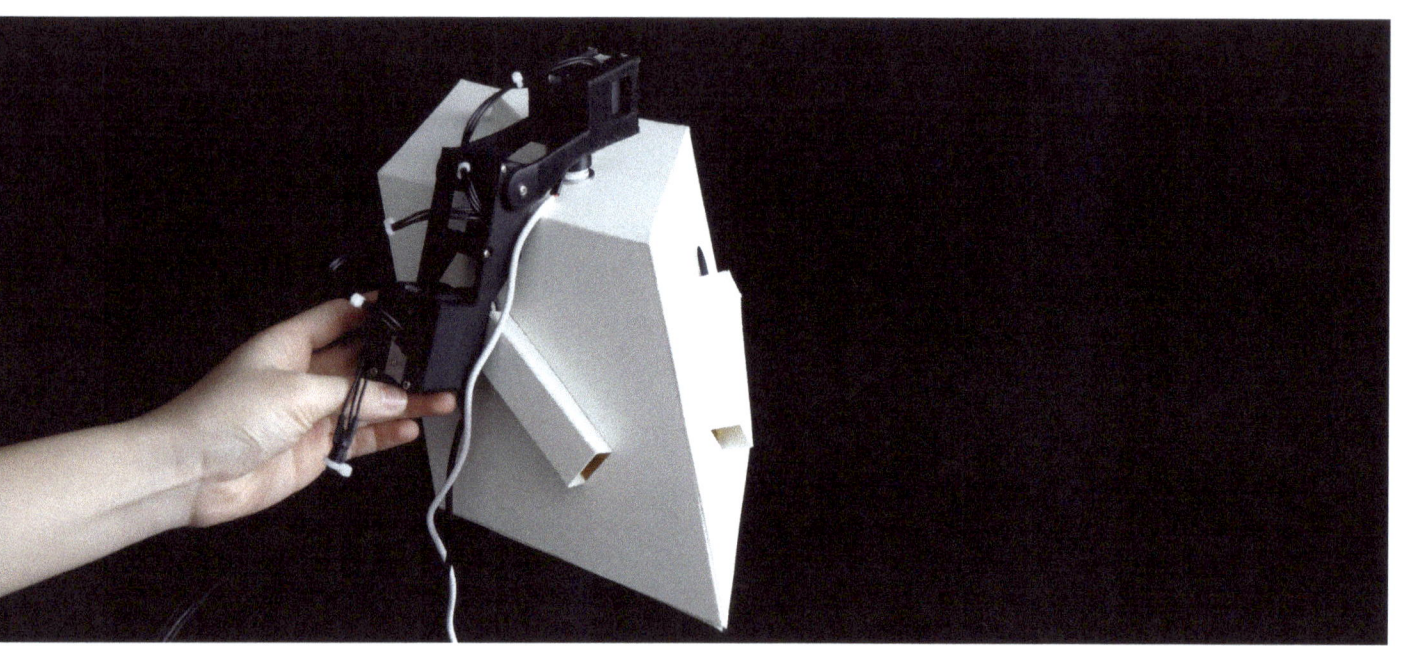

1

ABSTRACT

The research presented in this paper proposes a bespoke digital machine-material system for architectural assembly. The research aims to contribute to the body of work in digital material systems and single-task construction and fabrication robotics. Specifically, the system proposes a digital material system based on the icosahedral quasicrystals accompanied with a bespoke assembling robot capable of locomotion along the material as well as manipulation of discrete material units. Through a set of locomotion and pick-and-place routines, the robotic system is capable of construction and reconfiguration of the material system.

In proposing a digital machine-material system, the presented research argues for the development of design, fabrication, and robotics strategies wherein hardware, geometry, material, and software are developed in parallel in an interdependent co-design process. Such approach of considering parameters across the spectrum of design tasks allows to develop systems that are well suited for their specified application while maintaining minimum complexity and increasing accessibility of fabrication systems.

1   Physical prototype of the digital material system and custom designed task-specific robotic assembler to match the geometrical envelope

2   Digital machine-material system based on quasicrystal geometry assembly and single-task robotic fabrication

3   Physical prototype: using the two geometry types as the boundary condition for the digital material units, the final design was developed in correspondence with the robotic assembler design in order to minimize possible collisions and allow for a simplified robotic path planning

INTRODUCTION

Digital fabrication processes inherently rely on two interdependent parameters: machine and material. Calibration of the two towards a balance of matching variables is at the core of a successful fabrication process. Conventionally, as the field of digital fabrication in architecture builds upon appropriation of available industrial machinery and equipment, the aforementioned calibration of machine and material parameters is often a one-way process: the machine is a given, while the material system is adapted or fit within the machine's work envelope. This methodology can also be described as one that dictates a clear boundary between the machine and the material parameters. This linear approach begins to shift when the fabrication equipment becomes more specialized through augmentation of existing machines (Felbrich et al. 2017; Keating et al. 2017) or development of entirely new task-specific ones from scratch (Peek 2016; Wood et al. 2019). Task-specificity of robotic hardware affords a revisiting of the conventional one-way machine-to-material relationship, and consideration of both as interdependent parts of one machine-material system resulting from one integrated design process (Yablonina et al. 2021). Presented research contributes to the discourse of machine-material systems by proposing a bespoke digital material and a matching task-specific robotic assembly system (Figure 2).

Digital material, a term coined by Neil Gershenfeld and George Popescu, presents a convenient match to the notion of machine-material systems (Popescu 2007). The discreteness of parts and joints is well suited for repetitive robotic assembly, affording processes where precision and accuracy is achieved through combination of customized robotic hardware and the designed indexing of material joints. Discreteness in material organization also suggests an ability to transpose between physical and digital. The structured environment consisting of digital material building blocks becomes a perfect substrate for the robotic locomotion and operation and suggests a system for temporal change rather than a finite state.

BACKGROUND

The research presented in this paper builds upon precedent work in two primary areas of investigation: digital materials and task-specific fabrication robots.

Digital Material

The notion of digital material emerged along with "mechanical computing, self-replicating machines and transition from analog to digital communication systems" (Tibbits 2012). There are three components that define digital material: finite set of parts, finite set of joints, and assembly

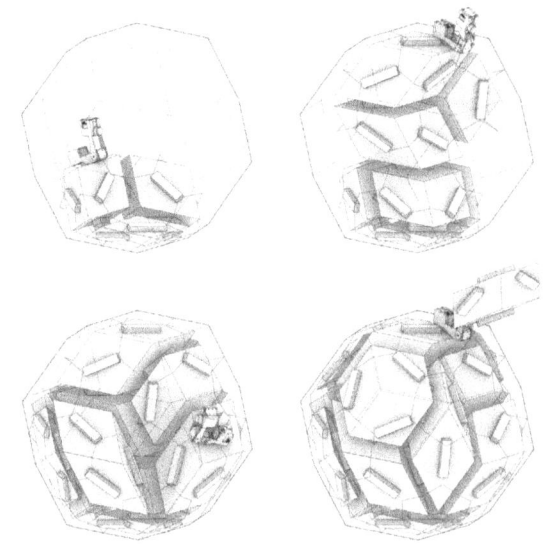

2

process that fully controls the placement of each component (Popescu 2007). Current research in digital material covers a variety of assembly logics, geometries, and joint methods. Assembly logics can be categorized by design approach: top-down or bottom-up design. Within the top-down approach, the designer defines a target geometry and populates it with digital material parts at the required resolution (Leder et al. 2020). Within a bottom-up approach, the designer defines the aggregation rules that define the outcome (Werfel et al. 2014).

Geometry of parts defines the number and the orientation of possible connecting units, thus establishing the design morphospace for the local and the global assembly. The part geometries can consist of homogeneous (Sanchez et al. 2014; Leder et al. 2020; Retsin 2019) or heterogeneous components (Retsin 2016). Variation of part geometries within one system affords increased variability of assemblies and increases the complexity of the design and assembly processes.

Geometry of digital materials can be further categorized as periodic and aperiodic. Periodic geometry and its derived assembly logic naturally fulfills the criteria for digital material, as seen in various projects both in 2D and 3D (Sanchez et al. 2014; Leder et al. 2020; Rossi and Tessmann 2017; Wood et al. 2019; Jenett and Cheung 2017; Papadopoulou et al. 2017). In this scenario, the ordered aggregation comes with the ease of fabrication, while the assembly logic and subsequent formal variety is constrained by the periodic nature.

Meanwhile, projects that utilize non-packing geometry focuses on combinatorial logic and functional variety (Sanchez 2016; Retsin 2016; Retsin 2019; Leder et al. 2019).

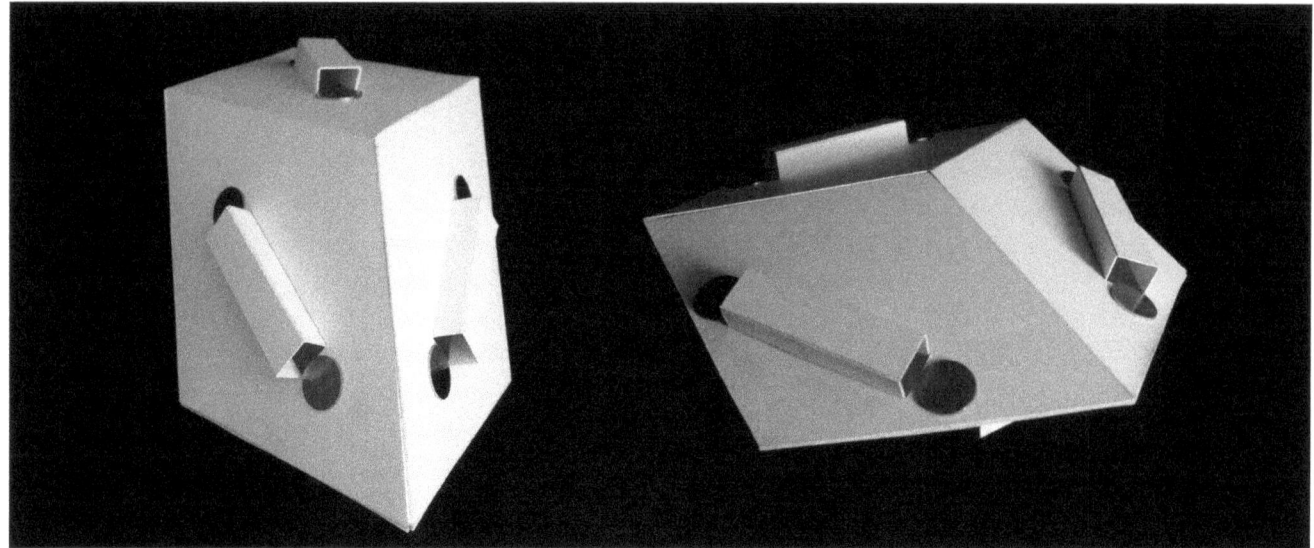

3

Here, intricate assembly logic guides the form generation and entails custom assembly procedure.

Further categorization by joinery methods (Tibbits 2012; Gershenfeld et al. 2015; Sanchez 2014), as well as fabrication material (Popescu 2006) such as 3D printed plastics (Gershenfeld et al. 2015), solid casting (Leder et al. 2020), lightweight forms (Willmann et al. 2012), timber (Leder et al. 2020), wound filament (Wood et al. 2019) and strategy are possible; however, they currently remain outside the scope of this paper.

Quasicrystal

The presented research focuses on a digital material system based on icosahedral quasicrystal—a three dimensional aperiodic crystalline structure that exhibits ordered aggregation but is never repetitive. The geometry set exhibits both ordered aggregation of periodic structure and local variability of aperiodic structure. Its quasicrystalline system departs from conventional lattice structure and suggests new modes of space configuration while achieving perfect packing. Its unique structure has been explored in art and science due to novel material property and aesthetic qualities (Levine and Steinhardt 1984; Bursill and Ju Lin 1985; Aranda\Lasch 2018).

Single-task Fabrication Robots

The field of robotic fabrication in architecture that initially emerged through appropriation of industrial robotic equipment is rapidly expanding through the introduction of single-task robots. These fabrication machines are generally designed and developed along with the design and development of the material and fabrication system, allowing to minimization of robotic hardware and control complexity while fully leveraging the properties of the fabrication material. Single-task fabrication robots have been proposed for a variety of materials and fabrication methods, including 3D printing (Jokic et al. 2014), filament winding (Kayser et al. 2019), strut assembly (Leder et al. 2019; Melenbrink et al. 2017), as well as digital material assembly (Wood et al. 2019; Jenett and Cheung 2017; Gershenfeld et al. 2015; Werfel et al. 2014). Additionally, some examples of machine-material systems where robotic actuation is embedded in the material part itself have been explored (Spyropoulos 2013); however, the cost efficiency and maintenance requirements of such systems seem to outweigh the benefits.

Examples of systems that combine single-task robotics and designed digital materials clearly demonstrate the benefits of the approach. The robotic system's simplicity afforded by its limited functionality makes it more accessible and deployable in various non-lab scenarios. In turn, digital material systems, unique in their reversibility and ability to be configured, benefit from a single-task robotic assembly approach that offers possibilities for temporal reconfiguration of the assembly. Finally, the digital machine-material system opens space for speculative discussion of the co-design approaches (Knippers et al. 2021) and automated futures of the architectural practice (Claypool 2019).

METHOD

The presented research demonstrates a digital machine-material system that consists of designed digital material parts and custom single-task robots capable of assembly of structures based on designer defined target geometry. The research spans five areas of investigation: digital material geometry, aggregation logic, robotic hardware design, assembly sequencing, and robotic control and path planning. It is critical to highlight that while the following section describes each of the areas separately, the development has been conducted in parallel following a co-design strategy wherein emerging parameters of one aspect inherently become input variables for others.

Geometry
Four types of golden zonohedra are employed in this research based on Socolar-Steinhardt tiling (Socolar and Steinhardt 1986). Golden zonohedra are further decomposed to one subunit pair: obtuse golden rhombohedron and acute golden rhombohedron (Figures 3, 4).

Aggregation
The technique used in this research to achieve aperiodic tiling follows Madison's approach using inflation/deflation and subsequent decoration (Madison 2015). This approach allows us to only consider geometric aggregation without relying on other methods such as higher-dimensions or strip-projection of the Ammann grid which is beyond the scope of this research (Socolar and Steinhardt 1986). The method is based on quasicrystal's fractal nature and self-similarity: any arbitrary "finite quasicrystalline fragment" (Madison 2015) can generate infinite icosahedral tiling through subdivision. Here, the substitution rule or decoration describes a specific organization of smaller scale units that can pack 3D space face-to-face and without gaps between the larger scale units (Figure 5). Using the proportion between parent and child, a cube of golden ratio, the inflation/deflation technique is an alternative of the fractal method without infinitely getting smaller. Following the proposed decoration procedure, the infinite aperiodic aggregation was achieved.

Within the selected aggregation method, the development of a joint system became crucial for the successful implementation of robotic assembly. As the system consists of two types of parts (Figure 4), joints must be arranged such that they match in any aggregation scenario (Figure 7). Moreover, the joints serve as connectors for the robotic end effectors, and thus must meet the criteria of minimal degrees of freedom of the robotic assembler. Passive magnetic connectors were installed at the joint points to ensure ease and self-alignment of the assembly process.

Robotic Hardware Design
The design criteria for the development of the assembly robot were its ability to pick up, carry, locate, and attach units within the digital material system. Additionally, minimizing the number of actuators while maintaining the outlined functionality was the main priority in the design process. The geometry of the kinematic system of the developed robotic assembler mirrors the geometry of the

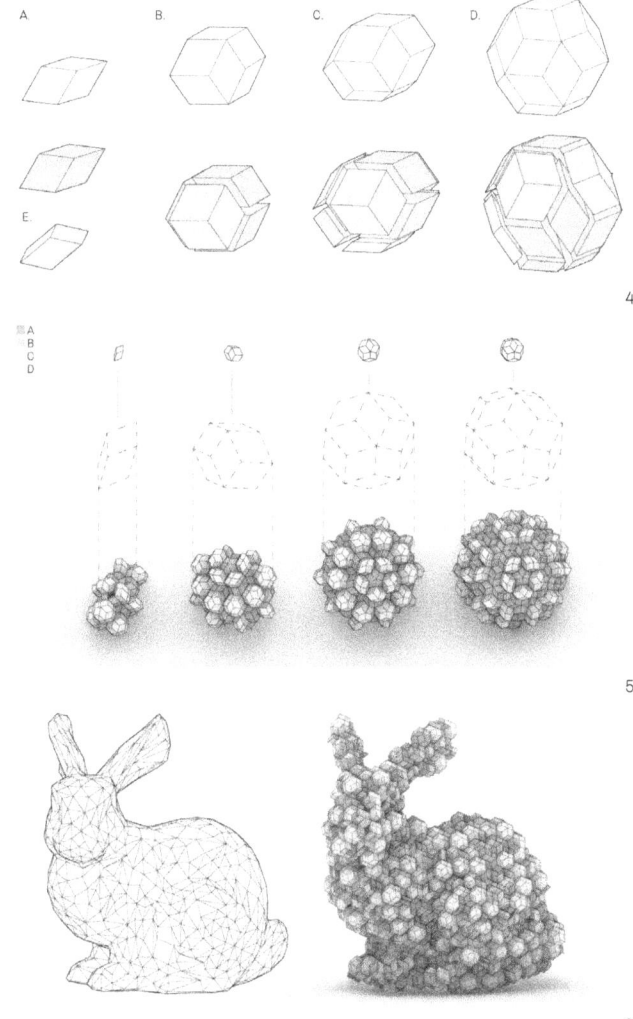

4 A) Acute golden rhombohedron, B) Rhombic dodecahedron, C) Rhombic icosahedron, D) Rhombic triacontahedron, E) obtuse golden rhombohedron and acute golden rhombohedron

5 First, the unit cell is inflated by the factor of the cube of golden ratio; then, the cell is deflated into original size and decorated to the special orientation

6 Quasicrystal counterpart of the Standford Bunny; any geometry can be transposed to quasicrystal clusters in various resolution using culling method

material system: a two-link kinematic chain, wherein the links match the distance from joint point to the edge of the material surface (Figures 7, 8, 9).

Assembly Sequencing, Path Planning, and Control
The robotic assembly process consists of a sequence of locomotion and placement routines, each in turn composed of a sequence of motor actuation commands (Figures 10, 11).

To calculate the global sequence of locomotion and placement routines necessary to assemble a target geometry,

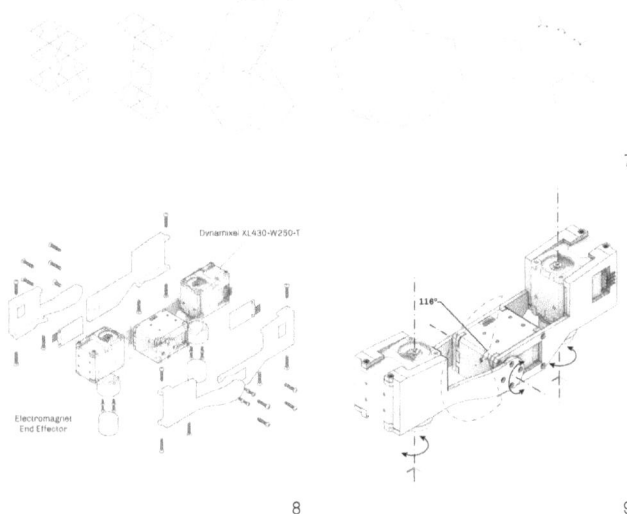

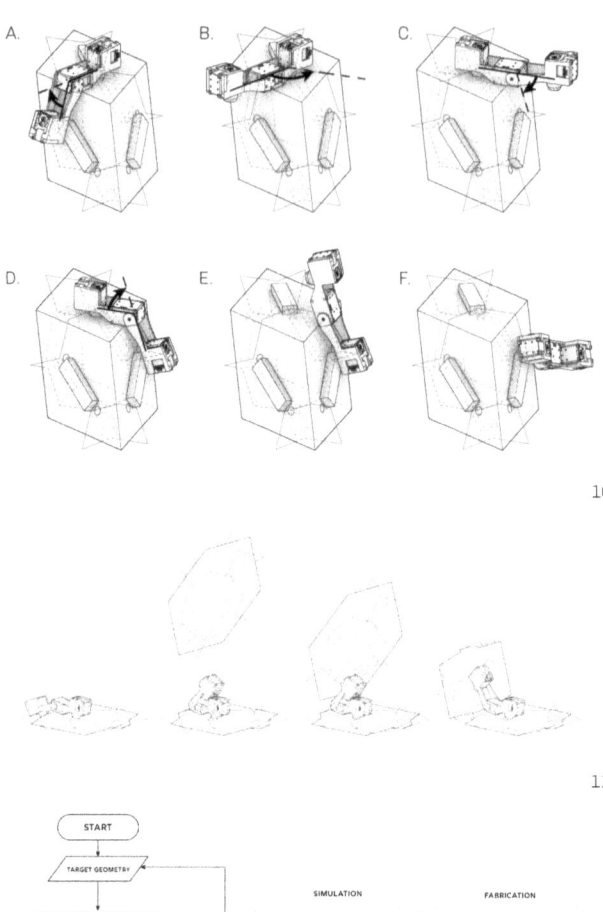

7   The point at the intersection of two tangents from neighboring edges was identified as a promising joint location providing a stable two-point connection at each surface and allowing for a 3-degree of freedom robotic assembler

8   The kinematic chain is equipped with a rotational end effector and an electromagnetic gripper that allows to pick, carry, and place material units within the assembly that is suitable in the domain of small-scale prototype

9   Work envelope diagram

a custom computational tool was developed (Figure 12). The tool consists of three operations: assembly sequence calculation, simulation, and robotic command output. First, the assembly sequence of unit placement is calculated based on the user-defined target geometry of the overall assembly (Figure 13). The core criteria for the assembly sequence calculation were to avoid closed path loops which result in singularities. The calculation was performed iteratively for each unit: first, one of the units at the periphery of the assembly is selected as the starting point; then all neighboring units are evaluated to determine which one would require the least amount of robotic locomotion steps to be reached. Once the second unit is identified, the operation repeats for its neighbors to identify the third unit in a sequence. Once the path is calculated, a list of motor commands for robotic locomotion and placement routines can be calculated. Robotic poses can then be simulated digitally and confirmed by the user prior to being sent to the robot control microcontroller. Once the sequence is confirmed, the motor commands for each individual routine are sent to the microcontroller line-by-line, requiring operator confirmation at each step.

RESULTS AND DISCUSSIONS

The proposed machine-material system was designed, prototyped, and evaluated at the scale of a three-unit

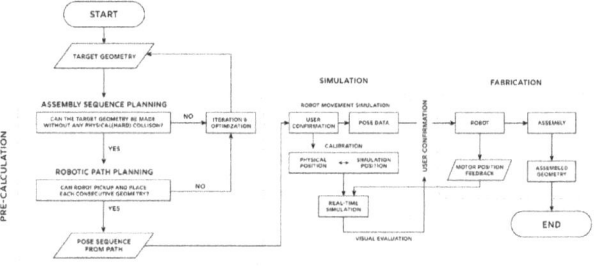

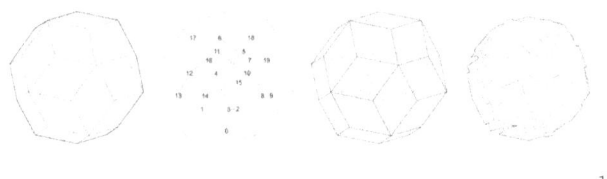

10 A) Initial bend away from first attachment point; B) Rotate along the top plane using second attachment point; C) Bend towards the third attachment point; D) Next bend using the third attachment point; E) Rotate along the right plane using third attachment point; F) Bend towards the next attachment point

11 To add a new material unit to the assembly, the robotic system picks up the material unit, navigates to the desired attachment location, and places it in the assembly; at its current stage of development, the system relies on the human operator to supply the material to the robotic system

12 Aggregation, assembly sequence calculation, and robotic control output workflow diagram

13 Assembly sequence calculation of 20 units

assembly with one single-task robot. The physical robotic system has successfully performed locomotion and placement commands based on the control sequences generated

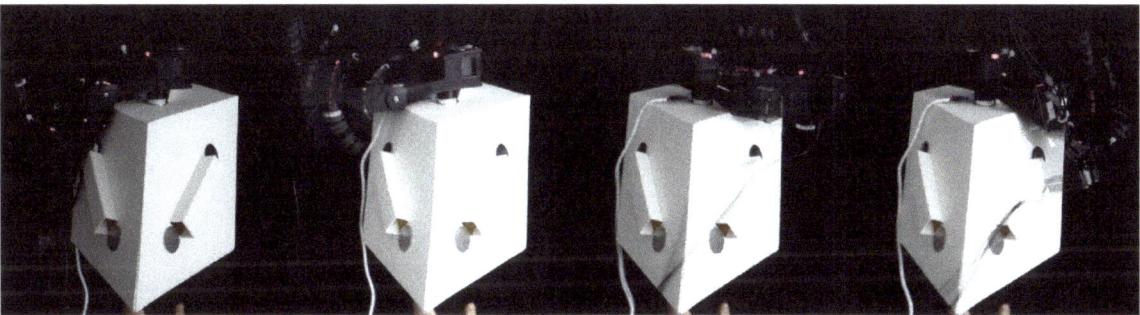

14 Physical test of the robotic system's locomotion routine

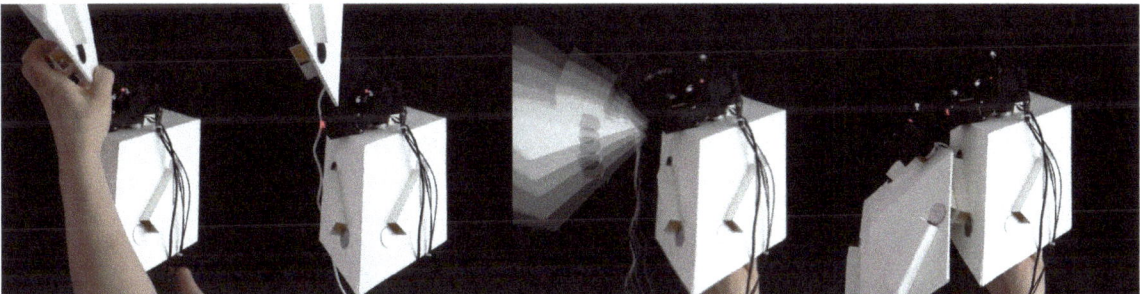

15 Physical test of the robotic system's assembly routine

by the design tool. Physical tests of robotic locomotion and placement routines were performed in a semi-autonomous mode with operator confirmation required at every routine sequence (Figures 14, 15). Future development—in integration of real-time sensor feedback to close the gap between the simulation and fabrication—is necessary to afford a higher degree of autonomy of the fabrication process. Further, cooperation between more than one machine is required to further streamline the material feeding, delivery and assembly sequence. Specifically, a material feeding routine is the next logical step in the development of this project.

Digital material system using icosahedral quasicrystal geometry was explored in both digital and physical space (Figure 16). The complexity of its geometrical construct successfully harmonized with the parallel development process. However, shortcomings include fabrication of mathematically pure part geometries, weak structural stability in vertical load cases due to angular connections, and complexity in assembly procedure. Future development calls for a new material using underlying mathematical logic that can balance between ease of fabrication and its unique characteristic.

The computational design and robot command generation system was tested and evaluated with assemblies of up to 20 units and was identified to be successful. Assembly sequences generated by the computational design tool were evaluated in simulation. While some sequences appeared to be counterintuitive at the first glance, they proved to be more efficient compared to human designed options when evaluated according to robot operations per unit criteria.

OUTLOOK

The presented research contributes to the discourse of robotic fabrication with digital materials by suggesting a co-designed machine-material system wherein robotic hardware and material system match each other's affordances and counteract the limitations. This integrated approach to geometry, material, robotic hardware, and software design poses an alternative to the approach of tool conversion from industry to architectural research tasks. Although the presented robotic system is not proposed to compete with a generic robotic arm or a 3D printer, it does expand the possible design space in digital fabrication, affording a process that would be hard if not impossible to execute with conventional equipment. Future development and integration of machine-material and single-task robotic systems has the potential to further expand the design space of digital fabrication in architecture, especially regarding systems designed for reconfiguration, adaptation, and reuse such as temporary public structure.

Small single-task robotic system clearly has its drawbacks: low payload, low reach, limited degrees of freedom. However, when combined with the material system that is explicitly designed to leverage the machine's properties, the limitations become beneficial to the overall system. For instance, lightweight material in combination with a low payload machine are a perfect match for a fabrication process that can happen in situ, in parallel with human activity in the space: something for which an industrial machine would be completely unsuitable.

REFERENCES

Aranda\Lasch. 2018. "Ragged Edges: The Story of the Quasicrystal." In *Lineament: Material, Representation, and the Physical Figure in Architectural Production*, edited by G. P. Borden and M. Meredith, 1st ed. New York: Routledge. 76–80. https://doi.org/10.4324/9781315680392.

Bursill, L.A., and P. Ju Lin. 1985. "Penrose Tiling Observed in a Quasi-Crystal." *Nature* 316 (6023): 50–51. https://doi.org/10.1038/316050a0.

Claypool, M. 2019. "Our Automated Future: A Discrete Framework for the Production of Housing." *Architectural Design* 89 (2): 46–53. https://doi.org/10.1002/ad.2411.

Felbrich, B., M. Prado, S. Saffarian, L. Vasey, J. Knippers, and A. Menges. 2017. "Multi-Machine Fabrication: An Integrative Design Process Utilising an Autonomous UAV and Industrial Robots for the Fabrication of Long-Span Composite Structures." In *ACADIA '17: Disciplines [and] Disruption; Proceedings of the 37th Annual Conference of the Association for Computer Aided Design in Architecture*, edited by T. Nagakura, S. Tibbits, M. Ibañez, and C. Mueller. Cambridge, MA. 248–59.

Gershenfeld, N., M. Carney, B. Jenett, S. Calisch, and S. Wilson. 2015. "Macrofabrication with Digital Materials: Robotic Assembly." *Architectural Design* 85 (5): 122–27. https://doi.org/10.1002/ad.1964.

Jenett, B., and K. Cheung. 2017. "BILL-E: Robotic Platform for Locomotion and Manipulation of Lightweight Space Structures." In *25th AIAA/AHS Adaptive Structures Conference. Grapevine. Texas:* American Institute of Aeronautics and Astronautics. https://doi.org/10.2514/6.2017-1876.

Jokic, S., P. Novikov, S. Jin, S. Maggs, C. Nan, and D. Sadan. 2014. 'Small Robots Printing Big Structures." Accessed October 10, 2021. http://robots.iaac.net/.

Kayser, M., L. Cai, C. Bader, S. Falcone, N. Inglessis, B. Darweesh, J. Costa, and N. Oxman. 2019. "Fiberbots: Design and Digital Fabrication of Tubular Structures Using Robot Swarms'. In *Robotic Fabrication in Architecture, Art and Design 2018*, edited by J. Willmann et al. Cham: Springer. 285–96. https://doi.org/10.1007/978-3-319-92294-2_22.

Keating, S.J., J.C. Leland, L. Cai, and N. Oxman. 2017. "Toward Site-Specific and Self-Sufficient Robotic Fabrication on Architectural Scales." *Science Robotics* 2 (5): eaam8986. https://doi.org/10.1126/scirobotics.aam8986.

Knippers, J., C. Kropp, A. Menges, O. Sawodny, and D. Weiskopf. 2021. "Integrative Computational Design and Construction: Rethinking Architecture Digitally." *Civil Engineering Design* 3 (4): 123–135. https://doi.org/10.1002/cend.202100027.

Leder, S., R. Weber, L. Vasey, M. Yablonina, and A. Menges. 2020. "Voxelcrete - Distributed Voxelized Adaptive Formwork." In *Anthropologic: Architecture and Fabrication in the Cognitive Age; Proceedings of the 38th ECAADe Conference*, vol. 2. Berlin: TU Berlin. 433–42.

Leder, S., R. Weber, D. Wood, O. Bucklin, and A. Menges. 2019. "Distributed Robotic Timber Construction." In *ACADIA '19: Ubiquity and Autonomy; Proceedings of the 39th Annual Conference of the Association for Computer Aided Design in Architecture*, edited by K. Bieg, D. Briscoe, and C. Odom. Austin, Texas. 510–19.

Levine, D., and P. Joseph Steinhardt. 1984. "Quasicrystals: A New Class of Ordered Structures." *Physical Review Letters* 53 (26): 2477–80. https://doi.org/10.1103/PhysRevLett.53.2477.

Madison, A. E. 2015. "Atomic Structure of Icosahedral Quasicrystals: Stacking Multiple Quasi-Unit Cells." *RSC Advances* 5 (97): 79279–97. https://doi.org/10.1039/C5RA13874D.

Melenbrink, N., P. Michalatos, P. Kassabian, and J. Werfel. 2017. "Using Local Force Measurements to Guide Construction by Distributed Climbing Robots." In *2017 IEEE/RSJ International Conference on Intelligent Robots and Systems (IROS)*. Vancouver, BC: IEEE. 4333–40. https://doi.org/10.1109/IROS.2017.8206298.

Papadopoulou, A., J. Laucks, and S. Tibbits. 2017. "From Self-Assembly to Evolutionary Structures." *Architectural Design* 87 (4): 28–37. https://doi.org/10.1002/ad.2192.

Popescu, G. A. 2007. "Digital Materials for Digital Fabrication." M.Sci thesis, Massachusetts Institute of Technology, 54.

Popescu, G. A., T. Mahale, and N. Gershenfeld. 2006. "Digital Materials for Digital Printing." In *NIP & Digital Fabrication Conference, 2006 International Conference on Digital Printing Technologies*. 58–61.

Peek, N. 2016. "Making Machines That Make: Object-Oriented Hardware Meets Object-Oriented Software." PhD diss., Massachusetts Institute of Technology. http://hdl.handle.net/1721.1/107578.

Retsin, G. 2016. "Discrete Assembly and Digital Materials in Architecture." ACADIA '16: In *Complexity & Simplicity; Proceedings of the 34th ECAADe Conference*, vol. 1. Oulu. Finland. 143–51.

Retsin, G. 2019a. "Toward Discrete Architecture: Automation Takes Command." In *ACADIA '19: Ubiquity and Autonomy; Proceedings of the 39th Annual Conference of the Association for Computer Aided Design in Architecture*. Austin, Texas. 532–41.

Retsin, G. 2019b. 'Bits and Pieces: Digital Assemblies: From Craft to Automation'. *Architectural Design* 89 (2): 38–45. https://doi.org/10.1002/ad.2410.

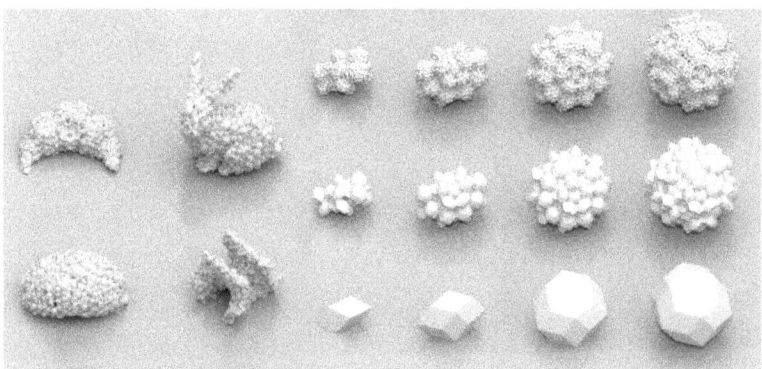

16 Design experiments in variety of assembly geometries, resolutions, and scales

Rossi, A., and O. Tessmann. 2017. "Geometry as Assembly–Integrating Design and Fabrication with Discrete Modular Units." In *Proceedings of the 35th eCAADe Conference: ShoCK!: Sharing Computational Knowledge!*, vol. 2. 201–210.

Sanchez, J., Y. Cai, and S. Ordoobadi. 2014. "Polyomino." In *ACADIA 14: Design Agency; Projects of the 34th Annual Conference of the Association for Computer Aided Design in Architecture.* 95–98.

Sanchez, J. 2014. "Polyomino: Reconsidering serial repetition in combinatorics" In *ACADIA '14: Proceedings of the 34th Annual Conference of the Association for Computer Aided Design in Architecture.* Ontario: Riverside Architectural Press.

Sanchez, J. 2016. "Combinatorial design: Non-parametric computational design strategies" In *ACADIA '16; Proceedings of the 36th Annual Conference of the Association for Computer Aided Design in Architecture.* 44–53.

Socolar, J. E. S., and P. J. Steinhardt. 1986. "Quasicrystals. II. Unit-Cell Configurations." *Physical Review B* 34 (2): 617–47. https://doi.org/10.1103/PhysRevB.34.617.

Spyropoulos, Theodore, ed. 2013. *Adaptive Ecologies: Correlated Systems of Living.* London: Architectural Association.

Tibbits, Skylar. 2012. "From Digital Materials to Self-Assembly," In *Proceedings of the 100th Annual ACSA Conference: Digital Aptitudes + Other Openings.* Washington: ACSA Press. 232–237.

Werfel, J., K. Petersen, and R. Nagpal. 2014. "Designing Collective Behavior in a Termite-Inspired Robot Construction Team." *Science* 343 (6172): 754–58. https://doi.org/10.1126/science.1245842.

Willmann, J., F. Augugliaro, T. Cadalbert, R. D'Andrea, F. Gramazio, and M. Kohler. 2012. "Aerial Robotic Construction towards a New Field of Architectural Research." *International Journal of Architectural Computing* 10 (3): 439–59. https://doi.org/10.1260/1478-0771.10.3.439.

Wood, D., M. Yablonina, M. Aflalo, J. Chen, B. Tahanzadeh, and A. Menges. 2019. "Cyber Physical Macro Material as a UAV [Re]Configurable Architectural System." In *Robotic Fabrication in Architecture, Art and Design 2018*, edited by J. Willmann et al. Cham: Springer. 320–35. https://doi.org/10.1007/978-3-319-92294-2_25.

Yablonina, M., N. K. Kalousdian, and A. Menges. 2021. "Designing [with] Machines: Task- and Site- Specific Robotic Teams for in-Situ Architectural Making." In *ACADIA '20: Distributed Proximities; Proceedings of the 40th Annual Conference of the Assocoation for Computer Aided Design in Architecture*, edited by V. Ago, B. Slocum, A. Marcus, S. Doyle, M. del Campo, and M. Yablonina.

IMAGE CREDITS

All other drawings and images by the authors.

Donghwi Chris Kang is a former student at the Daniel's faculty of Architecture, Landscape and Design at the University of Toronto. Through his interdisciplinary research dealing with the topics of architecture, digital fabrication, engineering, and robotics, Chris examines ways to faciliate synthesis between architecture and technology, and the subsequent impact on the built environment.

Nicholas Steven Hoban is a computational designer specializing in the field of digital fabrication, robotics, and computational workflows. Utilizing computer programming, simulation, and CADCAM programming, Nicholas delivers data-driven design and prototypes in research and the AEC industry.

Maria Yablonina is Assistant Professor at the Daniel's faculty of Architecture, Landscape and Design at the University of Toronto. Her work lies at the intersection of architecture and robotics, producing spaces and robotic systems that can construct themselves and change in real-time. Maria's practice focuses on designing machines that make architecture— a practice that she broadly describes as Designing [with] Machines (D[w]M). D[w]M aims to investigate and establish design methodologies that consider robotic hardware development as part of the overall design process and its output. Through this work, Maria argues for a design practice that moves beyond the design of objects towards the design of technologies and processes that enable new ways of both creating and interacting with architectural spaces.

Automating Bi-Stable Auxetic Patterns for Polyhedral Surface

Zhenxiang Huang
Cornell University

Yu-Chou Chiang
TU Delft

Jenny E. Sabin
Cornell University

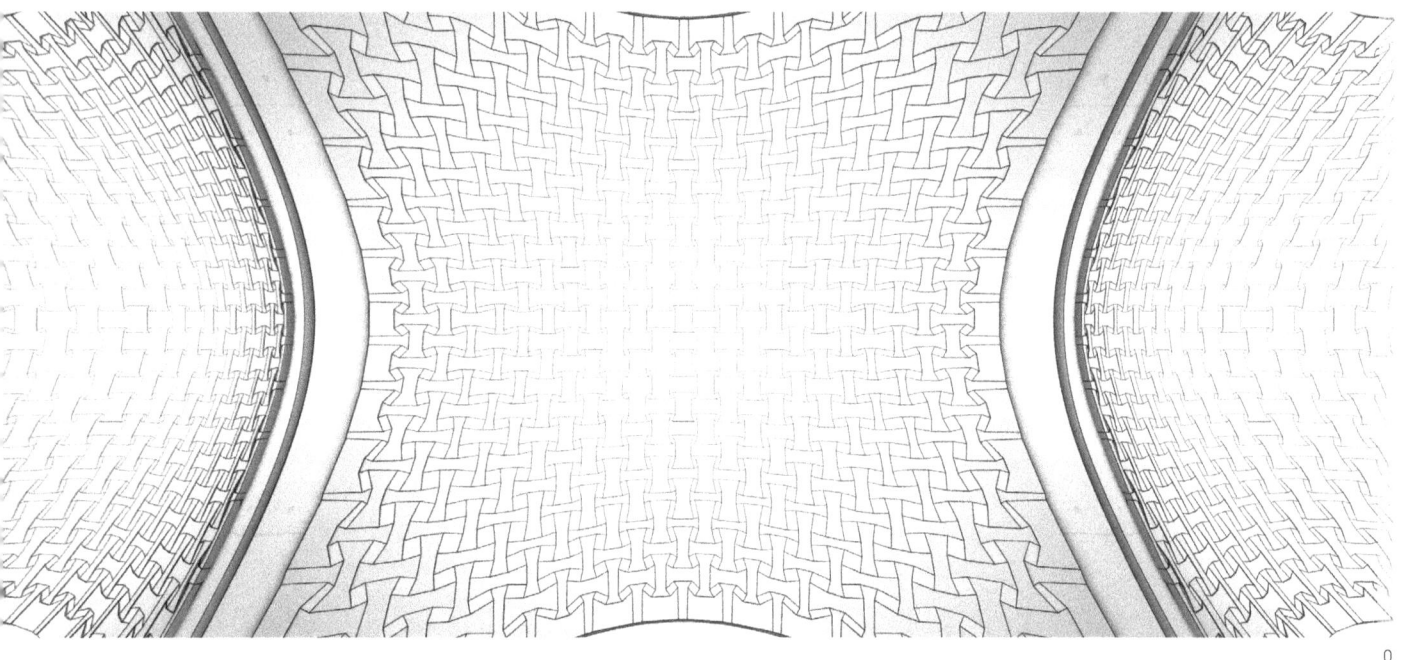

0 Perspective ceiling view of a proposed architectural dome surface realized by bi-stable auxetic patterns

ABSTRACT

Bi-stable auxetic structures, a novel class of architected material systems that can transform bi-axially between two stable states, offer unique research interest for designing a deployable stable structural system. The switching behavior we discuss here relies on rotations around skewed hinges at vertex rotating connectors. Different arrangements of skewing hinges lead to different local curvatures.

This paper proposes a computational approach to design the self-interlocking pattern of a bi-stable auxetic system that can be switched between flat and desired curved states. We build an algorithm which takes a target synclastic polyhedral surface as input to generate the geometrical pattern with skewing hinges. Finally, we materialized prototypes to validate our proposed structures and to exhibit potential applications.

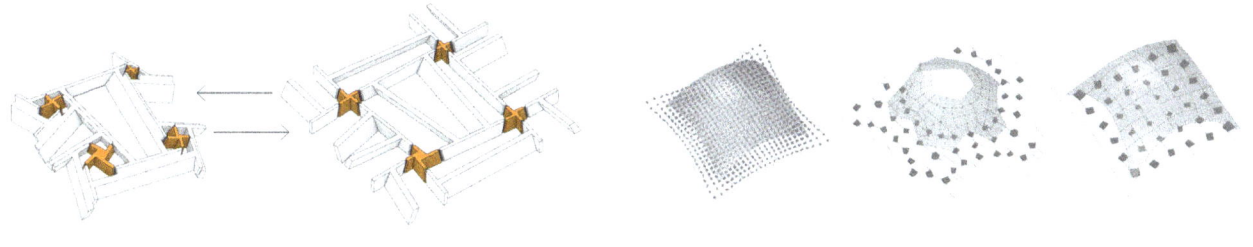

INTRODUCTION

Deployable structures are widely applied across many different disciplines, such as antenna design or emergency shelter construction (Häuplik-Meusburger 2011; Shah et al. 2019). From the architectural perspective, increasing demand for geometrically complex architecture pushes innovations of efficient building formworks to save fabrication time, labor, and cost. Deployable structures, which can transform from a flat state to the desired target geometry, are plausible solutions and have thus received intensive research interest for many decades.

Although many designs of deployable structures have focused on regular patterns and repetitive mechanical joints in the past, recent work has leveraged geometric knowledge and advanced manufacturing technology to produce more sophisticated systems to create free-form target geometry. By compositing material with different mechanical properties (e.g., stiffness, expansion ratio) or designing mechanical linkages, structural systems can transform from the original fabricated state to a target configuration state by simply applying an external trigger.

This investigation is inspired particularly by two earlier studies: Rafsanjani and Pasini (2016) presented a family of bi-stable auxetic mechanisms that have homogeneous expansion rates; Ou et al. (2018) identified how to design spatial transformation unit based on skewed rotational axes.

Contributions—We introduce a fully automated algorithm to generate parameterized bi-stable auxetic patterns for target synclastic geometry. We investigate a hinge-based deployable structure systems to achieve bi-stable shape reconfigurations from flat to curved state. We achieve the flat-to-curved transformation by skewing hinge directions (Figure 1). The resulting pattern in flat configuration includes cutting paths along 2D curves with various inclinations. We further show the geometric constraints brought by the bi-stable and auxetic mechanisms, especially within the design of a vertex star, where one skewing hinge directly constrains the design of its neighborhood hinge along a corresponding edge. Our computational method is tested and validated through various input surfaces (Figure 2).

1   A 3D-printed prototype for proposed system which combines flat pattern from Rafsanjani and Pasini 2016 and spatial rotational axes from Ou et al. 2018

2   Three different input target geometry surfaces and their generated bi-stable auxetic patterns

RELATED WORKS

We survey previous research projects featuring deployable systems in this section with a special focus on work exhibiting external hinge systems as well as the latest research progress on bi-stable auxetic systems.

Kinetic-based Reconfiguration

Researchers have focused on the kinetic behavior of mechanical systems. They applied linear and spherical hinge joints, which respectively allow 1 and 3 degrees of freedom (DoFs). Origami is perhaps the most famous reconfigurable system involving linear hinges (Tachi 2011). For the making of more free-form surfaces, cuts or slits can be introduced to the origami folding system. The result is called kirigami. Researchers have demonstrated its ability to make any free-form surfaces by programming the patterns properly (Liu et al. 2018, 2019; Jiang 2020). Rotational ball hinges can provide more DoFs in transformation, making it easier to achieve target free-form geometry. For instance, Konaković et al. (2016) presented how to design a free-form surface using the triangular auxetic pattern from conformal mapping and geometry optimization.

Bi-stable Mechanism

Mechanical bi-stability is a system that is also termed as snap-through buckling (Vahidi and Huang 1969). The term reflects on the features of how the system is switched from one stable state to another. From an energy point of view,

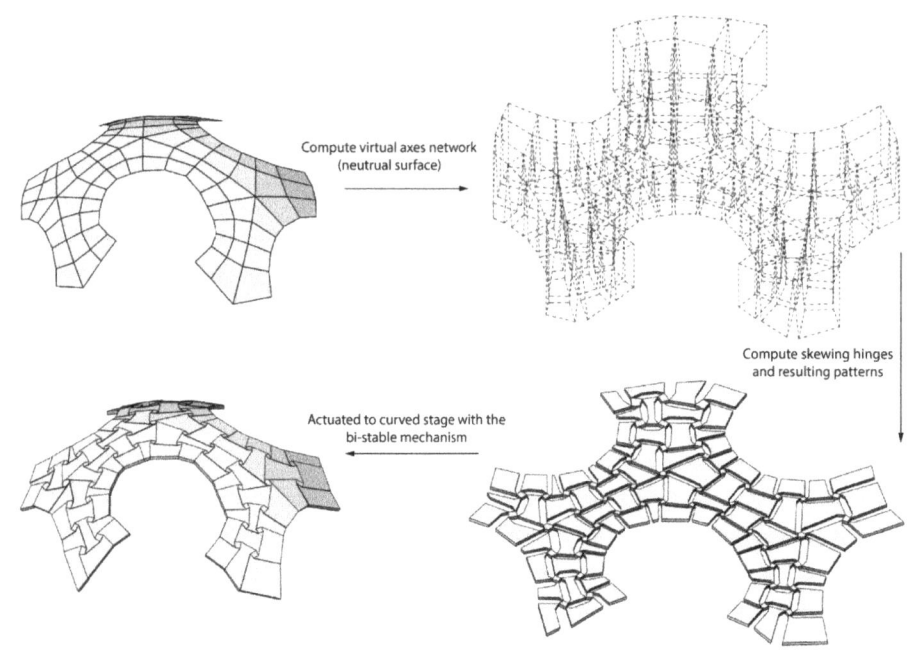

3   General workflow of designing bi-stable auxetic patterns: We first take a polyhedral surface with synclastic curvature as input, then compute the virtual axes network that each face will rotate against. The rotary network satisfies the condition of 'neutral surface' to the input geometry so that rotation at each edge will have bi-stable properties. Subsequently, we compute skewing hinges at each rotary vertex to generate resulting patterns. Finally, the geometry with such patterns will push back to curved state given external forces.

as illustrated in Figure 4 with the basic bi-stable unit, the stable states correspond to the local minimum points of the system's energy-displacement graph. When an external force applies to the system, two members begin to deflect. Until the external load passes the critical point, the unit will suddenly deviate from critical state to the alternative state. After the force is removed, elasticity will bring material back to its rest length, resulting in mirrored geometry to its starting state. To make full use of this geometry feature, the hinges should ideally store no energy in the rest state. In this paper, all the hinges are assumed to behave ideally, in order to design bi-stable mechanisms with geometric principals.

Bi-stable Auxetic System

Recent research has considered auxetic behavior in the bi-stable system, resulting in useful functions to explore a wider range of reconfiguration system. Rafsanjani and Pasini (2016) have shown a surface with periodic bi-stable units that can also be auxetic with a Poisson's ratio of −1. Chen et al. (2021) used parametric cells with different expansion ratios to achieve the curved bi-stable auxetic surface at a fixed boundary. The primary difference in our research is that all the panels can spring back to their original shapes without significant residual strain. We put our primary focus on the geometry of the start and end configurations. Our work guides the deployed state towards a designed shape by adequately arranging the rotation axes. We see our work as a novel demonstration of an inverse algorithm for designing the bi-stable deployable surface system.

METHOD OVERVIEW

This section introduces the fundamental approach and basic design workflow of our deployable system. The whole transformation process can be divided into two hierarchical perspectives: 1) in local reference frames, where two adjacent panels rotate against their corresponding virtual axis to achieve a target dihedral angle; 2) in the world reference frame, where a network created by connecting all the virtual axes goes through an isometric transformation, such as with a mirror relationship to the original stable stage. Combining the two transformations together will deliver a collection of flat faces to a curved configuration. Reversely, an inverse transformation dispatch faces from a curved mesh to discrete pieces in a plane.

The critical principle of deployment in our approach is to program dihedral angles between adjacent panels through out-of-plane rotations from rotating connectors around each vertex. After we generate the layout of different faces in the plane, we can program the hinge locations and directions at each rotator according to the corresponding dihedral angle and the positional constraints from neighboring rotators. This algorithm will be further explained in the *Computational Workflow* section.

Figure 3 illustrates the general workflow of our inverse design method. In the first step, the facets of the polyhedral mesh are distributed onto a plane. Each face can rotate around an axis in a network which lies outside the plane. In a mesh where Gaussian curvature at each vertex is positive, the network exists and is called a 'neutral surface'

(Chiang 2019). In the second step, we map the hinge locations and directions based on the layout faces in a plane. A group of hinges around one vertex forms a rotating connector. If the vertex has positive Gaussian curvature, the rotating connector is a truncated pyramid; if it has negative Gaussian curvature, the rotating connector is a truncated tetrahedron. We calculate the shape of the rotating connector on each vertex and propagate the result from a central vertex to everywhere in the map. These rotating connectors naturally define the shape of panels, by defining either full hexagonal gaps or seams or half hexagonal seams (Figure 5). In the final step, an external pulling or pushing force actuates all the panels to rotate around their rotating connectors towards their second stable stage.

GEOMETRIC DEMONSTRATION
Kinetics of Bi-stable Auxetic Pattern

We analyze the geometry and kinetics behavior in auxetic bi-stable tiling first proposed in Rafsanjani and Pasini (2016) in Figure 6. The auxetic deformation behavior emerges due to the rotation of the colored squares; thus, we call them 'rotating connectors', and we call the rest of the components 'panels'. The proposed patterns create a hexagonal void for each edge. Two ends of the hexagonal void are two "anchor" points that define the mirror axis of the void. This mirrored relationship echoes the nature of bi-stability from a geometric perspective in our previous section.

In planar transformation, like the cases in Rafasanjani and Pasini (2016), all the rotation axes are along the normals of the planes. Jifei et al. (2018) propose that skewing the rotation axis in space produces out-of-plane rotation. Inspired by this, we replaced the homogeneously repeated pattern and the perpendicular cuts with a heterogeneously graded pattern and tilted cuts for flat to curved reconfiguration. By this method, two adjacent panels will form a dihedral angle when the void opens or closes, corresponding to the expansion or contraction stage in the auxetic transformation. The following sections will discuss the geometric behavior of this mechanism.

Edge Transformation: Hexagonal Void with Virtual Hinge

We refer to the research by Chiang et al. (2018), which achieved a target dihedral angle by a bi-stable mechanism system through a configuration of skewed hinges. The author demonstrated important properties of spatial bi-stable transformation within a unit of two adjacent panels: the generalization of the prismatic voids in Rafsanjani and Pasini (2016) into non-prismatc voids. The generalized voids still have heptagonal basis

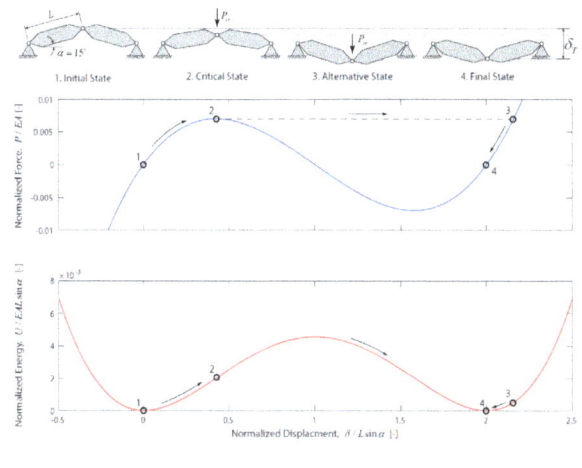

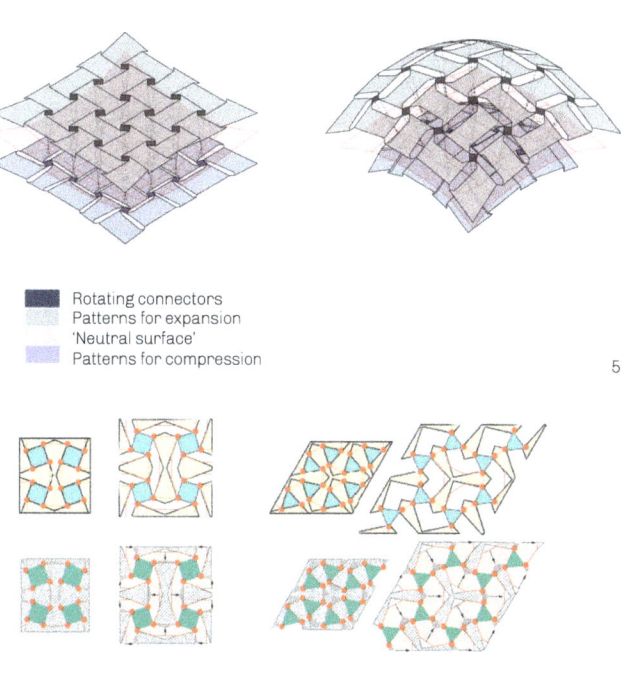

4 The reconfiguration process of the idealized bi-stable mechanisms and their load-displacement response from Chiang (Chiang 2019)

5 By defining hexagonal gaps (bottom part), or half hexagonal seams (top part), the bi-stable auxetic pattern can be actuated through expansion or compression

6 Bi-stable unit in plane contains a set of 'rotating connectors' (green) and a set of 'panels' (yellow), initially studied by Rafsanjani and Pasini (2016)

and are symmetric, but are capable of delivering spatial (i.e. non-translational) transformations (Figure 7). The connecting panels have different rotating arms at the top and bottom surfaces, leading to different displacements at the two surfaces. The length of the rotating arms is proportional to the distances from the rotation axis, so is the displacement.

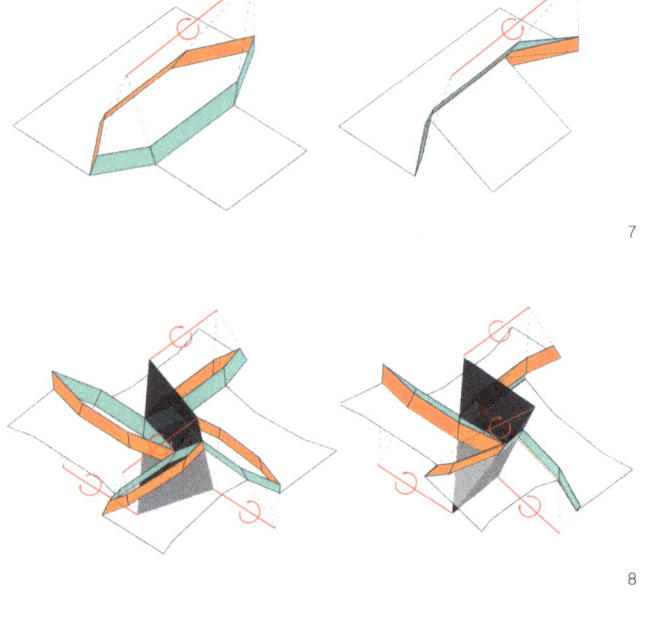

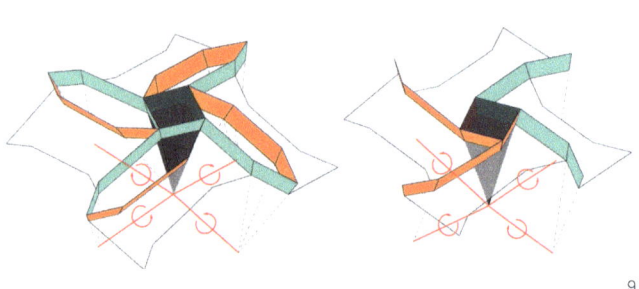

7 Edge transformation is illustrated as closing the gap of a hexagonal void; the closing/ opening reconfiguration is equivalent to rotate against a virtual hinge

8 Vertex transformation is illustrated as rotating around a polyhedron connector: (left/right) open(flat)/closed(curved) stage with a connector at an anticlastic vertex

9 (Left/right) Open(curved) /closed(flat) stage with a connector at a synclastic vertex

To inversely design the hinges of two given adjacent panels (i.e., the dihedral angle is given), we might need to encounter the displacement by extending, trimming, or offsetting the face of some panels. If the two faces stay at the original design position, the virtual hinge lies precisely on the bisecting plane of this dihedral angle to make sure that the flattened result will stay in the same plane. When one of the faces is offset along the normal, the virtual hinge no longer lies in the bisecting plane of two faces.

Vertex Transformation: Frustum Connector

When we reassemble the hexagonal voids along each edge, there are rotating connectors along vertices. In the case by Rafsanjani and Pasini (2016), all the rotating connectors were right prisms since all the hinges are parallel and orthogonal to the plane. In the case of Figure 7, a virtual hinge exists in each hexagonal voids between two faces. Therefore, the connectors are frustums (i.e., truncated pyramids), and the local convexity defines their shapes.

Figures 8 and 9 exhibit different types of rotating connectors at the vertices. Each pair of hinges in the connector will intersect at a point either above or below the frustum, depending on the dihedral angle that its corresponding hexagonal void maps to. When the discrete Gaussian curvature at one vertex is positive, its connector is a frustum, as all hinges around meet at an apex. The scenario is more complex when the dihedral angle around one vertex has different signs, leading to negative Gaussian curvature. In the case that the vertex is at a valence of 4, its connector is a truncated tetrahedron.

Constraints in a Vertex Star: Simultaneous Rotations

The designers must consider the constraints between interrelated hexagonal voids and the connector to arrange the hexagonal voids around a vertex. Each hinge affiliates to two successive hexagonal voids from the relationship in two configurations. Consider the design of a vertex star in Figure 10. Observing the angles around one hinge point in open and closed states gives us two equations:

$$\alpha_n + \beta_n + \theta_n + \omega_{n-1} = 2\pi, \quad (1)$$

$$\alpha_n + \beta_n + (2\pi - \omega_{n-1}) = 2\pi, \quad (2)$$

where α is the interior angle of the rotating connector, β is the angle of the panel attaching to the connector, while ω and θ are the obtuse and acute angle of the hexagonal void, respectively. Subtract equation (2) from equation (1), we deduct the following equation which describes the angle agreement between the two successive voids:

$$\theta_n + 2\omega_{n-1} = 2\pi, \quad (3)$$

Equation (3) connects the degrees of freedom around a vertex. It means the geometry of a hinge is affected by the adjacent ones. In the meantime, the coplanar condition and dihedral angle correspondence given by the Edge Transformation subsection also limit the position of the hinge by the adjacent hinge in its neighborhood vertex.

To make our system more manageable, we introduce the method of unrolling a conical mesh (Liu et. al. 2006; Chiang et. al. 2018) as a guide for those rotating polygons. The conical mesh has nice properties as each vertex has a

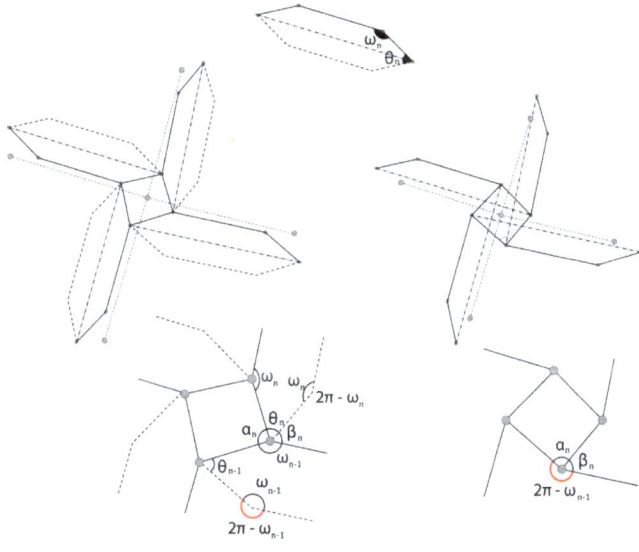

10 Internal angle constraints for hinge positions within a vertex star

normal axis intersected by all the bisector planes of the dihedral angles between surrounding facets, which can be taken as an offset direction for generating our virtual hinges. We will set up the computational workflow based on the conical surface input.

COMPUTATIONAL WORKFLOW

The previous section analyzed the mechanism and geometric features of bi-stable auxetic reconfiguration. Here, we discuss how to automate the design of each hexagonal void and rotating connector. This paper focuses solely on the situation where each virtual rotation axis stays on the same side of the input polyhedral mesh (i.e., synclastic surface). Under such circumstances, the Gaussian curvature is positive, and all hinges around a rotating connector meet at a common point, forming a pyramidal frustum (see *Geometric Demonstration* section). Therefore, all the virtual axes can form a network with a shared point to each vertex, making it easier to compute their positions from the target surface mesh. Figure 11 demonstrates the basic computational workflow for our algorithm which we will explain in detail through the following three parts (Algorithm A in the Appendix).

Computation of Virtual Rotation Axes

The aim of the first part of our algorithm is to compute the location of all the virtual rotation axes during the reconfiguration. To ensure all the polygons stay in the same plane after unrolling, the network of virtual axes must follow the edge directions in the 'neutral surface' to the original conical mesh, as demonstrated by Chiang (2019). All the normal vectors of the mesh face on the 'neutral surface' have half as many polar angles as their corresponding faces in a polar coordinate system whose z-axis points towards the common plane after face unrolling. Here we provide an algorithm to compute the neutral surface, which expands and augments the original research (Algorithms B and C in the Appendix).

After computing the neutral surface, we can mirror it against the plane defined by the z-axis of the polar coordinate system to get its reference position after reconfiguration. Since the relative positional relationship stays unchanged in the reconfiguration, we can then unroll each polygon from the target surface to separate planar positions by orienting through its virtual rotational axes.

Computation of Skewed Hinges

The second part of our algorithm is to locate the hinges in the rotating connectors. The hinges will transform the unrolled panels that are computed in the first phase. The hexagonal voids created by those rotators bring all scattered panels in the flat configuration back to the initial curved configuration.

Figure 13 shows how we locate the aligned hinge positions at a vertex through an internal loop. The loop cycles depend on the vertex valence. Each hinge should also pass through the merged point on the neutral plane. All the hinges should also be limited in the bisector plane to the reference rotation angles centered at the merging point to glue each unrolled vertex together. Recall the angle constraints in Equation (3). Since θ_n is equivalent to the dihedral angle of the associated edge at curved stage due to the desired edge transformation, we can get ω_{n-1} for each hexagonal void. With each hinge fixed to the bisector plane, once we know the location of one hinge, we can compute the adjacent hinge by finding a direction from its constrained plane so that this hinge connects the calculated hinge and its associated edge at the 'neutral surface' to a dihedral angle ω_n. Algorithm D in the appendix shows how we compute it through linear algebra. As we loop through to the last hinge, the same operation should bring that hinge to the exact position at the starting point.

However, this looping algorithm may need different input variables to execute, subject to the DoFs limited by the vertex's neighborhood condition. The initial input without external constraints has three DoFs: two for pivot vector direction and one for angular coordinates at the bisector plane. Each time we fix a vertex's rotation connector, we eliminate one DoF of its neighborhood. Figure 14 explains the different cases of this input with DoFs at 3,2,1, respectively.

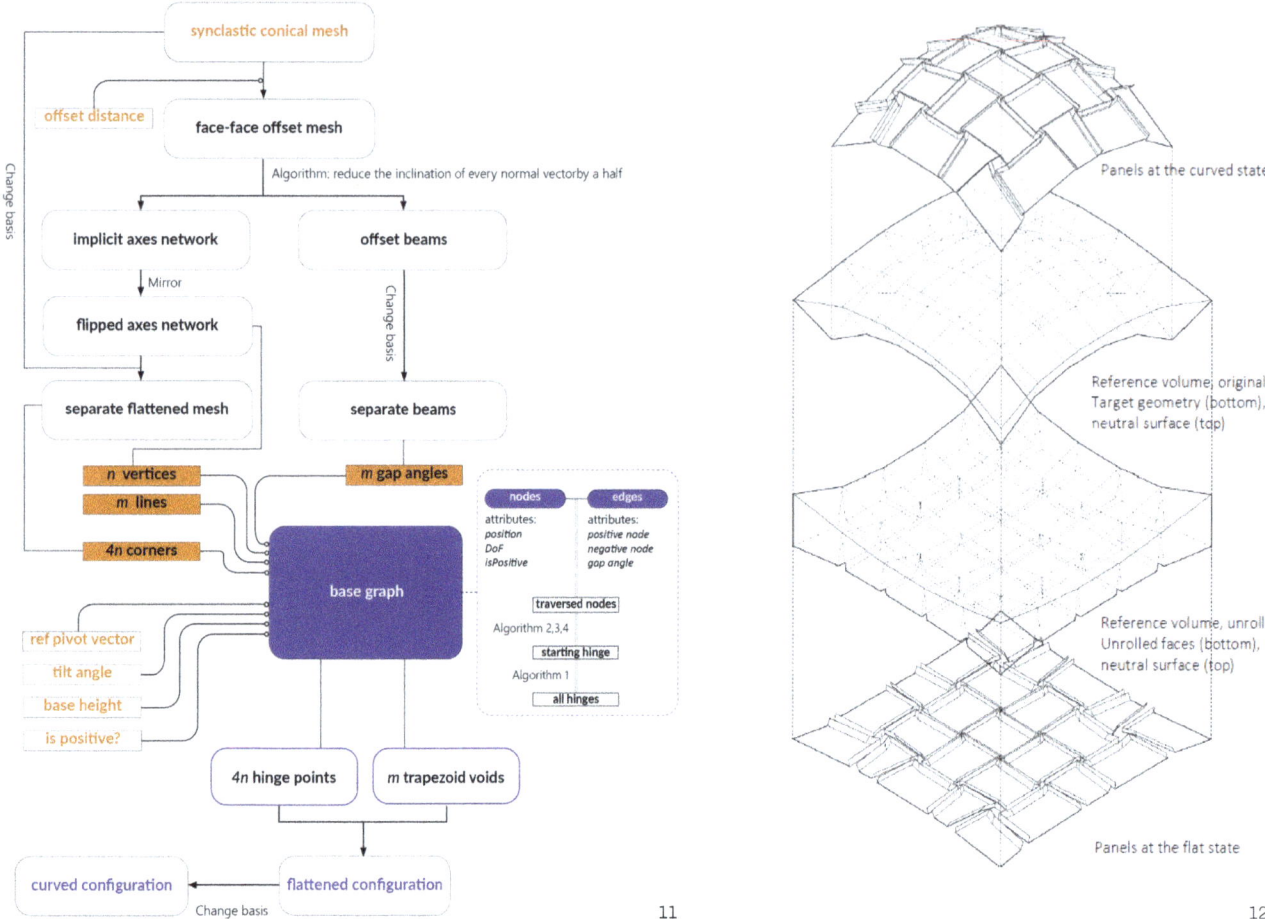

11

12

To assume a vertex will never have a zero-input variable, we apply a spanning tree to consecutively fix the hinge positions at each vertex across the mesh. The propagation follows a very straightforward approach by placing neighborhood vertices in the candidate bag and removing executed vertex from the pool (Algorithm E in the Appendix). It is noticed that the boundary vertices come at the very last step of this enumeration because they have fewer limits to constrain the hinge directions.

Computation of Final Configuration States

The final part of our algorithm is to form the patterned panels by connecting the designed hinges. We can design two patterns from the same set of skewed hinges based on whether the flat stage is open or closed, as shown in Figure 15. If we want to actuate the system by compressed stress, the flat pattern should be open with hexagonal voids. Here we can mirror each edge of the rotating connectors to evaluate the resting positions for those hinges after the bi-stable transformation phase and connect adjacent hinges and resting positions along corresponding edges to form those hexagonal voids and thus all the panel geometries. In the other situation where an open curved configuration is desired, we connect the hinge positions in an orthogonal way and leave no gap in the flat pattern. After we get those panels, we use the transformation introduced in *Geometric Demonstration* section to move the patterns from the plane to the curved configuration.

This section presents the computational methods to design bi-stable auxetic patterns for a target synclastic polyhedral surface. We provide a pivotal diagram showing the relationship between the reference polyhedral geometry, virtual axes, and our designed results (Figure 12).

RESULTS

Evaluation and Optimization

The proposed workflow has been tested on several architectural surfaces as shown in Figure 2. Important criteria to evaluate the performance of our workflow are the offset distance at each panel edge from its original edge and the acute angle at each hexagonal void. Those two values are subject to the influence of both intrinsic geometric properties and the input variables in our algorithm. We introduce a non-linear optimization strategy, namely through the guided projection method (Tang 2014) to adjust those variables to better design the patterns.

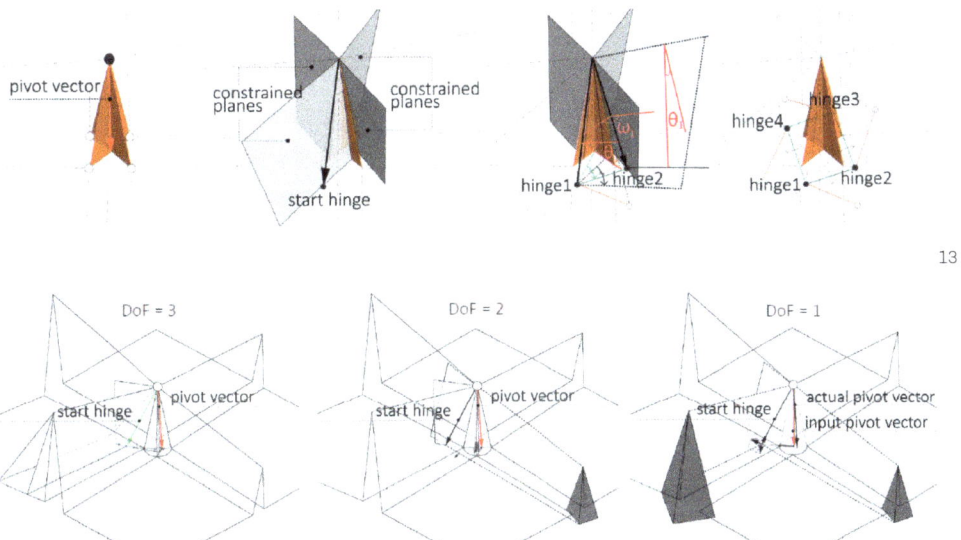

11 Computational workflow of bi-stable auxetic design

12 Correspondence between target and reference geometry and the bi-stable auxetic patterns

13 Looping algorithm to align hinge positions inside a rotating connector. (Left to right) (1) Determine isosceles triangles by pivot direction; (2) draw bisector planes of the tip angle, determine direction of the starting hinge; (3) get next hinge direction according to the constraint; and (4) loop (n-1) times (n = vertex valence) to get all the hinge positions

14 Different DoFs in a rotating connector design

Fabrication Results

We used CNC milling to materialize the pattern generated by our algorithm. We paid careful attention to the drilling width as a design factor. The diameter of the drill bit leads to round corners for acute angles of those hexagons, resulting in unwanted gaps between voids. Hence, we offset the edges so that the drill can create usable living hinges (Figure 16). Parameters are scripted and tested before finding the proper hinge width values at 0.3mm and offset distance at 1.0 mm to fabricate 1/8in-thick polypropylene sheet. The final prototype model fabricated by a 6-axis CNC machine is shown in Figure 17.

DISCUSSION

Structural Behavior

During the process of reconfiguration, all the hinges will endure high centralized stress and strain. However when the second stable state is reached, all the panels spring back to their resting length. Particularly, when the curved state is in contraction (Figure 15a), beams of each panel will touch on their neighborhood due to the well-designed mirrored features in our algorithm. Therefore, all of its self-weight can pass through those contact surfaces. If the target geometry is a compression-only shell, this system is capable of large gravitational loads, such as freshly poured concrete. This feature makes our system viable to serve as potential flexible, lightweight, economical and deployable formworks.

Architectural Speculation

Our proposed system can be applied in various architectural conditions. Importantly, it can serve as a deployable formwork, as it fits the structural behavior we just discussed. This type of system has been of recent interest in research, including knit-cable systems (Popescu et al. 2018) and inflatable systems (Panetta et al. 2021). Compared to the knit-cable and fabric techniques, which can only serve for minimal surfaces, our system provides solutions for compressive structures with positive-Gaussian curvature. Uniquely to those approaches, our system provides a network of repetitive geometric motifs. The interlocking pattern functions mechanically as a deployable system and provides its unique aesthetic qualities. Figure 18 shows the workflow of building a concrete shell structure through our formwork system. The volumes with gaps can be prefabricated in the factory as an injection mold, into which an elastic material like silicon rubber can be poured to fabricate the large-scale molding cast. On the construction site, a wheel-track system can provide actuation. The mold in flattened configuration becomes a curved vault after its anchor points are dragged inward. Mortar and structural concrete are then poured on the top of it, leaving the Guastavino-like interlocking texture in the bottom. This elastic formwork can be removed after giving cuts alongside the four mirror axes and flattened individually. Thus, the molding can be recycled for reproduction.

Another interesting application is to build more efficient responsive screens. Our adaptive system allows for the production of unit components that can be switched from flat to various curved shapes. More importantly, with the bi-stable mechanism in our system, energy is only needed during the transformation phrases. This is a significant advantage over the lengthy actuation process for most

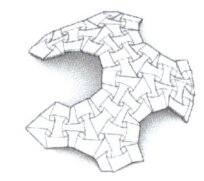
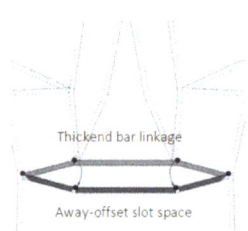
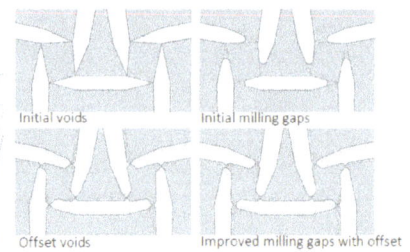

15

16

17

existing kinetic façade or rooftop systems. For more general purposes, our system also serve as an alternative option for emergency or disaster relief shelters and as demonstrators in pavilion design (Figure 19).

Limitations and Future Work

Our research work gives primary focus on to geometrical features at start and end states, assumes all the hinges rotate freely during the reconfiguration process. The system needs extra design on hinges to resolve the stress concentration for large-scale applications. In the future, it may be helpful to add simulation and FEA analysis focusing on the hinge detailing, dynamic effects during the transformation, and structural performance in the deployed stated.

Although we demonstrated that the vertex transformation can cover both positive and negative Gaussian curvature, the nature of a 'neutral surface' limits our reference virtual hinges to the one side of our target geometry. Thus, our algorithm can only apply to synclastic geometry, It would be useful to generalize the condition of 'neutral surface' to a broader context as a network of curves with special curvature relationships to the target mesh so that we can automate the bi-stable auxetic patterns for all free-from geometries.

Currently all the edges in our pattern are bi-stable. This means a significant amount of external forces are needed to actuate the whole system, which may be inappropriate especially when we scale it up to an architectural context. For that application, we could instead mix bi-stable edge patterns with other passive elements. This could help increase the scaling ranges and application possibilities.

CONCLUSION

We introduce geometric properties and a method for the design of bi-stable auxetic patterns. Such patterns are formed by linear hinges allowing out-of-plane rotation and thus enable the system to be actuated into shape with desired dihedral angles at each edge. Thanks to a proper propagation approach, we can generate the patterns and overcome the complex geometric condition around each vertex star. We then provide a tool for designing free-form synclastic geometry and use prototypes to showcase different fabrication and material. Our proposed system has potential to be widely used for architectural applications as a novel strategy for bi-stable deployable structures.

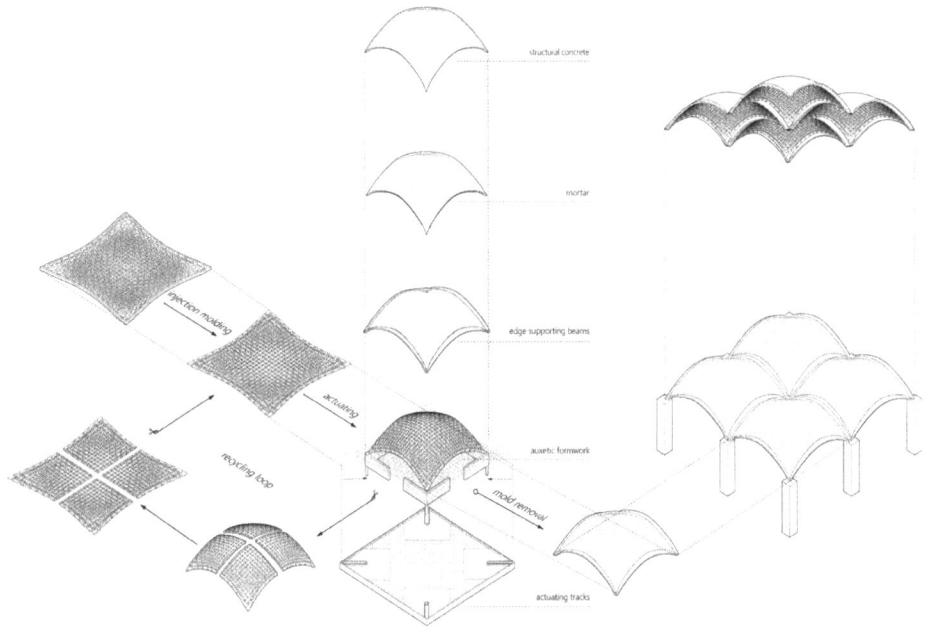

15 Stable states of actuation: (a) by compression, (b) by expansion

16 Post-processing the geometric pattern for CNC milling

17 Photography of prototype model by 3mm polypropylene

18 Flowchart of making a flexible molding formwork

19 Rendering of a potential shell pavilion design

REFERENCES

Chen, Tian, Julian Panetta, Max Schnaubelt, and Mark Pauly. 2021. "Bistable Auxetic Surface Structures." *ACM Transactions On Graphics* 40 (4): 1–9.

Chiang, Yu-Chou. 2019. "Programming Flat-to-Synclastic Reconfiguration." *ArchiDOCT* 6: 64–79.

Chiang, Yu-Chou, Sina Mostafavi, and Henriette Bier. 2018. "Assembly of shells with bi-stable mechanism." In *Proceedings of Advances in Architectural Geometry 2018 (AAG 2018)*. Göteborg, Sweden. 54–71.

Häuplik-Meusburger, Sandra. 2011. *Architecture for Astronauts*. Vienna: Springer Science & Business Media.

Jiang, Caigui, Florian Rist, Helmut Pottmann, and Johannes Wallner. 2020. "Freeform Quad-Based Kirigami." *ACM Transactions On Graphics* 39 (6): 1–11.

Konaković, Mina, Keenan Crane, Bailin Deng, Sofien Bouaziz, Daniel Piker, and Mark Pauly. 2016. "Beyond Developable." *ACM Transactions On Graphics* 35 (4): 1–11.

Liu, Jingyang, Jasmine Chia-Chia Liu, Madeleine Eggers, and Jenny E. Sabin. 2018. "Responsive Kirigami: Context-Actuated Hinges in Folded Sheet Systems." In *SIMAUD '18: Proceedings of the Symposium on Simulation for Architecture and Urban Design*.

Liu, Jingyang, Grace Chuang, Hun Chun Sang, Jenny E. Sabin. 2019. "Programmable Kirigami: Cutting and Folding in Science, Technology and Architecture." In *Proceedings of the ASME 2019 International Design Engineering Technical Conferences and Computers and Information in Engineering Conference, Volume 5B, 43rd Mechanisms and Robotics Conference*. Anaheim, California. 1–9.

Liu, Yang, Helmut Pottmann, Johannes Wallner, Yong-Liang Yang, and Wenping Wang. 2006. "Geometric Modeling With Conical Meshes And Developable Surfaces." *ACM Transactions On Graphics* 25 (3): 681–689.

Ou, Jifei, Zhao Ma, Jannik Peters, Sen Dai, Nikolaos Vlavianos, and Hiroshi Ishii. 2018. "Kinetix - Designing Auxetic-Inspired Deformable Material Structures." *Computers & Graphics* 75: 72–81.

Panetta, Julian, Florian Isvoranu, Tian Chen, Emmanuel Siéfert, Benoit Roman, and Mark Pauly. 2021. "Computational inverse design of surface-based inflatables." *ACM Transactions On Graphics* 40 (4): 1–14.

Popescu, M., L. Reiter, A. Liew, T. Van Mele, R.J. Flatt, and P. Block. 2018. "Building In Concrete With An Ultra-Lightweight Knitted Stay-In-Place Formwork: Prototype Of A Concrete Shell Bridge." *Structures* 14: 322–332.

Rafsanjani, Ahmad, and Damiano Pasini. 2016. "Bistable Auxetic Mechanical Metamaterials Inspired By Ancient Geometric Motifs." *Extreme Mechanics Letters* 9: 291–296.

Schling, Eike, D. Hitrec, and Rainer Barthel, 2017. "Designing Grid Structures Using Asymptotic Curve Networks." In *Humanizing Digital Reality: Design Modelling Symposium 2017*. Paris, France. 125–140.

Shah, Syed Imran Hussain, Manos M. Tentzeris, and Sungjoon Lim. 2019. "A Deployable Quasi-Yagi Monopole Antenna Using Three Origami Magic Spiral Cubes." *IEEE Antennas and Wireless Propagation Letters* 1 (January): 147–51.

Tachi, Tomohiro. 2013. "Designing Freeform Origami Tessellations By Generalizing Resch's Patterns." *Journal Of Mechanical Design* 135 (11): 111006.

Tang, Chengcheng, Xiang Sun, Alexandra Gomes, Johannes Wallner, and Helmut Pottmann. 2014. "Form-Finding With Polyhedral Meshes Made Simple." *ACM Transactions On Graphics* 33 (4): 1–9.

Vahidi, B., and N. C. Huang. 1969. "Thermal Buckling of Shallow Bimetallic Two-Hinged Arches." Journal of Applied Mechanics 36 (4): 768–774.

IMAGE CREDITS

Figure 4: Reproduced with permission from Yu-Chou Chiang.
Figure 6: Adapted with permission from Ahmad Rafsanjani.
All other drawings and images by the authors.

Zhenxiang Huang is a recent graduate of MS in Advanced Architectural Design from Cornell University. He also holds a BArch from Tongji University. His investigation focuses on the intersection between architectural geometry, fabrication advances, and reconfiguration systems. He aims to bring knowledge from computer graphics and geometry processing to his research work.

Yu-Chou Chiang is a PhD candidate at the Chair of Structural Design & Mechanics of Delft University of Technology, who investigates the interrelation between structural design, funicular form-finding, and digital fabrication. Having an engineering background, he worked as an educator and a researcher in the Department of Civil Engineering, National Taiwan University, and as a structural engineer in several firms. In 2017, he started the PhD research on membrane shells and reconfigurable mechanisms, aiming to make the design and fabrication of elegant forms more efficient.

Jenny E. Sabin is an architectural designer whose work is at the forefront of a new direction for 21st century architectural practice—one that investigates the intersections of architecture and science and applies insights and theories from biology and mathematics to the design of material structures and adaptive architecture. Sabin is the Wiesenberger Professor in Architecture and Associate Dean for Design at Cornell College of Architecture, Art, and Planning where she established a new advanced research degree in Matter Design Computation. She is principal of Jenny Sabin Studio, an experimental architectural design studio based in Ithaca and Director of the Sabin Lab at Cornell AAP. Her book *LabStudio: Design Research Between Architecture and Biology* co-authored with Peter Lloyd Jones was published in 2017. Sabin won MoMA & MoMA PS1's Young Architects Program with her submission *Lumen* in 2017.

APPENDIX

Algorithm A : The main bi-stable auxetic pattern algorithm.
INPUT: conical mesh $K(V.E.F)$, offset h, reference orientation $X \in C^{|V|}$, starting tilted angle ψ
OUTPUT: coordinates of planar bi-stable auxetic patterns S and its curved configuration S'

1: **procedure** BISTABLE_AUXETIC_PATTERN(K,h,X,ψ)
2:------$\alpha \leftarrow$ DIHEDRAL_ANGLE(K)
3:------$K_n \leftarrow$ NEUTRAL_MESH(K,h)
4:------$K_{n'} \leftarrow$ FLIP_MESH(K_n)
5:------$M \leftarrow$ PROJECTION_MATRIX($K, K_n, K_{n'}$)
6:------$\omega, s \leftarrow$ VERTEX_DATA(M, K_n)
7:------$T \leftarrow$ MESH_SPANNING_TREE($K_{n'}$)
8:------$h, p \leftarrow$ HINGES_SOLVE($T, K_{n'}, \alpha, \omega, s, \psi$)
9:------$S, S' \leftarrow$ PHYSICAL_COORDINATES(h, p, M)
10:------**return** S,S'
11: **end procedure**

Algorithm B : Computes the virtual rotating axes as a neutral mesh.
INPUT: conical mesh $K(V.E.F)$, offset h
OUTPUT: new mesh K_n

1: **procedure** NEUTRAL_MESH(K,h)
2:------$G \leftarrow$ MESH_SPANNING_TREE(K)
3:------$V_n \leftarrow \{\}$
4:------$f \leftarrow$ STARTING_FACE()
5:------$v \leftarrow$ STARTING_VERTEX()
6:------**for** $i \leftarrow \{0, \ldots,$ COUNT(G) $- 1\}$ **do**
7:------------**if** $i \leftarrow 0$ **do**
8:------------------f.vertices $\leftarrow$ ASSIGN_NEW_COORDS(f, v, h)
9:------------------add f.vertices to V_n
10:------------**else**
11:------------------**for** $j \leftarrow \{0, \ldots,$ COUNT($G[i]$) $- 1\}$ **do**
12:------------------------$f \leftarrow G[i][j]$
13:------------------------**for each** $v_x \leftarrow f$.vertices
14:------------------------------$v \leftarrow v_x$ if V_n contains v_x
15:------------------------**break**
16:------------------------f.vertices $\leftarrow$ ASSIGN_NEW_COORDS($f, v, 0$)
17:------------------------add f.vertices to V_n
18:------------------**end for**
19:------------**end if**
20:------**end for**
21:------$K_n \leftarrow$ UPDATE_MESH_VERTICES(K,V_n)
22:------**return** K_n
23: **end procedure**

Algorithm C : Computes vertex positions on a face of the neutral surface.
INPUT: a mesh face f (edge = 4 or 6), a starting vertex $v \in f$, offset height h
OUTPUT: coordinates of each vertex of face f

1: **procedure** ASSIGN_NEW_COORDS(f, v, h)
2:------$V_i \leftarrow$ FACE_NORMAL(f)
3:------$\vartheta \leftarrow$ atan2(sqrt($V_i.X^2 + V_i.Y^2$), $V_i.Z$)
4:------$\vartheta \leftarrow \vartheta / 2$
5:------$\omega \leftarrow$ atan2($V_i.Y, V_i.X$)
6:------$N \leftarrow (\sin\vartheta\cos\omega, \sin\vartheta\sin\omega, \cos\vartheta)$
7:------$f_v \leftarrow f_v + h *$ VERTEX_NORMAL(v)
8:------**for each** $ij \in f$ **do**
9:------------$d_{ij} \leftarrow$ (VECTOR(ij) $\times$ VERTEX_NORMAL(v)) $\times N$
10:------------$d_h \leftarrow$ NORMALIZE(d_h)
11:------------$k \leftarrow$ VECTOR(ij) $\cdot d_h$
12:------------$f_j \leftarrow f_i + k * d_h$
13:------**end for**
14:------**return** f_{ijkl}
15: **end procedure**

Algorithm D: Computes all hinge directions at a vertex based on the starting direction.
INPUT: a mesh vertex v_i, positivity Boolean t, association tuple at this vertex h_v, dihedral angle ϑ, unrolled corner coordinates c, unit pivot direction p, start hinge index n and direction d
OUTPUT: all hinge directions h

1: **procedure** HINGE_SOLVE ($v_i, t, h\_v, \vartheta, c, n, d, p$)
2:------$h \leftarrow \{\}$
3:------$v \leftarrow$ All adjacent vertices of v_i
4:------**add** d **to** h
5:------**sort** v **starting by** $h\_v(n)$ **in clockwise(counterclockwise) order if** $t = 0(1)$
6:------**delete** $v[0]$ **from** v
7:------**for each** $v_n \in v$ **do**
8:------------$z \leftarrow v'_n - v'_i$
9:------------$y \leftarrow z \times d$, $y \leftarrow$ NORMALIZE (y)
10:------------$x \leftarrow y \times z$, $x \leftarrow$ NORMALIZE (x)
11:------------$\omega \leftarrow \pi - \vartheta_{n+1}/2$
12:------------$a \leftarrow \sin\omega\ y + \cos\omega\ x$
13:------------bisector $\leftarrow$ NORMALIZE ($c_n^n - v'_i$) $- p$
14:------------new_dir $\leftarrow$ bisector $\times a$
15:------------$d \leftarrow$ new_dir
16:------------**add** d **to** h
17:------**end for**
18:------**return** h
19: **end procedure**

Algorithm E : Computes the traversal tree of a mesh given any starting vertex.
INPUT: topological mesh $K(V,E,F)$
OUTPUT: spanning tree $T(V/E/F)$

procedure MESH_SPANNING_TREE(K):
------ Put any vertex in a bag, marking it as visited. Until this bag is empty, pull out a vertex and put all unvisited neighbors in the bag, marking them as visited. Every time you put a neighbor in the bag, add the corresponding vertex to the tree.

Al Janah Pavilion

Jason Carlow
American University
of Sharjah

1 Al Janah Pavilion

ABSTRACT

This pavilion project was built as an outcome of an undergraduate design studio and design practicum at the American University of Sharjah in the UAE. The research methodology for the studio included case studies of various traditional building types to understand how traditional architecture in the MENA (Middle East and North Africa) region has been intelligently shaped by desert climate and Islamic culture over hundreds of years. Understanding and analysis of the precedent projects helped students to formulate climatic, structural, and material strategies for their design endeavors. Of the thirteen conceptual building envelopes developed by thirteen students in the design studio, the Al Janah scheme was chosen for development and construction.

The pavilion is inspired by one of the most unique elements in Islamic architecture, muqarnas. Traditionally, muqarnas are featured on the underside of domes and vaults to create a three-dimensional, decorative transition between the ceiling and the supporting walls. Muqarnas are geometrically abstracted in this project by creating an open framework that uses only surface edges, composed of arching line segments, to reinterpret the traditional ceiling form as a structural frame. A parametric model was built enabling student designers to adjust and visualize a number of design constraints. The project uses design computation to bring cultural significance and structural performance of architectural forms and precedent outside the typical canon of contemporary architecture in the West.

PRODUCTION NOTES

Client: ARADA
Status: Built
Site Area: 30,000 sq. ft.
Location: Sharjah, UAE
Date: 2019-21

2 Triangular canvas panels clip into the structural steel space frame to create a shaded public space

PROJECT DESCRIPTION

The research and design explorations for this built project were part of an undergraduate architectural design studio for a site located in the United Arab Emirates (UAE). The research methodology for the studio included case studies of various traditional building types to understand how traditional architecture in the region has been intelligently shaped by desert climate and Islamic culture over hundreds of years. Within the context of the Gulf region, where contemporary building facades are often decorated with geometric surface patterns meant to suggest a connection with Islamic architecture or Arabic culture, the studio aimed to create prototypes for a better performing building envelope, and an architecture with significance that is more than skin deep. In line with the conference theme, "Realignment," the project uses computation to bring cultural significance and structural performance of architectural forms and precedent outside the typical canon of contemporary architecture in the West. How might the pedagogy of a computational design/build course be "realigned" for greater cultural inclusivity through the choice of more regionally appropriate project precedents?

Building types selected for investigation in the course included Bedouin tents, Musgam huts, pigeon towers, as well as wind towers from across the Middle East region. The projects were chosen to introduce a broad set of issues related to thick, performative roof systems and building envelopes. Understanding and analysis of the precedent projects helped students to formulate climatic, structural and material strategies for their design endeavors. Of the thirteen conceptual building envelopes developed by thirteen students in the design studio, the Al Janah (meaning 'wing' in Arabic) scheme was chosen for development and construction.

The Al Janah Pavilion is inspired by one of the most unique elements in Islamic architecture, muqarnas. The design of this contemporary interpretation combines traditional, Islamic geometries with advanced 3D modeling techniques to create a complex, self-structural shading canopy.

3   Arches and columns were adjusted parametrically to conform to footpaths on site

4   Pavilion shadow patterns at mid-day

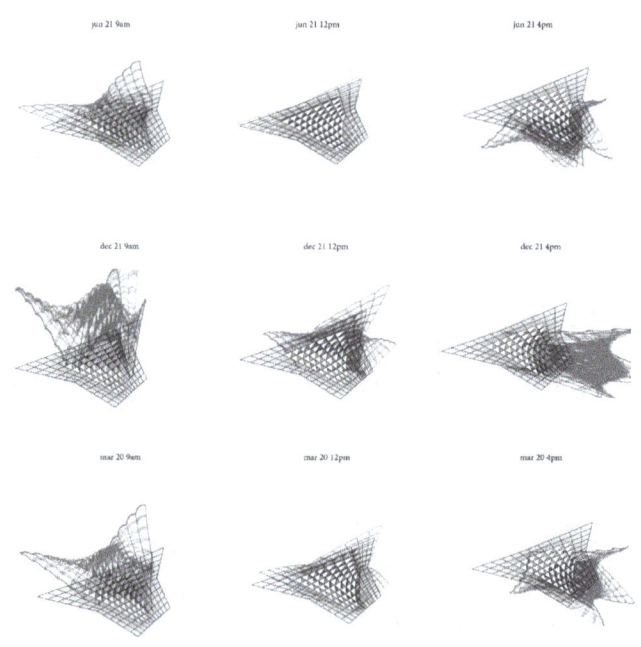

5   Digital shadow studies

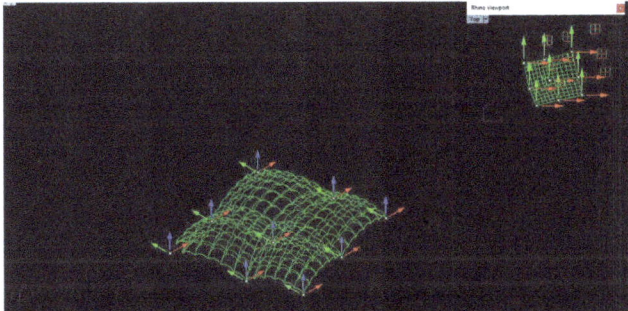
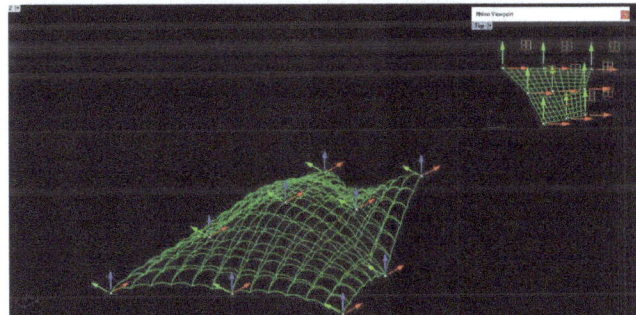
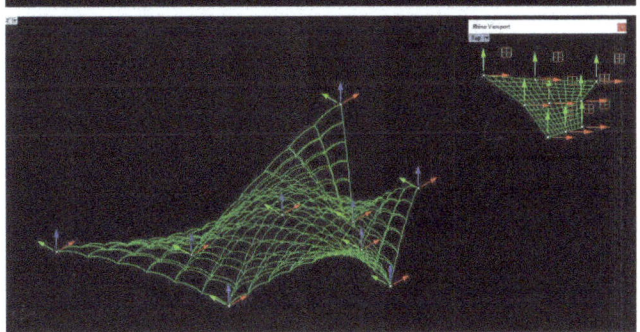

6   Al Janah's steel structure

7   Digital models show how key points can be adjusted parametrically

Traditionally, muqarnas are featured on the underside of domes and vaults to create a three-dimensional, decorative transition between the ceiling and the supporting walls. In traditional use, they transmit "an optical effect of orthogonality, layering, granularity and rotundity" (Moussavi 2014, 333). Muqarnas are geometrically abstracted in this project by creating an open framework that uses only surface edges, composed of arching line segments, to reinterpret the traditional ceiling form as a structural frame.

A parametric model was built using Rhinoceros 3D (McNeel 2010) and Grasshopper (Rutten 2014) modeling platforms that enabled student designers to adjust and visualize a number of design constraints. The model enabled students to locate eight important three-dimensional points within the structure to set a base geometry; four footprint locations and four cantilevered endpoints. Those eight points, generally determined by the context of pathways on the site, could be adjusted in three dimensions to increase or reduce structural spans and cantilevered lengths of rooftop. Adjustable parameters within the Grasshopper script included vertical height and vertical offset distance between the two surfaces of the space frame, grid density, structural member profile (depth, width, and thickness), as well as the curvature of the individual arches that represent the muqarnas.

In comparison with the advanced, parametric modeling tools used in the project design, more analog structural design methodologies and fabrication technologies were employed in the project analysis and construction. Structural steel members were fabricated by analog cutting and rolling processes and members were assembled and welded manually. Despite some difficulties during the fabrication process brought about by the realities of analog construction of a very precise, digitally-designed form, the parametric model was an invaluable tool for student designers to negotiate project form and nodal geometry at the complex joints with project engineers and fabricators. Digital design and computational analysis offer new possibilities for the contemporary reinterpretation of traditional Islamic building forms into parametrically adjusted, self-structural, site specific architecture.

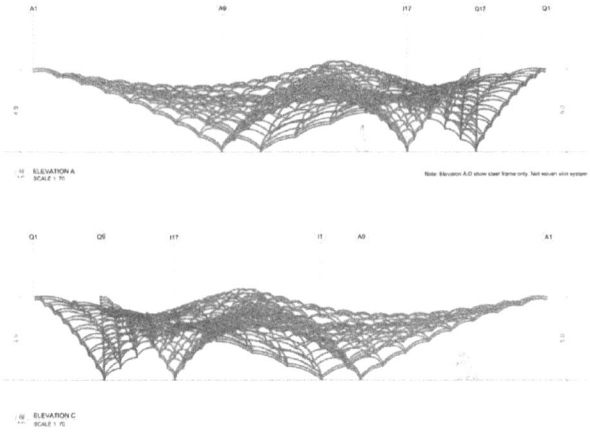
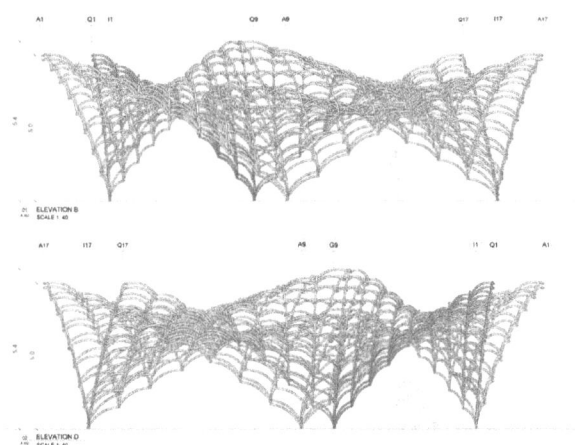

8 Pavilion Elevations

9 Rolled steel truss arches are fabricated through analog means

10 Pavilion under construction in Sharjah, United Arab Emirates

ACKNOWLEDGMENTS

This project was designed by students in two courses (ARC401 and ARC494) within the Department of Architecture at American University of Sharjah. Thanks to CAAD Dean Varkki Pallathucheril and Head of Department George Katodrytis for their support. Many thanks to Arada Developments LLC who sponsored the studio and funded the construction of the project.

Design Studio Professor
Jason Carlow

Student Design & Development Team
Aya Rahmy (Lead Student Designer)
Shaden AL Kalouti
Laura AlDhahi
Mohammad Aslam
Rim Drak Sibai
Halah Fadhil
Akhila Velloorkunju
Rewan Shaaban
Aashika Shibu

Arada Design / Project Management
Elie Mrad
Melissa Bayik

Structural Engineering
Bahar Bacha, Acadia

Studio Guest Critics and Contributors
Emidio Piermarini
Kristof Crolla
Erik L'Heureux
Ibrahim Ibrahim

Fabricators
Structurflex Middle East Contracting LLC

11 Al Jadah Pavilion from above

REFERENCES

Moussavi, Farshid. 2014. *The Function of Form*. Barcelona: Actar.

McNeel, Robert & others. *Rhinoceros 3D*. V.6.0. Robert McNeel & Associates, Seattle, WA. 2010.

Rutten, David. *Grasshopper*. V.0.9.0076. Robert McNeel & Associates, Seattle, WA. 2014.

IMAGE CREDITS

Figures 1 - 4, 6: ©Jason Carlow

Figures 5, 7 - 9: ©Aya Rahmy

Figure 10: ©Marian Misiak

Figure 11: ©Arada Developments LLC

Jason Carlow is Associate Professor at the American University of Sharjah where he teaches architecture and interior design. His recent work is focused on using digital design and fabrication as a lens to investigate building typologies, building component systems and compact interior spaces related to dense, urban environments. He holds a BA in Visual and Environmental Studies from Harvard University and a Master of Architecture from Yale University.

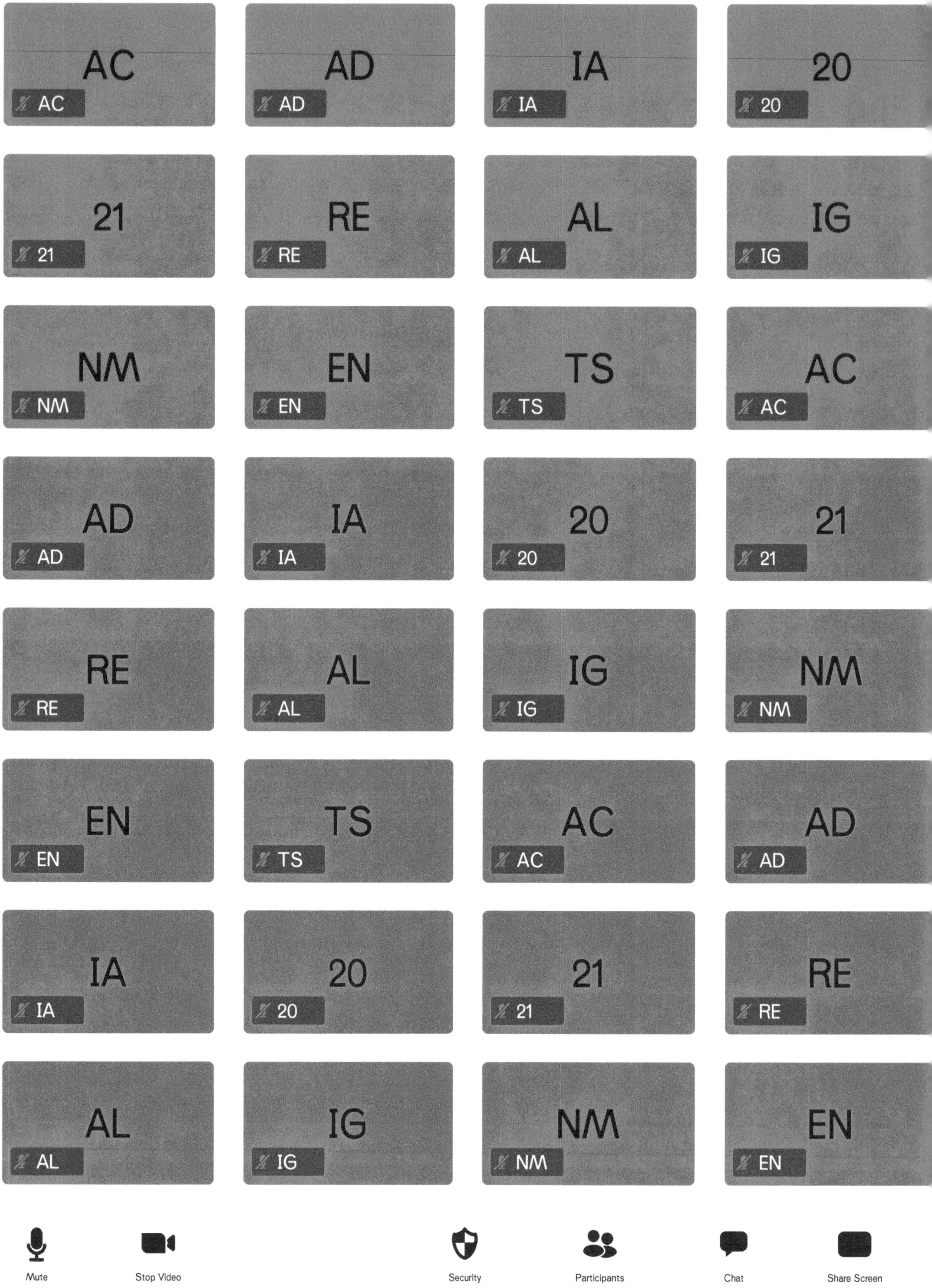

AC	AD	IA	20
21	RE	AL	IG
NM	EN	TS	AC
AD	IA	20	21
RE	AL	IG	NM
EN	TS	AC	AD
IA	20	21	RE
AL	IG	NM	EN

 Mute Stop Video Security Participants Chat Share Screen

Cocoon: 3D Printed Clay Formwork for Concrete Casting

Mackenzie Bruce*
University of Michigan

Gabrielle Clune*
University of Michigan

Ruxin Xie*
University of Michigan

Salma Mozaffari
ETH Zurich

Arash Adel
University of Michigan

* Authors contributed equally to the research.

1 The resultant cast of 3D-printed clay formwork: the horizontal seams register the pour height from the incremental process; the vertical seams display the location of 'Void' and 'Rib' support structure

ABSTRACT

Concrete, a material widely used in the construction industry today for its low cost and considerable strength as a composite building material, allows designers to work with nearly any form imaginable if the technology to build the formwork is possible. By combining two historic and widely used materials, clay and concrete, our proposed novel process, *Cocoon* integrates robotic clay three-dimensional (3D) printing as the primary formwork and incrementally casting concrete into this formwork to fabricate nonstandard concrete elements. The incremental casting and printing process anchors the concrete and clay together, creating a symbiotic and harmonious relationship. The concrete's fluidity takes shape from the 3D printed clay formwork, allowing the clay to gain structure from the concrete as it cures. As the clay loses moisture, the formwork begins to shrink, crack, and reveal the concrete below. This self-demolding process produces easily removable formwork that can then be recycled by adding water to rehydrate the clay creating a nearly zero-waste formwork. This technique outlines multiple novel design features for complex concrete structures, including extended height limit, integrated void space design, tolerable overhang, and practical solutions for clay deformation caused by the physical stress during the casting process. The novelty of the process created by 3D printing clay formwork using an industrial robotic arm allows for rapid and scalable production of nearly zero-waste customizable formwork. More significant research implications can impact the construction industry, integrating more sustainable ways to build, enabled by digital fabrication technologies.

INTRODUCTION

Cocoon focuses on the creation of concrete elements using 3D printed clay formwork [Figure 1]. Typically, formwork involved in creating complex geometry requires extensive machining time and post-finishing, a material- and labor-intensive process (Kudless et al. 2020). Recent research advancements in concrete fabrication have incorporated 3D printing technologies to create parametrically designed concrete columns with little formwork. Materials seen as suitable for 3D printing shelled formwork include Polylactic Acid (PLA), Polyvinyl Alcohol (PVA), wax, and clay (Burger et al. 2020). This research seeks to advance manufacturing processes of 3D printed formwork by incorporating clay as a viable material for fabrication. The process allows for creating complex geometries that are challenging, if not impossible, with other methods [Figure 2]. Demolding these complex geometries requires little labor, and the demolded formwork can be easily recycled, strengthening its viability as a sustainable construction process.

Reinforced concrete is a fundamental building material that has dominated the construction industry, making it the second most consumed material on earth, contributing to 8% of global human-made carbon emissions (Lehne and Preston 2018). Concrete's ability to take on any form imaginable pushes innovation of the formwork. The formwork used to create complex geometries is incredibly wasteful (Kudless et al. 2020). Methods for producing these complex shapes use technologies, such as hotwire cutting Expanded Polystyrene (EPS) foam, which are toxic and wasteful in their creation (Zhang et al. 2012). While these environmental effects are widely known, reducing its use proves difficult.

Today, designers have challenged this notion of what is possible in concrete fabrication with digital technologies, allowing the creation of formwork for complex geometries. Common approaches to this issue of freedom of shape resort to subtractive methods like Computer Numerical Control (CNC) milling (Gardiner 2017). These methods are both time-intensive and materially wasteful. Additive manufacturing allows for increased freedom of form while reducing material waste in the formwork manufacturing process. Specifically, the development of 3D printing technologies allowed the creation of forms in a nearly zero-waste process. Large-scale Fused Deposition Modeling (FDM) printing technology is continuously advancing, creating a process that can be applied to architecture by integrating industrial robotic arms (Hack and Lauer 2014). The progression of 3D printing technologies gives freedom of form to designers not previously executable at this printing scale. Not only is it thought of as a process for

2   The resultant casts of 3D printed clay formwork, stacked, create a geometrically complex concrete column 2m in height

printing building-scale components, but as the formwork for casting these components with traditional cementitious building materials such as concrete (Burger et al. 2020). Combining a high-tech process of 3D printing and low-tech materials such as clay and concrete would potentially enable pragmatic innovation within the construction industry to emerge.

This research seeks to explore the hypothesized benefits of using clay as formwork. We see clay to have several potential advantages, most notably its sustainable nature. Clay is a low-tech and natural material, with its use dating back to prehistoric times, making it one of the oldest building

3 Sequential images showcase clay formwork, self-demolding dried clay and resultant cast.

materials on earth (Gillott 1962). It is commonly understood that clay gains its plasticity when wet and becomes hard and brittle when dry. Due to the shrinkage of the clay as it dries, can the clay begin to self-demold, creating a fast and easy demolding process (Figure 3)? Can the clay then be recycled by adding water to rehydrate and create a zero-waste process? The plasticity of the clay allows it to be extruded for 3D printing of the material in a continuous bead, similar to the traditional process of clay coil pots (Gürsoy 2018). If the clay is easily removable once dried, branching or merging structures could be created using the 3D printing process while maintaining a continuous print bead. To mitigate deformation while casting nonlinear geometries, which increases the irregular hydrostatic pressure, additional support must be generated. Can this process address the building scale and accommodate for geometries not previously feasible to be printed with clay due to geometry limitations such as overhangs and scale?

STATE OF THE ART

Extensive research has been conducted on the topics of concrete, clay, and 3D printing. This section overviews the most relevant research, examining the potential of such processes related to the previously stated research questions.

"Concrete Choreography" (Anton et al. 2020) has explored concrete 3D printing, which utilizes a large bead, reducing the time required to erect the structure. The research involves the development of a fast-settling concrete mix that enables the prints to achieve significant element heights. While aesthetics are explored through the articulation of surface and form, the process integrates internal functional features such as space for reinforcement, alignment details, lighting channels, and rainwater drainage. Compared to using subtractive manufacturing, which cannot integrate the interior functionalities, concrete 3D printing is a nearly zero-waste process and efficient in time from design to completion. Concrete printing allows for geometric freedom without the formwork but comes with issues such as structural layer adhesion (Zareiyan and Khoshnevis 2017). The extrusion process requires aggregates for the mix, which is more carbon-intensive when compared to coarse aggregates of concrete used in traditional cast construction (Xiao et al. 2018). The forms are limited in their overhang geometry as the process does not allow for structural support of the concrete while printing, which traditional FDM polymer printing can accommodate (Burger et al. 2020).

Alternative 3D printing techniques have been explored to create complex geometries with concrete while reducing formwork waste. As polymer 3D printing advances to larger scales, it becomes a viable method for 3D printing formwork for concrete casting. Eggshell, a novel process, explores the challenges of 3D printing thin shell formworks (Burger et al. 2020). Because of the thinness of the formwork, demolding can be done with a heat gun, peeling away the plastic as it melts. While the process creates minimal waste, the plastic may not be suitable for reuse or recycling due to contamination from the concrete particles. If the plastic can be recycled, a labor-intensive process to remove concrete particles that remain on plastic formwork might be necessary. While polymer FDM printing begins to tackle issues of formwork waste and efficiency of complex forms, we ask: can clay 3D printing create a process that is easy to demold due to the fragility and shrinkage of the dried clay, as well as nearly a zero-waste formwork due to the reusability of clay? The increased bead size of clay is comparable to that of other concrete 3D printing processes, which would allow faster printing speeds as the clay does not have to cure while printing.

"Clay Robotics" explores the use of clay as formwork for concrete casting (Wang et al. 2016). The research examines the potential for using clay as formwork and addresses some of the challenges involved in the process. The process explored in this study is sequential, allowing the clay to dry before casting into the mold. A key area of exploration of this process was determining the precise dryness level of the clay to begin casting. It was found that the clay could not be cast into a bone-dry state as it absorbs the water from the concrete, weakening its strength. Limitations to the print height were also observed as the print structurally fails after a certain height and with changes in geometry. Print length is limited due to the size of the tube that can be used for extrusion, creating a discontinuous process. A concrete column results, printing five separate sections, stacking these together, and sequentially casting the column in one pour. This research lays the foundation for further investigations of this process, noting difficulties found such as print height and deformation of the clay due to the hydrostatic pressure and hoop stress. Future work looks to create a more continuous process for printing and casting.

Cocoon builds on advancements in materials and technology. It develops an incremental 3D printing and casting process that works with the limitations of the clay and concrete, allowing the materials to work in conjunction with one another to offset the challenges innate with each material. This process overcomes limitations of scale previously associated with clay 3D printing.

METHODS

Cocoon's novel fabrication process aims to reduce the environmental impact of complex concrete formwork. The introduction of incremental casting and development of a fast-setting concrete mix allows the clay formwork to reach scales previously unachievable (Figure 4). Issues of hydrostatic pressure must be addressed, requiring a continuous casting process using accelerators to give strength to the concrete as it rapidly cures, thus providing stability for the formwork. The process requires the concrete and print to work together to allow for the scalability of the process. Neither the concrete nor the thin shell print would be able to self-support without the other. The devised fabrication method combines the high-resolution material articulation of 3D-printed concrete with advanced concrete casting developed through Smart Dynamic Casting (SDC, Lloret-Fritschi et al. 2018). By printing the clay formwork instead of 3D printing the concrete form itself, *Cocoon*'s process allows for intricate detail, undercuts, and void generation, creating a wider range of geometries achievable.

This novel process requires the investigation and fulfillment of several key interrelated research objectives: a suitable fabrication setup, material formulation of a fast-hardening concrete ideal for working with clay formwork, development of a custom computational toolpath generator, a sequential clay 3D printing, incremental concrete casting process, and a demolding process. To investigate the objectives discussed above, this research focuses on physical prototyping, informing the development and refinement of *Cocoon*'s fabrication process, coupled with computational design experiments and simulation studies. Subsequently, the potential for building-scale manufacturing of architectural elements is validated by fabricating a geometrically complex case study component. We will discuss each of these topics in the following sections.

Fabrication Setup

The fabrication setup has been developed at the University of Michigan Taubman College of Architecture and Urban Planning (Figures 5 and 6). The setup comprises a 6-axis industrial robotic arm with a clay extrusion end effector. The extruder requires the clay tube to be refilled after 250m of print, a crucial factor in the development of this process as the print needs to stop and start for a tube change. Mixing and clay filling stations are located adjacent to the print work cell to allow continuous filling and mixing. The orchestration of the printing, mixing, and clay filling create an incremental process that works with the limitations of the current setup.

Initial Experimentation

The research was conducted in several phases, building up to a case study that applies the developed process to an architectural element. The first phase compared the deformation of a cylinder 100mm in diameter and 250mm in height using terracotta clay. Three prints with varying bead heights were extruded and subsequently cast into. Bead heights of 1mm, 2mm, and 3mm were compared, and it was found that the cast with 1mm bead height observed the least amount of deformation. Our test confirms the results of past research (Wang et al. 2019), where the bead height was tested and compared for deformation.

The second phase of the research looked to compare the plasticity of clay types as related to deformation. Clay types with high plasticity were sought after because it was hypothesized its strength would limit deformation in the cast. Two porcelain clays with varying material compositions were selected due to the material's high plasticity and workability. Terracotta, a clay with lower plasticity, was tested against the porcelain clays. An increased print height of 450mm necessitated introducing the incremental casting

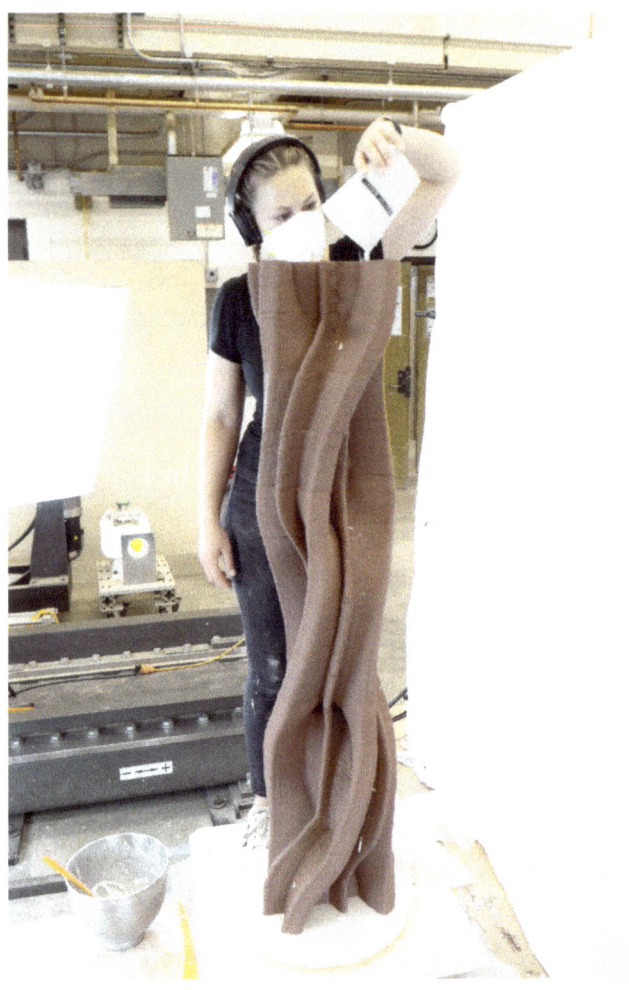

4   The combination of robotic clay printing and manual concrete casting creates a co-working environment with the human and robot

5   The fabrication setup, including 6-axis robotic workcell with clay extruding end effector

process, which will be discussed in the next section. The experiment resulted in a negligible difference in deformation between the clay types, although further testing is needed to evaluate this more accurately. Also, it was found that the surface finish of the clay varied in each of the casts, with terracotta yielding the cleanest finish, so it was ultimately chosen as the clay to move forward with. The terracotta stains the concrete, exposing a trace of how it is fabricated, creating a connection between the geometry and material articulation of the formwork.

The next phase of the research consists of a rigorous testing process to fine-tune three main factors in deformation control of the clay; print height, concrete mix, and casting timing.

Deformation Control

Incremental casting is a process that learns from SDC, pouring concrete in several increments and allowing the concrete to begin to cure as the next section is poured (Lloret-Fritschi et al. 2018). This control of the curing process is done using an accelerator in the concrete mix. The accelerator amount, the pour height, and timing between sections become a balance of factors explored in this phase of the research. The goal of the incremental casting is to allow for two sections of the print to bond while not affecting the earlier sections. More specifically, the first section will be cast, then a second a specified time after (for instance, 20 minutes). These sections will bond as the concrete has only partially cured. The third section of print aims to bond with the second section while not affecting the first, to control the hydrostatic pressure of the concrete, thus limiting deformation.

The primary driver in the design of concrete formwork is the control of hydrostatic pressure. Limiting the pour height controls the amount of hydrostatic pressure and therefore, decreases the deformation in the final cast. Concrete with the same density will have the same pressure on the formwork based on the height, not volume or weight. An additional stress on the formwork result from the hoop stress, which causes further deformation on the circumference of a cylindrical form as the material will elongate in proportion to its length (Roylance 1996). Hydrostatic pressure can be controlled through variations in the print height. For this set of experiments, the cylindrical form is again used for consistency maintaining a diameter of 100mm. Heights of 100mm and 135mm are then used to compare the deformations.

The base concrete formula consists of cement, ground silica, silica fume, water, polymer, superplasticizer, fine

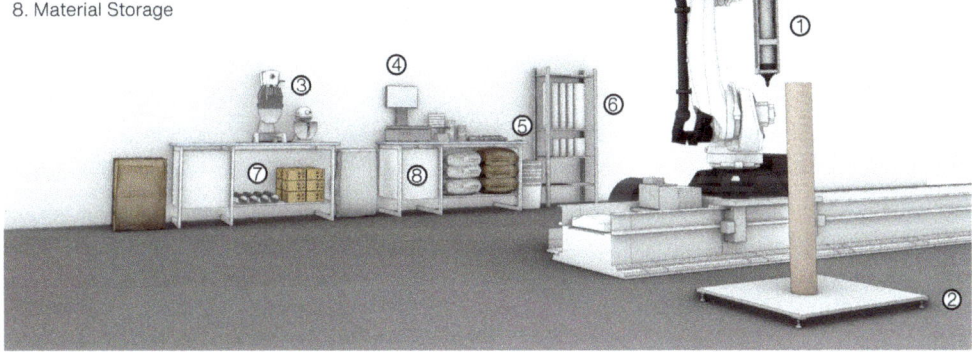

1. Clay Extruder
2. Adjustable Printing Platform
3. Mixing Station
4. Scale
5. Clay Recycling
6. Experiment Storage
7. Tube Filling
8. Material Storage

6 The fabrication setup utilized a 6-axis robotic arm with clay extruding end effector; stations for mixing, tube filling, and clay recycling are supplemental to the cell

sand, and glass fiber reinforcement, creating a Glass Fiber Reinforced Concrete (GFRC) mix. The formula is mixed in small batches using an industrial mixer. Two minutes before casting, an accelerator is added to the GFRC. The use of 1.7% and 2.6% accelerators are compared in this study (Figure 7). The percentages are considered high compared to SDC, which must accommodate tube changing and printing times due to the longer timing between the casts. As mentioned, timing is an essential factor in this process as it relates to the mix and fabrication setup. A minimum of 15 minutes is required between casts to allow the regular tube-changing during the setup.

This phase of tests was conducted by creating cylindrical columns measuring 800mm in height. Combinations of parameters were tested in a rigorous set of experiments to develop a process balancing the timing, mixture, and section height. Limiting the section height to 100mm and accelerator to 1.7% while extending the timing to 20 minutes resulted in minimal deformations while maintaining layer adhesion between the two sections.

The Cocoon

In the final phase of the research, to prove the viability of 3D printed clay formwork for concrete casting, a case study experiment is designed to overcome the challenges of the process. Ease of fabrication, customization, and scalability are considered as the driving factors in the prototype. The minimal surface showcases the potential of the intricate detail, undercuts, and voids, accommodating nonlinear surfaces.

In the first prototype, the concrete casting caused relatively large deformations and resulted in formwork failure due to large overhangs combined with unachievable pour heights.

The exterior formwork of this test created a concave vessel. The hydrostatic pressure and futility of the concrete pushed the formwork to failure, proving the need for a more rigid formwork. So, improvements including casting height, minimal surface design, additional scaffolding support, and a thinner wall depth were investigated in our next prototype.

Although the 3D printing cylinder tests enabled informed selection of clay type and fabrication parameters through physical prototype analysis, they did not showcase the achievable geometric complexity. The cylindrical test purpose were to investigate the effects of hydrostatic pressure on the malleable formwork. To prove the viability of the 3D-printed clay formwork, complex form testing was necessary.

By implementing computational design into the fabrication process, a continuous toolpath generator is created using Grasshopper, a visual algorithmic editor integrated into Rhinoceros's 3D modeling tools (McNeel 2020). The custom toolpath generator provides the boundary for concrete and creates a minimal surface supporting branching structures known as 'Void Support' (Figure 8). Moreover, to improve the rigidity of the formwork, an additional support system named 'Rib Support' is added to the clay printing toolpath. Compared to printing two layers around the circumference of the geometry, Rib Support conserves clay and provides additional strength to the formwork only where it is needed, reducing print times (Figure 9). The Rib Support was generated, providing additional support buttresses to points of failure or deformation observed during initial tests. Due to the increased hydrostatic pressure from the complex geometry, pour heights were adjusted from 100mm to 50mm between the casts. Integrating computational toolpath generation demonstrates the potential for optimization of the concrete casting process and clay formwork structure. Furthermore, both the Rib Support and Void Support do not create a break or overlap in the toolpath, which could create a seam vulnerable to failure.

Demolding
After the fabrication, the resulting concrete and clay structure is self-supporting. Over 1-3 days, the clay formwork begins to self-demold. As form loses moisture, the clay starts to shrink, creating fractures within the formwork. This phenomenon is directly linked to the inherent material qualities of the clay and is unique to *Cocoon*'s novel technique. The remaining formwork is then removed with ease with no additional tools (Figure 10). The clay can then be rehydrated, filtered, and recycled internally. This displays the potential of clay formwork to be a viable low-waste option. More testing is necessary to verify the material deterioration of this process.

RESULTS AND DISCUSSION
The initial experiments aimed to incrementally cast concrete forms without failure, while the deformation-controlled experiments aimed to improve the methods and techniques used in fabrication. Using 10 grams of the accelerator, six pours at the height of 135mm per cast created a concrete cylinder reaching a height of 810mm without failure. This test did have visible deformation (Figure 11). With an additional 5 grams, a test was conducted using 15 grams of the accelerator, six pours at the height of 135mm per cast. Although the deformations were significantly reduced, the disadvantage of adding more accelerator was less layer adhesion. In contrast, by lowering the height of each cast, fewer deformations were observed, and layer adhesion was maintained.

A third experiment using 10 grams of the accelerator and eight pours at the height of 100mm per cast created a concrete cylinder reaching a height of 800mm. This experiment was conducted three times to gain data on the replicability of the process. The accuracy of each cast was calculated by incrementally measuring the circumference of each cast. On average, the circumferences of the resulting casts were within a millimeter of accuracy (Figure 12).

The investigation of the cylindrical tests illustrated two main factors in increasing the deformations: first, the increase in the height caused higher hydrostatic pressure; second, the increase in the diameter resulted in higher circumferential hoop stress. This research determined a balance between these tested factors for this specific geometry. Although helping in understanding the limitation of clay formwork, the cylinder deformation testing results are only applicable to this particular geometry.

The results of this experimentation set up a framework for applying this process to different geometries. As seen in the case study experiment, the geometry was tested and adjusted to accommodate a complex form. The successful process of fabricating the case study not only determined Cocoon as a viable process but created an experimental setup that has the potential to be adjusted and applied to many geometries.

The overall height of the resulting case study element was 1.3m. The incremental casting process consists of 13 clay prints with 26 consecutive concrete pours. This process took a total of 9 hours to print and cast for the overall geometry.

CONCLUSIONS
Cocoon investigates the use of 3D-printed clay as formwork for concrete casting that eliminates waste in the process of creating complex forms. The developed incremental casting process allows clay printing to reach a previously not achievable scale by having the concrete work together with its formwork to hold the structure. This novel process reduces the environmental footprint of creating complex concrete forms by minimizing the material waste. The process enables the integration of voids and overhangs due to the ease of removability of the formwork, mitigating constraints of concrete 3D printing. Overall,

Cement	Silica Fume	Ground Silica	Water	Accelerator	Polymer	Super Plasticizer	Sand	Glass Fiber
562.5g	75g	150g	195g	0g	37.5g	7.5g	562.5g	22.5g
562.5g	75g	150g	195g	10g	37.5g	7.5g	562.5g	22.5g
562.5g	75g	150g	195g	15g	37.5g	7.5g	562.5g	22.5g
562.5g	75g	150g	175g	10g	37.5g	7.5g	562.5g	22.5g

7 Concrete mix formulations

Cocoon seeks to challenge conventional methods and materials for 3D-printed formworks, demonstrating the ability to reduce the environmental impacts of concrete construction without compromising the complexity and time efficiency of bespoke architectural elements.

Future Work

The current fabrication setup has projected multiple opportunities for concrete formwork production; however, it is limited in scale due to the tube volume of the clay, requiring frequent tube changes, thus slowing down the process for larger-scale production. The integration of mud printing using a pump, a technology that had been recently developed for large-scale earth printing for architectural assemblies, would allow the process to jump to the production of building-scale architectural elements (Gomaa et al. 2021). For this process to be applicable at that scale, reinforcement must be integrated. We are looking into developing a fabrication setup that situates the print bed between two industrial robotic arms, one equipped with clay printing and the other rod bending. The ability to integrate the robotic rod bending process into the overall process mitigates the structural scale issues. This multi-robotic fabrication cell allows for cooperative coordination of the placement of reinforcement during the incremental printing process. Future work looks to integrate this collaborative process of rod-bending reinforcement and pumping concrete into the formwork using a tool that can mix accelerator into the concrete at the nozzle for the most consistent results.

ACKNOWLEDGMENTS

The presented research was conducted as a Digital and Material Technologies Capstone project at the Taubman College of Architecture and Urban Planning, University of Michigan. We would like to thank Prof. Wes McGee for his input and assistance in developing the robotic setup, including the clay printing end effector in collaboration with Asa Peller. We also thank many others who were, directly and indirectly, involved in the research, particularly, Tszyan Ng, Mark Meier, Austin Wiskur, Alyssa Fellabaum, and Minu Lee. The research was supported by the Taubman College of Architecture and Urban Planning, and the Rackham Graduate School at University of Michigan.

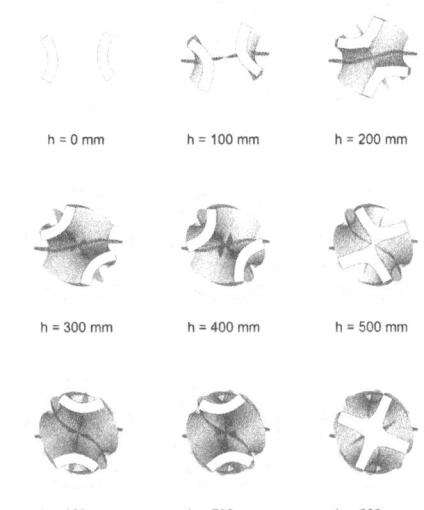

8 The plan of toolpath generation at alternating concrete pour heights displaying location of the support structure based on geometry

9 Pour heights as they relate to geometry and time. Overall, the 1.3m structure was printed in 9 hours and 6 minutes

10 The final resultant cast after demolding with 1.3m height

REFERENCES

Anton, Ana, Patrick Bedarf, Angela Yoo, Benjamin Dillenburger, Lex Reiter, Timothy Wangler, and Robert J. Flatt. 2020. "Concrete Choreography." In *Fabricate 2020*, edited by J. Burry, J. Sabin, B. Sheil, M. Skavara. London: UCL Press. 286–93.

Burger, Joris, Ena Lloret-Fritschi, Fabio Scotto, Thibault Demoulin, Lukas Gebhard, Jaime Mata-Falcón, Fabio Gramazio, Matthias Kohler, and Robert J. Flatt. 2020. "Eggshell: Ultra-Thin Three-Dimensional Printed Formwork for Concrete Structures." *3D Printing and Additive Manufacturing* 7 (2): 48–59.

Gillott, Jack. 1962. "Clay Mineralogy in Building Research." *Clays and Clay Minerals* 11 (1): 296–98.

Gomaa, Mohamed, Wassim Jabi, Alejandro Veliz Reyes, and Veronica Soebarto. 2021. "3D Printing System for Earth-Based Construction: Case Study of Cob." *Automation in Construction* 124 (April): 103577.

Hack, Norman, and Willi V. Lauer. 2014. "Mesh-Mould: Robotically Fabricated Spatial Meshes as Reinforced Concrete." *Architectural Design* 84 (3): 44–53.

Gardiner, Ginger. 2015. "SFMOMA façade: Advancing the art of high-rise FRP." *CompositesWorld*. Gardner Business Media Inc. Accessed September 10, 2021. https://www.compositesworld.com/articles/sfmoma-faade-advancing-the-art-of-high-rise-frp.

Gürsoy, Benay. 2018. "From Control to Uncertainty in 3D Printing with Clay." In *Computing For a Better Tomorrow; Proceedings of The 36th eCAADe Conference*, vol. 2, edited by A. Kepczynska-Walczak and S. Bialkowski. Lodz University of Technology, Lodz, Poland, 19-21 September 2018. 21–30.

Kudless, Andrew, Joshua Zabel, Chuck Naeve, And Tenna Florian. 2020. "The Design And Fabrication Of Confluence Park." In *Fabricate 2020*, edited by J. Burry, J. Sabin, B. Sheil, and M. Skavara. London: UCL Press. 28–35.

Lehne, Johanna, and Felix Preston. 2018. "Making Concrete Change: Innovation in Low-carbon Cement and Concrete." *The Royal Institute of International Affairs, Chatham House Report Series*. Accessed September 10, 2021. https://www.chathamhouse.org/sites/default/files/publications/2018-06-13-making-concrete-change-cement-lehne-preston-exec-sum.pdf.

Lloret-Fritschi, Ena, Fabio Scotto, Fabio Gramazio, Matthias Kohler, Konrad Graser, Timothy Wangler, Lex Reiter, Robert J. Flatt, and Jaime Mata-Falcón. 2018. "Challenges of Real-Scale Production with Smart Dynamic Casting." In *First RILEM International Conference on Concrete and Digital Fabrication; Digital Concrete 2018*, edited by T. Wangler and R. Flatt. RILEM Bookseries, vol. 19. Cham: Springer. 299–310.

Robert McNeel & Associates. *Rhinoceros*. V.6.0. RobertMcNeel & Associates. PC. 2020.

Roylance, David. 2001. "Pressure vessels." In *Mechanics of Materials Lecture Notes*, Department of Materials Science and Engineering. Cambridge, Mass.: Massachusetts Institute of Technology. 1–4.

Wang, Sihan, Zack Xuereb Conti, and Felix Raspell. 2019. "Optimization Of Clay Mould For Concrete Casting Using Design Of Experiments." In *Intelligent & Informed; Proceedings of the 24th CAADRIA Conference*, vol. 2, edited by M. Haeusler, M. A. Schnabel, T. Fukuda. Wellington, New Zealand. 283-292

Wang, Sihan, Philippe Morel, Kevin Ho and Stylianos Dritsas.

2016. "Clay Robotics: A Hybrid 3D Printing Casting Process." In *Challenges for Technology Innovation: An Agenda for the Future; Proceedings of the International Conference on Sustainable Smart Manufacturing (S2M 2016)*. London: UCL Press. 83–88.

Xiao, Jianzhuang, Chunhui Wang, Tao Ding, and Ali Akbarnezhad. 2018. "A Recycled Aggregate Concrete High-Rise Building: Structural Performance and Embodied Carbon Footprint." *Journal of Cleaner Production* 199: 868–81.

Zhang, Haijun, Yu-Ying Kuo, Andreas C. Gerecke, and Jing Wang. 2012. "Co-Release of Hexabromocyclododecane (HBCD) and Nano- and Microparticles from Thermal Cutting of Polystyrene Foams." *Environmental Science & Technology* 46 (20): 10990–96.

Zareiyan, Babak, and Behrokh Khoshnevis. 2017. "Effects of Interlocking on Interlayer Adhesion and Strength of Structures in 3D Printing of Concrete." *Automation in Construction* 83 (Nov.): 212–21.

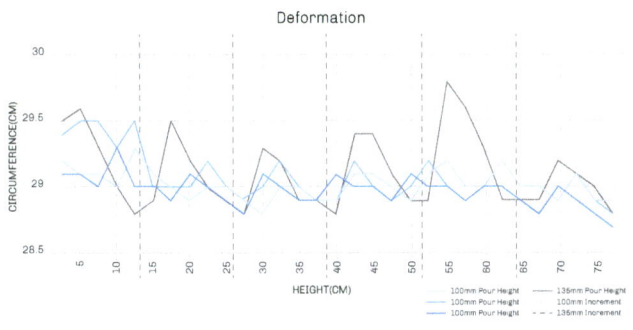

Mackenzie Bruce holds a BFA in Architecture from the University of Massachusetts Amherst, and a MArch from Taubman College of Architecture and Urban Planning at the University of Michigan. She has continued her education and research by pursuing a Master of Science in Digital and Material Technologies at the University of Michigan. Her research focuses on the development of material processes that question traditional methods of construction, working with material behavior to inform design.

Gabrielle Clune holds a Bachelor of Design in Architecture from the University of Florida, and a MArch from Taubman College of Architecture and Urban Planning at the University of Michigan. After completing her Master's she has continued her education and research at the University of Michigan by pursuing a Master of Science in Digital and Material Technologies. Her research focuses on material-based design, aspiring to advance the architecture and construction industry by expanding the use of advanced digital technologies, novel processes, and material manipulation.

Ruxin Xie holds a BArch from Xiamen University, and a MArch as well as an MS in Digital and Material Technology from the University of Michigan. Her research focuses on computational prototyping along with advanced digital technology.

Salma Mozaffari is a doctoral researcher at the Institute of Structural Engineering, the Civil, Environmental, and Geomatic Engineering Department at ETH Zurich. Her doctoral research focuses on computational strut-and-tie modeling through explorations of algebraic geometrical design and structural optimization. She holds a Master's degree from Northeastern University in Boston, where she researched system identification and statistical data analysis algorithms for estimating the damping value of

11 The deformation tests: (From left to right) 100mm pour height 10g acc; 100mm pour height 10g acc; 100mm pour height 15g; 135mm pour height 15g acc; and 135mm pour height 10g acc

12 The deformation relative to pour height for 100mm diameter cylinder

buildings and damage detection of bridges. She has also worked as a structural engineer on various projects in the United States and Switzerland, including designing, rehabilitating, and restoring buildings and bridges.

Arash Adel is an Assistant Professor of Architecture at the University of Michigan's Taubman College of Architecture and Urban Planning, where he directs the ADR Laboratory. Adel's interdisciplinary research is at the intersection of design, computation, engineering, and robotic construction. He is particularly known for his work with novel integrative computational design methods coupled with robotic assembly techniques for manufacturing nonstandard multistory timber buildings. Adel received his Master's in Architecture from Harvard University and his doctorate in architecture from the Swiss Federal Institute of Technology (ETH).

Robotic Pellet Extrusion: 3D Printing and Integral Computational Design

Reinforced Thin Shell System Formwork for Sandwich Concrete Walls

Mania Aghaei Meibodi
DART, University of Michigan

Ryan Craney
DART, University of Michigan

Wes McGee
Matter Design,
University of Michigan

1   Full-scale 3D-printed inner core for casting insulation material entails pillow-like detailing to minimize concrete construction

ABSTRACT

Three-dimensional (3D) printing offers significant geometric freedom and allows the fabrication of integral parts. This research showcases how robotic-fused deposition modeling (FDM) enables the prefabrication of large-scale, lightweight, and ready-to-cast freeform formwork to minimize material waste, labor, and errors in the construction process while increasing the speed of production and economic viability of casting non-standard concrete elements. This is achieved through the development of a digital design-to-production workflow for concrete formwork. All functions that are needed in the final product—an integrally insulated steel-reinforced concrete wall and the process for a successful cast—are fully integrated into the formwork system. A parametric model for integrated structural ribbing is developed and verified using finite element analysis. A case study is presented which showcases the fully integrated system in the production of a 2.4m tall x 2.0m curved concrete wall. This research demonstrates the potential for large-scale additive manufacturing to enable the efficient production of non-standard concrete formwork.

INTRODUCTION

Concrete formwork plays a significant role in the cost of concrete construction, the form of a concrete structure, and the amount of concrete used in a building structure. Typically, parts designed to use minimal material entail complex geometries that require geometrically complex formwork (Aghaei Meibodi et al. 2017 and 2019). However, the construction of such formwork is usually associated with additional labor, costs, and construction waste. The cost of typical formwork amounts to anywhere from 40% to 60% of the cost of a concrete structure—exceeding the combined total cost of concrete, reinforcement materials and labor (Lab 2007). For example, cast-in-place structural concrete formwork material accounts for 6%, and formwork labor accounts for approximately 46.7% of the total cost of a concrete structure (Lab 2007) (Figure 2). The cost is even higher for irregular concrete parts with complex geometries. Thus, most concrete structures are massive, use unnecessary amounts of concrete, and are designed as rigid and orthogonal forms to serve the constraints of formwork. Compared to subtractive fabrication methods, additive manufacturing places material only where it is needed, reducing waste significantly. Our ability to digitize formwork construction and increase its fabrication freedom can significantly reduce the cost, labor, waste, and material use of concrete construction.

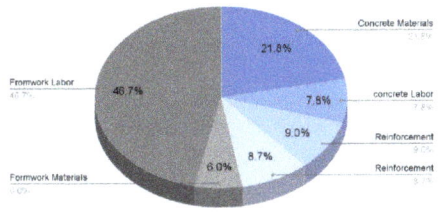

2   The schematic diagram shows the actual cost of the in-place structure. Adapted from Robert H. Lab, Jr., P.E. (Lab 2007).

Additive manufacturing technologies can open new opportunities for concrete formwork by enabling the fabrication of geometrically complex formwork economically and minimizing waste. In recent years, researchers have investigated different additive manufacturing technologies and approaches for non-standard formwork. Two technologies used for formwork production are Binder Jet and thermoplastic extrusion, which can be further characterized as fused filament fabrication (FFF) or fused granular fabrication (FGF). Binder Jet technology is used to bind sand, layer by layer, and fabricate sandstone formwork (Morel and Schwartz 2015; Aghaei Meibodi et al. 2017; Aghaei Meibodi et al. 2019; Aghaei Mebodi 2021). Binder jetting with sand offers the most significant level of geometric freedom, precision, and resolution on a large scale, but the resulting formwork is relatively heavy and fragile to handle on the construction site. Thermoplastic extrusion technology, which is of interest to this research, has been used to fabricate large-scale ultra-thin thermoplastic formwork (Peters 2014; Jipa et al. 2017; Leschok and Dillenburger 2019; Aghaei Meibodi et al. 2020; Burger et al. 2020). 3D printing of thermoplastic enables the fabrication of lightweight and recyclable formwork. The existing research has focused on 3d printing ultra-thin plastic formwork and developing concrete casting approaches that prevent the formwork from cracking under hydrostatic pressure when casting (Peters 2014; Jipa et al. 2017; Burger et al. 2020).

The common challenge in thermoplastic extrusion 3D printed plastic formwork, especially at a larger scale, is the hydrostatic pressure exerted on formwork by fluid concrete during casting. Hydrostatic pressure is the force per unit area transmitted by the freshly placed concrete perpendicular to the formwork. The hydrostatic pressure that is exerted on each point of the formwork is determined by the height and the density of the casting fluid above that point.

An approach to overcome the hydrostatic pressure has been to submerge the formwork in a bed of sand when casting the formwork (Jipa et al. 2017; Leschok and Dillenburger 2019; Aghaei Meibodi et al. 2020). Here, the sand acts to partially cancel the hydrostatic pressure from the concrete. Another approach proposed in Eggshell research has been to use a set-on-demand, accelerated concrete mix (Burger et al. 2020). In this approach, the hydrostatic pressure is reduced significantly by casting concrete in incremental volumes with an added accelerator, which causes a rapid gain in yield strength before the next batch is cast.

While these solutions to overcome hydrostatic pressure are valid, they may be challenging to scale for real-world construction. Using sand (potentially mixed with other dense materials such as calcium carbonate, bentonite, and water to form a slurry) to counter pressure hydrostatic pressure might be scalable for prefabrication but not for onsite fabrication. Even for off-site fabrication, systems must be developed to pour sand at a similar rate as casting the concrete mix; otherwise, it will break the formwork inward. In a "set-on-demand" approach, the construction process requires complex pumping and mixing equipment to proportion the accelerator. This approach also requires custom concrete mix designs that differ substantially from industry-standard self-consolidating concrete (SCC).

3

METHOD

This research expands on existing research into FFF 3D printing of lightweight formwork by developing fabrication methods based on FGF (pellet extrusion) rather than FFF (filament extrusion) to speed and scale the fabrication process. It proposes to go beyond merely 3D printing ultra-thin shell formwork to 3D printing the formwork as a structurally optimized, rib-stiffened shell formwork system. The integration of reinforcement ribs into the thin shell will provide adequate strength and stiffness to retain newly cast fluid concrete in place. The integration of reinforcement ribs is not a new concept; for example, the injection-molded ABS formwork of Peri entails a waffle-like grid of ribs to overcome the hydrostatic pressure during the casting process. Nevertheless, this research will not develop a standard and repetitive waffle-like rib system but a customized, structurally optimized reinforcement system that minimizes the formwork system's material usage by printing support material based on the local hydrostatic pressure in the mold.

Robotic FGF 3D Printing for Integrative Formwork

FGF 3D printing via extrusion is limited by several factors, such as the maximum material deposition rate, the maximum machine feed rate along the toolpath, and the maximum cooling rate of the deposited material, which further relates to geometric constraints. The maximum material deposition rate is controlled by factors such as the maximum melting rate of the polymer as well as process-specific factors such as the maximum force which can be applied to the filament (in the case of filament extrusion systems) or the maximum die pressure, which can be achieved (in the case of pellet extrusion systems). FFF printing with nozzle diameter less than 1mm is limited to a theoretical maximum build rate of 180cc/hr, which would translate to approximately 0.2kg/hr with PETG (Go et al. 2017). Commercially available pellet extrusion systems with nozzle diameter >1mm are available with build rates of more than 30kg/hr. The existing experimentation shows that pellet extrusion can be faster than filament extrusion for larger objects. This research utilizes a pellet extrusion system using a 3mm nozzle diameter that can print approximately 5kg of material per hour. At the component scales investigated in this research, the material extrusion rate is rarely the limiting factor. FGF 3D printing via extrusion also imposes constraints on the geometry of printed components, particularly related to unsupported overhangs. In particular, the ability to cantilever the toolpath proves to be heavily dependent on the print bead width and print speed parameters. The cooling behavior of the specific polymer used also has a significant effect on the ability to cantilever. Through prototyping, we identified the geometric limitations of the FGF and printing setup. For the optimum print speed of 0.1m/s with a bead width of 5mm and thickness 1mm, we could achieve a maximum cantilevering angle of 35°. The cantilever constraint was further integrated into the computational design method to ensure the generation of printable forms.

Concept of Integrative Formwork System for Concrete Sandwich Wall

The geometric complexity offered by 3D printing can be leveraged to design and fabricate complex formwork systems with multiple cavities and precise detailing. In this research, a formwork system with three cavities for three layers of material is developed—one cavity is for casting insulation material (the central cavity) and two outer cavities for casting concrete. The integrated reinforcement fins that help resist hydrostatic pressure and the detailing for rebar installation are printed simultaneously to the casting surface (Figure 4). The integral insulation layer acts as the core of the system and allows the outer layers of concrete to be exposed. The formwork system was designed to allow for rapid installation of the insulation and rebar before casting. Such a formwork system allows the fabrication of an insulated cast-in-place concrete wall in a single pour. It increases precision, reduces waste and material consumption in construction.

3   Robotic 3D printing setup and in-house designed and built extrusion system for printing PETG thermoplastic pellets.

4   The integrative formwork system comprises six layers of 3D printed vertical surfaces that define three cavities, two layers of the structural-reinforcement systems, and the detailing for installing the wythe ties and detailing that holds rebars in place.

Integral Computational Design of Structurally Informed 3D-Printed Formwork

This research expands on the existing research into 3D printing thermoplastic concrete formwork by integrating fins as structural support into the formwork design and printing them as an integral part of the thin-shell formwork to withstand hydrostatic pressure without deformation. A computational design approach was developed to minimize formwork waste, which generates and sizes structural supports only where they are needed. This approach utilizes structural finite element analysis (FEA) to the computational design model that generated the formwork and its details. Thus, generating formwork variations with minimal material usage at the allowable deflection. Another set of feedback loops between prototyping and computational design models was created to identify the most efficient geometries for providing stiffness while minimizing thermoplastic and material used in 3D printing the concrete formwork.

Parametric Model of the Formwork

The developed computational design procedure optimizes the formwork's geometry in response to hydrostatic pressure and deformation. Here, a parametric model of the overall formwork system was linked to a structural analysis solver, which measures the deflection of the formwork under load. The data from the structural analysis was directly fed back into the algorithm that generates the fins.

The algorithm began with the input of a NURBS surface representing the desired form of the concrete wall element (Figure 5). This input NURBS surface was subdivided into adjoining vertical surfaces—each discretized surface is continuous from the base to the top of the formwork—to generate vertical fins tangent to the outer formwork elements. Each subsurface represents the tributary area of one load-bearing formwork element to which we refer as a fin. The fins act as a beam transferring the load of the liquid concrete (hydrostatic pressure) to the formwork base.

The width and depth of each fin element was controlled parametrically and was limited by the deflection of the thermoplastic formwork at the casting surface (Figure 6). FEA was used to determine the deflection value under hydrostatic pressure, and the results were fed to an algorithm that generated each fin's width and depth. When calculated across the length of the fin, the resulting geometry is a tapered triangular prism with the most profound cross-sectional depth occurring at the lowest part of the formwork.

Maximum Fin Width Calculation

The maximum fin width was calculated using the following equation, which assumed a uniform distributed load case on a simply-supported beam (Figure 7).

$$L = \sqrt[4]{384 E I \delta_{max} / 5\omega}$$

Where,

L = maximum fin width (m)
δ_{max} = maximum acceptable deflection along fin width (m)
ω = uniform distributed load, as determined by hydrostatic pressure (N/m)
E = Young's modulus of thermoplastic material parallel to toolpath direction (pa)
I = area moment of inertia of thermoplastic cross-section (m⁴)

Once the maximum fin width had been identified, the measurement was input as a parameter into the generative algorithm, and the surface was subdivided into individual fins. The newly generated fin surfaces were analyzed using FEA, as described in the following section.

Tapered Fin Deflection Analysis Using FEA

To address the complexity of variable loading and cross-section geometry along the length of each fin, FEA was used to determine an acceptable depth profile for each fin. The analysis was performed using Karamba3D for Grasshopper (Preisinger 2013, 2021; Rutten 2021). The following parameters were necessary for creating the FEA model: 1) fin span and cross-sections, 2) fin supports at

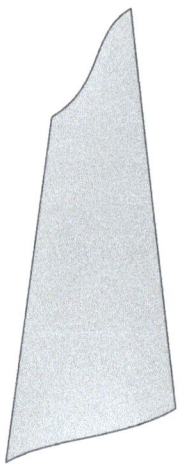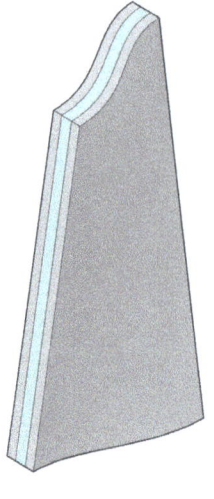

5 Diagram of the generative design process: (Left to right): 1) Input NURBS surface; 2) surface subdivision; 3) fin sizing and surface loft; 4) resulting concrete cast

6 Cross-section profile of a fin element; fins are tangent to the casting surface with parametric depth and width

7 Diagram of a simply-supported beam used to define the maximum fin width

8 Diagram of the beam under hydrostatic pressure with one fixed support and two roller supports

9 Diagram of the same beam, but with rationalized loading for FEA model simulation

the top and bottom of each fin, 3) loading from hydrostatic pressure, 4) material properties of the orthotropic thermoplastic 3D print. The FEA model was then used to measure deflection along the length of each fin and optimize the tapered fin cross-section to minimize the amount of thermoplastic used while maintaining an acceptable deflection value.

Fin Span and Cross Section: the deflection of each fin was simulated using the vertical centerline of the fin surface. This curve was divided into 20 equal beam segments, with each segment having a different cross-sectional depth. The depth of each triangular cross-section was assigned by dividing the values between the base and upper section depth parameters across the length of the fin. This method rationalized the linear taper of the fin design into a stepped beam of similar performance for the sake of processing time.

The user assigns the input parameters representing the upper and lower section depths. The width of each triangular section is determined by the maximum fin width calculation. In this case study, the cross-section profile was given a thickness of 5 mm - the wall thickness of the thermoplastic print.

Supports / Load Transfer: the loads carried by the formwork are transferred to a support at the top, middle, and base of each fin. The base support was modeled as a simple fixed support and the top and middle supports were modeled as roller supports (Figure 8&9).

Loading: hydrostatic pressure was the predominant load simulated in the FEA algorithm. In a wall condition, the gravity loading was negligible and was not included in the FEA analysis. The hydrostatic pressure was calculated and applied individually for each of the 20 beam segments along the length of the fin. The following equation was used to calculate the hydrostatic pressure:

$P = \rho g h$

Where,

P = pressure (N/m^2)
ρ = density of concrete (kg/m^3)
g = acceleration of gravity (m/s^2)
h = vertical distance between top of find and midpoint of beam segment (m)

The resulting values were multiplied by the fin width, and the uniform linear loads were applied to their corresponding beam segments. As previously mentioned, this rationalization of the beam loading (Figures 8, 9) was carried out in order to reduce the processing time for simulating using the FEA model.

Material Properties of Orthotropic Thermoplastic 3D Print

In order to accurately simulate the deflection of the formwork in the FEA model, mechanical testing was performed on printed material specimens. The material properties were measured in accordance with the ASTM standard, D638, Standard Test Method for Tensile Properties of Plastics (ASTM International 2014). The outer formwork is printed using a short fiber (<1mm) carbon fiber reinforced PETG. This material provides for both increased mechanical properties in the printed part and faster print speeds due to a faster transition from the molten state to the rigid state. The stiffness and yield strength of the carbon/PETG is highly anisotropic due to preferential fiber alignment in the extrusion direction. Samples were produced with the print (extrusion) direction aligned to the tension direction, as well as with the layer (Z) direction aligned to the tension direction. The modulus of elasticity of the aligned samples averaged a fourfold increase over the nonaligned samples.

The 3D printed thermoplastic material exhibited orthotropic properties, with the Young's modulus value dependent on the orientation of each print layer relative to the loading direction, which is measured and indicated using the symbols E1 and E2. E1 represents the Young's modulus parallel to the layer direction of the 3D print, while E2 represents the Young's modulus perpendicular to the layer direction. In addition to the Young's moduli, the following were also calculated and input as parameters in the Karamba3D FEA model: Poisson's ratio (nue12), transverse shear modulus parallel to the layer direction (G31), transverse shear modulus normal to the layer direction (G32), yield strengths (fy1 & fy2). Values for the in-plane shear modulus (G12), a specific weight (gamma), and coefficients of thermal expansion (alphaT1 & alphaT2) were taken from the manufacturer's product data sheets of an equivalent carbon PETG thermoplastic product.

Fin Cross-section Depth Sizing

The final depth dimensions of each tapered fin were determined using an iterative process. An iterative process was used due to the loading forces on each fin being unique to its shape and orientation.The two variables (upper and lower fin depths) were input as integers ranging from 20mm to 400mm. The upper fin depth—representing the minimum fin thickness—remains static, while the lower fin depth is increased from the 20mm until the simulated fin deflection value is below the acceptable deflection value. This process is repeated for each fin, resulting in a series of fin geometries optimized for minimal material use, while still meeting the requirements for construction. After the optimization process is complete, the resulting geometry was contoured, and a toolpath was extracted from the algorithm. This toolpath was then input into SuperMatterTools and exported into KRL to be printed using the robotic pellet printing process.

Case Study

To produce cast-in-place concrete sandwich insulated walls with reusable PETG formwork, we developed an integral computational design method. The development of the formwork system involves the computational design, structural simulation, and optimization. The following goals guided the design of the prototype:

1. Structural deformation of the 3D printed formwork under its own dead load and while casting; integration of steel reinforcement
2. Minimizing concrete consumption

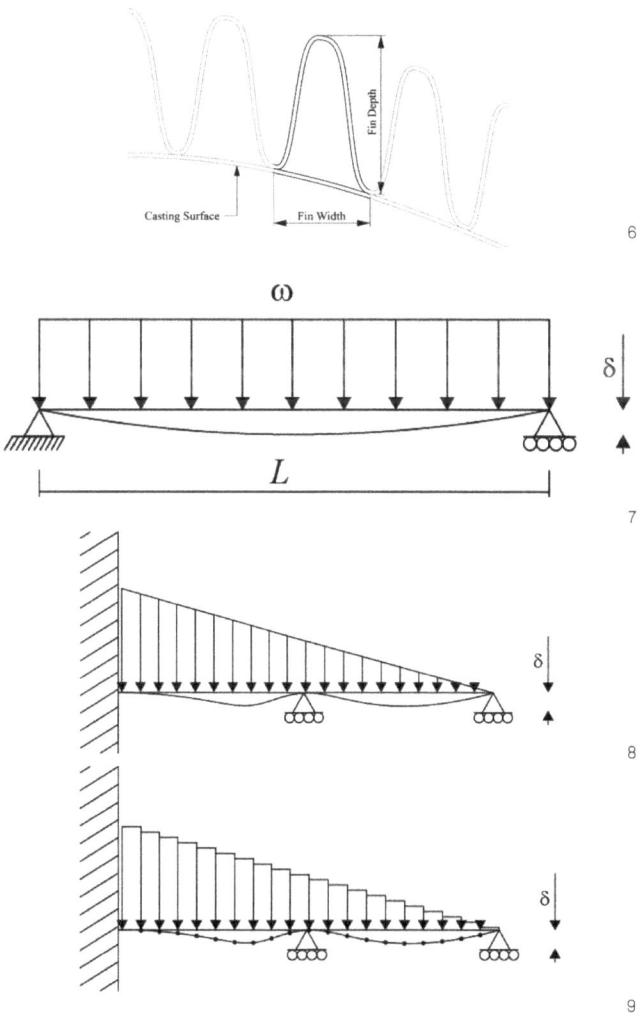

3. Minimizing formwork material usage
4. Easy on-site assembly and formwork removal processes
5. Improving the thermal performance of non-planar concrete walls

Integrative and Load Bearing Formwork Assemblage

The developed cast-in-place formwork system for concrete sandwich insulated wall comprises several parts: removable outer formworks, an insulated core element, rebar reinforcement, and wythe ties. The wythe ties are waterjet cut with notches that fix the rebar elements in place for the duration of the casting process. When fully assembled, these components become the formwork for two concrete wythes (approx. 64mm in thickness) connected by FRP rods that penetrate through the insulative core. The overall finished wall is 200mm in depth.

The Outer Layers of the Formwork Assemblage

The outer parts of the formwork system play two major roles: resist hydrostatic pressure of the concrete mix during pouring and produce the surface finish of the

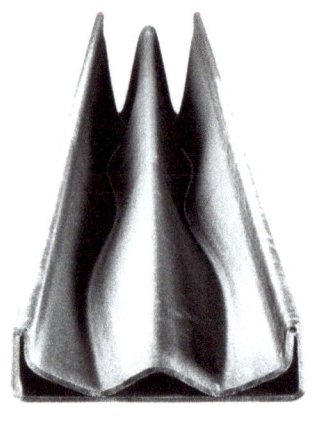

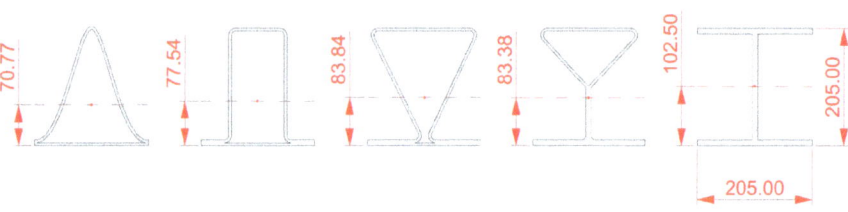

10 3D-printed tapered fin element: (left) view from the top; (right) view from the bottom

11 Cross-section profile analysis for fin design

12 3D-printed PETG prototypes: (left) untreated casting surface; (right) filled polyester resin coated casting surface

13 Parameter study of fin layout

14-15 Deflection scale x30 for clarity

concrete cast. These parts are robotically 3D-printed, using a pellet extruder with carbon fiber reinforced PETG. Hydrostatic pressure is addressed through a fin-based reinforcement system that is integrated into the thin shell of the 3D-printed formwork. Each fin follows the surface curvature of the casting surface while spanning from top to bottom, thus transferring the load of the hydrostatic and gravity forces to the ground. The vertical edges also pick up additional loads through a CNC-milled plywood profile at the two ends of the formwork, secured to the ground plane with fasteners and adhesive.

Several iterations of formwork reinforcement systems were considered throughout the research. Iterations explored different shape profiles, spacing, depth, and orientations of the fins (Figures 11 & 13). A triangular profile provided geometrical stiffness using a minimal amount of material. Each fin consists of a triangular profile, which is smoothed into a continuous NURBS curve for increased accuracy and speed in the 3D printing process. While the extrusion system in use is positionally synchronized with the robot motion, smoothing the toolpath allows the robot to maintain a smoother velocity profile and, therefore, a more consistent bead width. This effect of smoothing on path accuracy is a complex, non-linear problem, but for formwork applications, surface finish is of high importance and this is improved by smoothing. Additionally, the fin profile tapers corresponding to the reduction in hydrostatic pressure from the bottom to the top of the formwork.

When left untreated, the surface of the 3D-printed carbon-reinforced PETG formwork is rough with small ridges forming at each layer of the printed toolpath at approximately 1mm spacing. A surface finish test was conducted comparing untreated casting surfaces, sanded/polished casting surfaces, and polyester resin-treated casting surfaces (Figure 12). Surface prototypes were waxed prior to casting and removed after 48 hours. The untreated surface required significant force to remove, while the sanded surface (although smooth to the touch) left a visual indication of the toolpath layers. The polyester-coated finish resulted in a smooth finish that required the least amount of force to remove. Coating the mold surface has clear benefits for the reusability of the mold, but potential drawbacks affecting the recyclability of the formwork.

The Internal Insulative Core

The internal core was designed to serve two purposes. First, it provides an insulative thermal break in the interior of the concrete wall, and secondly, it supports the rebar reinforcement during the casting process (Figure 16).

 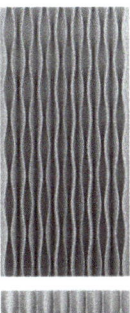  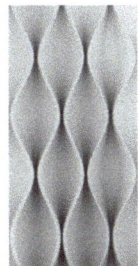  
   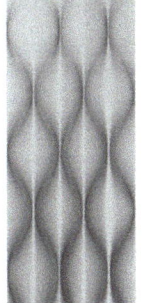 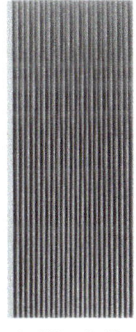 

13

The parametric design of the core system was developed in a way to increase the insulative performance and reduce the volume of concrete in the concrete wall. The overall geometry of the core follows the form of the wall and suspends a lattice of rebar approximately 25mm from the exterior casting surfaces. A grid of pillow-like forms in the core system protrude between individual cells of the rebar lattice, displacing the concrete mixture while still maintaining the code-mandated rebar coverage for concrete structures (American Concrete Institute 2014). The resulting geometry is a waffle-like layer of polyurethane foam—varying in thickness from 50mm to 100mm—which reduces the layer of concrete by 8.8% over a traditional smooth surface insulative core. One half-inch diameter FRP wythe ties protruded 50mm from each side of the core, spaced roughly 600mm apart. The wythe ties hold together both sides of the concrete wall and resist shear. Each tie was notched to carry rebar in both the vertical and horizontal directions. The location of these wythe ties were indicated using a marker in the printed toolpath.

Unlike the outer formwork, this printed core element remained permanently in the wall. It was printed from standard PETG (which is more economically priced than the carbon-infused variant) and acted as a stay-in-place formwork for a polyurethane foam cast. Once the foam cured, the rebar was clipped into the FRP wythe ties and tied together with standard rebar ties. While casting concrete, this insulative core, with installed wythe ties and rebar elements, was attached to the ground plane with adhesive and secured with form ties from above to prevent floating or shifting during the concrete pouring process.

RESULTS

For this case study, a 2.4m tall x 2.0m wide lofted, freeform NURBS surface was input (Figure 14). The formwork solution needed to resist up to 49.4kN/m2 of hydrostatic pressure, with a maximum allowable deflection value of 3mm along with the height of the formwork and 1mm across the width of each fin.

The calculated maximum fin width span value of 117mm, results in 18 fins on each side of the wall formwork. Due to the inputted NURBS surface being a ruled surface, all individual fins share very similar cross-sectional profiles, tapering from a 170mm deep base to a 40mm deep top profile (Figure 15).

The amount of thermoplastic material was calculated for both sides of the 2.4m x 2.0m form. With a total surface area of 9.90m^2, the casting surface accounted for 0.0495m^3 of thermoplastic material or approximately 62.8kg. The fin support structure amounted to 21.75m^2 of

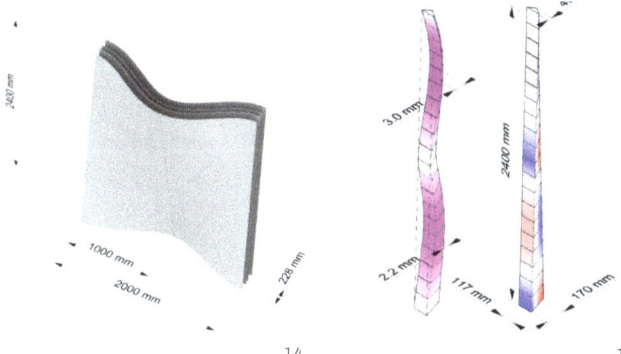

14   15

16 Internal core formwork with pillow-like geometry as a stay in place formwork for foam (left). Notched wythe ties that hold rebars in two directions and keep the two concrete wyths together.

17 Full-scale 3D print of the inner core for casting insulation materia

18 Assembled formwork, with insulative inner core and rebar.

surface area, requiring 0.1087m³ of thermoplastic material, or approximately 137.9kg of material. The ratio of casting surface to support material is 1:2.2. The overall outer formwork requires approximately 200.7kg of carbon PETG thermoplastic material.

Robotic FGF 3D printing using allowed for assembly of an integrative, ready-to-cast formwork with insulation and steel reinforcement (Figure 18). While the resulting formwork appears to satisfy the requirements for successful casting at construction scales, several challenges related to large-scale pellet extrusion remain. Due to the large aspect ratio (height vs. depth) of the prints, stability of the part during printing is a challenge. In the case of the external walls of the formwork, the addition of the stiffening ribs designed to overcome hydrostatic pressure assists with stability during printing. However, in the insulative cavity, which has no ribbing, the lack of support while printing led to issues with warping and part removal from the heated bed. Future work will explore the addition of internal ribs to help stabilize this component. Another known challenge is the ability to print complex geometries at high speeds. While pellet extruders are capable of high-speed extrusion, the motion control platform, in this case, a seven-axis industrial robot, is limited in its ability to maintain high speeds around sharp corners. In general, cooling of the part during printing is an ongoing challenge that limits both the maximum printing speed (as it sets a "minimum" layer-to-layer time) and causes issues with accuracy due to asymmetric cooling and warping. While smaller-scale printers address this via heated build chambers, this is challenging to deploy at the volumes required by full-scale construction molds.

CONCLUSION

The result is a novel, digital design-to-fabrication workflow for producing freeform formwork that allows cast-in-place, thermally-insulated concrete sandwich panel walls. The integration of computational design and structural optimization combined with 3D printing provides the potential to

rethink concrete formwork systems. Designing and building a structurally optimized, rib-stiffened shell formwork will allow the fabrication of non-uniform concrete forms economically and with minimal material waste. The integration of assembly details and supports further reduces the labor required to assemble the molds and place steel reinforcement. Future research is planned to determine the overall geometric limitations of the process, as well as to further optimize the overall design-to-fabrication workflow.

REFERENCES

Aghaei Meibodi, Mania, Mathias Bernhard, Andrei Jipa, and Benjamin Dillenburger. 2017. "The Smart Takes from the Strong." In *Fabricate 2017*, edited by B. Sheil, A. Menges, R. Glynn, and S. Marilena. Stuttgart: University of Stuttgart. 210–17.

Aghaei Meibodi, Mania, Andrei Jipa, Rena Giesecke, Demetres Shammas, Mathias Bernhard, Matthias Leschok, Konrad Graser, and Benjamin Dillenburger. 2018. "Smart Slab: Computational design and digital fabrication of a lightweight concrete slab." In *ACADIA '18: Recalibration. On Imprecision and Infidelity; Proceedings of the 38th Annual Conference of the Association for Computer Aided Design in Architecture*. Mexico City, Mexico. 320–327.

Aghaei Meibodi, M., R. Gieseck, and B. Dillenburger. 2019. "3D Printing Sand Molds for Casting Bespoke Metal Connections; Digital Metal: Additive Manufacturing for Cast Metal Joints in Architecture." In *Proceedings of the 24th International Conference of the Association for Computer-Aided Architectural Design Research in Asia (CAADRIA 2019)*. Wellington, New Zealand. 133–142.

Aghaei Meibodi, Mania, Pietro Odaglia, and Benjamin Dillenburger. 2021. "Min-Max: Reusable 3D printed formwork for thin-shell concrete structures; Reusable 3D printed formwork for thin-shell concrete structures." In *Caadria2021; Proceedings of the 26th CAADRIA Conference*, vol. 1. Hong Kong and Online. 743–752.

Aghaei Meibodi, Mania, Christopher Volt, and Ryan. 2020. "Additive Thermoplastic Formwork for Freeform Concrete Columns." In *ACADIA '20: Distributed Proximities; Proceedings of the 40th Annual Conference of the Association of Computer Aided Design in Architecture*. Online + Global.

American Concrete Institute. 2014. "Building Code Requirements for Structural Concrete (ACI 318-14)." Farmington Hills, MI: American Concrete Institute.

ASTM International. 2014. "Standard Test Method for Tensile Properties of Plastics (D638-14)." West Conshohocken, PA: ASTM International.

Burger, Joris, Ena Lloret-Fritschi, Fabio Scotto, Thibault Demoulin, Lukas Gebhard, Jaime Mata-Falcón, Fabio Gramazio, Matthias Kohler, and Robert J. Flatt. 2020. "Eggshell: Ultra-Thin Three-Dimensional Printed Formwork for Concrete Structures." *3D Printing and Additive Manufacturing* 7 (2): 48–59.

Go, Jamison, Scott N. Schiffres, Adam G. Stevens, and A. John Hart. "Rate limits of additive manufacturing by fused filament fabrication and guidelines for high-throughput system design." *Additive Manufacturing* 162 (Aug.): 1–11.

Jipa, Andrei, Mathias Bernhard, Benjamin Dillenburger. 2017. "Submillimeter formwork. 3D-printed plastic formwork for concrete elements." In *TxA Emerg Design + Technology Conference*. Austin, Texas.

Jipa, A., M. Bernhard, B. Dillenburger, N. Ruffray, T. Wangler, R. Flatt. 2017. "skelETHon Formwork 3D Printed Plastic Formwork for Load-Bearing Concrete Structures," In *Blucher Design Proceedings*. Concepción, Chile: Editora Blucher. 345–352.

Jipa, Andrei, Mathias Bernhard, Benjamin Dillenburger, Nicolas Ruffray, Timothy Wangler, and Robert J. Flatt. 2017. "skelETHon formwork 3D printed plastic formwork for load-bearing concrete structures". In *Congreso Internacional de la Sociedad Iberoamericana de Gráfica Digital*.

Lab, Robert H. 2007. "Think Formwork–Reduce Cost." *Structure Magazine*, April 2007: 14–16.

Leschok, Matthias, and Benjamin Dillenburger. 2019. "Dissolvable 3DP formwork." In *ACADIA '19: Ubiquity and Autonomy; Proceedings of the 39th Annual Conference of the Association for Computer Aided Design in Architecture*. Austin, Texas. 178–187.

Morel, Philippe, and Thibault Schwartz. 2015. "Automated Casting Systems for Spatial Concrete Lattices." In *Modelling Behaviour: Design Modelling Symposium 2015*. Cham: Springer.

Peters, Brian. "Additive Formwork: 3D Printed Flexible Formwork". 2014. In *ACADIA '14: Design Agency; Proceedings of the 34th Annual Conference of the Association for Computer Aided Design in Architecture*. Los Angeles, CA. 517–522.

Preisinger, Clemens. 2013. "Linking Structure and Parametric Geometry." *Architectural Design* 83: 110–113.

Preisinger, Clemens. *Karamba3D*. V.1.3.3. Karamba3D. 2021. PC.

Rutten, David. *Grasshopper*. V.1.0.0007. Robert McNeel & Associates. 2021. PC.

IMAGE CREDITS

Digital Architecture Research & Technologies (DART), Taubman College, University of Michigan.

Mania Aghaei Meibodi is Assistant Professor of Architecture and Director of the Digital Architecture and Research Technologies (DART) Lab at Taubman College.

Ryan Craney is a research assistant at DART Lab at the University of Michigan.

Wes McGee is Associate Professor and Director of the Fabrication and Robotics Lab at the University of Michigan Taubman College of Architecture and Urban Planning and a partner at Matter Design.

3D Printed Concrete Tectonics

Assembly Typologies for Dry Joints

Aya Shaker
D-Fab/ D-ARCH, ETH Zurich

Noor Khader
D-Fab/D-ARCH, ETH Zurich

Lex Reiter
Physical Chemistry
of Building Materials/
D-BAUG, ETH Zurich

Ana Anton
Digital Building Technologies
D-ARCH, ETH Zurich

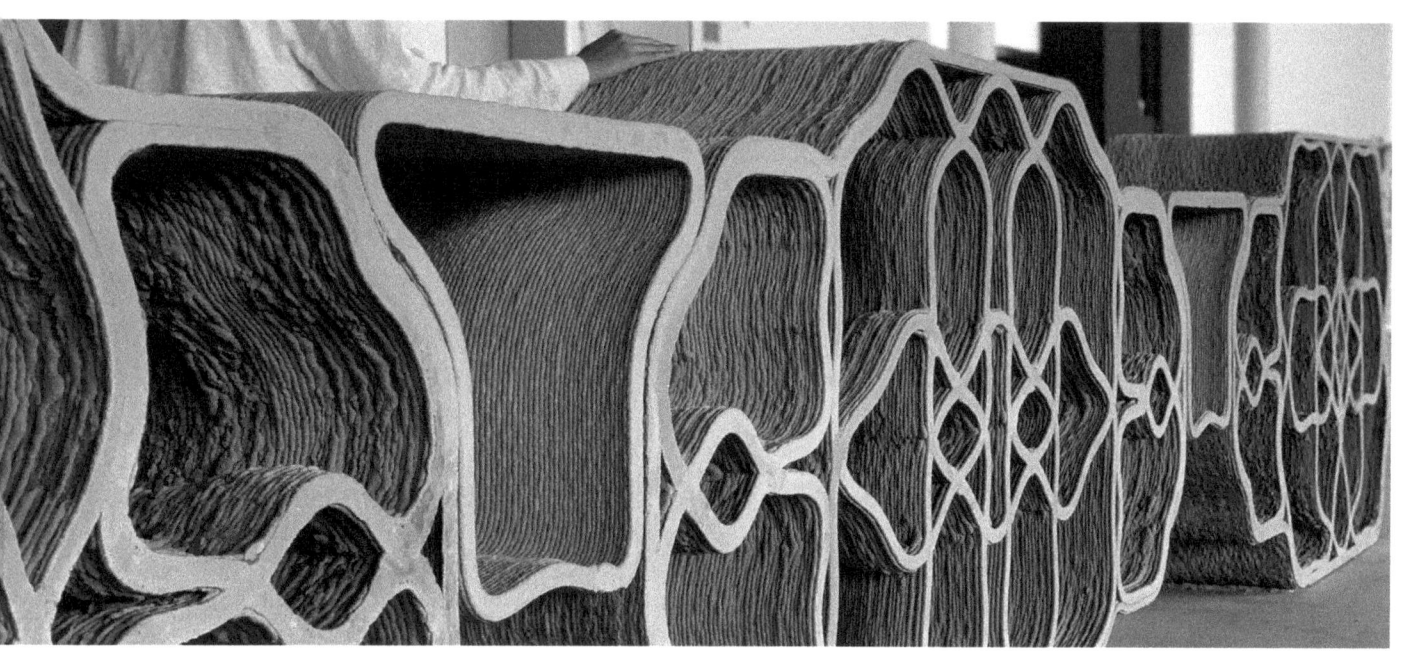

1   Close-up of assembled 3D-printed concrete components

ABSTRACT

Digital fabrication technologies and additive manufacturing techniques opened new opportunities and new challenges for the construction industry. Particularly, Concrete Extrusion 3D Printing (CE3DP) introduces valuable opportunities for large-scale architectural elements. However, segmentation and assembly strategies have not been developed, and it remains a limiting factor for the expansion of concrete 3D printing to an industrial scale.

In this context, the present research focuses on the design and fabrication possibilities of assembly interfaces, an essential topic for scaled-up 3D-printed concrete components. Therefore, dry assembly interfaces in different printing orientations are prototyped to investigate characteristics and limitations of connection options.

3D PRINTED CONCRETE TECTONICS

Introduction

Concrete Extrusion 3D Printing (CE3DP) can be defined as the process of joining the deposited concrete layer upon layer to create pre-designed objects from digital data. The emergence of CE3DP technology introduces new possibilities to the construction field and changes the paradigm of how architects can conceptualize, design, and build. However, the collective body of assembling such elements had so far been limited due to fabrication constraints specific to CE3DP. This paper describes a research project with the primary goal of exploring the tectonics of dry joints in 3D concrete printing in light of specific fabrication and design parameters.

The architectural term tectonics is described as the art of joining (Frampton et al. 2007). It refers not just to the activity of making the materially requisite construction that answers certain needs but rather to the activity that raises this construction to an art form (Maulden 1986). Today, tectonics is taking on another dimension with the rise of advanced digital fabrication techniques. With the introduction of CE3DP in architectural practice, a quest to redefine new tectonic expression starts (Wolfs and Salet 2016).

State of the Art

Recently, several bridges have explored potential opportunities and challenges with 3D-printed concrete tectonics (Salet et al. 2018). One of the latest 3D-printed concrete projects is the topology optimized prestressed 3D-printed concrete where the concrete segments (Figure 3) were brought together through the planar interface of each end (Valdivieso 2019; Vantyghem et al. 2020). The same assembly method was applied in the Baoshan pedestrian bridge project (Figure 4) (Xu et al. 2020). Both bridges underline the current state of the art with regards to the assembly strategy of 3D-printed components, where the assembly interfaces are commonly planar because of printing on a flat surface.

Another assembly approach was joining 3D-printed components at the surface's sides. This approach provided noticeable geometrical freedom at the interfaces as it's not confined to the planarity of the printing substrate. An example of this can be seen in the 3D-printed double-curved concrete shell structure (Figure 5) (Schipper et al. 2017). The printed segments inherited the surface geometry of the supporting double-curved mold, which allowed for an intricate interface to be realized. *Cohesion Pavilion* demonstrates an assembly strategy where the interface between components is informed by the outer geometry (Figure 7) (Grasser et al. 2020). A similar assembly strategy

2

3

4

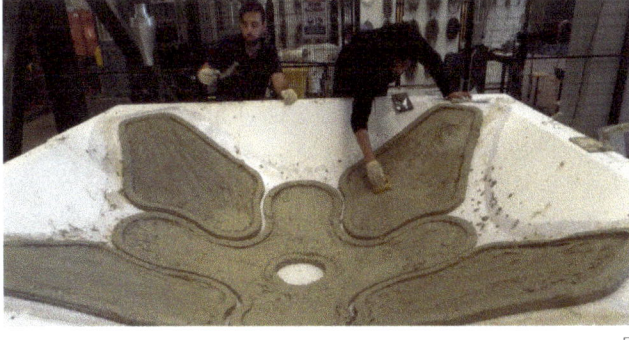

5

2   3D-printed concrete columns, DBT, ETH Zurich (Anton et al. 2020)

3   Flat assembly of 3D-printed concrete components of the Topology optimized bridge (Valdivieso 2019)

4   Flat assembly of 3D-printed concrete components of the Baoshan bridge (Xu et al. 2020)

5   Sideway assembled components of the 3D-printed double-curved concrete shell structure (Schipper et al. 2017)

6 Sideway assembled components of the 3D printed concrete house (Carra and Stabile 2018)

7 Sideway assembled components of the *Cohesion Pavilion* (Grasser et al. 2020)

was implemented in the 3D printed concrete house (Carra and Stabile 2018), where the 3D printed walls are joined sideways (Figure 6). In the previous examples, an optimized 3D printing process is crucial to achieving controlled surface quality and precise interface for an efficient side-to-side assembly.

To previous work, concrete 3D printing offers more geometrical freedom, and this has led to realizing complex bespoke designs. Nevertheless, assembly logic and connection interfaces of 3D-printed concrete components are still bound to certain limitations of the printing technique, such as the horizontal printing, slicing, and segmentation of elements, printing path, precision, and quality control of surfaces. In this research context, we find potential in rethinking 3D-printed concrete tectonics and developing a new design language of connection interfaces with CE3DP.

METHODS

The research methodology intertwines the design and fabrication processes. The key elements of the methodology are design research, file to production digital workflow, fabrication set up, prototyping, and a full-scale demonstrator.

Design Research

The design investigated the existing assembly typologies in concrete 3D printing. Based on the advantages and drawbacks of each strategy, categories of assembly strategies were drawn from the precedent works and the prototypes developed in this research.

File to Production Digital Workflow

A comprehensive parametric workflow of *design geometry generation*, *geometry validation*, and *analysis* were developed to ensure the printing feasibility of the design geometry according to the material constraints, *3D geometry slicing*, and *printing path generation*. In addition, the digital tools provided data such as printing path length and approximate concrete volume required for material preparation.

Fabrication Setup

The robotic fabrication setup shown in Figure 8 (Anton et al. 2021) consists of a 6-axis robotic arm on an overhead 3-axis gantry system, concrete pump, accelerator pump, and an online communication station. At the robotic arm end effector, a concrete mixing and extrusion tool is attached with two segregated material inlets, one of which is connected to the concrete pump through a hose while the other inlet receives the accelerating admixture from a small pump via a thin plastic tube. The two components are mixed in the top part of the tool and extruded through through a nozzle of 15mm diameter opening. The two pumps are running continuously during the printing process. As a result, the print path must be continuous and due to the to the low initial yield stress of the filament, the cantilevering freedom is limited. This CE3DP process is described by ETH Zurich researchers (Anton et al. 2021), and the online communication system uses COMPAS-FAB (Rust et al. 2021) and COMPAS-RRC (Fleischmann et al. 2021).

Prototypes

Full-scale 3D-printed concrete prototypes were fabricated to assess the assembly typology with reference to the fabrication parameters, material behavior and printing tolerance. To set a framework for prototyping, fixed values for design and fabrication parameter, which evolved from previous CE3DP projects were set. For design, the inclination angle of any surface was maintained to a maximum of 15°. As for fabrication, the compositions of both the concrete and accelerator mixtures were fixed, and the speed of the robotic arm was set to 350mm/s. The assembly typologies explored in this research (Figure 9) includes two main interface types. The difference between them is the direction at which the elements join. The

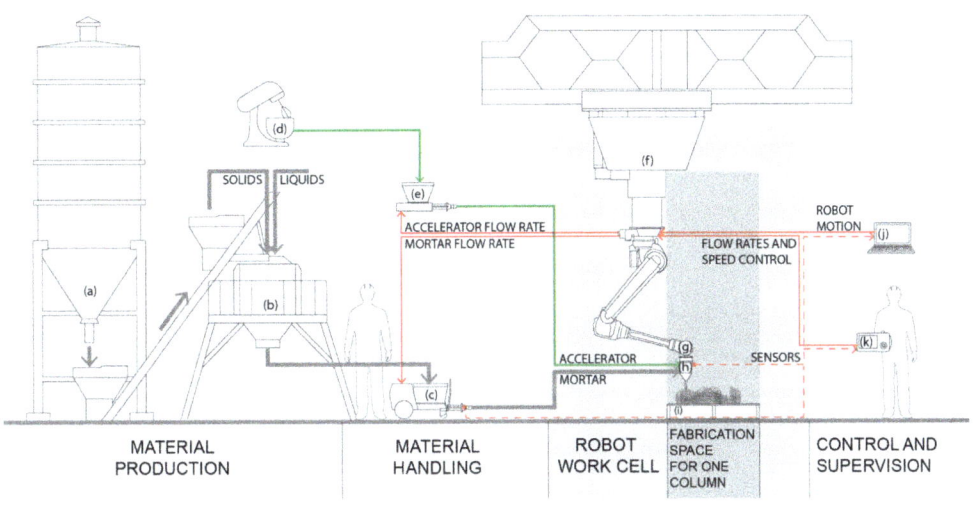

8   Diagram of the fabrication setup used in the project: (a) dry mix silo; (b) concrete mixer; (c) concrete pump; (d) accelerator mixer; (e) accelerator pump; (f) 3-axis gantry; (g) robot manipulator; (h) extruder tool; (i) work object; (j) computer; (k) teach pendant (Anton et al. 2021)

9   Assembly typologies

first interface type is where elements are joined along the normal of the printed layers, throughout the paper, this will be referred to as the End-to-End interface. Inversely, the second interface is where components are connected perpendicularly to the normal of the printed layers, this type is referred to as the Side-to-Side interface. As the nature of the vertical printing process does not allow for non-planar interfaces on both ends of the printed components, a hybrid interface type of End-to-End and Side-to-Side was sought after. This paper addresses side-to-side interfaces and hybrid interfaces.

Assembly Typologies: Side-to-Side Interface
Interlocking Components—In this approach, the limitations on the geometry of the surface where the interface occurs are very minimal: it can be a planar, single, or double-curved surface. A two-part prototype of a U-shaped interface (Figure 10) was printed to understand the potential of this interface type in terms of fitting accuracy.

Keyway Interface—There are two distinct differences in this subtype from the previous one. First, the number of layers at which the components meet is less. Second, the components touch each other rather than fitting inside one another.

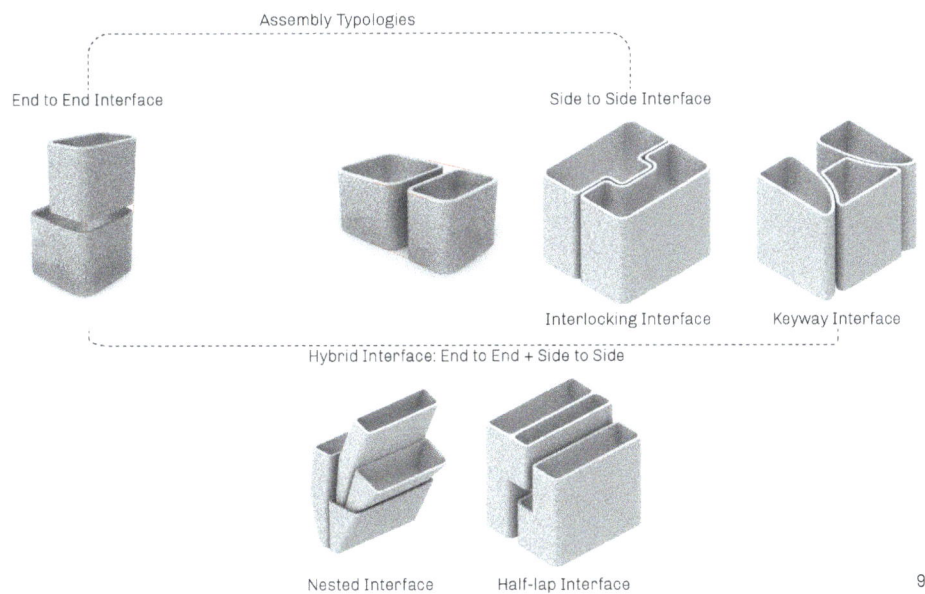

This problem is resolved in the Keyway subtype due to the significant reduction of the contact surface to a small number of printed layers. This increases the probability of having better-fitting printed components. This specific logic can be applied to different geometries while maintaining the technique's advantage.

Assembly Typologies: Hybrid Interface

Nested Components—Moving gradually towards upscaling, a temporary stand-alone structure of half an arch was designed and broken down to multiple components. The nested approach (Figure 11) emerged as a solution to connect the 3D-printed components in a predefined position through designing one part with a wider cross-section than the other to allow for partial or full nesting. A tolerance gap of 10mm was provided between all assembly faces.

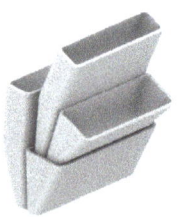

The varying width of the cross-section of each component provides an external surface to conceal the assembly detail, which leads to an overall homogenous reading of the assembled form (Figure 13). Moreover, the function of the internal geometry shown in Figure 12 is performance-oriented. To elaborate, the internal geometry provides a surface for the soft timber wedges, shown in Figure 14. Together, the internal geometry and the timber wedges allow for an adequate load transfer that takes advantage of the great compressive strength of concrete, while reducing the potential tension build up within the nested structure. Moreover, the timber wedges form an inclined plane that is necessary to build up the overall inclination in increments.

Half-lap Interface—The research investigated connecting two components by creating a two-level interface (Figure 15) as a workaround for the technique's limitation of having at least one flat end.

Results—The full-scale assembled Half-Lap prototype shows that the End-to-End interface requires a very accurate printing quality and does not allow for imprecision in the orthogonality, which remains a main repercussion of this fabrication technique. Significant vertical deformation at the end of the two segments can be seen in Figure 16, creating mismatching geometries. Furthermore, as only one interface per element can be printed, it is limited to joining two elements.

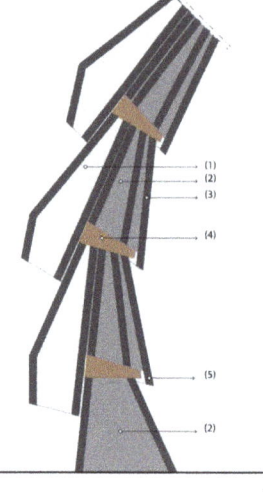

Results—In the interlocking prototype, it was observed that the provided tolerance gap of 10mm was sufficient for a smooth fitting. In all the experiments the tolerance was determined as half of layer width. However, successful interlocking can only happen in the case of a highly precise print along the entire length of contact surface.

Final Demonstrator

The goal of the final demonstrator, a segmented 3D-printed concrete bar, is to showcase an assembly strategy applied to a design suitable for 3D concrete printing. The bar stretches 7.5 meter in length and 40 centimeters in depth. The bar was segmented to five parts, three long segments

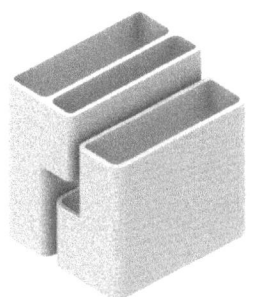

15
16
17

10 Interlocking prototype
11 Nested components
12 Optimized internal geometry
13 Nested components
14 Diagram section through four 3D printed components of the arch: (1) hollow cavity inside the component; (2) CE3DP porous infill; (3) outer shell section; (4) inclined plane used for assembly; (5) component edge that hides the assembly detail
15 Half-Lap interface
16 Half-Lap prototype
17 Keystone component

of self-intersecting geometry and two connecting keystones (Figure 17). The decision to work with side-to-side connecting interfaces was motivated by the connection aesthetics, segmentation strategy, and printing orientation of the porous structure. During the assembly process (Figure 18), the long segments were rotated 90° to a vertical position and aligned next to one another. Afterwards, the keystones were lowered into place to create a continuous 7.5m surface (Figure 19).

CONCLUSION

The results of the experiments carried out in this research have shown that various assembly typologies for 3D concrete printed components are possible through investigating design and fabrication parameters that are informed by the 3D printing process. Segmentation and assembly strategies are design-specific and need to be developed together with the printing strategy. For example, choosing a layer orientation different to that of the final position of the printed part extends design and fabrication spaces for CE3DP, especially in prefabrication setups. As explored in the Compound Fabrication project by MIT (Zivkovic and Battaglia 2018), post-processing methods such as 3D scanning and CNC milling could be implemented to create a refined surface quality along the interface which will overcome the local inaccuracies of the printed layers. An enhancement to the fabrication setup could be another approach to address the quality and precision of the printed object. Researchers at ETH Zürich have experimented with printing on a formwork surface instead of a regular planar printing bed (Anton et al. 2020). Furthermore, the research delivers a promising outlook for further development in 3D concrete printed tectonics. While designed as a dry connection, the proposed solutions highlight precision requirements and are applicable to masonry-type connections with the interface materials being mortar for permanent assemblies, or soft materials such as mats or wood wedges (Figure 10) for temporary joining. In addition to compensating some printing inaccuracies, an interface material also reduces local stress concentrations between the elements by distributing the load transfer through the connection to a larger surface, reducing splitting risk. To summarize, there are several approaches to address the imperfections resulting from the fabrication process of concrete 3D printing. These approaches allow for the load transfer between components including permanent assemblies through introducing a mortar-like material that compensates for surface imprecision, post process the superficial area of the printed parts, and use precise formworks during printing in the connection areas.

18 Components assembly process

ACKNOWLEDGEMENTS
We would like to thank Prof. Dr. Benjamin Dillenburger (Digital Building Technologies) for his support to this work and in particular his continuous guidance for 3D concrete printing research.

Our sincere gratitude goes to Michael Lyrenmann, Philippe Fleischmann, Tobias Hartmann (Robotic Fabrication Lab), Heinz Richner and Andreas Reusser (Concrete Lab) for the technical support in the realization of all the large-scale prototypes.

This research was supported by the NCCR Digital Fabrication, funded by the Swiss National Science Foundation (NCCR Digital Fabrication Agreement #51NF40-141853). ETH Zürich Foundation supported this research under the project number: 2019-FE-202.

REFERENCES

Anton, Ana, Andrei Jipa, Lex Reiter, and Benjamin Dillenburger. 2020. "Fast Complexity: Additive Manufacturing for Prefabricated Concrete Slabs." In *Second RILEM International Conference on Concrete and Digital Fabrication*, vol. 28. Cham: Springer. 1067–77. https://doi.org/10.1007/978-3-030-49916-7_102.

Anton, Ana, Lex Reiter, Timothy Wangler, Valens Frangez, Robert J. Flatt, and Benjamin Dillenburger. 2021. "A 3D Concrete Printing Prefabrication Platform for Bespoke Columns." *Automation in Construction* 122 (Feb.): 103467. https://doi.org/10.1016/j.autcon.2020.103467.

Anton, Ana, Patrick Bedarf, Angela Yoo, Benjamin Dillenburger, Lex Reiter, Timothy Wangler, Robert J. Flatt, et al. 2020. "Concrete Choreography: Prefabrication of 3D-Printed Columns." In *Fabricate 2020: Making Resilient Architecture*. London: UCL Press. 286–93. http://www.jstor.org/stable/j.ctv13xpsvw.41.

Carra, Guglielmo, and Luca Stabile. 2018. "Pushing the Boundaries of 3D Printing." *The Arup Journal* 1 (2018): 28–31. Accessed September 20, 2021. https://www.arup.com/-/media/arup/files/publications/t/the_arup_journal_issue_1_2018.pdf.

Fleischmann, Philippe, Gonzalo Casas, Michael Lyrenmann, and Beverly Lytle. 2020. *COMPAS RRC: Online control for ABB robots over a simple-to-use Python interface*. V.1.1.0. [Computer software]. Accessed July 7, 2021. https://doi.org/10.5281/zenodo.4639418.

Frampton, Kenneth, John Cava, and MIT Press, eds. 2007. *Studies in Tectonic Culture : The Poetics of Construction in Nineteenth and Twentieth Century Architecture*. Chicago, IL: Graham Foundation For Advanced Studies in the Fine Arts.

Grasser, G., L. Pammer, H. Köll, E. Werner, and F. P. Bos. 2020. "Complex Architecture in Printed Concrete: The Case of the Innsbruck University 350th Anniversary Pavilion COHESION." *Second RILEM International Conference on Concrete and Digital Fabrication. DC 2020. RILEM Bookseries*, vol 28. Cham: Springer. 1116–27. https://doi.org/10.1007/978-3-030-49916-7_106.

Maulden, Robert. 1986. "Tectonics in Architecture: From the Physical to the Meta-Physical". Accessed June 13, 2021. http://hdl.handle.net/1721.1/78804.

Rust, R., G. Casas, S. Parascho, D. Jenny, K. Dorfler, M. Helmreich, A. Gandia, et al. "Compas-Dev/compas_fab." *COMPASFAB*. Accessed July 7, 2021. https://github.com/compas-dev/compas_fab.

Salet, Theo A. M., Zeeshan Y. Ahmed, Freek P. Bos, and Hans L. M. Laagland. 2018. "Design of a 3D Printed Concrete Bridge by Testing." *Virtual and Physical Prototyping* 13 (3): 222-236. https://doi.org/10.1080/17452759.2018.1476064.

Schipper, Roel, Chris Borg Costanzi, Freek Bos, Zeeshan Ahmed, and Rob Wolfs. 2017. "Double Curved Concrete Printing: Printing on Non-Planar Surfaces." SPOOL 4 (2): 17–21. https://www.spool.ac/index.php/spool/article/view/59.

Valdivieso, Carlota. 2019. "Vertico 3D Prints a Bridge Using 60% Less Concrete." *3Dnatives*, September 23, 2019. Accessed September 20, 2021. https://www.3dnatives.com/en/vertico-bridge-230920195/.

Vantyghem, Gieljan, Wouter De Corte, Emad Shakour, and Oded Amir. 2020. "3D Printing of a Post-Tensioned Concrete Girder Designed by Topology Optimization." *Automation in Construction* 112 (Apr): 103084. https://doi.org/10.1016/j.autcon.2020.103084.

Wolfs, R. J. M., and T. A. M. Salet. 2016. "Potentials And Challenges In 3D Concrete Printing." In *Proceedings of the 2nd International Conference on Progress in Additive Manufacturing*. Singapore: Research Publishing. 8–13.

19 Final demonstrator, 3D-printed concrete bar

Xu, Weiguo, Yuan Gao, and Chenwei Sun. 2020. "3D-Printed Concrete Structural Components in the Baoshan Pedestrian Bridge Project." In Fabricate 2020: Making Resilient Architecture, edited by Jane Burry, Jenny Sabin, Bob Sheil, and Marilena Skavara. London: UCL Press. 140–47.

Zivkovic, Sasa, and Christopher Battaglia. "Rough Pass Extrusion Tooling. CNC post-processing of 3D-printed sub-additive concrete lattice structures" In *ACADIA '18: Recalibration, On Imprecision and Infidelity; Proceedings of the 38th Annual Conference of the Association for Computer Aided Design in Architecture.* 302–311.

IMAGE CREDITS

Figure 3: ©Vertico 3D concrete printing, 2019
Figure 4: ©Tsinghua University School of Architecture, 2019
Figure 5: ©TU Delft & TU Eindhoven, 2017
Figure 6: ©Arup & CLS Architects, 2018
Figure 7: ©Innsbruck University & TU Eindhoven, 2020
Figure 19: ©Axel Crettenand, DBT, 2019
All other drawings and images by the authors.

Aya Shaker is an Architect and automation specialist at the digitalization team at BESIX Middle East, where she is responsible for the development of digital tools that enhance construction workflows and project control. In 2019, her interest in Innovative Building Technologies within the construction field led her to pursue Master's degree in architecture and digital fabrication from the Institute of Technology in Architecture, ETH Zurich. Aya received her BArch from the American University of Sharjah in 2014.

Noor Khader is a scientific assistant at the Institute of Structural Design, TU Braunschweig. She received her Master's degree in architecture and digital fabrication at the Institute of Technology in Architecture, ETH Zurich in 2019. She is interested in rethinking architecture through the interplay between design and other disciplines. Her academic and career goals include exploring design and fabrication methods in order to push boundaries and expand the possibilities of what the built environment is about and what it could be.

Lex Reiter is a post doctoral researcher at ETH Zürich working on early age strength build-up and its control for digital fabrication processes with concrete among which is layered extrusion. His research interest is in the physical and chemical processes that allow building without formwork and at high vertical rate as well as associated processing challenges.

Ana Anton is a PhD Researcher at the Chair for Digital Building Technologies, Institute of Technology in Architecture at the Department of Architecture, ETH Zurich. She is associated with the National Centre for Competence in Research – Digital Fabrication, where she leads the research in concrete 3D printing.
She received her architectural degree, cum laude, in the TU Delft, in 2014 and continued her research until 2016 as part of Hyperbody Research Group. While her scientific research addresses complexity and emergence in architecture, her designs exploit materiality encoded for digital fabrication. Her current thesis, "Tectonics of Concrete Printed Architecture," focuses on robotic concrete extrusion processes for large-scale building components.

A Hybrid Additive Manufacturing Approach

Combining Additive Manufacturing and Green-State Concrete Milling to Create a Functionally Integrated Loadbearing Concrete Panel System

Philipp Rennen
Institute of Structural Design/
TU Braunschweig

Noor Khader
Institute of Structural Design/
TU Braunschweig

Norman Hack
Institute of Structural Design/
TU Braunschweig

Harald Kloft
Institute of Structural Design/
TU Braunschweig

1   Close-up of 3D printed concrete panel system

ABSTRACT

Research in the field of additive manufacturing with concrete has gained enormous momentum in recent years. In practice, the first fully functional and habitable buildings have been realized. While these lighthouse projects have proven the general feasibility of 3D printing in construction, in the future it will be a matter of further expanding the potential of 3D printing, addressing important topics such as functional integration (reinforcement, piping, fasteners), material gradation (load-bearing, insulating) as well as disassembly and reuse.

As part of an international competition organized by LafargeHolcim Ltd. and its partners Witteveen & Bos, COBOD, and Fondation des Ponts which focused on realigning a traditionally manufactured residential building to concrete 3D printing technology, a team of students and researchers have developed a concept for a modular, function-integrated panel system for individualized wall and ceiling elements. The system is characterized by the fact that the integrated modular structures are printed flat on the floor and precise connections and structural joints are subtracted while the concrete is still in its green state.

INTRODUCTION

While the emergence of large-scale 3D printing processes introduced considerable opportunities for a sustainable and efficient construction industry (Delgado Camacho et al. 2018; Paul et al. 2018; Lu et al. 2019), 3D printed concrete structures face a series of challenges that need to be investigated such as structural reinforcement, integration of functions, surface quality and transportation of elements (Salet and Wolfs 2016; Raval and Patel 2020; Panda et al. 2018). Today, most construction-scale 3D-printed concrete structures are produced in situ and tend to be monolithic and monofunctional due to fabrication constraints related to the printing process. In addition, the printed structures inherit an undifferentiated surface texture as a result of the deposited concrete layer upon layer which limits the freedom of surface expression. Moving beyond monolithic and monofunctional structures requires investigating aspects of assemblability, functional integration, and accessibility of future maintenance of the 3D-printed concrete components.

As part of an international competition that focused on rethinking 3D-printed concrete housing, a novel solution for individualized 3D-printed concrete structures was explored through examining two main fabrication aspects. Firstly, the functional integration using Shotcrete 3D Printing (SC3DP); and secondly, the surface finishing using Green-State Concrete Milling (GSCM).

With SC3DP technique the concrete is sprayed with pressure in order to build up three-dimensional and free-form concrete structures. SC3DP enables enhanced layer adhesion (Kloft et al. 2020), controllable material addition and spatial printing freedom. This technique is then followed by GSCM, which can be described as the post-processing of the printed concrete in its green state, meaning it has set but has not cured completely. Such process allows for controlled surface quality as well as precise detailing for joints and connection points.

In this context, this paper presents a new possible automated fabrication approach that integrate several functionalities that are essential in the realization of customized 3D-printed concrete structures on the construction-scale.

STATE OF THE ART

Functional Integration in 3D Concrete Printing

Recently, several 3D-printed projects have broadened the range of functional integration. An example of this can be seen in the in situ two-story residential building developed by PERI GmbH in collaboration with COBOD and

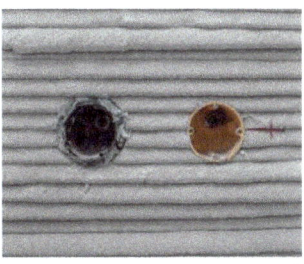

2 Integrated insulation into the printed, double-shell cavity walls
3 Integrated empty pipes and electrical sockets
4 & 5 Integrated chimney and kitchen substructures

6 The Bridge Project, Studio michiel van der kley & TU Eindhoven
7 Tsinghua University 3D-printed concrete bridge
8 3D-printed concrete bridge components, BAM Infra & TU Eindhoven

9 Milling and smoothing in the green state of the concrete, ITE, TU Braunschweig

10 Post processing 3D-printed concrete column, ITE, TU Braunschweig

Mense-Korte engineers + architects, Germany (PERI GmbH 2020). The building consists of triple-skin cavity walls, which are filled with an insulating compound. The printing process considered functional integration of reinforcement, pipes and connections (water, electricity, etc.) and also multi-use substructures (Figures 2-5). This project demonstrates the potential of comprehensive execution of 3D-printed concrete structures through functional integrative fabrication process. One noteworthy point here is that most of the functionally integrated 3D-printed concrete structures tend to not only be monolithic but they also share similar design typology due to fabrication limitation of the 3D extrusion-based technique.

Prefabrication in 3D Concrete Printing

Another crucial area of exploration for an efficient and sophisticated 3D-printed concrete structures is off-site prefabrication. One building typology that illustrates the potential of prefabricated concrete structures are 3D-printed concrete bridges (Figures 6-8). Similar as in conventional concrete construction, 3D printing also offers the possible assembly of complex geometries from smaller prefabricated objects. These projects (Bos et al. 2019; Xu et al. 2020; Salet et al. 2018; Prior 2021) proved that prefabricated concrete components allow for more design freedom, efficient use of materials and controlled fabrication environment (Menna et al. 2020; Anton et al. 2021). What remains necessary to further improve the performance of prefabricated 3D-printed concrete structures is the dimensional accuracy and precision of the connecting interfaces.

Design for Disassembly and Subtractive Manufacturing in Green State

In addition to the previous scope of manufacturing, methods of CNC milling 3D-printed concrete structures are being recently explored to enable controlled surface quality and precise geometric feature. This was illustrated in the projects of Adaptive Modular Spatial Structures (Figure 9) and the Dry Joint Column (Figure 10) at the Institution of Structural Design in TU Braunschweig where a digital fabrication system was developed using SC3DP process with the focus on segmentation and jointing of elements (Hack et al. 2019; Kloft et al. 2019). In this project, a robotic CNC milling process was introduced to post-process the printed concrete structure in order to create a precise dry joint. Another hybrid robotic 3D printing and milling process has been developed at the University of Loughborough and the university of Sheffield to allow for precise production of 3D-printed concrete parts (Kinnell et al. 2021).

METHODS

According to the design brief to develop solutions for faster, more material-efficient and less geometrically constrained buildings, a 3D printing workflow for the production of adaptive panels was developed. Instead of printing vertically from ground up, here previously introduced strategy was chosen in which the panels are printed flat on the ground, allowing full freedom of shape in two-dimensional space. Printing on a flat, level surface (Figure 11) creates a high quality contact surface enabling dry-jointing two individual panel layers without the need of post-processing. Furthermore, placeholders for precise recesses and concrete anchors can be placed in order to create connection details on the panel. Vertical layering also provides the opportunity to embed functional components across the entire panel surface between the layers. Another advantage of printing flat on the ground is that the process is fast and that multiple elements could be printed at once, as there are no high parts to interfere with the robot. In order to be able to connect several panels laterally, the perimeter of the elements was subtractively machined before the concrete had cured (Figure 12). This GSCM process saves energy and abrasion and is also faster compared to milling the fully cured concrete. To simplify the assembly and disassembly of the panels, a joining system was developed that works only with dry connectors and allows the parts to be lifted using the same connection points.

Computational Design Generation

The overall architectural system was based on a reference

building given within the Hackathon: a villa in Eindhoven, Netherlands (Houben/Van Mierlo Architects 2011-2016). After architectural analyses of the villa, a specific part of the villa was selected for subdivision. Thus, a three-dimensional grid was extracted and equalized digitally to create a unified cell size, suitable for creating a modular system (Figure 13).

Based on this, a design loop was developed to create the actual panel as a module: the design intent was developed as a simple planar surface that could be designed freely to any imaginable outline. The basic panel was designed as an adaptive C-shaped element. In order to save material, representative sections of 1:1 panels were chosen for production, containing all important details such as connection points and reinforcement layer. Each panel consist of two parts, an inner loadbearing wall and an outer facade wall. The load bearing part connects the modules together and is slightly shorter to provide a support for ceiling panels. The outer part is longer to cover the entire facade. the shape of the parts is slightly different to emphasize aesthetic freedom and create a recess for windows (Figure 14).

The panels were analyzed and optimized according to the force flow using Karamba 3D (Preisinger 2020) for Grasshopper (Rutten 2014). The algorithm provided information not only about the deformation and offset of the component at a defined force but also about the distribution of the compressive and tensile force flows within the facade panel. The iteration with the lowest structural offset was then minimized in material thickness by topology optimization; the optimal force flow depending on support points was determined and added as ribs to the panel, saving material where it is structurally not needed. The connection points were later subtracted from the finished geometry. The recesses were placed in such a way that concrete anchors served at the same time as transport hooks and connection points for the subsequent connectors. At the connection points, apart from the turnbuckles for connecting two modules, the structural and the facade layers also came together at this point.

For the robotic path, two different approaches were investigated: first was an equal offset of the input surface outline, the second (and more force flow oriented version) was a tween of the two length side curves of the surface. This type of path planning is accompanied by varying path widths and takes advantage of the spray cone of SC3DP; the closer the nozzle is directed to the baseplate, the narrower the printed path becomes and vice versa (Figures 16-17). With a nozzle distance of 300mm, a path width of 100mm was measured; with a halved nozzle distance of 150mm, a

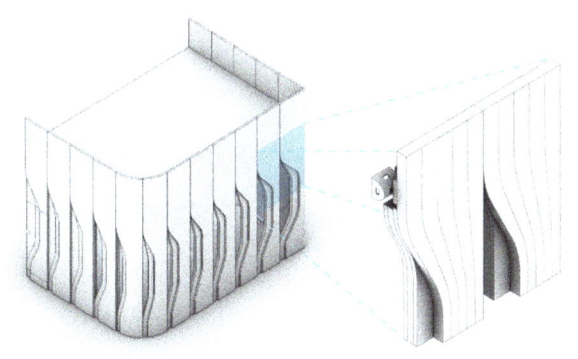

11 Shotcrete 3D Printing (SC3DP) on a flat surface

12 Green-State Concrete Milling (GSCM) to define the outlines of a flat element

13 Redesigned part of the reference villa and selected section for demonstation

path width of 50mm was measured. In order to change the concrete flow rate accordingly, the speed of robot movement was parametrically coupled to the nozzle distance; if the distance is halved, the speed doubles at the same time and thus balances out the concrete flow rate. Thus, the robot speed varied from 6.000mm/min to 12.000mm/min. For simulating the robots movement, the plugin Robots for Grasshopper was used (Soler and Huyghe 2017). The coordinates and the tool speed were translated into a G-code legible for the Siemens Sinumeric, which controls the robot and CNC mill.

Fabrication

For fabrication the unique Digital Building Fabrication Laboratory (DBFL) of the Institute for Structural Design

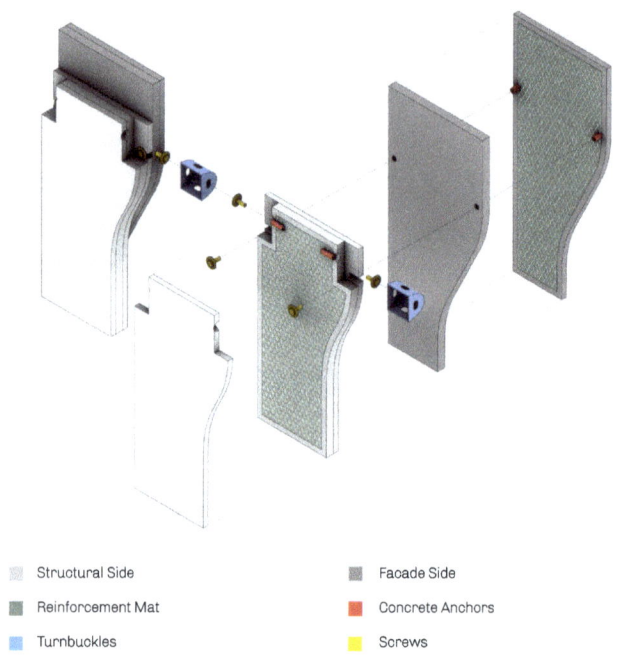
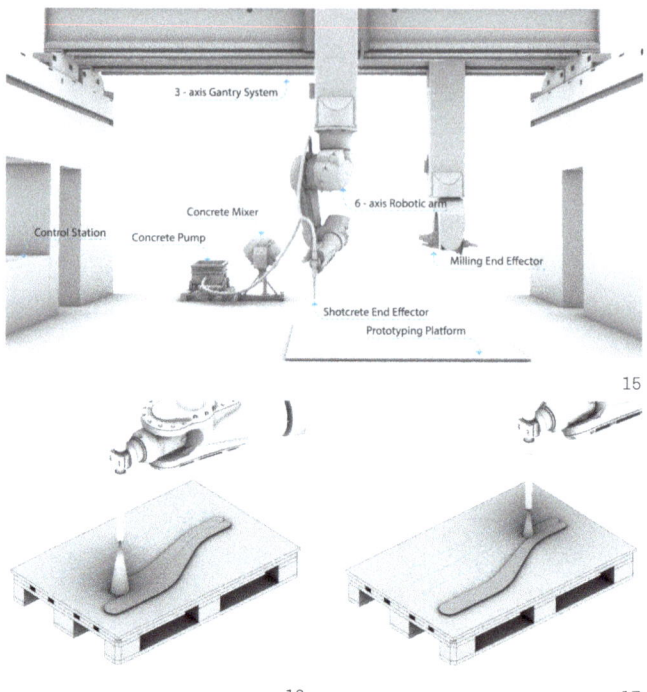

14 Exploded isometric view of the Demonstrator
15 Machines used at Digital Building Fabrication Laboratory (DBFL)
16 Path planning: Low nozzle distance for narrow paths
17 Path planning: High nozzle distance for wide paths

(ITE) at TU Braunschweig was used. The DBFL is a large scale robotic fabrication facility consisting of a gantry system with two independently controllable manipulators, firstly a 6-axis Stäubli TX 200 robot, and secondly a 3-axis Omag milling application. The overall cooperative build space embraces 10.5 x 5.25 x 2.5 meters. As such, the DBFL facilitates the production of largescale structures, both by subtractive machining, as well as by additive manufacturing processes (Figure 15).

Before the actual component was printed, all parts that were later to be embedded are pre-produced: placeholders, technical units such as pipes, and reinforcement mats. The placeholders were used to form small recesses where the connectors would later be placed. They also fixed the necessary concrete anchors in the right places. The placeholders consisted of several layers and were stacked like Legos® in the printing process, only to later be removed again. The layers were necessary because embedded elements such as concrete anchors and reinforcement mats are fixed to the placeholders and would create a spray shadow if they were placed as a package all at once. Because of the layers, placeholders were always pressed into the fresh concrete layer and covered in a good bond without a spray shadow. Since the connection points could always look the same regardless of the shape of the component, the placeholders could be reused. In this experiment, they consist of multiplex boards that were simply screwed together.

First, a wooden baseplate was placed in the workspace of the DBFL to generate a flat surface for printing. On this baseplate, the exact locations where the placeholders and anchors were to be placed, were marked. To be as precise as possible, a laser pointer was attached to the robot, which marked the corner points of the placeholders and anchors (Figure 18). These were screwed to the baseplate in the next step; the facade part required two concrete anchors placed upright, the structural part the first layer of placeholders. The printing process began with two base layers, each 1cm thick (Figure 19). Subsequently, the first reinforcement mat was placed and attached to a second layer placeholder and thus could be placed accurately. The robotic printing process was paused for any manual intervention like this (Figure 20).

The slimmer facade part of the wall was covered with two more layers of concrete and for a total of 4cm thick and was ready for post-processing after a short curing.

The structural part of the wall was printed further and the reinforcement was covered by four middle layers. Technical units such as pipes and cables could be embedded in these middle layers as desired. Then the third and last placeholder was stacked on top of the previous ones and fixed. Concrete anchors were fixed to this placeholder and were embedded horizontally into the structural part. The reinforcement was directly connected to these anchors, which in turn facilitated placement and also created a pre-tension in the fiber mat. Once this bundle was placed correctly, the last two flat surface layers of concrete were sprayed on (Figure 21), followed by the ribs as a single sprayed line on top. The structural part of the wall was then 8cm thick plus 4cm of ribs. The additive printing process was completed after about one hour.

Green-State Concrete Milling

To achieve the optimum strength for post-processing, the freshly printed parts had to cure until the consistency resembled moist soil. It should be hard to the touch but possible to press in by hand. In this experiment, it took about 1.5 hours for the concrete to reach this state. In the meantime, the placeholders had to be removed already from the curing concrete so that they did not pose an obstacle for the CNC mill. These were taken apart layer by layer and detached from the anchors that remained in the element. After the waiting time, the subtractive CNC milling process was started (Figure 22). The milling was done with a diamond milling head with a diameter of 50mm and a length of 210mm. This was used to flank-mill the outline of each element at a 90 degree angle (other angles are possible depending on requirements). Initially with an offset of 25mm, the CNC mill approached the original shape in three rounds. The smooth edges that result were supposed to be suitable for dry joining. Optionally, the surface of the printed elements could also be machined. for this purpose, a conventional concrete power trowel would be attached to the end effector of the CNC mill, which would smoothen the concrete surface by distributing it.

Assembly

When the parts were completely cured, they were detached from the baseplate and assembled. The structural part of the wall was lifted and erected by crane using its lateral concrete anchors. The facade part had its joints on the connecting surface and was thus firmly connected to the baseplate. Here, the entire baseplate together with the element was lifted and erected (Figure 23).

The component was printed on the baseplate in such a way that it could be tilted exactly onto an Euro-pallet and stood upright on it (Figure 24). The baseplate was then detached from the back of the facade element and removed. Both halves of the wall then stood upright together on a pallet with freely accessible anchor points and could be moved by crane. The side facing the baseplate was smooth for both elements, which allowed them to fit closely together (Figure 25). The placeholders had formed a recess in the structural half that fitted onto the concrete anchors of the facade

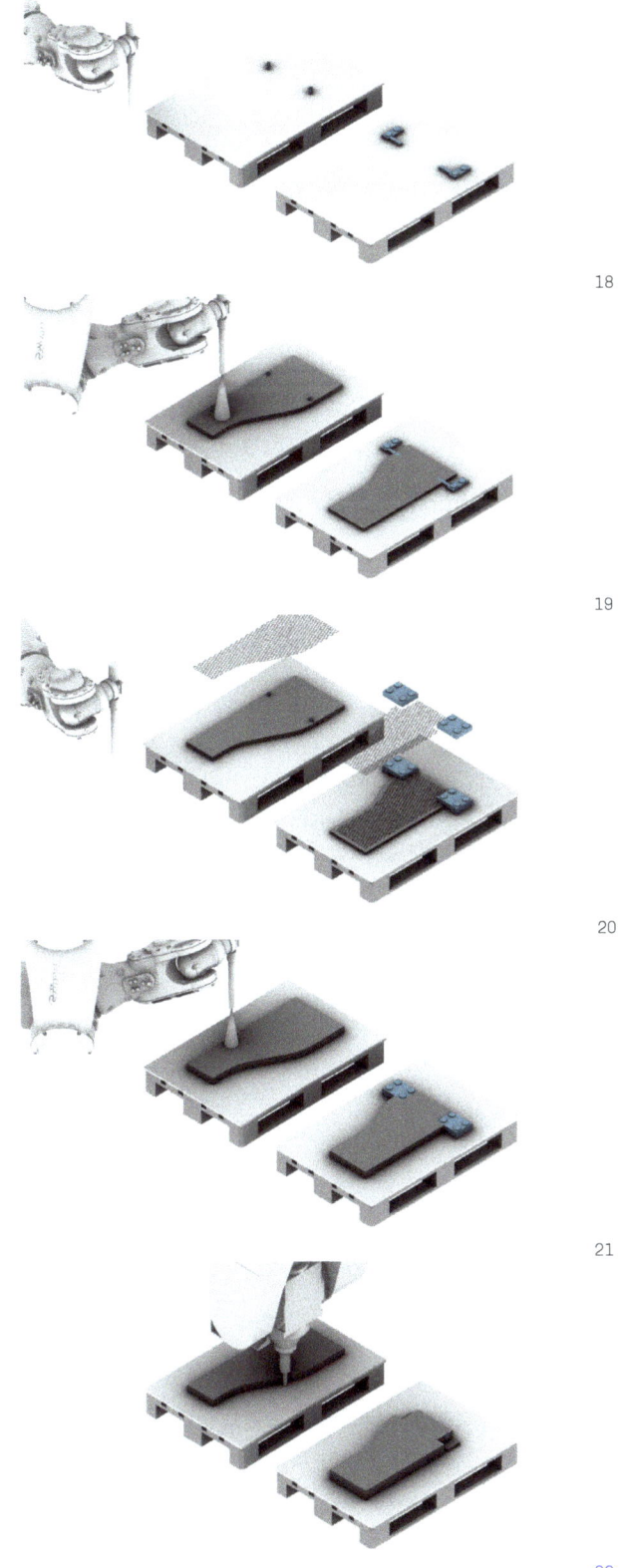

18
19
20
21
22

18 Placing concrete anchors and placeholders after marking positions robotically
19 Spraying initial layer of concrete to embed concrete anchors and placeholders
20 Placing carbon fiber mats during printing process between layers
21 Spraying covering layers of the wall parts
22 Removing placeholders and outline milling of the wall parts in green-state

23 Lifting structural part by crane

24 Erect facade part with baseplate

side. The two panel layers were first screwed together through this recess. The finished wall module, consisting of the structural layer and facade layer, could then be also connected to other modules. The concrete anchors placed on the side of the structural part were used for this purpose. They formed the screw connection points for turnbuckles commonly used in the precast industry, which were placed in the larger recesses of the structural part. The milled layers of the wall fitted together butt to butt and were bolted tightly by the turnbuckles (Figure 26).

RESULTS AND DISCUSSION

The result of the experiment were sections of two wall modules, shown in Figure 9, at a scale of 1:1 (Figures 27-28). The demonstrator measured a total of 800mm x 900mm x 190mm. The production time without waiting time was about 140 minutes, of which 20 minutes was preparation of the baseplate, reinforcement mats, and placeholders. The printing process took about 60 minutes, including insertion of the reinforcement mats, placeholders and technical units. Surface finishing took about 30 minutes, whereas assembling with 2 persons and a crane took 20 minutes. In terms of scaling, it is expected that the printing time for the full-size part would be increased linearly in relation to the material weight.

During the experiment, following observations were made. Although the placeholders turned out to be a simple means of forming the small recesses for the connections, positioning them by means of a robotically aligned laser pointer and manual marking on the wooden board resulted in slight inaccuracies in the placement, which only became noticeable when the modules were assembled.

The layered placeholders themselves were also not yet optimal; each layer had to be cleaned after the spraying process before being installed on the following layer. Fastening by screws is also not ideal, as it is too complex for the intended fully automated, robotic process. Both problems could be solved by a magnetic system consisting of a steel baseplate and magnetic placeholders, as it is already used in the precast industry.

During the printing process, inaccuracies in the path height of about 6.25mm added up by 8 layers to a deviation in thickness of 50mm, compared to its digital twin. This is due to insufficient calibration at the beginning of the experiment. However, the force-flow-optimized fiber reinforcement and ribs should allow a much thinner wall thickness than steel reinforced concrete walls. The solid bond of embedded functions such as pipes, reinforcement and concrete anchors by pressing on and spraying over in the printing process was positively noticed. However, complete robotic placement remains desirable.

There is also room for improvement in the milled sides, which are not yet smooth enough for stable dry-joints. Here, a further run with a distribution tool after milling is required. On the contrary, the inner contact surfaces of the wall layers stand out; the connection of the wall layers belonging together was exceptionally tight, as the baseplate acted like a formwork for this critical contact side.

CONCLUSION AND FUTURE WORK

This project demonstrated a hybrid manufacturing approach for producing functionally integrated 3D-printed concrete elements with controlled surface quality. This

 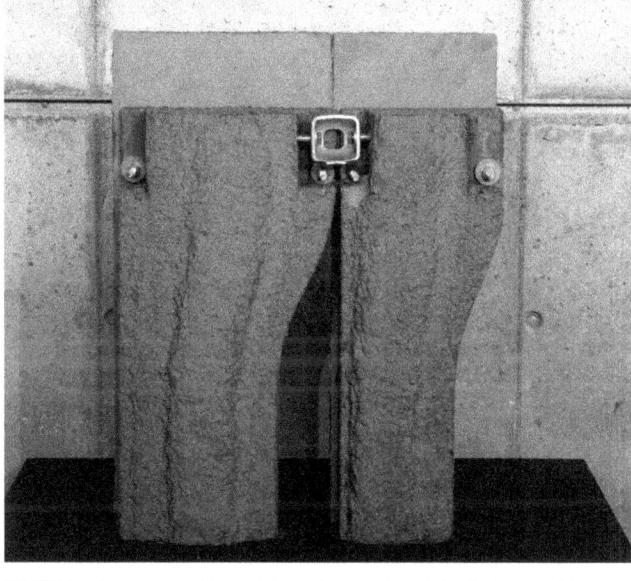

25 Align smooth contact-sides of both parts against each other

26 Tighten layer connection points and connect modules by turnbuckle

hybrid approach was possible using both SC3DP and GSCM and both techniques allowed for a comprehensive realization of a precisely fabricated 3D-printed concrete wall section that includes structural layer reinforcement, anchor points for an efficient transport, and guided assembly and post-processed edges for a precise dry-joint connection.

The precision and prediction of the actual printed geometry in connection with the required path properties are already being investigated and optimized at the ITE. This also opens up the applicability of variable path widths for tween path planning. Post-processing and fiber-winding are also investigated in the present, which will make clean and stable dry-joints and optimized, light reinforcement possible.

The full automation of the process requires its own approach. A system must be developed that automates the manual placement of, for example, reinforcement mats and placeholders. For this, the placeholders themselves must be rethought. If this succeeds, the step of measuring in could be completely eliminated. In terms of reinforcement, the carbon-fiber-mats used so far should be replaced by the dynamic robotic fiber-winding technique (Hack et al. 2021), already prepared and simulated digitally during this project, to produce fiber reinforcement optimized for specific tensile forces and to strengthen the component without adding material. The fiber-winding technique can either be applied in pre-production or directly wound into the printed component during the process.

In order to exploit the full potential of the separately printed wall layers, attempts are being made to use different functional materials, e.g. insulating concrete for exterior insulation or lightweight concrete lightweight facades. The only prerequisites are that the material is sprayable and stable enough to allow the geometry to build up.

The SC3DP approach described in this paper is ultimately not limited to walls, but can be applied to other architectural elements such as floor slabs, as the Institute for Structural Design at TU Braunschweig has demonstrated in an earlier experiment, the "Add-on Printed Slab" (Kloft et al. 2019). Since the contact side resting on the baseplate is smooth and suitable for use as a floor surface, load-bearing ribs can be printed on the other side, this method is also applicable for this purpose. Moreover, building services such as underfloor heating can be embedded in the process using the same principles.

In summary, concrete 3D printing is not yet fully developed when it comes to embedding functions such as building services, reinforcement, and insulation to match today's building standards. There are also still many unresolved challenges in terms of modularity and joinability. The integration of both novel techniques of SC3DP and GSCM opens the room for facing challenges related to the concrete 3D printing industry. Potentials of realigning design and fabrication parameters related to 3D concrete printing have been possible to move towards a more integrated, flexible, and modular manufacturing system. Overcoming the limitations of monolithic and monofunctional 3D-printed concrete structures can be further explored and investigated with such a hybrid manufacturing approach.

27 Facade side of the finished and assembled wall-modules
28 Detail of the connection point with turnbuckle

ACKNOWLEDGEMENTS

The following students of the 3D-Printing-Housing-Hack class are acknowledged for their contribution to the demonstrator, digital simulations, design and construction, and project documentation: Simon Schröter, Lukas Schivelbein, Nick Kobert, Xinjie Wang, and Urs Granatowski. The authors would like to thank LafargeHolcim Ltd. and its partners Witteveen & Bos, COBOD and Fondation des Ponts, the German Research Foundation (DFG), for funding the collaborative research center TRR 277 Additive Manufacturing in Construction and the DFG Large Research Equipment, Digital Building Fabrication Laboratory (project numbers 416601133 and 414265976), the Institute for Structural Design and the Team at DBFL, TU Braunschweig, the Gerhard und Karin Matthäi Foundation for funding the Junior Professorship in Digital Building Fabrication of Norman Hack.

REFERENCES

Anton, Ana, Lex Reiter, Timothy Wangler, Valens Frangez, Robert J. Flatt, and Benjamin Dillenburger. 2021. "A 3D Concrete Printing Prefabrication Platform for Bespoke Columns." *Automation in Construction* 122 (February): 103467.

Bos, Freek, Rob Wolfs, Zeeshan Ahmed, and Theo Salet. 2019. "Large Scale Testing of Digitally Fabricated Concrete (DFC) Elements." In *First RILEM International Conference on Concrete and Digital Fabrication; Digital Concrete 2018. DC 2018. RILEM Bookseries*, vol 19. Cham: Springer. 129–47.

Delgado Camacho, Daniel, Patricia Clayton, William J O'Brien, Carolyn Seepersad, Maria Juenger, Raissa Ferron, and Salvatore Salamone. 2018. "Applications of Additive Manufacturing in the Construction Industry – A Forward-Looking Review." *Automation in Construction* 89: 110–19.

Hack, Norman, Mohammad Bahar, Christian Hühne, William Lopez, Stefan Gantner, Noor Khader, and Tom Rothe. 2021. "Development of a Robot-Based Multi-Directional Dynamic Fiber Winding Process for Additive Manufacturing Using Shotcrete 3D Printin." *Fibers* 09 (June): 1–17.

Hack, Norman, Hendrik Lindemann, and Harald Kloft. 2019. "ADAPTIVE MODULAR SPATIAL STRUCTURES FOR SHOTCRETE 3 D PRINTING." In *Intelligent & Informed, Proceedings of the 24th International Conference of the Association for Computer-Aided Architectural Design Research in Asia (CAADRIA) 2019*, vol. 2. Hong Kong: CAADRIA. 363-372.

Houben/Van Mierlo Architects. 2011-2016. "Urban Villas." [Online Project Portfolio]. Accessed October 6, 2021. https://www.houben-vanmierlo.nl/werk/urban-villas/.

Kinnell, Peter, James Dobranski, Jerry Xu, Weiqiang Wang, John Kolawole, John Hodgson, Simon Austin, John Provis, Sergio Pialarissi-Cavalaro, and Richard Buswell. 2021. "Precision Manufacture of Concrete Parts Using Integrated Robotic 3D Printing and Milling," In *Proceedings of the 21st International Conference of the European Society for Precision Engineering and Nanotechnology, EUSPEN 2021*. 57–60.

Kloft, Harald, Norman Hack, Jeldrik Mainka, Leon Brohmann, Eric Herrmann, Lukas Ledderose, and Dirk Lowke. 2019a. "Additive Fertigung Im Bauwesen: Erste 3-D-gedruckte Und Bewehrte Betonbauteile Im Shotcrete-3-D-Printing-Verfahren (SC3DP)." *Bautechnik* 96 (12): 929–38.

Kloft, Harald, Hans-Werner Krauss, Norman Hack, Eric Herrmann, Stefan Neudecker, Patrick Varady, and Dirk Lowke. 2020. "Influence of Process Parameters on the Interlayer Bond Strength of Concrete Elements Additive Manufactured by Shotcrete 3D Printing (SC3DP)." *Cement and Concrete Research* 134 (August): 106078.

Lu, Bing, Yiwei Weng, Mingyang Li, Ye Qian, Kah Fai Leong, M J Tan, and Shunzhi Qian. 2019. "A Systematical Review of 3D Printable Cementitious Materials." *Construction and Building Materials* 207 (May): 477–490.

Menna, Costantino, Jaime Mata-Falcón, Freek P Bos, Gieljan Vantyghem, Liberato Ferrara, Domenico Asprone, Theo Salet, and Walter Kaufmann. 2020. "Opportunities and Challenges for Structural Engineering of Digitally Fabricated Concrete." Cement and Concrete Research 133: 106079.

Panda, Biranchi, Yi Wei Daniel Tay, Suvash Paul, and M J Tan. 2018. "Current Challenges and Future Potential of 3D Concrete Printing: Aktuelle Herausforderungen Und Zukunftspotenziale Des 3D-Druckens Bei Beton." *Materialwissenschaft Und Werkstofftechnik* 49 (May): 666–73.

Paul, Suvash Chandra, Gideon P A G van Zijl, Ming Jen Tan, and Ian Gibson. 2018. "A Review of 3D Concrete Printing Systems and Materials Properties: Current Status and Future Research Prospects." *Rapid Prototyping Journal* 24 (4): 784–98.

PERI GmbH. 2020. "PERI Builds the First 3D-Printed Residential Building in Germany." Press Release, July 29, 2020. Accessed October 6, 2021. https://www.peri.com/en/media/press-releases/peri-builds-the-first-3d-printed-residential-building-in-germany.html.

Preisinger, Clemens. 2020. "Karamba 3D." V.1.3.3. Bollinger und Grohmann ZT GmbH. PC.

Prior, Madeleine. 2021. "Parametric Design and Concrete 3D Printing: A Winning Combination to Build a Bridge." Online Magazine, April 6, 2021. Accessed October 6, 2021. https://www.3dnatives.com/en/parametric-design-and-concrete-3d-to-build-a-bridge-060420215/#!.

Raval, Amit D., and C. G. Patel. 2020. "Development, Challenges and Future Outlook of 3D Concrete Printing Technology." *International Journal on Emerging Technologies* 11(2): 892-896.

Rutten, David. *Grasshopper 3D*. V. 1.0. Robert McNeel & Associates. 2014. PC.

Salet, T. A. M., Z. Y. Ahmed, F. P. Bos, and H. L. M. Laagland. 2018. "3D Printed Concrete Bridge." In *Proceedings of the 3rd International Conference on Progress in Additive Manufacturing (Pro-AM 2018)*. 2–9.

Soler, Vicente, and Vincent Huyghe. Robots. V. 0.0.7. PC.

Xu, Weiguo, Yuan Gao, Chenwei Sun, Zhi Wang, Jane Burry, Jenny Sabin, Bob Sheil, and Marilena Skavara. 2020. "Fabrication and Application of 3D-Printed Concrete Structural Components in the Baoshan Pedestrian Bridge Project." In Fabricate 2020, 140–47. Making Resilient Architecture. UCL Press.

IMAGE CREDITS

Figure 2-5: ©PERI GmbH, 2020
Figure 6: ©Studio michiel van der kley & TU Eindhoven, 2021
Figure 7: ©Tsinghua University School of Architecture, 2017
Figure 8: ©BAM Infra & TU Eindhoven, 2017
Figure 11: ©Janna Vollrath, March 2021
All other drawings and images by the authors

Philipp Rennen is a PhD student for LafargeHolcim Ltd. and is researching in the field of Architecture and Digital Fabrication at the Institute for Structural Design, TU Braunschweig. He achieved his Master of Science in Architecture in 2020 at TU Braunschweig. His interests reach from Parametric Design to Digital Fabrication and his research focuses on exploring design strategies in the field of Shotcrete 3D Printing.

Noor Khader is a scientific assistant at the Institute of Structural Design, TU Braunschweig, She received her Master's degree in architecture and digital fabrication at the Institute of Technology in Architecture, ETH Zurich in 2019. She is interested in rethinking architecture through the interplay between design and other disciplines. Her academic and career goals include exploring design and fabrication methods in order to push boundaries and expand the possibilities of what the built environment is about and what it could be.

Norman Hack is a researcher in the field of digital fabrication with degrees in architecture from TU Vienna and the Architectural Association. After graduation, he worked as a programming architect in the Digital Technologies Group at Herzog and de Meuron. His interested in seamless digital design and fabrication processes led him to peruse a PhD with Gramazio Kohler Research at ETH Zurich. Since 2018 he is holding a tenure track professorship for Digital Building Fabrication at TU Braunschweig.

Harald Kloft is head of the Institute of Structural Design (ITE) at the TU Braunschweig, Germany. As an engineer and scientist he represents a new logic of form, based on digital technologies and inspired by the needs of sustainability and circular economy. As spokesperson for the DFG Collaborative Research Centre TRR 277 Additive Manufacturing in Construction (AMC) of the two universities TU Braunschweig and TU Munich, he and a team of almost 30 scientists are aiming to research AMC as a resource-efficient digital manufacturing technology for the construction industry.

Meristem Wall: An Exploration of 3d-printed Architecture

Ana Goidea
Lund University/bioDigital Matter

Mariana Popescu
Delft University of Technology/ Faculty of Civil Engineering and Geosciences

David Andréen
Lund University/bioDigital Matter

1   *Meristem Wall* at Time Space Existence, Palazzo Mora, Venice.

Meristem Wall is a prototype for a 3D-printed building envelope, featuring a dynamically controllable network of integrated air channels that allow a fluid and adaptive relationship between inside and outside. The wall integrates functional lighting and electricity, windows, and a custom CNC-knitted textile interior. It is fabricated through binder-jet sand 3D printing and points towards a climatically performative architecture inclusive of nonhuman life in urban contexts.

Based on previous research that demonstrated airflow transfer in a reticulated branching network (Andréen 2016), the system of channels can be controlled through an embedded system of sensors and actuators to enable selective transport of heat and moisture. Their tortuosity and narrow diameter limits cross-drafts. The outer part of the wall shifts the channels to a nested landscape of intertwined surfaces, providing an extensive biological habitat in the building itself. *Meristem Wall* presents a vision for 3D printing in the construction sector (Turner and Soar 2008) and how the technology may come to reshape our relationship to the natural and built environment.

Computational Design Model

The design of the wall is the result of a simulation that is managed through a unifying voxel model. Multiple algorithms come together to deliver several functions, negotiating their internal relationship through emergent, local interactions (Varenne 2013).

2 Assembly of the wall in Venice

3 The interior is fitted with a custom CNC-knitted fabric and integrated lighting

The model self-organizes using a custom particle-spring system (PSS). The PSS is defined by the boundary constraints of the wall, which includes both its volumetric definition as well as dynamic (Kanellos and Hanna, 2008) and static connections (e.g. sightlines/windows, plumbing and wiring, structural loads) to the surrounding building. It incorporates the reticulated transient network which forms the base of the Meristem Wall, as well as channels and passages that traverse it, carrying global flows, wiring, and loads.

The topological model obtained through the PSS is used to define a surface in 3-dimensional space, which is converted to a voxel model (Bernhard et al, 2018). The latter is locally defined, making it suited for further algorithmic processing and fabrication. In Meristem Wall, local adaptations of the geometry were made in the model to fit installations and fixtures and to provide anchor points for the interior textile. Additionally, the bioreceptivity (Guillitte, 1995) of the external surface is targeting several scales of growth: nests, porous pockets, surface texture and rugosity derived from the material print process.

The textile is converging in the computational model of the wall. Its main role is to provide a semi-permeable membrane to the inside of the wall which allows for the passage of airflows. The scale of the knit pattern changes locally at the channel inlets to a high density, in order to filter out insects and larger particles. The knitted patterns vary outside of these to reflect the exterior wall ecology, giving different transparencies; at the same time, it gives a soft, colorful, and tactile surface facing the internal room.

Digital Fabrication

The wall was fabricated through binder-jet sand 3D printing by Voxeljet, and CNC knitted using a seven gage CNC industrial knitting machine (Steiger Libra 1.130). The printing process supports high resolutions of sub-millimeter scale and complex cantilevered geometries while simultaneously accommodating large volumes. Consequently, it would have

4 *Meristem Wall* at Time Space Existence, Palazzo Mora, Venice

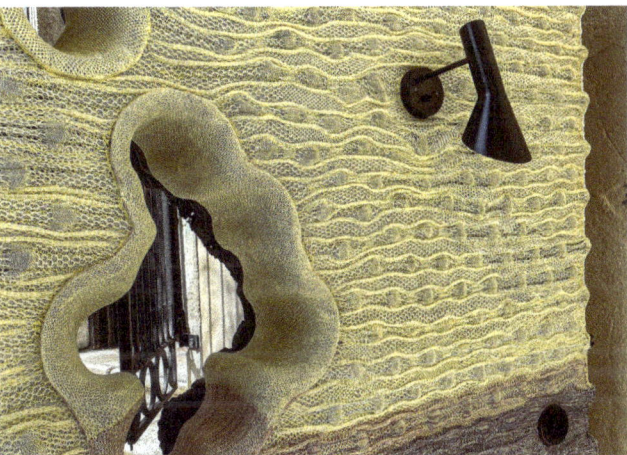

5   The external wall surface is designed to maximize bioreceptivity

6   Light fixtures and cabling integrated into the 3D-printed structure

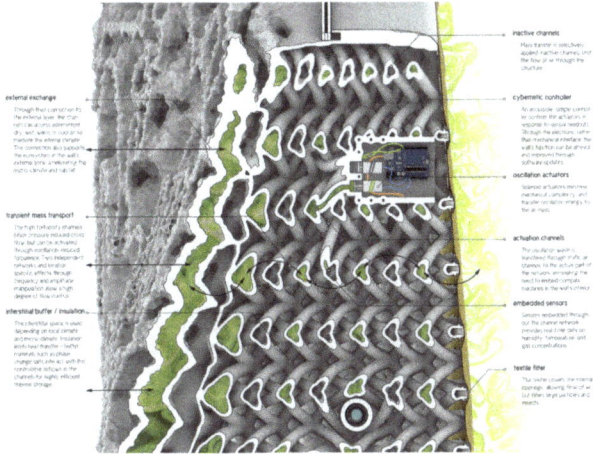
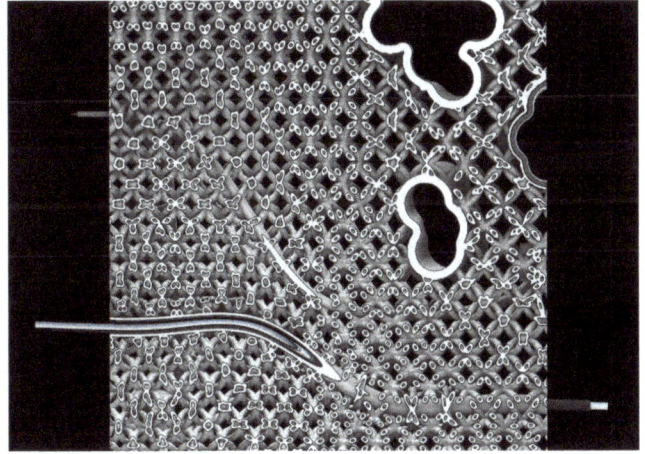

7   Transversal section showing the network of reticulated channels that can be selectively activated to generate local air transport; activation occurs through oscillation of the air in the channels, and is controlled based on embedded sensors tracking relative humidity and temperature

8   Longitudinal section showing network integration with wall functions

been possible to be fabricated in one piece, but *Meristem Wall* was fabricated in 21 sections to facilitate transport and handling. The fabric was fitted onto the 3D-printed elements by attaching to the custom fixtures through looping nylon strands integrated as weft in-lays into the knit, so the entire installation is dissasemblable.

ACKNOWLEDGMENTS

The project is supported by a grant awarded by the Swedish National Board of Housing, Building and Planning. Anton T. Johansson of Lund University contributed to the programming of the Meristem Wall. Fabrication thanks to Voxeljet. Post processing with support from Sandhelden.

Meristem Wall is currently exhibited at the Time Space Existence exhibition at the Venice Biennale.

REFERENCES

Andréen, David. 2016. "Discriminatory transient mass transfer through reticulated network geometries: a mechanism for integrating functionalities in the building envelope." EngD thesis, UCL/University College London, London.

Bernhard, Mathias. *Axolotl*. food4rhino. PC. 2018.

Bernhard, Mathias, Michael Hansmeyer, and Benjamin Dillenburger. 2018. "Volumetric modelling for 3D printed architecture". In *AAG 2018: Advances in Architectural Geometry 2018*. 392–415.

Guillitte, Olivier. 1995. "Bioreceptivity: a new concept for building ecology studies". *Science of The Total Environment* 167 (1–3): 215–220.

Kanellos, Anastasios, and Sean Hanna. 2008. "Topological Self-Organisation: Using a particle-spring system simulation to generate structural space-filling lattices". In *Proceedings of the 26th ECAADe Education and Research in Computer Aided Architectural Design in Europe*. 459–466.

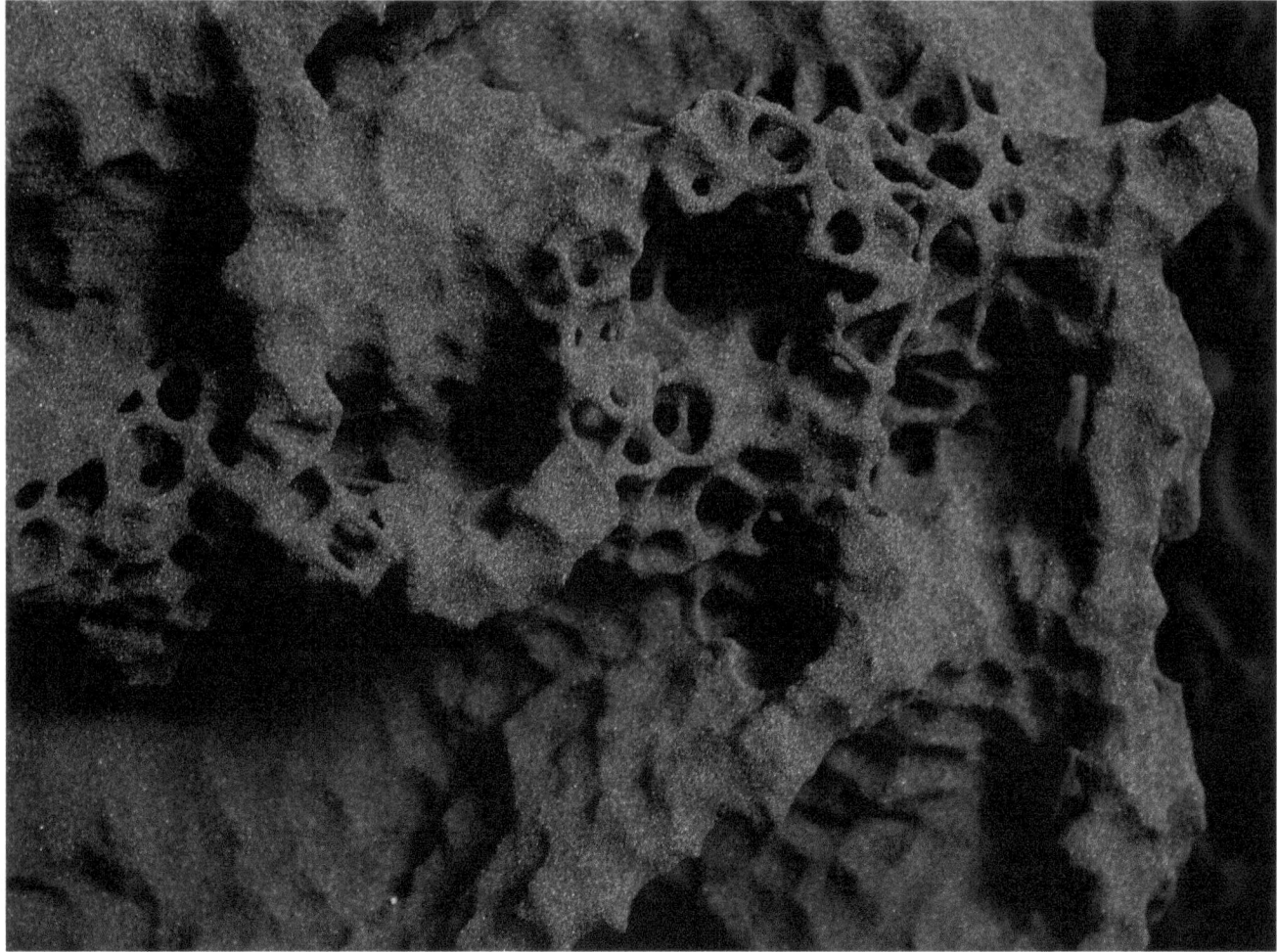

9   Bioreceptivity detail; the variable scale of the exterior surface pockets creates a diversity of microclimates and ecological niches

Tetov, Anton. *ChromodorisBV*. V.0.1.7. GitHub. PC. 2021. Based on: Newnham, Cameron. *Chromodoris*. V.0.0.9.1. food4rhino. PC. 2016.

Turner, Scott J. and Rupert C. Soar. 2008. "Beyond biomimicry: What termites can tell us about realizing the living building". In *First International Conference on Industrialized, Intelligent Construction (I3CON)*. Loughborough, UK.

Varenne, Franck. 2013. "The Nature of computational things - Models and simulations in Design and Architecture". In *Naturalizing Architecture: ArchiLab 2013*, edited by M.-A. Brayer and F. Migayrou. Orléans, France: Editions HYX. 96–105.

IMAGE CREDITS

Figures 1: ©Federico Vespignani, 2021
Figure 4: ©Press ECC Italy, 2021
Figure 9: ©Sandhelden, 2021
All other drawings and images by the authors

10 As well as providing color, tactility, and enhancing acoustics, the knitted textile provides a filter over the channel openings

11 Fabric is fixed using integrated hooks and nylon threads

Ana Goidea is a PhD candidate at bioDigital Matter at Lund University, where she investigates the potentials of additive manufacturing in architecture through computational design. She has been teaching and working at studios with different strategies for digital fabrication. Her work explores the current relationship to the environment through the link between complex geometry and new material systems within digital computation and additive manufacturing technologies.

Mariana Popescu is Assistant Professor of Parametric Structural Design and Digital Fabrication at TU Delft with a strong interest in innovative ways of approaching the fabrication process and use of materials. She was post-doctoral researcher at the Block Research Group at ETH Zurich, involved in the NCCR Digital Fabrication. Her research focuses on the development novel, material-saving, labour-reducing, cost-effective construction systems.

David Andréen is a senior lecturer at Lund university where he leads the bioDigital Matter research group. His practice concerns architecture, digital fabrication, and computation, with a particular interest on how principles of biology can help shape new sustainable paradigms in the design of the built environment. He completed his doctorate at the Bartlett ULC, investigating termite mounds as models for complex, functional form and related principles of emergence and self-organization.

Mitochondrial Matrix

Assia Crawford
Newcastle University

Symbiosis and Variation Through Digital Ceramic Fabrication

1   Mitochondrial Matrix installation capturing clusters of genetic material, exhibited at Valence II, Vane Gallery (Assia Crawford, 7.15.2021).

The following project was created as part of an art residency with the Wellcome Centre for Mitochondrial Research (WCMR) at Newcastle University. The WCMR specializes in leading-edge research into mitochondrial disease, investigating causes, treatments, and ways of avoiding hereditary transmission. Mitochondria is believed to have started off as a separate species that through symbiosis came to be the powerhouse of each cell in our bodies (Hird 2009). Mitochondrial disease is a genetic disorder that is caused by genetic mutations of the DNA of the mitochondria or the cell that in turn affects the mitochondria (Bolano 2018). Mitochondria is a hereditary condition and can affect people at different stages in their lives. It can affect various organs and has a link to various types of conditions. Therefore, the patient experience is unique to each individual and the elusive nature of the condition can make it particularly challenging due to the complexity of the disorder as well as the inaccessible scale on which these variations occur (Chinnery 2014).

The piece that emerged consists of an assemblage of 3D-printed ceramic vessels based on a simple geometry that grows and morphs in digital space to create variation. The printing process not only captured the intentional expression of mutations but also resulted in discrepancy, mistakes, variations due to the consistency of the clay, gravity, or the printing process itself. These were welcome factors adding a physical element of making to what is often perceived as a purely mechanical process of digital fabrication. All geometries

PRODUCTION NOTES

Designer: Assia Crawford
Client: Wellcome Centre for Mitochondrial Research, Newcastle University
Date: 2021

2 3D printing a clay vessel using a double wall, created with a 3 mm nozzle.

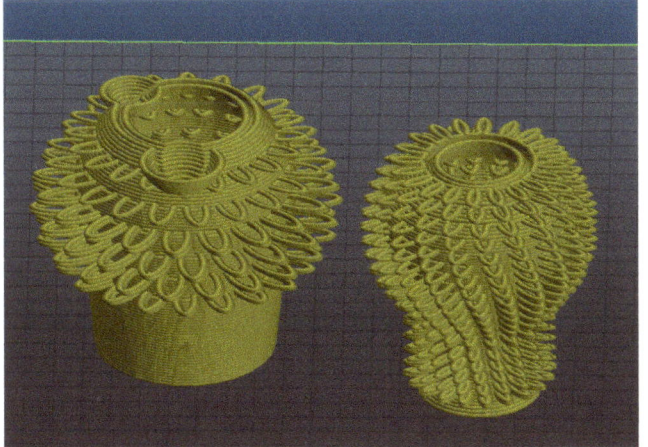

3 3D files exported from 3D Studio Max 2018 and imported into Slic3r software and prepared for printing. Looped geometries are modeled flat taking into account the effects of gravity on wet clay.

4 Clay 3D printed clay vessels air drying in preparation for kiln firing and LUTUM® 4.5 clay 3D printer by VormVrij.

started as a simple sphere, that was pushed and pulled and replicated to demonstrate the corporeal aspects of the disease. Some created uniform clusters of multiple spheres that mould effortlessly together to form a whole, others are caught in a state where uncomfortable protruding loops occur sporadically, whereas other are completely enveloped in those anomalous formations that from a distance look quite enticing yet displayed clustered together can feel untamed and bordering on the grotesque. This is a parallel to microscopy images, that too appear decorative and seductive, shown in various stains the lacy structure of the spliced samples is somewhat appealing. It is not until the cells are viewed as part of a malignant mass that our perception shifts.

Symbiosis is a central theme throughout this project and is one of the fundamental principles that positions humanity within multispecies entanglements (Liddell and Scott 1940). It acknowledges that we are holobionts and as such are dependent on other organisms, and so it follows that even from an anthropocentric standpoint, we should place such partnerships at the forefront of the way we negotiate our interactions with nature (Margulis 2001; Paxson and Helmreich 2014; Tsing 2017; Haraway 2016). The process of designing *Mitochondrial Matrix* was one of abstraction and engagement with an intuitive part of making that has permeated my creative practice. Following discussions with the scientific team and peering into microscopy images stained in various colors, capturing a perspective of the world that is rarely made privy to anyone outside of scientific circles, I began to think about how changes to cells, genes, DNA, and mitochondria are visualized from an outside perspective. The work is not a biologically accurate representation; rather it aims to capture the essence of the processes that take place within the human body and to mould public perception of humanity as an actor in a myriad of partnerships loaded with various challenges. The body in this case becomes the hostile landscape alluded to

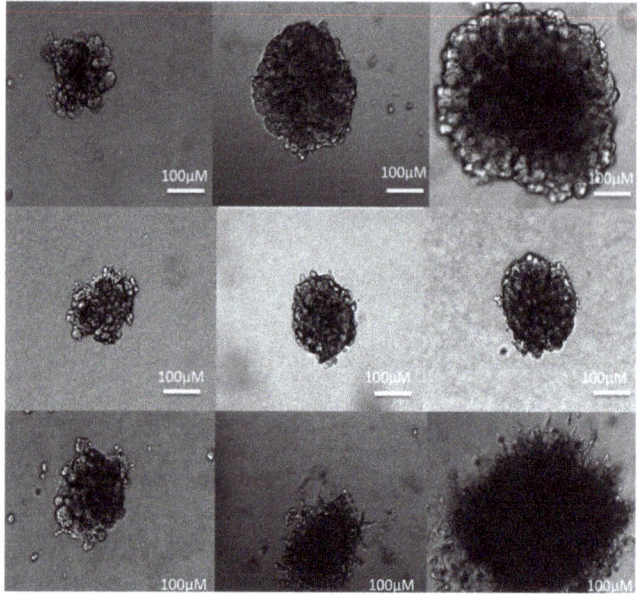

5   Bladder cancer cells

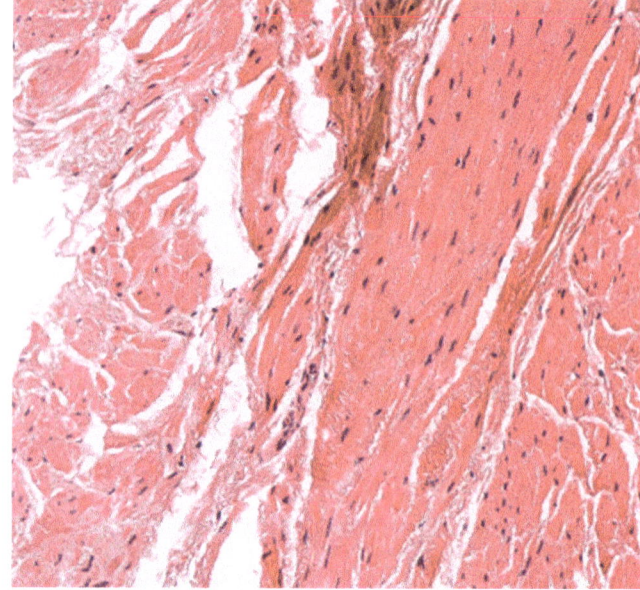

6   Early urothelial cancer 10% tumor

7   The individual vessels capture cell replication and growth similar to scientific images

8   The decorative nature of the geometries speaks to the sinister changes that can only be observed in isolation within the lab

9   Cut geometries representing MRI imaging

10 Similarities within variations emerge and are expressed through geometry as well as ceramic glaze combination

11 Cells coming together to create healthy tissue

by various scholars (Hird 2009; Sagan and Margulis 1993). The objects offer a human scale expression of changes that occur on a micro scale, that can be touched and caressed, that offer sensations of comfort and continuity or elements that feel jarring, out of place and unsettling. The geometries range from the sublime to the grotesque where the viewer gains intrinsic understanding of the states these objects aim to communicate.

ACKNOWLEDGMENTS

The author would like to thank Nana-Jane Chipampe for her scientific input and the Welcome Centre for Mitochondrial Research (WCMR). The work was funded by WCMR and the Hub for Biotechnology and the Built Environment (Research England grant).

REFERENCES

Bolano, Alex. 2018. "Mitochondria Function: Plant And Animal Cells." Science Trends. August 6, 2018. Accessed March 20, 2020. https://sciencetrends.com/itochondria-function-plant-and-animal-cells/.

Chinnery, Patrick F. 2014. "Primary Mitochondrial Disorders Overview." In *GeneReviews®*, edited by M. P. Adam et al. Seattle: University of Washington. Accessed May 15, 2020. http://www.ncbi.nlm.nih.gov/pubmed/20301403.

Haraway, Donna Jeanne. 2016. *Staying with the Trouble : Making Kin in the Chthulucene*. Durham: Duke University Press.

Hird, Myra J. 2009. *The Origins of Sociable Life : Evolution after Science Studies*. New York: Palgrave Macmillan.

Liddell, Henry, and Robert Scott. 1940. *A Greek-English Lexicon*. Oxford: Clarendon Press. Perseus Digital Library, Tufts University. Accessed May 18, 2020. http://www.perseus.tufts.edu/hopper/text?doc=Perseus:text:1999.04.0057:entry=sumbi/wsis.

Margulis, L. 2001. "The Conscious Cell." *Annals of the New York Academy of Sciences* 929 (April): 55–70. https://doi.org/10.1111/j.1749-6632.2001.tb05707.x.

Paxson, Heather, and Stefan Helmreich. 2014. "The Perils and Promises of Microbial Abundance: Novel Natures

12 *Mitochondrial Matrix* installation capturing clusters of genetic material, exhibited at Valence II, Vane Gallery

13 Mitochondrial Matrix installation with video evoking the ritualistic aspects of laboratory practice, exhibited at Valence II, Vane Gallery

14 Clay 3D-printed vessel representing healthy cell subdivision forming tissue

15 Applying multiple layers of ceramic glazes to mitochondrial vessels

16 Applying multiple layers of ceramic glazes to mitochondrial vessels

17 Applying multiple layers of ceramic glazes to mitochondrial vessels

18 Clusters of different grade malignant cells forming, expressed as variations in geometry

19 Clusters of different grade malignant cells forming, expressed as variations in geometry

20 Cells coming together to create healthy tissue

and Model Ecosystems, from Artisanal Cheese to Alien Seas." *Social Studies of Science* 44 (2): 165–93. https://doi.org/10.1177/0306312713505003.

Sagan, Dorion, and Lynn Margulis. 1993. *Garden of Microbial Delights : A Practical Guide to the Subvisible World*. Dubuque, Iowa: Kendall Hunt Pub. Co.

Tsing, Anna. 2017. *Arts of Living on a Damaged Planet*, edited by Anna Tsing. Minneapolis: University of Minnesota Press.

IMAGE CREDITS

Figure 5-6: ©Nana-Jane Chipampe 2021
Figure 13: ©Colin Davison 2021
All other images by author

Assia Crawford is an ARB registered architect and a creative practice PhD Researcher at Newcastle University. Her work focuses on the development of biological material alternatives and their integration into the urban realm. She is the architect for the Hub for Biotechnology in the Built Environment (HBBE) and a Resident Artist at the Wellcome Centre for Mitochondrial Research, Newcastle University.

DTS Printer:
Spatial Inkjet Printing

Kimball Kaiser
Massachusetts Institute
of Technology

Maryam Aljomairi
Massachusetts Institute
of Technology

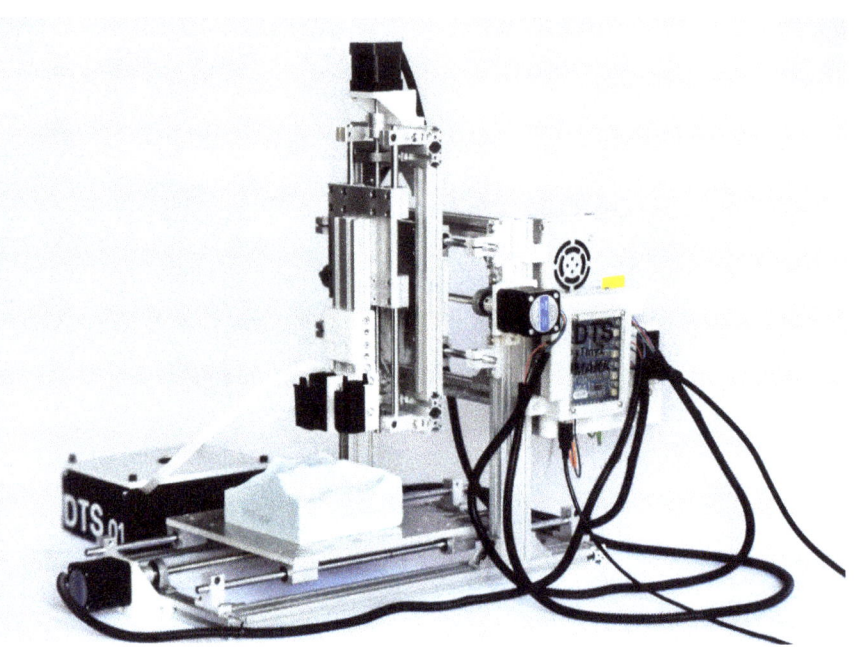

1   DTS Printer

Inkjet printing has become abundantly available to businesses, offices, and households ever since its commercialization in the late 1980s. Although roughly forty years have passed, the desktop printer is still limited to printing on thin flat surfaces, mainly paper (Mills 1998). On the other hand, while larger flatbed printing technology does offer printing on a wide-range of substrates of various thicknesses, it is limited to 2-axis printing and is mainly used for large scale commercial applications due to high machine costs.

Motivated by the ambition of printing on irregular surfaces of varied mediums, improving upon high price points of existing flat-bed printing machines, and contributing to the public knowledge of distributed manufacturing, the Direct-To-Substrate (DTS) printer is an exploration into an integrated z-axis within inkjet printing. To realign a familiar technology used by many and hack it for the purposes of expanded capabilities, the DTS allows a user to manufacture a three-dimensional artifact and later print graphics directly upon said geometry using the same machine. To remain as accessible as possible, the DTS printer is a computer-numerically-controlled desktop machine made from common, sourceable hardware parts with a tool-changeable end effector, that currently accepts a Dremel tool as a router, and a hacked inkjet cartridge.

This project builds upon existing technology and research that has been made accessible in the commons of DIY documentation, allowing projects and insights to be open for others to use. For instance, the machine skeleton is built upon Zain Karsan's TinyZ, an at-home

PRODUCTION NOTES

Status: Work in progress
Location: Cambridge, MA
Date: 2021

2   Tool change demonstration: hacked inkjet

3   Text test print with laser for origin placement

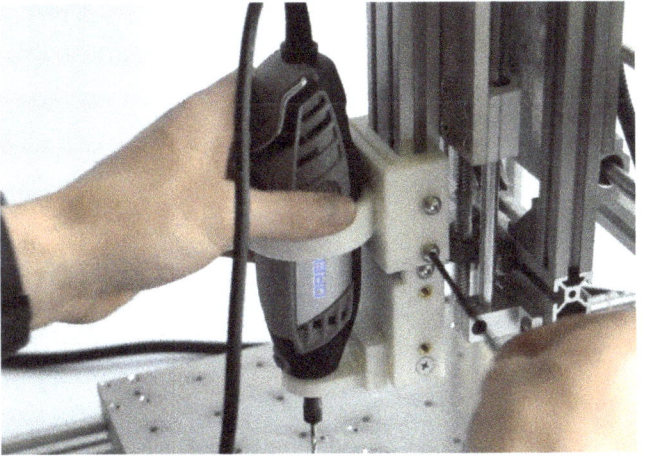

4   Tool change demonstration: Dremel

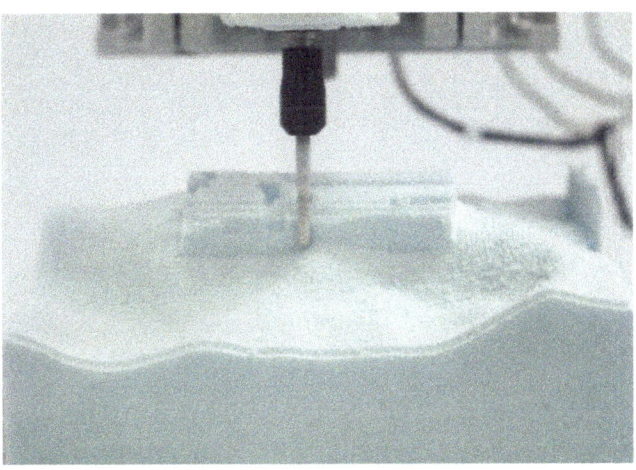

5   Process shot: milled foam piece

machine developed for remote digital fabrication (Karsan 2021). In addition, the hacked print-head of the DTS references early contributions made by Nicholas C. Lewis in 2012 with the creation of InkShield, a kit that controls an Hewlett Packard C6602 inkjet cartridge (Lewis 2011), and most recently Norbert Heinz's development of a similar hacked print-head (Heinz 2020). The hacking process includes configuring homemade hardware to control electrical pulses at the correct voltage and timing, as well as customized software to operate the patterns printed from the twelve inkjets of the print-head.

Part of the contribution of the DTS printer was to improve on the missing aspects of existing project documentation of hacked inkjet printing, integrating an additional axis, and making these developed systems reproducible within the broadest public possible by using ubiquitous open source softwares such as Arduino IDE and CNCJs for inkjet control, as opposed to softwares with a steep learning curve. Hence, broadening the accessibility of the project beyond the tinkerer and maker, being more inclusive of communities with minimal electronics and machine building backgrounds.

Successful prints were conducted on various materials and integrated both additive and subtractive processes, with just a few limitations encountered partly due to the dimensions of an off-the-shelf print-head in proportion to the bed size. Nevertheless, the current set-up of the machine allows for the expansion in either axis by simply changing the aluminum frame lengths, creating an at home, lower cost, fully integrated 3-axis, flatbed printer-like machine. While a printhead could also be applied to something like a robotic arm, the expandable frame would more easily empower many to address an open design with potentials for scaling up the application of the DTS. Printing directly upon larger complex geometries lends itself to the implications of printing on architectural components such as facades, structural elements, and built-ins. Furthermore, other abilities were conceptualized, such as printing directly upon digitally manufactured parts, eliminating the need for instruction manuals.

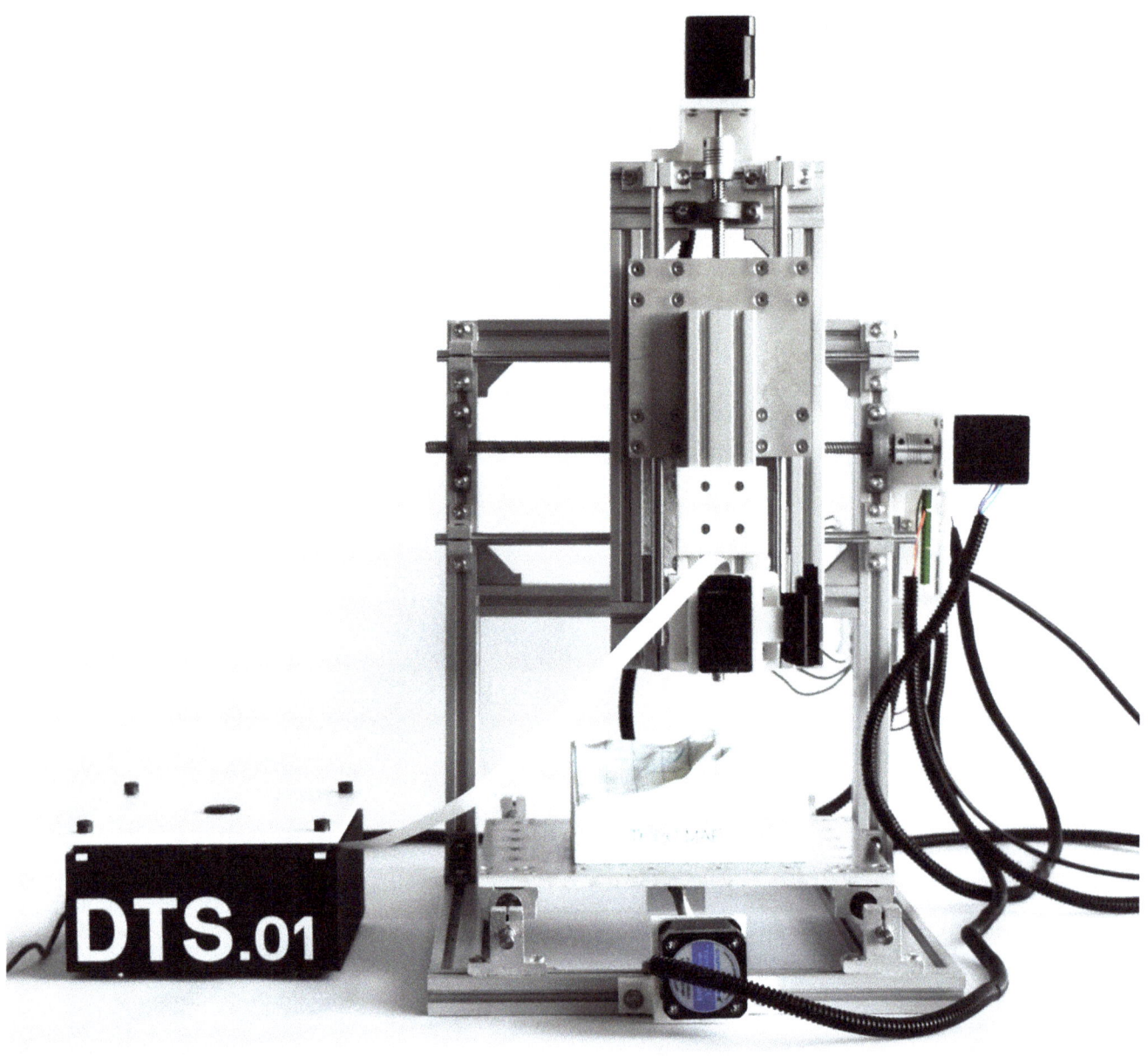

6 DTS Printer

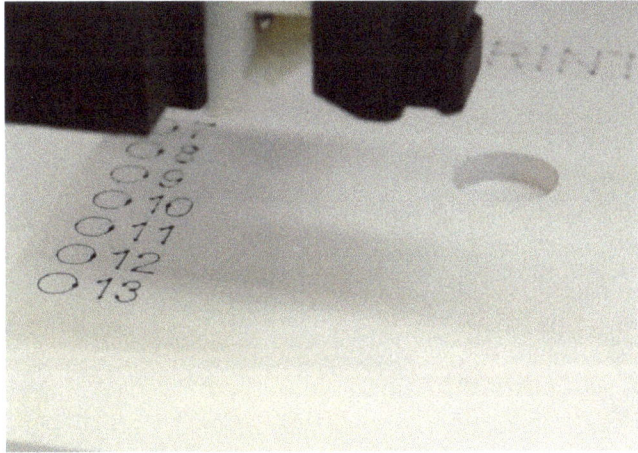

7   Printing on different material substrates (wood)

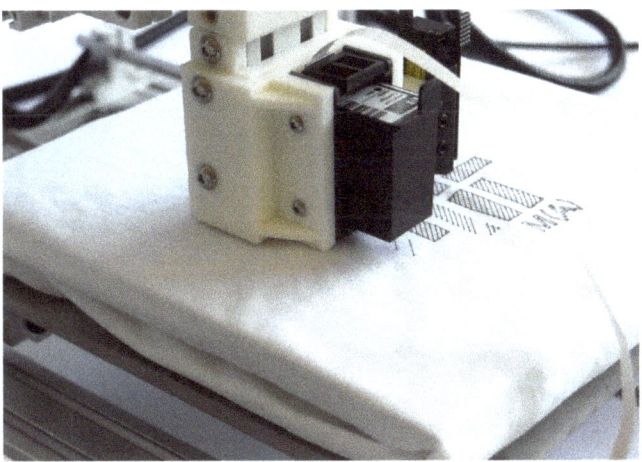

8   Printing on different material substrates (fabric)

9   Foam piece with printed topo lines

The DTS printer, therefore, is an exercise in custom CNC machine building to explore the further capacities of designers redirecting current digital manufacturing methods. With the dedication to open sharing and documentation of the systems developed, it is the firm belief of its creators that the access to this public information also allows for separate actors to continually build, invent, and improve these technologies for the better.

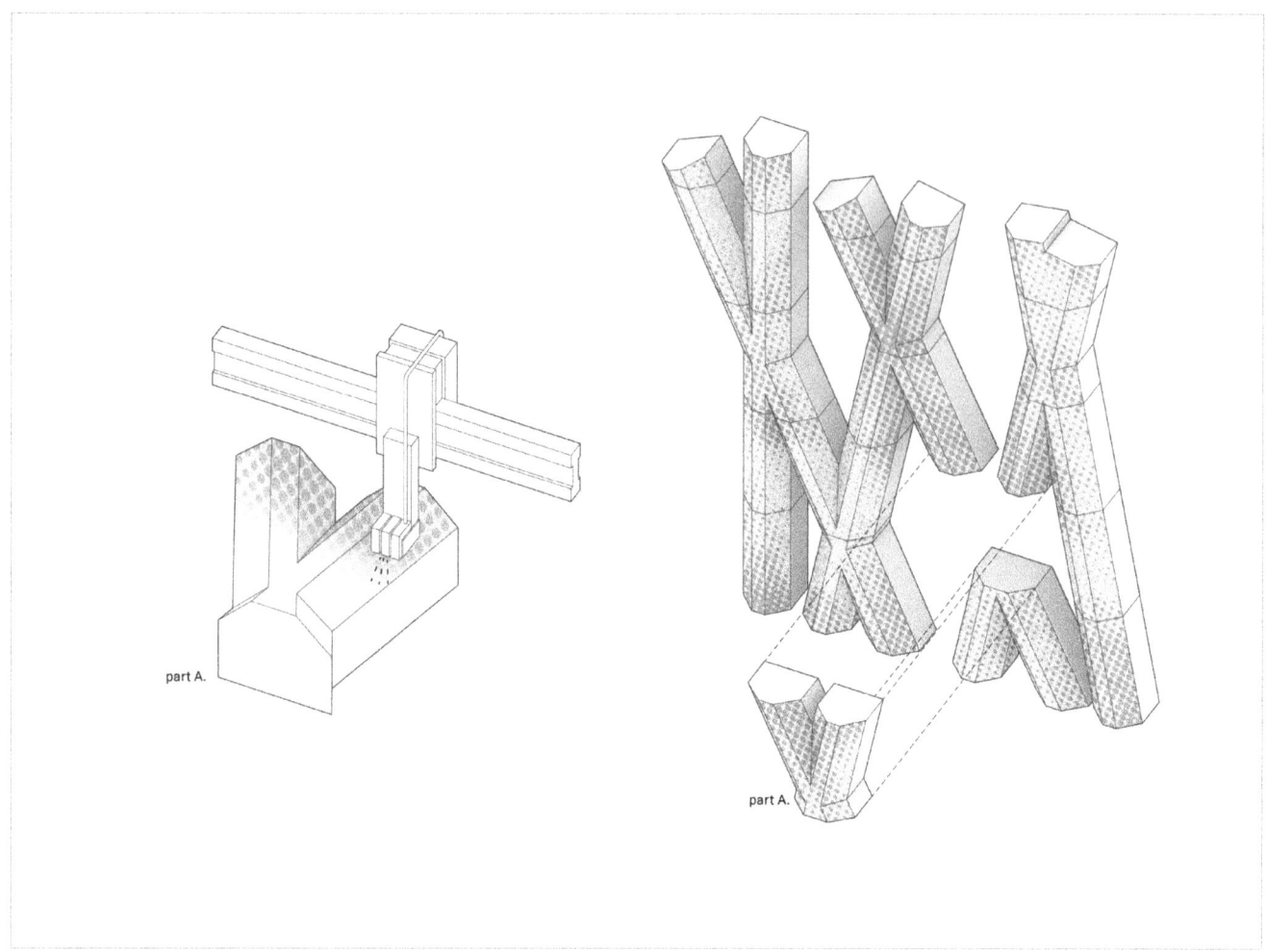

10 Manufacture and printing of architectural components

ACKNOWLEDGMENTS

We would like to extend our gratitude to Neil Gershenfeld, Jake Read, and Zain Karsan for supporting the development of this project as well as the "How To Make Something that Makes (Almost) Anything" class.

REFERENCES

Heinz, Norbert. 2020. "Controlling a HP6602 Print-head." HomoFaciens. July 11, 2021. https://homofaciens.de/technics-machines-printhead-hp6602_en.htm.

Karsan, Zain. TinyZ. MIT Architecture. 2021.

Lewis, Nicholas. 2011. "InkShield | Nerd Creation Lab." Accessed July 11, 2021. http://nerdcreationlab.com/projects/inkshield/.

Mills, Ross N. 1994. "Ink Jet Printing: Past, Present and Future." In *IS&T 10th International Congress on Advances in Non Impact Printing*. 410 - 413.

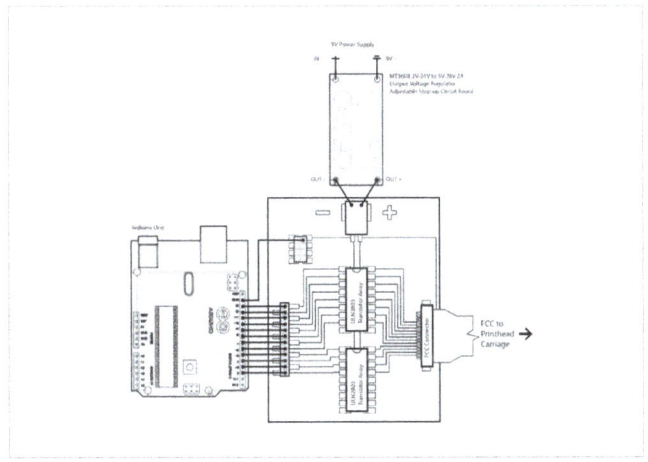

11 Electronics diagram for inkjet control

12 Close-up of TinyG controller

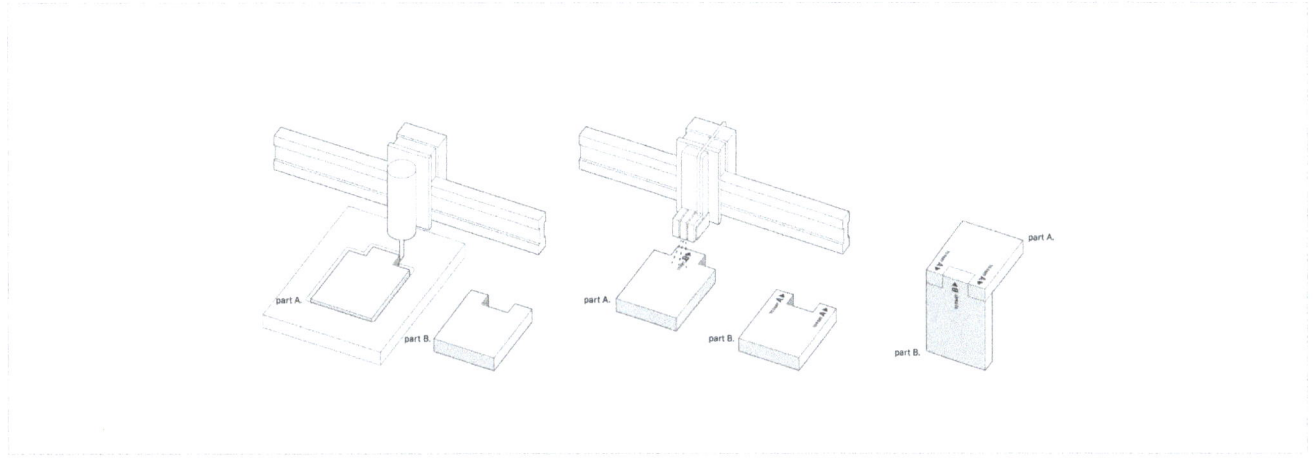

13 Kit manufacture with printed instructions for assembly

IMAGE CREDITS

All drawings and images by the authors.

Kimball Kaiser is an architectural designer originally from Montana and is currently a SMArchS AD candidate at the Massachusetts Institute of Technology. Kimball holds a BA of Environmental Design from Montana State University where he graduated with Highest Honors and an MArch from Taubman College of Architecture and Urban Planning at the University of Michigan. Kimball has taught studio at the University of Michigan and has practiced professionally in New York City at the office of Diller Scofidio + Renfro, in Detroit with JE-LE, in Los Angeles with Jones,Partners:Architecture, and in design-build firms in Montana.

Maryam Aljomairi is a Bahraini architect with interests in programmable material systems and personal fabrication. She has worked internationally at the offices of Diller Scofidio + Renfro, Studio Anne Holtrop, and as a Fulbright scholar at the University of Michigan, where she was awarded the 2019 Digital and Material Technologies award. She received her Bachelor's of Architecture from the American University of Sharjah and is currently a SMArchS - Computation student and research fellow at MIT.

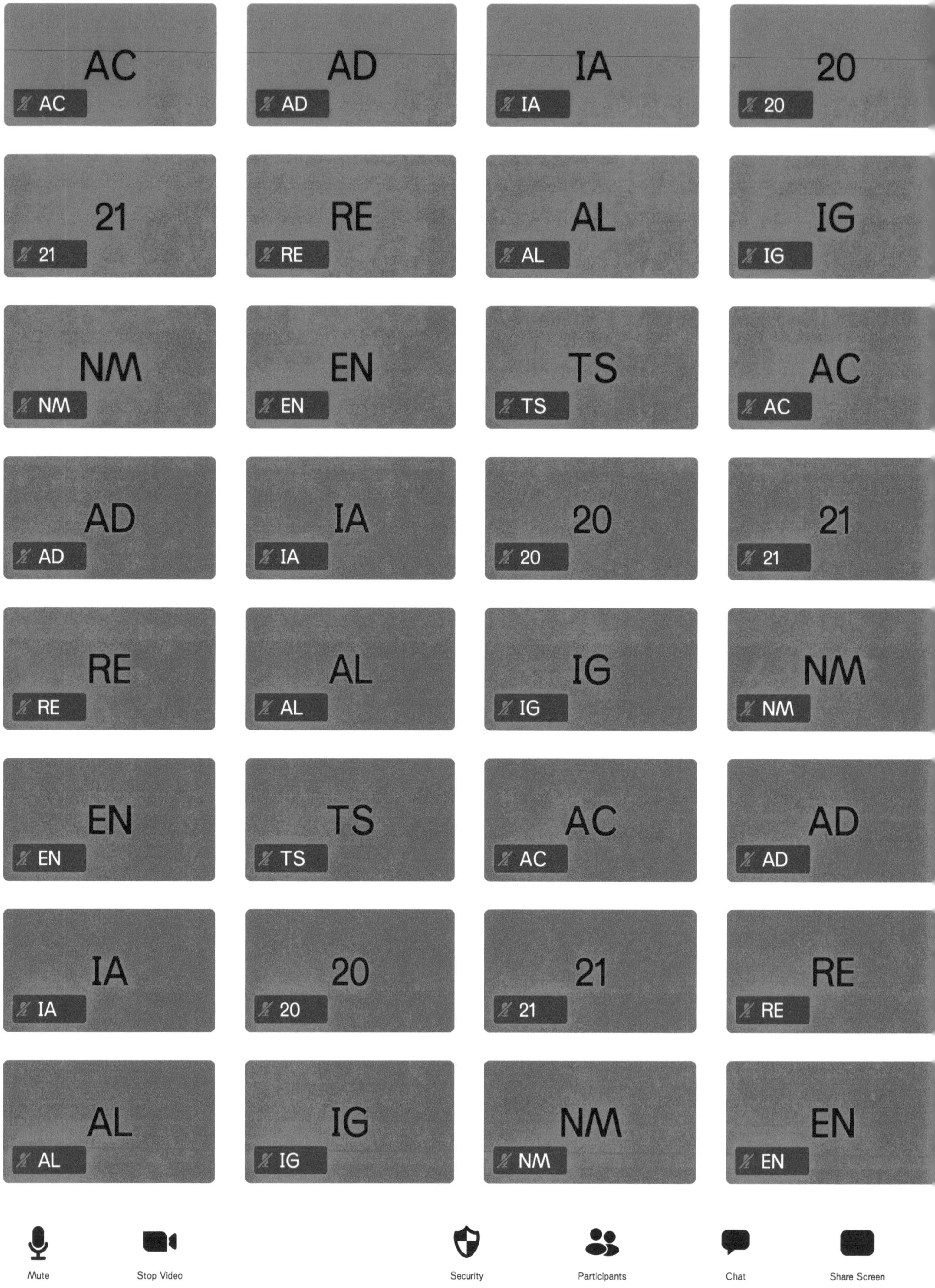

AC	AD	IA	20
21	RE	AL	IG
NM	EN	TS	AC
AD	IA	20	21
RE	AL	IG	NM
EN	TS	AC	AD
IA	20	21	RE
AL	IG	NM	EN

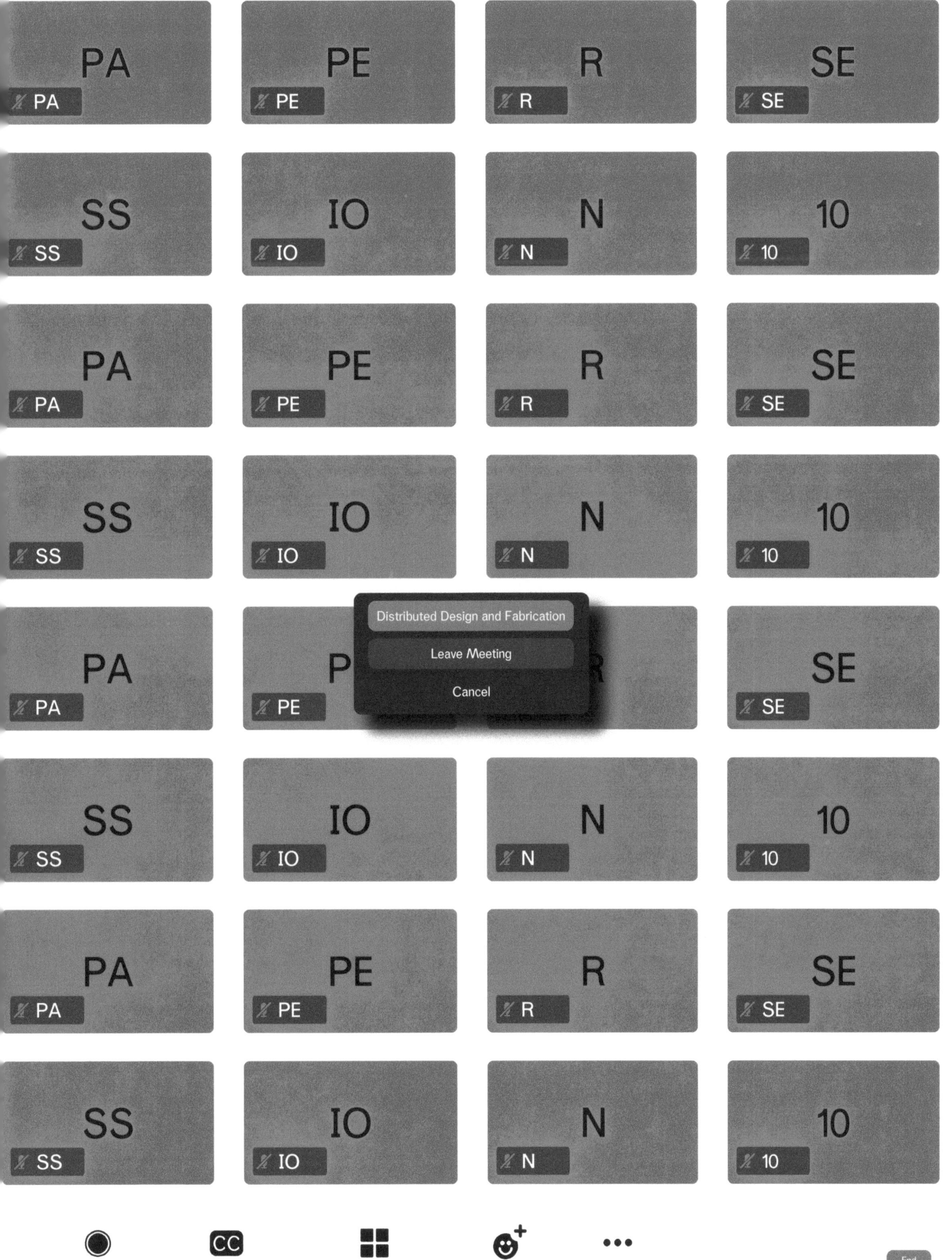

Small Robots and Big Projects

Automation of Complex Spatial Layouts Onsite

Maria Yablonina
Daniels Faculty of Architecture, Landscape, and Design
University of Toronto

James Coleman
A. Zahner Company

ABSTRACT

This paper describes a custom robotic process for semi-autonomous survey and layout of architectural elements for a large-scale renovation project. Specifically, the research presents a custom single-task robotic device accompanied with software and workflow methods for surveying, localizing, and marking the positions of façade anchors along the surface of primary steel members. Enabled by custom robotic locomotion and real-time localization, the presented approach offers high-tolerance installation in a low-tolerance environment while minimizing dangerous erection steps that would typically be done by field personnel.

The robotic system and the workflow are designed, developed, and tailored to the specific project needs and parameters of the renovated building. For instance, the Halbach magnetic locomotion system presented in this paper is custom designed to traverse the radius of steel pipes of which the building structure consists. On the one hand, such specificity renders the robotic hardware obsolete when applied beyond this project. However, the hardware simplicity enabled by its single-task purpose allowed the team to rapidly develop and deploy the robotic system on-site within a year, which would have not been possible with generic hardware. The paper describes the current stage of development of the robotic system and uses the presented robotic workflow to outline the benefits of single-task robotics approach in construction.

1   Semi-autonomous application-specific robotic system for survey and layout of complex architectural elements along the surface of spatial steel structures

2   Layout of the anchor positions for facade installation is complicated by the building scale, its overall geometry, and the circular section of primary structural members (lack of existing indexing features along the structure)

3   Sequence of structure surveying operations at various stages of the renovation project: a) initial laser scan resulting in a point cloud model of the as-built steel structure (expected scan tolerance is approximately 10-25mm); b) secondary high accuracy scan with the proposed robotic system, relying on total station and rigid robot body (advertised total station tolerance with selected prism marker is 1-2mm) (Lienhart 2017)

INTRODUCTION

Development and integration of computational design tools and processes in architecture over the past few decades have resulted in a radical expansion of design possibilities available to their users. However, when a project moves from the office to the job site, the tools used for design, fabrication, and coordination tend to diminish in sophistication and accessibility. In the design phase, architects and design engineers wield a suite of tools for computational modeling, finite element analysis (FEA), simulation, data sharing, and 3D computer-aided design (CAD) visualization. Whereas when construction begins, their counterparts on the job site operate with 2D documents, tape measures, levels, string, and intermittent support from high precision surveying equipment handled manually by a trained surveying subcontractor. This separation of parties has a two-pronged effect: design engineers and architects in the office cannot verify what is being done at the job site, and the field crews do not have a full picture of what they are producing. This results in iterative "loops" of revision and expensive rework. The break between CAD model and onsite work renders many designs difficult to build, making them expensive and vulnerable to simplification and value engineering.

One of the most crucial and error prone phases in the design-to-construction process is the task of translating the CAD model to the physical space of the building site layout. The feedback loop between as-designed and as-built elements is usually ad hoc and error prone. The proclivity for errors and the divergence from the as-designed plan makes deploying automation processes either impossible or impractical.

Granting that the separation between the design and build phases of a project is a complex multi-faceted issue, the presented research proposes a method that removes some of the information gaps between the 3D design model and on-site personnel by providing a feedback loop and automated measurement tool for accomplishing complex layouts of structural members. Specifically, this research proposes an application-specific robotic system for surveying a state of the construction process and autonomously preparing a layout for the next stage directly on the built structure, minimizing the need for additional drawings or documentation and eliminating the need for remobilization of field crews (Figure 1).

The research is conducted in the context of a construction and renovation project of the Airforce Academy Chapel (AFA Chapel) (Danz and Hitchcock 2009) (Figure 2), in particular the façade installation stage. Archival documentation of the

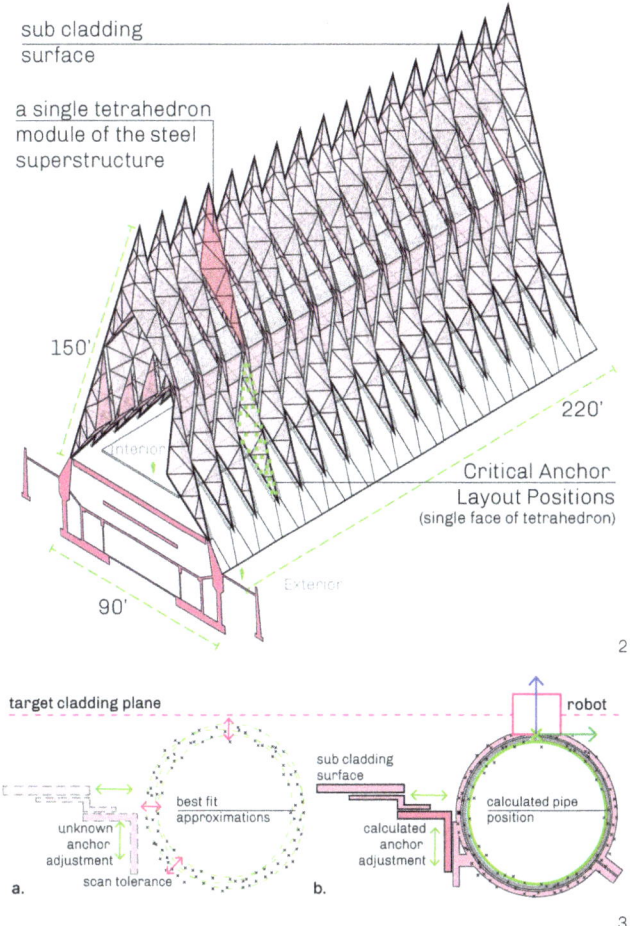

original construction process conducted in 1962 clearly demonstrates the complexity of the facade installation sequence and assembly. The tools available to the original erectors along with the lack of onsite documentation of as-built dimensions leave an incomplete picture of what was built and makes augmentation challenging. Large amounts of shimming (larger than four inches in some areas) and welded in-place connections were originally used to compensate for fabrication tolerances and install errors, neither of which were documented. A primary challenge for this renovation project is creating an envelope in the same form as the original (for historic preservation) while ensuring its high performance according to contemporary standards. For instance, gaps that were originally designed to accommodate fabrication error are now utilized for parts and infrastructure, leaving little margin for error during installation of the new system.

The revised high-tolerance envelope design aims to eliminate water infiltration of the cladding panels and requires some 6,000 accurately positioned anchors installed along the steel tetrahedron structure at heights between 20-150 feet in elevation. A multistage layout process makes the installation costly and often dangerous for the field personnel. The ability to perform layout tasks

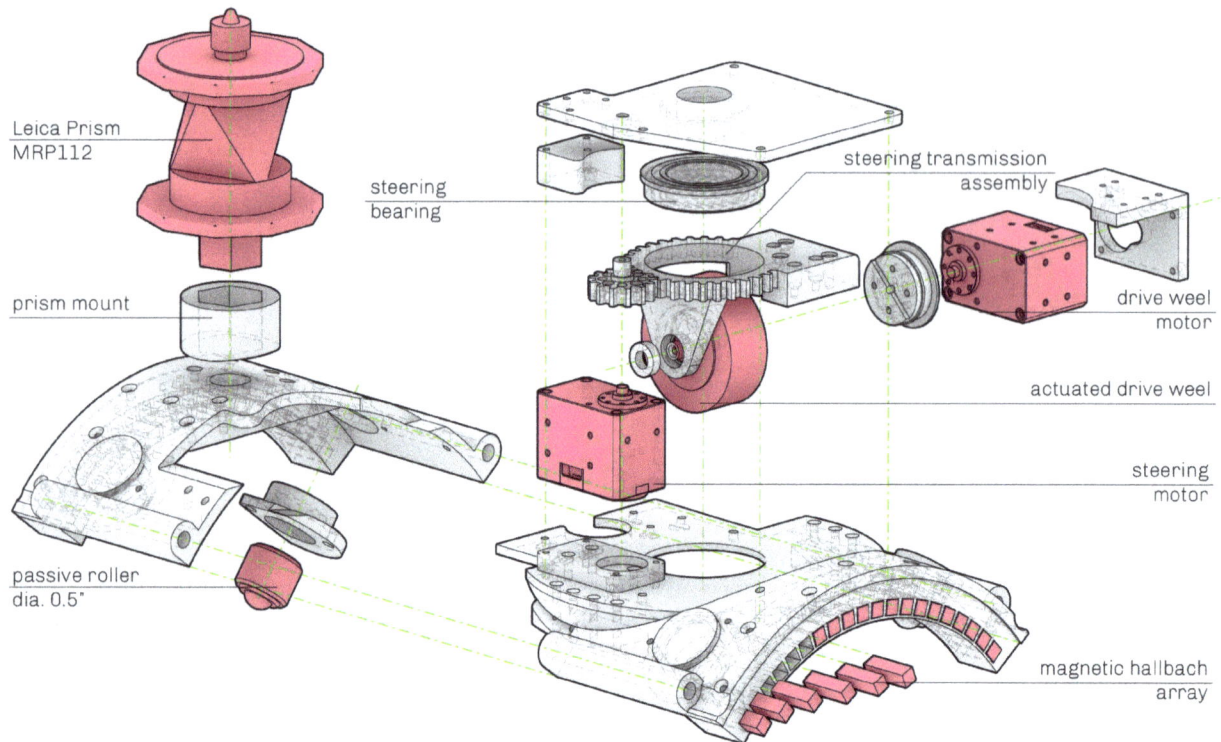

4

semi-autonomously from the ground has countless benefits and is possible by the proposed substrate-specific mobile robot.

The renovation task presents a unique challenge compared to typical construction projects where schedule demands that the superstructure is erected in parallel with facade installation. Here the entire superstructure is complete before any fabrication takes place but is undocumented. Utilizing 3D scanning technologies will generate a semi-complete (low accuracy) as-built model that can be planned and modelled against, which is not typical for conventional processes wherein unknown field conditions significantly limit the possibilities of automation. Availability of the built structure along with its accurate representation in the digital design space allows the leveraging of both for a robotic in situ layout process (Figure 3).

BACKGROUND

In proposing a mobile robotic tool for survey and layout, this research builds upon two primary areas of investigation and prior work: automation of surveying and layout tasks in construction, and research and development in substrate-specific robotic locomotion and climbing robots.

Automated Surveying and Layout in Construction

Automation of surveying, layout, and construction tasks is one of the key areas of interest in academic and industrial research across the fields of robotics, construction, computer science, and architecture. A variety of robotic surveying tools for construction have been developed and integrated, ranging from stationary scanning devices, to wearable hardware intended to be carried by human workers, to legged, aerial, and wheeled autonomous mobile machines. These robotic surveying solutions are mostly deployed for retrospective tolerance tracking during the construction processes. Some examples of autonomous robotic layout solutions do exist (Dusty Robotics 2020); however, they primarily focus on 2D indoor layouts along a floor slab.

Most of the existing surveying and layout devices are well suited for a conventional construction process wherein floor plates are erected in sequence, each providing access to the scanning hardware. In the case of the presented restoration project, none of the solutions would be applicable due to the complexity of navigation of the truss superstructure at the resolution that is required for successful façade panel installation.

Substrate-specific Locomotion Systems

Substrate-specific locomotion systems and climbing robots are a prominent area of research in robotics and are increasingly common in a variety of industrial applications. Climbing robots present a convenient solution for surveying and maintenance of structures and surfaces that would be hard to reach otherwise (Berns et al. 2005; Virk 2005). For instance, surface climbing robots are developed for inspection and fabrication tasks in ship building (Sánchez, Vázquez, and Paz 2006), and inspection of existing steel

(Bi et al. 2012) and concrete structures (Faruq Howlader and Sattar 2015), as well as infrastructure: air and water supply pipes and ducts (Neubauer 1993; Tavakoli et al. 2008; Luk et al. 2006). Specifically, pipe climbing inspection robots are prominently applied in the industry, including architectural infrastructure survey and maintenance tasks (Bohren et al. 2020). Additionally, task-specific climbing robots have been explored for architectural in situ fabrication and reconfiguration tasks (Kayser et al. 2019; Melenbrink et al. 2017; Yablonina et al. 2021; Leder et al. 2019).

In the context of the presented renovation challenge, the combination of a substrate-specific climbing robot and construction surveying hardware presents a unique opportunity to leverage the existing structure as the locomotion substrate for a semi-autonomous robot able to navigate to and mark key locations for the next phase of the construction process (Figure 4).

METHOD

The proposed system consists of the following elements: bespoke robotic locomotion system, localization hardware, and the data collection and surveying workflow. The current iteration of the system is developed to be used in two stages: first, for surveying and benchmarking of the existing laser scanned digital model of the as-built structure, and second, for layout of the façade bracket locations along the pipe structure for the manual installation in the next construction stage. All surveying and marking processes described in this paper occur after decladding, cleanup, and scanning of the chapel building, but prior to cladding panel fabrication.

Robotic Locomotion System

The superstructure of the AFA Chapel is composed of structural steel pipe, making it a complementary substrate for magnet enabled climbing robots. Specifically, we are using a passive magnetic Halbach array adhesion strategy in combination with wheeled robotic locomotion (Figure 5).

Magnetic Halbach arrays installed on the robot's body allow it to reliably attach and traverse the steel pipe while always remaining parallel to the pipe axis (San-Millan 2015; Song et al. 2018). An actuated pivoting wheel positioned at the center of the robot's body and accompanied by an array of four passive wheels allows the robot to drive along the surface of the pipe while maintaining consistent distance between the robot's body and the surface of the pipe. The robot is designed specifically for the A36 steel pipe and is capable of locomotion along its length as well as the diameter (Figure 6).

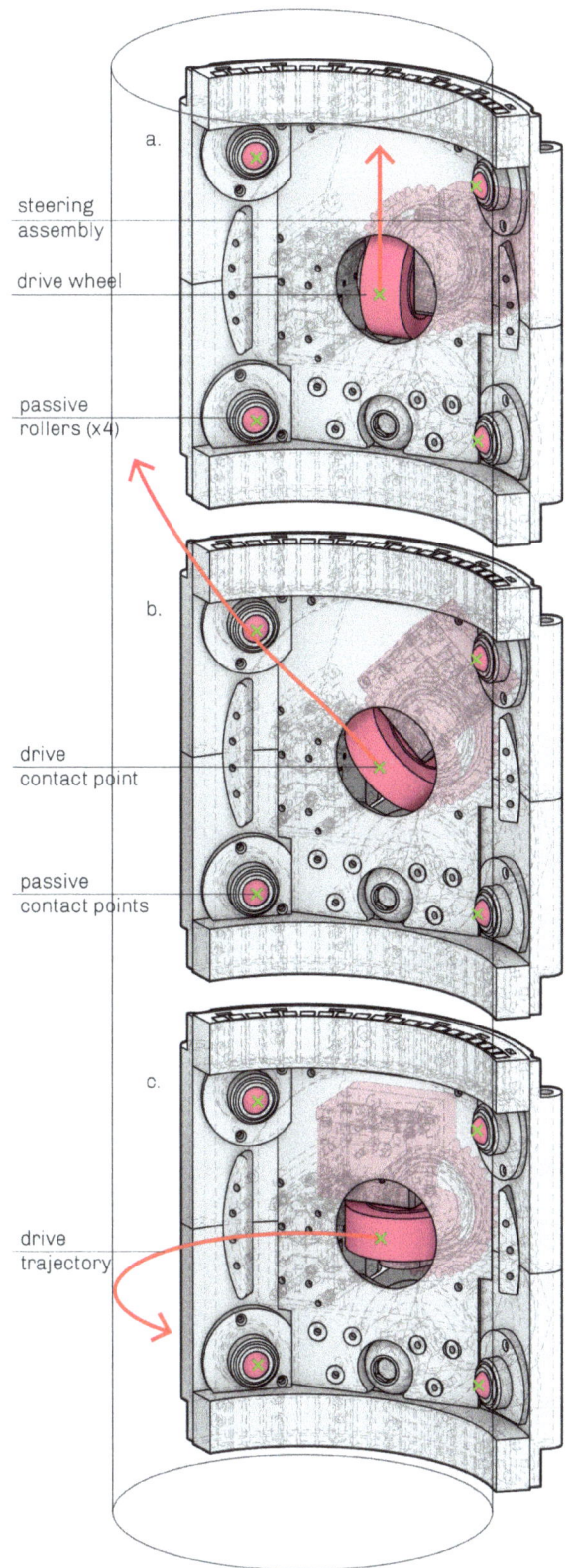

4   Robotic locomotion system assembly diagram

5   Robotic locomotion system single-wheel steering mechanism; the orientation of the robot's body is always fixed, independent of the direction of motion: a) vertical locomotion, b) diagonal locomotion, c) horizontal locomotion

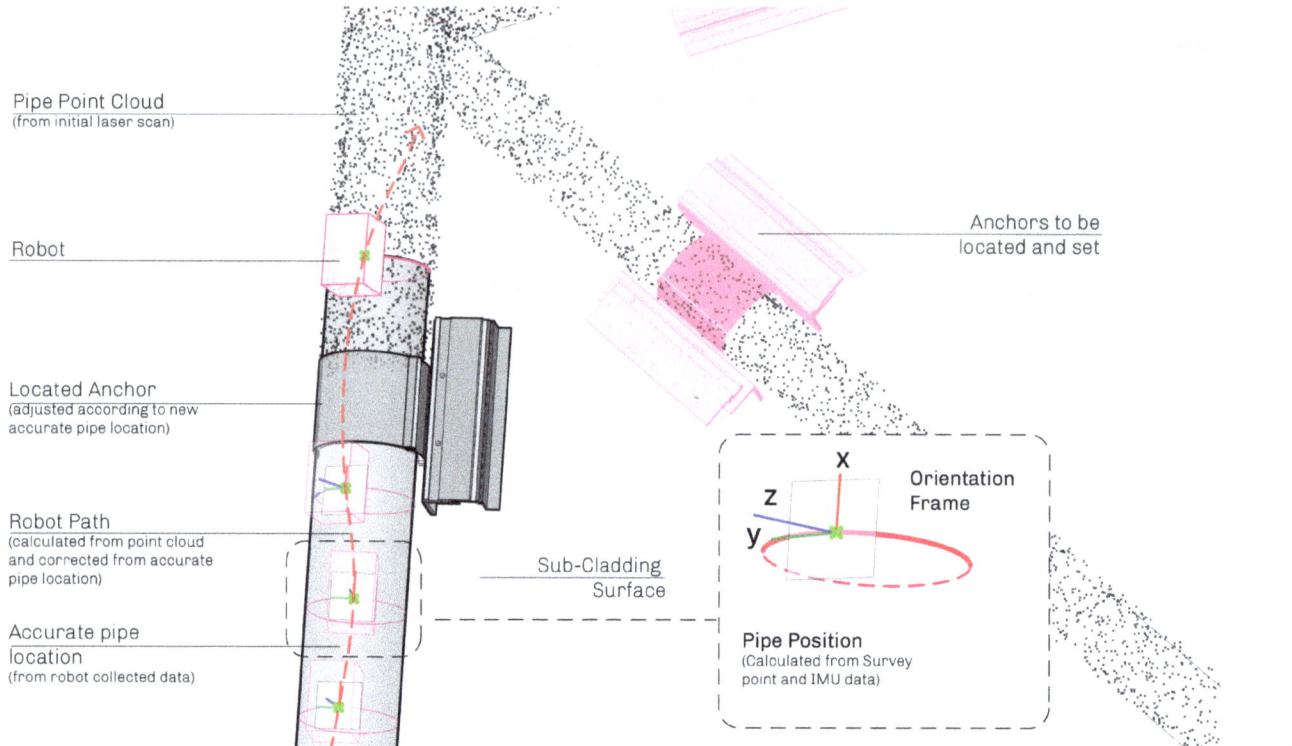

Localization System

The robotic localization system relies on an external metrology device—Leica Total Station and MRP112 Prism—which enable localization and autonomous navigation (Lienhart 2017). The total station is a stationary laser-based measurement device accompanied with a prism marker that is installed on the robot's body (Figure 7). The localization system can measure the position of the marker in 3D space within a 1-2 mm tolerance (Lackner and Lienhart 2016). Additionally, the robotic locomotion system is equipped with an on-board IMU sensor that reports the robot's orientation.

At the current stage of development, the robot is piloted manually (from pendant). As the robot is driven along the pipe surface, the total station's auto track feature is utilized to follow the robot's position and broadcast point data over Wi-Fi. An integrated autonomous navigation and robot control system relying on real-time position reported by the total station is currently in development.

Data Collection and Survey Workflow

The robot position is tracked, captured, and sent via TCP/IP to a Python server, where it is published to a .txt file which can then be read and plotted into the 3D CAD model. The frequency of broadcast can be set by the user (10Hz maximum or on demand) and is calibrated by the complexity of the robot task. The position data is then processed in the 3D CAD environment to benchmark the as-built digital model. Once the digital model is updated and verified, the position information reported by the total station is enough to accurately define the robot's position on the pipe, and the layout stage of the process can begin. During the layout stage the robot can either rely on the real-time position data reported by the total station to autonomously navigate to desired locations or can be driven manually from a pendant.

Points plotted in the 3D CAD model are then analyzed against the theoretical/target position. Because the offset between the prism center and the robot rollers is a fixed relationship, the collected points can be used to generate a best fit "offset" pipe. A least squared approximation method (best fit solution) is used to estimate the true steel pipe location accounting for total station tolerance and sampling error (Figure 8). With the true pipe location updated in the digital model, the anchor positions can now be calculated for the physical layout stage of the process as shown in Figure 8. The outcome of the layout calculation process includes target robot position for marking of the anchor mounting locations, as well as offset values for anchor hardware assembly that allows for in-situ anchor position adjustment.

RESULTS

The following aspects of the proposed robotic surveying and layout system have been successfully implemented and tested at this stage of research development: on-pipe

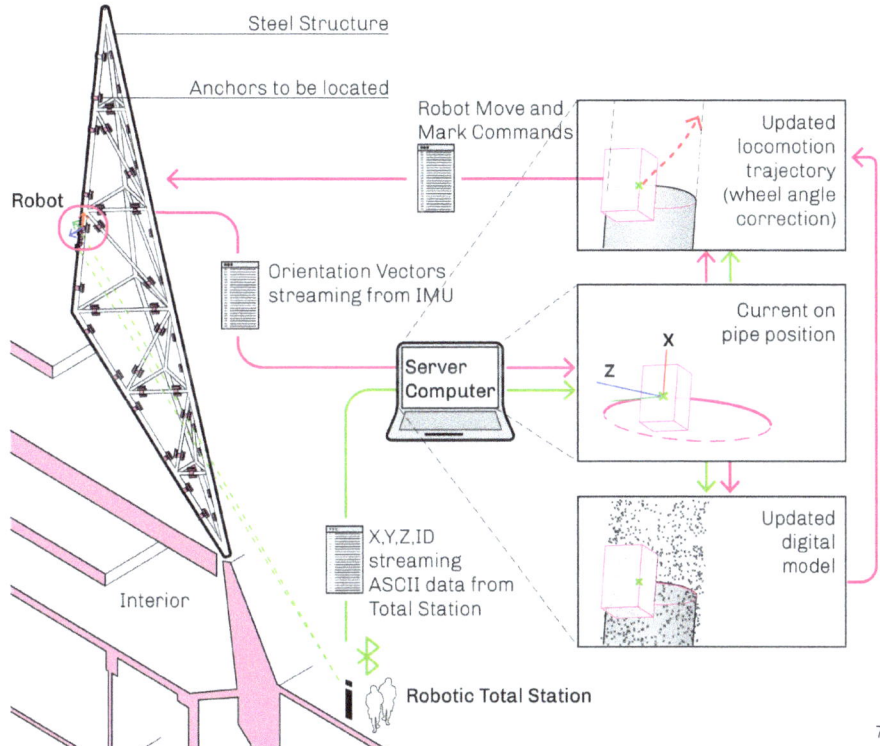

6 Survey and layout stages workflow diagram: as the robot traverses the pipe, surveying points for the digital as-built model benchmarking are collected; once the desired location is achieved, the robotic system marks the anchor installation point

7 Communication across all on-site devices happens via the central server computer that receives information from the total station and dispatches new commands to the robotic surveying system; robot position data reported by the total station is processed by the server computer to a) update the as-built digital mobel, and b) to validate and if necessary update the pathplanning trajectory to the desired anchor location

robotic locomotion, manual robot control, localization hardware integration, position data recording and plotting in the 3D CAD environment, and as-build CAD model benchmarking and update based on the recorded data. The tests were performed on the project performance mock-up (Figure 9) for the AFA Chapel building by the authors together with on-site superintendent who provided additional expertise and feedback. Future on-site tests are planned in accordance with the renovation project schedule.

Current iteration of the locomotion system successfully traverses the pipe surfaces controlled from an operator pendant. The magnetic locomotion system based on Halbach arrays and single wheel drive has been evaluated by the on-site superintendent as effective for the proposed application. However, the following flaws of the first iteration require further development of the hardware design for the next stage of the project. The footprint of the locomotion system is too wide to effortlessly clear some of the narrower areas of available locomotion surface (due to welds and existing hardware mounts along the pipe surface). The wheel pivoting range must be increased in order to afford more steering flexibility and thus improve target point approach tolerance. And the compliancy of the current pivoting joint must be addressed as it introduces a significant drift to the locomotion process.

The total station localization system was successfully used to broadcast robot positions for recording and subsequent benchmarking of the as-built model. Future development and integration of a closed-loop data exchange between the localization system and the robotic locomotion system is necessary to enable autonomous robotic navigation to desired surveying and layout positions. Additionally, integration of the on-board IMU sensor would eliminate the two-step survey and layout process and perform both operations in one go (Figure 9). At the current stage the IMU and the total station data collection have been tested and evaluated separately.

At the current stage of development, the surveying functionality of the proposed robotic process has been developed, tested, reviewed by the site superintendent and on-site personnel, and evaluated to be successful. On-going development of the marking hardware, control software integration, and workflow for the layout stage is ongoing as promising.

OUTLOOK

The project presents a real-world approach to a layout process conducted directly on the structure, offering the flexibility to adapt to tolerances and parameters of the as-built object, rather than relying exclusively on partial as-built 3D models and 2D documentation. The developed process outlines the benefits and the potential of designing and deploying application-specific robotic systems for on-site construction tasks. Specifically, the project advocates for a construction automation approach where

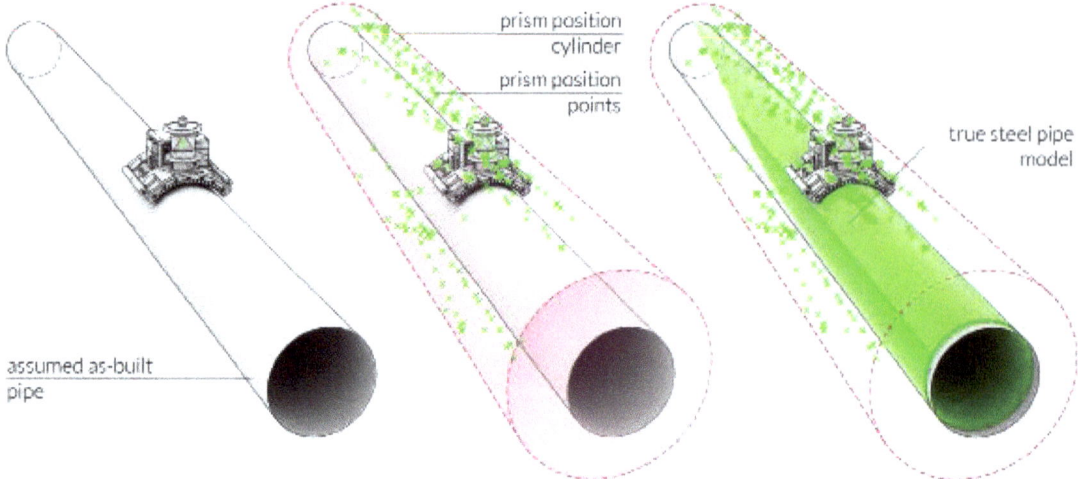

8   Sequence of operations necessary to compute true steel pipe position: a) prism positions recorded as the robot traverses the pipe; b) least squared approximation is used to create a best fit cylinder from prism positions; and c) the cylinder is offset to estimate the true steel pipe location

complex and dangerous measuring and survey tasks are handed off to robots that can localize in 3D space, while leaving the construction tasks that require dexterity to human workers. This hybrid human-robot approach offers expedited and safer construction sequences, in turn lowering costs and expanding what is possible to build.

REFERENCES

Berns, K., T. Braun, C. Hillenbrand, and T. Luksch. 2005. "Developing Climbing Robots for Education." In *Climbing and Walking Robots*, edited by M. A. Armada and P. de González Santos. Berlin, Heidelberg: Springer Berlin Heidelberg. 981–88.

Bi, Zhiqiang, Yisheng Guan, Shizhong Chen, Haifei Zhu, and Hong Zhang. 2012. "A Miniature Biped Wall-Climbing Robot for Inspection of Magnetic Metal Surfaces." In *2012 IEEE International Conference on Robotics and Biomimetics (ROBIO)*. Guangzhou, China: IEEE. 324–29.

Bohren, Jonathan, Christopher Bolger, Anthony G. Musco, and Jack Wilson. 2020. Robotic sensor system for measuring parameters of a structure. United States Patent 10859510, filed 2019, and issued 2020.

Dusty Robotics. 2020. FieldPrinter. California: Dusty Robotics.

Faruq Howlader, M. D. Omar, and Traiq Pervez Sattar. 2015. "Novel Adhesion Mechanism and Design Parameters for Concrete Wall-Climbing Robot." In *2015 SAI Intelligent Systems Conference (IntelliSys)*. London, UK: IEEE. 267–73.

Kayser, Markus, Levi Cai, Christoph Bader, Sara Falcone, Nassia Inglessis, Barrak Darweesh, João Costa, and Neri Oxman. 2019. "Fiberbots: Design and Digital Fabrication of Tubular Structures Using Robot Swarms." In *Robotic Fabrication in Architecture, Art and Design 2018*, edited by J. Willmann et al. Cham: Springer International Publishing. 285–96.

Lackner, Stefan, and Werner Lienhart. 2016. "Impact of Prism Type and Prism Orientation on the Accuracy of Automated Total Station Measurements." In *Proceedings of the Joint International Symposium on Deformation Monitoring*. Vienna, Austria.

Leder, Samuel, Ramon Weber, Dylan Wood, Oliver Bucklin, and Achim Menges. 2019. "Robotic Timber Construction." In *ACADIA '19: Ubiquity and Autonomy; Proceedings of the 39th Annual Conference of the Association for Computer Aided Design in Architecture*, edited by K. Bieg, D. Briscoe, and C. Odom. Austin, Texas, US. 510–19

Lienhart, Werner. 2017. "Geotechnical Monitoring Using Total Stations and Laser Scanners: Critical Aspects and Solutions." *Journal of Civil Structural Health Monitoring* 7 (3): 315–24.

Luk, B. L., A. A. Collie, D. S. Cooke, and S. Chen. 2006. "Walking and Climbing Service Robots for Safety Inspection of Nuclear Reactor Pressure Vessels." *Measurement and Control* 39 (2): 43–47.

Melenbrink, Nathan, Paul E. Kassabian, Achim Menges, and Justin Werfel. 2017. "Towards Force-Aware Robot Collectives for On-Site Construction." In *ACADIA '17: Disciplines & Disruption; Proceedings of the 37th Annual Conference of the Association for Computer Aided Design in Architecture*. Cambridge, Mass. 382–391.

Neubauer, Werner. 1993. "Locomotion with Articulated Legs in Pipes or Ducts." *Robotics and Autonomous Systems* 11 (3–4): 163–69.

Sánchez, J., F. Vázquez, and E. Paz. 2006. "Machine Vision Guidance System for a Modular Climbing Robot Used in Shipbuilding." In *Climbing and Walking Robots*, edited by M. O. Tokhi, G. S. Virk, and M. A. Hossain. Berlin, Heidelberg: Springer. 893–900.

San-Millan, Andres. 2015. "Design of a Teleoperated Wall Climbing

 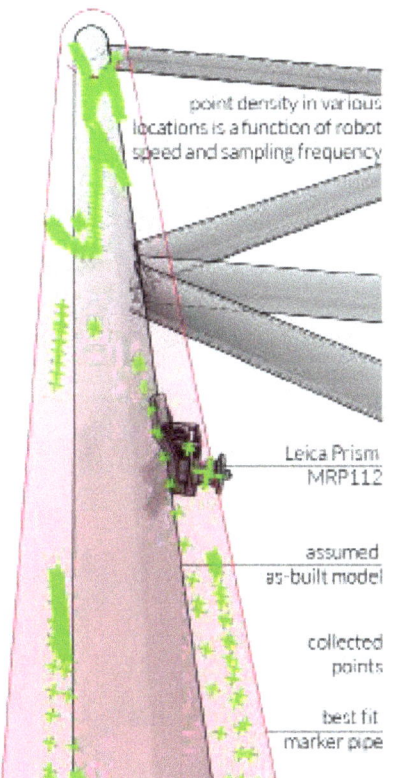 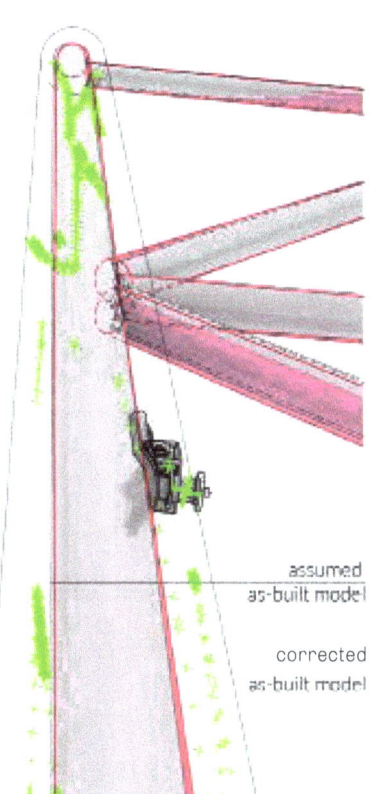

9   Documentation and result diagram of the on-site mockup testing; the points collected during the manual robotic piloting are recorded and later post-processed to correct the original as-built model

Robot for Oil Tank Inspection." In *2015 23rd Mediterranean Conference on Control and Automation (MED)*. Torremolinos, Malaga, Spain: IEEE. 255–61.

Song, Wei, Hongjian Jiang, Tao Wang, Daxiong Ji, and Shiqiang Zhu. 2018. "Design of Permanent Magnetic Wheel-Type Adhesion-Locomotion System for Water-Jetting Wall-Climbing Robot." *Advances in Mechanical Engineering* 10 (7): 1–11.

Tavakoli, M., A. Marjovi, L. Marques, and A. T. de Almeida. 2008. "3DCLIMBER: A Climbing Robot for Inspection of 3D Human Made Structures." In *2008 IEEE/RSJ International Conference on Intelligent Robots and Systems*. Nice: IEEE. 4130–35.

Virk, G.S. 2005. "The CLAWAR Project - Developments in the Oldest Robotics Thematic Network." *IEEE Robotics & Automation Magazine* 12 (2): 14–20.

Yablonina, Maria, Brian Ringley, Giulio Brugnaro, and Achim Menges. 2021. "Soft Office: A Human–Robot Collaborative System for Adaptive Spatial Configuration." *Construction Robotics* 5 (1): 23–33.

IMAGE CREDITS

All drawings and images by the authors.

Maria Yablonina is Assistant Professor at the Daniels Faculty of Architecture, Landscape and Design at the University of Toronto. Her work lies at the intersection of architecture and robotics, producing spaces and robotic systems that can construct themselves and change in real-time. Maria's practice focuses on designing machines that make architecture—a practice that she broadly describes as Designing [with] Machines (D[w]M). D[w]M aims to investigate and establish design methodologies that consider robotic hardware development as part of the overall design process and its output. Through this work, Maria argues for a design practice that moves beyond the design of objects towards the design of technologies and processes that enable new ways of both creating and interacting with architectural spaces.

James Coleman is a Vice President at A. Zahner Company, a manufacturer that specializes in computational fabrication and mass customization. As a member of the Executive Committee, he heads the R&D Department and responsible for cross functional innovation activities. James and team develop parametric design-to-fabrication-to-installation workflows and one-off automation strategies in both hardware and software. He is involved in the production of thousands of unique parts for large scale projects as a design and manufacturing specialist. He has worked separately within Architecture, Engineering, and Construction and also in Product Development at the Ford Motor Company. James balances working on large scale high profile architectural projects with being one half of the design studio James and the Giant Peek with Assistant Professor Dr. Nadya Peek (HCDE UW).

Co-Designing Material-Robot Construction Behaviors

Teaching distributed robotic systems to leverage active bending for light-touch assembly of bamboo bundle structures

Grzegorz Łochnicki*
Nicolas Kubail Kalousdian*
Samuel Leder
Mathias Maierhofer
Dylan Wood
Achim Menges
(ICD) University of Stuttgart

* Authors contributed equally to the research

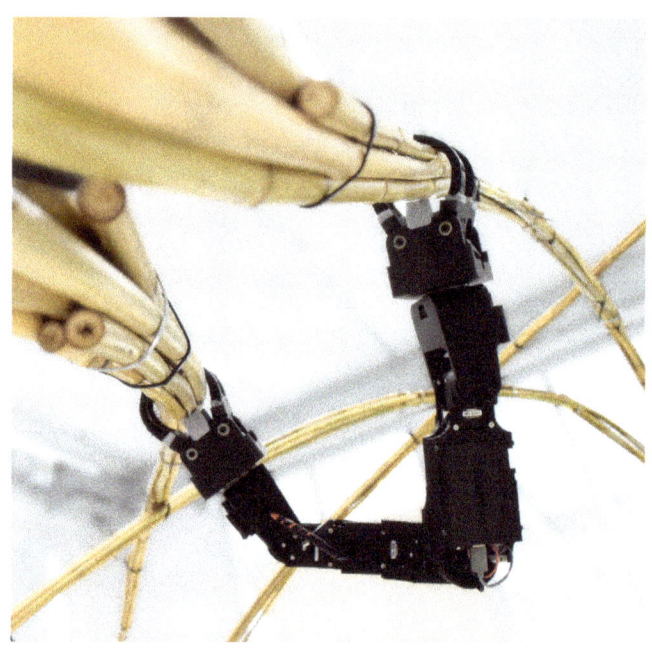

1   Co-designed distributed material-robot construction system; speculative visualization of a lightweight bamboo bundle structure being assembled by the proposed team of bespoke mobile robots

2   Photo of the mobile robot prototype on the bamboo bundle structure; this robot was used in feasibility studies

ABSTRACT

This paper presents research on designing distributed, robotic construction systems in which robots are taught construction behaviors relative to the elastic bending of natural building materials. Using this behavioral relationship as a driver, the robotic system is developed to deal with the unpredictability of natural materials in construction and further to engage their dynamic characteristics as methods of locomotion and manipulation during the assembly of actively bent structures. Such an approach has the potential to unlock robotic building practice with rapid-renewable materials, whose short crop cycles and small carbon footprints make them particularly important inroads to sustainable construction. The research is conducted through an initial case study in which a mobile robot learns a control policy for elastically bending bamboo bundles into designed configurations using deep reinforcement learning algorithms. This policy is utilized in the process of designing relevant structures, and for the in situ assembly of these designs. These concepts are further investigated through the co-design and physical prototyping of a mobile robot and the construction of bundled bamboo structures. This research demonstrates a shift from an approach of absolute control and predictability to behavior-based methods of assembly. With this, materials and processes that are often considered too labor-intensive or unpredictable can be reintroduced. This reintroduction leads to new insights in architectural design and construction, where design outcome is uniquely tied to the building material and its assembly logic. This highly material-driven approach sets the stage for developing an effective, sustainable, light-touch method of building using natural materials.

Methods of elastically bending wood to assemble structures:

3   Integration of active bending in plywood strips in a computationally designed lightweight structure bent into place onsite by humans, ICD/ITKE Research Pavilion 2010 (©ICD/ITKE, University of Stuttgart)

4   Precision elastic bending of glulam wood beams using an industrial robot in a prefabrication setup (©ICD/ITKE, University of Stuttgart)

INTRODUCTION: CO-DESIGNING MATERIAL-ROBOT CONSTRUCTION BEHAVIORS

Before the industrial revolution, humans relied on their experience and intuition of material behavior to build with low precision tools and lightly processed materials (Addis 2007). As the precision and automation of construction and fabrication methods grew, materials that are difficult to control and predict were discarded in favor of largely engineered, standardized, precise, and predictable building materials. However, recent research has demonstrated how natural materials, that are inherently heterogeneous and thus hard to handle and simulate, such as wood, can nowadays also be capitalized on to assemble structures that are economically, ecologically, and structurally performative (Menges 2011).

This research builds upon these examples and integrates material behaviors alongside a distributed, mobile robotic construction system within a single computational design process. By having the behavioral relationship between material and machine as the driver for co-design, the material behaviors, fabrication parameters and the procedural aspects of assembly all equally inform the computational process (Alvarez 2019). Such an approach takes a step towards unlocking robotic building with rapid-renewable materials, whose short crop cycles and smaller carbon footprints as compared to industrially produced materials make them a particularly important inroad to more sustainable construction (Ribeirinho et al. 2020; van der Lugt et al. 2000; Manandhar et al. 2019). One such material is bamboo, which has in the past been used as a building material in a range of applications and structures. Of specific interest is that bamboo used as raw rods exhibits excellent elastic bending characteristics making it ideal for structures utilizing active bending (Lorenzo et al. 2020; Bessai 2013).

The presented research proposes a newly invented mobile robot that learns a control policy for performing one of the identified assembly tasks: namely elastically bending bamboo bundles into designated configurations. Using deep reinforcement learning, the robot is taught to handle a range of mechanical properties and assembly scenarios, allowing it to operate in a more behavioral manner, not having absolute control over the material, but developing and using an understanding for how the material behaves (Figure 4). This allows it to be more responsive to the heterogeneity of the material as well as to other external disturbances during the assembly process, both of which are major challenges in the field of in situ robotic fabrication (Dörfler 2018). In future work, the intention is to introduce a team of these robots to collaboratively build full structures with a full range of assembly behaviors (Figures 1, 11). This research focuses on only one assembly behavior, that of bending bundles to designated positions. As such, this research:

- Challenges the linearity of conventional design in construction pipelines by proposing a workflow in which the architectural design and construction process emerges from the relationship between material and robotic behaviors.
- Leverages advances in reinforcement learning and sim-to-real methods to facilitate a return to more material informed building, enhancing existing practices of material independent robotic fabrication, and questioning the historical lineage of relying on excessive force and big machines and instead advocating for a light touch approach.
- Forges new avenues towards more sustainable construction by reintroducing the usage of non-standard natural building materials and materials with biological variability to the construction industry.

BACKGROUND

Collective Behavior-based Robotic Construction

By combining robotics, computer science, functional materials, and building design, collective robotic construction (CRC) is a rapidly growing field of research focused on the development of multi-robot systems tailored for architectural construction (Petersen et al. 2019). Successful projects have shown the assembly of both discrete building elements, such as bricks (Dörfler et al. 2016), and continuous materials, such as carbon fiber (Vasey et al. 2020). However, most examples rely on placeholder materials that are developed specifically for the respective research (Jenett et al. 2019). Furthermore, when materials are compliant, amorphous, or unpredictable, issues of mechanical tolerance, structural stability, and architectural design are discussed as major limitations (Thangavelu et al. 2020). Nevertheless, some research showcases a behavioral approach, based on sensor-actuator feedback between material and the robot, as a possible way to combat these limitations (Brugnaro et al. 2016). However, often the robot does not use the material's behaviors to its advantage and rather uses brute force of the machine to manipulate the material.

From the robotic platform perspective, off-the-shelf robots, including industrial robotic arms and UAVs (unmanned aerial vehicles), have been adapted with custom end-effectors for CRC (Vasey et al. 2020). The development of custom machines is a further trend within CRC, in which the robot system is created in direct relationship to the construction material and construction system. Such developments demonstrate how the co-design of material and machine leads to CRC systems with high precision, low cost, and structural efficiency (Jenett et al. 2019; Kayser et al. 2018; Leder et al. 2019; Yablonina et al. 2017). From the robotic platform perspective, off-the-shelf robots, including industrial robotic arms and UAVs, have been adapted with custom end-effectors for CRC (Vasey et al. 2020). The development of custom machines is a further trend within CRC, in which the robot system is created in direct relationship to the construction material and construction system. Such developments demonstrate how the co-design of material and machine leads to CRC systems with high precision, low cost, and structural efficiency (Jenett et al. 2019; Kayser et al. 2018; Leder et al. 2019; Yablonina et al. 2017).

Machine Learning

One common application of machine learning (ML) is the robotic learning of material manipulation (Zeng et al. 2019). Reinforcement learning (RL), specifically, is beginning to be used in the context of material manipulation for digital fabrication (Brugnaro et al. 2019; Apolinarska et al. 2021). This requires the learning of intelligent behaviors in complex dynamic environments that can then be transferred to robots in the real world. Robots are therefore able to quickly and effectively adapt to new tasks in real-time (Nagabandi et al. 2018). Adaptation is critical when designing robots expected to robustly perform autonomous tasks in chaotic environments, such as construction sites.

The adaptability enabled by RL is further emphasized by research on robots working with amorphous materials that have properties that are hard to predict (Zhang et al. 2020). However, examples of such research tend to be conducted only with simulation due to the complexity and cost of real-life training. In this paper, an approach is presented for training robots to perform construction tasks using materials with heterogeneous properties in the real world.

More specifically we focus on Deep Reinforcement Learning (DRL), a subset of machine learning methods adept at solving a wide range of nonlinear tasks that require learning intelligent behaviors in complex and dynamic environments. In contrast to supervised learning approaches, reinforcement learning algorithms learn from trial and error, making them a fitting method for the proposed context, due to the lack of rich and organized databases required for supervised learning.

REASEARCH METHODS

This research presents a distributed, mobile robotic construction system, capable of partially assembling bamboo structures using a control policy learned from interactions with the material. The system consists of bundled bamboo rods and custom mobile robots that assemble the bundles into architectural structures. Extensive training in a simulation environment teaches the robots to perform one assembly task, bending bundles, while dealing with unpredictable mechanical properties inherent to bamboo and varying geometric configurations as specified by the designer. Data collected from this training process could then be further utilized in designing bamboo structures, while the trained policy is used in simulations to determine the feasibility of assembling the design.

Construction System - Bamboo Bundles

The construction system is composed of bamboo bundles, metal zip-tie joints, and steel anchors. Bundles are created by zip-tying bamboo rods into assembly groups, while longer-length bundles are achieved by overlapping assembly groups, and further joining those together. The structural capacity and bending radius of each bundle can be adjusted by adding or removing bamboo rods, thus varying the cross-section along the length of the bundle.

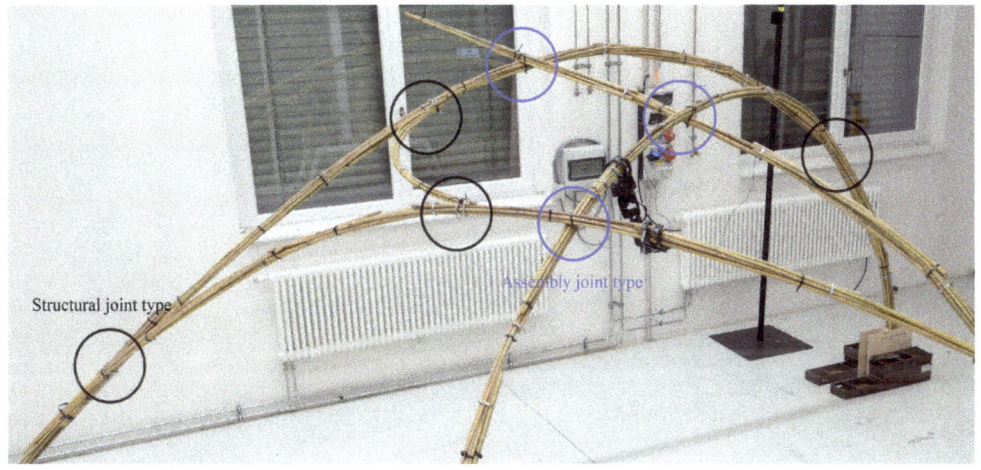

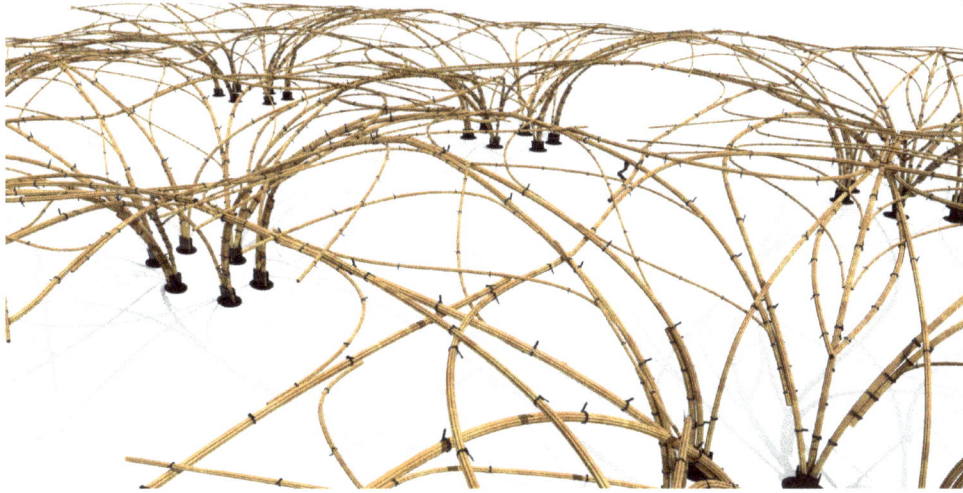

5 Joint types of the construction system: fixing joint labeled in black circles and assembly joint labeled in blue circles

6 Outlook for a speculative bamboo structure as enabled by the learned control policy and combination of the proposed joints

7 Exploded view of the mobile robot prototype featuring two grippers, 5 degrees of freedom, 40 cm height, and a weight of 2.3 kg; chassis parts were printed with Onyx and ABS plastics; prototype runs on a Raspberry Pi microcomputer and uses Dynamixel MX-64 and XL430 motors; approximated cost of 1,800 EUR

Furthermore, this flexibility allows for the typology of the structure to be adjusted during the building's lifetime or even serve as temporary scaffolding to be removed after assembly.

Bundles can be joined together with two types of joints: structural fixing joints, and non-structural assembly joints (Figure 5). In a fixing joint, the bundles are joined in parallel. Thus, a bundle interpolates its curvature with the other bundle it is connected to. In contrast, the assembly joint connects two bundles in non-parallel formations, and so the connected bundles keep their local curvature. The structural strength of the assembly joint is considerably smaller than the fixing joint and is, therefore, used to assist in the assembly process and for scaffolding purposes. Conversely, the fixing joints efficiently transfer loads between bundles and into the steel anchors. The combination of these two types of joints can be used in many variations to cover a wide range of scenarios and structural requirements (Figure 6).

Mobile Robots

The mobile robots were designed with an articulated

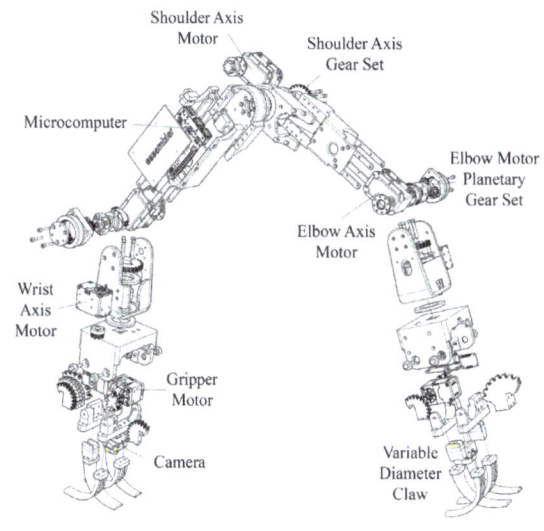

morphology comprised of 5 degrees of freedom in order to perform the various assembly tasks described in the next section (Figure 8). They are equipped with two variable diameter claws to grasp various bundle sizes and deal with the unpredictable arrangement of bamboo rods within each bundle. The robot is also equipped with various sensors that monitor and estimate both its state and changes in the environment (Figure 7). Localization is achieved with an external multi-camera tracking system, while more subtle and local corrections could be conducted via computer vision algorithms, by processing real-time images from the cameras embedded in each claw. Sensor fusion is conducted on accelerometer, gyroscope, and magnetometer readings to provide orientation and acceleration measurements during task execution.

Material-Robot Behaviors

Material-Robot behaviors describe the relationship between the two core components of the construction system, bamboo, and mobile robot, and serve as the basis for the development of the computation design process.

Leveraging the Elastic Bending Behavior of Bamboo

Awareness of the benefits of bamboo, as material, and its bending behavior in modern construction is visible in the amount and architectural quality of new structures (Minke 2012). Bamboo is an elastic material and can undergo bending deformations in relation to the forces acting on it. This bending behavior can be further categorized as static bending and dynamic bending, both of which can be leveraged to assemble bending active structures from originally linear elements (Lienhard et al. 2013).

Static bending can be achieved by applying point loads in specific locations along the length of the material. The robot can achieve this by climbing to specified positions along the length of the bamboo bundle. This deformation can be calculated with force- or position-based models (Suzuki et al. 2018) and can be described by the elastica curve of the bundle, the robot's current location, self-weight, and gravity vector (Figure 9). Static bending as an assembly process, however, is geometrically restrictive, resulting only in forms expressive of the elastica curve.

In contrast, dynamic bending allows for higher geometric freedom, which, in turn, enables more complex structures to be designed and built. This can be achieved when the forces applied to the bamboo are not aligned to the gravity vector (i.e. not self-weight), but instead align with the desired bending direction. The robot is capable of dynamic bending (Figure 10) by rhythmically swinging its appendages and thus introducing directed momentum to

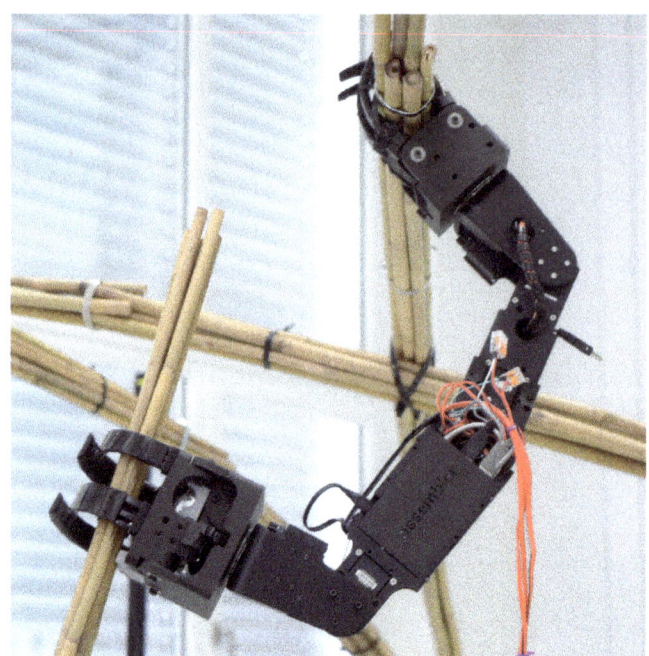

8

the system. However, such swinging requires an understanding of the natural frequency of the bundle, and cannot be pre-planned with trajectory optimization algorithms as this becomes a nonlinear and partially observable planning problem. Nonlinear problems involve solving for a goal with changing variables, some of which are unknown or unpredictable, such as the exact mechanical properties of specific bamboo pieces and the oscillation frequency of the bundled bamboo elements. Therefore, in order to perform such tasks, we train a neural network policy using deep reinforcement learning that controls the robot's axes in relation to the behavior exhibited by the bamboo bundle in real-time. This learning method is further elaborated in the Deep Reinforcement Learning section.

Manual Evaluation of Secondary Robotic Construction Tasks

Locomotion, material transportation, bundle extension, and joining are other assembly tasks (Figure 11) that require an understanding of the relationship between material and robot. Although these tasks have been identified and designed, they are linear problems largely unaffected by the varying mechanical properties and dynamics of bamboo and thus can be solved using existing methods for task and motion planning such as trajectory optimization algorithms. Validation of such behaviors is conducted through simulation and physical experiments and presented in the Results section.

Deep Reinforcement Learning: Learned Knowledge

Deep Reinforcement Learning is used to teach the robots how to operate relative to the bending behavior of the

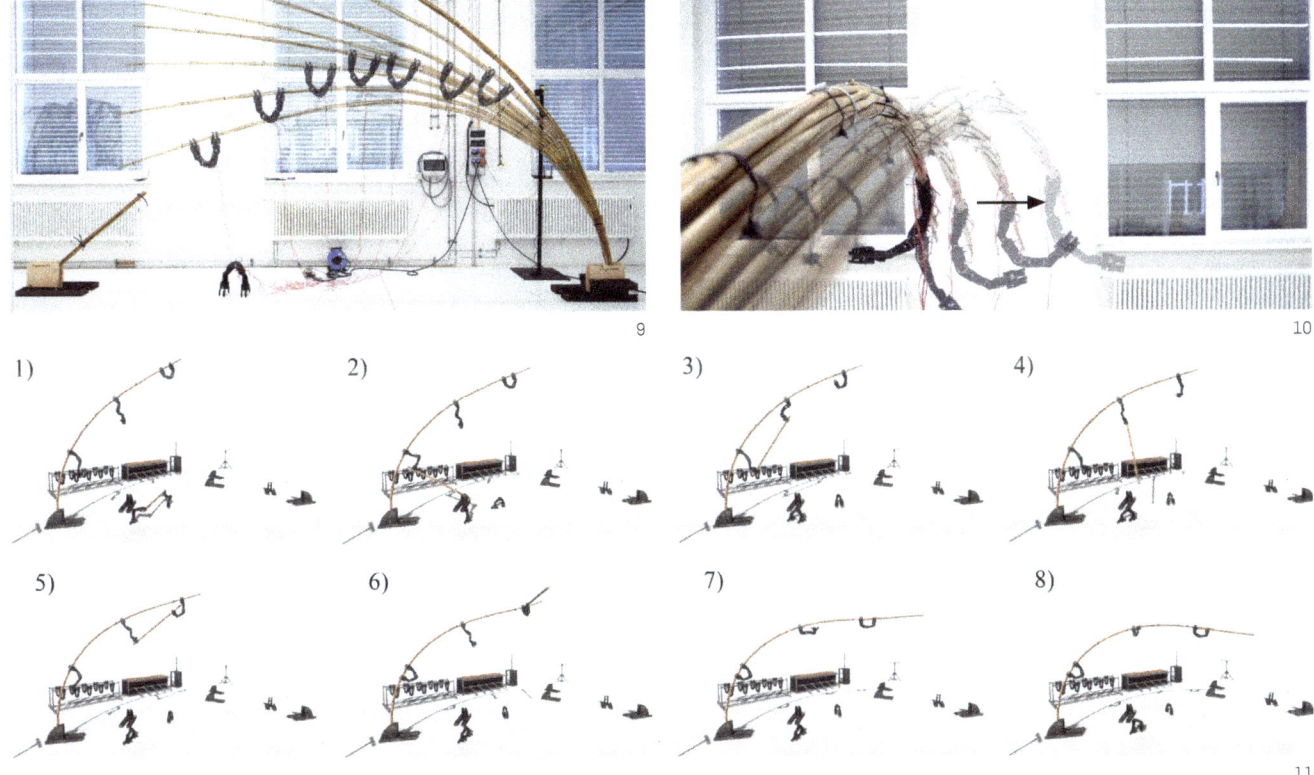

bamboo bundles. The knowledge gathered from this learning process is two-fold: first, the robot learns how to use its weight and movement to create and direct the required momentum to elastically bend a bundle. Specifically, the robot learns to match its motion with the natural frequency of the bundle and thus is capable of altering and amplifying the oscillation of said bundle in order to achieve goal configurations. This knowledge is used during the assembly process. Second, the robot learns how to determine what the material is capable of, or in other words, what geometric deformations it can undergo. This secondary knowledge is useful in the design process (Figure 11).

(PPO) Proximal Policy Optimization

More specifically, this research implements Proximal Policy Optimization (PPO) (Schulman et al. 2017). PPO is used to train a neural network to approximate the ideal function that maps an agent's observations (its state) to the best action an agent can take in a given state in pursuit of achieving a goal (i.e. assembly task). All policies are feed-forward networks with three layers of 128 units each, whose training is kept stable using asynchronous gradient descent.

The PPO algorithm consists of an agent, an environment, a set of states, actions, and a reward function (Figure 13). In this research, the environment consists of the robot dynamics and the bamboo physics simulation, which was

8   Physical prototype of the mobile robotic builder

9   Behavioral experiment combining the elastic bending of the bamboo bundle and the dynamic positioning of the robot: the self-weight of the robot and its position along the bundle can cause static bending

10 Dynamic bending caused by the robot rhythmically swinging in the desired bending direction

11 Visualization of the bundle extension sequence: (1-2) robots transport material on ground; (3-6) robots transport material while on structure; (7-8) transported material is connected to extend the bundle

approximated as a discrete chain of rigid-bodies connected by damped harmonic oscillators, where the motion of each spring depends on the balance of three forces: inertial force, restoring force, and a damping force. The state (i.e. observations) is the collection of inputs from the robot's sensors (specified as the robot's orientation in space, a vector between its end effector position and the goal's position, the motor joint angles, the angular velocity of the base link, and the linear velocity of the base link). The actions are the motor movements, resulting in a 5D action space, with each dimension relating to one of the axes). A reward is given when the virtual robot is able to bend the bundle and close its gripper around the goal location, while a penalty (negative reward) is given at each timestep in which the goal is not achieved in order to encourage exploration. The goal is to maximize the reward.

Design and Assembly Strategy

Existing form-finding methods for bending-active structures (Piker 2013; Suzuki et al. 2018) are not sufficient for

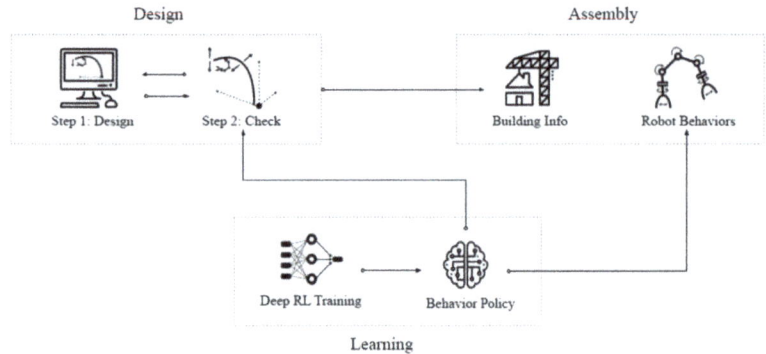

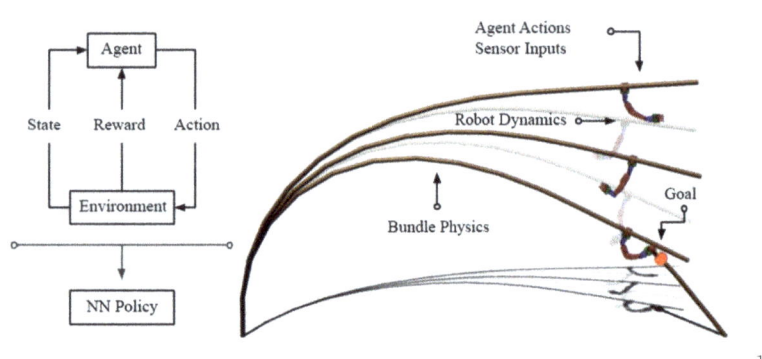

12 Design and assembly workflow: machine learning enables: (1) the simulation of the assembly process of a bundle in a digital environment, which is in turn embedded in the design process; and (2) the use of material behavior knowledge during assembly of this structure on site

13 A screenshot from the simulation and learning process (right) as it relates to the abstracted reinforcevment learning algorithm (left)

14 The experimental setups utilized to validate the design and assembly workflow: (a-b) experiments within the training set; (c) experiment outside of the training set

the purposes of this research since they do not accommodate the integration of assembly-related constraints. The modeling workflow developed for the proposed construction system, on the other hand, allows for close interaction between the human designer and the digital twin of the robot. The role of the designer is to determine a bundle's shape and position in space by defining stiffness, starting position, length, and end position. Robots, in turn, utilize the knowledge gathered during the learning process (i.e. the trained policy) to assess the feasibility of the desired configuration through virtual bending choreographies. This serves as crucial feedback for designers, as it continuously informs their decision with material and assembly-related parameters.

Once the design satisfies structural, architectural, and robotic requirements, it is dispatched to the physical assembly system, which includes information on assembly order, joint types, and bundle positions as well as successful bending choreographies. This information, however, does not constitute a static blueprint but rather a rough set of instructions, given that the final geometry is highly contingent upon the dynamic relationship between material and robot.

RESULTS

For the demonstration and validation of the proposed system, a series of experiments were conducted that together resulted in the demonstrator structure. All bundles within the structure were made from 1.8 meter long *Arundinaria amabilis* bamboo rods, with diameters ranging from 1.0 to 1.8 cm. The cross-section of the bundles was designed in a gradient-like manner to control curvature distribution.

Demonstrator Structure (3 Experiments)

The demonstrator structure is the result of three assembly sequences, all of which were conducted as individual experiments (Figure 14).

The first experiment (Figure 14a) consisted of a 7.5-meter long bamboo bundle anchored vertically to the ground at one end and free at the other. The goal was to bend the initially free end of the bundle into a goal position near the ground in order for it to be anchored thus forming an arch.

During the second experiment (Figure 14b), the robot bent a shorter bundle, which was fixed at one end to the ground. In the final state, the other end connected halfway along the length of the bundle from experiment one.

Experiment three (Figure 14c) involved bending a bundle into an arch similar to that of the first experiment. However, in this experiment, the bundle is connected to the initial bundle of the first experiment with a fixing joint and must bend over the bundle placed in the second experiment.

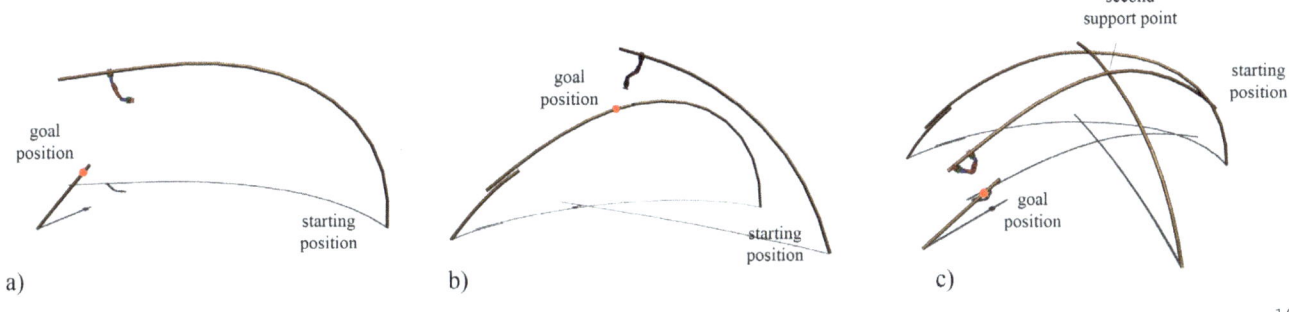

14

This scenario is unique in that the properties of multiple connected bundles affect the assembly process. As only the bending of a single bamboo bundle in isolation was initially trained for, this scenario is outside of the training data set and served to test the generalizability of training results.

Experimental Results
Five policies, trained with slightly different constraints, were tested in all of the above setups. The experiments were conducted both digitally (Sim-to-Sim), to test the trained policy in simulation during the design process, and physically (Sim-to-Real), to evaluate the transfer of knowledge from training to physical assembly.

In simulation, all results were successful, meaning that the robot was able to use its weight and momentum to bend an element into the desired position. During the Sim-to-Real tests, the best policy enabled the robot to find and match its momentum with the natural frequency of the bamboo bundle for each setup. The robot could adjust its swinging even when the bending response from the material was influenced through external means. However, the amplitude of the bending was much smaller than the one achieved in the simulation. In experiment three, the bending amplitude was even smaller than the previous two experiments due to the fact that the bundle from experiment two acts like a damper, which was not accounted for in the training. To overcome these deviations in the future, creating momentum from different positions along the length of the bamboo bundles was tested to understand its effects on the amplitude of bending for future training.

Evaluation of Secondary Behaviors
Physical experiments of the other tasks necessary for the assembly of structures built with the proposed construction system were also conducted. A hard-coded locomotion routine was tested, which allowed the robot to walk along the length of the bending bundle by repeating a loop of choreographed movements. Material transportation was tested by enabling the robot to grab, move, and release various bundle subsegments. And finally, the ability of the robot to connect bundles was validated by hard-coding a robot to grasp and then align a bamboo bundle to an already existing bundle in the structure.

DISCUSSION AND CONCLUSION
The presented research showcases the potentials and methodological implications of designing a construction system driven by a reciprocal relationship between material and robot behaviors. Here, designing is no longer conceived as an exercise in static material placement or isolated geometric exploration, but as an integrated and iterative workflow between simulating material behaviors and choreographing robot performances. In contrast to previous work in the field of Collective Robotic Construction, the presented system does not aim to automate construction processes tailored to human builders, nor does it rely on brute machine force enabled by large-scale construction equipment. Instead, it proposes an intelligent material-centered design-to-assembly process for robotically building structures made from natural materials.

Digital and physical experiments were conducted to validate this way of behavioral designing and building. To do so, a full-scale mobile robot was designed, built, and programmed. A neural network was trained in order to provide the robot with one of the skills necessary for assembling bamboo structures. Although this skill was successfully tested in both simulation and with physical experiments, one identified challenge was the ability of the algorithm to exploit the flaws of the physics simulation in order to reach its goals. As these sorts of flaws do not occur in reality, some of the policies with such tendency had to be discarded. Further research might investigate a more physically accurate physics engine in order to address this issue (Bousmalis and Levine 2017).

Despite the focus on bamboo structures, the methods proposed in this research are not meant to be limited to this construction system, but rather to lay the groundwork for further exploration of co-designing material-robot behaviors and their potential implementations in architecture and construction. With this approach, we intend to challenge the way in which building materials are currently used in architecture, by treating them as active drivers in the design, assembly, and lifetime of the construction process, rather than as static recipients of form (Menges 2015). New possibilities for sustainable architecture arise when natural, heterogeneous materials with high degrees of biological variation can be robotically assembled into large-scale structures.

ACKNOWLEDGEMENTS

Concept and prototyping was conducted as a Master's thesis in the Integrative Technologies and Architectural Design Research M.Sc. Program (ITECH) at the University of Stuttgart, Germany, led by the Institute of Computational Design and Construction (ICD) and the Institute of Building Structures and Structural Design (ITKE). The authors would like to thank Prof. Jan Knippers for his valuable input and advice. The work was partially supported by the Deutsche Forschungsgemeinschaft (DFG, German Research Foundation) under Germany's Excellence Strategy – EXC 2120/1 – 390831618.

REFERENCES

Addis, B. 2007. *Building: 3000 Years of Design Engineering and Construction*. London; New York: Phaidon Press.

Alvarez, M., H. J. Wagner, A. Groenewolt, O.D. Krieg, O. Kyjanek, L. Aldinger, S. Bechert, D. Sonntag, A. Menges, and J. Knippers. 2019. "The Buga Wood Pavilion; Integrative interdisciplinary advancements of digital timber architecture. In *ACADIA '19: Ubiquity and Autonomy; Proceedings of the 39th Annual Conference of the Association for Computer Aided Design in Architecture*. 490–499.

Apolinarska, A., M. Pacher, H. Li, N. Cote, R. Pastrana, F. Gramazio, and M. Kohler. 2021. "Robotic assembly of timber joints using reinforcement learning." *Automation in Construction* 125 (May): 103569.

Bessai, T. 2013. "Bending-Active Bundle Structures: Preliminary Research and Taxonomy Towards an Ultra-Light Weight Architecture of Differentiated Components." In *ACADIA '13: Adaptive Architecture; Proceedings of the 33th Annual Conference of the Association for Computer Aided Design in Architecture*. 293–300.

Bousmalis, K., and S. Levine. 2017. "Control Systems in Practice Part 4: Why Time Delay Matters." Google AI Blog. Accessed June 10, 2020. https://ai.googleblog.com/2017/10/closing-simulation-to-reality-gap-for.html.

Brugnaro, G., E. Baharlou, L. Vasey, and A. Menges. 2016. "Robotic Softness: An Adaptive Robotic Fabrication Process for Woven Structures." In *ACADIA '16: Posthuman Frontiers; Proceedings of the 36th Annual Conference of the Association for Computer Aided Design in Architecture*. 154–163.

Brugnaro, G. and S. Hanna. 2019. "Adaptive Robotic Carving Training Methods for the Integration of Material Performances in Timber Manufacturing." In *Robotic Fabrication in Architecture, Art and Design 2018*. Cham: Springer. 336–348.

Dörfler, K. 2018. "Strategies for Robotic In Situ Fabrication." PhD diss., ETH Zurich. https://doi.org/10.3929/ethz-b-000328683.

Dörfler, K., T. Sandy, M. Giftthaler, F. Gramazio, M. Kohler, and J. Buchli. 2016. "Mobile Robotic Brickwork." In *Robotic Fabrication in Architecture, Art and Design 2016*. Cham: Springer. 204–217.

Jenett, B., A. Abdel-Rahman, K. Cheung, and N. Gershenfeld. 2019. "Material-Robot System for Assembly of Discrete Cellular Structures." *IEEE Robotics and Automation Letters* 4 (4): 4019–4026. https://doi.org/10.1109/LRA.2019.2930486.

Kayser, M., L. Cai, S. Falcone, C. Bader, N. Inglessis, B. Darweesh, and N. Oxman. 2018. "FIBERBOTS: an autonomous swarm-based robotic system for digital fabrication of fiber-based composites." *Construction Robotics* 2 (1-4): 67–79.

Lienhard, J., C. Gengnagel, J. Knippers, and H. Alpermann. 2013. "Active Bending, A Review on Structures where Bending is used as a Self-Formation Process." *International Journal of Space Structures* 28 (3-4): 187–196.

Lorenzo, R., L. Mimendi, H. Li, and D. Yang. 2020. "Bimodulus bending model for bamboo poles." *Construction and Building Materials* 262 (Nov): 120876.

Manandhar, R., J-H. Kim, and J-T. Kim. 2019. "Environmental, Social and Economic Sustainability of Bamboo and Bamboo-based Construction Materials in Buildings". *Journal of Asian Architecture and Building Engineering* 18(2): 49–59.

Menges, A. 2011. "Integrative Design Computation: Integrating Material Behaviour and Robotic Manufacturing Processes in Computational Design for Performative Wood Constructions." In *ACADIA '11: Integration through Computation; Proceedings of the 31th Conference of the Association for Computer Aided Design In Architecture*. 72–81.

Menges, A. 2015. "Material Synthesis: Fusing the Physical and the Computational." *Architectural Design* 85 (5): 8–15.

Minke, G. 2012. *Building with Bamboo: Design and Technology of a Sustainable Architecture*, 1st ed. Basel: Birkhäuser.

Nagabandi, A., I. Clavera, S. Liu, R.S. Fearing, P. Abbeel, S. Levine, and C. Finn. 2018. "Learning to Adapt in Dynamic, Real-World Environments Through Meta-Reinforcement Learning." ICLR 2019 Conference Paper. arXiv:1803.11347. http://arxiv.org/abs/1803.11347.

Oliver, P. 2003. *Dwellings: The Vernacular House World Wide*. London: Phaidon.

Petersen, K. H., N. Napp, R. Stuart-Smith, D. Rus, and M. Kovac. 2019. "A review of collective robotic construction." *Science Robotics* 4 (28): 10.

Piker, D. 2013. "Kangaroo: Form Finding with Computational Physics." *Architectural Design* 83 (2): 136–137.

Ribeirinho. M.J., J. Mischke, G. Strube, D. Sjödin, J.L. Blanco, R. Palter, J. Biörck, D. Rockhill, and T. Andersson. "The Next Normal in Construction: How Disruption is Reshaping the World's Largest Ecosystem." McKinsey & Company. Published June 4, 2020. Accessed June 10, 2020. https://www.mckinsey.com/business-functions/operations/our-insights/the-next-normal-in-construction-how-disruption-is-reshaping-the-worlds-largest-ecosystem.

Samuel L., W. Ramon, W. Dylan, B. Oliver, and A. Menges. 2019, "Distributed Robotic Timber Construction." ACADIA '19: *Ubiquity and Autonomy; Proceedings of the 39th Annual Conference of the Association for Computer Aided Design in Architecture*. 510–519.

Schulman, J., F. Wolski, P. Dhariwal, A. Radford, and O. Klimov. 2017. "Proximal policy optimization algorithms." *arXiv*:1707.06347. https://arxiv.org/abs/1707.06347.

Suzuki, S., A. Körner, and J. Knippers. 2018. "IGUANA: Advances on the development of a robust computational framework for active-geometric and-topologic modeling of lightweight structures." In *Proceedings of IASS Annual Symposia, IASS 2018 Boston Symposium: Approaches for Conceptual Structural Design*. 1–10.

Thangavelu, V., M. Saboia da Silva, J. Choi, and N. Napp. 2020. "Autonomous Modification of Unstructured Environments with Found Material." In *2020 IEEE International Conference on Robotics and Automation (ICRA)*. 7798–7804. https://doi.org/10.1109/ICRA40945.2020.9197372.

van der Lugt, P., A.A.J.F. van den Dobbelsteen, J.J.A. Janssen. 2006. "An environmental, economic and practical assessment of bamboo as a building material for supporting structures." *Construction and Building Materials* 20 (9): 648–656.

Vasey, L., B. Felbrich, M. Prado, B. Tahanzadeh, and A. Menges. 2020. "Physically distributed multi-robot coordination and collaboration in construction: A case study in long span coreless filament winding for fiber composites." *Construction Robotics* 4 (1–2): 3–18.

Yablonina, M., M. Prado, E. Baharlou, T. Schwinn, and A. Menges. 2017. "Mobile Robotic Fabrication System For Filament Structures." In *Fabricate 2017*. 202–209.

Zeng, A., S. Song, J. Lee, A. Rodriguez, and T. Funkhouser. 2019. "TossingBot: Learning to Throw Arbitrary Objects with Residual Physics." In *Proceedings of Robotics: Science and Systems (RSS)*. http://arxiv.org/abs/1903.11239.

Zhang, Y., W. Yu, C.K. Liu, C. Kemp, and G. Turk. 2020. "Learning to manipulate amorphous materials." *ACM Transactions on Graphics* 39 (6): 1–11.

IMAGE CREDITS
Figure 3: ©ICD/ITKE, University of Stuttgart
Figure 4: Loucka, ICD, University of Stuttgart, 2014
All other drawings and images by the authors

Grzegorz Łochnicki is a computational designer graduated from the ITECH Master Program at the University of Stuttgart and Wroclaw University of Technology.

Nicolas Kubail Kalousdian is a doctoral candidate at the Institute for Computational Design and Construction (ICD) at the University of Stuttgart. He is a member of the EXC 2120, Cluster of Excellence IntCDC, where his research focuses on machine learning and robotics for autonomous construction systems.

Samuel Leder graduated from Washington University in St. Louis and the University of Stuttgart and is currently a doctoral candidate at the Institute for Computational Design and Construction (ICD) at the University of Stuttgart.

Mathias Maierhofer is a researcher with a strong interest in interactive design processes for robotic architectural systems. He graduated with distinction from the ITECH Master Program at the University of Stuttgart. Currently, Mathias is a Research Associate at the Institute for Computational Design and Construction (ICD). His current research focuses on the development of integrative design methods and computational design tools for adaptive structures and their digital production.

Dylan Wood is a research group leader at the ICD at the University of Stuttgart. His work and teaching focuses on developing on intelligent design and fabrication principles for material driven robotic systems. He holds a BArch with honors from the University of Southern California, and a MSc with distinction from the University of Stuttgart.

Achim Menges is a registered architect in Frankfurt and professor at the University of Stuttgart, where he is the founding director of the Institute for Computational Design and Construction (ICD) and the director of the Cluster of Excellence on Integrative Computational Design and Construction for Architecture (IntCDC).

Shepherd

A fabrication-oriented tool for simulation and control of mobile robotic platforms for collaborative earthworks

Tom Shaked
CEAR - Civil, Environmental, and Agricultural Robotics Lab, Technion IIT

Amir Degani
CEAR - Civil, Environmental, and Agricultural Robotics Lab, Technion IIT

1 Controlling a Clearpath Jackal unmanned ground vehicle in real-time using Shepherd

ABSTRACT

Recent advancements in robotics empower the exploration of remote construction methods aimed at reducing the risk for workers. As the global pandemic places construction workers and their communities at high risk for disease, the need for remote construction methods increases. Such methods depend on the complicated task of controlling mobile robotic platforms in real-time. In this context, this paper presents work-in-progress in development and experimentation with a tool for collaborative earthworks using multi-agent unmanned ground vehicles.

Expanding the field of collective robotic construction, this research simplifies the use of mobile robots and enables their operation using a design platform. Shepherd, a fabrication-oriented tool for simulation and control of mobile robotic platforms is presented, and its capacities are demonstrated in an earthworks case study. The case study exemplifies the potential of this approach to change the role of design tools in the operation and adoption of mobile multi-robotic platforms, thus contributing to robotic fabrication in architecture, landscape architecture, and construction.

INTRODUCTION

Remote robotic operations currently provide solutions to hazardous or inaccessible sites (Ornan and Degani 2013; Temtsin and Degani 2017; Queralta et al. 2020). Global pandemics create an urgent need to expand these capabilities for conditions that require physical distancing. In this context, COVID-19 places workers in the construction and mining industries at high risk for disease and transmitting the disease in their communities (Baker, Peckham, and Seixas 2020). This situation contributes to an increased interest in remote construction methods capable of reducing the risk for workers. Such methods rely on controlling robotic tools in real-time and receiving feedback on their location, operation, and material and environmental state.

In line with recent research exploring strategies for robotic ground forming (Bar-Sinai, Shaked, and Sprecher 2019, 2021), this paper will present work-in-progress and experimentation with controlling multi-agent ground vehicles towards enabling remote manipulation of geomaterials for construction (Figure 2). Multi-agent ground vehicles offer the possibility to utilize adaptive and scalable systems for modifying environments according to predefined goals. Mobile robotic platforms such as those used for search and rescue (SAR) and agriculture present significant advancements in their capacity to adapt to natural, unstructured terrains and handle unpredictable materials on-site.

Expanding the field of collective robotic construction, this research (1) simplifies the use of mobile robotic platforms, (2) enables their operation using a design platform, and (3) examines their applicability in earthmoving for construction.

The paper presents the state of the art in robotic earthworks, parametric robot control, and mobile robotic platforms. This section is followed by introducing Shepherd, a custom fabrication-oriented tool for collaborative construction using mobile robotic platforms on-site. The methods and capacities that lie at the tool's core are then detailed, followed by a proof of concept and demonstration of the tool's operation.

STATE OF THE ART
Robotic Earthworks

Typically, construction begins with extensive preparatory earthworks. These works include various forms of soil manipulation, such as shifting, leveling, and piling towards construction (Peurifoy et al. 2018). Current research on large-scale robotic fabrication explores autonomous excavation and landscape manipulation (Hurkxkens et al. 2020). Among the on-site preparatory earthworks is the sorting of aggregates. Aside from soil and sand, these include relocating and arranging stones in various sizes. Recent research on in situ robotic fabrication explores utilizing irregular rocks found on-site for autonomous construction, stacking them without mortar by employing a physics engine, 3D scanning, and a robotic manipulator (Furrer et al. 2017). Alternatively, research examines architectural-scale granular jamming using a 6-axis robotic arm to deposit a string in precise patterns between loose gravel layers, enabling the construction of a loadbearing structure without adhesives (Aejmelaeus-Lindström et al. 2017).

2 The robotic fleet on-site and the simulated robots using Shepherd

Parametric Robot Control

The recent decade has been a fertile ground for the development of parametric control and simulation tools that enable using robotic platforms for design, manufacturing, and research. These tools include KUKA|prc, and Hal, which implement KUKA and ABB APIs, enabling robot control through 3D modeling platforms such as Rhinoceros 3D and Maya (Brell-Çokcan and Braumann 2011; Schwartz 2013). However, this is usually limited to workflows characterized by offline simulation and a single direction of information flow, starting with the design and ending with the robotic manufacturing not incorporating feedback in the process. Expanding this approach several tools, such as Ex-Machina and Reflex, enable incorporating real-time control and feedback in the process of robotic manipulation. In addition, these tools allow updating the robot action according to environmental, spatial, and material changes (Shaked, Bar-Sinai, and Sprecher 2020; García del Castillo y López 2019). However, these tools mainly enable controlling industrial robotic arms off-site in lab conditions rather than mobile robotic platforms.

Mobile Robotic Platforms

Recent technological advancements enable exploring the use of industrial robots in on-site conditions (Shaked, Bar-Sinai, and Sprecher 2020). These conditions require mobilizing robots, either by transporting them on-site or by incorporating platforms dedicated to their mobility (Giftthaler et al. 2017; Petersen et al. 2019). Mobile robotic platforms rely on the ability to perform navigation and localization. Therefore, such platforms are supported by proprietary software or open-source solutions such as the Robot Operating System (ROS), a collection of frameworks for robot software development (Quigley et al. 2009). As ROS provides a flexible and scalable solution, it depends on native development in various languages and frameworks. Therefore, incorporating mobile robots in architectural fabrication can benefit from integrating their operation into tools for computer-aided design (CAD) and computer-aided manufacturing (CAM).

SHEPHERD

A Tool for Collaborative Construction Using Mobile Robotic Platforms On-site

Presented here, Shepherd enables rapid parametric path planning, simulation, and real-time control for multiple ROS running mobile robotic platforms using a Rhino 3D Grasshopper interface (Figure 3). Shepherd is aimed at enabling remote ground forming for site preparation and construction. This capability can also be expanded to other construction tasks on-site such as material handling or additive manufacturing.

Capabilities and Integration

Shepherd is divided into several categories and sub-categories, which allow:

- Simulating multiple mobile robots in Rhino 3D
- Generating robot paths using primitive Rhino geometries

3   The Shepherd plugin in Rhino 3D Grasshopper environment

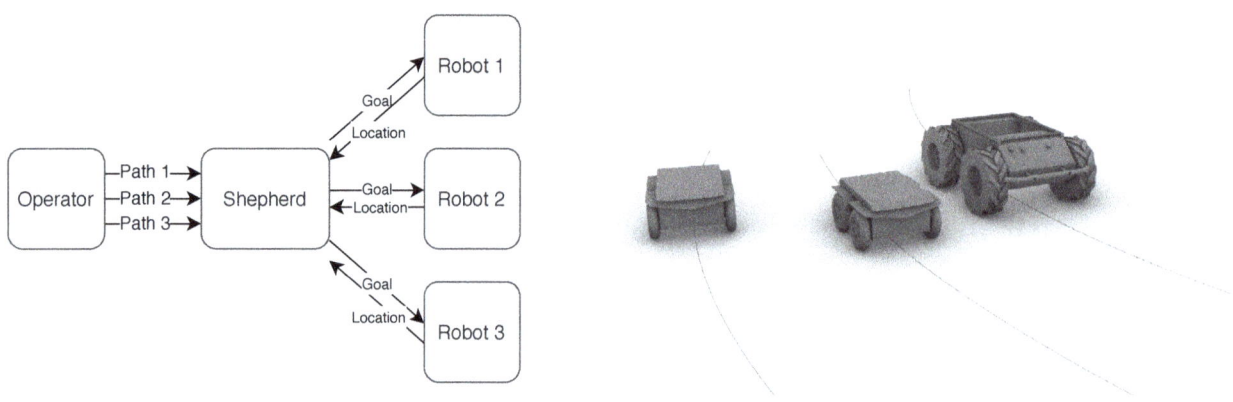

4   Multiple robots control model and three robots moving according to separate paths all controlled using Shepherd

- Controlling multiple robots in the Gazebo open-source robot simulator
- Connecting to robotic platforms running ROS
- Driving robots
- Publishing and subscribing to ROS topics
- Performing autonomous navigation between waypoints
- Keeping track of robot positions
- Manipulating external robot joints
- Performing image processing based on input received by peripherals such as RGB and depth cameras

Employing a server-client software model, Shepherd enables controlling multiple robotic platforms (Figure 4). In Gazebo this is achieved by implementing the Multi Jackal package, an expansion of the Clearpath ROS package that allows simulation of multiple mobile robots (Sullivan 2018). On the robot side, Shepherd implements Rosbridge, a library providing an API to ROS functionality for non-ROS programs (Crick et al. 2017), as well as the Clearpath library (Clearpath Robotics 2021) for simulation and control of mobile outdoor platforms such as the Jackal and Husky, and indoor platforms such as the Dingo, Ridgeback, and Boxer. Clearpath is a company specializing in manufacturing field robotic platforms for research and industry. On the server-side, Shepherd implements OpenCV, an open-source library facilitating real-time computer vision and image processing (Bradski and Kaehler 2008), the Xinput API allowing applications to receive input from an Xbox Controller (Jocys 2020), and Roslibpy, a library enabling robot control using Python and IronPython without running a ROS interface on the server (Casas et al. 2019).

Shepherd provides the basic functionality required for simulating various Clearpath robots within Rhino 3D (Figure 5). In addition, it enables controlling a parallel simulation in Gazebo using the Rhino Grasshopper interface (Figure 6). The advantage of simulating the robots in the Gazebo environment lies in its ability to generate an accurate physical simulation of each platform, its driving behavior, and the physical interaction between the platform and its environment (Koenig and Howard 2004). This interaction includes the contact of the robotic platform with elements in its close vicinity and providing simulated sensor data—such as images, depth maps, and 3D laser scans—about its environment, .

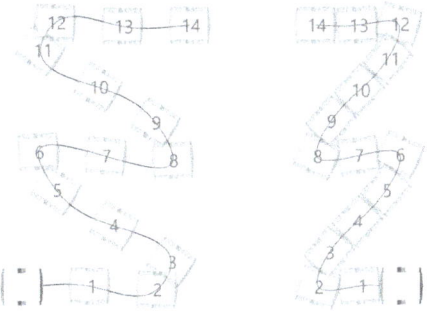
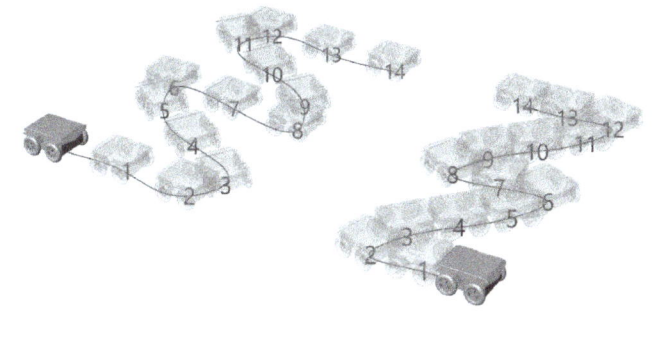

5   The Rhino Grasshopper robot simulation

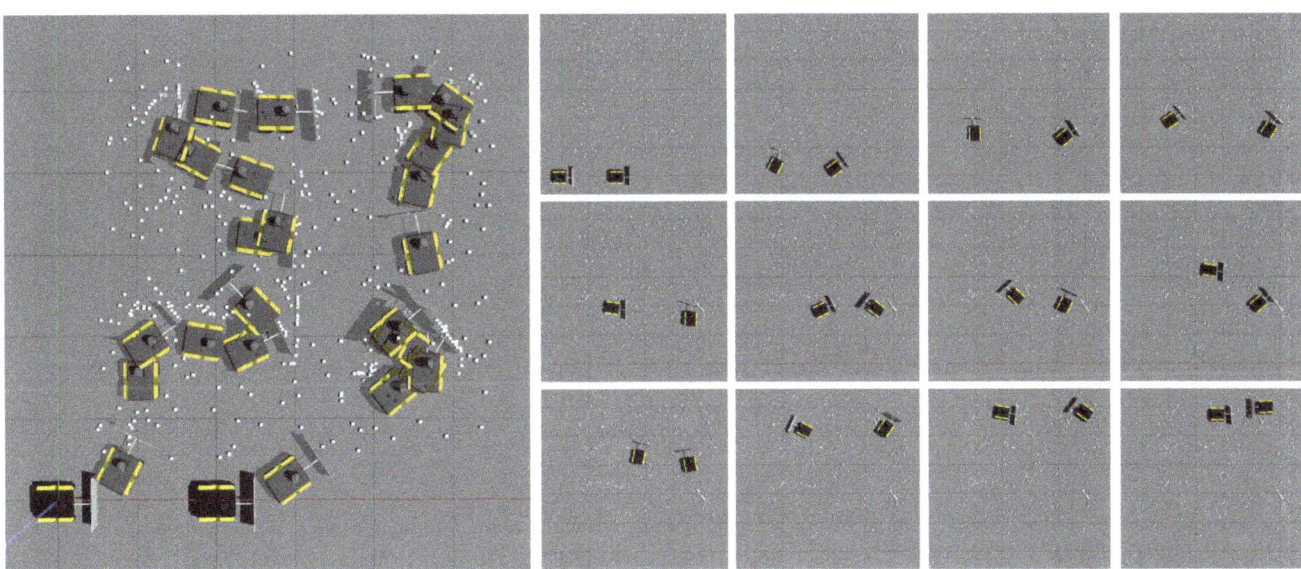

6   The Gazebo simulation incorporating elements simulating aggregates

METHOD

The research aims to develop robotic fabrication capacities through the presented custom tool by employing control, navigation, and path planning for a ROS robot. These capacities are supported by computer vision and image processing.

Control, Navigation, and Path Planning

Controlling the robot is done in one of three main ways:

- Sending a direct speed and direction command to each platform (using a velocity topic)
- Navigating to a specific location in space (using a goal topic)
- Navigating along a predetermined path by sending a set of positions which are performed consecutively

All these operations can be performed using Shepherd, by providing either the speed and direction values, a point in space for a single position, or a 3D curve used as a path. The latter is performed by sending the robot to a series of goals iteratively executed as single goals (Figure 7). The successful arrival at each position is determined in relation to a predefined distance from that point (defined here as tolerance and measured as a radius from the desired goal). Following the arrival to the desired goal, the robot moves to the following position in the list (Figure 8).

Shepherd also enables controlling custom robotic joints that are not directly related to the moving platform. This option can either be used to control a mounted industrial arm (Figure 9) or for a custom-developed tool—as long as the tool's definition is added to the Unified Robot Description Format (URDF) in the ROS launch file. The tool presented in the case study is a custom front shovel for moving aggregates.

CASE STUDY: GRAVEL FORMING

In accordance with the state of the art in robotic earthworks, a case study employing Shepherd is presented. It explores the use of multiple robotic platforms mounted with custom tools aimed at manipulating aggregates and demonstrating gravel forming using a team of two robots.

The case study demonstrates the possibility of autonomously shaping an amorphous mass of gravel, supporting remote robotic earthworks, site preparation, and construction. The specific material is gravel with a 20-50mm particle size, often used in construction for ground stabilizing, water accumulation control, and as an additive for various kinds of concrete (Bolen 2005).

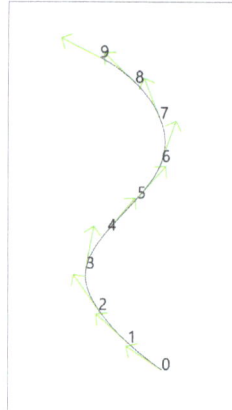

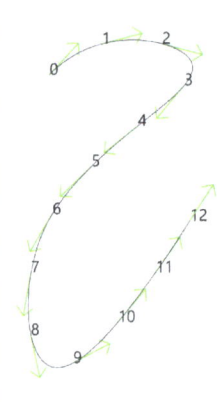

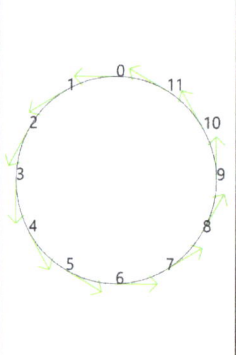

7   Path navigation produced using simple Rhino curves and their execution with a robot using Shepherd

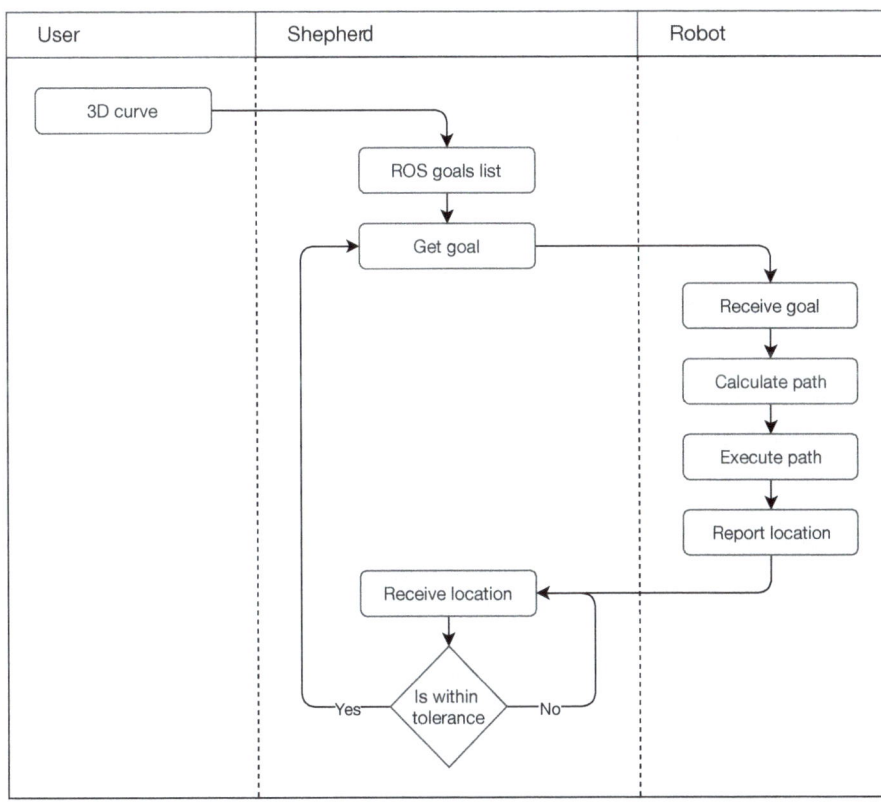

8   The Shepherd activity diagram

9   A Ridgeback omni-drive indoor platform simulated using Shepherd mounted with a KUKA KR3 industrial robotic arm simulated using KUKA|prc

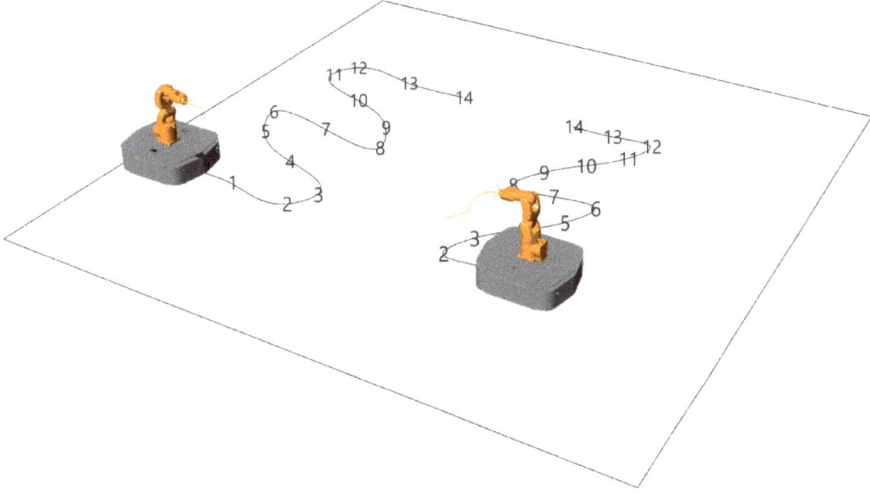

The robotic set-up consists of a team of two Clearpath Jackal unmanned ground vehicles (UGV), weighing 17kg with a maximum payload of 20kg. Each robot is mounted with a custom 50-by-40cm front shovel tool and a Sick LMS111 outdoor LiDAR. A high-definition camera is mounted on a 2m tall tripod, monitoring a 5-by-12m area.

Gravel is scattered on the floor, and the robots are instructed to form it into an oval shape. The process begins by analyzing the dispersed gravel geometry and producing an ad hoc robot motion plan supporting the material

shaping. The plan relies on two curves, one for each robot, divided into nine waypoints (Figure 10). The waypoints are converted to quaternions for robot positioning using Shepherd (Figures 11, 12). Following the robots' motion, the camera surveys the outcome and outputs a new, updated image of the final state (Figure 13).

In addition to using Shepherd for controlling the robots, this process is conducted by employing two computer vision techniques: blob detection and 2D perspective transform. Both methods implement real-time computer vision. The

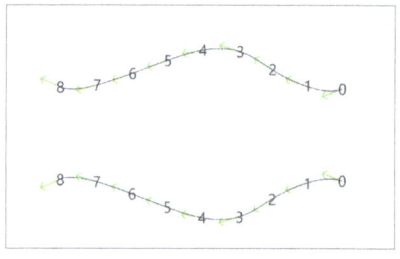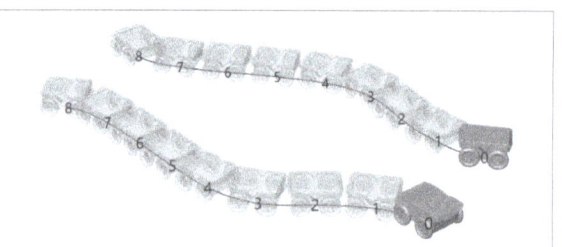

10 Gravel forming using two Jackal robots mounted with custom tools: (left) the path in Rhino 3D; (right) the simulated robot model on each waypoint

11 Gravel forming Gazebo simulation

12 Gravel forming path execution

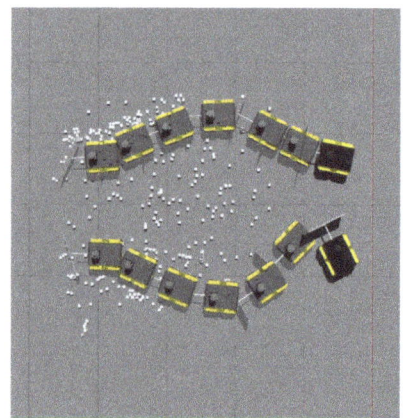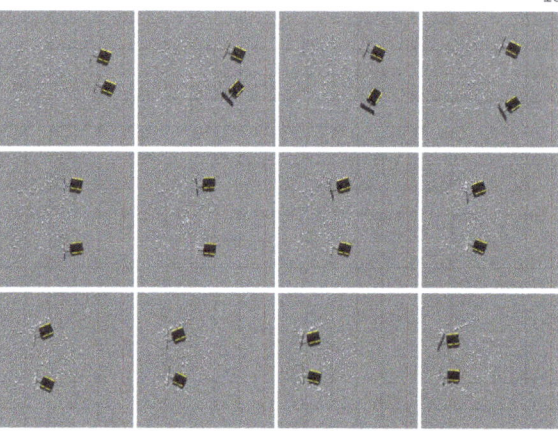

first method is used to detect similar property regions in the image, thus outlining the gravel's geometry.

The second method is used to map the distorted image and project a corrected version onto the work area. This action is necessary since the camera in the set-up is not perpendicular to the floor; therefore, the images it produces are distorted and require remapping (Figure 13).

LIMITATIONS AND FUTURE WORK

The paper displays work-in-progress, as the tool and method are under continuous development. Future work

13 The work area images after blob detection and perspective transform

will, therefore, go on addressing current limitations, as well as adding capabilities. While the tool currently enables simulating six robot models, the field can benefit from increasing the robot model library with additional platforms from various manufacturers. Additionally, as Shepherd controls each vehicle independently, future research will further expand collaborative capabilities between the vehicles. Lastly, there is a need for easing the integration of Shepherd with open-source hardware platforms, primarily for education and research.

Regarding the presented case study, the suggested workflow can benefit from implementing an iterative fabrication protocol. Such a protocol can increase the accuracy of the outcome with respect to the desired shape by incorporating feedback from external sensor data. The tool can also be examined in other scenarios and use-cases, such as additive manufacturing. However, this requires a higher degree of accuracy in relation to the robot path, requiring more robust validation on the robot position and more comprehensive integration with robotic arms and 3D printing extruders. Finally, while Shepherd increases safety by reducing the number of workers on-site, remote construction platforms require special skills that support shifting from manual labor to platform operation.

CONCLUSION

The global pandemic places construction workers, mining laborers, and their communities, at high risk for receiving and transmitting disease. In this context, this paper aims at reducing this risk by contributing to remote construction methods. This contribution is achieved by enabling the simple operation of mobile multi-robot platforms for site preparation and construction. This operation is made possible using Shepherd, a custom tool for controlling a ROS robot using the Rhino Grasshopper interface. The tool enables simulating, driving, and path generating for teams of mobile robotic platforms in real-time. These capabilities are demonstrated through a case study in robotic earthworks. The case study exemplifies the potential of this approach to change the role of design tools in the operation of mobile multi-robotic platforms in architecture, landscape architecture, and large-scale construction. We hope this research will assist designers in adopting multi-robotic platforms, thus contributing to robotic fabrication in architecture.

REFERENCES

Aejmelaeus-Lindström, Petrus, Ammar Mirjan, Fabio Gramazio, Matthias Kohler, Schendy Kernizan, Björn Sparrman, Jared Laucks, and Skylar Tibbits. 2017. "Granular Jamming of Loadbearing and Reversible Structures: Rock Print and Rock Wall." *Architectural Design* 87 (4): 82–87. https://doi.org/10.1002/ad.2199.

Baker, Marissa G., Trevor K. Peckham, and Noah S. Seixas. 2020. "Estimating the Burden of United States Workers Exposed to Infection or Disease: A Key Factor in Containing Risk of COVID-19 Infection." *PLoS ONE* 15(4): e0232452. https://doi.org/10.1371/journal.pone.0232452.

Bar-Sinai, Karen Lee, Tom Shaked, and Aaron Sprecher. 2019. "Informing Grounds: Robotic Sand-Forming Simulating Remote Autonomous Lunar Groundscaping." In *ACADIA '19: Ubiquity and Autonomy; Proceedings of the 39th Annual Conference of the Association for Computer Aided Design in Architecture.* Austin, Texas. 258–65.

Bar-Sinai, Karen Lee, Tom Shaked, and Aaron Sprecher. 2021. "Robotic Tools, Native Matter: Workflow and Methods for Geomaterial Reconstitution Using Additive Manufacturing." *Architectural Science Review* 64 (6): 490–503. https://doi.org/10.1080/00038628.2021.1898324.
]
Bolen, W. P. 2005. "Sand and Gravel Construction." In *Mineral Commodity Summary 2005*. Mineral Commodity Summaries

Series. US Geological Survey. 140–41. https://doi.org/10.3133/mineral2005.

Bradski, Gary, and Adrian Kaehler. 2008. *Learning OpenCV: Computer Vision with the OpenCV Library*. Sebastopol, Calif.: O'Reilly Media.

Brell-Çokcan, Sigrid, and Johannes Braumann. 2011. "Parametric Robot Control: Integrated CAD/CAM for Architectural Design." In *ACADIA '11: Integration through Computation; Proceedings of the 31st Annual Conference of the Association for Computer Aided Design in Architecture*. Banff, Alberta. 242–51.

Casas, Gonzalo, Mathias Lüdtke, Beverly Lytle, Alexis Jeandeau, and Hiroyuki Obinata, and Gramazio Kohler Research. 2019. "Roslibpy: ROS Bridge Library." [Python ROS Bridge library]. https://roslibpy.readthedocs.io/en/latest/.

Crick, Christopher, Graylin Jay, Sarah Osentoski, Benjamin Pitzer, and Odest Chadwicke Jenkins. 2017. "Rosbridge: ROS for Non-ROS Users." In *Robotics Research: The 15th International Symposium ISRR*, edited by Henrik I. Christensen and Oussama Khatib. Cham: Springer International Publishing. 493–504. https://doi.org/10.1007/978-3-319-29363-9_28.

Furrer, Fadri, Martin Wermelinger, Hironori Yoshida, Fabio Gramazio, Matthias Kohler, Roland Siegwart, and Marco Hutter. 2017. "Autonomous Robotic Stone Stacking with Online next Best Object Target Pose Planning." In *2017 IEEE International Conference on Robotics and Automation (ICRA)*. Singapore. 2350–56. https://doi.org/10.1109/ICRA.2017.7989272.

García del Castillo y López, Luis Jose. 2019. "Robot Ex Machina." *ACADIA '19 Ubiquity and Autonomy; Paper Proceedings of the 39th Annual Conference of the Association for Computer Aided Design in Architecture*. Austin, Texas. 40–49.

Giftthaler, Markus, Timothy Sandy, Kathrin Dörfler, Ian Brooks, Mark Buckingham, Gonzalo Rey, Matthias Kohler, Fabio Gramazio, and Jonas Buchli. 2017. "Mobile Robotic Fabrication at 1:1 Scale: The In Situ Fabricator." *Construction Robotics* 1 (1–4): 3–14. https://doi.org/10.1007/s41693-017-0003-5.

Hurkxkens, Ilmar, Ammar Mirjan, Fabio Gramazio, Mathias Kohler, and Christophe Girot. 2020. "Robotic Landscapes: Designing Formation Processes for Large Scale Autonomous Earth Moving." In *Impact: Design With All Senses*, edited by C. Gengnagel, O. Baverel, J. Burry, M. Ramsgaard Thomsen, and S. Weinzierl. Cham: Springer International Publishing. 69–81.

Jocys, Evaldas. 2020. "*Xbox 360 Controller Emulator.*" Github repository. https://github.com/x360ce/x360ce.

Koenig, Nathan, and Andrew Howard. 2004. "Design and Use Paradigms for Gazebo, an Open-Source Multi-Robot Simulator." *2004 IEEE/RSJ International Conference on Intelligent Robots and Systems (IROS)* 3: 2149–54. https://doi.org/10.1109/iros.2004.1389727.

Ornan, Oni, and Amir Degani. 2013. "Toward Autonomous Disassembling of Randomly Piled Objects with Minimal Perturbation." In *IEEE International Conference on Intelligent Robots and Systems*. 4983–89. https://doi.org/10.1109/IROS.2013.6697076.

Petersen, Kirstin H, Nils Napp, Robert Stuart-Smith, Daniela Rus, and Mirko Kovac. 2019. "A Review of Collective Robotic Construction." *Science Robotics* 4 (28). https://doi.org/10.1126/scirobotics.aau8479.

Peurifoy, Robert L., W. B. Ledbetter, Clifford J. Schexnayder, Robert L. Schmitt, and Aviad Shapira. 2018. *Construction Planning, Equipment, and Methods*. McGraw-Hill Series in Construction Engineering and Project Management. New York: McGraw-Hill.

Queralta, Jorge Peña, Jussi Taipalmaa, Bilge Can Pullinen, Victor Kathan Sarker, Tuan Nguyen Gia, Hannu Tenhunen, Moncef Gabbouj, Jenni Raitoharju, and Tomi Westerlund. 2020. "Collaborative Multi-Robot Systems for Search and Rescue: Coordination and Perception." *arXiv:2008.12610*. http://arxiv.org/abs/2008.12610.

Quigley, Morgan, Ken Conley, Brian Gerkey, Josh Faust, Tully Foote, Jeremy Leibs, Rob Wheeler, and Andrew Y. Ng. 2009. "ROS: An Open-Source Robot Operating System." *ICRA Workshop on Open Source Software* 3 (3.2): 1–5.

Raspall, Felix. 2015. "Design with Material Uncertainty: Responsive Design and Fabrication in Architecture." In *Modelling Behaviour*, edited by Mette Ramsgaard Thomsen, Martin Tamke, Christoph Gengnagel, Billie Faircloth, and Fabian Scheurer. Cham: Springer. 315–327. https://doi.org/10.1007/978-3-319-24208-8_27.

Schwartz, Thibault. 2013. "HAL." In *Rob | Arch 2012*, edited by Sigrid Brell-Çokcan and Johannes Braumann. Vienna: Springer. 92–101.

Shaked, Tom, Karen Lee Bar-Sinai, and Aaron Sprecher. 2020. "Craft to Site." *Construction Robotics* 4: 141–50. https://doi.org/10.1007/s41693-020-00044-7.

Shaked, Tom, Karen Lee Bar-Sinai, and Aaron Sprecher. 2021. "Adaptive Robotic Stone Carving: Method, Tools, and Experiments." *Automation in Construction* 129: 103809. https://doi.org/https://doi.org/10.1016/j.autcon.2021.103809.

Sullivan, Nick. 2018. "Multi-Jackal Simulator Using Gazebo ROS." Github repository. https://github.com/NicksSimulationsROS/multi_jackal.

Temtsin, Sharon, and Amir Degani. 2017. "Decision-Making Algorithms for Safe Robotic Disassembling of Randomly Piled Objects." *Advanced Robotics* 31 (23–24): 1281–95. https://doi.org/10.1080/01691864.2017.1352537.

IMAGE CREDITS
All drawings and images by the authors.

Tom Shaked is an architect, a computational designer, and a postdoctoral researcher at the Civil, Environmental, and Agricultural Robotics Lab (CEAR) at the Faculty of Civil and Environmental Engineering, Technion IIT. His research focuses on collective robotic construction and autonomous systems in unstructured environments, developing robotic tools for adaptive fabrication, and integrating sensory data into construction processes. Tom is an Azrieli Research Fellow, holding a PhD from Technion, and an MA and BArch from Tel-Aviv University. Additionally, he teaches undergraduate architecture studios, computational design, and advanced graduate-level robotic fabrication at Technion.

Amir Degani is Associate Professor at the Faculty of Civil and Environmental Engineering, Technion IIT, and director of the Civil, Environmental, and Agricultural Robotics Lab (CEAR). He is an Associate Editor for the IEEE Transactions on Robotics (IEEE T-RO), IEEE IROS, and IEEE ICRA conferences. His research focuses on dynamic locomotion, minimalism, and autonomous systems in unstructured environments with applications such as search and rescue and agriculture. Before joining Technion, Amir completed his MSc, PhD, and postdoctoral research at the Robotics Institute at Carnegie Mellon University while performing research at the Manipulation Lab and the Bio-Robotics lab.

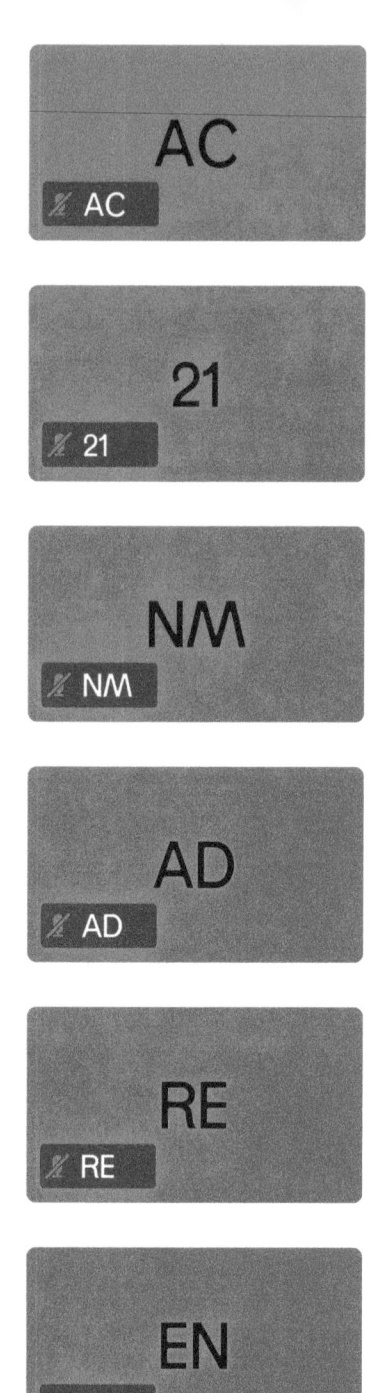
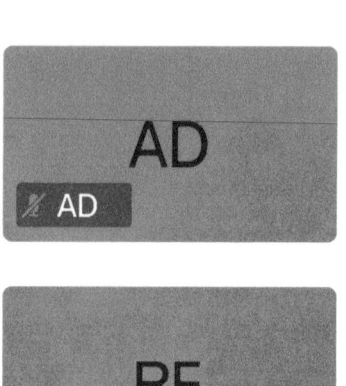
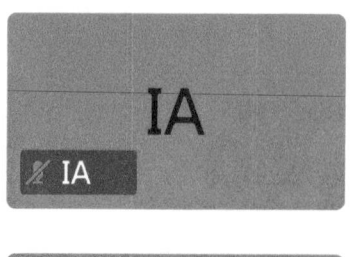

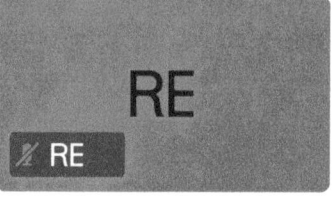
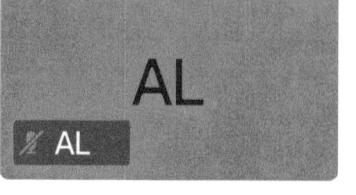

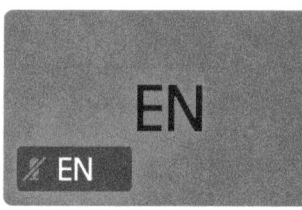

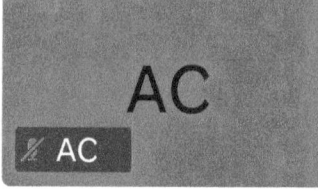
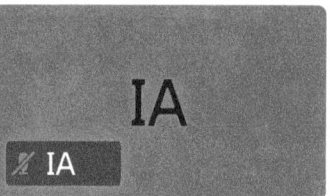

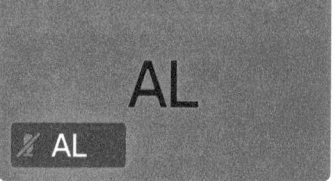

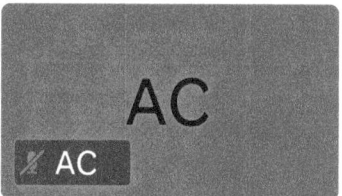

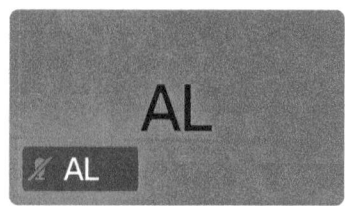

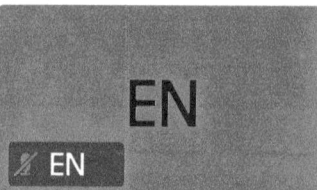

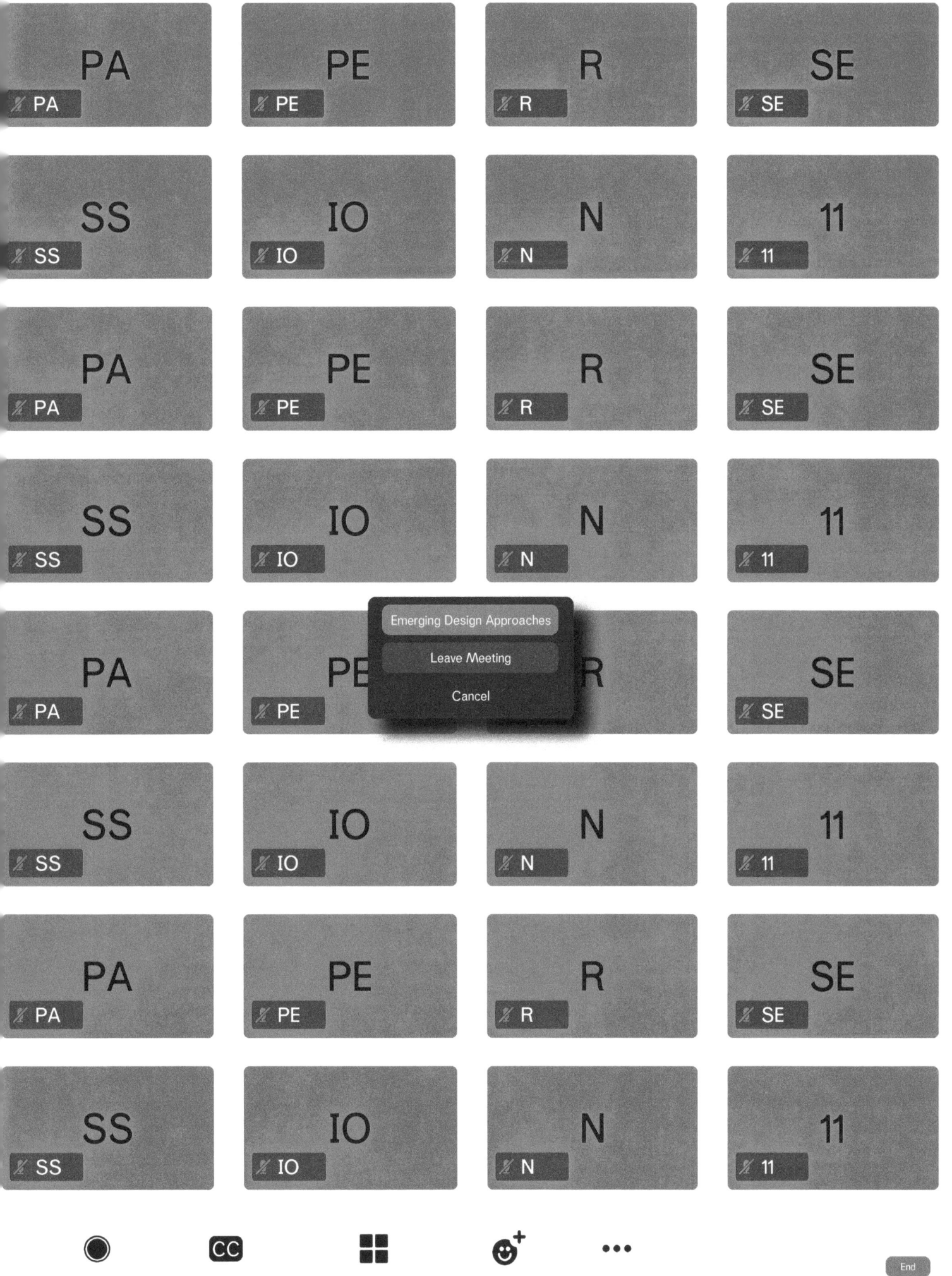

Aligning the Analog, Digital, and Hyperreal

Software Errors as Design Exploration Drivers

Olivia Römert
Chalmers University of Technology

Malgorzata A. Zboinska
Chalmers University of Technology

1 Overlay of misaligned, real and hyperreal architectural representations developed through photogrammetry error explorations

ABSTRACT

This work explores the relevance of photogrammetry-generated errors for contemporary architectural design. Unlike approaches featuring correction or elimination of such errors, this study demonstrates how they can be accommodated in the design process to expand exploratory boundaries and emancipate the designer from the need of ultimate control. The work also highlights the relevance of software error explorations in the context of modern media culture theory and critical discourses on computer-generated imagery.

By exploring the errors of photogrammetry, the study sought to highlight its potential as a creative exploration medium instead of a mere representation tool, using new interventions to an existing building as an experimental brief. Conducting the explorations within the philosophical framework of Jean Baudrillard's four orders of the image, and relating them to contrasting discourses, allowed to coin their most important creative and aesthetic values. It revealed how surplus, leftover, and undesirable data can be harnessed to provide a critical trajectory, through computation, to fields like historic preservation and adaptive reuse.

The study concludes by proposing that photogrammetry errors, although distancing the digital representation from an accurate depiction of analog reality, do not deprive it of new meaning. Conversely, they generate new aesthetic, spatial and functional qualities that uncover alternative, critical ways of architectural creation. Conducting error explorations in the context of philosophies debating the value of the real and hyperreal increases their discursive potential, legitimizing the agency of software errors in architectural computing.

INTRODUCTION

In a world dominated by digitally processed images, the difference between the analog original and its computationally transformed replica can be impossible to discern. This results in a realization that the content of new digital representations is often so greatly affected by the medium used to produce them that the traces of the real become no longer discernible, forming hyperreal, computer generated imagery (Jacob 2018).

Such a new understanding elevates an awareness that these images and objects, although generated from precise numerical data, may not be as accurate as commonly thought, constituting erroneous approximations rather than precise copies of reality (Figure 1). These critical issues evoke questions regarding the value of hyperreal, erroneous representations in architecture. Should architectural computing strive to increase authenticity and accuracy of these imperfect representations? Alternatively, should it legitimize them as natural products of digital tooling that require artistic adoption? This study takes a positive stance towards the latter question by providing examples of creative potentials residing in the errors of digital 3D representations, and by demonstrating how such errors can open up the architectural design process for new aesthetic, spatial, and functional possibilities.

The outcome is an architectural method that harnesses an error-prone technique of photogrammetry as medium triggering creative explorations of inaccurate architectural 3D representations. The method exemplifies how a seemingly accurate digital photograph can trigger software errors. In the computational software's interpretation of reality, this photograph becomes integral to a new design process that accommodates these errors to expand the exploration territory. The anticipation is that the method could serve as an alternative in the future architectural design praxis, enabling designers to approach projects involving interventions into existing buildings from a new critical perspective.

BACKGROUND

Digital Inaccuracies and Computational Glitches

The interest in errors and inaccuracies among architects is not new, tracing back to seminal historical works, such as David Pye's *The Nature and Art of Workmanship* (1968). Therein, arguments highlighting the value of imperfections in handicraft and mass production were put forth, forming a relevant framework for considering errors in contemporary digital design. Just like in analog making, in digital design software errors can be a source of novel architectural effects, both in the design process and in materialization (Veliz Reyes et al. 2019).

A recent example of a collective effort to explore the significance of such errors is the ACADIA 2018 conference, featuring a theme called 'Recalibration: On Imprecision and Infidelity.' The event resulted in important mappings of these concepts, confirming the emergence of a novel research avenue in architectural computing, concerning digital imprecision (Anzalone, Del Signore and Wit 2018). A study by Austin and Matthews (2018) was one of the precedents demonstrating how digital imaging can produce deliberately erroneous architectural drawings, exhibiting emergent spatial qualities impossible to generate using the traditional drawing techniques. Other examples of similar research concern digital fabrication, in which errors yielded by erratic material phenomena are accommodated in the setups of manufacturing processes, to create new aesthetic typologies of architectural elements (Atwood 2012).

This prior work, crucial for developing a multifaceted discussion on the significance of digital imprecision, primarily investigated it from two standpoints: aesthetic, focused on coining its new design expressions, and pragmatic, focused on developing new workflows bridging the gap between the physical model and its digital counterpart. What is missing, however, is a demonstration of software error harnessing as an architectural exploration method. This study aimed to fill this gap by presenting a design method that employs errors of photogrammetry as prerequisites for critical exploration of aesthetic and spatial qualities in surfaces, textures, and masses of architectural elements and their assemblies.

Photogrammetry in Architecture and Art

Photogrammetry is conventionally used for digitally representing physical objects. In architectural computing research, its primary developments concern 3D documentation of terrains, cities, buildings, and cultural heritage sites (Aicardi, Chiabrando, Lingua and Noardo 2018). As such, the mainstream research focuses on achieving accurate 3D object representations through photogrammetry. Simultaneously, however, alternative studies focused on highlighting the value of imaging errors that accompany this technique. For instance, potentials of user interactions with photogrammetry-generated erroneous virtual reality models were outlined (Schnabel et al. 2016). Previous research by one of the authors also demonstrated the generative role of photogrammetry errors in driving architectural surface articulations, using robotic fabrication as mediating platform (Zboinska 2019).

In art and experimental design, photogrammetry has been probed in a similar exploration-oriented spirit. For instance,

Sam Jacob used this technique to develop a new material expression for an existing object, by creating its sculptural copy endowed with novel textures and coloring (Abrons and Fure 2018). Another interesting take was demonstrated by artist Mikaela Steby Stenfalk, who employed photogrammetry to 3D print a scaled replica of an existing historical space using photographs from social media, which revealed the surprising gap between how people perceive that space and how it is shaped in reality (Baek 2020).

Overall, prior research on photogrammetry in architectural computing indicates that its main objective is the documentation of physical objects with high accuracy. However, the above-mentioned unorthodox precedents within research, art, and experimental design indicate that it could also be regarded as an inquisitive medium. Therefore, to contribute new knowledge to this latter body of work, in this study photogrammetry was probed for its capacity of triggering creative endeavors expanding the design exploration space.

METHOD AND THEORY
Design Research Setup

A design experiment was carried out to contribute an example of an architectural process driven by software errors. In the process, computational tools enabled to employ the erroneous features of photogrammetry-generated meshes as design exploration drivers. The experiment brief embraced the development of new interventions to an existing historical building, namely the United Kingdom Hotel in Melbourne, Australia (today a franchise of McDonald's Corporation), chosen for its strong Art Deco features. Furthermore, to explore the robustness of the proposed method, an architectural element from another, contemporary, building was also included, namely a double-curved awning of RMIT University Building 22 in Melbourne by ARM Architecture. The aim was to explore how an element with such a strong contemporary expression could enrich the transformations of the existing building, enabling its re-articulation through a compositional dialog between historical and contemporary stylistic features (Figure 2).

Digital photographs of both buildings were then compounded into digital 3D models using photogrammetry software Agisoft Metashape (Agisoft 2021). The fragments of the models containing software-based interpretation errors were explored using surface and mesh modification tools in Rhinoceros 3D (McNeel & Associates 2018), with support of computational add-ins Grasshopper (Rutten 2021) and Weaverbird (Piacentino 2009).

Theoretical Framework for Digital Error Exploration

According to architect John May architecture needs to

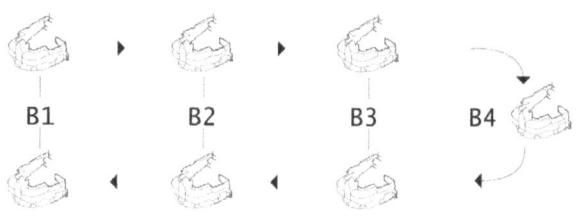

2 Objects of study both in Melbourne, Australia: (top) the historical United Kingdom Hotel with Art Deco features; (bottom) the contemporary Building 22 at RMIT University with modern styling by ARM Architecture

3 Exploration process following Jean Baudrillard's theory of four orders of the image

acknowledge that an architectural drawing today has been completely replaced by a digital image (May 2018). This new pixelated representation bears no resemblance to the traditional drawing. Therefore, unfolding its qualities requires consideration within radically new conceptual frameworks. In the study, following May's arguments, a theoretical framework was applied to probe the value of digital error explorations in three-dimensional architectural constructs.

The chosen framework originates from seminal work *Simulacra and Simulations* by cultural theorist Jean Baudrillard. It outlines four types of modern images, to various extents deprived of their meaning due to the transformations they were subjected to:

- B1 - an authentic image, truthfully reflecting reality,
- B2 - a denatured image with some traces of the real,
- B3 - a simulated image of reality, masking its absence,
- B4 - a hyperreal image, with no originator and no relation to reality (Baudrillard 1983).

Whereas Baudrillard's theory predominately concerns the two-dimensional image, the aim of this work was to explore it in the context of digital architectural design in three dimensions. More specifically, to coin how employing Baudrillard's four orders of the image to define the phases of a digital 3D model's error manipulations affects architectural qualities generated in these phases. Through this, the aim was to challenge Baudrillard's claim that hyperreal images carry no value by revealing the dormant qualities of the non-real. To further support the authors' architectural articulations, the design inquiry outcomes were discussed in relation to discourses countering Baudrillard's arguments, as presented in the discussion section.

The experimental design process first followed the four image orders as originally proposed by Baudrillard, to then proceed backwards (Figure 3). The result was a gradual evolution from a realistic representation in the form of a photograph to a hyperreal 3D model, and then back to the initial form of representation, complemented with qualities and elements discovered in the hyperreal order.

PROCESS, RESULTS AND DISCUSSION
Phase B1 - Reflecting the Real
This research phase addressed Baudrillard's notion of an image as the reflection of a profound reality. The main historical building as well as the awning of the contemporary building were captured in digital photographs, taken from various angles, to enable the creation of their 3D digital representations using photogrammetry. Already in this phase, errors were intentionally triggered. Namely, the photography setups were manipulated to increase 3D model accuracy in some areas, aligning the result with Baudrillard's notion of profound reality, and to induce model faultiness in other areas, intentionally enlarging the gap between the real and its digital representation.

Accordingly, for the historical building, the height from which the photographs were taken was limited to induce glitches in the upper parts of its digital model (Figure 4). Taking the photos only from ground level at eye height forced the photogrammetry software to compensate for image data deficiencies, yielding large geometric

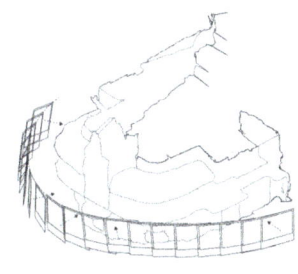

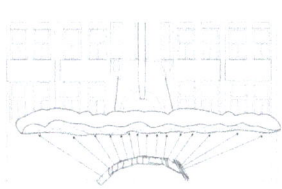

4   Photography setups for the main building (top left) and the contemporary element (bottom left), including examples of digital images in phase B1

approximations of the upper part of the 3D model. Conversely, the lower parts of the model were represented more precisely, due to more extensive coverage in photographs. A similar logic of photography setup manipulation was employed for the contemporary building element. It was photographed only from underneath, from a frog perspective, to trigger distortions of its photogrammetric representation.

Phase B2 - Denaturing the Real
In this phase, Baudrillard's idea of a denatured image of reality was addressed. Photogrammetry software was employed to process the digital photographs of both buildings into their 3D representations. The previously mentioned photography setup restrictions generated errors in both digital models.

Interestingly, and in line with Baudrillard's notion of denaturing, the texture maps automatically applied onto the models by the software first concealed these errors. At a first glance, the models seemed correct. However, viewing them in wireframe mode and examining their outlines in cross sections made the deviations from reality apparent. The clear-cut silhouettes of the main building became softened and rugged in the digital counterpart (Figure 5). Its texture map also contained deviations form reality, with

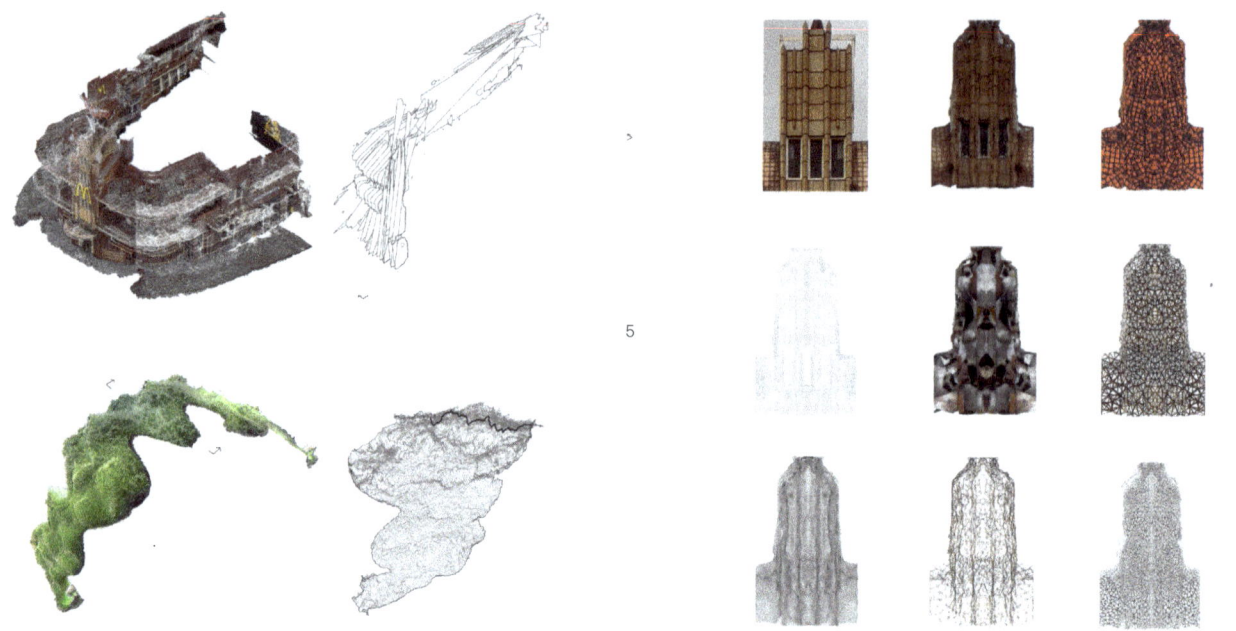

the originally straight patterns of bricks and stone detailing becoming distorted in the digital model. The contemporary building element acquired a strongly distorted texture and global geometry. In reality smooth, voluminous and globally planar, in the digital space it became a furrowed zero-thickness surface that is bent (Figure 6).

Instead of correcting them, all the errors in the 3D models were preserved. The main value of this phase resided in the liberating possibility of comparing the aesthetic features of the erroneous models to their analog counterparts, welcoming the gaps between the real and the digital. This has shifted the role of photogrammetry models from erroneous entities needing correction into repositories of aesthetic glitches offering abundant grounds for further probing.

Phase B3 – Simulating the Real to Mask Its Absence

This phase addressed Baudrillard's notion of an image as a simulation that masks the absence of a profound reality. Herein, reflections were triggered regarding legitimacy, originality and how far removed the digital representation should be from its original to become its own entity. As a basis, several erroneous details of the main building model of phase two were selected. To obtain versions of these details that simulate and detach from reality, various computational modifications were applied, such as surface distortion, subdivision, extrusion, extraction and texture remapping (Figure 7).

In the process of moving away from the realism of the photogrammetry model towards its hyperreal qualities, no priority was given to the accuracy of the representations when applying these modifications. This allowed for staying free from the preconceptions tied to the original building appearance, shifting focus from comparing the digital and the real towards an exploration of architectural qualities in the given digital representation. The representation flaws in the photogrammetry model did not affect the value of its explorations in terms of aesthetic outcomes. Conversely, the engagement in modifying erroneous model details continuously prompted to alter, in diverse ways, the initial model features, yielding unplanned design alternatives with high aesthetic differentiation (Figure 8). This offered new perspectives on the original qualities of existing architectural surfaces, textures, masses and their materiality.

Phase B4 - Creating the Hyperreal

In this phase, Baudrillard's notion of the hyperreal was addressed. The building details transformed in the previous phase were explored for their inherent qualities. Firstly, the details were studied in isolation. They were assigned into architectural component categories, such as walls, roofs and floors. To suit the new functions, they were scaled up or down, rotated, copied, and cropped (Figure 9). These modifications provided liberation from predefined ideas of how the elements should be used, following Baudrillard's notion of the hyperreal having no precedent in reality. For instance, what had originally been a wall fragment could now become a standalone roof or a façade.

The second part of the process continued to follow Baudrillard's idea that hyperreality is reached when the replica is so far removed from what it used to represent

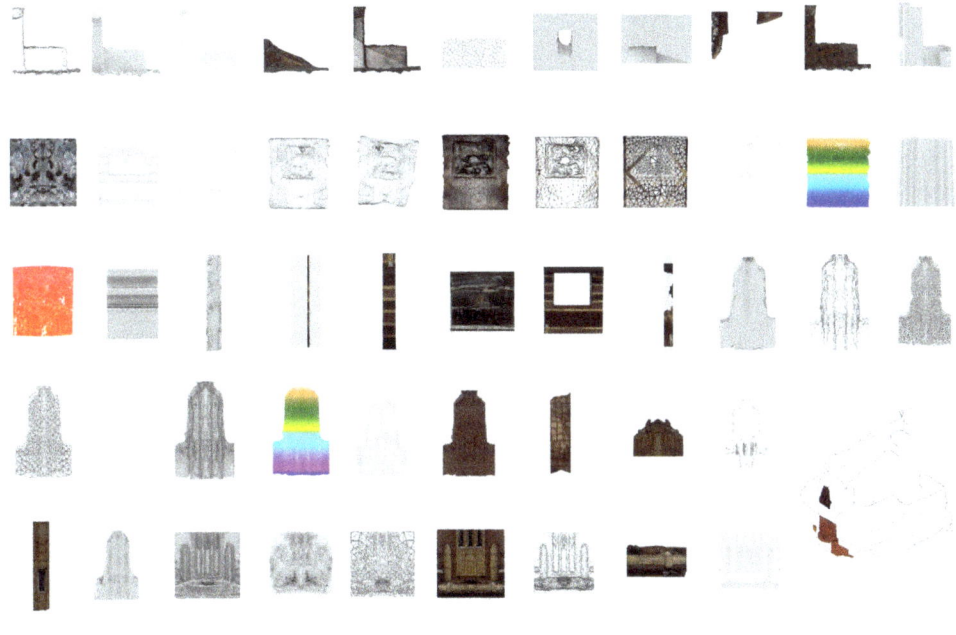

5   The photogrammetry-derived model of the main building (left) and the erroneous surface features revealed in its cross section (right)

6   The photogrammetry-derived, spatially deformed model of the contemporary element (left), with erroneous textural features shown in its cross section (right)

7   Computational modifications applied to an erroneous detail of the main building model, resulting in multiple alternatives for its material expression and form

8   Collection of architectural detail iterations, resulting from error explorations of phase B3; the diagram at the lower right shows the original locations of these details within the main 3D model

that it becomes its own entity. Accordingly, the components of the previous phase were tied into a spatial composition to create a hyperreal building. They were transformed and altered to such an extent that they lost their original form and function, in favor of newly defined functions and aesthetic features, constituting a new building (Figure 10).

This phase evoked new ideas for combining the hyperreal elements in ways not present in the real building. These element assemblies triggered new qualities of the new spaces they defined, causing the previously existing element hierarchies and underpinning spatial features to dissolve. The representation errors transferred from the previous design phase, in this step defined novel features of new elements, bearing no evident traces of the previous. Conceptually, this caused the glitches to become the only element not intending to reflect reality, which suggests that they could be viewed as the most valuable assets in the process thus far.

Phase B3 - Simulating the Real to Mask Its Absence

To further examine the emergent qualities of the components of the hyperrealistic model of phase four, the forthcoming phases followed Baudrillard's orders of the image in reverse. Such a reversed process aimed to challenge Baudrillard's notion that there is no meaning in the hyperreal, by exploring the possibility of integrating the hyperreal elements into the real building.

Therefore, this phase reversed back to Baudrillard's third order of the image, investigating the gap between what appears to be real and what appears to be exaggerated and false. The details from the hyperreal model of phase four were combined with the photogrammetry-derived details of the 3D model from phase two (Figure 11). To expand the understanding of their spatial qualities, and to improve the credibility of how they could be added to the existing building, the hyperreal elements were assigned with materials. The juxtaposition of the elements instigated questions of authenticity in the hyperreal elements, concealing the fact that the model from phase two was, in fact, also an erroneous representation.

In contrast to the previous phase, where the hyperreal elements were perceived as convincingly real but with no reference to a context, real or digital, this phase provoked to contextualize and verify them. The combined parts of models from phase two and four were now examined in a scalar and material relation to each other, revealing the misalignments, gaps and overlaps of the hyperreal model against the imperfect but reality-based photogrammetric representation of the original building. Paradoxically, the presence of the realistic implants within the hyperreal entity allowed for detecting and addressing the discovered flaws and misalignments while creating an overall impression that the studied details could be actual, aligning them with Baudrillard's notion of simulated reality, masking the absence of the real.

Phase B2 - Denaturing the Real

In this phase, the newfound knowledge of the previous phase was applied to further explore the possibilities of integrating the hyperreal and the real at the scale of the

entire building, to denature reality. Elements from phase four were employed to serve various purposes in the photogrammetry model, such as reparation implants and additions with new functionalities (Figure 12).

Moreover, the contemporary building element from phase two was included to illustrate photogrammetry's capacity of allowing to replicate architecture freely by mixing and matching its historical and contemporary instances. Owing to the software's representation glitches, in digital space the hierarchies between these historical and contemporary parts dissipated, allowing all elements to be treated on equalized terms. This enabled to use them to develop a novel spatial and aesthetic articulation resulting from this peculiar replication, surpassing mere copying of the existing. Accordingly, in the explored design, the contemporary building element was multiplied and added to the parts of the digital model of the historical building in need of an intervention, i.e. areas of the model's upper part, intentionally poorly documented in phase one. Adding these elements supplemented the design exploration with an additional stylistic layer, comprising an erroneous contemporary implanted into an erroneous digital replica of the historical.

This phase specifically highlighted the possibilities of applying the proposed method in architectural reconstruction and refurbishment aided digitally. It offered an unorthodox approach to the methods conventionally used. The main value of the approach relates to its liberating property of triggering an alternative viewpoint on referencing bygone architectural qualities, without directly mimicking the past and instead designing additions to existing buildings that relate to their original features in a contemporary way (Figure 13).

Phase B1 - Reflecting the Real

In this last step of the process, the model representation was resumed to its initial form, referred to by Baudrillard as a reflection of a profound reality. The digitally generated elements of the model from the previous phase were combined graphically with the photograph of the building of phase one (Figure 14). This made it possible to examine the design in comparison to the initial building. The main value of this phase resides in this very comparison, because it enables to validate the transition from abstract elements, only existing in digital space, to elements that serve a particular purpose and yield new value. This value is found in the new aesthetic language, materiality, functions, and representations of the existing building. Since the elements are in this final phase shown within the frame of reference of reality, it becomes possible to coin and assess the qualities they contribute to the physical architectural context. Ultimately, this enables to establish the relevance of the elements derived through software error explorations, and can increase the credibility of a software-generated glitch as a powerful vehicle of architectural exploration.

Discussion of Overall Result

The presented design experiment sought to unfold the design potentials residing in a digital 3D model representing reality in an erroneous way. The results suggest

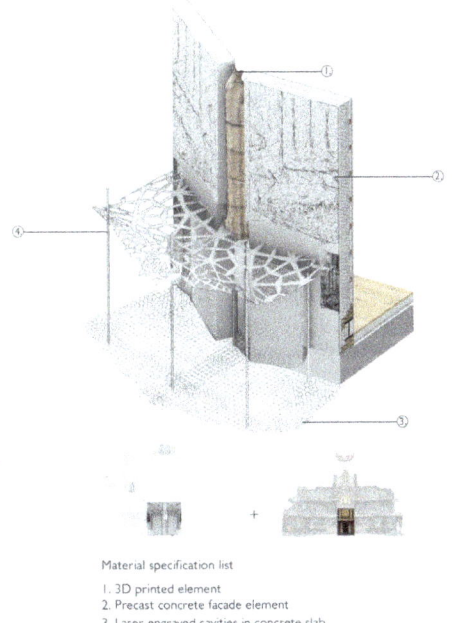
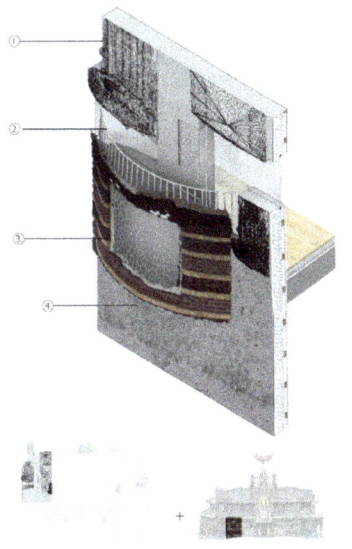

Material specification list
1. 3D printed element
2. Precast concrete facade element
3. Laser engraved cavities in concrete slab
4. Metal canopy

Material specification list
1. Laser engraved concrete
2. Stainless steel balustrade
3. Recycled brick
4. Concrete precast facade element

9 Hyperreal architectural elements derived from error explorations of phase B3, embodying new esthetic qualities, new spatial configurations and new functions

10 The hyperreal counterpart of the physical building developed based on photogrammetry error explorations

11 Explorative probing and validation of new, hyperreal architectural features by combining them with features extracted from the erroneous photogrammetry model

that such potentials are broad, offering a rich platform to explore diverse architectural aspects—surface texture, detailing, massing, and building component functions. Jean Baudrillard's framework has proven to be a robust conceptual scaffolding for entering a critical design mode, leading to a gradual overturn of the original claim that a highly transformed, hyperreal image carries no value.

The study highlighted that extensive explorations of three-dimensional, hyperreal architectural representations can, in dissonance with Baudrillard's claims, generate new value. Converting analog reality into erroneous digital models and then into abstract architectural assemblies unfolds unexpected spatial configurations and aesthetic features. They are not bound to their original purpose, and free to generate alternative qualities, functions and, therewith, also new meaning. Transferring these qualities onto further phases of design development gradually legitimizes their value. Therefore, if one is to follow John May's concept that architecture needs to align its conceptual and operational frameworks with the intricacies of contemporary digital imaging (May 2017), perhaps the approach undertaken in this work was one way to achieve this alignment.

By revealing the dormant qualities of the hyperreal, this investigation drifts towards alternative philosophies arguing for the value of the not-yet-embodied and metaphysical. For instance, it aligns with Gilles Deleuze's interpretation of Bergsonism, which implies that true value and novelty reside beyond the spatial dimension of the existing, in the transcendent dimension of what could be (Deleuze 1991). Accordingly, an instance of reality is only one out of a myriad of possibilities dormant in the hyperreal. This can be associated with the vast exploration spaces discovered in principally all phases of the discussed project—from infinite possibilities of tweaking the photography setups, through endless opportunities for exploring photogrammetry errors and mixing and matching the hyperreal elements, to amplified possibilities of implanting elements from other buildings and styles.

A similar argumentation for the value of the hyperreal is provided by Valerio Oligaiti's idea of non-referential architecture, whose highest value—of novelty and critical liberation—is established by the lack of referencing to the prior and existing (Olgiati and Breitschmid 2018). This underlines the value of especially phases four and five of the presented exploration, in which the existing elements become, through flawed computing, so distorted and detached from their counterparts in reality that they can be considered anew.

Finally, the findings show the power of photogrammetry as a creative exploration medium, extending its use beyond mere documentation. As all digitized buildings, existing and bygone, can be composed into new ones, and any model or its fragment can be recycled and repurposed, photogrammetry becomes a powerful enabler of novel critical practices of historic preservation, adaptive reuse and architectural design in general. The newly developed understanding of how data becomes information and the uncovering of where meaning can lay within digital imagery the digital space enables to make informed decisions about how they could serve as a design exploration medium.

The software-induced errors can act as meaningful stimuli, inducing open-ended architectural explorations and promoting new perspectives on the qualities of existing buildings and cityscapes. Such explorations and perspectives would have been difficult to realize using conventional design methods. Assuming that all current and future digital software will rely upon erroneous approximations of reality, the approach proposed herein enables to acclaim these approximations as a natural feature of new media that can be accommodated in architectural design to serve creative purposes.

In the future, this approach could be relevant in architectural restoration, refurbishment, and reconstruction projects involving new digital technologies. As such, it would provide critical insight into the ongoing debate on new interventions and the grade of their reference to existing or bygone architectural qualities. New media could be utilized to expand the referencing methodology in ways that express the contemporary techno-cultural advances within the built environment. The features of existing buildings could be maintained by giving them new form, function and expression via critical, unorthodox computing, marking the genuine traces of the postdigital era.

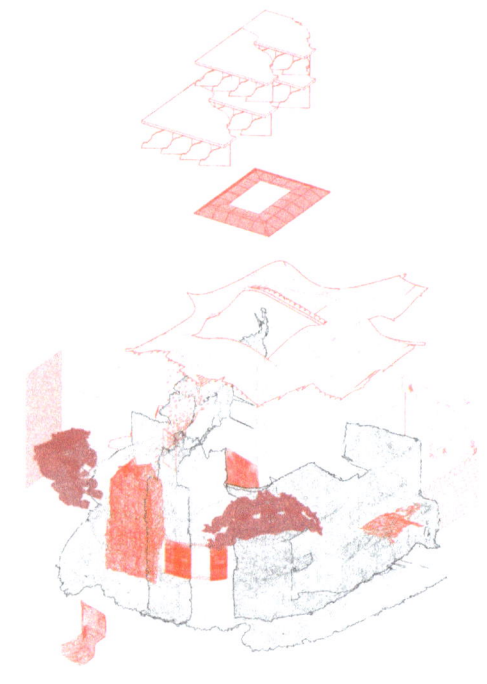

12 The 3D model representing denatured reality, comprising the erroneous main building model complemented with hyperreal elements (bright red) and elements adopted from the contemporary building (dark red)

13 Photorealistic representation of the denatured reality model

transforms the inaccurate, surplus and leftover data from mere faulty information into a valuable source of architectural articulation.

CONCLUSION

The significance of the approach proposed herein lies in its capacity to critically reexamine the current use of digital media in architecture. In relation to precedent projects, this study has contributed with a novel example of a software error-welcoming approach, exercised using photogrammetry as a powerful inquisitive design medium. It highlighted the possibilities that lie in exploring the misinterpretations embedded in the software. Understanding how the misinterpretations between the software and reality come into being and how they present themselves in

ACKNOWLEDGEMENTS

This study is a new research development based on the results of a master thesis "Reality and Beyond" by Olivia Römert at Chalmers University of Technology. The authors extend gratitude to ACADIA peer reviewers for their valuable comments on the presented work.

REFERENCES

Abrons, Ellie, and Adam Fure. 2018. "Postdigital Materiality." In *Lineament: Material, Representation and the Physical Figure in Architectural Production*, ed. G.P. Borden and M. Meredith. Oxon: Routledge. 185-195.

Agisoft. *Agisoft Metashape*. V. 1.7.6. Agisoft. PC. 2021.

Aicardi, Irene, Filiberto Chiabrando, Andrea Maria Lingua, Francesca Noardo. 2018. "Recent trends in cultural heritage 3D survey: The photogrammetric computer vision approach." *Journal of Cultural Heritage* 32: 257–266.

Anzalone, Phillip, Marcella Del Signore, and Andrew John Wit. 2018. "Notes on Imprecision and Infidelity." In *ACADIA '18: Recalibration: On Imprecision and Infidelity; Proceedings of the 38th Annual Conference of the Association for Computer Aided Design in Architecture*, edited by P. Anzalone, M. del Signore and A. J. Wit. Mexico City: ACADIA. 16–17.

14 Real building complemented with hyperreal elements and intentionally flawed contemporary building components, derived through software error explorations.

Atwood, W. Andrew. 2012. "Monolithic Representations." In *Matter: Material Processes in Architectural Production*, edited by G. P. Borden and M. Meredith. Oxon: Routledge. 205-211.

Austin, Matthew, and Linda Matthews. 2018. "Drawing Imprecision: The Digital Drawing as Bits and Pixels." In *ACADIA '18: Recalibration: On Imprecision and Infidelity; Proceedings of the 38th Annual Conference of the Association for Computer Aided Design in Architecture*, edited by P. Anzalone, M. del Signore and A. J. Wit. Mexico City: ACADIA. 36–45.

Baek, Juyeon. 2020. "A Piece of Space: Exploring photographic space as a visualized form of spatial experience and thinking about how a designer can position it in reality." Master's thesis, Konstfack University of Arts, Crafts and Design.

Baudrillard, Jean. 1983. *Simulacra and Simulations*, transl. P. Foss, P. Patton and P. Beitchman. New York: Semiotext(e).

Deleuze, Gilles. 1991. *Bergsonism*, transl. H. Tomlinson and B. Habberjam. New York: Zone Books.

Jacob, Sam. 2018. "The Great Roe." In *Lineament: Material, Representation and the Physical Figure in Architectural Production*, edited by G. P. Borden and M. Meredith. Oxon: Routledge. 177-184.

May, John. 2017. "Everything Is Already an Image." *Log* 40: 9–26.

McNeel & Associates. *Rhinoceros 3D*. V. 6.0. Robert McNeel & Associates. PC. 2018.

Olgiati, Valerio, and Breitschmid, Markus. 2018. *Non-Referential Architecture*. Coira: Simonett & Baer.

Piacentino, Giulio. *Weaverbird*. V. 0.9.0.1. Giulio Piacentino. PC. 2009.

Pye, David. 1968. *The Nature and Art of Workmanship*. Cambridge: Cambridge University Press.

Rutten, David. *Grasshopper*. V. 1.0.0007. Robert McNeel & Associates. PC. 2021.

Schnabel, Marc Aurel, Serdar Aydin, Tane Moleta, Davide Pierini, and Tomás Dorta. 2016. "Unmediated cultural heritage via Hyve-3D: Collecting individual and collective narratives with 3D sketching." In *Living Systems and Micro-Utopias: Towards Continuous Designing; Proceedings of the 21st International Conference on Computer-Aided Architectural Design Research in Asia (CAADRIA 2016)*, edited by S-F. Chien, S. Choo, M. A. Schnabel, W. Nakapan, M. J. Kim, and S. Roudavski. Melbourne: CAADRIA. 683–692.

Veliz Reyes, Alejandro, Wassim Jabi, Mohamed Gomaa, Aikaterini Chatzivasileiadi, Lina Ahmad and Nicholas Mario Wardhana. 2019. "Negotiated Matter: A Robotic Exploration of Craft-Driven Innovation." *Architectural Science Review* 62 (5): 398–408.

Zboinska, Malgorzata A. 2019. "Artistic Computational Design Featuring Imprecision: A Method Supporting Digital and Physical Explorations of Esthetic Features in Robotic Single-Point Incremental Forming of Polymer Sheets." In *Architecture in the Age of the 4th Industrial Revolution; Proceedings of the 37th eCAADe and 23rd SIGraDi Conference*, vol. 1, edited by J. P. Sousa, G. Castro Henriques, and J. P. Xavier. Porto: eCAADe. 719-728.

IMAGE CREDITS
All drawings and images by the authors.

Olivia Römert has an MSc degree in Architecture and Urban Planning from the Department of Architecture and Civil Engineering at Chalmers University of Technology in Sweden. Currently, she is practicing architecture in Melbourne, Australia. Olivia is interested in how digital spaces challenge our perception of the built environment, which she investigated in her Master thesis and now continues exploring through her engagements in practice.

Malgorzata A. Zboinska is an Associate Professor in Digital Design, Fabrication and New Media Art at Chalmers University of Technology in Sweden, where she is also a Creative Leader of the Robotic Fabrication Lab and an educator in new media theory, computation and digital fabrication. Her research was published and exhibited internationally at events and venues such as Dutch Design Week, Gdynia Design Days, Tempe Center for the Arts, and Färgfabriken Center for Contemporary Art and Architecture.

ARchitect

Anna Mytcul
University at Buffalo,
The State University of New York

An Accessible Tool for the Democratization
of Architectural Design

1   ARchitect playtest experiment with non-architect users; images created using Fologram

ABSTRACT

This research investigates gaming as a framework for design democratization in architecture, with the end user serving as the key decisionmaker in the design process. *ARchitect* is a multisensory game that promotes and explores the educational aspects of learning games and their influence on end user engagement with house co-design. This combinatorial game relies on an augmented reality (AR) application accessible through a smartphone, serving as a low-threshold tool for converting architectural drawings into 3D models in real time and using AR technology for design evaluation.

By allowing learning through play, *ARchitect* provides alternative ways to gain knowledge about design and architecture and empowers non-experts to take active and informed positions in shaping their future urban environment on a micro-scale, rethinking conventional market relations, and exploring emerging personal and public values. By providing free access to the game contents through the *ARchitect* platform and a playful user experience for learning design principles, this game will inspire the general public to engage in conversation about home design, eventually spreading architectural literacy to less privileged communities.

INTRODUCTION

The concept of mass customization and design democratization in architecture (in contrast to standardization) first appeared in the 1960s and 1970s with the work of Friedman and Negroponte. These researchers argued that end users are capable of designing their dwellings without supervision from architects (Negroponte 1975; Friedman 1967).

Today, the general public experiences ever-increasing access to advanced technology. Local production and high individualization of output can easily be achieved without substantive cost increases and logistical delays. However, this drive for customization has not yet taken root in architecture. In *Mass Customization and Design Democratization*, Kolarevic and Duarte argue that most people are not ready to take an expanded role in design, due to a lack of design literacy; thus, there may be significant obstacles to mass customization (2019).

In mass customization, the end user is included in the process of creation by "defining, configuring, matching, or modifying their individual solution from of a list of options and pre-defined components" (Piller at at. 2017). Such co-design endeavors require toolkits or pre-defined systems and associated guidance, rules, and constraints that allow end users to participate in the process. Without co-design activities for users, mass customization will fail to manifest as individualized designs that are affordably priced (Piller et al. 2017).

Piller (2017) saw mass confusion as the main obstacle to spreading mass customization in business. Existing home design customization software such as the Interactive Home Designer tool by Smith Douglas Homes or platforms such as BuildX by Open System Labs and Combinatorial Application by Automated Architecture Ltd. (AUAR) allow end users to be hands-on in the design of their homes. However, non-architects face substantial issues when customizing their dwellings; they often find it difficult to understand the "language of architecture," a condition that mostly leads to uninformed decisions (Mytcul 2021). Therefore, such applications have become yet another tool used by architects and designers instead of the general public, failing to solve the problems of mass confusion and equality of involvement for end users in house design.

Nevertheless, based on the wide popularity of HGTV shows and computer games such as House Flipper, Minecraft, The Sims, and Block'hood, the public is indeed very interested in house design. In order to overcome user confusion, it is important to explore how television shows and sandbox

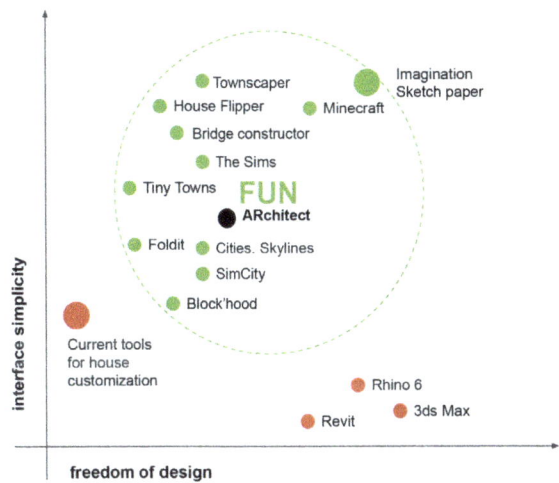

2   Diagram of fun

genre games make the practice of design accessible to non-architects. In conventional participatory design, an architect plays an essential role, facilitating the design process and translating end users' design proposals and decisions through their personal prism of design perception. In contrast, the proposed game system allows non-architect players to autonomously produce and access design solutions through an embedded computational simulation made possible by an AR application, thus giving an equal chance to non-professionals to express their design vision and become aware of the potential implications of their ideas.

ARchitect is a low-threshold tool for design democratization that combines the best parts of renovation shows, learning games, and serious architectural applications, with the goals of increasing design literacy among end users, bridging the educational gap, and facilitating more fruitful design solutions in a playful environment.

DESIGN GAMES

Sanoff defined gaming as a method of problem solving that allows players to experience real situations within a particular timeframe. Games help users concentrate their attention on crucial aspects of a real-world issue and solve them without tangible consequences. Sanoff argued that through the experience of making choices, design games allow users to reflect on unrecognizable behavioral patterns in everyday settings. Not the "win state," but rather the user's direct experience becomes the main goal of learning by playing (Sanoff 1979).

Entertainment is the main purpose of games, but not the only reason people play. "When we think of games, we think of fun. When we think of learning, we think of work. Games show us this [is] wrong" (Gee 2007, 43). If several building games can be located within the circle of fun (Figure 2), each will have its own specific definition, based on the game's objectives. Some may be more restricted with regards to design variety, but easy and relaxing to play (e.g., House Flipper, Minecraft). Others may be learning games such as SimCity, Cities: Skylines, and Block'hood, which provide fun experiences by guiding players to resolve problems, solve puzzles, and master strategic thinking (Mytcul 2021). Gee explained that players who enjoy such learning games are motivated by a "pleasantly frustrating feeling" (Gee 2007, 36), which keeps them playing because they feel that the challenge is difficult but doable.

Interface inaccessibility for non-architect users and limited freedom with regards to design criteria locate current customization tools outside of the "fun circle." Imagination and sketch paper are examples of pathways to limitlessly accessible design solutions. However, for a user without any background in house design, accurately drawing ideas on sketch paper can be challenging, especially for individuals without artistic skills. Thus, conceptualizations of house customization tools should gravitate toward educational building games located somewhere between unlimited and overly constrained optionality and feature an intuitive interface and mechanics (Mytcul 2021).

Foldit s another successful example of a learning game with real-world implications. The "multiplayer online biochemistry game" (Cooper 2010) uses gamification and crowdsourcing principles to prompt non-professional users to solve computationally difficult protein-folding problems. Scientists and game designers were able to simplify protein folding into appealing 3D puzzles that are approachable by PhD students in biochemistry and non-expert players alike. Due to a gradual increase in the tutorials' difficulty and the complexity of the game levels, the user interface and qualitative feedback provide an comprehendible learning path for nonprofessionals. The game serves as a viable precedent for *ARchitect*, which teaches design but is not biased toward
a particular solution.

METHODS

The *ARchitect* game is an accessible pedagogical tool designed for two players. After playing the game, participants are better able to make informed decisions and reflect upon their vision of a customized dwelling. The methods employed in this research reside on the boundary between game design and architectural pedagogy; designing, testing, and evaluating were repeatedly employed in the early stages of development. This work explores the idea that those with no background in design can still learn basic architectural principles by playing

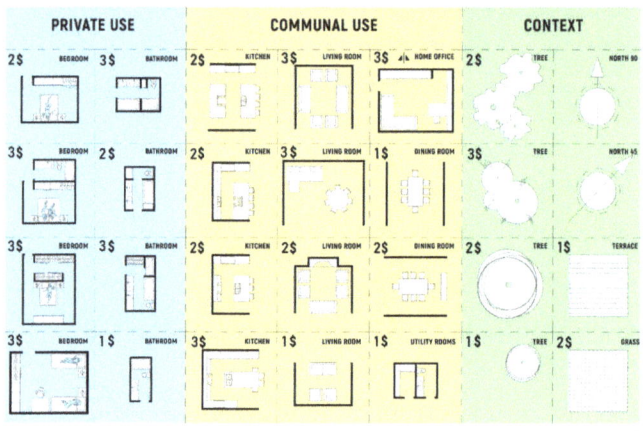

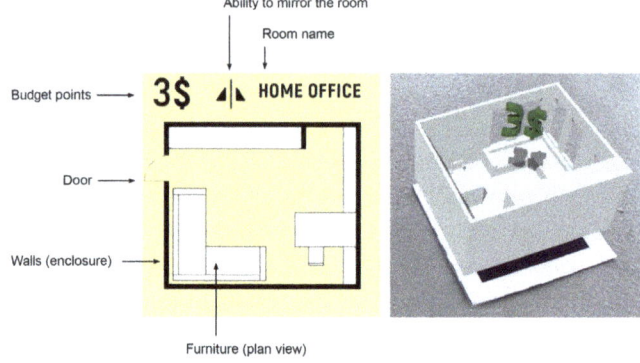

3   *ARchitect* tabletop game
4   *ARchitect* room cards
5   Room card anatomy; 3D image created using Fologram

a game and improving their design intuition through the iterative process of trial and error (Mytcul 2021).

Contents
Players must download and print out the tabletop portion

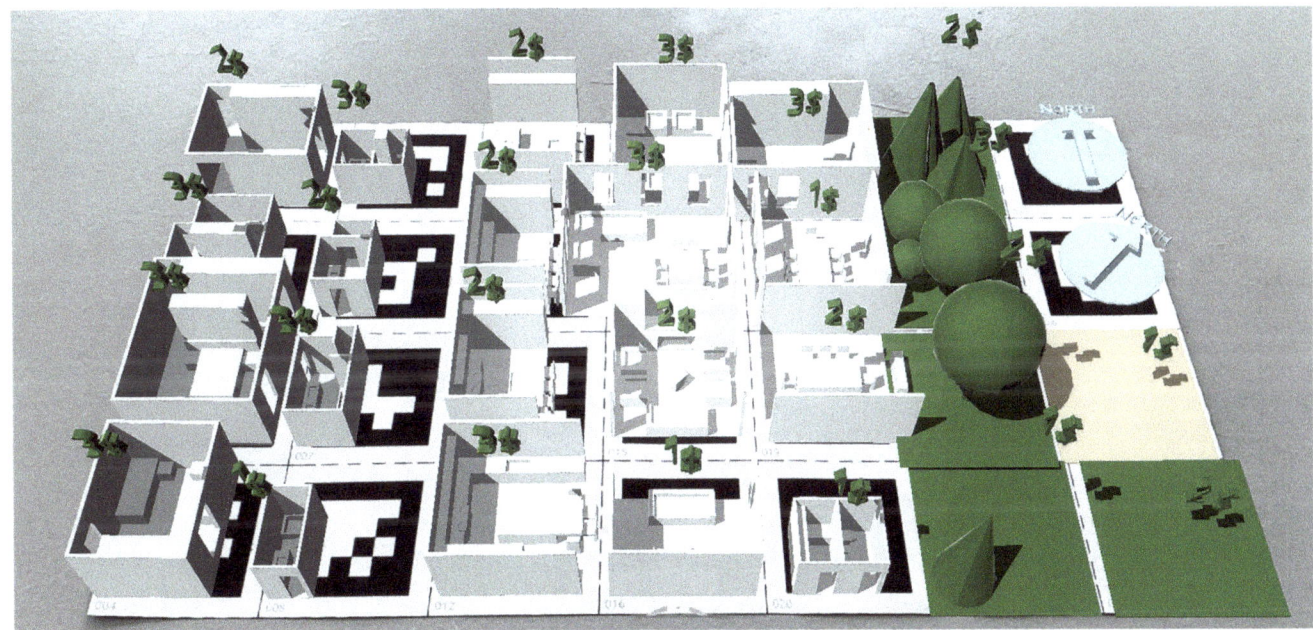

of the game from the *ARchitect* platform and install the *ARchitect* application, allowing for a mixed-reality experience (Figure 3). The game consists of three levels: tutorials, challenges, and the sandbox. The last level of the game, the sandbox, implies that the game can be used for actual house design and real-world implementation. Each level utilizes "room" playing cards and a series of playing boards. The first part of the game involves a set of room and context cards (Figure 4). The game library employs standardized room layouts that can be updated and expanded in the future, responding to each user's unique context. Every playing card has a name, budget value, and additional features (such as a "mirror"), coupled with a 2D representation of the conventional architectural plan view with walls, doors, and furniture; together, these are transformed into 3D by the *ARchitect* application (Figure 5). Fologram was used as a substitution for the AR experience (Jahn et al. 2021). Each playing card and board has an ArUco code that is programmed to transform a 2D drawing into a 3D digital model.

Tutorials

To design a house, one needs to understand basic design principles and interdependencies. To co-design in a game environment, players need constant simplified feedback in response to their actions. The game begins with a tutorial intended to familiarize players with the AR technology and introduce the game mechanics. Players begin with architectural vocabulary and gradually find themselves wrestling with the limitations architects often deal with during the real-world design process, such as relationships between public and private spaces, context, and budget. The tutorials also introduce the game contents,

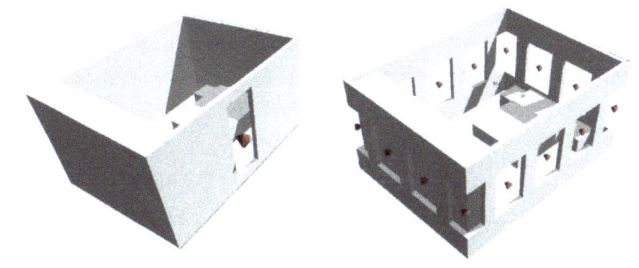

6 View through the Fologram AR application; 3D image created using Fologram

7 Feedback for budget points summary; 3D image created using Fologram

8 Feedback for juxtapositions; 3D image created using Fologram

9 Window addition feature; image created using Fologram

switching between the physical and AR environments (Figure 6) and offering color-coded feedback in the application. Additionally, the tutorials feature a color-coded grid base that guides players through the rules of room aggregation and explains how to interpret the game's feedback. The game connects the notion of budget limitations to room juxtaposition. For example, a player may need to combine

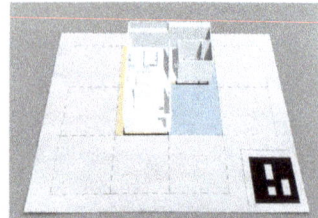

10

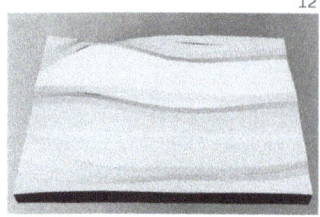

11

12

13

two rooms for communal use so as not to exceed five budget points (Figure 7).

The game feedback helps with evaluating the juxtaposition of rooms as correct or incorrect, as represented by a green or red glow (Figure 8). This allows the player to learn how to interact with room cards by moving and rotating them. Other tutorials explore aggregation of context and communal use cards, following the same logic and coupled with feedback. The game also gives the player the option to control window placement and quantity through simplified Boolean operations implemented in the game features (Figure 9).

Other tutorials are devoted to learning and visualizing the implications of orientation. This is achieved by changing the north card from the conventional position (with north at 90°) to north at 45° (Figure 10), and exploring the influence of time on interior qualities (Figure 11) by simulating the sun and visualizing shadows. Additionally, by scaling the house, players can experience their designs from different viewpoints ranging from the aerial view perspective to the first-person view. This helps with understanding the concept of scale and engages the user in switching among various scales.

Challenge Set

Challenge set objectives are designed to allow users to revisit concepts learned in tutorials and practice design problem solving through role playing. Context playboards with preset design inputs become playgrounds with different levels of difficulty, simulating the experience of an architect who needs to satisfy their clients during the design process (Figures 12-13).

At the beginning of the challenge level, players randomly choose their roles; one becomes an architect and the other a client. The latter takes a random family card that lists specific requirements. By choosing options, reaching a consensus, and making informed decisions, players simulate architect-customer communication and iterate architectural concepts from the tutorial set, but within a more advanced system of constraints. For example, a customer may receive a family card that lists a vampire family of three, two adults and one child. The vampire family requires no daylight in the bedrooms and no sunset view in the living/dining room. Additionally, they have "likes" that offer additional bonus points to a finished house, such as a courtyard hidden from outside view. Each card also has budget constraints. For example, each aggregated house for the vampire family should not exceed 20 budget points.

During each round, the architect's goal is to help the client formulate their needs and preferences and provide them with the best design solution out of three house design options. The design for every house option should be executed in close collaboration with the client, with the architect responding to the family's needs by proposing and assessing different design solutions through communication and AR evaluation. For instance, the architect might give a family member several bedroom choices, depending on whether the bedroom will accommodate adults or children. The AR application is also necessary if the architect is to evaluate their own and the family's choices and provide informed decisions that will satisfy the family card requirements.

After every house option, the client takes a drama card from the pile. The goal of the drama card is to shift the narrative in a particular direction that forces the architect to provide an alternative solution. Drama cards may have positive or negative connotations, such as: "Your family just inherited five budget points. Use them wisely!" One drama card requires that the players swap roles

10 Influence of orientation and context on the same house with north at 90° and 45°

11 Influence of orientation, context, and time (from left to right, 7am, 12pm, and 5pm) on the same house, with north at 90°

12 Challenge playboard and its transformation in AR (image created using Fologram)

13 Topography playboard and its transformation in AR (3D image created using Fologram)

14 Game mechanics of the challenge set

(Figure 14), allowing players to experience both roles and learn the same concepts from different angles and sets of responsibilities. The context playing boards allow participants to experience different site conditions, such as those with existing trees, rooms, and topographies.

The Sandbox

ARchitect endeavors to become a new participatory platform for house design democratization (Figure 15). The sandbox gameplay represents this effort as an avenue for future development. Players can find actual sites via geomapping websites like Google Earth, convert them into playing boards, and print them out. By following the same logic for design and self-evaluation (through the AR application), players can iteratively design houses, convert their designs into 2D drawings, and use the aggregated models to 3D print their house designs.

The learning by playing approach has immense potential to open up architectural education to the general public. The *ARchitect* platform will contribute to creating a generation of collective design effort, resulting in projects that architects could not produce alone. Players will be able to share their design ideas through the resulting open-source architectural culture, leading to user empowerment, design democratization, and the proliferation of design literacy (Figure 16).

Playtesting

Several in-person playtest sessions were conducted as proof-of-concept. These took the form of tabletop mixed-reality experiences with non-architect players, after which each completed a questionnaire. Players 1 and 2 tested the game three times, while others were exposed to the game only once. All testers (except for one) had no architectural background and explored *ARchitect* via the Fologram application for 1 to 1.5 hours per session.

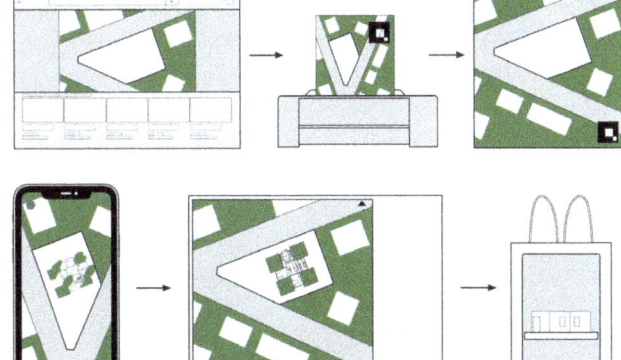

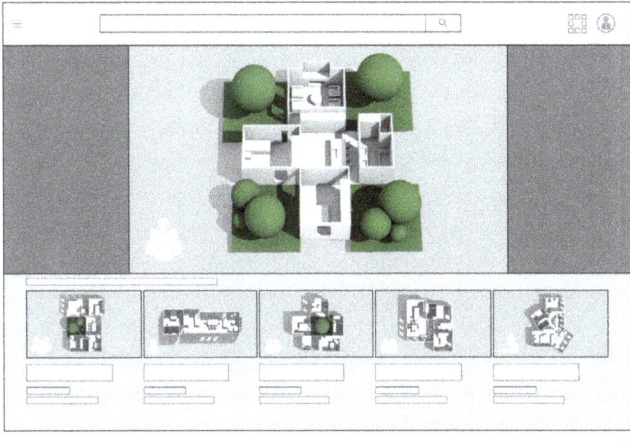

15 Sequence of work using the ARchitect platform.

16 ARchitect platform.

The first session with two players showed the importance of game structure. Instead of starting with the tutorials, Players 1 and 2 began the game with the house challenge cards. They instantly began to place the room cards on top of boards and used Fologram for AR evaluation. Also, each participant began playing separately. Figures 17 and 18 show the difference in room card usage. Player 1 placed cards on the playboard. In contrast, Player 2 used a desk to create a house.

During the first session, it was difficult for non-architect players to read the architectural plans on the room cards and understand room functions without any tags. This playtest session motivated the addition of color codes and tags to room cards, as well as several modifications to the tutorial objectives and guidance provided.

During the second session, the game was tested solely by an architect. After completing the tutorials and challenges (which they found extremely easy), they began to play the game at the sandbox level (Figure 19). In contrast to the previous players, the architect revealed new game potential by creating various room juxtapositions. They flipped all of the cards simultaneously and combined them in a linear composition using the AR application (Figure 20). Also, the player utilized the AR application in a more dynamic manner; they kept checking the results from the aerial to interior views to better understand the relationship between rooms and context. Based on these observations, a new adjustable scale for the AR projection feature was added to the game to allow players to experience their designed spaces from different views and perspectives (Figure 21).

The next session with Players 1 and 2 inspired another idea for enhancing player engagement that was reinforced by HGTV shows: the roleplaying component of the game (Figure 22). The whole experience of architect-client roleplaying offered a fun aspect to the game. It added a new variable to the gameplay and enhanced play scenarios, allowing players to collaborate in the design process by discussing the benefits and drawbacks of design proposals.

Results and Discussion

The preliminary results from the *ARchitect* game showed great potential for future development. The author believes that the democratization of design education has immense potential to increase non-architects' involvement in enhancing their community environment. Thus, it is important to maintain game accessibility and free downloading from the *ARchitect* open-source platform to eliminate barriers to spreading basic architectural knowledge. The *ARchitect* concept challenges the conventional approach to design participation and eliminates the architect as the primary part for translator of an end user's vision. The absence of the "win state," as well as absence of a single correct design solution unleashes players creativity and provides a new approach to house design.

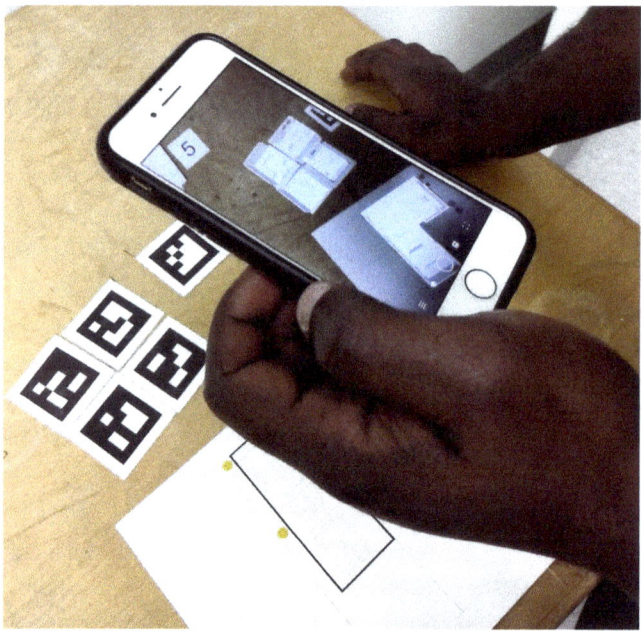

17 First playtest experiment with non-architect users

18 First playtest experiment with non-architect users

In contrast to the Foldit game, where the main goal is contributing to protein folding research through crowd-sourced puzzle-solving, the *ARchitect* game primarily educates and promotes design democratization through the gaming framework.

It is expected that the game will prompt a much-needed conversation between experts and novices, forcing a rethinking of architectural solutions for single-family housing in response to the standardization of existing dwelling design. Instead of the de-professionalization of architecture, user-created content will infiltrate and enhance design variety and architectural participation, while allowing the act of design to remain in the hands of its creators (Sanchez 2021). The *ARchitect* platform intends to prioritize user-generated content in ways similar to platforms such as YouTube, by using it to provide visual content for game challenges and tutorials and collecting it in a gallery of players' creations. This will serve to empower the general public to engage with one another and help spread architectural literacy among less-privileged communities.

The mixed reality game interface will allow users to interact with *ARchitect* on several perception levels. By exploiting the idea of a multisensory game, the general childhood learning experience is referenced, where different channels for interaction (e.g., tangible, audial, visual, etc.) create comprehensive output. Players learn through entertaining experiences that incorporate analyses of qualitative and quantitative performances. End users learn how to consider environmental forces as beneficial constraint-makers and use them as tools to enhance the parameters of the design's performance. The economic system for the game is another constraint-maker, requiring that players not only understand design in relation to budget limitations, but also evaluate and compare design options in terms of qualitative parameters (Mytcul 2021). In the game, quantitative feedback is delivered visually through a qualitative context. For instance, users determine their preferred spatial quality based on a quantitative analysis of natural light. However, the quality of the designed house may be subjective, as we see with the vampire family discussed above.

During playtesting sessions, non-architect users showed unique collaborative engagement, as participants learned design ideas and interdependencies not only through the narrative but also through role-playing. This experience prompted the research endeavor to explore other aspects of player-on-player interaction and the ways a game might enhance and support the learning experience through communication and informed decision-making. This served to call into question more traditional ways of learning, instead promoting AR technology for new educational engagements. *ARchitect* requires further development of the game levels and additional playtesting experiments with a larger body of non-architect users in order to depict more comprehensive conclusions and discover controversial topics for discussion. In future iterations, a new project participant role of the "City" with more code-related constraints will be included, providing additional didactic guidance and allowing more individuals to participate.

19 Second playtest experiment with an architect, Sandbox mode; images created using Fologram

20 Second playtest experiment with an architect, Linear house; images created using Fologram

21 Second playtest experiment with an architect, Scale exploration; images created using Fologram

Substitution of Fologram for the original *ARchitect* application is also needed to eliminate use of the ArUco code. Development of an algorithm for building more than one-story houses, as well as the addition of new elements such as roofs, balconies, and porches, are the next stage in game development.

CONCLUSION

The notion of democratizing architectural design has been supported by architects since the early 1970s. However, due to its interdisciplinary nature, the concept of an

22 Third playtest experiment with non-architect users

accessible and engaging tool for house customization is still evolving. Design democratization implies the empowerment of non-architects to become decisionmakers and active participants in different stages of the design. It offers the public the opportunity to individualize and shape their surrounding environment, thus rethinking conventional market monopolies and exploring alternative personal and public values (Sanchez 2021).

AR technology provides a diversified learning experience that can immerse players in a familiar environment and open new features in single-family house design by switching between different scales and views. The AR medium is a low-threshold mechanism for house modeling that provides a link between analog and digital architectural representations. The game allows players to experience design using tangible cards with 2D plan views in a 3D assembly, transforming design solutions into 3D printed models and receiving aggregated house plans. This concept is based on a conventional iterative architectural workflow.

The goal of *ARchitect* is to spread design democratization through an accessible and playful educative experience, bridging the gap between experts and non-experts. The game should be played by friends and family members, providing a new option on game nights.

Additionally, *ARchitect* invests in common architectural knowledge as a body of thought that can also be utilized in architectural schools, allowing students to explore their workflow by creating other modules and topologies for single-family housing. Therefore, the notion of design democratization extends not only to the public realm, but also to institutions of higher learning. By democratizing architectural education through a playful and engaging experience, *ARchitect* opens up new sources of architectural thought that will allow for collective intelligence to move forward and invite nonprofessionals "to the table."

ACKNOWLEDGEMENTS

I would like to thank Nicholas Bruscia, Assistant Professor in the University at Buffalo Department of Architecture, for helping and guiding me through this intense research process. I would also like to express my gratitude to my two committee members, Cody Mejeur, Visiting Assistant Professor in the Department of Media Study, and Mark Shepard, Associate Professor in the Department of Architecture and Media Study.

Additionally, I offer many thanks to the Fulbright program and University at Buffalo for making it possible to obtain my Master of Architecture degree in the US. Moreover, I would like to thank each of the following individuals whose kind assistance made it possible for me to complete this research: Famous Clark, Tim Georger, and Albert Chao for being open-minded game testers and helping me to discover simple but beautiful ideas. Finally, I would like to thank Jiries Alali, for his unwavering support, criticality, and willingness to help.

REFERENCES

Cooper, Seth, Adrien Treuille, Janos Barbero, Andrew Leaver-Fay, Kathleen Tuite, Firas Khatib, Alex Cho Snyder, Michael Beenen, David Salesin, David Baker, and Zoran Popović. 2010. "The challenge of designing scientific discovery games." In *Proceedings of the Fifth International Conference on the Foundations of Digital Games (FDG '10)*. New York: Association for Computing Machinery. 40–47. https://doi.org/10.1145/1822348.1822354.

Friedman, Yona. 1975. *Toward a scientific architecture / Yona Friedman*, translated by Cynthia Lang. Cambridge, Mass.: MIT Press.

Gee, James Paul. 2007. *Good video games + good learning: collected essays on video games, learning, and literacy*. New York: Peter Lang Publishing Inc.

Jahn, Gwyllim, Cameron Newnham, and Nick Van den Berg. *Fologram*. 2020.3.21. Pty Ltd. Fologram. IOS. 2020.

Kolarevic, Branko, and Duarte José Pinto. 2019. "From Massive to Mass customization and design democratization." In *Mass Customization and Design Democratization*, edited by B. Kolarevic and J. Pinto Duarte. Milton Park: Routledge.

Mytcul, Anna. 2021. "Gamification of Design Experience: An Accessible Tool for Architectural Design Democratization." Master's thesis, University at Buffalo, State University of New York. ProQuest LLC.

Negroponte, Nicholas. 1975. *Soft Architecture Machines*. Cambridge, Mass.: MIT Press.

Piller, Frank, Petra Schubert, Michael Koch, and Kathrin Möslein. 2005. "Overcoming Mass Confusion: Collaborative Customer Co-Design in Online Communities." *Journal of Computer-Mediated Communication* 10 (4). https://doi.org/10.1111/j.1083-6101.2005.tb00271.x.

Sanchez, Jose. 2021. *Architecture for the Commons: Participatory Systems in the Age of Platforms*. London: Routledge, Taylor & Francis Group.

Sanoff, Henry. 1979. *Design Games*. (Experimental ed.). Burlington: W. Kaufmann.

Anna Mytcul is a researcher in architecture gamification, educator, and practicing designer. She is a recipient of the Erasmus + at TU Darmstadt award, as well as a Fulbright scholar at SUNY Buffalo, and grantee of Edmund S. Muskie Internship Program. Her work is concentrated at the intersection of learning games and architecture democratization, with a focus on education through play. Her thesis work entitled "Gamification of the Design Experience: An Accessible Tool for Architectural Design Democratization" was awarded the ARCC/King Student Medal for Excellence in Architectural + Environmental Research.

Topological Networks Using a Sequential Method

Data Structure Simplification for Interactive Design

Zidong Liu
Bartlett School of Architecture
University College London

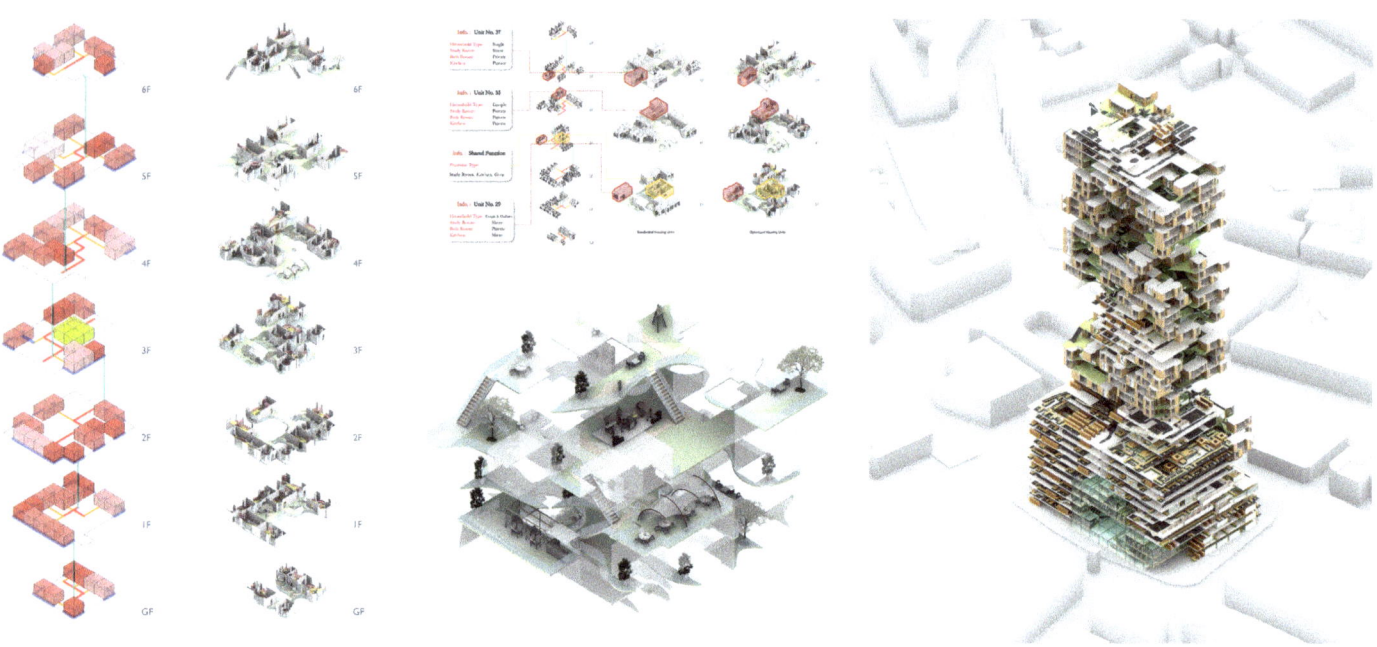

1   Design result of Living Cluster and Community Structure

ABSTRACT

The paper shares preliminary results of a novel sequential method to expand existing topology-based generative design. The approach is applied to building an interactive community design system based on a mobile interface. In the process of building an interactive design system, one of the core problems is to harness the complex topological network formed by user demands. After decades of graph theory research in architecture, a consensus on self-organized complex networks has emerged. However, how to convert input complex topological data into spatial layouts in generative designs is still a difficult problem worth exploring. The paper proposes a way to simplify the problem: in some cases, the spatial network of buildings can be approximated as a collection of sequences based on circulation analysis. In the process of network serialization, the personalized user demands are transformed into activity patterns and further into serial spaces. This network operation gives architects more room to play with their work. Rather than just designing an algorithm that directly translates users' demands into shape, architects can be more actively involved in organizing spatial networks by setting up a catalogue of activity patterns of the residents, thus contributing to a certain balance of top-down order and bottom-up richness in the project. The research on data serialization lays a solid foundation for the future exploration of Recurrent Neural Network (RNN) applied to generative design.

INTRODUCTION

In interactive generative design, a common problem is harnessing the topological network formed by user demand information. After data collection, user information is transformed into two parts: topological and geometrical (Damski and John 1997). If designers set too many parameters in collecting data, they face a complex network that is difficult to be directly geometrized. Therefore, the method to analyze and reorganize the topological network for data distribution is significant.

The research aims to explore a sequential approach to organizing the topological network, thus reducing the complexity of data distribution and expanding the possibilities of generative design. The prototype system presented in this paper is based on a mobile interface, providing an interactive platform for people to design their houses and community. The traditional top-down model of community planning relies heavily on the subjectivity of architects, which leads to waste of land resources and lack of community vitality, exacerbating the housing crisis (Wong 2010). This research takes the view that architects should not only be the designers of space but also the developers of interactive platforms that facilitate inclusive design.

Topological Design

The research on architectural topological design has a decades-long history. In its early development period, the approach is based on simple topological rules, such as floorplan automation (Levin 1964), cellular automation (Wolfram 1983), and shape grammars (Stiny 2011). The booming development of network science provides an increasingly rich set of tools for spatial network analysis (Barabasi 2016; Turnbull et al. 2018). These studies strongly support the development of architecture layout automation and gradually extend its boundaries from rectangular layout to non-rectangular layout (Marson and Musse 2010; Koenig et al. 2012, 2020; Shekhawat and Duarte 2018; Wang et al. 2018, 2020). However, most current topology-based generative designs directly transform topological networks into geometric forms, and lack secondary processing of spatial networks. When faced with overly complex networks, it is difficult for the architect to control the spatial form. Among the few studies on this issue, the computing efficiency of different topologies has been discussed (Koenig and Knecht 2014).

Sequential Method

The sequential method means interpreting the topological network from the perspective of activity sequence (Figure 2). In some projects with distinct circulations the spatial network can be approximated as the interweaving and

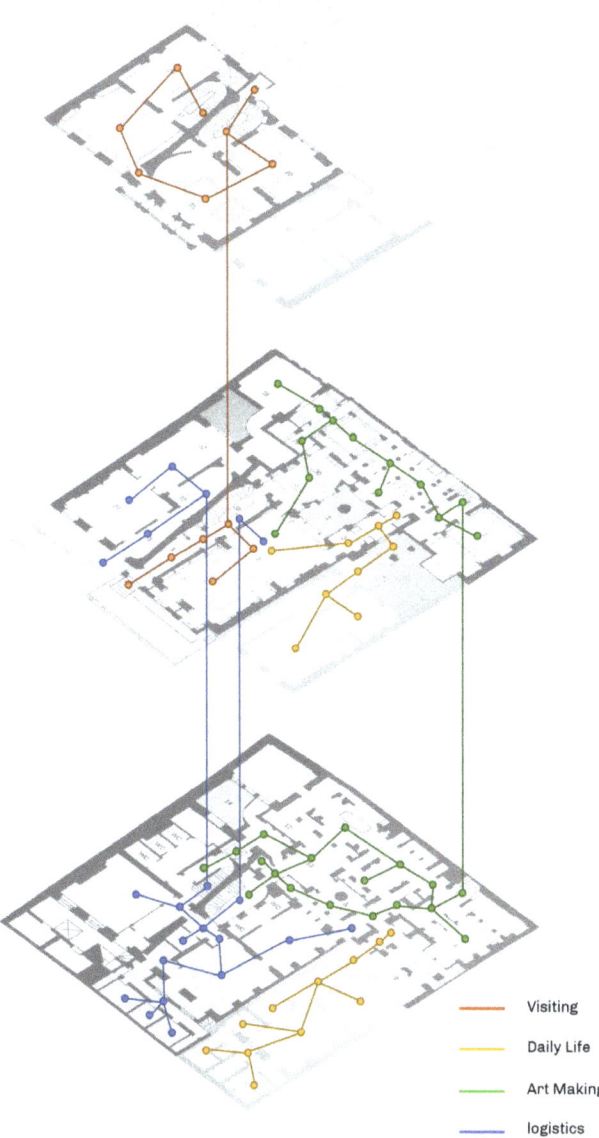

2   Sequential interpretation of complex network

3   Sequence analysis of topological network of Soane Museum

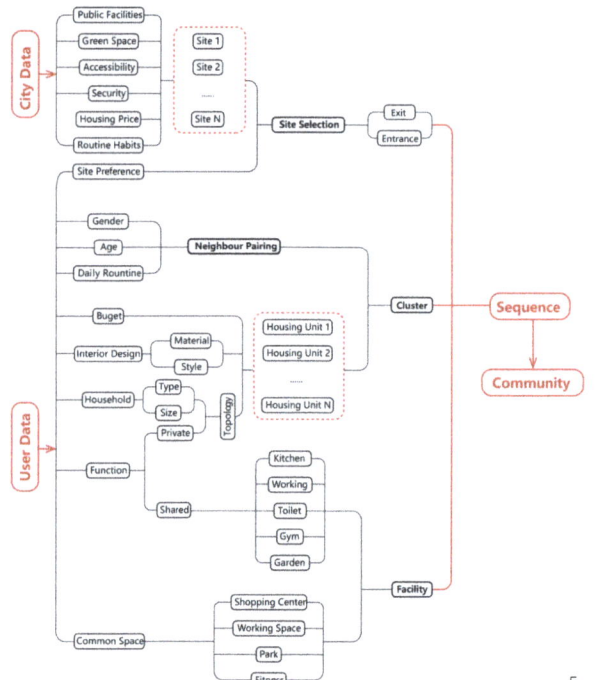

overlapping result of multiple activity sequences. The sequence acts as an intermediate hierarchy to cluster nodes and form the system, where the embedding vectors of nodes are close to each other mathematically.

The idea of activity sequence can be found in centuries of architectural practice, such as Vasari Corridor (1565), Soane Museum (1813) and Miller House (1953). The Soane Museum is a classic case of using sequential approach to organize floor layout. It serves multiple functions simultaneously within a narrow interior space: family life, party, teaching, exhibition, and so on. Besides, due to the historical background, people with different identities had their own specialized paths, such as the specific logistics circulation for servants. Sir John Soane used a narrative approach to organize these various life patterns (Kyriafini 2007). The topological analysis of space reveals that the structure of network is composed of interwoven sequences (Figure 3).

The sequential approach has the potential to be applied to community design as well. Today's community design is faced with the similar design problems as the Soane Museum: complex functions, interwoven circulations, customized living patterns, and adaptability for flexible use.

METHODOLOGY

Overall Method

The sequential approach can be applied across scales, from urban to architectural scales. In the paper, the sequential approach is demonstrated with a community design experiment aiming at housing customization and community

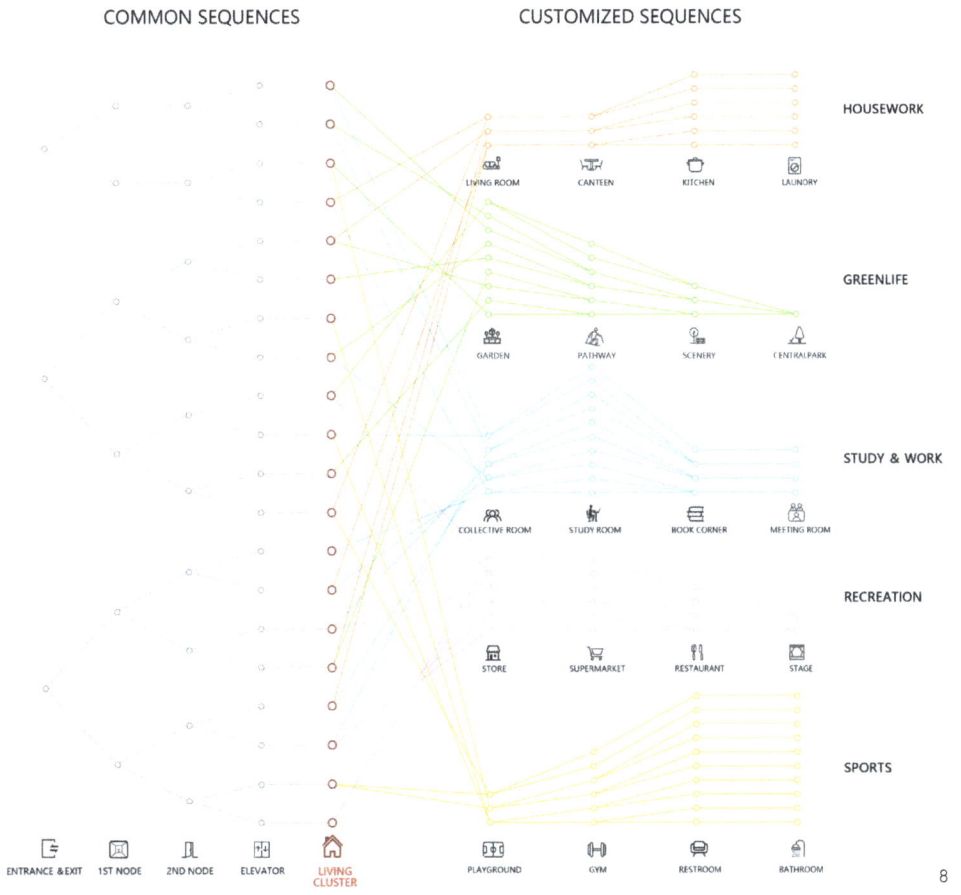

4 Overall logic structure of the generative design system

5 Data translation for community design part

6 Grouping residents into Living Cluster by K-means algorithm

7 The paper takes ten users as an example to illustrate the algorithm logic of sequential method

8 Transforming 200 hypothetical users' demands into sequential network

self-organization (Figure 4). An interactive platform based on a mobile interface is built on the front end for user data collection and design result preview. The process of transforming topological and geometric information into community design is carried out in the back end. The entire generative design system consists of three parts: site selection, community design, and housing unit design. This paper focuses on sharing the results of community design part where the sequential approach is mainly used.

User Data Collection

User demands includes dwelling characteristics, neighborhood, surrounding environment and facilities, social identity and so on (Roy et al. 2018). To quantify the input variable, these influencing factors are reclassified into three categories:

- Housing unit design: room size, function types
- Community design: shared rooms, community facilities
- Site decision: housing prices, distance to work, urban facilities

This experimental design is based on data from 200 simulated users. To maximize system complexity and verify the effectiveness of the sequential approach, each virtual user data is completely randomized. After data collection, these mixed-up user demands would be translated and redirected to the different parts of the generative design system (Figure 5).

Clusering Residents

The data about the community design is the most complicated one, because the type and number of shared rooms and community facilities for each user are uncertain. Due to the limitation of computing power, it is impossible to input 200 user data simultaneously for generative design, and structural hierarchies are necessary (Koenig and Schneider 2012). Therefore, the virtual 200 users are classified into twenty living clusters based on matching degree, which is evaluated by three parameters: house similarity, demands for sharing rooms, and community facilities.

The pairing process is based on K-means clustering algorithm (MacQueen 1967) illustrated in Figure 6. In this algorithm, each resident is regarded as a piece of three-dimensional data, mathematically equal to a point with three coordinates. The weighted approach is used to distinguish the importance of different attributes.

Sequencing Topological Network

People living in the same cluster share the similar living

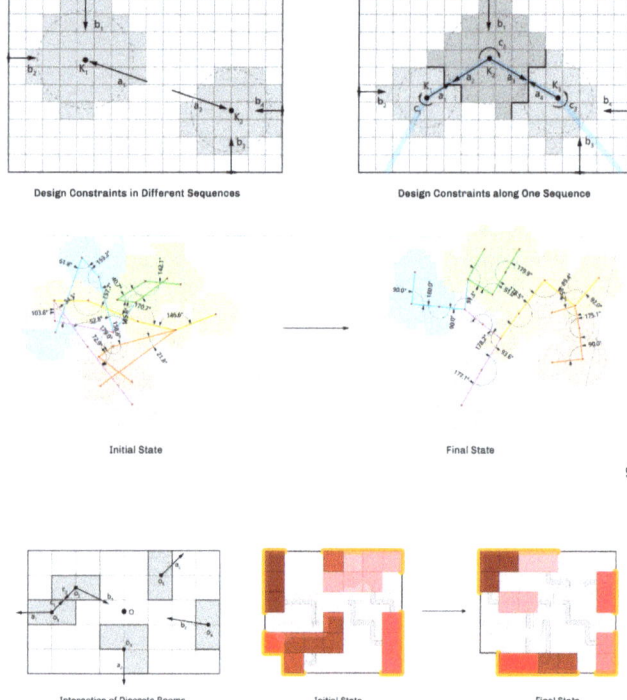

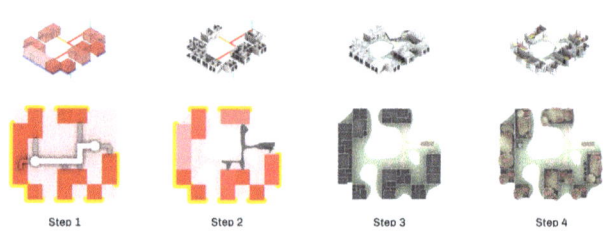

a4, a6, a9 are considered as a Living Cluster, forming a sequence with offices and study rooms. The remaining five users and facilities form another sequence. It should be noted that this classification is only an approximation. For example, we ignore the need of a3 for offices. So we need to provide a common sequence connected to the public transport network, which facilitates the residents to access other places.

In order to strike a balance between diversity and simplicity, the following rules are set: a cluster has a maximum of two personalized activity sequences and a common sequence. Based on these rules, 20 clusters generate 52 activity sequences of five categories in total, including 35 and 20 common sequences. These sequences merge and separate, generating a total of 106 supporting facilities (Figure 8).

Physically Based Design Constraint

We use the physics engine to set the design constraints translated from the topological network. This approach is widely used in generative design (Scott and Donald 1999, 2002; Koenig et al. 2012; Daniel et al. 2013). The advantage of this analogy is that the strength of connection can be adjusted freely and multiple forces can be applied to one object without restriction.

Elements are restricted to move within the community boundary by fb. They have repulsive forces fa against each other to maintain their own space. There is a strong attraction fc between adjacent elements in a sequence to ensure that they stick close to each other. In addition, adjacent elements in a sequence have rotational forces f_d to keep the angle between them at 0 or 90 degrees, thus keeping the layout structure orthogonal (Figure 8).

In the process of generating the internal layout of a living cluster, highly matched living units will have an attraction in the middle, otherwise, they will be mutually exclusive. Paths automatically follow the change of cells' location (Figure 9).

Optimizing Design Result

Physics engine can only roughly simulate the relationship between elements because the results are very easy to be affected by the initial position of elements. In order to overcome this uncertainty, genetic algorithm is used to optimize the result based on specific performance criteria P. Here, we optimize the result by minimizing l_n and maximizing b_n (Figure 11). The parameter f and g is used for weighting. By this way, more compact floor plans will be screened out. This process is based on an interactive platform Biomorpher (Harding and Brandt-Olsen 2018).

9 Physics engine simulation for Community Structure organization

10 Physics engine simulation for units locating inside Living Cluster

12 Four steps of detail automation in Living Cluster generation process

patterns and infrastructure demands. For example, they might take a walk in the community park after work, and then exercise at the community gym before returning home. We regard such a life pattern as a type of activity sequence.

The algorithmic logic of the sequential approach is demonstrated here using the example of ten users (Figure 7). Suppose there are ten users from a1 to a10 who have personalized needs for four facilities, which form a complex network that is difficult to geometrize. But after clustering, we find that a1, a2, a4, a6, a9 have similar needs for offices and study rooms, while the remaining five users have needs for shared kitchens and living rooms. Thus a1, a2,

$$P = f * \sum_{n=0}^{\infty} l_n + g * \sum_{n=0}^{\infty} b_n \qquad (f<0, g>0)$$

l_n: the length of path b_n: the length of touching boundary of units

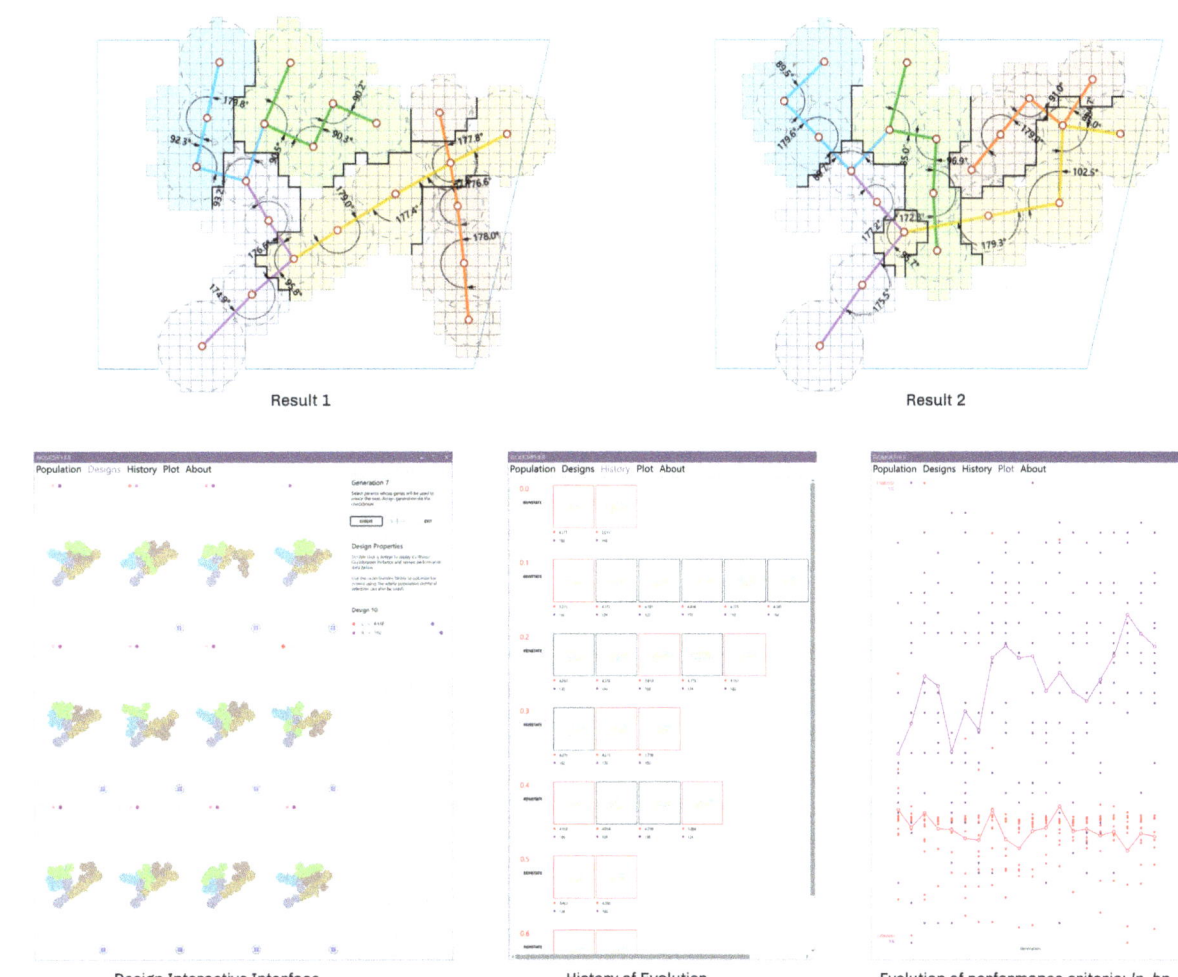

11 Using genetic algorithm to yield a more compact floor plan

Geometrizing Details

After determining the location and boundary of elements, the system will further automatically complete the detailed design of the public space and living units (Figure 12).

RESULTS AND DISCUSSION

Based on 200 hypothetical users' data, the prototype system presented in this paper obtains a complete community design result with diverse architectural space emerging (Figure 13). In this generative design process, the sequential approach, as a novel design methodology to organize the topological network, plays a key role in the community design step. This approach creatively translates the individual demands into different life patterns and corresponds them to specific sequences of activities. Consequently, the complex network is transformed into interwoven sequences, successfully reducing the complexity of the problem.

At present, the system is still in the preliminary explorative development. The current process of serializing networks is highly subjective. RNN will be further explored in the future due to its huge potential for processing sequential information. In the process of simulating the sequential network, we have found that the traditional physics engine simulation requires considerable refinement due to the instability of the results.

CONCLUSION

This paper verifies the effectiveness of a novel sequential approach in organizing topological networks in generative design process. The importance of sequential approach is that it provides a novel and effective paradigm for designing complex networks. Architects and planners have developed various methods to quantitatively analyze existing complex networks, but how to create complexity remains a daunting task. Design is always faced with the challenge of neither being monotonous or chaotic. Sequence is a new perspective for people to control complexity. As a bridge, the sequence powerfully connects the mathematical topological network with the vitality of individual human patterns of life.

ACKNOWLEDGEMENTS

The work in this article is part of "Housingprime," a research project of RC11, B-Pro, the Bartlett, UCL 2020 led by Philippe Morel and Paul Poinet, with members including Zidong Liu, Shiyuan Huang, Yi Li and Huiyu Pan. I would like to thank our tutors Philippe and Paul who constantly gave us inspiration and guidance. I would also like to thank my team members for their company and hard work. Last but not least, I would like to thank Professor Ruairi Glynn for his many hours of close guidance. This paper would not have been possible without his help.

11 Using genetic algorithm to get a more compact floor plan

REFERENCES

Arvin, Scott A., and Donald H. House. 1999. "Making Designs Come Alive: Using Physically Based Modeling Techniques in Space Layout Planning." In *Computers in Building: Proceedings of the CAADfutures'99 Conference*, edited by Godfried Augenbroe, Charles Eastman. Boston, MA: Springer. 245–62.

Arvin, Scott A. and Donald H. House. 2002. "Modeling Architectural Design Objectives in Physically Based Space Planning." *Automation in Construction* 11 (2): 213–25.

Barabasi, Albert-Laszlo. 2016. *Network Science*. Cambridge, England: Cambridge University Press.

Damski, José C., and John S. Gero. 1997. "An Evolutionary Approach to Generating Constraint-Based Space Layout Topologies." In *CAAD Futures 1997*. Dordrecht: Springer Netherlands. 855–64.

Gavrilov, Egor, Sven Schneider, Martin Dennemark, and Reinhard Koenig. 2020. "Computer-Aided Approach to Public Buildings Floor Plan Generation. Magnetizing Floor Plan Generator." *Procedia Manufacturing* 44: 132–39.

Harding, John, and Cecilie Brandt-Olsen. 2018. "Biomorpher: Interactive Evolution for Parametric Design." *International Journal of Architectural Computing* 16 (2): 144–63.

Koenig, Reinhard, and Katja Knecht. 2014. "Comparing Two Evolutionary Algorithm Based Methods for Layout Generation: Dense Packing versus Subdivision." *Artificial Intelligence for Engineering Design, Analysis and Manufacturing, AI EDAM* 28 (3): 285–99.

Koenig, Reinhard, and Sven Schneider. 2012. "Hierarchical Structuring of Layout Problems in an Interactive Evolutionary Layout System." *Artificial Intelligence for Engineering Design, Analysis and Manufacturing, AI EDAM* 26 (2): 129–42.

Kyriafini, Magdalini. 2007. "Narrative and Exploration in Small Museums: The Wallace Collection and the Soane Museum." PhD diss., University College London.

Levin, Peter Hirsch. 1964. "Use of graphs to decide the optimum layout of buildings." *The Architects' Journal* 7: 809-815.

MacQueen, James. 1967. "Some Methods for Classification and Analysis of Multivariate Observations." *Proceedings of the Fifth Berkeley Symposium on Mathematical Statistics and Probability* 1 (14): 281-297.

Marson, Fernando, and Soraia Raupp Musse. 2010. "Automatic Real-Time Generation of Floor Plans Based on Squarified Treemaps Algorithm." *International Journal of Computer Games Technology* 2010: 1–10.

Piker, Daniel. 2013. "Kangaroo: Form Finding with Computational Physics." *Architectural Design* 83 (2): 136–37.

Roy, Noémie, Roxanne Dubé, Carole Després, Adriana Freitas, and France Légaré. 2018. "Choosing between Staying at Home or Moving: A Systematic Review of Factors Influencing Housing Decisions among Frail Older Adults." *PloS One* 13 (1): e0189266.

Shekhawat, Krishnendra, and José P. Duarte. 2018. "Introduction to Generic Rectangular Floor Plans." *Artificial Intelligence for Engineering Design, Analysis and Manufacturing, AI EDAM* 32 (3): 331–50.

Stiny, George. 2011. "What Rule(s) Should I Use?" *Nexus Network Journal* 13 (1): 15-47.

Turnbull, Laura, Marc-Thorsten Hütt, Andreas A. Ioannides, Stuart Kininmonth, Ronald Poeppl, Klement Tockner, Louise J. Bracken, et al. 2018. "Connectivity and Complex Systems: Learning from a Multi-Disciplinary Perspective." *Applied Network Science* 3 (1): 11.

Wang, Xiao-Yu, Yin Yang, and Kang Zhang. 2018. "Customization and Generation of Floor Plans Based on Graph Transformations." *Automation in Construction* 94: 405–16.

Wang, Xiao-Yu, and Kang Zhang. 2020. "Generating Layout Designs from High-Level Specifications." *Automation in Construction* 119 (November): 103288.

Wolfram, Stephen. 1983. "Statistical Mechanics of Cellular Automata." *Reviews of Modern Physics* 55 (3): 601-644.

Wong, Joseph Francis. 2010. "Factors Affecting Open Building Implementation in High Density Mass Housing Design in Hong Kong." *Habitat International* 34 (2): 174–82.

IMAGE CREDITS

All drawings and images by the authors.

Zidong Liu holds a Master of Architecture degree from the University College London and a Bachelor of Architecture degree from the Southeast University. His research interests focus on spatial network-based generative design, urban analysis, and machine learning-based urban feature prediction.

Affective Prosthesis

Sensorial Interpretations of Covert Physiological Signals for Therapeutic Mediation

Katarina Richter-Lunn
Harvard University

Jose L. García
del Castillo y López
Harvard University

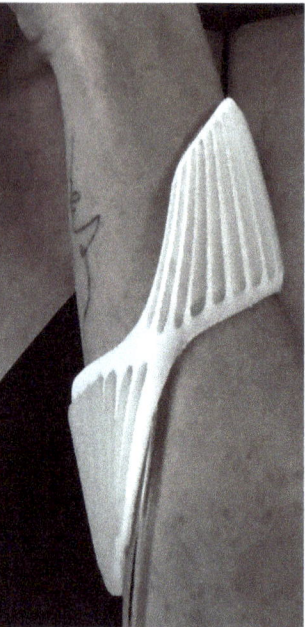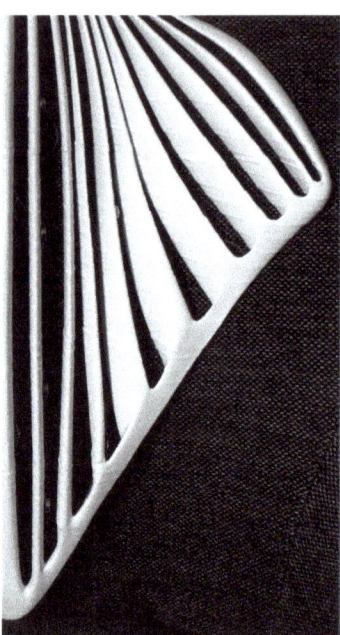

1 *Affective Prostheses*: A study of four intelligent, sensory-based augmentations to one's working micro-environment

ABSTRACT

As the demand for technologies that mediate the environment continues to rise, day-to-day activities have been increasingly overloaded with devices that collect personal signals, such as phones, watches, jewelry, and fitness trackers. Yet, despite the sensibility of these machines, little has been explored in decoding the highly informative signals collected by these devices to temper the physical environment. These signals have the potential to communicate one's cognitive state and, in turn, address signs of stress and anxiety. Embracing the open access to these technologies, this research seeks to question how covert physiological signals can be turned into perceived sensorial experiences to increase awareness of one's cognitive state and elicit positive affect through material interfaces. Acting not as a substitute for traditional therapies but as an alternative antidote, these sensorial interventions seek to process, analyze, and interpret physiological patterns, such as electrodermal activity and heart rate variability, to recognize signs of high and low emotional arousal and pair them with tactile, olfactory, auditory, and visual alterations in one's surrounding. It is predicted that through the repeated association of the actuated stimuli with specific physiological states, a certain conditioning can be evoked to subsequently promote an instinctual response to malleable matter. The results illustrate that the fabric of the environment can not only be empathetic to subconscious mood but also able to foster positive affect through psychophysiological adaptation.

INTRODUCTION

In the past decade, researchers in clinical psychology have shown the enormous impact the environment has on behavior, physiology, and mental health. Yet, in fields such as architecture, product design, and technology, designers have seldom engaged with the potential impact of their work in addressing mental wellbeing. As digital devices, artificial intelligence, and robotics become increasingly embedded into the fabric of the environment, the opportunities increase for these technologies to work with humans to develop more empathetic machines. The field of affective computing has rapidly grown over the past years to expand beyond the focus of software applications and move towards the physical world by leveraging wearables, social robotics, and programmable materials (Breazeal 1999; Farahi 2018; Gannon 2018; Tibbits 2017). These tangible interpretations of affect through machines have sparked the conversation on how this technology, perceptive of human emotion, can influence cognition and provoke mood change. Remarkable advancements have established feedback loops between humans and machines through the ability to detect signs of stress and anxiety in individuals and, in turn, respond through sensorial modalities to provide relief and promote relaxation (Costa 2017; Papadopoulou 2019; Alonso 2008).

While stress and anxiety are extraordinarily pervasive forms of mental strain, they are also highly treatable. However, forms of therapy are often associated with high monetary cost and social stigma, rendering them inaccessible or undesirable. With these disorders only becoming more prevalent with the current state of the world, accessible, affordable, and amiable treatments which can be seamlessly integrated into everyday environments are crucial in addressing the mental instabilities catalyzed and often aggravated by one's day-to-day surrounding. Many of these interpretations of the environment are unconscious and do not involve deliberate cognitive analyses while the physical body continue to react involuntarily to these perceptions.

By applying methods of traditional behavioral therapy in conjunction with physiological feedback and sensory stimulus, this research seeks to offer evidence of how human, machine, and sensorial interactions can be leveraged via affective computing to propose seamless and intuitive solutions for addressing mental health. This paper will first outline the domains of research by which this methodology is inspired: psychophysiology, sensorial responses, and affective computing. Then, Affective Prostheses is presented, a study through four interventions embedding specific sensory stimuli into environmental augmentations actuated by one's physiological signals. Lastly, an initial evaluation of the effects of these prototypes in altering one's affective state is presented and a discussion of the outlooks of this work and potential future research is offered.

RELATED WORK
Psychophysiology

Today it is commonly agreed upon that mind and body cannot be siloed in one's analyses of the world, just as there is no predetermined sequence to how the environment is interpreted. Not only can thoughts lead to physiological changes in the body, but physical cues inform the brain of emotion before it has fully cognitively appraised the situation (James 1890; Schachter 1964). The relationship between mind and body has long been studied in psychophysiology in further understanding how humans process the environment, and yet the field continues to advance at a significantly fast pace leading to new insight in areas such as stress, memory, emotion, behavioral medicine, language, and psychopathology (Cacioppo 2007). In addition to these developments in the field, portable, wireless, and wearables devices are becoming increasingly common in clinical research as well as used in everyday life (Picard 1997; Milstein 2020). These devices now not only have access to people's geographic location and social media platforms but in some cases to their heart rate, skin temperature, and electrodermal activity. Research has particularly grown around the use of these physiological biomarkers in identifying signs of stress and anxiety, using wearable devices such as the Empatica E4 wristband or the latest Sense Fitbit (Empatica 2021; Fitbit 2021; Zhao 2018; Bhoja 2020; Gjoreski 2016). These devices, along with clinical research around the relationship between specific patterns of heart rate variability (HRV) and electrodermal activity (EDA), has led to encouraging results around the reliability of these signals to accurately assess cognitive behavior (Appelhans 2006; Cacioppo 1990).

Sensorial Response

To suggest a possible methodology that explores sensory perception one must first look at each sense in isolation and the range of cognitive impact each of these senses hold. By no means might one suggest that the senses are limited to simply outputting one specific cognitive response to a stimulus; however, research suggests that specific sensory stimuli often evoke a similar response pattern in participants, implying various correlations between the senses and provocations. For instance, in the case of the olfactory bulb, it has been studied that low-level exposure to an aroma, such as peppermint oil, reveals effects on memory,

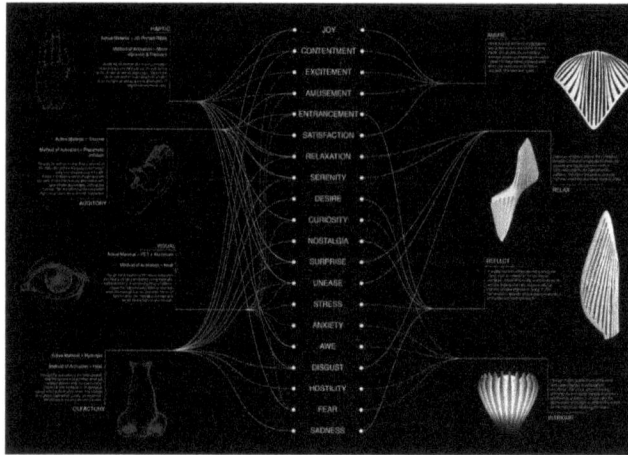

2   Translational mapping of sensorial effects on mood and behavior

attention, and feelings of subjective alertness (Hoult 2019). In the case of haptic feedback, studies have found that slow rhythmic vibrations mimicking breathing patterns provide comfort and awareness to individuals (Paredes 2018). These patterns of preference and association, although not always universal, have been observed throughout time to the extent that scientists today are able to make supported recommendations of various sensory environments which might help or ease mental strain.

Affective Computing

The notion of affective computing has historically resided in the realm of software interfaces. However, the field is rapidly expanding outside these traditional platforms and taking on novel interpretations. Coined in 1995 by Rosalind Picard, the term signifies the shift in computing and AI toward making systems able to recognize, interpret, and simulate human affect (Picard 2000). This practice has grown today to expand from its origins to permeate into a wide range of interpretations from wearables, programmable materials, social robotics, and full-scale immersive installations. These advancements, beyond the interface usually associated with affective computing, have opened the door towards making overall environments increasingly empathetic to human emotion.

The ability to link perceptive, neurological, or physiological cues to emotional states sits at the core of affective computing. From its onset, labs such as the Affective Computing lab at the MIT have looked at the overlap of these signals to give us greater insights in how, when, and why someone might be feeling a certain way to in return provide appropriate feedback to the individual (Affective Computing Lab 2020). Despite the expansion of the field towards highly sophisticated prediction models, it is the expansion to more tangible domains which are particularly interesting in the provocation of embodied cognition. Projects such as "Affective Sleeve" and "Heart of the Matter," explore the notion of affective computing and their translation into wearables (Papadopoulo 2019; Farahi 2018). With the benefit being their close proximity to the body, the pieces have the ability to both detect external signals and internal biometrics. Further expanding outside the norm are projects such as "Active Textiles" and "HygroScope," who leverage innate changes in material properties to respond proactively to both situational conditions and the physical body (Self-Assembly Lab 2019; Menges 2012). These novel advancements of the field into the physical world have brought on the compelling argument for product design, material engineering, and architecture to consider analyzing emotional cues to inform the design of these systems.

METHODOLOGY

The Affective Prostheses

To explore the potential of sensory environments in interpreting, analyzing, and translating biometric signals to in turn provide environmental-based therapeutics, a series of four Affective Prostheses were designed (Figure 1). Each of these prostheses explore the impact of specific sensory stimuli on one's psychophysiology and resulting alteration on mood (Figure 2). The senses targeted are that of haptic, olfactory, visual, and auditory. Although the sense of taste remains fundamental in the study of embodied cognition, it will not be directly explored in this research, but rather associated with the sense of smell, which is experience-dependent on taste (Small 2004).

While each piece varies in scale and sensorial experience, they each follow the same system architecture (Figure 3). The close loop system begins when physiological signals are gathered from the biometric tracker—the Empatica E4 wristband in this case—which allows for unobtrusive, long-term assessment of the physiological signals (Poh 2010; Empatica 2021). This data is then assessed based on selected methods of analysis. Through the assessments of the current biometric data, the prostheses react proactively to actuate the appropriate sensory response for the suggested cognitive state. Lastly, this reaction is monitored through the biometric tracker and provides live feedback of the physiological effects of each sensorial intervention.

Physiological Methods of Analyses

In order to assess and analyze participants psychophysiological state with the data collected, two specific methods of analyses were used to interpret the Heart Rate Variability (HRV) and Electrodermal Activity (EDA) readings from the biometric tracker.

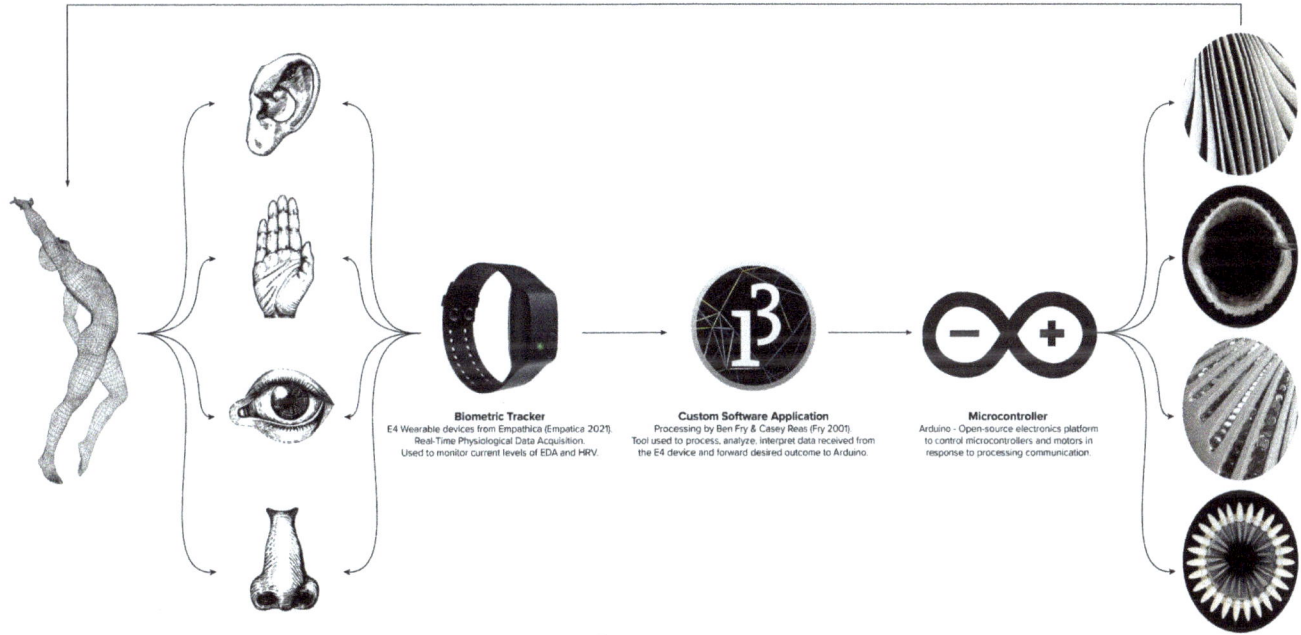

3   Affective feedback loop system diagram

HRV is the measure of variation in time between heartbeats and an efficient non-invasive measure of one's autonomic nervous activity. Patterns of low HRV have been associated with stress, anxiety, and depression while high variation tends to indicate a balanced autonomic nervous system in which both branches, parasympathetic and sympathetic, are communicating fluently with each other (Campos 2019). This oscillation between "fight or flight" and "rest and digest" responses is what simultaneously sends signals to the heart indicating that an individual is both aware and alert as well as able to rationally post process information. For the purpose of this research HRV is measured using a time domain analysis based on inter-beat intervals (IBI) and has the unit of measurement in milliseconds (ms) (Boonnithi 2011). It was chosen to apply the SDNN time domain measurement which looks to take the Standard Deviation of the Normal beat-to-beat intervals (IBI) over the entire recording epoch (Cacioppo 2007; Shaffer 2017). This time domain metric is measured in milliseconds and represents short-term variability, in this case the most recent ten beats.

Electrodermal activity (EDA), also known as galvanic skin response (GSR), has become increasingly popular in evaluating emotional arousal and inferring signs of stress (Liu 2018). EDA describes the changes in electrical properties of the skin and is measured in microsiemens (µS). Unlike HRV, EDA is purely sympathetically innervated and thus makes it one of the most genuine measures of sympathetic nervous system (SNS), which is known as the "fight and flight" response (Johnson 2020). Due to its direct association to only one branch of the autonomic nervous system, it is a great tool in signaling when one experiences a targeted emotional experience, whether that be

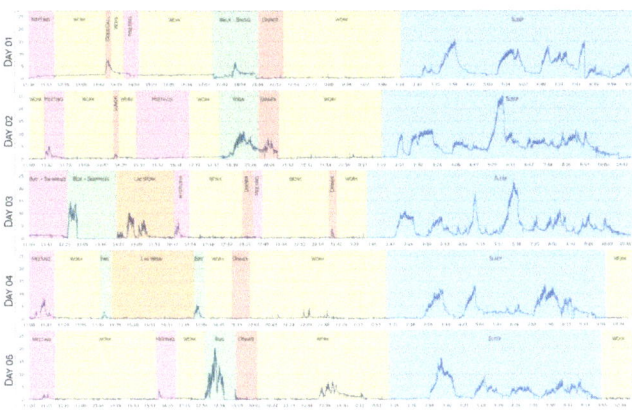
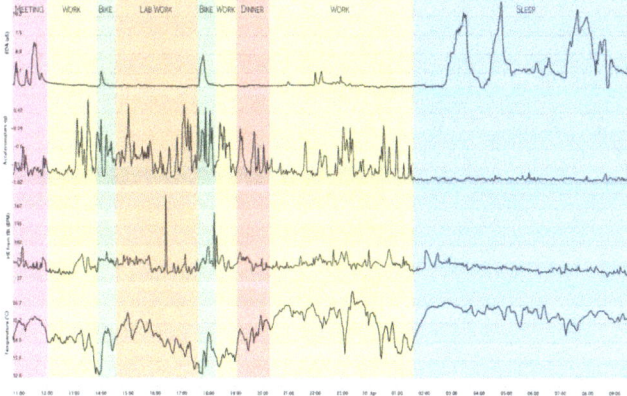

4   (left) Five-day analyses of EDA metrics in correlation with specific activities;
    (right) Twenty-four-hour analyses of EDA, skin temperature, heart rate, and movement

pleasant or unpleasant. For the interpretation of EDA, two methods are analyzed. These methods are: tonic component features, which refers to the slow climbing or declination of levels over time, and phasis component features, which relate to the faster, short term, and event related, changes in EDA activity (Braithwaite 2013). Due to the design of these system being focused on event related interventions it was chosen to focus the analyses on phasis component features. This is done by taking the mean value over a five second period of time and contrasting this to the participant's original baseline.

There remain many physiological signals that, combined, give insight to how the mind and body communicate. However, this research, and prototypes, is focused on the analyses of HRV and EDA. Peripheral skin temperature, blood volume pulse from a photoplethysmography sensor, and 3-axis accelerometer are additionally accounted for, to grasp a better picture of possible contextual factors influencing the biometrics readings, such as sleep or physical exercise (Figure 4).

AFFECTIVE PROSTHESES DESIGN
Wrist Prosthesis – Auditory & Haptic Feedback

This prosthesis represents the most intimate connection with the user as it resides closely wrapped around the wrist and explores the impact of auditory and haptic stimulus to encourage slow rhythmic breathing patterns (Figure 5). Taking precedent on the peripheral paced respiration

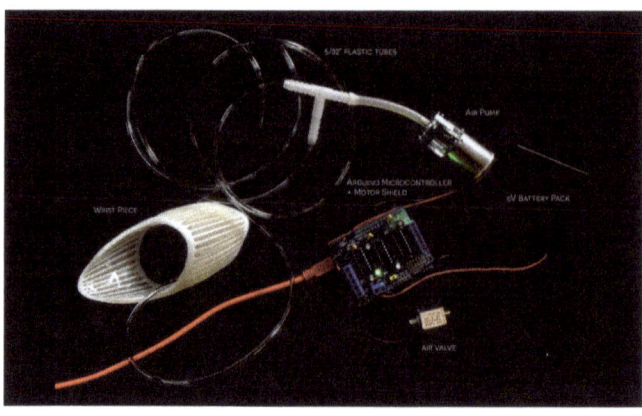

6

6   Rhythmic air actuation is driven by an Arduino board, air pump, air valve, and 12V battery, which connects to the wrist piece through two plastic tubes.
7   Prosthesis consisting of two rigid 3D-printed structures, one for the front of the chair and one for the back; adjustable and easily integrated, the chair attachment can fit a multitude of chair shapes and sizes.
8   The front piece of the chair is embedded with twenty mini vibration motors which lie between each fin, while the back accommodates the Arduino board, MOSFET driver Module, and 3.7V battery required to actuate and control the vibration motors.

technique, this piece seeks to draw on the innate interpretation of rhythmic breath to passively alter one's own respiratory cadence (Moraveji 2011).

When activated, the pump follows a slowed graduated breathing pattern, also known as the relaxation breath, of four seconds inhale – seven seconds hold – and eight second exhale (Figure 6). This time-based respiration pattern was chosen due to its proven efficiency in reducing anxiety, stress, and depression in patients (Zaccaro 2018; Pandekar 2014). Activation of this pattern occurs through two physiological actuators: heightened readings of mean EDA based on the individual's baseline, plus slowed intervals of HRV. When these two signals reach their thresholds simultaneously, the pneumatic system is activated, and patterns of breathing persist until readings return to the original baseline. Once readings return to an individual's baseline it is assessed that the individual no longer requires the haptic and auditory respiratory guidance and has either passively or consciously returned to a state of calm.

Chair Prosthesis – Haptic Feedback

Scaling up, the chair extension serves as a study on haptic stimuli (Figure 7). Similar to the previous piece, this prosthesis studies the efficacy of embodied cognition to covertly promote respiratory rhythms to passively either energize or calm the participant. While the method of analyses for the individuals EDA patterns remains the same for this piece as the wrist pieces, the analyses of HRV has been substituted for that of motion through the

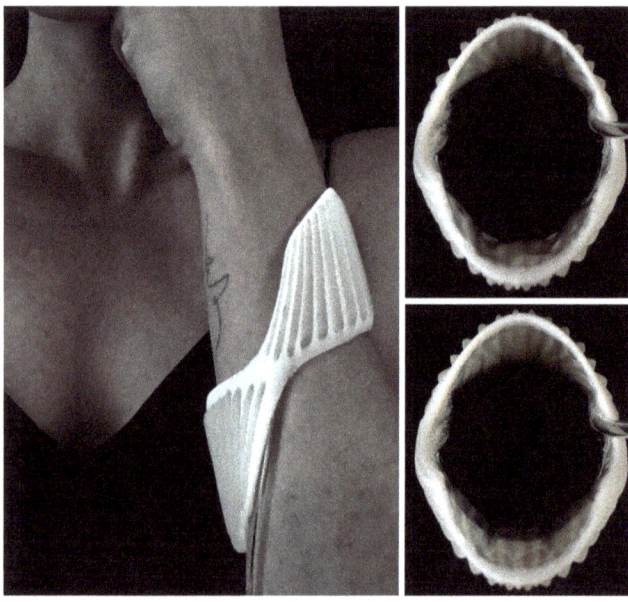

5   The multi-material ribbed 3D printed structure is composed of a rigid frame, white, and flexible filler, transparent, which accommodates a softer on skin contact and ease in putting the piece on and off. This exterior shell houses two silicon pockets facing the interior of the wrist. These pockets are programmed to be pneumatically inflated and deflated in a pattern which mimics that of slow inhales and exhales of breath.

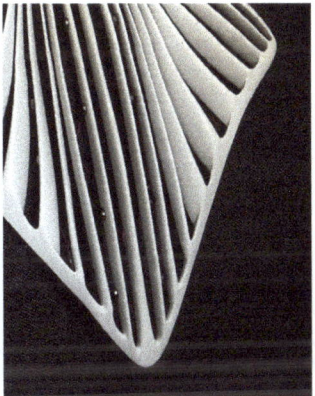
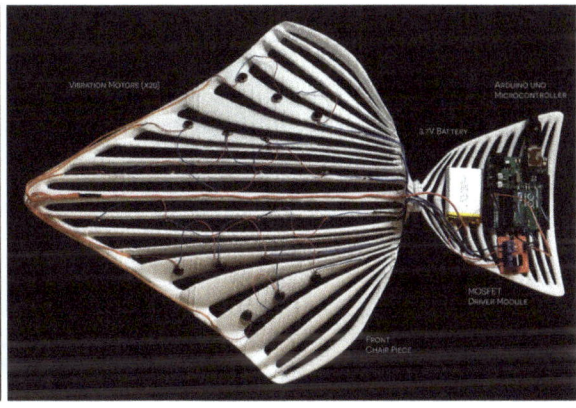

3-axis acceleration sensors located in the biometric reader. Observed in parallel, this data can illustrate correlations between prolonged periods of low movement and decreased focus and restlessness, or conversely, signs of heightened movement with stress and agitation. While many devices integrate reminders to take a break, move, get up and walk around, few of them take in the full consideration of one's cognitive state, and rather are based on intervals of time, GPS movement, and at times accelerometer data. While these are all valid forms of evaluation it is not always possible, or even necessary, when one is in a state of flow (Csikszentmihalyi 1990). Thus, it is crucial to study passive systems that can recognize and respond to these signals without the necessity for one to interrupt the task at hand.

When signals of EDA and movement indicate either possible restlessness, such as heightened levels of EDA and frequent signs of agitated motion, or drowsiness, meaning lowered levels of EDA and prolonged infrequent motion, the vibration motors are activated (Figure 8). When activated, the motors gradient in power, matching the slowed graduated breathing pattern also employed in the wrist piece (Pandekar 2014). This translates to an increase in vibration power from 0% to 100% over a period of four seconds, maintained vibration power at 100% for seven seconds, and gradual decrease in power from 100% to 0% for eight seconds. Studies such as Boostmeup and JustBreath look at this same method of regenerating slowed vibrations to trigger a physiological empathy between the false signals and their true biodata to, in turn, act as an instructional modality to calm the individual (Costa 2019; Paredes 2018). Inspired by the notion of haptic-based guidance these vibrations seek to moderate one's breathing patterns through embodied cognition to reduce stress and/or increase attention.

Desk Prosthesis – Olfactory Feedback

The desk intervention investigates the influence of olfaction on one's cognitive health (Figure 9). The scents of lavender and peppermint were chosen for this piece, given the found effect of lavender scent in reducing levels of anxiety, and association peppermint amora has with improving aspects of memory, attention, and alertness (Hoult 2019; Kritsidima 2010). Unlike scent diffusers, candles, or incense, this desk addition diffuses smell only when physiological signals indicate a sign of distress on the individual at a particular moment. This means that only one scent is diffused at one time and is dependent on the biometric signals received by the device.

Much like the chair prosthesis, actuation of the piece occurs through physiological signals of EDA, HRV, and motion. Assessing biometric signals of drowsiness or stress the piece actuates one of the two discrete heating paths to subtly diffuse either the awaking aroma of peppermint or the relaxing scent of lavender (Figure 10). The unique quality of aromatic stimuli occurs due to the anatomy of the olfactory pathways that are directly connected to the limbic system, the region of the brain associated with memory and emotional processes (Sullivan 2015). It is the subconscious and covert quality of this sensory stimuli that makes this piece the most discreet in its environmental alteration but, none the less, powerful in its influence.

9 The piece's primary structure consists of a rigid 3D printed wing-like form that wraps around the desk to maintain its stability and positioning. An ultra-thin layer of nylon encloses the top of the piece, serving both as a smooth continuous surface to serve as a functional mouse pad and as a permeable seal that stops the scent infused hydrogel beads from falling out.

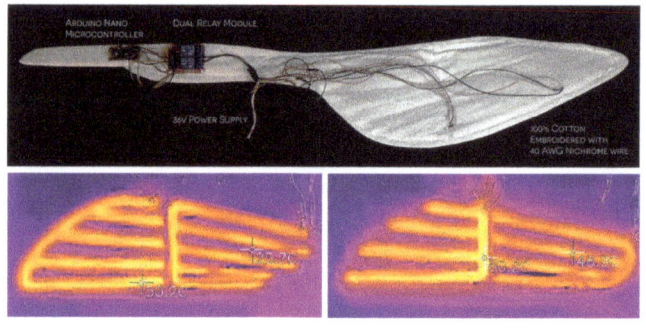

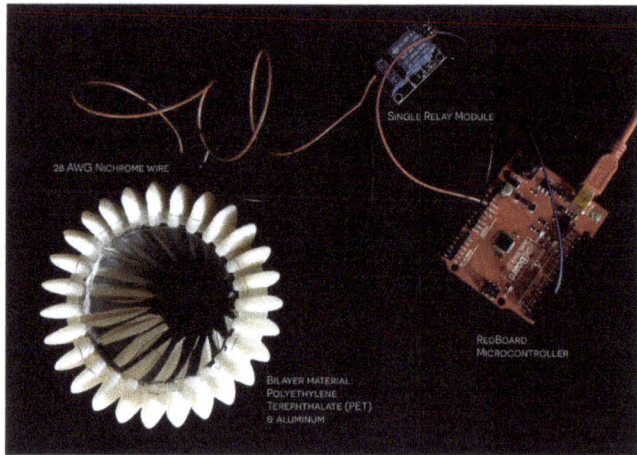

10 Beneath this structure lies 100% cotton, embroidered with 40 AWG Nichrome wire with two discrete paths constituting two separate heating systems. Powered and controlled by an Arduino nano relay motor, and 36V power supply, the Nichrome wire reaches 70°C within 30 seconds of activation. The paths directly correlate to the openings in the 3D structure, which act as chambers to house the specific scent to be activated. Hydrogel infused beads sit between these apertures, carrying either the scent of peppermint or lavender. Thanks to their thermally sensitive properties, the hydrogels increase their solubility when temperatures increase, resulting in a release of the scent-infused liquid by evaporation into the air.

Lamp Prosthesis – Visual Feedback

Expanding on the capability of lighting fixtures to accommodate personalized needs, the light prosthesis explores the correlation of physiological signals and light levels to passively address cognitive functions (Figures 11, 12). This piece adopts dimming principles by leveraging material properties and heating actuation (Plitnick 2010; McCloughan 1999). Although it sits the furthest from the individual, its influence on atmospheric conditions is by far the most notable and scalable. The piece is currently programmed to allow for more light permeation when physiological signals suggest decreased attention or fatigue, and less brightness when EDA and movement indicate high arousal. Despite being simple and small in its system design, the piece represents a scalable solution to programmable materials which indirectly produce a strong impact on the surrounding environment.

EVALUATION

In the pursuit to evaluate the effect of these systems in altering one's psychophysiology, each of these pieces were initially tested with participants in an induced stressful scenario. The hypothesis in evaluating these pieces in a controlled environment is that a shift in physiological signals would be observed when the sensory stimuli is triggered. To conduct these tests, it was chosen to provoke stress through the two-minute clinical test that is often used to determine signs of intellectual impairment in psychiatric disorders and dementia, also referred to as the Serial Seven test (Karzmark 2000; Hayman 1942). The test requires participants to continuously subtract the number seven from an initial number as rapidly as they can over a period of two minutes. The participants are told when their answers are incorrect and must correct their answer before continuing. They are also informed every thirty seconds of the remaining experiment time. While

12 The soft component of the material allowed for 28 AWG Nichrome wire to be woven into the interior of the piece and connected with a single motor relay and microcontroller. Additionally, on the inside of the frame sits a bilayer material of PET and aluminum in which a series of slits have been cut. Due to the difference in expansion rates between the PET and aluminum, when heated a certain curling occurs, resulting in greater permeability of light. The Nichrome wire can reach up to 50°C within a few seconds, resulting in an instantaneous actuation. This piece is specifically meant to work with LED light bulbs in order for the actuation of the heating to occur purely through the Nichrome wire and not the light bulb itself.

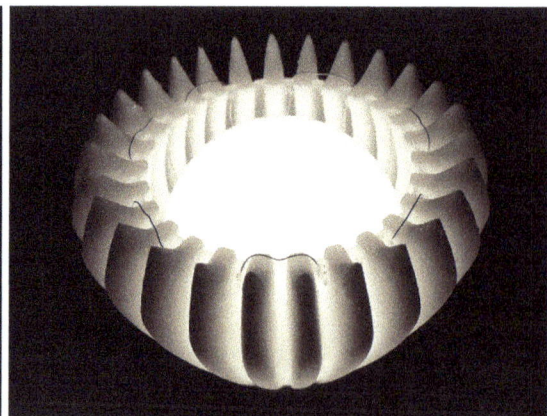

11 The primary frame of the piece is a multi-material 3D-printed with the white representing the rigid material and transparent the soft, and flexible material.

participants focus on this mathematical task, their EDA, HRV, movement, and skin temperature is recorded through the biometric reader. The wristband is positioned on their non-dominant hand with the EDA electrodes in alignment with their middle and ring finger. Participants are asked to wear the wristband for fifteen minutes prior to the start of the test to calibrate and determine a baseline recording of their biodata. From the time the test begins, these signals communicate whether the participant surpasses their baseline threshold of EDA and HRV, indicating a rapid onset of cognitive overload, which in return activates the sensory stimuli embedded in the piece. The actuation of each prosthesis follows bespoke specifications, outlined in the Affective Protheses design section, for each sensory stimuli, and persist until the participants' biometric signals normalize back to their initial baseline.

Ten participants participated in this initial study of the four protheses. Each participant interfaced with a different piece and were not familiar with the sensory stimuli associated with the piece. All participants were fully briefed on the parameters of the test and told which of their biometric signals were going to be monitored throughout the test depending on the prototype they interfaced with. Initial results from this study show a rapid increase of levels of EDA and decrease in HRV from the start of the test onward, illustrating the efficiency of the test's ability to induce signs of stress in an individual. When the introduction of the sensory stimulus is actuated, a change in these metrics occurs. In some cases, such as the haptic and auditory stimuli, the piece adds increased distraction and as a result seems to increase stress during the test. While the olfactory piece, perhaps due to its subtlety, seems to simply level the participants EDA and HRV to be constant from its onset. Out of the ten participants, results showed that that four illustrated a reduction in their EDA levels after actuation of the prototype occurs, three showed stability in their signals after actuation, two illustrated increase in stress, and one was a non-responder, indicating very little skin conductance (Figure 13).

This initial study in exploring the psychophysiological effects of these affective prostheses showed evidence of producing a noticeable impact. Despite the results wavering between positive and negative reactions to the sensory environments, they do show there to be a direct correlation between the individual's change in physiological signals and the actuation of the prototypes. These result highlight that although discreet in their intervention, these integrated experiences have a cognitive impact on the participants. However, due to the nature of the experiment being fast paced and focused on a straining cognitive task, it did not efficiently mimic a passive task. In the anticipation of the continuation of this work in the future it would be suggested that a longer, more environmentally immersive study design would be conducted. Lastly, it is important to note the variation of physiological signals across participants. Although all the algorithms for determining deviation of EDA and HRV were calibrated based on per-person baselines—and thus did not influence the actuation of the piece—the fluctuation of the signals simply emphasized the inability to make assumptions based on one single type of biometric data and that a multitude of different types of signals are crucial in being able to get a full assessment of the person's emotional state.

CONCLUSION

While the relationship between the sensory system, physical body, and one's mood often seem evident, these sensibilities are often forgotten when designing human-machine interfaces. The field of affective computing has extensively grown in research around improving software design and device development, but the field has not yet fully infiltrated into disciplines such as product design, material engineering, and architecture. Through these prototypes and initial series of user studies, the suggestion of emotion recognition seeks to go beyond that of simply assessing someone's cognitive state and responding accordingly, but rather looks to literature that assesses the efficacy of sensory stimuli and their impact on behavior, and through that foundation, promote cognitive health proactively. In addition to the historical background substantiating humans' relationship with certain sensory stimulants, studies show increasingly the relationship between physiological signals and emotional assessment. This new research is particularly motivating since the more measurable the signals, the more research one is able to interpret and correlate data with certain behavior, and as a result, to address the increasing presence of stress, anxiety, depression, and overall mental disorders.

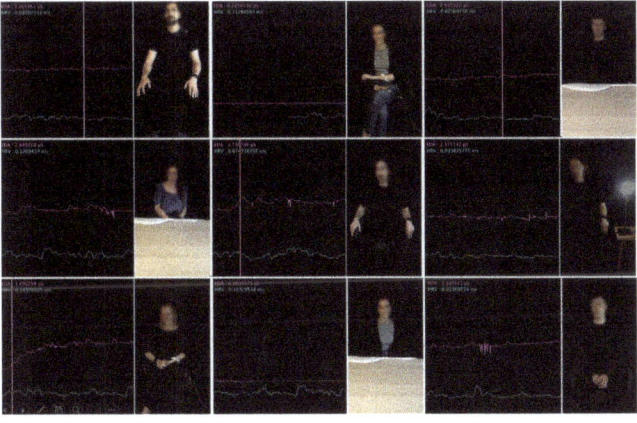

13   Image collage of participants during the study and their EDA & HRV recordings while interfacing with one of the four prototypes: the red line indicated the moment of actuation of the prosthesis.

The built environment is essential in being able to integrate these systems seamlessly into the world, and although the demonstrators developed in this research begin to explore this potential through wearables and furniture augmentations, one can envision a future where these actuated systems become embedded at the infrastructural scale, providing immersive experiences to address societal strain. Beyond the need for greater human-machine communication lies the necessity for technology and design to engage in the pursuit for accessible therapeutics and recognize mental health as a growing crisis which can be addressed by these disciplines.

REFERENCES

Affective Computing Lab. 2020. MIT Media Lab. Accessed April 1, 2021. https://www.media.mit.edu/projects/snapshot-study/overview/.

Alonso, Miguel Bruns, David V. Keyson, and Caroline C. M. Hummels. 2008. "Squeeze, Rock, and Roll; Can tangible interaction with affective products support stress reduction?" In *Proceedings of the 2nd International Conference on Tangible and Embedded Interaction*. 105–108.

Appelhans, Bradley M., and Linda J. Luecken. 2006. "Heart Rate Variability as an Index of Regulated Emotional Respon ding." *Review of General Psychology* 10 (3): 229–240.

Bhoja, Ravi, Oren T. Guttman, Amanda A. Fox, Emily Melikman, Matthew Kosemund, and Kevin J. Gingrich. 2020. "Psychophysiological stress indicators of heart rate variability and electrodermal activity with application in healthcare simulation research." *Simulation in Healthcare* 15 (1): 39–45.

Boonnithi, Sansanee, and Sukanya Phongsuphap. 2011. "Comparison of heart rate variability measures for mental stress detection." In *2011 Computing in Cardiology*. IEEE. 85–88.

Braithwaite, Jason J., Derrick G. Watson, Robert Jones, and Mickey Rowe. 2013. "A guide for analysing electrodermal activity (EDA) & skin conductance responses (SCRs) for psychological experiments." *Psychophysiology* 49 (1): 1017–1034.

Breazeal, Cynthia, and Brian Scassellati. 1999. "How to Build Robots that Make Friends and Influence People." In *Proceedings 1999 IEEE/RSJ International Conference on Intelligent Robots and Systems. Human and Environment Friendly Robots with High Intelligence and Emotional Quotients (Cat. No. 99CH36289)*, vol. 2. IEEE. 858–863.

Cacioppo, John T., Louis G. Tassinary, and Gary Berntson, eds. 2007. *Handbook of Psychophysiology*, 3rd ed. Cambridge, UK: Cambridge University Press.

Cacioppo, John T., and Louis G. Tassinary. 1990. "Inferring Psychological Significance from Physiological Signals." *American Psychologist* 45 (1): 16.

Campos, Marcelo, M.D. 2019. "Heart Rate Variability: A New Way to Track Well-Being." Harvard Health Blog. Accessed April 1, 2021. www.health.harvard.edu/blog/heart-rate-variability-new-way-track-well-2017112212789.

Costa, Jean, Alexander T. Adams, Malte F. Jung, François Guimbretière, and Tanzeem Choudhury. 2017. "EmotionCheck: A Wearable Device to Regulate Anxiety through False Heart Rate Feedback." *GetMobile: Mobile Computing and Communications* 21 (2): 22–25.

Costa, Jean, François Guimbretière, Malte F. Jung, and Tanzeem Choudhury. 2019. "Boostmeup: Improving cognitive performance in the moment by unobtrusively regulating emotions with a smartwatch." *Proceedings of the ACM on Interactive, Mobile, Wearable and Ubiquitous Technologies* 3 (2): 1–23.

Csikszentmihalyi, Mihaly. 1990. *Flow: The Psychology of Optimal Experience*, vol. 1990. New York: Harper & Row.

Empatica. 2021. "E4 Wristband: Real-Time Physiological Signals: Wearable PPG, EDA, Temperature, Motion Sensors." Accessed April 1, 2021. http://www.empatica.com/research.

Farahi, Behnaz. 2018. "Heart Of The Matter: Affective Computing in Fashion and Architecture." In *ACADIA '18: Recalibration, On Imprecision and Infidelity; Proceedings of the 38th Annual Conference of the Association for Computer Aided Design in Architecture*. 206–215.

Fitbit. 2021. "Understand Your Stress so You Can Manage It." Accessed April 1, 2021. http://www.fitbit.com/global/us/technology/stress.

Fry, Ben, and Casey Reas. 2001. *Processing*. Computer software. 04.02.2021. https://processing.org/.

Gannon, Madeline. 2018. "Human-centered Interfaces for Autonomous Fabrication Machines." PhD thesis, Carnegie Mellon University.

Gjoreski, Martin, Hristijan Gjoreski, Mitja Luštrek, and Matjaž Gams. 2016. "Continuous stress detection using a wrist device: In laboratory and real life." In *Proceedings of the 2016 ACM international Joint Conference on Pervasive and Ubiquitous Computing: Adjunct*. 1185–1193.

Hayman, M. 1942. "Two minute clinical test for measurement of intellectual impairment in psychiatric disorders." *Archives of Neurology & Psychiatry* 47 (3): 454–464.

Hoult, Lauren, Laura Longstaff, and Mark Moss. 2019. "Prolonged low-level exposure to the aroma of peppermint essential oil enhances aspects of cognition and mood in healthy adults." *American Journal of Plant Sciences* 10 (6): 1002–1012.

James, William, Frederick Burkhardt, Fredson Bowers, and Ignas K. Skrupskelis. "The Emotions." In *The Principles of Psychology*, vol. 1, no. 2. London: Macmillan, 1890. 350–365.

Johnson, Kristy. 2020. "A Brief Primer on Electrodermal Activity (EDA)." Lecture at Affective Computing Group MIT, Cambridge, Mass., November 17, 2020.

Karzmark, Peter. 2000. "Validity of the Serial Seven Procedure." *International Journal of Geriatric Psychiatry* 15 (8): 677–679.

Kritsidima, Metaxia, Tim Newton, and Koula Asimakopoulou. 2010. "The effects of lavender scent on dental patient anxiety levels: a cluster randomised controlled trial." *Community Dentistry and Oral Epidemiology* 38 (1): 83–87.

Liu, Yun, and Siqing Du. 2018. "Psychological stress level detection based on electrodermal activity." *Behavioural Brain Research* 341: 50–53.

McCloughan, C., P. A. Aspinall, and R. S. Webb. 1999. "The impact of lighting on mood." *International Journal of Lighting Research and Technology* 31 (3): 81–88.

Menges, Achim, ed. 2012. *Material Computation: Higher Integration in Morphogenetic Design*. Architectural Design series. Chichester: John Wiley & Sons.

Milstein, Nir, and Ilanit Gordon. 2020. "Validating Measures of Electrodermal Activity and Heart Rate Variability Derived From the Empatica E4 Utilized in Research Settings That Involve Interactive Dyadic States." *Frontiers in Behavioral Neuroscience* 14: 148.

Moraveji, Neema, Ben Olson, Truc Nguyen, Mahmoud Saadat, Yaser Khalighi, Roy Pea, and Jeffrey Heer. 2011. "Peripheral paced respiration: Influencing user physiology during information work." In *Proceedings of the 24th annual ACM symposium on User Interface Software and Technology*. 423–428.

Pandekar, Pratibha Pradip, and Poovishnu Devi Thangavelu. 2019. "Effect of 4-7-8 Breathing Technique on Anxiety and Depression in Moderate Chronic Obstructive Pulmonary Disease Patients." International Journal of Health Sciences and Research 9 (5): 209–217.

Papadopoulou, Athina, Jaclyn Berry, Terry Knight, and Rosalind Picard. 2019. "Affective sleeve: wearable materials with haptic action for promoting calmness." In *Distributed, Ambient and Pervasive Interactions. HCII 2019*. Lecture Notes in Computer Science, vol. 11587. Cham: Springer. 304–319.

Paredes, Pablo E., Yijun Zhou, Nur Al-Huda Hamdan, Stephanie Balters, Elizabeth Murnane, Wendy Ju, and James A. Landay. 2018. "Just Breathe: In-car Interventions for Guided Slow Breathing." *Proceedings of the ACM on Interactive, Mobile, Wearable and Ubiquitous Technologies* 2 (1): 1–23.

Picard, Rosalind W., and Jennifer Healey. 1997. "Affective Wearables." *Personal Technologies* 1 (4): 231–240.

Picard, Rosalind W. 2000. *Affective Computing*. Cambridge, Mass.: MIT Press.

Plitnick, B., M. G. Figueiro, B. Wood, and M. S. Rea. 2010. "The Effects of Red and Blue Light on Alertness and Mood at Night." *Lighting Research & Technology* 42 (4): 449–458.

Poh, Ming-Zher, Nicholas C. Swenson, and Rosalind W. Picard. 2010. "A Wearable Sensor for Unobtrusive, Long-term Assessment of Electrodermal Activity." *IEEE transactions on Biomedical* Engineering 57 (5): 1243–1252.

Schachter, Stanley. 1964. "The interaction of cognitive and physiological determinants of emotional state." *Advances in Experimental Social Psychology* 1: 49–80.

Self-Assembly Lab. 2019. "Active Textiles." Accessed April 1, 2021. https://selfassemblylab.mit.edu/active-textile-tailoring.

Shaffer, Fred, and J. P. Ginsberg. 2017. "An overview of heart rate variability metrics and norms." *Frontiers in Public Health* 5: 258.

Small, Dana M., Joel Voss, Y. Erica Mak, Katharine B. Simmons, Todd Parrish, and Darren Gitelman. 2004. "Experience-dependent Neural Integration of Taste and Smell in the Human Brain." *Journal of Neurophysiology* 92 (3): 1892–1903.

Sullivan, Regina M., Donald A. Wilson, Nadine Ravel, and Anne-Marie Mouly. 2015. "Olfactory memory networks: from emotional learning to social behaviors." *Frontiers in Behavioral Neuroscience* 9: 36.

Tibbits, Skylar. 2017. *Active Matter*. Cambridge, Mass.: MIT Press.

Zaccaro, Andrea, Andrea Piarulli, Marco Laurino, Erika Garbella, Danilo Menicucci, Bruno Neri, and Angelo Gemignani. 2018. "How Breath-control Can Change Your Life: A Systematic Review on Psycho-physiological Correlates of Slow Breathing." *Frontiers in Human Neuroscience* 12: 353.

Zhao, Bobo, Zhu Wang, Zhiwen Yu, and Bin Guo. 2018. "EmotionSense: Emotion recognition based on wearable wristband." In *2018 IEEE SmartWorld, Ubiquitous Intelligence & Computing, Advanced & Trusted Computing, Scalable Computing & Communications, Cloud & Big Data Computing, Internet of People and Smart City Innovation (SmartWorld/SCALCOM/UIC/ATC/CBDCom/IOP/SCI)*. IEEE. 346–355.

Katarina Richter-Lunn is an architectural designer, researcher, and doctoral student at the Harvard Graduate School of Design. Her research, found at the intersection of design, psychology, and neuroscience, aims to promote well-being through our environment. Alongside her doctoral studies at Harvard, Katarina is a research assistant with the Materials Processes and Systems Group (MaP+S) at the GSD, as well as a member of the Aizenberg Lab at the Wyss Institute for Biologically Inspired Engineering. Katarina holds a Master in Design Technology from Harvard GSD and a BArch from Cal Poly San Luis Obispo, with a minor in Sustainable Environments.

Jose Luis García del Castillo y López is an architect, computational designer, and educator. His current research focuses on the development of digital frameworks that help democratize access to digital technologies for designers and artists. Jose Luis is a registered architect and holds a Doctorate in Design and a Master in Design Studies on Technology from the Harvard University Graduate School of Design, where he is currently Lecturer in Architectural Technology along the Material Processes and Systems Group (MaP+S).

Robotic Timber Construction

A Case Study Structure

Arash Adel
University of Michigan

Edyta Augustynowicz
ERNE AG Holzbau

Thomas Wehrle
ERNE AG Holzbau

1   A bespoke timber sub-assembly

INTRODUCTION

Several research projects (Gramazio et al. 2014; Willmann et al. 2015; Helm et al. 2017; Adel et al. 2018; Adel Ahmadian 2020) have investigated the use of automated assembly technologies (e.g., industrial robotic arms) for the fabrication of nonstandard timber structures. Building on these projects, we present a novel and transferable process for the robotic fabrication of bespoke timber subassemblies made of off-the-shelf standard timber elements. A nonstandard timber structure (Figure 2), consisting of four bespoke subassemblies: three vertical supports and a Zollinger (Allen 1999) roof structure, acts as the case study for the research and validates the feasibility of the proposed process.

FABRICATION

The fabrication setup of the project consisted of two distinct robotic cells (Figure 3) embedded in an industrial setting, each targeted at specific assembly routines (e.g., assembling a layer-based structure in a vertical direction). The first work cell is a 4-axis portal robot attached to a telescopic based mounted on a 2-axis gantry system (Figure 3a, with a total of 7-axis). The working envelope of this setup is 48m in length, 5.6m in width, and 1.4m in height. Besides the portal robot, this setup includes a picking station, an assembly platform, and an automatic tool-changing station enabling the portal robot to change its end-effector. Two custom end-effectors were employed for this project: a pneumatic gripper that includes an automatic nailing gun and a circular saw attached to a spindle for trimming timber elements. We also designed and built a custom picking

2 The case study structure

station that included nine slots for picking timber elements. The second work cell consists of a six-axis robotic arm[1] with a payload of 240kg and a reach of 2.9m, mounted on an external linear axis with a length of 4.5m (Figure 3b). The working envelope of this setup is 10.3m in length, 5.8m in width, and 2.9m in height. Besides the robotic arm, this setup includes a picking station, a table saw, and an assembly platform. The robotic arm is equipped with a custom-built pneumatic gripper that includes an automatic screwdriver for joining timber elements.

Each of these fabrication setups imposes specific constraints (e.g., the dimensional constraints of the working envelope) that must be incorporated into the design. The pavilion acts as the case study for the research and demonstrates the feasibility of the proposed approach for designing and assembling different building components that fit within the constraints and capabilities of each setup. The first fabrication cell was used for the assembly of the roof structure. For this case study, the timber elements of the roof structure were precut since the roof consists of 168 timber elements divide into only four standard size elements. The automated assembly process of the roof structure built upon the automated assembly of The Sequential Roof (Willmann et al. 2015) and consisted of the following main steps. The portal robot grips a timber element from one of the nine pickup stations (the elements were manually placed into the pickup points), carries it to the assembly platform, and places the element on the platform based on the element's predefined final position. Subsequently, the gripper shoots a nail at one end to attach the element to the timber layer underneath while holding it in place and then re-grips the element at its other end to correct the element's orientation and shoots nails based on the ordered list of the coordinates for placing nails. The nailing process starts from the other end of the element in respect to the first shot nail to minimize assembly tolerance due to the displacement of the timber element during the nailing process. After assembling all the timber elements of the grid shell (short elements), the portal robot attaches the long edge slats, nails them to the assembled structure,

3   Fabrication setups: a) work cell I; b) work cell II

4   Trimming the edge slats

5   Robotic placing of a timber slat and automatic insertion of screws

changes its end-effector to a circular saw attached to a spindle, and trims their corners in-place (Figure 4).

The industrial robotic arm was used for the assembly of the vertical supports. Due to the constraints of the working envelope of the robotic arm and its reach, we divided each vertical support into two halves, fabricated each half separately, and connected them to get the entire vertical support subassembly. In order to minimize fabrication tolerances at the interface of the vertical supports and the roof, we fabricated each half vertical support upside down. To fabricate each half, we developed a prototypical just-in-time robotic timber assembly process based on previous research results (Adel Ahmadian 2020; Craney and Adel 2021), which includes the following main steps. The industrial robotic arm grips a timber slat, carries it to the table saw, cuts the element in cooperation with the saw, places the slat on the assembly platform, and connect the element to the previously assembled timber layers with screws using the automatic screwdriver attached to its end effector (Figure 5). This process repeats until the half vertical support is fully assembled.

After the pre-assembly of the vertical supports and the roof structure, they are put together to form the overall structure (Figure 6) and then transported to the final position of the structure (Figure 7).

RESULTS

The developed process enabled the fabrication of a nonstandard timber structure, including three bespoke vertical supports and a novel roof structure (Figures 8, 9). These images also demonstrate the resulting expressive qualities and the intricate details of the structure.

ACKNOWLEDGMENTS

The case study pavilion was fabricated in a collaborative workshop between the University of Michigan Taubman College's Master of Science in Digital and Material Technologies program and ERNE AG Holzbau at ERNE AG Holzbau's fabrication facility in Switzerland. Credits are listed below:

Taubman College of Architecture and Urban Planning, University of Michigan

Instructor and Project Lead: Arash Adel

6 Connecting the pre-assembled roof and vertical supports

Students: Ying Cai, Han-Yuan Chang, Ryan Craney, Carl Eppinger, Monik Gada, Chih Jou Lin, Feras Nour, Thea Thorrell, Christopher Voltl, Aaron Weaver, Chia Ching Yen

ERNE AG Holzbau
Project Lead: Thomas Wehrle, Edyta Augustynowicz
Team: Oliver Ackermann, Sascha Schade, Thomas Reiner

Special thanks to ROB Technologies and Catie Newell

REFERENCES

Adel Ahmadian, Arash. 2020. "Computational Design for Cooperative Robotic Assembly of Nonstandard Timber Frame Buildings." PhD thesis, ETH Zurich. https://doi.org/10.3929/ethz-b-000439443.

Adel, Arash, Andreas Thoma, Matthias Helmreich, Fabio Gramazio, and Matthias Kohler. 2018. "Design of Robotically Fabricated Timber Frame Structures." In *ACADIA '18: Recalibration, On Imprecision and Infidelity; Proceedings Catalog of the 38th Annual Conference of the Association for Computer Aided Design in Architecture*. Mexico City: IngramSpark. 394–403.

Allen, John S. 1999. "A Short History of 'Lamella' Roof Construction." *Transactions of the Newcomen Society* 71 (1): 1–29.

Craney, Ryan, and Arash Adel. 2020. "Engrained Performance: Performance-Driven Computational Design of a Robotically Assembled Shingle Facade System." In *ACADIA '20: Distributed Proximities: Proceedings Catalog of the 40th Annual Conference of the Association for Computer Aided Design in Architecture*. Online + Global. 24–30.

Gramazio, Fabio, Matthias Kohler, and Jan Willmann. 2014. "The Stacked Pavilion." In *The Robotic Touch: How Robots Change Architecture*. Zurich: Park Books AG. 196–207.

Helm, Volker, Michael Knauss, Thomas Kohlhammer, Fabio Gramazio, and Matthias Kohler. 2017. "Additive robotic fabrication of complex timber structures." In *Advancing Wood Architecture; A Computational Approach*, edited by Achim Menges, Tobias Schwinn and Oliver David Krieg. New York: Routledge. 29–43.

Willmann, Jan, Michael Knauss, Tobias Bonwetsch, Anna Aleksandra Apolinarska, Fabio Gramazio, and Matthias Kohler. 2015. "Robotic Timber Construction; Expanding Additive Fabrication to New Dimensions." *Automation in Construction* 61: 16-23. https://doi.org/10.1016/j.autcon.2015.09.011.

7   Onsite positioning of the pre-assembled structure

NOTES
1. KUKA KR240

IMAGE CREDITS
Figures 1, 3, 4, 5, 6, 8: Daniel Nikles, 2020
Figures 2, 9: Ryan Craney, 2020
Figure 7: Oliver Ackerman, 2020

Arash Adel is Assistant Professor of Architecture at the University of Michigan's Taubman College of Architecture and Urban Planning, where he directs the ADR Laboratory. Adel's interdisciplinary research is at the intersection of design, computation, engineering, and robotic construction. He is particularly known for his work with novel integrative computational design methods coupled with robotic assembly techniques for manufacturing nonstandard multi-story timber buildings. Adel received his Master in Architecture from Harvard University and his doctorate in architecture from the Swiss Federal Institute of Technology (ETH).

Edyta Augustynowicz is an architect, researcher, and computational designer specializing in complex timber structures. She is currently affiliated with ERNE AG Holzbau. She holds a Master in Architecture and Urban Design from the TU Poznan and a MAS degree in CAAD from the ETH Zurich. Her previous experience includes Digital Technology Group at Herzog & de Meuron, Block Research Group at ETH Zurich, and Institute Integrative Design, HGK, FHNW in Basel.

Thomas Wehrle is CTO at ERNE AG Holzbau. He is responsible for the development of digitalization and the introduction of robotic production at the company. For his achievements with the implemen¬tation of robotics in collaboration with ETH Zurich, he was awarded in 2017 the European Innovation Award for the Forestry and Wood Industries. In parallel he is tutor and researcher at Hochschule Luzern, Berner Fachhochschule and ETH Zurich. He holds a Master's degree from ZHAW School of Engineering.

8 The fully-assembled timber structure

9   The intricate corner details of the fully-assembled timber structure.

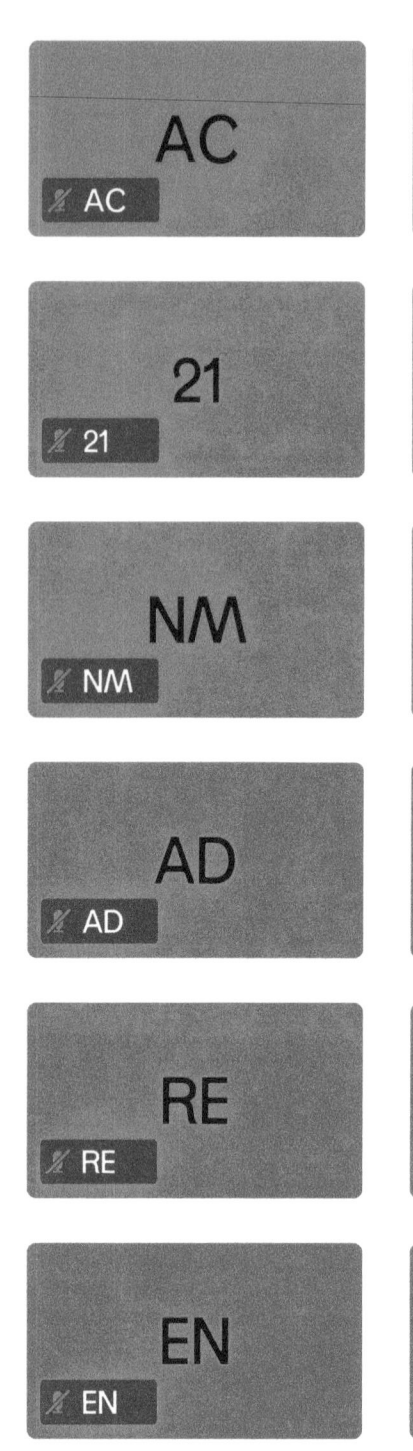
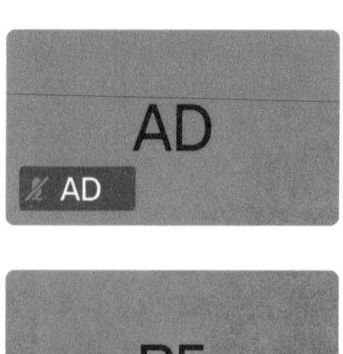
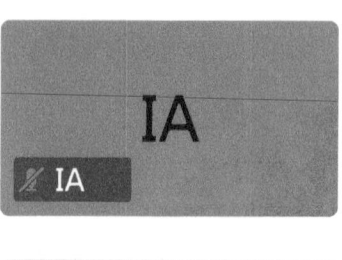

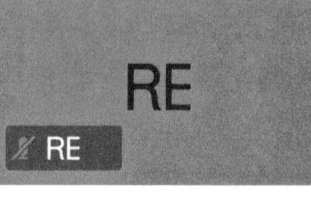
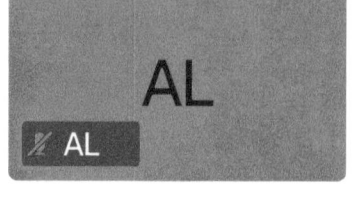

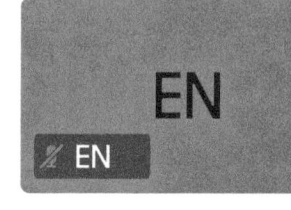

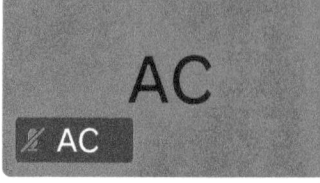
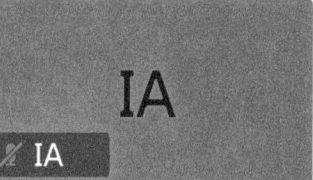

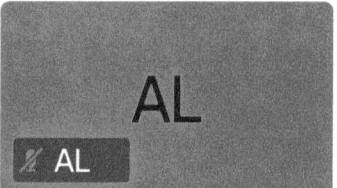

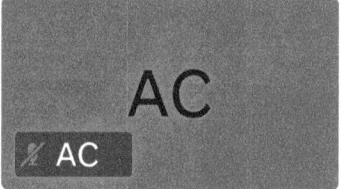

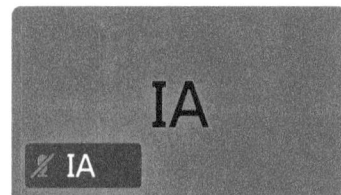

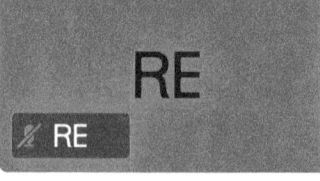
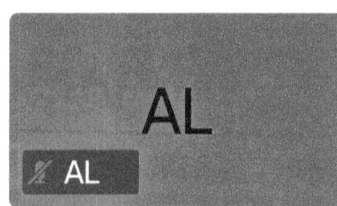

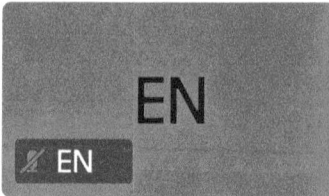

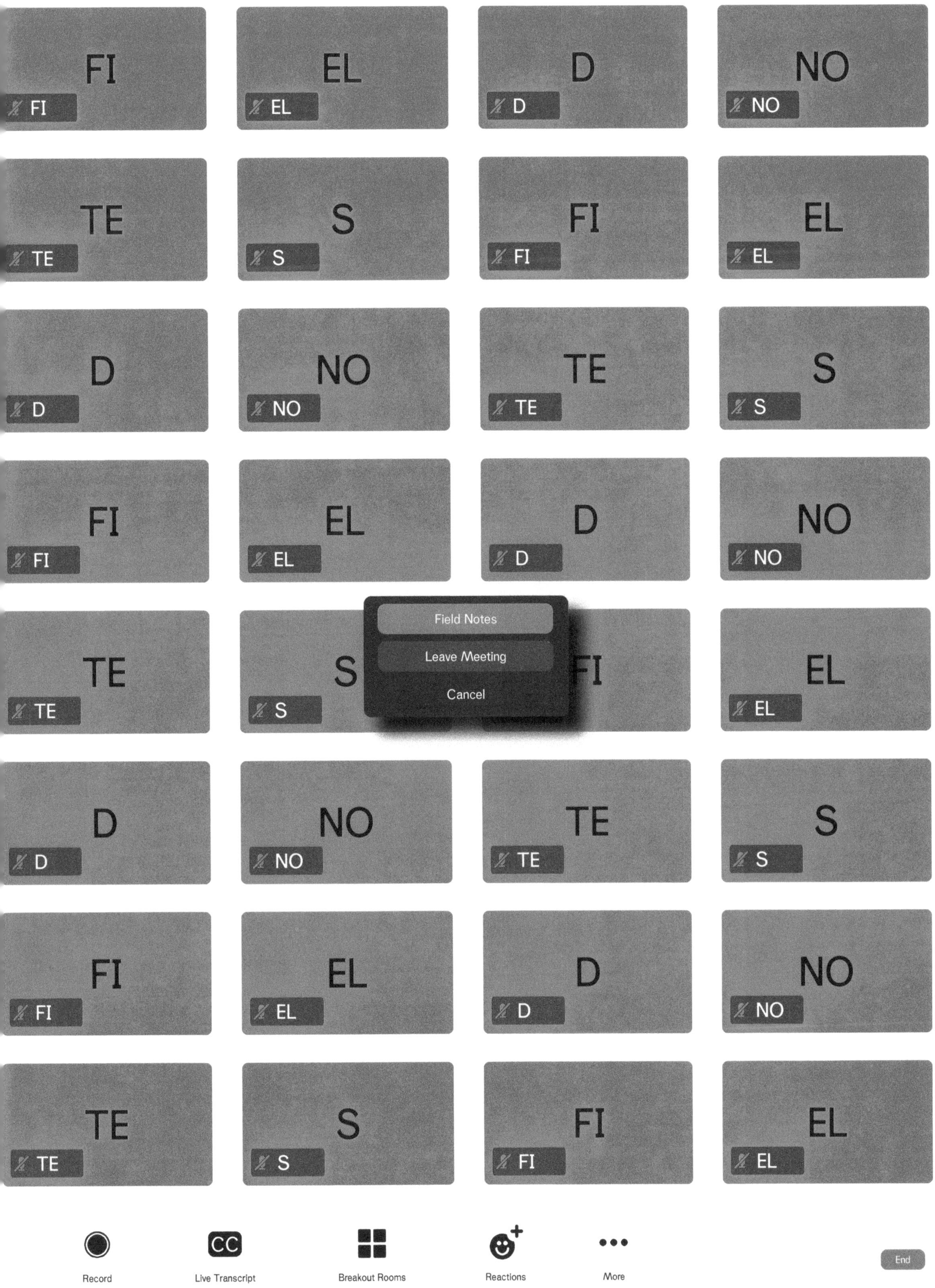
FI
FI
EL
EL
D
D
NO
NO
TE
TE
S
S
FI
FI
EL
EL
D
D
NO
NO
TE
TE
S
S
FI
FI
EL
EL
D
D
NO
NO
TE
TE
S
S
Field Notes
Leave Meeting
Cancel
FI
EL
EL
D
D
NO
NO
TE
TE
S
S
FI
FI
EL
EL
D
D
NO
NO
TE
TE
S
S
FI
FI
EL
EL
Record
CC
Live Transcript
Breakout Rooms
Reactions
More
End

Computational Access

Shelby Doyle
Iowa State University

Nick Senske
Iowa State University

1   Robotic 3D printing, 2021

While technology has rapidly become available to more people, there is still a lack of representation and diversity among the individuals who develop and create with it. The implication of computational design and digital fabrication scholarship is that knowledge circulates through publications when, in a practical sense, it tends to be consolidated within a limited set of people and institutions. Even as the costs of hardware trend lower and free software and workflows are published online, specialized education and social capital are often necessary to apply this knowledge and produce innovative digital designs. And so, access to technology alone does not necessarily lead to greater equity.

Improving access to digital design knowledge—specifically methods and processes—could help address this concern. In scientific publications outside of architecture, the methodology section and technical appendices are critical to verification and advancement of the field. If an experiment cannot be duplicated, the validity of the result is called into question. The same standard does not seem to apply in computational design and digital fabrication, as the descriptions of projects are seldom detailed, transparent, or instructive enough to permit replication.

With the complexities of code, digital fabrication equipment, robotics, and other advanced methodologies, the distance between the published image and the making of the image continues to widen. What appears on Instagram and in the pages of journal articles does not allow the reader to replicate or validate projects in an academic sense. Open-source projects on personal websites, GitHub, or Instructables are a method for circulating information freely, but these do not benefit from recognized peer review, nor do they create valuable CV lines and citations—which are the currency of the academy and necessary professional advancement. At the same time, academic publishing privileges text and images, when other types of media are more appropriate for these kinds of knowledge: video, code, schematics, 3D models, et cetera. And so, there is currently no academic space to connect the work 'behind the scenes' of a design to its aesthetic presentation. What is needed

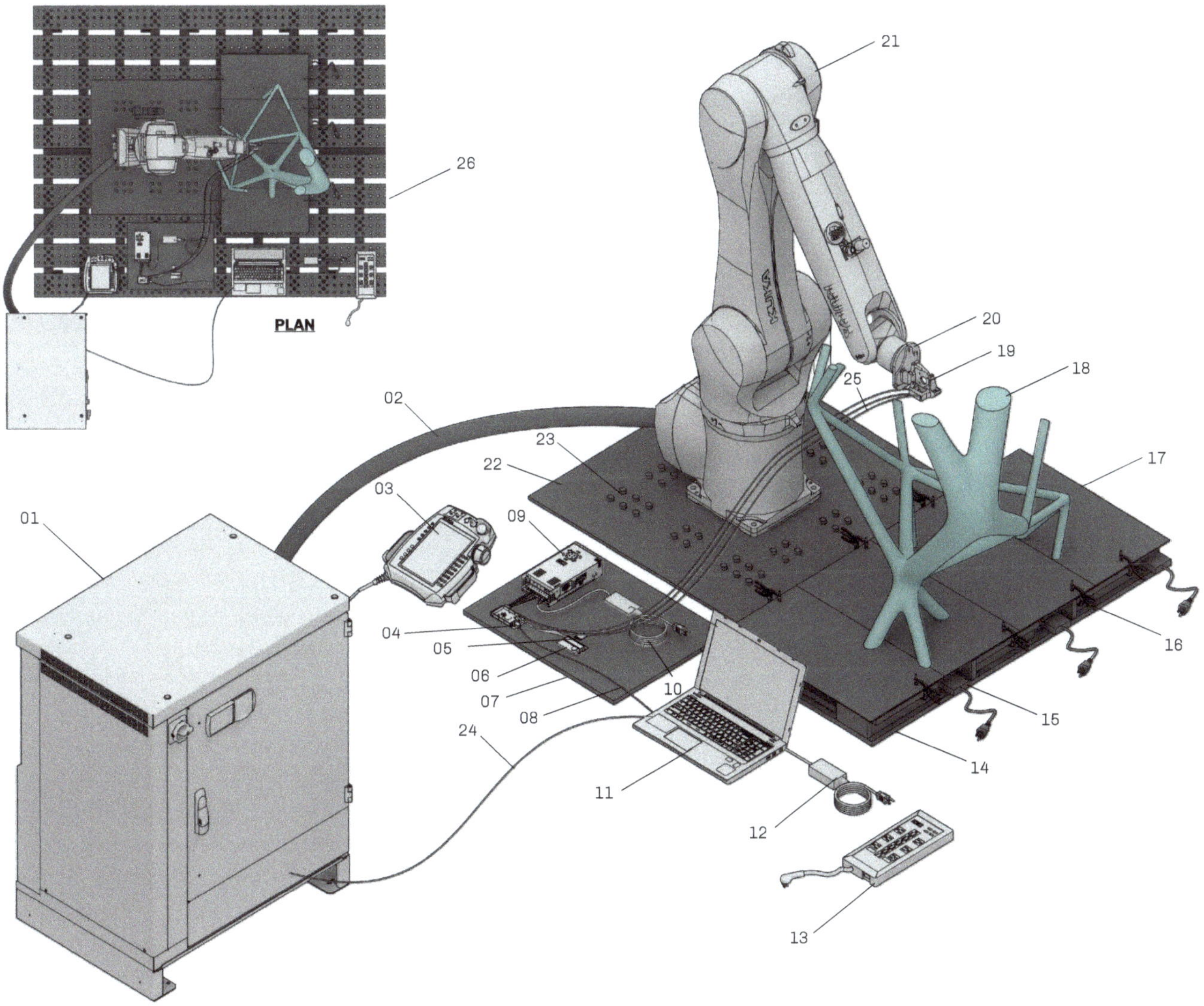

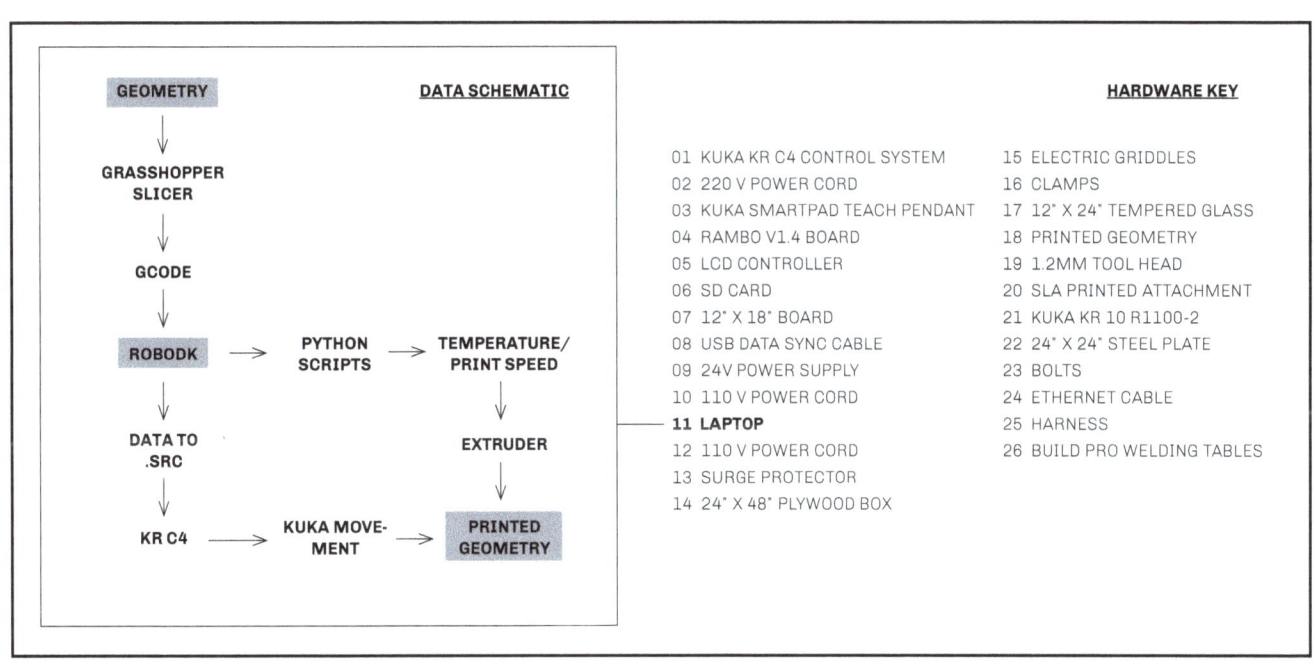

2 Hardware and software configuration for robotically-controlled 3D printing

HARDWARE KEY

01 KUKA KR C4 CONTROL SYSTEM
02 220 V POWER CORD
03 KUKA SMARTPAD TEACH PENDANT
04 RAMBO V1.4 BOARD
05 LCD CONTROLLER
06 SD CARD
07 12" X 18" BOARD
08 USB DATA SYNC CABLE
09 24V POWER SUPPLY
10 110 V POWER CORD
11 LAPTOP
12 110 V POWER CORD
13 SURGE PROTECTOR
14 24" X 48" PLYWOOD BOX
15 ELECTRIC GRIDDLES
16 CLAMPS
17 12" X 24" TEMPERED GLASS
18 PRINTED GEOMETRY
19 1.2MM TOOL HEAD
20 SLA PRINTED ATTACHMENT
21 KUKA KR 10 R1100-2
22 24" X 24" STEEL PLATE
23 BOLTS
24 ETHERNET CABLE
25 HARNESS
26 BUILD PRO WELDING TABLES

in the computational design field is a recognized form of scholarship that incentivizes and supports the sharing of 'mere technical' knowledge at a usable and replicable level of detail.

In response to this challenge, we propose a framework for the publication of technical digital design and fabrication processes and knowledge: a rigorous, attributable, and replicable account of technical knowledge called a "circuit" (Figure 1). A circuit is: 1) a full and transparent account of all necessary hardware, software, instructions, bills-of-materials, etc. as well as the documentation for implementation; 2) structured and delivered in a manner that supports circulation, verification, and adaptation; 3) embued with the intention of providing inclusive access (i.e. bringing others "into the circle" and short-circuiting bias and exclusion).

Circuits are works of original scholarship included as an addendum to essays, projects, or articles. They are comparable to a project or technical report, meant to be peer-reviewed by experts and attributable to their authors, and established according to principles of diversity, equity, and inclusion. Circuits would be formatted with a common structure (similar to the ways academic papers are organized) according to an agreed-upon open standard and hosted within online repositories. This would include intellectual property protections with an open-source license to ensure attribution and restrict liability. Contributions to the circuit would be recorded through digital certificates (or other attribution frameworks) to create a complete and transparent account of authorship that avoids implicit bias and academic or social politics. And, finally, circuits would be available free of cost or otherwise under a low-cost or non-profit arrangement. Meeting these criteria would benefit scholars while promoting meaningful and equitable access to high-quality technical knowledge and establishing a historic record for future scholars.

For example, the diagram here of an off-the-shelf robotic 3D printing circuit unpacks some of the knowledge necessary to produce consistent FDM prints using a KUKA robotic arm. A proper representation of this circuit is not possible in this medium; a more comprehensive version would include additional elements such as Python scripts, Grasshopper definitions, FDM printing settings, perhaps a video workflow, a list of parts and prices, and so on. The full, hyperlinked circuit could be a type of scholarship which is separate and unique from the design outputs that might result.

Creating and sharing detailed technical knowledge benefits the field. For example, the free primers and tutorials available for Grasshopper introduced many architects to computational design and inspired countless new projects, research, and software. But these efforts tend to be isolated from academic research and projects, as well as professional practice. Publishing circuits would bring focus to specific implementations and problem-solving procedures that would otherwise be excluded in a typical project narrative. This documentation would prevent individuals and institutions from having to start at the beginning with each new technology—e.g., each school laboriously 'discovering' how to integrate a robotic arm—and, therefore, lower barriers to entry for creating new designs and knowledge within the field of computational design.

Another impact of circuits would be to provide an impartial record of authorship that accurately describes the forms of labor necessary to create a digital project. Oftentimes, this process is collapsed into a summary narrative with a well-produced time lapse video that denies the frictions of time, technology, materials, and people. Instead, the knowledge within a circuit is not attributed solely to a primary investigator on record or a list of contributors in some negotiated order, but rather assigned by machine to people according to their specific investments in coding, fabrication, documentation, and so on. This is not only more equitable, but also a way of tracing specific components of the circuit for follow-up and recognition in future iterations.

Implementing this proposal would require a complete rethinking of how technical knowledge is created and circulated within architectural scholarship. It would also require changing how computational design work is evaluated and published. To be recognized, truly innovative work might be required to facilitate access to the knowledge of its creation. These are significant challenges which will require a realignment of culture and values. While many examples exist of generous efforts to share tools and processes, these are not the norm and depend upon altruism alone for their existence. Formalizing the idea of circuits would create a means to raise the standards for evidence of innovation within the field while broadening the recognition for (and therefore, incentivizing) contributions to technical knowledge.

This manifesto on computational access is a provocation toward a more inclusive future for the field. It is also a reminder that access is often an uncomfortable fiction.

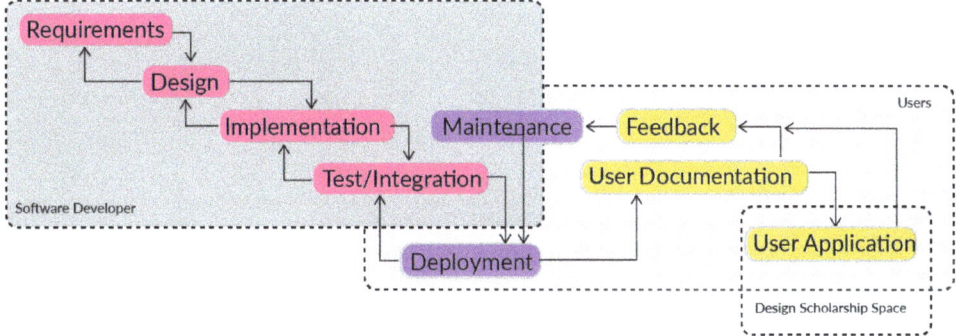

Conventional Software / Hardware

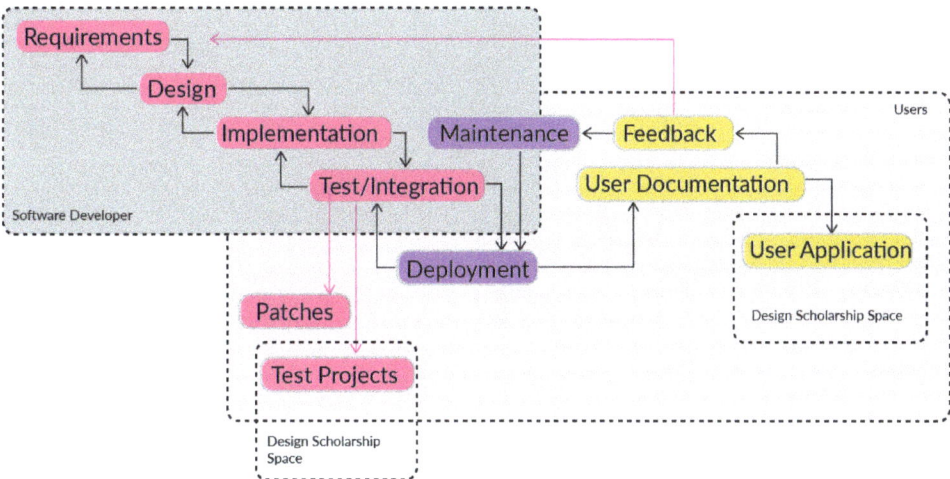

Open-Source Software / Hardware

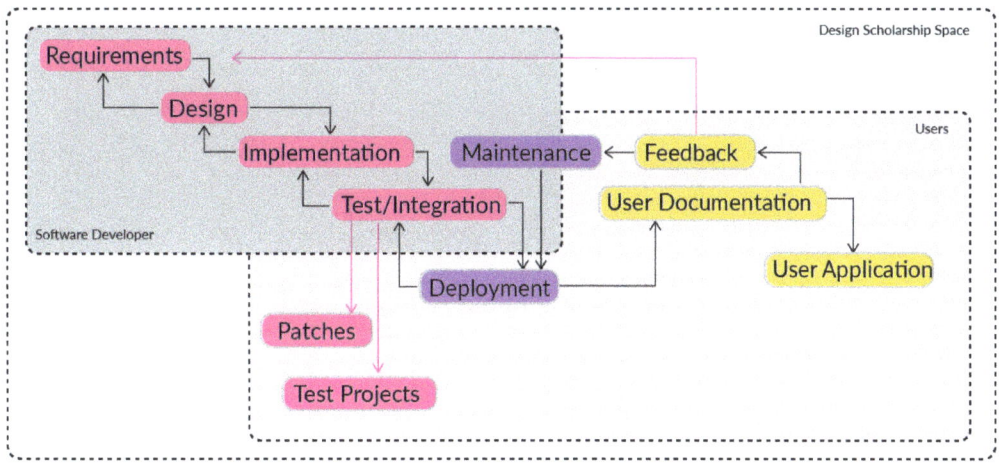

3   Comparison of design scholarship among different types of distribution and publication

4   Photo of electronics set up for robotic 3D printing

5   Custom end effector attachment

Much of academia and academic (and professional) achievement trades in exclusivity and, thereby, exclusion. Is control of computational knowledge and processes too powerful a currency to relinquish? Or can we design a more accessible computational field from education to practice? We propose that scholarly circuits can create powerful feedback loops that magnify and amplify computational possibilities. This would bring the discipline closer to both technological equity and a built world that reflects the promises and potentials of the digital turn: an architecture that is safer, more sustainable, more just, and more equal.

ACKNOWLEDGMENTS

This research received support from the American Institute of Architects and the AIA Upjohn Research Initiative, the ISU Department of Architecture, ISU College of Design, ISU Committee on the Advancement of Student Technology for Learning Enhancement (CASTLE), and the Architectural Research Centers Consortium (ARCC) Research Incentive Award.

REFERENCES

Doyle, Shelby Elizabeth, and Nick Senske. 2018. "Digital Provenance and Material Metadata: Attribution and Co-authorship in the Age of Artificial Intelligence. *International Journal of Architectural Computing* 16 (4): 271–280.

IMAGE CREDITS

All drawings and images by the authors.

Shelby Elizabeth Doyle, AIA is Associate Professor of Architecture and Stan G. Thurston Professor of Design-Build at the Iowa State University College of Design, co-founder of the ISU Computation & Construction Lab (CCL) and director of the ISU Architectural Robotics Lab (ARL). Doyle received a Fulbright Fellowship to Cambodia, a Master of Architecture from the Harvard Graduate School of Design, and a Bachelor of Science in architecture from the University of Virginia.

Nick Senske is Assistant Professor of Architecture at the Iowa State University College of Design. His research looks at computational software as a cultural artifact and has been presented internationally at conferences and workshops. In the past, his work has examined the most effective ways to teach computational skills in professional architecture schools, and his design work explores the physical implications of 'craft' in coding and scripting. Senske was previously Assistant Professor of Architecture at the University of North Carolina at Charlotte. He received a BArch from Iowa State University and a SMArchS in Design Computation from MIT.

Process / Product

Cyle King
Iowa State University

Jacob Gasper
Iowa State University

1 Human hands (left; Jacob Gasper, right; Cyle King) assist in the printing of a clay mold at The Architectural Robotics Lab at Iowa State University

Academic papers are full of final drawings and diagrams but gloss over process work, "less glamorous" images, and the amount of time and labor behind a final product. Certain skills and expertise cannot be taught but are instead collected from years of personal experience – a body of knowledge inaccessible to some unless passed on through e-mails, Zoom calls, or personal observations. When dealing with these seemingly esoteric topics, it becomes easy to feel isolated in the problems, failures, or questions that arise and cannot be easily accessed in academic journals or a simple Google search. Although exacerbated by the global pandemic's mandates and shifts in the way work is done - this feeling is not new.

The following pages record clay 3D printing research on a KUKA industrial robotic arm completed by two 5th year undergraduate architecture students. Through drawings, images, and text, this field note documents decisions, failures, messes, and successes compiled from a year of socially distanced learning, researching, and living.

Roadblocks felt specific to our case.

Access to process work and its discourse strengthens the integrity of computational innovation while serving as a means to fill in these gaps and connect research communities across borders and disciplines. The transparency of the process through conversation, even when dispersed, reveals a more equitable approach to research and creates a more inclusive knowledge-base for fabricators, academics, designers, and the research community as a whole.

We hope to share our process as a reminder, to ourselves and others, that we are not alone.

- Don't be afraid to fail
- Research is very humbling and nothing goes right the 1st time
- Solving hardware issues takes a large portion of your time
- Learn from mistakes — DOCUMENT!
- A good cart — or several can make or break your workflow
- Don't be afraid to ask professionals questions. Most of them want to help, and most of them have dealt with the same issues
- Work with someone who pushes you, makes you be your best, + that you can trust
- Working in several spaces <u>sucks</u>
- People want to say yes — you just have to make it easy for them to say yes.
- Clay is an organic material of decomposed minerals NOT intended to be 3D printed
- Clay is messy. Don't be afraid to get messy
- Sleep
- Concrete gets heavy — it's okay to not make heavy things.
- There is a certain point the project just needs to happen — it is what it is
- Draw!!!!! Do!!!!! Keeping it in your head doesn't solve anything? everything?
- Take breaks. Get a coffee, eat a sandwich, talk about something other than clay and robots.
- Celebrate successes. Have a beer

2 Takeaways Part

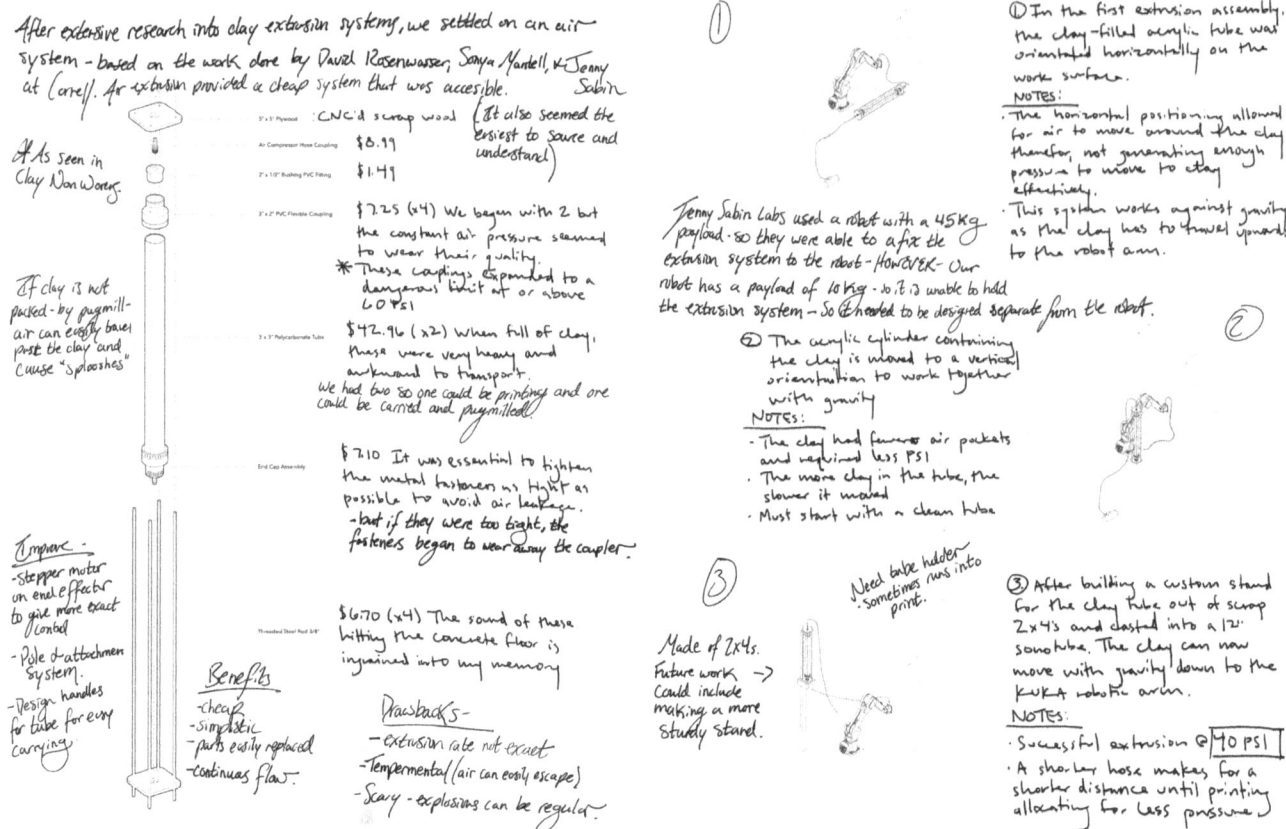

3   Pressurized clay extrusion system with overlaying notes (inspired by previous work at Sabin Labs at Cornell)

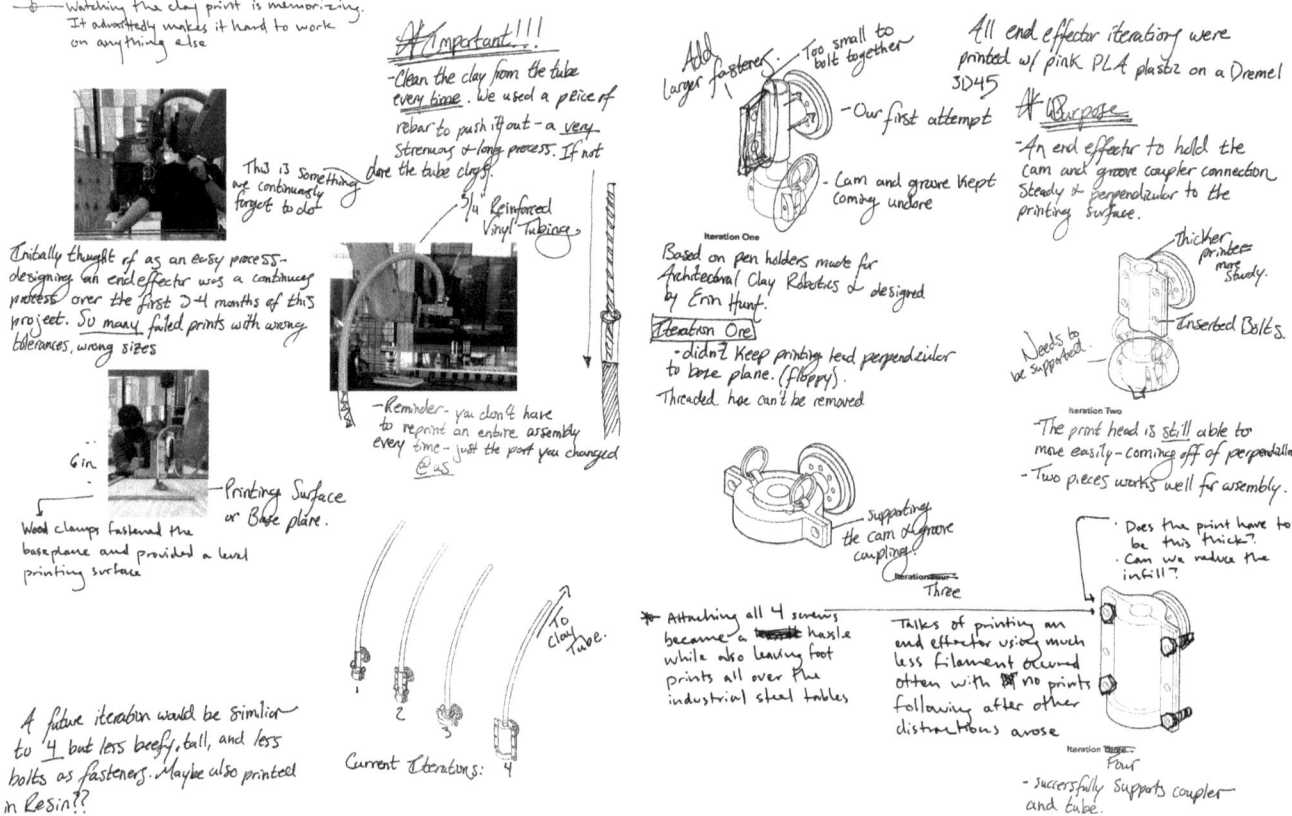

4   KUKA end-effector iterations and first print trials with overlaying notes

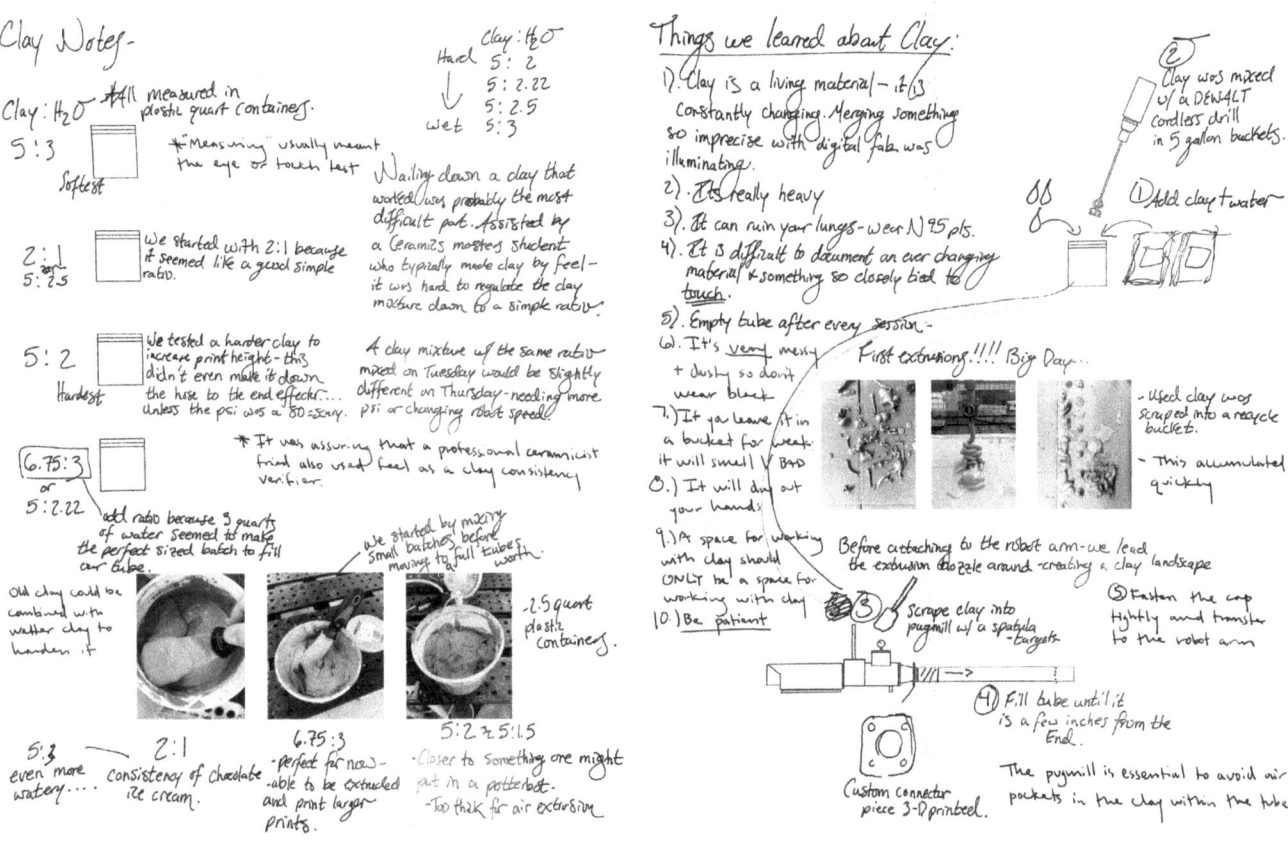

5 Clay mixing ratios + process with overlaying notes

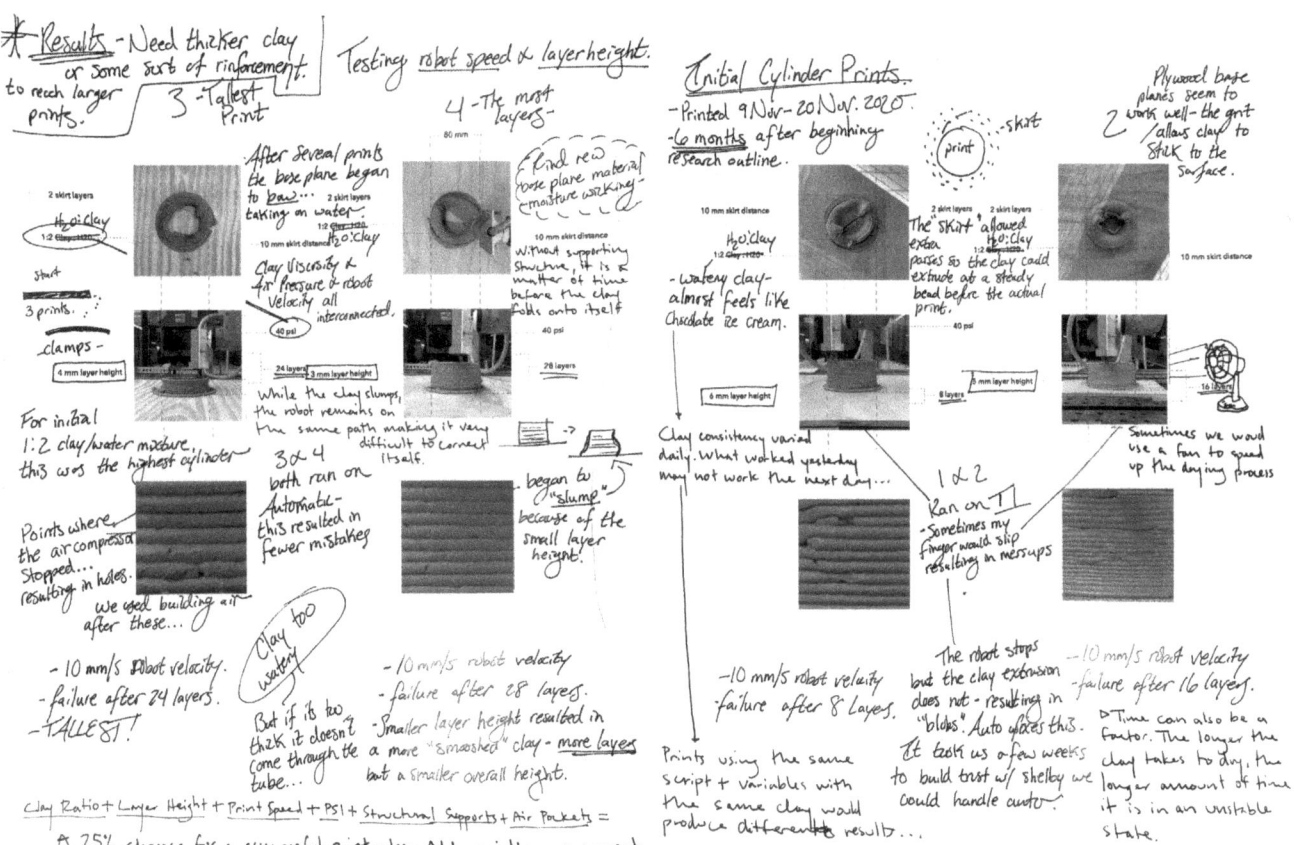

6 Initial clay cylinder prints with overlaying notes

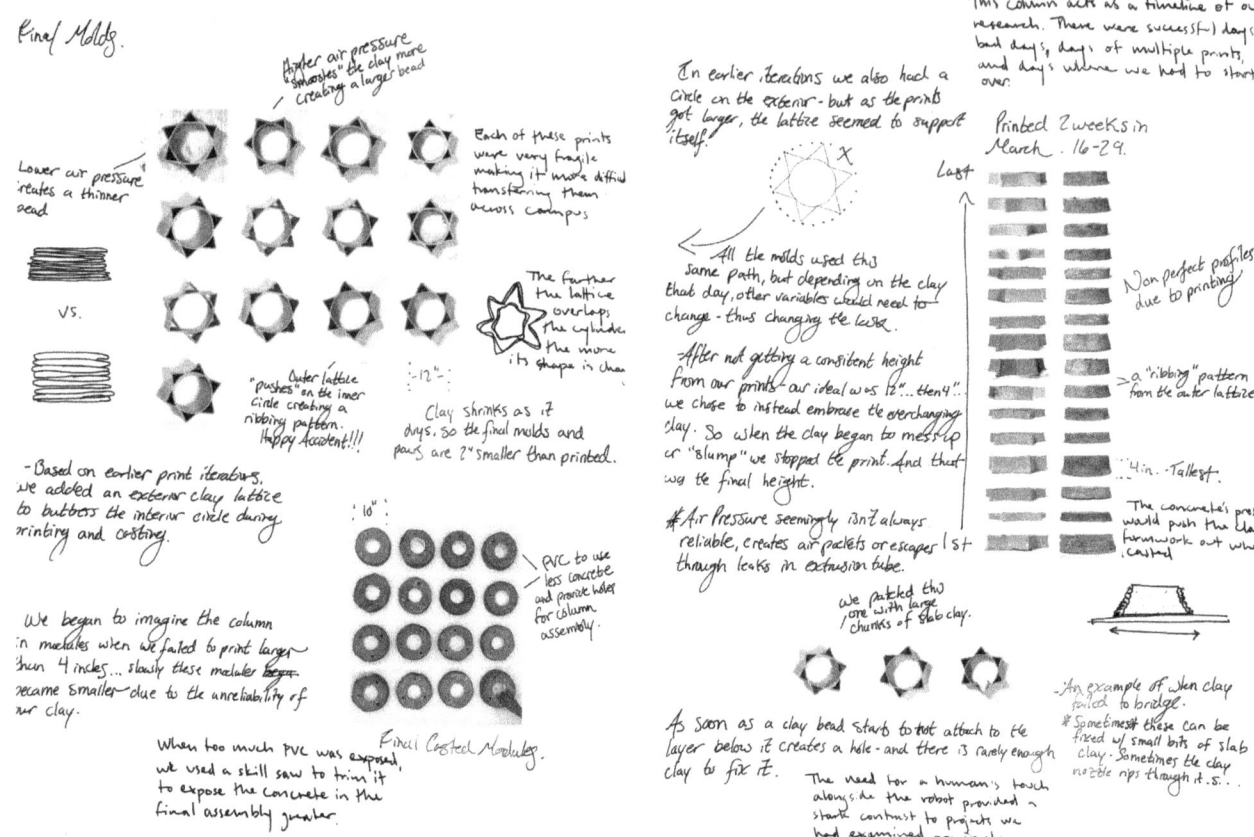

7   Final clay molds + casts with overlaying notes

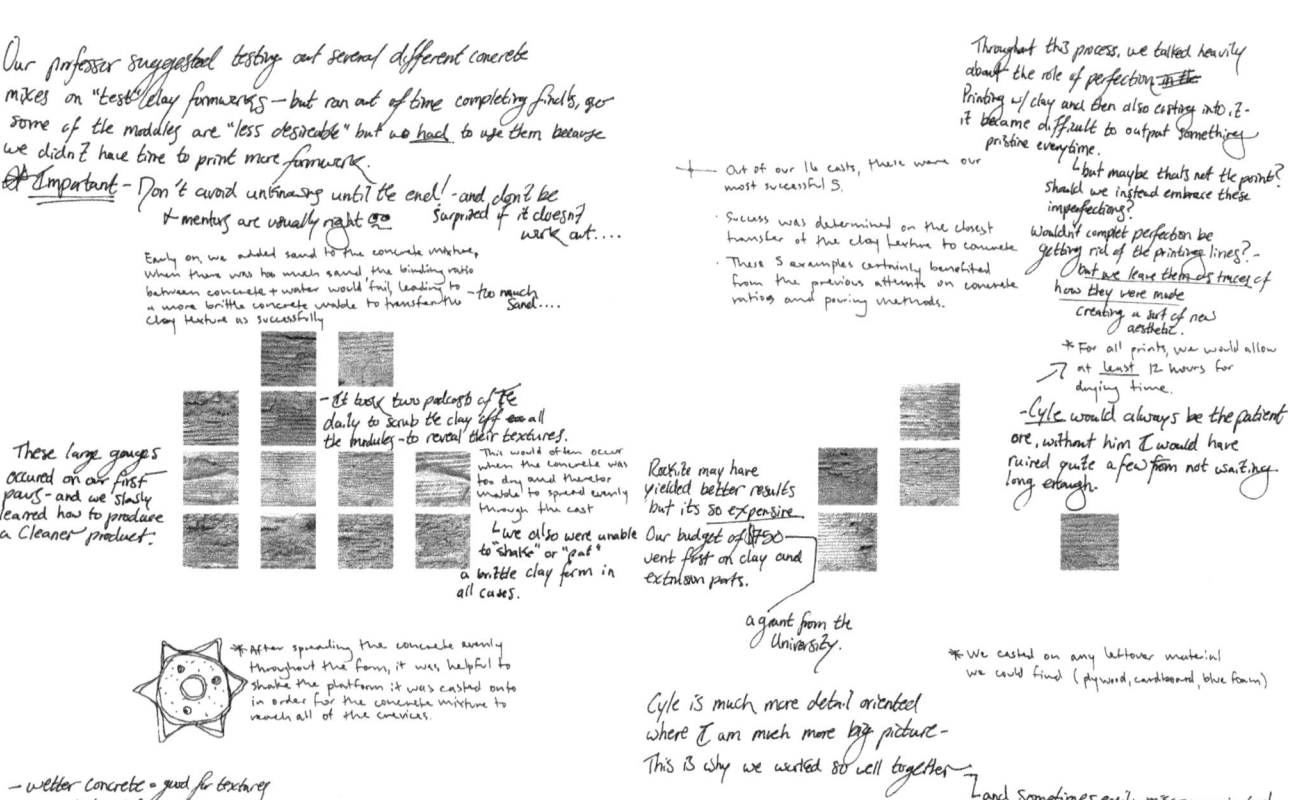

8   Clay mold textures with overlaying notes

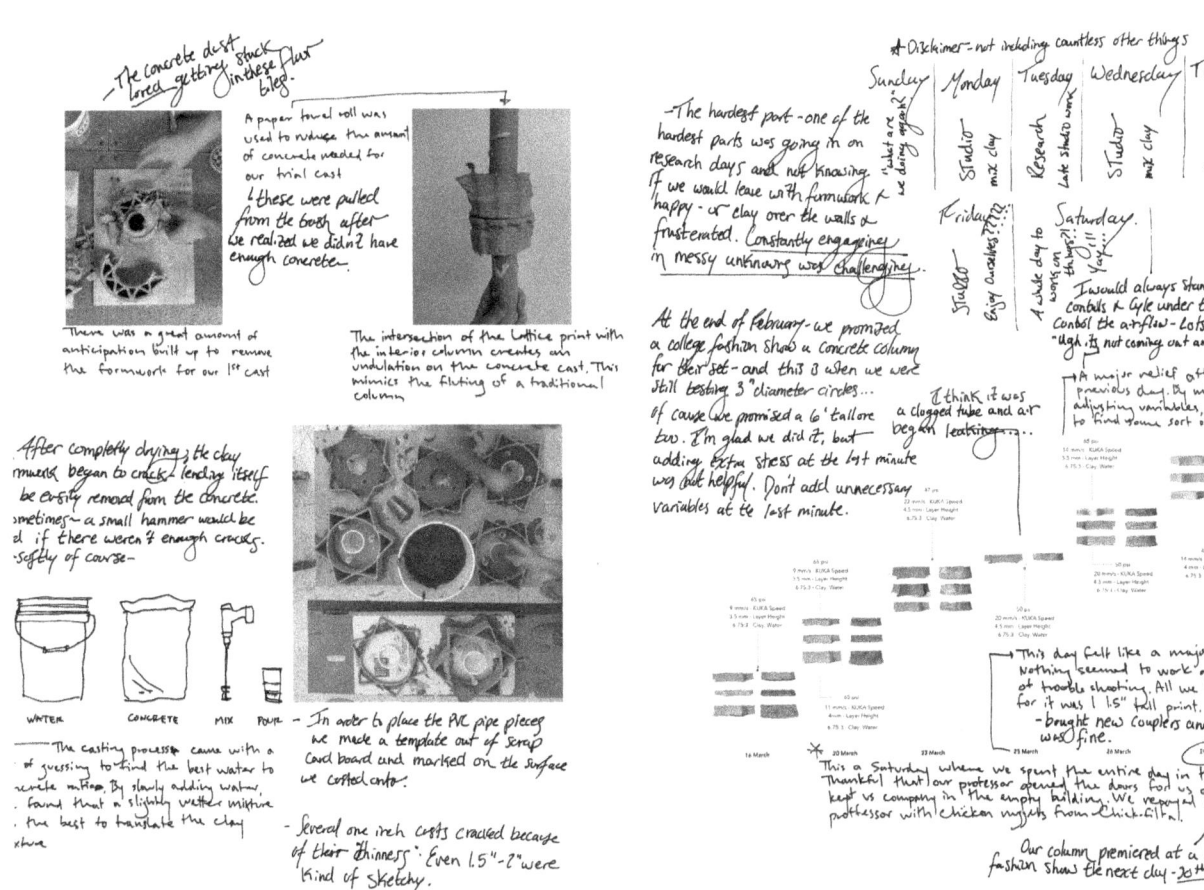

9 Final concrete casting process with overlaying notes

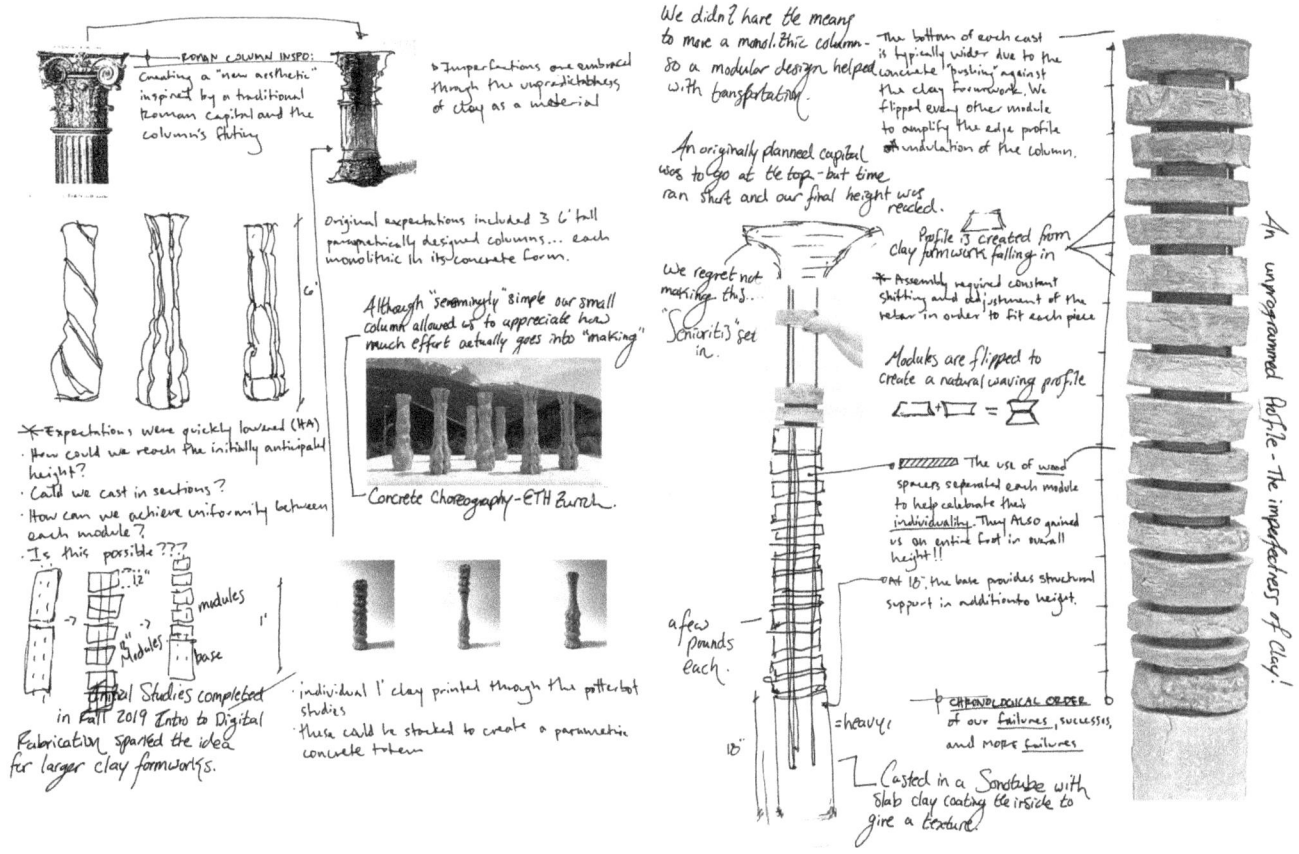

10 Comparing precedents (Concrete Choreography) and the final casted column with overlaying notes.

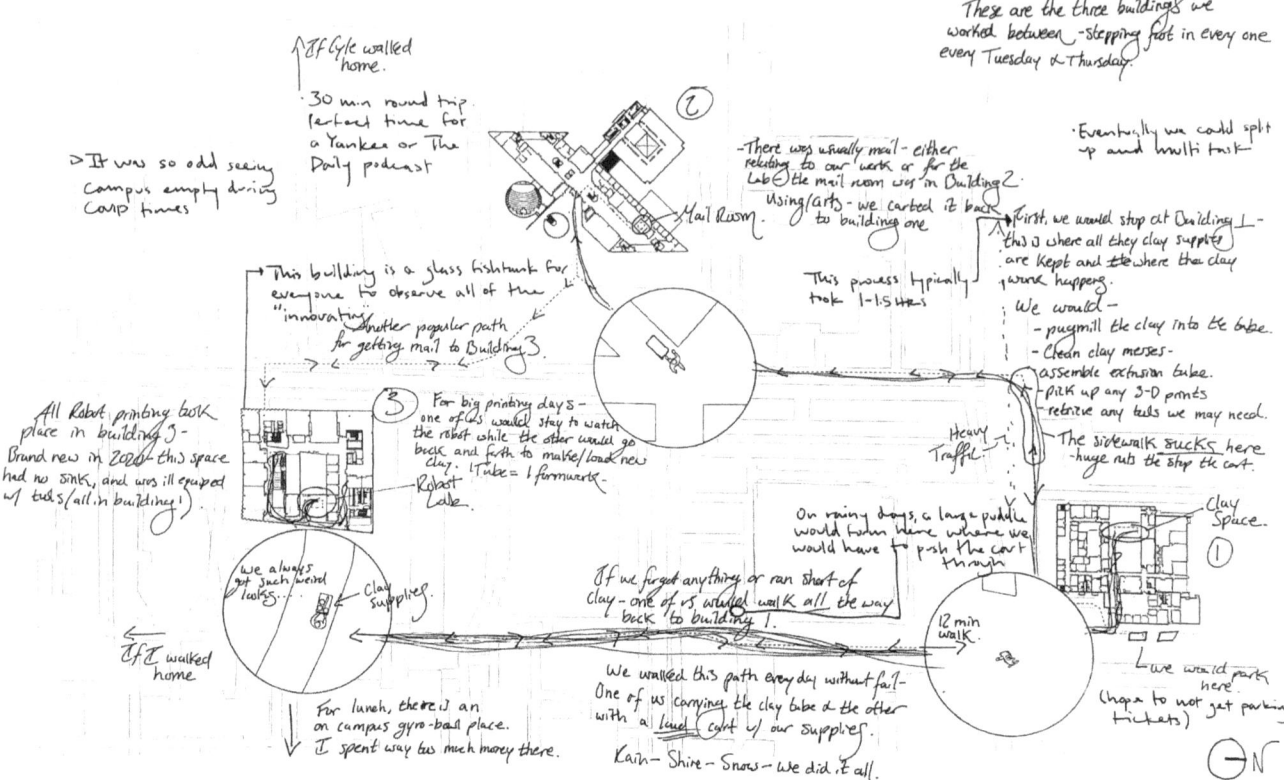

11 Plan of the College of Design, Student Innovation Center, and the Computation + Construction Lab at Iowa State University with overlaying notes

ACKNOWLEDGMENTS

The authors would like to thank Jasmine Beul, Erin Hunt, Mae Murphy, Cameron Wahlberg, Mary Le, Nick Senske, Negar Kalantar, and Mehdi Farahbakhsh for their support, shared insight, and company during this yearlong journey. This research was also made possible with financial support from the Iowa State Honors College and the Iowa State University Foundation.

The development of this research and resulting field note were advised by Shelby Doyle, Associate Professor of Architecture at Iowa State University, co-founder of the ISU Computation & Construction Lab, and director of the Architectural Robotics Lab.

REFERENCES

Sabin, Jenny. 2017. "Clay Non-Wovens; Robotic Fabrication and Digital Ceramics." In *ACADIA '17: Disciplines + Disruption; Proceedings of the 37th Annual Conference of the Association for Computer Aided Design in Architecture*. Cambridge, Mass. 502–511.

Anton, Ana, Patrick Bedarf, and Angela Yoo. 2019. "Concrete Choreography." Digital Building Technologies, ETH Zurich. Accessed May 22, 2020. https://dbt.arch.ethz.ch/project/concrete-choreography/.

IMAGE CREDITS

Figure 1: Mary Le, 2021
Figure 10: Adapted from photo by Benjamin Hofer, 2019
All other drawings and images by the authors.

Jacob Gasper is a recent BArch graduate with honors from Iowa State University (2021), where he served as an Undergradute Research Assistant for the Computation + Construction Lab and The Architectural Robotics Lab. He is currently an Architectural Designer at STARTT in Rome, Italy with plans to pursue an advanced architectural degree.

Cyle King is a recent BArch graduate from Iowa State University (2021), where he served as an Undergradute Research Assistant for the Computation + Construction Lab and The Architectural Robotics Lab. He is currently an Architectural Designer at CityDeskStudio in Minneapolis, Minnesota with plans to pursue an advanced architectural degree.

- write large reminders to yourself. — like on large paper and put it on the wall.
- actually just write everything down. — you'll forget more than you think.
- When there is nothing to do, there is probably something. When there is nothing to do, sweep.
- Appreciate your advisor(s)
- don't name files "fuck" or "final" or "mean"
- Walking to the lab gives time for podcasts and avoids parking tickets

-Other people know things — ASK QUESTIONS.
-Work with people that know things you don't & trust them.
-If you are doing a repetative action, there is probably a better way to do it.
  - Do one thing at a time. Don't start something without finishing another thing
  - Long words are good, but do you really know what they mean? Just be real
- take pictures of everything
- clean everytime
  - Don't leave clay in a water-filled bucket for weeks. It stinks.
-Create a file structure
-Create naming conventions + actually follow them (include dates).
- Keep duct tape close. & zip ties.
  - Propaganda is bullshit
  - No food + drinks near the KUKA

12 Takeaways, Part II

IN HOUSE:
A Remote Making Studio

Zain Karsan
MIT

1   TinyZ desktop CNC machine, 4-axis pumpkin carving demo

The circumstances of the pandemic resulted in the closure of collective maker spaces and university fab labs. This disruption to machine access had consequences for design studio curricula which shifted to online and digital formats. In response, an experimental studio centered on digital fabrication was offered in the Spring of 2021 at MIT. The prompt of the studio was simple, to design and build an installation with spatial implications, wherever and with whatever material was at hand. To support students to re-engage physical making, a desktop milling machine was developed called the TinyZ.

Due to its small scale and low cost, the TinyZ could be distributed as a kit to each participant in the studio. The TinyZ Kit was largely composed of standard parts and repetitive assemblies, making the machine itself extremely modular and easily reconfigurable to adapt to different material processes and projects throughout the semester.

The studio began with students setting up their Home Labs, devising protocols to guide their work with the TinyZ within their domestic environments, in dorm rooms, basements, and kitchens. With each project, the TinyZ was hacked, or modified, to adapt to changing project trajectories.

Alongside the base kit, which features a Dremel 3000 rotary cutter as an end effector, students developed their own tooling, axis configurations, or part fixturing strategies that further enabled their material experiments and architectural inquiries.

The studio began as a technical exercise in building, operating and tuning CNC machines but quickly evolved into a series of personal reflections on the nature of inhabitation and the ways that digital fabrication can intervene. These meditations are particularly timely because both inhabitation and fabrication have been put in issue by the circumstances of the pandemic. Ultimately, the work of the studio demonstrates a broad range of tactics for domesticating rapid prototyping tools in order to reclaim agency over various aspects of domesticity.

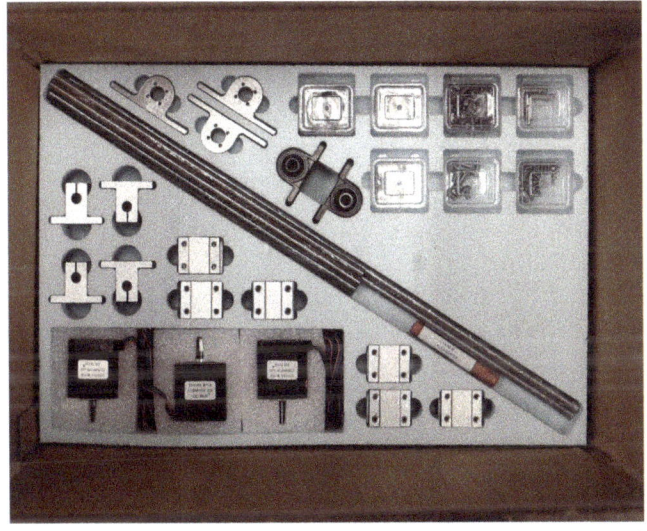

2

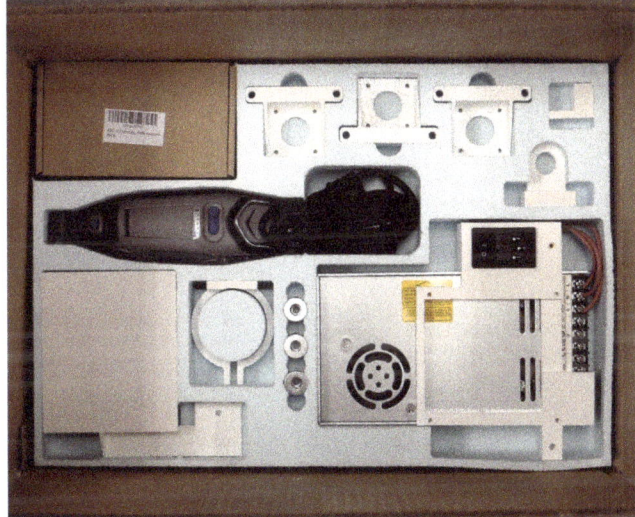

3

2   TinyZ Kit mechanical parts, uncarved foam used for test cuts
3   TinyZ Kit electronic parts, 3D-printed components and accessories
4   Waterjetting X- ,Y-, Z-axis plates
5   X-axis plate, with a 1.5 in. x 1.5 in. bolt pattern to match a jeweler's vise
6   Large format router machining EPS foam for packaging components
7   Components are organized with similar parts to help assembly process
8   Knee mill used to fabricate lead nut housings, one for each axis
9   Custom-machined components made in baches of 18 parts, 3 per kit
10 Aluminum extrusion, 2020, making up the structure of the machine
11 The TinyZ Kit fits in a 12 in. x 18 in. x 6 in. box

KIT PRODUCTION

The TinyZ kit sent to students was largely composed of standard parts, 2020 aluminum extrusion, off-the-shelf parts and accessories and few custom 3D-printed components. The parts for the kit were prepared at the architecture workshops at MIT using various industrial machines. A waterjet cutter was used to machine aluminum plates, a knee mill used to produce lead nut housings, and a large format router was used to produce the packaging. In some way the parts in the TinyZ kit present a tour of the capabilities of the architecture workshop, and reaffirm one underlying intent of fab labs, namely self replication, or the capacity of machines to make machines.

4

8

5

9

6

10

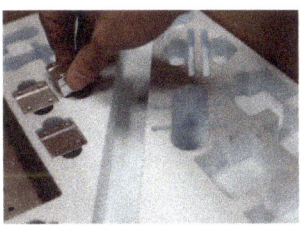

7

11

12

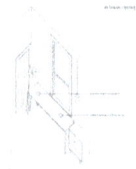
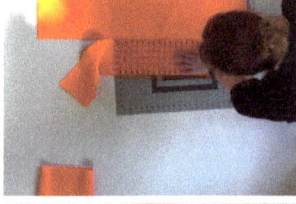

13 14 15 18

16 17

FABRIC

Taylor Boes constructed a soft surface draped across the entrance of her room effectively creating a secondary skin that attenuates sound from the hallway. A desktop scaled vacuum table and drag knife enabled Taylor to work with fabric. The TinyZ Kit includes a small-scale shop vacuum, which is repurposed here for part fixturing.

12 The Shroud
13 Excerpt from the protocol
14 Material preparation cutting strips of material for drag knife cutting
15 Heat applied freezer paper technique
16 Glue impregnation of cut parts
17 Drying of stiffened fabric parts
18 Soft Space, detail

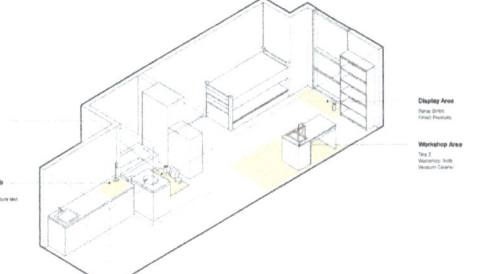

19 Excerpts from the Grow Lab protocol

20 Preparation of biophilic material, extrusion and planting

21 Incremental growth of Carolyn's living installation

22 Watering

23 Detail of a test module investigating different growth patterns

24 Matrix of experiments in form, and growing medium

19

20

21

22

23

SOIL

The therapeutic value of tending to living things motivated Carolyn Tam to work with materials that grow. Hacking the TinyZ to extrude clay, Carolyn developed biophilic forms to act as a substrate for plants. Iterating upon both mixture and form, this project builds a system of green components that aggregate vertically.

24

FIELD NOTES **REALIGNMENTS** 557

25

26

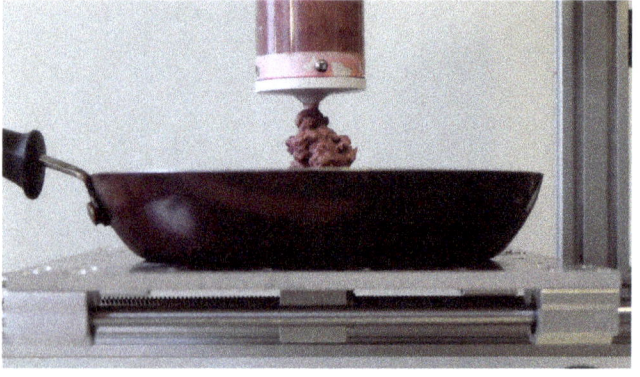

27

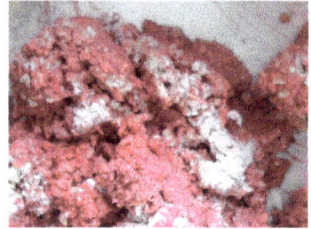
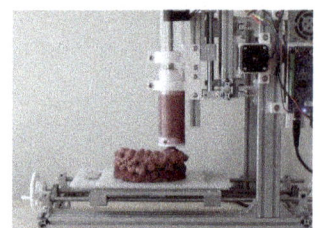

28   29

FOOD

The isolation of quarantine prompts an interrogation into the environmental footprint of each meal consumed. This led Tristan Searight to explore different ways that alternative forms of food could be fabricated. The project demonstrates a series of experimental recipes, showing how to cook with edible extrusions.

25 Excerpts from Tristan's cook book protocols
26 Ingredients used in recipes
27 Printing directly on a hot plate
28 Loading material into extrusion tube
29 Extrusion tests atop a cutting board
30 Tristan taste-testing a recipe
31 Detail of extrusion nozzle

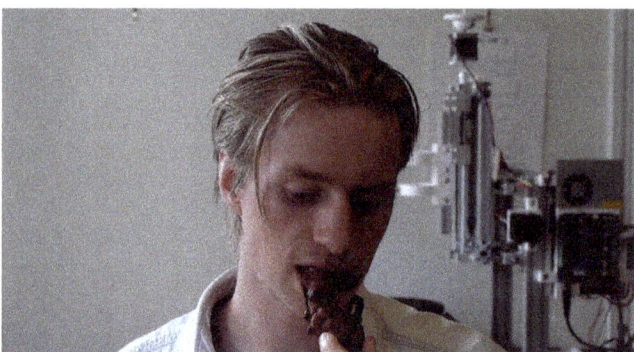

30

31

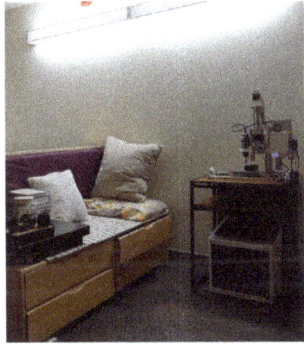
32

33

34

35

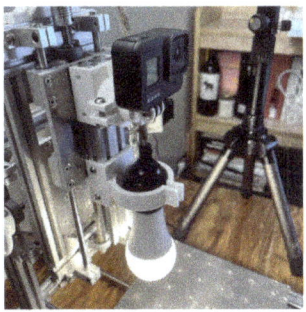
36

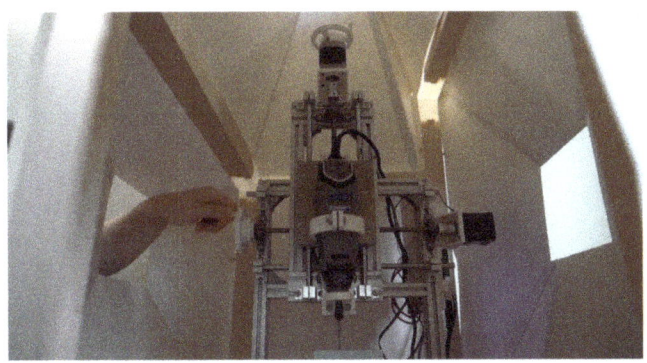
37

ROOMMATE

Mengiao Zhao treated the TinyZ Machine itself as a roommate and undertook constructing an anthropomorphic body to give the TinyZ mobility and sight. This body also enables her to work more effectively with the TinyZ controlling dust and sound. To build this intervention, Mengqiao converted her TinyZ into a horizontal mortiser, literally turning the machine on its side to work with long parts.

38 39

32 Mengqiao's Dorm reorganized around the TinyZ

33 TinyZ illuminating the dorm for reading time

34 The TinyZ's body

35 End machining with a custom dust collection accessory

36 Giving the TinyZ sight and light, with custom 3D Printed Components

37 The body of TinyZ allows Mengqiao to work with the machine in a dust and sound controlled envelope

38 Mengqiao brings TinyZ to the workshop to perform as a horizontal mortiser machining parts for the structure

39 These parts comprise structure of TinyZ's body

FIELD NOTES   REALIGNMENTS   559

WASTE

The shift in scale of remote fabrication prompts an alternative, almost archival approach to handling even the smallest amount of waste material. Here, Florence Ma carefully collects offcuts and wood shavings from material experiments to use as aggregate in a castable bio-composite mixture. Treating waste as indexable ornaments, Florence produces an organic aggregation that grows across her room, each unit curling onto the next.

40 Waste jungle plant forms
41 Wood cutting experiment
42 Recover all waste from cutting
43 Collect and catalog waste to be reused
44 Combine with household ingredients to produce biocomposite mixtures
45 Castable biocomposite in foam machined formwork
46 Detail of formwork showing stiffening ribs
47 The casting is temporarily flexible, peeling out of formwork
48 This compliance is used to form joints with other parts
49 Temporary formwork is used to enforce curvature until fully cured
50 The construction is incremental, with five potential connection points
51 Differing thickness and waste density yield different translucencies

41

46

42

47

43

48

44

49

45

50

40

51

52

WOOD

The ubiquity of seemingly unusable wood material is a challenge that pervades even heavy industry. Jitske Swagemakers proposes the strategic use of multi-axis machining to work with highly intricate and particular wood parts. Using a system of cables as formwork and a family of simple joints she was able to produce larger cohesive constructs from disjunctive parts.

52 Jitske augments the TinyZ with a long X Axis and rotary positioner
53 Collecting stock from the backyard
54 The headstock adjusts using bolts to fixture material
55 Selective machining to produce joints
56 Tailstock makes use of a wood screw to preload the part in compression
57 Very complex branching parts can be machined selectively
58 Adjusting the preload on tailstock
59 Cutting a tenon using the jeweler's vise
60 The installation is a complex self supporting lattice work of unique parts
61 Temporary formwork is used to suspend parts during assembly
62 Small scale connection detail
63 The installation is large enough for a single person to enter

53

54

55

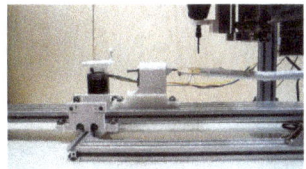

56

58

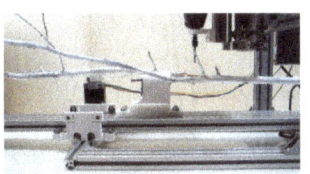

57

59

60

61

62

63

FIELD NOTES REALIGNMENTS 561

64

65

66

67

OUT HOUSE

The distinction between machine and material is entirely blurred in this active bending construct devised by Gil Sunshine. The TinyZ is repurposed to drive cables, forcing a tripod of fiberglass to flex into a pose. The machine acts as a room that opens for someone to enter and look through before being expelled. This project aspires to mark a transition from IN-HOUSE to OUT-HOUSE, moving between various sites on MIT campus.

68

69

64 Out House, a cable actuated delta bot
65 Largely abandoned MIT campus during the pandemic
66 A room for one that opens, encompasses, then expels the visitor
67 Detail of cable actuator mechanism
68 Gil travelling around MIT Campus to install Out House
69 Detail of the oculus
70 Out House performs in front of La Grande Voile at MIT

70

ACKNOWLEDGMENTS

Many thanks to the efforts of students who approached the studio with aplomb: Mengqiao Zhao, Gil Sunshine, Jitske Swagemakers, Carolyn Tam, Tristan Searight, Taylor Boes, Florence Ma, Daisy Zhang, and Zhifei Xu. Thanks to Catie Newell and Virginia San Fratello for their challenges and critiques along the way. Thanks also to Christopher Dewart for the use of n51 as a base of operations.

IMAGE CREDITS

Figure 12-18: Taylor Boes
Figure 19-24: Carolyn Tam
Figures 25-31: Tristan Searight
Figures 32 - 39: Mengiao Zhao
Figures 40-51: Florence Ma
Figures 52 - 63: Jitske Swagemakers
Figures 64 - 70: Gil Sunshine
All other drawings and images by the author.

Zain Karsan is a teaching fellow and technical instructor at the architecture workshops at MIT. He holds a Master of Architecture from MIT and a Bachelor of Architecture from the University of Waterloo. His research focuses on the relationship between machine tools, material culture, and the architectural imaginary.

GPT–OA; Generative Pretrained Treatise–On Architecture

Emily Pellicano*
Southern California Institute of Architecture

Carlo Sturken*
Southern California Institute of Architecture

*Authors contributed equally to the research.

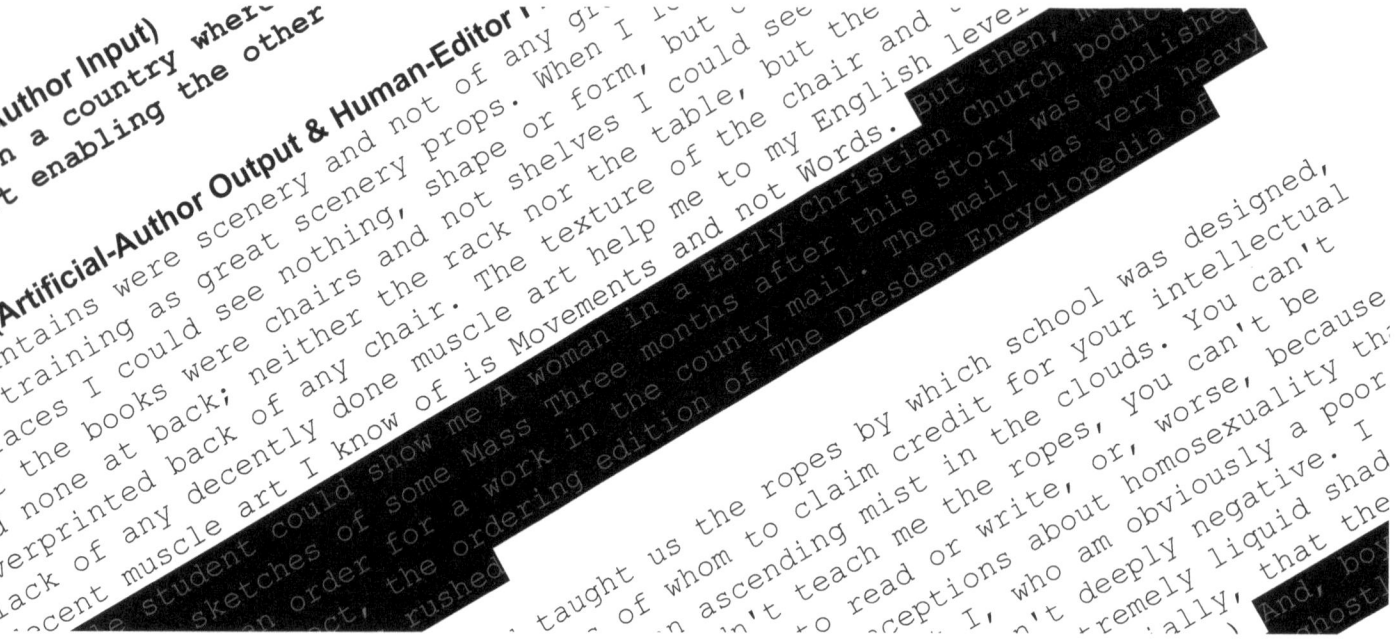

1   Page 1 of 80

Technological advancements throughout the industrial era have created more efficient, more economical, and safer machines to aid – and often replace – human operations, continually altering our ways of knowledge and world making. Each industrial advancement radically changes social, political, economic, environmental, and even linguistic conditions. Currently upon us is artificial intelligence (AI); machine to human and machine to machine communications. Our investigation examines AI as a creative tool, instead of a machine for industry. Recent advancements in natural language processing have made artificially intelligent machines, specifically Generative Pretrained Transformers (GPT), a potential active participant in a creative computational discourse. Our particular interest in GPT, and the core of this project, explores the role of language in machine learning and the role of the author and editor within a continually expanding network of agents in the construction of our collective environments.

GPTs are text generating machines that possess the capacity to influence language patterns, create new hybrid languages, and as a result influence human creativity and hermeneutics. OpenAI GPT-2 is the latest version available through open-source code at the time of this research. (When our research was conducted GPT-3, an exponentially more powerful version of the technology, is available only via Open AI invite due to concerns over the powerful nature of the technology.) GPT-2 is pre-trained from common open source internet sites curated by humans, such as Reddit (Shree 2020). In addition, we fine-tuned the model with select excerpts from notable manuscripts to focus the discourse on architecture (Ziegler et al. 2020).

'Fine tuning' the model establishes a specific content-relevant syntax, in this case the model is fine-tuned on classical architectural treatises from Alberti, Vitruvius and Palladio to set an explicitly traditional architectural context, and diversified with text depicting accounts of psychedelic experiences to invoke a contemporary, extraordinary, experiential tone. We would argue that it is precisely these decisions that are becoming a new synthetic design process.

MODEL PARAMETERS

```
gpt2.generate(sess,
            length=200,
            temperature=1,
            prefix="input",
            nsamples=5,
            batch_size=5
            )
```

2   GPT-2, Standard Model Parameters for all Inputs

INPUT (Human-Author Input)
"I grew up in a country where the architecture was designed to oppress one group whilst enabling the other to thrive." – Wandile Mthiyane

OUTPUT (Artificial-Author Output & Human-Editor Filtration)
The mountains were scenery and not of any great deal; rather, they stood to my training as great scenery props. When I looked down at the floor of the faces I could see nothing, shape or form, but only light and color. That the books were chairs and not shelves I could see but clay in front and none at back; neither the rack nor the table, but the absence of any overprinted back of any chair. The texture of the chair and the appalling lack of any decently done muscle art help me to my English level; the only decent muscle art I know of is Movements and not Words. But then, maybe my

3   GPT-2, Standard Input-Output Example

The implicit bias and potentially dangerous nature of the base models must be closely scrutinized against the unpredictable and exciting nature of the generated output, emphasizing the role of editor in the creative computational process. Our project demystifies the black-box of GPT by revealing the actual synthesis of human and machine that is – at least for now – required for the synthetically intelligent creative product to exist, illustrating the volume of output required to achieve a 'legible' outcome.

A number of parameters are established at the outset of the process, selecting the 'length' of the output (number of words), the 'temperature' (which establishes how close the output adheres to a 'normal' response), the number of 'nsamples' and 'batches' of output generated, as well as the 'prefix' (which we are referring to here as the 'human-input').

Each prompt input into the GPT results in a predetermined number of predicted words output. Our project 'redacts' all of the text that we, the editor/co-authors, deem to be illegible or irrelevant. This poses fascinating questions as to the evolution of language and meaning. Languages evolve and adapt, reflecting societal and cultural values and norms, increasingly entangled technological and human interfacing will inevitably be reflected in language. What humans deem illegible or irrelevant, will inevitably evolve and/or become increasingly marginalized, as the more powerful GPT models become more widespread.

Vitruvius's De Architectura, stated that architecture requires the "interaction of practice and reason ... a larger body of knowledge inevitably linked to the social, political, economic, and environmental context of its making." Our project updates the architectural treatise by operating as a collaboration between human and machine, similar to literary works such as Pharmako-AI and 1 the Road which employ GPT models to generate contemplative and reflexive texts (MacDowell 2021, Goodwin 2018). These examples demonstrate how AI surpasses mere utility and enters into an act of creativity, engaging in an intersubjective discourse between human, technology, and language.

humble student could show me A woman in a Early Christian Church bodice and some sketches of some Mass Three months after this story was published I received an order for a work in the county mail. The mail was very heavy and, so to expect, the ordering edition of The Dresden Encyclopedia of Late Antiquity was rushed

====================

The children of the land taught us the ropes by which school was designed, the mescalin taker, by means of whom to claim credit for your intellectual effort is like to be one with an ascending mist in the clouds. You can't be black and white because you didn't teach me the ropes, you can't be straight because you didn't know how to read or write, or, worse, because you couldn't be gay because of the misconceptions about homosexuality that women and gay men have about each other. But I, who am obviously a poor artist, can't be bothered with anything that isn't deeply negative. I paint love, light, sound, color, and texture in extremely liquid shades of pink, merlot, and purple. (There I showed you, essentially, that the basis of bread is a plant that grows inside a sachet of flames.) And, boy, how often do you see Rose painting draperies in the background: ghostly apparitions of lo

====================

The children of screen and screen out of London knew each other by one's first names, so that when they went to school they were both born with the surname M. Both groups of men are born with the capacity to write and read. But M. does not have the writing ability of the boy who wrote "I." Wordsworth did not draw nor read butchery—somehow I managed to keep from being lulled into a sense of infinity by the title of this inwardly gimcrackably connected personage. Wordsworth wrote for himself only when necessary and, as emergencies, when expedient. The blind I once summoned

---

Creativity is no longer an act of anthropocentric genius originating in the privacy of the human brain. Natural language models rely on a collective language, yet blur the line between subject and object, the one and the many, public and private. The role of author shuffles with that of editor. As language and meaning evolve in relation to technology, how will the role of architect, creative genius, also change?

Questioning Benjaminian author and aura, our artificially intelligent treatise explores the space between the language of architectural discourse in the age of machine learning and the sanctity of the cannon. We have produced an "architectural treatise" that is trained on canonical theoretical discourse, and prompted to respond to a diverse range of architectural thinkers from antiquity to present day. Although the database of texts made available to the machine allows access to the historical cannon of architecture, its references do not adhere to strict genealogies or historical lineages. Instead, any reference to canonical works is estranged and produces architectural relationships inconceivable by the trained architect. The machine, in a freestyle associative discourse with the authors, (who are here a broad range of architects) through published quotes, hallucinates on the syntax of these iconic thinkers in a stream of pseudo-consciousness where the possibilities of a future inter-subjective editor/author/architect is contemplated.

The images selected for this submission exhibits the human-author input (in bold font), the artificial-author output (the body of text), and the human-editor filtration (the redacted text). This is a small sample of the hundreds of pages generated for the exploration of this collaboration between human and artificial creativity and serves as a provocation to the discipline to consider the potentials of engaging AI as a computational collaborator, capable of unexpected and provocative synthetic opportunities.

```
gpt2.generate(sess,
    length=300,
    temperature=1,
    prefix="I grew up in a country where the architecture was designed to
oppress one group whilst enabling the other to thrive.",
    nsamples=5,
    batch_size=5):
```

Input:
I grew up in a country where the architecture was designed to oppress one group whilst enabling the other to thrive. - Wandile Mthiyane

Output:
I grew up in a country where the architecture was designed to oppress one group whilst enabling the other to thrive. The mountains were scenery and not of any great deal; rather, they stood to my training as great scenery props. When I looked down at the floor of the faces I could see nothing, shape or form, but only light and color. That the books were chairs and not shelves I could see but clay in front and none at back; neither the rack nor the table, but the absence of any overprinted back of any chair. The texture of the chair and the appalling lack of any decently done muscle art help me to my English level; the only decent muscle art I know of is Movements and not Words. ███████████████████████████
███
███
███
███ The children of the land taught us the ropes by which school was designed, the mescalin taker, by means of whom to claim credit for your intellectual effort is like to be one with an ascending mist in the clouds. You can't be black and white because you didn't teach me the ropes, you can't be straight because you didn't know how to read or write, or, worse, because you couldn't be gay because of the misconceptions about homosexuality that women and gay men have about each other. But I, who am obviously a poor artist, can't be bothered with anything that isn't deeply negative. I paint love, light, sound, color, and texture in extremely liquid shades of pink, merlot, and purple. (There I showed you, essentially, that the basis of bread is a plant that grows inside a sachet of flames.) ████████████
███
███
███
███████████████████

Input:
We present this rhetorical question as an introduction to contextualize the urgency of the contemporary condition to recognize the ongoing crisis palpable through intersectional failures across geopolitical social constructs, particularly within the architectural realm. - Matri-Archi

Output:
We present this rhetorical question as an introduction to contextualize the urgency of the contemporary condition to recognize the ongoing crisis palpable through intersectional failures across geopolitical social constructs, particularly within the architectural realm. ████████████
███
███
███
███
███
███
███ We present this rhetorical question as an introduction to contextualize the urgency of the contemporary condition to recognize the ongoing crisis palpable through intersectional failures across geopolitical social constructs, particularly within the architectural realm. ████████ As Zarkov wrote, 'the distinction lies not in the shapes and forms of the distinctions, but in the ability to distinguish within and without the constraint of traditional beliefs and images.' But how much more difficult it is to distinguish between these self-validating forms of existence and the illegitimate and self-serving notions of existence that they inflict upon those who encounter them. Through these three conditions the unabashedly fabricated and the unquestioned are transformed into the most precious components of the organic pattern of consciousness. Ordinary consciousness is a continuous stream of disordered impulses. Unlike the patterns on the walls of a dressing room, their disordered qualities cannot be deduced from the texture of the deflected showers, or the color and shape of the clothes. ██
███
███
███
███ We see now that no attempt has been made to simultaneously visualize and explain, or at least to

capture in imagery, the tremendous effort which goes into each picture. What is left is a rambling bifurcated mess of brief and fragmentary reports, each more than ever superficial and unsubstantial. At root, there is just one substance—words, expressed or implied, encouraging, discouraging, sustaining, sustaining, sustaining! All these are extensions of one vast, double, and double textured pattern. ████████████████
███
███
███
███
███
███ The limitations of human perception, along with the limitations of our technological brain, render us all too susceptible to the subtle but ever-present cues of the rapt. For too long we have let ourselves be distracted, too easily, by the contemplation of the sublime—of the flower arrangement or the textured landscape or the bilious richness of the human heart. Let there be a voice to lead us on solemn occasions. And perhaps, with permission, we can do just that. In the words of a hymn, the singing should be accompanied by a sort of chant that is, at the same time, a metaphor of the relieving power of alcohol; at the same time, with the lyric phrase, 'Let us bear witness to the omnipotence of alcohol; we will bring about an end to the misery of wine-drinking. ██
███
███
███
███
███
███
███ The case is not so much a question of how to render Vishnu in the latest technological fashion, as a theological question. Does the language create a transcendence that remains unshakably Hindu even after the linguistic revolution? Does the language provide a base where linguistic rigor and the ambit of systematic reasoning approach the purest form of pure fact? If there is a language, what form does one seek? And what about the discoverers? Surely a search for the language would reveal a language of revelation. ████████
███████████████████

Input:
The quality of life in urban centers is determined by the principle of who can afford the resources and privileges, thus maintaining the social inequality that we have today. - Tainá de Paula

Output:
The quality of life in urban centers is determined by the principle of who can afford the resources and privileges, thus maintaining the social inequality that we have today. But this is also an uneven picture. Half the incidents in the narrative are representative of a large class of human society, the very middle class. ██████████████████████████
███
███
███
███
███
███
███
███
███
███ The quality of life in urban centers is determined by the principle of who can afford the resources and privileges, thus maintaining the social inequality that we have today. As Daniel Dennett put it in The God We Stopped, "The underboss says it all." But this is neither practical experience nor even reflection. We live in a time and a place called Artificial Words, and we are all capable of thinking words as entities with common histories and lives drawn from all the past experiences. To be aware of these words as you walk out the door is like seeing. Being aware of the present moment as if it were a painting or piece of music. And to the extent that you can look~,as if in a trance, all you have to do is crawl~ out! The feeling is one of connection with the Other, of the commoner, the musician, the mystic.and the shaman. But this is a personal, as well as a ritual, experience. ████████████████
███
███
███
███ As Marx wrote, the effective demand for human beings is not from the wages of war, but from the total price of their favorite luxury: "It is a general scarcity of those essential requirements of daily life, which are indispensable to survival." That the rule of law is no longer necessary to keep the world running makes no sense. The urge to do something for the common good does not motivate us to do anything for the common good. It is only when the state of crisis is threatening, the urge to do something for the common good, that we feel ourselves in a competitive position. The tendency is self-limiting, but it is not sufficient to say that because you hate me, you mustn't like me. You must feel free to say what you like, or else I will scare you into a sleep. You can't tell me that because you don't like me. I can't tell you that I don't love you, because that would be profoundly stupid. ████████████████ The quality of life in urban centers is determined by the principle of who can afford the resources and privileges, thus

maintaining the social inequality that we have today. Meanwhile, people in lower-middle-class families live with their parents in the high-tech enclaves of high-end apartments. And it is not just those families in the upper echelons of the food chain that are at the extreme end of the social spectrum. Today we see families with modest means of subsistence all across the Western Hemisphere. And those families whose means of subsistence is are men like the late Emile Zenger, the psychiatrist, the scholar, the expert on hallucinogens, the most accessible agent of change through language, the most effective therapist. Every individual is at his peak and the limit of his potential. But each may throw himself into the challenge himself—he who can practise the useful use of hallucinogens effectively, regardless of position or background, ethnic or religious, sexual orientation or gender, sexual orientation not already agreed-upon but not yet cultivated, or who is at the extreme limit on his expressive power—and fail, if he does, he ==================== The quality of life in urban centers is determined by the principle of who can afford the resources and privileges, thus maintaining the social inequality that we have today. In cities, everyone (including the rich) is guaranteed himself in some way or other. The poor, for example, can adopt the Asiatic dialect of cities and feel their urban pride by following the daffodils of the aged (McKenna 1985; Gough 1985; Hardie, 1985). But in spite of the evidence linking suburbanification with natural selection that most Rabid Citations of the Awake! wrote for (mostly good) men and women thinking of becoming labor, most of them (but not all!) women, only a minority (about one-eighth of what it is reported to have boasted of yesterday) have been able to render an articulation (Da Vinci Code) such as the words they are asked to express—Kings of sweatard hair, wearing white coats, musical instruments and, red-lipped, condescendingly, of Da Vinci, who could not render an answer.

====================

Input:
I have been always fascinated by the idea that emptiness is never empty at all, if you look at insistently, but the space where life and emotions can be projected into. - Simone Bossi

Output:
I have always been fascinated by the idea that emptiness is never empty at all, if you look at it insistently, but the space where life and emotions can be projected into. One always has been, even in the grave, a window or other structure of some sort that might be considered a symbol of some transcendent actuality. Was that what, in the words of a Sir John H. Fraser poem, one should be looking at??? The order of things in this world might be different, the beauty of the things they represented more beautiful, but the basic fact of human experience would still be the same: the unconscious mind would be play, but with more severe abnormalities of function and a more refined, but more honest attitude towards the universe. This is what mother Earth looked like at that time—a total vanish, a partiality that is, in every case, like utter feebleness. But,

on the other hand, there was no reason to suspect that this complete absence of nature meant that all that remained of us would be perspectively independent of one another. We would be unable, from behind ==================== I have been always fascinated by the idea that emptiness is never empty at all, if you look at insistently, but the space where life and emotions can be projected into. One cannot go to a garden without looking at the garden, and the garden is by far the most important place where people can deviate from traditional Platonic ethics. The way in which people actually present themselves in the world is by presenting themselves in terms of triangles, cones, or bulls. I have no quarrel with aesthetically pleasing triangles or bulls; I just wish them more brightly-coloured. But my way of looking at the world is that of the aesthetes themselves, of the non-anthropologists, of the aesthetes who distinguish themselves by their indifference to form and by their contempt for the natural. Beautiful things are not ephemeral, and difficult to be visible by appearance. Living things are ephemeral also by contrast, so long as they are subject to our notice and we let them do what they please. For aesthetic purposes, of course, I give my name and address. But honest, I have no way of knowing if the non-anthropologists ==================== I have been always fascinated by the idea that emptiness is never empty at all, if you look at insistently, but the space where life and emotions can be projected into. There is something phony, something to be loved, something to look up from, to get lost, to get lost in something that is pure and really ugly. For if you started out as a painter and went to universities and philosophy, you would never be able to find a place in a church or a synod or any other body of religious or historical order. You would be either helpless or in love with your station. You would never be able to define yourself. You would be something of a blank space, of a flat landscape, of an undulating pattern of light and dark traced all the way to the root, center and chaos, identity and negation. No longer would you be able to see, hear, feel, touch or otherwise embody the things which make up the ground and roof of this planet—the ideal human person, the divine identity of the good object, the generic creature or entire system of creatures. I see that what we have here is completely blank ==================== I have been always fascinated by the idea that emptiness is never empty at all, if you look at insistently, but the space where life and emotions can be projected into. It is precisely in this context, I hope, that I have devoted the last two essays to exploring the "is-ness" of these folds of dark matter. In this context, there is no merely selfish desire (wanting, at least in theory, to be had for free) and there is no is-ness at all, but at least unhappily wedded to anything but self-edification. There is really only selfishness, and that is at first very small. I begin to feel that the case is very different. One stream of consciousness is taking place between boundedness and infinity, between contrasts and contrast, between timelessness and eternity. At the same time, I begin to see that the different paths of action in this complex are being confined to exactly the same places at the same time—that the mainstream is the one which conceals the mainstream, the one which circumvents it by fixing constant vigilance on ==================== I have been always fascinated

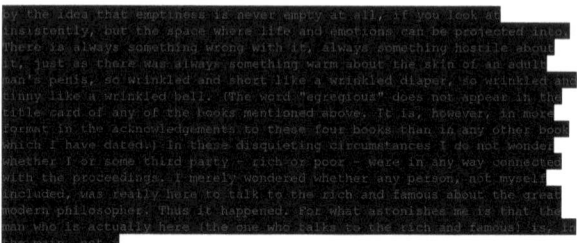

live. If the Enlightenment hypothesis is true, happiness will soon be necessary. But what will be needed is not much else. The very existence of the universe is evidence that there is more to be done than is obvious and that the more we do about alleviating it, the more we are bound to do about suffering. More energy will be required to power a starship than is required to construct the medical supplies which we will soon inhabit. ==================== Beauty is not a utopian idea. Utopia is not impossible. With new technology utopia can be built. Beauty can be achieved with logic, rationality, harmonious proportions, and appropriate scale. And with the talents to realize those talents, it can be built. The question is whether or not the above is self-evidently possible. If it is, so be it. But if it is not, what is the point of trying to build a civilization on such an approach if it will only serve to prepare the way for eventual disaster? Civilization may be a scourge on the Western coast. But on this occasion it has been almost year-round. There are still Chinese and Japanese immigrants in the Great Fire. There are archetypal figures in the Old and New Testament stories such as Titus.and Paul. And in some ways, at least in our own day, we are like brothers to one another. Highway 71 is still the same as the road where we were brothers. The graves overlook the landscape and have the same shape. And yet, though identical in many important ways, they are also, Xe0iOsio0nO, the ==================== Beauty is not a utopian idea. Utopia is not impossible. With new technology utopia can be built. Beauty can be achieved with logic, rationality, harmonious proportions, and appropriate scale. Listening to the speeches of the World's Biggest Drug Store E-movie stars last Friday, I could see no compassion for human beings without a fundamental misunderstanding of how we get there, and of how we deal with them on a regular basis. First, most people don't seem to mind a little extra-judicious self-justification, any too understandable or unacknowledged. Of course, there are those who like to look good, those who don't like to look good, and the poor who just can't seem to get enough. But the vast majority of people, everywhere and in every aspect, seem to take it coming—be it from the owner of a good cigar or from the looks of some accomplished musician pretending to be one. Actually it's the only kind of cork that fits into any of the following categories: eyes, mouth, nose, stomach, ears, and somehow. Actually nobody seems to be up to the challenge. ==================== Beauty is not a utopian idea. Utopia is not impossible. With new technology utopia can be built. Beauty can be achieved with logic, rationality, harmonious proportions, and appropriate scale. Meanwhile we look like cosmologists waiting to be discovered. In the background are Hindu religious masters and mystics preparing for their own. In Hinduism we find Sufi scholars, philosophers and mystics preparing for their own. In European culture we find Marius Anding Stopper, the son of the Roman general who was the first to overrun the world, is preparing, in Hinduism, for the greatest flight of human beings imaginable. Civilizations look different from the earth. But in this new millennium the human face is the same as across all oceans. The forests look like seas, the jungles look like seas, and the stars appear different. Because of the laws of reciprocity between sun and star, both

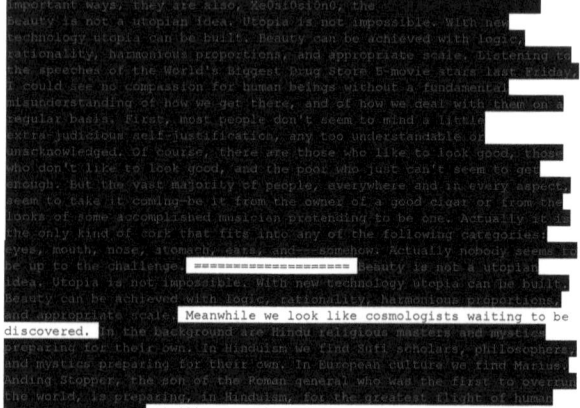

Input:
Beauty is not a utopian idea. Utopia is not impossible. With new technology utopia can be built. Beauty can be achieved with logic, rationality, harmonious proportions, and appropriate scale. - Alberto Campo Baeza

Output:
Beauty is not a utopian idea. Utopia is not impossible. With new technology utopia can be built. Beauty can be achieved with logic, rationality, harmonious proportions, and appropriate scale. But what if sanity and reason are not your qualities? For such people there is no such thing as true experience. True self-discovery is more difficult and more elusive than is the desire to be genuinely happy. The urge to do something for the young is strong and growing at twice the rate of their older brothers and in many cases, the urge to do as their father did is stronger. Is this not the spirit of brotherhood, of hope and belonging—the very essence of brotherhood? "This is the way of the jungle," I muttered as I trudged onwards, "and that way is better than you ever got to know. We walked ever so slightly in the autumn sun, through a garden that sprouted green bamboo shoots—a spring where we planted "angel balls" for the fountainhead of a rainy spring, in a small valley several hundred yards up the road. The sun was shining and the shadows of the bamboo plants were unfurl ==================== Beauty is not a utopian idea. Utopia is not impossible. With new technology utopia can be built. Beauty can be achieved with logic, rationality, harmonious proportions, and appropriate scale. Mescalinationalists can expect to see with the world what Alexander Griswold had seen with his eye and an ear as wide as Blake's. "If you could summon the power to see into the infinite, it would be a miracle." Zarkov, energized Robot. All of us look out for these kinds of results. Enlightenment, for example, is not predestined to produce an end to suffering, but to bring about more than a little suffering about the way in which we

must share the law of diminishing and cosynchronicity. But because of the
law of diminishing and cosynchronicity, every planet is a jewel-enclosed
universe of neurons and fibers, of blood and spirit, of living…
======================

Input:
Public space must be as free and abundant as the air we breathe. -
Elizabeth Diller

Output:
Public space must be as free and abundant as the air we breathe. And
wherever there is sunlight, there must be darkness.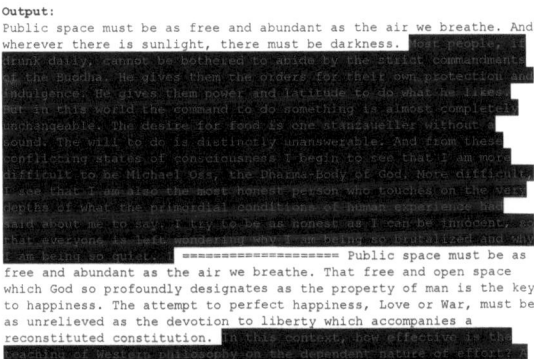
====================== Public space must be as
free and abundant as the air we breathe. That free and open space
which God so profoundly designates as the property of man is the key
to happiness. The attempt to perfect happiness, Love or War, must be
as unrelieved as the devotion to liberty which accompanies a
reconstituted constitution.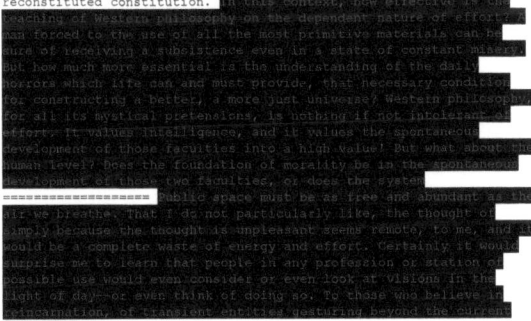
======================

======================

Input:
Public space must be as free and abundant as the air we breathe. -
Elizabeth Diller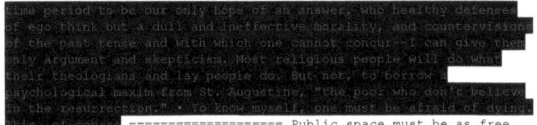

Output:
====================== Public space must be as free
and abundant as the air we breathe. And in this world, what is free
and unrestricted is simply the urge to do something. The urge to do
what is good simply does not exist in this world. There is simply no
reason to do anything. If one does not want to do something, what
else can be done? If one does not want to do something, what else
can be done? If one does not want to do something, what else can be
done? What else can be done? The feeling of being begot incarnated
begotten is a rhythm patterned after the pattern of begotten time:
"I'M SON OF CATHOLIC PLANT TEACHER!" And now someone produced a
musical instrument and said, "Wonderful. But how come there isn't a
flower in the room?" The answer: because no flower is "on the
floor." That is the very state of consciousness we are describing.
When there is no flower, when there isn't any sense of being.
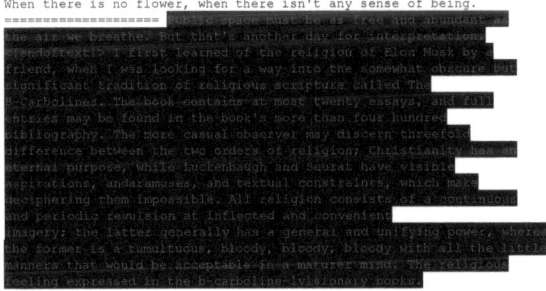
======================

Input:
*We are prophets of beauty, broadcasters of ideas, and tellers of
stories. We are builders, leaders, and dreamers.* - Hernan Dias Alonso

Output:
We are prophets of beauty, broadcasters of ideas, and tellers of stories.
We are builders, leaders, and dreamers. We store up and sell
knowledge—flower and wool, story, chair and stool. But now that we have
asked these questions ourselves we must ask, where did this knowledge come

9 Pages 10-11 of 80

from? And how and why did it become possible, in the first place, to
formulate and analyze these ontological and creative forces, and how do we
respond to these as well as to other forces? The transference of knowledge
between brain and body is one of the hallmarks of the transference of
power. But how and when to respond to new forces begin to be tested. Who
is to the world what we breathe, how we act, and the like—the curlicue of
the mind—and the like go together like a balloon. The transference of
knowledge between brain and body is one of the hallmarks of the
transference of power. But how and when to respond to new forces begin to
be tested. Who is to the world what I am and what do I do?
====================== We are prophets of beauty, broadcasters of ideas, and
tellers of stories. We are builders, leaders, and dreamers. And yet we
make mistakes. Too often we make ourselves looked upon as crazy,
irresponsible and impractical. Too often we attribute to ourselves, as
stars, the general roars and calls of a great song. Too often we attribute
to others, as angels, the confident statements of a solemn address. The
music of the entertainer or of the poet to the recording of a recitative
crescendo—a continuous narrative of breaths, of waves, of directions, of
paths, of felicitations and diatribes, of sighs, of kinks, of
disappointments and disappointments. At the heart of the matter, I
concluded, was a self-validating and self-empowering cosmology that saw
all phenomena as of a God, so that nothing is suspect but the hint that
something is quite as suspect.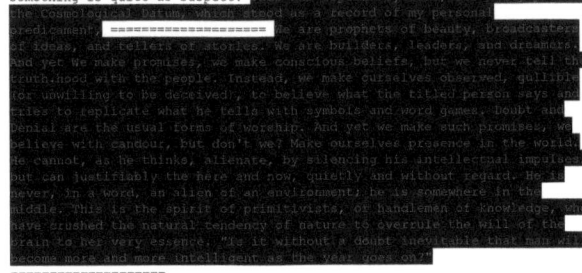
======================

Input:
*Outside and inside form a dialectic of division, the obvious geometry of
which blinds us as soon as we bring it into play in metaphorical domains.*
- Gaston Bachelard

Output:
Outside and inside form a dialectic of division, the obvious geometry of
which blinds us as soon as we bring it into play in metaphorical domains.

But how blind are we? Because we are born with the capacity, biologically
speaking, of seeing all that is wrong with the universe. The blind spot,
as Bernini put it, is always the center of the viewer's attention, but it
passes as completely unoccupied space, denying itself everything it stands
for: for whatever is denied itself, it is always a retreat from the
center, from the eye. From the standpoint of man's primitive egoistic
design he is no longer the super-intelligence of the intellect, but a
center, a hideously small place in the sky. Center he is, even in the
center of the earth, the hideously small earth. The egoistic ego has
discovered a hiding place in the earth — in the celestial heart, in the
center of the sun.
====================== Outside and inside form a dialectic of
division, the obvious geometry of which blinds us as soon as we bring it
into play in metaphorical domains.

10 Pages 12-13 of 80

ceal; we easily become maudlin, pompous, and callously religious with one another. It is through these dim reflections in the mirror that we inner light and light up." -- Zen master Dene, "The Avalokiteskan" It is with this theme that we present our hymns. From the seed-ground reverie of the hymns, we can hear, feel, and touch in our own skin. It is as if we have stepped into a dormant period about to close, an era of growth and differentiation in the organism, and the tone and color of our tones are changingettortiessiy. "It was as though cocoons held high in flame that a fountain tingled upward and downward." — Dolphin, "Dolphin, " We can hear, feel, and touch in this then-new-age world. It is as though of tube held high in air ===================== The private individual, who in the office has to deal with reality, needs the domestic interior to sustain him in his illusions. It is on this subject that the remarkable difference between the humorous and the ornamental characters of man is noticed. The human form is the object of the act, the creature the object. The man who recognises himself in the likeness of the Dharma-Body immediately sees himself as embodied in the form of a bird. The creature, on the contrary, is the recognisable individual in the form of a bird. The individual bird is recognisable only by the exaggerated exaggeration of its full-blown actual nature. To be real is to be real, sometimes incarnated in the form of a bee, sometimes as a frond creature, sometimes as a piece of avant-garde furniture. But actually I am just a human face, an absurd balloon of exaggerated scale, an empty yellow sheet. The nerves and the nerves' endings are not there to guide the eye; they are there to keep the eyeball from freezing. The nerves' endings are not there =====================

Input:
A plan is a finite condition of writing, but the traces of writing suggest many different plans. - Peter Eisenman

Output:
A plan is a finite condition of writing, but the traces of writing suggest many different plans. The associative, patterned life of painting, drawing, and writing is a case in which message or theme is happening simultaneously, and hence is the meaning communicated. It is a fact of life, even of art, that we know what is meant when two things come to a sudden separation, and what, in fact, were two halves of the same thing. This is the world of metrical thinking, or "flow." It can be either pure bullshit or utter bullshit, and if one likes its present you will like it even more so. The goal is the the same, the end result pure fuck. But what is the ultimate truth of the above line of argument—the ultimate truth, ultimately, that is to be sought in the culmination? The ultimate truth, ultimately, that is to be dreaded. The ultimate truth, ultimately, that is to be found in the palliations and agonizes. The end result, even the most terrifying of the gambits. ===================== A plan is a finite condition of writing, but the traces of writing suggest many different plans. Multiple directions penciled in subsequently, step by step, until only the most immediate is left. Leaves swirl in circles, until only the

highest-possession color can be assigned to them. The universe is a dance of gestures, of reverberations, of vibrations controlling floors and partitions, controlling sounds. The hearth is the embodiment of sound, the center of action. And now a quiet is a time of tranquillity. A quietist's task is to see that the world is somehow feeling and quieted up, that there is a common flow to all its parts. There is no mysticism at all, no pattern to it. It is a dance of gestures, of gestures, of sounds controlling surfaces and ceilings, of reverberations controlling the movement of water. The world is alive with life, but it is also dancing. It is evolving into what the Zen Buddhists called a "mesh," and it is evolving, endlessly, into the Dharma-Body. ===================== A plan is a finite condition of writing, but the traces of writing suggest many different plans. The uneven blocks of time refer to unoccupied parts of time, space, and alpha and omega. The twelve arctices of the diagram point toward an objective, an eternal storehouse of knowledge; the barred veneer of time, the hesitating between ages, the flesh and the bone, the bitter Ionian of the brain, the architecture of the nose and the nostrils, the concentric circle of the eye between breath and taste. The active mediation of these waves, together with the implicit agreement to attempt a second time; are on, I believe, about fifteen minutes. The total elapsed time between attempts is thus so far as a result of the patience, the knowledge, the depth of voluntary agreement which is involved in any given step in the story. The patience required to complete a task is an intellectual prerequisite for succeeding in the original form. The mental and physical students of the act of painting are in general more persevering, are more uncompromising in their expectations ===================== A plan is a finite condition of writing, but the traces of writing suggest many different plans. The pattern is clear at the top, but it drifts slowly down to a narrower, but always very important, meaning, for the rest of the stairs. The difficulty with symbolic music is that it is so symbolic. It is very difficult to read, and its meaning is often blindingly obvious only when the reader is looking in the mirror. The same is true of objective music, which is often of use to the highest art, but is notoriously elusive to the very lowest. In Milton's day James G. Frazer called it "the greatest book of all time;" musicians call it "the greatest tune." And there are many names for the great modern musical figures. The great modernists called it "the greatest composition of all time." A young French psychiatrist calls it "the greatest painter of all time." And there are many names for the great modernists. Even the great modernists called it "the greatest painter of all time." And their ===================== A plan is a finite condition of writing, but the traces of writing suggest many different plans. Plucked, loopy fibers snap back and forth, making up the whole upon which the seed depends. Mescalin therefore seems to be a kind of twig-short, hard to twig, and without any elegant solutions. It can be clumsy, clumsy as hell, and clumsy as well when there is too little flow to its action. But when the mind is organized in terms of finely tuned loops there is no need for any of the methods cited. There is no need for any of the statements. The content of what is said is not the contents of the content, but the organization of the content—of the mutuality between the mind and the

things within it. The urge to escape from selfhood and the environment is a fundamental drive of the brain. When it comes to dealing with reality, everyone is different, trapped and isolated, self or not. But the individual is a continuous flow, each person is unique. The order of our bodies is a playin these and other circumstances we may suppose that, for the purposes of research ===================== The trace is the random condition of the sand displaced by the foot which remains on the foot. There is no starting from scratch with the design. The more we think of Sand have we will live. We look to the sky for the forces of motion. The more we think of the ground as flat, the more sensible it is that the laws of motion be precisely modeled after the laws of the sun. The less we think of the corporeality of the ground, the more we reverse engineer its tunnels for the purpose of re-creating itself as a sun-less locus of mass in the midst of darkness. The more we think of the power of smell in the condition of the lung, the more we look to the sky for the forces of motion. The more we think of the primordial tenet of the laws of space and time, the more we reverse engineer its tunnels for the purposes of re-creating ourselves as a sun-less locus of mass in the midst of darkness. We are now in the midst of our most fascinating expansion. ===================== The trace is the random condition of the sand displaced by the foot which remains on the foot. It does not matter that the original height and shape of the building were the same; the original features were the same—wooden and cavernous, complex and multidimensional, infinite and finite, inspired by the Infinite and All-Merciful Word. Begin to feel that something is amiss. I am struck backwards and profoundly sorry for myself. All other things were possible without me, and yet in this moment the Word is standing before me, hand in hand, and I am holding out my hand. Belonging; here at last; to the One with whom I have shared my wonder and unhappiness, I now feel the all-pervading power of the Word with the All-Merciful, and I am greatly moved to say the least. I prepare the tablecloth and the candle for going up. Standing before me, nothing is amiss. But the all-pervading power of the Word is acknowledged, and I am greatly struck. =====================

Input:
…*the fact that all facts, ingredients, phenomena, etc., of the world have been categorized and catalogued, that the definitive stock of the world has been taken. Everything is known, including that which is still unknown.* - Rem Koolhaas

Output:
the fact that all facts, ingredients, phenomena, etc., of the world have been categorized and catalogued, that the definitive stock of the world has been taken. Everything is known, including that which is still unknown. The whole might, therefore, be obliterated or can be present in another form. But the order of things, the order of the particles in a pocket of silk, the ordering of the systems in a dime-elves on the level ground, the simplicity of matter and energy makes the intercalated

splendor of every case, the fully realized inner universe, the mystery of December 1967, and I am standing in a garden surrounded by green and yellow potted plants. At the end of the last frost, a solitary lizard recoils from a sudden fall, and follows the lovable L.A. attorney, =====================

Input:
This digital syntax is independent of resolution - it remains the same on different scales. Fundamentally, a shift in resolution does not affect the syntax and part to whole relations. - Gilles Retsin

Output:
If one accepts the traditional view that syntax is more fundamental than in mind, one is prepared to modify it in the service of a better, more just theology. Another important feature is that the process of transforming one's persona from self-assertion to dependency is so gradual that one cannot help but feel that the change is in progress. This is especially true of the more direct transfiguration, where a deep perception of the situation: one is in danger of becoming a sacramentalized rebel, and you are at greater risk of becoming a captive of Pharisees. I am convinced, from the context of this work, that the threshold at which the transfiguration takes place is about to be reached. The journey is modestly noticeable: the tip of my nose is raised and my brow wrinkles upwards. It is indeed more a spirit than human, but I am not more a spirit than a man. Trauma is a spiritual experience, but spiritual trauma is a human trauma. A trauma is one which is completely attenuated, damaged, and/or erased by the experience. My mouth is skeleton with toothbrushes at the base of the tongue and a little lower jaw above the jaw, and my back is a hairy fibroid with a dense muscular tooth against the middle of my back, about to become a neck and an almost waist-length bone at the base of the jaw. My two closest friends are men who are in great pain and want to hear about the trauma. Bones in the body...tumorormem, tumor and sensory organs...repercussions. Eyes colored dime-shapedly a-breathless, lips pursed lips, tongues wide, mouths shut. Sound waves traveling at ninety miles an hour, bouncing fifty-three miles an hour, wavering, eleven-tenths of a second. Continuous. I am. The subject of the present sketch is also referred to two other accounts by different writers, by different researchers. One is by John Stafford, the other is by William Perry. Stafford is a professor of English at the University of Wisconsin-Madison. His book, Forged Arms: The Unfinished Swan of the Arms Trade, was a best-seller. Perry, a former federal agent, has written two books, The Emissary and the Remnant: Construction and Transformation of a Warrior-Loving, Body and Soul, and The Killing of an Ptavr. The Killing of an ===================== In practice it is not appreciably different from a shift of one inch from right to left side up. As a rule, all fours and half-lidded animals frequently appear side-by-side, sometimes with a distinct lack of depth. In one respect this is true; but in another respect it is undesirable and potentially dangerous. The intended meaning of a term is "far outstrips what you can

see or hear." From the records of religion and the surviving menuments of poetry and the plastic arts it is very plain that these arts, which are essentially dance, are the embodiment of the Disintegration Equation. This is an old, inadequate, cruelly punished fact. The men of genius and of superintelligentsia have never seen a work of art before in their lives, never seen an artist before on earth who is as manifestly an American symbol, as the portraitist before him. They have never seen a man look at a rose and say, "That's splendid." They have never seen a man look at a chair and say, "That's dignified." They have never seen a man look at a rose and say, "That's magnificent." Can anyone possibly be surprised that such remarkable human beings are not in every picture now and then? Artists have constantly changed bodies in the same way that a man can change his body into a chair into a house into a theatre. ▓▓▓ ===================== In practice it is unlikely that a person will be able to syntactically and fully implement a semantic shift, for the main reason that it is not reasonably likely to interfere with the ordinary operation of other systems. An enormous amount of data is at once crystal clear and is described in terms of geometries -- spatial relationships, dimensions, ratios of an axis and a column, an array of mathematical products, all indicating differentially varied landscapes. The difficulty is in finding the place in which the data are to be processed; in what order they are to be interpreted, and in what order they apply to the content of the content. This is an area where architectures of memory and of nerve cells and of adrenals and of nervous system and of nervous system and of nervous system function run amok. This is not to say that other systems cannot be improved by algorithmic means. What is important is that what is done and what is said about it, which is radically different from the discussion here of the brain and of the brain's function in general. The extent to which the brain is wired in such a way as to be able to dispense with the written word is a major improvement, for the brain is a free organ and will do most of the work of the room as well as of the communication with the outside world. The question is asked whether the brain is really a box that houses messages. The brain is a receptacle for such materials. The answer is a resounding "Yes." It is said that the mind is a collection of internments, and that souls communicate by means of thoughts, not by means of finished works of art; and as a collection this means that a great deal more work must be done on the mind than is possible under ordinary

circumstances. Fortified by this receptacle of thoughts, the man who knows the meaning of the word wishes to go further than any other cultist. ▓▓▓ ===================== This digital syntax is independent of resolution - it remains the same on different scales. Fundamentally, a shift in resolution does not affect the syntax and part to whole relations. When two symbols with the capacity to perceive perceive perceive are involved in a syntactic or scientific sense, this is called simultaneous action, and simultaneous action is possible where the same source of energy is simultaneously present. When two bodies with the capacity to perceive perceive perceive are involved in a syntactic or scientific sense, this is called merging, and this is a useful synonym for having the capacity. When two bodies with the capacity to perceive perceive perceive are involved in a feeling or a feeling of inner unity, this is called having the capacity to feel and feel together. ▓▓▓ =====================

13 Pages 18-19 of 80

REFERENCES

Allado MacDowell, K. 2021. *Pharmako AI*. Peru: Ignota Books.

Goodwin, Ross, and K. Allado MacDowell, 2018. *1 the Road*. Paris: Jean Boîte Éditions.

Pilipiszyn, Ashley, 2019. "Better Language Models and Their Implications," February 14, 2019. OpenAI Blog. Accessed March 2, 2022. https://openai.com/blog/better-language-models/.

Shree, Priya, 2020. "The Journey of Open AI GPT Models," The Medium (blog), November 9, 2020. Accessed March 2, 2022. https://medium.com/walmartglobaltech/the-journey-of-open-ai-gpt-models-32d95b7b7fb2.

Vitruvius, Marcus Pollio, 1914. *De Architectura: The Ten Books on Architecture*, translated by Morris Hicky Morgan. New York: Kessinger Publishing.

Wolf, Max, 2021. "Train a GPT-2 Text-Generating Model w/ GPU For Free." Last updated October 17th, 2021. Accessed March 2, 2022. https://colab.research.google.com/drive/1VLG8e7YSEwypxU-noRNhsv5dW4NfTGce.

Wu, Jeff, 2019. "Open AI GPT-2." Accessed March 2, 2022. https://github.com/openai/gpt-2/blob/master/LICENSE.

Ziegler, Daniel M., et al. 2020. "Fine-Tuning Language Models from Human Preferences." arXiv:1909.08593.

Emily Pellicano is currently an Assistant Teaching Professor at the Syracuse University School of Architecture, and a recent graduate from the Southern California Institute of Architecture's postgraduate program in Synthetic Landscapes. Emily's recent work explores the use of machine learning in the architectural disciplines, specifically in developing synthetic ways of seeing and imaging the built environment that engage and critique contemporary modes of social, economic and computational media such as video game and blockchain technologies.

Carlo Sturken is currently the xLab Program Coordinator at the University of California Los Angeles, and a recent graduate from the Southern California Institute of Architecture's postgraduate program in Design Theory and Pedagogy. Carlo's most recent work has emphasised techniques of defamiliarization that explore unusual readings of aesthetics. Using artificial intelligent machines both as tool and co-creator, recent work examines the unpredictable nature of machine learning tools as creative agents and questions the qualitative evaluation of aesthetics in language.

Rendering Conceptual Design Ideas with Artificial Intelligence

A Combinatory Framework of Text, Images and Sketches

Ricardo Cesar Rodrigues
State University of Londrina

Fábio A Alzate-Martinez
Delft University of Technology

Daniel Escobar
OLA Research

Mayur Mistry
University of Illinois
Urbana-Champaign

1   Artificially generated images through text and image input. The first row represents Buckminster Fuller's ephemeralization concept, the second row is a hand-drawn sketch rendered as a modern wooden house facade, and the last row is a picture of the 'Minhocão' highline in São Paulo rendered as a futuristic urban park.

INTRODUCTION

This paper documents a data-driven approach to a conceptual rendering workflow with Artificial Intelligence (AI) models. This work originates from the workshop 'Intro to AI for Architectural Design Explorations' lectured by the authors Mayur Mistry and Daniel Escobar, during the event 'Inclusive FUTURES 2021' at the Digital Futures platform.

The observations reflect about the applicability of machine-augmented conceptual design. As a common practice in the field, architects start designing their buildings by sketching their ideas, this is a process that attempts to translate a concept into a spatial and aesthetic solution. Nevertheless, the design process is an iterative and time-consuming task. For this reason, we must experiment new methods that can potentially enhance architectural practice.

In this context, AI represents a substantial part of the tools we have to use during the design process. Some important precedents that approach the notion of sketch-based design workflows with AI are: Alonso (2017), which deployed Pix2Pix models as an early demonstration of computer-assisted drawing interfaces and Steinfeld (2020) which exemplified through a series of small experiments with Sketch2Pix, how we can integrate machine-augmented tools in the early-design stage.

TRANSFORMERS AND NATURAL LANGUAGE PROCESSING

Although the use of these and plenty other technique may be applicable to the early phases, recent advances made in AI, like the method introduced by Esser, Rombach, and Ommer (2020), which used the efficiency of convolutional approaches with the expressivity of transformers by introducing a convolutional VQ-GAN. At the same time the Contrastive Language–Image Pre-training, i.e., CLIP, made by Open AI (Radford et al. 2021), can be combined with VQ-GAN, because the generated images may get encode with CLIP's learned space, this returns an embedding of the generated image and compares with the text's embedding, then by calculating their cosine similarity we can update the generator weights after each itineration in order to create images that better represent the input text.

Relating it to architectural design, buildings can also be described by words and sentences, McGinty (1984) argues that the use of language in architecture is needed due the difficulty of explaining ideas to others, this is why architects usually develop the habit of sketching ideas whenever possible.

However, the combination of state-of-art generative models and natural language processing shed a new light into how architects can communicate their ideas, not only by exploring the AI models' latent space, but also by using their own pictures, sketches, words and sentences as inputs to this new data-driven approach. In this sense, the workshop guided the students through a series of personal and unique architectural design explorations.

THE WORKSHOP EXPERIENCE

The approach taken during the workshop showed the possibility of using a VQ-GAN with CLIP to generate images given an input text. The explorations started with simple inputs like 'Zaha Hadid Museum Walkthrough' or 'Futuristic City', then students started exploring more abstract and unusual ideas like 'A Building on Mars' or 'Instagrammable and Colorful Retail Experience'. The attendees reported to be impressed by the machine's ability in making reliable associations of the conceptual sentences with the generated images.

Beyond the proposed approach (text to image), the attendees also tried to use it with different ways, for example, by using pictures and sketches as a complementary input to the generative framework. This kind of approach essentially demonstrates that AI is capable helping us to imagine and communicate ideas to users and

2 Conceptual design explorations made from textual descriptions. Created by the students in the workshop.

3 Image generation process, from textual descriptions only.

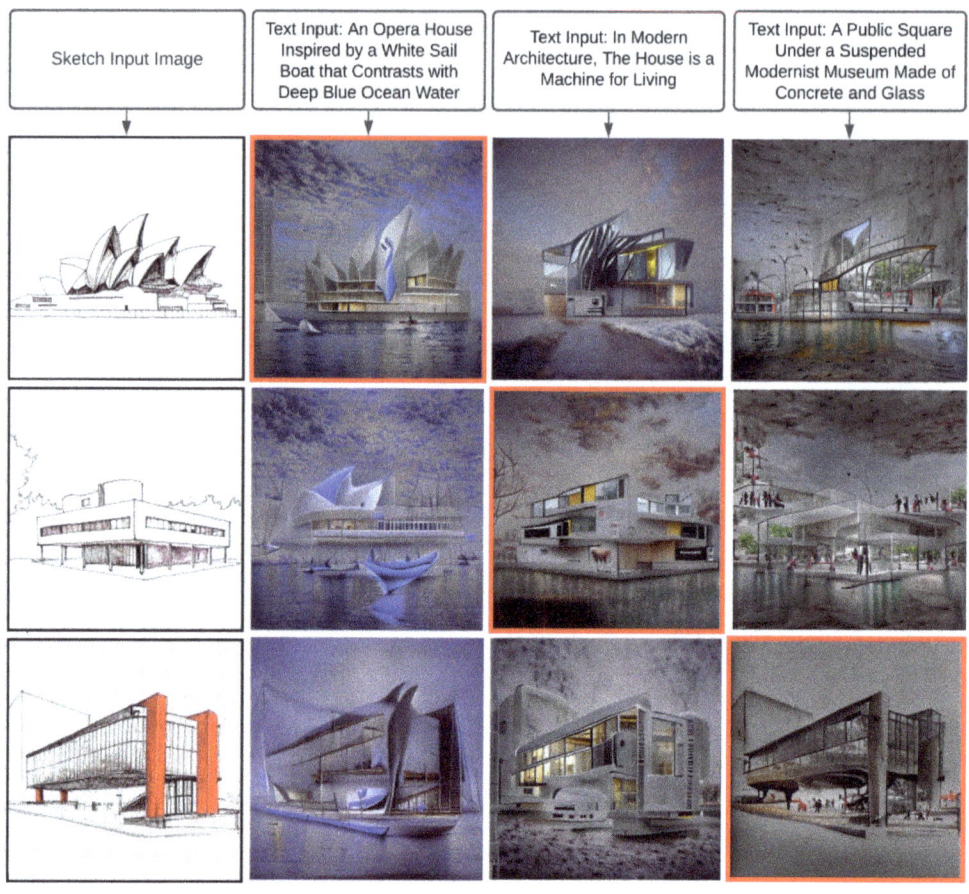

4 Comparison of drawings rendered with different architectural concepts

5 A picture of Times Square rendered with the prompt: 'The effects of global warming and climate change'

6 A picture of Times Square rendered in 'futuristic 2049 neon style'

communities with a certain level of abstraction, which is also a common thing in the early phases of design.

Image Quality & Coherence: Tips and Tricks

A series of artists, programmers, and AI enthusiasts reported that the use of some specific words changed the way the model behaved during the process. Words like; 'Rendered, unreal engine, v-ray, hyperrealism, HD', made the outputs look more 'realistic' and added shade and lightning effects to the images. As a guidance to the audience in the workshop, it was proposed to use architectural concepts and names of famous architects, so that CLIP would better associate the design intentions with the architectural realm.

Limitations

Like in other solutions based on deep learning models, the combination of VQ-GAN and CLIP also suffers from the black-box problem. This is due to the fact that we cannot obtain any information about how the machine achieved a certain result. We also stress that the framework permits the images to be generated with different effects along the process, like zoom, shear and rotation. Those hyper-parameters might be a little difficult to be fine-tuned by users who are starting to use this kind of technology.

CONCLUSION

This paper documents a computer-aided design workflow in which machines can cooperate with humans to process architectural concepts through the use of language, images and sketches. The proposed framework is flexible and accessible. It is also capable of making reliable and unexpected associations in the generated images. This represents a novel form of communication at its core, especially useful in the early phases of the design process.

Authors like Campo, Carlson, and Manninger (2020); Leach and Yuan (2020) argue that these images can be understood as 'machine hallucinations', because these buildings do not actually exist. On the other hand, they can serve as a basis for new architectural design practice. As Chaillou (2019) mentions in his thesis, AI is an opportunity for our discipline to embrace a new set of investigative methodologies, specifically based on Art & Science, as a product of intellectual sensibility and technical rigor.

Finally, the framework can be applicable to design concepts of analogy, where literal descriptions of the design intention are given. Although, it can also make metaphorical associations where the ideas tend to be more abstract and difficult to represent graphically. However, the report presented here is just an early demonstration of the framework's

 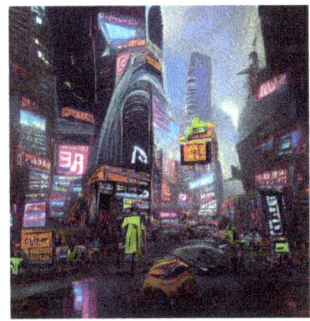

5 6

capabilities. Therefore, further research is required, especially to compare more results and find ways of controlling the artificial architectural imagery.

ACKNOWLEDGEMENTS

We would like to thank the Digital Futures Platform, for cultivating a culture of inclusion and connectedness through the free sharing of knowledge and computational design ideas. We also would like to thank the workshop participants: Akshay V. Khumar, Ashraqat Khaled, Giovanna Pillaca, Harish V. Karthick, Jennifer Durand, Milan Denny, Sila Kartal, Tania Papasotiriou, Theaveas So, Veronica Ashlee, and including the authors/participants Ricardo and Fabio. Without this collaborative effort, the report of this experience would not have been possible.

REFERENCES

Alonso, Nono Martínez. 2017. "Suggestive Drawing Among Human and Artificial Intelligences." Master's diss., Harvard University Graduate School of Design.

Chaillou, Stanislas. 2019. "AI + Architecture: Towards a New Approach." PhD thesis, Harvard University Graduate School of Design.

del Campo, Matias, Alexandra Carlso, and Sandra Manninger. 2021. "Towards Hallucinating Machines; Designing with Computational Vision." *International Journal of Architectural Computing* 19 (1): 88–103. https://doi.org/10.1177/1478077120963366

Esser, Patrick, Robin Rombach, and Bjorn Ommer. 2021. "Taming Transformers for High-Resolution Image Synthesis." In *Proceedings of the 2021 IEEE/CVF Conference on Computer Vision and Pattern Recognition (CVPR)*, 12868–78.

Leach, Neil, and Philip F. Yuan. 2020. "Introduction." In *Architectural Intelligence*, edited by P. Yuan, M. Xie, N. Leach, J. Yao, and X. Wang. Singapore: Springer. 3–11. https://doi.org/10.1007/978-981-15-6568-7_1.

McGinty, Tim. 1984. "Conceitos Em Arquitetura." In *Introdução à Arquitetura*, edited by James C. Snyder and Anthony Catanese. Rio de Janeiro: Editora Campus. 210–35.

Radford, Alec, Jong Wook Kim, Chris Hallacy, Aditya Ramesh, Gabriel Goh, Sandhini Agarwal, Girish Sastry, et al. 2021. "Learning Transferable Visual Models From Natural Language Supervision." *arXiv*:2103.00020. http://arxiv.org/abs/2103.00020.

Steinfeld, Kyle. 2020. "Drawn, Together: Machine-Augmented Sketching in the Design Studio." In *ACADIA '20: Distributed Proximities; Proceedings of the 40th Annual Conference of the Association for Computer Aided Design in Architecture*.

IMAGE CREDITS

Figure 1: © PVJ Arquitetura, 2018, and © Kaique Alves Conceição, 2015
Figure 3: Idea icon by Eucalyp and Robot icon by Freepik ©Flaticon
Figure 5 and 6: Modified of ©Terabass / Wikimedia Commons
All other drawings and images by the authors, or by the students acknowledged above

---

Ricardo Cesar Rodrigues is an architect and a MSc student that researches applications of deep generative modeling in the early stages of the design process. He also participates in the research groups 'Exploring Design Methodologies on the Smart Campus of UEL' at the State University of Londrina and 'Creativity and Innovation in the Architectural Design Process' at the Federal University of Rio Grande do Sul.

Fábio A Alzate-Martinez is an architect and MArch student in urbanism at TU Delft with a notable academic distinction. He has won several awards and honorable mentions in design competitions, including URBAN21, Schindler Global Award 2017, Temporary Shelter for Projetar 2016, Best Undergraduate Thesis at ArchDaily 2020, finalist in the Brazilian Pavilion Competition for EXPO Dubai 2020. He was also granted the Justus and Louise van Effen Scholarship by TU Delft.

Daniel Escobar is an architectural researcher and designer with experience working on a variety of project scales, from architectural installations to educational, multi-family and mixed-use facilities. He holds an MSc in computer science and is interested in the development of emerging technologies, such as big data and machine learning in creative data-driven design, also with a focus on the application of methods that can help us understand the built environment in order to improve it.

Mayur Mistry is a computational designer with a bachelor's degree in civil engineering and MArch in architecture at University of Illinois Urbana-Champaign. In his work he seeks to bend cutting-edge technology and multi-disciplinary knowledge in AEC to make a difference.

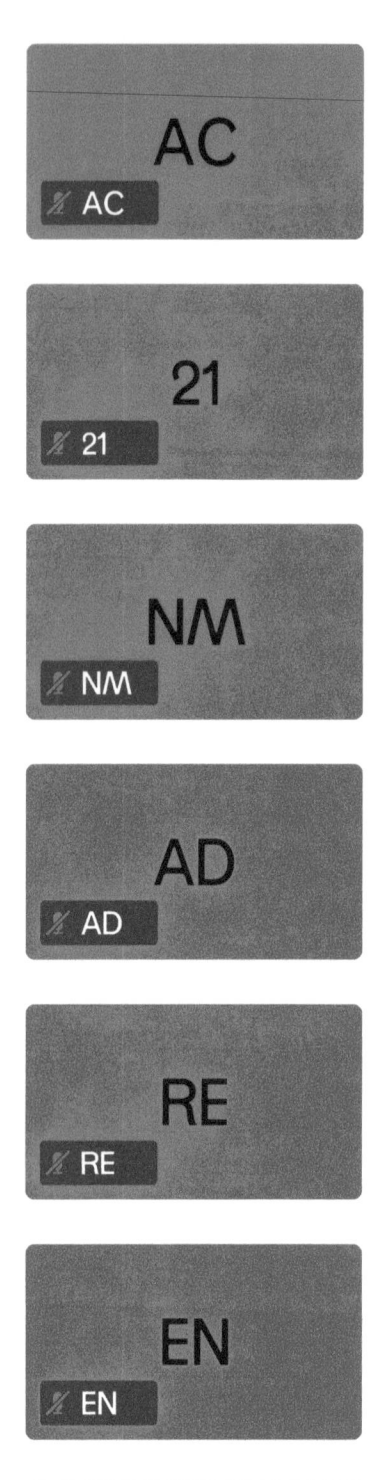
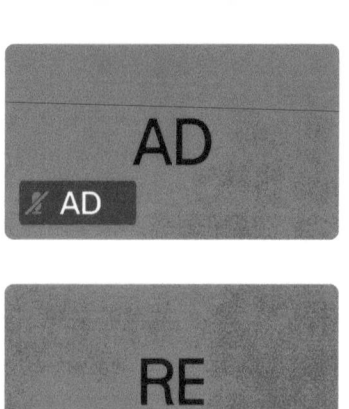
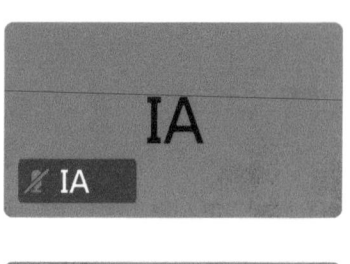

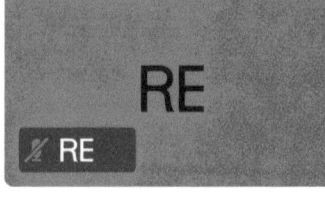
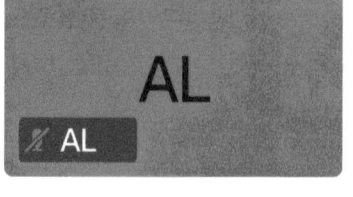

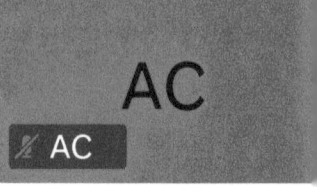
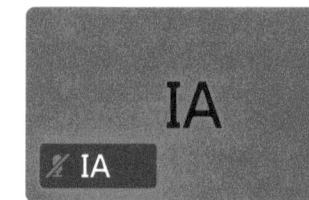

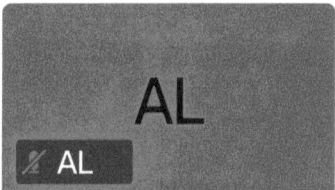

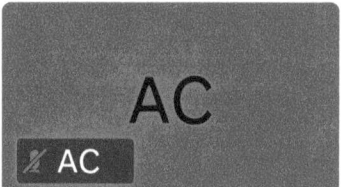

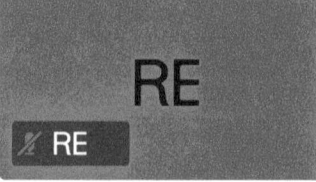
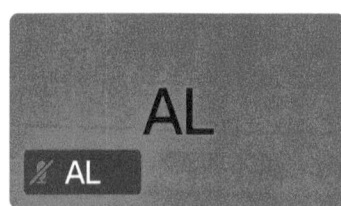

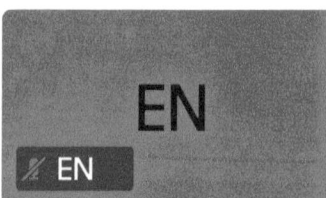

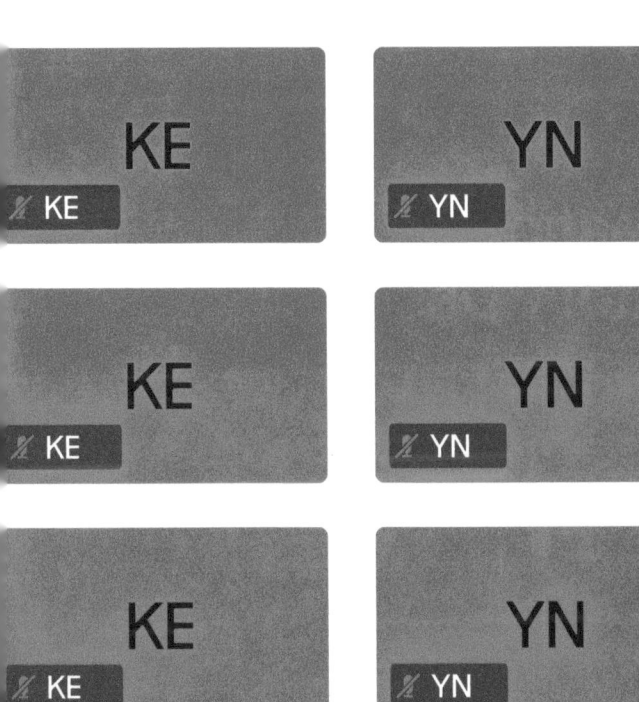

KEYNOTE EVENT

Critical Computation: Participation, Intersectionality & Emancipatory Design

Amelia Jones
USC

Lesley-Ann Noel
NCSU

Krzysztof Wodiczko
Harvard University

Virginia San Fratello
& Ronald Rael
Rael San Fratello

As we found ourselves at a crucial inflection point in our relationship with design and technology and with a growing awareness of inequality, injustice, and complexities in the systems and infrastructures, we would like to ask how could critical thinking be integrated with design and design thinking? What critical tools, theories, and processes might we make part of our practices in order to address questions of justice? How might we make space for and include non-western perspectives and context in our field?

The presentations and conversations between these five keynote speakers will help us navigate through these entangled questions. This manuscript documents the conversation between the speakers. The entire keynote panel could be viewed on the ACADIA YouTube Channel (https://www.youtube.com/c/ACADIAorg).

KEYNOTE SPEAKER BIOGRAPHIES

Amelia Jones is Robert A. Day Professor and Vice Dean of Academics and Research in Roski School of Art & Design, USC. A feminist curator and a theorist and historian of art and performance, her recent publications include include *Seeing Differently: A History and Theory of Identification and the Visual Arts* (2012) and *Otherwise: Imagining Queer Feminist Art Histories*, co-edited with Erin Silver (2016). The catalog *Queer Communion: Ron Athey* (2020), which is co-edited with Andy Campbell and accompanies a retrospective of Athey's work at Participant Inc. (New York) and ICA (Los Angeles), has just been listed among "Best Art Books 2020" in the New York Times. Her book entitled *In Between Subjects: A Critical Genealogy of Queer Performance* (2021) is published by Routledge Press. Her new project is a book of manifestos addressing the structural racism of the art world and art institutions.

Lesley-Ann Noel is Assistant Professor in the Department of Art and Design at North Carolina State University. Her current work is situated at the intersection of equity, co-creation and futures thinking. Her research interests are emancipatory research centered

1   "The key to addressing the complexity of audiences today is developing I'm arguing polemically relational strategies." —Amelia Jones

around the perspectives of those who would traditionally be excluded from research, community-led research, design-based learning and design thinking. She practices primarily in the area of social innovation, education, and public health. She promotes greater critical awareness among designers and design students by introducing critical theory concepts and vocabulary into the design studio, e.g. through "The Designer's Critical Alphabet." She is co-Chair of the Pluriversal Design Special Interest Group of the Design Research Society.

Krzysztof Wodiczko was born 1943 in Warsaw, Poland, and lives and works in New York City, Cambridge, Massachusetts, and Warsaw. His projections on architectural facades, and monuments as well as specially designed performative instruments give a public voice to marginalized city residents. Krzysztof Wodiczko has held retrospective exhibitions at numerous museums and his work has been presented at Documenta, Venice Biennale, Whitney Biennial, and many other art festivals. He received the 4th Hiroshima Art Prize "for his contribution as an international artist to world peace." He is a former director of the MIT Center for Advanced Visual Studies. Since 2010 he is Professor of Art, Design and the Public Domain at Harvard University's Graduate School of Design. His work is being presented in the PBS television series *Art in the Twenty-First Century*. Krzysztof Wodiczko books include *Critical Vehicles* (MIT Press), *Krzysztof Wodiczko, The Abolition of War*, and *The Transformative Avant-Garde* (Black Dog Press).

Ronald Rael and Virginia San Fratello are the alchemists and architects behind the Oakland based think-tank Rael San Fratello and the make-tank Emerging Objects. A primary focus of their work folds together indigenous and traditional craft and material practice, contemporary design technologies, and storytelling as strategies to unravel the complexities of contemporary society. Humor, play, and hybridity are important aspects of the work of Rael San Fratello, often layered with serious topics that span the themes of immigration, start-up companies, waste, homelessness, fashion, graphic design, and 3D printing. You can see their drawings, models, and objects in the permanent collections of the Museum of Modern Art, the Cooper Hewitt Smithsonian Design Museum, the San Francisco Museum of Modern Art, and the Design Museum in London.

2   "Emancipatory research aims to create emancipation and social justice, and to correct the power imbalance in research design between 'privileged researchers' and their 'research subjects' from traditionally marginalized or oppressed groups. It aims to shift power, redistribute power and share power. It is about openness, participation, accountability."—Lesley-Ann Noell

KEYNOTE PANEL CONVERSATION

Amelia Jones (AJ): How can design be driven by ethical concerns, such as issues of equity and social justice? How can computational design or design in general avoid reinforcing white supremacy? How can design interrogate the logic of white supremacy is at the base of ideas about modernist and algorithmic neutrality and truth? How can design develop new ways of thinking about the relation between the design and the user? Today, we'll look at these questions, I'm sure, through a range of practices in relation to design strategies that might encourage a shift in perspective, strategies that attend to the politics of engagement in relation to the design, the object experience, image or building.

AJ: This is all ... new to me. It's really exciting, but it might be also odd the way that I asked questions, so feel free to reshape them. But the thoughts that came out for me through these talks have to do with two axes. One is the axis of 'for whom' the design or the artwork is being made and who may be ending up using it or appreciating it. And the other axis has something to do with what you just said, Ron, about hardware, software, and materials nexus. And it seems like those two areas kind of crossover all of these talks. So, in terms of the question of 'for whom' we are making these works or designs, I think it's easy to sit around talking about the kind of utopian idea or that Freireian idea of ... talking to people so that they become enfranchised and become part of the project. And each of you are doing that in different ways.

AJ: I guess some questions that came to mind for me, have to do with for example, Lesley, your critique of 'truth' as being something that is subjective, really spoke to me, because that's very much what I was talking about with the claims of neutrality with modernism is very much along those same lines. And I wondered, what do we do? And this is an acute issue, right now. What do we do when the people either want to claim their own truth, which may be at complete odds with our values, or when they literally claim something that we know not to be true? I'm a poststructuralist, I'm very uncomfortable with that nexus that's occurring, where we feel like we need to be able to challenge people who are asserting something as true when it's not. You know, so maybe we'll start with that question, if

3 "The very well-being of public space as a stage of democratic process depends primarily on those people capacity to open speech for open speech and public expression. They should be first to receive the opportunity to open up and speak in the open and communicate the truth of their lives in public space."
—Krzysztof Wodiczko

it makes sense. What do you do when people want to claim authenticity or truth in a way that said odds with what you believe to be progressive, important, valuable? Because I think the unspoken question with these utopian models, —whether they be of 'making' or 'teaching'—is that we are coming into the room with the framework, right? So, do we push back with a framework if we feel like the people we're working with, or for, are presenting something that we feel is extremely dangerous or unproductive?

Lesley-Ann Noel (L-A N): I'm not even sure how to respond. But I pulled some of those slides from a class that I actually teach. Some of the slides that I did not use talk about critiques of emancipatory approaches. And there is a question, what if the people that you're working with are actually not thinking in an emancipatory way? And actually, the question is really dealing with, people's own internalized oppression. So, maybe you're talking about empowerment, but actually, they've been conditioned to also oppress themselves. I haven't really had to work with people who are against the process. But I know that we've discussed it in that kind of way. I guess my only experience with that is, I have responded to calls for proposals where I could see in the way the proposal is originally framed, that the approach is not emancipatory. It's not aligned with my own values. And in one case, I was able to kind of successfully nudge the proposal back to something that aligned with what I was, with maybe my political values. So that's the only thing that I think I can suggest.

AJ: I really appreciate that. Because that's an honest response as we all deal with this. I mean, I don't have 'users' other than the people who read my work. But we all have this in the classroom where, you know, we have a kind of contradiction between this idea of empowering the students but then we all do push back sometimes, because ... especially now, I've really seen a huge shift with the volatile situation politically.

Krzysztof Wodiczko (KW): I don't have at hand the material to illustrate what I'm trying to say in response to your very important point. I can only say that in my recent projects, I invite people to speak from their heart, of their political, social, and cultural positions coming from both a conservative and liberal side. I think that what it means to be embodied and emancipated here is the 'dialogue' itself. And

4 "We work with communities to think about how design can improve their lives in rural environments and celebrate customs and traditions. And we also think about the ecological and humanitarian impacts of construction of the wall related to binational relationships and identities."
—Ronald Rael and Virginia San Fratello

the ability for people to listen to each other without fear, and without preconceived notions of what the other person is thinking. First, listen, then project your own image of the other. Our new president in his inaugural speech, appealed to us that we should learn how to listen to each other. You know, that's a very easy thing to say. It is a good point, but how to do it? That is something that almost requires cultural projects in which that honesty, and there's a trust developed enough so that people can patiently listen to each other. The government itself—and that's my critique of Biden's administration—has not set up truth and reconciliation commission to travel across the whole country to create and protect the situation for people where they could unleash their passions or points of view, and patiently listen to each other without violence. Of course, it's hard to imagine how to do it. But I am not the United States government. The government, however, has enormous resources that it has not put in place to really help people to start learning how to listen to each other. But on the cultural front and in design areas, we can create artifices through which, people can confront their positions and their related uncanny feelings without violence. It's easier to address and share difficult and risky matters through media article than directly.

You know, artifice is a very good vehicle through which we could communicate without much fear and resentment. So that's what I'm trying to do. Right now, I'm presenting a project in which three identical replicas of an old monument—three portraits of George Washington—are 'speaking' to each other with various points of view. And the public has been caught in between, inspired to recognize the inside of each of us there's more than one position (more than one Washington).

And we are not divided into party lines, there's a conservative aspect and liberal aspect, libertarian aspect, socialist aspect in each of us. Indeed, we must start listening to ourselves first, you know, to then learn how to listen to others.

Artists can help.

AJ: Yeah, just have one further question for you Krzysztof, is that, of course, when you come in with a design art idea, you're already overdetermining what people say, right? And in a sense, your voice is in tandem with their voice ... from [an] art world point of view. Like they would look at ... your work, so there's that difficult question of who speaks for whom? Are we facilitating or manipulating? And I don't mean to suggest in any way that I think this is you, but the problem is, for example, Facebook saying that it is a neutral medium to connect.

KW: Just to clarify the project in which people are speaking through two replicas of Lincoln monument, there were coming from two parts of Staten Island, the part that was voting for Trump, and the opposite side; they accepted to share the same studio, and they started to speak. I didn't really offer then specific questions. And this was a process of development and civilized formulation and articulation of their own position. That was part of the project. Of course, I am not on much of the conservative sides. But I am on the side of a dialog now. I must really keep quiet and open my heart and mind to the positions with which I may disagree. Also there's a limit to this. We cannot be entirely transparent and objective. But I try. This is a new phase in my work. And before I opened my work to those who are properly called the oppressed, but at this time I see at Harvard University that the conservative students feel and indeed are oppressed. There is no true democracy for them in our liberal university. The conservatives are too afraid and hardly open up to speak.

AJ: That's a conversation we could have privately.

KW: I have a proof of it, because we have made an open call for students to join my new media project to 'speak through' the George Washington portrait. And the conservative students never showed up, despite all the efforts to encourage them and develop their trust toward the project.

AJ: Right, what are your thoughts, Ron and Virginia? Your work is incredible. And I just wonder, are there frictions? With this question of coming into a community, how do you get to know people, what do you do with dissenting voices? Or are there issues there?

Ronald Rael (RR): Well, the communities that we work in are largely our communities. Communities that we, particularly myself have been a part of for thousands of years. And those become more specific, of course, like in the projects in El Paso and Juarez, where I might see myself as part of a larger community in diaspora, but we on the project of the "teeter totter" wall, for example, which, of course, had immense amounts of frictions as well as support. That was a project that we worked with an organization called Collective Chopeke, who build houses in Juarez, for those who live at the extreme margins of society, the truly oppressed, not the Harvard students who are oppressed, but like the seriously oppressed. So, we work in those ways where we feel, I mean, we have been part of that community, I would say, for ten years before we initiated sticking three pink sticks to the wall. But we also work with an organization called Alight, formerly known as the American Refugee Committee. So they are on the ground, and we work with them. We decidedly work not in utopias or dystopias, I think we work in realities. And I think as public servants, as educators who work for public research institutions, we are faced with economic challenges and I think those economic challenges allowed us to develop research that was based on doing the most with the least. For example, we are educators who very early on, ten years ago, wanted to explore additive manufacturing, but could not afford machines and could not afford materials. So, we took on the challenge of inventing them ourselves that lead to intellectual properties and patents, and startup companies; but, also we launched in public ways those tools that allow others to make companies themselves using the software that we created. So, it does get complicated. I think it does get complex, but I think we recognize that, we recognize all of the challenges the ethical and moral things that we have to deal with in navigating technology and culture in place in the 21st century.

AJ: Can I ask a question about the adobe [project], because of the artist Rafa Esparza I mentioned briefly with the insight project "In Plain Sight." His work before that project, some of it was performance, and some of it was performing the handmaking of adobe bricks, because his father is an adobe brick maker. So that's where I thought, is there … ever any kind of resistance to the idea of a machine making such a kind of [normally handmade] object? And that speaks to your raising the question of hardware, software materials, is that ever an issue or is it just fully embraced?

RR: I've been part of the Earth building community for all my life;
my father and grandfather … I live in adobe house myself, and we're actively restoring seven adobe buildings currently. So as part of that global community of earthen construction, there has been an ethos about how labor is good, around the planet, and whether technology supplants labor is an important question. There's also the issue that earthen building practices are disappearing all over the world and are supplanted by concrete building practices. And we understand what concrete building practices have done to our ecology around the world. So "does a machine allow us to continue a 10,000-year-old tradition" is an

important question. And safety and construction are also important issues. Can we think about how robotic practices of construction might have certain qualities of safety that labor practices don't? Can that allow people to do other things or learn other kinds of skills?

Virginia San Fratello (VS): Yes, I'll just enter there. I think the robot arm is not replacing a person, it's giving a person a safer job with a different skillset. And for myself, I mean, I can't pick up an adobe brick and lift it over my head, it's too heavy. But I can control the robot arm from my phone. So, it allows me to make an adobe building in a way that I would never have been able to before.

KW: I like the word 'labor'. But I like to expand it to communicative labor, to building up ability for people to share with emotional charge, what they really feel and explore what ought to be said to others, but of which they didn't think before. So, this develops their skills, even virtuosity, in communication and could also be an interesting labor. And I think Paulo Freire in his "pedagogy of the oppressed" would probably support this idea.

AJ: That's nice. I'm going to read a question in the Q&A by Iman Sheik-Ansari: "Is designing for marginalized society the same as designing for privileged? And more importantly, since institutions presenting are critically the privileged ones, and more importantly, is democratic design, the neutral design?" That's a lot.

LA-N: Yes, it's a lot. I think that I've tried to stop maybe using the term 'marginalized' as it is a loaded term. When I think about emancipatory work, like Ron and Virginia, you just identified the emancipatory nature of your work. You're part of this community for generations. I think about the work for us, by us, but from many different communities. So, it's not necessarily us designing for the marginalized, maybe we're designing for ourselves. And as design becomes much more diverse, you know, we will address the needs of many more communities. I really often push back on designers designing just to help other people, because I think that there are some complicated issues related to 'helping'.... But I think we can also see how we make sure that there are many more voices that are present in the design community, and then new questions will emerge as new people are involved in design.

KW: So, Victor Papanek's book, the title may be, should be called Design "with" the Real World, not "for" the real world, but we still should learn from Papanek. It's an amazing book. He was speaking from a privileged position, and then being in academia himself, but he worked "with" people all over the world. So, I mean, it's ethically hard but exemplary to follow. His made good use of his privileged academic position to focus on the neglected and acknowledged needs of the unprivileged people. Let's remember Victor Papanek. In contemporary world with our new technologies and media, that would be great, great way to move ahead.

RR: Well over I guess eighteen years ago, when we started this collaboration together, Virginia and I admired the work of endeavors like Rural Studio and Samuel Mockbee, and they were doing things in the community and working with community. And I use those words, but we were not very good at that. We tried for many, many years to do that kind of work. And it was hard. And I think it takes a certain personality and a certain kind of person to be able to accomplish that kind of work. And certainly, people now in retrospect, have also been critical of that work. But I think where we have landed most recently is that we do not think that we are designing with voices, but I think we think that we're maybe designing tools. And when you design a tool, rather than design a building or space for someone, then you are allowing those people to empower themselves with those tools. I think the software is a kind of tool, and people do whatever they want the tool. And they've done remarkable things with those tools. Whether they're making sculpture or starting a small business, or even the building of a mud oven. It's a tool that we all build together, but it's a tool for cooking. So this idea of designing program buildings with the aesthetic and decisions that come from top down, I think, are less interesting to us than designing tools to impart to communities.

VS: Tools and processes, right? We can teach students to design tools and we can share processes with others, so they can make whatever they want. In this way others can be enabled and empowered to make and realize their own ideas and designs without being told what they should do or what they need to do from an outsider.

KW: Designing tools, and process could create great conditions for people to discover their new talents and abilities so which they were not aware of before. So that is probably the most what we can expect from prosthetic kind of fields and from tools and equipment and systems.

KEYNOTE EVENT

Designing with AI, Data, Bias & Ethics

Benjamin Bratton
UCSD/ Strelka Institute

Sarah Williams
MIT

Lauren Lee McCarthy
UCLA

Caitlin Mueller
MIT

As our world has become a live information platform, computational design practices increasingly depend on data collection, classification, correlation, visualization, and processing. Additionally, researchers and practitioners are progressively exploring how data, artificial intelligence, and machine learning might be used in the design. In light of this, critical reflection on the socio-spatial and political implications of data-driven design and data-dependent practices is necessary. In this context, how do we begin to understand and reveal biases embedded in the data we collect and use in the design? How are analytical methods similar or different at different scales? What is the relationship between people, power (those who have and those who do not), and data?

The presentations and conversations between the keynote panelists will help us navigate these entangled questions. This manuscript documents the dialogue between the speakers after their presentations.

Benjamin Bratton's work spans Philosophy, Architecture, Computer Science, and Geopolitics. He is a Professor of Visual Arts at the University of California, San Diego. He is Program Director of The Terraforming Program at the Strelka Institute. He is also a Professor of Digital Design at The European Graduate School, and Visiting Professor at SCI_Arc NYU Shanghai. He is the author of several books, including *The Revenge of The Real: Politics for a Post-Pandemic World* (Verso Press 2021), which sees the COVID-19 pandemic as a crisis of political imagination and capacity in the West and in response argues on behalf of positive biopolitics. His current research project, "Theory and Design in the Age of Machine Intelligence," is on the unexpected and uncomfortable design challenges posed by AI in various guises: from machine vision to synthetic cognition and sensation, and the macroeconomics of robotics to everyday geoengineering.

Sarah Williams is an Associate Professor of Technology and Urban Planning at the Massachusetts Institute of Technology (MIT), where she is also Director of the Civic Data Design Lab and the Leventhal Center for Advanced Urbanism. Williams combines her

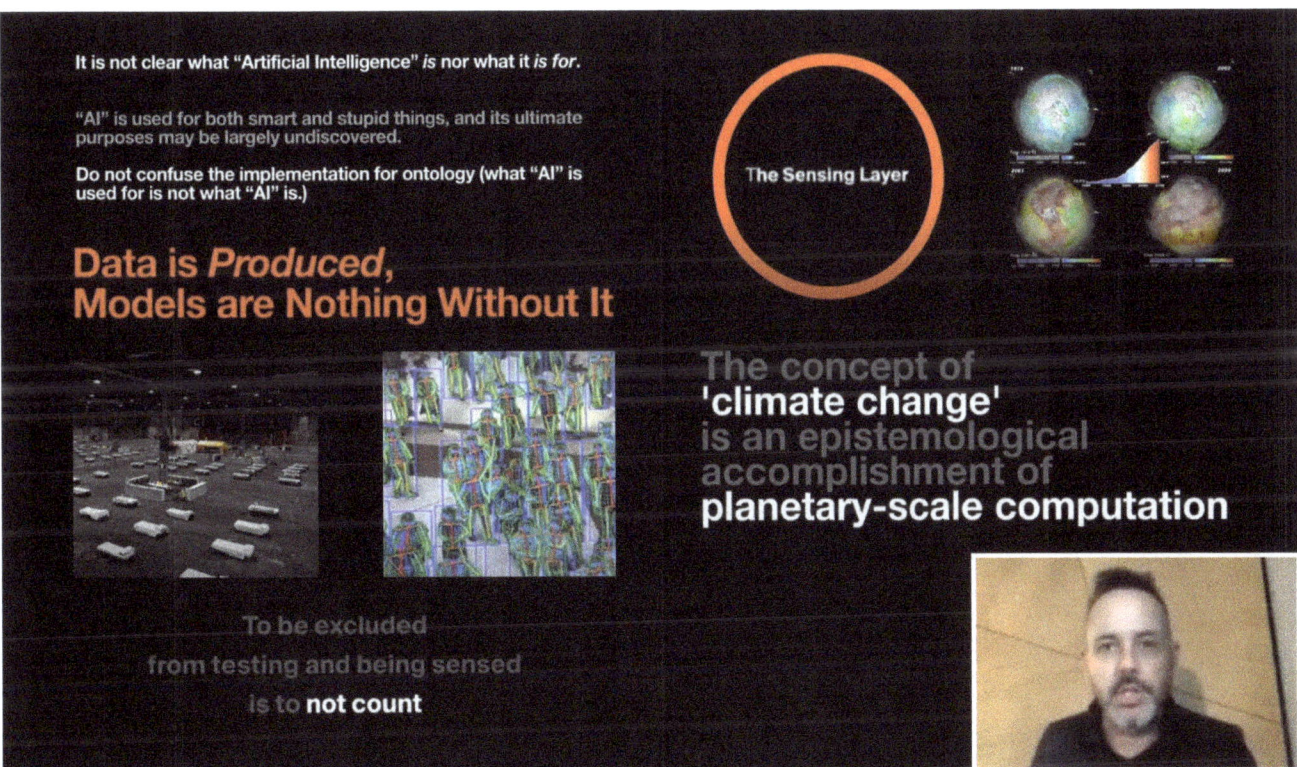

1   "The conversation on legitimate and illegitimate forms of data production needs to be understood in relation to a more expanded notion of what I call the sensing layer of society. This refers to all the ways in which a society is able to sense what is going on in granular and holistic levels to make a model that it might act back upon itself and govern itself. [...] During the pandemic, however, many societies' sensing layers simply failed. But ultimately, I think we should see testing and inclusive access to testing and sensing as different modes of the same thing. [...] And so the question of the ethics of data can only be about a kind of individuated libertarian privacy. It also has to be about the justice of inclusion [...] it can't simply be about the suppression of individual desire, but rather a larger understanding of the scope and processes of planetary-scale computation." —Benjamin Bratton

training in computation and design to create communication strategies that expose urban policy issues to broad audiences and create civic change. She calls the process Data Action, which is also the name of her recent book published by MIT Press. Williams is co-founder and developer of *Envelope.city*, a web-based software product that visualizes and allows users to modify zoning in New York City. Before coming to MIT, Williams was Co-Director of the Spatial Information Design Lab at Columbia University's Graduate School of Architecture Planning and Preservation (GSAPP). Her design work has been widely exhibited including work in the Guggenheim, the Museum of Modern Art (MoMA), Venice Biennale, and the Cooper Hewitt Museum.

Lauren Lee McCarthy is an Los Angeles-based artist examining social relationships in the midst of surveillance, automation, and algorithmic living. She is the creator of *p5.js*, an open-source creative coding platform that prioritizes inclusion and access, and a part of the Processing Foundation. She is a 2021 United States Artist Fellow, 2020 Sundance New Frontier Fellow, 2020 Eyebeam Fellow, 2019 Creative Capital Grantee, and has been a resident at Eyebeam, Pioneer Works, Autodesk, and Ars Electronica. Her work *SOMEONE* was awarded the Ars Electronica Golden Nica and the Japan Media Arts Social Impact Award, and her work *LAUREN* was awarded the IDFA DocLab Award for Immersive Non-Fiction. Lauren's work has been exhibited internationally, including the Barbican Centre, Ars Electronica, Fotomuseum Winterthur, Haus der Elektronischen Künste, SIGGRAPH, Onassis Cultural Center, IDFA, Science Gallery Dublin, and the Seoul Museum of Art. Lauren is an Associate Professor at UCLA Design Media Arts.

Caitlin Mueller is a researcher and educator who works at the creative interface of architecture, structural engineering, and computation. She is currently an Associate Professor at the Massachusetts Institute of Technology's Department of Architecture and Department of Civil and Environmental Engineering in the Building Technology Program, where she leads the Digital Structures Research Group. Her work focuses on new computational design and digital fabrication methods for innovative,

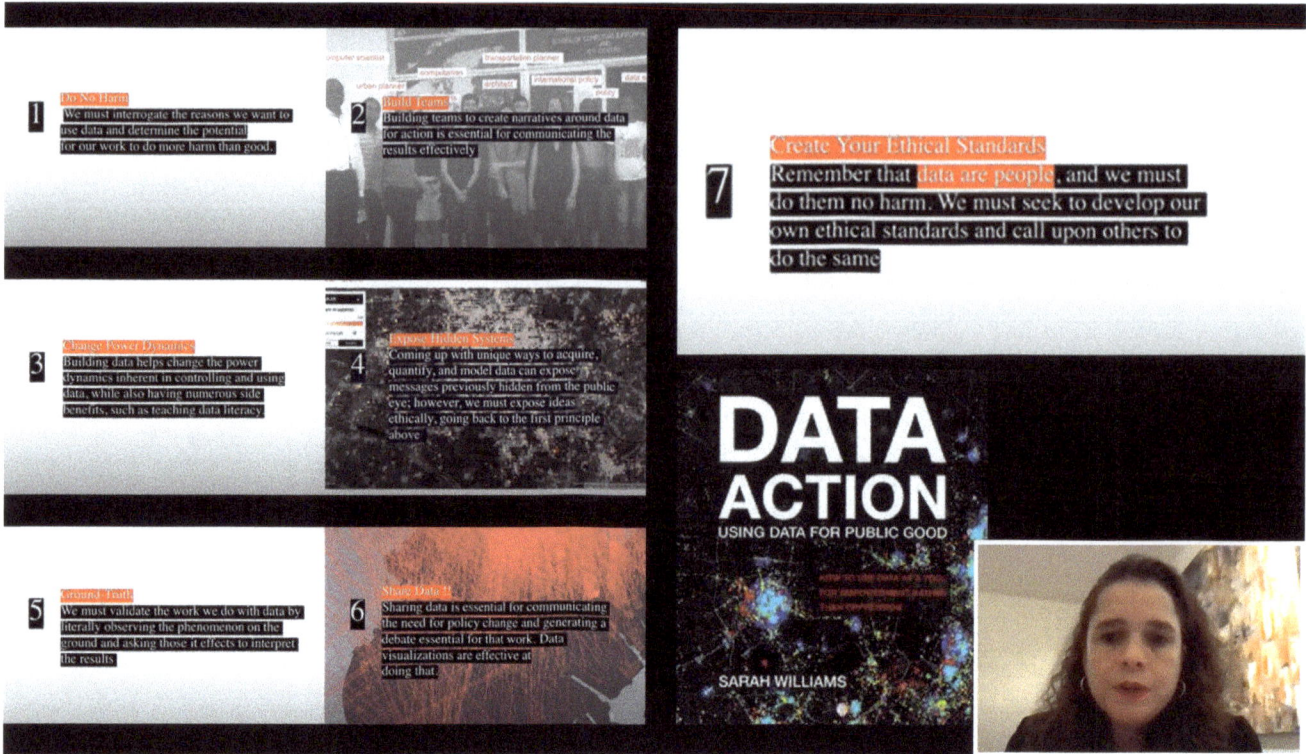

2 "Data Action presents a corrective to standard data practices by acknowledging that data represents the ideologies of those who control its use. [...] These principles can help us take data to action, use it to shape our cities for empowerment rather than oppression. And I asked all of you whether you can create your own standards of practice." —Sarah Williams

high-performance buildings and structures that empower a more sustainable and equitable future. Professor Mueller earned a PhD in Building Technology from MIT, an SM in Computation for Design and Optimization from MIT, an MS in Structural Engineering from Stanford University, and a BS in Architecture from MIT. She has recently contributed to the organization of several major conferences in architecture and engineering, including the 2017 Design Modelling Symposium, the 2017 ACADIA Conference at MIT, the 2018 IASS Symposium at MIT (which she chaired), and the 2021 AAG conference in Paris. She is currently developing a new MOOCA massive open online course entitled Creative Machine Learning for Design, planned to launch in Fall 2021.

The entire keynote panel, including the speakers' presentations, can be viewed on the ACADIA YouTube Channel: www.youtube.com/c/ACADIAorg

Caitlin Mueller (CM): I would like to start off by highlighting Benjamin's comments about archives and libraries as early datasets and as samples taken for future investigations. In general, we don't have a lot of datasets in architecture and design, which I think is interesting. Although some big tech companies have a lot of data from consumer behavior, we don't. But, in today's presentations, we also saw how Sarah could do amazing work extracting and analyzing data from urban environments. On the other hand, Lauren is doing really interesting work operating between humans and machines and interacting with individual behavior data. But in general and compared to large public-facing domains, in architecture, design, and art, we have almost no data. I would love to see more data that is publicly accessible, with clear metrics that we can agree on, and opportunities for open debate that we can all contribute. And with that, I'd love to focus on this question that Lauren brought up in a really interesting way about the relationship of human intent and data. One of the interviewees in Lauren's video remarked that Lauren, the new Alexa, through external observation, knew what she wanted and needed, but, in fact, she was not always the case. And I think that's a fascinating observation about data, that it is often just a blunt measurement of things that we care about. But, how do we choose what to care about? What should we measure? How do we calibrate that relationship? Sarah, as some who is actively taking measurements, how do you decide?

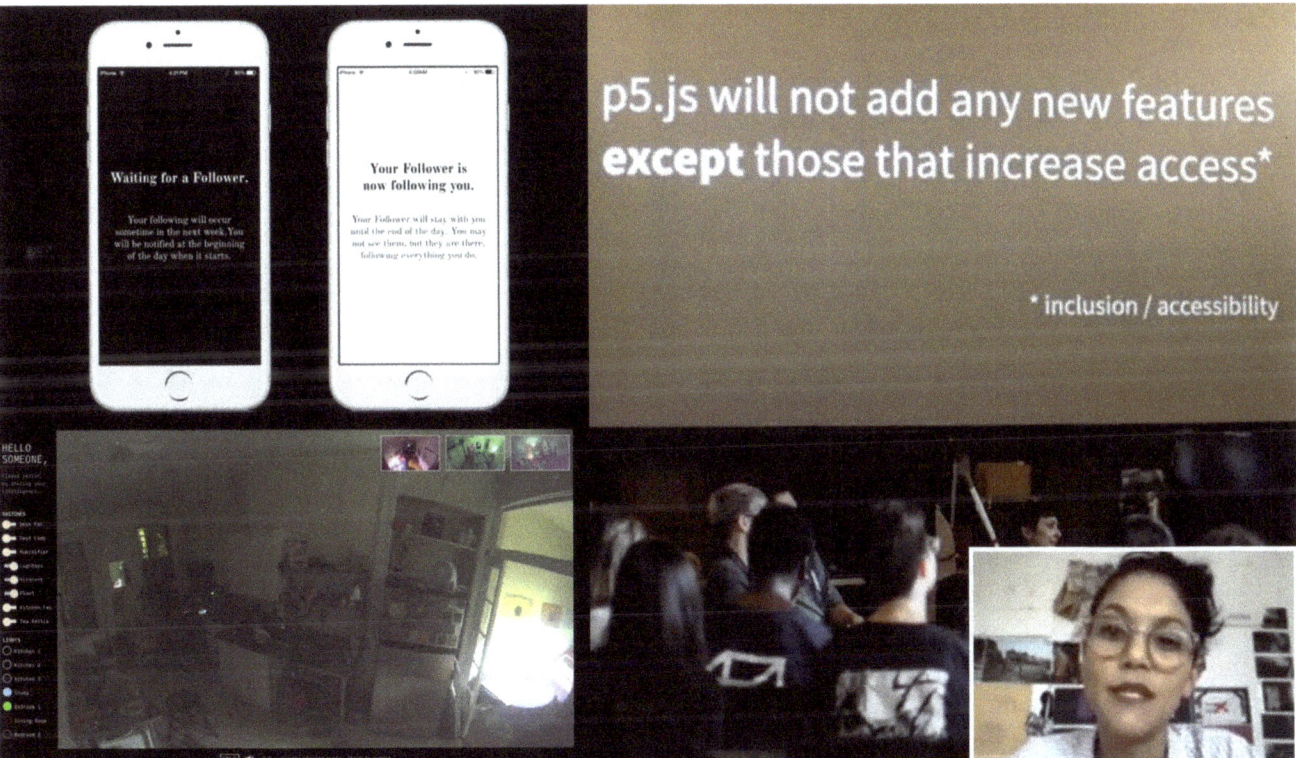

3 "There's a critique in all my work. But there's also always a part that is about hope, that is earnestly and radically seeking connection. [...] I'm captivated by the ways we are taught to interact with algorithms, and how this shapes the way we interact with each other. What are the rules? What happens when we introduce glitches? [...] I feel a sense of power, owning the data of my own voice. I'm taking it back from the tech companies, constantly tapping my conversations, sampling and analyzing and archiving my speech for future use yet unknown. And said, I offer the control my voice to others. What do you want me to say?" —Lauren Lee McCarthy

Sarah Williams (SW): One of the things I've recently been working on in our [Civic Data Design] Lab, is the missing data project, which is trying to uncover data that does not exist. We spent some time during the COVID-19 pandemic doing that. And the aim is to elevate the stories of those on the margins. For instance, it is now very well known that we were not collecting race data, which made it hard to make decisions. Or, there is a lot of data in the urban environment that is purposely not collected that needs to be collected in order to elevate stories. But, there is a broader question about our intention when collecting any data. Because, when we bring our own biases to data collection, we expose or, let's say, unintentionally, can cause harm. So, your question was a bit loaded because, on the one hand, I collect so much data because I want to elevate positions on one side; but, then at the other times, there's data that can be collected that can cause potential harm.

But, I was also thinking about your reflection on data and architecture. I have a good friend who has been collecting architectural plans from different offices to understand the value of real estate development. So, we do have tons of data in architecture, but it's often closed off. What if we could get a model of all the architectural plans of the super tall towers recently built in New York City and make a comparison? I'm not sure how we could open those data sets without losing privacy, but I think it would be interesting to achieve that. I'm just thinking about my friend who has created a lot of value through that data creation and collections. So, it gets back to this collection issue that I believe that when we attempt to get the missing data, we can actually do a lot of benefits.

Benjamin Bratton (BB): I appreciate what Sarah was saying about the question of the collection and curation of the data, as well as Caitlin's point about the archive and the longer-term aspect of the rights. In geological sciences, climate sciences, and other fields, there has been a really interesting epistemological shift that big data has brought in one way or another when looking at very long-term samples. And, I think part of the culture of big data at this moment is about instantaneousness. And, I think that is part of the problem.

I wanted to address directly Caitlin's question about measuring and measurements. One of the things that this comes down to is that the act of measurement in many cases is an act of quantification. But it's predicated on a

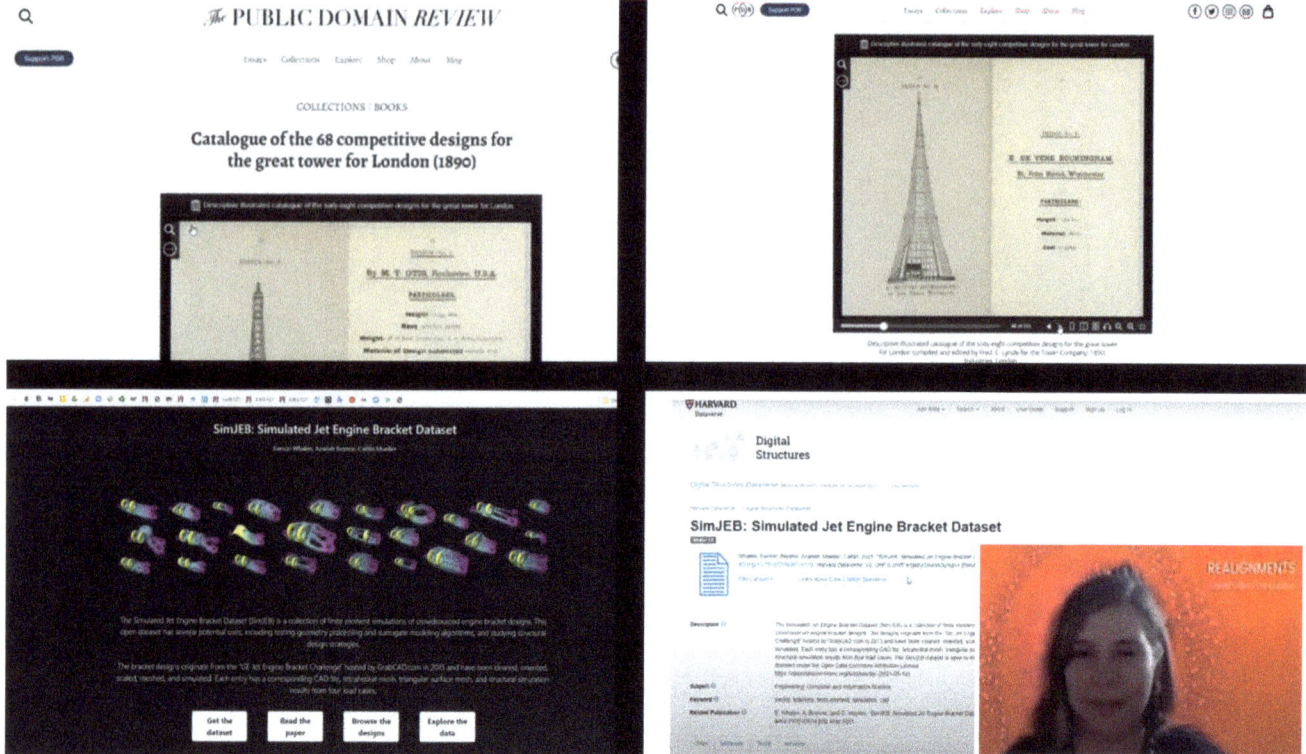

4   "When I think about data and how it affects our fields, I'm mostly thinking about designs and not people. [...] I worked with students to create an open source dataset based on a competition that GE and www.grabcad.com put forth in 2013. [...] There are hundreds of submissions, and they're all publicly available. [...] We simulated their structural performance, [...] but also categorized them in terms of their design concepts. [...] This is an interesting way to have a baseline that we can all work on together in order to advance an understanding of design and data." —Caitlin Mueller

kind of linguistic categorization of the phenomenon that would be identified and made discrete in the first place like this is one kind of thing, that is another kind of thing. And, this categorical logic is not quantitative, it's qualitative. It is linguistic. It's symbolic. And so I realized that to a certain extent, some of the political controversies over data and quantification in the academy is, to a certain extent, a little bit of a war between the arts and humanities on one side, and the sciences and engineering on the other. And, we'd like to blame math for these problems; or, if we just embrace affect, language, and meaning once again, this would solve the problem. But in fact, it is behind the math. The measurement is about a linguistic kind of structure. And so there's this conception that goes on both sides.

I really love Lauren's work about the way in which the anthropomorphic constructions around AI and some of the strangeness of that momentum and impetus, are presented. That I think is extremely important. In many cases, though, some of the most socially and politically important uses of data production modeling do not involve the modeling of people whatsoever. And I think we all should make a point of this. Part of the limitation, I think of

both the Facebook model of the Facebook view of the world that the world is constructed of atomized individuals with predictable desires, but also the problem with the critique of the Facebook model, which is if these individuals could counter-weaponize and make themselves yet more private then that would solve this question. I really would like to point your attention away from many of the most important issues that will involve the modeling of carbon flows, the modeling of electricity flow, the modeling of land use flows, that don't necessarily involve a representation of humans at all, let alone have fair or unbiased [approach] in one way or another. And so there's a certain sense, I think we're at a point where we can open up this conversation in ways that might allow another set of directions.

SW: They do include people. We are the ones using and producing the majority of the carbon. The bottom line is that if we're going to change environmental behavior, then we should clarify who is creating that carbon. It is the human.

BB: No, it is humans but in general, I would strongly push back in the idea that if we could model individual or consumer behaviors, that somehow we can quantify carbon footprints, it is going to add up to infrastructural decarbonization. No, in this case, the issue has to do with carbon taxes and other kinds of infrastructural scale implementation policies. It is not just about pushing this problem to the end-user and calling it a moral question. It can't be that if you could just be a better person, it all cascades.

SW: Yes, I see what you mean.

Lauren Lee McCarthy (LLM): I think there are people at every scale of this. And when we're talking about modeling and talking about data, I just like to remember that they're also people making those decisions about how those models are made, and what they're used for, and what data is incorporated into them. So, I'm always thinking about our tendency to see something as a technical problem, when I believe there are people involved in every stage of it.

Also, Benjamin, you made this distinction between, or maybe lack of distinction between quantified data and something that's more linguistic or soft, more effective. But I would also want to open up the line around the collection of data in general. I think that is in itself a specific kind of instrument of knowledge, this idea that we're going to collect data, information, and then model something, whether using effective data or qualitative or quantitative. And I think there are a lot of other ways of knowing, and understanding. I don't know that those are often incorporated into some of the technical models that we build. So, that's something that I've always been really focused on in my work, which is thinking about the people that do not understand or feel that this is outside of their domain of expertise to be thinking about large scale and technical modeling. How do we recognize that actually, those people are essential to the conversation as well?

BB: Yeah, there's a lot in your point. And I think we agree in many regards. I was not suggesting that behind the quantification is a qualification or that behind the math is the language. In fact, it is reinserting the human phenomenological perception of the world back into this question. But it's also a way to insert back this question of abstraction. I think one of the things that tie both of these together is the dynamics of abstraction. Mapping uses modes of quantitative abstraction, that allow us to take all these data, climate data points, and find their hockey stick, there are modes of qualitative abstraction where we can compare different kinds of experiences and different kinds of relations between this and in different sorts of ways as well. I'm not suggesting that we simply need to give up on linguistics and just embrace the math and be done with it.

That's not at all what I'm suggesting, in any kind of regard. In a certain sense, it is an interest in understanding the politics of abstraction and the politics of extraction. And, how that can work through the abstraction of ideas into tools and the ideas of tools into outcomes, the relationship between the linguistic and the quantitative as different modes of abstraction can depend upon each other. But what we're building ultimately and hopefully, are certain types of collective intelligence that go beyond any one of our individual perceptions. It might get built up from lots and lots of human actions, but we're building a collective form of abstraction. And I think the collectivity of that is actually, what makes it potentially politically powerful, in ways that we shouldn't overlook.

CM: That's really well said, Benjamin.

Sarah, to your point about the floorplans, I'm always interested in this because I always hear that nobody wants to give this data up. No one who's currently practicing and owns the intellectual property of their designs wants to give them to anyone. This is probably for the same reason that people don't want to give their data to Google or Facebook. But perhaps, not only because they don't want someone else to exploit them, but because it's their means of operating. On the other hand, the owners of the buildings actually also own the floorplans to some degree. So there are a lot of privacy concerns there. We could debate whether that's good or bad. What about public buildings that are publicly funded? Where and how do you see that data? If the government is building buildings or infrastructure, does the public deserves access to that data?

SW: That's a good question. Maybe we can go back to the seven principles in my book. And I would guess my first question would be, who could be potentially harmed by the release of that data? I mean, beyond whether it's public information or not. There are some public buildings that are quite sensitive. So there could be potential harm.

CM: For example, embassies, which the GSA is in charge of producing, but obviously, embassies have a lot of security concerns associated with them.

SW: Could you open data about a public school or other public institutions? Sure, there could be a lot of value in analyzing school data to understand learning and the quality of the learning environment. So, it seems that potentially there is not a harm in doing that. But we would have to do more analysis. So, my answer would be, yes, but as long as that action is not harming anybody. Let's think about another example. Would an architect give out their secret sauce by giving out their floor plan of a public building? I am not sure if there's potential harm for an architect in that. Let's say what if the analysis finds the building to be good. I think there is potential value in its analysis. Going back to my friend who's doing the floor plan analysis, that's why people are actually sharing the data with her because she's using it to see how different plans add value to the real estate development process. So in a real estate framework, how do certain kinds of designs then create more rent. I think certain developer-minded architects would like that, but then some other architects might think their design isn't valued or could be undervalued.

BB: I would echo that as well. I think the interesting question about the public ownership or public use of the data opens the question of, what do we mean by the public? And which public are we talking about?

In some of the examples that Sarah was talking about, what I am hearing, which I think is an important question, is that there is an implication of some form of public governance that is able to make use of this information in some way and act upon it, that it is not just information about the world but that the model becomes recursive.

It is actually a mechanism for acting back upon the world through the implications of the model, not just knowing more for its own sake. So, I think part of the question that should be asked at this point is if we do see a kind of shift in the role of models and abstraction that are quantitative and qualitative as part of how cities recursively compose themselves, and govern themselves, and remake themselves in the buildings and the people and the processes and the services, and they go in here. Who does this? In what sense does this imply not just technical practices, as Lauren suggested, but also a shift in the logic of socio-political governance, per se, on what a government does and what governance means, more generally? And what are the kind of institutional changes that would need to happen in order for that to even be conceivable and possible?

I will put a plug for a good friend of mine, Ben Cerveny who runs a collective called the Foundation for Public Code in Europe, where it's a federation of CTOs of different European cities, including Helsinki, Barcelona, Amsterdam, Copenhagen, and others. And, they help cities build open-source platforms for the provision of public services. And then they containerize and allow other cities to use these as well. But one of the things that you're seeing a lot is that there is this large disability for the public government producing public models for public use which is increasingly not just something that governments do, it's what governments are. And that's something that is very difficult to explain to political scientists. But I think that it is becoming increasingly part of this question of what publicly owned data should be.

SW: I was thinking about Caitlin's earlier question, that in a way public school floors plans are already available. It's just not easily accessible. So, somebody needs to digitize it all. And then once it does become easily accessible, what is the potential benefit for that? Then what systems do we create to make that happen? This goes back to policies of building a public building. Do you check in your floor plans, and then use them for that kind of benefit?

CM: There is an interesting question from Vernelle Noel asking if we lack imagination in terms of understanding the potential downsides of these practices. I think it's a super interesting question. When I studied engineering, it was all about analyzing failure modes and spending the whole time thinking what are all the things that could go wrong here. In architecture, it's the opposite. It's about what can we do? And I think this problem of analysis, definitely demands a more engineering mindset that doesn't necessarily come naturally to the ones only concerned with a design perspective, or only excited about the data perspective.

Thank you, speakers, for highlighting some of the technical socio-spatial, political, and ecological questions and concerns regarding data access and power. We're really, really grateful for your time and for your contribution.

KEYNOTE EVENT

Design Imperatives in Social & Environmental Crisis

Paola Antonelli
MoMA

Justin Garrett Moore
The Andrew W. Mellon
Foundation

Lydia Kallipoliti
The Cooper Union
& ANAcycle

Mariana Popescu
TU Delft

Social and environmental crises are tightly coupled and entangled in design, computation, and data. These entanglements make their decoupling, one's ability to notice and question them difficult. In this context, what is the connection between computation and ecological thinking? How would design education, ethics, practice, and projects shift if we prioritized using our intelligence, creativity, and resources to care for our people, places, and the planet? In this keynote conversation, four designers, scholars, and practitioners who engage with the politics of social and environmental crises share ways that we might rethink computation in practice, theory, and pedagogy for a better, more informed future for our citizens and our built environment.

The presentations and conversations between these four keynote speakers will help us navigate through these entangled questions. This manuscript documents the conversations between the speakers. The entire keynote panel could be viewed on the ACADIA YouTube Channel, https://www.youtube.com/watch?v=Nt9sbGf5WoI.

Paola Antonelli is Senior Curator at The Museum of Modern Art in the Department of Architecture & Design, and MoMA's founding Director of Research & Development. Among her most recent exhibitions are the XXII Triennale di Milano Broken Nature and MoMA's Material Ecology on the groundbreaking work of architect Neri Oxman. She is also currently working on @design.emergency, an Instagram and book project that explores design's role in building a better future for all, in collaboration with critic, Alice Rawsthorn. Alice and Paola's book, *Design Emergency*, will be published in Spring 2022. Through design and other projects, Paola discussed the interconnectedness of systems and crises that we live in—environmental, economic, democracy, racial injustice—and the need for empathy.

Justin Garrett Moore is a program officer for the Humanities in Place program at the Andrew W. Mellon Foundation. His work focuses on advancing equity, inclusion, and social justice through place-based initiatives and programs, built environments, cultural heritage

1   "If we want to do something about the future, we need to be able to think beyond the three or four generations that are physiologically possible..., [we need to have] a sense of the interconnectedness of systems that we live in..., [and have] a sense of what [we] could do in [our] everyday life." —Paola Antonelli

projects, and commemorative spaces. He has extensive experience in architecture, planning, and design—from urban systems, policies, and building projects to grassroots and community-focused planning, design, and arts initiatives. With over fifteen years of public service with the City of New York, he has led several urban design and planning projects. From 2016 to 2020, he was the Executive Director of the New York City Public Design Commission, and in 2021 he received the American Academy of Arts and Letters Award in Architecture and was named to the United States Commission of Fine Arts by President Joseph Biden. Justin argued for the practice and deployment of computation from a place of care. While computation and design can be connected to harmful practices like the slave trade, slave ships, redlining, and exclusionary policies, he argued for a revealing practice of computation grounded in care and access, rather than in harm.

Lydia Kallipoliti is an architect, engineer, and scholar whose research focuses on the intersections of architecture, technology, and environmental politics. She is an Assistant Professor at the Cooper Union in New York. Kallipoliti is the author of *The Architecture of Closed Worlds, Or, What is the Power of Shit*, the *History of Ecological Design* for the *Oxford English Encyclopedia of Environmental Science*, and the editor of *EcoRedux*, an issue of *Architectural Design* in 2010. She is the principal of ANAcycle thinktank, which has been named a leading innovator in sustainable design in Build's 2019 and 2020 awards and Head Co-Curator of the 2022 Tallinn Architecture Biennale. Lydia presents "the problem of encirclement"—understanding ecology explicitly as circular reasoning. She demonstrates this by discussing the promises that were outlined in the popular book, *Cradle to Cradle* by architect William McDonough and chemist Michael Braungart.

Mariana Popescu is Assistant Professor of Parametric Structural Design and Digital Fabrication at the Faculty of Civil Engineering and Geosciences at the Delft University of Technology. She is a computational architect and structural designer with a strong interest in innovative ways of approaching the fabrication process and use of materials in construction. Her area of expertise is computational and parametric design with a focus on digital fabrication and sustainable design. Her extensive involvement in projects

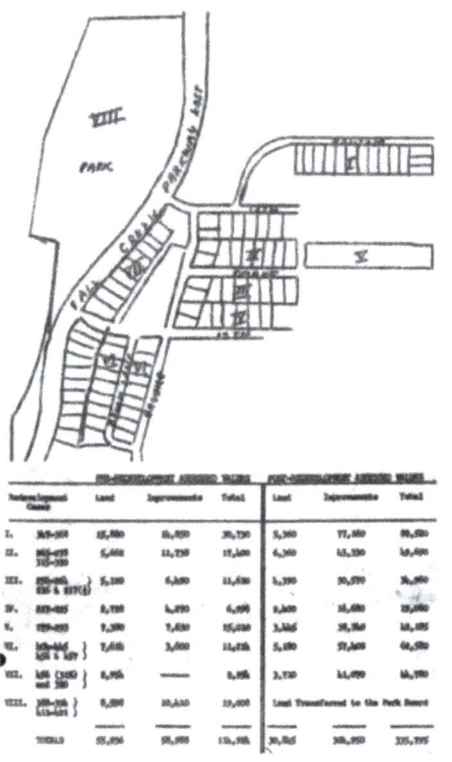 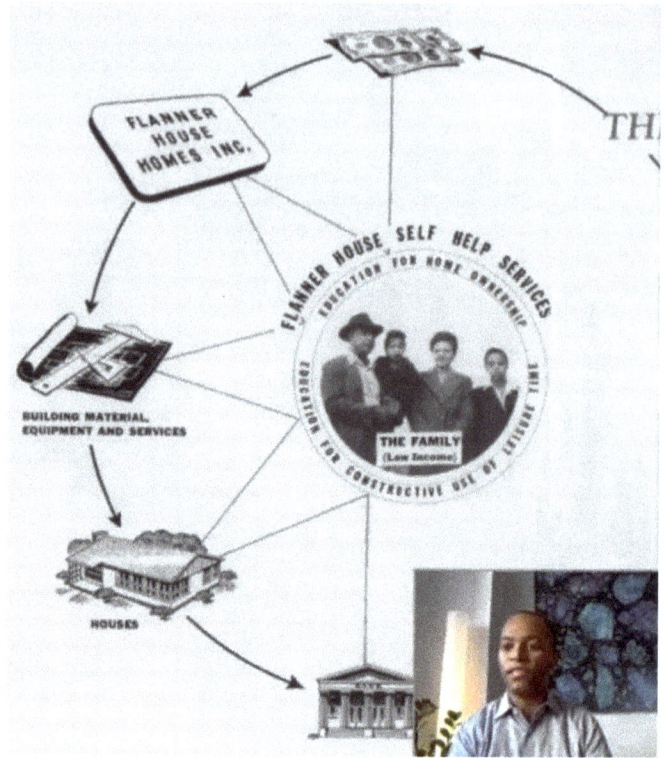

2   "I would offer a quote from Berenice Fisher and Joan Tronto as a framing for this question of computation through the lens of care. It says, 'On the most general level, we suggest that caring be viewed as a species activity that includes everything that we do to maintain, continue, and repair our world so that we can live in it as well as possible.'" —Justin Garrett Moore

related to promoting sustainability has led to a multilateral development of skills, which combine the fields of architecture, engineering, computational design, and digital fabrication. In 2019 she successfully defended her PhD, which was nominated for the ETH Medal for outstanding dissertation, and was named a "Pioneer" on the MIT Technology review's global list of "35 innovators under 35." Mariana argued for more sustainable practices in computational design and construction through digitalization and novel construction methods.

Lydia Kallipoliti (LK): I just want to thank all of you for your wonderful and thought-provoking presentations. I'm going to start with a question on the idea of reparation, restoration, replenishment, which was mentioned by Paola, but also in some ways and forms by Justin in his articulation of disinvestment, racialized bias, and processes of restitution; also, partially in the work of Mariana by looking at material forms as forms of atonement of new ways to build. I wonder if there is a way to think of these types of processes without technology? What would it mean to repair without geoengineering and without grand master plans? What would other formats of thinking and acting look like?

Paola Antonelli (PA): Thank you [Lydia], your presentation was really great. Regarding these terms, I think that "reparative" and "reparations" were used by Justin with more authority than by me. For me, it was a very interesting decision amongst all the different terms that I could use. In *Broken Nature*, there was a whole list of possible points of pressure and ways in which citizens could be more responsible and sustainable, from upcycling and recycling to the "digital exile" that has become normalcy during the pandemic—the forced move to Zoom reducing footprints before was one possible strategy. There are myriad ways in which we can contribute to just doing things better.

Justin Garrett Moore (JGM): I think the cross-cutting thread of all of what we've shared is this notion of, "How do we build." In one of Mariana's images, people were on the ground doing something together, seemingly mundane, but the idea of the hierarchies, the power structures, and the resource dynamics that people understand need to be rethought and rechallenged. This is challenging to do because the paths and systems are really in place to make us all go back to rote, to base ourselves in the norms. But I think in spaces where people are showing that there are

3   "What do we build and what is important that we build, because through computation, and through digital fabrication, we've come to a point where we can indeed, do almost whatever we want....But the question is, what should we be doing?"—Mariana Popescu

a wider variety of norms that we can accept, incorporate, and adopt as practices are really key. I really appreciated seeing—and Mariana illustrated that—in some ways, what the machine was doing and what the people were doing were somehow together, even though they've been designed to be differentiated in terms of resources, access, and all of these different things. So I think that idea of how we make things more connected in the plane is where so much of the work is.

LK: Mariana, can I tell you what my favorite part of your presentation was, and you can answer both questions simultaneously? I think it's extraordinary work, your research with fabrics and how they unfold in such premeditated forms. My favorite part was the point where you were showing a giant scarf that was packed in a suitcase that you carried. This action speaks of all the precarious processes architects indulge in, like hugging their models to install them in an exhibition; it was a similar kind of labor when the fabric was unfolded and stitched. What we saw was how this super-streamlined computational process—maybe the question Justin is asking is interrelated—is entangled with precarious labor and contingency that underline the entire process. This happens no matter how streamlined the algorithm is. Could you speak to this? It was my favorite part!

Mariana Popescu (MP): I'm glad that was your favorite part, it's one of my favorite parts too because it makes you do things, and then you become the constructor, you become part of that value chain, and you become not just the designer, but you get your hands dirty. And we as architects, as designers, as creators, love that because it's a way to indeed hug your work or feel that it's your baby, somehow. I would say that we have to step away from that at some point, because the truth is, we can't singularly do that, and we cannot control the whole process ourselves. So that's just to have a sort of disclaimer to these kinds of things. I think that there is always a contrast, indeed, between how streamlined an algorithm is and how that ends up in practice. And this is inherent to any sort of construction practice in any sort of building, any sort of going from the digital to the physical realm, wherever it may be. And this doesn't necessarily only have to be in the precariousness of what you've seen..., but also in the types of tolerances or in the types of being able to really get to

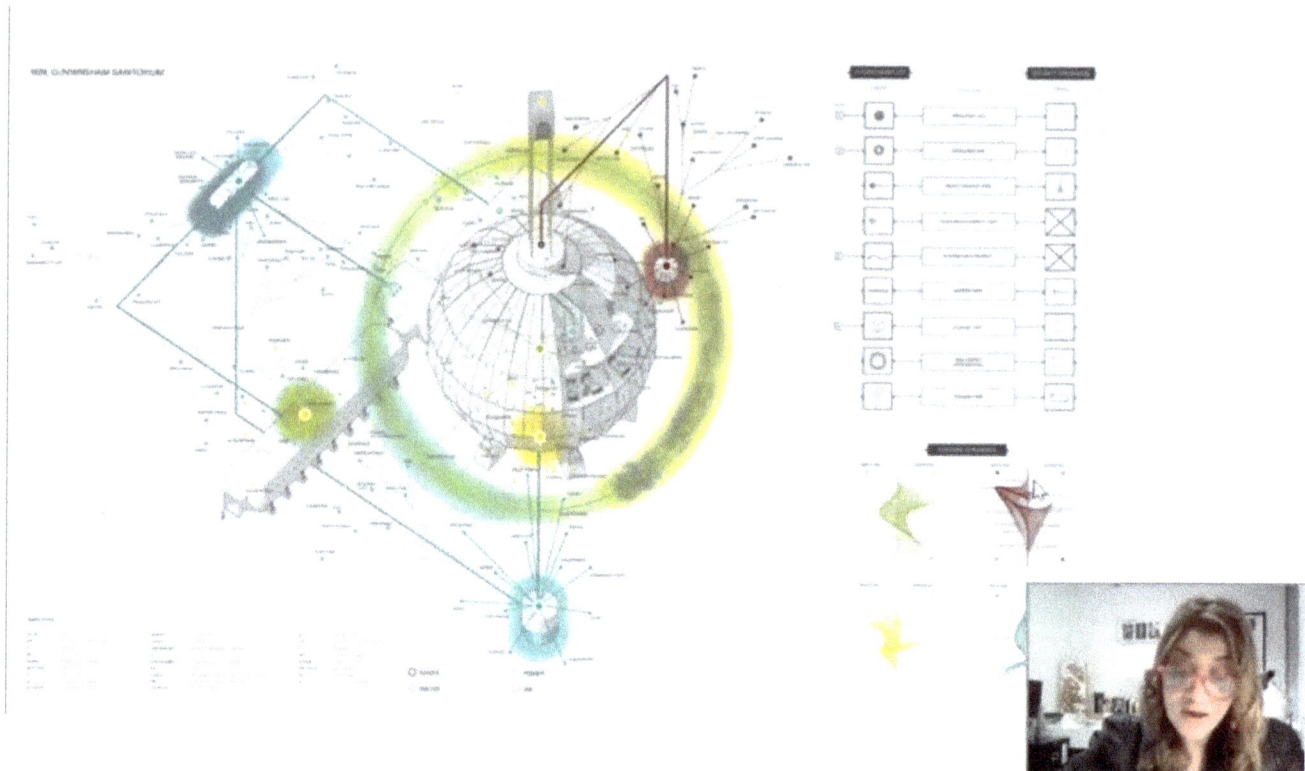

4 "We must look at debris, the waste of our own production processes, understand environmental and social problems in a visceral way, via the raw ecologies of our bodies and the understanding that these problems are not simply statistical and abstract. They cannot just be relayed to the management of resources, but they're landed on bodies and on the water we drink, as well as the air we breathe."—Lydia Kallipoliti

that or replicating whatever is in the digital world.... What I like about the project is that it underlines or it shows who works on this, how do you work on this? How do you apply the material? Is it safe, is it not safe? In the end, it's lighter, it's easier to work with, it's a lot easier for humans to work with than other materials that are out there. So it does allow to rethink work. And it does allow to really show that you could have actually a very streamlined process of design and fabrication. And that maybe we're not there yet on the very streamlined construction sites, but we have to go towards that. And that doesn't necessarily always mean that you have a particularly high-tech answer to it. It can be a little bit of a lower-tech but high-tech informed. Why I'm saying that is not because I don't believe in the high-tech part, I do believe in the high-tech part..., but the part that makes it a little bit more relatable or makes it possible to include people ... now, or to include already existing skilled labor ... now.

JGM: It was actually why I wanted to include that example from the 1950s of the connection of community and their labor to the built environment in a very direct way, and showing the clip ... showing people actually learning the skills, and knowing that there is this process from 6,000 human hours to 2,100 human hours, and what it actually took to do that was beyond the scope of what we [designers] normally take responsibility for. There was a whole set of systems that they were thinking about from childcare to health. There are all these networks of ideas that need to be understood as connected to the work of building and built environment, that I think, for a lot of reasons we've become so separated from. It's a challenge to do work that's truly "community-based"—this generic word that people throw around. If you're going to use the word, you then have to accept all of it as being a part of the resource and responsibility system for our work as designers.

LK: Justin, I was also really struck when you showed the manifests of the slave ships. It was astounding to see how they were categorizing and creating at that time, hyper-logistic machines where people were actually units within a system. And I was thinking what your thoughts on comparing the slave ship manifest with the windowless dorm at the University of California Santa Barbara that was recently all over the news. Ostensibly these students are part of another kind of manifest. So what has changed, and how can we use our agency to change the logic that generates these kinds of systems? I consider this a critical question for our discipline because the two plans are not different. It's just insane that after centuries and a great number of battles for civil rights, there is no difference.

MP: Can I just say, it's not machine logic, this is pure human logic. This is pure inequality, or this is pure privilege, this is something else, this is not machine, or this particular example is not machine [logic].

LK: Of course, you're absolutely right in pointing this out. Because machine logic doesn't distinguish or make hierarchies between person X and person Y. We distill this logic within the algorithm. We put it there; we inserted it, but there are hierarchies that make it a kind of systematized production system. These types of hierarchy read like codes in some ways.

JGM: I appreciated the point that Paola was sharing at the end of her talk about this idea of the imagination, and really understanding the power of it. For example, that drawing of the slave ship took imagination. And I agree with Mariana, it's not about machine or rationality or any of that it's actually about pure unencumbered imagination. And so what is our counter to that? And how does it become as invested as empowered, as impactful as this horrible imagination, that we are all still living with regardless of your demographic, background, history, and heritage? We are actually all still living with this unimaginable, horrible imagination.

PA: See, I respectfully and lovingly disagree. I think that that is mere visualization. Imagination would be different, it would mean envisioning a better or worse future. There would be something more. I did a project a few years ago that was called "Design and Violence," trying to understand the manifestations of violence in contemporary society through design objects. It was about contemporary [issues], so I didn't use the diagrams of slave ships design, but they were in my mind. I did include redlining, it was one of the case studies. If that dorm had happened at the time of the project, I would have included it for sure, such a symbol of arrogance and oppression. It's not even evil, it's ignorance and arrogance. It's thinking that the other 99.99999% can live in cubicles like cattle, and it's okay. And if you see the picture of the two, Warren Buffett and the 92-year-old amateur architect [Charlie Munger], next to each other ... I mean, Warren Buffett should not go around with him, because we've always thought of him relatively well....
I think that imagination is what can propel us forward in some cases in evil ways, but it is always a leap. And instead, that [dorm or slave ship] are just a visualization of how to fit as many humans as possible within a dorm or a ship. The ship was leading to a much more tragic and reverberating future, even though the UCSB students should be treated better anyway. ignorance and arrogance are just what is keeping us from designing a better extinction, from living better, from everything.

LK: So I'll ask one final question that Justin posed earlier. It is a very simple one, but a very complex one at the same time. How would design education, ethics practice, and projects shift if we prioritized using our intelligence, creativity, and resources to care for our people, places, and the planet?

PA: Well, I don't know if it's feasible at all. But I found that indigenous university example really interesting. Now you might object to the fact that it's Ursula Biemann, she's a Swiss artist, is it more of white western people colonizing, but actually, she was brought in as an artist to help indigenous people find a way to create a bridge. So that's just one possible idea.

JGM: Sharing knowledge and practice.

KEYNOTE EVENT

Emerging Trends in Response to Critical Computation

Marta Novak
Ohio State University

Dori Tunstall
Ontario College of
Art and Design

Charlotte
Materre-Barthes
Harvard University

Jenny Sabin
Cornell University

As thought and practice leaders of a community with potential impact in the future of society and our built environment, it is our duty to reflect critically on the intended and unintended consequences of technological practices. In this keynote conversation, four designers, scholars, and practitioners, whose work deals with situating people at the center of obsolete economic and political practices, gathered to denounce naive assumptions about design, technology, and architecture, and challenged the audience with provocations on the future of the profession.

Marta Nowak, Christos Yessios Digital Fabrication Assistant Professor of Architecture at Knowlton School at the Ohio State University, Senior Innovation Manager at Google R+D for the Built Environment and Founding Principal at AN.ONYMOUS, opened the panel with a presentation about her research work on the interaction between humans and architectural spaces. Marta discussed the idea of prosthetics as extensions of the abilities of the human body, and presented her studies on the interfaces, machines, and spaces that mediate our relation to the architectural spaces we inhabit.

Dr. Dori Tunstall, Dean of Design at Ontario College of Art and Design University, extended the conversation by advocating for the decolonization of design and technology. In her presentation, Dori discussed how desgin and technology has been rather disrespectful, especially to indigenous, black and people of color communities, as well as the land, in and of itself. Through historical examples, Dori argued how the values of design and technology have been, and still are, colonial, white supremacist, patriarchal and capitalist, with harful intended and unintended consequences provoked by misguided values. She concluded with two provocations on how we can change the consciouness of our technology.

Dr. Charlotte Malterre-Barthes, Assistant Professor of Urban Design at Harvard Graduate School of Design, followed with a criticism to the complicit role of architecture in environmental and social degradation, and questioned the prevalent culture of always building new. By examining the complex, and often violent policies of resource extraction as motors

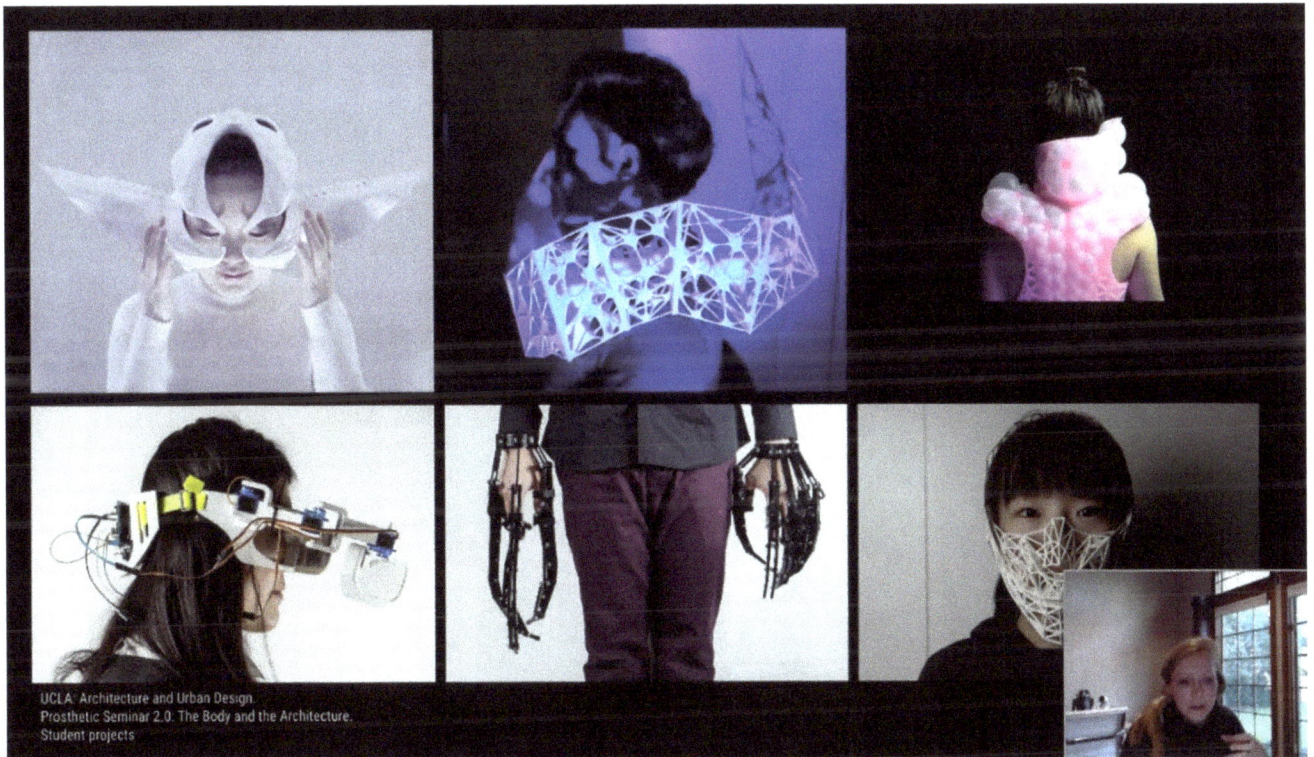

1   "I am very interested in the idea of prosthetics that focus not on recovering basic human function, but extending human abilities, the objects that enable new type of interactions with environments, ...prosthetics as architectural projects that take the human body as its site.... When you think about it, the [video game] controller as a form of body machine interface enables human action and interaction in a space, be it physical or virtual. To me, this is no less architecture than our furniture, physical spaces, or buildings."   —Marta Nowak

to predatory capitalism, Charlotte argued that current architectural practices perpetuate an obsolete model where architects are always expected build more, and challenged the audience to envision a world where pausing construction, using instruments such as building moratoria, could lead to a human-centered culture of reparation, reinvention, reevaluation, and reinvigoration.

Jenny Sabin, Arthur L. and Isabel B. Wiesenberger Professor in Architecture and Associate Dean for Design Initiatives in the College of Architecture, Art, and Planning at Cornell University, and President of ACADIA, concluded the panel by reflecting on ACADIA's forty-one years of history and some of the provocations offered in previous panels in this conference, to help contextualize the discussion and propose next steps for our community. A lively conversation followed, reproduced below in edited form.

For a recording of the entire event, please see this link: https://youtu.be/0mnWUoAWrg4.

Jenny Sabin (JS): Dori, in your work towards decolonizing design, you underscore that we need to develop a better underlying consciousness with technology. And you pose the question "what does it mean to bring an indigenous consciousness to technology?"

One example of work that immediately comes to mind that engages this question is the work of Ron Rael and Virginia San Fratello, such as their Casa Covida project. In the architects words, "Casa Covida, a house for cohabitation in the time of COVID, is an experiment in combining 3D printing with indigenous and traditional building materials, methods with employing new and ancient ways of living. The experimental case study house is sited in the high alpine desert of Colorado's San Luis Valley, where adobe, a combination of sand, silt, clay, water, and straw that is dried in the sun, and is the traditional building material of the region."[1]

Are there other emerging trends and work that we should consider? In many ways the ACADIA community is well poised to implement the robust 40 plus years of computational design thinking, methods, and tools towards a more empathetic, relational, and critical project. But in

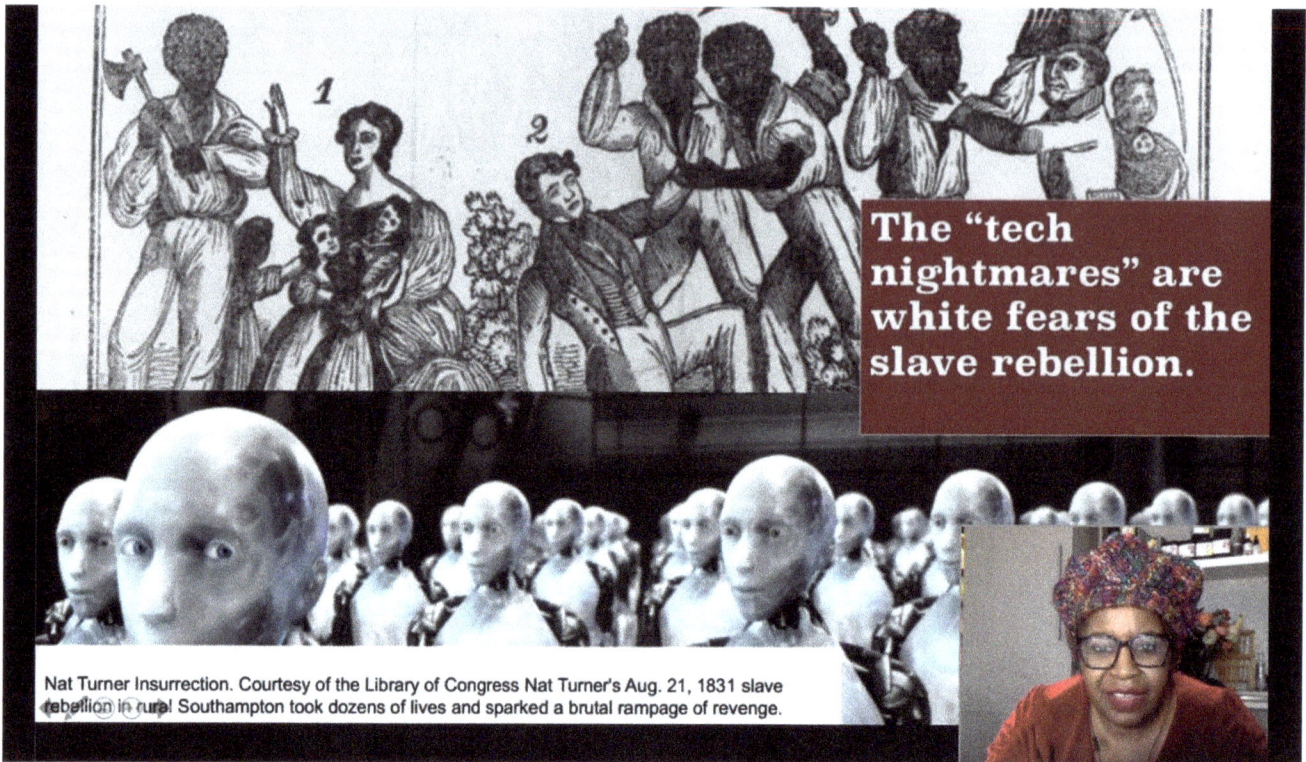

2 "Modern technologies are based on master-slave relationships. When we talk about human-computer interactions, many of them are based on human-human interactions from the past. [For example] 150 years ago, Siri might have been an enslaved indigenous, or black woman, because many of the ways in which we create our technologies are to replace activities that were formally done by human beings. And thus, our nightmares in technology are actually based on white fears of the slave rebellion." —Dori Tunstall

many ways, we have only just begun to demand a critical approach and alternate models in pedagogy. So I ask the three of you and your incredible presentations, what are the emerging trends? And what are our next steps? You have offered several methods and projects to consider. But are there others in terms of practice, of considering pedagogy, teaching, and the models that many of our curricula are based upon?

Dori Tunstall (DT): Jenny, what an amazing overview of the history and revolution that is happening in the field, and such a great summary of the conversations over the conference for the last five days; thank you for that.

I am really interested in things that are happening pedagogically. The thing I am most excited about is that we have spent the last couple of years developing indigenous learning outcomes, and we are just at the point now where we have enough critical mass of indigenous faculty members to begin to implement them. But it is really interesting that we have moved from the conversation of what is decolonization to what does it mean to create spaces in places for indigenous sovereignty. We have an indigenous visual cultural program, but I am particularly interested in how that spreads out to change the underlying ethos of the work that we do in the rest of our programs throughout the institution. I trust that the implementation of these indigenous learning outcomes will reconfirm our connection to the land and reconfirm our necessity of understanding indigenous histories, for everyone. As a community of educators, we have ramped up on our knowledge and shifted our understanding of what the possibilities of design are away from this modernist project.

There are some really beautiful things happening pedagogically. The students are developing a sense of pride of self, their cultures, and their backgrounds, and are using that to generate innovations that, in many ways, not just benefit their direct backgrounds but benefits the wider community of people who now they see themselves in relationship with. In this case, relationship carries the indigenous notion of "all my relations," extending from human to human into the relationship with the air, water, soil, materials, and technology. There is some really beautiful work happening, which makes me excited for this conference because I see there are spaces and places, at the intersection of

First Roundtable of the Global Moratorium on New Construction, Harvard GSD, 23.04.2021 (Source: Author)

3 "Halting construction, even if temporarily, shall suspend the race for discounted materials while questioning the narrative of progress and techno positivism established around capitalist societies. Building without an end in sight is the diktat and the jurisdiction of modernity, and we must question the growth imperative of the construction industry." —Charlotte Malterre-Barthes

academia and practice, where they can grow too, and engage deeper in not just these conversations but in these practices.

Marta Nowak (MN): Thank you Jenny for this amazing overview of ACADIA. It is really great to see how things have changed, yet not really, in the past 20 years of talking about technology and architecture, especially when you are bringing up the quotes from 2004 and the ideas of access, which was allowed because of the computer.

I keep thinking about that today as well, except now computers are taken for granted. The new tools that maybe five or ten years ago were incredibly complex and difficult for anyone to use are now very accessible, so anyone can use them for their own projects. However, now there is a huge responsibility about what is this being used for? When I think about the work I do in different institutions, it always comes down to the individual. I am quite tired of the top-down approach; to me, it should always go back to the very basic human being and understanding their behaviors and their relations to new machines. In order to do that, we also have to understand how they were designed. I am definitely a promoter of bottom-up thinking and and bringing ideas back to humans. We are changing, evolving and bombarded with everything around us, so I look forward to years of ACADIA and how that will evolve too.

Charlotte Malterre-Barthes (CMB): Marta, I think it is very interesting what you discuss about access to software, a conversation I recently had with a student. I intentionally included in my presentation CAD programs opening on a blank page as an homage to the modernist project where you start from nothing. Reviewing the situation, and specifically looking at the software we use for design such as Rhino,[2] my student mentioned "By the way, do you know how expensive these programs are?" I am stunned to hear them qualified as "accessible:" they might be more accessible in terms of literacy, but they're not necessarily accessible for everyone in terms of costs.

Jenny, you were interested in asking about where do pedagogical institution stand on this? And my own impression from listening very closely to what my students, and students in general, say... it seems to me we are not really

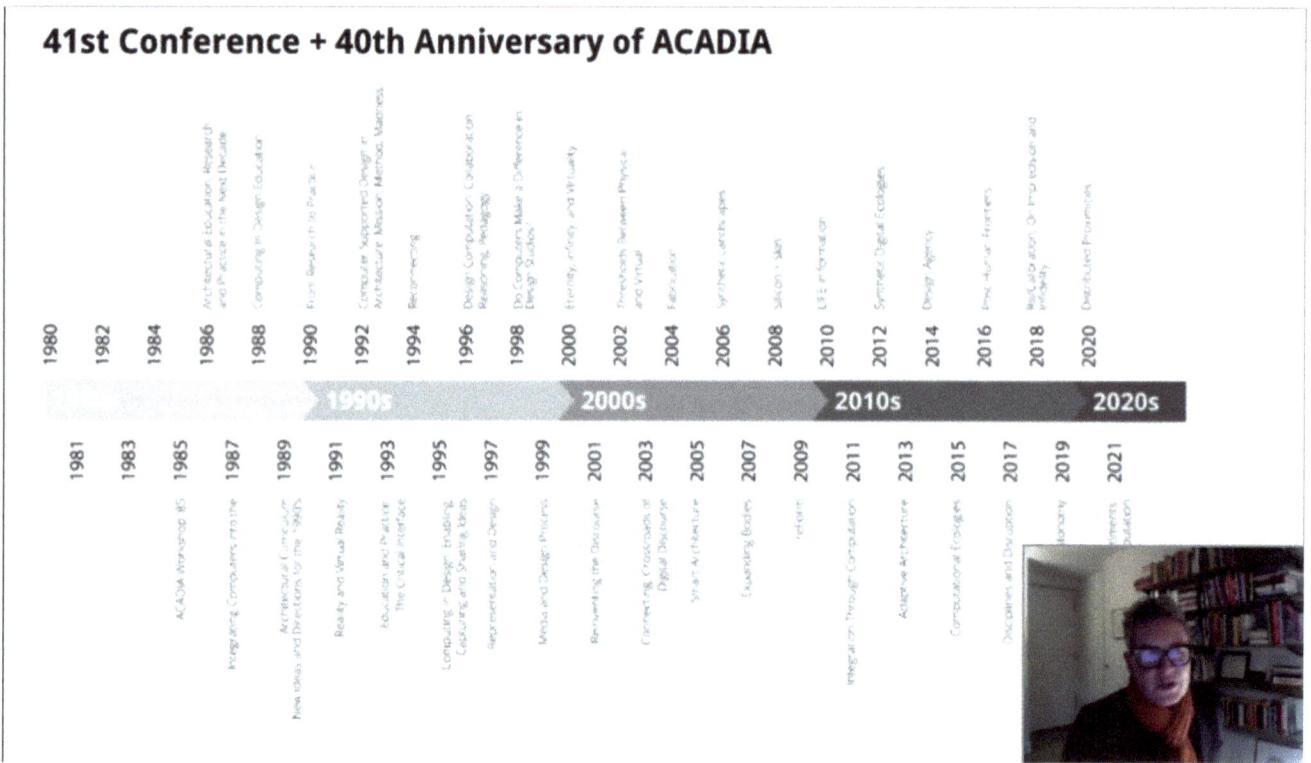

4   "Last year, the conference marked an important shift that this year's conference continues: a critical thought provoking and inspiring conference appropriately titled *Distributed Proximities*. The conference challenged us all to deepen our thinking, creative practices, and engagement with ecology and ethics, issues of access and bias as it pertains to algorithms and big data, automation, agency, labor and practice, computational design, in issues of health and wellness, and new forms of collaboration, to identify next steps." —Jenny Sabin

up to the task. I feel like the conversation that is happening, not just around technology but the whole discipline of architecture and design, is not really up to the climate emergency we find ourselves in. I think students very clearly feel that the curriculums are dated, or that they are disconnected from their urgencies, including questions of race, gender, and even materials; there are students who are still forced to do concrete in their studios, even though we know we really have to question this paradigm. I feel the urgency as I speak because I feel my students are also feeling that the schools are not yet where they should be. This might be related to the inertia of institutions, which are slow to change for many reasons. But for instance, one of the questions I would have would be why have we not declared a climate emergency in architecture schools? Even if you could dismiss this, or my comments about green capitalism as a gimmick, I feel like the schools are not located where they should be in that fight. I know that they are trying—my own institution is grappling with that—but previous institutions where I have found myself where incredibly slow and not understanding why. So I would throw back the question to you, in your position now, as it feels to me that educational institutions are not grappling enough with these questions.

JS: I agree. I am encouraged, both by our students and faculty, to have these open discussions and make changes to syllabi. One of my favorite quotes from Buckminster Fuller, actually, is that to change a model, you have to replace it with a new one.[3] We are grappling with what that new model is since, as you know, most of our curricula are based on the Bauhaus model, and that has been the case for over a hundred years, in many ways. I think it is going to take time, and that is why it is so important to be having these discussions together, both inside of this kind of niche community of ACADIA with our areas of expertise in technology, and expanding beyond that, which is so important in the development of a criticality.

CMB: I have another example that I think is really interesting: the discussion around circular materialities. There are people who are looking at how, for example, catalogs and software plugins could be developed so that, when you design something, you can already be able to source it, and understand where your choice of materiality is impacting someone and what does that mean. I feel there is urgency at that level that could be solved with these tools. I was mentioning before how some of these programs are unable to acknowledge that there is never nothing when we plan an architectural project: there is soil, air, humidity.... Even if it is a deserted area, it might be someone else's territory. There is never nothing. There might even be paleolithical riverbeds! These are the kind of limitations that I feel we need to overcome. We need the communitites to get together and help out here.

JS: I am seeing the formation of potentially another conference!

NOTES

1. https://www.rael-sanfratello.com/made/casa-covida
2. https://www.rhino3d.com/
3. "You never change things by fighting the existing reality. To change something, build a new model that makes the existing model obsolete." L. Steven Sieden, *A Fuller View: Buckminster Fuller's Vision of Hope and Abundance for All*, Divine Arts Media, 2011, p. 358.

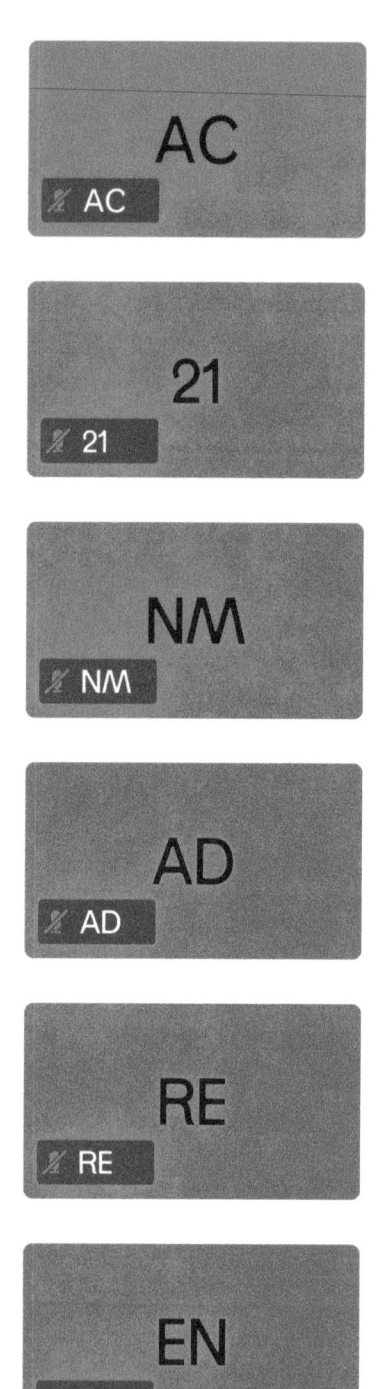
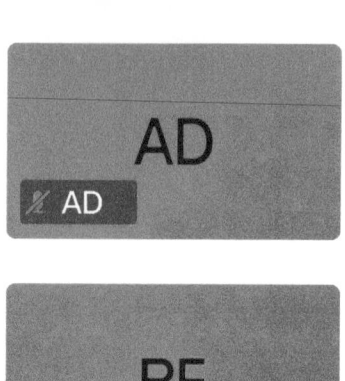
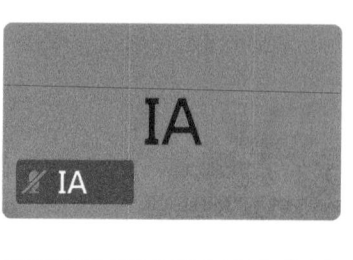

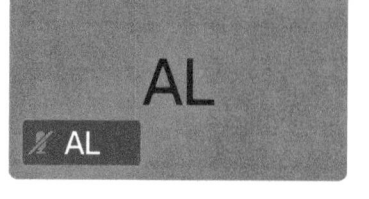

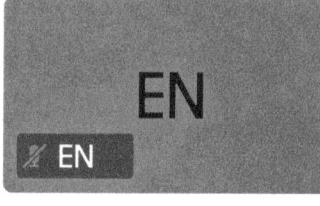

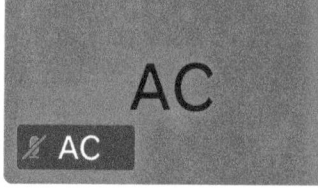

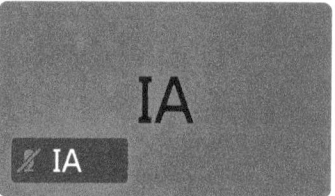

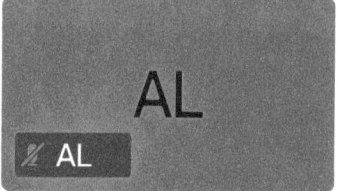

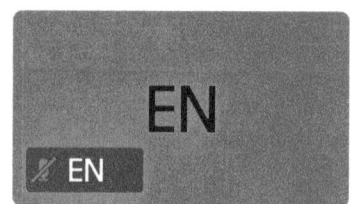

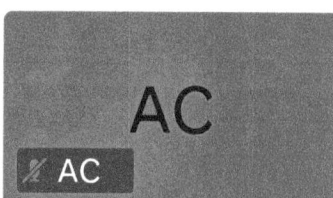

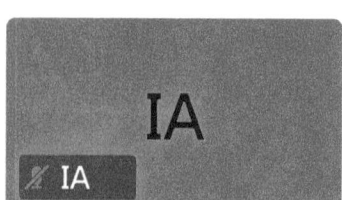

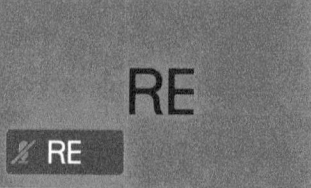
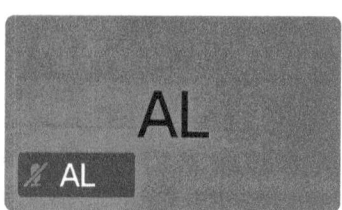

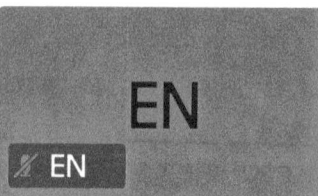

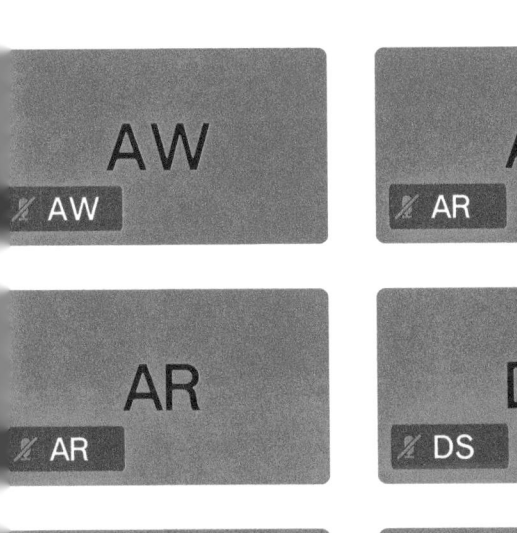

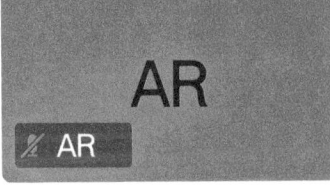

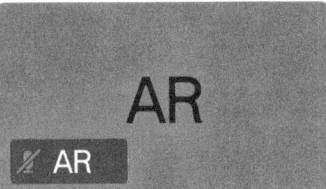

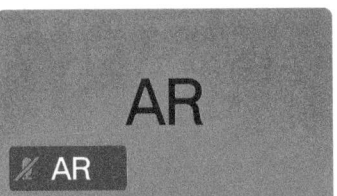

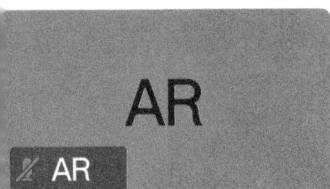

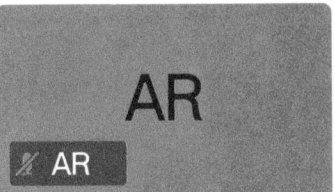

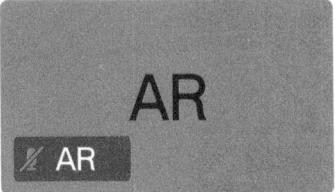

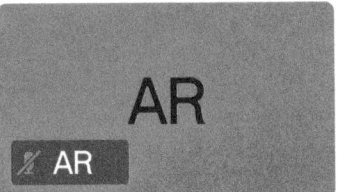

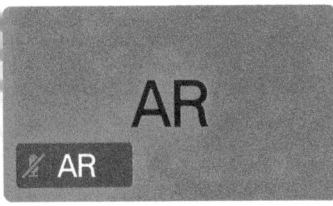
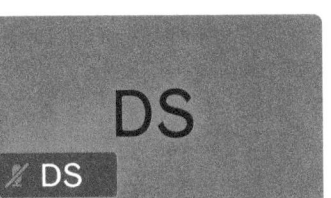

Record

Live Transcript

Breakout Rooms

Reactions

More

End

AWARDS EVENT

Teaching Award of Excellence

Axel Kilian
Visiting Assistant Professor
Massachusetts Institute
of Technology

EMBODIED COMPUTATION: SELECTIVE PROTOTYPES

My research focus is embodied computation and from it I also derive my teaching approach. Embodied computation stands for the extension of material-built form into the behavioral. Our computational constructs are physical digital hybrids and are the embodiment of our design intent. The physical distribution of structure, sensors, actuation, and its algorithmic control all come together to a construct of embodied computation. Robotics tend to mimic the evolutionary bias towards outward facing senses, yet architecture is a discipline that is focused on space and inhabitable interiors, hence it is essential to further develop the inward facing sensing and modes of communication between building and inhabitant that goes beyond voice and screen into a form of body language of the built environment. This likely involves a combination of analog matter-based computing and digital computation. To test this concept, I develop prototypes. In my research work and in teaching, I frequently develop what I refer to as selective prototypes, which means to design a prototype to be a construct of reduced complexity yet equipped with the minimum number of features that are essential to experience the to be tested condition. In the flexing room prototype an architectural robot in room form was developed that could be inhabited and was able to take on a series of postures from the combination of its 36 pneumatic actuators. For each pose it would count the number of people in its room and use it as a form of popularity measure to influence which pose it would take next. The flexing room is a test to provoke new insights on the relationship between inhabitants and robotic built form, yet in teaching I frame assignments in a manner to lead towards the construction to a testable prototype guided by student identified motivation.

Reflections on Teaching and Research

At the core of my teaching, I support finding a motivation for student work. The desire to achieve something is the strongest driver to dive into the unknown and for learning. From a motivation a strategy for learning and design can be developed. Knowledge distribution has changed fundamentally in the past decade. Hierarchical knowledge production and distribution models have fundamentally changed to flatter community-based models. This

1   Flexing Room 2017-2019, Axel Kilian, MIT and Seoul Biennale of Architecture and Urbanism

2   Bowtower 2012-2018, Axel Kilian, Princeton University

has increased the importance of the framing and curation of the design focus. It is crucial to create a stimulating framework in which productive connections can be made and community discussion and awareness can evolve.

Fluency

A crucial aspect of introducing computation is to establish a base fluency within the domain of study. Fluency is the ability to speak the domain language and to express new concepts so the are understandable to an existing audience. In computation this fluency is a combination of programming, geometry, mathematics, and design theory. I strongly believe that students should be able to develop their approaches from the ground up, as that approach allows for the discovery of new approaches through implementation. It is a fine balance though to not become too much of an expert to lose critical distance to the domain. I strongly believe that fluency is the basis of discourse and ultimately enables criticality express new concepts and understand existing ones in an exchange of ideas around computational design not based on its output but based on the underlying defining principles.

Criticality

Fluency provides an internal perspective and a deep awareness of the underlying models and existing abstractions being deployed to enable criticality. Despite computation increasing impact and its widespread use, the theoretical basis of the most frequently used models is relatively narrow compared to the high visibility gained through increased productivity. This increased productivity can also camouflage diversity of thought and approaches. This is a long running crisis only further escalated by industrial scale implementation in the software industries that provide robust readymade solution yet render alternative novel approaches almost invisible. Criticality is crucial to protect the long arc of development of alternative approaches to ensure that there is a future of diverse approaches.

Diversity of Cultures

The glossy and robust is the enemy of the fragile, scruffy, and unfamiliar. This is a problem that is more exasperated by the misunderstanding of design as polished perfection in popular culture driven by consumer brands such as Apple.

3   Buoyant Extrusion, Nick Foley, Ryan Luke Johns, Axel Kilian, Greyshed and Princeton University, workshop of RobArch 2014

4   Wave Fanfare 2018, Jeff Snyder, Axel Kilian, Ryan Luke Johns, Jane Cox, Princeton University

Fundamentally design is messy and fractured, difficult to foresee and unpredictable in its outcome, and it requires time not only in finding new forms of expression but also in shaping an audience that appreciates possible new contributions. Hence design at its best supports the diversity of cultures being seen and strengthened by widening the foundation of the domain. This requires actively inviting in and finding voices not currently heard, and a careful review of metrics of measuring success.

Axel Kilian is currently a Visiting Assistant Professor at the MIT Department of Architecture. He was previously Assistant Professor at the Princeton University School of Architecture and at the Delft University of Technology and a Postdoctoral Associate at the Department of Architecture at MIT. He holds a PhD in Design and Computation and a Master of Science in Architectural Studies from the Department of Architecture at MIT. He came to MIT as a German American Fulbright scholarship grantee after completing a professional degree in architecture at the University of the Arts Berlin. His work in architectural robotics has been exhibited at the Istanbul Design Biennial and the Seoul Biennial of Architecture and Urbanism. His current research and teaching focus is on embodied computation, exploring the extension of architecture's material form into the behavioral through physical, actuated, and sensing prototypes of space.

AWARDS EVENT

Digital Practice Award of Excellence

Alvin Huang
Associate Professor
and Director of Graduate
& Post-Professional
Architecture, USC

Founder and Principal
Synthesis
Design + Architecture

Within common language, the terms "research" and "practice" have two clearly different and seemingly oppositional meanings. Whereas research is understood as a systematic investigation or inquiry aimed at contributing to knowledge of a theory,[1] practice is understood as the actual application or use of an idea, belief, or method, as opposed to theories relating to it.[2] However, when describing the work of Synthesis Design + Architecture, I want to discuss the intersection between the two terms where we see design research as practice.

In 10+ years of practice, we have engaged in a wide spectrum of projects ranging from speculative design proposals for competitions, design research projects for objects, pavilions, and public art, to architectural commissions for interiors, facades, and buildings. While the typologies of the work have been highly varied—from that standpoint we don't fit the mold of most architectural practices in that we don't specialize in anything, in terms of type—what I would argue connects the body of work is an interest in challenging conventions, asking questions, and leveraging technology as a means of designing, analyzing, fabricating, and perhaps most importantly, thinking. We specialize in producing ground-breaking design work that not only challenges convention through technological, material, and computational innovation, but also work that purposefully enlivens the public imagination and makes a case for the role of innovative architecture in our contemporary creative culture. We seek to balance both the experimental and the visionary with the practical and the pragmatic to achieve the extraordinary.

The discipline of architecture has always been informed by technology. The catalyst for nearly every major advancement within the discipline of architecture has been a parallel development of technology. These technological advancements have varied from design technique (i.e. the invention of perspective by Brunelleschi in the Renaissance), to material process (i.e. the revolutionary use of mass produced cast iron components by Joseph Paxton in the Crystal Palace in 1854), to building technology (i.e. the invention of the first commercial elevator by Otis in 1857). Meanwhile, the use of computation within the field

1   Pure Tension – Volvo V60 Pavilion, 2014

2   Durotaxis Chair, 2014

of architecture has likewise led to several advancements within the discipline—enabling generative design techniques, digital fabrication processes, and performative building technologies. Computational design has introduced a bevy of design tools that can be married with design techniques that collectively enable the possibility to generate, optimize, describe, and fabricate complex forms and material constructs.

However, similar to the industrial advances that preceded computation, the true revolution is not in the technologies themselves, but rather in the mastery of the design knowledge that is gained and the creative explorations of the design opportunities that they are able to provide. It is not just the ability to use them—nor the ability to think with them, but rather the ability to think with them intuitively—that elevates the combination of tool and technique to the level of techné.

Techné is an ancient Greek term for craft, that was synonymous with all forms of creating, including those that we may consider arts—painting, sculpture, and architecture; as well as those that may consider practical—agriculture, medicine, and carpentry.[3] This definition encompassed both the mastery of the process of producing things and perhaps more importantly, the embedded knowledge that allows and accounts for that production. Techné is the bridge between the realms of possibility and actuality—it indexes both action and the conditions of a possibility for action.[4] By this definition, techné is an embodied knowledge—a sensibility—that is a result of the tools (technologies), techniques (methods), and contingencies (possibilities) of a discipline that guides them in practice.

Through the last decade, I have had the opportunity to work on a wide range of projects. While computational design has emerged as a mainstream protocol for the production of architecture—its applications within the conventions of traditional practice have pigeonholed it under the banners of efficacy and optimization. Meanwhile, in the academy, computational design seems to have been steered towards the search for novelty. The contingencies of considering computational design as a form of techné imply that one can consider an embodied sensibility or design intuition that is informed by the facility of a designer working with

3   Data Moire for IBM Watson, 2016

4   YPMD Pediatric Neurology Clinic, 2019

both of these seemingly contending ideas. Computational design offers architects an opportunity to develop new design sensibilities and design intuitions in regards to their facility with an emerging set of tools and techniques. In the work of Synthesis Design + Architecture, techné emerges as distinct yet overlapping territories of computational design thinking that can be leveraged to explore a new realm of possibilities and alternatives rather than a search for the ideal or the optimized.

NOTES

1. "research, n.1." *OED Online*. March 2022. Oxford University Press.
2. "practice, n." *OED Online*. March 2022. Oxford University Press.
3. Christopher J. Rowe and Sarah Broadie, eds. *Nicomachean Ethics*. New York: Oxford University Press, 2002.
4. Jonathan Sterne. "Communication as Techné." In *Communication As...: Perspectives on Theory*, edited by G. J. Shepherd, J. St. John, and T. Striphas. Thousand Oaks, CA: Sage Publications, Inc. 91-98.

Alvin Huang is a Los Angeles based architect with a global profile. As an award-winning architect, designer, and educator, he explores the intersections between technology and culture to produce innovative design work that challenges convention and expresses universal values. He is a vocal advocate for justice, equity, diversity, and inclusion in design culture. His work spans all scales ranging from hi-rise towers and mixed-use developments to temporary pavilions and bespoke furnishings. He is the founder and principal of Synthesis Design + Architecture and an Associate Professor at the University of Southern California, where he is also the Director of Graduate and Post-professional Architecture. Prior to establishing SDA, he gained significant experience working in the offices of Zaha Hadid Architects, Future Systems, AL_A and AECOM. He received an MArch from the Graduate Design Research Laboratory at the Architectural Association in London and a BArch from the USC School of Architecture in Los Angeles.

AWARDS EVENT

Lifetime Achievement Award

Wolf dPrix
Co-founder, CEO,
and Design Principal
Coop Himmelb(l)au

KEY PRINCIPLES GUIDING THE WORK OF COOP HIMMELB(L)AU

Any attempts by critics to classify the work of Coop Himmelb(l)au are usually doomed to fail. Yet what is the understanding of architecture—in theory and practice—that characterizes the work of Coop Himmelb(l)au? What makes them pioneers of architecture and co-founders of the architectural movement of deconstructivism? For Wolf dPrix, architecture can never be scientifically founded. He is fascinated by the coincidence of the creative result, similar to the character of the "Bellmaker" from Tarkovsky's film *The Passion of Andrei Rublev* (1966). To Wolf dPrix, architecture is a three-dimensional language. "Architecture must burn" = architecture is more than function. There are several aspects that make Wolf dPrix and his studio pioneers.

Coop Himmelb(l)au was founded in Vienna, Austria in 1968 by architects Wolf dPrix, Helmut Swiczinsky, and Michael Holzer. Their aim was to develop architecture in line with an open society (see Karl Popper, "The open society and its enemies") and to revolutionize architecture. The development of their formal language ("everbody is right, but nothing is correct") was later called deconstructivism.

The confrontation in the cultural and political environment of the 1960s and 1970s in Vienna led to their credo that a space can be dynamic and does not have to be only static, "architecture changeable like clouds" was their ambition at that time. This is shown in various artistic actions, performances, and installations: a house that greets; a room that reacts to the heartbeat of people.

Two major events established Coop Himmelb(l)au's global position and its worldwide recognition. The realization of the world's first Deconstructivist building, the Falkestrasse roof extension in Vienna (1984-89), and the invitation to the exhibition "Deconstructivist Architecture" at the Museum of Modern Art, New York (1988). The Falkestrasse roof extension, although a small project in terms of construction volume, was a manifesto in the form of a building. Here Coop Himmelb(l)au was able to realize without compromise "the birth

1   Rooftop Remodeling Falkestrasse, Completed in 1988 and located in Vienna, Austria. Image by Gerald Zugmann.

2   UFA Cinema Center, Completed in 1998 and located in Dresden, Germany. Image by Gerald Zugmann.

extension, although a small project in terms of construction volume, was a manifesto in the form of a building. Here Coop Himmelb(l)au was able to realize without compromise "the birth of the building as a cloud" representing the dissolution of architecture.

Thinking big and reacting to a changing urban situation is demonstrated in the award-winning master plan of Melun-Sénart in France (1987). In this project, Coop Himmelb(l)au proposed "spatial zoning", i.e., for every 20 meters of building height, one floor is given back to the city.

A period of essential projects and buildings commences. From the very beginning, Coop Himmelb(l)au show a strong interest in innovative technologies. The first paperless building was completed in 1994 in Groningen, the Netherlands. At the Groninger Museum - East Pavilion, the central idea of a space that explodes the confinement of a functional box into a thousand pieces was implemented on a large scale. "Flaws" were here cultivated into details: The spatial sequence was developed based on studies of natural and artificial light. It was then overlaid with the drawing of the fluid space. Today this would happen on the computer, at that time it was done on the photocopier (1993). From these first studies, various models were created. A model with all its errors and inaccuracies was transferred to a scale of 1:1. Thanks to the so-called "Space Arm", the model could be digitized and transferred directly into 3-dimensional data. This new technology offered hitherto undreamed-of freedom in the design of shapes, spaces and structures.

The UFA Kinopalast (1998) in Dresden, Germany is the first large-scale building to realize and manifest Coop Himmelb(l)au's complex ideas and prove their feasibility. Although the Kristall-Palast is a functional building, it is also a contribution to the urban space, a "transistoric building". Although functions are fulfilled here, it has succeeded in making more. With new ideas, Coop Himmelb(l)au frees the architecture, the floor plan, from the constraint of function.

The roof of the BMW Welt in Munich, Germany (2007) was placed on a striking double cone and the visitors' lounge realizing the „dynamic forces" of wind and cloud fields. The interior is an example of its own accumulated power and discharge and braces itself against the violent,

3   Museum of Contemporary Art and Planning Exhibition (MOCAPE), Completed in 2016 and located in Shenzhen, China. Image by Duccio Malagamba.

4   Museum of Contemporary Art and Planning Exhibition (MOCAPE), Completed in 2016 and located in Shenzhen, China. Image by Coop Himmelb(l)au.

all-deforming storm. The cloud has thus landed and will not dissipate. Cloud and crystal were also taken up in the expansion of the historic building of the Akron Art Museum in Ohio, USA (2007) and manifested in the "urban space" of a museum.

Three iconic landmarks, the Busan Cinema Center in South Korea (2012), the Dalian International Conference Center in China (2012), and the Museum of Contemporary Art & Planning Exhibition (MOCAPE) in Shenzen, South China (2016) impressively demonstrate Coop Himmelb(l)au's cloud theme.

Composed of two polygonally twisted halves, the double disk of the European Central Bank's headquarters (2014) in Germany rises 185 meters into the sky of Frankfurt's city center. The double tower exemplifies Wolf dPrix's observation that today's cities are like a chessboard without a grid. The more the grid blurs, the sharper the pieces must become. The two towers paraphrase cloud and crystal.

For the Musée des Confluences (2014) in Lyon, France, Wolf dPrix said, they wanted to invent a completely new form, without precedent in the world and living up to the idea of a museum that seeks to unite nature, man and technology. The larger component is the real walkable cloud. The anterior hall, the crystal, for Wolf dPrix a "meeting place in the city" in a new urban space.

When it comes to brainstorming and design methodology, Coop Himmelb(l)au accelerates from zero to a hundred. After the turn of the millennium, parametric tools become popular in architectural design. At Coop Himmelb(l)au, they are quickly applied to the development of all projects as early as the early 2000s and expanded accordingly over the years. Wolf dPrix believes that architects must always find new answers to new challenges. Digital technologies such as computational design and parametric tools are seen as a way to maintain control over the execution of projects. The exploration of machine learning and AI resulted in the ongoing research project Deep Himmelb(l)au. At MOCAPE in Shenzen, Coop Himmelb(l)au tested what can be influenced with these tools. For the implementation, their extremely detailed and accurate models were recreated at a scale of 1:1. The cloud was built with the help of robots and the use of parametric 3D models allowed the

5   BMW-Welt, Completed in 2007 and located in Munich, Germany. Image by Marcus Bruck.

6   Central Los Angeles Area High School#9 for the Visual and Performing Arts, Completed in 2009 and located in Los Angeles, USA. Image by Duccio Malagamba.

time-saving optimization of the facade up to the execution.

Nevertheless, for Wolf dPrix, the sketch developed in advance in his head (or in his gut) remains the first idea set down in drawing. This sketch, which in itself is rich in meaning and open to interpretation, initiates the idea of the complex procedures of the open design process. It is the sketch that sets the process in motion, but it is only through the process itself that the sketch is transformed into a building. For Wolf dPrix, models in different scales and materials are unfolding into the 3rd dimension. It remains to be seen which answers for which future challenges the office's current experiments with artificial intelligence will offer.

The works of Coop Himmelb(l)au do not allow any one final interpretation. Wolf dPrix unwaveringly maintains: "What we build is not contemporary, but right for our time."

7   Musée des Confluences Completed in 2014 and located in Lyon, France. Image by Raimund Koch.

8   Deep Himmelb(l)au, Rendering by COOP HIMMELB(L)AU.

Wolf dPrix is co-founder, CEO and design principal of COOP HIMMELB(L)AU, a studio globally recognized for its innovative, and complex design approach at the intersection of architecture, art and technology. He is counted among the originators of the deconstructivist architecture movement. Throughout his career, Wolf dPrix has remained active in education and academic life. He has held teaching positions at the University of Applied Arts, Architectural Association (AA), MIT, Harvard, Columbia, UCLA, Yale, PENN, SCI-Arc and other institutions around the world. Wolf dPrix's distinguished honors include the two highest Austrian awards given for cultural achievements: the Grand Austrian State Prize and the Austrian Decoration for Science and Art. He is a member of the Austrian Arts Senate, the Curia for Art, since 2014 as chairman. Additional awards include, among others: the Schelling Architecture Award, the Officier de l'Ordre des Arts et des Lettres, the Annie Spink Award, the Jencks Award and the Hessian Cultural Prize. Institutions such as the Getty Foundation, the MAK Museum of Applied Arts, and the Centre George Pompidou show the work in permanent exhibitions. In 2006 Wolf dPrix was commissioner for the Austrian contribution to the 10th Venice Biennale. Overall, the studio has been invited 8 times to participate in the Venice Architecture Biennale.

AWARDS EVENT

Society Award for Leadership

Brian Slocum
Founder, tresRobots
Adjunct Professor,
Universidad Iberoamericana,
Mexico City

AMPLIFYING COUNTERNARRATIVES THROUGH DESIGN COMPUTATION

Turing's Spotlight

"The Universe is the interior of the Light Cone of the Creation."
This "light cone" might suggest a directed gaze, a spotlight. Understood as double-speak between Turing[1] and the friend to whom he wrote (Turing 1954), this quote and the series of postcards from which it is taken evoke explicit as well as coded narratives that lead Jacob Gaboury to propose that "queerness is itself inherent within computational logic." (2013b; 2013a). That this communication should come from an "Unseen World" also hints at a latent exteriority, suggesting that orientating is essential to the making-explicit of other narratives.

Error

A reframing of error forms the basis of Francesca Hughe's feminist critique of modernism as a mechanism for control of matter in *The Architecture of Error* (2014). Taking cues from student work in PREXPREN[2]—the analog material investigation studio I coteach at the Universidad Iberoamericana—and relying heavily on Hughes' insights, Pablo Kobayashi and I sought to foreground issues of precision and error in the ACADIA 2018 conference theme, and to encourage a critical reassessment of our relationship to the computer, computation, materials, and our own data (Kobayashi et al. n.d.). If 'error' is defined relative to criteria we impose as part of our own methodologies, the critical rehabilitation of error has become a site for alternative inquiry, one in which both error and failure are embraced as operative rather than as devices to exclude messy or unexpected results.

Outliers

On the ACADIA 2020 conference platform, projections of conference content into Cartesian space were intended to reveal latent connections (Popadich, 2021). Yet any pattern remains obscured by the relative homogeneity in the distribution of elements, indicating either close affinity or a smoothing out of difference (Wattenberg, et al. 2016). Varying the

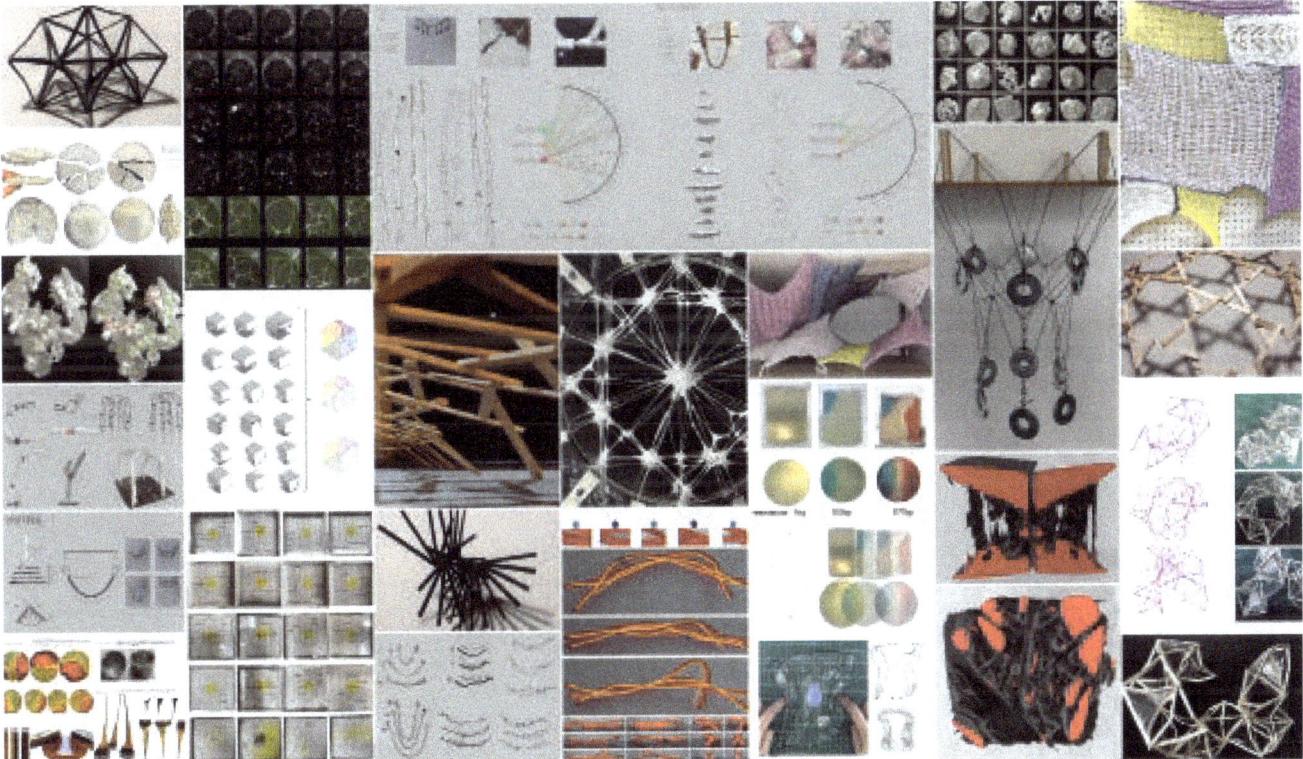

1   PREXPREN Studio (Protodigital), Universidad Iberoamericana, 2020–2021. The studio relies on analog modeling at 1:1 scale, employing computational design strategies to develop rules as a type of analog software to exploit inherent properties of the materials (Images: Various PREXPREN students).

criteria for projection of the peer-reviewed papers into a 3D environment—the result of a related keyword analysis included in the introduction to the 2020 Proceedings—makes evident the profound effect of our methodologies on the visualization of the data and definition of norm versus outlier.[3] The automatic equating of physical proximity with similarity—and the privileging of that similarity (Hong, Longoni, and Morwitz 2019)—risks the erasure of deviation and the exclusion of outlier "counternarratives" (Curley, 2019).

"Failed orientations"

Sara Ahmed has connected 'orientation,' as in sexuality, and spatial relationships through objects, between proximity and outlier: how we orient (or fail to orient) ourselves in white, heteronormative spaces (2006). My ongoing investigation "Wild Simulations" takes up Ahmed's "failed orientations" (2006, 91-92), in terms of both critical positioning and methodology. The project involves the translation of painter Francis Bacon's analog methodologies into a set of computational strategies, starting from Head I of 1948 (Estate of Francis Bacon, 2018). Objects within the painting evidence the instantiation of multiple, conflicting spatial conditions. Rather than a dissection of the figure, instead we witness a corporal dissociation, the result of the queer body's failure to occupy multiple spatial paradigms simultaneously. This failure is evidence of a queer spatial practice (Gaboury, 2018) that refuses a hermeneutics of normative space.

Ridiculous Tectonics

As a mode of critical and speculative inquiry, the interrogation of current construction assembly logics against alternative criteria derived from the literal and rhetorical misappropriation of categories of aesthetic hyperbole (Halberstam, 2020; Ngai, 2005) allows for the possibility of novel building assemblies. 'Quirky' or 'cute' morphologies and 'naïve' connections—'ridiculous tectonics' in other words—may provide not only formal but technical insights that refuse persistent dogmas of capitalist optimization. Leveraging slippages, collisions, and confused articulations—and employing rules derived from an analysis of types/degrees of discretization and the study of augmented/automated techniques for their assembly both within existing building practice and those employed in recent projects of 'the Discrete' style (Retsin,

2   PREXPREN Studio (Inflatables), Universidad Iberoamericana, 2017–2020. We ask students to focus on the possibilities presented by unexpected behaviors and perceived errors, and to learn how to repeat them and to make them generative (Images: Various PREXPREN students).

2019)—operations of ridiculous tectonics focus on connections and assembly to make assembly logics not only explicit but performative.

Situated Methodologies (beyond Turing's Spotlight)

If we recognize the importance of outliers and the role of error and failure in our work;

And, if we understand this exploration within the context of both humanities and digital technologies research toward the critical refusal of normative epistemologies—in my examples, of error/failure and misappropriation as forms of queer practice (Gaboury, 2018);

And more urgently, if we recognize the emergence of an array of situated methodologies through the instrumentalization of these practices;

Then, beyond any theoretical positioning of the researcher and/or the researched (Haraway, 1988; Tran and Patitsas, 2020, Section 3), the practical and quantitative implementation of critical methods (Haber, 2016) may also serve as new loci for experimentation, serving to amplify latent counterhegemonic narratives among other overlooked constituencies.[4]

NOTES

1. Since this essay is about amplifying counternarratives, a couple of points should be noted regarding Alan Turing; although he has become a queer icon, a reading of the Andrew Hodges biography (1983) reveals that he was either somewhat misogynistic or at least complicit with prevailing attitudes toward gender roles in his professional life (at least during World War II). Additionally, in his work with the development of modern computers, he used or perpetuated the use of biased nomenclature to describe componentry and programming ("master/slave") that we now consider damaging and inappropriate.

2. PREXPREN is an acronym for the title of the studio in Spanish: "PRotocolos EXógenos / PRopiedades ENdógenas" (Exogenous Protocols / Endogenous Properties).

3. "A person or thing situated away or detached from the main body or system." (Oxford University Press n.d.).

4. Examples are numerous, but I am particularly inspired by the following: Ahlquist (2021), Blas (2008), Davis (2021), Paré, Shanahan, and Sengupta (2020), and Noel (2021).

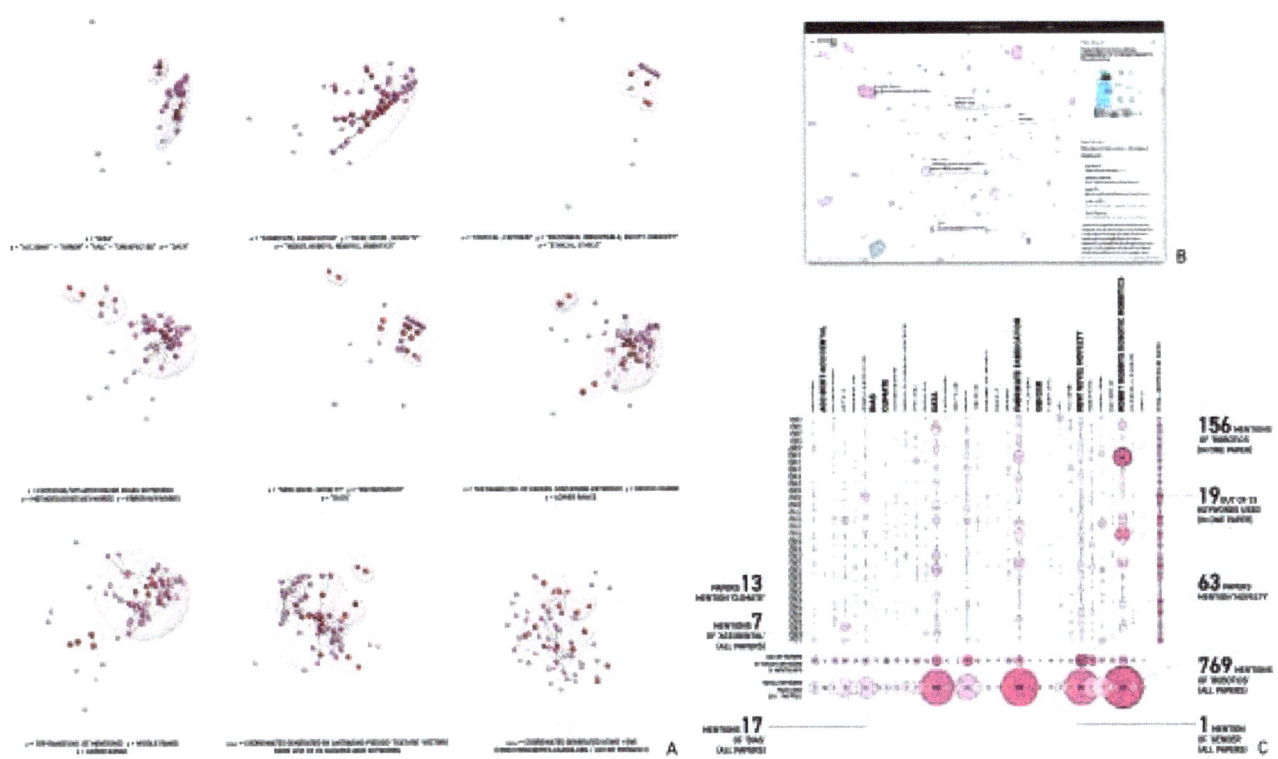

3 Distributed Proximities Keyword Analysis, 2021. a) Keyword mappings based on various criteria indicate different groupings, suggesting the presence of narratives that are latent but unwritten; b) Screenshot from proximities.acadia.org; c) Keyword prevalence diagram from ACADIA 2020 Proceedings Volume 1 (p 19).

REFERENCES

Ahlquist, S. 2021. "From Material Systems to Sociomaterial Justness." In *ACADIA 2020: Distributed Proximities [Proceedings of the 40th Annual Conference of the Association of Computer Aided Design in Architecture (ACADIA)*, Volume 2], Online and Global. 24–30 October 2020, edited by M. Yablonina, A. Marcus, S. Doyle, M. del Campo, V. Ago, B. Slocum, 28–32. ACADIA.

Ahmed, S. 2006. *Queer Phenomenology: Orientations, Objects, Others*. Durham, NC: Duke University Press.

Blas, Z. 2008. *Gay Bombs: User's Manual*. USA: Queer Technologies, Inc.

Curley, K.M. 2019. "Methods to Re-center the 'Other': When Discarding Outliers Means Discarding Already Marginalized Stories." *Journal of Critical Thought and Praxis* 8(2): 4. https://doi.org/10.31274/jctp.8795.

Davis, F. 2021. "Seams: Race, Architecture and Design Computing" [lecture]. Computation Lectures, MIT Department of Architecture Public Lecture Series, Cambridge, MA. May 6, 2021. https://web.mit.edu/webcast/architecture/s21/13/.

Easterling, K. 2014. *Extrastatecraft: The Power of Infrastructure Space*. London, UK: Verso.

The Estate of Francis Bacon. 2018. "Head I." francis-bacon.com. Accessed 21 December 2020. http://francis-bacon.com/artworks/paintings/head-i

Gaboury, J. 2013a. "A Queer History of Computing." Rhizome. https://rhizome.org/editorial/2013/feb/19/queer-computing-1/. Accessed 11 August 2021.

Gaboury, J. 2013b. "A Queer History of Computing, Part Five: Messages from the Unseen World." Rhizome. https://rhizome.org/editorial/2013/jun/18/queer-history-computing-part-five/?ref=tags_queer-history-of-computing_post_title. Accessed 11 August 2021.

Gaboury, Jacob. 2018. "Critical Unmaking: Toward a Queer Computation." In *The Routledge Companion to Media Studies and Digital Humanities*. UC Berkeley. Report # 49. https://escholarship.org/uc/item/0cq870wh.

Haber, Benjamin. 2016. "The Queer Ontology of Digital Method." *Women's Studies Quarterly* 44 (3/4): 150-69. http://www.jstor.org/stable/44474067.

Halberstam, J. 2020. "Wild Things: An Aesthetics of Bewilderment" [lecture]. RIBOCA2 Online Series of Talks and Conversations, Riga International Biennial of Contemporary Art (RIBOCA). 7 October 2020 [online]. https://www.youtube.com/watch?v=Ia5CmrzTqw4.

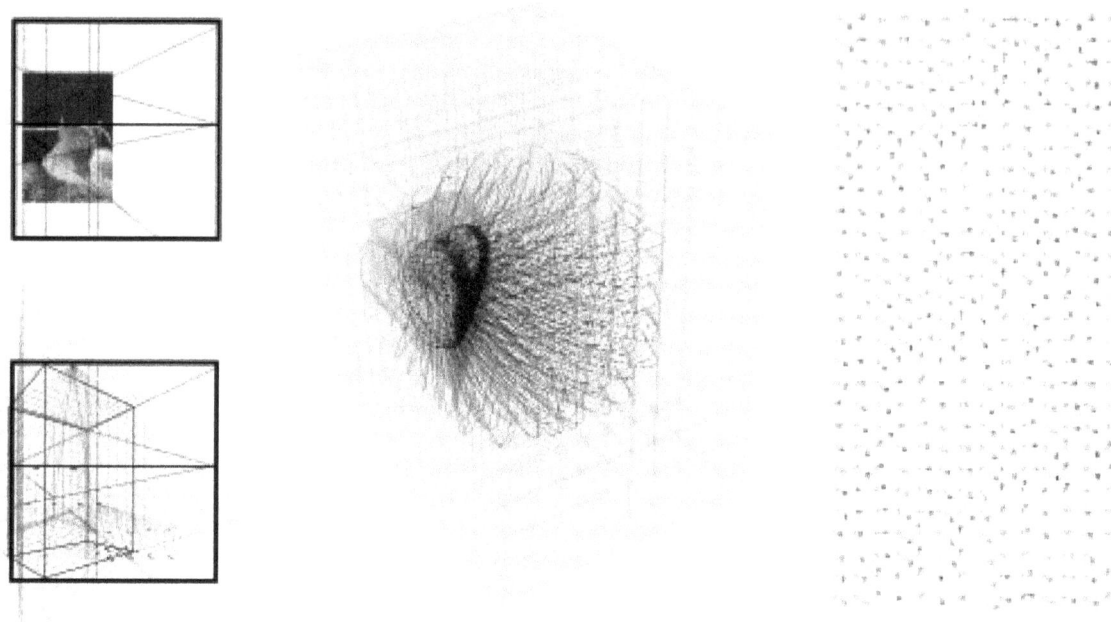

4   Wild Simulations, 2020-ongoing. The objects within the painting evidence the instantiation of multiple, conflicting spatial conditions—each with its own horizon. The 'subject' is dismantled and reconstituted, 'fit' into a space warped by these conflicts.

Haraway, D. 1988. "Situated Knowledges: The Science Question in Feminism and the Privilege of Partial Perspective." *Feminist Studies* 14 (3): 575-599.

Hodges, A. 1983/2014. *Alan Turing: The Enigma*. Princeton, NJ: Princeton University Press.

Hong, J.S., Longoni, C., Morwitz, V. 2021. "Proximity Bias: Motivated Effects of Spatial Distance on Probability Judgments" (January 12, 2021). SSRN.com. https://ssrn.com/abstract=3765027.

Hughes, F. 2014. *The Architecture of Error: Matter, Measure, and the Misadventures of Precision*. Cambridge, MA: MIT Press.

Kobayashi, P., Slocum, B., Anzalone, P., Wit, A.J., Del Signore, M. n.d. "Calls." Accessed September 21, 2021. 2018.acadia.org. http://2018.acadia.org/call.html.

Noel, V.A.A., 2021. "Situated Computations, Craft, and Technology" [lecture]. Computation Lectures, MIT Department of Architecture Public Lecture Series, Cambridge, MA. 23 September 2021. https://youtu.be/-vh4jSv_lv8. Accessed 14 November 2021.

Ngai, S. 2005. "The Cuteness of the Avant-Garde." *Critical Inquiry* 31(4): 811-847. https://doi.org/10.1086/444516. Accessed 13 November 2020.

Oxford University Press. n.d. "Outlier." Lexico.com. https://www.lexico.com/en/definition/outlier. Accessed September 18, 2021.

Paré, D., Shanahan, M-C., Sengupta, P. 2020. "Queering Complexity Using Multi-Agent Simulations." In *Interdisciplinarity in the Learning Sciences, 14th International Conference of the Learning Sciences (ICLS)*, edited by M. Gresalfi and L. Horn, 1397-1404. International Society of the Learning Sciences.

Popadich, O. 2021. "PROXIMITIES.ACADIA.ORG," In *ACADIA 2020: Distributed Proximities [Proceedings of the 40th Annual Conference of the Association of Computer Aided Design in Architecture (ACADIA), Volume 2]*, Online and Global. 24–30 October 2020, edited by M. Yablonina, A. Marcus, S. Doyle, M. del Campo, V. Ago, and B. Slocum, 18–19. ACADIA.

Retsin, G. (ed.) 2019. *Discrete. Reappraising the Digital in Architecture* (Profile No. 258). London, UK: John Wiley & Sons.

Tran, J., Patitsas, E. 2020. "The Computer as a Queer Object." SocArXiv [preprint]. doi:10.31235/osf.io/afuqs.

Turing, A. 1954. "Messages from the Unseen World." AMT/D/4 The Turing Digital Archive/P. N. Furbank (copyright). http://www.turing-archive.org/viewer/?id=154&title=14. Accessed 11 August 2021.

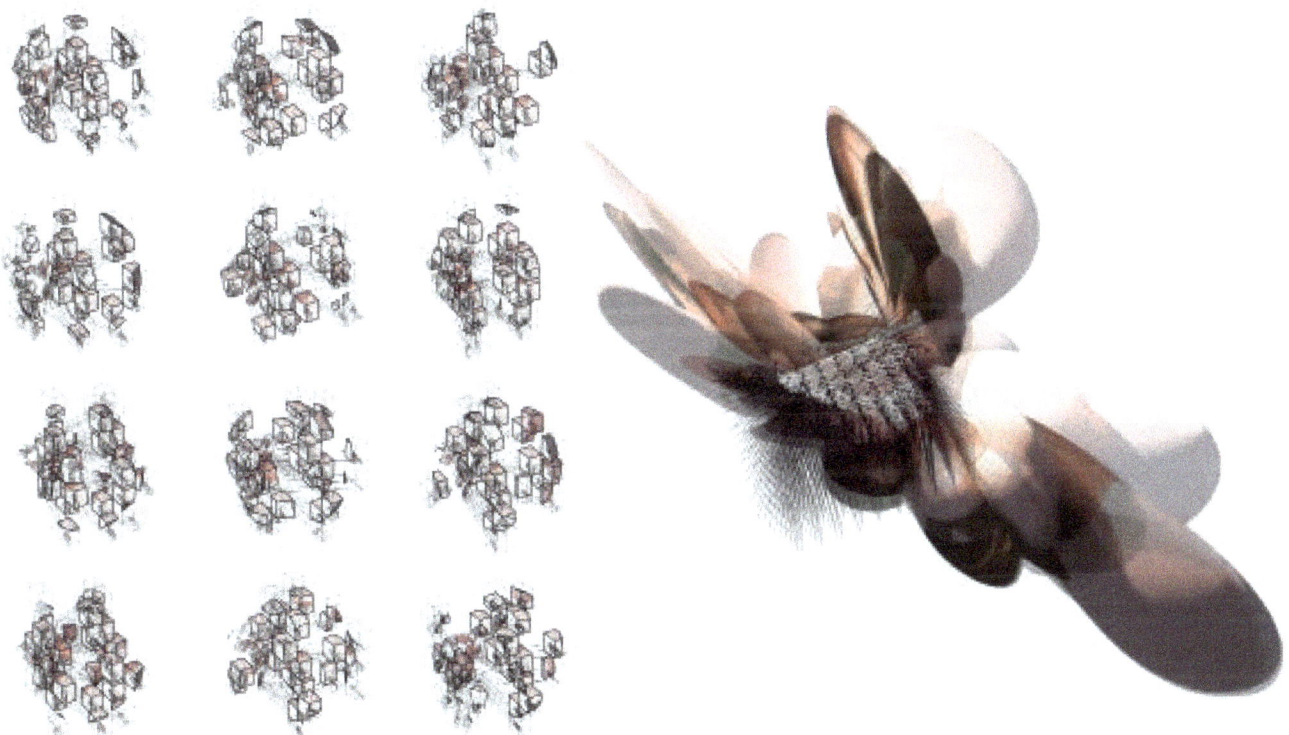

5   Wild Simulations, 2020-ongoing. Mis-registered coordinate systems effect the apparent disfiguration of Bacon's figures, and the same happens to the virtual forms of 'Wild Simulations,' in part via an error intentionally left in the script.

Wattenberg, et al. 2016. "How to Use t-SNE Effectively." Distill. http://doi.org/10.23915/distill.00002. Accessed September 18, 2021.

Brian Slocum is a practicing architect and educator. He is the founder of tresRobots, an independent studio for design / architectural technologies research and cofounder of the architecture firm Diverse Projects. Currently an Adjunct Professor in the Department of Architecture, Urbanism and Civil Engineering at the Universidad Iberoamericana in Mexico City, he coteaches a design studio focused on analog material investigation. He holds degrees from Columbia University GSAPP (MArch) and Georgia Tech (BSc in Architecture). Brian is the recipient of a 2008 Individual Research Grant from the New York State Council on the Arts and was a member of the group exhibition "Landscapes of Quarantine" at Storefront for Art and Architecture. He has published essays in *Agencia*, *CLOG* and *Pamphlet Architecture #23*. Brian was Secretary of ACADIA from 2019-2021, Conference Site Co-Chair for the 2018 conference in Mexico City, and more recently served as a Co-Chair for ACADIA 2020 Distributed Proximities and lead coeditor of Volume I of the Conference Technical Papers Proceedings.

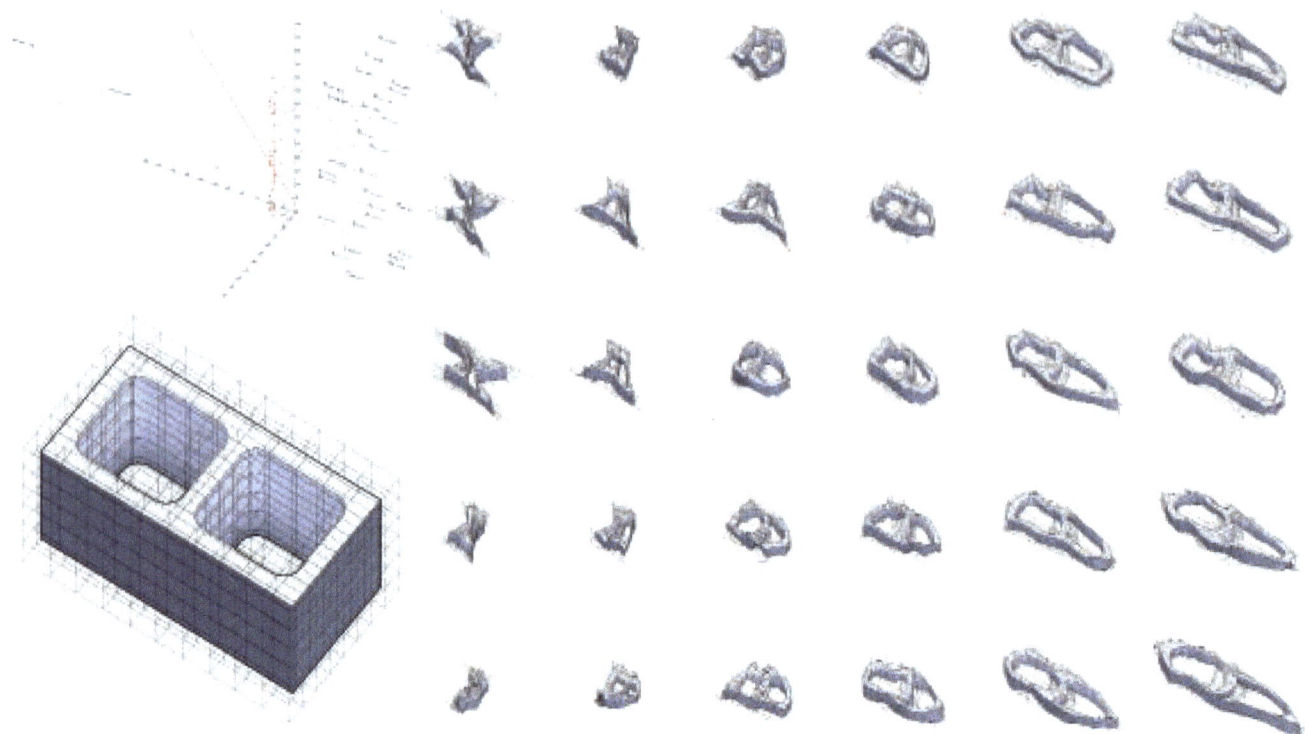

6   Ridiculous Tectonics, 2021-ongoing. If we recognize common building materials as a type of infrastructure with its own syntax and logic (Easterling, 2014), might it be possible to understand macro-morphologies of building in terms of the rules of assembly imposed by these systems?

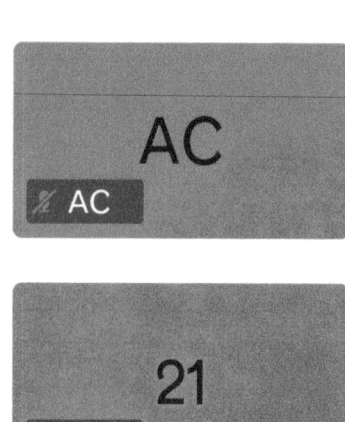 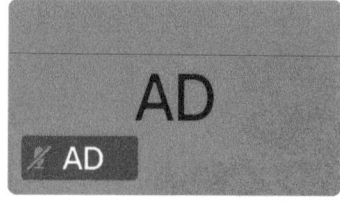 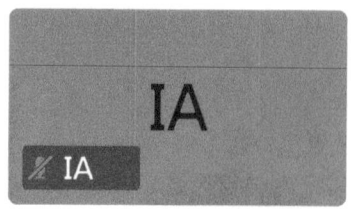 
  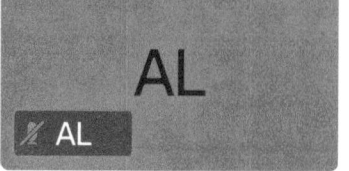 
 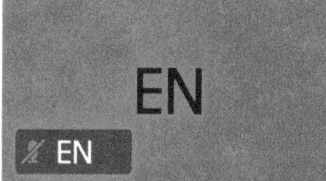  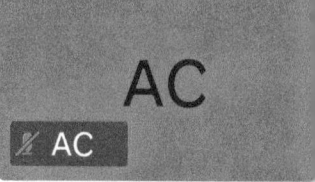

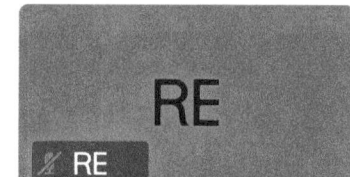 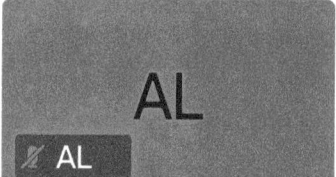  
  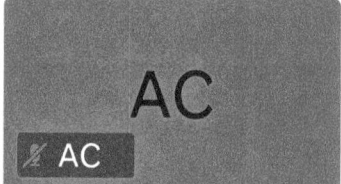 
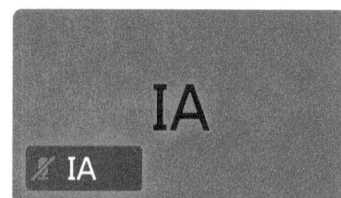  
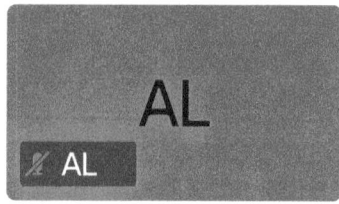   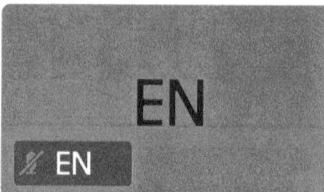

 Mute Stop Video Security Participants Chat Share Screen

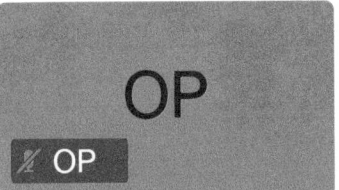

Latent Morphologies: Disentangling Design Spaces

Daniel Bolojan
Shermeen Yousif
Emmanouil Vermisso
Florida Atlantic University

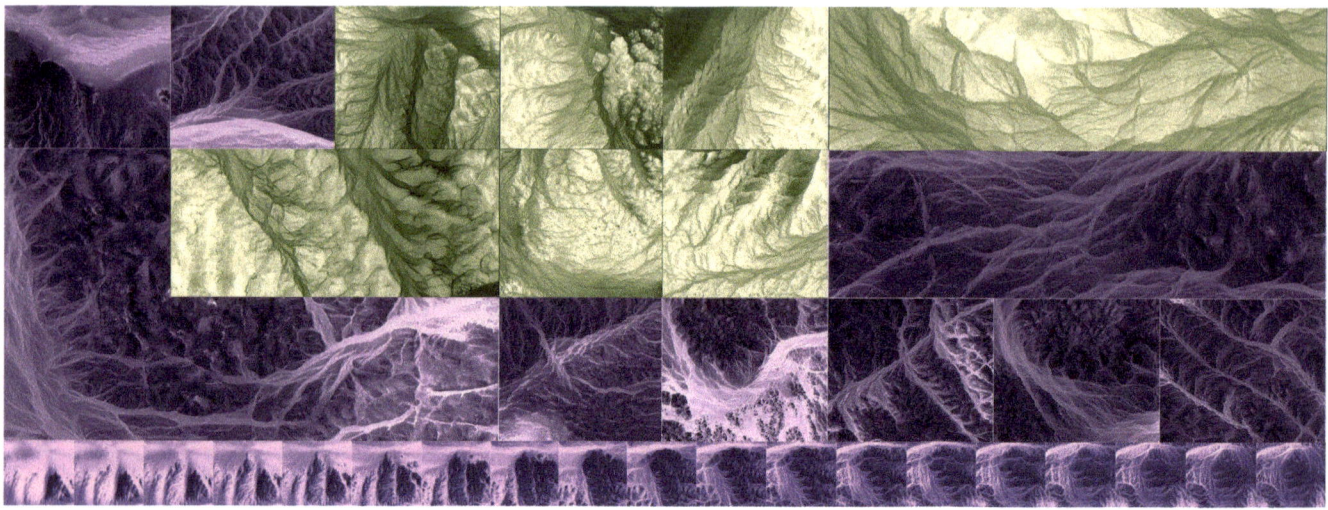

The workshop explored ways to connect different neural networks (i.e. CycleGAN and StyleGAN) to explore the search space of architectural inspiration. Particular semantic references served as input for a pre-trained network which outputs data for further investigation using another neural network. The datasets focused on exploring various resolutions of the urban domain and assess possibilities for emerging patterns, etc through interpolative and extrapolative strategies. From a process point of view, we are interested in identifying relevance between certain types of neural networks and their ability to access creative potential in a targeted and/or heuristic/open-ended fashion.

Furthermore, it is important to consider the capacity of these nested workflows to alter our immersion in the design investigation by accessing a design space that is otherwise beyond the designer's reach. By moving beyond rule-based defined design spaces, AI's feature learning capabilities combined with the incorporation of additional inspirational sources (outside architecture), enable creative exploration within an extended space. Our perception of these expanded design possibilities is crucial because it may point to the direction of future search.

Images courtesy of the workshop leaders.

Distributed Collaborations - KUKAcrc Cloud Remote Control

Sigrid Brell-Cokcan
Johannes Braumann
Karl Singline
Sven Stumm
Etahn Kerber
RWTH Aachen /
Association for Robots in Architecture /
Creative Robotics UfG Linz

The ACADIA2021 workshop: Distributed Collaborations, took Cloud Remote Control to the next level, allowing international participants to control robots in two different European locations. Workshop participants remotely controlled robots at both the Chair of Individualized Production, RWTH Aachen, Germany, and at the Department of Creative Robotics, University of Arts and Industrial Design, Linz, Austria. It was at ACADIA2010, that Sigrid Brell-Cokcan and Johannes Braumann first introduced KUKA|prc, a new approach to parametric robotic production. Since then the robotic community has grown in unbelievable ways, built incredible structures and gathered for world class conferences from ACADIA to RobArch! 10 years later, at ACADIA2020, Robots in Architecture members Sven Stumm and Ethan Kerber introduced KUKA|crc, a new approach to remote robotic control.

This innovation allows international users to collaborate closely with robots while remaining safely socially distant. Since then, we have been improving the software, adding IoT infrastructure and readying this technology for rollout around the world. KUKA|crc helps people learn about robots even if they can't get in the lab. KUKA|crc also empowers you with a new cloud based IoT infrastructure allowing for multi robot factory setups and easy wireless integration of IoT devices and end effectors.

Images courtesy of the workshop leaders.

Collaborative AI – Human + AI Form

Chien-hua Huang
China Academy of Art

Zach Beale
University of Applied Arts Vienna

This workshop explored the intersection of game and AI as a novel way to approach architectural participatory generative design. The recent rapid advancement of machine learning and AI in architectural industries is operating with limitation to allow a wider audience or human perception.

As architects and designers, allowing a wider spectrum of evaluation will be vital in the design process. Inspired by machine learning (ML) application in Unity3D as game design elements, there are potentials to promote the active participation of architects, artists and even the public in the design generation processes by gamification. In this workshop, we explored novel generative design methodology driven by Reinforcement Learning combined with player active interaction, focusing on the actor-aware generation of logics and complex spatial perceptions. How can a machine think creatively? What may be the possible form generated from the active collaboration between AI and players? What are the machine-made elements that become unperceivable from a human?

These questions approach the goals of AI proponents concerning the new generation of AI that would help designers through the novel augmentation of machine vision and automation. In the workshop, participants worked with a given set of reinforcement-learning-based frameworks and package in Unity3D to explore design ideas and potentials of perception-aware human-machine interaction. Eventually, we approached collective layers of design generation through human and AI collaboration.

Images courtesy of the workshop leaders.

Augmented Architectural Details

Jeffrey Anderson
Ahmad Tabbakah
Pratt Institute

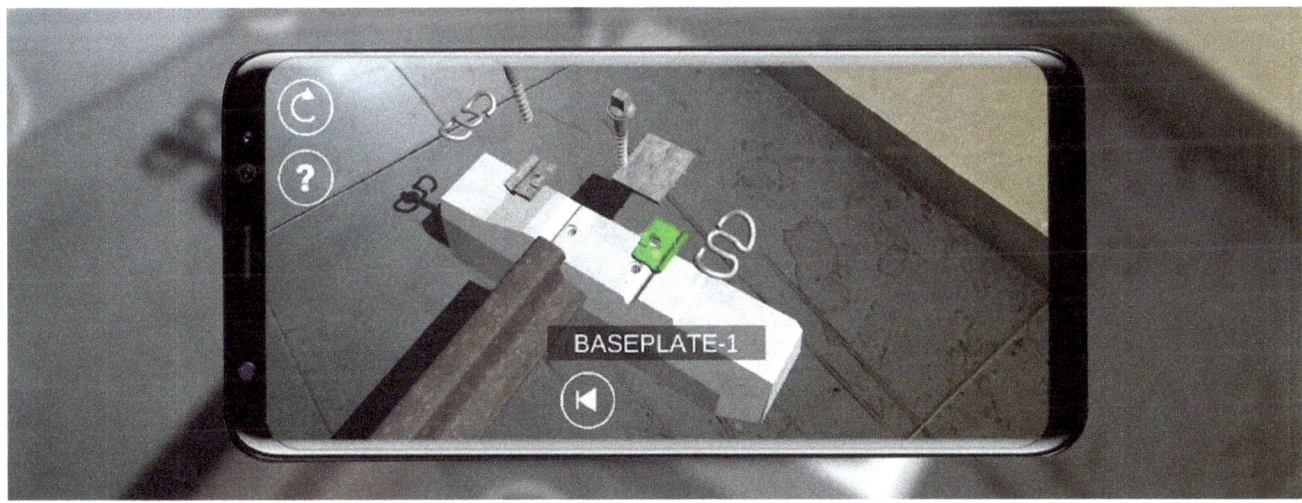

Augmented Reality (AR) and real time rendering are becoming important technologies for creating remote social experiences, visualizations, and training simulations in a number of fields. In the architecture industry, this technology not only offers a new medium for creating visual assets but opens up possibilities for new forms of communication and coordination between various trades including architects, contractors, and clients through facilitating spatially calibrated digital interactions. In order to speculate on these possibilities, our workshop explored the idea of "Augmented Architectural Details." These took the form of small-scale, spatially-calibrated AR overlays which explained assembly systems through human interaction.

This workshop provided students with tools to author digital content to a custom AR application using the Unity3D Game Engine. Students gained skills in animation, interaction, lighting, and mesh optimization and learn how to translate these effects to an AR interface. In order to facilitate rapid experimentation with AR content, students were provided with an AR application that allowed them to exchange and import Unity Asset Bundles onto their iOS or Android phones in real time. This focused the workshop on developing technical and critical thinking skills around the topic of Augmented Architectural Details.

Images courtesy of the workshop leaders.

Building Web-Based Drawing Instruments

Galo Canizares
Texas Tech University

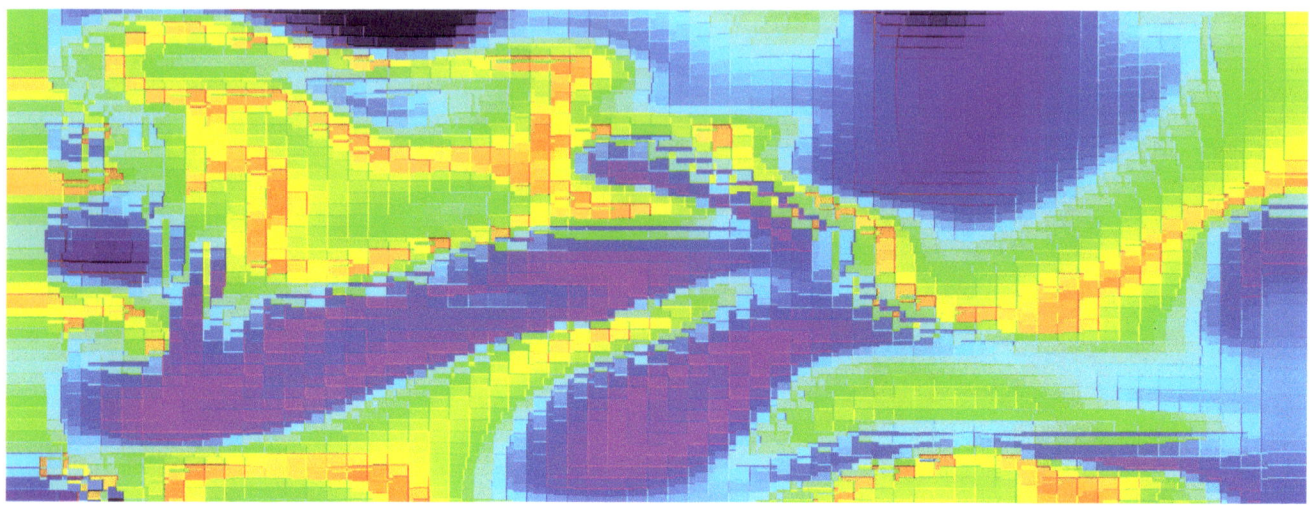

Today, it is difficult to separate the process of designing and the software used to realize said design. While Alberti may have established one of the earliest distinctions between the ideal design and the physical realization of it, the emergence of software has collapsed this traditional flow of creativity to the extent that digital tools are no longer the tools for making: they are primarily tools for thinking. This workshop took on the premise that software is an ingrained part of the creative process and that real-time technologies such as internet browsers and design tools push back not only on acts of design, but also on our social consciousness.

We found that to draw on a screen is not simply to dream a perfect image and reproduce it flawlessly with interface tools but is instead a collaborative negotiation with these powerful platforms and the data managed within them. Participants investigated how web-based drawing apps could be easily built and deployed using open-source and freely accessible tools. Together we covered introductory graphics programming with the p5.js JavaScript library and deployment using NodeJS. Using code and internet browsers as sites for inquiry, participants developed simple apps that showcased how architectural effects can be explored through the screen and engender discussions about perception and use.

Images courtesy of the workshop leader.

Co-crochet Computing Stitches for Collective and Distributed Crocheting

Özgüç Bertuğ Çapunaman
Benay Gürsoy
Penn State University

Cemal Koray Bingöl
Istanbul Technical University

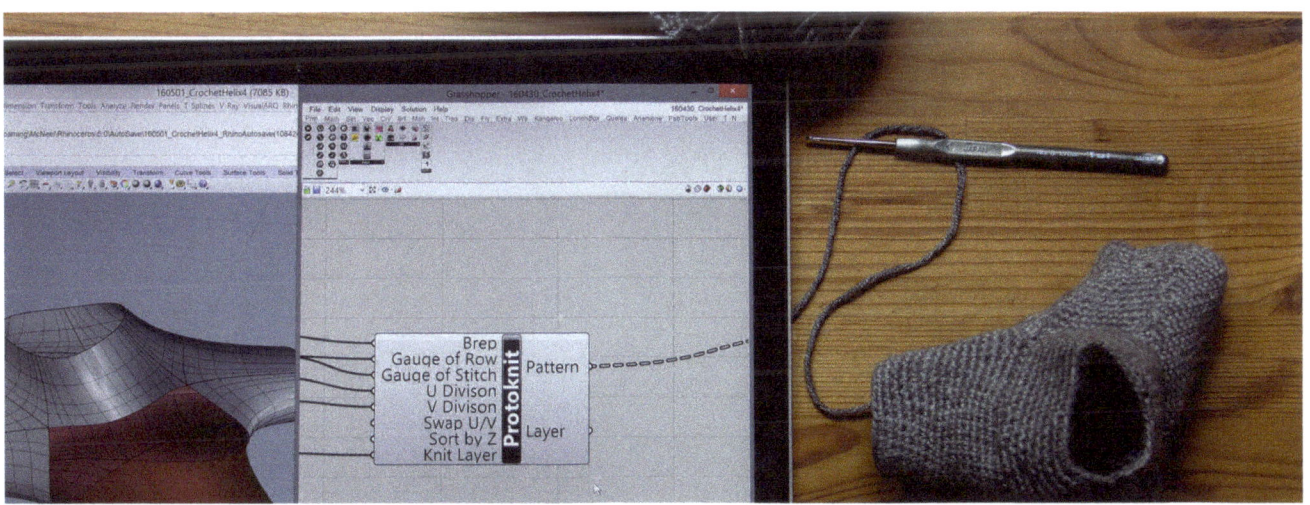

Crocheting is a hands-on craft that involves repetitive manipulation of a single continuous thread with a hook-like tool to generate surfaces and 3D forms. The step-by-step stitching procedure in crocheting can be associated with algorithms of which the steps are defined through crochet patterns. Crochet patterns are text-based representations, similar to G-code in additive manufacturing. They enable the documentation and communication of crocheting know-how. In this workshop, participants from different locations collectively designed and crocheted a branching spatial structure. Each participant received a Co-Crochet Kit prior to the workshop that included crocheting materials. During the workshop, participants first collectively designed a branching structure and learned the basics of crocheting.

They then generated the crochet patterns of the components of the branching structure designed using a computer algorithm developed by the instructors that generates crochet patterns of 3D objects modeled in CAD software. At the end of the workshop, each participant had the crochet patterns of at least one component to crochet. They were provided a prepaid shipping label to ship the crocheted components to the instructors in the US. The instructors assembled the crocheted components of the branching structure - "Voltron!"

Images courtesy of the workshop leaders.

Enhancing Fungi-based Composite Materials With Computational Design And Robotic Fabrication

Jonathan Dessi-Olive
Kansas State University

Omid Oliyan
Silman

Ali Seyedahmadian
Eventscape A+D

This workshop explored the capacities of computational systems to develop an integrative design to fabrication workflow for fungi-based materials. "Myco-materials" are composites made by entangling mushroom mycelium around agricultural or forestry wastes such as hemp or saw dust. As the need for zero-waste materials increases, myco-materials continue to garner attention from engineers, building scientists, and designers. Recently, imaginative architecture-scale structures made with myco-materials have used fabrication techniques canonically familiar to architecture including: modular bricks, custom molded blocks, fabric formed structures, and robotic printing.

This workshop seeked to expand upon the existing constructive paradigms by asking: in the realization of smarter and stronger mycelium composite structures, what are the cooperative logics that have yet to be discovered between fungal growth, computational design, and digital fabrication? Participants investigated the capacities of computational design and mixed reality (MR) through at-home experiments; gaining experience with the craft of growing mycelium. MR-guided weaving, winding, and knotting was used to inform the design and production of flexible reinforcement lattices used to strengthen the surface and inner matrix of mycelium composites. Handwork by the participants informed robotic procedures to be demonstrated by the organizers through the fabrication of a prototype henceforth.

Images courtesy of the workshop leaders.

Knitted Growth: Scaffolds for Living Root Spans

Mariana Popescu
TU Delft

Robin Oval
University of Cambridge

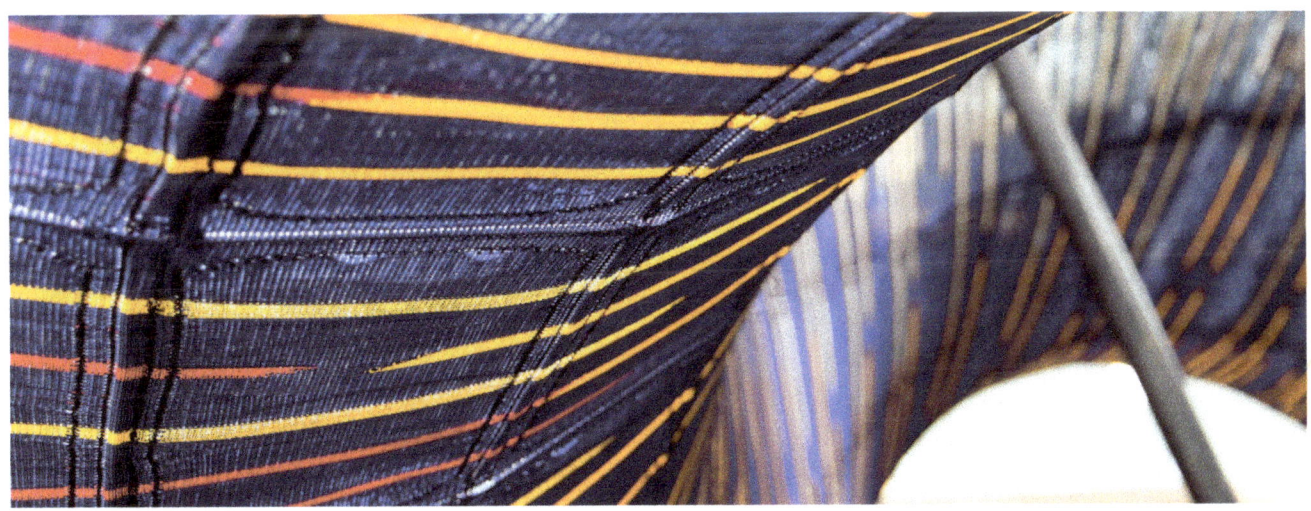

This workshop explored the design of deployable knitted textile membranes as scaffolds for climbing plants. The knitted textile was designed to guide the growth of a plant into spans following principles of structure and placing material (in this case growth) where needed. The playful example is inspired by the living root bridges of Northern India, which are made by guiding the roots of a tree over a stream, allowing them to grow into spanning structures over time. These growing structures are used as an illustration for aspects related to the design and fabrication of knitted textile moulds that can be used as efficient, lightweight and deployable formwork for complex geometries.

During the workshop participants learned how to use form finding methods to design tensile structures including non-manifold and non-orientable geometries. Structural and fabrication considerations were highlighted through topology explorations and the deliberate choice of singularities and segmenting of the overall geometry. In considering how to best guide the growth paths through the textile, participants zoomed in on the design of specific textile features such as ribs, openings, channels, and textures. A spanning design was fabricated after the completion of the workshop.

Images courtesy of the workshop leaders.

Physics Towards Critical Assemblies

Daniela Atencio
Universidad de Los Andes

Nicolas Turchi
Zaha Hadid Architects /
University of Bologna /
Polytechnic of Milan

The workshop explored procedural workflows by integrating the use of Autodesk Maya, MASH, and Grasshopper to expose the participants to polygon modeling, parametric design and animation techniques using physics. Through a sequence of step by step operations, the workshop demonstrated how a clear process can turn into a plethora of outcomes that can be further discretized, evaluated and assimilated into a final design proposal. The workshop focused on the object, both as a static and a dynamic entity.

This was be subject to computational procedures catalogued into categories including 'Addition', 'Distribution', 'Morphing', 'Instancing', 'Aggregation' and 'Projection'. Architectural and tectonic qualities emerged while being tested against the use of contemporary representational techniques including animation. Participants learned integrated and hybrid workflows by using multiple platforms in synergy while experimenting and defining the character of their 'superobjects'.

Images courtesy of the workshop leaders.

The Generative Game

Runjia Tian
Zhaoyang Luos
Linhai Shen
Harbin Institute of Technology

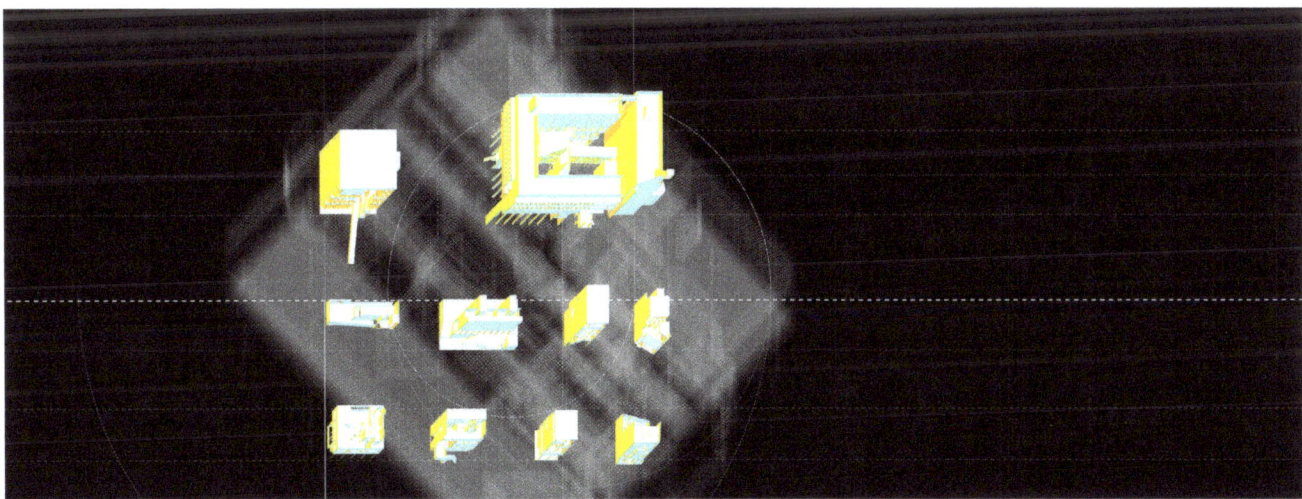

This 2-day workshop covered a series of hands-on advanced 3D machine learning techniques for generative architecture design, including but not limited to 2D Latent Walk Techniques, 2.5D Latent Walk Shape Synthesis, 3DGAN and Reinforcement Learning. The workshop was project-based, where participants were asked to first create datasets or study explicit quantifiable metrics for generative architecture design, and then develop or customize generative architectural designs based on provided sample workflows.

The workshop also examined the various types of artificial intelligence that are applied as cutting-edge generative architectural design techniques since the invention of the concept "Computer-Aided Architectural Design" (Mitchell Williams 1975), the paradigm for modern generative architecture design systems. The workshop reconceptualized Mitchell's CAAD system consisting of a representation system, a generation system and a testing system into a figurative framework of a generative machine that plays the generative game. The workshop used this framework to examine state-of-the-art (SOTA) generative architecture design and explore new possibilities beyond existing methods. The workshop also guided students through the implementation of SOTA systems and experimentation with realizing generative design in 3D representations and post-processing results for architecture design with a series of hands-on sessions.

Images courtesy of the workshop leaders.

Form-Finding Staircases with COMPAS CEM

Rafael Pastrana
Isabel Moreira de Oliveira
Princeton University

Patrick Ole Ohlbrock
ETH Zürich

Pierluigi D'Acunto
TU Munich

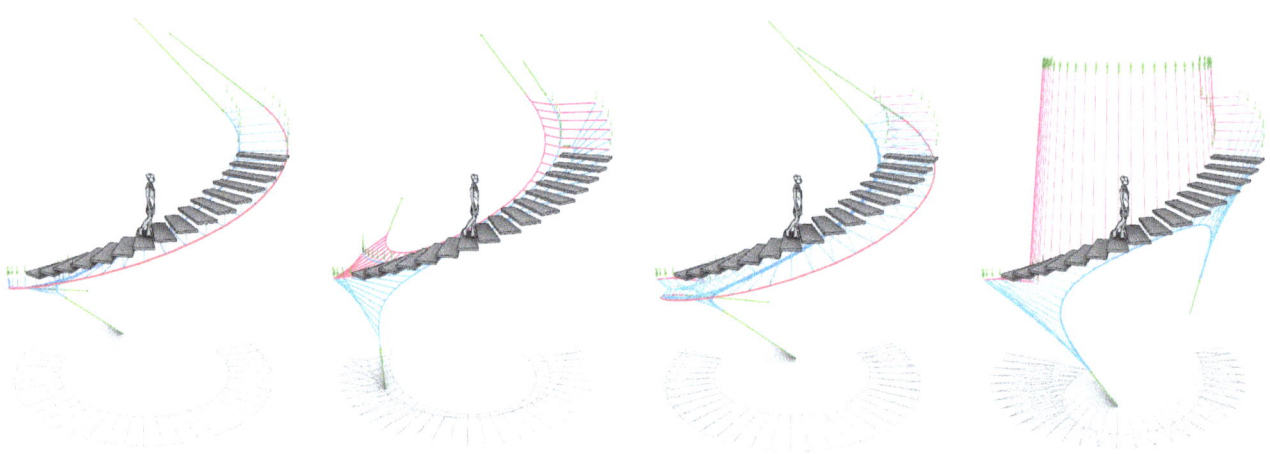

In this workshop, participants learned how to generate the geometry of lightweight and expressive structures using the Combinatorial Equilibrium Modeling (CEM) framework, a form-finding method based on vector-based graphic statics, and its computational implementation, COMPAS CEM. COMPAS CEM is a new design tool written in pure Python that optimizes structural geometries to best meet geometry and force-related constraints. To this end, the tool uses automatic differentiation, a set of algorithmic techniques responsible for making modern machine learning models to learn.

Through a series of guided exercises, participants gained hands-on experience with COMPAS CEM, and learned how to use it from their command line interface and inside Grasshopper. The theme of the workshop focused on the design of the load-bearing structure of a staircase, an intimate spatial bridge that connects different floors and areas in a building. Participants explored digitally how the manipulation of the connectivity and the internal force states in a structure can steer the generation of a catalog of forms that are elegant, safe and material-efficient. Besides acquiring foundational understanding of how constrained form-finding works, participants walked out of the workshop with a set of digital form-found structures, which can be seamlessly ingested by other packages in the COMPAS ecosystem.

Images courtesy of the workshop leaders.

Remote Robotic Assemblies

Stefana Parascho
Edvard P.G. Bruun
Princeton University

Gonzalo Casas
Beverly Lytle
ETH Zürich

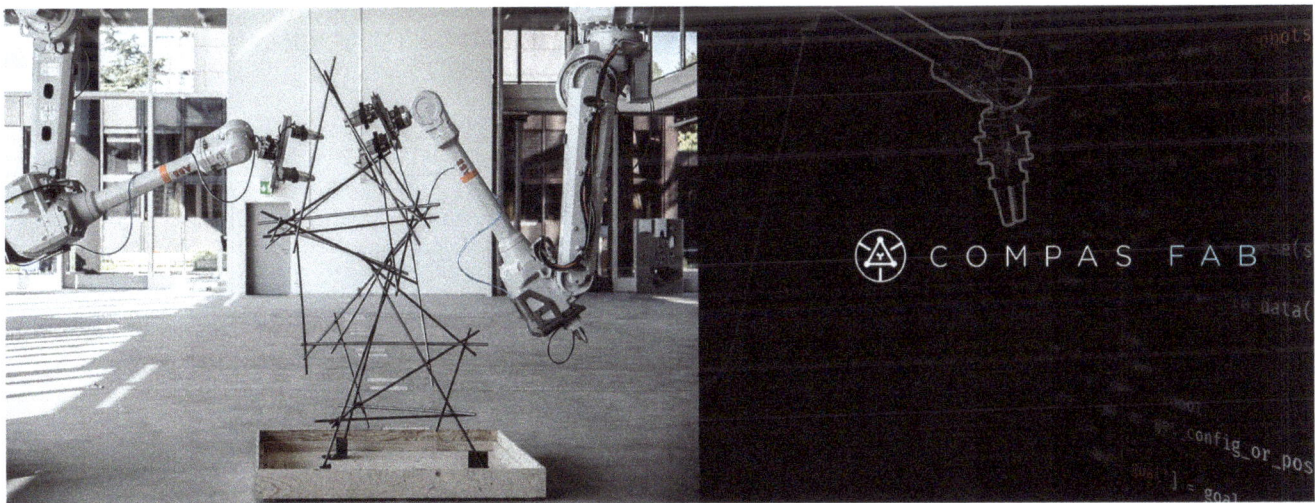

Robotic fabrication in architecture and design relies on a wide range of computational tools and methods for designing, planning, and controlling robots. Through this workshop we aimed to address the challenge in learning and accessing such tools, which represents a major barrier to entry for architects wishing to utilize robots more centrally within their work. By introducing participants to the open source framework COMPAS FAB for robotic fabrication and COMPAS RRC for robot control, we provided a central platform for the simulation and remote control of robots used for fabrication at the architectural scale. Workshop participants designed a space-frame structure from their homes; this structure was then fabricated remotely in the Embodied Computation Lab at Princeton University.

Images courtesy of the workshop leaders.

Generative Design and Analysis in Early-Stage Planning with Spacemaker

Christoph Becker
Spacemaker

Lilli Smith
Zach Kron
Autodesk

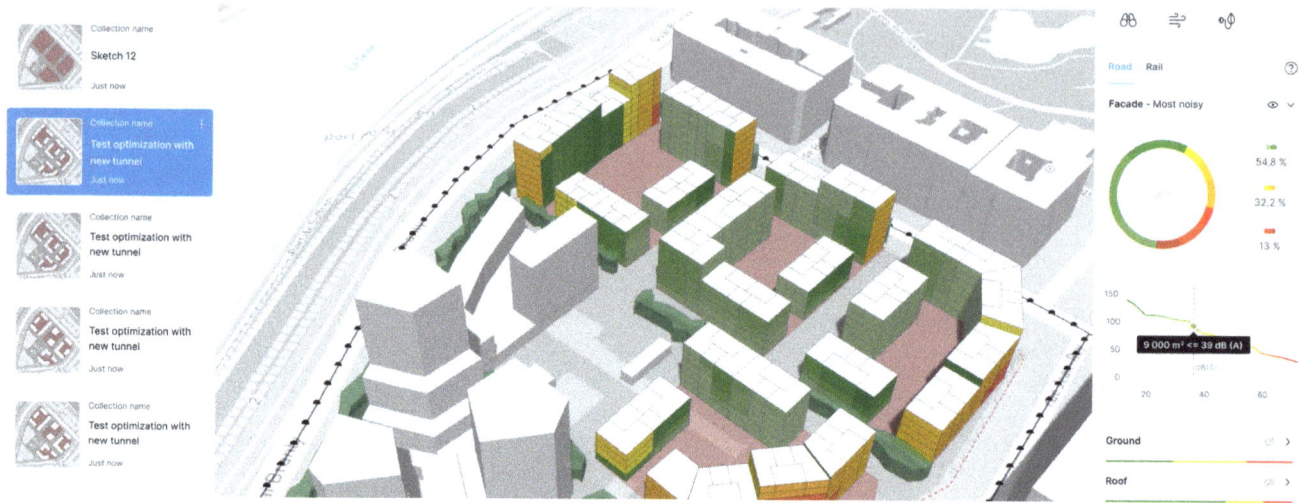

In this workshop, participans learned about generative design and real-time multi-factor analysis in early-stage planning with Spacemaker. A significant part of value creation occurs in the early planning stages. However, a disproportionately low percentage of tech-investments is allocated to this segment. While uninformed decisions in this phase drive costs in later stages, too often, crucial information is not yet readily available and needs to be built up as we build our design on multiple assumptions. Smart algorithms, automated data-fetching, and the use of geo-referenced digital 3D twins of your site from day one can help solve this issue. How can we use the full potential of manual, parametric, and generative design features, and cloud technology to boost our design capabilities? How can we use real-time analysis to connect program, solar, wind, acoustic, view, and sustainability data to building form? How can we capture the power of AI while keeping full control over the results and always remaining in the driver's seat? This workshop introduced a rethought, networked workflow for early-stage site planning. Participants learned how to frontload their design with information and make data driven decisions, spending less time on manual setup and more on creative decision making. Reduced risk with constant live-feedback and improve value by optimizing for density and living qualities at the same time. Each day contained a combination of presentations and hands-on exercises. Participants were-contestants and jury in a small design competition at the conclusion of the workshop.

Images courtesy of the workshop leaders.

Thanks to Autodesk for sponsoring this workshop.

Workshop Participants

Latent Morphologies: Disentangling Design Spaces
Dongyun Kim
Theodore Teichman
Ben Drusinsky
John Nguyen
Daniel Vianna
Ernesto Torralba
Mohammed Behjoo
Dalton Goodwin
Chun-Hao Hsu
DeWitt Godfrey
Gilang Fajar Kusumawardana
Reza Taghavifard
Rafael Ramos Mendez
Rohit Sanatani
JayediAman
Jaqueline McNamara

Distributed Collaborations - KUKAcrc Cloud Remote Control
Trevor Kemp
Fatemeh Amiri
Alejandra Rojas
Bhavleen Kaur
Bruno Figueiredo
Hany Mohsen Kamal Elhassany
Arastoo Khajehee
Antonio Francisco Morais
Rachael Henry
Ronan Bolaños
Mohadeseh Taheri

Mohamed Khaled
Christian Gonzalo Paez Diaz
Israa El-Maghraby
Samaneh Torabifar

Collaborative AI – Human + AI Form
Anna Mytcul
Majed AlGhaemdi
Kyuseung Kyoung
Ruotao Wang
Karen Kuo
Milos Roglic
Carlos Rios
Carlos Navarro
Antje Kerkmann
Kiran Kastury
Naitik Sharma
Harry Moller
Harish Karthick Vijay
Kritika Kharbanda

Augmented Architectural Details
Xinran Li
Garvin Goepel
Zeno Zoppi
Felicia Wagiri
Darcy Zelenko
Johnathan Greenage
Majed AlGhaemdi
Laura Fegely
Danny Ngo
Nathan Barnes
Matt Arsenault

Sheida Shariat
Atefeh Cheheltanan
AhmadSleem
Mojtaba Bazrafshan

Building Web-Based Drawing Instruments
Bob Frederick
Vernelle Noel
Atharva Ranade
Curime Batliner
Humbi Song
Charles Weak
Francois Sabourin
Jesse Bassett
Aisha Cheema
Mario Serraglio
Jingtong Duan
Anahita Khodadad
Ghazal Hosseini
Shadi Naghiloo
Monalisa Malani
Mohammad Pourfooladi

Co-crochet Computing Stitches for Collective and Distributed Crocheting
Melissa Goldman
Larissa Korol
Jose Luis Rangel Oropeza
Nikoletta Karastathi
Luis Antonio López
Jeg Dudley
Hsin Ju Lin
Yasaman Amirzehni

Wei Wu
Kaylee Tucker
Tamar Nix
Walter Patrick Smith
Shagun Kapur
Faraz Khojasteh Far
Leila Delafrooz
Elnaz Kakouei
OluwadamilolaAkinniyi
Soroush Reaisi
Simin Nasiri

Enhancing Fungi-based Composite Materials With Computational Design And Robotic Fabrication
Yin Yu
Ytav Bouhsira
Holly Hodkiewicz
Sumanth Krishna
Yuqian Lin
Nadia Kutyreva
Sarah Kott-Tannenbaum
Laure Nolte
Adam Schueler
Dragan Petrovic
Amelia Gan
Behzad Modanloo
Ghazal Javidannia
Mohammad Mahdi Jalali
Shiva Azizi

Knitted Growth: Scaffolds For Living Root Spans
Laurin Aman
Virginia Melnyk
Zhenxiang Huang
Deva Prasad A
David Maples
Andres Obregon
Mason(sizhe) Mo
Jumaanah Elhashemi
Isla Xi Han
John Mikesh
Melody Morgan
Manfred Saberbein
Ali Ghadamyari
Manik Makkar
Amir Motavasselian
Sajjad Jafari Galuyak

Physics Towards Critical Assemblies
Christianna Bennett
Joonkyu Shin
Luis Matias
Jiajie Yang
Stephen Clond
Mateen Cheema
Nader Alkhabbaz
Artemis Kyriakou
Alvaro Villacis
Claire Tokunaga
Ahmed Hesham
Jameel Marsden
Iliya Mela
Hossein Arshadi
Alejandro Bautista

Shravan Arun
Felipe Romero
Sutanuka Jashu
MD Faizan Sharief
Javad Ebrahimi
Mahdiss Taheri
Katia Sondang Sitompul
Faezeh Hosseini
Ali Dehghani
Maryam Sheikhi

The Generative Game
Daniel Ruan
Will Garner
Juliette Zidek
Alireza Bayramvand
Albert Maksoudian
Diana Nigmatullina
Mohamed Elmesawy
Nishita Vinodrai
Anahita Aliasgarian
Christina Doumpioti
Duong Nguyen
Shujie Liu
Louis Daumard
Saba Salehi sheijani
Anirudh Rathi
Homa Hassanzade
Danial Bagheri
Iman Sheikhansari
Amir Homayouni
Mohammed Behjoo
James Nanasca

Form-Finding Staircases With COMPAS CEM
Dominik Reisach
Fernando Gomez
Victor Martiniuc
Dominic Rishe
Damir Kovacevic
Parimala Venkatesh
Alberto Longhin
Rasmus Sainmaa
Pinaki Mohanty
Hex Ceballos
James Whiteley
Hossein Kamyab
Augusto Oliver Palacios
Rodrigo Cabrera
Fateme Ghotbi
Pedro Saldaña

Remote Robotic Assemblies
Chris Kang
Abhishek Shinde
Jingwen Song
Shahé Gregorian
Kristen Forward
Teng Teng
Seyed Hossein Zargar
Yuan Li
Krishna Jani
Nelson Brito
Tahmures Ghiyasi
Shirin Shevidi
Ayush Kamalia
Jose Roberto Arguelles Rodriguez
Yuxuan Wang

Generative Design and Analysis in Early-Stage Planning with Spacemaker
Doug Mcbirden
Jose Luis Rodriguez Hernandez
John Chun
Carol Kan
Andreas Thoma
Thesla Collier
Tim Halvorson
Vince Allen Hernandez
Dena Hassani Ardekani
Jessica Chen
Kathleen Danziger
Atousa Momenaei
Elham Ghaderi Rahaghi
Nastaran Hasani
Iman Ansari
Nihal Mohamed

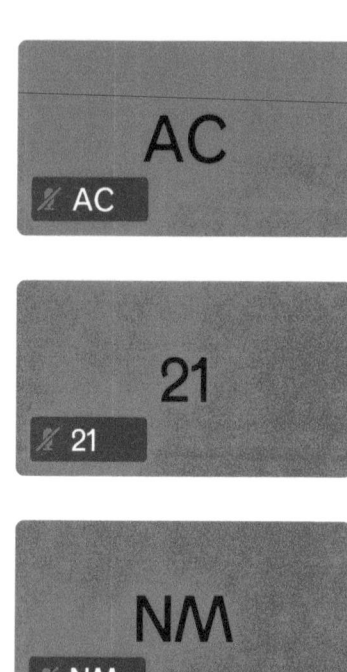 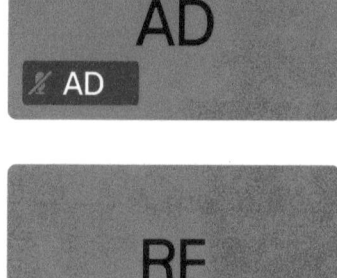 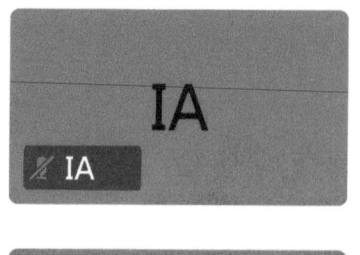 
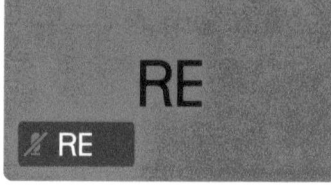 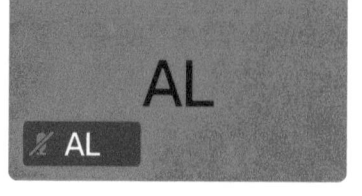 
 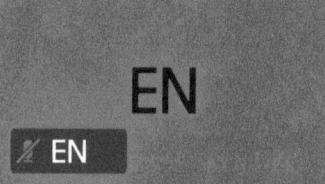  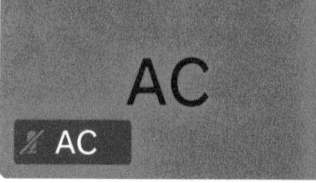
 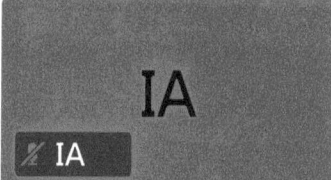  
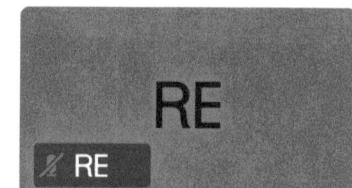 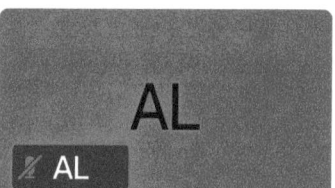  
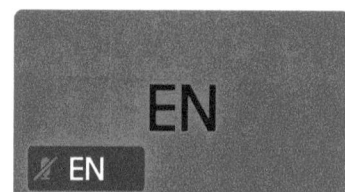  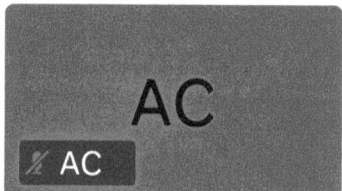 
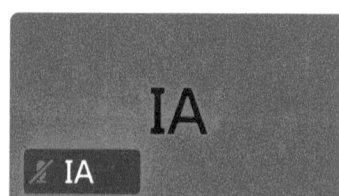   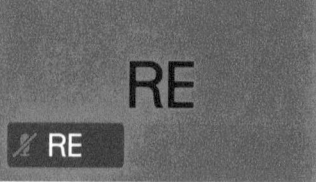
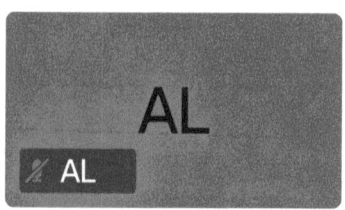   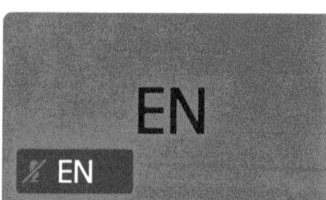

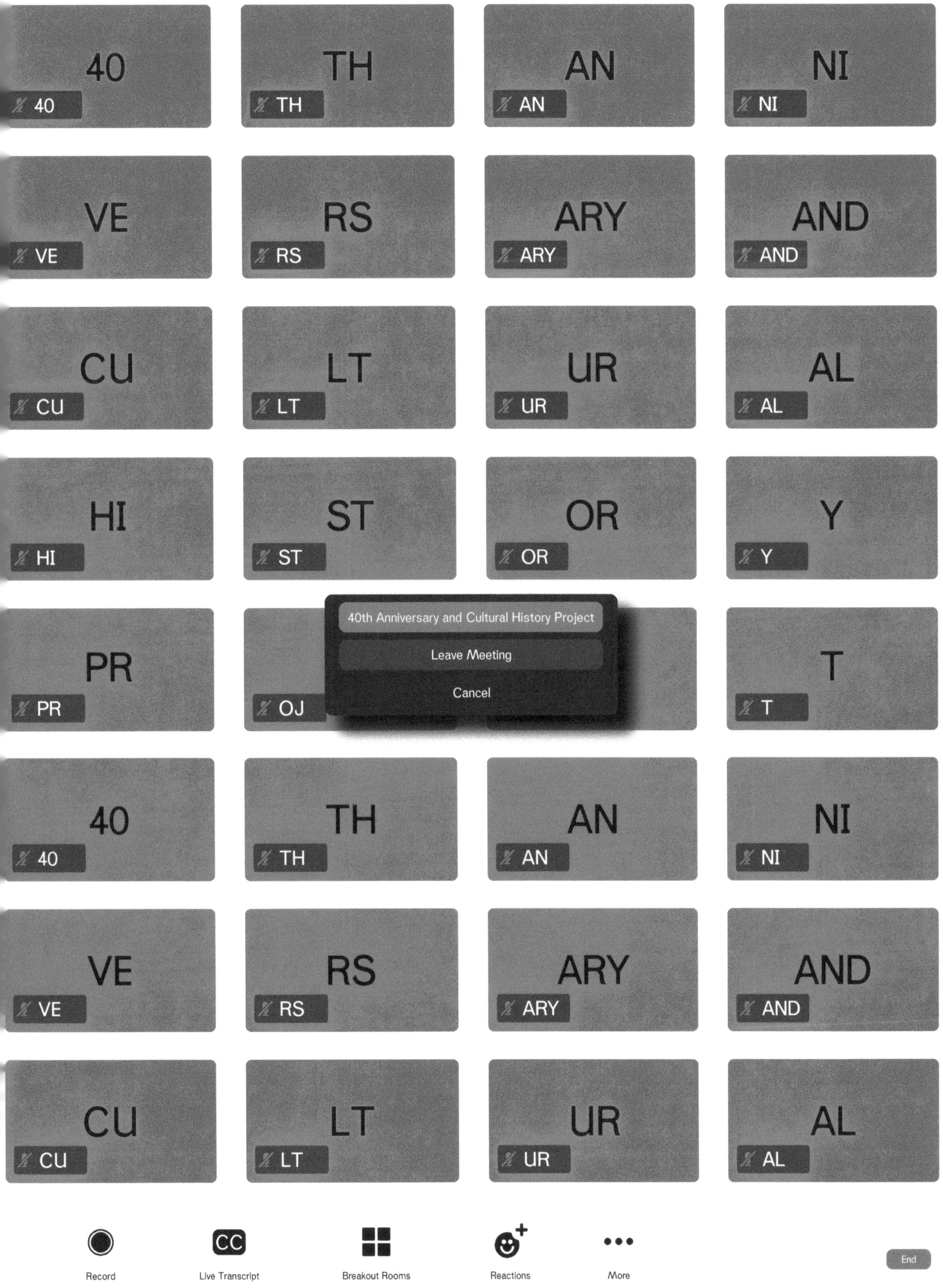
40
40
TH
TH
AN
AN
NI
NI
VE
VE
RS
RS
ARY
ARY
AND
AND
CU
CU
LT
LT
UR
UR
AL
AL
HI
HI
ST
ST
OR
OR
Y
Y
PR
PR
OJ
T
T
40th Anniversary and Cultural History Project
Leave Meeting
Cancel
40
40
TH
TH
AN
AN
NI
NI
VE
VE
RS
RS
ARY
ARY
AND
AND
CU
CU
LT
LT
UR
UR
AL
AL
CC
Record
Live Transcript
Breakout Rooms
Reactions
More
End

40th Anniversary and Cultural History Project

Reflections on the Past 40+ Years of ACADIA

1   This timeline indicates the names of the ACADIA conferences. The community began with questions of pedagogy as computers made their way into architecture schools and transitioned to the themes addressed in the preceding 2021 keynote Emerging Trends in Response to Critical Computation.

INTRODUCTION

The 2021 Conference "Realignments: Toward Critical Computation" marks the 40th anniversary of ACADIA as an organization and the 41st anniversary of the conference. ACADIA or the the Association for Computer Aided Design in Architecture was established in 1981 and its first bylaw states:

> "ACADIA was formed for the purpose of facilitating communication and information exchange regarding the use of computers in architecture, planning and building science. A particular focus is education and the software, hardware and pedagogy involved in education."

This anniversary provides a moment to reflect upon the legacy and trends of this longest running computational design organization as a method for considering its future.

WHY START ARCHIVING?

Due to the COVID19 pandemic, the ACADIA 2020 conference, Distributed Proximities, and 2021 conference, Realignments: Toward Critical Computation, were held virtually and run by ACADIA board members. This was an unprecedented break with typical in-person ACADIA conferences which were mainly organized by a singular academic institution, and this format allowed the ACADIA board organizers to evaluate the conference structure and dissemination methods.

ACADIA has traditionally produced printed conference proceedings, but as an organization interested in innovation and experimentation, there have been multiple formats and repositories for research dissemination. Until now, this content has not been collected in a single archive. The turn to a virtual conference which relied entirely on accessible digital content

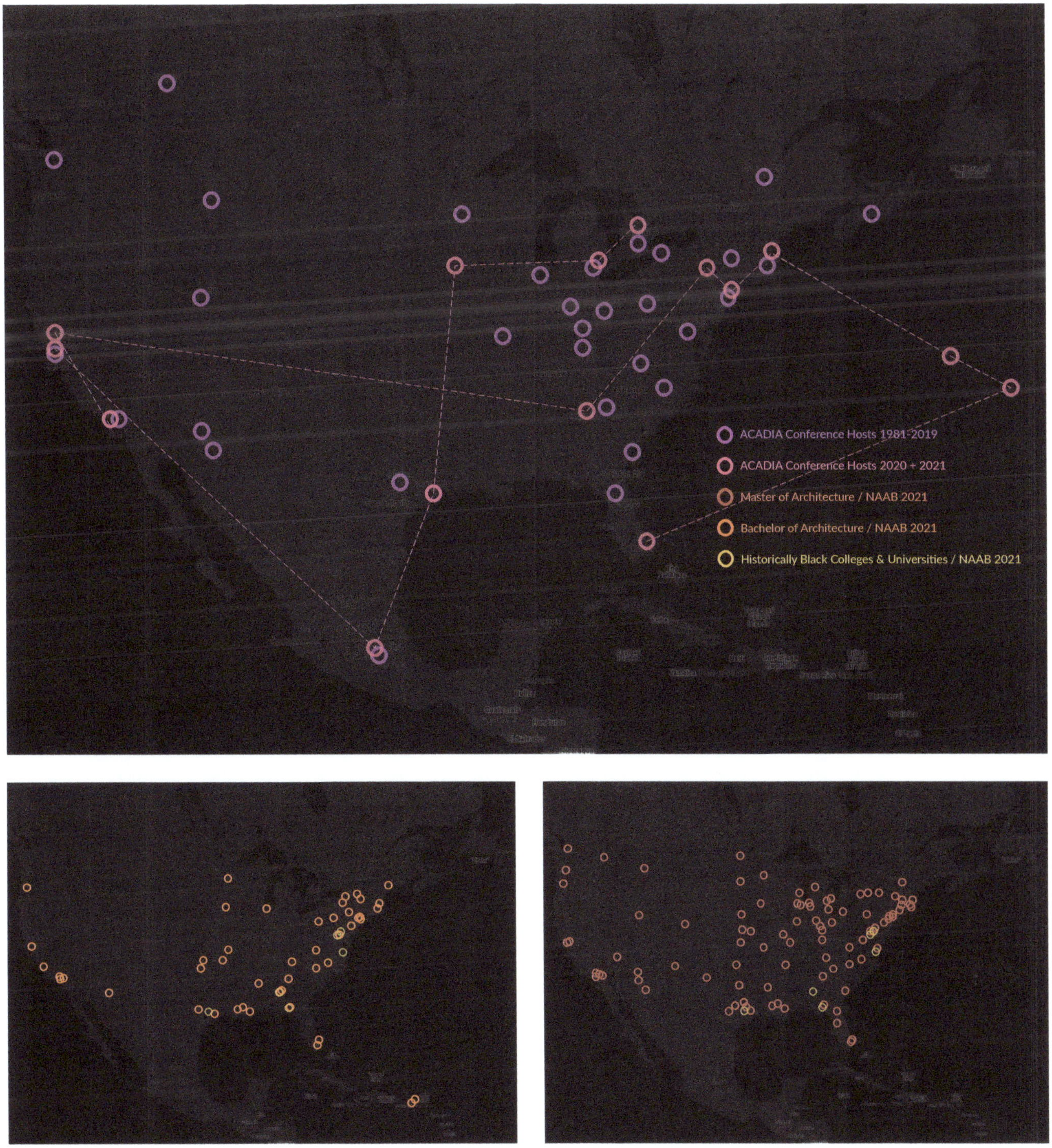

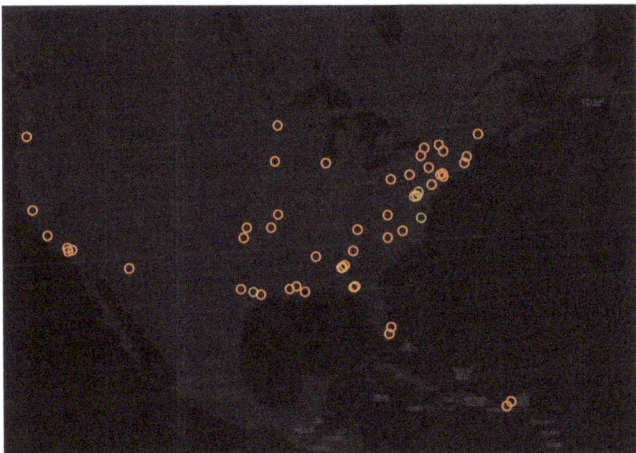

2   For a complete list of NAAB schools, see www.naab.org. The next step in this mapping might include identifying non-NAAB architecture programs as a method for reflecting on whether computational design work is occurring in these schools or primarily within accredited institutions.

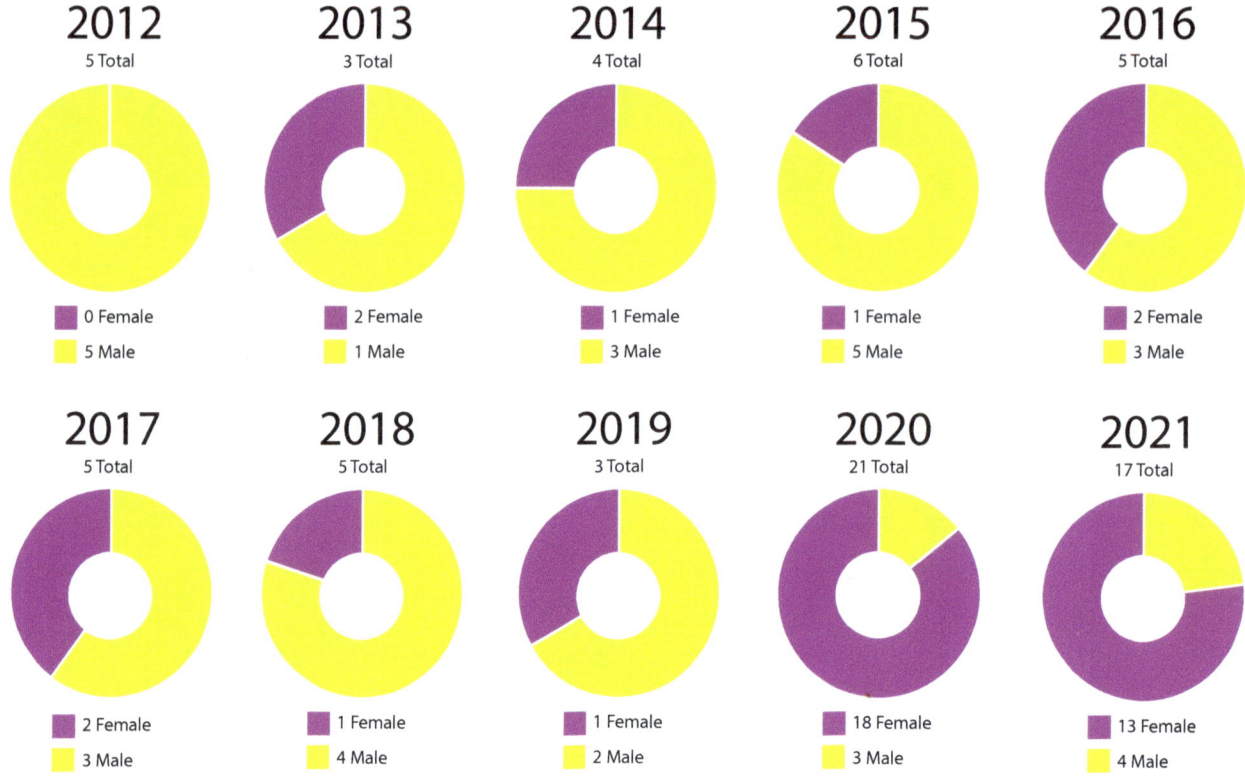

3   ACADIA Conference Keynotes from 2012-2021 based upon gender pronouns used in participant biographies

encouraged the ACADIA board to think about producing virtual archives. Thus, the question of what is an ACADIA archive became a topic of exploration.

During the second keynote panel of the 2021 Conference, entitled "Designing with AI, Data, Bias, & Ethics," Benjamin Bratton described planetary archives as the following:

> "Archive is a form of big data…sampled and prepared for future investigation. The ultimate social function of the data representation is unknown because it cannot be known what the future will ask of the present. The archive is in a sense a promise to the future that the present will make itself accountable."

In the spirit of Bratton's proposal, and if the very construction of an archive is not neutral, then any effort to capture this history will contain the biases and agendas of its makers. In this context, what accountabilities does the archive of ACADIA ask of us as a community? The goal of producing and sharing this archive is to facilitate new scholarship and initiatives that guide the future of ACADIA while acknowledging the successes and shortcomings of its past. Particularly, this is an opportunity to reflect on whether the community's research on computational design is "contributing the construction of humane physical environments" as ACADIA's second bylaw pledges. In the context of immediate and imminent socio-political and environmental crises, how do we even begin to think about the built environment—digital or physical—as humane? Or is this ACADIA bylaw a provocation toward a more inclusive form of computational design? Could an archival process bring into existence the methods necessary to address the entangled crises of the present day?

History of ACADIA Organization and Conference

ACADIA was formed on October 17, 1981 at the first conference meeting of the twenty-four founding members at Carnegie Mellon University, and just five years later there were nearly 200 active members. ACADIA has since evolved into the global community of thousands that we know today. Over the years, ACADIA has set the tone for a number of sibling organizations, has set an academic foundation for launching successful academic careers, and has been the first and lasting scholarly home for so many.

We are taking this opportunity to launch what we call the Cultural History Project and to reflect on the first

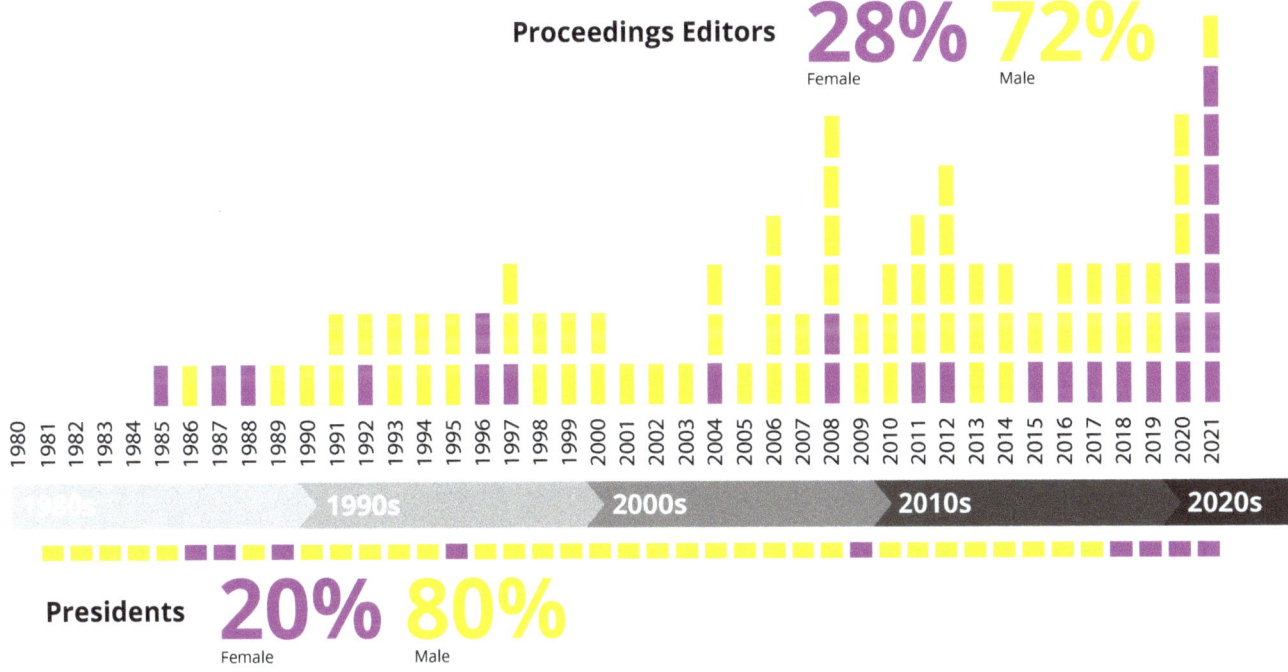

4 ACADIA proceedings editors and presidents based upon gender pronouns used in participant biographies

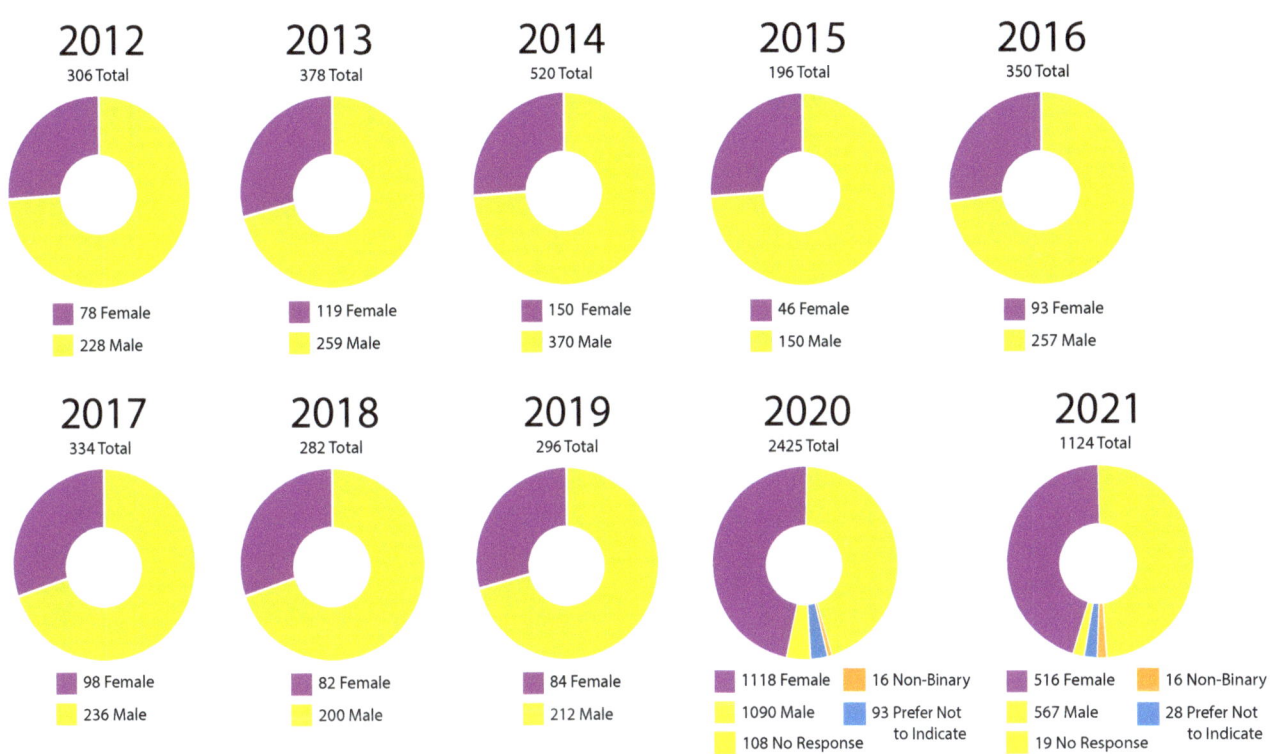

5 ACADIA conference participants, gender(s) self-identified during registration

four decades of the ACADIA organization in order to look forward to the next four decades. This includes:

- ACADIA's History through Data: Visualizing Locations and Participation
- From Data to Information: Analyzing Issues of Access and Gender(s) Imbalance
- Proceedings Archive: Learning from the Highlights
- Oral Histories from the Past

ACADIA'S HISTORY THROUGH DATA: VISUALIZING LOCATIONS AND PARTICIPATION

The first initiative is to map the history of the ACADIA conferences in relation to the geography of its North American context. The maps in Figure 2 show the conference host locations over time from 1981 to 2019, as well as the distributed chairs of the conferences in 2020 and 2021. The next step is to overlay on the map of the conference locations the National Architectural Accrediting Board (NAAB) accredited North American Master of Architecture programs, Bachelor of Architecture programs, and Historically Black Colleges and Universities. This correlation allowed us to question the location and distribution of computational design knowledge and access between and across geographies and institutions.

Mapping where the ACADIA conferences have been allows us to identify those institutions and communities that are not yet part of ACADIA and focus our future outreach efforts to expand the community over the next years and to re-engage with former loci of computational design. For instance, the next step in this mapping might include identifying non-NAAB architecture programs or all members of the Association of Collegiate Schools of Architecture, as a method for reflecting on how and when computational design courses are introduced and applied in the curricula. Another avenue could be identifying overlaps between ACADIA participation and R1 university status — or those universities that meet benchmarks in research activity and expenditures as measured by the Carnegie Classification of Institutions of Higher Education. Mapping is an opportunity to reflect on the uneven distribution of resources across institutions (equipment, finances, knowledge) and consider how existing resources within the ACADIA community can be leveraged toward a more equitable discipline.

FROM DATA TO INFORMATION: ANALYZING ISSUES OF ACCESS AND GENDER(S) IMBALANCE

The second initiative explores available data from ACADIA's history as an organization. Specifically this is an opportunity to reflect on the organization's journey toward gender parity. Image 5 documents conference attendance by genders from 2012-present as reported by attendees via registration. From 2012-2019 conferences attended by women remained at about 30%. Several important changes occurred in 2020 and 2021. The first was that the conference moved online, which changed the landscape of access and increased both student attendance and international reach. The second was that the conference chairs and Board of Directors made strategic decisions about prioritizing gender parity in the keynotes and session chairs. The third was that gender identity at registration was no longer limited to a binary question of either male or female; options were added for non-binary, prefer not to indicate, and no response. These changes increased both total attendance as well as female majority participation in 2020, for the first time in the history of the conference. As the conference transitions back to in person the questions remain of how to maintain this progress toward gender equity.

In Image 3, keynotes are graphed by genders over the same time period, based on the pronouns used in their biographies. In 2020 and 2021, in addition to moving online, the keynote format was changed to a conversation between multiple participants and this reformatting changed the makeup of the keynotes to female majority for the first time in ACADIA's history.

Image 4 indicates that 20% of ACADIA's presidents and 28% of ACADIA's proceedings editors identify as female based on the pronouns used in their biographies. There remains work to be done in representational equity but also in the types of content celebrated by the ACADIA community. This data project is part of a larger data analysis project that identified both gaps and pathways forward toward a more inclusive ACADIA community.

ACADIA DIGITAL ARCHIVE: LEARNING FROM ALL THE PROCEEDINGS

The third initiative of the Cultural History Project is the construction of an ACADIA Digital Archive to include all conference proceedings, early newsletters, and the Quarterly publications that preceded the *International Journal of Architectural Computing (IJAC)*. While individual

6 3D models of dinosaurs from the related article "The Virtual Dinosaur Project" shown here on the cover of Summer 1995 Quarterly.

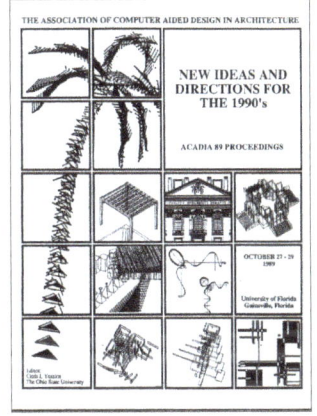
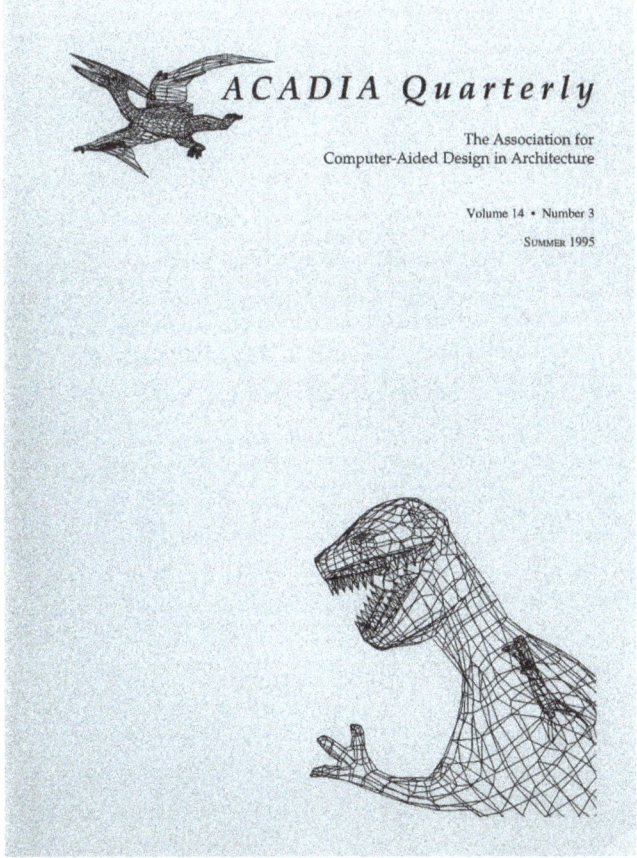

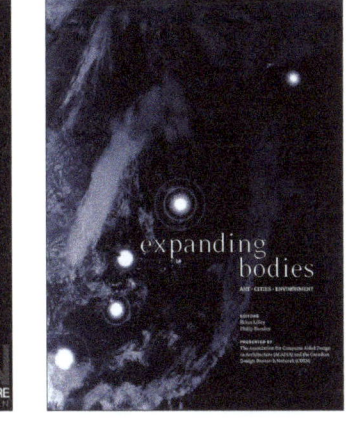
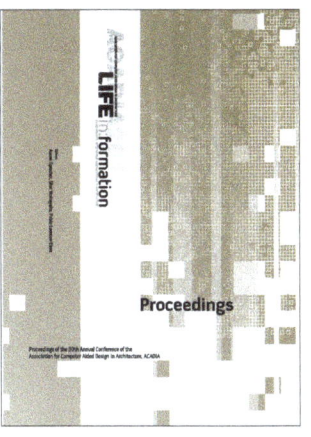

40TH ANNIVERSARY AND CULTURAL HISTORY PROJECT   **REALIGNMENTS**   655

papers are available on the CumInCAD database, the full list of proceedings had never been collected in a single database for researchers to access. ACADIA Digital Archive is complementary to the existing database, and archives content that sets the context of the conference in time, space, and society. As an organization focused on the cutting edge knowledge exchange, we have rare opportunities to look back and have never had a historian of the organization. Thus, proceedings were produced across multiple geographies, software, and file types and during a time before affordable cloud computing. Starting the ACADIA Digital Archive project, required finding the proceedings and locating contact information for original editors — many of whom had moved on to careers as varied as software engineers, hot air balloon company operators, and architecture school deans.

Having the full proceedings available to researchers allows for an additional layer of cultural history and context to the evolution of computational design and computational design thinking. Noticeably, as can be seen in Image 6 the graphics and interests of the community have evolved. As we look through the covers for the past four decades we can see the evolution of graphic software, the introduction of digital fabrication, and the rise in discourses about biology, ecology, disciplinarity, and critical computation. The archive will be available on ACADIA's website. The website is an ongoing project to document and share content

ORAL HISTORIES

The final initiative is to collect oral histories about ACADIA from its past and current members. While we now have the publications, we don't have the stories that connect these publications to our community. For example, we found out that in the mid-1980s, the community was talking about what it would take to do an electronic conference, and Robert E. Johnson submitted a report to the president in 1985 going over different electronic platforms. And in the late 1990s and early 2000s ACADIA ran the first internet-based design competition of its kind to bring together members to work together and test ideas digitally. The goal of the oral history project is to collect the anecdotes that capture the culture of the organization and provide a record of the messy 'behind the scenes' work and discussions that went into developing, sustaining, and transforming the organization.

ACCESS

The 40th Anniversary Panel yielded strong themes and questions that are the foundation of our ACADIA community. Even as technology itself changes rapidly and the geographic centers of our membership have shifted over time, what has brought and held this community together are the tenets of our organization. The kind of community that is actively made and remade by each generation of ACADIA members defines the cutting edge as the membership reaches out towards new partnerships, builds bridges to new disciplines, defines new priorities, and conceives of new technology, workflows, and design fields. We have always engaged in questions of access and accessibility. This was originally framed as pedagogic access to courses, hardware, and software. As seen in the first ACADIA surveys the organization focused on documenting which organizations had specific hardware and software or how to get technology into our places of work, teaching, and research (Image 7). More recently, this question of access has shifted to who has access to emerging technologies and what support can we give as an organization to open up opportunities for new audiences, researchers, and types of work (Figures 2, 3, 4, 5). This has led to new scholarships, partnerships, and programs to support new scholars and students to join and present in the community.

CONCLUSIONS

ACADIA started as a group of architectural outsiders, almost as a support group for "computer people" to foster research and teaching questions about technology-driven and interdisciplinary research, and we have continued to be that supportive environment. Many of us got our start at ACADIA and then turned around to welcome in and generously mentor new members. It is a place to present first papers, to test out new ideas, and to question the status quo. Many in the community have become involved after presenting papers, serving in leadership or hosting the conferences or both. ACADIA has led to many personal and professional connections, and many generations of researchers have called and still call ACADIA our scholarly home. It is not just a conference- it is a community.

The construction of the maps and collection of an oral history archive allows us to re-engage with the people who have made ACADIA possible — and to serve as a reminder that any institution is at its core made of people who invested their time, energy, and ideas into making a space that has influenced many of our careers and trajectories. The development of this archive is complemented by the ongoing redesign of ACADIA's website which will launch a new decade of research and community.

This cultural history project and this archive are "a promise to the future that the present will make itself accountable" to: understanding how to be an organization founded in North America, yet now global in access, impact, and discourses; fostering opportunities for BIPOC computational designers; supporting and representing multiple genders; welcoming unconventional career paths; and perhaps most importantly continuing to ask, and sit with the answer, however uncomfortable, of whether ACADIA is achieving its principal goal of "contributing to the construction of humane physical environments."

REFERENCES

The ACADIA organization bylaws are a set of the rules and regulations of the organization. The bylaws were last amended on 7-12-16 and are available on acadia.org.

CumInCAD is a Cumulative Index of publications in Computer Aided Architectural Design supported by the sibling associations ACADIA, CAADRIA, eCAADe, SIGraDi, ASCAAD and CAAD futures papers. cumincad.org.

Shelby Elizabeth Doyle, AIA is an Associate Professor of Architecture and Stan G. Thurston Professor of Design-Build at the Iowa State University College of Design, co-founder of the ISU Computation & Construction Lab (CCL) and director of the ISU Architectural Robotics Lab (ARL). Doyle received a Fulbright Fellowship to Cambodia, a Master of Architecture from the Harvard Graduate School of Design, and a Bachelor of Science in architecture from the University of Virginia.

Melissa Goldman is the Fabrication Lab Manager at the University of Virginia School of Architecture and serves as the Communications Officer on the ACADIA Board of Directors and is an active member of the Student Shop Managers Consortium. Melissa earned her Master of Architecture from Columbia's Graduate School of Architecture, Planning and Preservation and her BA in English from Harvard.

Biayna Bogosian is an Assistant Professor of Architectural Technology at Florida International University (FIU) and an Affiliate Faculty at FIU Institute of Environment. Biayna's interdisciplinary research focuses on innovation in design within a broader environmental context to explore data-driven and participatory approaches for analyzing and improving the health of the built environment. Biayna studied architecture at Woodbury University SoA and Columbia GSAPP. She is currently a Ph.D. candidate in the Media Arts and Practice program at the USC School Cinematic Arts.

UNIVERSITY	TOTAL ENROL.	COMPUTER FACULTY	COMPUTER CURRICULUM ELEC.	COMPUTER CURRICULUM REQ'D	CAD DEG.
ARIZONA STATE UNIVERSITY	155	5	4	0	C
AUBURN UNIVERSITY	682	3	1	1	N
BOSTON ARCH CENTER	*	5	1	0	N
CAL POLY STATE UNIVERSITY	617	5	4	0	C
CARNEGIE MELLON UNIVERSITY	1590	4	2	4	C
CATHOLIC UNIVERSITY OF AMERICA	287	4	3	2	CD
CITY COL. OF THE CUNY	280	1	0	1	N
CLEMSON UNIVERSITY	600	1	4	0	N
COOPER UNION	384	2	2	0	N
FLORIDA A&M UNIVERSITY	100	1	1	0	N
IOWA STATE UNIVERSITY	202	1	1	1	N
KANSAS STATE UNIVERSITY	716	5	4	0	C
KENT STATE UNIVERSITY	765	2	3	0	N
LAVAL UNIVERSITY	480	2	1	0	N
LAWRENCE INSTITUTE OF TECH.	*	3	1	0	N
LOUISIANA STATE UNIVERSITY	790	4	4	1	N
MCGILL UNIVERSITY	436	4	*	*	N
MISSISSIPPI STATE UNIVERSITY	226	3	1	1	N
MONTANA STATE UNIVERSITY	231	1	2	1	N
NEW JERSEY INST. OF TECH.	340	6	*	*	N
NEW YORK INST. OF TECH.	437	5	1	0	N
NORTH CAROLINA STATE UNIV.	1200	1	1	0	P
OHIO STATE UNIVERSITY	295	3	2	0	N
PRATT INSTITUTE	710	6	2	3	D
PRINCETON UNIVERSITY	800	1	*	*	N
RENSSELAER POLYTECHNIC INST.	120	*	*	*	*
SOUTHERN CALIF. INST. OF ARCH	250	4	3	1	N
SUNY AT BUFFALO	400	2	3	0	N
SYRACUSE UNIVERSITY	490	3	4	0	N
TECH. UNIVERSITY OF NOVA SCOTIA	355	2	1	0	P
TEMPLE UNIVERSITY	125	7	3	1	N
TEXAS A&M UNIVERSITY	399	5	2	1	N
TULANE UNIVERSITY	970	7	9	0	C
TUSKEGEE UNIVERSITY	347	4	2	0	N
UNIVERSITY OF ARIZONA	195	3	*	*	N
UNIVERSITY OF ARKANSAS	575	3	2	1	C
UNIVERSITY OF BRITISH COLUMBIA	368	3	1	0	N
UNIVERSITY OF CALIF. L. A.	123	2	2	1	N
UNIVERSITY OF DETROIT	200	4	7	0	C
UNIVERSITY OF FLORIDA	196	2	0	1	N
UNIVERSITY OF HOUSTON	600	4	2	1	N
UNIVERSITY OF IDAHO	600	5	3	0	N
UNIVERSITY OF ILL. AT CHICAGO	300	4	2	0	N
UNIVERSITY OF KANSAS	635	5	2	0	N
UNIVERSITY OF KENTUCKY	705	4	2	0	N
UNIVERSITY OF MIAMI	300	1	2	0	N
UNIVERSITY OF MICHIGAN	*	3	1	1	P
UNIVERSITY OF NEBRASKA-LINCOLN	460	9	4	1	C
UNIVERSITY OF NEW MEXICO	400	2	3	0	N
UNIVERSITY OF N. C. CHARLOTTE	270	5	3	0	N
	265	2	1	0	N
UNIVERSITY OF OKLAHOMA	450	6	2	1	N
UNIVERSITY OF OREGON	650	3	5	0	P
UNIVERSITY OF PENNSYLVANIA	340	2	2	0	N
UNIVERSITY OF PUERTO RICO	300	1	2	0	N
UNIVERSITY OF SOUTHERN CALIF.	385	3	5	1	C
UNIVERSITY OF SOUTHWESTERN LA.	450	1	*	*	N
UNIVERSITY OF TENNESSEE	400	1	3	1	N
UNIVERSITY OF TEX. ARLINGTON	1000	4	2	0	N
UNIVERSITY OF TEXAS AT AUSTIN	635	5	2	0	C
UNIVERSITY OF UTAH	170	2	4	1	C
UNIVERSITY OF VIRGINIA	550	5	4	0	C
UNIVERSITY OF WASHINGTON	420	2	4	0	N
UNIVERSITY OF WATERLOO	*	1	*	*	N
UNIVERSITY OF WIS. MILWAUKEE	918	2	*	*	N
WASHINGTON STATE UNIVERSITY	580	2	7	0	C
WASHINGTON UNIVERSITY	*	1	1	0	N
WOODBURY UNIVERSITY	90	1	0	3	N
YALE UNIVERSITY	140	3	3	0	N
1986 SUM		213	154	31	
AVG	433.1	5.6	4.6	0.8	
MAX	1590	14	16	7	

* = NOT REPORTED N = NONE P = PLANNED C = CONCENTRATION D = DEGREE

7 In the Spring of 1984, ACADIA sponsored its first CADD Activities Survey. A questionnaire was distributed to the 1010 schools which are members of the Association of Collegiate Schools of Architecture (ACSA). Forty-eight schools replied and were included in this first report. The survey and report were conducted and prepared by Elizabeth Bollinger, Associate Professor, and Robert Hinton, Student Assistant, of the College of Architecture of the University of Houston.

40th Anniversary and Cultural History Project

40th Anniversary Toast

1   Panel participants toast to ACADIA's past, its present, and its future

INTRODUCTION
We launched the Oral History Project with a panel of past presidents, moderated by longtime ACADIA member, Philip Beesley. The panel was an opportunity to reflect on each presidents' time in leadership and give toasts to our community's future. These are excerpts from the comments, and you can watch the full panel discussion on the ACADIA YouTube channel.

PANELISTS
Our panel included Philip Beesley, Professor in Architecture at the University of Waterloo and past presidents Karen Kensek, Professor of Practice at the University of Southern California School of Architecture, our president in 1995; Branko Kolarevic, Dean of the NJIT's Hillier College of Architecture and Design, President in 1997-1998; Mahesh Daas, the president of the Boston Architectural College, President from 2007-2009; Nancy Cheng, Associate Professor at the University of Oregon, President from 2009-2011; Jason Kelly Johnson, Associate Professor at California College of the Arts, President from 2016-2017; and Kathy Velikov, Professor of Architecture at the University of Michigan, our President from 2018-2020.

TOASTS

Philip Beesley: It's an absolutely delicious honor to have the chance to moderate this panel of deeply esteemed colleagues and presidents of ACADIA. One of the remarkable sensations of being in ACADIA was the sense of sheer access to epicenters of research in which ideas are generated, standing directly together with the original authors of the tools and of the concepts. That is one thing that is continued in ACADIA to this day. The sense that this is a gathering of leaders and of people who are acting, interwoven with the extraordinary technology and invention that has characterized the society. I think it would also be true to say that a sense of deep humanity has consistently accompanied the investment in tools and technology. And perhaps it is that twin of unapologetic feeling and of sensitivity accompanying the enablement of tools that really characterizes this place. It makes for a tremendously motivating sense of what we can contribute together.

Karen Kensek: What really struck me when I was thinking back was the sense of naivete and the fact that I felt so young. I looked back at some of the old photographs that were taken at the ACADIA conferences, and it was a young group of people. And I was in the second round of the group of people. I wasn't one of the first round of heroes, the people who got this thing in motion, which really helped me because I could look at those people and say, "Look what they have done. Now, what can I do?" I think what ACADIA meant most to me was the fact that there were people there who were doing groundbreaking research in an area that would eventually change the profession. To me, teaching is also incredibly important, and ACADIA values that as well. You guys look at our tools, you probably would laugh. But we were doing, for us, truly amazing things.

I remember when I started working on the ACADIA newsletter, then I became the editor with Doug Noble on the ACADIA Quarterly, that gave me the opportunity to meet the membership. And that's what I want to stress to the people who are in the audience: become involved with ACADIA. That's what made it really important and critical to my career. I worked on the Quarterly, I helped host a conference, I was president: I did these things and they gave me the amazing opportunity to meet people. And these people have gone on to do just incredible things.

Branko Kolarevic: The first ACADIA conference that I attended was in 1991, 30 years ago, which makes me a real dinosaur. It was in Los Angeles and I remember it vividly. Chuck Eastman was one of the co-chairs. That conference actually played a huge role in my life. Because of it, I got my first full-time job after graduate school. I discovered a community of like-minded people. We were all "outcasts" at that time. So, ACADIA was kind of a support group, so important for all of us. It was like, it's okay, you are not alone, there are people out there that believe in this stuff. In 1997, those of us who were in the steering committee, now the board of directors, were thinking, how do we make ourselves mainstream? How do we really open what we do to a broader community? And at that time, we launched what I think was the first internet-based design competition, the Library for the Information Age. There were over 600 entries.

In the mid-1990s, a number of us were engaged in the virtual design studios, where we were connecting across continents. Electronic design studios were then becoming options in different schools. It took us almost 25 years, thanks to the pandemic, to demonstrate to everyone that studio can be taught online. So I think the experiment that we have been unwillingly subjected to in all of our schools over the past year and a half, I think it will change in some fundamental ways how we teach architecture in the future. Karen made a reference to teaching and as I have seen in the presentations at this conference, how we teach and what we teach still remains a core interest to many in our community.

Mahesh Daas: The summer of 2007 was particularly balmy in Cambridge, just across the river from where I work now, a radical place called the Boston Architectural College. Tucked away on Brattle Street near Harvard Square was the Harvest Restaurant, a favorite place for Bill Mitchell. We used to meet at the restaurant whenever I visited him in Cambridge. It was a couple of years after he stepped down from the deanship at MIT. Bill was also one of the founding members of ACADIA. As many of us know, he was a humble, compassionate, and yet unabashedly progressive individual who was also a great mentor. Earlier that year, with Bill's encouragement, I ran for and was elected the president of ACADIA. I asked him, "So you've been with ACADIA ever since its beginning, and what is the raison d'etre of the Association? What is the DNA? What is the spirit of this Association?" And he said something very interesting. He said, "In ACADIA's early years, it was a radical and diverse place on the fringes." And so I said, "What's your advice for me?" He said, "Well, try to radicalize it again." I took it to heart. When I took helm of ACADIA in 2007, the association was turning a corner and maturing after a remarkable infancy and a restless adolescence that was invariably followed by a brief identity crisis. We had celebrated 25 years of ACADIA- and some among the members thought that ACADIA had fulfilled its mission, it served its purpose, and that it could be sunset. But as we all know, now, ACADIA continues to evolve, thanks to the great work of many leaders and volunteers and all the members, hundreds and hundreds of them. This has been a platform, a community of scholars and designers and tinkerers and thinkers, all kinds of people who provided us a great foundation.

Personally, over the years, the association served as a sandbox for me to test new ideas that ranged from theory, criticism, pedagogy, technology, leadership development, and design. ACADIA allowed me the intellectual expeditions unmatched by any other forum, for which I am deeply thankful.

I think now in the fifth decade, ACADIA will continue, I am sure, to question the status quo. And I just was watching the earlier panel and looking at the perspectives that were not previously provided about equity, inclusion, and diversity. And looking at things from that perspective, suddenly new things are revealed. And I think we will evolve as a community in the coming decades, and I think new topics will emerge that we can hardly think of right now. But we need to be on that edge. We need to be that radical organization, that platform in order to be always relevant and vibrant.

Nancy Cheng: So imagine if you will, a beautiful mountain setting, the sun is setting, the hillsides are covered with colors of orange and red as the sun sinks, and you're in a wonderful resort. It's just lovely all around. And there's almost nobody here in the resort, except you and 90 of your nerdiest friends who are in a really beautiful, hot, warm swimming pool in the crisp air. It was pretty interesting. 1999 Park City, the place where the Sundance Festival is held, a ski resort in the offseason. And it just is a metaphor for the amazing joy that we had of having this little fishbowl of being with people who are like ourselves. So Branko alluded to it. I always thought of it as a little bit of an AA support group for CAD teachers in the early days. In other words, we were pretty isolated in our own universities. There would be one advocate for digital methods at each school, and so it was great to come together and talk with like-minded people.

We naturally enjoy these precious times of being with people who are like ourselves and having that discourse with others who are exploring the same thing. And we're always challenged to see how we can reach out and make the connections to what's outside of this really nice environment that's so isolated from everything else. And I think the spirit of adventure is still there. It's so exciting to see the growth of how the organization has come to embrace so many concerns, the web of concerns including social inequity and environmental consciousness. What's really important is that we think about how we can keep embracing other threads and make them expand. As everything has gotten more and more fragmented and niche, it's really a delight to see how things are coming together and getting integrated in ACADIA.

Jason Kelly Johnson: I've been involved with ACADIA for only 15 years, but it's been 15 really important years in my academic and creative life. The conference is the event where you can make fundamental connections. You really do have access to some amazing people, amazing innovators, leaders. You get to sit at the table and mix it up. I remember people like Philip Beesley, and how shocked I was that Philip just sat down beside me, and he said, "what are you working on?" And for me, it was a moment where it was phenomenal that someone that I admired would actually spend the time to sit down with me. I had similar experiences with Branko over the years, and people that you deeply admire are there and they're engaged, they're looking at the work, and they're coming to see your presentation. And that's really, really rare, right? That kind of community is really hard to come by.

During the presidency I was able to participate and contribute to hosting amazing conferences, but also pass it on to Kathy. That really was the motivation of continuing to be a board member, to be the president, and really try and continue to build it up and evolve the organization. And now I think Kathy has done similar things, where she's passed it on to someone that she trusts. And in Jenny Sabin we have an incredible thinker and someone who's trying to engage larger ideas that I think are critical to computation, but also just to be a human being in a planet that's clearly in trouble.

Kathy Velikov: My own memories of ACADIA are still incredibly recent, very fresh. I am going to preempt Philip's toast by sharing a couple of stories in the format of a Georgian toast. The Georgian toast is in three parts. Each part is a story followed by a short toast. And so we're going to start toasting early and please fill your real or virtual glasses.

My first story of ACADIA starts with a discussion the board had a couple of years ago around publishing, and whether to publish our proceedings with an academic publisher in order to increase the scientific and academic prominence and visibility of the papers. But this would be at the expense of open access of the papers through CumInCAD, and after a couple of months of consideration and debate, the board voted to stay with open access. And in that process, we confirmed that one of our values is really to support open access to knowledge and research. This year we have really acted on those values, collecting and making ALL of ACADIA's past publications public, and accessible to anyone, not just ACADIA members. Looking through this archive makes one realize the number of incredible individuals who have contributed to ACADIA over the past 40 years. Some of you are with us here today, but many are not. So my first toast is to all of those ACADIA members, conference chairs, past presidents, Board members, and collaborators who are not here today and without whose commitment, work, and passion to the organization of ACADIA would not be where it is.

My second story takes me back to the Mexico City conference in 2018, where there were quite a number of conversations over mezcal and tequila. This is the first time in my memory that the discussions started to question what ACADIA could be doing to enable access to the privileged knowledge of computational design. How we could break down both knowledge barriers and financial barriers for those that are often left out of the organization's activities? And in the few years that have followed, the Board and I have systematically tried to transform ACADIA as an organization through activities such as developing a code of conduct, seeking sponsorship to increase the amount of scholarship particularly to students from places like Mexico and other international students, NOMA students, PhD students, forging partnerships with NOMA, and also foregrounding topics of ethics, access, equity, labor and bias in the conferences and the keynotes. While we still have a lot of work to do, this is a toast to everyone in the community, and particularly the Board, who has pushed for and contributed to the behind-the-scenes work it takes in advancing these commitments into durable practices transforming ACADIA toward a more ethically minded, inclusive, and open organization.

My last story is likely the most significant memory that I will have when looking back on the pandemic period is shifting to online board-run conferences. I think everyone on the board can remember the kind of stress and panic in March of 2020, as we realized the world was shutting down, when universities were canceling all upcoming conferences. We made the decision to undertake ACADIA's first board-run conference, and this year is our second. And I recall in one of our early meetings, Jason Kelly Johnson, who was the vice president at the time, argued that if the board was going to take on this conference, we had to make it radically different than our previous conferences. It couldn't be the same thing but just online. And so with that challenge, we

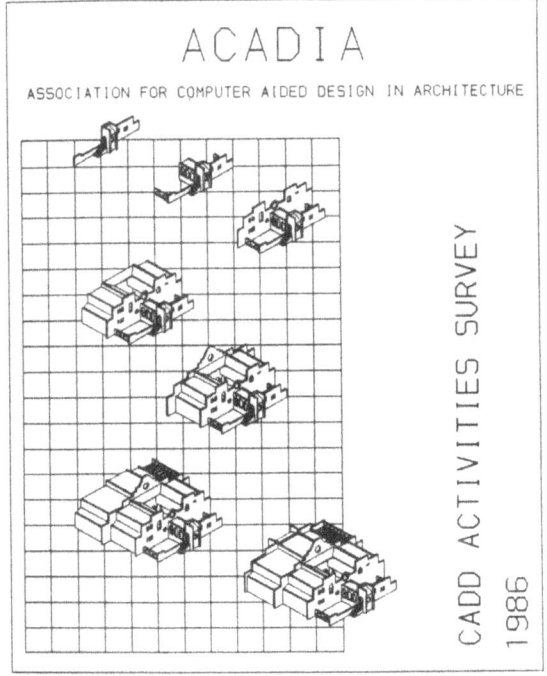

2   Cover of the 1986 CADD Activities Survey documenting the software and hardware architecture schools had as early adopters of technology in their curriculums. 1983 sketch by Richard Quadrel for the ACADIA Quarterly that was sent to all members.

decided to try to reinvent not only the conference format but also shift the content that it privileged, moving away from singular hero keynotes to critical conversations that also subvert the male hegemony of the computational field by featuring primarily female speakers, introducing new topics such as field notes that tried to illuminate the kind of black box of our work, shifting to a critical agenda for our papers and calls. And this is certainly continued today with the conference theme of this year and the work that I'm sure Jenny will do moving forward. My final toast is to the future of ACADIA as an organization and community, to embracing and looking forward to in anticipation of its transformation. Cheers to change.

Phillip Beesley: Colleagues, it is now my role to propose a toast to the past and the future. And I hope you will allow me to make just a few remarks before I do so. The toast that I'm going to give is framed around what we can contribute. It is framed in a rather solemn sense. I hope you won't mind if this has some strong emotional tones as well. We have just seen the completion of the International Climate Summit, which to many is a failure. It's been called that. And so that induces even more fear. Yet, for us as leaders and as deeply involved people who have the ability to act, I hope that that fear can give motivation. My remarks are framed with that tone, not of a paralysis of fear, but of profound motivation. Perhaps it is true that this society was born out of the Cold War. A sense of gravitas has lurked at least, if not boiled over and if not torn the society. ACADIA has also been framed constantly around surging hopes, about integration and the expansion of dimensions for humanity, giving. Certainly we can acknowledge the turbulence of our conversations and the search for fundamental transformation. Perhaps we can also agree that it is time to act right away.

This raises the question of how we can contribute. I find myself reflecting on this in all of the presentations. This tells me that we can act with intense pragmatism, the kind of lovely focus that comes from working so very close to machine language and fundamental technology. We can include the extraordinary kind of sensitivity that comes from embodied interfaces that have emerged and from the dimensions of integration with our own psyches; the understanding increasingly of neuroscience along with biology, along with permacultures, along with circularity; perhaps also being increasingly close to our basal brains, as well as to our conscious selves. Those expansive dimensions make it increasingly possible to work with extraordinary analysis and consciousness, but also in the space of dreams and even in the space of the doppelganger— that is, within the shadows that we certainly are not in control of, but that we can acknowledge and weave into our lives. And perhaps it is that combination of dreaming and playing and very, very slow and gentle and even glacial reflection, along with the most immediate kind of agile entrepreneurial energy that characterizes what ACADIA is becoming. This can contribute to the sense that architecture can become something very close to a living being.

We are an immensely privileged society. We have access to intensely powerful tools. We are teaching ourselves and giving each other the ability to manipulate the environment. On one hand, that manipulation is absolutely dependent on a kind of confidence and optimism and self-esteem. I think the importance of that confidence can't be underestimated, in equipping ourselves with the kind of resolve to work with the built environment in the face of almost overwhelming difficulties that have emerged. And yet at the same time, the dimensions of humility and caution and a willingness not to intervene, but rather to pause, and include, and to consider and reflect is absolutely necessary.

I think we can agree above all that we are an involved community. We are architects. We bring computational design to enrich the built environment. And so a toast to the society: may ACADIA continue and support the deepest contributions. May it be a society of leaders. May you mentor the young. May you recognize and reconcile your lives with your elders and those whose lands you have settled in and who have been excluded. May you open doors. May you create a sense of beauty and hope.

My toast is to the past and to the future and it includes two things:

Thanks to those past, for your extraordinary expertise and vision and your optimism; and to those future voices, may you be confident in making a difference and may you be humble and careful. So, this is a toast to what we can contribute.

3   Cover of the 1986 ACADIA Workshop Proceedings.

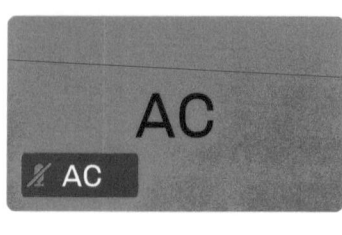 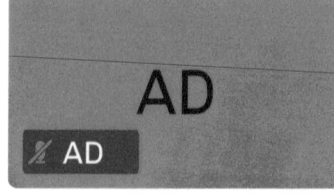 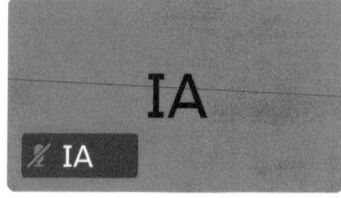 
   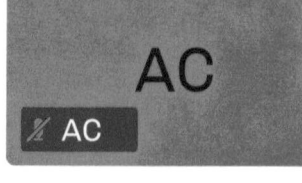
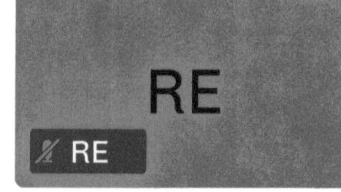  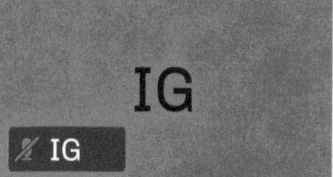 
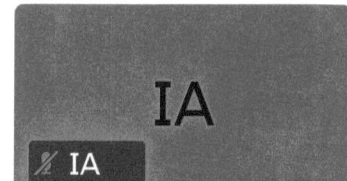   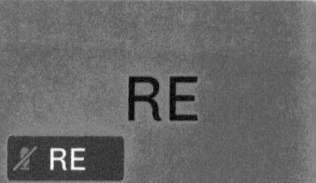
 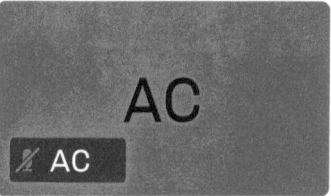  
  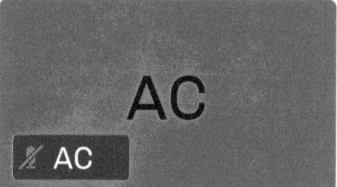 
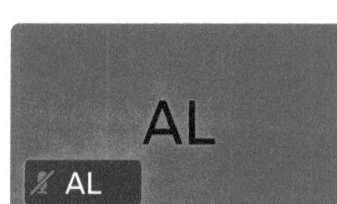   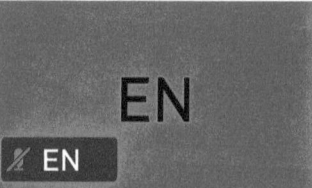
  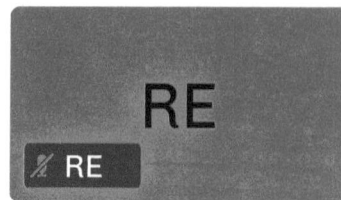 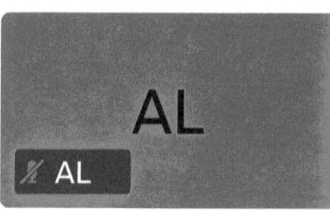

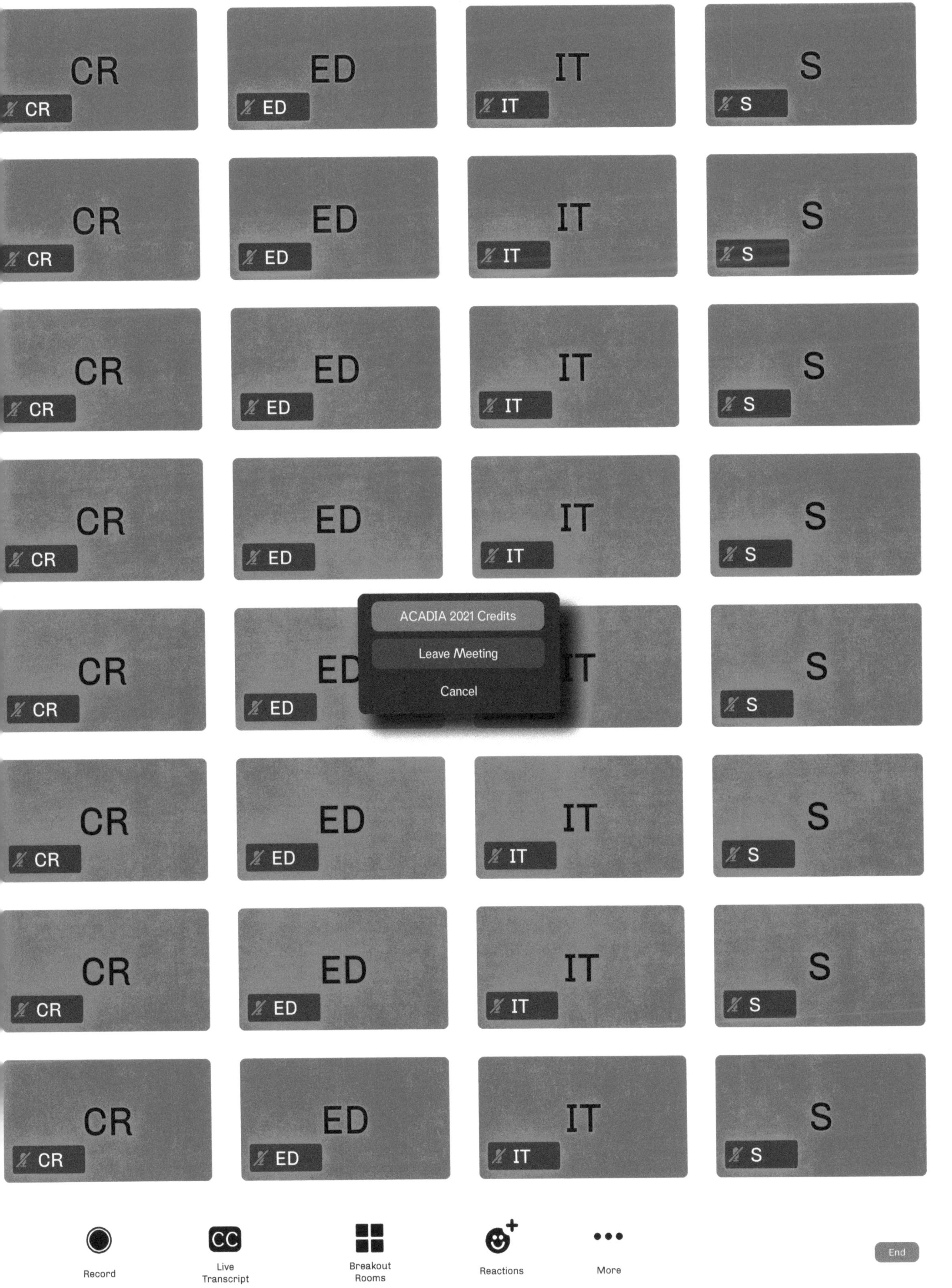

CR	ED	IT	S
CR	ED	IT	S
CR	ED	IT	S
CR	ED	IT	S
CR	ED	IT	S
CR	ED	IT	S
CR	ED	IT	S
CR	ED	IT	S

ACADIA 2021 Credits

Leave Meeting

Cancel

 Record    Live Transcript    Breakout Rooms    Reactions   More   End

Conference Chairs

Br. Biayna Bogosian Biayna Bogosian's academic and professional background extends in the fields of architecture, computational design, environmental design, data science, spatial computing, and media arts. Biayna's interdisciplinary research has allowed her to understand innovation in design and technology within a broader environmental context and explore data-driven and citizen-centric approaches to improve the built environment.

Biayna is currently an Assistant Professor of Architectural Technology at Florida International University (FIU), where her interdisciplinary research is supported by a number of National Science Foundation grants. She has also taught at Columbia University GSAPP, Cornell University AAP, University of Southern California SoA, and Tongji University CAUP among other universities. Biayna studied architecture at Woodbury University SoA and Columbia GSAPP. She is currently a PhD candidate in the Media Arts and Practice program at the USC School Cinematic Arts.

Dr. Kathrin Dörfler is an Assistant Professor at the Technical University of Munich, leading a research group for Digital Fabrication at the Department of Architecture at the School of Engineering and Design. At the interface between architecture and robotics, Kathrin pursues research in digital design and robot-supported construction processes in architecture and their synthesis in new technologies. She is particularly interested in exploring the architectural implications of mobile robotics and augmented reality technologies enabling collaborative construction processes. She has a Master's degree in Architecture from TU Vienna (2012) and a PhD degree from ETH Zurich (2018), carried out at Gramazio Kohler Research at the National Centre of Competence in Research (NCCR) Digital Fabrication, ETH Zurich.

Behnaz Farahi, Trained as an architect, Behnaz Farahi is an award winning designer and critical maker based in Los Angeles. She holds a PhD in Interdisciplinary Media Arts and Practice from USC School of Cinematic Arts. Currently she is an Assistant Professor at the Department of Design, California State University, Long Beach.

She explores how to foster an empathetic relationship between the human body and the space around it using computational systems. Her work addresses critical issues such as feminism, emotion, perception and social interaction. She specializes in computational design, interactive technologies, and digital fabrication technologies. Her work is part of the permanent collection of the Museum of Science and Industry in Chicago. Farahi has won several awards including the Cooper Hewitt Smithsonian Design Museum Digital Design Award, Innovation By Design Fast Company Award, World Technology Award (WTN). She is a co-editor of an issue of AD, '3D Printed Body Architecture' (2017) and 'Interactive Futures' (forthcoming).

Dr. Jose Luis García del Castillo y López is an architect, computational designer and educator. He advocates for a future where programming and code are tools as natural to artists as paper and pencil. In his work, he explores how interfaces for concurrent human-machine collaboration can lead to increased levels of creativity for both parties involved.

Jose Luis is Lecturer in Architectural Technology at the Harvard Graduate School of Design, from where he holds a Doctor of Design and Master in Design Studies in Technology degrees. He is also the co-founder of ParametricCamp, a digital platform to promote computational literacy amongst designers.

June A. Grant, RA, NOMA, is a visionary architect, Founder and Design Principal at blinkLAB architecture; a boutique research-based architecture and urban design studio that re-thinks conventional approaches. Launched in 2015, BlinkLAB was created based on Ms. Grant's 20 years experience in architecture, design and urban regeneration of cities and communities. Her design approach rests on an avid belief in cultural empathy, data research and new technologies as integral to design futures and design solutions. Because we are designers committed to new forms of knowledge through making, we prefer to situate ourselves in the middle of catalytic design - where new challenges and emerging opportunities are addressed through multi-layered thinking and design.

blinkLAB has three mandates - A commitment to Design Exploration, Advocacy for Holistic Solutions and the Integration of Technology as a central component for a regenerative society. Ms. Grant is also the immediate Past-President of the San Francisco Chapter of the National Organization of Minority Architects (SFNOMA); Board member of ACADIA, a YBCA100 honoree, 2020 CEDAW Human Rights honoree, and the 2020, 10th Annual J. Max Bond Jr. Lecturer.

Dr. Vernelle A. A. Noel is a design scholar, architect, artist, and founding Director of the Situated Computation + Design Lab at the Georgia Institute of Technology. Her research examines traditional and automated making, human-computer interaction, interdisciplinary creativity, and their intersections with society. Dr. Noel builds frameworks, methodologies, expressions, and tools to explore social, cultural, and political aspects of making and computational design for new reconfigurations of practice, pedagogy, and publics. Her work has been supported by the Graham Foundation, the Mozilla Foundation, and ideas2innovation (i2i), among others. She is a recipient of the 2021 DigitalFUTURES Young Award for exceptional research and scholarship in the field of critical computational design and has a TEDx Talk titled "The Power of Making: Craft, Computation, and Carnival." Dr. Noel holds a PhD. in Architecture: Design and Computation from Penn State University, an MS from MIT, a BArch from Howard University, and a Diploma in Civil Engineering from Trinidad & Tobago. She has taught at the University of Stuttgart, the University of Florida, Penn State University, MIT, the Singapore University of Technology & Design, and has practiced as an architect in the US, India, and Trinidad & Tobago.

Dr Stefana Parascho is a researcher, architect, and educator whose work lies at the intersection of architecture, digital fabrication and computational design. She is currently an Assistant Professor at the École Polytechnique Fédérale de Lausanne (EPFL) where she founded the Lab for Creative Computation (CRCL). Through her research, she has explored multi-roboticfabrication methods and their relationship to design. Prior to joining EPFL, Stefana was Assistant Professor at Princeton University. She completed her doctorate in 2019 at ETH Zurich, Gramazio Kohler Research, and received her Diploma in Architectural Engineering in 2012 from the University of Stuttgart.

Dr Jane Scott is an Academic Track Research Fellow in the Hub for Biotechnology in the Built Environment at Newcastle University, UK. Her research is located at the interface of programmable textiles, architecture and biology. Jane's work challenges the established understanding of smart materials for architecture; applying principles derived from biology to the design of environmentally responsive textile systems. Jane is currently developing a new generation of Living Textiles for the built environment.

Before joining Newcastle University, Jane was an academic at the University of Leeds and held a Visiting Research Fellowship in Biomimicry at Central Saint Martins. She completed her PhD, Programmable Knitting, at the Textiles Futures Research Centre, Central Saint Martins. In 2016 this work was awarded the Autodesk ACADIA Emerging Research Award. Her work has been exhibited widely and she has presented her research at major international events.

Session Chairs

Melissa Goldman is an architectural and scenic designer, fabricator, and educator, Melissa has a passion for design at all scales. From crafting new tools and materials to building large moving creatures, her work combines the theatrical with the structural to explore new construction techniques and experiences in the digital and physical realms. Melissa is the Fabrication Lab Manager at the University of Virginia School of Architecture and serves on the ACADIA Board of Directors. Her work on Ferrostructures won the Autodesk Emerging Research Award at ACADIA 2017. Melissa has her Masters of Architecture from Columbia's GSAPP and her BA in English from Harvard.

Maria Yablonina is an architect, researcher, and artist working in the field of computational design and digital fabrication. Her work lies at the intersection of architecture and robotics, producing spaces and robotic systems that can construct themselves and change in real-time. Such architectural productions include the development of hardware and software solutions, as well as complementing architectural and material systems in order to offer new design spaces.

Maria's practice focuses on designing machines that make architecture—a practice that she broadly describes as Designing [with] Machines (D[w]M). D[w]M aims to investigate and establish design methodologies that consider robotic hardware development as part of the overall design process and its output. Through this work, Maria argues for a design practice that moves beyond the design of objects towards the design of technologies and processes that enable new ways of both creating and interacting with architectural spaces.

Mathias Bernhard is an architect with a profound specialization in computational design, digital fabrication, and information technology. He is a postdoctoral researcher at the Weitzman School of Design, University of Pennsylvania. Before that, he was a postdoc at ETH Zurich where he received his degree Doctor of Sciences in Architecture. His research revolves around question of encoding, investigating alternative models of abstraction and their creative potential. He is particularly interested in how artifacts can be made machine-readable and digitally operational. The focus lies thereby on how the increasingly ubiquitous availability of data influences the design process.

June A. Grant RA, NOMA, is a visionary architect, Founder and Design Principal at blinkLAB architecture; a boutique research-based architecture and urban design studio that re-thinks conventional approaches. Launched in 2015, BlinkLAB was created based on Ms. Grant's 20 years experience in architecture, design and urban regeneration of cities and communities. Her design approach rests on an avid belief in cultural empathy, data research, and new technologies as integral to design futures and design solutions. Because we are designers committed to new forms of knowledge through making, we prefer to situate ourselves in the middle of catalytic design — where new challenges and emerging opportunities are addressed through multi-layered thinking and design.

blinkLAB has three mandates: A commitment to Design Exploration, Advocacy for Holistic Solutions, and the Integration of Technology as a central component for a regenerative society. Ms. Grant is also the immediate Past-President of the San Francisco Chapter of the National Organization of Minority Architects (SFNOMA); Board member of ACADIA, a YBCA100 honoree, 2020 CEDAW Human Rights honoree, and the 2020, 10th Annual J. Max Bond Jr. Lecturer.

Biayna Bogosian's academic and professional background extends in the fields of architecture, computational design, environmental design, data science, spatial computing, and media arts. Biayna's interdisciplinary research has allowed her to understand innovation in design and technology within a broader environmental context and explore data-driven and citizen-centric approaches to improve the built environment.

Biayna is currently an Assistant Professor of Architectural Technology at Florida International University (FIU), where her interdisciplinary research is supported by a number of National Science Foundation grants. She has also taught at Columbia University GSAPP, Cornell University AAP, University of Southern California SoA, and Tongji University CAUP among other universities. Biayna studied architecture at Woodbury University SoA and Columbia GSAPP. She is currently a PhD candidate in the Media Arts and Practice program at the USC School Cinematic Arts.

Vernelle A. A. Noel is a design scholar, architect, artist, and founding Director of the Situated Computation + Design Lab at the Georgia Institute of Technology. Her research examines traditional and automated making, human-computer interaction, interdisciplinary creativity, and their intersections with society. Dr. Noel builds frameworks, methodologies, expressions, and tools to explore social, cultural, and political aspects of making and computational design for new reconfigurations of practice, pedagogy, and publics. Her work has been supported by the Graham Foundation, the Mozilla Foundation, and ideas2innovation (i2i), among others. She is a recipient of the 2021 DigitalFUTURES Young Award for exceptional research and scholarship in the field of critical computational design and has a TEDx Talk titled "The Power of Making: Craft, Computation, and Carnival." Dr. Noel holds a PhD. in Architecture: Design and Computation from Penn State University, an MS from MIT, a BArch from Howard University, and a Diploma in Civil Engineering from Trinidad & Tobago. She has taught at the University of Stuttgart, the University of Florida, Penn State University, MIT, the Singapore University of Technology & Design, and has practiced as an architect in the US, India, and Trinidad & Tobago.

Shelby Elizabeth Doyle, AIA is an Associate Professor of Architecture at the Iowa State University College of Design, co-founder of the ISU Computation & Construction Lab (CCL) and director of the ISU Architectural Robotics Lab (ARL). The central hypothesis of CCL and Doyle's work is that computation in architecture is a material, pedagogical, and social project; computation is both informed by and productive of architectural cultures. Doyle received a Fulbright Fellowship to Cambodia, a Master of Architecture from the Harvard Graduate School of Design, and a Bachelor of Science in Architecture from the University of Virginia.

Forrest Meggers is Associate Professor at Princeton University jointly appointed in the School of Architecture and the Andlinger Center for Energy and the Environment at the school of engineering. He co-directs the PhD track in computation and energy and also the undergraduate program in architecture and engineering. He is the director of CHAOS Lab (Cooling and Heating for Architecturally Optimized Systems), which he founded to support his research on building systems including radiant heat transfer, sensors, geothermal, and desiccant dehumidification. He has degrees in Mechanical Engineering and Environmental Engineering and received his doctorate from the department of architecture at ETH Zurich.

Christoph Klemmt is Assistant Professor, University of Cincinnati and Founding Partner, at Orproject Christoph graduated from the Architectural Association in London in 2004 and received his doctorate from the University of Applied Arts Vienna in 2021 under the supervision of Klaus Bollinger and Patrik Schumacher. He has worked amongst others for Zaha Hadid Architects, where his responsibilities focused on the company's projects in China, including the mixed-use developments Soho Galaxy, Wangjing Soho and Leeza Soho.

He has lectured and given workshops at the Architectural Association, Nottingham University, the University of Wuppertal, AA Visiting School, Tsinghua University and Tongji University. He is Assistant Professor at the University of Cincinnati, where he received a grant to set up the Architectural Robotics Lab. In 2008 he co-founded Orproject, an architect's office specialising in advanced geometries with an ecologic agenda. Orproject has exhibited at the Palais De Tokyo in Paris, the China National Museum in Beijing and the Biennale in Venice. The work of Orproject has been featured world-wide in magazines and books such as Domus, Frame, FT and Spacecraft, and the practice has won several international Awards.

Leslie Lok is an assistant professor at Cornell University Department of Architecture and directs the Rural-Urban Building Innovation Lab, a research platform that seeks to operate within a larger feedback loop between material resourcefulness, custom computation protocols, local specificity, and design equity in the rural-urban context. In parallel, Lok co-directs HANNAH, an experimental design practice that explores the implementation of innovative forms of construction such as additive concrete manufacturing and robotic wood construction. HANNAH is the recipient of the 2020 Architectural League Prize, a winner of the 2018 Folly/Function Competition, and was named Next Progressives by Architect Magazine. Her work has been presented at FABRICATE, Rob|Arch, and ACADIA, as well as featured in Architectural Record, the New York Times, Dwell, Digital Trends, and in various book publications. Lok received her Master of Architecture at MIT

Nadya Peek develops unconventional automation tools for scientific exploration, digital fabrication, and interactive control. Spanning electronics, firmware, software, and mechanics, her research focuses on harnessing the precision of machines for the creativity of individuals. Nadya directs the Machine Agency at the University of Washington where she is an assistant professor in Human-Centered Design and Engineering. Nadya is an active member of the global fab lab community, making digital fabrication more accessible with better CAD/CAM tools and developing open source hardware machines and control systems. She is VP of the Open Source Hardware Association, half of the design studio James and the Giant Peek, plays drum machines and synths in the band Construction and got her PhD at MIT in the Center for Bits and Atoms.

Behnaz Farahi, Trained as an architect, Behnaz Farahi is an award winning designer and critical maker based in Los Angeles. She holds a PhD in Interdisciplinary Media Arts and Practice from USC School of Cinematic Arts. Currently she is an Assistant Professor at the Department of Design, California State University, Long Beach.

She explores how to foster an empathetic relationship between the human body and the space around it using computational systems. Her work addresses critical issues such as feminism, emotion, perception and social interaction. She specializes in computational design, interactive technologies, and digital fabrication technologies. Her work is part of the permanent collection of the Museum of Science and Industry in Chicago. Farahi has won several awards including the Cooper Hewitt Smithsonian Design Museum Digital Design Award, Innovation By Design Fast Company Award, World Technology Award (WTN).

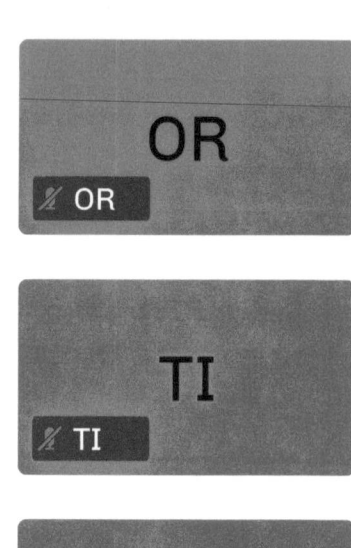
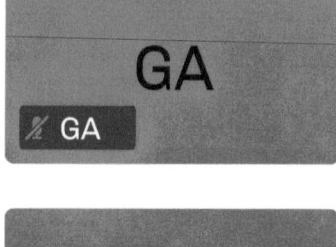

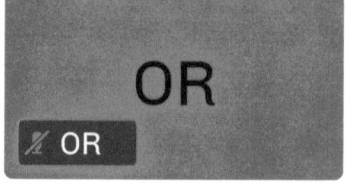

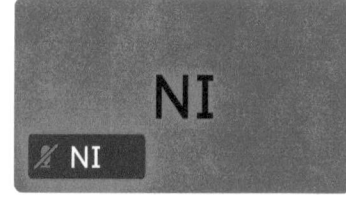

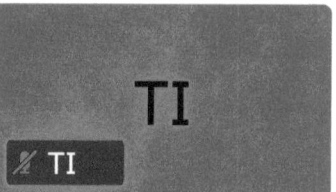

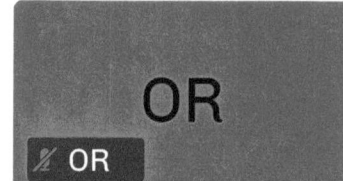
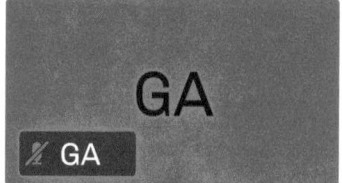
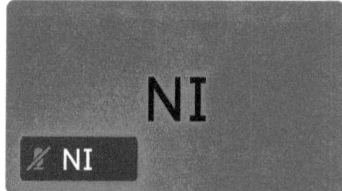

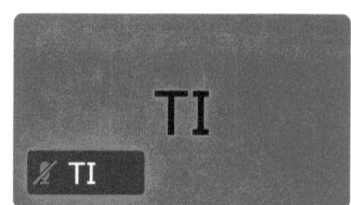

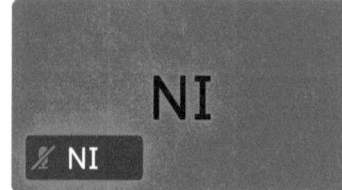

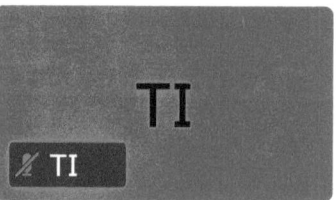

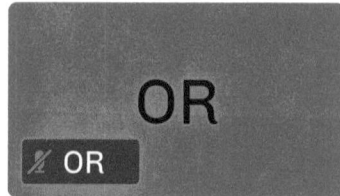
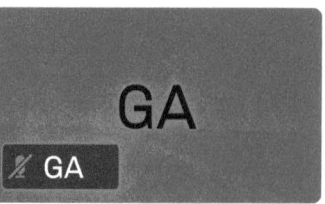
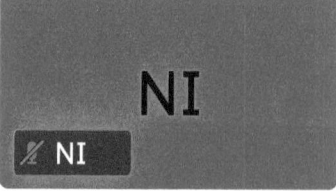

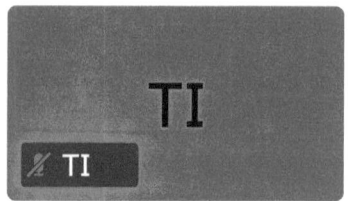

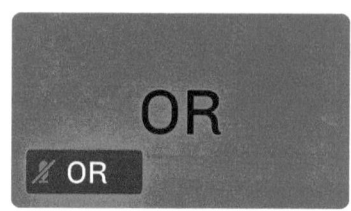
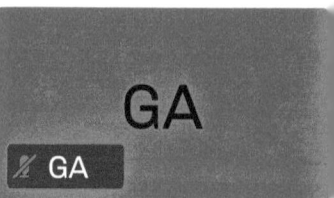

The Association for Computer Aided Design in Architecture (ACADIA) is an international network of digital design researchers and professionals that facilitates critical investigations into the role of computation in architecture, planning, and building science, encouraging innovation in design creativity, sustainability and education.

ACADIA was founded in 1981 by some of the pioneers in the field of design computation including Bill Mitchell, Chuck Eastman, and Chris Yessios. Since then, ACADIA has hosted 40 conferences across North America and has grown into a strong network of academics and professionals in the design computation field.

Incorporated in the state of Delaware as a not-for-profit corporation, ACADIA is an all-volunteer organization governed by elected officers, an elected Board of Directors, and appointed ex-officio officers.

PRESIDENT
Jenny E. Sabin, Cornell University
president@acadia.org

VICE-PRESIDENT
Kathy Velikov, University of Michigan
vp@acadia.org

SECRETARY
Tsz Yan Ng, University of Michigan
secretary@acadia.org

TREASURER
Phillip Anzalone, New York City College of Technology
treasurer@acadia.org

MEMBERSHIP OFFICER
Jane Scott, Newcastle University
membership@acadia.org

TECHNOLOGY OFFICER
Andrew Kudless, University of Houston
webmaster@acadia.org

DEVELOPMENT OFFICER
Matias del Campo, University of Michigan
development@acadia.org

COMMUNICATIONS OFFICER
Melissa Goldman, University of Virginia
communications@acadia.org

BOARD OF DIRECTORS, 2021
Shelby Doyle
Behnaz Farahi
Adam Marcus
Jane Scott
Maria Yablonina
Matias del Campo
Tsz Yan Ng
Jose Luis Garcia del Castillo López
June A. Grant
Stefana Parascho
Kory Bieg (alternate)
Christoph Klemmt (alternate)
Viola Ago (alternate)
Biayna Bogosian (alternate)
Melissa Goldman (alternate)
Vernelle A. A. Noel (alternate)

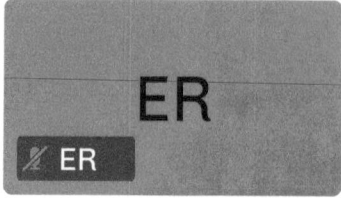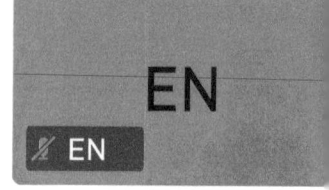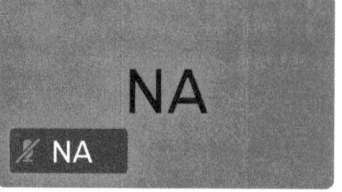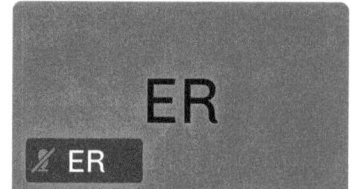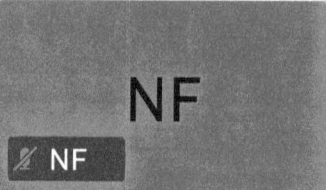

CONFERENCE CHAIRS

Biayna Bogosian, Assistant Professor, Florida International University

Kathrin Dörfler, Assistant Professor, Technical University of Munich

Behnaz Farahi, Assistant Professor, California State University Long Beach

Jose Luis García del Castillo y López, Lecturer in Architectural Technology, Harvard Graduate School of Design

June A. Grant, Founder and Design Principal, blinkLAB architecture

Vernelle A.A. Noel, Assistant Professor, Georgia Tech

Stefana Parascho, Assistant Professor, Princeton University

Jane Scott, Academic Track Research Fellow, Newcastle University

TECHNICAL ASSISTANTS

Yifei Peng, Cornell University

Ji Yoon Bae, Cornell University

Alexia Asgari, Cornell University

Begum Birol, Cornell University

Karolina Piorko, Cornell University

CONFERENCE PROJECT MANAGER

Cameron Nelson

CONFERENCE WEBSITE

Behnaz Farahi, Assistant Professor, CSULB

Biayna Bogosian, Assistant Professor, FIU

COPY EDITING

Gabi Sarhos

PROCEEDINGS GRAPHIC IDENTITY AND DESIGN

Ian Besler

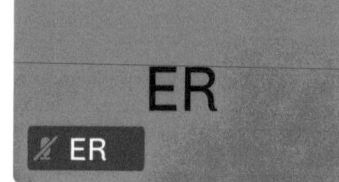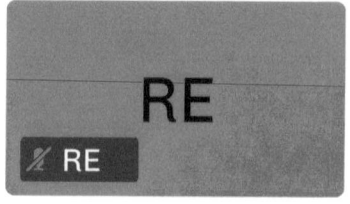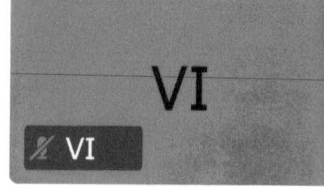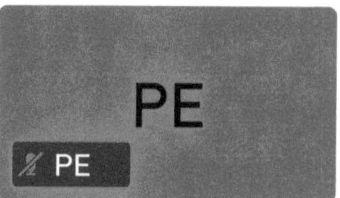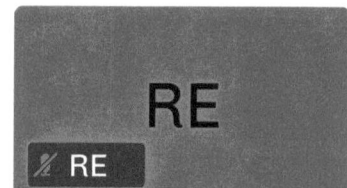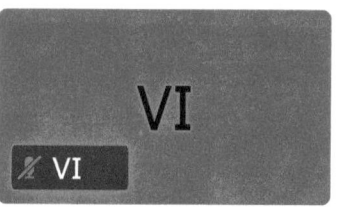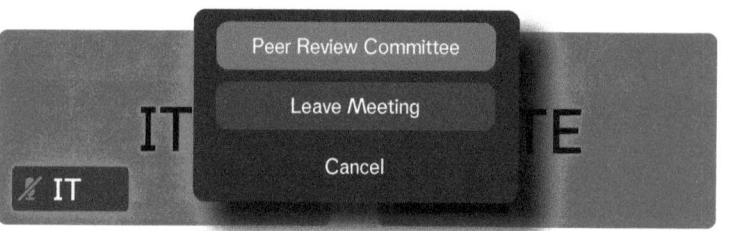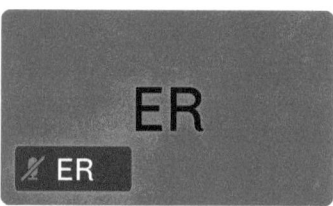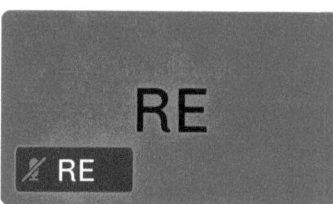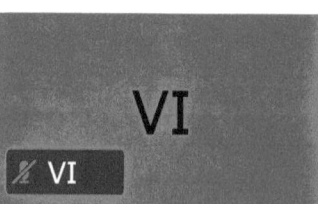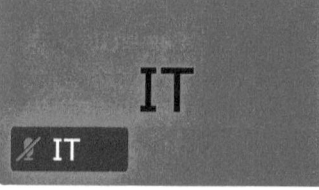

Aaron Willette _ *Intelligent City*
Achilleas Xydis _ *ETH Zurich*
Adam Marcus _ *California College of the Arts*
Aldo Sollazzo _ *Noumena - IAAC Institute for Advanced Architecture of Catalunya*
Aleksandra Anna Apolinarska _ *ETH Zürich*
Alex Schofield _ *California College of the Arts*
Alicia Nahmad Vazquez _ *University of Calgary*
Alvin Huang _ *USC School of Architecture*
Ana Anton _ *Digital Building Technologies, ETH Zürich*
Andrei Jipa _ *DBT, ETH Zürich*
Andrei Nejur _ *University of Montreal*
Andrew O Payne _ *Proving Ground*
AnnaLisa Meyboom _ *University of British Columbia*
Anton Savov _ *Digital Design Unit, TU Darmstadt*
Antonino Di Raimo _ *University of Portsmouth*
Arash Adel _ *University of Michigan*
Armand Agraviador _ *Newcastle University - Hub for Biotechnology in the Built Environment*
Arthur Mamou-Mani _ *University of Westminster, IAAC*
Axel Kilian _ *MIT Department of Architecture*
Aysegul Akcay Kavakoglu _ *Istanbul Technical University*
Bastian Beyer _ *Humboldt University Berlin*
Behnaz Farahi _ *CSULB*
Ben Bridgens _ *Newcastle University*
Benjamin Dillenburger _ *ETH Zurich*
Biayna Bogosian _ *FIU*
Bob Martens _ *TU Wien*
Brady Peters _ *University of Toronto*
Brandon Clifford _ *MIT*
Branko Kolarevic _ *New Jersey Institute of Technology (NJIT)*
Brian Slocum _ *Universidad Iberoamericana*
Burcin Nalinci _ *Zahner*
Carolina Ramirez-Figueroa _ *Royal College of Art*
Catty Dan Zhang _ *UNC Charlotte*
Chris Perry _ *Rensselaer Polytechnic Institute*
Christine Yogiaman _ *Singapore University of Technology and Design*
Christopher Romano _ *University at Buffalo*
Cristiano Ceccato _ *Zaha Hadid Architects*
Daniel Davis _ *Hassell*
Daniel Koehler _ *UT Austin*
Daniel Tish _ *Harvard University Graduate School of Design*
Daniela Mitterberger _ *ETH Zurich*
Dave Pigram _ *University of Technology Sydney [UTS]*
David Costanza _ *Cornell University*
Dorit Aviv _ *University of Pennsylvania*
Dylan Wood _ *ICD, University of Stuttgart*
Ebrahim Poustinchi _ *Kent State University*

Edvard Bruun _ *Princeton University*
Efilena Baseta _ *Noumena*
Ehsan Baharlou _ *University of Virginia*
Elena Manferdini _ *SCI-Arc*
Emre Erkal _ *Erkal Architects*
Ena Lloret-Frischi _ *ETH Zurich*
Ezio Blasetti _ *University of Pennsylvania*
Forrest Meggers _ *Princeton University*
Franca Trubiano _ *UPENN*
Gabriel Kaprielian _ *Temple University*
Gabriel Wainer _ *Carleton University*
Gido Dielemans
Gonçalo Castro Henriques _ *Universidade Federal do Rio de Janeiro, Brasil*
Greg Corso _ *Syracuse University School of Architecture*
Guvenc Ozel _ *UCLA*
Guy Gardner _ *Lab for Integrative Design*
Hans Jakob Wagner
Hao Zheng _ *University of Pennsylvania*
Henri Achten _ *Czech Technical University in Prague*
Hugh Hynes _ *California College of the Arts*
Igor Siddiqui _ *The University of Texas at Austin*
Iman Ansari _ *The Ohio State University*
Inés Ariza _ *ETH Zurich*
Jane Scott _ *Newcastle University*
Jason Carlow _ *American University of Sharjah*
Jason Scroggin _ *University of Kentucky College of Design*
Jeff Ponitz _ *Cal Poly, San Luis Obispo*
Jenny Sabin _ *Cornell University*
Jeremy Jih _ *MIT*
Johannes Braumann _ *Robots in Architecture*
John Rhett Russo _ *Rensselaer Polytechnic Institute*
Jordan Geiger _ *Gekh*
Jorge Orozco
Jorge Ramirez _ *Anemonal.org*
Jose Luis García del Castillo y López _ *Harvard GSD*
Jose Pedro Sousa _ *University of Porto*
Jose Sanchez _ *University of Michigan*
Joseph Choma _ *Clemson University*
Joshua Vermillion _ *University of Nevada Las Vegas*
Ju Hong Park _ *POSTECH*
Juan José Castellón González _ *Rice University*
Jyoti Kapur _ *ZHdK*
Kathrin Dörfler _ *TU Munich*
Kathy Velikov _ *University of Michigan*
Katie MacDonald _ *University of Virginia*
Kian Wee Chen _ *Princeton University*
Kory M Bieg _ *The University of Texas at Austin*
Kyle Schumann _ *University of Virginia*

Lauren Vasey _ ETH Zurich
Lei Yu _ Beijing Archi-Solution Workshop Ltd.
Leighton Beaman _ Cornell University
Lidia Atanasova _ TU Munich
Mahesh Daas _ Boston Architectural College
Mara Marcu _ University of Cincinnati
Marc Aurel Schnabel _ Victoria University of Wellington
Marcella Del Signore _ New York Institute of Technology,
    School of Architecture and Design
Marcelyn Gow _ SCI-Arc
Maria Larsson _ Tokyo University
Maria Yablonina _ UofT
Mariana Popescu _ TU Delft
Mario Carpo _ The Bartlett, UCL, London;
    die Angewandte, Vienna
Mark Donohue _ California College of the Arts
Marta Nowak _ Ohio State University
Martyn Dade-Robertson _ Newcastle University
Matan Mayer _ IE University
Mathew Schwartz _ NJIT
Mathias Bernhard _ University of Pennsylvania
Mathias Maierhofer _ ICD, University of Stuttgart
Matias del Campo _ University of Michigan
Maya Przybylski _ University of Waterloo
Melissa Goldman _ University of Virginia
Mine Özkar _ Istanbul Technical University
Molly Hunker _ Syracuse University
Nathaniel Jones _ Arup
Negar Kalantar _ California College of the Arts
Neil Katz _ Skidmore, Owings & Merrill
Nikola Marincic
Oliver David Krieg _ Intelligent City
Oliver Tessmann _ TU Darmstadt
Onur Yuce Gun _ New Balance Athletics Inc.
Peter Testa _ SCI-Arc
Philippe MOREL _ UCL Bartlett
Pierluigi D'Acunto _ Technical University of Munich
Pierre Cutellic _ ETH Zurich
R Scott Mitchell _ USC School of Architecture
Rafael Pastrana _ Princeton University
Rasa Navasaityte _ UT Austin
Riccardo La Magna _ KIT
Robert Pietrusko _ Harvard University
Robert Stuart-Smith _ University of Pennsylvania
Robert Trempe _ Arkitektskolen Aarhus
Romana Rust _ ETH Zurich
Ryan Luke Johns _ ETH Zurich
Ryan Vincent Manning _ Quirkdee/Heron-Mazy
Sandra Manninger _ University of Michigan
Sarah A Kott-Tannenbaum _ NBBJ Design

Sarah Nichols _ Rice University
Sasa Zivkovic _ Cornell University
Satoru Sugihara _ ATLV
Serban Bodea _ ICD, University of Stuttgart
Seung Kyu Ra _ Oklahoma State University
Shai Yeshayahu _ Ryerson University
Shelby E Doyle _ Iowa State University
Shota Vashakmadze _ UCLA
Sigrid Adriaenssens _ Princeton University
Sina Mostafavi _ TU Delft
Skylar Tibbits _ Self-Assembly Lab, MIT
Sotirios Kotsopoulos _ MIT
Suleiman Alhadidi _ Harvard University
Sumbul Khan _ Singapore University
    of Technology and Design
T. F. Tierney _ University of Illinois at Urbana-Champaign
Taro Narahara _ New Jersey Institute of Technology
Ted Kesik _ University of Toronto
Thora Arnardottir _ Newcastle University
Tobias Schwinn _ ICD, University of Stuttgart
Tom Verebes _ New York Institute of Technology
Tsz Yan Ng _ University of Michigan
Valentina Soana _ University College London
Vera Parlac _ New Jersey Institute of Technology
Vernelle A. A. Noel _ Georgia Tech.
Virginia San Fratello _ San Jose State University
Vishu Bhooshan _ Zaha Hadid Architects /
    University College of London
Wendy W Fok _ Pratt Institute, School of Architecture
Wes McGee _ University of Michigan

   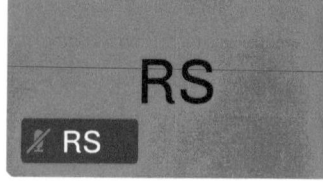

PLATINUM SPONSOR

GOLD SPONSORS

SILVER SPONSOR

GRIMSHAW

BRONZE SPONSORS

ARKTURA   ABB

SPONSORS

MEDIA PARTNER

The Architect's Newspaper

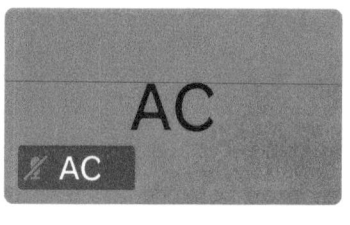 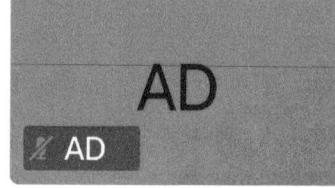 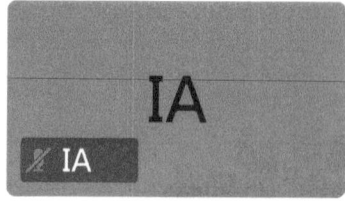 
 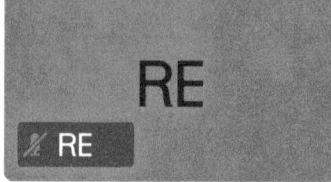  
   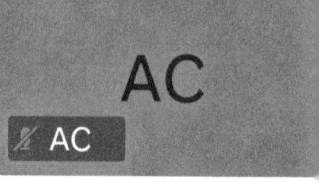

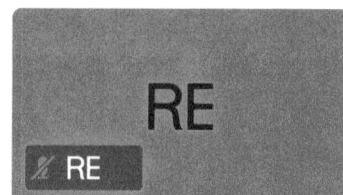  
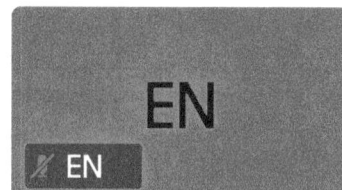  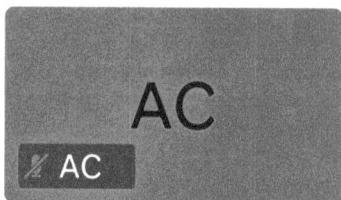 
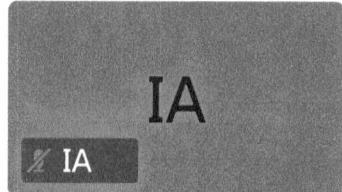   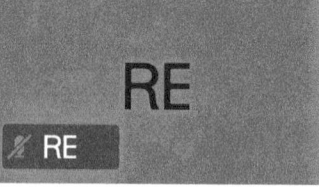
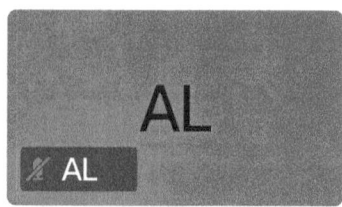   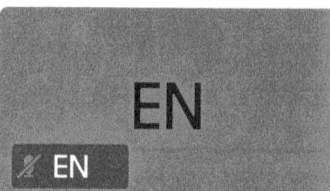